D0151373

Manual of
Steel Construction

Allowable Stress
Design

Manual of

STEEL CONSTRUCTION

Allowable Stress Design

NINTH EDITION

American Institute of Steel Construction, Inc.

1 East Wacker Drive, Suite 3100

Chicago, Illinois 60601

Printed in the United States of America
First Revision of the Ninth Edition (1/91)

FOREWORD

The American Institute of Steel Construction, founded in 1921, is the non-profit technical specifying and trade organization for the fabricated structural steel industry in the United States. Executive and engineering headquarters of AISC are maintained in Chicago.

The Institute is supported by three classes of membership: Active Members totaling about 400 companies engaged in the fabrication and erection of structural steel; Associate Members who are allied product manufacturers; and Professional Members who are individuals or firms engaged in the practice of architecture or engineering. Professional Members also include architectural and engineering educators. The continuing financial support and active participation of Active Members in the engineering, research and development activities of the Institute make possible the publishing of this *Allowable Stress Design Manual of Steel Construction*.

The Institute's objectives are to improve and advance the use of fabricated structural steel through research and engineering studies to develop the most efficient and economical design of structures. It also conducts programs to improve product quality.

To accomplish these objectives the Institute publishes manuals, textbooks, specification and technical booklets. Best known and most widely used are the *Manuals of Steel Construction* which hold a highly respected position in engineering literature. Outstanding among AISC standards are the *Specification for Structural Steel Buildings* and the *Code of Standard Practice for Steel Buildings and Bridges*.

The Institute also assists designers, contractors, educators and others by publishing technical information and timely articles on structural applications through two quarterly publications: *Engineering Journal* and *Modern Steel Construction*. In addition, public appreciation of aesthetically designed structures is encouraged through its annual award programs: Prize Bridges, Architectural Awards of Excellence and student Fellowship Awards.

The Ninth Edition Manual was produced under the guidance of the AISC Manual Committee, made up of experienced and knowledgeable engineers from industry and member fabricator companies. The valuable assistance furnished by the American Iron and Steel Institute and the Welded Steel Tube Institute in assembling data and in generating many of the tables by electronic computers is gratefully acknowledged.

PREFACE TO THE FIRST REVISED PRINTING

In Part 1, domestic availability of W shapes has been updated.

In Part 4, design procedure for column base plates has been revised.

American Institute of Steel Construction

January 1991

PART 1
Dimensions and Properties

STRUCTURAL STEELS

PRODUCT AVAILABILITY

Section A3.1 of the *Specification for Structural Steel Buildings Allowable Stress Design and Plastic Design,* (from here on referred to as the ASD Specification), lists 16 ASTM specifications for structural steel approved for use in building construction.

Six of these steels are available in hot-rolled structural shapes, plates and bars. Two steels, ASTM A514 and A852, are only available in plates. Table 1 shows five groups of shapes and 11 ranges of thicknesses of plates and bars available in the various minimum yield stress* and tensile strength levels afforded by the eight steels. For complete information on each steel, reference should be made to the appropriate ASTM specification. A listing of the shape sizes included in each of the five groups follows in Table 2, corresponding to the groupings given in Table A of ASTM Specification A6.

Seven additional grades of steel, other than those covering hot-rolled shapes, plates and bars, are listed in Sect. A3.1. These steels cover pipe, cold- and hot-formed tubing and cold- and hot-rolled sheet and strip.

For additional information on availability of structural tubing, refer to separate discussion beginning on pg. 1-91. For additional information on availability and classification of structural steel plates and bars, refer to separate discussion beginning on pg. 1-105.

Space does not permit inclusion in the listing of shapes and plates in Part 1 of this Manual of all rolled shapes or plates of greater thickness that are occasionally used in construction. For such products, refer to the various producers' catalogs.

To obtain an economical structure, it is often advantageous to minimize the number of different sections. Cost per sq. ft. often can be reduced by designing this way.

SELECTION OF THE APPROPRIATE STRUCTURAL STEEL

ASTM A36 is the all-purpose carbon grade steel widely used in building and bridge construction. ASTM A529 structural carbon steel, ASTM A572 high-strength, low-alloy structural steel, ASTM A242 and A588 atmospheric corrosion-resistant high-strength low-alloy structural steel, ASTM A514 quenched and tempered alloy structural steel plate and ASTM A852 quenched and tempered low-alloy structural steel plate may each have certain advantages over ASTM A36 structural carbon steel, depending on the application. These high-strength steels have proven economical choices where lighter members, resulting from use of higher allowable stresses, are not penalized because of instability, local buckling, deflection or other similar reasons. They are frequently used in tension members, beams in continuous and composite construction where deflections can be minimized, and columns having low slenderness ratios. The reduction of dead load, and associated savings in shipping costs, can be significant. However, higher strength steels are not to be used indiscriminately. Effective use of all steels depends on thorough cost and engineering analysis.

With suitable procedures and precautions, all steels listed in the AISC Specification are suitable for welded fabrication.

ASTM A242 and A588 atmospheric corrosion-resistant, high-strength low-alloy

---

*As used in the AISC Specification, "yield stress" denotes either the specified minimum yield point (for those steels that have a yield point) or specified minimum yield strength (for those steels that do not have a yield point).

steels can be used in the bare (uncoated) condition in most atmospheres. Where boldly exposed under such conditions, exposure to the normal atmosphere causes a tightly adherent oxide to form on the surface which protects the steel from further atmospheric corrosion. To achieve the benefits of the enhanced atmospheric corrosion resistance of these bare steels, it is necessary that design, detailing, fabrication, erection and maintenance practices proper for such steels be observed. Designers should consult with the steel producers on the atmospheric corrosion-resistant properties and limitations of these steels prior to use in the bare condition. When either A242 or A588 steel is used in the coated condition, the coating life is typically longer than with other steels. Although A242 and A588 steels are more expensive than other high-strength, low-alloy steels, the reduction in maintenance resulting from the use of these steels usually offsets their higher initial cost.

ASTM A852 and A514 Types E, F, P, and Q are higher strength atmospheric corrosion-resistant steels suitable for use in the bare (uncoated) condition in most atmospheres.

BRITTLE FRACTURE CONSIDERATIONS IN STRUCTURAL DESIGN
General Considerations

As the temperature decreases, an increase is generally noted in the yield stress, tensile strength, modulus of elasticity and fatigue strength of structural steels. In contrast, the ductility of these steels, as measured by reduction in area or by elongation, decreases with decreasing temperature. Furthermore, there is a temperature below which a structural steel subjected to tensile stresses may fracture by cleavage,* with little or no plastic deformation, rather than by shear,* which is usually preceded by a considerable amount of plastic deformation or yielding.

Fracture that occurs by cleavage at a nominal tensile stress below the yield stress is commonly referred to as brittle fracture. Generally, a brittle fracture can occur in a structural steel when there is a sufficiently adverse combination of tensile stress, temperature, strain rate and geometrical discontinuity (notch) present. Other design and fabrication factors may also have an important influence. Because of the interrelation of these effects, the exact combination of stress, temperature, notch and other conditions that will cause brittle fracture in a given structure cannot be calculated readily. Consequently, designing against brittle fracture often consists mainly of (1) avoiding conditions that tend to cause brittle fracture and (2) selecting a steel appropriate for the application. A discussion of these factors is given in the following sections. Refs. 1 through 5 cover the subject in much more detail.

Conditions Causing Brittle Fracture

It has been established that plastic deformation can occur only in the presence of shear stresses. Shear stresses are always present in a uniaxial or biaxial state-of-stress. However, in a triaxial state-of-stress, the maximum shear stress approaches zero as the principal stresses approach a common value. Thus, under equal triaxial tensile stresses, failure occurs by cleavage rather than by shear. Consequently, triaxial tensile stresses tend to cause brittle fracture and should be avoided. A triaxial state-of-stress can result from a uniaxial loading when notches or geometrical discontinuities are present.

---

*Shear and cleavage are used in the metallurgical sense (macroscopically) to denote different fracture mechanisms. Ref. 2, as well as most elementary textbooks on metallurgy, discusses these mechanisms.

Increased strain rates tend to increase the possibility of brittle behavior. Thus structures that are loaded at fast rates are more susceptible to brittle fracture. However, a rapid strain rate or impact load is not a required condition for a brittle fracture.

Cold work, and the strain aging that normally follows, generally increases the likelihood of brittle fracture. This behavior usually is attributed to the previously mentioned reduction in ductility. The effect of cold work that occurs in cold forming operations can be minimized by selecting a generous forming radius, thus limiting the amount of strain. The amount of strain that can be tolerated depends on both the steel and the application.

When tensile residual stresses are present, such as those resulting from welding, they add to any applied tensile stress and thus, the actual tensile stress in the member will be greater than the applied stress. Consequently, the likelihood of brittle fracture in a structure that contains high residual stresses may be minimized by a post-weld heat treatment. The decision to use a post-weld heat treatment should be made with assurance the anticipated benefits are needed and will be realized, and that possible harmful effects can be tolerated. Many modern steels for welded construction are designed to be used in the less costly as-welded condition when possible. The soundness and mechanical properties of welded joints in some steels may be adversely affected by a post-weld heat treatment.

Welding may also contribute to the problem of brittle fracture by introducing notches and flaws into a structure and by causing an unfavorable change in microstructure of the base metal. However, properly designed welds, care in selecting their location and the use of good welding practice, can minimize such detrimental effects. The proper electrode must be selected so that the weld metal will be as resistant to brittle fracture as the base metal.

Selecting a Steel

The best guide in selecting a steel appropriate for a given application is experience with existing and past structures. The A36 steel has been used successfully in a great number of applications, such as buildings, transmission towers, transportation equipment and bridges, even at the lowest atmospheric temperatures encountered in the U.S. Therefore, it appears that any of the structural steels, when designed and fabricated in an appropriate manner, could be used for similar applications with little likelihood of brittle fracture. Consequently, brittle fracture is not usually experienced in such structures unless unusual temperature, notch and stress conditions are present. Nevertheless, it is always desirable to avoid or minimize the previously cited adverse conditions that increase the susceptibility to brittle fracture.

In applications where notch toughness is considered important, it usually is required that steels must absorb a certain amount of energy, 15 ft-lb. or higher (Charpy V-notch test), at a given temperature. The test temperature may be higher than the lowest operating temperature depending on the rate of loading.[5] For example, the toughness requirements for A709 steels are based on the loading rate for bridges.[6]

LAMELLAR TEARING

The information on strength and ductility presented in the previous sections generally pertains to loadings applied in the planar direction (longitudinal or transverse orientation) of the steel plate or shape. It should be noted that elongation and area reduction values may well be significantly lower in the through-thickness direction than in the planar direction. This inherent directionality is of small consequence in many applications, but does become important in the design and fabrication of structures containing massive members with highly restrained welded joints.

With the increasing trend toward heavy welded-plate construction, there has been a broader recognition of occurrences of lamellar tearing in some highly restrained joints of welded structures, especially those using thick plates and heavy structural shapes. The restraint induced by some joint designs in resisting weld deposit shrinkage can impose tensile strain sufficiently high to cause separation or tearing on planes parallel to the rolled surface of the structural member being joined. The incidence of this phenomenon can be reduced or eliminated through greater understanding by designers, detailers and fabricators of (1) the inherent directionality of constructional forms of steel, (2) the high restraint developed in certain types of connections and (3) the need to adopt appropriate weld details and welding procedures with proper weld metal for through-thickness connections. Further, steels can be specified to be produced by special practices and/or processes to enhance through-thickness ductility and thus assist in reducing the incidence of lamellar tearing. Steels produced by such practices are available from several producers. However unless precautions are taken in both design and fabrication, lamellar tearing may still occur in thick plates and heavy shapes of such steels at restrained through-thickness connections. Some guidelines in minimizing potential problems have been developed.[7]

JUMBO SHAPES AND HEAVY WELDED BUILT-UP SECTIONS

Although Group 4 and 5 W-shapes, commonly referred to as jumbo shapes, generally are contemplated as columns or compression members, their use in non-column applications has been increasing. These heavy shapes have been known to exhibit segregation and a coarse grain structure in the mid-thickness region of the flange and the web. Because these areas may have low toughness, cracking might occur as a result of thermal cutting or welding.[8] Similar problems may also occur in welded built-up sections. To minimize the potential of brittle failure, the current AISC ASD Specification (see Manual, Part 5) includes provisions for material toughness requirements, methods of splicing and fabrication methods for Group 4 and 5 hot-rolled or welded built-up cross sections with an element of the cross section more than 2 in. in thickness intended for tension applications.

REFERENCES

1. *Brockenbrough, R.L. and B.G. Johnson* U.S.S. Steel Design Manual *1981, U.S. Steel.*
2. *Parker, E.R.* Brittle Behavior of Engineering Structures *John Wiley & Sons, 1957, New York, N.Y.*
3. *Welding Research Council* Control of Steel Construction to Avoid Brittle Failure *1957.*
4. *Lightner, M.W. and R.W. Vanderbeck* Factors Involved in Brittle Fracture *Regional Technical Meetings, American Iron and Steel Institute, 1956.*
5. *Rolfe, S.T. and J.M. Barsom* Fracture and Fatigue Control in Structures—Applications of Fracture Mechanics *Prentice-Hall, Inc., 1977, Englewood Cliffs, N.J.*
6. *Rolfe, S.T.* Fracture and Fatigue Control in Steel Structures *AISC Engineering Journal, 1st Qtr. 1977, New York, N.Y. (pg. 2).*
7. *American Institute of Steel Construction, Inc.* Commentary in Highly Restrained Welded Connections *AISC Engineering Journal, 3rd Qtr. 1973, New York, N.Y. (pg. 61).*
8. *Fisher, John W. and Alan W. Pense* Experience with Use of Heavy W Shapes in Tension *AISC Engineering Journal, 2nd Qtr. 1987, Chicago, Ill. (pp. 63–77).*

TABLE 1
Availability of Shapes, Plates and Bars According to ASTM Structural Steel Specifications

Steel Type	ASTM Designation		F_y Minimum Yield Stress (ksi)	F_u Tensile Stress[a] (ksi)	Shapes — Group per ASTM A6					Plates and Bars										
					[b]1	2	3	4	5	To ½″ Incl.	Over ½″ to ¾″ Incl.	Over ¾″ to 1¼″ Incl.	Over 1¼″ to 1½″ Incl.	Over 1½″ to 2″ Incl.	Over 2″ to 2½″ Incl.	Over 2½″ to 4″ Incl.	Over 4″ to 5″ Incl.	Over 5″ to 6″ Incl.	Over 6″ to 8″ Incl.	Over 8″
Carbon	A36		32	58–80																
			36	58–80[c]																
	A529		42	60–85																
High-strength Low-alloy	A441		40	60																
			42	63																
			46	67																
			50	70																
	A572 Grade	42	42	60																
		50	50	65																
		60	60	75																
		65	65	80																
Corrosion-resistant High-strength Low-alloy	A242		42	63																
			46	67																
			50	70																
	A588		42	63																
			46	67																
			50	70																
Quenched & Tempered Low-alloy	A852[d]		70	90–110																
Quenched & Tempered Alloy	A514[d]		90	100–130																
			100	110–130																

[a]Minimum unless a range is shown.
[b]Includes bar-size shapes.
[c]For shapes over 426 lbs./ft, minimum of 58 ksi only applies.
[d]Plates only.
▨ Available.
☐ Not available.

TABLE 2
Structural Shape Size Groupings for Tensile Property Classification

Structural Shapes	Group 1	Group 2	Group 3	Group 4	Group 5
W shapes	W 24×55, 62 W 21×44 to 57 incl. W 18×35 to 71 incl. W 16×26 to 57 incl. W 14×22 to 53 incl. W 12×14 to 58 incl. W 10×12 to 45 incl. W 8×10 to 48 incl. W 6×9 to 25 incl. W 5×16, 19 W 4×13	W 44×198, 224 W 40×149 to 268 incl. W 36×135 to 210 incl. W 33×118 to 152 incl. W 30×90 to 211 incl. W 27×84 to 178 incl. W 24×68 to 162 incl. W 21×62 to 147 incl. W 18×76 to 143 incl. W 16×67 to 100 incl. W 14×61 to 132 incl. W 12×65 to 106 incl. W 10×49 to 112 incl. W 8×58, 67	W 44×248, 285 W 40×277 to 328 incl. W 36×230 to 300 incl. W 33×201 to 291 incl. W 30×235 to 261 incl. W 27×194 to 258 incl. W 24×176 to 229 incl. W 21×166 to 223 incl. W 18×158 to 192 incl. W 14×145 to 211 incl. W 12×120 to 190 incl.	W 40×362 to 655 incl. W 36×328 to 798 incl. W 33×318 to 619 incl. W 30×292 to 581 incl. W 27×281 to 539 incl. W 24×250 to 492 incl. W 21×248 to 402 incl. W 18×211 to 311 incl. W 14×233 to 550 incl. W 12×210 to 336 incl.	W 36×848 W 14×605 to 730 incl.
M Shapes	to 37.7 lb./ft incl.				
S Shapes	to 35 lb./ft incl.				
HP Shapes		to 102 lb./ft incl.	over 102 lb./ft		
American Standard Channels (C)	to 20.7 lb./ft incl.	over 20.7 lb./ft			
Miscellaneous Channels (MC)	to 28.5 lb./ft incl.	over 28.5 lb./ft			
Angles (L) Structural Bar-size	to ½ in. incl.	over ½ to ¾ in. incl.	over ¾ in.		

Notes: Structural tees from W, M and S shapes fall into the same group as the structural shape from which they are cut.

Group 4 and Group 5 shapes are generally contemplated for application as columns or compression components. When used in other applications (e.g., trusses) and when thermal cutting or welding is required, special material specification and fabrication procedures apply to minimize the possibility of cracking. (See Part 5, Specification Sects. A3.1, J1.7, J1.8, J2.7, and M2.2 and corresponding Commentary sections.)

DIMENSIONS AND PROPERTIES
W Shapes
M Shapes
S Shapes
HP Shapes
American Standard Channels (C)
Miscellaneous Channels (MC)
Angles (L)

STRUCTURAL SHAPES—
DESIGNATIONS, DIMENSIONS AND PROPERTIES

The hot rolled shapes shown in Part 1 of this Manual are published in ASTM Specification A6/A6M, *Standard Specification for General Requirements for Rolled Steel Plates, Shapes, Sheet Piling, And Bars For Structural Use.*

W shapes have essentially parallel flange surfaces. The profile of a W shape of a given nominal depth and weight available from different producers is essentially the same except for the size of fillets between the web and flange.

HP bearing pile shapes have essentially parallel flange surfaces and equal web and flange thicknesses. The profile of an HP shape of a given nominal depth and weight available from different producers is essentially the same.

American Standard beams (S) and American Standard channels (C) have a slope of approximately 16⅔% (2 in 12 in.) on their inner flange surfaces. The profiles of S and C shapes of a given nominal depth and weight available from different producers are essentially the same.

The letter M designates shapes that cannot be classified as W, HP or S shapes. Similarly, MC designates channels that cannot be classified as C shapes. Because many of the M and MC shapes are only available from a limited number of producers, or are infrequently rolled, their availability should be checked prior to specifying these shapes. They have various slopes on their inner flange surfaces, dimensions for which may be obtained from the respective producing mills.

The flange thickness given in the tables for S, M, C and MC shapes is the *average* flange thickness.

In calculating the theoretical weights, properties and dimensions of the rolled shapes listed in Part 1 of this Manual, fillets and roundings have been included for all shapes except angles. The properties of these rolled shapes are based on the *smallest* theoretical size fillets produced; dimensions for detailing are based on the *largest* theoretical size fillets produced. These properties and dimensions are either exact or slightly conservative for all producers who offer them.

Equal leg and unequal leg angle (L) shapes of the same nominal size available from different producers have profiles which are essentially the same, except for the size of fillet between the legs and the shape of the ends of the legs. The *k* distance given in the tables for each angle is based on the largest theoretical size fillet available. Availability of certain angles is subject to rolling accumulation and geographical location, and should be checked with material suppliers.

AMERICAN INSTITUTE OF STEEL CONSTRUCTION

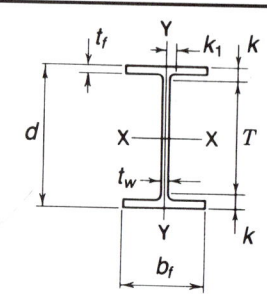

W SHAPES
Dimensions

Desig-nation	Area A	Depth d	Web Thickness t_w	$\frac{t_w}{2}$	Flange Width b_f	Thickness t_f	Distance T	k	k_1
	In.²	In.	In.	In.	In.	In.	In.	In.	In.
W 44×285	83.8	44.02	44	1.024	1	½	11.811	11¾	1.772	1¾	38⅝	2¹¹/₁₆	1⅜
×248	72.8	43.62	43⅝	0.865	⅞	⁷/₁₆	11.811	11¾	1.575	1⁹/₁₆	38⅝	2½	1⁵/₁₆
×224	65.8	43.31	43¼	0.787	¹³/₁₆	⁷/₁₆	11.811	11¾	1.416	1⁷/₁₆	38⅝	2⁵/₁₆	1⁵/₁₆
×198	58.0	42.91	42⅞	0.709	¹¹/₁₆	⅜	11.811	11¾	1.220	1¼	38⅝	2⅛	1¼
W 40×328	96.4	40.00	40	0.910	¹⁵/₁₆	½	17.910	17⅞	1.730	1¾	33¾	3⅛	1¹¹/₁₆
×298	87.6	39.69	39¾	0.830	¹³/₁₆	⁷/₁₆	17.830	17⅞	1.575	1⁹/₁₆	33¾	3	1⅝
×268	78.8	39.37	39⅜	0.750	¾	⅜	17.750	17¾	1.415	1⁷/₁₆	33¾	2¹³/₁₆	1⁹/₁₆
×244	71.7	39.06	39	0.710	¹¹/₁₆	⅜	17.710	17¾	1.260	1¼	33¾	2⅝	1⁹/₁₆
×221	64.8	38.67	38⅝	0.710	¹¹/₁₆	⅜	17.710	17¾	1.065	1¹/₁₆	33¾	2⁷/₁₆	1⁹/₁₆
×192	56.5	38.20	38¼	0.710	¹¹/₁₆	⅜	17.710	17¾	0.830	¹³/₁₆	33¾	2¼	1⁹/₁₆
W 40×655ᵃ	192.0	43.62	43⅝	1.970	2	1	16.870	16⅞	3.540	3⁹/₁₆	33¾	4¹⁵/₁₆	2¼
×593ᵃ	174.0	42.99	43	1.790	1¹³/₁₆	¹⁵/₁₆	16.690	16¾	3.230	3¼	33¾	4⅝	2⅛
×531ᵃ	156.0	42.34	42⅜	1.610	1⅝	¹³/₁₆	16.510	16½	2.910	2¹⁵/₁₆	33¾	4⁵/₁₆	2
×480ᵃ	140.0	41.81	41¾	1.460	1⁷/₁₆	¾	16.360	16⅜	2.640	2⅝	33¾	4	2
×436ᵃ	128.0	41.34	41⅜	1.340	1⁵/₁₆	¹¹/₁₆	16.240	16¼	2.400	2⅜	33¾	3¹³/₁₆	1¹⁵/₁₆
×397ᵃ	116.0	40.95	41	1.220	1¼	⅝	16.120	16⅛	2.200	2³/₁₆	33¾	3⅝	1⅞
×362ᵃ	106.0	40.55	40½	1.120	1⅛	⁹/₁₆	16.020	16	2.010	2	33¾	3⅜	1¹³/₁₆
×324	95.3	40.16	40⅛	1.000	1	½	15.905	15⅞	1.810	1¹³/₁₆	33¾	3³/₁₆	1¾
×297	87.4	39.84	39⅞	0.930	¹⁵/₁₆	½	15.825	15⅞	1.650	1⅝	33¾	3¹/₁₆	1¹¹/₁₆
×277	81.3	39.69	39¾	0.830	¹³/₁₆	⁷/₁₆	15.830	15⅞	1.575	1⁹/₁₆	33¾	3	1⅝
×249	73.3	39.38	39⅜	0.750	¾	⅜	15.750	15¾	1.420	1⁷/₁₆	33¾	2¹³/₁₆	1⁹/₁₆
×215	63.3	38.98	39	0.650	⅝	⁵/₁₆	15.750	15¾	1.220	1¼	33¾	2⅝	1⁹/₁₆
×199	58.4	38.67	38⅝	0.650	⅝	⁵/₁₆	15.750	15¾	1.065	1¹/₁₆	33¾	2⁷/₁₆	1⁹/₁₆
W 40×183ᵇ	53.7	38.98	39	0.650	⅝	⁵/₁₆	11.810	11¾	1.220	1¼	33¾	2⅝	1⁹/₁₆
×167	49.1	38.59	38⅝	0.650	⅝	⁵/₁₆	11.810	11¾	1.025	1	33¾	2⁷/₁₆	1⁹/₁₆
×149	43.8	38.20	38¼	0.630	⅝	⁵/₁₆	11.810	11¾	0.830	¹³/₁₆	33¾	2¼	1½

ᵃFor application refer to Notes in Table 2.
ᵇHeavier shapes in this series are available from some producers.
Shapes in shaded rows are not available from domestic producers.

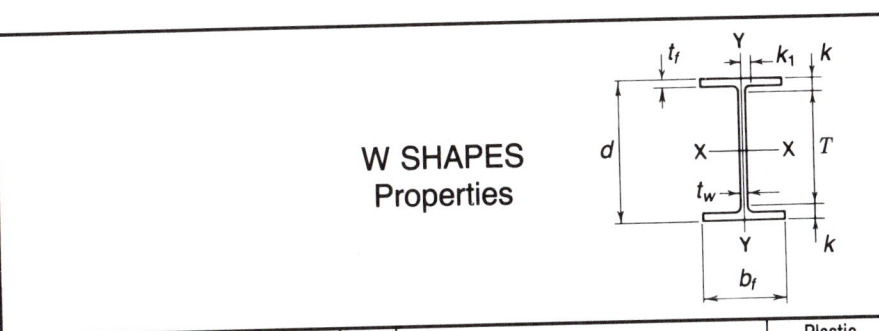

W SHAPES
Properties

Nom-inal Wt. per Ft Lb.	Compact Section Criteria $\frac{b_f}{2t_f}$	F_y' Ksi	$\frac{d}{t_w}$	F_y''' Ksi	r_T In.	$\frac{d}{A_f}$	Elastic Properties Axis X-X I In.⁴	S In.³	r In.	Axis Y-Y I In.⁴	S In.³	r In.	Plastic Modulus Z_x In.³	Z_y In.³
285	3.3	—	43.0	35.7	2.95	2.10	24600	1120	17.1	490	83.0	2.42	1310	135
248	3.7	—	50.4	26.0	2.96	2.34	21400	983	17.2	435	74.0	2.44	1150	118
224	4.2	—	55.0	21.8	2.96	2.59	19200	889	17.1	391	66.0	2.44	1030	105
198	4.8	—	60.5	18.0	2.94	2.98	16700	776	16.9	336	57.0	2.41	902	90.0
328	5.2	—	44.0	34.2	4.73	1.29	26800	1340	16.7	1660	185	4.15	1510	286
298	5.7	—	47.8	28.9	4.70	1.41	24200	1220	16.6	1490	167	4.12	1370	257
268	6.3	—	52.5	24.0	4.67	1.57	21500	1090	16.5	1320	149	4.09	1220	229
244	7.0	—	55.0	21.8	4.63	1.75	19200	983	16.4	1170	132	4.04	1100	203
221	8.3	61.1	54.5	22.3	4.56	2.05	16600	858	16.0	988	112	3.90	967	172
192	10.7	37.1	53.8	22.8	4.43	2.60	13500	708	15.5	770	87.0	3.69	807	135
655	2.4	—	22.1	—	4.43	0.73	56500	2590	17.2	2860	339	3.86	3060	541
593	2.6	—	24.0	—	4.38	0.80	50400	2340	17.0	2520	302	3.81	2750	481
531	2.8	—	26.3	—	4.33	0.88	44300	2090	16.9	2200	266	3.75	2450	422
480	3.1	—	28.6	—	4.28	0.97	39500	1890	16.8	1940	237	3.72	2180	374
436	3.4	—	30.9	—	4.24	1.06	35400	1710	16.6	1720	212	3.67	1980	334
397	3.7	—	33.6	58.6	4.21	1.15	32000	1560	16.6	1540	191	3.65	1790	300
362	4.0	—	36.2	50.4	4.17	1.26	28900	1420	16.5	1380	173	3.61	1630	270
324	4.4	—	40.2	41.0	4.14	1.40	25600	1280	16.4	1220	153	3.57	1460	239
297	4.8	—	42.8	36.0	4.11	1.53	23200	1170	16.3	1090	138	3.54	1330	215
277	5.0	—	47.8	28.9	4.13	1.59	21900	1100	16.4	1040	132	3.58	1250	204
249	5.5	—	52.5	24.0	4.10	1.76	19500	992	16.3	926	118	3.56	1120	182
215	6.5	—	60.0	18.4	4.09	2.03	16700	858	16.2	796	101	3.54	963	156
199	7.4	—	59.5	18.7	4.04	2.31	14900	769	16.0	695	88.2	3.45	868	137
183	4.8	—	60.0	18.4	2.98	2.71	13300	682	15.7	336	56.9	2.50	781	89.6
167	5.8	—	59.4	18.7	2.91	3.19	11600	599	15.3	283	47.9	2.40	692	76.0
149	7.1	—	60.6	18.0	2.84	3.90	9780	512	14.9	229	38.8	2.29	597	62.2

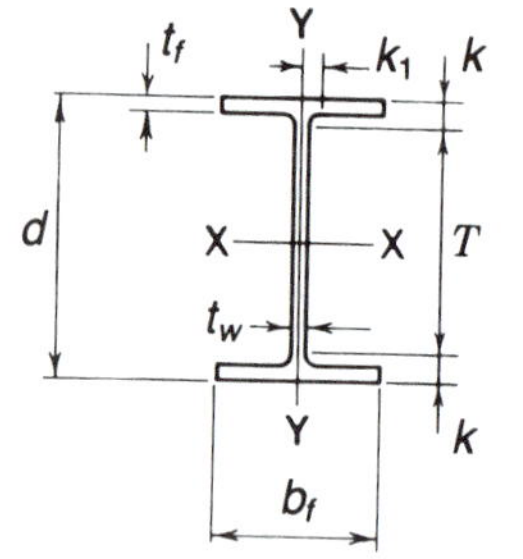

W SHAPES
Dimensions

Designation	Area A	Depth d		Web Thickness t_w		$\frac{t_w}{2}$	Flange Width b_f		Flange Thickness t_f		Distance T	Distance k	Distance k_1
	In.^{2}	In.		In.		In.	In.		In.		In.	In.	In.
W 36×848[a]	249.0	42.45	42 1/2	2.520	2 1/2	1 1/4	18.130	18 1/8	4.530	4 1/2	31 1/8	5 11/16	2 1/4
×798[a]	234.0	41.97	42	2.380	2 3/8	1 3/16	17.990	18	4.290	4 5/16	31 1/8	5 7/16	2 3/16
×720[a]	211.0	41.19	41 1/4	2.165	2 3/16	1 1/8	17.775	17 3/4	3.900	3 7/8	31 1/8	5 1/16	2 1/16
×650[a]	190.0	40.47	40 1/2	1.970	2	1	17.575	17 5/8	3.540	3 9/16	31 1/8	4 11/16	2
×588[a]	172.0	39.84	39 7/8	1.790	1 13/16	15/16	17.400	17 3/8	3.230	3 1/4	31 1/8	4 3/8	1 7/8
×527[a]	154.0	39.21	39 1/4	1.610	1 5/8	13/16	17.220	17 1/4	2.910	2 15/16	31 1/8	4 1/16	1 3/4
×485[a]	142.0	38.74	38 3/4	1.500	1 1/2	3/4	17.105	17 1/8	2.680	2 11/16	31 1/8	3 13/16	1 3/4
×439[a]	128.0	38.26	38 1/4	1.360	1 3/8	11/16	16.965	17	2.440	2 7/16	31 1/8	3 9/16	1 5/8
×393[a]	115.0	37.80	37 3/4	1.220	1 1/4	5/8	16.830	16 7/8	2.200	2 3/16	31 1/8	3 5/16	1 5/8
×359[a]	105.0	37.40	37 3/8	1.120	1 1/8	9/16	16.730	16 3/4	2.010	2	31 1/8	3 1/8	1 9/16
×328[a]	96.4	37.09	37 1/8	1.020	1	1/2	16.630	16 5/8	1.850	1 7/8	31 1/8	3	1 1/2
×300	88.3	36.74	36 3/4	0.945	15/16	1/2	16.655	16 5/8	1.680	1 11/16	31 1/8	2 13/16	1 1/2
×280	82.4	36.52	36 1/2	0.885	7/8	7/16	16.595	16 5/8	1.570	1 9/16	31 1/8	2 11/16	1 1/2
×260	76.5	36.26	36 1/4	0.840	13/16	7/16	16.550	16 1/2	1.440	1 7/16	31 1/8	2 9/16	1 1/2
×245	72.1	36.08	36 1/8	0.800	13/16	7/16	16.510	16 1/2	1.350	1 3/8	31 1/8	2 1/2	1 7/16
×230	67.6	35.90	35 7/8	0.760	3/4	3/8	16.470	16 1/2	1.260	1 1/4	31 1/8	2 3/8	1 7/16
W 36×256	75.4	37.43	37 3/8	0.960	1	1/2	12.215	12 1/4	1.730	1 3/4	32 1/8	2 5/8	1 5/16
×232	68.1	37.12	37 1/8	0.870	7/8	7/16	12.120	12 1/8	1.570	1 9/16	32 1/8	2 1/2	1 1/4
×210	61.8	36.69	36 3/4	0.830	13/16	7/16	12.180	12 1/8	1.360	1 3/8	32 1/8	2 5/16	1 1/4
×194	57.0	36.49	36 1/2	0.765	3/4	3/8	12.115	12 1/8	1.260	1 1/4	32 1/8	2 3/16	1 3/16
×182	53.6	36.33	36 3/8	0.725	3/4	3/8	12.075	12 1/8	1.180	1 3/16	32 1/8	2 1/8	1 3/16
×170	50.0	36.17	36 1/8	0.680	11/16	3/8	12.030	12	1.100	1 1/8	32 1/8	2	1 3/16
×160	47.0	36.01	36	0.650	5/8	5/16	12.000	12	1.020	1	32 1/8	1 15/16	1 1/8
×150	44.2	35.85	35 7/8	0.625	5/8	5/16	11.975	12	0.940	15/16	32 1/8	1 7/8	1 1/8
×135	39.7	35.55	35 1/2	0.600	5/8	5/16	11.950	12	0.790	13/16	32 1/8	1 11/16	1 1/8

[a] For application refer to Notes in Table 2.
Shapes in shaded rows are not available from domestic producers.

W SHAPES
Properties

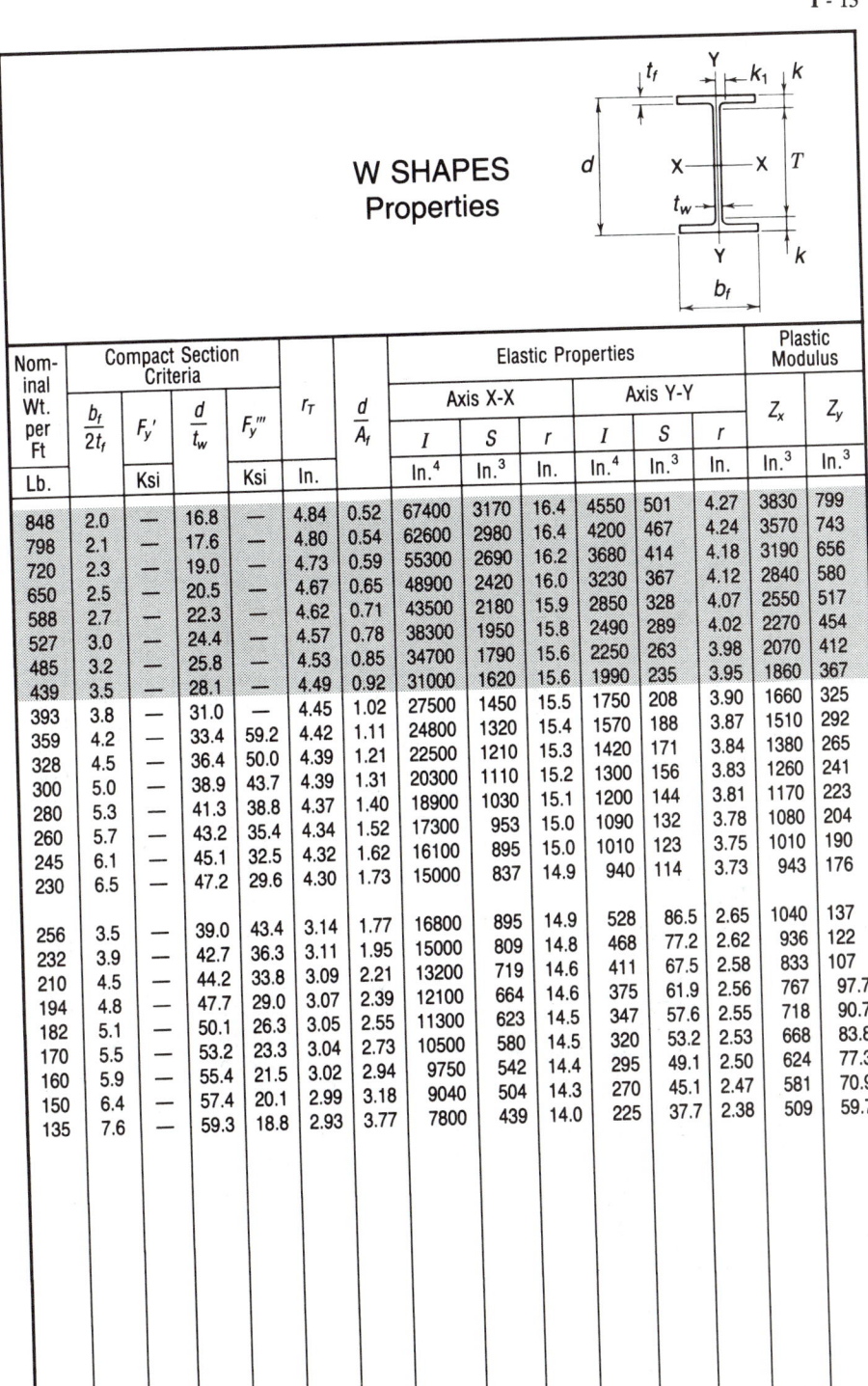

Nom-inal Wt. per Ft	Compact Section Criteria				r_T	d/A_f	Elastic Properties						Plastic Modulus	
	$\frac{b_f}{2t_f}$	F_y'	$\frac{d}{t_w}$	F_y'''			Axis X-X			Axis Y-Y			Z_x	Z_y
							I	S	r	I	S	r		
Lb.		Ksi		Ksi	In.		In.^{4}	In.^{3}	In.	In.^{4}	In.^{3}	In.	In.^{3}	In.^{3}
848	2.0	—	16.8	—	4.84	0.52	67400	3170	16.4	4550	501	4.27	3830	799
798	2.1	—	17.6	—	4.80	0.54	62600	2980	16.4	4200	467	4.24	3570	743
720	2.3	—	19.0	—	4.73	0.59	55300	2690	16.2	3680	414	4.18	3190	656
650	2.5	—	20.5	—	4.67	0.65	48900	2420	16.0	3230	367	4.12	2840	580
588	2.7	—	22.3	—	4.62	0.71	43500	2180	15.9	2850	328	4.07	2550	517
527	3.0	—	24.4	—	4.57	0.78	38300	1950	15.8	2490	289	4.02	2270	454
485	3.2	—	25.8	—	4.53	0.85	34700	1790	15.6	2250	263	3.98	2070	412
439	3.5	—	28.1	—	4.49	0.92	31000	1620	15.6	1990	235	3.95	1860	367
393	3.8	—	31.0	—	4.45	1.02	27500	1450	15.5	1750	208	3.90	1660	325
359	4.2	—	33.4	59.2	4.42	1.11	24800	1320	15.4	1570	188	3.87	1510	292
328	4.5	—	36.4	50.0	4.39	1.21	22500	1210	15.3	1420	171	3.84	1380	265
300	5.0	—	38.9	43.7	4.39	1.31	20300	1110	15.2	1300	156	3.83	1260	241
280	5.3	—	41.3	38.8	4.37	1.40	18900	1030	15.1	1200	144	3.81	1170	223
260	5.7	—	43.2	35.4	4.34	1.52	17300	953	15.0	1090	132	3.78	1080	204
245	6.1	—	45.1	32.5	4.32	1.62	16100	895	15.0	1010	123	3.75	1010	190
230	6.5	—	47.2	29.6	4.30	1.73	15000	837	14.9	940	114	3.73	943	176
256	3.5	—	39.0	43.4	3.14	1.77	16800	895	14.9	528	86.5	2.65	1040	137
232	3.9	—	42.7	36.3	3.11	1.95	15000	809	14.8	468	77.2	2.62	936	122
210	4.5	—	44.2	33.8	3.09	2.21	13200	719	14.6	411	67.5	2.58	833	107
194	4.8	—	47.7	29.0	3.07	2.39	12100	664	14.6	375	61.9	2.56	767	97.7
182	5.1	—	50.1	26.3	3.05	2.55	11300	623	14.5	347	57.6	2.55	718	90.7
170	5.5	—	53.2	23.3	3.04	2.73	10500	580	14.5	320	53.2	2.53	668	83.8
160	5.9	—	55.4	21.5	3.02	2.94	9750	542	14.4	295	49.1	2.50	624	77.3
150	6.4	—	57.4	20.1	2.99	3.18	9040	504	14.3	270	45.1	2.47	581	70.9
135	7.6	—	59.3	18.8	2.93	3.77	7800	439	14.0	225	37.7	2.38	509	59.7

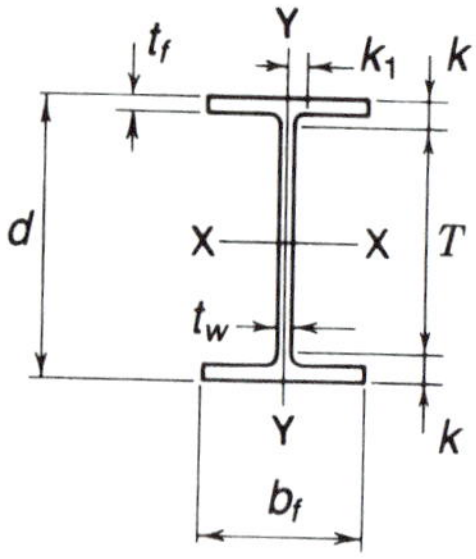

W SHAPES
Dimensions

Designation	Area A	Depth d		Web Thickness t_w		$\frac{t_w}{2}$	Flange Width b_f		Flange Thickness t_f		Distance T	Distance k	Distance k_1
	In.²	In.		In.		In.	In.		In.		In.	In.	In.
W 33×619[a]	181.0	38.47	38½	1.970	2	1	16.910	16⅞	3.540	3 9/16	29¾	4⅜	1¾
×567[a]	166.0	37.91	37⅞	1.810	1 13/16	15/16	16.750	16¾	3.270	3¼	29¾	4 1/16	1 11/16
×515[a]	151.0	37.36	37⅜	1.650	1⅝	13/16	16.590	16⅝	2.990	3	29¾	3 13/16	1⅝
×468[a]	137.0	36.81	36¾	1.520	1½	¾	16.455	16½	2.720	2¾	29¾	3½	1 9/16
×424[a]	124.0	36.34	36⅜	1.380	1⅜	11/16	16.315	16⅜	2.480	2½	29¾	3 5/16	1 7/16
×387[a]	113.0	35.95	36	1.260	1¼	⅝	16.200	16¼	2.280	2¼	29¾	3⅛	1⅜
×354[a]	104.0	35.55	35½	1.160	1 3/16	⅝	16.100	16⅛	2.090	2 1/16	29¾	2⅞	1⅜
×318[a]	93.5	35.16	35⅛	1.040	1 1/16	9/16	15.985	16	1.890	1⅞	29¾	2 11/16	1 5/16
×291	85.6	34.84	34⅞	0.960	1	½	15.905	15⅞	1.730	1¾	29¾	2 9/16	1¼
×263	77.4	34.53	34½	0.870	⅞	7/16	15.805	15¾	1.570	1 9/16	29¾	2⅜	1 3/16
×241	70.9	34.18	34⅛	0.830	13/16	7/16	15.860	15⅞	1.400	1⅜	29¾	2 3/16	1 3/16
×221	65.0	33.93	33⅞	0.775	¾	⅜	15.805	15¾	1.275	1¼	29¾	2 1/16	1 3/16
×201	59.1	33.68	33⅝	0.715	11/16	⅜	15.745	15¾	1.150	1⅛	29¾	1 15/16	1⅛
W 33×169[b]	49.5	33.82	33⅞	0.670	11/16	⅜	11.500	11½	1.220	1¼	29¾	2 1/16	1⅛
×152	44.7	33.49	33½	0.635	⅝	5/16	11.565	11⅝	1.055	1 1/16	29¾	1⅞	1⅛
×141	41.6	33.30	33¼	0.605	⅝	5/16	11.535	11½	0.960	15/16	29¾	1¾	1 1/16
×130	38.3	33.09	33⅛	0.580	9/16	5/16	11.510	11½	0.855	⅞	29¾	1 11/16	1 1/16
×118	34.7	32.86	32⅞	0.550	9/16	5/16	11.480	11½	0.740	¾	29¾	1 9/16	1 1/16
W 30×581[a]	170.0	35.39	35⅜	1.970	2	1	16.200	16¼	3.540	3 9/16	26¾	4 5/16	1 11/16
×526[a]	154.0	34.76	34¾	1.790	1 13/16	1	16.020	16	3.230	3¼	26¾	4	1⅝
×477[a]	140.0	34.21	34¼	1.630	1⅝	13/16	15.865	15⅞	2.950	3	26¾	3¾	1 9/16
×433[a]	127.0	33.66	33⅝	1.500	1½	¾	15.725	15¾	2.680	2 11/16	26¾	3 7/16	1½
×391[a]	114.0	33.19	33¼	1.360	1⅜	11/16	15.590	15⅝	2.440	2 7/16	26¾	3¼	1 7/16
×357[a]	104.0	32.80	32¾	1.240	1¼	⅝	15.470	15½	2.240	2¼	26¾	3	1⅜
×326[a]	95.7	32.40	32⅜	1.140	1⅛	9/16	15.370	15⅜	2.050	2 1/16	26¾	2 13/16	1 5/16
×292[a]	85.7	32.01	32	1.020	1	½	15.255	15¼	1.850	1⅞	26¾	2⅝	1¼
×261	76.7	31.61	31⅝	0.930	15/16	½	15.155	15⅛	1.650	1⅝	26¾	2 7/16	1 3/16
×235	69.0	31.30	31¼	0.830	13/16	7/16	15.055	15	1.500	1½	26¾	2¼	1⅛
×211	62.0	30.94	31	0.775	¾	⅜	15.105	15⅛	1.315	1 5/16	26¾	2⅛	1⅛
×191	56.1	30.68	30⅝	0.710	11/16	⅜	15.040	15	1.185	1 3/16	26¾	1 15/16	1 1/16
×173	50.8	30.44	30½	0.655	⅝	5/16	14.985	15	1.065	1 1/16	26¾	1⅞	1 1/16

[a]For application refer to Notes in Table 2.
[b]Heavier shapes in this series are available from some producers.
Shapes in shaded rows are not available from domestic producers.

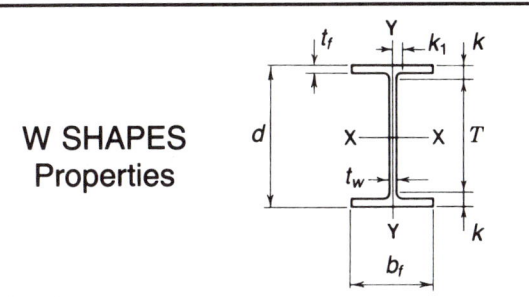

W SHAPES
Properties

Nom-inal Wt. per Ft	Compact Section Criteria				r_T	d/A_f	Elastic Properties						Plastic Modulus	
	$\frac{b_f}{2t_f}$	F_y'	$\frac{d}{t_w}$	F_y'''			Axis X-X			Axis Y-Y			Z_x	Z_y
							I	S	r	I	S	r		
Lb.		Ksi		Ksi	In.		In.^{4}	In.^{3}	In.	In.^{4}	In.^{3}	In.	In.^{3}	In.^{3}
619	2.4	—	19.5	—	4.51	0.64	41800	2170	15.2	2870	340	3.98	2560	537
567	2.6	—	20.9	—	4.46	0.69	37700	1990	15.1	2580	308	3.94	2330	485
515	2.8	—	22.6	—	4.42	0.75	33700	1810	14.9	2290	276	3.89	2110	433
468	3.0	—	24.2	—	4.37	0.82	30100	1630	14.8	2030	247	3.85	1890	387
424	3.3	—	26.3	—	4.33	0.90	26900	1480	14.7	1800	221	3.81	1700	345
387	3.6	—	28.5	—	4.30	0.97	24300	1350	14.7	1620	200	3.79	1550	312
354	3.8	—	30.6	—	4.27	1.06	21900	1230	14.5	1460	181	3.74	1420	282
318	4.2	—	33.8	57.8	4.24	1.16	19500	1110	14.4	1290	161	3.71	1270	250
291	4.6	—	36.3	50.1	4.21	1.27	17700	1010	14.4	1160	146	3.69	1150	226
263	5.0	—	39.7	41.9	4.18	1.39	15800	917	14.3	1030	131	3.66	1040	202
241	5.7	—	41.2	38.9	4.17	1.54	14200	829	14.1	932	118	3.63	939	182
221	6.2	—	43.8	34.5	4.15	1.68	12800	757	14.1	840	106	3.59	855	164
201	6.8	—	47.1	29.8	4.12	1.86	11500	684	14.0	749	95.2	3.56	772	147
169	4.7	—	50.5	25.9	2.95	2.41	9290	549	13.7	310	53.9	2.50	629	84.4
152	5.5	—	52.7	23.7	2.94	2.74	8160	487	13.5	273	47.2	2.47	559	73.9
141	6.0	—	55.0	21.8	2.92	3.01	7450	448	13.4	246	42.7	2.43	514	66.9
130	6.7	—	57.1	20.3	2.88	3.36	6710	406	13.2	218	37.9	2.39	467	59.5
118	7.8	—	59.7	18.5	2.84	3.87	5900	359	13.0	187	32.6	2.32	415	51.3
581	2.3	—	18.0	—	4.34	0.62	33000	1870	13.9	2530	312	3.86	2210	492
526	2.5	—	19.4	—	4.29	0.67	29300	1680	13.8	2230	278	3.80	1990	438
477	2.7	—	21.0	—	4.24	0.73	26100	1530	13.7	1970	249	3.75	1790	390
433	2.9	—	22.4	—	4.20	0.80	23200	1380	13.5	1750	222	3.71	1610	348
391	3.2	—	24.4	—	4.16	0.87	20700	1250	13.5	1550	198	3.68	1430	310
357	3.5	—	26.5	—	4.12	0.95	18600	1140	13.4	1390	179	3.65	1300	279
326	3.7	—	28.4	—	4.09	1.03	16800	1030	13.2	1240	162	3.61	1190	252
292	4.1	—	31.4	—	4.06	1.13	14900	928	13.2	1100	144	3.58	1060	223
261	4.6	—	34.0	57.2	4.02	1.26	13100	827	13.1	959	127	3.54	941	196
235	5.0	—	37.7	46.4	4.00	1.39	11700	746	13.0	855	114	3.52	845	175
211	5.7	—	39.9	41.4	3.99	1.56	10300	663	12.9	757	100	3.49	749	154
191	6.3	—	43.2	35.4	3.97	1.72	9170	598	12.8	673	89.5	3.46	673	138
173	7.0	—	46.5	30.6	3.94	1.91	8200	539	12.7	598	79.8	3.43	605	123

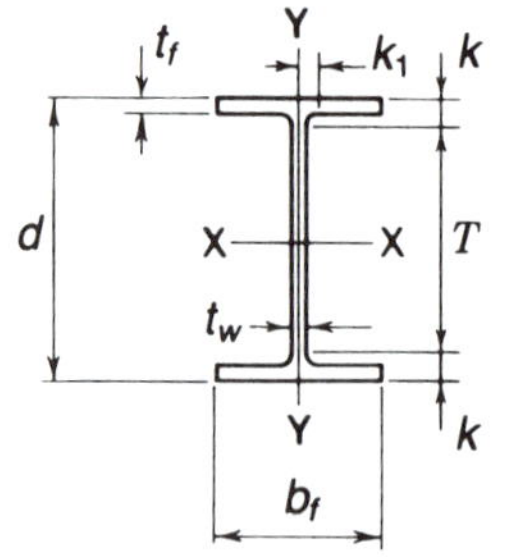

W SHAPES
Dimensions

Designation	Area A	Depth d		Web Thickness t_w		$\frac{t_w}{2}$	Flange Width b_f		Flange Thickness t_f		Distance T	Distance k	Distance k_1
	In.^{2}	In.		In.		In.	In.		In.		In.	In.	In.
W 30×148[b]	43.5	30.67	30 5/8	0.650	5/8	5/16	10.480	10 1/2	1.180	1 3/16	26 3/4	2	1
×132	38.9	30.31	30 1/4	0.615	5/8	5/16	10.545	10 1/2	1.000	1	26 3/4	1 3/4	1 1/16
×124	36.5	30.17	30 1/8	0.585	9/16	5/16	10.515	10 1/2	0.930	15/16	26 3/4	1 11/16	1
×116	34.2	30.01	30	0.565	9/16	5/16	10.495	10 1/2	0.850	7/8	26 3/4	1 5/8	1
×108	31.7	29.83	29 7/8	0.545	9/16	5/16	10.475	10 1/2	0.760	3/4	26 3/4	1 9/16	1
× 99	29.1	29.65	29 5/8	0.520	1/2	1/4	10.450	10 1/2	0.670	11/16	26 3/4	1 7/16	1
× 90	26.4	29.53	29 1/2	0.470	1/2	1/4	10.400	10 3/8	0.610	9/16	26 3/4	1 5/16	1
W 27×539[a]	158.0	32.52	32 1/2	1.970	2	1	15.255	15 1/4	3.540	3 9/16	24	4 1/4	1 5/8
×494[a]	145.0	31.97	32	1.810	1 13/16	15/16	15.095	15 1/8	3.270	3 1/4	24	4	1 9/16
×448[a]	131.0	31.42	31 3/8	1.650	1 5/8	13/16	14.940	15	2.990	3	24	3 11/16	1 1/2
×407[a]	119.0	30.87	30 7/8	1.520	1 1/2	3/4	14.800	14 3/4	2.720	2 3/4	24	3 7/16	1 7/16
×368[a]	108.0	30.39	30 3/8	1.380	1 3/8	11/16	14.665	14 5/8	2.480	2 1/2	24	3 3/16	1 5/16
×336[a]	98.7	30.00	30	1.260	1 1/4	5/8	14.545	14 1/2	2.280	2 1/4	24	3	1 5/16
×307[a]	90.2	29.61	29 5/8	1.160	1 3/16	5/8	14.445	14 1/2	2.090	2 1/16	24	2 13/16	1 1/4
×281[a]	82.6	29.29	29 1/4	1.060	1 1/16	9/16	14.350	14 3/8	1.930	1 15/16	24	2 5/8	1 3/16
×258	75.7	28.98	29	0.980	1	1/2	14.270	14 1/4	1.770	1 3/4	24	2 1/2	1 1/8
×235	69.1	28.66	28 5/8	0.910	15/16	1/2	14.190	14 1/4	1.610	1 5/8	24	2 5/16	1 1/8
×217	63.8	28.43	28 3/8	0.830	13/16	7/16	14.115	14 1/8	1.500	1 1/2	24	2 3/16	1 1/16
×194	57.0	28.11	28 1/8	0.750	3/4	3/8	14.035	14	1.340	1 5/16	24	2 1/16	1
×178	52.3	27.81	27 3/4	0.725	3/4	3/8	14.085	14 1/8	1.190	1 3/16	24	1 7/8	1 1/16
×161	47.4	27.59	27 5/8	0.660	11/16	3/8	14.020	14	1.080	1 1/16	24	1 13/16	1
×146	42.9	27.38	27 3/8	0.605	5/8	5/16	13.965	14	0.975	1	24	1 11/16	1
W 27×129[b]	37.8	27.63	27 5/8	0.610	5/8	5/16	10.010	10	1.100	1 1/8	24	1 13/16	15/16
×114	33.5	27.29	27 1/4	0.570	9/16	5/16	10.070	10 1/8	0.930	15/16	24	1 5/8	15/16
×102	30.0	27.09	27 1/8	0.515	1/2	1/4	10.015	10	0.830	13/16	24	1 9/16	15/16
× 94	27.7	26.92	26 7/8	0.490	1/2	1/4	9.990	10	0.745	3/4	24	1 7/16	15/16
× 84	24.8	26.71	26 3/4	0.460	7/16	1/4	9.960	10	0.640	5/8	24	1 3/8	15/16

[a]For application refer to Notes in Table 2.
[b]Heavier shapes in this series are available from some producers.
Shapes in shaded rows are not available from domestic producers.

W SHAPES
Properties

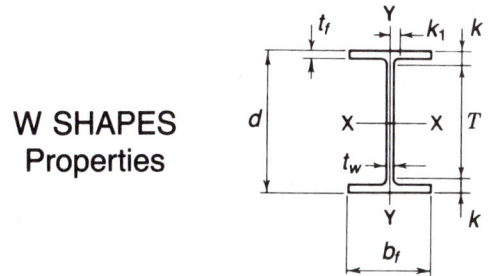

Nom-inal Wt. per Ft	Compact Section Criteria				r_T	d/A_f	Elastic Properties						Plastic Modulus	
	$\frac{b_f}{2t_f}$	F_y'	$\frac{d}{t_w}$	F_y'''			Axis X-X			Axis Y-Y			Z_x	Z_y
							I	S	r	I	S	r		
Lb.		Ksi		Ksi	In.		In.^{4}	In.^{3}	In.	In.^{4}	In.^{3}	In.	In.^{3}	In.^{3}
148	4.4	—	47.2	29.7	2.70	2.48	6680	436	12.4	227	43.3	2.28	500	68.0
132	5.3	—	49.3	27.2	2.68	2.87	5770	380	12.2	196	37.2	2.25	437	58.4
124	5.7	—	51.6	24.8	2.66	3.09	5360	355	12.1	181	34.4	2.23	408	54.0
116	6.2	—	53.1	23.4	2.64	3.36	4930	329	12.0	164	31.3	2.19	378	49.2
108	6.9	—	54.7	22.0	2.61	3.75	4470	299	11.9	146	27.9	2.15	346	43.9
99	7.8	—	57.0	20.3	2.57	4.23	3990	269	11.7	128	24.5	2.10	312	38.6
90	8.5	58.1	62.8	16.7	2.56	4.65	3620	245	11.7	115	22.1	2.09	283	34.7
539	2.2	—	16.5	—	4.10	0.60	25500	1570	12.7	2110	277	3.66	1880	437
494	2.3	—	17.7	—	4.05	0.65	22900	1440	12.6	1890	250	3.61	1710	394
448	2.5	—	19.0	—	4.01	0.70	20400	1300	12.5	1670	224	3.57	1530	351
407	2.7	—	20.3	—	3.96	0.77	18100	1170	12.3	1480	200	3.52	1380	313
368	3.0	—	22.0	—	3.93	0.84	16100	1060	12.2	1310	179	3.48	1240	279
336	3.2	—	23.8	—	3.89	0.90	14500	970	12.1	1170	161	3.45	1130	252
307	3.5	—	25.5	—	3.86	0.98	13100	884	12.0	1050	146	3.42	1020	227
281	3.7	—	27.6	—	3.84	1.06	11900	811	12.0	953	133	3.40	933	206
258	4.0	—	29.6	—	3.81	1.15	10800	742	11.9	859	120	3.37	850	187
235	4.4	—	31.5	—	3.78	1.25	9660	674	11.8	768	108	3.33	769	168
217	4.7	—	34.3	56.3	3.76	1.34	8870	624	11.8	704	99.8	3.32	708	154
194	5.2	—	37.5	47.0	3.74	1.49	7820	556	11.7	618	88.1	3.29	628	136
178	5.9	—	38.4	44.9	3.72	1.66	6990	502	11.6	555	78.8	3.26	567	122
161	6.5	—	41.8	37.8	3.70	1.82	6280	455	11.5	497	70.9	3.24	512	109
146	7.2	—	45.3	32.2	3.68	2.01	5630	411	11.4	443	63.5	3.21	461	97.5
129	4.5	—	45.3	32.2	2.59	2.51	4760	345	11.2	184	36.8	2.21	395	57.6
114	5.4	—	47.9	28.8	2.58	2.91	4090	299	11.0	159	31.5	2.18	343	49.3
102	6.0	—	52.6	23.9	2.56	3.26	3620	267	11.0	139	27.8	2.15	305	43.4
94	6.7	—	54.9	21.9	2.53	3.62	3270	243	10.9	124	24.8	2.12	278	38.8
84	7.8	—	58.1	19.6	2.49	4.19	2850	213	10.7	106	21.2	2.07	244	33.2

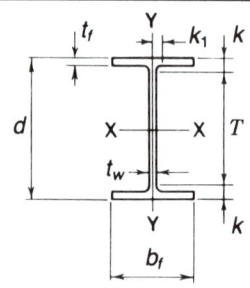

W SHAPES
Dimensions

Desig-nation	Area A	Depth d		Web Thickness t_w		$\dfrac{t_w}{2}$	Flange Width b_f		Flange Thickness t_f		Distance T	k	k_1
	In.²	In.		In.		In.	In.		In.		In.	In.	In.
W 24×492ᵃ	144.0	29.65	29⅝	1.970	2	1	14.115	14⅛	3.540	3⁹⁄₁₆	21	4⁵⁄₁₆	1⁹⁄₁₆
×450ᵃ	132.0	29.09	29⅛	1.810	1¹³⁄₁₆	¹⁵⁄₁₆	13.955	14	3.270	3¼	21	4¹⁄₁₆	1½
×408ᵃ	119.0	28.54	28½	1.650	1⅝	¹³⁄₁₆	13.800	13¾	2.990	3	21	3¾	1⅜
×370ᵃ	108.0	27.99	28	1.520	1½	¾	13.660	13⅝	2.720	2¾	21	3½	1⁵⁄₁₆
×335ᵃ	98.4	27.52	27½	1.380	1⅜	¹¹⁄₁₆	13.520	13½	2.480	2½	21	3¼	1¼
×306ᵃ	89.8	27.13	27⅛	1.260	1¼	⅝	13.405	13⅜	2.280	2¼	21	3¹⁄₁₆	1³⁄₁₆
×279ᵃ	82.0	26.73	26¾	1.160	1³⁄₁₆	⅝	13.305	13¼	2.090	2¹⁄₁₆	21	2⅞	1⅛
×250ᵃ	73.5	26.34	26⅜	1.040	1¹⁄₁₆	⁹⁄₁₆	13.185	13⅛	1.890	1⅞	21	2¹¹⁄₁₆	1⅛
×229	67.2	26.02	26	0.960	1	½	13.110	13⅛	1.730	1¾	21	2½	1
×207	60.7	25.71	25¾	0.870	⅞	⁷⁄₁₆	13.010	13	1.570	1⁹⁄₁₆	21	2⅜	1
×192	56.3	25.47	25½	0.810	¹³⁄₁₆	⁷⁄₁₆	12.950	13	1.460	1⁷⁄₁₆	21	2¼	1
×176	51.7	25.24	25¼	0.750	¾	⅜	12.890	12⅞	1.340	1⁵⁄₁₆	21	2⅛	¹⁵⁄₁₆
×162	47.7	25.00	25	0.705	¹¹⁄₁₆	⅜	12.955	13	1.220	1¼	21	2	1¹⁄₁₆
×146	43.0	24.74	24¾	0.650	⅝	⁵⁄₁₆	12.900	12⅞	1.090	1¹⁄₁₆	21	1⅞	1¹⁄₁₆
×131	38.5	24.48	24½	0.605	⅝	⁵⁄₁₆	12.855	12⅞	0.960	¹⁵⁄₁₆	21	1¾	1¹⁄₁₆
×117	34.4	24.26	24¼	0.550	⁹⁄₁₆	⁵⁄₁₆	12.800	12¾	0.850	⅞	21	1⅝	1
×104	30.6	24.06	24	0.500	½	¼	12.750	12¾	0.750	¾	21	1½	1
W 24×103ᵇ	30.3	24.53	24½	0.550	⁹⁄₁₆	⁵⁄₁₆	9.000	9	0.980	1	21	1¾	¹³⁄₁₆
× 94	27.7	24.31	24¼	0.515	½	¼	9.065	9⅛	0.875	⅞	21	1⅝	1
× 84	24.7	24.10	24⅛	0.470	½	¼	9.020	9	0.770	¾	21	1⁹⁄₁₆	¹⁵⁄₁₆
× 76	22.4	23.92	23⅞	0.440	⁷⁄₁₆	¼	8.990	9	0.680	¹¹⁄₁₆	21	1⁷⁄₁₆	¹⁵⁄₁₆
× 68	20.1	23.73	23¾	0.415	⁷⁄₁₆	¼	8.965	9	0.585	⁹⁄₁₆	21	1⅜	¹⁵⁄₁₆
W 24× 62	18.2	23.74	23¾	0.430	⁷⁄₁₆	¼	7.040	7	0.590	⁹⁄₁₆	21	1⅜	¹⁵⁄₁₆
× 55	16.2	23.57	23⅝	0.395	⅜	³⁄₁₆	7.005	7	0.505	½	21	1⁵⁄₁₆	¹⁵⁄₁₆

ᵃFor application refer to Notes in Table 2.
ᵇHeavier shapes in this series are available from some producers.

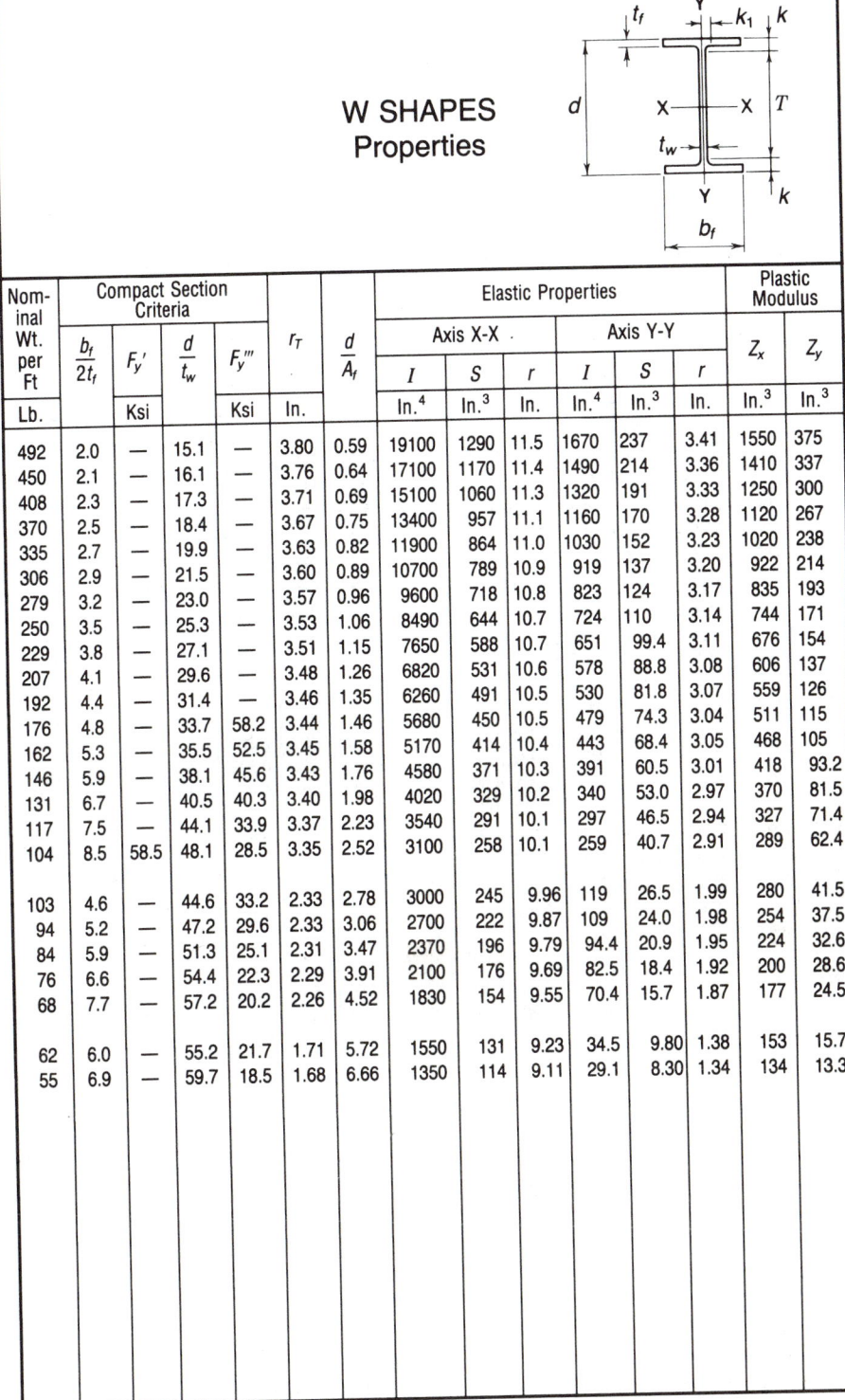

W SHAPES
Properties

Nom-inal Wt. per Ft	Compact Section Criteria				r_T	$\dfrac{d}{A_f}$	Elastic Properties						Plastic Modulus	
	$\dfrac{b_f}{2t_f}$	F_y'	$\dfrac{d}{t_w}$	F_y'''			Axis X-X			Axis Y-Y			Z_x	Z_y
							I	S	r	I	S	r		
Lb.		Ksi		Ksi	In.		In.^{4}	In.^{3}	In.	In.^{4}	In.^{3}	In.	In.^{3}	In.^{3}
492	2.0	—	15.1	—	3.80	0.59	19100	1290	11.5	1670	237	3.41	1550	375
450	2.1	—	16.1	—	3.76	0.64	17100	1170	11.4	1490	214	3.36	1410	337
408	2.3	—	17.3	—	3.71	0.69	15100	1060	11.3	1320	191	3.33	1250	300
370	2.5	—	18.4	—	3.67	0.75	13400	957	11.1	1160	170	3.28	1120	267
335	2.7	—	19.9	—	3.63	0.82	11900	864	11.0	1030	152	3.23	1020	238
306	2.9	—	21.5	—	3.60	0.89	10700	789	10.9	919	137	3.20	922	214
279	3.2	—	23.0	—	3.57	0.96	9600	718	10.8	823	124	3.17	835	193
250	3.5	—	25.3	—	3.53	1.06	8490	644	10.7	724	110	3.14	744	171
229	3.8	—	27.1	—	3.51	1.15	7650	588	10.7	651	99.4	3.11	676	154
207	4.1	—	29.6	—	3.48	1.26	6820	531	10.6	578	88.8	3.08	606	137
192	4.4	—	31.4	—	3.46	1.35	6260	491	10.5	530	81.8	3.07	559	126
176	4.8	—	33.7	58.2	3.44	1.46	5680	450	10.5	479	74.3	3.04	511	115
162	5.3	—	35.5	52.5	3.45	1.58	5170	414	10.4	443	68.4	3.05	468	105
146	5.9	—	38.1	45.6	3.43	1.76	4580	371	10.3	391	60.5	3.01	418	93.2
131	6.7	—	40.5	40.3	3.40	1.98	4020	329	10.2	340	53.0	2.97	370	81.5
117	7.5	—	44.1	33.9	3.37	2.23	3540	291	10.1	297	46.5	2.94	327	71.4
104	8.5	58.5	48.1	28.5	3.35	2.52	3100	258	10.1	259	40.7	2.91	289	62.4
103	4.6	—	44.6	33.2	2.33	2.78	3000	245	9.96	119	26.5	1.99	280	41.5
94	5.2	—	47.2	29.6	2.33	3.06	2700	222	9.87	109	24.0	1.98	254	37.5
84	5.9	—	51.3	25.1	2.31	3.47	2370	196	9.79	94.4	20.9	1.95	224	32.6
76	6.6	—	54.4	22.3	2.29	3.91	2100	176	9.69	82.5	18.4	1.92	200	28.6
68	7.7	—	57.2	20.2	2.26	4.52	1830	154	9.55	70.4	15.7	1.87	177	24.5
62	6.0	—	55.2	21.7	1.71	5.72	1550	131	9.23	34.5	9.80	1.38	153	15.7
55	6.9	—	59.7	18.5	1.68	6.66	1350	114	9.11	29.1	8.30	1.34	134	13.3

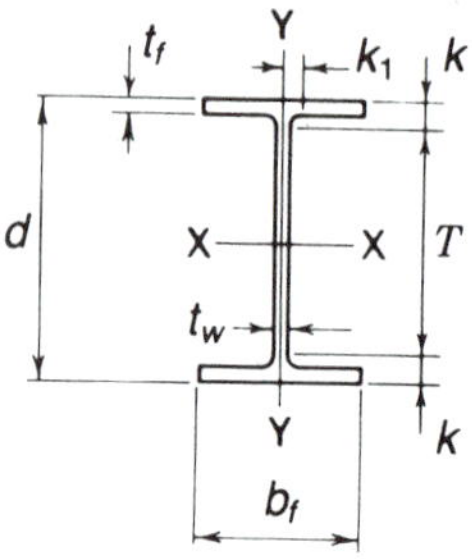

W SHAPES
Dimensions

Designation	Area A	Depth d		Web Thickness t_w		Web $\frac{t_w}{2}$	Flange Width b_f		Flange Thickness t_f		Distance T	Distance k	Distance k_1
	In.^{2}	In.		In.		In.	In.		In.		In.	In.	In.
W 21×402[a]	118.0	26.02	26	1.730	1 3/4	7/8	13.405	13 3/8	3.130	3 1/8	18 1/4	3 7/8	1 7/16
×364[a]	107.0	25.47	25 1/2	1.590	1 9/16	13/16	13.265	13 1/4	2.850	2 7/8	18 1/4	3 5/8	1 3/8
×333[a]	97.9	25.00	25	1.460	1 7/16	3/4	13.130	13 1/8	2.620	2 5/8	18 1/4	3 3/8	1 5/16
×300[a]	88.2	24.53	24 1/2	1.320	1 5/16	11/16	12.990	13	2.380	2 3/8	18 1/4	3 1/8	1 1/4
×275[a]	80.8	24.13	24 1/8	1.220	1 1/4	5/8	12.890	12 7/8	2.190	2 3/16	18 1/4	3	1 3/16
×248[a]	72.8	23.74	23 3/4	1.100	1 1/8	9/16	12.775	12 3/4	1.990	2	18 1/4	2 3/4	1 1/8
×223	65.4	23.35	23 3/8	1.000	1	1/2	12.675	12 5/8	1.790	1 13/16	18 1/4	2 9/16	1 1/16
×201	59.2	23.03	23	0.910	15/16	1/2	12.575	12 5/8	1.630	1 5/8	18 1/4	2 3/8	1
×182	53.6	22.72	22 3/4	0.830	13/16	7/16	12.500	12 1/2	1.480	1 1/2	18 1/4	2 1/4	1
×166	48.8	22.48	22 1/2	0.750	3/4	3/8	12.420	12 3/8	1.360	1 3/8	18 1/4	2 1/8	15/16
×147	43.2	22.06	22	0.720	3/4	3/8	12.510	12 1/2	1.150	1 1/8	18 1/4	1 7/8	1 1/16
×132	38.8	21.83	21 7/8	0.650	5/8	5/16	12.440	12 1/2	1.035	1 1/16	18 1/4	1 13/16	1
×122	35.9	21.68	21 5/8	0.600	5/8	5/16	12.390	12 3/8	0.960	15/16	18 1/4	1 11/16	1
×111	32.7	21.51	21 1/2	0.550	9/16	5/16	12.340	12 3/8	0.875	7/8	18 1/4	1 5/8	15/16
×101	29.8	21.36	21 3/8	0.500	1/2	1/4	12.290	12 1/4	0.800	13/16	18 1/4	1 9/16	15/16
W 21× 93	27.3	21.62	21 5/8	0.580	9/16	5/16	8.420	8 3/8	0.930	15/16	18 1/4	1 11/16	1
× 83	24.3	21.43	21 3/8	0.515	1/2	1/4	8.355	8 3/8	0.835	13/16	18 1/4	1 9/16	15/16
× 73	21.5	21.24	21 1/4	0.455	7/16	1/4	8.295	8 1/4	0.740	3/4	18 1/4	1 1/2	15/16
× 68	20.0	21.13	21 1/8	0.430	7/16	1/4	8.270	8 1/4	0.685	11/16	18 1/4	1 7/16	7/8
× 62	18.3	20.99	21	0.400	3/8	3/16	8.240	8 1/4	0.615	5/8	18 1/4	1 3/8	7/8
W 21× 57	16.7	21.06	21	0.405	3/8	3/16	6.555	6 1/2	0.650	5/8	18 1/4	1 3/8	7/8
× 50	14.7	20.83	20 7/8	0.380	3/8	3/16	6.530	6 1/2	0.535	9/16	18 1/4	1 5/16	7/8
× 44	13.0	20.66	20 5/8	0.350	3/8	3/16	6.500	6 1/2	0.450	7/16	18 1/4	1 3/16	7/8

[a]For application refer to Notes in Table 2.
Shapes in shaded rows are not available from domestic producers.

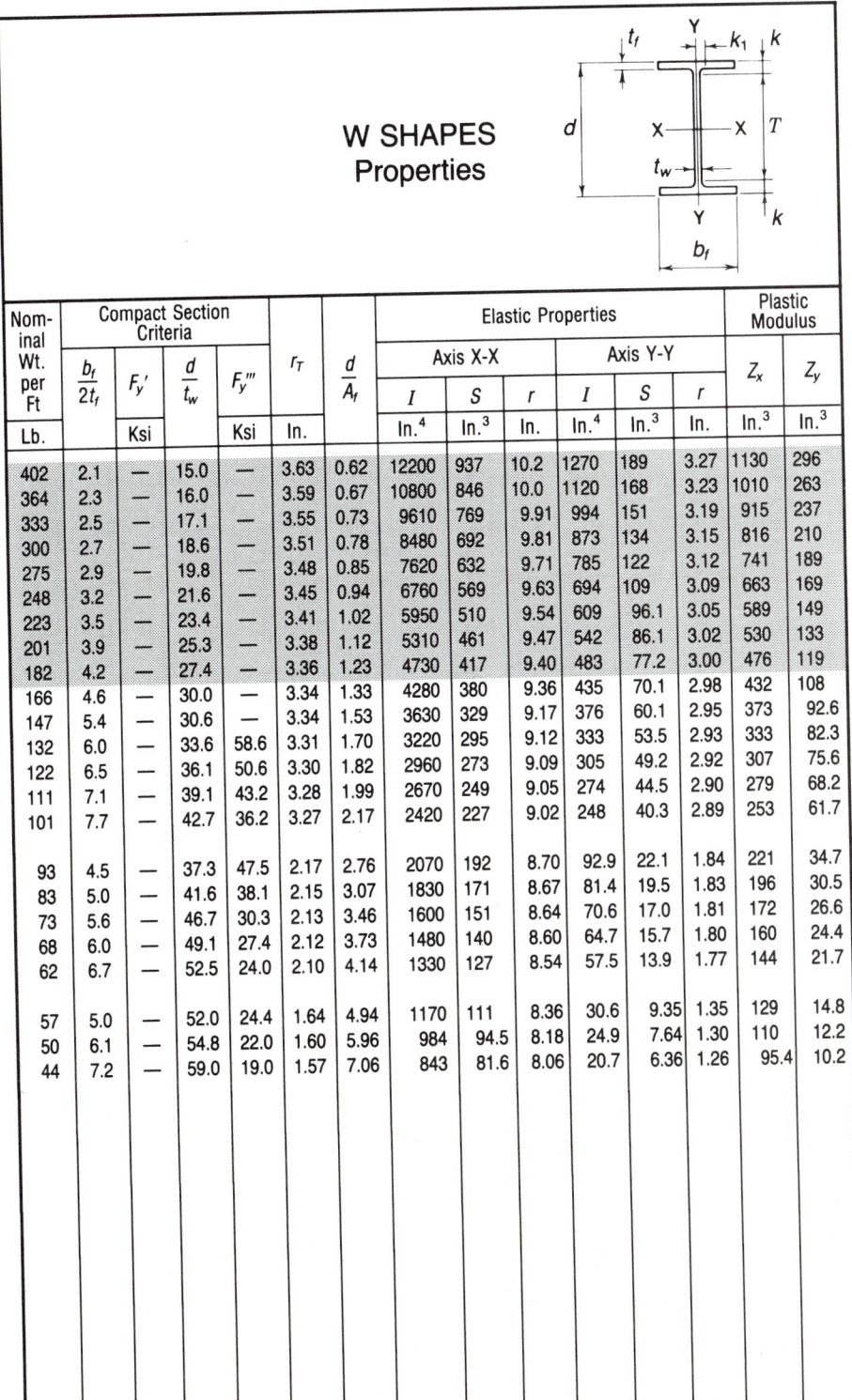

W SHAPES
Properties

Nominal Wt. per Ft	Compact Section Criteria				r_T	d/A_f	Elastic Properties						Plastic Modulus	
							Axis X-X			Axis Y-Y			Z_x	Z_y
	$\dfrac{b_f}{2t_f}$	F_y'	$\dfrac{d}{t_w}$	F_y'''			I	S	r	I	S	r		
Lb.		Ksi		Ksi	In.		In.^{4}	In.^{3}	In.	In.^{4}	In.^{3}	In.	In.^{3}	In.^{3}
402	2.1	—	15.0	—	3.63	0.62	12200	937	10.2	1270	189	3.27	1130	296
364	2.3	—	16.0	—	3.59	0.67	10800	846	10.0	1120	168	3.23	1010	263
333	2.5	—	17.1	—	3.55	0.73	9610	769	9.91	994	151	3.19	915	237
300	2.7	—	18.6	—	3.51	0.78	8480	692	9.81	873	134	3.15	816	210
275	2.9	—	19.8	—	3.48	0.85	7620	632	9.71	785	122	3.12	741	189
248	3.2	—	21.6	—	3.45	0.94	6760	569	9.63	694	109	3.09	663	169
223	3.5	—	23.4	—	3.41	1.02	5950	510	9.54	609	96.1	3.05	589	149
201	3.9	—	25.3	—	3.38	1.12	5310	461	9.47	542	86.1	3.02	530	133
182	4.2	—	27.4	—	3.36	1.23	4730	417	9.40	483	77.2	3.00	476	119
166	4.6	—	30.0	—	3.34	1.33	4280	380	9.36	435	70.1	2.98	432	108
147	5.4	—	30.6	—	3.34	1.53	3630	329	9.17	376	60.1	2.95	373	92.6
132	6.0	—	33.6	58.6	3.31	1.70	3220	295	9.12	333	53.5	2.93	333	82.3
122	6.5	—	36.1	50.6	3.30	1.82	2960	273	9.09	305	49.2	2.92	307	75.6
111	7.1	—	39.1	43.2	3.28	1.99	2670	249	9.05	274	44.5	2.90	279	68.2
101	7.7	—	42.7	36.2	3.27	2.17	2420	227	9.02	248	40.3	2.89	253	61.7
93	4.5	—	37.3	47.5	2.17	2.76	2070	192	8.70	92.9	22.1	1.84	221	34.7
83	5.0	—	41.6	38.1	2.15	3.07	1830	171	8.67	81.4	19.5	1.83	196	30.5
73	5.6	—	46.7	30.3	2.13	3.46	1600	151	8.64	70.6	17.0	1.81	172	26.6
68	6.0	—	49.1	27.4	2.12	3.73	1480	140	8.60	64.7	15.7	1.80	160	24.4
62	6.7	—	52.5	24.0	2.10	4.14	1330	127	8.54	57.5	13.9	1.77	144	21.7
57	5.0	—	52.0	24.4	1.64	4.94	1170	111	8.36	30.6	9.35	1.35	129	14.8
50	6.1	—	54.8	22.0	1.60	5.96	984	94.5	8.18	24.9	7.64	1.30	110	12.2
44	7.2	—	59.0	19.0	1.57	7.06	843	81.6	8.06	20.7	6.36	1.26	95.4	10.2

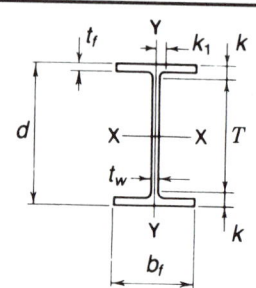

W SHAPES
Dimensions

Desig- nation	Area A	Depth d		Web Thickness t_w		$\frac{t_w}{2}$	Flange Width b_f		Flange Thickness t_f		Distance T	k	k_1
	In.²	In.		In.		In.	In.		In.		In.	In.	In.
W 18×311ᵃ	91.5	22.32	22 3/8	1.520	1 1/2	3/4	12.005	12	2.740	2 3/4	15 1/2	3 7/16	1 3/16
×283ᵃ	83.2	21.85	21 7/8	1.400	1 3/8	11/16	11.890	11 7/8	2.500	2 1/2	15 1/2	3 3/16	1 3/16
×258ᵃ	75.9	21.46	21 1/2	1.280	1 1/4	5/8	11.770	11 3/4	2.300	2 5/16	15 1/2	3	1 1/8
×234ᵃ	68.8	21.06	21	1.160	1 3/16	5/8	11.650	11 5/8	2.110	2 1/8	15 1/2	2 3/4	1
×211ᵃ	62.1	20.67	20 5/8	1.060	1 1/16	9/16	11.555	11 1/2	1.910	1 15/16	15 1/2	2 9/16	1
×192	56.4	20.35	20 3/8	0.960	1	1/2	11.455	11 1/2	1.750	1 3/4	15 1/2	2 7/16	15/16
×175	51.3	20.04	20	0.890	7/8	7/16	11.375	11 3/8	1.590	1 9/16	15 1/2	2 1/4	7/8
×158	46.3	19.72	19 3/4	0.810	13/16	7/16	11.300	11 1/4	1.440	1 7/16	15 1/2	2 1/8	7/8
×143	42.1	19.49	19 1/2	0.730	3/4	3/8	11.220	11 1/4	1.320	1 5/16	15 1/2	2	13/16
×130	38.2	19.25	19 1/4	0.670	11/16	3/8	11.160	11 1/8	1.200	1 3/16	15 1/2	1 7/8	13/16
W 18×119	35.1	18.97	19	0.655	5/8	5/16	11.265	11 1/4	1.060	1 1/16	15 1/2	1 3/4	15/16
×106	31.1	18.73	18 3/4	0.590	9/16	5/16	11.200	11 1/4	0.940	15/16	15 1/2	1 5/8	15/16
× 97	28.5	18.59	18 5/8	0.535	9/16	5/16	11.145	11 1/8	0.870	7/8	15 1/2	1 9/16	7/8
× 86	25.3	18.39	18 3/8	0.480	1/2	1/4	11.090	11 1/8	0.770	3/4	15 1/2	1 7/16	7/8
× 76	22.3	18.21	18 1/4	0.425	7/16	1/4	11.035	11	0.680	11/16	15 1/2	1 3/8	13/16
W 18× 71	20.8	18.47	18 1/2	0.495	1/2	1/4	7.635	7 5/8	0.810	13/16	15 1/2	1 1/2	7/8
× 65	19.1	18.35	18 3/8	0.450	7/16	1/4	7.590	7 5/8	0.750	3/4	15 1/2	1 7/16	7/8
× 60	17.6	18.24	18 1/4	0.415	7/16	1/4	7.555	7 1/2	0.695	11/16	15 1/2	1 3/8	13/16
× 55	16.2	18.11	18 1/8	0.390	3/8	3/16	7.530	7 1/2	0.630	5/8	15 1/2	1 5/16	13/16
× 50	14.7	17.99	18	0.355	3/8	3/16	7.495	7 1/2	0.570	9/16	15 1/2	1 1/4	13/16
W 18× 46	13.5	18.06	18	0.360	3/8	3/16	6.060	6	0.605	5/8	15 1/2	1 1/4	13/16
× 40	11.8	17.90	17 7/8	0.315	5/16	3/16	6.015	6	0.525	1/2	15 1/2	1 3/16	13/16
× 35	10.3	17.70	17 3/4	0.300	5/16	3/16	6.000	6	0.425	7/16	15 1/2	1 1/8	3/4
W 16×100	29.4	16.97	17	0.585	9/16	5/16	10.425	10 3/8	0.985	1	13 5/8	1 11/16	15/16
× 89	26.2	16.75	16 3/4	0.525	1/2	1/4	10.365	10 3/8	0.875	7/8	13 5/8	1 9/16	7/8
× 77	22.6	16.52	16 1/2	0.455	7/16	1/4	10.295	10 1/4	0.760	3/4	13 5/8	1 7/16	7/8
× 67	19.7	16.33	16 3/8	0.395	3/8	3/16	10.235	10 1/4	0.665	11/16	13 5/8	1 3/8	13/16
W 16× 57	16.8	16.43	16 3/8	0.430	7/16	1/4	7.120	7 1/8	0.715	11/16	13 5/8	1 3/8	7/8
× 50	14.7	16.26	16 1/4	0.380	3/8	3/16	7.070	7 1/8	0.630	5/8	13 5/8	1 5/16	13/16
× 45	13.3	16.13	16 1/8	0.345	3/8	3/16	7.035	7	0.565	9/16	13 5/8	1 1/4	13/16
× 40	11.8	16.01	16	0.305	5/16	3/16	6.995	7	0.505	1/2	13 5/8	1 3/16	13/16
× 36	10.6	15.86	15 7/8	0.295	5/16	3/16	6.985	7	0.430	7/16	13 5/8	1 1/8	3/4

ᵃFor application refer to Notes in Table 2.

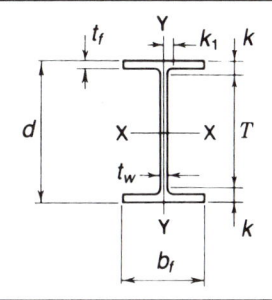

W SHAPES
Properties

Nom- inal Wt. per Ft	Compact Section Criteria				r_T	$\dfrac{d}{A_f}$	Elastic Properties						Plastic Modulus	
							Axis X-X			Axis Y-Y				
	$\dfrac{b_f}{2t_f}$	F_y'	$\dfrac{d}{t_w}$	F_y'''			I	S	r	I	S	r	Z_x	Z_y
Lb.		Ksi		Ksi	In.		In.^{4}	In.^{3}	In.	In.^{4}	In.^{3}	In.	In.^{3}	In.^{3}
311	2.2	—	14.7	—	3.26	0.68	6960	624	8.72	795	132	2.95	753	207
283	2.4	—	15.6	—	3.23	0.74	6160	564	8.61	704	118	2.91	676	185
258	2.6	—	16.8	—	3.19	0.79	5510	514	8.53	628	107	2.88	611	166
234	2.8	—	18.2	—	3.16	0.86	4900	466	8.44	558	95.8	2.85	549	149
211	3.0	—	19.5	—	3.13	0.94	4330	419	8.35	493	85.3	2.82	490	132
192	3.3	—	21.2	—	3.10	1.02	3870	380	8.28	440	76.8	2.79	442	119
175	3.6	—	22.5	—	3.07	1.11	3450	344	8.20	391	68.8	2.76	398	106
158	3.9	—	24.3	—	3.05	1.21	3060	310	8.12	347	61.4	2.74	356	94.8
143	4.2	—	26.7	—	3.03	1.32	2750	282	8.09	311	55.5	2.72	322	85.4
130	4.6	—	28.7	—	3.01	1.44	2460	256	8.03	278	49.9	2.70	291	76.7
119	5.3	—	29.0	—	3.02	1.59	2190	231	7.90	253	44.9	2.69	261	69.1
106	6.0	—	31.7	—	3.00	1.78	1910	204	7.84	220	39.4	2.66	230	60.5
97	6.4	—	34.7	54.7	2.99	1.92	1750	188	7.82	201	36.1	2.65	211	55.3
86	7.2	—	38.3	45.0	2.97	2.15	1530	166	7.77	175	31.6	2.63	186	48.4
76	8.1	64.2	42.8	36.0	2.95	2.43	1330	146	7.73	152	27.6	2.61	163	42.2
71	4.7	—	37.3	47.4	1.98	2.99	1170	127	7.50	60.3	15.8	1.70	145	24.7
65	5.1	—	40.8	39.7	1.97	3.22	1070	117	7.49	54.8	14.4	1.69	133	22.5
60	5.4	—	44.0	34.2	1.96	3.47	984	108	7.47	50.1	13.3	1.69	123	20.6
55	6.0	—	46.4	30.6	1.95	3.82	890	98.3	7.41	44.9	11.9	1.67	112	18.5
50	6.6	—	50.7	25.7	1.94	4.21	800	88.9	7.38	40.1	10.7	1.65	101	16.6
46	5.0	—	50.2	26.2	1.54	4.93	712	78.8	7.25	22.5	7.43	1.29	90.7	11.7
40	5.7	—	56.8	20.5	1.52	5.67	612	68.4	7.21	19.1	6.35	1.27	78.4	9.95
35	7.1	—	59.0	19.0	1.49	6.94	510	57.6	7.04	15.3	5.12	1.22	66.5	8.06
100	5.3	—	29.0	—	2.81	1.65	1490	175	7.10	186	35.7	2.51	198	54.9
89	5.9	—	31.9	64.9	2.79	1.85	1300	155	7.05	163	31.4	2.49	175	48.1
77	6.8	—	36.3	50.1	2.77	2.11	1110	134	7.00	138	26.9	2.47	150	41.1
67	7.7	—	41.3	38.6	2.75	2.40	954	117	6.96	119	23.2	2.46	130	35.5
57	5.0	—	38.2	45.2	1.86	3.23	758	92.2	6.72	43.1	12.1	1.60	105	18.9
50	5.6	—	42.8	36.1	1.84	3.65	659	81.0	6.68	37.2	10.5	1.59	92.0	16.3
45	6.2	—	46.8	30.2	1.83	4.06	586	72.7	6.65	32.8	9.34	1.57	82.3	14.5
40	6.9	—	52.5	24.0	1.82	4.53	518	64.7	6.63	28.9	8.25	1.57	72.9	12.7
36	8.1	64.0	53.8	22.9	1.79	5.28	448	56.5	6.51	24.5	7.00	1.52	64.0	10.8

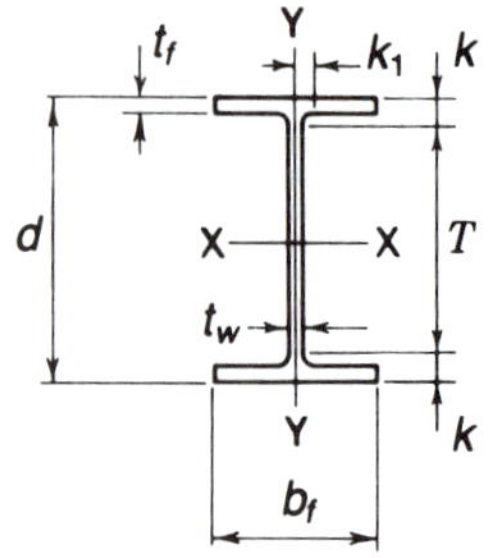

W SHAPES
Dimensions

Designation	Area A	Depth d		Web Thickness t_w		Web $\frac{t_w}{2}$	Flange Width b_f		Flange Thickness t_f		Distance T	Distance k	Distance k_1
	In.^2	In.		In.		In.	In.		In.		In.	In.	In.
W 16× 31	9.12	15.88	15 7/8	0.275	1/4	1/8	5.525	5 1/2	0.440	7/16	13 5/8	1 1/8	3/4
× 26	7.68	15.69	15 3/4	0.250	1/4	1/8	5.500	5 1/2	0.345	3/8	13 5/8	1 1/16	3/4
W 14×730[a]	215.0	22.42	22 3/8	3.070	3 1/16	1 9/16	17.890	17 7/8	4.910	4 15/16	11 1/4	5 9/16	2 3/16
×665[a]	196.0	21.64	21 5/8	2.830	2 13/16	1 7/16	17.650	17 5/8	4.520	4 1/2	11 1/4	5 3/16	2 1/16
×605[a]	178.0	20.92	20 7/8	2.595	2 5/8	1 5/16	17.415	17 3/8	4.160	4 3/16	11 1/4	4 13/16	1 15/16
×550[a]	162.0	20.24	20 1/4	2.380	2 3/8	1 3/16	17.200	17 1/4	3.820	3 13/16	11 1/4	4 1/2	1 13/16
×500[a]	147.0	19.60	19 5/8	2.190	2 3/16	1 1/8	17.010	17	3.500	3 1/2	11 1/4	4 3/16	1 3/4
×455[a]	134.0	19.02	19	2.015	2	1	16.835	16 7/8	3.210	3 3/16	11 1/4	3 7/8	1 5/8
W 14×426[a]	125.0	18.67	18 5/8	1.875	1 7/8	15/16	16.695	16 3/4	3.035	3 1/16	11 1/4	3 11/16	1 9/16
×398[a]	117.0	18.29	18 1/4	1.770	1 3/4	7/8	16.590	16 5/8	2.845	2 7/8	11 1/4	3 1/2	1 1/2
×370[a]	109.0	17.92	17 7/8	1.655	1 5/8	13/16	16.475	16 1/2	2.660	2 11/16	11 1/4	3 5/16	1 7/16
×342[a]	101.0	17.54	17 1/2	1.540	1 9/16	13/16	16.360	16 3/8	2.470	2 1/2	11 1/4	3 1/8	1 3/8
×311[a]	91.4	17.12	17 1/8	1.410	1 7/16	3/4	16.230	16 1/4	2.260	2 1/4	11 1/4	2 15/16	1 5/16
×283[a]	83.3	16.74	16 3/4	1.290	1 5/16	11/16	16.110	16 1/8	2.070	2 1/16	11 1/4	2 3/4	1 1/4
×257[a]	75.6	16.38	16 3/8	1.175	1 3/16	5/8	15.995	16	1.890	1 7/8	11 1/4	2 9/16	1 3/16
×233[a]	68.5	16.04	16	1.070	1 1/16	9/16	15.890	15 7/8	1.720	1 3/4	11 1/4	2 3/8	1 3/16
×211	62.0	15.72	15 3/4	0.980	1	1/2	15.800	15 3/4	1.560	1 9/16	11 1/4	2 1/4	1 1/8
×193	56.8	15.48	15 1/2	0.890	7/8	7/16	15.710	15 3/4	1.440	1 7/16	11 1/4	2 1/8	1 1/16
×176	51.8	15.22	15 1/4	0.830	13/16	7/16	15.650	15 5/8	1.310	1 5/16	11 1/4	2	1 1/16
×159	46.7	14.98	15	0.745	3/4	3/8	15.565	15 5/8	1.190	1 3/16	11 1/4	1 7/8	1
×145	42.7	14.78	14 3/4	0.680	11/16	3/8	15.500	15 1/2	1.090	1 1/16	11 1/4	1 3/4	1

[a]For application refer to Notes in Table 2.

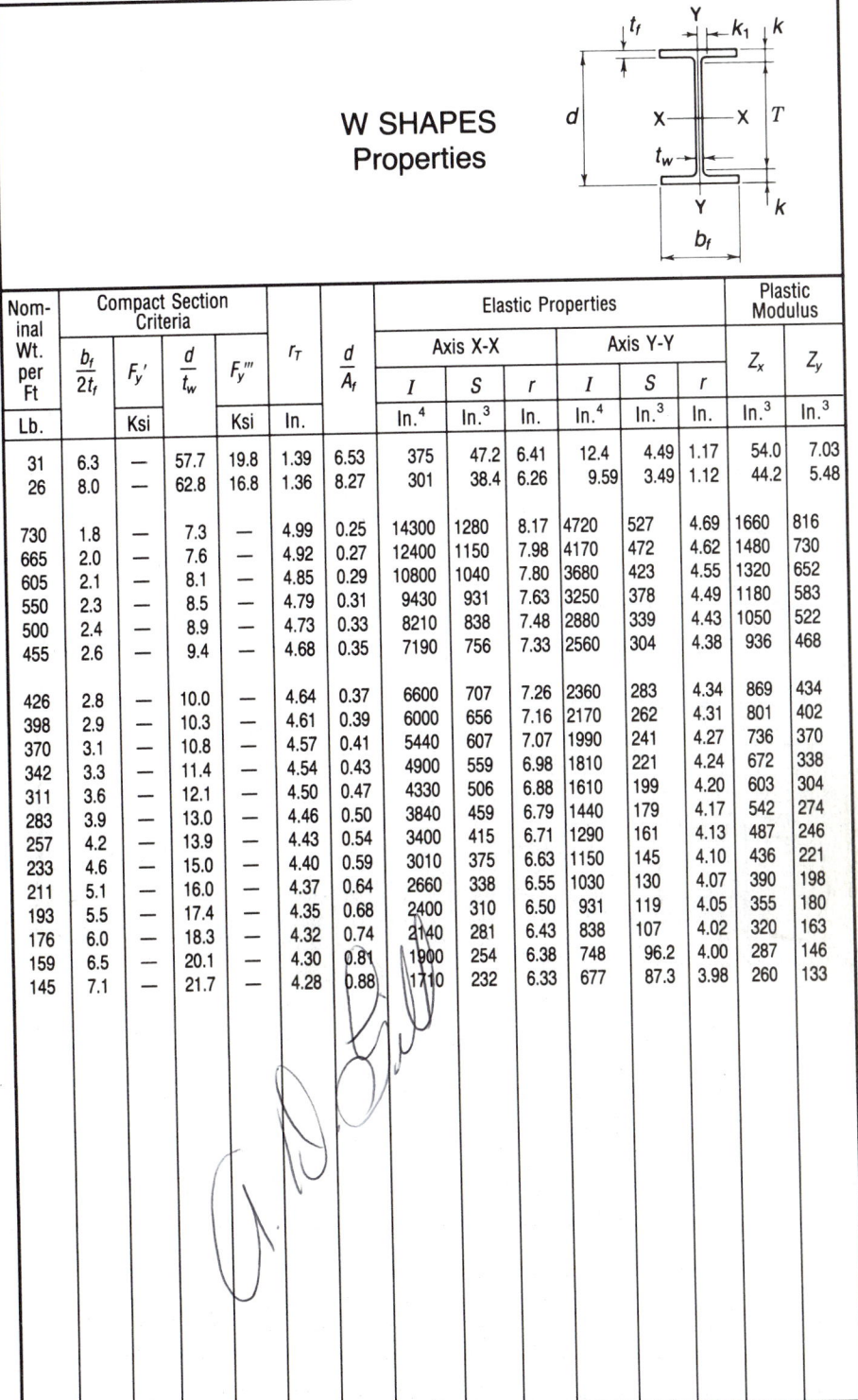

W SHAPES
Properties

Nominal Wt. per Ft	Compact Section Criteria				r_T	d/A_f	Elastic Properties						Plastic Modulus	
	$\frac{b_f}{2t_f}$	F_y'	$\frac{d}{t_w}$	F_y'''			Axis X-X			Axis Y-Y			Z_x	Z_y
							I	S	r	I	S	r		
Lb.		Ksi		Ksi	In.		In.^{4}	In.^{3}	In.	In.^{4}	In.^{3}	In.	In.^{3}	In.^{3}
31	6.3	—	57.7	19.8	1.39	6.53	375	47.2	6.41	12.4	4.49	1.17	54.0	7.03
26	8.0	—	62.8	16.8	1.36	8.27	301	38.4	6.26	9.59	3.49	1.12	44.2	5.48
730	1.8	—	7.3	—	4.99	0.25	14300	1280	8.17	4720	527	4.69	1660	816
665	2.0	—	7.6	—	4.92	0.27	12400	1150	7.98	4170	472	4.62	1480	730
605	2.1	—	8.1	—	4.85	0.29	10800	1040	7.80	3680	423	4.55	1320	652
550	2.3	—	8.5	—	4.79	0.31	9430	931	7.63	3250	378	4.49	1180	583
500	2.4	—	8.9	—	4.73	0.33	8210	838	7.48	2880	339	4.43	1050	522
455	2.6	—	9.4	—	4.68	0.35	7190	756	7.33	2560	304	4.38	936	468
426	2.8	—	10.0	—	4.64	0.37	6600	707	7.26	2360	283	4.34	869	434
398	2.9	—	10.3	—	4.61	0.39	6000	656	7.16	2170	262	4.31	801	402
370	3.1	—	10.8	—	4.57	0.41	5440	607	7.07	1990	241	4.27	736	370
342	3.3	—	11.4	—	4.54	0.43	4900	559	6.98	1810	221	4.24	672	338
311	3.6	—	12.1	—	4.50	0.47	4330	506	6.88	1610	199	4.20	603	304
283	3.9	—	13.0	—	4.46	0.50	3840	459	6.79	1440	179	4.17	542	274
257	4.2	—	13.9	—	4.43	0.54	3400	415	6.71	1290	161	4.13	487	246
233	4.6	—	15.0	—	4.40	0.59	3010	375	6.63	1150	145	4.10	436	221
211	5.1	—	16.0	—	4.37	0.64	2660	338	6.55	1030	130	4.07	390	198
193	5.5	—	17.4	—	4.35	0.68	2400	310	6.50	931	119	4.05	355	180
176	6.0	—	18.3	—	4.32	0.74	2140	281	6.43	838	107	4.02	320	163
159	6.5	—	20.1	—	4.30	0.81	1900	254	6.38	748	96.2	4.00	287	146
145	7.1	—	21.7	—	4.28	0.88	1710	232	6.33	677	87.3	3.98	260	133

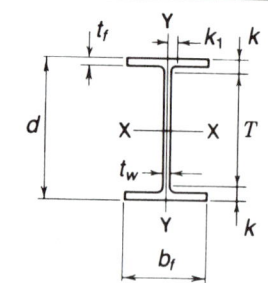

W SHAPES
Dimensions

Desig-nation	Area A	Depth d	Web Thickness t_w		$\frac{t_w}{2}$	Flange Width b_f		Thickness t_f		Distance T	k	k_1
	In.²	In.	In.		In.	In.		In.		In.	In.	In.
W 14×132	38.8	14.66	14⅝	0.645	⅝	5/16	14.725	14¾	1.030	1	11¼	1 11/16	15/16
×120	35.3	14.48	14½	0.590	9/16	5/16	14.670	14⅝	0.940	15/16	11¼	1⅝	15/16
×109	32.0	14.32	14⅜	0.525	½	¼	14.605	14⅝	0.860	⅞	11¼	1 9/16	⅞
× 99	29.1	14.16	14⅛	0.485	½	¼	14.565	14⅝	0.780	¾	11¼	1 7/16	⅞
× 90	26.5	14.02	14	0.440	7/16	¼	14.520	14½	0.710	11/16	11¼	1⅜	⅞
W 14× 82	24.1	14.31	14¼	0.510	½	¼	10.130	10⅛	0.855	⅞	11	1⅝	1
× 74	21.8	14.17	14⅛	0.450	7/16	¼	10.070	10⅛	0.785	13/16	11	1 9/16	15/16
× 68	20.0	14.04	14	0.415	7/16	¼	10.035	10	0.720	¾	11	1½	15/16
× 61	17.9	13.89	13⅞	0.375	⅜	3/16	9.995	10	0.645	⅝	11	1 7/16	15/16
W 14× 53	15.6	13.92	13⅞	0.370	⅜	3/16	8.060	8	0.660	11/16	11	1 7/16	15/16
× 48	14.1	13.79	13¾	0.340	5/16	3/16	8.030	8	0.595	⅝	11	1⅜	⅞
× 43	12.6	13.66	13⅝	0.305	5/16	3/16	7.995	8	0.530	½	11	1 5/16	⅞
W 14× 38	11.2	14.10	14⅛	0.310	5/16	3/16	6.770	6¾	0.515	½	12	1 1/16	⅝
× 34	10.0	13.98	14	0.285	5/16	3/16	6.745	6¾	0.455	7/16	12	1	⅝
× 30	8.85	13.84	13⅞	0.270	¼	⅛	6.730	6¾	0.385	⅜	12	15/16	⅝
W 14× 26	7.69	13.91	13⅞	0.255	¼	⅛	5.025	5	0.420	7/16	12	15/16	9/16
× 22	6.49	13.74	13¾	0.230	¼	⅛	5.000	5	0.335	5/16	12	⅞	9/16

W SHAPES
Properties

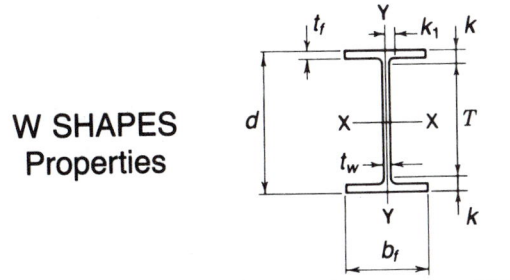

Nom-inal Wt. per Ft	Compact Section Criteria				r_T	$\dfrac{d}{A_f}$	Elastic Properties						Plastic Modulus	
	$\dfrac{b_f}{2t_f}$	F_y'	$\dfrac{d}{t_w}$	F_y'''			Axis X-X			Axis Y-Y			Z_x	Z_y
							I	S	r	I	S	r		
Lb.		Ksi		Ksi	In.		In.^{4}	In.^{3}	In.	In.^{4}	In.^{3}	In.	In.^{3}	In.^{3}
132	7.1	—	22.7	—	4.05	0.97	1530	209	6.28	548	74.5	3.76	234	113
120	7.8	—	24.5	—	4.04	1.05	1380	190	6.24	495	67.5	3.74	212	102
109	8.5	58.6	27.3	—	4.02	1.14	1240	173	6.22	447	61.2	3.73	192	92.7
99	9.3	48.5	29.2	—	4.00	1.25	1110	157	6.17	402	55.2	3.71	173	83.6
90	10.2	40.4	31.9	—	3.99	1.36	999	143	6.14	362	49.9	3.70	157	75.6
82	5.9	—	28.1	—	2.74	1.65	882	123	6.05	148	29.3	2.48	139	44.8
74	6.4	—	31.5	—	2.72	1.79	796	112	6.04	134	26.6	2.48	126	40.6
68	7.0	—	33.8	57.7	2.71	1.94	723	103	6.01	121	24.2	2.46	115	36.9
61	7.7	—	37.0	48.1	2.70	2.15	640	92.2	5.98	107	21.5	2.45	102	32.8
53	6.1	—	37.6	46.7	2.15	2.62	541	77.8	5.89	57.7	14.3	1.92	87.1	22.0
48	6.7	—	40.6	40.2	2.13	2.89	485	70.3	5.85	51.4	12.8	1.91	78.4	19.6
43	7.5	—	44.8	32.9	2.12	3.22	428	62.7	5.82	45.2	11.3	1.89	69.6	17.3
38	6.6	—	45.5	31.9	1.77	4.04	385	54.6	5.87	26.7	7.88	1.55	61.5	12.1
34	7.4	—	49.1	27.4	1.76	4.56	340	48.6	5.83	23.3	6.91	1.53	54.6	10.6
30	8.7	55.3	51.3	25.1	1.74	5.34	291	42.0	5.73	19.6	5.82	1.49	47.3	8.99
26	6.0	—	54.5	22.2	1.28	6.59	245	35.3	5.65	8.91	3.54	1.08	40.2	5.54
22	7.5	—	59.7	18.5	1.25	8.20	199	29.0	5.54	7.00	2.80	1.04	33.2	4.39

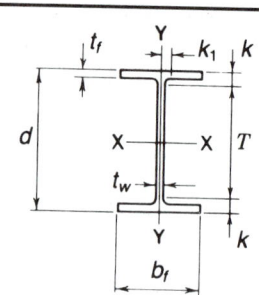

W SHAPES
Dimensions

Desig-nation	Area A	Depth d		Web Thickness t_w		$\dfrac{t_w}{2}$	Flange Width b_f		Flange Thickness t_f		Distance T	k	k_1
	In.²	In.		In.		In.	In.		In.		In.	In.	In.
W 12×336[a]	98.8	16.82	16⅞	1.775	1¾	⅞	13.385	13⅜	2.955	2¹⁵⁄₁₆	9½	3¹¹⁄₁₆	1½
×305[a]	89.6	16.32	16⅜	1.625	1⅝	¹³⁄₁₆	13.235	13¼	2.705	2¹¹⁄₁₆	9½	3⁷⁄₁₆	1⁷⁄₁₆
×279[a]	81.9	15.85	15⅞	1.530	1½	¾	13.140	13⅛	2.470	2½	9½	3³⁄₁₆	1⅜
×252[a]	74.1	15.41	15⅜	1.395	1⅜	¹¹⁄₁₆	13.005	13	2.250	2¼	9½	2¹⁵⁄₁₆	1⁵⁄₁₆
×230[a]	67.7	15.05	15	1.285	1⁵⁄₁₆	¹¹⁄₁₆	12.895	12⅞	2.070	2¹⁄₁₆	9½	2¾	1¼
×210[a]	61.8	14.71	14¾	1.180	1³⁄₁₆	⅝	12.790	12¾	1.900	1⅞	9½	2⅝	1¼
×190	55.8	14.38	14⅜	1.060	1¹⁄₁₆	⁹⁄₁₆	12.670	12⅝	1.735	1¾	9½	2⁷⁄₁₆	1³⁄₁₆
×170	50.0	14.03	14	0.960	¹⁵⁄₁₆	½	12.570	12⅝	1.560	1⁹⁄₁₆	9½	2¼	1⅛
×152	44.7	13.71	13¾	0.870	⅞	⁷⁄₁₆	12.480	12½	1.400	1⅜	9½	2⅛	1¹⁄₁₆
×136	39.9	13.41	13⅜	0.790	¹³⁄₁₆	⁷⁄₁₆	12.400	12⅜	1.250	1¼	9½	1¹⁵⁄₁₆	1
×120	35.3	13.12	13⅛	0.710	¹¹⁄₁₆	⅜	12.320	12⅜	1.105	1⅛	9½	1¹³⁄₁₆	1
×106	31.2	12.89	12⅞	0.610	⅝	⁵⁄₁₆	12.220	12¼	0.990	1	9½	1¹¹⁄₁₆	¹⁵⁄₁₆
× 96	28.2	12.71	12¾	0.550	⁹⁄₁₆	⁵⁄₁₆	12.160	12⅛	0.900	⅞	9½	1⅝	⅞
× 87	25.6	12.53	12½	0.515	½	¼	12.125	12⅛	0.810	¹³⁄₁₆	9½	1½	⅞
× 79	23.2	12.38	12⅜	0.470	½	¼	12.080	12⅛	0.735	¾	9½	1⁷⁄₁₆	⅞
× 72	21.1	12.25	12¼	0.430	⁷⁄₁₆	¼	12.040	12	0.670	¹¹⁄₁₆	9½	1⅜	⅞
× 65	19.1	12.12	12⅛	0.390	⅜	³⁄₁₆	12.000	12	0.605	⅝	9½	1⁵⁄₁₆	¹³⁄₁₆
W 12× 58	17.0	12.19	12¼	0.360	⅜	³⁄₁₆	10.010	10	0.640	⅝	9½	1⅜	¹³⁄₁₆
× 53	15.6	12.06	12	0.345	⅜	³⁄₁₆	9.995	10	0.575	⁹⁄₁₆	9½	1¼	¹³⁄₁₆
W 12× 50	14.7	12.19	12¼	0.370	⅜	³⁄₁₆	8.080	8⅛	0.640	⅝	9½	1⅜	¹³⁄₁₆
× 45	13.2	12.06	12	0.335	⁵⁄₁₆	³⁄₁₆	8.045	8	0.575	⁹⁄₁₆	9½	1¼	¹³⁄₁₆
× 40	11.8	11.94	12	0.295	⁵⁄₁₆	³⁄₁₆	8.005	8	0.515	½	9½	1¼	¾
W 12× 35	10.3	12.50	12½	0.300	⁵⁄₁₆	³⁄₁₆	6.560	6½	0.520	½	10½	1	⁹⁄₁₆
× 30	8.79	12.34	12⅜	0.260	¼	⅛	6.520	6½	0.440	⁷⁄₁₆	10½	¹⁵⁄₁₆	½
× 26	7.65	12.22	12¼	0.230	¼	⅛	6.490	6½	0.380	⅜	10½	⅞	½
W 12× 22	6.48	12.31	12¼	0.260	¼	⅛	4.030	4	0.425	⁷⁄₁₆	10½	⅞	½
× 19	5.57	12.16	12⅛	0.235	¼	⅛	4.005	4	0.350	⅜	10½	¹³⁄₁₆	½
× 16	4.71	11.99	12	0.220	¼	⅛	3.990	4	0.265	¼	10½	¾	½
× 14	4.16	11.91	11⅞	0.200	³⁄₁₆	⅛	3.970	4	0.225	¼	10½	¹¹⁄₁₆	½

[a]For application refer to Notes in Table 2.

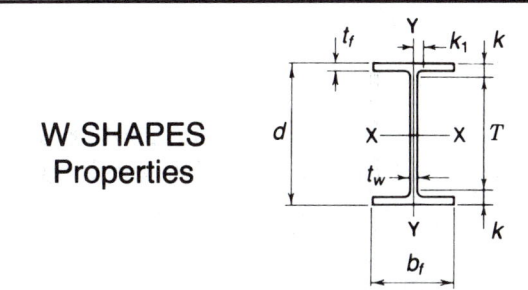

W SHAPES
Properties

Nominal Wt. per Ft	Compact Section Criteria				r_T	$\dfrac{d}{A_f}$	Elastic Properties						Plastic Modulus	
	$\dfrac{b_f}{2t_f}$	F_y'	$\dfrac{d}{t_w}$	F_y'''			Axis X-X			Axis Y-Y			Z_x	Z_y
							I	S	r	I	S	r		
Lb.		Ksi		Ksi	In.		In.^{4}	In.^{3}	In.	In.^{4}	In.^{3}	In.	In.^{3}	In.^{3}
336	2.3	—	9.5	—	3.71	0.43	4060	483	6.41	1190	177	3.47	603	274
305	2.4	—	10.0	—	3.67	0.46	3550	435	6.29	1050	159	3.42	537	244
279	2.7	—	10.4	—	3.64	0.49	3110	393	6.16	937	143	3.38	481	220
252	2.9	—	11.0	—	3.59	0.53	2720	353	6.06	828	127	3.34	428	196
230	3.1	—	11.7	—	3.56	0.56	2420	321	5.97	742	115	3.31	386	177
210	3.4	—	12.5	—	3.53	0.61	2140	292	5.89	664	104	3.28	348	159
190	3.7	—	13.6	—	3.50	0.65	1890	263	5.82	589	93.0	3.25	311	143
170	4.0	—	14.6	—	3.47	0.72	1650	235	5.74	517	82.3	3.22	275	126
152	4.5	—	15.8	—	3.44	0.79	1430	209	5.66	454	72.8	3.19	243	111
136	5.0	—	17.0	—	3.41	0.87	1240	186	5.58	398	64.2	3.16	214	98.0
120	5.6	—	18.5	—	3.38	0.96	1070	163	5.51	345	56.0	3.13	186	85.4
106	6.2	—	21.1	—	3.36	1.07	933	145	5.47	301	49.3	3.11	164	75.1
96	6.8	—	23.1	—	3.34	1.16	833	131	5.44	270	44.4	3.09	147	67.5
87	7.5	—	24.3	—	3.32	1.28	740	118	5.38	241	39.7	3.07	132	60.4
79	8.2	62.6	26.3	—	3.31	1.39	662	107	5.34	216	35.8	3.05	119	54.3
72	9.0	52.3	28.5	—	3.29	1.52	597	97.4	5.31	195	32.4	3.04	108	49.2
65	9.9	43.0	31.1	—	3.28	1.67	533	87.9	5.28	174	29.1	3.02	96.8	44.1
58	7.8	—	33.9	57.6	2.72	1.90	475	78.0	5.28	107	21.4	2.51	86.4	32.5
53	8.7	55.9	35.0	54.1	2.71	2.10	425	70.6	5.23	95.8	19.2	2.48	77.9	29.1
50	6.3	—	32.9	60.9	2.17	2.36	394	64.7	5.18	56.3	13.9	1.96	72.4	21.4
45	7.0	—	36.0	51.0	2.15	2.61	350	58.1	5.15	50.0	12.4	1.94	64.7	19.0
40	7.8	—	40.5	40.3	2.14	2.90	310	51.9	5.13	44.1	11.0	1.93	57.5	16.8
35	6.3	—	41.7	38.0	1.74	3.66	285	45.6	5.25	24.5	7.47	1.54	51.2	11.5
30	7.4	—	47.5	29.3	1.73	4.30	238	38.6	5.21	20.3	6.24	1.52	43.1	9.56
26	8.5	57.9	53.1	23.4	1.72	4.95	204	33.4	5.17	17.3	5.34	1.51	37.2	8.17
22	4.7	—	47.3	29.5	1.02	7.19	156	25.4	4.91	4.66	2.31	0.847	29.3	3.66
19	5.7	—	51.7	24.7	1.00	8.67	130	21.3	4.82	3.76	1.88	0.822	24.7	2.98
16	7.5	—	54.5	22.2	0.96	11.3	103	17.1	4.67	2.82	1.41	0.773	20.1	2.26
14	8.8	54.3	59.6	18.6	0.95	13.3	88.6	14.9	4.62	2.36	1.19	0.753	17.4	1.90

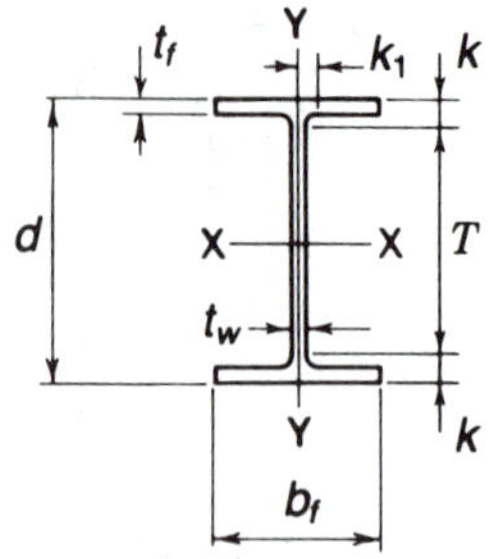

W SHAPES
Dimensions

Designation	Area A	Depth d		Web Thickness t_w		Web $\frac{t_w}{2}$	Flange Width b_f		Flange Thickness t_f		Distance T	Distance k	Distance k_1
	In.^{2}	In.		In.		In.	In.		In.		In.	In.	In.
W 10×112	32.9	11.36	11 3/8	0.755	3/4	3/8	10.415	10 3/8	1.250	1 1/4	7 5/8	1 7/8	15/16
×100	29.4	11.10	11 1/8	0.680	11/16	3/8	10.340	10 3/8	1.120	1 1/8	7 5/8	1 3/4	7/8
× 88	25.9	10.84	10 7/8	0.605	5/8	5/16	10.265	10 1/4	0.990	1	7 5/8	1 5/8	13/16
× 77	22.6	10.60	10 5/8	0.530	1/2	1/4	10.190	10 1/4	0.870	7/8	7 5/8	1 1/2	13/16
× 68	20.0	10.40	10 3/8	0.470	1/2	1/4	10.130	10 1/8	0.770	3/4	7 5/8	1 3/8	3/4
× 60	17.6	10.22	10 1/4	0.420	7/16	1/4	10.080	10 1/8	0.680	11/16	7 5/8	1 5/16	3/4
× 54	15.8	10.09	10 1/8	0.370	3/8	3/16	10.030	10	0.615	5/8	7 5/8	1 1/4	11/16
× 49	14.4	9.98	10	0.340	5/16	3/16	10.000	10	0.560	9/16	7 5/8	1 3/16	11/16
W 10× 45	13.3	10.10	10 1/8	0.350	3/8	3/16	8.020	8	0.620	5/8	7 5/8	1 1/4	11/16
× 39	11.5	9.92	9 7/8	0.315	5/16	3/16	7.985	8	0.530	1/2	7 5/8	1 1/8	11/16
× 33	9.71	9.73	9 3/4	0.290	5/16	3/16	7.960	8	0.435	7/16	7 5/8	1 1/16	11/16
W 10× 30	8.84	10.47	10 1/2	0.300	5/16	3/16	5.810	5 3/4	0.510	1/2	8 5/8	15/16	1/2
× 26	7.61	10.33	10 3/8	0.260	1/4	1/8	5.770	5 3/4	0.440	7/16	8 5/8	7/8	1/2
× 22	6.49	10.17	10 1/8	0.240	1/4	1/8	5.750	5 3/4	0.360	3/8	8 5/8	3/4	1/2
W 10× 19	5.62	10.24	10 1/4	0.250	1/4	1/8	4.020	4	0.395	3/8	8 5/8	13/16	1/2
× 17	4.99	10.11	10 1/8	0.240	1/4	1/8	4.010	4	0.330	5/16	8 5/8	3/4	1/2
× 15	4.41	9.99	10	0.230	1/4	1/8	4.000	4	0.270	1/4	8 5/8	11/16	7/16
× 12	3.54	9.87	9 7/8	0.190	3/16	1/8	3.960	4	0.210	3/16	8 5/8	5/8	7/16

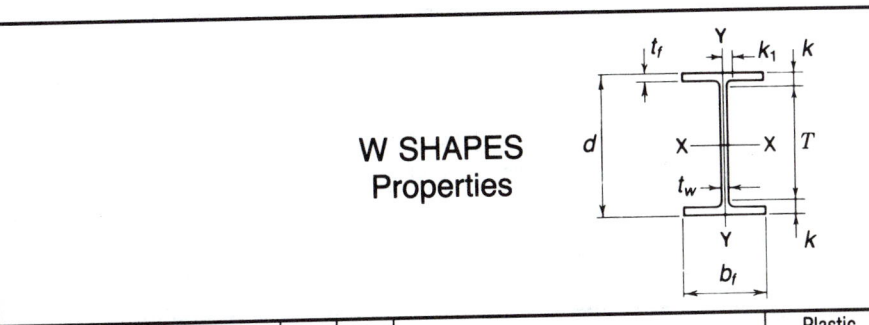

W SHAPES
Properties

Nom-inal Wt. per Ft	Compact Section Criteria				r_T	d/A_f	Elastic Properties						Plastic Modulus	
	$\frac{b_f}{2t_f}$	F_y'	$\frac{d}{t_w}$	F_y'''			Axis X-X			Axis Y-Y			Z_x	Z_y
							I	S	r	I	S	r		
Lb.		Ksi		Ksi	In.		In.⁴	In.³	In.	In.⁴	In.³	In.	In.³	In.³
112	4.2	—	15.0	—	2.88	0.87	716	126	4.66	236	45.3	2.68	147	69.2
100	4.6	—	16.3	—	2.85	0.96	623	112	4.60	207	40.0	2.65	130	61.0
88	5.2	—	17.9	—	2.83	1.07	534	98.5	4.54	179	34.8	2.63	113	53.1
77	5.9	—	20.0	—	2.80	1.20	455	85.9	4.49	154	30.1	2.60	97.6	45.9
68	6.6	—	22.1	—	2.79	1.33	394	75.7	4.44	134	26.4	2.59	85.3	40.1
60	7.4	—	24.3	—	2.77	1.49	341	66.7	4.39	116	23.0	2.57	74.6	35.0
54	8.2	63.5	27.3	—	2.75	1.64	303	60.0	4.37	103	20.6	2.56	66.6	31.3
49	8.9	53.0	29.4	—	2.74	1.78	272	54.6	4.35	93.4	18.7	2.54	60.4	28.3
45	6.5	—	28.9	—	2.18	2.03	248	49.1	4.32	53.4	13.3	2.01	54.9	20.3
39	7.5	—	31.5	—	2.16	2.34	209	42.1	4.27	45.0	11.3	1.98	46.8	17.2
33	9.1	50.5	33.6	58.7	2.14	2.81	170	35.0	4.19	36.6	9.20	1.94	38.8	14.0
30	5.7	—	34.9	54.2	1.55	3.53	170	32.4	4.38	16.7	5.75	1.37	36.6	8.84
26	6.6	—	39.7	41.8	1.54	4.07	144	27.9	4.35	14.1	4.89	1.36	31.3	7.50
22	8.0	—	42.4	36.8	1.51	4.91	118	23.2	4.27	11.4	3.97	1.33	26.0	6.10
19	5.1	—	41.0	39.4	1.03	6.45	96.3	18.8	4.14	4.29	2.14	0.874	21.6	3.35
17	6.1	—	42.1	37.2	1.01	7.64	81.9	16.2	4.05	3.56	1.78	0.844	18.7	2.80
15	7.4	—	43.4	35.0	0.99	9.25	68.9	13.8	3.95	2.89	1.45	0.810	16.0	2.30
12	9.4	47.5	51.9	24.5	0.96	11.9	53.8	10.9	3.90	2.18	1.10	0.785	12.6	1.74

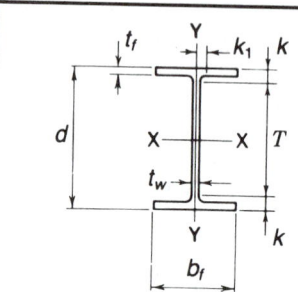

W SHAPES
Dimensions

Desig-nation	Area A	Depth d		Web Thickness t_w		$\dfrac{t_w}{2}$	Flange Width b_f		Flange Thickness t_f		Distance T	k	k_1
	In.²	In.		In.		In.	In.		In.		In.	In.	In.
W 8×67	19.7	9.00	9	0.570	9/16	5/16	8.280	8¼	0.935	15/16	6⅛	1 7/16	11/16
×58	17.1	8.75	8¾	0.510	½	¼	8.220	8¼	0.810	13/16	6⅛	1 5/16	11/16
×48	14.1	8.50	8½	0.400	⅜	3/16	8.110	8⅛	0.685	11/16	6⅛	1 3/16	⅝
×40	11.7	8.25	8¼	0.360	⅜	3/16	8.070	8⅛	0.560	9/16	6⅛	1 1/16	⅝
×35	10.3	8.12	8⅛	0.310	5/16	3/16	8.020	8	0.495	½	6⅛	1	9/16
×31	9.13	8.00	8	0.285	5/16	3/16	7.995	8	0.435	7/16	6⅛	15/16	9/16
W 8×28	8.25	8.06	8	0.285	5/16	3/16	6.535	6½	0.465	7/16	6⅛	15/16	9/16
×24	7.08	7.93	7⅞	0.245	¼	⅛	6.495	6½	0.400	⅜	6⅛	⅞	9/16
W 8×21	6.16	8.28	8¼	0.250	¼	⅛	5.270	5¼	0.400	⅜	6⅝	13/16	½
×18	5.26	8.14	8⅛	0.230	¼	⅛	5.250	5¼	0.330	5/16	6⅝	¾	7/16
W 8×15	4.44	8.11	8⅛	0.245	¼	⅛	4.015	4	0.315	5/16	6⅝	¾	½
×13	3.84	7.99	8	0.230	¼	⅛	4.000	4	0.255	¼	6⅝	11/16	7/16
×10	2.96	7.89	7⅞	0.170	3/16	⅛	3.940	4	0.205	3/16	6⅝	⅝	7/16
W 6×25	7.34	6.38	6⅜	0.320	5/16	3/16	6.080	6⅛	0.455	7/16	4¾	13/16	7/16
×20	5.87	6.20	6¼	0.260	¼	⅛	6.020	6	0.365	⅜	4¾	¾	7/16
×15	4.43	5.99	6	0.230	¼	⅛	5.990	6	0.260	¼	4¾	⅝	⅜
W 6×16	4.74	6.28	6¼	0.260	¼	⅛	4.030	4	0.405	⅜	4¾	¾	7/16
×12	3.55	6.03	6	0.230	¼	⅛	4.000	4	0.280	¼	4¾	⅝	⅜
× 9	2.68	5.90	5⅞	0.170	3/16	⅛	3.940	4	0.215	3/16	4¾	9/16	⅜
W 5×19	5.54	5.15	5⅛	0.270	¼	⅛	5.030	5	0.430	7/16	3½	13/16	7/16
×16	4.68	5.01	5	0.240	¼	⅛	5.000	5	0.360	⅜	3½	¾	7/16
W 4×13	3.83	4.16	4⅛	0.280	¼	⅛	4.060	4	0.345	⅜	2¾	11/16	7/16

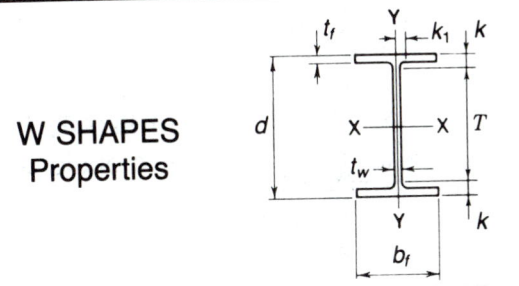

W SHAPES
Properties

Nominal Wt. per Ft	Compact Section Criteria				r_T	$\frac{d}{A_f}$	Axis X-X			Axis Y-Y			Plastic Modulus	
	$\frac{b_f}{2t_f}$	F_y'	$\frac{d}{t_w}$	F_y'''			I	S	r	I	S	r	Z_x	Z_y
Lb.		Ksi		Ksi	In.		In.^{4}	In.^{3}	In.	In.^{4}	In.^{3}	In.	In.^{3}	In.^{3}
67	4.4	—	15.8	—	2.28	1.16	272	60.4	3.72	88.6	21.4	2.12	70.2	32.7
58	5.1	—	17.2	—	2.26	1.31	228	52.0	3.65	75.1	18.3	2.10	59.8	27.9
48	5.9	—	21.3	—	2.23	1.53	184	43.3	3.61	60.9	15.0	2.08	49.0	22.9
40	7.2	—	22.9	—	2.21	1.83	146	35.5	3.53	49.1	12.2	2.04	39.8	18.5
35	8.1	64.4	26.2	—	2.20	2.05	127	31.2	3.51	42.6	10.6	2.03	34.7	16.1
31	9.2	50.0	28.1	—	2.18	2.30	110	27.5	3.47	37.1	9.27	2.02	30.4	14.1
28	7.0	—	28.3	—	1.77	2.65	98.0	24.3	3.45	21.7	6.63	1.62	27.2	10.1
24	8.1	64.1	32.4	63.0	1.76	3.05	82.8	20.9	3.42	18.3	5.63	1.61	23.2	8.57
21	6.6	—	33.1	60.2	1.41	3.93	75.3	18.2	3.49	9.77	3.71	1.26	20.4	5.69
18	8.0	—	35.4	52.7	1.39	4.70	61.9	15.2	3.43	7.97	3.04	1.23	17.0	4.66
15	6.4	—	33.1	60.3	1.03	6.41	48.0	11.8	3.29	3.41	1.70	0.876	13.6	2.67
13	7.8	—	34.7	54.7	1.01	7.83	39.6	9.91	3.21	2.73	1.37	0.843	11.4	2.15
10	9.6	45.8	46.4	30.7	0.99	9.77	30.8	7.81	3.22	2.09	1.06	0.841	8.87	1.66
25	6.7	—	19.9	—	1.66	2.31	53.4	16.7	2.70	17.1	5.61	1.52	18.9	8.56
20	8.2	62.1	23.8	—	1.64	2.82	41.4	13.4	2.66	13.3	4.41	1.50	14.9	6.72
15	11.5	31.8	26.0	—	1.61	3.85	29.1	9.72	2.56	9.32	3.11	1.46	10.8	4.75
16	5.0	—	24.2	—	1.08	3.85	32.1	10.2	2.60	4.43	2.20	0.966	11.7	3.39
12	7.1	—	26.2	—	1.05	5.38	22.1	7.31	2.49	2.99	1.50	0.918	8.30	2.32
9	9.2	50.3	34.7	54.8	1.03	6.96	16.4	5.56	2.47	2.19	1.11	0.905	6.23	1.72
19	5.8	—	19.1	—	1.38	2.38	26.2	10.2	2.17	9.13	3.63	1.28	11.6	5.53
16	6.9	—	20.9	—	1.37	2.78	21.3	8.51	2.13	7.51	3.00	1.27	9.59	4.57
13	5.9	—	14.9	—	1.10	2.97	11.3	5.46	1.72	3.86	1.90	1.00	6.28	2.92

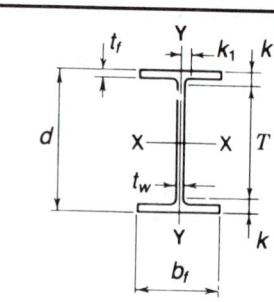

M SHAPES
Dimensions

Desig-nation	Area A	Depth d		Web Thickness t_w		$\dfrac{t_w}{2}$	Flange Width b_f		Thickness t_f		Distance T	k	Grip	Max. Flge. Fastener
	In.²	In.		In.		In.	In.		In.		In.	In.	In.	In.
M 14×18	5.10	14.00	14	0.215	3/16	1/8	4.000	4	0.270	1/4	12¾	5/8	1/4	3/4
M 12×11.8	3.47	12.00	12	0.177	3/16	1/8	3.065	3⅛	0.225	1/4	10⅞	9/16	1/4	—
M 12×10.8	3.18	11.97	12	0.160	3/16	1/16	3.065	3⅛	0.210	1/4	11	1/2	1/4	1/2
M 12×10	2.94	11.97	12	0.149	3/16	1/16	3.250	3¼	0.180	3/16	11	1/2	3/16	1/2
M 10×9	2.65	10.00	10	0.157	3/16	1/8	2.690	2¾	0.206	3/16	8⅞	9/16	3/16	—
M 10×8	2.35	9.95	10	0.141	3/16	1/16	2.690	2¾	0.182	3/16	9⅛	7/16	3/16	3/8
M 10×7.5	2.21	9.99	10	0.130	1/8	1/16	2.690	2¾	0.173	3/16	9⅛	7/16	3/16	3/8
M 8×6.5	1.92	8.00	8	0.135	1/8	1/16	2.281	2¼	0.189	3/16	7	1/2	3/16	—
M 6×4.4	1.29	6.00	6	0.114	1/8	1/16	1.844	1⅞	0.171	3/16	5⅛	7/16	3/16	—
M 5×18.9	5.55	5.00	5	0.316	5/16	3/16	5.003	5	0.416	7/16	3¼	7/8	7/16	7/8

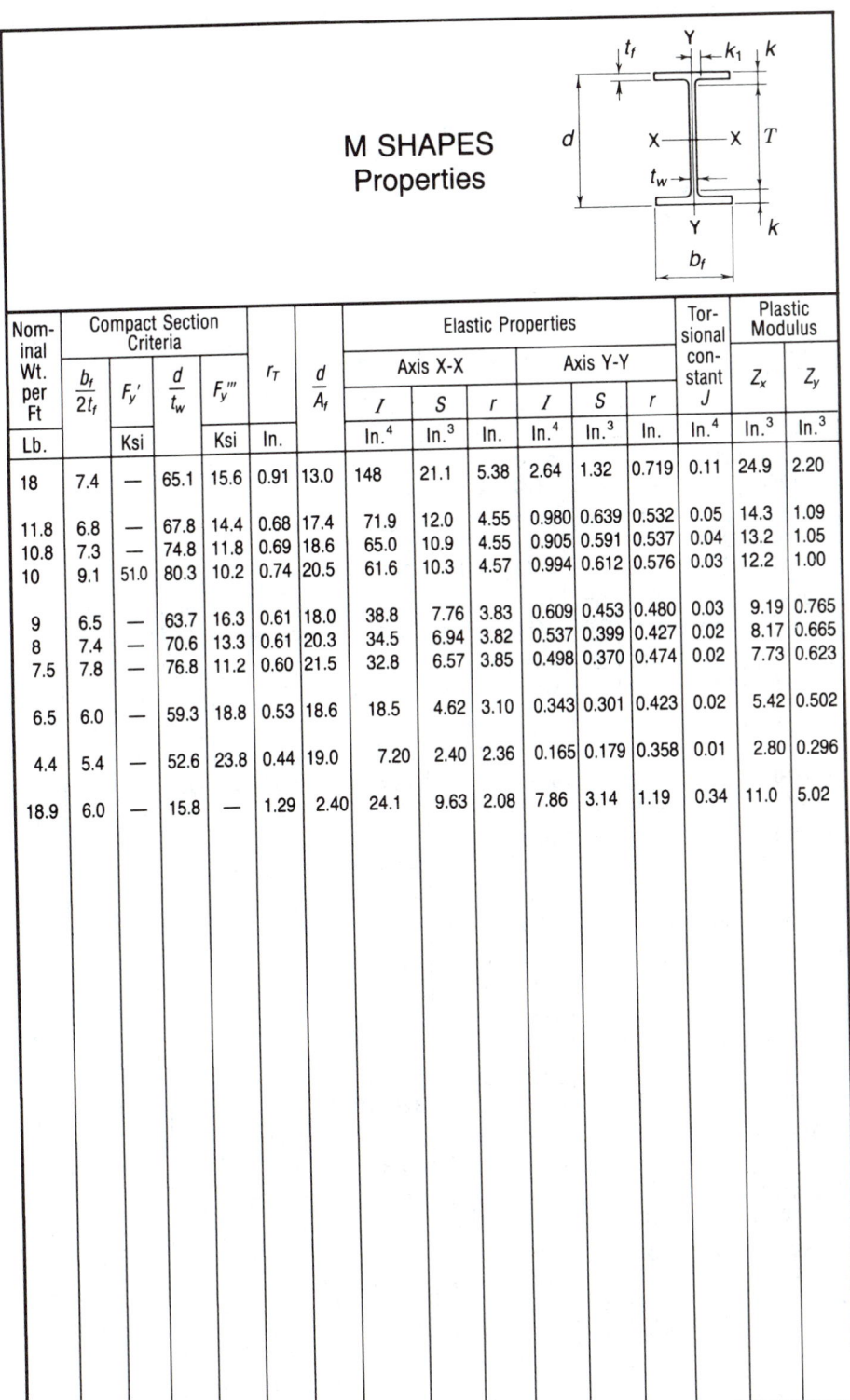

M SHAPES
Properties

Nominal Wt. per Ft Lb.	Compact Section Criteria				r_T In.	$\dfrac{d}{A_f}$	Elastic Properties						Torsional constant J In.⁴	Plastic Modulus	
	$\dfrac{b_f}{2t_f}$	F_y' Ksi	$\dfrac{d}{t_w}$	F_y''' Ksi			Axis X-X			Axis Y-Y				Z_x In.³	Z_y In.³
							I In.⁴	S In.³	r In.	I In.⁴	S In.³	r In.			
18	7.4	—	65.1	15.6	0.91	13.0	148	21.1	5.38	2.64	1.32	0.719	0.11	24.9	2.20
11.8	6.8	—	67.8	14.4	0.68	17.4	71.9	12.0	4.55	0.980	0.639	0.532	0.05	14.3	1.09
10.8	7.3	—	74.8	11.8	0.69	18.6	65.0	10.9	4.55	0.905	0.591	0.537	0.04	13.2	1.05
10	9.1	51.0	80.3	10.2	0.74	20.5	61.6	10.3	4.57	0.994	0.612	0.576	0.03	12.2	1.00
9	6.5	—	63.7	16.3	0.61	18.0	38.8	7.76	3.83	0.609	0.453	0.480	0.03	9.19	0.765
8	7.4	—	70.6	13.3	0.61	20.3	34.5	6.94	3.82	0.537	0.399	0.427	0.02	8.17	0.665
7.5	7.8	—	76.8	11.2	0.60	21.5	32.8	6.57	3.85	0.498	0.370	0.474	0.02	7.73	0.623
6.5	6.0	—	59.3	18.8	0.53	18.6	18.5	4.62	3.10	0.343	0.301	0.423	0.02	5.42	0.502
4.4	5.4	—	52.6	23.8	0.44	19.0	7.20	2.40	2.36	0.165	0.179	0.358	0.01	2.80	0.296
18.9	6.0	—	15.8	—	1.29	2.40	24.1	9.63	2.08	7.86	3.14	1.19	0.34	11.0	5.02

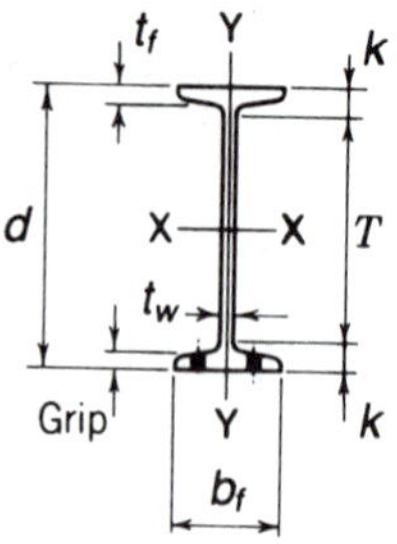

S SHAPES
Dimensions

Designation	Area A	Depth d		Web Thickness t_w		Web $\frac{t_w}{2}$	Flange Width b_f		Flange Thickness t_f		Distance T	Distance k	Grip	Max. Flge. Fastener
	In.^{2}	In.		In.		In.	In.		In.		In.	In.	In.	In.
S 24×121	35.6	24.50	24½	0.800	13/16	7/16	8.050	8	1.090	1 1/16	20½	2	1⅛	1
×106	31.2	24.50	24½	0.620	⅝	5/16	7.870	7⅞	1.090	1 1/16	20½	2	1⅛	1
S 24×100	29.3	24.00	24	0.745	¾	⅜	7.245	7¼	0.870	⅞	20½	1¾	⅞	1
×90	26.5	24.00	24	0.625	⅝	5/16	7.125	7⅛	0.870	⅞	20½	1¾	⅞	1
×80	23.5	24.00	24	0.500	½	¼	7.000	7	0.870	⅞	20½	1¾	⅞	1
S 20×96	28.2	20.30	20¼	0.800	13/16	7/16	7.200	7¼	0.920	15/16	16¾	1¾	15/16	1
×86	25.3	20.30	20¼	0.660	11/16	⅜	7.060	7	0.920	15/16	16¾	1¾	15/16	1
S 20×75	22.0	20.00	20	0.635	⅝	5/16	6.385	6⅜	0.795	13/16	16¾	1⅝	13/16	⅞
×66	19.4	20.00	20	0.505	½	¼	6.255	6¼	0.795	13/16	16¾	1⅝	13/16	⅞
S 18×70	20.6	18.00	18	0.711	11/16	⅜	6.251	6¼	0.691	11/16	15	1½	11/16	⅞
×54.7	16.1	18.00	18	0.461	7/16	¼	6.001	6	0.691	11/16	15	1½	11/16	⅞
S 15×50	14.7	15.00	15	0.550	9/16	5/16	5.640	5⅝	0.622	⅝	12¼	1⅜	9/16	¾
×42.9	12.6	15.00	15	0.411	7/16	¼	5.501	5½	0.622	⅝	12¼	1⅜	9/16	¾
S 12×50	14.7	12.00	12	0.687	11/16	⅜	5.477	5½	0.659	11/16	9⅛	1 7/16	11/16	¾
×40.8	12.0	12.00	12	0.462	7/16	¼	5.252	5¼	0.659	11/16	9⅛	1 7/16	⅝	¾
S 12×35	10.3	12.00	12	0.428	7/16	¼	5.078	5⅛	0.544	9/16	9⅝	1 3/16	½	¾
×31.8	9.35	12.00	12	0.350	⅜	3/16	5.000	5	0.544	9/16	9⅝	1 3/16	½	¾
S 10×35	10.3	10.00	10	0.594	⅝	5/16	4.944	5	0.491	½	7¾	1⅛	½	¾
×25.4	7.46	10.00	10	0.311	5/16	3/16	4.661	4⅝	0.491	½	7¾	1⅛	½	¾
S 8×23	6.77	8.00	8	0.441	7/16	¼	4.171	4⅛	0.426	7/16	6	1	7/16	¾
×18.4	5.41	8.00	8	0.271	¼	⅛	4.001	4	0.426	7/16	6	1	7/16	¾
S 7×20	5.88	7.00	7	0.450	7/16	¼	3.860	3⅞	0.392	⅜	5⅛	15/16	⅜	⅝
×15.3	4.50	7.00	7	0.252	¼	⅛	3.662	3⅝	0.392	⅜	5⅛	15/16	⅜	⅝
S 6×17.25	5.07	6.00	6	0.465	7/16	¼	3.565	3⅝	0.359	⅜	4¼	⅞	⅜	⅝
×12.5	3.67	6.00	6	0.232	¼	⅛	3.332	3⅜	0.359	⅜	4¼	⅞	⅜	—
S 5×14.75	4.34	5.00	5	0.494	½	¼	3.284	3¼	0.326	5/16	3⅜	13/16	5/16	—
×10	2.94	5.00	5	0.214	3/16	⅛	3.004	3	0.326	5/16	3⅜	13/16	5/16	—
S 4×9.5	2.79	4.00	4	0.326	5/16	3/16	2.796	2¾	0.293	5/16	2½	¾	5/16	—
×7.7	2.26	4.00	4	0.193	3/16	⅛	2.663	2⅝	0.293	5/16	2½	¾	5/16	—
S 3×7.5	2.21	3.00	3	0.349	⅜	3/16	2.509	2½	0.260	¼	1⅝	11/16	¼	—
×5.7	1.67	3.00	3	0.170	3/16	⅛	2.330	2⅜	0.260	¼	1⅝	11/16	¼	—

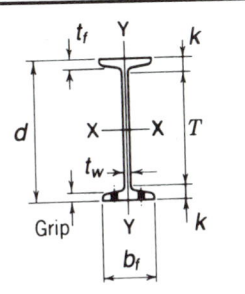

S SHAPES
Properties

Nominal Wt. per Ft	Compact Section Criteria				r_T	d/A_f	Elastic Properties						Torsional constant	Plastic Modulus	
							Axis X-X			Axis Y-Y					
	$\frac{b_f}{2t_f}$	F_y'	$\frac{d}{t_w}$	F_y'''			I	S	r	I	S	r	J	Z_x	Z_y
Lb.		Ksi		Ksi	In.		In.^{4}	In.^{3}	In.	In.^{4}	In.^{3}	In.	In.^{4}	In.^{3}	In.^{3}
121	3.7	—	30.6	—	1.86	2.79	3160	258	9.43	83.3	20.7	1.53	12.8	306	36.2
106	3.6	—	39.5	42.3	1.86	2.86	2940	240	9.71	77.1	19.6	1.57	10.1	279	33.2
100	4.2	—	32.2	63.6	1.59	3.81	2390	199	9.02	47.7	13.2	1.27	7.58	240	23.9
90	4.1	—	38.4	44.8	1.60	3.87	2250	187	9.21	44.9	12.6	1.30	6.04	222	22.3
80	4.0	—	48.0	28.7	1.61	3.94	2100	175	9.47	42.2	12.1	1.34	4.88	204	20.7
96	3.9	—	25.4	—	1.63	3.06	1670	165	7.71	50.2	13.9	1.33	8.39	198	24.9
86	3.8	—	30.8	—	1.63	3.13	1580	155	7.89	46.8	13.3	1.36	6.64	183	23.0
75	4.0	—	31.5	—	1.43	3.94	1280	128	7.62	29.8	9.32	1.16	4.59	153	16.7
66	3.9	—	39.6	42.1	1.44	4.02	1190	119	7.83	27.7	8.85	1.19	3.58	140	15.3
70	4.5	—	25.3	—	1.36	4.17	926	103	6.71	24.1	7.72	1.08	4.15	125	14.4
54.7	4.3	—	39.0	43.3	1.37	4.34	804	89.4	7.07	20.8	6.94	1.14	2.37	105	12.1
50	4.5	—	27.3	—	1.26	4.28	486	64.8	5.75	15.7	5.57	1.03	2.12	77.1	9.97
42.9	4.4	—	36.5	49.6	1.26	4.38	447	59.6	5.95	14.4	5.23	1.07	1.54	69.3	9.02
50	4.2	—	17.5	—	1.25	3.32	305	50.8	4.55	15.7	5.74	1.03	2.82	61.2	10.3
40.8	4.0	—	26.0	—	1.24	3.46	272	45.4	4.77	13.6	5.16	1.06	1.76	53.1	8.85
35	4.7	—	28.0	—	1.16	4.34	229	38.2	4.72	9.87	3.89	0.980	1.08	44.8	6.79
31.8	4.6	—	34.3	56.2	1.16	4.41	218	36.4	4.83	9.36	3.74	1.00	0.90	42.0	6.40
35	5.0	—	16.8	—	1.10	4.12	147	29.4	3.78	8.36	3.38	0.901	1.29	35.4	6.22
25.4	4.7	—	32.2	63.9	1.09	4.37	124	24.7	4.07	6.79	2.91	0.954	0.60	28.4	4.96
23	4.9	—	18.1	—	0.95	4.51	64.9	16.2	3.10	4.31	2.07	0.798	0.55	19.3	3.68
18.4	4.7	—	29.5	—	0.94	4.70	57.6	14.4	3.26	3.73	1.86	0.831	0.34	16.5	3.16
20	4.9	—	15.6	—	0.88	4.63	42.4	12.1	2.69	3.17	1.64	0.734	0.45	14.5	2.96
15.3	4.7	—	27.8	—	0.87	4.88	36.7	10.5	2.86	2.64	1.44	0.766	0.24	12.1	2.44
17.25	5.0	—	12.9	—	0.81	4.69	26.3	8.77	2.28	2.31	1.30	0.675	0.37	10.6	2.36
12.5	4.6	—	25.9	—	0.79	5.02	22.1	7.37	2.45	1.82	1.09	0.705	0.17	8.47	1.85
14.75	5.0	—	10.1	—	0.74	4.66	15.2	6.09	1.87	1.67	1.01	0.620	0.32	7.42	1.88
10	4.6	—	23.4	—	0.72	5.10	12.3	4.92	2.05	1.22	0.809	0.643	0.11	5.67	1.37
9.5	4.8	—	12.3	—	0.65	4.88	6.79	3.39	1.56	0.903	0.646	0.569	0.12	4.04	1.13
7.7	4.5	—	20.7	—	0.64	5.13	6.08	3.04	1.64	0.764	0.574	0.581	0.07	3.51	0.964
7.5	4.8	—	8.6	—	0.59	4.60	2.93	1.95	1.15	0.586	0.468	0.516	0.09	2.36	0.826
5.7	4.5	—	17.6	—	0.57	4.95	2.52	1.68	1.23	0.455	0.390	0.522	0.04	1.95	0.653

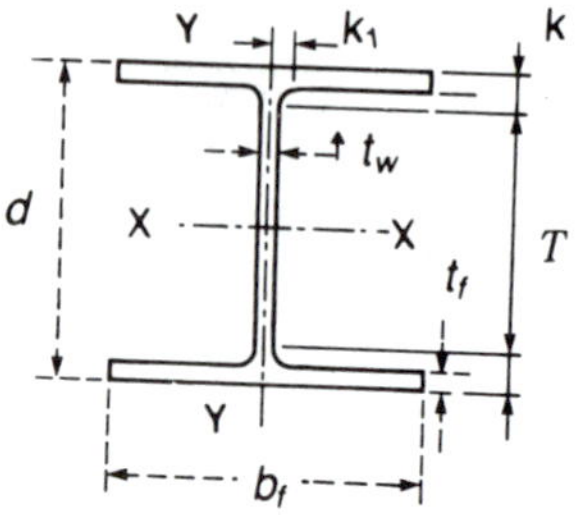

HP SHAPES
Dimensions

Designation	Area A	Depth d		Web Thickness t_w		$\frac{t_w}{2}$	Flange Width b_f		Flange Thickness t_f		Distance T	Distance k	Distance k_1
	In.^{2}	In.		In.		In.	In.		In.		In.	In.	In.
HP 14×117	34.4	14.21	14¼	0.805	13/16	7/16	14.885	14⅞	0.805	13/16	11¼	1½	1 1/16
×102	30.0	14.01	14	0.705	11/16	⅜	14.785	14¾	0.705	11/16	11¼	1⅜	1
× 89	26.1	13.83	13⅞	0.615	⅝	5/16	14.695	14¾	0.615	⅝	11¼	1 5/16	15/16
× 73	21.4	13.61	13⅝	0.505	½	¼	14.585	14⅝	0.505	½	11¼	1 3/16	⅞
HP 13×100	29.4	13.15	13⅛	0.765	¾	⅜	13.205	13¼	0.765	¾	10¼	1 7/16	1
× 87	25.5	12.95	13	0.665	11/16	⅜	13.105	13⅛	0.665	11/16	10¼	1⅜	15/16
× 73	21.6	12.75	12¾	0.565	9/16	5/16	13.005	13	0.565	9/16	10¼	1¼	15/16
× 60	17.5	12.54	12½	0.460	7/16	¼	12.900	12⅞	0.460	7/16	10¼	1⅛	⅞
HP 12× 84	24.6	12.28	12¼	0.685	11/16	⅜	12.295	12¼	0.685	11/16	9½	1⅜	1
× 74	21.8	12.13	12⅛	0.605	⅝	5/16	12.215	12¼	0.610	⅝	9½	1 5/16	15/16
× 63	18.4	11.94	12	0.515	½	¼	12.125	12⅛	0.515	½	9½	1¼	⅞
× 53	15.5	11.78	11¾	0.435	7/16	¼	12.045	12	0.435	7/16	9½	1⅛	⅞
HP 10× 57	16.8	9.99	10	0.565	9/16	5/16	10.225	10¼	0.565	9/16	7⅝	1 3/16	13/16
× 42	12.4	9.70	9¾	0.415	7/16	¼	10.075	10⅛	0.420	7/16	7⅝	1 1/16	¾
HP 8× 36	10.6	8.02	8	0.445	7/16	¼	8.155	8⅛	0.445	7/16	6⅛	15/16	⅝

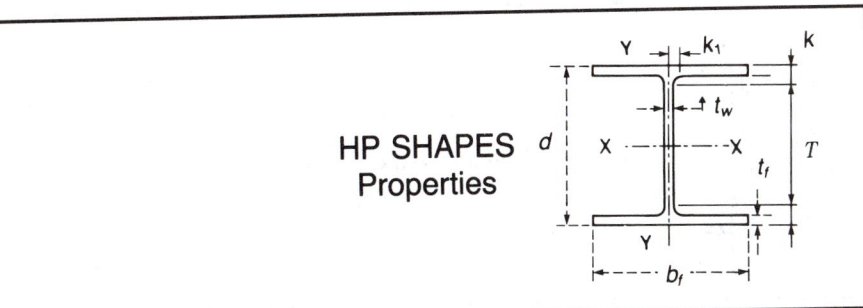

HP SHAPES
Properties

Nom-inal Wt. per Ft Lb.	Compact Section Criteria $\frac{b_f}{2t_f}$	F_y' Ksi	$\frac{d}{t_w}$	F_y''' Ksi	r_T In.	$\frac{d}{A_f}$	Elastic Properties Axis X-X I In.^{4}	S In.^{3}	r In.	Axis Y-Y I In.^{4}	S In.^{3}	r In.	Tor-sional con-stant J In.^{4}	Plastic Modulus Z_x In.^{3}	Z_y In.^{3}
117	9.2	49.4	17.7	—	4.00	1.19	1220	172	5.96	443	59.5	3.59	8.02	194	91.4
102	10.5	38.4	19.9	—	3.97	1.34	1050	150	5.92	380	51.4	3.56	5.40	169	78.8
89	11.9	29.6	22.5	—	3.94	1.53	904	131	5.88	326	44.3	3.53	3.60	146	67.7
73	14.4	20.3	27.0	—	3.90	1.85	729	107	5.84	261	35.8	3.49	2.01	118	54.6
100	8.6	56.7	17.2	—	3.54	1.30	886	135	5.49	294	44.5	3.16	6.25	153	68.6
87	9.9	43.5	19.5	—	3.51	1.49	755	117	5.45	250	38.1	3.13	4.12	131	58.5
73	11.5	31.9	22.6	—	3.47	1.74	630	98.8	5.40	207	31.9	3.10	2.54	110	48.8
60	14.0	21.5	27.3	—	3.43	2.11	503	80.3	5.36	165	25.5	3.07	1.39	89.0	39.0
84	9.0	52.5	17.9	—	3.29	1.46	650	106	5.14	213	34.6	2.94	4.24	120	53.2
74	10.0	42.1	20.0	—	3.26	1.63	569	93.8	5.11	186	30.4	2.92	2.98	105	46.6
63	11.8	30.5	23.2	—	3.23	1.91	472	79.1	5.06	153	25.3	2.88	1.83	88.3	38.7
53	13.8	22.0	27.1	—	3.20	2.25	393	66.8	5.03	127	21.1	2.86	1.12	74.0	32.2
57	9.0	51.6	17.7	—	2.74	1.73	294	58.8	4.18	101	19.7	2.45	1.97	66.5	30.3
42	12.0	29.4	23.4	—	2.69	2.29	210	43.4	4.13	71.7	14.2	2.41	0.81	48.3	21.8
36	9.2	50.3	18.0	—	2.18	2.21	119	29.8	3.36	40.3	9.88	1.95	0.77	33.6	15.2

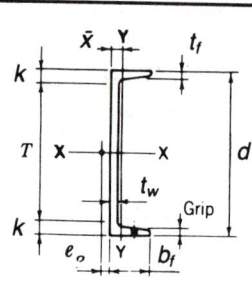

CHANNELS
AMERICAN STANDARD
Dimensions

Designation	Area A	Depth d	Web Thickness t_w	Web $\frac{t_w}{2}$	Flange Width b_f	Flange Average thickness t_f	Distance T	Distance k	Grip	Max. Flge. Fastener
	In.²	In.	In.	In.	In.	In.	In.	In.	In.	In.
C 15×50	14.7	15.00	0.716 11/16	3/8	3.716 3¾	0.650 5/8	12⅛	1 7/16	5/8	1
×40	11.8	15.00	0.520 ½	¼	3.520 3½	0.650 5/8	12⅛	1 7/16	5/8	1
×33.9	9.96	15.00	0.400 3/8	3/16	3.400 3⅜	0.650 5/8	12⅛	1 7/16	5/8	1
C 12×30	8.82	12.00	0.510 ½	¼	3.170 3⅛	0.501 ½	9¾	1⅛	½	7/8
×25	7.35	12.00	0.387 3/8	3/16	3.047 3	0.501 ½	9¾	1⅛	½	7/8
×20.7	6.09	12.00	0.282 5/16	⅛	2.942 3	0.501 ½	9¾	1⅛	½	7/8
C 10×30	8.82	10.00	0.673 11/16	5/16	3.033 3	0.436 7/16	8	1	7/16	¾
×25	7.35	10.00	0.526 ½	¼	2.886 2⅞	0.436 7/16	8	1	7/16	¾
×20	5.88	10.00	0.379 3/8	3/16	2.739 2¾	0.436 7/16	8	1	7/16	¾
×15.3	4.49	10.00	0.240 ¼	⅛	2.600 2⅝	0.436 7/16	8	1	7/16	¾
C 9×20	5.88	9.00	0.448 7/16	¼	2.648 2⅝	0.413 7/16	7⅛	15/16	7/16	¾
×15	4.41	9.00	0.285 5/16	⅛	2.485 2½	0.413 7/16	7⅛	15/16	7/16	¾
×13.4	3.94	9.00	0.233 ¼	⅛	2.433 2⅜	0.413 7/16	7⅛	15/16	7/16	¾
C 8×18.75	5.51	8.00	0.487 ½	¼	2.527 2½	0.390 3/8	6⅛	15/16	3/8	¾
×13.75	4.04	8.00	0.303 5/16	⅛	2.343 2⅜	0.390 3/8	6⅛	15/16	3/8	¾
×11.5	3.38	8.00	0.220 ¼	⅛	2.260 2¼	0.390 3/8	6⅛	15/16	3/8	¾
C 7×14.75	4.33	7.00	0.419 7/16	3/16	2.299 2¼	0.366 3/8	5¼	7/8	3/8	5/8
×12.25	3.60	7.00	0.314 5/16	3/16	2.194 2¼	0.366 3/8	5¼	7/8	3/8	5/8
× 9.8	2.87	7.00	0.210 3/16	⅛	2.090 2⅛	0.366 3/8	5¼	7/8	3/8	5/8
C 6×13	3.83	6.00	0.437 7/16	3/16	2.157 2⅛	0.343 5/16	4⅜	13/16	5/16	5/8
×10.5	3.09	6.00	0.314 5/16	3/16	2.034 2	0.343 5/16	4⅜	13/16	3/8	5/8
× 8.2	2.40	6.00	0.200 3/16	⅛	1.920 1⅞	0.343 5/16	4⅜	13/16	5/16	5/8
C 5× 9	2.64	5.00	0.325 5/16	3/16	1.885 1⅞	0.320 5/16	3½	¾	5/16	5/8
× 6.7	1.97	5.00	0.190 3/16	⅛	1.750 1¾	0.320 5/16	3½	¾	—	—
C 4× 7.25	2.13	4.00	0.321 5/16	3/16	1.721 1¾	0.296 5/16	2⅝	11/16	5/16	5/8
× 5.4	1.59	4.00	0.184 3/16	1/16	1.584 1⅝	0.296 5/16	2⅝	11/16	—	—
C 3× 6	1.76	3.00	0.356 3/8	3/16	1.596 1⅝	0.273 ¼	1⅝	11/16	—	—
× 5	1.47	3.00	0.258 ¼	⅛	1.498 1½	0.273 ¼	1⅝	11/16	—	—
× 4.1	1.21	3.00	0.170 3/16	1/16	1.410 1⅜	0.273 ¼	1⅝	11/16	—	—

CHANNELS
AMERICAN STANDARD
Properties

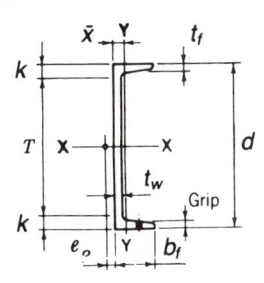

Nominal Wt. per Ft	$\bar{x}$	Shear Center Location e_o	$\dfrac{d}{A_f}$	Axis X-X			Axis Y-Y		
				I	S	r	I	S	r
Lb.	In.	In.		In.⁴	In.³	In.	In.⁴	In.³	In.
50	0.798	0.583	6.21	404	53.8	5.24	11.0	3.78	0.867
40	0.777	0.767	6.56	349	46.5	5.44	9.23	3.37	0.886
33.9	0.787	0.896	6.79	315	42.0	5.62	8.13	3.11	0.904
30	0.674	0.618	7.55	162	27.0	4.29	5.14	2.06	0.763
25	0.674	0.746	7.85	144	24.1	4.43	4.47	1.88	0.780
20.7	0.698	0.870	8.13	129	21.5	4.61	3.88	1.73	0.799
30	0.649	0.369	7.55	103	20.7	3.42	3.94	1.65	0.669
25	0.617	0.494	7.94	91.2	18.2	3.52	3.36	1.48	0.676
20	0.606	0.637	8.36	78.9	15.8	3.66	2.81	1.32	0.692
15.3	0.634	0.796	8.81	67.4	13.5	3.87	2.28	1.16	0.713
20	0.583	0.515	8.22	60.9	13.5	3.22	2.42	1.17	0.642
15	0.586	0.682	8.76	51.0	11.3	3.40	1.93	1.01	0.661
13.4	0.601	0.743	8.95	47.9	10.6	3.48	1.76	0.962	0.669
18.75	0.565	0.431	8.12	44.0	11.0	2.82	1.98	1.01	0.599
13.75	0.553	0.604	8.75	36.1	9.03	2.99	1.53	0.854	0.615
11.5	0.571	0.697	9.08	32.6	8.14	3.11	1.32	0.781	0.625
14.75	0.532	0.441	8.31	27.2	7.78	2.51	1.38	0.779	0.564
12.25	0.525	0.538	8.71	24.2	6.93	2.60	1.17	0.703	0.571
9.8	0.540	0.647	9.14	21.3	6.08	2.72	0.968	0.625	0.581
13	0.514	0.380	8.10	17.4	5.80	2.13	1.05	0.642	0.525
10.5	0.499	0.486	8.59	15.2	5.06	2.22	0.866	0.564	0.529
8.2	0.511	0.599	9.10	13.1	4.38	2.34	0.693	0.492	0.537
9	0.478	0.427	8.29	8.90	3.56	1.83	0.632	0.450	0.489
6.7	0.484	0.552	8.93	7.49	3.00	1.95	0.479	0.378	0.493
7.25	0.459	0.386	7.84	4.59	2.29	1.47	0.433	0.343	0.450
5.4	0.457	0.502	8.52	3.85	1.93	1.56	0.319	0.283	0.449
6	0.455	0.322	6.87	2.07	1.38	1.08	0.305	0.268	0.416
5	0.438	0.392	7.32	1.85	1.24	1.12	0.247	0.233	0.410
4.1	0.436	0.461	7.78	1.66	1.10	1.17	0.197	0.202	0.404

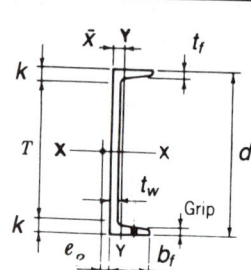

CHANNELS
MISCELLANEOUS
Dimensions

Desig-nation	Area A	Depth d	Web Thickness t_w		$\frac{t_w}{2}$	Flange Width b_f		Average Thickness t_f		Distance T	k	Grip	Max. Flge. Fastener
	In.²	In.	In.		In.	In.		In.		In.	In.	In.	In.
MC 18×58	17.1	18.00	0.700	¹¹⁄₁₆	⅜	4.200	4¼	0.625	⅝	15¼	1⅜	⅝	1
×51.9	15.3	18.00	0.600	⅝	⁵⁄₁₆	4.100	4⅛	0.625	⅝	15¼	1⅜	⅝	1
×45.8	13.5	18.00	0.500	½	¼	4.000	4	0.625	⅝	15¼	1⅜	⅝	1
×42.7	12.6	18.00	0.450	⁷⁄₁₆	¼	3.950	4	0.625	⅝	15¼	1⅜	⅝	1
MC 13×50	14.7	13.00	0.787	¹³⁄₁₆	⅜	4.412	4⅜	0.610	⅝	10¼	1⅜	⅝	1
×40	11.8	13.00	0.560	⁹⁄₁₆	¼	4.185	4⅛	0.610	⅝	10¼	1⅜	⁹⁄₁₆	1
×35	10.3	13.00	0.447	⁷⁄₁₆	¼	4.072	4⅛	0.610	⅝	10¼	1⅜	⁹⁄₁₆	1
×31.8	9.35	13.00	0.375	⅜	³⁄₁₆	4.000	4	0.610	⅝	10¼	1⅜	⁹⁄₁₆	1
MC 12×50	14.7	12.00	0.835	¹³⁄₁₆	⁷⁄₁₆	4.135	4⅛	0.700	¹¹⁄₁₆	9⅜	1⁵⁄₁₆	¹¹⁄₁₆	1
×45	13.2	12.00	0.712	¹¹⁄₁₆	⅜	4.012	4	0.700	¹¹⁄₁₆	9⅜	1⁵⁄₁₆	¹¹⁄₁₆	1
×40	11.8	12.00	0.590	⁹⁄₁₆	⁵⁄₁₆	3.890	3⅞	0.700	¹¹⁄₁₆	9⅜	1⁵⁄₁₆	¹¹⁄₁₆	1
×35	10.3	12.00	0.467	⁷⁄₁₆	¼	3.767	3¾	0.700	¹¹⁄₁₆	9⅜	1⁵⁄₁₆	¹¹⁄₁₆	1
×31	9.12	12.00	0.370	⅜	³⁄₁₆	3.670	3⅝	0.700	¹¹⁄₁₆	9⅜	1⁵⁄₁₆	¹¹⁄₁₆	1
MC 12×10.6	3.10	12.00	0.190	³⁄₁₆	⅛	1.500	1½	0.309	⁵⁄₁₆	10⅝	¹¹⁄₁₆	—	—
MC 10×41.1	12.1	10.00	0.796	¹³⁄₁₆	⅜	4.321	4⅜	0.575	⁹⁄₁₆	7½	1¼	⁹⁄₁₆	⅞
×33.6	9.87	10.00	0.575	⁹⁄₁₆	⁵⁄₁₆	4.100	4⅛	0.575	⁹⁄₁₆	7½	1¼	⁹⁄₁₆	⅞
×28.5	8.37	10.00	0.425	⁷⁄₁₆	³⁄₁₆	3.950	4	0.575	⁹⁄₁₆	7½	1¼	⁹⁄₁₆	⅞
MC 10×25	7.35	10.00	0.380	⅜	³⁄₁₆	3.405	3⅜	0.575	⁹⁄₁₆	7½	1¼	⁹⁄₁₆	⅞
×22	6.45	10.00	0.290	⁵⁄₁₆	⅛	3.315	3⅜	0.575	⁹⁄₁₆	7½	1¼	⁹⁄₁₆	⅞
MC 10× 8.4	2.46	10.00	0.170	³⁄₁₆	¹⁄₁₆	1.500	1½	0.280	¼	8⅝	¹¹⁄₁₆	—	—
MC 10× 6.5	1.91	10.00	0.152	⅛	¹⁄₁₆	1.127	1⅛	0.202	³⁄₁₆	9⅛	⁷⁄₁₆	—	—

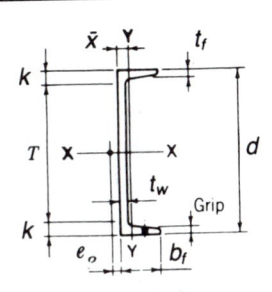

CHANNELS
MISCELLANEOUS
Properties

Nominal Wt. per Ft	$\bar{x}$	Shear Center Location e_o	$\dfrac{d}{A_f}$	Axis X-X			Axis Y-Y		
				I	S	r	I	S	r
Lb.	In.	In.		In.^{4}	In.^{3}	In.	In.^{4}	In.^{3}	In.
58	0.862	0.695	6.86	676	75.1	6.29	17.8	5.32	1.02
51.9	0.858	0.797	7.02	627	69.7	6.41	16.4	5.07	1.04
45.8	0.866	0.909	7.20	578	64.3	6.56	15.1	4.82	1.06
42.7	0.877	0.969	7.29	554	61.6	6.64	14.4	4.69	1.07
50	0.974	0.815	4.83	314	48.4	4.62	16.5	4.79	1.06
40	0.963	1.03	5.09	273	42.0	4.82	13.7	4.26	1.08
35	0.980	1.16	5.23	252	38.8	4.95	12.3	3.99	1.10
31.8	1.00	1.24	5.33	239	36.8	5.06	11.4	3.81	1.11
50	1.05	0.741	4.15	269	44.9	4.28	17.4	5.65	1.09
45	1.04	0.844	4.27	252	42.0	4.36	15.8	5.33	1.09
40	1.04	0.952	4.41	234	39.0	4.46	14.3	5.00	1.10
35	1.05	1.07	4.55	216	36.1	4.59	12.7	4.67	1.11
31	1.08	1.18	4.67	203	33.8	4.71	11.3	4.39	1.12
10.6	0.269	0.284	25.9	55.4	9.23	4.22	0.382	0.310	0.351
41.1	1.09	0.864	4.02	158	31.5	3.61	15.8	4.88	1.14
33.6	1.08	1.06	4.24	139	27.8	3.75	13.2	4.38	1.16
28.5	1.12	1.21	4.40	127	25.3	3.89	11.4	4.02	1.17
25	0.953	1.03	5.11	110	22	3.87	7.35	3.00	1.00
22	0.990	1.13	5.25	103	20.5	3.99	6.50	2.80	1.00
8.4	0.284	0.332	23.8	32.0	6.40	3.61	0.328	0.270	0.365
6.5	0.180	0.167	43.8	22.1	4.42	3.40	0.112	0.118	0.242

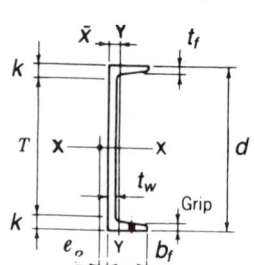

CHANNELS
MISCELLANEOUS
Dimensions

Desig-nation	Area A	Depth d	Web Thickness t_w		Web $\frac{t_w}{2}$	Flange Width b_f		Flange Average Thickness t_f		Distance T	Distance k	Grip	Max. Flge. Fastener
	In.²	In.	In.		In.	In.		In.		In.	In.	In.	In.
MC 9×25.4	7.47	9.00	0.450	7/16	1/4	3.500	3½	0.550	9/16	6⅝	1 3/16	9/16	⅞
×23.9	7.02	9.00	0.400	⅜	3/16	3.450	3½	0.550	9/16	6⅝	1 3/16	9/16	⅞
MC 8×22.8	6.70	8.00	0.427	7/16	3/16	3.502	3½	0.525	½	5⅝	1 3/16	½	⅞
×21.4	6.28	8.00	0.375	⅜	3/16	3.450	3½	0.525	½	5⅝	1 3/16	½	⅞
MC 8×20	5.88	8.00	0.400	⅜	3/16	3.025	3	0.500	½	5¾	1⅛	½	⅞
×18.7	5.50	8.00	0.353	⅜	3/16	2.978	3	0.500	½	5¾	1⅛	½	⅞
MC 8× 8.5	2.50	8.00	0.179	3/16	1/16	1.874	1⅞	0.311	5/16	6½	¾	5/16	⅝
MC 7×22.7	6.67	7.00	0.503	½	1/4	3.603	3⅝	0.500	½	4¾	1⅛	½	⅞
×19.1	5.61	7.00	0.352	⅜	3/16	3.452	3½	0.500	½	4¾	1⅛	½	⅞
MC 6×18	5.29	6.00	0.379	⅜	3/16	3.504	3½	0.475	½	3⅞	1 1/16	½	⅞
×15.3	4.50	6.00	0.340	5/16	3/16	3.500	3½	0.385	⅜	4¼	⅞	⅜	⅞
MC 6×16.3	4.79	6.00	0.375	⅜	3/16	3.000	3	0.475	½	3⅞	1 1/16	½	¾
×15.1	4.44	6.00	0.316	5/16	3/16	2.941	3	0.475	½	3⅞	1 1/16	½	¾
MC 6×12	3.53	6.00	0.310	5/16	1/8	2.497	2½	0.375	⅜	4⅜	13/16	⅜	⅝

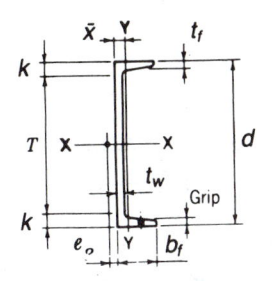

CHANNELS
MISCELLANEOUS
Properties

Nom-inal Wt. per Ft	$\bar{x}$	Shear Center Loca-tion e_o	$\dfrac{d}{A_f}$	Axis X-X			Axis Y-Y		
				I	S	r	I	S	r
Lb.	In.	In.		In.^{4}	In.^{3}	In.	In.^{4}	In.^{3}	In.
25.4	0.970	0.986	4.68	88.0	19.6	3.43	7.65	3.02	1.01
23.9	0.981	1.04	4.74	85.0	18.9	3.48	7.22	2.93	1.01
22.8	1.01	1.04	4.35	63.8	16.0	3.09	7.07	2.84	1.03
21.4	1.02	1.09	4.42	61.6	15.4	3.13	6.64	2.74	1.03
20	0.840	0.843	5.29	54.5	13.6	3.05	4.47	2.05	0.872
18.7	0.849	0.889	5.37	52.5	13.1	3.09	4.20	1.97	0.874
8.5	0.428	0.542	13.7	23.3	5.83	3.05	0.628	0.434	0.501
22.7	1.04	1.01	3.89	47.5	13.6	2.67	7.29	2.85	1.05
19.1	1.08	1.15	4.06	43.2	12.3	2.77	6.11	2.57	1.04
18	1.12	1.17	3.60	29.7	9.91	2.37	5.93	2.48	1.06
15.3	1.05	1.16	4.45	25.4	8.47	2.38	4.97	2.03	1.05
16.3	0.927	0.930	4.21	26.0	8.68	2.33	3.82	1.84	0.892
15.1	0.940	0.982	4.29	25.0	8.32	2.37	3.51	1.75	0.889
12	0.704	0.725	6.41	18.7	6.24	2.30	1.87	1.04	0.728

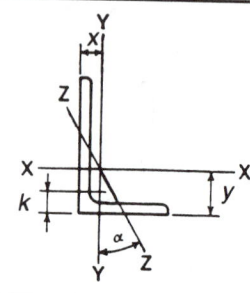

ANGLES
Equal legs and unequal legs
Properties for designing

Size and Thickness	k	Weight per Ft	Area	AXIS X-X				AXIS Y-Y				AXIS Z-Z	
				I	S	r	y	I	S	r	x	r	Tan
In.	In.	Lb.	In.²	In.⁴	In.³	In.	In.	In.⁴	In.³	In.	In.	In.	α
L 9×4× ⅝	1⅛	26.3	7.73	64.9	11.5	2.90	3.36	8.32	2.65	1.04	0.858	.847	0.216
9/16	1 1/16	23.8	7.00	59.1	10.4	2.91	3.33	7.63	2.41	1.04	0.834	.850	0.218
½	1	21.3	6.25	53.2	9.34	2.92	3.31	6.92	2.17	1.05	0.810	.854	0.220
L 8×8×1⅛	1¾	56.9	16.7	98.0	17.5	2.42	2.41	98.0	17.5	2.42	2.41	1.56	1.000
1	1⅝	51.0	15.0	89.0	15.8	2.44	2.37	89.0	15.8	2.44	2.37	1.56	1.000
⅞	1½	45.0	13.2	79.6	14.0	2.45	2.32	79.6	14.0	2.45	2.32	1.57	1.000
¾	1⅜	38.9	11.4	69.7	12.2	2.47	2.28	69.7	12.2	2.47	2.28	1.58	1.000
⅝	1¼	32.7	9.61	59.4	10.3	2.49	2.23	59.4	10.3	2.49	2.23	1.58	1.000
9/16	1 3/16	29.6	8.68	54.1	9.34	2.50	2.21	54.1	9.34	2.50	2.21	1.59	1.000
½	1⅛	26.4	7.75	48.6	8.36	2.50	2.19	48.6	8.36	2.50	2.19	1.59	1.000
L 8×6×1	1½	44.2	13.0	80.8	15.1	2.49	2.65	38.8	8.92	1.73	1.65	1.28	0.543
⅞	1⅜	39.1	11.5	72.3	13.4	2.51	2.61	34.9	7.94	1.74	1.61	1.28	0.547
¾	1¼	33.8	9.94	63.4	11.7	2.53	2.56	30.7	6.92	1.76	1.56	1.29	0.551
⅝	1⅛	28.5	8.36	54.1	9.87	2.54	2.52	26.3	5.88	1.77	1.52	1.29	0.554
9/16	1 1/16	25.7	7.56	49.3	8.95	2.55	2.50	24.0	5.34	1.78	1.50	1.30	0.556
½	1	23.0	6.75	44.3	8.02	2.56	2.47	21.7	4.79	1.79	1.47	1.30	0.558
7/16	15/16	20.2	5.93	39.2	7.07	2.57	2.45	19.3	4.23	1.80	1.45	1.31	0.560
L 8×4×1	1½	37.4	11.0	69.6	14.1	2.52	3.05	11.6	3.94	1.03	1.05	0.846	0.247
¾	1¼	28.7	8.44	54.9	10.9	2.55	2.95	9.36	3.07	1.05	0.953	0.852	0.258
9/16	1 1/16	21.9	6.43	42.8	8.35	2.58	2.88	7.43	2.38	1.07	0.882	0.861	0.265
½	1	19.6	5.75	38.5	7.49	2.59	2.86	6.74	2.15	1.08	0.859	0.865	0.267
L 7×4× ¾	1¼	26.2	7.69	37.8	8.42	2.22	2.51	9.05	3.03	1.09	1.01	0.860	0.324
⅝	1⅛	22.1	6.48	32.4	7.14	2.24	2.46	7.84	2.58	1.10	0.963	0.865	0.329
½	1	17.9	5.25	26.7	5.81	2.25	2.42	6.53	2.12	1.11	0.917	0.872	0.335
⅜	⅞	13.6	3.98	20.6	4.44	2.27	2.37	5.10	1.63	1.13	0.870	0.880	0.340

ANGLES
Equal legs and unequal legs
Properties for designing

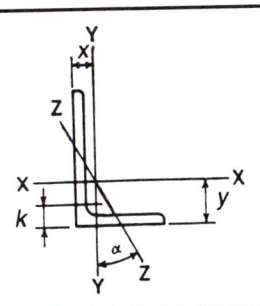

Size and Thickness In.	k In.	Weight per Ft Lb.	Area In.^{2}	AXIS X-X				AXIS Y-Y				AXIS Z-Z	
				I In.^{4}	S In.^{3}	r In.	y In.	I In.^{4}	S In.^{3}	r In.	x In.	r In.	Tan α
L 6×6 ×1	1½	37.4	11.0	35.5	8.57	1.80	1.86	35.5	8.57	1.80	1.86	1.17	1.000
⅞	1⅜	33.1	9.73	31.9	7.63	1.81	1.82	31.9	7.63	1.81	1.82	1.17	1.000
¾	1¼	28.7	8.44	28.2	6.66	1.83	1.78	28.2	6.66	1.83	1.78	1.17	1.000
⅝	1⅛	24.2	7.11	24.2	5.66	1.84	1.73	24.2	5.66	1.84	1.73	1.18	1.000
⁹⁄₁₆	1¹⁄₁₆	21.9	6.43	22.1	5.14	1.85	1.71	22.1	5.14	1.85	1.71	1.18	1.000
½	1	19.6	5.75	19.9	4.61	1.86	1.68	19.9	4.61	1.86	1.68	1.18	1.000
⁷⁄₁₆	¹⁵⁄₁₆	17.2	5.06	17.7	4.08	1.87	1.66	17.7	4.08	1.87	1.66	1.19	1.000
⅜	⅞	14.9	4.36	15.4	3.53	1.88	1.64	15.4	3.53	1.88	1.64	1.19	1.000
⁵⁄₁₆	¹³⁄₁₆	12.4	3.65	13.0	2.97	1.89	1.62	13.0	2.97	1.89	1.62	1.20	1.000
L 6×4 × ⅞	1⅜	27.2	7.98	27.7	7.15	1.86	2.12	9.75	3.39	1.11	1.12	0.857	0.421
¾	1¼	23.6	6.94	24.5	6.25	1.88	2.08	8.68	2.97	1.12	1.08	0.860	0.428
⅝	1⅛	20.0	5.86	21.1	5.31	1.90	2.03	7.52	2.54	1.13	1.03	0.864	0.435
⁹⁄₁₆	1¹⁄₁₆	18.1	5.31	19.3	4.83	1.90	2.01	6.91	2.31	1.14	1.01	0.866	0.438
½	1	16.2	4.75	17.4	4.33	1.91	1.99	6.27	2.08	1.15	0.987	0.870	0.440
⁷⁄₁₆	¹⁵⁄₁₆	14.3	4.18	15.5	3.83	1.92	1.96	5.60	1.85	1.16	0.964	0.873	0.443
⅜	⅞	12.3	3.61	13.5	3.32	1.93	1.94	4.90	1.60	1.17	0.941	0.877	0.446
⁵⁄₁₆	¹³⁄₁₆	10.3	3.03	11.4	2.79	1.94	1.92	4.18	1.35	1.17	0.918	0.882	0.448
L 6×3½× ½	1	15.3	4.50	16.6	4.24	1.92	2.08	4.25	1.59	0.972	0.833	0.759	0.344
⅜	⅞	11.7	3.42	12.9	3.24	1.94	2.04	3.34	1.23	0.988	0.787	0.767	0.350
⁵⁄₁₆	¹³⁄₁₆	9.8	2.87	10.9	2.73	1.95	2.01	2.85	1.04	0.996	0.763	0.772	0.352
L 5×5 × ⅞	1⅜	27.2	7.98	17.8	5.17	1.49	1.57	17.8	5.17	1.49	1.57	0.973	1.000
¾	1¼	23.6	6.94	15.7	4.53	1.51	1.52	15.7	4.53	1.51	1.52	0.975	1.000
⅝	1⅛	20.0	5.86	13.6	3.86	1.52	1.48	13.6	3.86	1.52	1.48	0.978	1.000
½	1	16.2	4.75	11.3	3.16	1.54	1.43	11.3	3.16	1.54	1.43	0.983	1.000
⁷⁄₁₆	¹⁵⁄₁₆	14.3	4.18	10.0	2.79	1.55	1.41	10.0	2.79	1.55	1.41	0.986	1.000
⅜	⅞	12.3	3.61	8.74	2.42	1.56	1.39	8.74	2.42	1.56	1.39	0.990	1.000
⁵⁄₁₆	¹³⁄₁₆	10.3	3.03	7.42	2.04	1.57	1.37	7.42	2.04	1.57	1.37	0.994	1.000

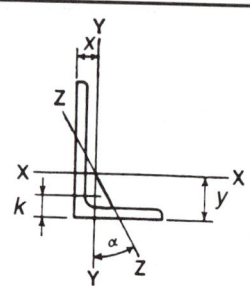

ANGLES
Equal legs and unequal legs
Properties for designing

Size and Thickness In.	k In.	Weight per Ft Lb.	Area In.²	AXIS X-X I In.⁴	S In.³	r In.	y In.	AXIS Y-Y I In.⁴	S In.³	r In.	x In.	AXIS Z-Z r In.	Tan α
L 5×3½× ¾	1¼	19.8	5.81	13.9	4.28	1.55	1.75	5.55	2.22	0.977	0.996	0.748	0.464
⅝	1⅛	16.8	4.92	12.0	3.65	1.56	1.70	4.83	1.90	0.991	0.951	0.751	0.472
½	1	13.6	4.00	9.99	2.99	1.58	1.66	4.05	1.56	1.01	0.906	0.755	0.479
⁷⁄₁₆	¹⁵⁄₁₆	12.0	3.53	8.90	2.64	1.59	1.63	3.63	1.39	1.01	0.883	0.758	0.482
⅜	⅞	10.4	3.05	7.78	2.29	1.60	1.61	3.18	1.21	1.02	0.861	0.762	0.486
⁵⁄₁₆	¹³⁄₁₆	8.7	2.56	6.60	1.94	1.61	1.59	2.72	1.02	1.03	0.838	0.766	0.489
¼	¾	7.0	2.06	5.39	1.57	1.62	1.56	2.23	0.830	1.04	0.814	0.770	0.492
L 5×3 × ⅝	1	15.7	4.61	11.4	3.55	1.57	1.80	3.06	1.39	0.815	0.796	0.644	0.349
½	1	12.8	3.75	9.45	2.91	1.59	1.75	2.58	1.15	0.829	0.750	0.648	0.357
⁷⁄₁₆	¹⁵⁄₁₆	11.3	3.31	8.43	2.58	1.60	1.73	2.32	1.02	0.837	0.727	0.651	0.361
⅜	⅞	9.8	2.86	7.37	2.24	1.61	1.70	2.04	0.888	0.845	0.704	0.654	0.364
⁵⁄₁₆	¹³⁄₁₆	8.2	2.40	6.26	1.89	1.61	1.68	1.75	0.753	0.853	0.681	0.658	0.368
¼	¾	6.6	1.94	5.11	1.53	1.62	1.66	1.44	0.614	0.861	0.657	0.663	0.371
L 4×4 × ¾	1⅛	18.5	5.44	7.67	2.81	1.19	1.27	7.67	2.81	1.19	1.27	0.778	1.000
⅝	1	15.7	4.61	6.66	2.40	1.20	1.23	6.66	2.40	1.20	1.23	0.779	1.000
½	⅞	12.8	3.75	5.56	1.97	1.22	1.18	5.56	1.97	1.22	1.18	0.782	1.000
⁷⁄₁₆	¹³⁄₁₆	11.3	3.31	4.97	1.75	1.23	1.16	4.97	1.75	1.23	1.16	0.785	1.000
⅜	¾	9.8	2.86	4.36	1.52	1.23	1.14	4.36	1.52	1.23	1.14	0.788	1.000
⁵⁄₁₆	¹¹⁄₁₆	8.2	2.40	3.71	1.29	1.24	1.12	3.71	1.29	1.24	1.12	0.791	1.000
¼	⅝	6.6	1.94	3.04	1.05	1.25	1.09	3.04	1.05	1.25	1.09	0.795	1.000
L 4×3½× ½	¹⁵⁄₁₆	11.9	3.50	5.32	1.94	1.23	1.25	3.79	1.52	1.04	1.00	0.722	0.750
⁷⁄₁₆	⅞	10.6	3.09	4.76	1.72	1.24	1.23	3.40	1.35	1.05	0.978	0.724	0.753
⅜	¹³⁄₁₆	9.1	2.67	4.18	1.49	1.25	1.21	2.95	1.17	1.06	0.955	0.727	0.755
⁵⁄₁₆	¾	7.7	2.25	3.56	1.26	1.26	1.18	2.55	0.994	1.07	0.932	0.730	0.757
¼	¹¹⁄₁₆	6.2	1.81	2.91	1.03	1.27	1.16	2.09	0.808	1.07	0.909	0.734	0.759

ANGLES
Equal legs and unequal legs
Properties for designing

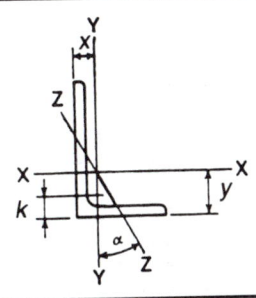

Size and Thickness In.	k In.	Weight per Ft Lb.	Area In.²	AXIS X-X				AXIS Y-Y				AXIS Z-Z	
				I In.⁴	S In.³	r In.	y In.	I In.⁴	S In.³	r In.	x In.	r In.	Tan α
L 4 ×3 × ½	¹⁵⁄₁₆	11.1	3.25	5.05	1.89	1.25	1.33	2.42	1.12	0.864	0.827	0.639	0.543
⁷⁄₁₆	⅞	9.8	2.87	4.52	1.68	1.25	1.30	2.18	0.992	0.871	0.804	0.641	0.547
⅜	¹³⁄₁₆	8.5	2.48	3.96	1.46	1.26	1.28	1.92	0.866	0.879	0.782	0.644	0.551
⁵⁄₁₆	¾	7.2	2.09	3.38	1.23	1.27	1.26	1.65	0.734	0.887	0.759	0.647	0.554
¼	¹¹⁄₁₆	5.8	1.69	2.77	1.00	1.28	1.24	1.36	0.599	0.896	0.736	0.651	0.558
L 3½×3½× ½	⅞	11.1	3.25	3.64	1.49	1.06	1.06	3.64	1.49	1.06	1.06	0.683	1.000
⁷⁄₁₆	¹³⁄₁₆	9.8	2.87	3.26	1.32	1.07	1.04	3.26	1.32	1.07	1.04	0.684	1.000
⅜	¾	8.5	2.48	2.87	1.15	1.07	1.01	2.87	1.15	1.07	1.01	0.687	1.000
⁵⁄₁₆	¹¹⁄₁₆	7.2	2.09	2.45	0.976	1.08	0.990	2.45	0.976	1.08	0.990	0.690	1.000
¼	⅝	5.8	1.69	2.01	0.794	1.09	0.968	2.01	0.794	1.09	0.968	0.694	1.000
L 3½×3 × ½	¹⁵⁄₁₆	10.2	3.00	3.45	1.45	1.07	1.13	2.33	1.10	0.881	0.875	0.621	0.714
⁷⁄₁₆	⅞	9.1	2.65	3.10	1.29	1.08	1.10	2.09	0.975	0.889	0.853	0.622	0.718
⅜	¹³⁄₁₆	7.9	2.30	2.72	1.13	1.09	1.08	1.85	0.851	0.897	0.830	0.625	0.721
⁵⁄₁₆	¾	6.6	1.93	2.33	0.954	1.10	1.06	1.58	0.722	0.905	0.808	0.627	0.724
¼	¹¹⁄₁₆	5.4	1.56	1.91	0.776	1.11	1.04	1.30	0.589	0.914	0.785	0.631	0.727
L 3½×2½× ½	¹⁵⁄₁₆	9.4	2.75	3.24	1.41	1.09	1.20	1.36	0.760	0.704	0.705	0.534	0.486
⁷⁄₁₆	⅞	8.3	2.43	2.91	1.26	1.09	1.18	1.23	0.677	0.711	0.682	0.535	0.491
⅜	¹³⁄₁₆	7.2	2.11	2.56	1.09	1.10	1.16	1.09	0.592	0.719	0.660	0.537	0.496
⁵⁄₁₆	¾	6.1	1.78	2.19	0.927	1.11	1.14	0.939	0.504	0.727	0.637	0.540	0.501
¼	¹¹⁄₁₆	4.9	1.44	1.80	0.755	1.12	1.11	0.777	0.412	0.735	0.614	0.544	0.506
L 3 ×3 × ½	¹³⁄₁₆	9.4	2.75	2.22	1.07	0.898	0.932	2.22	1.07	0.898	0.932	0.584	1.000
⁷⁄₁₆	¾	8.3	2.43	1.99	0.954	0.905	0.910	1.99	0.954	0.905	0.910	0.585	1.000
⅜	¹¹⁄₁₆	7.2	2.11	1.76	0.833	0.913	0.888	1.76	0.833	0.913	0.888	0.587	1.000
⁵⁄₁₆	⅝	6.1	1.78	1.51	0.707	0.922	0.865	1.51	0.707	0.922	0.865	0.589	1.000
¼	⁹⁄₁₆	4.9	1.44	1.24	0.577	0.930	0.842	1.24	0.577	0.930	0.842	0.592	1.000
³⁄₁₆	½	3.71	1.09	0.962	0.441	0.939	0.820	0.962	0.441	0.939	0.820	0.596	1.000

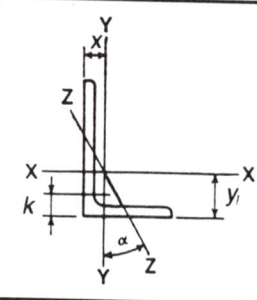

ANGLES
Equal legs and unequal legs
Properties for designing

Size and Thickness	k	Weight per Ft	Area	AXIS X-X				AXIS Y-Y				AXIS Z-Z	
				I	S	r	y	I	S	r	x	r	Tan
In.	In.	Lb.	In.^{2}	In.^{4}	In.^{3}	In.	In.	In.^{4}	In.^{3}	In.	In.	In.	α
L 3 ×2½× ½	7/8	8.5	2.50	2.08	1.04	0.913	1.00	1.30	0.744	0.722	0.750	0.520	0.667
7/16	13/16	7.6	2.21	1.88	0.928	0.920	0.978	1.18	0.664	0.729	0.728	0.521	0.672
3/8	3/4	6.6	1.92	1.66	0.810	0.928	0.956	1.04	0.581	0.736	0.706	0.522	0.676
5/16	11/16	5.6	1.62	1.42	0.688	0.937	0.933	0.898	0.494	0.744	0.683	0.525	0.680
1/4	5/8	4.5	1.31	1.17	0.561	0.945	0.911	0.743	0.404	0.753	0.661	0.528	0.684
3/16	9/16	3.39	0.996	0.907	0.430	0.954	0.888	0.577	0.310	0.761	0.638	0.533	0.688
L 3 ×2 × ½	13/16	7.7	2.25	1.92	1.00	0.924	1.08	0.672	0.474	0.546	0.583	0.428	0.414
7/16	3/4	6.8	2.00	1.73	0.894	0.932	1.06	0.609	0.424	0.553	0.561	0.429	0.421
3/8	11/16	5.9	1.73	1.53	0.781	0.940	1.04	0.543	0.371	0.559	0.539	0.430	0.428
5/16	5/8	5.0	1.46	1.32	0.664	0.948	1.02	0.470	0.317	0.567	0.516	0.432	0.435
1/4	9/16	4.1	1.19	1.09	0.542	0.957	0.993	0.392	0.260	0.574	0.493	0.435	0.440
3/16	1/2	3.07	0.902	0.842	0.415	0.966	0.970	0.307	0.200	0.583	0.470	0.439	0.446
L 2½×2½× ½	13/16	7.7	2.25	1.23	0.724	0.739	0.806	1.23	0.724	0.739	0.806	0.487	1.000
3/8	11/16	5.9	1.73	0.984	0.566	0.753	0.762	0.984	0.566	0.753	0.762	0.487	1.000
5/16	5/8	5.0	1.46	0.849	0.482	0.761	0.740	0.849	0.482	0.761	0.740	0.489	1.000
1/4	9/16	4.1	1.19	0.703	0.394	0.769	0.717	0.703	0.394	0.769	0.717	0.491	1.000
3/16	1/2	3.07	0.902	0.547	0.303	0.778	0.694	0.547	0.303	0.778	0.694	0.495	1.000
L 2½×2 × 3/8	11/16	5.3	1.55	0.912	0.547	0.768	0.831	0.514	0.363	0.577	0.581	0.420	0.614
5/16	5/8	4.5	1.31	0.788	0.466	0.776	0.809	0.446	0.310	0.584	0.559	0.422	0.620
1/4	9/16	3.62	1.06	0.654	0.381	0.784	0.787	0.372	0.254	0.592	0.537	0.424	0.626
3/16	1/2	2.75	0.809	0.509	0.293	0.793	0.764	0.291	0.196	0.600	0.514	0.427	0.631
L 2 ×2 × 3/8	5/8	4.7	1.36	0.479	0.351	0.594	0.636	0.479	0.351	0.594	0.636	0.389	1.000
5/16	9/16	3.92	1.15	0.416	0.300	0.601	0.614	0.416	0.300	0.601	0.614	0.390	1.000
1/4	1/2	3.19	0.938	0.348	0.247	0.609	0.592	0.348	0.247	0.609	0.592	0.391	1.000
3/16	7/16	2.44	0.715	0.272	0.190	0.617	0.569	0.272	0.190	0.617	0.569	0.394	1.000
1/8	3/8	1.65	0.484	0.190	0.131	0.626	0.546	0.190	0.131	0.626	0.546	0.398	1.000

ANGLES
Equal legs and unequal legs
Properties for designing

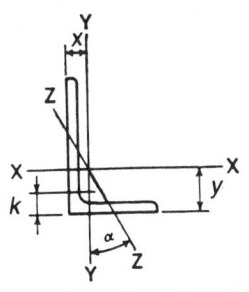

Size and Thickness	k	Weight per Ft	Area	AXIS X-X				AXIS Y-Y				AXIS Z-Z	
				I	S	r	y	I	S	r	x	r	Tan
In.	In.	Lb.	In.^{2}	In.^{4}	In.^{3}	In.	In.	In.^{4}	In.^{3}	In.	In.	In.	α
L 1¾×1¾×¼	½	2.77	0.813	0.227	0.227	0.529	0.529	0.227	0.227	0.529	0.529	0.341	1.000
×³⁄₁₆	⁷⁄₁₆	2.12	0.621	0.179	0.144	0.537	0.506	0.179	0.144	0.537	0.506	0.343	1.000
L 1½×1½×¼	⁷⁄₁₆	2.34	0.688	0.139	0.134	0.449	0.466	0.139	0.134	0.449	0.466	0.292	1.000
×³⁄₁₆	⅜	1.80	0.527	0.110	0.104	0.457	0.444	0.110	0.104	0.457	0.444	0.293	1.000
L 1¼×1¼×¼	⁷⁄₁₆	1.92	0.563	0.077	0.091	0.369	0.403	0.077	0.091	0.369	0.403	0.243	1.000
×³⁄₁₆	⅜	1.48	0.434	0.061	0.071	0.377	0.381	0.061	0.071	0.377	0.381	0.244	1.000
L 1⅛×1⅛×⅛	⁷⁄₃₂	0.900	0.266	0.032	0.040	0.345	0.327	0.032	0.040	0.345	0.327	0.221	1.000
L 1 ×1 ×⅛	¼	0.800	0.234	0.022	0.031	0.304	0.296	0.022	0.031	0.304	0.296	0.196	1.000

	USUAL GAGES FOR ANGLES, INCHES														CRIMPS
Leg	8	7	6	5	4	3½	3	2½	2	1¾	1½	1⅜	1¼	1	$b=t+1\frac{1}{2}$ Min.$=2$
g	4½	4	3½	3	2½	2	1¾	1⅜	1⅛	1	⅞	⅞	¾	⅝	
g_1	3	2½	2¼	2											
g_2	3	3	2½	1¾											

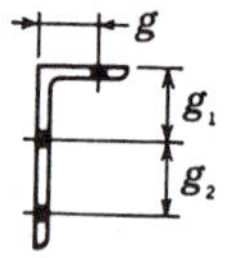

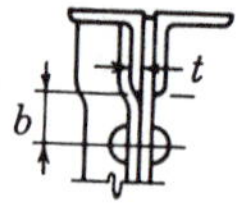

Other gages are permitted to suit specific requirements subject to clearances and edge distance limitations.

STRUCTURAL TEES

Dimensions and Properties

Structural tees are obtained by splitting the webs of various beams, generally with the aid of rotary shears, and straightening to meet established tolerances listed in Standard Mill Practice, Part 1 of this Manual.

Although structural tees may be obtained by off-center splitting, or by splitting on two lines, as specified on order, the Dimensions and Properties for Designing are based on a depth of tee equal to ½ the published beam depth. The table shows properties and dimensions for these full-depth tees. Values of Q_s and C_c' are given for F_y = 36 ksi and F_y = 50 ksi, for those tees having stems which exceed the noncompact section criteria of AISC ASD Specification Sect. B5.1. Since the cross section is comprised entirely of unstiffened elements, $Q_a = 1.0$ and $Q = Q_s$, for all tee sections. The flexural-torsional properties table also lists the dimensional values ($\bar{r}_o$ and H) and cross section constant J needed for checking torsional and flexural-torsional buckling.

USE OF TABLE

The table may be used as follows for checking allowable stresses for (1) flexural buckling and (2) torsional or flexural-torsional buckling.

(1) Flexural Buckling

Where no value of Q_s is shown, the allowable compressive stress is given by AISC ASD Specification Sect. E2. Where a value of Q_s is shown, the strength must be reduced in accordance with Appendix B5.

(2) Torsional or Flexural-torsional Buckling

The allowable stresses for torsional or flexural-torsional buckling can be determined from the AISC Load and Resistance Factor Design (LRFD) Specification Appendix E3. This involves calculations with J, $\bar{r}_o$, and H tabulated in Part 1 of this Manual. For further discussion see Part 3 of this Manual.

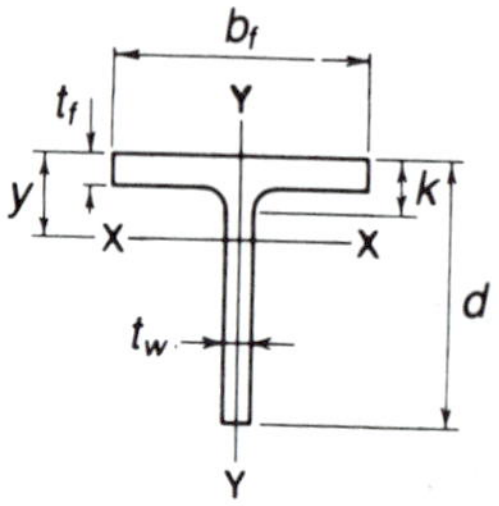

STRUCTURAL TEES
Cut from W shapes
Dimensions

Designation	Area	Depth of Tee d		Stem Thickness t_w		Stem $\frac{t_w}{2}$	Area of Stem	Flange Width b_f		Flange Thickness t_f		Distance k
	In.^{2}	In.		In.		In.	In.^{2}	In.		In.		In.
WT 18 ×179.5	52.7	18.700	18 11/16	1.120	1 1/8	9/16	20.9	16.730	16 3/4	2.010	2	3 1/8
×164	48.2	18.540	18 9/16	1.020	1	1/2	18.9	16.630	16 5/8	1.850	1 7/8	3
×150	44.1	18.370	18 3/8	0.945	15/16	1/2	17.4	16.655	16 5/8	1.680	1 11/16	2 13/16
×140	41.2	18.260	18 1/4	0.885	7/8	7/16	16.2	16.595	16 5/8	1.570	1 9/16	2 11/16
×130	38.2	18.130	18 1/8	0.840	13/16	7/16	15.2	16.550	16 1/2	1.440	1 7/16	2 9/16
×122.5	36.0	18.040	18	0.800	13/16	7/16	14.4	16.510	16 1/2	1.350	1 3/8	2 1/2
×115	33.8	17.950	18	0.760	3/4	3/8	13.6	16.470	16 1/2	1.260	1 1/4	2 3/8
WT 18 ×128	37.7	18.710	18 11/16	0.960	1	1/2	18.0	12.215	12 1/4	1.730	1 3/4	2 5/8
×116	34.1	18.560	18 9/16	0.870	7/8	7/16	16.1	12.120	12 1/8	1.570	1 9/16	2 1/2
×105	30.9	18.345	18 3/8	0.830	13/16	7/16	15.2	12.180	12 1/8	1.360	1 3/8	2 5/16
× 97	28.5	18.245	18 1/4	0.765	3/4	3/8	14.0	12.115	12 1/8	1.260	1 1/4	2 3/16
× 91	26.8	18.165	18 1/8	0.725	3/4	3/8	13.2	12.075	12 1/8	1.180	1 3/16	2 1/8
× 85	25.0	18.085	18 1/8	0.680	11/16	3/8	12.3	12.030	12	1.100	1 1/8	2
× 80	23.5	18.005	18	0.650	5/8	5/16	11.7	12.000	12	1.020	1	1 15/16
× 75	22.1	17.925	17 7/8	0.625	5/8	5/16	11.2	11.975	12	0.940	15/16	1 7/8
× 67.5	19.9	17.775	17 3/4	0.600	5/8	5/16	10.7	11.950	12	0.790	13/16	1 11/16
WT 16.5×177	52.1	17.775	17 3/4	1.160	1 3/16	5/8	20.6	16.100	16 1/8	2.090	2 1/16	2 7/8
×159	46.7	17.580	17 9/16	1.040	1 1/16	9/16	18.3	15.985	16	1.890	1 7/8	2 11/16
×145.5	42.8	17.420	17 7/16	0.960	1	1/2	16.7	15.905	15 7/8	1.730	1 3/4	2 9/16
×131.5	38.7	17.265	17 1/4	0.870	7/8	7/16	15.0	15.805	15 3/4	1.570	1 9/16	2 3/8
×120.5	35.4	17.090	17 1/8	0.830	13/16	7/16	14.2	15.860	15 7/8	1.400	1 3/8	2 3/16
×110.5	32.5	16.965	17	0.775	3/4	3/8	13.1	15.805	15 3/4	1.275	1 1/4	2 1/16
×100.5	29.5	16.840	16 7/8	0.715	11/16	3/8	12.0	15.745	15 3/4	1.150	1 1/8	1 15/16
WT 16.5× 84.5	24.8	16.910	16 15/16	0.670	11/16	3/8	11.3	11.500	11 1/2	1.220	1 1/4	2 1/16
× 76	22.4	16.745	16 3/4	0.635	5/8	5/16	10.6	11.565	11 5/8	1.055	1 1/16	1 7/8
× 70.5	20.8	16.650	16 5/8	0.605	5/8	5/16	10.1	11.535	11 1/2	0.960	15/16	1 3/4
× 65	19.2	16.545	16 1/2	0.580	9/16	5/16	9.60	11.510	11 1/2	0.855	7/8	1 11/16
× 59	17.3	16.430	16 3/8	0.550	9/16	5/16	9.04	11.480	11 1/2	0.740	3/4	1 9/16
WT 15 ×117.5	34.5	15.650	15 5/8	0.830	13/16	7/16	13.0	15.055	15	1.500	1 1/2	2 1/4
×105.5	31.0	15.470	15 1/2	0.775	3/4	3/8	12.0	15.105	15 1/8	1.315	1 5/16	2 1/8
× 95.5	28.1	15.340	15 3/8	0.710	11/16	3/8	10.9	15.040	15	1.185	1 3/16	1 15/16
× 86.5	25.4	15.220	15 1/4	0.655	5/8	5/16	9.97	14.985	15	1.065	1 1/16	1 7/8
WT 15 × 74	21.7	15.335	15 5/16	0.650	5/8	5/16	9.96	10.480	10 1/2	1.180	1 3/16	2
× 66	19.4	15.155	15 1/8	0.615	5/8	5/16	9.32	10.545	10 1/2	1.000	1	1 3/4
× 62	18.2	15.085	15 1/8	0.585	9/16	5/16	8.82	10.515	10 1/2	0.930	15/16	1 11/16
× 58	17.1	15.005	15	0.565	9/16	5/16	8.48	10.495	10 1/2	0.850	7/8	1 5/8
× 54	15.9	14.915	14 7/8	0.545	9/16	5/16	8.13	10.475	10 1/2	0.760	3/4	1 9/16
× 49.5	14.5	14.825	14 7/8	0.520	1/2	1/4	7.71	10.450	10 1/2	0.670	11/16	1 7/16

STRUCTURAL TEES
Cut from W shapes
Properties

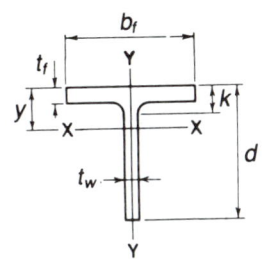

$$C_c' = \sqrt{\frac{2\pi^2 E}{Q_s Q_a F_y}}, \quad Q_a = 1.0$$

Nom-inal Wt. per Ft Lb.	$\frac{d}{t_w}$	Axis X-X I In.⁴	S In.³	r In.	y In.	Axis Y-Y I In.⁴	S In.³	r In.	$F_y = 36$ Q_s	C_c'	$F_y = 50$ Q_s	C_c'
179.5	16.7	1500	104.0	5.33	4.33	786	94.0	3.86	—	—	—	—
164	18.2	1350	94.1	5.29	4.21	711	85.5	3.84	—	—	—	—
150	19.4	1230	86.1	5.27	4.13	648	77.8	3.83	—	—	0.927	111
140	20.6	1140	80.0	5.25	4.07	599	72.2	3.81	—	—	0.867	115
130	21.6	1060	75.1	5.26	4.05	545	65.9	3.78	0.981	127	0.816	118
122.5	22.6	995	71.0	5.26	4.03	507	61.4	3.75	0.943	130	0.770	122
115	23.6	934	67.0	5.25	4.01	470	57.1	3.73	0.896	133	0.715	127
128	19.5	1200	87.4	5.66	4.92	264	43.2	2.65	—	—	0.927	111
116	21.3	1080	78.5	5.63	4.82	234	38.6	2.62	0.994	126	0.831	117
105	22.1	985	73.1	5.65	4.87	206	33.8	2.58	0.960	129	0.791	120
97	23.8	901	67.0	5.62	4.80	187	30.9	2.56	0.887	134	0.705	127
91	25.1	845	63.1	5.62	4.77	174	28.8	2.55	0.831	138	0.635	134
85	26.6	786	58.9	5.61	4.73	160	26.6	2.53	0.767	144	0.565	142
80	27.7	740	55.8	5.61	4.74	147	24.6	2.50	0.720	149	0.521	148
75	28.7	698	53.1	5.62	4.78	135	22.5	2.47	0.677	153	0.486	153
67.5	29.6	636	49.7	5.66	4.96	113	18.9	2.38	0.634	158	0.457	158
177	15.3	1320	96.8	5.03	4.16	729	90.6	3.74	—	—	—	—
159	16.9	1160	85.8	4.99	4.02	645	80.7	3.71	—	—	—	—
145.5	18.1	1050	78.3	4.97	3.94	581	73.1	3.69	—	—	0.993	107
131.5	19.8	943	70.2	4.94	3.84	517	65.5	3.66	—	—	0.907	112
120.5	20.6	871	65.8	4.96	3.85	466	58.8	3.63	—	—	0.867	115
110.5	21.9	799	60.8	4.96	3.81	420	53.2	3.59	0.968	128	0.801	120
100.5	23.6	725	55.5	4.95	3.78	375	47.6	3.56	0.896	133	0.715	127
84.5	25.2	649	51.1	5.12	4.21	155	27.0	2.50	0.827	139	0.630	135
76	26.4	592	47.4	5.14	4.26	136	23.6	2.47	0.775	143	0.574	141
70.5	27.5	552	44.7	5.15	4.29	123	21.3	2.43	0.728	148	0.529	147
65	28.5	513	42.1	5.18	4.36	109	18.9	2.39	0.685	152	0.492	153
59	29.9	469	39.2	5.20	4.47	93.6	16.3	2.32	0.621	160	0.447	160
117.5	18.9	674	55.1	4.42	3.42	427	56.8	3.52	—	—	0.952	110
105.5	20.0	610	50.5	4.43	3.40	378	50.1	3.49	—	—	0.897	113
95.5	21.6	549	45.7	4.42	3.35	336	44.7	3.46	0.981	127	0.816	118
86.5	23.2	497	41.7	4.42	3.31	299	39.9	3.43	0.913	132	0.735	125
74	23.6	466	40.6	4.63	3.84	113	21.7	2.28	0.896	133	0.715	127
66	24.6	421	37.4	4.66	3.90	98.0	18.6	2.25	0.853	137	0.664	131
62	25.8	396	35.3	4.66	3.90	90.4	17.2	2.23	0.801	141	0.601	138
58	26.6	373	33.7	4.67	3.94	82.1	15.7	2.19	0.767	144	0.565	142
54	27.4	349	32.0	4.69	4.01	73.0	13.9	2.15	0.733	147	0.533	147
49.5	28.5	322	30.0	4.71	4.09	63.9	12.2	2.10	0.685	152	0.492	153

Where no value of C_c' or Q_s is shown, the Tee complies with the noncompact section criteria of Specification Sect. B5.

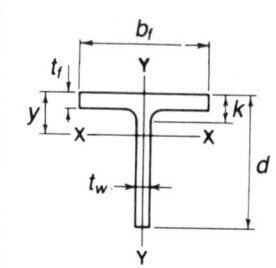

STRUCTURAL TEES
Cut from W shapes
Dimensions

Designation	Area	Depth of Tee d		Stem Thickness t_w	$\frac{t_w}{2}$	Area of Stem	Flange Width b_f		Flange Thickness t_f		Distance k
	In.²	In.		In.	In.	In.²	In.		In.		In.
WT 13.5×108.5	31.9	14.215	14³/₁₆	0.830	¹³/₁₆	⁷/₁₆	11.8	14.115	14⅛	1.500	1½	2³/₁₆
× 97	28.5	14.055	14¹/₁₆	0.750	¾	⅜	10.5	14.035	14	1.340	1⁵/₁₆	2¹/₁₆
× 89	26.1	13.905	13⅞	0.725	¾	⅜	10.1	14.085	14⅛	1.190	1³/₁₆	1⅞
× 80.5	23.7	13.795	13¾	0.660	¹¹/₁₆	⅜	9.10	14.020	14	1.080	1¹/₁₆	1¹³/₁₆
× 73	21.5	13.690	13¾	0.605	⅝	⁵/₁₆	8.28	13.965	14	0.975	1	1¹¹/₁₆
WT 13.5× 64.5	18.9	13.815	13¹³/₁₆	0.610	⅝	⁵/₁₆	8.42	10.010	10	1.100	1⅛	1¹³/₁₆
× 57	16.8	13.645	13⅝	0.570	⁹/₁₆	⁵/₁₆	7.78	10.070	10⅛	0.930	¹⁵/₁₆	1⅝
× 51	15.0	13.545	13½	0.515	½	¼	6.98	10.015	10	0.830	¹³/₁₆	1⁹/₁₆
× 47	13.8	13.460	13½	0.490	½	¼	6.60	9.990	10	0.745	¾	1⁷/₁₆
× 42	12.4	13.355	13⅜	0.460	⁷/₁₆	¼	6.14	9.960	10	0.640	⅝	1⅜
WT 12 × 88	25.8	12.625	12⅝	0.750	¾	⅜	9.46	12.890	12⅞	1.340	1⁵/₁₆	2⅛
× 81	23.9	12.500	12½	0.705	¹¹/₁₆	⅜	8.81	12.955	13	1.220	1¼	2
× 73	21.5	12.370	12⅜	0.650	⅝	⁵/₁₆	8.04	12.900	12⅞	1.090	1¹/₁₆	1⅞
× 65.5	19.3	12.240	12¼	0.605	⅝	⁵/₁₆	7.41	12.855	12⅞	0.960	¹⁵/₁₆	1¾
× 58.5	17.2	12.130	12⅛	0.550	⁹/₁₆	⁵/₁₆	6.67	12.800	12¾	0.850	⅞	1⅝
× 52	15.3	12.030	12	0.500	½	¼	6.01	12.750	12¾	0.750	¾	1½
WT 12 × 51.5	15.1	12.26	12¼	0.550	⁹/₁₆	⁵/₁₆	6.74	9.000	9	0.980	1	1¾
× 47	13.8	12.155	12⅛	0.515	½	¼	6.26	9.065	9⅛	0.875	⅞	1⅝
× 42	12.4	12.050	12	0.470	½	¼	5.66	9.020	9	0.770	¾	1⁹/₁₆
× 38	11.2	11.960	12	0.440	⁷/₁₆	¼	5.26	8.990	9	0.680	¹¹/₁₆	1⁷/₁₆
× 34	10.0	11.865	11⅞	0.415	⁷/₁₆	¼	4.92	8.965	9	0.585	⁹/₁₆	1⅜
WT 12 × 31	9.11	11.870	11⅞	0.430	⁷/₁₆	¼	5.10	7.040	7	0.590	⁹/₁₆	1⅜
× 27.5	8.10	11.785	11¾	0.395	⅜	³/₁₆	4.66	7.005	7	0.505	½	1⁵/₁₆
WT 10.5× 83	24.4	11.240	11¼	0.750	¾	⅜	8.43	12.420	12⅜	1.360	1⅜	2⅛
× 73.5	21.6	11.030	11	0.720	¾	⅜	7.94	12.510	12½	1.150	1⅛	1⅞
× 66	19.4	10.915	10⅞	0.650	⅝	⁵/₁₆	7.09	12.440	12½	1.035	1¹/₁₆	1¹³/₁₆
× 61	17.9	10.840	10⅞	0.600	⅝	⁵/₁₆	6.50	12.390	12⅜	0.960	¹⁵/₁₆	1¹¹/₁₆
× 55.5	16.3	10.755	10¾	0.550	⁹/₁₆	⁵/₁₆	5.92	12.340	12⅜	0.875	⅞	1⅝
× 50.5	14.9	10.680	10⅝	0.500	½	¼	5.34	12.290	12¼	0.800	¹³/₁₆	1⁹/₁₆
WT 10.5× 46.5	13.7	10.810	10¾	0.580	⁹/₁₆	⁵/₁₆	6.27	8.420	8⅜	0.930	¹⁵/₁₆	1¹¹/₁₆
× 41.5	12.2	10.715	10¾	0.515	½	¼	5.52	8.355	8⅜	0.835	¹³/₁₆	1⁹/₁₆
× 36.5	10.7	10.620	10⅝	0.455	⁷/₁₆	¼	4.83	8.295	8¼	0.740	¾	1½
× 34	10.0	10.565	10⅝	0.430	⁷/₁₆	¼	4.54	8.270	8¼	0.685	¹¹/₁₆	1⁷/₁₆
× 31	9.13	10.495	10½	0.400	⅜	³/₁₆	4.20	8.240	8¼	0.615	⅝	1⅜
WT 10.5× 28.5	8.37	10.530	10½	0.405	⅜	³/₁₆	4.26	6.555	6½	0.650	⅝	1⅜
× 25	7.36	10.415	10⅜	0.380	⅜	³/₁₆	3.96	6.530	6½	0.535	⁹/₁₆	1⁵/₁₆
× 22	6.49	10.330	10⅜	0.350	⅜	³/₁₆	3.62	6.500	6½	0.450	⁷/₁₆	1³/₁₆

STRUCTURAL TEES
Cut from W shapes
Properties

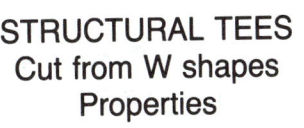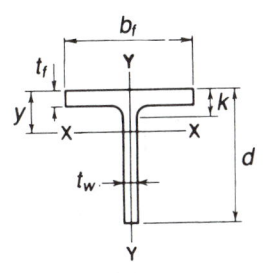

$$C_c' = \sqrt{\frac{2\pi^2 E}{Q_s Q_a F_y}}, \quad Q_a = 1.0$$

Nom-inal Wt. per Ft Lb.	$\frac{d}{t_w}$	Axis X-X I	S	r	y	Axis Y-Y I	S	r	$F_y = 36$ Q_s	C_c'	$F_y = 50$ Q_s	C_c'
		In.^{4}	In.^{3}	In.	In.	In.^{4}	In.^{3}	In.	Q_s	C_c'	Q_s	C_c'
108.5	17.1	502	45.2	3.97	3.11	352	49.9	3.32	—	—	—	—
97	18.7	444	40.3	3.95	3.03	309	44.1	3.29	—	—	0.963	109
89	19.2	414	38.2	3.98	3.05	278	39.4	3.26	—	—	0.937	111
80.5	20.9	372	34.4	3.96	2.99	248	35.4	3.24	—	—	0.851	116
73	22.6	336	31.2	3.95	2.95	222	31.7	3.21	0.938	130	0.765	122
64.5	22.6	323	31.0	4.13	3.39	92.2	18.4	2.21	0.938	130	0.765	122
57	23.9	289	28.3	4.15	3.42	79.4	15.8	2.18	0.883	134	0.700	128
51	26.3	258	25.3	4.14	3.37	69.6	13.9	2.15	0.780	143	0.578	141
47	27.5	239	23.8	4.16	3.41	62.0	12.4	2.12	0.728	148	0.529	147
42	29.0	216	21.9	4.18	3.48	52.8	10.6	2.07	0.664	155	0.476	155
88	16.8	319	32.2	3.51	2.74	240	37.2	3.04	—	—	—	—
81	17.7	293	29.9	3.50	2.70	221	34.2	3.05	—	—	—	—
73	19.0	264	27.2	3.50	2.66	195	30.3	3.01	—	—	0.947	110
65.5	20.2	238	24.8	3.52	2.65	170	26.5	2.97	—	—	0.887	114
58.5	22.1	212	22.3	3.51	2.62	149	23.2	2.94	0.960	129	0.791	120
52	24.1	189	20.0	3.51	2.59	130	20.3	2.91	0.874	135	0.690	129
51.5	22.3	204	22.0	3.67	3.01	59.7	13.3	1.99	0.951	129	0.781	121
47	23.6	186	20.3	3.67	2.99	54.5	12.0	1.98	0.896	133	0.715	127
42	25.6	166	18.3	3.67	2.97	47.2	10.5	1.95	0.810	140	0.610	137
38	27.2	151	16.9	3.68	3.00	41.3	9.18	1.92	0.741	146	0.541	145
34	28.6	137	15.6	3.70	3.06	35.2	7.85	1.87	0.681	153	0.489	153
31	27.6	131	15.6	3.79	3.46	17.2	4.90	1.38	0.724	148	0.525	148
27.5	29.8	117	14.1	3.80	3.50	14.5	4.15	1.34	0.626	159	0.450	160
83	15.0	226	25.5	3.04	2.39	217	35.0	2.98	—	—	—	—
73.5	15.3	204	23.7	3.08	2.39	188	30.0	2.95	—	—	—	—
66	16.8	181	21.1	3.06	2.33	166	26.7	2.93	—	—	—	—
61	18.1	166	19.3	3.04	2.28	152	24.6	2.92	—	—	0.993	107
55.5	19.6	150	17.5	3.03	2.23	137	22.2	2.90	—	—	0.917	112
50.5	21.4	135	15.8	3.01	2.18	124	20.2	2.89	0.990	127	0.826	118
46.5	18.6	144	17.9	3.25	2.74	46.4	11.0	1.84	—	—	0.968	109
41.5	20.8	127	15.7	3.22	2.66	40.7	9.75	1.83	—	—	0.856	116
36.5	23.3	110	13.8	3.21	2.60	35.3	8.51	1.81	0.908	132	0.730	125
34	24.6	103	12.9	3.20	2.59	32.4	7.83	1.80	0.853	137	0.664	131
31	26.2	93.8	11.9	3.21	2.58	28.7	6.97	1.77	0.784	142	0.583	140
28.5	26.0	90.4	11.8	3.29	2.85	15.3	4.67	1.35	0.793	142	0.592	139
25	27.4	80.3	10.7	3.30	2.93	12.5	3.82	1.30	0.733	147	0.533	147
22	29.5	71.1	9.68	3.31	2.98	10.3	3.18	1.26	0.638	158	0.460	158

Where no value of C_c' or Q_s is shown, the Tee complies with the noncompact section criteria of Specification Sect. B5.1

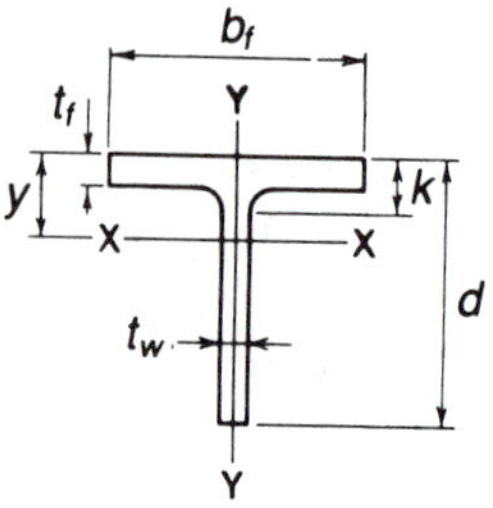

STRUCTURAL TEES
Cut from W shapes
Dimensions

Designation	Area	Depth of Tee d		Stem			Area of Stem	Flange				Dis-tance k
				Thickness t_w		$\frac{t_w}{2}$		Width b_f		Thickness t_f		
	In.^{2}	In.		In.		In.	In.^{2}	In.		In.		In.
WT 9×71.5	21.0	9.745	9¾	0.730	¾	⅜	7.11	11.220	11¼	1.320	1 5/16	2
×65	19.1	9.625	9⅝	0.670	11/16	⅜	6.45	11.160	11⅛	1.200	1 3/16	1⅞
×59.5	17.5	9.485	9½	0.655	⅝	5/16	6.21	11.265	11¼	1.060	1 1/16	1¾
×53	15.6	9.365	9⅜	0.590	9/16	5/16	5.53	11.200	11¼	0.940	15/16	1⅝
×48.5	14.3	9.295	9¼	0.535	9/16	5/16	4.97	11.145	11⅛	0.870	⅞	1 9/16
×43	12.7	9.195	9¼	0.480	½	¼	4.41	11.090	11⅛	0.770	¾	1 7/16
×38	11.2	9.105	9⅛	0.425	7/16	¼	3.87	11.035	11	0.680	11/16	1⅜
WT 9×35.5	10.4	9.235	9¼	0.495	½	¼	4.57	7.635	7⅝	0.810	13/16	1½
×32.5	9.55	9.175	9⅛	0.450	7/16	¼	4.13	7.590	7⅝	0.750	¾	1 7/16
×30	8.82	9.120	9⅛	0.415	7/16	¼	3.78	7.555	7½	0.695	11/16	1⅜
×27.5	8.10	9.055	9	0.390	⅜	3/16	3.53	7.530	7½	0.630	⅝	1 5/16
×25	7.33	8.995	9	0.355	⅜	3/16	3.19	7.495	7½	0.570	9/16	1¼
WT 9×23	6.77	9.030	9	0.360	⅜	3/16	3.25	6.060	6	0.605	⅝	1¼
×20	5.88	8.950	9	0.315	5/16	3/16	2.82	6.015	6	0.525	½	1 3/16
×17.5	5.15	8.850	8⅞	0.300	5/16	3/16	2.65	6.000	6	0.425	7/16	1⅛
WT 8×50	14.7	8.485	8½	0.585	9/16	5/16	4.96	10.425	10⅜	0.985	1	1 11/16
×44.5	13.1	8.375	8⅜	0.525	½	¼	4.40	10.365	10⅜	0.875	⅞	1 9/16
×38.5	11.3	8.260	8¼	0.455	7/16	¼	3.76	10.295	10¼	0.760	¾	1 7/16
×33.5	9.84	8.165	8⅛	0.395	⅜	3/16	3.23	10.235	10¼	0.665	11/16	1⅜
WT 8×28.5	8.38	8.215	8¼	0.430	7/16	¼	3.53	7.120	7⅛	0.715	11/16	1⅜
×25	7.37	8.130	8⅛	0.380	⅜	3/16	3.09	7.070	7⅛	0.630	⅝	1 5/16
×22.5	6.63	8.065	8⅛	0.345	⅜	3/16	2.78	7.035	7	0.565	9/16	1¼
×20	5.89	8.005	8	0.305	5/16	3/16	2.44	6.995	7	0.505	½	1 3/16
×18	5.28	7.930	7⅞	0.295	5/16	3/16	2.34	6.985	7	0.430	7/16	1⅛
WT 8×15.5	4.56	7.940	8	0.275	¼	⅛	2.18	5.525	5½	0.440	7/16	1⅛
×13	3.84	7.845	7⅞	0.250	¼	⅛	1.96	5.500	5½	0.345	⅜	1 1/16

STRUCTURAL TEES
Cut from W shapes
Properties

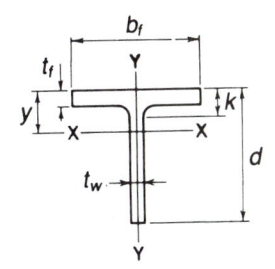

Nom- inal Wt. per Ft	$\dfrac{d}{t_w}$	Axis X-X				Axis Y-Y			$C_c' = \sqrt{\dfrac{2\pi^2 E}{Q_s Q_a F_y}}$, $Q_a = 1.0$			
		I	S	r	y	I	S	r	$F_y = 36$ ksi		$F_y = 50$ ksi	
Lb.		In.⁴	In.³	In.	In.	In.⁴	In.³	In.	Q_s	C_c'	Q_s	C_c'
71.5	11.0	142	18.5	2.60	2.09	156	27.7	2.72	—	—	—	—
65	11.9	127	16.7	2.58	2.02	139	24.9	2.70	—	—	—	—
59.5	14.5	119	15.9	2.60	2.03	126	22.5	2.69	—	—	—	—
53	15.9	104	14.1	2.59	1.97	110	19.7	2.66	—	—	—	—
48.5	17.4	93.8	12.7	2.56	1.91	100	18.0	2.65	—	—	—	—
43	19.2	82.4	11.2	2.55	1.86	87.6	15.8	2.63	—	—	0.937	111
38	21.4	71.8	9.83	2.54	1.80	76.2	13.8	2.61	0.990	127	0.826	118
35.5	18.7	78.2	11.2	2.74	2.26	30.1	7.89	1.70	—	—	0.963	109
32.5	20.4	70.7	10.1	2.72	2.20	27.4	7.22	1.69	—	—	0.877	114
30	22.0	64.7	9.29	2.71	2.16	25.0	6.63	1.69	0.964	128	0.796	120
27.5	23.2	59.5	8.63	2.71	2.16	22.5	5.97	1.67	0.913	132	0.735	125
25	25.3	53.5	7.79	2.70	2.12	20.0	5.35	1.65	0.823	139	0.625	135
23	25.1	52.1	7.77	2.77	2.33	11.3	3.72	1.29	0.831	138	0.635	134
20	28.4	44.8	6.73	2.76	2.29	9.55	3.17	1.27	0.690	152	0.496	152
17.5	29.5	40.1	6.21	2.79	2.39	7.67	2.56	1.22	0.638	158	0.460	158
50	14.5	76.8	11.4	2.28	1.76	93.1	17.9	2.51	—	—	—	—
44.5	16.0	67.2	10.1	2.27	1.70	81.3	15.7	2.49	—	—	—	—
38.5	18.2	56.9	8.59	2.24	1.63	69.2	13.4	2.47	—	—	0.988	108
33.5	20.7	48.6	7.36	2.22	1.56	59.5	11.6	2.46	—	—	0.861	115
28.5	19.1	48.7	7.77	2.41	1.94	21.6	6.06	1.60	—	—	0.942	110
25	21.4	42.3	6.78	2.40	1.89	18.6	5.26	1.59	0.990	127	0.826	118
22.5	23.4	37.8	6.10	2.39	1.86	16.4	4.67	1.57	0.904	133	0.725	126
20	26.2	33.1	5.35	2.37	1.81	14.4	4.12	1.57	0.784	142	0.583	140
18	26.9	30.6	5.05	2.41	1.88	12.2	3.50	1.52	0.754	145	0.553	144
15.5	28.9	27.4	4.64	2.45	2.02	6.20	2.24	1.17	0.668	154	0.479	155
13	31.4	23.5	4.09	2.47	2.09	4.80	1.74	1.12	0.563	168	0.406	168

Where no value of C_c' or Q_s is shown, the Tee complies with the noncompact section criteria of Specification Sect. B5.1

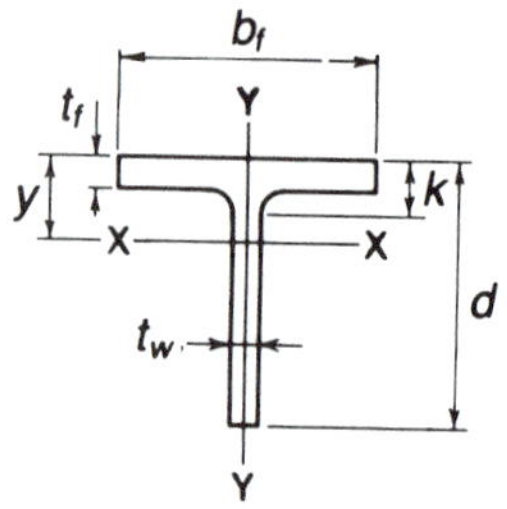

STRUCTURAL TEES
Cut from W shapes
Dimensions

Designation	Area	Depth of Tee d		Stem Thickness t_w		Stem $\frac{t_w}{2}$	Area of Stem	Flange Width b_f		Flange Thickness t_f		Distance k
	In.^{2}	In.		In.		In.	In.^{2}	In.		In.		In.
WT 7×365	107	11.210	11¼	3.070	3$\frac{1}{16}$	1$\frac{9}{16}$	34.4	17.890	17⅞	4.910	4$\frac{15}{16}$	5$\frac{9}{16}$
×332.5	97.8	10.820	10⅞	2.830	2$\frac{13}{16}$	1$\frac{7}{16}$	30.6	17.650	17⅝	4.520	4½	5$\frac{3}{16}$
×302.5	88.9	10.460	10½	2.595	2⅝	1$\frac{5}{16}$	27.1	17.415	17⅜	4.160	4$\frac{3}{16}$	4$\frac{13}{16}$
×275	80.9	10.120	10⅛	2.380	2⅜	1$\frac{3}{16}$	24.1	17.200	17¼	3.820	3$\frac{13}{16}$	4½
×250	73.5	9.800	9¾	2.190	2$\frac{3}{16}$	1⅛	21.5	17.010	17	3.500	3½	4$\frac{3}{16}$
×227.5	66.9	9.510	9½	2.015	2	1	19.2	16.835	16⅞	3.210	3$\frac{3}{16}$	3⅞
×213	62.6	9.335	9⅜	1.875	1⅞	$\frac{15}{16}$	17.5	16.695	16¾	3.035	3$\frac{1}{16}$	3$\frac{11}{16}$
×199	58.5	9.145	9⅛	1.770	1¾	⅞	16.2	16.590	16⅝	2.845	2⅞	3½
×185	54.4	8.960	9	1.655	1⅝	$\frac{13}{16}$	14.8	16.475	16½	2.660	2$\frac{11}{16}$	3$\frac{5}{16}$
×171	50.3	8.770	8¾	1.540	1$\frac{9}{16}$	$\frac{13}{16}$	13.5	16.360	16⅜	2.470	2½	3⅛
×155.5	45.7	8.560	8½	1.410	1$\frac{7}{16}$	¾	12.1	16.230	16¼	2.260	2¼	2$\frac{15}{16}$
×141.5	41.6	8.370	8⅜	1.290	1$\frac{5}{16}$	$\frac{11}{16}$	10.8	16.110	16⅛	2.070	2$\frac{1}{16}$	2¾
×128.5	37.8	8.190	8¼	1.175	1$\frac{3}{16}$	⅝	9.62	15.995	16	1.890	1⅞	2$\frac{9}{16}$
×116.5	34.2	8.020	8	1.070	1$\frac{1}{16}$	$\frac{9}{16}$	8.58	15.890	15⅞	1.720	1¾	2⅜
×105.5	31.0	7.860	7⅞	0.980	1	½	7.70	15.800	15¾	1.560	1$\frac{9}{16}$	2¼
× 96.5	28.4	7.740	7¾	0.890	⅞	$\frac{7}{16}$	6.89	15.710	15¾	1.440	1$\frac{7}{16}$	2⅛
× 88	25.9	7.610	7⅝	0.830	$\frac{13}{16}$	$\frac{7}{16}$	6.32	15.650	15⅝	1.310	1$\frac{5}{16}$	2
× 79.5	23.4	7.490	7½	0.745	¾	⅜	5.58	15.565	15⅝	1.190	1$\frac{3}{16}$	1⅞
× 72.5	21.3	7.390	7⅜	0.680	$\frac{11}{16}$	⅜	5.03	15.500	15½	1.090	1$\frac{1}{16}$	1¾

STRUCTURAL TEES
Cut from W shapes
Properties

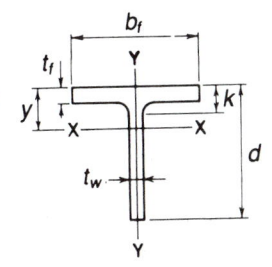

Nom-inal Wt. per Ft Lb.	$\dfrac{d}{t_w}$	Axis X-X				Axis Y-Y			$C_c' = \sqrt{\dfrac{2\pi^2 E}{Q_s Q_a F_y}}$, $Q_a = 1.0$			
									$F_y = 36$ ksi		$F_y = 50$ ksi	
		I	S	r	y	I	S	r	Q_s	C_c'	Q_s	C_c'
		In.⁴	In.³	In.	In.	In.⁴	In.³	In.				
365	3.7	739	95.4	2.62	3.47	2360	264	4.69	—	—	—	—
332.5	3.8	622	82.1	2.52	3.25	2080	236	4.62	—	—	—	—
302.5	4.0	524	70.6	2.43	3.05	1840	211	4.55	—	—	—	—
275	4.3	442	60.9	2.34	2.85	1630	189	4.49	—	—	—	—
250	4.5	375	52.7	2.26	2.67	1440	169	4.43	—	—	—	—
227.5	4.7	321	45.9	2.19	2.51	1280	152	4.38	—	—	—	—
213	5.0	287	41.4	2.14	2.40	1180	141	4.34	—	—	—	—
199	5.2	257	37.6	2.10	2.30	1090	131	4.31	—	—	—	—
185	5.4	229	33.9	2.05	2.19	994	121	4.27	—	—	—	—
171	5.7	203	30.4	2.01	2.09	903	110	4.24	—	—	—	—
155.5	6.1	176	26.7	1.96	1.97	807	99.4	4.20	—	—	—	—
141.5	6.5	153	23.5	1.92	1.86	722	89.7	4.17	—	—	—	—
128.5	7.0	133	20.7	1.88	1.75	645	80.7	4.13	—	—	—	—
116.5	7.5	116	18.2	1.84	1.65	576	72.5	4.10	—	—	—	—
105.5	8.0	102	16.2	1.81	1.57	513	65.0	4.07	—	—	—	—
96.5	8.7	89.8	14.4	1.78	1.49	466	59.3	4.05	—	—	—	—
88	9.2	80.5	13.0	1.76	1.43	419	53.5	4.02	—	—	—	—
79.5	10.1	70.2	11.4	1.73	1.35	374	48.1	4.00	—	—	—	—
72.5	10.9	62.5	10.2	1.71	1.29	338	43.7	3.98	—	—	—	—

Where no value of C_c' or Q_s is shown, the Tee complies with the noncompact section criteria of Specification Sect. B5.1

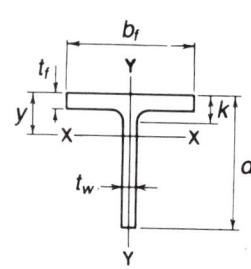

STRUCTURAL TEES
Cut from W shapes
Dimensions

Designation	Area	Depth of Tee d		Stem Thickness t_w		$\frac{t_w}{2}$	Area of Stem	Flange Width b_f		Flange Thickness t_f		Distance k
	In.²	In.		In.		In.	In.²	In.		In.		In.
WT 7×66	19.4	7.330	7⅜	0.645	⅝	⁵⁄₁₆	4.73	14.725	14¾	1.030	1	1¹¹⁄₁₆
×60	17.7	7.240	7¼	0.590	⁹⁄₁₆	⁵⁄₁₆	4.27	14.670	14⅝	0.940	¹⁵⁄₁₆	1⅝
×54.5	16.0	7.160	7⅛	0.525	½	¼	3.76	14.605	14⅝	0.860	⅞	1⁹⁄₁₆
×49.5	14.6	7.080	7⅛	0.485	½	¼	3.43	14.565	14⅝	0.780	¾	1⁷⁄₁₆
×45	13.2	7.010	7	0.440	⁷⁄₁₆	¼	3.08	14.520	14½	0.710	¹¹⁄₁₆	1⅜
WT 7×41	12.0	7.155	7⅛	0.510	½	¼	3.65	10.130	10⅛	0.855	⅞	1⅝
×37	10.9	7.085	7⅛	0.450	⁷⁄₁₆	¼	3.19	10.070	10⅛	0.785	¹³⁄₁₆	1⁹⁄₁₆
×34	9.99	7.020	7	0.415	⁷⁄₁₆	¼	2.91	10.035	10	0.720	¾	1½
×30.5	8.96	6.945	7	0.375	⅜	³⁄₁₆	2.60	9.995	10	0.645	⅝	1⁷⁄₁₆
WT 7×26.5	7.81	6.960	7	0.370	⅜	³⁄₁₆	2.58	8.060	8	0.660	¹¹⁄₁₆	1⁷⁄₁₆
×24	7.07	6.895	6⅞	0.340	⁵⁄₁₆	³⁄₁₆	2.34	8.030	8	0.595	⅝	1⅜
×21.5	6.31	6.830	6⅞	0.305	⁵⁄₁₆	³⁄₁₆	2.08	7.995	8	0.530	½	1⁵⁄₁₆
WT 7×19	5.58	7.050	7	0.310	⁵⁄₁₆	³⁄₁₆	2.19	6.770	6¾	0.515	½	1¹⁄₁₆
×17	5.00	6.990	7	0.285	⁵⁄₁₆	³⁄₁₆	1.99	6.745	6¾	0.455	⁷⁄₁₆	1
×15	4.42	6.920	6⅞	0.270	¼	⅛	1.87	6.730	6¾	0.385	⅜	¹⁵⁄₁₆
WT 7×13	3.85	6.995	7	0.255	¼	⅛	1.77	5.025	5	0.420	⁷⁄₁₆	¹⁵⁄₁₆
×11	3.25	6.870	6⅞	0.230	¼	⅛	1.58	5.000	5	0.335	⁵⁄₁₆	⅞

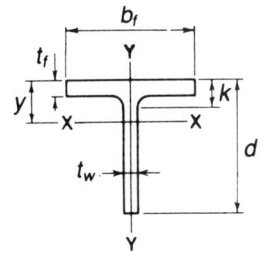

STRUCTURAL TEES
Cut from W shapes
Properties

Nom-inal Wt. per Ft	$\dfrac{d}{t_w}$	Axis X-X				Axis Y-Y			$C_c' = \sqrt{\dfrac{2\pi^2 E}{Q_s Q_a F_y}}$, $Q_a = 1.0$			
		I	S	r	y	I	S	r	$F_y = 36$ ksi		$F_y = 50$ ksi	
Lb.		In.^{4}	In.^{3}	In.	In.	In.^{4}	In.^{3}	In.	Q_s	C_c'	Q_s	C_c'
66	11.4	57.8	9.57	1.73	1.29	274	37.2	3.76	—	—	—	—
60	12.3	51.7	8.61	1.71	1.24	247	33.7	3.74	—	—	—	—
54.5	13.6	45.3	7.56	1.68	1.17	223	30.6	3.73	—	—	—	—
49.5	14.6	40.9	6.88	1.67	1.14	201	27.6	3.71	—	—	—	—
45	15.9	36.4	6.16	1.66	1.09	181	25.0	3.70	—	—	—	—
41	14.0	41.2	7.14	1.85	1.39	74.2	14.6	2.48	—	—	—	—
37	15.7	36.0	6.25	1.82	1.32	66.9	13.3	2.48	—	—	—	—
34	16.9	32.6	5.69	1.81	1.29	60.7	12.1	2.46	—	—	—	—
30.5	18.5	28.9	5.07	1.80	1.25	53.7	10.7	2.45	—	—	0.973	108
26.5	18.8	27.6	4.94	1.88	1.38	28.8	7.16	1.92	—	—	0.958	109
24	20.3	24.9	4.48	1.87	1.35	25.7	6.40	1.91	—	—	0.882	114
21.5	22.4	21.9	3.98	1.86	1.31	22.6	5.65	1.89	0.947	130	0.775	122
19	22.7	23.3	4.22	2.04	1.54	13.3	3.94	1.55	0.934	130	0.760	123
17	24.5	20.9	3.83	2.04	1.53	11.7	3.45	1.53	0.857	136	0.669	131
15	25.6	19.0	3.55	2.07	1.58	9.79	2.91	1.49	0.810	140	0.610	137
13	27.3	17.3	3.31	2.12	1.72	4.45	1.77	1.08	0.737	147	0.537	146
11	29.9	14.8	2.91	2.14	1.76	3.50	1.40	1.04	0.621	160	0.447	160

Where no value of C_c' or Q_s is shown, the Tee complies with the noncompact section criteria of Specification Sect. B5.1

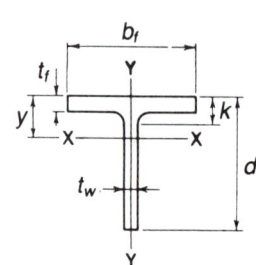

STRUCTURAL TEES
Cut from W shapes
Dimensions

Designation	Area	Depth of Tee d	Stem Thickness t_w		$\frac{t_w}{2}$	Area of Stem	Flange Width b_f		Flange Thickness t_f		Distance k
	In.²	In.	In.		In.	In.²	In.		In.		In.
WT 6×168	49.4	8.410	8⅜	1.775 1¾	⅞	14.9	13.385 13⅜		2.955 2¹⁵/₁₆		3¹¹/₁₆
×152.5	44.8	8.160	8⅛	1.625 1⅝	¹³/₁₆	13.3	13.235 13¼		2.705 2¹¹/₁₆		3⁷/₁₆
×139.5	41.0	7.925	7⅞	1.530 1½	¾	12.1	13.140 13⅛		2.470 2½		3³/₁₆
×126	37.0	7.705	7¾	1.395 1⅜	¹¹/₁₆	10.7	13.005 13		2.250 2¼		2¹⁵/₁₆
×115	33.9	7.525	7½	1.285 1⁵/₁₆	¹¹/₁₆	9.67	12.895 12⅞		2.070 2¹/₁₆		2¾
×105	30.9	7.355	7⅜	1.180 1³/₁₆	⅝	8.68	12.790 12¾		1.900 1⅞		2⅝
× 95	27.9	7.190	7¼	1.060 1¹/₁₆	⁹/₁₆	7.62	12.670 12⅝		1.735 1¾		2⁷/₁₆
× 85	25.0	7.015	7	0.960 ¹⁵/₁₆	½	6.73	12.570 12⅝		1.560 1⁹/₁₆		2¼
× 76	22.4	6.855	6⅞	0.870 ⅞	⁷/₁₆	5.96	12.480 12½		1.400 1⅜		2⅛
× 68	20.0	6.705	6¾	0.790 ¹³/₁₆	⁷/₁₆	5.30	12.400 12⅜		1.250 1¼		1¹⁵/₁₆
× 60	17.6	6.560	6½	0.710 ¹¹/₁₆	⅜	4.66	12.320 12⅜		1.105 1⅛		1¹³/₁₆
× 53	15.6	6.445	6½	0.610 ⅝	⁵/₁₆	3.93	12.220 12¼		0.990 1		1¹¹/₁₆
× 48	14.1	6.355	6⅜	0.550 ⁹/₁₆	⁵/₁₆	3.50	12.160 12⅛		0.900 ⅞		1⅝
× 43.5	12.8	6.265	6¼	0.515 ½	¼	3.23	12.125 12⅛		0.810 ¹³/₁₆		1½
× 39.5	11.6	6.190	6¼	0.470 ½	¼	2.91	12.080 12⅛		0.735 ¾		1⁷/₁₆
× 36	10.6	6.125	6⅛	0.430 ⁷/₁₆	¼	2.63	12.040 12		0.670 ¹¹/₁₆		1⅜
× 32.5	9.54	6.060	6	0.390 ⅜	³/₁₆	2.36	12.000 12		0.605 ⅝		1⁵/₁₆
WT 6× 29	8.52	6.095	6⅛	0.360 ⅜	³/₁₆	2.19	10.010 10		0.640 ⅝		1⅜
× 26.5	7.78	6.030	6	0.345 ⅜	³/₁₆	2.08	9.995 10		0.575 ⁹/₁₆		1¼
WT 6× 25	7.34	6.095	6⅛	0.370 ⅜	³/₁₆	2.26	8.080 8⅛		0.640 ⅝		1⅜
× 22.5	6.61	6.030	6	0.335 ⁵/₁₆	³/₁₆	2.02	8.045 8		0.575 ⁹/₁₆		1¼
× 20	5.89	5.970	6	0.295 ⁵/₁₆	³/₁₆	1.76	8.005 8		0.515 ½		1¼
WT 6× 17.5	5.17	6.250	6¼	0.300 ⁵/₁₆	³/₁₆	1.88	6.560 6½		0.520 ½		1
× 15	4.40	6.170	6⅛	0.260 ¼	⅛	1.60	6.520 6½		0.440 ⁷/₁₆		¹⁵/₁₆
× 13	3.82	6.110	6⅛	0.230 ¼	⅛	1.41	6.490 6½		0.380 ⅜		⅞
WT 6× 11	3.24	6.155	6⅛	0.260 ¼	⅛	1.60	4.030 4		0.425 ⁷/₁₆		⅞
× 9.5	2.79	6.080	6⅛	0.235 ¼	⅛	1.43	4.005 4		0.350 ⅜		¹³/₁₆
× 8	2.36	5.995	6	0.220 ¼	⅛	1.32	3.990 4		0.265 ¼		¾
× 7	2.08	5.995	6	0.200 ³/₁₆	⅛	1.19	3.970 4		0.225 ¼		¹¹/₁₆

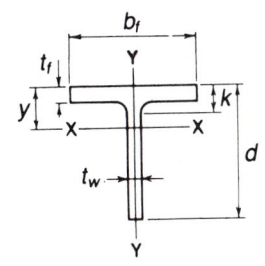

STRUCTURAL TEES
Cut from W shapes
Properties

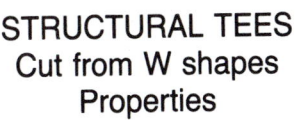

Nom-inal Wt. per Ft Lb.	$\frac{d}{t_w}$	Axis X-X				Axis Y-Y			$C_c' = \sqrt{\frac{2\pi^2 E}{Q_s Q_a F_y}}$, $Q_a = 1.0$			
									$F_y = 36$ ksi		$F_y = 50$ ksi	
		I	S	r	y	I	S	r	Q_s	C_c'	Q_s	C_c'
Lb.		In.⁴	In.³	In.	In.	In.⁴	In.³	In.				
168	4.7	190	31.2	1.96	2.31	593	88.6	3.47	—	—	—	—
152.5	5.0	162	27.0	1.90	2.16	525	79.3	3.42	—	—	—	—
139.5	5.2	141	24.1	1.86	2.05	469	71.3	3.38	—	—	—	—
126	5.5	121	20.9	1.81	1.92	414	63.6	3.34	—	—	—	—
115	5.9	106	18.5	1.77	1.82	371	57.5	3.31	—	—	—	—
105	6.2	92.1	16.4	1.73	1.72	332	51.9	3.28	—	—	—	—
95	6.8	79.0	14.2	1.68	1.62	295	46.5	3.25	—	—	—	—
85	7.3	67.8	12.3	1.65	1.52	259	41.2	3.22	—	—	—	—
76	7.9	58.5	10.8	1.62	1.43	227	36.4	3.19	—	—	—	—
68	8.5	50.6	9.46	1.59	1.35	199	32.1	3.16	—	—	—	—
60	9.2	43.4	8.22	1.57	1.28	172	28.0	3.13	—	—	—	—
53	10.6	36.3	6.91	1.53	1.19	151	24.7	3.11	—	—	—	—
48	11.6	32.0	6.12	1.51	1.13	135	22.2	3.09	—	—	—	—
43.5	12.2	28.9	5.60	1.50	1.10	120	19.9	3.07	—	—	—	—
39.5	13.2	25.8	5.03	1.49	1.06	108	17.9	3.05	—	—	—	—
36	14.2	23.2	4.54	1.48	1.02	97.5	16.2	3.04	—	—	—	—
32.5	15.5	20.6	4.06	1.47	0.985	87.2	14.5	3.02	—	—	—	—
29	16.9	19.1	3.76	1.50	1.03	53.5	10.7	2.51	—	—	—	—
26.5	17.5	17.7	3.54	1.51	1.02	47.9	9.58	2.48	—	—	—	—
25	16.5	18.7	3.79	1.60	1.17	28.2	6.97	1.96	—	—	—	—
22.5	18.0	16.6	3.39	1.58	1.13	25.0	6.21	1.94	—	—	0.998	107
20	20.2	14.4	2.95	1.57	1.08	22.0	5.51	1.93	—	—	0.887	114
17.5	20.8	16.0	3.23	1.76	1.30	12.2	3.73	1.54	—	—	0.856	116
15	23.7	13.5	2.75	1.75	1.27	10.2	3.12	1.52	0.891	134	0.710	127
13	26.6	11.7	2.40	1.75	1.25	8.66	2.67	1.51	0.767	144	0.565	142
11	23.7	11.7	2.59	1.90	1.63	2.33	1.16	0.847	0.891	134	0.710	127
9.5	25.9	10.1	2.28	1.90	1.65	1.88	0.939	0.822	0.797	141	0.596	139
8	27.2	8.70	2.04	1.92	1.74	1.41	0.706	0.773	0.741	146	0.541	146
7	29.8	7.67	1.83	1.92	1.76	1.18	0.594	0.753	0.626	159	0.450	159

Where no value of C_c' or Q_s is shown, the Tee complies with the noncompact section criteria of Specification Sect. B5.1

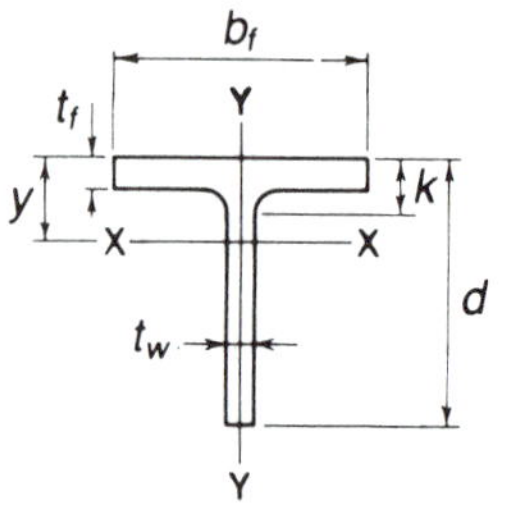

STRUCTURAL TEES
Cut from W shapes
Dimensions

Designation	Area	Depth of Tee d		Stem Thickness t_w		Stem $\frac{t_w}{2}$	Area of Stem	Flange Width b_f		Flange Thickness t_f		Distance k
	In.^{2}	In.		In.		In.	In.^{2}	In.		In.		In.
WT 5×56	16.5	5.680	5⅝	0.755	¾	⅜	4.29	10.415	10⅜	1.250	1¼	1⅞
×50	14.7	5.550	5½	0.680	11/16	⅜	3.77	10.340	10⅜	1.120	1⅛	1¾
×44	12.9	5.420	5⅜	0.605	⅝	5/16	3.28	10.265	10¼	0.990	1	1⅝
×38.5	11.3	5.300	5¼	0.530	½	¼	2.81	10.190	10¼	0.870	⅞	1½
×34	9.99	5.200	5¼	0.470	½	¼	2.44	10.130	10⅛	0.770	⅜	1⅜
×30	8.82	5.110	5⅛	0.420	7/16	¼	2.15	10.080	10⅛	0.680	11/16	1 5/16
×27	7.91	5.045	5	0.370	⅜	3/16	1.87	10.030	10	0.615	⅝	1¼
×24.5	7.21	4.990	5	0.340	5/16	3/16	1.70	10.000	10	0.560	9/16	1 3/16
WT 5×22.5	6.63	5.050	5	0.350	⅜	3/16	1.77	8.020	8	0.620	⅝	1¼
×19.5	5.73	4.960	5	0.315	5/16	3/16	1.56	7.985	8	0.530	½	1⅛
×16.5	4.85	4.865	4⅞	0.290	5/16	3/16	1.41	7.960	8	0.435	7/16	1 1/16
WT 5×15	4.42	5.235	5¼	0.300	5/16	3/16	1.57	5.810	5¾	0.510	½	15/16
×13	3.81	5.165	5⅛	0.260	¼	⅛	1.34	5.770	5¾	0.440	7/16	⅞
×11	3.24	5.085	5⅛	0.240	¼	⅛	1.22	5.750	5¾	0.360	⅜	¾
WT 5× 9.5	2.81	5.120	5⅛	0.250	¼	⅛	1.28	4.020	4	0.395	⅜	13/16
× 8.5	2.50	5.055	5	0.240	¼	⅛	1.21	4.010	4	0.330	5/16	¾
× 7.5	2.21	4.995	5	0.230	¼	⅛	1.15	4.000	4	0.270	¼	11/16
× 6	1.77	4.935	4⅞	0.190	3/16	⅛	0.938	3.960	4	0.210	3/16	⅝

STRUCTURAL TEES
Cut from W shapes
Properties

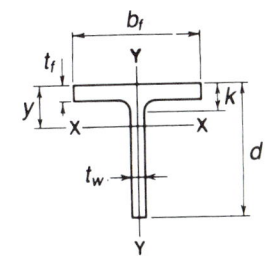

Nom-inal Wt. per Ft Lb.	$\dfrac{d}{t_w}$	Axis X-X				Axis Y-Y			$C_c' = \sqrt{\dfrac{2\pi^2 E}{Q_s Q_a F_y}}$, $Q_a = 1.0$			
									$F_y = 36$ ksi		$F_y = 50$ ksi	
		I	S	r	y	I	S	r	Q_s	C_c'	Q_s	C_c'
		In.^{4}	In.^{3}	In.	In.	In.^{4}	In.^{3}	In.				
56	7.5	28.6	6.40	1.32	1.21	118	22.6	2.68	—	—	—	—
50	8.2	24.5	5.56	1.29	1.13	103	20.0	2.65	—	—	—	—
44	9.0	20.8	4.77	1.27	1.06	89.3	17.4	2.63	—	—	—	—
38.5	10.0	17.4	4.04	1.24	0.990	76.8	15.1	2.60	—	—	—	—
34	11.1	14.9	3.49	1.22	0.932	66.8	13.2	2.59	—	—	—	—
30	12.2	12.9	3.04	1.21	0.884	58.1	11.5	2.57	—	—	—	—
27	13.6	11.1	2.64	1.19	0.836	51.7	10.3	2.56	—	—	—	—
24.5	14.7	10.0	2.39	1.18	0.807	46.7	9.34	2.54	—	—	—	—
22.5	14.4	10.2	2.47	1.24	0.907	26.7	6.65	2.01	—	—	—	—
19.5	15.7	8.84	2.16	1.24	0.876	22.5	5.64	1.98	—	—	—	—
16.5	16.8	7.71	1.93	1.26	0.869	18.3	4.60	1.94	—	—	—	—
15	17.4	9.28	2.24	1.45	1.10	8.35	2.87	1.37	—	—	—	—
13	19.9	7.86	1.91	1.44	1.06	7.05	2.44	1.36	—	—	0.902	113
11	21.2	6.88	1.72	1.46	1.07	5.71	1.99	1.33	0.999	126	0.836	117
9.5	20.5	6.68	1.74	1.54	1.28	2.15	1.07	0.874	—	—	0.872	115
8.5	21.1	6.06	1.62	1.56	1.32	1.78	0.888	0.844	—	—	0.841	117
7.5	21.7	5.45	1.50	1.57	1.37	1.45	0.723	0.810	0.977	128	0.811	119
6	26.0	4.35	1.22	1.57	1.36	1.09	0.551	0.785	0.793	142	0.592	139

Where no value of C_c' or Q_s is shown, the Tee complies with the noncompact section criteria of Specification Sect. B5.1.

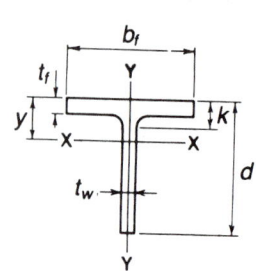

STRUCTURAL TEES
Cut from W shapes
Dimensions

Designation	Area	Depth of Tee d		Stem Thickness t_w		$\frac{t_w}{2}$	Area of Stem	Flange Width b_f		Flange Thickness t_f		Distance k
	In.²	In.		In.		In.	In.²	In.		In.		In.
WT 4 ×33.5	9.84	4.500	4½	0.570	9/16	5/16	2.56	8.280	8¼	0.935	15/16	1 7/16
×29	8.55	4.375	4⅜	0.510	½	¼	2.23	8.220	8¼	0.810	13/16	1 5/16
×24	7.05	4.250	4¼	0.400	⅜	3/16	1.70	8.110	8⅛	0.685	11/16	1 3/16
×20	5.87	4.125	4⅛	0.360	⅜	3/16	1.48	8.070	8⅛	0.560	9/16	1 1/16
×17.5	5.14	4.060	4	0.310	5/16	3/16	1.26	8.020	8	0.495	½	1
×15.5	4.56	4.000	4	0.285	5/16	3/16	1.14	7.995	8	0.435	7/16	15/16
WT 4 ×14	4.12	4.030	4	0.285	5/16	3/16	1.15	6.535	6½	0.465	7/16	15/16
×12	3.54	3.965	4	0.245	¼	⅛	0.971	6.495	6½	0.400	⅜	⅞
WT 4 ×10.5	3.08	4.140	4⅛	0.250	¼	⅛	1.03	5.270	5¼	0.400	⅜	13/16
× 9	2.63	4.070	4⅛	0.230	¼	⅛	0.936	5.250	5¼	0.330	5/16	¾
WT 4 × 7.5	2.22	4.055	4	0.245	¼	⅛	0.993	4.015	4	0.315	5/16	¾
× 6.5	1.92	3.995	4	0.230	¼	⅛	0.919	4.000	4	0.255	¼	11/16
× 5	1.48	3.945	4	0.170	3/16	⅛	0.671	3.940	4	0.205	3/16	⅝
WT 3 ×12.5	3.67	3.190	3¼	0.320	5/16	3/16	1.02	6.080	6⅛	0.455	7/16	13/16
×10	2.94	3.100	3⅛	0.260	¼	⅛	0.806	6.020	6	0.365	⅜	¾
× 7.5	2.21	2.995	3	0.230	¼	⅛	0.689	5.990	6	0.260	¼	⅝
WT 3 × 8	2.37	3.140	3⅛	0.260	¼	⅛	0.816	4.030	4	0.405	⅜	¾
× 6	1.78	3.015	3	0.230	¼	⅛	0.693	4.000	4	0.280	¼	⅝
× 4.5	1.34	2.950	3	0.170	3/16	⅛	0.502	3.940	4	0.215	3/16	9/16
WT 2.5× 9.5	2.77	2.575	2⅝	0.270	¼	⅛	0.695	5.030	5	0.430	7/16	13/16
× 8	2.34	2.505	2½	0.240	¼	⅛	0.601	5.000	5	0.360	⅜	¾
WT 2 × 6.5	1.91	2.080	2⅛	0.280	¼	⅛	0.582	4.060	4	0.345	⅜	11/16

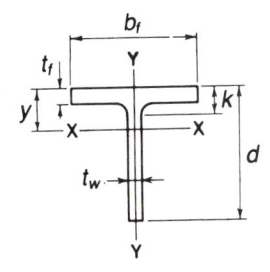

STRUCTURAL TEES
Cut from W shapes
Properties

Nominal Wt. per Ft Lb.	$\dfrac{d}{t_w}$	Axis X-X				Axis Y-Y			$C_c' = \sqrt{\dfrac{2\pi^2 E}{Q_s Q_a F_y}}$, $Q_a = 1.0$			
		I	S	r	y	I	S	r	$F_y = 36$ ksi		$F_y = 50$ ksi	
		In.⁴	In.³	In.	In.	In.⁴	In.³	In.	Q_s	C_c'	Q_s	C_c'
33.5	7.9	10.9	3.05	1.05	0.936	44.3	10.7	2.12	—	—	—	—
29	8.6	9.12	2.61	1.03	0.874	37.5	9.13	2.10	—	—	—	—
24	10.6	6.85	1.97	0.986	0.777	30.5	7.52	2.08	—	—	—	—
20	11.5	5.73	1.69	0.988	0.735	24.5	6.08	2.04	—	—	—	—
17.5	13.1	4.81	1.43	0.967	0.688	21.3	5.31	2.03	—	—	—	—
15.5	14.0	4.28	1.28	0.968	0.667	18.5	4.64	2.02	—	—	—	—
14	14.1	4.22	1.28	1.01	0.734	10.8	3.31	1.62	—	—	—	—
12	16.2	3.53	1.08	0.999	0.695	9.14	2.81	1.61	—	—	—	—
10.5	16.6	3.90	1.18	1.12	0.831	4.89	1.85	1.26	—	—	—	—
9	17.7	3.41	1.05	1.14	0.834	3.98	1.52	1.23	—	—	—	—
7.5	16.6	3.28	1.07	1.22	0.998	1.70	0.849	0.876	—	—	—	—
6.5	17.4	2.89	0.974	1.23	1.03	1.37	0.683	0.843	—	—	—	—
5	23.2	2.15	0.717	1.20	0.953	1.05	0.532	0.841	0.913	132	0.735	125
12.5	10.0	2.28	0.886	0.789	0.610	8.53	2.81	1.52	—	—	—	—
10	11.9	1.76	0.693	0.774	0.560	6.64	2.21	1.50	—	—	—	—
7.5	13.0	1.41	0.577	0.797	0.558	4.66	1.56	1.45	—	—	—	—
8	12.1	1.69	0.685	0.844	0.676	2.21	1.10	0.966	—	—	—	—
6	13.1	1.32	0.564	0.861	0.677	1.50	0.748	0.918	—	—	—	—
4.5	17.4	0.950	0.408	0.842	0.623	1.10	0.557	0.905	—	—	—	—
9.5	9.5	1.01	0.485	0.605	0.487	4.56	1.82	1.28	—	—	—	—
8	10.4	0.845	0.413	0.601	0.458	3.75	1.50	1.27	—	—	—	—
6.5	7.4	0.526	0.321	0.524	0.440	1.93	0.950	1.00	—	—	—	—

Where no value of C_c' or Q_s is shown, the Tee complies with the noncompact section criteria of Specification Sect. B5.1.

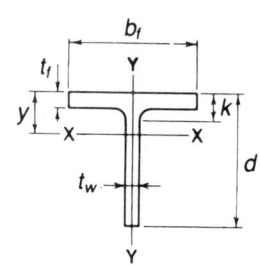

STRUCTURAL TEES
Cut from M shapes
Dimensions

Designation	Area	Depth of Tee d		Stem			Area of Stem	Flange				Dis-tance k	Grip	Max Flge Fas-ten-er
				Thickness t_w		$\frac{t_w}{2}$		Width b_f		Thickness t_f				
	In.²	In.		In.		In.	In.²	In.		In.		In.	In.	In.
MT 7 ×9	2.55	7.000	7	0.215	3/16	1/8	1.50	4.000	4	0.270	1/4	5/8	1/4	3/4
MT 6 ×5.9	1.73	6.000	6	0.177	3/16	1/8	1.06	3.065	3 1/8	0.225	1/4	9/16	1/4	—
MT 5 ×4.5	1.32	5.000	5	0.157	3/16	1/8	0.785	2.690	2 3/4	0.206	3/16	9/16	3/16	—
MT 4 ×3.25	0.958	4.000	4	0.135	1/8	1/16	0.540	2.281	2 1/4	0.189	3/16	1/2	3/16	—
MT 3 ×2.2	0.646	3.000	3	0.114	1/8	1/16	0.342	1.844	1 7/8	0.171	3/16	7/16	3/16	—
MT 2.5×9.45	1.78	2.500	2 1/2	0.316	5/16	3/16	0.790	5.003	5	0.416	7/16	7/8	7/16	7/8

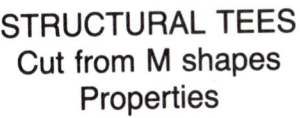
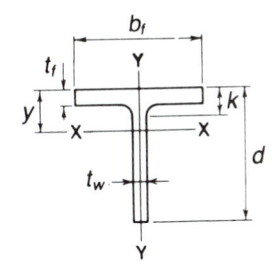

STRUCTURAL TEES
Cut from M shapes
Properties

Nom-inal Wt. per Ft	$\dfrac{d}{t_w}$	Axis X-X				Axis Y-Y			$C_c' = \sqrt{\dfrac{2\pi^2 E}{Q_s Q_a F_y}}$, $Q_a = 1.0$			
		I	S	r	y	I	S	r	$F_y = 36$ ksi		$F_y = 50$ ksi	
Lb.		In.^{4}	In.^{3}	In.	In.	In.^{4}	In.^{3}	In.	Q_s	C_c'	Q_s	C_c'
9	32.6	13.1	2.69	2.27	2.12	1.32	0.660	0.719	0.523	174	0.376	174
5.9	33.9	6.60	1.60	1.95	1.89	0.490	0.320	0.532	0.483	181	0.348	181
4.5	31.8	3.46	0.997	1.62	1.53	0.305	0.227	0.480	0.549	170	0.396	170
3.25	29.6	1.57	0.556	1.28	1.17	0.172	0.150	0.423	0.634	158	0.457	158
2.2	26.3	0.577	0.267	0.945	0.836	0.083	0.090	0.358	0.780	143	0.578	141
9.45	7.9	1.05	0.527	0.615	0.511	3.93	1.57	1.19	—	—	—	—

Where no value of C_c' or Q_s is shown, the Tee complies with the noncompact section criteria of Specification Sect. B5.1.

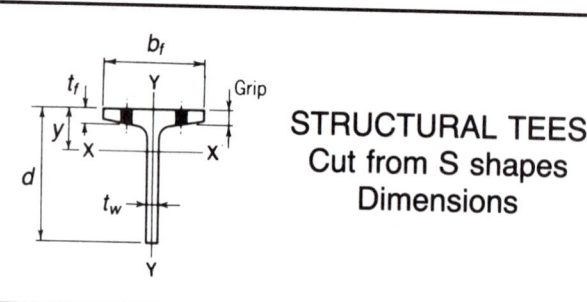

STRUCTURAL TEES
Cut from S shapes
Dimensions

Designation	Area	Depth of Tee d		Stem Thickness t_w		$\frac{t_w}{2}$	Area of Stem	Flange Width b_f		Flange Thickness t_f		Distance k	Grip	Max Flge Fastener
	In.²	In.		In.		In.	In.²	In.		In.		In.	In.	In.
ST 12 ×60.5	17.8	12.250	12¼	0.800	13/16	7/16	9.80	8.050	8	1.090	1 1/16	2	1 1/8	1
12 ×53	15.6	12.250	12¼	0.620	5/8	5/16	7.59	7.870	7 7/8	1.090	1 1/16	2	1 1/8	1
ST 12 ×50	14.7	12.000	12	0.745	3/4	3/8	8.94	7.245	7 1/4	0.870	7/8	1 3/4	7/8	1
12 ×45	13.2	12.000	12	0.625	5/8	5/16	7.50	7.125	7 1/8	0.870	7/8	1 3/4	7/8	1
12 ×40	11.7	12.000	12	0.500	1/2	1/4	6.00	7.000	7	0.870	7/8	1 3/4	7/8	1
ST 10 ×48	14.1	10.150	10 1/8	0.800	13/16	7/16	8.12	7.200	7 1/4	0.920	15/16	1 3/4	15/16	1
10 ×43	12.7	10.150	10 1/8	0.660	11/16	3/8	6.70	7.060	7	0.920	15/16	1 3/4	15/16	1
ST 10 ×37.5	11.0	10.000	10	0.635	5/8	5/16	6.35	6.385	6 3/8	0.795	13/16	1 5/8	13/16	7/8
10 ×33	9.70	10.000	10	0.505	1/2	1/4	5.05	6.255	6 1/4	0.795	13/16	1 5/8	13/16	7/8
ST 9 ×35	10.3	9.000	9	0.711	11/16	3/8	6.40	6.251	6 1/4	0.691	11/16	1 1/2	11/16	7/8
9 ×27.35	8.04	9.000	9	0.461	7/16	1/4	4.15	6.001	6	0.691	11/16	1 1/2	11/16	7/8
ST 7.5×25	7.35	7.500	7 1/2	0.550	9/16	5/16	4.13	5.640	5 5/8	0.622	5/8	1 3/8	9/16	3/4
7.5×21.45	6.31	7.500	7 1/2	0.411	7/16	1/4	3.08	5.501	5 1/2	0.622	5/8	1 3/8	9/16	3/4
ST 6 ×25	7.35	6.000	6	0.687	11/16	3/8	4.12	5.477	5 1/2	0.659	11/16	1 7/16	11/16	3/4
6 ×20.4	6.00	6.000	6	0.462	7/16	1/4	2.77	5.252	5 1/4	0.659	11/16	1 7/16	5/8	3/4
ST 6 ×17.5	5.15	6.000	6	0.428	7/16	1/4	2.57	5.078	5 1/8	0.545	9/16	1 3/16	1/2	3/4
6 ×15.9	4.68	6.000	6	0.350	3/8	3/16	2.10	5.000	5	0.544	9/16	1 3/16	1/2	3/4
ST 5 ×17.5	5.15	5.000	5	0.594	5/8	5/16	2.97	4.944	5	0.491	1/2	1 1/8	1/2	3/4
5 ×12.7	3.73	5.000	5	0.311	5/16	3/16	1.55	4.661	4 5/8	0.491	1/2	1 1/8	1/2	3/4
ST 4 ×11.5	3.38	4.000	4	0.441	7/16	1/4	1.76	4.171	4 1/8	0.425	7/16	1	7/16	3/4
4 × 9.2	2.70	4.000	4	0.271	1/4	1/8	1.08	4.001	4	0.425	7/16	1	7/16	3/4
ST 3.5×10	2.94	3.500	3 1/2	0.450	7/16	1/4	1.57	3.860	3 7/8	0.392	3/8	15/16	3/8	5/8
3.5× 7.65	2.25	3.500	3 1/2	0.252	1/4	1/8	0.882	3.662	3 5/8	0.392	3/8	15/16	3/8	5/8
ST 3 × 8.625	2.53	3.000	3	0.465	7/16	1/4	1.39	3.565	3 5/8	0.359	3/8	7/8	3/8	5/8
3 × 6.25	1.83	3.000	3	0.232	1/4	1/8	0.696	3.332	3 3/8	0.359	3/8	7/8	3/8	
ST 2.5× 7.375	2.17	2.500	2 1/2	0.494	1/2	1/4	1.23	3.284	3 1/4	0.326	5/16	13/16	5/16	—
2.5× 5	1.47	2.500	2 1/2	0.214	3/16	1/8	0.535	3.004	3	0.326	5/16	13/16	5/16	—
ST 2 × 4.75	1.40	2.000	2	0.326	5/16	3/16	0.652	2.796	2 3/4	0.293	5/16	3/4	5/16	—
2 × 3.85	1.13	2.000	2	0.193	3/16	1/8	0.386	2.663	2 5/8	0.293	5/16	3/4	5/16	—
ST 1.5× 3.75	1.10	1.500	1 1/2	0.349	3/8	3/16	0.523	2.509	2 1/2	0.260	1/4	11/16	1/4	—
1.5× 2.85	0.835	1.500	1 1/2	0.170	3/16	1/8	0.255	2.330	2 3/8	0.260	1/4	11/16	1/4	—

STRUCTURAL TEES
Cut from S shapes
Properties

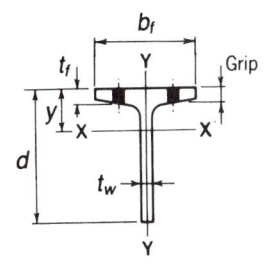

Nominal Wt. per Ft	d/t_w	Axis X-X				Axis Y-Y			$C_c' = \sqrt{\dfrac{2\pi^2 E}{Q_s Q_a F_y}}$, $Q_a = 1.0$			
		I	S	r	y	I	S	r	$F_y = 36$ ksi		$F_y = 50$ ksi	
Lb.		In.^{4}	In.^{3}	In.	In.	In.^{4}	In.^{3}	In.	Q_s	C_c'	Q_s	C_c'
60.5	15.3	259	30.1	3.82	3.63	41.7	10.4	1.53	—	—	—	—
53	19.8	216	24.1	3.72	3.28	38.5	9.80	1.57	—	—	0.907	112
50	16.1	215	26.3	3.83	3.84	23.8	6.58	1.27	—	—	—	—
45	19.2	190	22.6	3.79	3.60	22.5	6.31	1.30	—	—	0.937	111
40	24.0	162	18.7	3.72	3.29	21.1	6.04	1.34	0.878	135	0.695	128
48	12.7	143	20.3	3.18	3.13	25.1	6.97	1.33	—	—	—	—
43	15.4	125	17.2	3.14	2.91	23.4	6.63	1.36	—	—	—	—
37.5	15.7	109	15.8	3.15	3.07	14.9	4.66	1.16	—	—	—	—
33	19.8	93.1	12.9	3.10	2.81	13.8	4.43	1.19	—	—	0.907	112
35	12.7	84.7	14.0	2.87	2.94	12.1	3.86	1.08	—	—	—	—
27.35	19.5	62.4	9.61	2.79	2.50	10.4	3.47	1.14	—	—	0.922	111
25	13.6	40.6	7.73	2.35	2.25	7.85	2.78	1.03	—	—	—	—
21.45	18.2	33.0	6.00	2.29	2.01	7.19	2.61	1.07	—	—	0.988	108
25	8.7	25.2	6.05	1.85	1.84	7.85	2.87	1.03	—	—	—	—
20.4	13.0	18.9	4.28	1.78	1.58	6.78	2.58	1.06	—	—	—	—
17.5	14.0	17.2	3.95	1.83	1.64	4.94	1.95	0.980	—	—	—	—
15.9	17.1	14.9	3.31	1.78	1.51	4.68	1.87	1.00	—	—	—	—
17.5	8.4	12.5	3.63	1.56	1.56	4.18	1.69	0.901	—	—	—	—
12.7	16.1	7.83	2.06	1.45	1.20	3.39	1.46	0.954	—	—	—	—
11.5	9.1	5.03	1.77	1.22	1.15	2.15	1.03	0.798	—	—	—	—
9.2	14.8	3.51	1.15	1.14	0.941	1.86	0.932	0.831	—	—	—	—
10	7.8	3.36	1.36	1.07	1.04	1.59	0.821	0.734	—	—	—	—
7.65	13.9	2.19	0.816	0.987	0.817	1.32	0.720	0.766	—	—	—	—
8.625	6.5	2.13	1.02	0.917	0.914	1.15	0.648	0.675	—	—	—	—
6.25	12.9	1.27	0.552	0.833	0.691	0.911	0.547	0.705	—	—	—	—
7.375	5.1	1.27	0.740	0.764	0.789	0.833	0.507	0.620	—	—	—	—
5	11.7	0.681	0.353	0.681	0.569	0.608	0.405	0.643	—	—	—	—
4.75	6.1	0.470	0.325	0.580	0.553	0.451	0.323	0.569	—	—	—	—
3.85	10.4	0.316	0.203	0.528	0.448	0.382	0.287	0.581	—	—	—	—
3.75	4.3	0.204	0.191	0.430	0.432	0.293	0.234	0.516	—	—	—	—
2.85	8.8	0.118	0.101	0.376	0.329	0.227	0.195	0.522	—	—	—	—

Where no value of C_c' or Q_s is shown, the Tee complies with the noncompact section criteria of Specification Sect. B5.1.

AMERICAN INSTITUTE OF STEEL CONSTRUCTION

DOUBLE ANGLES
Properties of Sections

Properties of double angles in contact and separated are listed in the following tables. Each table shows properties of double angles in contact, and the radius of gyration about the Y-Y axis when the legs of the angles are separated. Values of Q_s are given for $F_y = 36$ ksi and $F_y = 50$ ksi, for those angles exceeding the limiting width-thickness ratio for a noncompact section given in AISC ASD Specification Sect. B5.1. Since the cross section is comprised entirely of unstiffened elements, $Q_a = 1.0$ and $Q = Q_s$ for all angle sections. The flexural-torsional properties table also lists the dimensional values ($\bar{r}_o$ and H) needed for checking torsional and flexural-torsional buckling.

USE OF TABLE

The table may be used as follows for checking allowable stresses for (1) flexural buckling and (2) torsional or flexural-torsional buckling.

(1) Flexural Buckling

Where no value of Q_s is shown, the allowable compressive strength is given by AISC ASD Specification Sect. E2. Where a value of Q_s is shown, the strength must be reduced in accordance with Appendix B5.

(2) Torsional or Flexural-torsional Buckling

The allowable stresses for torsional or flexural-torsional buckling can be determined from design equations in the AISC Load and Resistance Factor Design Specification Appendix E3. This involves calculations with J, $\bar{r}_o$, and H. For further discussion see Part 3 of this Manual.

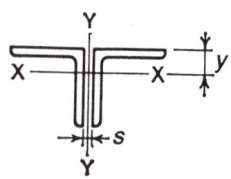

DOUBLE ANGLES
Two equal leg angles
Properties of sections

Designation	Wt. per Ft 2 Angles	Area of 2 Angles	AXIS X-X				AXIS Y-Y			Q_s*			
			I	S	r	y	Radii of Gyration Back to Back of Angles, In.			Angles in Contact		Angles Separated	
	Lb.	In.²	In.⁴	In.³	In.	In.	0	³⁄₈	¾	$F_y =$ 36 ksi	$F_y =$ 50 ksi	$F_y =$ 36 ksi	$F_y =$ 50 ksi
L 8×8×1⅛	113.8	33.5	195.0	35.1	2.42	2.41	3.42	3.55	3.69	—	—	—	—
1	102.0	30.0	177.0	31.6	2.44	2.37	3.40	3.53	3.67	—	—	—	—
⅞	90.0	26.5	159.0	28.0	2.45	2.32	3.38	3.51	3.64	—	—	—	—
¾	77.8	22.9	139.0	24.4	2.47	2.28	3.36	3.49	3.62	—	—	—	—
⅝	65.4	19.2	118.0	20.6	2.49	2.23	3.34	3.47	3.60	—	—	.997	.935
½	52.8	15.5	97.3	16.7	2.50	2.19	3.32	3.45	3.58	.995	.921	.911	.834
L 6×6×1	74.8	22.0	70.9	17.1	1.80	1.86	2.59	2.73	2.87	—	—	—	—
⅞	66.2	19.5	63.8	15.3	1.81	1.82	2.57	2.70	2.85	—	—	—	—
¾	57.4	16.9	56.3	13.3	1.83	1.78	2.55	2.68	2.82	—	—	—	—
⅝	48.4	14.2	48.3	11.3	1.84	1.73	2.53	2.66	2.80	—	—	—	—
½	39.2	11.5	39.8	9.23	1.86	1.68	2.51	2.64	2.78	—	—	—	.961
⅜	29.8	8.72	30.8	7.06	1.88	1.64	2.49	2.62	2.75	.995	.921	.911	.834
L 5×5× ⅞	54.4	16.0	35.5	10.3	1.49	1.57	2.16	2.30	2.45	—	—	—	—
¾	47.2	13.9	31.5	9.06	1.51	1.52	2.14	2.28	2.42	—	—	—	—
½	32.4	9.50	22.5	6.31	1.54	1.43	2.10	2.24	2.38	—	—	—	—
⅜	24.6	7.22	17.5	4.84	1.56	1.39	2.09	2.22	2.35	—	—	.982	.919
⁵⁄₁₆	20.6	6.05	14.8	4.08	1.57	1.37	2.08	2.21	2.34	.995	.921	.911	.834
L 4×4× ¾	37.0	10.9	15.3	5.62	1.19	1.27	1.74	1.88	2.03	—	—	—	—
⅝	31.4	9.22	13.3	4.80	1.20	1.23	1.72	1.86	2.00	—	—	—	—
½	25.6	7.50	11.1	3.95	1.22	1.18	1.70	1.83	1.98	—	—	—	—
⅜	19.6	5.72	8.72	3.05	1.23	1.14	1.68	1.81	1.95	—	—	—	—
⁵⁄₁₆	16.4	4.80	7.43	2.58	1.24	1.12	1.67	1.80	1.94	—	—	.997	.935
¼	13.2	3.88	6.08	2.09	1.25	1.09	1.66	1.79	1.93	.995	.921	.911	.834

* Where no value of Q_s is shown, the angles comply with the noncompact section criteria of Specification Sect. B5.1 and may be considered fully effective.
For F_y = 36 ksi: $C'_c = 126.1/\sqrt{Q_s}$
For F_y = 50 ksi: $C'_c = 107.0/\sqrt{Q_s}$

DOUBLE ANGLES
Two equal leg angles
Properties of sections

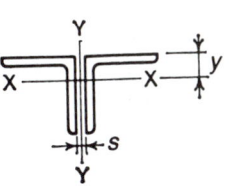

Designation	Wt. per Ft 2 Angles (Lb.)	Area of 2 Angles (In.²)	AXIS X-X I (In.⁴)	S (In.³)	r (In.)	y (In.)	AXIS Y-Y Radii of Gyration Back to Back of Angles, In. 0	3/8	3/4	Q_s^* Angles in Contact $F_y=$36 ksi	$F_y=$50 ksi	Angles Separated $F_y=$36 ksi	$F_y=$50 ksi
L 3½×3½×⅜	17.0	4.97	5.73	2.30	1.07	1.01	1.48	1.61	1.75	—	—	—	—
⁵⁄₁₆	14.4	4.18	4.90	1.95	1.08	.990	1.47	1.60	1.74	—	—	—	.986
¼	11.6	3.38	4.02	1.59	1.09	.968	1.46	1.59	1.73	—	.982	.965	.897
L 3 ×3 ×½	18.8	5.50	4.43	2.14	.898	.932	1.29	1.43	1.59	—	—	—	—
⅜	14.4	4.22	3.52	1.67	.913	.888	1.27	1.41	1.56	—	—	—	—
⁵⁄₁₆	12.2	3.55	3.02	1.41	.922	.865	1.26	1.40	1.55	—	—	—	—
¼	9.8	2.88	2.49	1.15	.930	.842	1.26	1.39	1.53	—	—	—	.961
³⁄₁₆	7.42	2.18	1.92	.882	.939	.820	1.25	1.38	1.52	.995	.921	.911	.834
L 2½×2½×⅜	11.8	3.47	1.97	1.13	.753	.762	1.07	1.21	1.36	—	—	—	—
⁵⁄₁₆	10.0	2.93	1.70	.964	.761	.740	1.06	1.20	1.35	—	—	—	—
¼	8.2	2.38	1.41	.789	.769	.717	1.05	1.19	1.34	—	—	—	—
³⁄₁₆	6.14	1.80	1.09	.606	.778	.694	1.04	1.18	1.32	—	—	.982	.919
L 2 ×2 ×⅜	9.4	2.72	.958	.702	.594	.636	.870	1.01	1.17	—	—	—	—
⁵⁄₁₆	7.84	2.30	.832	.681	.601	.614	.859	1.00	1.16	—	—	—	—
¼	6.38	1.88	.695	.494	.609	.592	.849	.989	1.14	—	—	—	—
³⁄₁₆	4.88	1.43	.545	.381	.617	.569	.840	.977	1.13	—	—	—	—
⅛	3.30	.960	.380	.261	.626	.546	.831	.965	1.11	.995	.921	.911	.834

* Where no value of Q_s is shown, the angles comply with the noncompact section criteria of Specification Sect. B5.1 and may be considered fully effective.

For $F_y = 36$ ksi: $C_c' = 126.1/\sqrt{Q_s}$

For $F_y = 50$ ksi: $C_c' = 107.0/\sqrt{Q_s}$

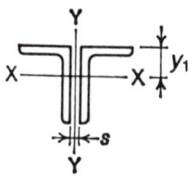

DOUBLE ANGLES
Two unequal leg angles
Properties of sections
Long legs back to back

Designation	Wt. per Ft 2 Angles	Area of 2 Angles	AXIS X-X				AXIS Y-Y Radii of Gyration Back to Back of Angles, In.			Q_s* Angles in Contact		Q_s* Angles Separated	
			I	S	r	y	0	$3/8$	$3/4$	$F_y = $ 36 ksi	$F_y = $ 50 ksi	$F_y = $ 36 ksi	$F_y = $ 50 ksi
	Lb.	In.^{2}	In.^{4}	In.^{3}	In.	In.							
L 8×6 ×1	88.4	26.0	161.0	30.2	2.49	2.65	2.39	2.52	2.66	—	—	—	—
¾	67.6	19.9	126.0	23.3	2.53	2.56	2.35	2.48	2.62	—	—	—	—
½	46.0	13.5	88.6	16.0	2.56	2.47	2.32	2.44	2.57	—	—	.911	.834
L 8×4 ×1	74.8	22.0	139.0	28.1	2.52	3.05	1.47	1.61	1.75	—	—	—	—
¾	57.4	16.9	109.0	21.8	2.55	2.95	1.42	1.55	1.69	—	—	—	—
½	39.2	11.5	77.0	15.0	2.59	2.86	1.38	1.51	1.64	—	—	.911	.834
L 7×4 × ¾	52.4	15.4	75.6	16.8	2.22	2.51	1.48	1.62	1.76	—	—	—	—
½	35.8	10.5	53.3	11.6	2.25	2.42	1.44	1.57	1.71	—	—	.965	.897
⅜	27.2	7.97	41.1	8.88	2.27	2.37	1.43	1.55	1.68	—	—	.839	.750
L 6×4 × ¾	47.2	13.9	49.0	12.5	1.88	2.08	1.55	1.69	1.83	—	—	—	—
⅝	40.0	11.7	42.1	10.6	1.90	2.03	1.53	1.67	1.81	—	—	—	—
½	32.4	9.50	34.8	8.67	1.91	1.99	1.51	1.64	1.78	—	—	—	.961
⅜	24.6	7.22	26.9	6.64	1.93	1.94	1.50	1.62	1.76	—	—	.911	.834
L 6×3½× ⅜	23.4	6.84	25.7	6.49	1.94	2.04	1.26	1.39	1.53	—	—	.911	.834
⁵⁄₁₆	19.6	5.74	21.8	5.47	1.95	2.01	1.26	1.38	1.51	—	—	.825	.733
L 5×3½× ¾	39.6	11.6	27.8	8.55	1.55	1.75	1.40	1.53	1.68	—	—	—	—
½	27.2	8.00	20.0	5.97	1.58	1.66	1.35	1.49	1.63	—	—	—	—
⅜	20.8	6.09	15.6	4.59	1.60	1.61	1.34	1.46	1.60	—	—	.982	.919
⁵⁄₁₆	17.4	5.12	13.2	3.87	1.61	1.59	1.33	1.45	1.59	—	—	.911	.834
L 5×3 × ½	25.6	7.50	18.9	5.82	1.59	1.75	1.12	1.25	1.40	—	—	—	—
⅜	19.6	5.72	14.7	4.47	1.61	1.70	1.10	1.23	1.37	—	—	—	—
⁵⁄₁₆	16.4	4.80	12.5	3.77	1.61	1.68	1.09	1.22	1.36	—	—	.982	.919
¼	13.2	3.88	10.2	3.06	1.62	1.66	1.08	1.21	1.34	—	—	.911	.834
												.804	.708

* Where no value of Q_s is shown, the angles comply with the noncompact section criteria of Specification Sect. B5.1 and may be considered fully effective.

For $F_y = $ 36 ksi: $C_c' = 126.1/\sqrt{Q_s}$
For $F_y = $ 50 ksi: $C_c' = 107.0/\sqrt{Q_s}$

DOUBLE ANGLES
Two unequal leg angles
Properties of sections
Long legs back to back

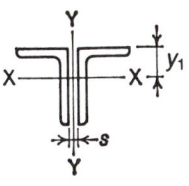

Designation	Wt. per Ft 2 Angles (Lb.)	Area of 2 Angles (In.²)	AXIS X-X I (In.⁴)	AXIS X-X S (In.³)	AXIS X-X r (In.)	AXIS X-X y (In.)	AXIS Y-Y Radii of Gyration Back to Back of Angles, In. 0	AXIS Y-Y 3/8	AXIS Y-Y 3/4	Q_s* Angles in Contact $F_y=$ 36 ksi	$F_y=$ 50 ksi	Q_s* Angles Separated $F_y=$ 36 ksi	$F_y=$ 50 ksi
L 4 ×3½×½	23.8	7.00	10.6	3.87	1.23	1.25	1.44	1.58	1.72	—	—	—	—
3/8	18.2	5.34	8.35	2.99	1.25	1.21	1.42	1.56	1.70	—	—	—	—
5/16	15.4	4.49	7.12	2.53	1.26	1.18	1.42	1.55	1.69	—	—	.997	.935
1/4	12.4	3.63	5.83	2.05	1.27	1.16	1.41	1.54	1.67	—	.982	.911	.834
L 4 ×3 ×½	22.2	6.50	10.1	3.78	1.25	1.33	1.20	1.33	1.48	—	—	—	—
3/8	17.0	4.97	7.93	2.92	1.26	1.28	1.18	1.31	1.45	—	—	—	—
5/16	14.4	4.18	6.76	2.47	1.27	1.26	1.17	1.30	1.44	—	—	.997	.935
1/4	11.6	3.38	5.54	2.00	1.28	1.24	1.16	1.29	1.43	—	—	.911	.834
L 3½×3 ×3/8	15.8	4.59	5.45	2.25	1.09	1.08	1.22	1.36	1.50	—	—	—	—
5/16	13.2	3.87	4.66	1.91	1.10	1.06	1.21	1.35	1.49	—	—	—	.986
1/4	10.8	3.13	3.83	1.55	1.11	1.04	1.20	1.33	1.48	—	—	.965	.897
L 3½×2½×3/8	14.4	4.22	5.12	2.19	1.10	1.16	.976	1.11	1.26	—	—	—	—
5/16	12.2	3.55	4.38	1.85	1.11	1.14	.966	1.10	1.25	—	—	—	.986
1/4	9.8	2.88	3.60	1.51	1.12	1.11	.958	1.09	1.23	—	—	.965	.897
L 3 ×2½×3/8	13.2	3.84	3.31	1.62	.928	.956	1.02	1.16	1.31	—	—	—	—
1/4	9.0	2.63	2.35	1.12	.945	.911	1.00	1.13	1.28	—	—	—	.961
3/16	6.77	1.99	1.81	.859	.954	.888	.993	1.12	1.27	—	—	.911	.834
L 3 ×2 ×3/8	11.8	3.47	3.06	1.56	.940	1.04	.777	.917	1.07	—	—	—	—
5/16	10.0	2.93	2.63	1.33	.948	1.02	.767	.903	1.06	—	—	—	—
1/4	8.2	2.38	2.17	1.08	.957	.993	.757	.891	1.04	—	—	—	.961
3/16	6.1	1.80	1.68	.830	.966	.970	.749	.879	1.03	—	—	.911	.834
L 2½×2 ×3/8	10.6	3.09	1.82	1.09	.768	.831	.819	.961	1.12	—	—	—	—
5/16	9.0	2.62	1.58	.932	.776	.809	.809	.948	1.10	—	—	—	—
1/4	7.2	2.13	1.31	.763	.784	.787	.799	.935	1.09	—	—	—	—
3/16	5.5	1.62	1.02	.586	.793	.764	.790	.923	1.07	—	—	.982	.919

* Where no value of Q_s is shown, the angles comply with the noncompact section criteria of Specification Sect. B5.1 and may be considered fully effective.

For F_y = 36 ksi: $C_c' = 126.1/\sqrt{Q_s}$

For F_y = 50 ksi: $C_c' = 107.0/\sqrt{Q_s}$

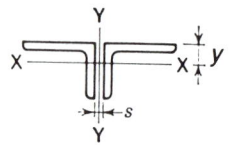

DOUBLE ANGLES
Two unequal leg angles
Properties of sections
Short legs back to back

Designation	Wt. per Ft 2 Angles Lb.	Area of 2 Angles In.²	AXIS X-X I In.⁴	S In.³	r In.	y In.	AXIS Y-Y Radii of Gyration Back to Back of Angles, In. 0	3/8	3/4	Q_s* Angles in Contact F_y = 36 ksi	F_y = 50 ksi	Angles Separated F_y = 36 ksi	F_y = 50 ksi
L 8×6 ×1	88.4	26.0	77.6	17.8	1.73	1.65	3.64	3.78	3.92	—	—	—	—
¾	67.6	19.9	61.4	13.8	1.76	1.56	3.60	3.74	3.88	—	—	—	—
½	46.0	13.5	43.4	9.58	1.79	1.47	3.56	3.69	3.83	.995	.921	.911	.834
L 8×4 ×1	74.8	22.0	23.3	7.88	1.03	1.05	3.95	4.10	4.25	—	—	—	—
¾	57.4	16.9	18.7	6.14	1.05	.953	3.90	4.05	4.19	—	—	—	—
½	39.2	11.5	13.5	4.29	1.08	.859	3.86	4.00	4.14	.995	.921	.911	.834
L 7×4 × ¾	52.4	15.4	18.1	6.05	1.09	1.01	3.35	3.49	3.64	—	—	—	—
½	35.8	10.5	13.1	4.23	1.11	.917	3.30	3.44	3.59	—	.982	.965	.897
⅜	27.2	7.97	10.2	3.26	1.13	.870	3.28	3.42	3.56	.926	.838	.839	.750
L 6×4 × ¾	47.2	13.9	17.4	5.94	1.12	1.08	2.80	2.94	3.09	—	—	—	—
⅝	40.0	11.7	15.0	5.07	1.13	1.03	2.78	2.92	3.06	—	—	—	—
½	32.4	9.50	12.5	4.16	1.15	.987	2.76	2.90	3.04	—	—	—	.961
⅜	24.6	7.22	9.81	3.21	1.17	.941	2.74	2.87	3.02	.995	.921	.911	.834
L 6×3½× ⅜	23.4	6.84	6.68	2.46	.988	.787	2.81	2.95	3.09	.995	.921	.911	.834
5/16	19.6	5.74	5.70	2.08	.996	.763	2.80	2.94	3.08	.912	.822	.825	.733
L 5×3½× ¾	39.6	11.6	11.1	4.43	.977	.996	2.33	2.48	2.63	—	—	—	—
½	27.2	8.00	8.10	3.12	1.01	.906	2.29	2.43	2.57	—	—	—	—
⅜	20.8	6.09	6.37	2.41	1.02	.861	2.27	2.41	2.55	—	—	.982	.919
5/16	17.4	5.12	5.44	2.04	1.03	.838	2.26	2.39	2.54	.995	.921	.911	.834
L 5×3 × ½	25.6	7.50	5.16	2.29	.829	.750	2.36	2.50	2.65	—	—	—	—
⅜	19.6	5.72	4.08	1.78	.845	.704	2.34	2.48	2.63	—	—	.982	.919
5/16	16.4	4.80	3.49	1.51	.853	.681	2.33	2.47	2.61	.995	.921	.911	.834
¼	13.2	3.88	2.88	1.23	.861	.657	2.32	2.46	2.60	.891	.797	.804	.708

* Where no value of Q_s is shown, the angles comply with the noncompact section criteria of Specification Sect. B5.1 and may be considered fully effective.
For F_y = 36 ksi: $C'_c = 126.1/\sqrt{Q_s}$
For F_y = 50 ksi: $C'_c = 107.0/\sqrt{Q_s}$

AMERICAN INSTITUTE OF STEEL CONSTRUCTION

DOUBLE ANGLES
Two unequal leg angles
Properties of sections
Short legs back to back

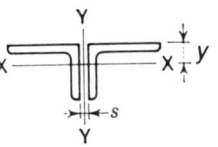

Designation	Wt. per Ft 2 Angles (Lb.)	Area of 2 Angles (In.²)	AXIS X-X I (In.⁴)	S (In.³)	r (In.)	y (In.)	AXIS Y-Y Radii of Gyration Back to Back of Angles, In. 0	3/8	3/4	Q_s* Angles in Contact F_y=36 ksi	F_y=50 ksi	Angles Separated F_y=36 ksi	F_y=50 ksi
L 4 ×3½×½	23.8	7.00	7.58	3.03	1.04	1.00	1.76	1.89	2.04	—	—	—	—
⅜	18.2	5.34	5.97	2.35	1.06	.955	1.74	1.87	2.01	—	—	—	—
5/16	15.4	4.49	5.10	1.99	1.07	.932	1.73	1.86	2.00	—	—	.997	.935
¼	12.4	3.63	4.19	1.62	1.07	.909	1.72	1.85	1.99	.995	.921	.911	.834
L 4 ×3 ×½	22.2	6.50	4.85	2.23	.864	.827	1.82	1.96	2.11	—	—	—	—
⅜	17.0	4.97	3.84	1.73	.879	.782	1.80	1.94	2.08	—	—	—	—
5/16	14.4	4.18	3.29	1.47	.887	.759	1.79	1.93	2.07	—	—	.997	.935
¼	11.6	3.38	2.71	1.20	.896	.736	1.78	1.92	2.06	.995	.921	.911	.834
L 3½×3 ×⅜	15.8	4.59	3.69	1.70	.897	.830	1.53	1.67	1.82	—	—	—	—
5/16	13.2	3.87	3.17	1.44	.905	.808	1.52	1.66	1.80	—	—	—	.986
¼	10.8	3.13	2.61	1.18	.914	.785	1.52	1.65	1.79	—	.982	.965	.897
L 3½×2½×⅜	14.4	4.22	2.18	1.18	.719	.660	1.60	1.74	1.89	—	—	—	—
5/16	12.2	3.55	1.88	1.01	.727	.637	1.59	1.73	1.88	—	—	—	.986
¼	9.8	2.88	1.55	.824	.735	.614	1.58	1.72	1.86	—	.982	.965	.897
L 3 ×2½×⅜	13.2	3.84	2.08	1.16	.736	.706	1.33	1.47	1.62	—	—	—	—
¼	9.0	2.63	1.49	.808	.753	.661	1.31	1.45	1.60	—	—	—	.961
3/16	6.77	1.99	1.15	.620	.761	.638	1.30	1.44	1.58	.995	.921	.911	.834
L 3 ×2 ×⅜	11.8	3.47	1.09	.743	.559	.539	1.40	1.55	1.70	—	—	—	—
5/16	10.0	2.93	.941	.634	.567	.516	1.39	1.53	1.68	—	—	—	—
¼	8.20	2.38	.784	.520	.574	.493	1.38	1.52	1.67	—	—	—	.961
3/16	6.14	1.80	.613	.401	.583	.470	1.37	1.51	1.66	.995	.921	.911	.834
L 2½×2 ×⅜	10.6	3.09	1.03	.725	.577	.581	1.13	1.28	1.43	—	—	—	—
5/16	9.00	2.62	.893	.620	.584	.559	1.12	1.26	1.42	—	—	—	—
¼	7.24	2.13	.745	.509	.592	.537	1.11	1.25	1.40	—	—	—	—
3/16	5.50	1.62	.583	.392	.600	.514	1.10	1.24	1.39	—	—	.982	.919

* Where no value of Q_s is shown, the angles comply with the noncompact section criteria of Specification Sect. B5.1 and may be considered fully effective.

For F_y = 36 ksi: $C_c' = 126.1/\sqrt{Q_s}$

For F_y = 50 ksi: $C_c' = 107.0/\sqrt{Q_s}$

COMBINATION SECTIONS

Standard rolled shapes are frequently combined to produce efficient and economical structural members for special applications. Experience has established a demand for certain combinations. When properly sized and connected to satisfy the design and specification criteria, these members may be used as struts, lintels, eave struts and light crane and trolley runways.

Properties of several combined sections are tabulated for those combinations that experience has proven to be in popular demand. For properties, dimensions and discussion of other combination shapes, see *Light and Heavy Industrial Buildings*, Fisher, J. M. and D. R. Buettner, American Institute of Steel Construction, Inc., Chicago, IL 1979.

COMBINATION SECTIONS
S shapes and channels
Properties of sections

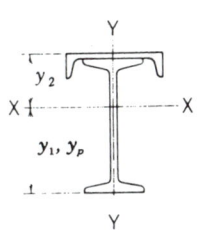

Beam	Channel	Total Wt. per Ft	Total Area	AXIS X-X					AXIS Y-Y			r_T
				I	$S_1 = I/y_1$	$S_2 = I/y_2$	r	y_1	I	S	r	
		Lb.	In.^{2}	In.^{4}	In.^{3}	In.^{3}	In.	In.	In.^{4}	In.^{3}	In.	In.
S 10× 25.4	C 8×11.5	36.9	10.84	176	27.2	46.6	4.02	6.45	39.4	9.85	1.91	2.44
× 25.4	C 10×15.3	40.7	11.95	186	27.6	52.9	3.94	6.73	74.2	14.8	2.49	3.16
S 12× 31.8	C 8×11.5	43.3	12.73	299	39.8	63.2	4.84	7.50	42.0	10.5	1.82	2.38
× 31.8	C 10×15.3	47.1	13.84	316	40.4	71.4	4.78	7.82	76.8	15.4	2.36	3.06
× 40.8	C 10×15.3	56.1	16.49	377	50.1	80.0	4.78	7.53	81.0	16.2	2.22	2.94
S 15× 42.9	C 8×11.5	54.4	15.98	585	64.9	94.2	6.05	9.01	47.0	11.8	1.71	2.29
× 42.9	C 10×15.3	58.2	17.09	616	65.8	105	6.01	9.37	81.8	16.4	2.19	2.94
S 18× 54.7	C 10×15.3	70.0	20.59	1070	98.0	145	7.20	10.88	88.2	17.6	2.07	2.83
× 54.7	C 12×20.7	75.4	22.19	1130	99.8	164	7.15	11.36	150	25.0	2.60	3.52
S 20× 66	C 10×15.3	81.3	23.89	1530	130	181	8.00	11.81	95.1	19.0	2.00	2.74
× 66	C 12×20.7	86.7	25.49	1620	132	203	7.97	12.29	157	26.1	2.48	3.41
× 86	C 12×20.7	106.7	31.39	2050	170	240	8.08	12.04	176	29.3	2.37	3.27
S 24× 80	C 10×15.3	95.3	27.99	2610	188	252	9.66	13.86	110	21.9	1.98	2.69
× 80	C 12×20.7	100.7	29.59	2750	191	278	9.65	14.38	171	28.5	2.41	3.31
×106	C 12×20.7	126.7	37.29	3660	258	345	9.90	14.18	206	34.4	2.35	3.17

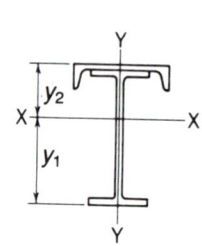

COMBINATION SECTIONS
W shapes and channels
Properties of sections

Beam	Channel	Total Wt. per Ft	Total Area	AXIS X-X					AXIS Y-Y			
				I	$S_1 = I/y_1$	$S_2 = I/y_2$	r	y_1	I	S	r	r_T
		Lb.	In.^{2}	In.^{4}	In.^{3}	In.^{3}	In.	In.	In.^{4}	In.^{3}	In.	In.
W 12×26	C 10×15.3	41.3	12.14	299	36.3	70.5	4.96	8.22	84.7	16.9	2.64	3.23
W 14×30	C 10×15.3	45.3	13.34	420	46.1	84.6	5.61	9.12	87.0	17.4	2.55	3.05
×43	C 12×20.7	63.7	18.69	601	67.4	120	5.67	8.92	174	29.0	3.05	3.73
×61	C 15×33.9	94.9	27.86	923	99.4	185	5.76	9.29	422	56.3	3.89	4.65
W 16×36	C 12×20.7	56.7	16.69	670	62.8	123	6.34	10.67	154	25.6	3.03	3.83
×67	C 15×33.9	100.9	29.66	1360	126	229	6.78	10.78	434	57.9	3.83	4.63
×67	MC 18×42.7	109.7	32.3	1430	128	225	6.65	11.18	673	74.8	4.56	5.53
W 18×50	C 12×20.7	70.7	20.79	1120	97.4	166	7.34	11.51	169	28.2	2.85	3.66
×50	C 15×33.9	83.9	24.66	1250	100	211	7.11	12.47	355	47.3	3.79	4.74
×76	C 15×33.9	109.9	32.26	1860	158	273	7.60	11.80	467	62.3	3.80	4.62
×76	MC 18×42.7	118.7	34.90	1950	159	304	7.48	12.24	706	78.4	4.50	5.49
W 21×62	C 12×20.7	82.7	24.39	1800	138	218	8.59	13.01	187	31.1	2.77	3.22
×62	C 15×33.9	95.9	28.26	2000	142	272	8.41	14.06	373	49.7	3.63	4.24
×68	C 12×20.7	88.7	26.09	1960	152	232	8.68	12.93	194	32.3	2.72	3.53
×68	C 15×33.9	101.9	29.96	2180	156	287	8.52	13.95	380	50.6	3.56	4.57
×101	MC 18×42.7	143.7	42.4	3370	245	416	8.91	13.73	802	89.1	4.35	5.35
W 24×62	C 12×20.7	82.7	24.29	2150	146	232	9.41	14.74	164	27.3	2.59	3.56
×62	C 15×33.9	95.9	28.16	2410	151	293	9.25	15.93	350	46.6	3.52	4.67
×68	C 12×20.7	88.7	26.19	2450	168	258	9.67	14.53	199	33.2	2.76	3.60
×68	C 15×33.9	101.9	30.06	2720	173	321	9.50	15.67	385	51.4	3.58	4.63
×84	C 12×20.7	104.7	30.79	3040	212	303	9.93	14.35	223	37.2	2.69	3.48
×84	C 15×33.9	117.9	34.66	3340	217	368	9.82	15.40	409	54.6	3.44	4.45
×104	MC 18×42.7	146.7	43.2	4320	280	475	10.0	15.41	813	90.3	4.34	5.38

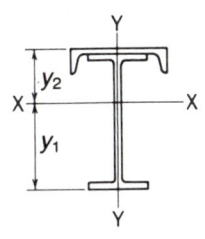

COMBINATION SECTIONS
W shapes and channels
Properties of sections

Beam	Channel	Total Wt. per Ft	Total Area	AXIS X-X					AXIS Y-Y			r_T
				I	$S_1 = I/y_1$	$S_2 = I/y_2$	r	y_1	I	S	r	
		Lb.	In.^{2}	In.^{4}	In.^{3}	In.^{3}	In.	In.	In.^{4}	In.^{3}	In.	In.
W 27×84	C 15×33.9	117.9	34.76	4050	237	404	10.8	17.07	421	56.1	3.48	4.53
×114	C 15×33.9	147.9	43.46	5450	327	495	11.2	16.68	474	63.2	3.30	4.30
×146	MC 18×42.7	188.7	55.50	7360	441	661	11.5	16.70	997	111	4.24	5.23
W 30×99	C 15×33.9	132.9	39.06	5540	300	480	11.9	18.51	443	59.1	3.37	4.47
×99	MC 18×42.7	141.7	41.70	5830	304	533	11.8	19.18	682	75.8	4.04	5.36
×116	C 15×33.9	149.9	44.16	6590	360	544	12.2	18.30	479	63.9	3.29	4.34
×116	MC 18×42.7	158.7	46.80	6900	365	599	12.1	18.93	718	79.8	3.92	5.19
×132	C 15×33.9	165.9	48.86	8660	371	542	13.3	23.34	511	68.1	3.23	4.43
×173	MC 18×42.7	215.7	63.40	10400	574	819	12.8	18.16	1150	128	4.26	5.23
W 33×118	C 15×33.9	151.9	44.66	7900	395	596	13.3	20.01	502	66.9	3.35	4.42
×118	MC 18×42.7	160.7	47.30	8280	400	656	13.2	20.69	741	82.3	3.96	5.26
×141	C 15×33.9	174.9	51.56	9580	484	689	13.6	19.79	561	74.8	3.30	4.29
×141	MC 18×42.7	183.7	54.20	10000	490	751	13.6	20.42	800	88.9	3.84	5.07
×152	MC 18×42.7	194.7	57.30	10800	531	793	13.7	20.33	827	91.9	3.80	5.01
W 36×150	C 15×33.9	183.9	54.16	11500	546	765	14.6	21.15	585	78.0	3.29	4.30
×150	MC 18×42.7	192.7	56.80	12100	553	832	14.6	21.81	824	91.6	3.81	5.07
×170	MC 18×42.7	212.7	62.60	13700	631	911	14.8	21.64	874	97.1	3.74	4.95
×194	MC 18×42.7	236.7	69.60	15400	717	995	14.9	21.47	929	103	3.65	4.84

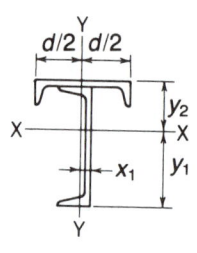

COMBINATION SECTIONS
Two channels
Properties of sections

Vertical Channel	Horizontal Channel	Total Wt. per Ft	Total Area	AXIS X-X					AXIS Y-Y			
				I	$S_1 = I/y_1$	$S_2 = I/y_2$	r	y_1	I	S	r	x_1
		Lb.	In.^{2}	In.^{4}	In.^{3}	In.^{3}	In.	In.	In.^{4}	In.^{3}	In.	In.
C 3× 4.1	C 4× 5.4	9.5	2.80	3.0	1.4	3.0	1.04	2.20	4.0	2.0	1.20	0.44
C 4× 5.4	C 4× 5.4	10.8	3.18	6.5	2.3	4.9	1.43	2.86	4.2	2.1	1.14	0.46
	C 5× 6.7	12.1	3.56	6.9	2.3	5.5	1.39	2.94	7.8	3.1	1.48	0.46
C 5× 6.7	C 5× 6.7	13.4	3.94	12.8	3.5	8.0	1.80	3.60	8.0	3.2	1.42	0.48
	C 6× 8.2	14.9	4.37	13.4	3.6	8.9	1.75	3.70	13.6	4.5	1.76	0.48
	C 7× 9.8	16.5	4.84	14.0	3.7	9.8	1.70	3.79	21.8	6.2	2.12	0.48
C 6× 8.2	C 5× 6.7	14.9	4.37	21.5	5.1	10.9	2.22	4.22	8.2	3.3	1.37	0.51
	C 6× 8.2	16.4	4.80	22.5	5.2	12.1	2.16	4.34	13.8	4.6	1.70	0.51
	C 7× 9.8	18.0	5.27	23.4	5.2	13.3	2.11	4.45	22.0	6.3	2.04	0.51
	C 8×11.5	19.7	5.78	24.3	5.3	14.5	2.05	4.55	33.3	8.3	2.40	0.51
	C 9×13.4	21.6	6.34	25.2	5.4	15.8	1.99	4.64	48.6	10.8	2.77	0.51
	C 10×15.3	23.5	6.89	26.0	5.5	16.9	1.94	4.70	68.1	13.6	3.14	0.51
C 7× 9.8	C 6× 8.2	18.0	5.27	35.3	7.1	15.7	2.59	4.95	14.1	4.7	1.63	0.54
	C 7× 9.8	19.6	5.74	36.7	7.2	17.3	2.53	5.08	22.3	6.4	1.97	0.54
	C 8×11.5	21.3	6.25	38.0	7.3	18.8	2.47	5.20	33.6	8.4	2.32	0.54
	C 9×13.4	23.2	6.81	39.3	7.4	20.5	2.40	5.31	48.9	10.9	2.68	0.54
	C 10×15.3	25.1	7.36	40.5	7.5	21.9	2.34	5.39	68.4	13.7	3.05	0.54
C 8×11.5	C 6× 8.2	19.7	5.78	52.4	9.5	19.6	3.01	5.53	14.4	4.8	1.58	0.57
	C 7× 9.8	21.3	6.25	54.5	9.6	21.6	2.95	5.68	22.6	6.5	1.90	0.57
	C 8×11.5	23.0	6.76	56.4	9.7	23.6	2.89	5.82	33.9	8.5	2.24	0.57
	C 9×13.4	24.9	7.32	58.4	9.8	25.6	2.82	5.95	49.2	10.9	2.59	0.57
	C 10×15.3	26.8	7.87	60.0	9.9	27.5	2.76	6.06	68.7	13.7	2.95	0.57
	C 12×20.7	32.2	9.47	64.4	10.2	32.6	2.61	6.30	130	21.7	3.71	0.57

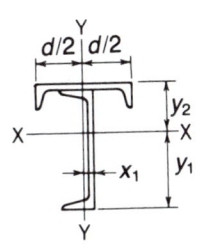

COMBINATION SECTIONS
Two channels
Properties of sections

Vertical Channel	Horizontal Channel	Total Wt. per Ft	Total Area	AXIS X-X					AXIS Y-Y			
				I	$S_1 = I/y_1$	$S_2 = I/y_2$	r	y_1	I	S	r	x_1
		Lb.	In.^{2}	In.^{4}	In.^{3}	In.^{3}	In.	In.	In.^{4}	In.^{3}	In.	In.
C 9×13.4	C 7× 9.8	23.2	6.81	77.7	12.4	26.3	3.38	6.26	23.1	6.6	1.84	0.60
	C 8×11.5	24.9	7.32	80.5	12.6	28.7	3.32	6.42	34.4	8.6	2.17	0.60
	C 9×13.4	26.8	7.88	83.3	12.7	31.2	3.25	6.57	49.7	11.0	2.51	0.60
	C 10×15.3	28.7	8.43	85.6	12.8	33.5	3.19	6.69	69.2	13.8	2.86	0.60
	C 12×20.7	34.1	10.03	91.7	13.1	39.8	3.02	6.98	131	21.8	3.61	0.60
C 10×15.3	C 8×11.5	26.8	7.87	110	15.8	34.2	3.75	7.00	34.9	8.7	2.11	0.63
	C 9×13.4	28.7	8.43	114	15.9	37.2	3.68	7.16	50.2	11.2	2.44	0.63
	C 10×15.3	30.6	8.98	117	16.1	39.9	3.61	7.30	69.7	13.9	2.79	0.63
	C 12×20.7	36.0	10.58	126	16.4	47.5	3.45	7.64	131	21.9	3.52	0.63
	C 15×33.9	49.2	14.45	141	17.3	63.7	3.13	8.18	317	42.3	4.69	0.63
C 12×20.7	C 9×13.4	34.1	10.03	207	25.2	51.4	4.54	8.21	51.8	11.5	2.27	0.70
	C 10×15.3	36.0	10.58	213	25.4	55.0	4.48	8.38	71.3	14.3	2.60	0.70
	C 12×20.7	41.4	12.18	228	25.9	65.3	4.32	8.79	133	22.1	3.30	0.70
	C 15×33.9	54.6	16.05	256	27.0	87.8	4.00	9.48	319	42.5	4.46	0.70
C 15×33.9	C 10×15.3	49.2	14.45	474	48.8	85.6	5.72	9.71	75.5	15.1	2.29	0.79
	C 12×20.7	54.6	16.05	509	49.9	99.8	5.63	10.19	137	22.9	2.92	0.79
	C 15×33.9	67.8	19.92	575	52.0	132	5.37	11.06	323	43.1	4.03	0.79
	MC 18×42.7	76.6	22.56	608	53.1	152	5.19	11.45	562	62.5	4.99	0.79
MC 18×42.7	C 12×20.7	63.4	18.69	860	72.9	133	6.78	11.80	143	23.9	2.77	0.88
	C 15×33.9	76.6	22.56	975	76.1	174	6.57	12.80	329	43.9	3.82	0.88
	MC 18×42.7	85.4	25.20	1030	77.6	200	6.40	13.29	568	63.2	4.75	0.88

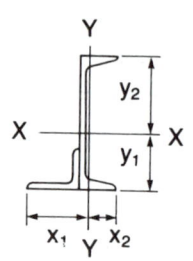

COMBINATION SECTIONS
Channels and angles
Properties of sections
Long leg of angle turned out

Channel	Angle	Total Wt. per Ft	Total Area	AXIS X-X					AXIS Y-Y				
				I	$S_1 = I/y_1$	$S_2 = I/y_2$	r	y_1	I	$S_1 = I/x_1$	$S_2 = I/x_2$	r	x_1
		Lb.	In.^{2}	In.^{4}	In.^{3}	In.^{3}	In.	In.	In.^{4}	In.^{3}	In.^{3}	In.	In.
C 6× 8.2	L 2½×2½×¼	12.3	3.59	17.9	8.0	4.8	2.24	2.24	2.6	1.0	1.4	0.85	2.60
	L 3 ×2½×¼	12.7	3.71	18.5	8.5	4.8	2.23	2.17	3.6	1.2	1.9	0.98	3.01
	L 3½×3 ×¼	13.6	3.96	19.0	8.9	4.9	2.19	2.13	4.9	1.4	2.4	1.11	3.40
	×⁵⁄₁₆	14.8	4.33	19.8	9.8	5.0	2.14	2.02	5.7	1.7	2.7	1.14	3.31
	L 4 ×3 ×¼	14.0	4.09	19.5	9.5	5.0	2.19	2.06	6.5	1.7	3.1	1.26	3.79
C 7× 9.8	L 2½×2½×¼	13.9	4.06	28.5	10.6	6.6	2.65	2.68	3.0	1.1	1.6	0.86	2.67
	L 3 ×2½×¼	14.3	4.18	29.3	11.2	6.7	2.65	2.61	4.0	1.3	2.0	0.98	3.09
	L 3½×3 ×¼	15.2	4.43	30.0	11.8	6.7	2.60	2.54	5.4	1.6	2.6	1.10	3.48
	×⁵⁄₁₆	16.4	4.80	31.2	12.9	6.8	2.55	2.42	6.3	1.8	2.9	1.14	3.40
	L 4 ×3 ×¼	15.6	4.56	30.8	12.4	6.8	2.60	2.48	7.1	1.8	3.2	1.25	3.88
	×⁵⁄₁₆	17.0	4.96	32.0	13.7	6.9	2.54	2.35	8.3	2.2	3.6	1.29	3.78
C 8×11.5	L 3 ×2½×¼	16.0	4.69	43.9	14.3	8.9	3.06	3.07	4.6	1.4	2.2	0.99	3.16
	L 3½×3 ×¼	16.9	4.94	44.9	15.1	9.0	3.02	2.98	6.0	1.7	2.7	1.10	3.56
	×⁵⁄₁₆	18.1	5.31	46.7	16.4	9.0	2.97	2.84	6.9	2.0	3.0	1.14	3.48
	L 4 ×3 ×¼	17.3	5.07	46.0	15.8	9.0	3.01	2.91	7.8	2.0	3.4	1.24	3.97
	×⁵⁄₁₆	18.7	5.47	47.8	17.3	9.1	2.96	2.76	9.0	2.3	3.8	1.28	3.87
	L 5 ×3½×⁵⁄₁₆	20.2	5.94	49.9	18.9	9.3	2.90	2.64	14.7	3.2	5.6	1.57	4.64
C 9×13.4	L 3 ×2½×¼	17.9	5.25	63.1	17.8	11.6	3.47	3.54	5.2	1.6	2.3	0.99	3.22
	L 3½×3 ×¼	18.8	5.50	64.6	18.8	11.6	3.43	3.45	6.7	1.8	2.9	1.10	3.64
	×⁵⁄₁₆	20.0	5.87	67.1	20.4	11.7	3.38	3.29	7.7	2.2	3.2	1.14	3.55
	L 4 ×3 ×¼	19.2	5.63	66.0	19.6	11.7	3.42	3.37	8.5	2.1	3.6	1.23	4.05
	×⁵⁄₁₆	20.6	6.03	68.7	21.4	11.8	3.37	3.20	9.9	2.5	4.0	1.28	3.96
	L 5 ×3½×⁵⁄₁₆	22.1	6.50	71.4	23.4	12.0	3.31	3.06	15.8	3.3	5.9	1.56	4.74
C 10×15.3	L 3½×3 ×¼	20.7	6.05	89.3	22.8	14.7	3.84	3.91	7.4	2.0	3.1	1.11	3.70
	×⁵⁄₁₆	21.9	6.42	92.7	24.8	14.8	3.80	3.74	8.5	2.3	3.4	1.15	3.62
	L 4 ×3 ×¼	21.1	6.18	91.1	23.8	14.8	3.84	3.83	9.4	2.3	3.8	1.23	4.12
	×⁵⁄₁₆	22.5	6.58	94.7	25.9	14.9	3.79	3.65	10.8	2.7	4.2	1.28	4.03
	L 5 ×3½×⁵⁄₁₆	24.0	7.05	98.4	28.2	15.1	3.74	3.49	16.9	3.5	6.1	1.55	4.83
	×³⁄₈	25.7	7.54	102	30.6	15.2	3.67	3.33	19.2	4.1	6.7	1.60	4.73

COMBINATION SECTIONS
Channels and angles
Properties of sections
Long leg of angle turned out

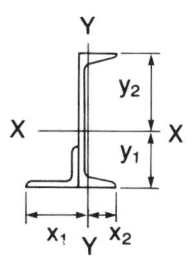

Channel	Angle		Total Wt. per Ft	Total Area	AXIS X-X					AXIS Y-Y				
					I	$S_1 = I/y_1$	$S_2 = I/y_2$	r	y_1	I	$S_1 = I/x_1$	$S_2 = I/x_2$	r	x_1
			Lb.	In.²	In.⁴	In.³	In.³	In.	In.	In.⁴	In.³	In.³	In.	In.
C 12×20.7	L 3½×3	×¼	26.1	7.65	164	33.2	23.2	4.63	4.94	9.5	2.5	3.7	1.12	3.84
		×⁵⁄₁₆	27.3	8.02	170	35.8	23.5	4.61	4.75	10.7	2.8	4.0	1.16	3.77
	L 4 ×3	×¼	26.5	7.78	167	34.4	23.4	4.63	4.86	11.6	2.7	4.4	1.22	4.28
		×⁵⁄₁₆	27.9	8.18	173	37.2	23.6	4.60	4.66	13.2	3.2	4.8	1.27	4.20
	L 5 ×3½×	⁵⁄₁₆	29.4	8.65	180	40.2	23.9	4.56	4.47	19.9	4.0	6.8	1.52	5.02
		×⅜	31.1	9.14	186	43.4	24.1	4.51	4.29	22.5	4.6	7.5	1.57	4.93
	L 6 ×4	×⅜	33.0	9.70	192	46.6	24.3	4.45	4.12	33.2	5.8	10.3	1.85	5.72
		×½	36.9	10.84	202	53.2	24.7	4.32	3.80	40.6	7.3	11.9	1.93	5.52
C 12×25	L 3½×3	×¼	30.4	8.91	180	35.4	26.1	4.50	5.09	10.2	2.6	3.8	1.07	3.87
		×⁵⁄₁₆	31.6	9.28	187	38.0	26.4	4.49	4.92	11.4	3.0	4.2	1.11	3.81
	L 4 ×3	×¼	30.8	9.04	183	36.6	26.3	4.50	5.02	12.3	2.8	4.5	1.17	4.32
		×⁵⁄₁₆	32.2	9.44	190	39.3	26.6	4.49	4.84	13.9	3.3	5.0	1.22	4.25
	L 5 ×3½×	⁵⁄₁₆	33.7	9.91	197	42.3	26.9	4.46	4.67	20.8	4.1	7.0	1.45	5.09
		×⅜	35.4	10.40	204	45.4	27.2	4.43	4.49	23.5	4.7	7.7	1.50	5.00
	L 6 ×4	×⅜	37.3	10.96	211	48.7	27.5	4.39	4.33	34.5	5.9	10.7	1.77	5.81
		×½	41.2	12.10	223	55.3	28.0	4.29	4.03	42.3	7.5	12.4	1.87	5.63
C 15×33.9	L 4 ×3	×¼	39.7	11.65	383	58.7	45.1	5.73	6.52	16.8	3.7	5.8	1.20	4.49
		×⁵⁄₁₆	41.1	12.05	395	62.4	45.6	5.73	6.33	18.7	4.2	6.3	1.25	4.43
	L 5 ×3½×	⁵⁄₁₆	42.6	12.52	408	66.5	46.1	5.71	6.14	26.2	4.9	8.5	1.45	5.30
		×⅜	44.3	13.01	421	70.8	46.5	5.69	5.94	29.3	5.6	9.2	1.50	5.23
	L 6 ×4	×⅜	46.2	13.57	434	75.4	46.9	5.65	5.76	41.3	6.8	12.4	1.75	6.06
		×½	50.1	14.71	458	84.8	47.7	5.58	5.40	50.3	8.5	14.3	1.85	5.89

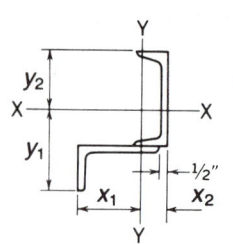

COMBINATION SECTIONS
Channels and angles
Properties of sections
Short leg of angle turned down

Channel	Angle	Total Wt. per Ft	Total Area	AXIS X-X				AXIS Y-Y			
				$S_1 = I/y_1$	$S_2 = I/y_2$	r	y_1	$S_1 = I/x_1$	$S_2 = I/x_2$	r	x_1
		Lb.	In.^{2}	In.^{3}	In.^{3}	In.	In.	In.^{3}	In.^{3}	In.	In.
C 6× 8.2	L 3×2½×¼	12.7	3.71	6.0	5.9	2.61	4.21	2.4	4.4	1.22	2.25
	L 3×3 ×¼	13.1	3.84	6.1	6.2	2.68	4.56	2.8	4.6	1.26	2.18
	L 4×3 ×¼	14.0	4.09	6.4	6.2	2.63	4.46	3.8	6.6	1.64	2.85
	L 5×3 ×⁵⁄₁₆	16.4	4.80	7.5	6.4	2.55	4.16	6.0	9.3	2.05	3.33
	L 6×3½×⁵⁄₁₆	18.0	5.27	7.7	6.8	2.56	4.45	8.4	12.1	2.47	3.82
C 7× 9.8	L 3×2½×¼	14.3	4.18	8.0	7.8	3.00	4.70	2.6	5.0	1.19	2.32
	L 3×3 ×¼	14.7	4.31	8.0	8.2	3.07	5.05	2.9	5.2	1.23	2.25
	L 4×3 ×¼	15.6	4.56	8.5	8.2	3.03	4.93	3.9	7.5	1.60	2.95
	L 5×3 ×⁵⁄₁₆	18.0	5.27	10.0	8.5	2.95	4.60	6.1	10.5	2.01	3.47
	L 6×3½×⁵⁄₁₆	19.6	5.74	10.3	8.9	2.96	4.87	8.6	13.6	2.44	3.98
C 8×11.5	L 3×2½×¼	16.0	4.69	10.4	10.2	3.39	5.20	2.7	5.6	1.16	2.37
	L 3×3 ×¼	16.4	4.82	10.4	10.6	3.45	5.55	3.0	5.8	1.20	2.31
	L 4×3 ×¼	17.3	5.07	10.9	10.6	3.42	5.42	4.0	8.3	1.55	3.03
	L 5×3 ×⁵⁄₁₆	19.7	5.78	12.9	11.0	3.36	5.06	6.3	11.7	1.97	3.58
	L 6×3½×⁵⁄₁₆	21.3	6.25	13.3	11.4	3.36	5.31	8.7	15.2	2.40	4.13
C 9×13.4	L 3×2½×¼	17.9	5.25	13.1	12.9	3.78	5.71	2.8	6.2	1.14	2.40
	L 3×3 ×¼	18.3	5.38	13.1	13.4	3.84	6.07	3.2	6.5	1.18	2.35
	L 4×3 ×¼	19.2	5.63	13.8	13.5	3.81	5.93	4.2	9.2	1.51	3.10
	L 5×3 ×⁵⁄₁₆	21.6	6.34	16.2	13.9	3.76	5.54	6.4	12.9	1.92	3.68
	L 6×3½×⁵⁄₁₆	23.2	6.81	16.7	14.4	3.77	5.78	8.9	16.9	2.36	4.26
C 10×15.3	L 3×2½×¼	19.8	5.80	16.2	16.0	4.17	6.22	3.0	6.8	1.12	2.42
	L 3×3 ×¼	20.2	5.93	16.1	16.5	4.22	6.58	3.4	7.1	1.16	2.37
	L 4×3 ×¼	21.1	6.18	17.0	16.6	4.20	6.43	4.3	10.0	1.48	3.15
	L 5×3 ×⁵⁄₁₆	23.5	6.89	19.9	17.1	4.17	6.02	6.5	14.0	1.88	3.76
	L 6×3½×⁵⁄₁₆	25.1	7.36	20.5	17.7	4.18	6.25	9.0	18.3	2.31	4.36
C 12×20.7	L 3×2½×¼	25.2	7.40	24.3	24.7	4.90	7.32	3.6	8.6	1.10	2.47
	L 3×3 ×¼	25.6	7.53	24.0	25.3	4.95	7.69	4.0	8.9	1.13	2.43
	L 4×3 ×¼	26.5	7.78	25.3	25.5	4.95	7.54	4.7	12.2	1.40	3.25
	L 5×3 ×⁵⁄₁₆	28.9	8.49	29.2	26.3	4.94	7.11	6.9	17.0	1.78	3.92
	L 6×3½×⁵⁄₁₆	30.5	8.96	30.1	27.1	4.97	7.33	9.3	22.4	2.19	4.59
C 15×33.9	L 3×3 ×¼	38.8	11.40	42.7	47.2	5.95	9.45	5.6	13.5	1.10	2.48
	L 4×3 ×¼	39.7	11.65	44.5	47.7	5.96	9.31	5.9	17.2	1.30	3.35
	L 5×3 ×⁵⁄₁₆	42.1	12.36	50.1	49.1	6.01	8.91	7.8	23.4	1.61	4.12
	L 6×3½×⁵⁄₁₆	43.7	12.83	51.4	50.3	6.05	9.15	10.2	30.7	1.97	4.88

STEEL PIPE AND STRUCTURAL TUBING
DIMENSIONS AND PROPERTIES

GENERAL

When designing and specifying steel pipe or tubing as compression members, refer to comments in the notes for Columns, Steel Pipe and Structural Tubing, p. 3-35. For standard mill practices and tolerances, refer to p. 1-155. For material specifications and availability, see Table 3, p. 1-92.

STRUCTURAL TUBING

The tables of dimensions and properties of square and rectangular structural tubing (unfilled) list a selected range of frequently used sizes. For dimensions and properties of other sizes, refer to manufacturers' catalogs.

The tables are based on an outside corner radius equal to two times the specified wall thickness. Material specifications stipulate that the outside corner radius may vary up to three times the specified wall thickness. This variation should be considered in those details where a close match or fit is important.

STEEL PIPE

The tables of dimensions and properties of steel pipe (unfilled) list a selected range of sizes of Standard, Extra Strong and Double-extra Strong pipe. For a complete range of sizes manufactured, refer to manufacturers' catalogs. The properties and dimensions table also shows the relationship between Standard, Extra Strong and Double-extra Strong pipe with pipe ordered by Schedule Number (see ASTM A53 for a more complete listing of pipe diameters).

TABLE 3
Availability of Steel Pipe and Structural Tubing According to ASTM Material Specifications

Steel	ASTM Specification	Grade	F_y Minimum Yield Stress (ksi)	F_u Minimum Tensile Stress (ksi)	Shape Round	Shape Square & Rectangular	Availability
Electric-Resistance Welded	A53 Type E	B	35	60	▨		Note 3
Seamless	Type S	B	35	60	▨		Note 3
Cold Formed	A500	A	33	45	▨		Note 1
		B	42	58	▨		Note 1
		C	46	62	▨		Note 1
		A	39	45		▨	Note 1
		B	46	58		▨	Note 2
		C	50	62		▨	Note 1
Hot Formed	A501	—	36	58	▨		Note 1
High-strength Low-alloy	A618	I	50	70	▨		Note 1
		II	50	70	▨	▨	Note 1
		III	50	65	▨		Note 1

Notes:
1. Available in mill quantities only; consult with producers.
2. Normally stocked in local steel service centers.
3. Normally stocked by local pipe distributors.
▨ Available.
☐ Not available.

PIPE
Dimensions and properties

Nominal Diameter In.	Outside Diameter In.	Inside Diameter In.	Wall Thickness In.	Weight per Ft Lbs. Plain Ends	A In.^{2}	I In.^{4}	S In.^{3}	r In.	Schedule No.
	Dimensions			Weight	Properties				
Standard Weight									
½	.840	.622	.109	.85	.250	.017	.041	.261	40
¾	1.050	.824	.113	1.13	.333	.037	.071	.334	40
1	1.315	1.049	.133	1.68	.494	.087	.133	.421	40
1¼	1.660	1.380	.140	2.27	.669	.195	.235	.540	40
1½	1.900	1.610	.145	2.72	.799	.310	.326	.623	40
2	2.375	2.067	.154	3.65	1.07	.666	.561	.787	40
2½	2.875	2.469	.203	5.79	1.70	1.53	1.06	.947	40
3	3.500	3.068	.216	7.58	2.23	3.02	1.72	1.16	40
3½	4.000	3.548	.226	9.11	2.68	4.79	2.39	1.34	40
4	4.500	4.026	.237	10.79	3.17	7.23	3.21	1.51	40
5	5.563	5.047	.258	14.62	4.30	15.2	5.45	1.88	40
6	6.625	6.065	.280	18.97	5.58	28.1	8.50	2.25	40
8	8.625	7.981	.322	28.55	8.40	72.5	16.8	2.94	40
10	10.750	10.020	.365	40.48	11.9	161	29.9	3.67	40
12	12.750	12.000	.375	49.56	14.6	279	43.8	4.38	—
Extra Strong									
½	.840	.546	.147	1.09	.320	.020	.048	.250	80
¾	1.050	.742	.154	1.47	.433	.045	.085	.321	80
1	1.315	.957	.179	2.17	.639	.106	.161	.407	80
1¼	1.660	1.278	.191	3.00	.881	.242	.291	.524	80
1½	1.900	1.500	.200	3.63	1.07	.391	.412	.605	80
2	2.375	1.939	.218	5.02	1.48	.868	.731	.766	80
2½	2.875	2.323	.276	7.66	2.25	1.92	1.34	.924	80
3	3.500	2.900	.300	10.25	3.02	3.89	2.23	1.14	80
3½	4.000	3.364	.318	12.50	3.68	6.28	3.14	1.31	80
4	4.500	3.826	.337	14.98	4.41	9.61	4.27	1.48	80
5	5.563	4.813	.375	20.78	6.11	20.7	7.43	1.84	80
6	6.625	5.761	.432	28.57	8.40	40.5	12.2	2.19	80
8	8.625	7.625	.500	43.39	12.8	106	24.5	2.88	80
10	10.750	9.750	.500	54.74	16.1	212	39.4	3.63	80
12	12.750	11.750	.500	65.42	19.2	362	56.7	4.33	—
Double-Extra Strong									
2	2.375	1.503	.436	9.03	2.66	1.31	1.10	.703	—
2½	2.875	1.771	.552	13.69	4.03	2.87	2.00	.844	—
3	3.500	2.300	.600	18.58	5.47	5.99	3.42	1.05	—
4	4.500	3.152	.674	27.54	8.10	15.3	6.79	1.37	—
5	5.563	4.063	.750	38.55	11.3	33.6	12.1	1.72	—
6	6.625	4.897	.864	53.16	15.6	66.3	20.0	2.06	—
8	8.625	6.875	.875	72.42	21.3	162	37.6	2.76	—

The listed sections are available in conformance with ASTM Specification A53 Grade B or A501. Other sections are made to these specifications. Consult with pipe manufacturers or distributors for availability.

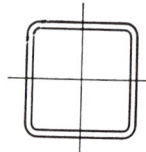

STRUCTURAL TUBING
Square
Dimensions and properties

Nominal* Size	Wall Thickness		Weight per Ft	Area	I	S	r	J	Z
In.	In.		Lb.	In.^{2}	In.^{4}	In.^{3}	In.	In.^{4}	In.^{3}
16×16	0.6250	⅝	127.37	37.4	1450	182	6.23	2320	214
	0.5000	½	103.30	30.4	1200	150	6.29	1890	175
	0.3750	⅜	78.52	23.1	931	116	6.35	1450	134
	0.3125	5⁄16	65.87	19.4	789	98.6	6.38	1220	113
14×14	0.6250	⅝	110.36	32.4	952	136	5.42	1530	161
	0.5000	½	89.68	26.4	791	113	5.48	1250	132
	0.3750	⅜	68.31	20.1	615	87.9	5.54	963	102
	0.3125	5⁄16	57.36	16.9	522	74.6	5.57	812	86.1
12×12	0.6250	⅝	93.34	27.4	580	96.7	4.60	943	116
	0.5000	½	76.07	22.4	485	80.9	4.66	777	95.4
	0.3750	⅜	58.10	17.1	380	63.4	4.72	599	73.9
	0.3125	5⁄16	48.86	14.4	324	54.0	4.75	506	62.6
	0.2500	¼	39.43	11.6	265	44.1	4.78	410	50.8
	0.1875	3⁄16	29.84	8.77	203	33.8	4.81	312	38.7
10×10	0.6250	⅝	76.33	22.4	321	64.2	3.78	529	77.6
	0.5625	9⁄16	69.48	20.4	297	59.4	3.81	485	71.3
	0.5000	½	62.46	18.4	271	54.2	3.84	439	64.6
	0.3750	⅜	47.90	14.1	214	42.9	3.90	341	50.4
	0.3125	5⁄16	40.35	11.9	183	36.7	3.93	289	42.8
	0.2500	¼	32.63	9.59	151	30.1	3.96	235	34.9
	0.1875	3⁄16	24.73	7.27	116	23.2	3.99	179	26.6
9× 9	0.6250	⅝	67.82	19.9	227	50.4	3.37	377	61.5
	0.5625	9⁄16	61.83	18.2	211	46.8	3.40	347	56.6
	0.5000	½	55.66	16.4	193	42.9	3.43	315	51.4
	0.3750	⅜	42.79	12.6	154	34.1	3.49	246	40.3
	0.3125	5⁄16	36.10	10.6	132	29.3	3.53	209	34.3
	0.2500	¼	29.23	8.59	109	24.1	3.56	170	28.0
	0.1875	3⁄16	22.18	6.52	83.8	18.6	3.59	130	21.4

*Outside dimensions across flat sides.
**Properties are based upon a nominal outside corner radius equal to two times the wall thickness.

STRUCTURAL TUBING
Square
Dimensions and properties

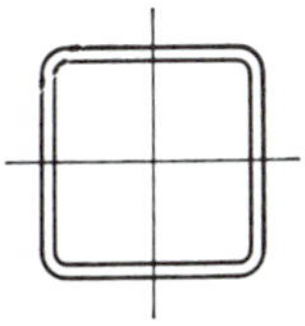

Dimensions				Properties**					
Nominal* Size	Wall Thickness		Weight per Ft	Area	*I*	*S*	*r*	*J*	*Z*
In.	In.		Lb.	In.^{2}	In.^{4}	In.^{3}	In.	In.^{4}	In.^{3}
8×8	0.6250	5/8	59.32	17.4	153	38.3	2.96	258	47.2
	0.5625	9/16	54.17	15.9	143	35.7	3.00	238	43.6
	0.5000	1/2	48.85	14.4	131	32.9	3.03	217	39.7
	0.3750	3/8	37.69	11.1	106	26.4	3.09	170	31.3
	0.3125	5/16	31.84	9.36	90.9	22.7	3.12	145	26.7
	0.2500	1/4	25.82	7.59	75.1	18.8	3.15	118	21.9
	0.1875	3/16	19.63	5.77	58.2	14.6	3.18	90.6	16.8
7×7	0.5625	9/16	46.51	13.7	91.4	26.1	2.59	154	32.3
	0.5000	1/2	42.05	12.4	84.6	24.2	2.62	141	29.6
	0.3750	3/8	32.58	9.58	68.7	19.6	2.68	112	23.5
	0.3125	5/16	27.59	8.11	59.5	17.0	2.71	95.6	20.1
	0.2500	1/4	22.42	6.59	49.4	14.1	2.74	78.3	16.5
	0.1875	3/16	17.08	5.02	38.5	11.0	2.77	60.2	12.7
6×6	0.5625	9/16	38.86	11.4	54.1	18.0	2.18	92.9	22.7
	0.5000	1/2	35.24	10.4	50.5	16.8	2.21	85.6	20.9
	0.3750	3/8	27.48	8.08	41.6	13.9	2.27	68.5	16.8
	0.3125	5/16	23.34	6.86	36.3	12.1	2.30	58.9	14.4
	0.2500	1/4	19.02	5.59	30.3	10.1	2.33	48.5	11.9
	0.1875	3/16	14.53	4.27	23.8	7.93	2.36	37.5	9.24
5×5	0.5000	1/2	28.43	8.36	27.0	10.8	1.80	46.8	13.7
	0.3750	3/8	22.37	6.58	22.8	9.11	1.86	38.2	11.2
	0.3125	5/16	19.08	5.61	20.1	8.02	1.89	33.1	9.70
	0.2500	1/4	15.62	4.59	16.9	6.78	1.92	27.4	8.07
	0.1875	3/16	11.97	3.52	13.4	5.36	1.95	21.3	6.29

*Outside dimensions across flat sides.

**Properties are based upon a nominal outside corner radius equal to two times the wall thickness.

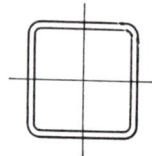

STRUCTURAL TUBING
Square
Dimensions and properties

Dimensions			Properties**					
Nominal* Size	Wall Thickness		Weight per Ft	Area	I	S	r	J	Z
In.	In.		Lb.	In.^{2}	In.^{4}	In.^{3}	In.	In.^{4}	In.^{3}
4.5×4.5	0.2500	¼	13.91	4.09	12.1	5.36	1.72	19.7	6.43
	0.1875	³⁄₁₆	10.70	3.14	9.60	4.27	1.75	15.4	5.03
4×4	0.5000	½	21.63	6.36	12.3	6.13	1.39	21.8	8.02
	0.3750	⅜	17.27	5.08	10.7	5.35	1.45	18.4	6.72
	0.3125	⁵⁄₁₆	14.83	4.36	9.58	4.79	1.48	16.1	5.90
	0.2500	¼	12.21	3.59	8.22	4.11	1.51	13.5	4.97
	0.1875	³⁄₁₆	9.42	2.77	6.59	3.30	1.54	10.6	3.91
3.5×3.5	0.3125	⁵⁄₁₆	12.70	3.73	6.09	3.48	1.28	10.4	4.35
	0.2500	¼	10.51	3.09	5.29	3.02	1.31	8.82	3.70
	0.1875	³⁄₁₆	8.15	2.39	4.29	2.45	1.34	6.99	2.93
3×3	0.3125	⁵⁄₁₆	10.58	3.11	3.58	2.39	1.07	6.22	3.04
	0.2500	¼	8.81	2.59	3.16	2.10	1.10	5.35	2.61
	0.1875	³⁄₁₆	6.87	2.02	2.60	1.73	1.13	4.28	2.10
2.5×2.5	0.3125	⁵⁄₁₆	8.45	2.48	1.87	1.50	0.868	3.32	1.96
	0.2500	¼	7.11	2.09	1.69	1.35	0.899	2.92	1.71
	0.1875	³⁄₁₆	5.59	1.64	1.42	1.14	0.930	2.38	1.40
2×2	0.3125	⁵⁄₁₆	6.32	1.86	0.880	0.880	0.690	1.49	1.11
	0.2500	¼	5.41	1.59	0.766	0.766	0.694	1.36	1.00
	0.1875	³⁄₁₆	4.32	1.27	0.668	0.668	0.726	1.15	0.840

*Outside dimensions across flat sides.
**Properties are based upon a nominal outside corner radius equal to two times the wall thickness.

STRUCTURAL TUBING
Rectangular
Dimensions and properties

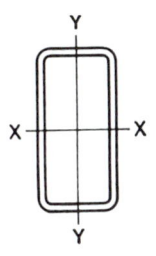

Dimensions			Properties**									
				X-X Axis				Y-Y Axis				
Nominal* Size	Wall Thickness	Weight per Ft	Area	I_x	S_x	Z_x	r_x	I_y	S_y	Z_y	r_y	J
In.	In.	Lb.	In.²	In.⁴	In.³	In.³	In.	In.⁴	In.³	In.³	In.	In.⁴
20×12	0.5000 ½	103.30	30.4	1650	165	201	7.37	750	125	141	4.97	1650
	0.3750 ⅜	78.52	23.1	1280	128	154	7.45	583	97.2	109	5.03	1270
	0.3125 ⁵⁄₁₆	65.87	19.4	1080	108	130	7.47	495	82.5	91.8	5.06	1070
20× 8	0.5000 ½	89.68	26.4	1270	127	162	6.94	300	75.1	84.7	3.38	806
	0.3750 ⅜	68.31	20.1	988	98.8	125	7.02	236	59.1	65.6	3.43	625
	0.3125 ⁵⁄₁₆	57.36	16.9	838	83.8	105	7.05	202	50.4	55.6	3.46	529
20× 4	0.5000 ½	76.07	22.4	889	88.9	123	6.31	61.6	30.8	36.0	1.66	205
	0.3750 ⅜	58.10	17.1	699	69.9	95.3	6.40	50.3	25.1	28.5	1.72	165
	0.3125 ⁵⁄₁₆	48.86	14.4	596	59.6	80.8	6.44	43.7	21.8	24.3	1.74	143
18× 6	0.5000 ½	76.07	22.4	818	90.9	119	6.05	141	47.2	53.9	2.52	410
	0.3750 ⅜	58.10	17.1	641	71.3	92.2	6.13	113	37.6	42.1	2.57	322
	0.3125 ⁵⁄₁₆	48.86	14.4	546	60.7	78.1	6.17	97.0	32.3	35.8	2.60	274
16×12	0.6250 ⅝	110.36	32.4	1160	145	175	5.98	742	124	144	4.78	1460
	0.5000 ½	89.68	26.4	962	120	144	6.04	618	103	118	4.84	1200
	0.3750 ⅜	68.31	20.1	748	93.5	111	6.11	482	80.3	91.3	4.90	922
	0.3125 ⁵⁄₁₆	57.36	16.9	635	79.4	93.8	6.14	409	68.2	77.2	4.93	777
16× 8	0.5000 ½	76.07	22.4	722	90.2	113	5.68	244	61.0	69.7	3.30	599
	0.3750 ⅜	58.10	17.1	565	70.6	87.6	5.75	193	48.2	54.2	3.36	465
	0.3125 ⁵⁄₁₆	48.86	14.4	481	60.1	74.2	5.79	165	41.2	45.9	3.39	394
16× 4	0.5000 ½	62.46	18.4	481	60.2	82.2	5.12	49.3	24.6	29.0	1.64	157
	0.3750 ⅜	47.90	14.1	382	47.8	64.2	5.21	40.4	20.2	23.0	1.69	127
	0.3125 ⁵⁄₁₆	40.35	11.9	327	40.9	54.5	5.25	35.1	17.6	19.7	1.72	110
14×10	0.6250 ⅝	93.34	27.4	728	104	127	5.15	431	86.2	101	3.96	885
	0.5000 ½	76.07	22.4	608	86.9	105	5.22	361	72.3	83.6	4.02	730
	0.3750 ⅜	58.10	17.1	476	68.0	81.5	5.28	284	56.8	64.8	4.08	564
	0.3125 ⁵⁄₁₆	48.86	14.4	405	57.9	69.0	5.31	242	48.4	54.9	4.11	477

*Outside dimensions across flat sides.
**Properties are based upon a nominal outside corner radius equal to two times the wall thickness.

AMERICAN INSTITUTE OF STEEL CONSTRUCTION

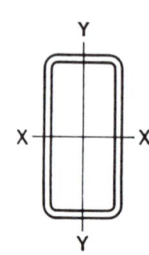

STRUCTURAL TUBING
Rectangular
Dimensions and properties

Nominal* Size In.	Wall Thickness In.		Weight per Ft Lb.	Area In.²	X-X Axis				Y-Y Axis				J In.⁴
					I_x In.⁴	S_x In.³	Z_x In.³	r_x In.	I_y In.⁴	S_y In.³	Z_y In.³	r_y In.	
14×6	0.5000	½	62.46	18.4	426	60.8	78.3	4.82	111	37.1	42.9	2.46	296
	0.3750	⅜	47.90	14.1	337	48.1	61.1	4.89	89.1	29.7	33.6	2.52	233
	0.3125	⁵⁄₁₆	40.35	11.9	288	41.2	51.9	4.93	76.7	25.6	28.7	2.54	199
	0.2500	¼	32.63	9.59	237	33.8	42.3	4.97	63.4	21.1	23.4	2.57	162
14×4	0.5000	½	55.66	16.4	335	47.8	64.8	4.52	43.1	21.5	25.5	1.62	134
	0.3750	⅜	42.79	12.6	267	38.2	50.8	4.61	35.4	17.7	20.3	1.68	108
	0.3125	⁵⁄₁₆	36.10	10.6	230	32.8	43.3	4.65	30.9	15.4	17.4	1.71	93.1
	0.2500	¼	29.23	8.59	189	27.0	35.4	4.69	25.8	12.9	14.3	1.73	77.0
12×8	0.6250	⅝	76.33	22.4	418	69.7	87.1	4.32	222	55.3	65.6	3.14	481
	0.5625	⁹⁄₁₆	69.48	20.4	387	64.5	79.9	4.35	205	51.3	60.3	3.17	442
	0.5000	½	62.46	18.4	353	58.9	72.4	4.39	188	46.9	54.7	3.20	401
	0.3750	⅜	47.90	14.1	279	46.5	56.5	4.45	149	37.3	42.7	3.26	312
	0.3125	⁵⁄₁₆	40.35	11.9	239	39.8	47.9	4.49	128	32.0	36.3	3.28	265
	0.2500	¼	32.63	9.59	196	32.6	39.1	4.52	105	26.3	29.6	3.31	216
	0.1875	³⁄₁₆	24.73	7.27	151	25.1	29.8	4.55	81.1	20.3	22.7	3.34	165
12×6	0.6250	⅝	67.82	19.9	337	56.2	72.9	4.11	112	37.2	44.5	2.37	286
	0.5625	⁹⁄₁₆	61.83	18.2	313	52.2	67.1	4.15	104	34.7	41.0	2.39	264
	0.5000	½	55.66	16.4	287	47.8	60.9	4.19	96.0	32.0	37.4	2.42	241
	0.3750	⅜	42.79	12.6	228	38.1	47.7	4.26	77.2	25.7	29.4	2.48	190
	0.3125	⁵⁄₁₆	36.10	10.6	196	32.6	40.6	4.30	66.6	22.2	25.1	2.51	162
	0.2500	¼	29.23	8.59	161	26.9	33.2	4.33	55.2	18.4	20.6	2.53	132
	0.1875	³⁄₁₆	22.18	6.52	124	20.7	25.4	4.37	42.8	14.3	15.8	2.56	101
12×4	0.6250	⅝	59.32	17.4	257	42.8	58.6	3.84	41.8	20.9	25.8	1.55	127
	0.5625	⁹⁄₁₆	54.17	15.9	240	39.9	54.2	3.88	39.6	19.8	24.0	1.58	119
	0.5000	½	48.85	14.4	221	36.8	49.4	3.92	36.9	18.5	22.0	1.60	110
	0.3750	⅜	37.69	11.1	178	29.6	39.0	4.01	30.5	15.2	17.6	1.66	89.0
	0.3125	⁵⁄₁₆	31.84	9.36	153	25.5	33.3	4.05	26.6	13.3	15.1	1.69	76.9
	0.2500	¼	25.82	7.59	127	21.1	27.3	4.09	22.3	11.1	12.5	1.71	63.6
	0.1875	³⁄₁₆	19.63	5.77	98.2	16.4	21.0	4.13	17.5	8.75	9.63	1.74	49.3

*Outside dimensions across flat sides.
**Properties are based upon a nominal outside corner radius equal to two times the wall thickness.

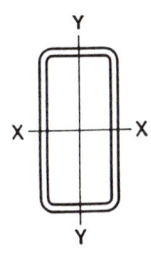

STRUCTURAL TUBING
Rectangular
Dimensions and properties

Dimensions			Properties**									
				X-X Axis				Y-Y Axis				
Nominal* Size	Wall Thickness	Weight per Ft	Area	I_x	S_x	Z_x	r_x	I_y	S_y	Z_y	r_y	J
In.	In.	Lb.	In.²	In.⁴	In.³	In.³	In.	In.⁴	In.³	In.³	In.	In.⁴
12×2	0.2500 ¼	22.42	6.59	92.2	15.4	21.4	3.74	4.62	4.62	5.38	0.837	15.9
	0.1875 ³⁄₁₆	17.08	5.02	72.0	12.0	16.6	3.79	3.76	3.76	4.24	0.865	12.8
10×8	0.6250 ⅝	67.82	19.9	266	53.2	65.9	3.65	187	46.8	56.4	3.07	367
	0.5625 ⁹⁄₁₆	61.83	18.2	247	49.3	60.6	3.68	174	43.5	52.0	3.09	337
	0.5000 ½	55.66	16.4	226	45.2	55.1	3.72	160	39.9	47.2	3.12	306
	0.3750 ⅜	42.79	12.6	180	35.9	43.1	3.78	127	31.8	37.0	3.18	239
	0.3125 ⁵⁄₁₆	36.10	10.6	154	30.8	36.7	3.81	109	27.3	31.5	3.21	203
	0.2500 ¼	29.23	8.59	127	25.4	30.0	3.84	90.2	22.5	25.8	3.24	166
	0.1875 ³⁄₁₆	22.18	6.52	97.9	19.6	23.0	3.87	69.7	17.4	19.7	3.27	127
10×6	0.6250 ⅝	59.32	17.4	211	42.2	54.2	3.48	93.5	31.2	37.7	2.32	221
	0.5625 ⁹⁄₁₆	54.17	15.9	197	39.3	50.0	3.51	87.5	29.2	34.9	2.34	204
	0.5000 ½	48.85	14.4	181	36.2	45.6	3.55	80.8	26.9	31.9	2.37	187
	0.3750 ⅜	37.69	11.1	145	29.0	35.9	3.62	65.4	21.8	25.2	2.43	147
	0.3125 ⁵⁄₁₆	31.84	9.36	125	25.0	30.7	3.65	56.5	18.8	21.5	2.46	126
	0.2500 ¼	25.82	7.59	103	20.6	25.1	3.69	46.9	15.6	17.7	2.49	103
	0.1875 ³⁄₁₆	19.63	5.77	79.8	16.0	19.3	3.72	36.5	12.2	13.6	2.51	79.1
10×5	0.6250 ⅝	55.06	16.2	183	36.7	48.3	3.37	60.0	24.0	29.3	1.93	157
	0.5625 ⁹⁄₁₆	50.34	14.8	171	34.3	44.7	3.40	56.5	22.6	27.2	1.95	146
	0.5000 ½	45.45	13.4	158	31.6	40.8	3.44	52.5	21.0	25.0	1.98	134
	0.3750 ⅜	35.13	10.3	128	25.5	32.3	3.51	42.9	17.1	19.9	2.04	107
	0.3125 ⁵⁄₁₆	29.72	8.73	110	22.0	27.6	3.55	37.2	14.9	17.0	2.07	91.5
	0.2500 ¼	24.12	7.09	91.2	18.2	22.7	3.59	31.1	12.4	14.0	2.09	75.2
	0.1875 ³⁄₁₆	18.35	5.39	70.8	14.2	17.4	3.62	24.3	9.71	10.8	2.12	58.0
10×4	0.5625 ⁹⁄₁₆	46.51	13.7	146	29.3	39.4	3.27	32.9	16.4	20.1	1.55	93.8
	0.5000 ½	42.05	12.4	136	27.1	36.1	3.31	30.8	15.4	18.5	1.58	86.9
	0.3750 ⅜	32.58	9.58	110	22.0	28.7	3.39	25.5	12.8	14.9	1.63	70.4
	0.3125 ⁵⁄₁₆	27.59	8.11	95.5	19.1	24.6	3.43	22.4	11.2	12.8	1.66	60.8
	0.2500 ¼	22.42	6.59	79.3	15.9	20.2	3.47	18.8	9.39	10.6	1.69	50.4
	0.1875 ³⁄₁₆	17.08	5.02	61.7	12.3	15.6	3.51	14.8	7.39	8.20	1.72	39.1

*Outside dimensions across flat sides.
**Properties are based upon a nominal outside corner radius equal to two times the wall thickness.

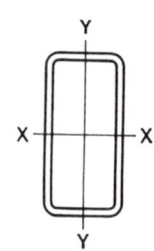

STRUCTURAL TUBING
Rectangular
Dimensions and properties

Nominal* Size	Wall Thickness		Weight per Ft	Area	X-X Axis				Y-Y Axis				J
					I_x	S_x	Z_x	r_x	I_y	S_y	Z_y	r_y	
In.	In.		Lb.	In.²	In.⁴	In.³	In.³	In.	In.⁴	In.³	In.³	In.	In.⁴
10×2	0.3750	⅜	27.48	8.08	75.4	15.1	21.5	3.06	4.85	4.85	6.05	0.775	16.5
	0.3125	⁵⁄₁₆	23.34	6.86	66.1	13.2	18.5	3.10	4.42	4.42	5.33	0.802	14.9
	0.2500	¼	19.02	5.59	55.5	11.1	15.4	3.15	3.85	3.85	4.50	0.830	12.8
	0.1875	³⁄₁₆	14.53	4.27	43.7	8.74	11.9	3.20	3.14	3.14	3.56	0.858	10.3
9×7	0.6250	⅝	59.32	17.4	183	40.6	51.0	3.24	123	35.1	42.8	2.66	248
	0.5625	⁹⁄₁₆	54.17	15.9	170	37.9	47.1	3.27	115	32.8	39.5	2.69	229
	0.5000	½	48.85	14.4	157	34.8	42.9	3.30	106	30.2	36.1	2.71	209
	0.3750	⅜	37.69	11.1	126	27.9	33.8	3.37	85.1	24.3	28.4	2.77	164
	0.3125	⁵⁄₁₆	31.84	9.36	108	24.0	28.8	3.40	73.5	21.0	24.3	2.80	140
	0.2500	¼	25.82	7.59	89.4	19.9	23.6	3.43	60.8	17.4	19.9	2.83	114
	0.1875	³⁄₁₆	19.63	5.77	69.2	15.4	18.1	3.46	47.2	13.5	15.3	2.86	87.7
9×6	0.6250	⅝	55.06	16.2	161	35.8	45.8	3.15	84.5	18.2	34.4	2.28	189
	0.5625	⁹⁄₁₆	50.34	14.8	150	33.4	42.3	3.19	79.2	26.5	31.9	2.31	175
	0.5000	½	45.45	13.4	139	30.8	38.7	3.22	73.2	24.4	29.1	2.34	160
	0.3750	⅜	35.13	10.3	112	24.8	30.6	3.29	59.4	19.8	23.1	2.40	127
	0.3125	⁵⁄₁₆	29.72	8.73	96.4	21.4	26.1	3.32	51.4	17.1	19.8	2.43	108
	0.2500	¼	24.12	7.09	79.8	17.7	21.4	3.36	42.7	14.2	16.2	2.46	88.8
	0.1875	³⁄₁₆	18.35	5.39	61.9	13.8	16.5	3.39	33.3	11.1	12.5	2.48	68.2
9×5	0.5625	⁹⁄₁₆	46.51	13.7	130	29.0	37.6	3.09	50.9	20.4	24.7	1.93	126
	0.5000	½	42.05	12.4	121	26.8	34.4	3.12	47.4	18.9	22.7	1.96	115
	0.3750	⅜	32.58	9.58	97.8	21.7	27.3	3.20	38.8	15.5	18.1	2.01	92.2
	0.3125	⁵⁄₁₆	27.59	8.11	84.6	18.8	23.4	3.23	33.8	13.5	15.6	2.04	79.2
	0.2500	¼	22.42	6.59	70.3	15.6	19.3	3.27	28.2	11.3	12.8	2.07	65.2
	0.1875	³⁄₁₆	17.08	5.02	54.7	12.1	14.8	3.30	22.1	8.84	9.90	2.10	50.2
9×3	0.5000	½	35.24	10.4	84.4	18.8	25.9	2.86	13.7	9.11	11.3	1.15	41.6
	0.3750	⅜	27.48	8.08	69.9	15.5	20.9	2.94	11.7	7.79	9.29	1.20	34.9
	0.3125	⁵⁄₁₆	23.34	6.86	61.0	13.6	18.0	2.98	10.4	6.92	8.08	1.23	30.5
	0.2500	¼	19.02	5.59	51.1	11.4	14.9	3.02	8.84	5.90	6.73	1.26	25.6
	0.1875	³⁄₁₆	14.53	4.27	40.1	8.91	11.5	3.06	7.06	4.70	5.26	1.29	20.1

*Outside dimensions across flat sides.
**Properties are based upon a nominal outside corner radius equal to two times the wall thickness.

STRUCTURAL TUBING
Rectangular
Dimensions and properties

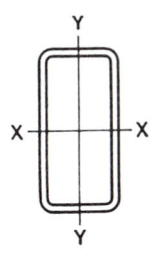

Dimensions			Properties**									
				X-X Axis				Y-Y Axis				
Nominal* Size	Wall Thickness	Weight per Ft	Area	I_x	S_x	Z_x	r_x	I_y	S_y	Z_y	r_y	J
In.	In.	Lb.	In.^{2}	In.^{4}	In.^{3}	In.^{3}	In.	In.^{4}	In.^{3}	In.^{3}	In.	In.^{4}
8×6	0.5625 9/16	46.51	13.7	112	27.9	35.2	2.86	70.8	23.6	28.8	2.28	147
	0.5000 1/2	42.05	12.4	103	25.8	32.2	2.89	65.7	21.9	26.4	2.31	135
	0.3750 3/8	32.58	9.58	83.7	20.9	25.6	2.96	53.5	17.8	21.0	2.36	107
	0.3125 5/16	27.59	8.11	72.4	18.1	21.9	2.99	46.4	15.5	18.0	2.39	91.3
	0.2500 1/4	22.42	6.59	60.1	15.0	18.0	3.02	38.6	12.9	14.8	2.42	74.9
	0.1875 3/16	17.08	5.02	46.8	11.7	13.9	3.05	30.1	10.0	11.4	2.45	57.6
8×4	0.5625 9/16	38.86	11.4	80.5	20.1	26.9	2.65	26.2	13.1	16.2	1.51	69.0
	0.5000 1/2	35.24	10.4	75.1	18.8	24.7	2.69	24.6	12.3	15.0	1.54	64.1
	0.3750 3/8	27.48	8.08	61.9	15.5	19.9	2.77	20.6	10.3	12.2	1.60	52.2
	0.3125 5/16	23.34	6.86	53.9	13.5	17.1	2.80	18.1	9.05	10.5	1.62	45.2
	0.2500 1/4	19.02	5.59	45.1	11.3	14.1	2.84	15.3	7.63	8.72	1.65	37.5
	0.1875 3/16	14.53	4.27	35.3	8.83	11.0	2.88	12.0	6.02	6.77	1.68	29.1
8×3	0.5000 1/2	31.84	9.36	61.0	15.3	21.0	2.55	12.1	8.05	10.1	1.14	35.7
	0.3750 3/8	24.93	7.33	51.0	12.7	17.0	2.64	10.4	6.92	8.31	1.19	29.9
	0.3125 5/16	21.21	6.23	44.7	11.2	14.7	2.68	9.25	6.16	7.24	1.22	26.3
	0.2500 1/4	17.32	5.09	37.6	9.40	12.2	2.72	7.90	5.26	6.05	1.25	22.1
	0.1875 3/16	13.25	3.89	29.6	7.40	9.49	2.76	6.31	4.21	4.73	1.27	17.3
8×2	0.3750 3/8	22.37	6.58	40.1	10.0	14.2	2.47	3.85	3.85	4.83	0.765	12.6
	0.3125 5/16	19.08	5.61	35.5	8.87	12.3	2.51	3.52	3.52	4.28	0.792	11.4
	0.2500 1/4	15.62	4.59	30.1	7.52	10.3	2.56	3.08	3.08	3.63	0.819	9.84
	0.1875 3/16	11.97	3.52	23.9	5.97	8.02	2.60	2.52	2.52	2.88	0.847	7.94
7×5	0.5000 1/2	35.24	10.4	63.5	18.1	23.1	2.48	37.2	14.9	18.2	1.90	79.9
	0.3750 3/8	27.48	8.08	52.2	14.9	18.5	2.54	30.8	12.3	14.6	1.95	64.2
	0.3125 5/16	23.34	6.86	45.5	13.0	15.9	2.58	26.9	10.8	12.6	1.98	55.3
	0.2500 1/4	19.02	5.59	38.0	10.9	13.2	2.61	22.6	9.04	10.4	2.01	45.6
	0.1875 3/16	14.53	4.27	29.8	8.50	10.2	2.64	17.7	7.10	8.10	2.04	35.3

*Outside dimensions across flat sides.
**Properties are based upon a nominal outside corner radius equal to two times the wall thickness.

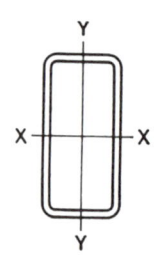

STRUCTURAL TUBING
Rectangular
Dimensions and properties

Nominal* Size In.	Wall Thickness In.		Weight per Ft Lb.	Area In.^{2}	X-X Axis				Y-Y Axis				J In.^{4}
					I_x In.^{4}	S_x In.^{3}	Z_x In.^{3}	r_x In.	I_y In.^{4}	S_y In.^{3}	Z_y In.^{3}	r_y In.	
7×4	0.5000	½	31.84	9.36	52.9	15.1	19.8	2.38	21.5	10.8	13.3	1.52	53.0
	0.3750	⅜	24.93	7.33	44.0	12.6	16.0	2.45	18.1	9.06	10.8	1.57	43.3
	0.3125	⁵⁄₁₆	21.21	6.23	38.5	11.0	13.8	2.49	16.0	7.98	9.36	1.60	37.5
	0.2500	¼	17.32	5.09	32.3	9.23	11.5	2.52	13.5	6.75	7.78	1.63	31.2
	0.1875	³⁄₁₆	13.25	3.89	25.4	7.26	8.91	2.55	10.7	5.34	6.06	1.66	24.2
7×3	0.5000	½	28.43	8.36	42.3	12.1	16.6	2.25	10.5	6.99	8.84	1.12	29.8
	0.3750	⅜	22.37	6.58	35.7	10.2	13.5	2.33	9.08	6.05	7.32	1.18	25.1
	0.3125	⁵⁄₁₆	19.08	5.61	31.5	9.00	11.8	2.37	8.11	5.41	6.40	1.20	22.0
	0.2500	¼	15.62	4.59	26.6	7.61	9.79	2.41	6.95	4.63	5.36	1.23	18.5
	0.1875	³⁄₁₆	11.97	3.52	21.1	6.02	7.63	2.45	5.57	3.71	4.21	1.26	14.6
7×2	0.2500	¼	13.91	4.09	20.9	5.98	8.10	2.26	2.69	2.69	3.19	0.812	8.36
	0.1875	³⁄₁₆	10.70	3.14	16.7	4.77	6.36	2.31	2.21	2.21	2.54	0.839	6.74
6×5	0.5000	½	31.84	9.36	42.9	14.3	18.1	2.14	32.1	12.8	16.0	1.85	62.9
	0.3750	⅜	24.93	7.33	35.6	11.9	14.7	2.21	26.8	10.7	12.9	1.91	50.9
	0.3125	⁵⁄₁₆	21.21	6.23	31.2	10.4	12.7	2.24	23.5	9.40	11.2	1.94	43.9
	0.2500	¼	17.32	5.09	26.2	8.74	10.5	2.27	19.8	7.91	9.26	1.97	36.3
	0.1875	³⁄₁₆	13.25	3.89	20.6	6.87	8.15	2.30	15.6	6.23	7.20	2.00	28.1
6×4	0.5000	½	28.43	8.36	35.3	11.8	15.4	2.06	18.4	9.21	11.5	1.48	42.1
	0.3750	⅜	22.37	6.58	29.7	9.90	12.5	2.13	15.6	7.82	9.44	1.54	34.6
	0.3125	⁵⁄₁₆	19.08	5.61	26.2	8.72	10.9	2.16	13.8	6.92	8.21	1.57	30.1
	0.2500	¼	15.62	4.59	22.1	7.36	9.06	2.19	11.7	5.87	6.84	1.60	25.0
	0.1875	³⁄₁₆	11.97	3.52	17.4	5.81	7.06	2.23	9.32	4.66	5.34	1.63	19.5
6×3	0.3750	⅜	19.82	5.83	23.8	7.92	10.4	2.02	7.78	5.19	6.34	1.16	20.3
	0.3125	⁵⁄₁₆	16.96	4.98	21.1	7.03	9.11	2.06	6.98	4.65	5.56	1.18	17.9
	0.2500	¼	13.91	4.09	17.9	5.98	7.62	2.09	6.00	4.00	4.67	1.21	15.1
	0.1875	³⁄₁₆	10.70	3.14	14.3	4.76	5.97	2.13	4.83	3.22	3.68	1.24	11.9

*Outside dimensions across flat sides.
**Properties are based upon a nominal outside corner radius equal to two times the wall thickness.

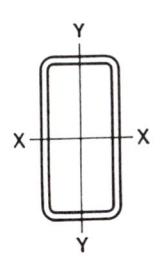

STRUCTURAL TUBING
Rectangular
Dimensions and properties

Dimensions			Properties**									
				X-X Axis				Y-Y Axis				J
Nominal* Size In.	Wall Thickness In.	Weight per Ft Lb.	Area In.^{2}	I_x In.^{4}	S_x In.^{3}	Z_x In.^{3}	r_x In.	I_y In.^{4}	S_y In.^{3}	Z_y In.^{3}	r_y In.	In.^{4}
6×2	0.3750 ⅜	17.27	5.08	17.8	5.94	8.33	1.87	2.84	2.84	3.61	0.748	8.72
	0.3125 ⁵⁄₁₆	14.83	4.36	16.0	5.34	7.33	1.92	2.62	2.62	3.22	0.775	7.94
	0.2500 ¼	12.21	3.59	13.8	4.60	6.18	1.96	2.31	2.31	2.75	0.802	6.88
	0.1875 ³⁄₁₆	9.42	2.77	11.1	3.70	4.88	2.00	1.90	1.90	2.20	0.829	5.56
5×4	0.3750 ⅜	19.82	5.83	18.7	7.50	9.44	1.79	13.2	6.58	8.08	1.50	26.3
	0.3125 ⁵⁄₁₆	16.96	4.98	16.6	6.65	8.24	1.83	11.7	5.85	7.05	1.53	22.9
	0.2500 ¼	13.91	4.09	14.1	5.65	6.89	1.86	9.98	4.99	5.90	1.56	19.1
	0.1875 ³⁄₁₆	10.70	3.14	11.2	4.49	5.39	1.89	7.96	3.98	4.63	1.59	14.9
5×3	0.5000 ½	21.63	6.36	16.9	6.75	9.20	1.63	7.33	4.88	6.35	1.07	18.2
	0.3750 ⅜	17.27	5.08	14.7	5.89	7.71	1.70	6.48	4.32	5.35	1.13	15.6
	0.3125 ⁵⁄₁₆	14.83	4.36	13.2	5.27	6.77	1.74	5.85	3.90	4.72	1.16	13.8
	0.2500 ¼	12.21	3.59	11.3	4.52	5.70	1.77	5.05	3.37	3.99	1.19	11.7
	0.1875 ³⁄₁₆	9.42	2.77	9.06	3.62	4.49	1.81	4.08	2.72	3.15	1.21	9.21
5×2	0.3125 ⁵⁄₁₆	12.70	3.73	9.74	3.90	5.31	1.62	2.16	2.16	2.70	0.762	6.24
	0.2500 ¼	10.51	3.09	8.48	3.39	4.51	1.66	1.92	1.92	2.32	0.789	5.43
	0.1875 ³⁄₁₆	8.15	2.39	6.89	2.75	3.59	1.70	1.60	1.60	1.86	0.816	4.40
4×3	0.3125 ⁵⁄₁₆	12.70	3.73	7.45	3.72	4.75	1.41	4.71	3.14	3.88	1.12	9.89
	0.2500 ¼	10.51	3.09	6.45	3.23	4.03	1.45	4.10	2.74	3.30	1.15	8.41
	0.1875 ³⁄₁₆	8.15	2.39	5.23	2.62	3.20	1.48	3.34	2.23	2.62	1.18	6.67
4×2	0.3125 ⁵⁄₁₆	10.58	3.11	5.32	2.66	3.60	1.31	1.71	1.71	2.17	0.743	4.58
	0.2500 ¼	8.81	2.59	4.69	2.35	3.09	1.35	1.54	1.54	1.88	0.770	4.01
	0.1875 ³⁄₁₆	6.87	2.02	3.87	1.93	2.48	1.38	1.29	1.29	1.52	0.798	3.26
3.5×2.5	0.2500 ¼	8.81	2.59	3.97	2.27	2.88	1.24	2.33	1.86	2.28	0.948	4.99
	0.1875 ³⁄₁₆	6.87	2.02	3.26	1.86	2.31	1.27	1.93	1.54	1.83	0.977	4.02
3×2	0.2500 ¼	7.11	2.09	2.21	1.47	1.92	1.03	1.15	1.15	1.44	0.742	2.63
	0.1875 ³⁄₁₆	5.59	1.64	1.86	1.24	1.57	1.06	0.977	0.977	1.18	0.771	2.16

*Outside dimensions across flat sides.
**Properties are based upon a nominal outside corner radius equal to two times the wall thickness.

AMERICAN INSTITUTE OF STEEL CONSTRUCTION

BARS AND PLATES
Product availability

Plates and bars are available in eight of the structural steel specifications listed in Sect. A3.1 of the AISC ASD Specification. These are ASTM A36, A242, A441, A529, A572, A588, A514 and A852. Bars are available in all of these steels except A514 and A852. Table 1, p. 1-7 shows the availability of each steel in terms of plate thickness.

The Manual user is referred to the discussion on Selection of the Appropriate Structural Steel, p. 1-3, for guidance in selection of both plate and structural shapes. For additional information designers should consult the steel producers.

CLASSIFICATION

Bars and plates are generally classified as follows:

Bars: 6 in. or less in width, .203 in. and over in thickness.
Over 6 in. to 8 in. in width, .230 in. and over in thickness.
Plates: Over 8 in. to 48 in. in width, .230 in. and over in thickness.
Over 48 in. in width, .180 in. and over in thickness.

BARS

Bars are available in various widths, thicknesses, diameters and lengths. The preferred practice is to specify widths in ¼-in. increments and thickness and diameter in ⅛-in. increments.

PLATES

Defined according to rolling procedure:

Sheared plates are rolled between horizontal rolls and trimmed (sheared or gas cut) on all edges.
Universal (UM) plates are rolled between horizontal and vertical rolls and trimmed (sheared or gas cut) on ends only.
Stripped plates are furnished to required widths by shearing or gas cutting from wider sheared plates.

Sizes

Plate mills are located in various districts, but the sizes of plates produced differ greatly and the catalogs of individual mills should be consulted for detail data. The extreme width of UM plates currently rolled is 60 in. and for sheared plates 200 in., but their availability together with limiting thickness and lengths should be checked with the mills before specifying. The preferred increments for width and thickness are:

Widths: Various. The catalogs of individual mills should be consulted to determine the most economical widths.
Thickness: ¹⁄₃₂ in. increments up to ½ in.
¹⁄₁₆ in. increments over ½ to 1 in.
⅛ in. increments over 1 in. to 3 in.
¼ in. increments over 3 in.

AMERICAN INSTITUTE OF STEEL CONSTRUCTION

Ordering

Plate thickness may be specified in inches or by weight per square foot, but no decimal edge thickness can be assured by the latter method. Separate tolerance tables apply to each method.

"*Sketch*" plates (i.e., plates whose dimensions and cuts are detailed), exclusive of those with re-entrant cuts, can be supplied by most mills by shearing or gas cutting, depending on thickness.

"*Full circles*" are also available, either by shearing up to 1 in. thickness, or by gas cutting for heavier gages.

Invoicing

Standard practice is to invoice plates to the fabricator at theoretical weight at point of shipment. Permissible variations in weight are in accordance with the tables of ASTM Specification A6.

All sketch plates, including circles, are invoiced at theoretical weight and, except as noted, are subject to the same weight variations as apply to rectangular plates. Odd shapes in most instances require gas cutting, for which gas cutting extras are applicable.

All plates ordered gas cut for whatever reason, or beyond published shearing limits, take extras for gas cutting in addition to all other extras. Rolled steel bearing plates are often gas cut to prevent distortion due to shearing but would also take the regular extra for the thickness involved.

Extras for thickness, width, length, cutting, quality and quantity, etc., which are added to the base price of plates, are subject to revision, and should be obtained by inquiry to the producer. The foregoing general statements are made as a guide toward economy in design.

FLOOR PLATES

Floor plates having raised patterns are available from several mills, each offering their own style of surface projections and in a variety of widths, thicknesses and lengths. A maximum width of 96 in. and a maximum thickness of 1 in. are available, but availability of matching widths, thicknesses and lengths should be checked with the producer. Floor plates are generally not specified to chemical composition limits or mechanical property requirements; a commercial grade of carbon steel is furnished. However, when strength or corrosion resistance is a consideration, raised pattern floor plates are procurable in any of the regular steel specifications. As in the case of plain plates, the individual manufacturers should be consulted for precise information. The nominal or ordered thickness is that of the flat plate, exclusive of the height of raised pattern. For Table of Loads-Floor Plate, see p. 2-145. The usual weights are as follows:

THEORETICAL WEIGHTS OF ROLLED FLOOR PLATES WITH RAISED PATTERN

Gauge No.	Theoretical Weight per Sq. Ft, Lb.	Nominal Thickness, In.	Theoretical Weight per Sq. Ft, Lb.	Nominal Thickness, In.	Theoretical Weight per Sq. Ft, Lb.
18	2.40	1/8	6.16	1/2	21.47
16	3.00	3/16	8.71	9/16	24.02
14	3.75	1/4	11.26	5/8	26.58
13	4.50	5/16	13.81	3/4	31.68
12	5.25	3/8	16.37	7/8	36.78
		7/16	18.92	1	41.89

Note: Thickness is measured near the edge of the plate, exclusive of raised pattern. The plate thickness listed is exclusive of raised pattern height, see manufacturer's catalog.

SQUARE AND ROUND BARS
Weight and area

Size In.	Weight Lb. per Ft ■	Weight Lb. per Ft ●	Area Sq. In ▨	Area Sq. In ◎	Size In.	Weight Lb. per Ft ■	Weight Lb. per Ft ●	Area Sq. In ▨	Area Sq. In ◎
0					**3**	30.63	24.05	9.000	7.069
1/16	0.013	0.010	0.0039	0.0031	1/16	31.91	25.07	9.379	7.366
1/8	0.053	0.042	0.0156	0.0123	1/8	33.23	26.10	9.766	7.670
3/16	1.120	0.094	0.0352	0.0276	3/16	34.57	27.15	10.160	7.980
1/4	0.213	0.167	0.0625	0.0491	1/4	35.94	28.23	10.563	8.296
5/16	0.332	0.261	0.0977	0.0767	5/16	37.34	29.32	10.973	8.618
3/8	0.479	0.376	0.1406	0.1105	3/8	38.76	30.44	11.391	8.946
7/16	0.651	0.512	0.1914	0.1503	7/16	40.21	31.58	11.816	9.281
1/2	0.851	0.668	0.2500	0.1963	1/2	41.68	32.74	12.250	9.621
9/16	1.077	0.846	0.3164	0.2485	9/16	43.19	33.92	12.691	9.968
5/8	1.329	1.044	0.3906	0.3068	5/8	44.71	35.12	13.141	10.321
11/16	1.608	1.263	0.4727	0.3712	11/16	46.27	36.34	13.598	10.680
3/4	1.914	1.503	0.5625	0.4418	3/4	47.85	37.58	14.063	11.045
13/16	2.246	1.764	0.6602	0.5185	13/16	49.46	38.85	14.535	11.416
7/8	2.605	2.046	0.7656	0.6013	7/8	51.09	40.13	15.016	11.793
15/16	2.991	2.349	0.8789	0.6903	15/16	52.76	41.43	15.504	12.177
1	3.403	2.673	1.0000	0.7854	**4**	54.44	42.76	16.000	12.566
1/16	3.841	3.017	1.1289	0.8866	1/16	56.16	44.11	16.504	12.962
1/8	4.307	3.382	1.2656	0.9940	1/8	57.90	45.47	17.016	13.364
3/16	4.796	3.769	1.4102	1.1075	3/16	59.67	46.86	17.535	13.772
1/4	5.317	4.176	1.5625	1.2272	1/4	61.46	48.27	18.063	14.186
5/16	5.862	4.604	1.7227	1.3530	5/16	63.28	49.70	18.598	14.607
3/8	6.433	5.053	1.8906	1.4849	3/8	65.13	51.15	19.141	15.033
7/16	7.032	5.523	2.0664	1.6230	7/16	67.01	52.63	19.691	15.466
1/2	7.656	6.013	2.2500	1.7671	1/2	68.91	54.12	20.250	15.904
9/16	8.308	6.525	2.4414	1.9175	9/16	70.83	55.63	20.816	16.349
5/8	8.985	7.067	2.6406	2.0739	5/8	72.79	57.17	21.391	16.800
11/16	9.690	7.610	2.8477	2.2365	11/16	74.77	58.72	21.973	17.257
3/4	10.421	8.185	3.0625	2.4053	3/4	76.78	60.30	22.563	17.721
13/16	11.179	8.780	3.2852	2.5802	13/16	78.81	61.90	23.160	18.190
7/8	11.963	9.396	3.5156	2.7612	7/8	80.87	63.51	23.766	18.665
15/16	12.774	10.032	3.7539	2.9483	15/16	82.96	65.15	24.379	19.147
2	13.611	10.690	4.0000	3.1416	**5**	85.07	66.81	25.000	19.635
1/16	14.475	11.369	4.2539	3.3410	1/16	87.21	68.49	25.629	20.129
1/8	15.366	12.068	4.5156	3.5466	1/8	89.38	70.20	26.266	20.629
3/16	16.283	12.788	4.7852	3.7583	3/16	91.57	71.92	26.910	21.135
1/4	17.227	13.530	5.0625	3.9761	1/4	93.79	73.66	27.563	21.648
5/16	18.197	14.292	5.3477	4.2000	5/16	96.04	75.43	28.223	22.166
3/8	19.194	15.075	5.6406	4.4301	3/8	98.31	77.21	28.891	22.691
7/16	20.217	15.879	5.9414	4.6664	7/16	100.61	79.02	29.566	23.221
1/2	21.267	16.703	6.2500	4.9087	1/2	102.93	80.84	30.250	23.758
9/16	22.344	17.549	6.5664	5.1572	9/16	105.29	82.69	30.941	24.301
5/8	23.447	18.415	6.8906	5.4119	5/8	107.67	84.56	31.641	24.850
11/16	24.577	19.303	7.2227	5.6727	11/16	110.07	86.45	32.348	25.406
3/4	25.734	20.211	7.5625	5.9396	3/4	112.50	88.36	33.063	25.967
13/16	26.917	21.140	7.9102	6.2126	13/16	114.96	90.29	33.785	26.535
7/8	28.126	22.090	8.2656	6.4918	7/8	117.45	92.24	34.516	27.109
15/16	29.362	23.061	8.6289	6.7771	15/16	119.96	94.22	35.254	27.688
					6	122.50	96.21	36.000	28.274

AMERICAN INSTITUTE OF STEEL CONSTRUCTION

SQUARE AND ROUND BARS
Weight and area

Size In.	Weight Lb. per Ft ■	Weight Lb. per Ft ●	Area Sq. In ▨	Area Sq. In ◯	Size In.	Weight Lb. per Ft ■	Weight Lb. per Ft ●	Area Sq. In ▨	Area Sq. In ◯
6	122.50	96.21	36.000	28.274	9	275.63	216.48	81.000	63.617
1/16	125.07	98.23	36.754	28.866	1/16	279.47	219.49	82.129	64.504
1/8	127.66	100.26	37.516	29.465	1/8	283.33	222.53	83.266	65.397
3/16	130.28	102.32	38.285	30.069	3/16	287.23	225.59	84.410	66.296
1/4	132.92	104.40	39.063	30.680	1/4	291.15	228.67	85.563	67.201
5/16	135.59	106.49	39.848	31.296	5/16	295.10	231.77	86.723	68.112
3/8	138.29	108.61	40.641	31.919	3/8	299.07	234.89	87.891	69.029
7/16	141.02	110.75	41.441	32.548	7/16	303.07	238.03	89.066	69.953
1/2	143.77	112.91	42.250	33.183	1/2	307.10	241.20	90.250	70.882
9/16	146.55	115.10	43.066	33.824	9/16	311.15	244.38	91.441	71.818
5/8	149.35	117.30	43.891	34.472	5/8	315.24	247.59	92.641	72.760
11/16	152.18	119.52	44.723	35.125	11/16	319.34	250.81	93.848	73.708
3/4	155.04	121.77	45.563	35.785	3/4	323.48	254.06	95.063	74.662
13/16	157.92	124.03	46.410	36.450	13/16	327.64	257.33	96.285	75.622
7/8	160.83	126.32	47.266	37.122	7/8	331.82	260.61	97.516	76.589
15/16	163.77	128.63	48.129	37.800	15/16	336.04	263.92	98.754	77.561
7	166.74	130.95	49.000	38.485	10	340.28	267.25	100.000	78.540
1/16	169.73	133.30	49.879	39.175	1/16	344.54	270.60	101.254	79.525
1/8	172.74	135.67	50.766	39.871	1/8	348.84	273.98	102.516	80.516
3/16	175.79	138.06	51.660	40.574	3/16	353.16	277.37	103.785	81.513
1/4	178.86	140.48	52.563	41.282	1/4	357.50	280.78	105.063	82.516
5/16	181.96	142.91	53.473	41.997	5/16	361.88	284.22	106.348	83.525
3/8	185.08	145.36	54.391	42.718	3/8	366.28	287.67	107.641	84.541
7/16	188.23	147.84	55.316	43.445	7/16	370.70	291.15	108.941	85.562
1/2	191.41	150.33	56.250	44.179	1/2	375.16	294.65	110.250	86.590
9/16	194.61	152.85	57.191	44.918	9/16	379.64	298.17	111.566	87.624
5/8	197.84	155.38	58.141	45.664	5/8	384.14	301.70	112.891	88.664
11/16	201.10	157.94	59.098	46.415	11/16	388.67	305.26	114.223	89.710
3/4	204.38	160.52	60.063	47.173	3/4	393.23	308.84	115.563	90.763
13/16	207.69	163.12	61.035	47.937	13/16	397.82	312.45	116.910	91.821
7/8	211.03	165.74	62.016	48.707	7/8	402.43	316.07	118.266	92.886
15/16	214.39	168.38	63.004	49.483	15/16	407.07	319.71	119.629	93.956
8	217.78	171.04	64.000	50.265	11	411.74	323.38	121.000	95.033
1/16	221.19	173.73	65.004	51.054	1/16	416.43	327.06	122.379	96.116
1/8	224.64	176.43	66.016	51.849	1/8	421.15	330.77	123.766	97.205
3/16	228.11	179.15	67.035	52.649	3/16	425.89	334.49	125.160	98.301
1/4	231.60	181.90	68.063	53.456	1/4	430.66	338.24	126.563	99.402
5/16	235.12	184.67	69.098	54.269	5/16	435.46	342.01	127.973	100.510
3/8	238.67	187.45	70.141	55.088	3/8	440.29	345.80	129.391	101.623
7/16	242.25	190.26	71.191	55.914	7/16	445.14	349.61	130.816	102.743
1/2	245.85	193.09	72.250	56.745	1/2	450.02	353.44	132.250	103.869
9/16	249.48	195.94	73.316	57.583	9/16	454.92	357.30	133.691	105.001
5/8	253.13	198.81	74.391	58.426	5/8	459.85	361.17	135.141	106.139
11/16	256.82	201.70	75.473	59.276	11/16	464.81	365.06	136.598	107.284
3/4	260.53	204.62	76.563	60.132	3/4	469.80	368.98	138.063	108.434
13/16	264.26	207.55	77.660	60.994	13/16	474.81	372.91	139.535	109.591
7/8	268.02	210.50	78.766	61.862	7/8	479.84	376.87	141.016	110.753
15/16	271.81	213.48	79.879	62.737	15/16	484.91	380.85	142.504	111.922
					12	490.00	384.85	144.000	113.097

AREA OF RECTANGULAR SECTIONS
Square inches

Width In.	Thickness, Inches													
	3/16	1/4	5/16	3/8	7/16	1/2	9/16	5/8	11/16	3/4	13/16	7/8	15/16	1
1/4	0.047	0.063	0.078	0.094	0.109	0.125	0.141	0.156	0.172	0.188	0.203	0.219	0.234	0.250
1/2	0.094	0.125	0.156	0.188	0.219	0.250	0.281	0.313	0.344	0.375	0.406	0.438	0.469	0.500
3/4	0.141	0.188	0.234	0.281	0.328	0.375	0.422	0.469	0.516	0.563	0.609	0.656	0.703	0.750
1	0.188	0.250	0.313	0.375	0.438	0.500	0.563	0.625	0.688	0.750	0.813	0.875	0.938	1.00
1 1/4	0.234	0.313	0.391	0.469	0.547	0.625	0.703	0.781	0.859	0.938	1.02	1.09	1.17	1.25
1 1/2	0.281	0.375	0.469	0.563	0.656	0.750	0.844	0.938	1.03	1.13	1.22	1.31	1.41	1.50
1 3/4	0.328	0.438	0.547	0.656	0.766	0.875	0.984	1.09	1.20	1.31	1.42	1.53	1.64	1.75
2	0.375	0.500	0.625	0.750	0.875	1.00	1.13	1.25	1.38	1.50	1.63	1.75	1.88	2.00
2 1/4	0.422	0.563	0.703	0.844	0.984	1.13	1.27	1.41	1.55	1.69	1.83	1.97	2.11	2.25
2 1/2	0.469	0.625	0.781	0.938	1.09	1.25	1.41	1.56	1.72	1.88	2.03	2.19	2.34	2.50
2 3/4	0.516	0.688	0.859	1.03	1.20	1.38	1.55	1.72	1.89	2.06	2.23	2.41	2.58	2.75
3	0.563	0.750	0.938	1.13	1.31	1.50	1.69	1.88	2.06	2.25	2.44	2.63	2.81	3.00
3 1/4	0.609	0.813	1.02	1.22	1.42	1.63	1.83	2.03	2.23	2.44	2.64	2.84	3.05	3.25
3 1/2	0.656	0.875	1.09	1.31	1.53	1.75	1.97	2.19	2.41	2.63	2.84	3.06	3.28	3.50
3 3/4	0.703	0.938	1.17	1.41	1.64	1.88	2.11	2.34	2.58	2.81	3.05	3.28	3.52	3.75
4	0.750	1.00	1.25	1.50	1.75	2.00	2.25	2.50	2.75	3.00	3.25	3.50	3.75	4.00
4 1/4	0.797	1.06	1.33	1.59	1.86	2.13	2.39	2.66	2.92	3.19	3.45	3.72	3.98	4.25
4 1/2	0.844	1.13	1.41	1.69	1.97	2.25	2.53	2.81	3.09	3.38	3.66	3.94	4.22	4.50
4 3/4	0.891	1.19	1.48	1.78	2.09	2.38	2.67	2.97	3.27	3.56	3.86	4.16	4.45	4.75
5	0.938	1.25	1.56	1.88	2.19	2.50	2.81	3.13	3.44	3.75	4.06	4.38	4.69	5.00
5 1/4	0.984	1.31	1.64	1.97	2.30	2.63	2.95	3.28	3.61	3.94	4.27	4.59	4.92	5.25
5 1/2	1.03	1.38	1.72	2.06	2.41	2.75	3.09	3.44	3.78	4.13	4.47	4.81	5.16	5.50
5 3/4	1.08	1.44	1.80	2.16	2.52	2.88	3.23	3.59	3.95	4.31	4.67	5.03	5.39	5.75
6	1.13	1.50	1.88	2.25	2.63	3.00	3.38	3.75	4.13	4.50	4.88	5.25	5.63	6.00
6 1/4	1.17	1.56	1.95	2.34	2.73	3.13	3.52	3.91	4.30	4.69	5.08	5.47	5.86	6.25
6 1/2	1.22	1.63	2.03	2.44	2.84	3.25	3.66	4.06	4.47	4.88	5.28	5.69	6.09	6.50
6 3/4	1.27	1.69	2.10	2.53	2.95	3.38	3.80	4.22	4.64	5.06	5.48	5.91	6.33	6.75
7	1.31	1.75	2.19	2.63	3.06	3.50	3.94	4.38	4.81	5.25	5.69	6.13	6.56	7.00
7 1/4	1.36	1.81	2.27	2.72	3.17	3.63	4.08	4.53	4.98	5.44	5.89	6.34	6.80	7.25
7 1/2	1.41	1.88	2.34	2.81	3.28	3.75	4.22	4.69	5.16	5.63	6.09	6.56	7.03	7.50
7 3/4	1.45	1.94	2.42	2.91	3.39	3.88	4.36	4.84	5.33	5.81	6.30	6.78	7.27	7.75
8	1.50	2.00	2.50	3.00	3.50	4.00	4.50	5.00	5.50	6.00	6.50	7.00	7.50	8.00
8 1/2	1.59	2.13	2.66	3.19	3.72	4.25	4.78	5.31	5.84	6.38	6.91	7.44	7.97	8.50
9	1.69	2.25	2.81	3.38	3.94	4.50	5.06	5.63	6.19	6.75	7.31	7.88	8.44	9.00
9 1/2	1.78	2.38	2.97	3.56	4.16	4.75	5.34	5.94	6.53	7.13	7.72	8.31	8.91	9.50
10	1.88	2.50	3.13	3.75	4.38	5.00	5.63	6.25	6.88	7.50	8.13	8.75	9.38	10.00
10 1/2	1.97	2.63	3.28	3.94	4.59	5.25	5.91	6.56	7.22	7.88	8.53	9.19	9.84	10.50
11	2.06	2.75	3.44	4.13	4.81	5.50	6.19	6.88	7.56	8.25	8.94	9.63	10.31	11.00
11 1/2	2.16	2.88	3.59	4.31	5.03	5.75	6.47	7.19	7.91	8.63	9.34	10.06	10.78	11.50
12	2.25	3.00	3.75	4.50	5.25	6.00	6.75	7.50	8.25	9.00	9.75	10.50	11.25	12.00

WEIGHT OF RECTANGULAR SECTIONS
Pounds per linear foot

Width In.	3/16	1/4	5/16	3/8	7/16	1/2	9/16	5/8	11/16	3/4	13/16	7/8	15/16	1
1/4	0.16	0.21	0.27	0.32	0.37	0.43	0.48	0.53	0.58	0.64	0.69	0.74	0.80	0.85
1/2	0.32	0.43	0.53	0.64	0.74	0.85	0.96	1.06	1.17	1.28	1.38	1.49	1.60	1.70
3/4	0.48	0.64	0.80	0.96	1.12	1.28	1.44	1.60	1.75	1.91	2.07	2.23	2.39	2.55
1	0.64	0.85	1.06	1.28	1.49	1.70	1.91	2.13	2.34	2.55	2.76	2.98	3.19	3.40
1 1/4	0.80	1.06	1.33	1.60	1.86	2.13	2.39	2.66	2.92	3.19	3.46	3.72	3.99	4.25
1 1/2	0.96	1.28	1.60	1.91	2.23	2.56	2.87	3.19	3.51	3.83	4.15	4.47	4.79	5.10
1 3/4	1.12	1.49	1.86	2.23	2.61	2.98	3.35	3.72	4.09	4.47	4.84	5.21	5.58	5.95
2	1.28	1.70	2.13	2.55	2.98	3.40	3.83	4.25	4.68	5.10	5.53	5.95	6.38	6.81
2 1/4	1.44	1.91	2.39	2.87	3.35	3.83	4.31	4.79	5.26	5.74	6.22	6.70	7.18	7.66
2 1/2	1.60	2.13	2.66	3.19	3.72	4.25	4.79	5.32	5.85	6.38	6.91	7.44	7.96	8.51
2 3/4	1.75	2.34	2.92	3.51	4.09	4.68	5.26	5.85	6.43	7.02	7.60	8.19	8.77	9.36
3	1.91	2.55	3.19	3.83	4.47	5.10	5.74	6.38	7.02	7.66	8.29	8.93	9.57	10.2
3 1/4	2.07	2.76	3.46	4.15	4.84	5.53	6.22	6.91	7.60	8.29	8.99	9.68	10.4	11.1
3 1/2	2.23	2.98	3.72	4.47	5.21	5.95	6.70	7.44	8.19	8.93	9.68	10.4	11.2	11.9
3 3/4	2.39	3.19	3.99	4.79	5.58	6.38	7.18	7.98	8.77	9.57	10.4	11.2	12.0	12.8
4	2.55	3.40	4.25	5.10	5.95	6.81	7.66	8.51	9.36	10.2	11.1	11.9	12.8	13.6
4 1/4	2.71	3.62	4.52	5.42	6.33	7.23	8.13	9.04	9.94	10.8	11.8	12.7	13.6	14.5
4 1/2	2.87	3.83	4.79	5.74	6.70	7.66	8.61	9.57	10.5	11.5	12.4	13.4	14.4	15.3
4 3/4	3.03	4.04	5.05	6.06	7.07	8.08	9.09	10.1	11.1	12.1	13.1	14.1	15.2	16.2
5	3.19	4.25	5.32	6.38	7.44	8.51	9.57	10.6	11.7	12.8	13.8	14.9	16.0	17.0
5 1/4	3.35	4.47	5.58	6.70	7.82	8.93	10.0	11.2	12.3	13.4	14.5	15.6	16.7	17.9
5 1/2	3.51	4.68	5.85	7.02	8.19	9.36	10.5	11.7	12.9	14.0	15.2	16.4	17.5	18.7
5 3/4	3.67	4.89	6.11	7.34	8.56	9.78	11.0	12.2	13.5	14.7	15.9	17.1	18.3	19.6
6	3.83	5.10	6.38	7.66	8.93	10.2	11.5	12.8	14.0	15.3	16.6	17.9	19.1	20.4
6 1/4	3.99	5.32	6.65	7.98	9.30	10.6	12.0	13.3	14.6	16.0	17.3	18.6	19.9	21.3
6 1/2	4.15	5.53	6.91	8.29	9.68	11.1	12.4	13.8	15.2	16.6	18.0	19.4	20.7	22.1
6 3/4	4.31	5.74	7.18	8.61	10.0	11.5	12.9	14.4	15.8	17.2	18.7	20.1	21.5	23.0
7	4.47	5.95	7.44	8.93	10.4	11.9	13.4	14.9	16.4	17.9	19.4	20.8	22.3	23.8
7 1/4	4.63	6.17	7.71	9.25	10.8	12.3	13.9	15.4	17.0	18.5	20.0	21.6	23.1	24.7
7 1/2	4.79	6.38	7.98	9.57	11.2	12.8	14.4	16.0	17.5	19.1	20.7	22.3	23.9	25.5
7 3/4	4.94	6.59	8.24	9.89	11.5	13.2	14.8	16.5	18.1	19.8	21.4	23.1	24.7	26.4
8	5.10	6.81	8.51	10.2	11.9	13.6	15.3	17.0	18.7	20.4	22.1	23.8	25.5	27.2
8 1/2	5.42	7.23	9.04	10.8	12.7	14.5	16.3	18.1	19.9	21.7	23.5	25.3	27.1	28.9
9	5.74	7.66	9.57	11.5	13.4	15.3	17.2	19.1	21.1	23.0	24.9	26.8	28.7	30.6
9 1/2	6.06	8.08	10.1	12.1	14.1	16.2	18.2	20.2	22.2	24.2	26.3	28.3	30.3	32.3
10	6.38	8.51	10.6	12.8	14.9	17.0	19.1	21.3	23.4	25.5	27.6	29.8	31.9	34.0
10 1/2	6.70	8.93	11.2	13.4	15.6	17.9	20.1	22.3	24.6	26.8	29.0	31.3	33.5	35.7
11	7.02	9.36	11.7	14.0	16.4	18.7	21.1	23.4	25.7	28.1	30.4	32.8	35.1	37.4
11 1/2	7.34	9.78	12.2	14.7	17.1	19.6	22.0	24.5	26.9	29.3	31.8	34.2	36.7	39.1
12	7.66	10.2	12.8	15.3	17.9	20.4	23.0	25.5	28.1	30.6	33.2	35.7	38.3	40.8

CRANE RAILS

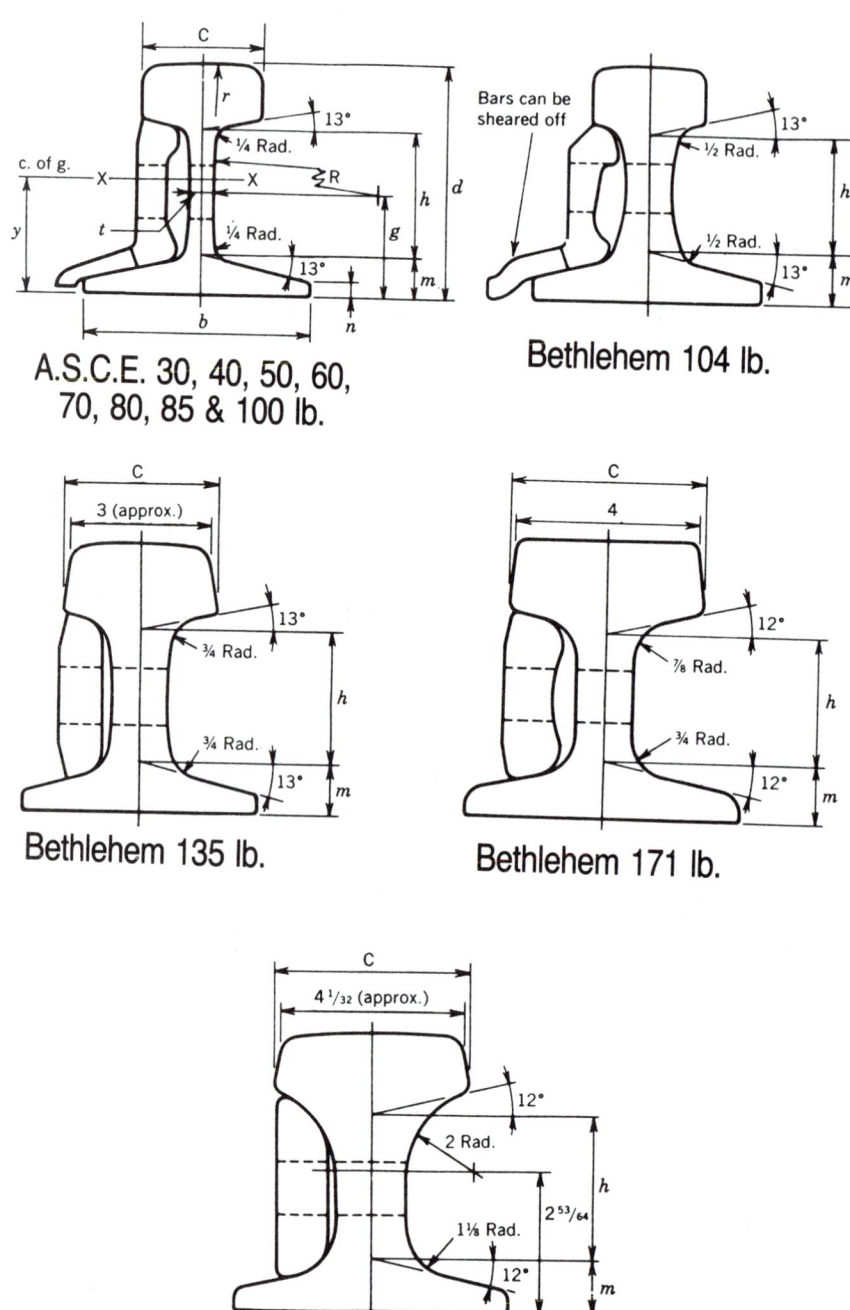

A.S.C.E. 30, 40, 50, 60, 70, 80, 85 & 100 lb.

Bethlehem 104 lb.

Bethlehem 135 lb.

Bethlehem 171 lb.

Bethlehem 175 lb.

Nomenclature of sketch for A.S.C.E. rails also applies to the other sections.

CRANE RAILS

GENERAL NOTES

The ASCE rails and the 104- to 175-lb. crane rails listed below are recommended for crane runway use. For complete details and for profiles and properties of rails not listed, consult manufacturers' catalogs.

Rails should be arranged so that joints on opposite sides of the crane runway will be staggered with respect to each other and with due consideration to the wheelbase of the crane. Rail joints should not occur at crane girder splices. Light 40-lb. rails are available in 30-ft lengths, 60-lb. rails in 30-, 33- or 39-ft lengths, standard rails in 33- or 39- ft lengths and crane rails up to 60 ft. Consult manufacturer for availability of other lengths. Odd lengths, which must be included to complete a run or obtain the necessary stagger, should be not less than 10 ft long. For crane rail service, 40-lb. rails are furnished to manufacturers' specifications and tolerances. 60- and 85-lb. rails are furnished to manufacturers' specifications and tolerances, or to ASTM A1. Crane rails are furnished to ASTM A759. Rails will be furnished with standard drilling (see p. 1-115) in both standard and odd lengths unless stipulated otherwise on order. For controlled cooling, heat treatment and rail end preparation, see manufacturers' catalogs. Purchase orders for crane rails should be noted "For crane service."

DIMENSIONS AND PROPERTIES

Type	Clas-sifi-ca-tion	Nom-inal Wt. per Yd.	d	Gage g	Base b	m	n	Head c	r	Web t	h	R	Area	I_x	S Hd.	Bse.	y
		Lb.	In.	In.	In.	In.	In.	In.	In.	In.	In.	In.	In.^{2}	In.^{4}	In.^{3}	In.^{3}	In.
ASCE	Light	30	3⅛	1²⁵/₆₄	3⅛	¹⁷/₃₂	¹¹/₆₄	1¹¹/₁₆	12	²¹/₆₄	1²³/₃₂	12	3.00	4.10	2.55	—	—
ASCE	Light	40	3½	1⁷¹/₁₂₈	3½	⅝	⁷/₃₂	1⅞	12	²⁵/₆₄	1⁵⁵/₆₄	12	3.94	6.54	3.59	3.89	1.68
ASCE	Light	50	3⅞	1²³/₃₂	3⅞	¹¹/₁₆	¼	2⅛	12	⁷/₁₆	2¹/₁₆	12	4.90	10.1	5.10	—	1.88
ASCE	Light	60	4¼	1¹¹⁵/₁₂₈	4¼	⁴⁹/₆₄	⁹/₃₂	2⅜	12	³¹/₆₄	2¹⁷/₆₄	12	5.93	14.6	6.64	7.12	2.05
ASCE		70	4⅝	2³/₆₄	4⅝	¹³/₁₆	⁹/₃₂	2⁷/₁₆	12	³³/₆₄	2¹⁵/₃₂	12	6.81	19.7	8.19	8.87	2.22
ASCE		80	5	2³/₁₆	5	⅞	¹⁹/₆₄	2½	12	³⁵/₆₄	2⅝	12	7.86	26.4	10.1	11.1	2.38
ASCE	Std.	85	5³/₁₆	2¹⁷/₆₄	5³/₁₆	⁵⁷/₆₄	¹⁹/₆₄	2⁹/₁₆	12	⁹/₁₆	2¾	12	8.33	30.1	11.1	12.2	2.47
ASCE	Std.	100	5¾	2⁶⁵/₁₂₈	5¾	³¹/₃₂	¹⁰/₃₂	2¾	12	⁹/₁₆	3⁵/₆₄	12	9.84	44.0	14.6	16.1	2.73
Bethlehem	Crane	104	5	2⁷/₁₆	5	1¹/₁₆	½	2½	12	1	2⁷/₁₆	3½	10.3	29.8	10.7	13.5	2.21
Bethlehem	Crane	135	5¾	2¹⁵/₃₂	5³/₁₆	1¹/₁₆	¹⁵/₃₂	3⁷/₁₆	14	1¼	2¹³/₁₆	12	13.3	50.8	17.3	18.1	2.81
Bethlehem	Crane	171	6	2⅝	6	1¼	⅝	4.3	Flat	1¼	2¾	Vert.	16.8	73.4	24.5	24.4	3.01
Bethlehem	Crane	175	6	2²¹/₃₂	6	1⁹/₆₄	½	4¼	18	1½	3⁷/₆₄	Vert.	17.1	70.5	23.4	23.6	2.98

For maximum wheel loadings see manufacturers' catalogs.

CRANE RAILS
Splices

BOLTED SPLICES

It is often more desirable to use properly installed and maintained bolted splice bars in making up rail joints for crane service than welded splice bars.

Standard rail drilling and joint bar punching, as furnished by manufacturers of light standard rails for track work, include round holes in rail ends and slotted holes in joint bars to receive standard oval neck track bolts. Holes in rails are oversize and punching in joint bars is spaced to allow 1/16 to 1/8 in. clearance between rail ends (see manufacturers' catalogs for spacing and dimensions of holes and slots). Although this construction is satisfactory for track and light crane service, its use in general crane service may lead to joint failure.

For best service in bolted splices, it is recommended that tight joints be stipulated for all rails for crane service. This will require rail ends to be finished by milling or grinding, and the special rail drilling and joint bar punching tabulated below. Special rail drilling is accepted by some mills, or rails may be ordered blank for shop drilling. End finishing of standard rails can be done at the mill; light rails must be end-finished in the fabricating shop or ground at the site prior to erection. In the crane rail range, from 104 to 175 lbs. per yard, rails and joint bars are manufactured to obtain a tight fit and no further special end finishing, drilling or punching is required. Because of cumulative tolerance variations in holes, bolt diameters and rail ends, a slight gap may sometimes occur in the so-called tight joints. Conversely, it may sometimes be necessary to ream holes through joint bar and rail to permit entry of bolts.

Joint bars for crane service are provided in various sections to match the rails. Joint bars for light and standard rails may be purchased blank for special shop punching to obtain tight joints. See Bethlehem Steel Corp. Booklet 3351 for dimensions, material specifications and the identification necessary to match the crane rail section.

Joint bar bolts, as distinguished from oval neck track bolts, have straight shanks to the head and are manufactured to ASTM A449 specifications. Nuts are manufactured to ASTM A563 Gr. B specifications. ASTM A325 bolts and nuts may be used. Bolt assembly includes an alloy steel spring washer, furnished to AREA specification.

After installation, bolts should be retightened within 30 days and every three months thereafter.

WELDED SPLICES

When welded splices are specified, consult the manufacturer for recommended rail end preparation, welding procedure and method of ordering. Although joint continuity, made possible by this method of splicing, is desirable, it should be noted that the careful control required in all stages of the welding operation may be difficult to meet during crane rail installation.

Rails should not be attached to structural supports by welding. Rails with holes for joint bar bolts should not be used in making splices.

CRANE RAILS
Splices for tight joints

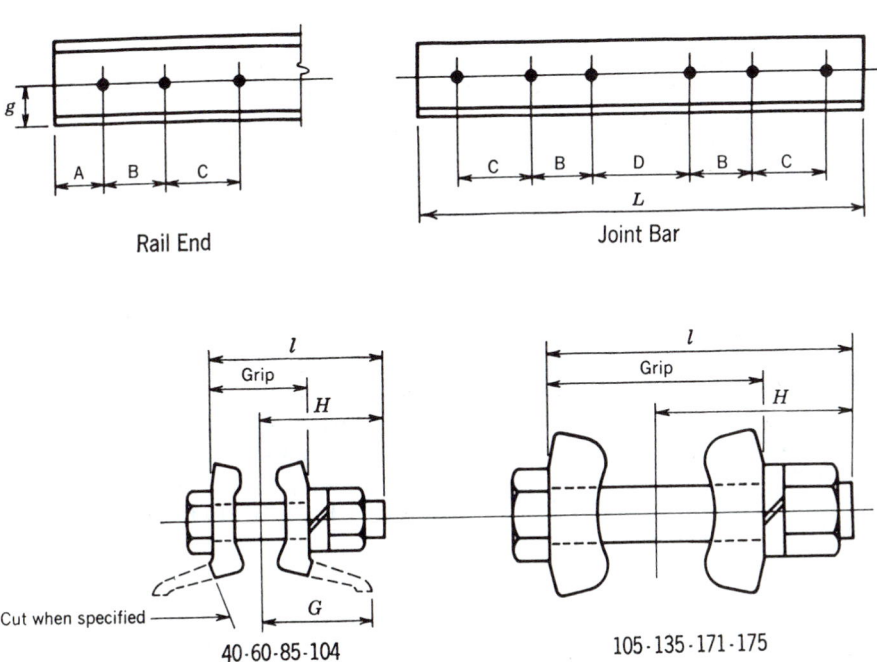

Rail End

Joint Bar

Cut when specified

40·60·85·104

105·135·171·175

Rail						Joint Bar						Bolt				Washer		Wt. 2 Bars Bolts, Nuts Washers	
	Drilling					Punching													
Wt. per Yard	g	Hole Dia.	A	B	C	Hole Dia.	D	B	C	L	G	Dia.	Grip	l	H	Inside Dia.	Thickness and Width	With Fig.	Less Fig.
Lb.	In.	In.	In.	In.	In.	In.	In.	In.	In.	In.	In.	In.	In.	In.	In.	In.	In.	Lb.	Lb.
40	1 71/128	13/16*	2½	5	...	13/16*	4 15/16*	5	...	20	2 3/16	¾	1 15/16	3½	2½	13/16	7/16x3/8	20.0	16.5
60	1 115/128	13/16*	2½	5	...	13/16*	4 15/16*	5	...	24	2 11/16	¾	2 19/32	4	2 11/16	13/16	7/16x3/8	36.5	29.6
85	2 17/64	15/16*	2½	5	...	15/16*	4 15/16*	5	...	24	3 11/32	7/8	3 5/32	4¾	3 3/16	15/16	7/16x3/8	56.6	45.3
104	2 7/16	1 1/16	4	5	6	1 1/16	7 15/16	5	6	34	3½	1	3½	5¼	3½	1 1/16	7/16x½	73.5	55.4
135	2 15/32	1 3/16	4	5	6	1 3/16	7 15/16	5	6	34	...	1 1/8	3 5/8	5½	3 11/16	1 3/16	7/16x½	...	75.3
171	2 5/8	1 3/16	4	5	6	1 3/16	7 15/16	5	6	34	...	1 1/8	4 7/16	6¼	4 1/16	1 3/16	7/16x½	...	90.8
175	2 21/32	1 3/16	4	5	6	1 3/16	7 15/16	5	6	34	...	1 1/8	4 1/8	6¼	3 15/16	1 3/16	7/16x½	...	87.7

*Special rail drilling and joint bar punching.

CRANE RAILS
Fastenings

HOOK BOLTS

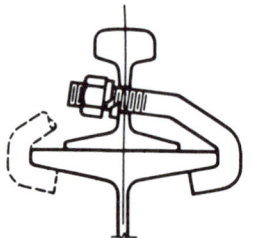

Hook bolts are used primarily with light rails when attached to beams with flanges too narrow for clamps. Rail adjustment to ±½ in. is inherent in the threaded shank. Hook bolts are paired alternately 3 to 4 in. apart, spaced at about 24-in. centers. The special rail drilling required must be done at the fabricator's shop.

RAIL CLAMPS

Although a variety of satisfactory rail clamps are available from track accessory manufactures, the two frequently recommended for crane runway use are the fixed and floating types illustrated below. These are available in forgings or pressed steel, either for single bolts or for double bolts as shown. The fixed-type features adjustment through eccentric punching of fillers and positive attachment of rail to support. The floating-type permits longitudinal and controlled tansverse movement through clamp clearances and filler adjustment, useful in allowing for thermal expansion and contraction of rails and possible misalignment of supports. Both types should be spaced 3 ft or less apart.

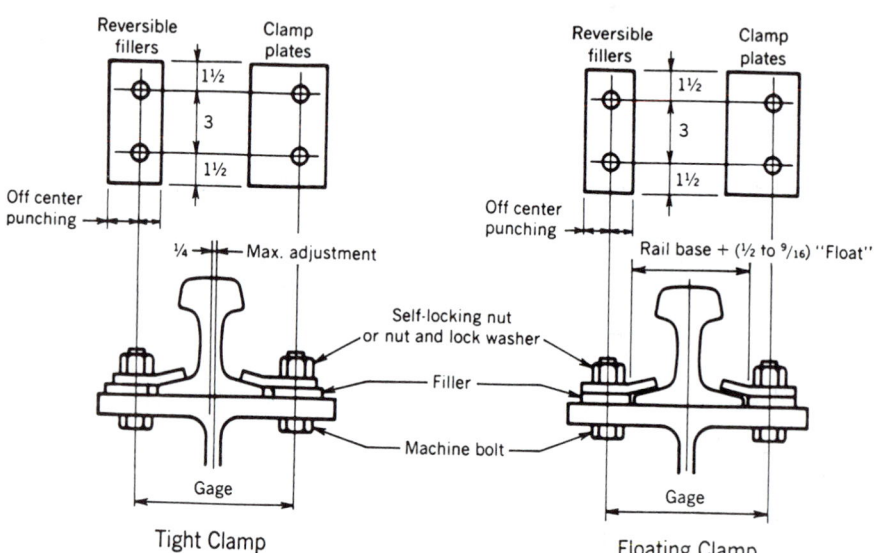

Dimensions shown above are suggested. See manufacturers' catalogs for recommended gages, bolt sizes and detail dimensions not shown.

TORSION PROPERTIES
W Shapes

An analysis for torsional shear is not required for the routine design of most structural steel members. When torsional analysis is required, the table of Torsion Properties will be of assistance in utilizing current analysis methods. The reader is referred to the AISC publication, *Torsional Analysis of Steel Members,* for additional information and appropriate design aids.

Torsion Properties are also required to determine the allowable torsional buckling stresses as specified in the AISC LRFD Specification Appendix E3.

NOMENCLATURE

C_w Warping constant for a section, in.^{6}

E Modulus of elasticity of steel (29,000 ksi)

G Shear modulus of elasticity of steel (11,200 ksi)

H Flexural constant

J Torsional constant for a section, in.^{4}

Q_f Statical moment for a point in the flange directly above the vertical edge of the web, in.^{3}

Q_w Statical moment at mid-depth of the section, in.^{3}

$\bar{r}_o$ Polar radius of gyration about the shear center, in.

S_w Warping statical moment at a point in the section, in.^{4}

W_{no} Normalized warping function at a point at the flange edge, in.^{2}

TORSION PROPERTIES
W shapes

Designation	Torsional Constant J	Warping Constant C_w	$\sqrt{\dfrac{EC_w}{GJ}}$	Normalized Warping Constant W_{no}	Warping Statical Moment S_w	Statical Moment	
						Q_f	Q_w
	In.^{4}	In.^{6}	In.	In.^{2}	In.^{4}	In.^{3}	In.^{3}
W 44×285	60.0	219000	97.2	125	653	204	657
×248	40.7	192000	111	124	577	184	574
×224	30.0	172000	122	124	517	166	517
×198	20.1	146000	137	123	443	144	450
W 40×328	74.2	607000	145	171	1330	287	755
×298	56.3	540000	158	170	1190	261	684
×268	41.1	475000	173	168	1060	234	612
×244	30.4	417000	188	167	934	208	550
×221	21.2	349000	206	166	785	176	483
×192	13.7	268000	225	165	608	137	404
W 40×655	596	1140000	70.4	169	2520	534	1530
×593	451	989000	75.4	166	2240	484	1380
×531	329	848000	81.7	163	1950	433	1230
×480	245	739000	88.4	160	1730	391	1090
×436	186	649000	95.1	158	1540	354	992
×397	142	577000	103	156	1380	323	894
×362	109	511000	110	154	1240	294	813
×324	79.4	446000	121	152	1100	264	730
×297	61.2	397000	130	151	986	240	665
×277	51.1	378000	138	151	940	230	624
×249	37.7	333000	151	149	836	208	560
×215	24.4	283000	173	149	714	179	481
×199	18.1	245000	187	148	621	157	434
W 40×183	19.6	119000	126	111	402	134	391
×167	14.0	99300	136	111	336	113	346
×149	9.62	79600	146	110	270	92.0	299
W 36×848	1270	1620000	57.5	172	3530	674	1910
×798	1070	1480000	59.8	169	3270	634	1790
×720	804	1270000	64.0	166	2870	571	1590
×650	600	1090000	68.6	162	2520	513	1420
×588	453	950000	73.7	159	2240	465	1280
×527	330	816000	80.0	156	1960	415	1130
×485	260	727000	85.2	154	1770	380	1040
×439	195	637000	91.9	152	1570	344	928
×393	143	554000	100	150	1390	309	830
×359	109	493000	108	148	1240	281	757
×328	84.5	441000	117	146	1130	258	691
×300	64.2	398000	127	146	1020	235	628
×280	52.6	366000	134	145	944	219	585
×260	41.5	330000	144	144	858	200	538
×245	34.6	306000	151	143	799	187	505
×230	28.6	282000	160	143	740	175	472
W 36×256	53.3	168000	90.6	109	576	176	520
×232	39.8	148000	98.3	108	512	159	468
×210	28.0	128000	109	108	446	138	416
×194	22.2	116000	116	107	407	128	383
×182	18.4	107000	123	106	378	120	359
×170	15.1	98500	130	105	349	111	334
×160	12.4	90200	137	105	321	103	312
×150	10.1	82200	145	105	294	95.1	291
×135	6.99	68100	159	104	245	79.9	255
W 33×619	567	870000	63.0	148	2210	463	1280
×567	444	768000	66.9	145	1990	425	1170
×515	338	672000	71.8	143	1770	385	1060
×468	256	587000	77.1	140	1570	348	947
×424	193	514000	83.1	138	1400	315	852
×387	149	458000	89.2	136	1260	288	773
×354	115	408000	96.1	135	1130	263	709
×318	84.4	357000	105	133	1000	237	634
×291	65.0	319000	113	132	906	216	577
×263	48.5	281000	123	130	808	195	519
×241	35.8	250000	135	130	721	174	469
×221	27.5	224000	145	129	650	158	428
×201	20.5	198000	158	128	580	142	386

TORSION PROPERTIES
W shapes

Designation	Torsional Constant J	Warping Constant C_w	$\sqrt{\dfrac{EC_w}{GJ}}$	Normalized Warping Constant W_{no}	Warping Statical Moment S_w	Statical Moment	
						Q_f	Q_w
	In.^{4}	In.^{6}	In.	In.^{2}	In.^{4}	In.^{3}	In.^{3}
W 33×169	17.7	82400	110	93.7	329	109	314
×152	12.4	71700	123	93.8	286	95.1	279
×141	9.70	64400	131	93.3	258	86.5	257
×130	7.37	56600	141	92.8	228	76.9	233
×118	5.30	48300	154	92.2	196	66.6	207
W 30×581	537	636000	55.4	129	1850	402	1110
×526	405	550000	59.3	126	1630	364	993
×477	307	480000	63.6	124	1450	329	896
×433	231	417000	68.3	122	1280	297	805
×391	174	364000	73.6	120	1140	268	716
×357	134	323000	79.1	118	1020	245	650
×326	103	286000	85.0	117	919	223	595
×292	74.9	249000	92.8	115	812	200	530
×261	53.8	215000	102	114	710	177	470
×235	40.0	190000	111	112	633	160	422
×211	27.9	166000	124	112	556	141	374
×191	20.6	146000	135	111	494	126	337
×173	15.3	129000	148	110	439	113	303
×148	14.6	49400	93.9	77.3	239	86.8	250
×132	9.72	42100	106	77.3	204	74.0	219
×124	7.99	38600	112	76.9	188	68.8	204
×116	6.43	34900	119	76.5	171	62.8	189
×108	4.99	30900	127	76.1	152	56.1	173
× 99	3.77	26800	136	75.7	133	49.5	156
× 90	2.94	24000	146	75.2	119	45.1	142
W 27×539	499	440000	47.8	111	1490	342	940
×494	391	386000	50.6	108	1340	313	856
×448	297	336000	54.1	106	1190	283	766
×407	225	291000	57.9	104	1050	255	688
×368	169	254000	62.3	102	930	231	620
×336	131	225000	66.8	101	836	211	564
×307	101	199000	71.5	99.4	750	192	511
×281	78.8	178000	76.5	98.2	680	176	466
×258	61.0	159000	82.2	97.1	613	161	424
×235	46.3	140000	88.4	96.0	548	146	384
×217	37.0	128000	94.7	95.0	503	135	354
×194	26.5	111000	104	93.9	442	120	314
×178	19.5	98300	114	93.7	393	107	284
×161	14.7	87300	124	92.9	352	96.6	256
×146	10.9	77200	135	92.2	314	87.0	231
×129	11.2	32500	86.9	66.4	183	69.5	197
×114	7.33	27600	98.7	66.4	155	59.2	171
×102	5.29	24000	108	65.7	137	52.7	153
× 94	4.03	21300	117	65.4	122	47.3	139
× 84	2.81	17900	129	64.9	103	40.6	122
W 24×492	456	283000	40.1	92.1	1150	281	774
×450	357	247000	42.4	90.1	1030	257	703
×408	271	214000	45.2	88.1	909	233	626
×370	205	184000	48.2	86.3	802	209	562
×335	154	160000	51.8	84.6	709	189	509
×306	119	141000	55.4	83.3	636	173	461
×279	91.7	125000	59.4	82.0	570	157	418
×250	67.3	108000	64.4	80.6	502	141	372
×229	51.8	95800	69.2	79.6	451	128	338
×207	38.6	83900	75.0	78.5	401	116	303
×192	31.0	76200	79.7	77.7	367	107	280
×176	24.1	68400	85.9	77.0	333	97.8	255
×162	18.5	62600	93.6	77.0	304	89.4	234
×146	13.4	54600	103	76.3	268	79.5	209
×131	9.50	47100	113	75.6	233	69.7	185
×117	6.72	40800	125	74.9	204	61.5	164
×104	4.72	35200	139	74.3	178	54.1	144

I TORSION PROPERTIES
W shapes

Designation	Torsional Constant J	Warping Constant C_w	$\sqrt{\dfrac{EC_w}{GJ}}$	Normalized Warping Constant W_{no}	Warping Statical Moment S_w	Statical Moment	
						Q_f	Q_w
	In.^{4}	In.^{6}	In.	In.^{2}	In.^{4}	In.^{3}	In.^{3}
W 24×103	7.11	16600	77.8	53.0	117	49.4	140
× 94	5.26	15000	85.8	53.1	105	44.4	127
× 84	3.70	12800	94.8	52.6	91.3	39.0	112
× 76	2.68	11100	104	52.2	79.8	34.4	100
× 68	1.87	9430	114	51.9	68.0	29.5	88.3
W 24× 62	1.71	4620	83.7	40.7	42.3	23.2	76.6
× 55	1.18	3870	92.0	40.4	35.7	19.8	67.1
W 21×402	297	165000	37.9	76.7	805	210	564
×364	225	142000	40.4	75.0	709	189	505
×333	174	124000	42.9	73.5	632	172	457
×300	130	107000	46.2	71.9	556	154	408
×275	101	94100	49.1	70.7	499	141	370
×248	75.2	81800	53.0	69.5	441	127	331
×223	54.9	70600	57.6	68.3	388	113	295
×201	41.3	61800	62.1	67.3	345	102	265
×182	31.0	54300	67.2	66.4	307	92.3	238
×166	23.9	48500	72.6	65.6	277	84.4	216
×147	15.4	41100	83.0	65.4	235	71.4	187
×132	11.3	36000	90.9	64.7	208	64.0	167
×122	8.98	32700	97.1	64.2	191	59.2	154
×111	6.83	29200	105	63.7	172	53.7	139
×101	5.21	26200	114	63.2	155	49.0	127
W 21× 93	6.03	9940	65.3	43.6	85.3	38.2	110
× 83	4.34	8630	71.8	43.0	75.0	34.2	98.0
× 73	3.02	7410	79.7	42.5	65.2	30.3	86.2
× 68	2.45	6760	84.6	42.3	59.9	28.0	79.9
× 62	1.83	5960	91.8	42.0	53.2	25.1	72.2
W 21× 57	1.77	3190	68.3	33.4	35.6	20.9	64.3
× 50	1.14	2570	76.2	33.1	28.9	17.2	55.0
× 44	0.77	2110	84.3	32.8	24.0	14.5	47.7
W 18×311	177	75700	33.2	58.8	483	141	376
×283	135	65600	35.4	57.5	427	127	338
×258	104	57400	37.7	56.4	382	116	306
×234	79.7	49900	40.2	55.2	339	105	274
×211	59.3	43200	43.4	54.2	299	94.3	245
×192	45.2	37900	46.5	53.3	267	85.7	221
×175	34.2	33200	50.1	52.5	237	77.2	199
×158	25.4	28900	54.2	51.6	210	69.4	178
×143	19.4	25700	58.6	51.0	189	63.2	161
×130	14.7	22700	63.4	50.4	169	57.1	145
×119	10.6	20300	70.3	50.4	151	50.6	131
×106	7.48	17400	77.7	49.8	131	44.6	115
× 97	5.86	15800	83.5	49.4	120	41.2	105
× 86	4.10	13600	92.7	48.9	104	36.3	92.8
× 76	2.83	11700	103	48.4	90.7	31.9	81.4
W 18× 71	3.48	4700	59.1	33.7	52.1	25.8	72.7
× 65	2.73	4240	63.4	33.4	47.5	23.8	66.6
× 60	2.17	3850	67.8	33.1	43.5	22.1	61.4
× 55	1.66	3430	73.1	32.9	39.0	19.9	55.9
× 50	1.24	3040	79.6	32.6	34.9	18.0	50.4
W 18× 46	1.22	1710	60.3	26.4	24.2	15.3	45.3
× 40	0.81	1440	67.9	26.1	20.6	13.3	39.2
× 35	0.51	1140	76.6	25.9	16.5	10.7	33.2

TORSION PROPERTIES
W shapes

Designation	Torsional Constant J	Warping Constant C_w	$\sqrt{\dfrac{EC_w}{GJ}}$	Normalized Warping Constant W_{no}	Warping Statical Moment S_w	Statical Moment	
						Q_f	Q_w
	In.^{4}	In.^{6}	In.	In.^{2}	In.^{4}	In.^{3}	In.^{3}
W 16×100	7.73	11900	63.1	41.7	107	39.0	99.0
×89	5.45	10200	69.8	41.1	93.3	34.4	87.3
×77	3.57	8590	78.9	40.6	79.3	29.7	75.0
×67	2.39	7300	88.9	40.1	68.2	25.9	64.9
W 16×57	2.22	2660	55.8	28.0	35.6	19.0	52.6
×50	1.52	2270	62.1	27.6	30.8	16.7	46.0
×45	1.11	1990	68.0	27.4	27.2	15.0	41.1
×40	0.79	1730	75.2	27.1	23.9	13.4	36.5
×36	0.54	1460	83.2	26.9	20.2	11.4	32.0
W 16×31	0.46	739	64.5	21.3	13.0	9.17	27.0
×26	0.26	565	74.8	21.1	10.0	7.20	22.1
W 14×730	1450	362000	25.4	78.3	1720	319	831
×665	1120	305000	26.5	75.5	1510	287	740
×605	870	258000	27.7	73.0	1320	259	660
×550	670	219000	29.1	70.6	1160	233	588
×500	514	187000	30.7	68.5	1020	209	524
×455	395	160000	32.4	66.5	899	189	468
W 14×426	331	144000	33.6	65.3	827	176	434
×398	273	129000	35.1	64.1	756	163	401
×370	222	116000	36.7	62.9	689	151	368
×342	178	103000	38.6	61.6	623	138	336
×311	136	89100	41.2	60.3	553	125	301
×283	104	77700	44.0	59.1	493	113	271
×257	79.1	67800	47.1	57.9	438	102	243
×233	59.5	59000	50.7	56.9	389	91.7	218
×211	44.6	51500	54.7	55.9	345	82.3	195
×193	34.8	45900	58.4	55.1	312	75.4	177
×176	26.5	40500	63.0	54.4	279	68.0	160
×159	19.8	35600	68.3	53.7	248	61.3	143
×145	15.2	31700	73.6	53.0	224	55.8	130
W 14×132	12.3	25500	73.2	50.2	190	49.9	117
×120	9.37	22700	79.2	49.7	171	45.3	106
×109	7.12	20200	85.8	49.1	154	41.2	95.9
×99	5.37	18000	93.1	48.7	138	37.2	86.6
×90	4.06	16000	101	48.3	125	33.7	78.3
W 14×82	5.08	6710	58.5	34.1	73.8	28.1	69.3
×74	3.88	5990	63.2	33.7	66.6	25.7	62.8
×68	3.02	5380	68.0	33.4	60.4	23.5	57.3
×61	2.20	4710	74.5	33.1	53.3	21.0	51.1

I	TORSION PROPERTIES W shapes

Designation	Torsional Constant J	Warping Constant C_w	$\sqrt{\dfrac{EC_w}{GJ}}$	Normalized Warping Constant W_{no}	Warping Statical Moment S_w	Statical Moment	
						Q_f	Q_w
	In.^{4}	In.^{6}	In.	In.^{2}	In.^{4}	In.^{3}	In.^{3}
W 14×53	1.94	2540	58.1	26.7	35.5	17.3	43.6
×48	1.46	2240	63.1	26.5	31.6	15.6	39.2
×43	1.05	1950	69.3	26.2	27.8	13.9	34.8
W 14×38	0.80	1230	63.2	23.0	20.0	11.5	30.7
×34	0.57	1070	69.6	22.8	17.5	10.2	27.3
×30	0.38	887	77.7	22.6	14.7	8.59	23.6
W 14×26	0.36	405	54.1	16.9	8.94	6.96	20.1
×22	0.21	314	62.6	16.8	7.02	5.58	16.6
W 12×336	243	57000	24.7	46.4	459	119	301
×305	185	48600	26.1	45.0	403	107	269
×279	143	42000	27.6	44.0	357	96.3	241
×252	108	35800	29.3	42.8	313	86.4	214
×230	83.8	31200	31.1	41.8	279	78.4	193
×210	64.7	27200	33.0	41.0	249	71.1	174
×190	48.8	23600	35.3	40.1	220	64.1	156
×170	35.6	20100	38.3	39.2	192	56.9	137
×152	25.8	17200	41.5	38.4	168	50.4	121
×136	18.5	14700	45.3	37.7	146	44.5	107
×120	12.9	12400	50.0	37.0	126	38.9	93.2
×106	9.13	10700	55.0	36.4	110	34.6	81.9
×96	6.86	9410	59.6	35.9	98.2	31.3	73.6
×87	5.10	8270	64.8	35.5	87.2	28.0	66.0
×79	3.84	7330	70.3	35.2	78.1	25.3	59.5
×72	2.93	6540	76.0	34.9	70.3	22.9	53.9
×65	2.18	5780	82.9	34.5	62.7	20.6	48.4
W 12×58	2.10	3570	66.4	28.9	46.3	18.2	43.2
×53	1.58	3160	72.0	28.7	41.2	16.3	39.0
W 12×50	1.78	1880	52.3	23.3	30.2	14.7	36.2
×45	1.31	1650	57.0	23.1	26.7	13.1	32.4
×40	0.95	1440	62.5	22.9	23.6	11.8	28.8
W 12×35	0.74	879	55.4	19.6	16.8	9.86	25.6
×30	0.46	720	63.9	19.4	13.9	8.30	21.6
×26	0.30	607	72.4	19.2	11.8	7.15	18.6
W 12×22	0.29	164	38.1	12.0	5.13	4.87	14.7
×19	0.18	131	43.4	11.8	4.14	4.01	12.4
×16	0.10	96.9	49.4	11.7	3.09	3.04	10.0
×14	0.07	80.4	54.4	11.6	2.59	2.59	8.72

TORSION PROPERTIES
W shapes

Designation	Torsional Constant J	Warping Constant C_w	$\sqrt{\dfrac{EC_w}{GJ}}$	Normalized Warping Constant W_{no}	Warping Statical Moment S_w	Statical Moment	
						Q_f	Q_w
	In.^{4}	In.^{6}	In.	In.^{2}	In.^{4}	In.^{3}	In.^{3}
W 10×112	15.1	6020	32.2	26.3	85.7	30.8	73.7
×100	10.9	5150	35.0	25.8	74.7	27.2	64.9
×88	7.53	4330	38.6	25.3	64.2	23.8	56.4
×77	5.11	3630	42.9	24.8	54.9	20.7	46.8
×68	3.56	3100	47.4	24.4	47.6	18.1	42.6
×60	2.48	2640	52.6	24.0	41.2	15.9	37.3
×54	1.82	2320	57.4	23.8	36.6	14.3	33.3
×49	1.39	2070	62.1	23.6	33.0	13.0	30.2
W 10×45	1.51	1200	45.3	19.0	23.6	11.5	27.5
×39	0.98	992	51.3	18.7	19.8	9.77	23.4
×33	0.58	790	59.2	18.5	16.0	7.98	19.4
W 10×30	0.62	414	41.5	14.5	10.7	7.09	18.3
×26	0.40	345	47.1	14.3	9.05	6.08	15.6
×22	0.24	275	54.5	14.1	7.30	4.95	13.0
W 10×19	0.23	104	34.0	9.89	3.93	3.76	10.8
×17	0.16	85.1	37.6	9.80	3.24	3.13	9.33
×15	0.10	68.3	41.2	9.72	2.62	2.56	8.00
×12	0.05	50.9	49.1	9.56	1.99	2.00	6.32
W 8×67	5.06	1440	27.2	16.7	32.3	14.7	35.1
×58	3.34	1180	30.3	16.3	27.2	12.5	29.9
×48	1.96	931	35.0	15.8	22.0	10.4	24.5
×40	1.12	726	40.9	15.5	17.5	8.42	19.9
×35	0.77	619	45.6	15.3	15.2	7.39	17.3
×31	0.54	530	50.5	15.1	13.1	6.46	15.2
W 8×28	0.54	312	38.8	12.4	9.43	5.64	13.6
×24	0.35	259	44.0	12.2	7.94	4.83	11.6
W 8×21	0.28	152	37.3	10.4	5.47	4.03	10.2
×18	0.17	122	42.8	10.3	4.44	3.31	8.52
W 8×15	0.14	51.8	31.3	7.82	2.47	2.39	6.78
×13	0.09	40.8	34.8	7.74	1.97	1.93	5.70
×10	0.04	30.9	43.4	7.57	1.53	1.56	4.43
W 6×25	0.46	150	29.0	9.01	6.23	3.92	9.46
×20	0.24	113	34.9	8.78	4.82	3.10	7.45
×15	0.10	76.5	44.2	8.58	3.34	2.18	5.39
W 6×16	0.22	38.2	21.0	5.92	2.42	2.28	5.84
×12	0.09	24.7	26.6	5.75	1.61	1.55	4.15
×9	0.04	17.7	33.7	5.60	1.19	1.19	3.12
W 5×19	0.31	50.8	20.4	5.94	3.21	2.44	5.81
×16	0.19	40.6	23.4	5.81	2.62	2.02	4.82
W 4×13	0.15	14.0	15.5	3.87	1.36	1.27	3.14

TORSION PROPERTIES
M shapes

Designation	Torsional Constant J	Warping Constant C_w	$\sqrt{\dfrac{EC_w}{GJ}}$	Normalized Warping Constant W_{no}	Warping Statical Moment S_w	Statical Moment	
						Q_f	Q_w
	In.^{4}	In.^{6}	In.	In.^{2}	In.^{4}	In.^{3}	In.^{3}
M 14×18	.11	124	54.0	13.7	3.71	3.60	12.4
M 12×11.8	.05	34.0	42.0	9.02	1.56	1.98	7.14
M 12×10.8	.04	31.3	45.0	9.01	1.45	1.86	6.58
M 12×10	.03	34.5	54.6	9.58	1.40	1.72	6.07
M 10× 9	.03	14.6	35.5	6.59	.91	1.32	4.60
M 10× 8	.02	12.8	40.7	6.57	.80	1.18	4.06
M 10× 7.5	.02	12.0	39.4	6.60	.77	1.13	3.86
M 8× 6.5	.02	5.23	26.0	4.45	.48	.82	2.72
M 6× 4.4	.01	1.40	19.0	2.69	.21	.45	1.40
M 5×18.9	.34	41.3	17.7	5.73	2.98	2.28	5.53

TORSION PROPERTIES
S shapes

I

Designation	Torsional Constant J	Warping Constant C_w	$\sqrt{\dfrac{EC_w}{GJ}}$	Normalized Warping Constant W_{no}	Warping Statical Moment S_w	Statical Moment	
						Q_f	Q_w
	In.^{4}	In.^{6}	In.	In.^{2}	In.^{4}	In.^{3}	In.^{3}
S 24×121	12.8	11400	48.0	47.1	103	47.1	154
×106	10.1	10600	52.0	46.1	98.8	47.1	141
S 24×100	7.58	6380	46.7	41.9	66.0	33.5	121
×90	6.04	6000	50.7	41.2	63.8	33.5	112
×80	4.88	5640	54.7	40.5	61.6	33.5	103
S 20×96	8.39	4710	38.1	34.9	57.8	29.2	99.7
×86	6.64	4390	41.4	34.2	55.5	29.2	92.5
S 20×75	4.59	2750	39.4	30.7	38.9	22.6	77.0
×66	3.58	2550	43.0	30.0	37.3	22.6	70.5
S 18×70	4.15	1800	33.6	27.0	29.2	17.1	63.0
×54.7	2.37	1560	41.3	26.0	26.9	17.1	52.9
S 15×50	2.12	811	31.5	20.3	17.8	11.8	39.0
×42.9	1.54	744	35.4	19.8	16.9	11.8	35.1
S 12×50	2.82	505	21.5	15.5	14.0	9.30	31.0
×40.8	1.75	437	25.4	14.9	12.9	9.30	26.9
S 12×35	1.08	324	27.9	14.5	10.0	7.48	22.7
×31.8	0.90	307	29.7	14.3	9.74	7.48	21.3
S 10×35	1.29	189	19.5	11.8	7.13	5.24	17.9
×25.4	0.60	153	25.7	11.1	6.34	5.24	14.4
S 8×23	0.55	61.8	17.1	7.90	3.50	3.10	9.74
×18.4	0.34	53.5	20.3	7.58	3.22	3.10	8.38
S 7×20	0.45	34.6	14.1	6.38	2.41	2.29	7.26
×15.3	0.24	28.8	17.6	6.05	2.17	2.29	6.12
S 6×17.25	0.37	18.4	11.3	5.03	1.61	1.63	5.35
×12.5	0.17	14.5	14.9	4.70	1.41	1.63	4.30
S 5×14.75	0.32	9.12	8.59	3.84	1.03	1.11	3.72
×10	0.11	6.66	12.3	3.51	0.86	1.11	2.88
S 4× 9.5	0.12	3.10	8.17	2.59	0.53	0.70	2.05
× 7.7	0.07	2.62	9.64	2.47	0.48	0.70	1.79
S 3× 7.5	0.09	1.10	5.59	1.72	0.28	0.40	1.20
× 5.7	0.04	0.85	7.08	1.60	0.24	0.40	1.00

FLEXURAL-TORSIONAL PROPERTIES
Channels

Designation	Torsional Constant J	Warping Constant C_w	Polar Radius of Gyration $\bar{r}_o$	Flexural Constant H^*
	In.⁴	In.⁶	In.	No Units
C 15×50	2.67	492	5.49	.937
×40	1.46	411	5.72	.927
×33.9	1.02	358	5.94	.920
C 12×30	0.87	151	4.55	.919
×25	0.54	130	4.72	.909
×20.7	0.37	112	4.93	.899
C 10×30	1.23	79.3	3.63	.921
×25	0.69	68.4	3.75	.912
×20	0.37	56.9	3.93	.900
×15.3	0.21	45.6	4.19	.883
C 9×20	0.43	39.5	3.46	.899
×15	0.21	31.0	3.69	.882
×13.4	0.17	28.2	3.79	.874
C 8×18.75	0.44	25.1	3.06	.894
×13.75	0.19	19.2	3.27	.874
×11.5	0.13	16.5	3.42	.862
C 7×14.75	0.27	13.1	2.75	.875
×12.25	0.16	11.2	2.87	.862
× 9.8	0.10	9.18	3.02	.846
C 6×13	0.24	7.22	2.37	.858
×10.5	0.13	5.95	2.49	.843
× 8.2	0.08	4.72	2.65	.824
C 5× 9	0.11	2.93	2.10	.814
× 6.7	0.06	2.22	2.26	.790
C 4× 7.25	0.08	1.24	1.75	.768
×5.4	0.04	0.92	1.89	.741
C 3× 6	0.07	0.46	1.39	.689
× 5	0.04	0.38	1.45	.674
× 4.1	0.03	0.31	1.53	.656

FLEXURAL-TORSIONAL PROPERTIES
Channels

$\Bigl[$

Designation	Torsional Constant J	Warping Constant C_w	Polar Radius of Gyration $\bar{r}_o$	Flexural Constant H *
	In.^{4}	In.^{6}	In.	No Units
MC 18×58	2.81	1070	6.56	.944
×51.9	2.03	986	6.70	.939
×45.8	1.45	897	6.88	.933
×42.7	1.23	852	6.97	.930
MC 13×50	2.98	558	5.07	.875
×40	1.57	463	5.33	.860
×35	1.14	413	5.50	.849
×31.8	0.94	380	5.64	.842
MC 12×50	3.24	411	4.77	.859
×45	2.35	374	4.87	.851
×40	1.70	336	5.01	.842
×35	1.25	297	5.18	.832
×31	1.01	268	5.34	.821
MC 12×10.6	0.06	11.7	4.27	.983
MC 10×41.1	2.27	270	4.26	.790
×33.6	1.21	224	4.47	.771
×28.5	0.79	194	4.68	.752
×25	0.64	125	4.46	.802
×22	0.51	111	4.63	.790
MC 10× 8.4	0.04	7.01	3.68	.972
MC 10× 6.5	0.02	2.45	3.43	.990
MC 9×25.4	0.69	104	4.08	.770
×23.9	0.60	98.2	4.15	.763
MC 8×22.8	0.57	75.3	3.85	.716
×21.4	0.50	70.9	3.91	.709
MC 8×20	0.44	47.9	3.59	.780
×18.7	0.38	45.1	3.65	.773
MC 8× 8.5	0.06	8.22	3.24	.910
MC 7×22.7	0.63	58.5	3.53	.662
×19.1	0.41	49.4	3.71	.638
MC 6×18	0.38	34.6	3.46	.562
×15.3	0.22	30.1	3.41	.581
MC 6×16.3	0.34	22.1	3.11	.643
×15.1	0.29	20.6	3.18	.634
MC 6×12	0.15	11.2	2.80	.740

FLEXURAL-TORSIONAL PROPERTIES
Single Angles

Designation	Torsional Constant J In.^{4}	Polar Radius of Gyration $\bar{r}_o$ In.	Flexural Constant H No Units	Designation	Torsional Constant J In.^{4}	Polar Radius of Gyration $\bar{r}_o$ In.	Flexural Constant H No Units
L 9×4× ⅝	1.06	4.37	—	L 6×6 ×1	3.68	3.19	.637
⁹⁄₁₆	.782	4.38	—	⅞	2.51	3.22	.632
½	.557	4.39	—	¾	1.61	3.26	.629
				⅝	.954	3.29	.628
L 8×8×1⅛	7.13	4.31	.632	⁹⁄₁₆	.704	3.31	.627
1	5.08	4.35	.630	½	.501	3.32	.627
⅞	3.46	4.37	.629	⁷⁄₁₆	.340	3.34	.627
¾	2.21	4.41	.627	⅜	.218	3.36	.626
⅝	1.30	4.45	.627	⁵⁄₁₆	.129	3.38	.625
⁹⁄₁₆	.960	4.47	.627				
½	.682	4.48	.624	L 6×4 × ⅞	2.07	2.83	—
				¾	1.33	2.86	—
L 8×6×1	4.35	3.89	—	⅝	.792	2.89	—
⅞	2.96	3.93	—	⁹⁄₁₆	.585	2.90	—
¾	1.90	3.96	—	½	.417	2.92	—
⅝	1.12	3.99	—	⁷⁄₁₆	.284	2.94	—
⁹⁄₁₆	.822	4.01	—	⅜	.183	2.96	—
½	.584	4.02	—	⁵⁄₁₆	.108	2.97	—
⁷⁄₁₆	.396	4.04	—				
				L 6×3½× ½	.396	2.88	—
L 8×4×1	3.68	3.77	—	⅜	.174	2.92	—
¾	1.61	3.82	—	⁵⁄₁₆	.103	2.93	—
⁹⁄₁₆	.704	3.86	—				
½	.501	3.88	—	L 5×5 × ⅞	2.07	2.65	.634
				¾	1.33	2.68	.634
L 7×4× ¾	1.47	3.33	—	⅝	.792	2.71	.630
⅝	.873	3.36	—	½	.417	2.74	.630
½	.459	3.38	—	⁷⁄₁₆	.284	2.77	.629
⅜	.200	3.42	—	⅜	.183	2.79	.627
				⁵⁄₁₆	.108	2.81	.626

C_w can be taken conservatively as zero for single angles.

FLEXURAL-TORSIONAL PROPERTIES
Single Angles

Designation	Torsional Constant J In.^{4}	Polar Radius of Gyration $\bar{r}_o$ In.	Flexural Constant H No Units	Designation	Torsional Constant J In.^{4}	Polar Radius of Gyration $\bar{r}_o$ In.	Flexural Constant H No Units
L 5×3½×¾	1.11	2.37	—	L 4 ×3 ×½	.281	1.95	—
⅝	.660	2.40	—	⁷⁄₁₆	.192	1.96	—
½	.348	2.44	—	⅜	.123	1.98	—
⁷⁄₁₆	.238	2.45	—	⁵⁄₁₆	.0731	2.00	—
⅜	.153	2.47	—	¼	.0386	2.01	—
⁵⁄₁₆	.0905	2.49	—				
¼	.0479	2.50	—	L 3½×3½×½	.281	1.89	.631
				⁷⁄₁₆	.192	1.91	.629
L 5×3 ×⅝	.610	2.36	—	⅜	.123	1.91	.628
½	.322	2.39	—	⁵⁄₁₆	.0731	1.93	.627
⁷⁄₁₆	.219	2.41	—	¼	.0386	1.95	.626
⅜	.141	2.42	—				
⁵⁄₁₆	.0832	2.43	—	L 3½×3 ×½	.260	1.76	—
¼	.0438	2.45	—	⁷⁄₁₆	.178	1.77	—
				⅜	.114	1.79	—
L 4×4 ×¾	1.02	2.11	.639	⁵⁄₁₆	.0680	1.81	—
⅝	.610	2.14	.631	¼	.0360	1.83	—
½	.322	2.17	.632				
⁷⁄₁₆	.219	2.19	.631	L 3½×2½×½	.234	1.67	—
⅜	.141	2.20	.625	⁷⁄₁₆	.160	1.68	—
⁵⁄₁₆	.0832	2.22	.623	⅜	.103	1.70	—
¼	.0438	2.23	.627	⁵⁄₁₆	.0611	1.72	—
				¼	.0322	1.73	—
L 4×3½×½	.301	2.04	—				
⁷⁄₁₆	.206	2.06	—	L 3 ×3 ×½	.234	1.60	.634
⅜	.132	2.08	—	⁷⁄₁₆	.160	1.61	.632
⁵⁄₁₆	.0782	2.09	—	⅜	.103	1.63	.629
¼	.0412	2.11	—	⁵⁄₁₆	.0611	1.65	.628
				¼	.0322	1.66	.627
				³⁄₁₆	.0142	1.68	.626

C_w can be taken conservatively as zero for single angles.

FLEXURAL-TORSIONAL PROPERTIES
Single Angles

Designation	Torsional Constant J In.^{4}	Polar Radius of Gyration $\bar{r}_o$ In.	Flexural Constant H No Units	Designation	Torsional Constant J In.^{4}	Polar Radius of Gyration $\bar{r}_o$ In.	Flexural Constant H No Units
L 3 ×2½×½	.213	1.47	—	L 2½×2 ×⅜	.0728	1.22	—
⁷⁄₁₆	.146	1.49	—	⁵⁄₁₆	.0432	1.24	—
⅜	.0943	1.50	—	¼	.0227	1.25	—
⁵⁄₁₆	.0560	1.52	—	³⁄₁₆	.00990	1.27	—
¼	.0296	1.54	—				
³⁄₁₆	.0131	1.55	—	L 2 ×2 ×⅜	.0640	1.05	.637
				⁵⁄₁₆	.0381	1.07	.633
L 3 ×2 ×½	.192	1.40	—	¼	.0201	1.09	.630
⁷⁄₁₆	.132	1.41	—	³⁄₁₆	.00880	1.10	.628
⅜	.0855	1.43	—	⅛	.00274	1.12	.626
⁵⁄₁₆	.0509	1.45	—				
¼	.0270	1.46	—				
³⁄₁₆	.0120	1.48	—				
L 2½×2½×½	.185	1.31	.639				
⅜	.0816	1.34	.632				
⁵⁄₁₆	.0483	1.36	.630				
¼	.0253	1.37	.628				
³⁄₁₆	.0110	1.39	.627				

C_w can be taken conservatively as zero for single angles.

FLEXURAL-TORSIONAL PROPERTIES
Structural Tees

Designation	Torsional Constant J In.⁴	Polar Radius of Gyration $\bar{r}_o$ In.	Flexural Constant H No Units	Designation	Torsional Constant J In.⁴	Polar Radius of Gyration $\bar{r}_o$ In.	Flexural Constant H No Units
WT 18 ×179.5	54.3	7.38	0.797	WT 13.5×108.5	18.5	5.72	0.830
×164	42.1	7.32	0.799	× 97	13.2	5.66	0.826
×150	32.0	7.30	0.797	× 89	9.74	5.70	0.815
×140	26.2	7.27	0.796	× 80.5	7.31	5.67	0.813
×130	20.7	7.28	0.791	× 73	5.44	5.65	0.810
×122.5	17.3	7.28	0.788				
×115	14.3	7.27	0.784	WT 13.5× 64.5	5.60	5.48	0.731
				× 57	3.65	5.54	0.716
WT 18 ×128	26.6	7.43	0.703	× 51	2.64	5.52	0.714
×116	19.8	7.40	0.703	× 47	2.01	5.57	0.703
×105	13.9	7.49	0.687	× 42	1.40	5.63	0.685
× 97	11.1	7.45	0.687				
× 91	9.19	7.45	0.686	WT 12 × 88	12.0	5.09	0.835
× 85	7.51	7.44	0.684	× 81	9.22	5.09	0.831
× 80	6.17	7.46	0.678	× 73	6.70	5.08	0.827
× 75	5.04	7.50	0.670	× 65.5	4.74	5.09	0.818
× 67.5	3.48	7.65	0.644	× 58.5	3.35	5.08	0.813
				× 52	2.35	5.07	0.809
WT 16.5×177	57.2	7.00	0.802				
×159	42.1	6.94	0.803	WT 12 × 51.5	3.54	4.88	0.733
×145.5	32.4	6.90	0.801	× 47	2.62	4.89	0.727
×131.5	24.2	6.86	0.802	× 42	1.84	4.89	0.721
×120.5	17.9	6.91	0.792	× 38	1.34	4.93	0.709
×110.5	13.7	6.90	0.788	× 34	.932	4.99	0.692
×100.5	10.2	6.89	0.784				
				WT 12 × 31	.850	5.13	0.619
WT 16.5× 84.5	8.83	6.74	0.714	× 27.5	.588	5.18	0.606
× 76	6.16	6.82	0.700				
× 70.5	4.84	6.85	0.691	WT 10.5× 83	11.9	4.69	0.861
× 65	3.67	6.93	0.678	× 73.5	7.69	4.64	0.847
× 59	2.64	7.02	0.659	× 66	5.62	4.61	0.845
				× 61	4.47	4.58	0.846
WT 15 ×117.5	19.9	6.25	0.817	× 55.5	3.40	4.56	0.846
×105.5	13.9	6.27	0.809	× 50.5	2.60	4.54	0.846
× 95.5	10.3	6.25	0.806				
× 86.5	7.61	6.25	0.802	WT 10.5× 46.5	3.01	4.37	0.729
				× 41.5	2.16	4.33	0.732
WT 15 × 74	7.27	6.10	0.716	× 36.5	1.51	4.31	0.732
× 66	4.85	6.19	0.698	× 34	1.22	4.31	0.727
× 62	3.98	6.20	0.693	× 31	.513	4.31	0.722
× 58	3.21	6.24	0.683				
× 54	2.49	6.31	0.669	WT 10.5× 28.5	.884	4.36	0.665
× 49.5	1.88	6.38	0.654	× 25	.570	4.44	0.640
				× 22	.383	4.49	0.623

C_w can be taken conservatively as zero for tees.

T FLEXURAL-TORSIONAL PROPERTIES
Structural Tees

Designation	Torsional Constant J In.^{4}	Polar Radius of Gyration $\bar{r}_o$ In.	Flexural Constant H No Units	Designation	Torsional Constant J In.^{4}	Polar Radius of Gyration $\bar{r}_o$ In.	Flexural Constant H No Units
WT 9×71.5	9.69	4.03	0.874	WT 7×365	714	5.47	0.966
×65	7.31	3.99	0.874	×332.5	555	5.36	0.966
×59.5	5.30	4.03	0.862	×302.5	430	5.25	0.966
×53	3.73	4.00	0.860	×275	331	5.15	0.967
×48.5	2.92	3.97	0.862	×250	255	5.06	0.967
×43	2.04	3.95	0.860	×227.5	196	4.98	0.967
×38	1.41	3.92	0.862	×213	164	4.92	0.968
				×199	135	4.87	0.968
WT 9×35.5	1.74	3.72	0.751	×185	110	4.81	0.968
×32.5	1.36	3.69	0.755	×171	88.3	4.77	0.968
×30	1.08	3.67	0.756	×155.5	67.5	4.71	0.968
×27.5	0.829	3.68	0.749	×141.5	51.8	4.66	0.969
×25	0.613	3.66	0.748	×128.5	39.3	4.61	0.969
				×116.5	29.6	4.56	0.970
WT 9×23	0.609	3.67	0.694	×105.5	22.2	4.52	0.970
×20	0.403	3.65	0.692	× 96.5	17.3	4.49	0.971
×17.5	0.252	3.74	0.662	× 88	13.2	4.46	0.971
				× 79.5	9.84	4.42	0.971
WT 8×50	3.85	3.62	0.877	× 72.5	7.56	4.40	0.971
×44.5	2.72	3.60	0.877				
×38.5	1.78	3.56	0.877				
×33.5	1.19	3.53	0.879				
WT 8×28.5	1.10	3.30	0.770				
×25	0.760	3.28	0.770				
×22.5	0.655	3.27	0.767				
×20	0.396	3.24	0.769				
×18	0.271	3.30	0.745				
WT 8×15.5	0.229	3.26	0.695				
×13	0.130	3.32	0.667				

C_w can be taken conservatively as zero for tees.

FLEXURAL-TORSIONAL PROPERTIES
Structural Tees

Designation	Torsional Constant J In.^{4}	Polar Radius of Gyration $\bar{r}_o$ In.	Flexural Constant H No Units	Designation	Torsional Constant J In.^{4}	Polar Radius of Gyration $\bar{r}_o$ In.	Flexural Constant H No Units
WT 7×66	6.13	4.21	0.966	WT 6×168	120	4.07	0.958
×60	4.67	4.18	0.966	×152.5	92.0	4.00	0.959
×54.5	3.55	4.16	0.968	×139.5	70.9	3.94	0.957
×49.5	2.68	4.14	0.968	×126	53.5	3.88	0.958
×45	2.03	4.12	0.968	×115	41.6	3.84	0.958
				×105	32.2	3.79	0.959
WT 7×41	2.53	3.25	0.912	× 95	24.3	3.74	0.959
×37	1.94	3.21	0.917	× 85	17.7	3.69	0.960
×34	1.51	3.19	0.915	× 76	12.8	3.65	0.960
×30.5	1.10	3.18	0.915	× 68	9.22	3.61	0.960
				× 60	6.43	3.58	0.959
WT 7×26.5	0.970	2.89	0.868	× 53	4.55	3.54	0.961
×24	0.726	2.87	0.866	× 48	3.42	3.51	0.961
×21.5	0.524	2.85	0.866	× 43.5	2.54	3.49	0.960
				× 39.5	1.92	3.46	0.960
WT 7×19	0.398	2.87	0.800	× 36	1.46	3.45	0.961
×17	0.284	2.86	0.793	× 32.5	1.09	3.43	0.960
×15	0.190	2.90	0.772				
				WT 6× 29	1.05	3.01	0.944
WT 7×13	0.179	2.82	0.713	× 26.5	0.788	3.00	0.940
×11	0.104	2.86	0.691				
				WT 6× 25	0.889	2.67	0.899
				× 22.5	0.656	2.64	0.898
				× 20	0.476	2.62	0.901
				WT 6× 17.5	0.369	2.56	0.835
				× 15	0.228	2.55	0.830
				× 13	0.150	2.54	0.826
				WT 6× 11	0.146	2.52	0.683
				× 9.5	0.0899	2.54	0.663
				× 8	0.0511	2.62	0.624
				× 7	0.0350	2.64	0.610

C_w can be taken conservatively as zero for tees.

T FLEXURAL-TORSIONAL PROPERTIES
Structural Tees

Designation	Torsional Constant J In.^{4}	Polar Radius of Gyration $\bar{r}_o$ In.	Flexural Constant H No Units	Designation	Torsional Constant J In.^{4}	Polar Radius of Gyration $\bar{r}_o$ In.	Flexural Constant H No Units
WT 5×56	7.50	3.04	.963	WT 4 ×33.5	2.52	2.41	.962
×50	5.41	3.00	.964	×29	1.66	2.39	.961
×44	3.75	2.98	.964	×24	0.979	2.34	.966
×38.5	2.55	2.93	.964	×20	0.559	2.31	.961
×34	1.78	2.92	.965	×17.5	0.385	2.29	.963
×30	1.23	2.89	.965	×15.5	0.268	2.29	.961
×27	0.909	2.87	.966				
×24.5	0.693	2.85	.966	WT 4 ×14	0.268	1.97	.935
				×12	0.173	1.96	.936
WT 5×22.5	0.753	2.44	.940				
×19.5	0.487	2.42	.936	WT 4 ×10.5	0.141	1.80	.877
×16.5	0.291	2.40	.927	× 9	0.0855	1.81	.863
WT 5×15	0.310	2.17	.848	WT 4 × 7.5	0.0679	1.72	.762
×13	0.201	2.15	.848	× 6.5	0.0433	1.74	.732
×11	0.119	2.17	.831	× 5	0.0212	1.69	.748
WT 5× 9.5	0.116	2.08	.728	WT 3 ×12.5	0.229	1.76	.952
× 8.5	0.0776	2.12	.702	×10	0.120	1.73	.952
× 7.5	0.0518	2.16	.672	× 7.5	0.0504	1.71	.937
× 6	0.0272	2.16	.662				
				WT 3 × 8	0.111	1.37	.880
				× 6	0.0449	1.37	.846
				× 4.5	0.0202	1.34	.852
				WT 2.5× 9.5	0.154	1.44	.964
				× 8	0.0930	1.43	.962
				WT 2 × 6.5	0.0750	1.16	.947

C_w can be taken conservatively as zero for tees.

FLEXURAL-TORSIONAL PROPERTIES
Structural Tees

Designation	Torsional Constant J	Polar Radius of Gyration $\bar{r}_o$	Flexural Constant H	Designation	Torsional Constant J	Polar Radius of Gyration $\bar{r}_o$	Flexural Constant H
	In.^{4}	In.	No Units		In.^{4}	In.	No Units
MT 7 × 9	.0599	3.10	.590	ST 12 ×60.5	6.38	5.14	.640
				12 ×53	5.04	4.88	.685
MT 6 × 5.9	.0307	2.69	.564				
				ST 12 ×50	3.76	5.28	.584
MT 5 × 4.5	.0213	2.21	.584	12 ×45	3.01	5.11	.616
				12 ×40	2.43	4.88	.657
MT 4 × 3.25	.0146	1.73	.611				
				ST 10 ×48	4.15	4.36	.625
MT 3 × 2.2	.00993	1.26	.645	10 ×43	3.30	4.21	.661
MT 2.5× 9.45	.165	1.37	.951	ST 10 ×37.5	2.28	4.29	.612
				10 ×33	1.78	4.10	.655
				ST 9 ×35	2.05	4.02	.583
				9 ×27.35	1.18	3.71	.662
				ST 7.5×25	1.05	3.22	.637
				7.5×21.45	.767	3.05	.689
				ST 6 ×25	1.39	2.60	.663
				6 ×20.4	.872	2.42	.733
				ST 6 ×17.5	.538	2.49	.697
				6 ×15.9	.449	2.39	.731
				ST 5 ×17.5	.633	2.23	.653
				5 ×12.7	.300	1.98	.768
				ST 4 ×11.5	.271	1.73	.707
				4 × 9.2	.167	1.59	.789
				ST 3.5×10	.221	1.55	.703
				3.5× 7.65	.120	1.40	.802
				ST 3 × 8.625	.182	1.36	.706
				3 × 6.25	.0838	1.21	.820
				ST 2.5× 7.375	.155	1.17	.712
				2.5× 5	.0568	1.02	.842
				ST 2 × 4.75	.0589	0.909	.800
				2 × 3.85	.0364	0.841	.872
				ST 1.5× 3.75	.0440	0.736	.832
				1.5× 2.85	.0220	0.673	.913

C_w can be taken conservatively as zero for tees.

FLEXURAL-TORSIONAL PROPERTIES
Double Angles

| Designation | Long Legs Vertical Back to Back of Angles, In. | | | | | | Short Legs Vertical Back to Back of Angles, In. | | | | | |
| | 0 | | 3/8 | | 3/4 | | 0 | | 3/8 | | 3/4 | |
	$\bar{r}_o$	H	$\bar{r}_o$	H	$\bar{r}_o$	H	$\bar{r}_o$	H	$\bar{r}_o$	H	$\bar{r}_o$	H
L 8×8 ×1⅛	4.58	.837	4.68	.844	4.79	.851	4.58	.837	4.68	.844	4.79	.851
1	4.58	.833	4.68	.840	4.79	.847	4.58	.833	4.68	.840	4.79	.847
⅞	4.58	.831	4.68	.838	4.78	.845	4.58	.831	4.68	.838	4.78	.845
¾	4.58	.828	4.68	.835	4.78	.842	4.58	.828	4.68	.835	4.78	.842
⅝	4.58	.825	4.68	.832	4.78	.839	4.58	.825	4.68	.832	4.78	.839
½	4.59	.822	4.69	.829	4.78	.836	4.59	.822	4.69	.829	4.78	.836
L 8×6 ×1	4.07	.721	4.15	.731	4.23	.742	4.19	.925	4.31	.929	4.44	.933
¾	4.08	.714	4.16	.724	4.24	.735	4.17	.919	4.29	.924	4.41	.928
½	4.11	.708	4.18	.718	4.26	.728	4.17	.914	4.28	.919	4.40	.923
L 8×4 ×1	3.87	.566	3.93	.578	3.99	.591	4.12	.982	4.26	.983	4.41	.984
¾	3.89	.562	3.94	.573	4.00	.586	4.08	.980	4.22	.981	4.36	.982
½	3.93	.558	3.97	.568	4.03	.580	4.05	.977	4.19	.979	4.33	.980
L 7×4 × ¾	3.42	.609	3.48	.623	3.55	.637	3.58	.968	3.71	.971	3.85	.973
½	3.45	.604	3.50	.616	3.57	.629	3.55	.965	3.68	.967	3.82	.969
⅜	3.46	.602	3.51	.614	3.57	.627	3.54	.963	3.67	.965	3.80	.968
L 6×6 ×1	3.43	.843	3.54	.852	3.65	.861	3.43	.843	3.54	.852	3.65	.861
⅞	3.43	.838	3.54	.847	3.65	.856	3.43	.838	3.54	.847	3.65	.856
¾	3.44	.833	3.54	.842	3.65	.852	3.44	.833	3.54	.842	3.65	.852
⅝	3.44	.830	3.54	.839	3.64	.848	3.44	.830	3.54	.839	3.64	.848
½	3.44	.827	3.54	.836	3.64	.845	3.44	.827	3.54	.836	3.64	.845
⅜	3.44	.822	3.54	.831	3.64	.841	3.44	.822	3.54	.831	3.64	.841
L 6×4 × ¾	2.98	.672	3.05	.687	3.13	.704	3.10	.948	3.23	.952	3.36	.956
⅝	2.98	.668	3.05	.683	3.13	.699	3.09	.946	3.21	.950	3.34	.954
½	3.00	.663	3.06	.678	3.14	.693	3.08	.943	3.20	.947	3.34	.951
⅜	3.01	.661	3.07	.675	3.15	.690	3.07	.940	3.19	.944	3.32	.948
L 6×3½× ⅜	2.97	.610	3.02	.624	3.09	.640	3.05	.961	3.17	.964	3.31	.967
5/16	2.97	.610	3.02	.624	3.09	.639	3.03	.960	3.16	.963	3.29	.966
L 5×5 × ⅞	2.87	.844	2.97	.855	3.09	.865	2.87	.844	2.97	.855	3.09	.865
¾	2.85	.839	2.96	.850	3.07	.861	2.85	.839	2.96	.850	3.07	.861
½	2.86	.830	2.96	.841	3.07	.852	2.86	.830	2.96	.841	3.07	.852
⅜	2.87	.824	2.96	.835	3.07	.846	2.87	.824	2.96	.835	3.07	.846
5/16	2.87	.821	2.97	.833	3.07	.844	2.87	.821	2.97	.833	3.07	.844

FLEXURAL-TORSIONAL PROPERTIES
Double Angles

Designation	Long Legs Vertical						Short Legs Vertical					
	Back to Back of Angles, In.						Back to Back of Angles, In.					
	0		³⁄₈		³⁄₄		0		³⁄₈		³⁄₄	
	$\bar{r}_o$	H	$\bar{r}_o$	H	$\bar{r}_o$	H	$\bar{r}_o$	H	$\bar{r}_o$	H	$\bar{r}_o$	H
L 5 ×3½×¾	2.50	.697	2.58	.715	2.67	.734	2.61	.943	2.74	.948	2.87	.953
½	2.51	.685	2.59	.703	2.67	.722	2.59	.936	2.71	.941	2.84	.947
³⁄₈	2.52	.682	2.59	.699	2.67	.717	2.58	.932	2.70	.938	2.83	.943
⁵⁄₁₆	2.53	.679	2.60	.695	2.68	.713	2.58	.930	2.70	.936	2.82	.942
L 5 ×3 ×½	2.45	.626	2.52	.645	2.59	.665	2.55	.962	2.69	.965	2.82	.969
³⁄₈	2.46	.623	2.52	.641	2.60	.661	2.54	.959	2.67	.963	2.80	.966
⁵⁄₁₆	2.47	.621	2.53	.638	2.60	.657	2.54	.957	2.67	.961	2.80	.965
¼	2.48	.618	2.54	.634	2.61	.653	2.53	.956	2.66	.960	2.79	.964
L 4 ×4 ×¾	2.29	.847	2.40	.861	2.52	.873	2.29	.847	2.40	.861	2.52	.873
⁵⁄₈	2.29	.839	2.40	.853	2.51	.867	2.29	.839	2.40	.853	2.51	.867
½	2.29	.834	2.39	.848	2.50	.862	2.29	.834	2.39	.848	2.50	.862
³⁄₈	2.29	.827	2.39	.841	2.50	.855	2.29	.827	2.39	.841	2.50	.855
⁵⁄₁₆	2.29	.824	2.39	.838	2.50	.852	2.29	.824	2.39	.838	2.50	.852
¼	2.29	.823	2.39	.837	2.49	.850	2.29	.823	2.39	.837	2.49	.850
L 4 ×3½×½	2.15	.783	2.24	.801	2.34	.818	2.17	.881	2.29	.892	2.41	.903
³⁄₈	2.15	.774	2.24	.792	2.34	.809	2.17	.875	2.28	.887	2.40	.898
⁵⁄₁₆	2.15	.774	2.24	.791	2.34	.808	2.17	.872	2.28	.884	2.40	.895
¼	2.16	.770	2.24	.787	2.34	.805	2.17	.870	2.28	.882	2.39	.893
L 4 ×3 ×½	2.04	.719	2.12	.740	2.22	.762	2.10	.924	2.22	.933	2.35	.940
³⁄₈	2.04	.714	2.12	.735	2.21	.757	2.09	.919	2.21	.928	2.34	.935
⁵⁄₁₆	2.05	.710	2.13	.731	2.22	.752	2.09	.917	2.21	.925	2.33	.933
¼	2.06	.706	2.13	.726	2.22	.747	2.09	.914	2.20	.923	2.33	.931
L 3½×3½×³⁄₈	2.00	.831	2.10	.847	2.22	.862	2.00	.831	2.10	.847	2.22	.862
⁵⁄₁₆	2.01	.827	2.10	.843	2.21	.858	2.01	.827	2.10	.843	2.21	.858
¼	2.01	.824	2.10	.839	2.21	.855	2.01	.824	2.10	.839	2.21	.855
L 3½×3 ×³⁄₈	1.86	.771	1.95	.791	2.06	.812	1.89	.884	2.00	.897	2.13	.909
⁵⁄₁₆	1.87	.766	1.96	.787	2.06	.807	1.89	.881	2.00	.894	2.12	.906
¼	1.87	.762	1.96	.782	2.06	.803	1.89	.878	2.00	.891	2.12	.903
L 3½×2½×³⁄₈	1.76	.696	1.84	.721	1.94	.748	1.82	.932	1.94	.941	2.08	.948
⁵⁄₁₆	1.77	.691	1.85	.716	1.94	.742	1.81	.930	1.94	.938	2.07	.946
¼	1.77	.691	1.85	.715	1.93	.740	1.81	.927	1.93	.936	2.06	.944
L 3 ×3 ×½	1.72	.842	1.83	.860	1.95	.877	1.72	.842	1.83	.860	1.95	.877
³⁄₈	1.72	.834	1.82	.852	1.94	.869	1.72	.834	1.82	.852	1.94	.869
⁵⁄₁₆	1.72	.830	1.82	.848	1.93	.866	1.72	.830	1.82	.848	1.93	.866
¼	1.72	.825	1.82	.844	1.93	.862	1.72	.825	1.82	.844	1.93	.862
³⁄₁₆	1.72	.822	1.82	.841	1.93	.858	1.72	.822	1.82	.841	1.93	.858

FLEXURAL-TORSIONAL PROPERTIES
Double Angles ⌐L

Designation	Long Legs Vertical						Short Legs Vertical					
	Back to Back of Angles, In.						Back to Back of Angles, In.					
	0		⅜		¾		0		⅜		¾	
	$\bar{r}_o$	H	$\bar{r}_o$	H	$\bar{r}_o$	H	$\bar{r}_o$	H	$\bar{r}_o$	H	$\bar{r}_o$	H
L 3 ×2½×⅜	1.58	.763	1.67	.789	1.78	.813	1.61	.896	1.73	.910	1.86	.922
¼	1.59	.754	1.67	.779	1.78	.804	1.61	.889	1.72	.903	1.84	.916
³⁄₁₆	1.59	.750	1.67	.775	1.77	.800	1.61	.885	1.72	.899	1.84	.912
L 3 ×2 ×⅜	1.49	.672	1.57	.704	1.66	.737	1.55	.949	1.68	.956	1.82	.963
⁵⁄₁₆	1.50	.667	1.57	.698	1.66	.730	1.55	.946	1.68	.954	1.82	.961
¼	1.50	.664	1.57	.694	1.66	.726	1.54	.943	1.67	.951	1.80	.958
³⁄₁₆	1.50	.661	1.57	.690	1.66	.721	1.54	.940	1.66	.949	1.80	.956
L 2½×2½×⅜	1.43	.839	1.54	.861	1.66	.880	1.43	.839	1.54	.861	1.66	.880
⁵⁄₁₆	1.43	.834	1.54	.856	1.66	.876	1.43	.834	1.54	.856	1.66	.876
¼	1.43	.829	1.53	.851	1.65	.871	1.43	.829	1.53	.851	1.65	.871
³⁄₁₆	1.43	.825	1.53	.847	1.65	.867	1.43	.825	1.53	.847	1.65	.867
L 2½×2 ×⅜	1.29	.752	1.39	.785	1.50	.816	1.33	.912	1.45	.927	1.59	.939
⁵⁄₁₆	1.30	.746	1.39	.779	1.50	.810	1.33	.908	1.45	.923	1.58	.935
¼	1.30	.741	1.39	.773	1.50	.804	1.33	.903	1.45	.919	1.58	.932
³⁄₁₆	1.31	.736	1.39	.767	1.49	.798	1.32	.899	1.44	.915	1.57	.928
L 2 ×2 ×⅜	1.15	.846	1.26	.873	1.39	.896	1.15	.846	1.26	.873	1.39	.896
⁵⁄₁₆	1.15	.840	1.26	.867	1.38	.890	1.15	.840	1.26	.867	1.38	.890
¼	1.15	.834	1.25	.861	1.38	.885	1.15	.834	1.25	.861	1.38	.885
³⁄₁₆	1.15	.828	1.25	.855	1.37	.880	1.15	.828	1.25	.855	1.37	.880
⅛	1.15	.822	1.25	.850	1.37	.875	1.15	.822	1.25	.850	1.37	.875

SURFACE AREAS AND BOX AREAS
W shapes
Square feet per foot of length

Designation	Case A	Case B	Case C	Case D	Designation	Case A	Case B	Case C	Case D
W 44×285	10.0	11.0	8.32	9.31	W 36×256	9.02	10.0	7.26	8.27
×248	9.97	10.9	8.25	9.24	×232	8.96	9.97	7.20	8.21
×224	9.93	10.9	8.20	9.19	×210	8.91	9.93	7.13	8.14
×198	9.87	10.9	8.14	9.12	×194	8.88	9.89	7.09	8.10
					×182	8.85	9.85	7.06	8.07
W 40×328	10.8	12.3	8.16	9.65	×170	8.82	9.82	7.03	8.03
×298	10.8	12.3	8.10	9.59	×160	8.79	9.79	7.00	8.00
×268	10.7	12.2	8.04	9.52	×150	8.76	9.76	6.97	7.97
×244	10.7	12.1	7.99	9.46	×135	8.71	9.70	6.92	7.92
×221	10.6	12.1	7.92	9.40					
×192	10.5	12.0	7.84	9.32	W 33×619	10.2	11.6	7.82	9.23
					×567	10.1	11.5	7.71	9.11
W 40×655	11.0	12.4	8.68	10.1	×515	10.0	11.4	7.61	8.99
×593	10.9	12.3	8.56	9.95	×468	9.89	11.3	7.51	8.88
×531	10.7	12.1	8.43	9.81	×424	9.80	11.2	7.42	8.78
×480	10.6	12.0	8.33	9.70	×387	9.73	11.1	7.34	8.69
×436	10.6	11.9	8.24	9.60	×354	9.66	11.0	7.27	8.61
×397	10.5	11.8	8.17	9.51	×318	9.58	10.9	7.19	8.52
×362	10.4	11.7	8.09	9.43	×291	9.52	10.8	7.13	8.46
×324	10.3	11.7	8.02	9.34	×263	9.46	10.8	7.07	8.39
×297	10.3	11.6	7.96	9.28	×241	9.42	10.7	7.02	8.34
×277	10.3	11.6	7.93	9.25	×221	9.38	10.7	6.97	8.29
×249	10.2	11.5	7.88	9.19	×201	9.33	10.6	6.93	8.24
×215	10.2	11.5	7.81	9.12					
×199	10.1	11.4	7.76	9.07	W 33×169	8.30	9.26	6.60	7.55
					×152	8.27	9.23	6.55	7.51
W 40×183	9.17	10.2	7.48	8.47	×141	8.23	9.19	6.51	7.47
×167	9.11	10.1	7.42	8.40	×130	8.20	9.15	6.47	7.43
×149	9.05	10.0	7.35	8.34	×118	8.15	9.11	6.43	7.39
W 36×848	11.1	12.6	8.59	10.1	W 30×581	9.52	10.9	7.25	8.60
×798	11.0	12.5	8.49	9.99	×526	9.40	10.7	7.13	8.46
×720	10.8	12.3	8.35	9.83	×477	9.30	10.6	7.02	8.35
×650	10.7	12.1	8.21	9.67	×433	9.20	10.5	6.92	8.23
×588	10.6	12.0	8.09	9.54	×391	9.11	10.4	6.83	8.13
×527	10.4	11.9	7.97	9.41	×357	9.03	10.3	6.76	8.05
×485	10.3	11.8	7.88	9.31	×326	8.96	10.2	6.68	7.96
×439	10.3	11.7	7.79	9.20	×292	8.88	10.2	6.61	7.88
×393	10.2	11.6	7.70	9.10	×261	8.81	10.1	6.53	7.79
×359	10.1	11.5	7.63	9.02	×235	8.75	10.0	6.47	7.73
×328	10.0	11.4	7.57	8.95	×211	8.71	9.97	6.42	7.67
×300	9.99	11.4	7.51	8.90	×191	8.66	9.92	6.37	7.62
×280	9.95	11.3	7.47	8.85	×173	8.62	9.87	6.32	7.57
×260	9.90	11.3	7.42	8.80					
×245	9.87	11.2	7.39	8.77	W 30×148	7.53	8.40	5.99	6.86
×230	9.84	11.2	7.36	8.73	×132	7.49	8.37	5.93	6.81

Case A: Shape perimeter, minus one flange surface.
Case B: Shape perimeter.
Case C: Box perimeter, equal to one flange surface plus twice the depth.
Case D: Box perimeter, equal to two flange surfaces plus twice the depth.

SURFACE AREAS AND BOX AREAS
W shapes
Square feet per foot of length

Desig-nation	Case A	Case B	Case C	Case D	Desig-nation	Case A	Case B	Case C	Case D
W 30×124	7.47	8.34	5.90	6.78	W 24×103	6.18	6.93	4.84	5.59
×116	7.44	8.31	5.88	6.75	× 94	6.16	6.92	4.81	5.56
×108	7.41	8.28	5.84	6.72	× 84	6.12	6.87	4.77	5.52
× 99	7.37	8.25	5.81	6.68	× 76	6.09	6.84	4.74	5.49
× 90	7.35	8.22	5.79	6.66	× 68	6.06	6.80	4.70	5.45
W 27×539	8.82	10.1	6.69	7.96	W 24× 62	5.57	6.16	4.54	5.13
×494	8.72	9.97	6.59	7.84	× 55	5.54	6.13	4.51	5.10
×448	8.61	9.86	6.48	7.73					
×408	8.51	9.74	6.38	7.61	W 21×402	7.32	8.44	5.45	6.57
×368	8.42	9.64	6.29	7.51	×364	7.22	8.32	5.35	6.46
×336	8.34	9.55	6.21	7.42	×333	7.13	8.22	5.26	6.36
×307	8.27	9.47	6.14	7.34	×300	7.04	8.12	5.17	6.25
×281	8.21	9.40	6.08	7.27	×275	6.96	8.04	5.10	6.17
×258	8.15	9.34	6.02	7.21	×248	6.89	7.95	5.02	6.09
×235	8.09	9.27	5.96	7.14	×223	6.82	7.87	4.95	6.00
×217	8.04	9.22	5.91	7.09	×201	6.75	7.80	4.89	5.93
×194	7.98	9.15	5.85	7.02	×182	6.69	7.74	4.83	5.87
×178	7.95	9.12	5.81	6.98	×166	6.65	7.68	4.78	5.82
×161	7.91	9.08	5.77	6.94	×147	6.61	7.66	4.72	5.76
×146	7.87	9.03	5.73	6.89	×132	6.57	7.61	4.67	5.71
					×122	6.54	7.57	4.65	5.68
W 27×129	6.92	7.75	5.44	6.27	×111	6.51	7.54	4.61	5.64
×114	6.88	7.72	5.39	6.23	×101	6.48	7.50	4.58	5.61
×102	6.85	7.68	5.35	6.18					
× 94	6.82	7.65	5.32	6.15	W 21× 93	5.54	6.24	4.31	5.01
× 84	6.78	7.61	5.28	6.11	× 83	5.50	6.20	4.27	4.96
					× 73	5.47	6.16	4.23	4.92
W 24×492	8.07	9.25	6.12	7.29	× 68	5.45	6.14	4.21	4.90
×450	7.96	9.13	6.01	7.17	× 62	5.42	6.11	4.19	4.87
×408	7.86	9.01	5.91	7.06					
×370	7.75	8.89	5.80	6.94	W 21× 57	5.01	5.56	4.06	4.60
×335	7.66	8.79	5.71	6.84	× 50	4.97	5.51	4.02	4.56
×306	7.59	8.71	5.64	6.76	× 44	4.94	5.48	3.99	4.53
×279	7.51	8.62	5.56	6.67					
×250	7.44	8.54	5.49	6.59	W 18×311	6.41	7.41	4.72	5.72
×229	7.38	8.47	5.43	6.52	×283	6.32	7.31	4.63	5.62
×207	7.32	8.40	5.37	6.45	×258	6.24	7.23	4.56	5.54
×192	7.27	8.35	5.32	6.40	×234	6.17	7.14	4.48	5.45
×176	7.23	8.31	5.28	6.36	×211	6.10	7.06	4.41	5.37
×162	7.22	8.30	5.25	6.33	×192	6.03	6.99	4.35	5.30
×146	7.17	8.24	5.20	6.27	×175	5.97	6.92	4.29	5.24
×131	7.12	8.19	5.15	6.22	×158	5.92	6.86	4.23	5.17
×117	7.08	8.15	5.11	6.18	×143	5.87	6.81	4.18	5.12
×104	7.04	8.11	5.07	6.14	×130	5.83	6.76	4.14	5.07
					×119	5.81	6.75	4.10	5.04
					×106	5.77	6.70	4.05	4.99
					× 97	5.74	6.67	4.03	4.96

Case A: Shape perimeter, minus one flange surface.
Case B: Shape perimeter.
Case C: Box perimeter, equal to one flange surface plus twice the depth.
Case D: Box perimeter, equal to two flange surfaces plus twice the depth.

SURFACE AREAS AND BOX AREAS
W shapes
Square feet per foot of length

Desig-nation	Case A	Case B	Case C	Case D	Desig-nation	Case A	Case B	Case C	Case D
W 18× 86	5.70	6.62	3.99	4.91	W 14×132	5.93	7.16	3.67	4.90
× 76	5.67	6.59	3.95	4.87	×120	5.90	7.12	3.64	4.86
					×109	5.86	7.08	3.60	4.82
W 18× 71	4.85	5.48	3.71	4.35	× 99	5.83	7.05	3.57	4.79
× 65	4.82	5.46	3.69	4.32	× 90	5.81	7.02	3.55	4.76
× 60	4.80	5.43	3.67	4.30					
× 55	4.78	5.41	3.65	4.27	W 14× 82	4.75	5.59	3.23	4.07
× 50	4.76	5.38	3.62	4.25	× 74	4.72	5.56	3.20	4.04
					× 68	4.69	5.53	3.18	4.01
W 18× 46	4.41	4.91	3.52	4.02	× 61	4.67	5.50	3.15	3.98
× 40	4.38	4.88	3.48	3.99					
× 35	4.34	4.84	3.45	3.95	W 14× 53	4.19	4.86	2.99	3.66
					× 48	4.16	4.83	2.97	3.64
W 16×100	5.28	6.15	3.70	4.57	× 43	4.14	4.80	2.94	3.61
× 89	5.24	6.10	3.66	4.52					
× 77	5.19	6.05	3.61	4.47	W 14× 38	3.93	4.50	2.91	3.48
× 67	5.16	6.01	3.57	4.43	× 34	3.91	4.47	2.89	3.45
					× 30	3.89	4.45	2.87	3.43
W 16× 57	4.39	4.98	3.33	3.93					
× 50	4.36	4.95	3.30	3.89	W 14× 26	3.47	3.89	2.74	3.16
× 45	4.33	4.92	3.27	3.86	× 22	3.44	3.86	2.71	3.12
× 40	4.31	4.89	3.25	3.83					
× 36	4.28	4.87	3.23	3.81	W 12×336	5.77	6.88	3.92	5.03
					×305	5.67	6.77	3.82	4.93
W 16× 31	3.92	4.39	3.11	3.57	×279	5.59	6.68	3.74	4.83
× 26	3.89	4.35	3.07	3.53	×252	5.50	6.58	3.65	4.74
					×230	5.43	6.51	3.58	4.66
W 14×730	7.61	9.10	5.23	6.72	×210	5.37	6.43	3.52	4.58
×665	7.46	8.93	5.08	6.55	×190	5.30	6.36	3.45	4.51
×605	7.32	8.77	4.94	6.39	×170	5.23	6.28	3.39	4.43
×550	7.19	8.62	4.81	6.24	×152	5.17	6.21	3.33	4.36
×500	7.07	8.49	4.68	6.10	×136	5.12	6.15	3.27	4.30
×455	6.96	8.36	4.57	5.98	×120	5.06	6.09	3.21	4.24
					×106	5.02	6.03	3.17	4.19
W 14×426	6.89	8.28	4.50	5.89	× 96	4.98	5.99	3.13	4.15
×398	6.81	8.20	4.43	5.81	× 87	4.95	5.96	3.10	4.11
×370	6.74	8.12	4.36	5.73	× 79	4.92	5.93	3.07	4.08
×342	6.67	8.03	4.29	5.65	× 72	4.89	5.90	3.05	4.05
×311	6.59	7.94	4.21	5.56	× 65	4.87	5.87	3.02	4.02
×283	6.52	7.86	4.13	5.48					
×257	6.45	7.78	4.06	5.40	W 12× 58	4.39	5.22	2.87	3.70
×233	6.38	7.71	4.00	5.32	× 53	4.37	5.20	2.84	3.68
×211	6.32	7.64	3.94	5.25					
×193	6.27	7.58	3.89	5.20	W 12× 50	3.90	4.58	2.71	3.38
×176	6.22	7.53	3.84	5.15	× 45	3.88	4.55	2.68	3.35
×159	6.18	7.47	3.79	5.09	× 40	3.86	4.52	2.66	3.32
×145	6.14	7.43	3.76	5.05					

Case A: Shape perimeter, minus one flange surface.
Case B: Shape perimeter.
Case C: Box perimeter, equal to one flange surface plus twice the depth.
Case D: Box perimeter, equal to two flange surfaces plus twice the depth.

SURFACE AREAS AND BOX AREAS
W shapes
Square feet per foot of length

Desig-nation	Case A	Case B	Case C	Case D	Desig-nation	Case A	Case B	Case C	Case D
W 12× 35	3.63	4.18	2.63	3.18	W 8×10	2.23	2.56	1.64	1.97
× 30	3.60	4.14	2.60	3.14	W 6×25	2.49	3.00	1.57	2.08
× 26	3.58	4.12	2.58	3.12	×20	2.46	2.96	1.54	2.04
W 12× 22	2.97	3.31	2.39	2.72	×15	2.42	2.92	1.50	2.00
× 19	2.95	3.28	2.36	2.69	W 6×16	1.98	2.31	1.38	1.72
× 16	2.92	3.25	2.33	2.66	×12	1.93	2.26	1.34	1.67
× 14	2.90	3.23	2.32	2.65	× 9	1.90	2.23	1.31	1.64
W 10×112	4.30	5.17	2.76	3.63	W 5×19	2.04	2.45	1.28	1.70
×100	4.25	5.11	2.71	3.57	×16	2.01	2.43	1.25	1.67
× 88	4.20	5.06	2.66	3.52	W 4×13	1.63	1.96	1.03	1.37
× 77	4.15	5.00	2.62	3.47					
× 68	4.12	4.96	2.58	3.42					
× 60	4.08	4.92	2.54	3.38					
× 54	4.06	4.89	2.52	3.35					
× 49	4.04	4.87	2.50	3.33					
W 10× 45	3.56	4.23	2.35	3.02					
× 39	3.53	4.19	2.32	2.98					
× 33	3.49	4.16	2.28	2.95					
W 10× 30	3.10	3.59	2.23	2.71					
× 26	3.08	3.56	2.20	2.68					
× 22	3.05	3.53	2.17	2.65					
W 10× 19	2.63	2.96	2.04	2.38					
× 17	2.60	2.94	2.02	2.35					
× 15	2.58	2.92	2.00	2.33					
× 12	2.56	2.89	1.98	2.31					
W 8× 67	3.42	4.11	2.19	2.88					
× 58	3.37	4.06	2.14	2.83					
× 48	3.32	4.00	2.09	2.77					
× 40	3.28	3.95	2.05	2.72					
× 35	3.25	3.92	2.02	2.69					
× 31	3.23	3.89	2.00	2.67					
W 8× 28	2.87	3.42	1.89	2.43					
× 24	2.85	3.39	1.86	2.40					
W 8× 21	2.61	3.05	1.82	2.26					
× 18	2.59	3.03	1.79	2.23					
W 8× 15	2.27	2.61	1.69	2.02					
× 13	2.25	2.58	1.67	2.00					

Case A: Shape perimeter, minus one flange surface.
Case B: Shape perimeter.
Case C: Box perimeter, equal to one flange surface plus twice the depth.
Case D: Box perimeter, equal to two flange surfaces plus twice the depth.

AMERICAN INSTITUTE OF STEEL CONSTRUCTION

STANDARD MILL PRACTICE
General Information

Rolling structural shapes and plates involves such factors as roll wear, subsequent roll dressing, temperature variations, etc., which cause the finished product to vary from published profiles. Such variations are limited by the provisions of the American Society for Testing and Materials Specification A6. Contained in this section is a summary of these provisions, not a reproduction of the complete specification. In its entirety, A6 covers a group of common requirements, which, unless otherwise specified in the purchase order or in an individual specification, shall apply to rolled steel plates, shapes, sheet piling and bars.

In accordance with Table 1, *carbon steel* refers to ASTM Designations A36 and A529; *high-strength, low-alloy steel* refers to Designations A242, A572, and A588; *alloy steel* refers to Designation A514; and low-alloy steel refers to A852.

For further information on mill practices, including permissible variations for rolled tees, zees and bulb angles in structural and bar sizes, pipe, tubing, sheets and strip, and for other grades of steel, see ASTM A6, A53, A500, A568 and A618, and the AISI Steel Products Manuals and Producers' Catalogs.

The data on spreading rolls to increase areas and weights, and mill cambering of beams, is not a part of A6.

Additional material on mill practice is included in the descriptive material preceding the "Dimensions and Properties" tables for shapes and plates.

Letter symbols representing dimensions on sketches shown herein are in accordance with ASTM A6, AISI and mill catalogs and *not necessarily as defined by the general nomenclature of this manual.*

STANDARD MILL PRACTICE
Methods of increasing areas and weights by spreading rolls

W SHAPES

To vary the area and weight within a given nominal size, the flange width, the flange thickness and the web thickness are changed, as shown in Fig. 1.

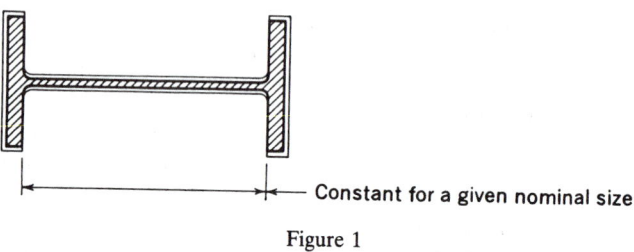

Constant for a given nominal size

Figure 1

S SHAPES AND AMERICAN STANDARD CHANNELS

To vary the area and weight within a given nominal size, the web thickness and the flange width are changed by an equal amount, as shown in Figs. 2 and 3.

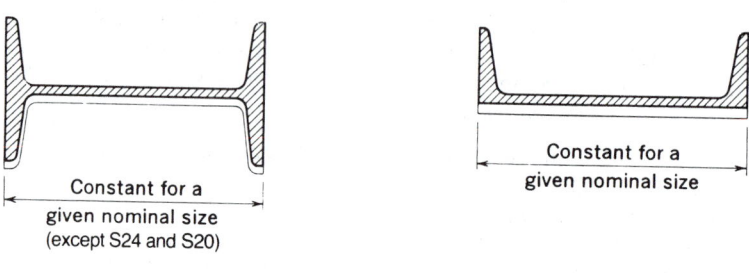

Constant for a
given nominal size
(except S24 and S20)

Figure 2

Constant for a
given nominal size

Figure 3

ANGLES

To vary area and weight for a given leg length, the thickness of each leg is changed. Note that leg length is changed slightly by this method (Fig. 4).

Figure 4

AMERICAN INSTITUTE OF STEEL CONSTRUCTION

STANDARD MILL PRACTICE
Cambering of rolled beams

All beams are straightened after rolling to meet permissible variations for sweep and camber listed hereinafter for W shapes and S shapes. The following data refers to the subsequent cold cambering of beams to produce a predetermined dimension.

The maximum lengths that can be cambered depend on the length to which a given section can be rolled, with a maximum of 100 ft. The following table outlines the maximum and minimum induced camber of W shapes and S shapes.

MAXIMUM AND MINIMUM INDUCED CAMBER

Sections Nominal Depth In.	Specified Length of Beam, Ft				
	Over 30 to 42, incl.	Over 42 to 52, incl.	Over 52 to 65, incl.	Over 65 to 85, incl.	Over 85 to 100, incl.
	Max. and Min. Camber Acceptable, In.				
W shapes, 24 and over	1 to 2, incl.	1 to 3, incl.	2 to 4, incl.	3 to 5, incl.	3 to 6, incl.
W shapes, 14 to 21, incl. and S shapes, 12 in. and over	¾ to 2½, incl.	1 to 3, incl.	2 to 4, incl.	2½ to 5, incl.	Inquire

Consult the producer for specific camber and/or lengths outside the above listed available lengths and sections.

Mill camber in beams of less depth than tabulated should not be specified.

A single minimum value for camber, within the ranges shown above for the length ordered, should be specified.

Camber is measured at the mill and will not necessarily be present in the same amount in the section of beam as received due to release of stress induced during the cambering operation. In general, 75% of the specified camber is likely to remain.

Camber will approximate a simple regular curve nearly the full length of the beam, or between any two points specified.

Camber is ordinarily specified by the ordinate at the mid-length of the portion of the beam to be curved. Ordinates at other points should not be specified.

Although mill cambering to achieve reverse or other compound curves is not considered practical, fabricating shop facilities for cambering by heat can accomplish such results as well as form regular curves in excess of the limits tabulated above. Refer to Effect of Heat on Steel, Part 6 of this Manual, for further information.

PERMISSIBLE VARIATIONS FOR CAMBER ORDINATE

Lengths	Plus Variation	Minus Variation
50 ft and less	½ inch	0
Over 50 ft	½ in. plus ⅛ in. for each 10 ft or fraction thereof in excess of 50 ft	0

STANDARD MILL PRACTICE
Positions for measuring camber and sweep

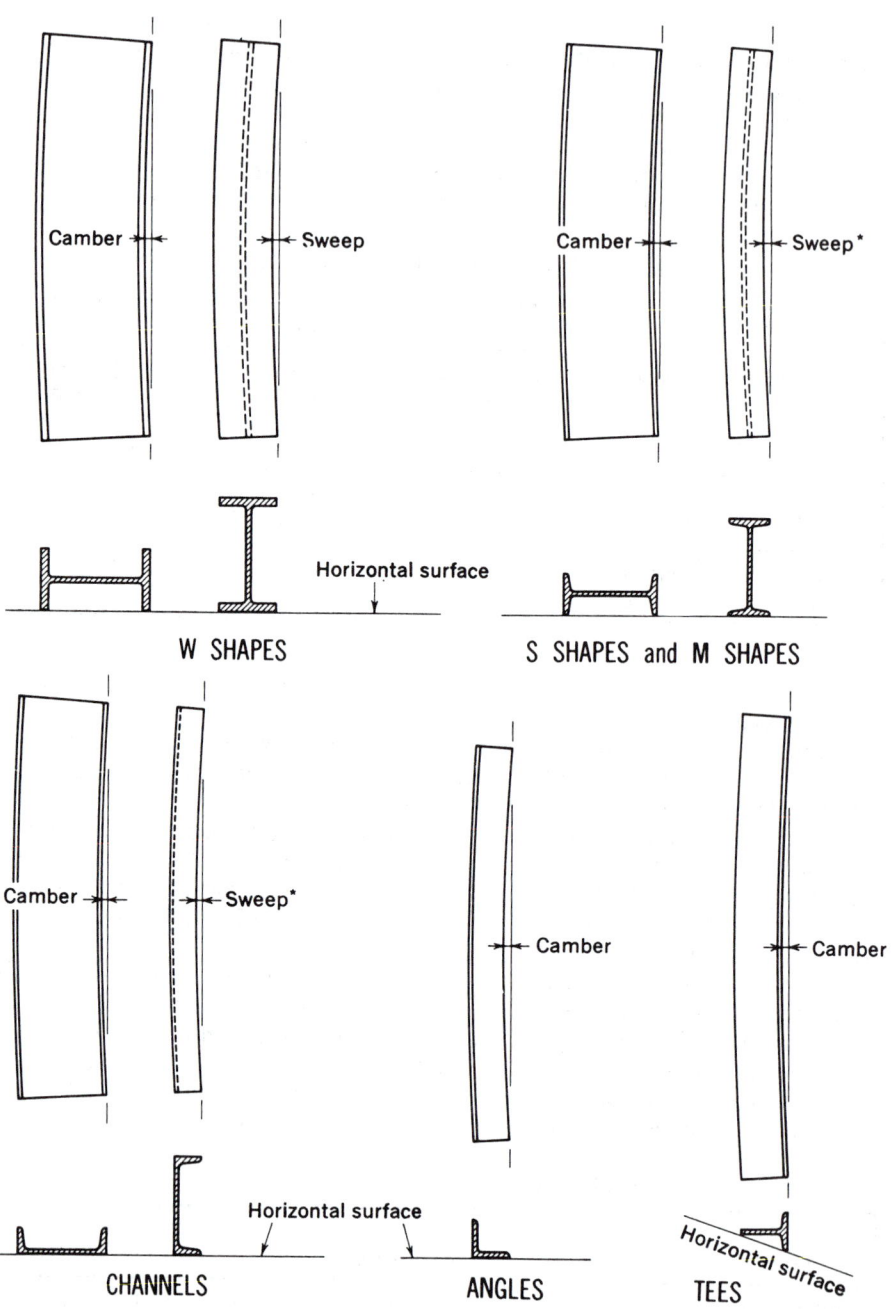

* Due to the extreme variations in flexibility of these shapes, straightness tolerances for sweep are subject to negotiations between manufacturer and purchaser for individual sections involved.

STANDARD MILL PRACTICE

W Shapes
HP Shapes

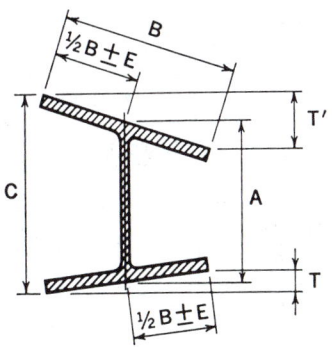 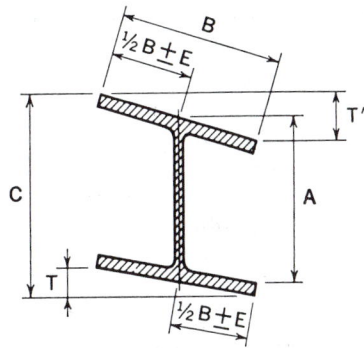

PERMISSIBLE VARIATIONS IN CROSS SECTION

Section Nominal Size, In.	A, Depth, In.		B, Flg. Width, In.		T + T', Flanges, Out of Square, Max, In.	E,[a] Web off Center, Max, In.	C, Max, Depth at any Cross-Section over Theoretical Depth, In.
	Over Theo-retical	Under Theo-retical	Over Theo-retical	Under Theo-retical			
To 12, incl.	1/8	1/8	1/4	3/16	1/4	3/16	1/4
Over 12	1/8	1/8	1/4	3/16	5/16	3/16	1/4

[a]Variation of 5/16-in. max. for sections over 426 lb./ft.

PERMISSIBLE VARIATIONS IN LENGTH

W Shapes	Variations from Specified Length for Lengths Given, In.			
	30 ft and Under		Over 30 ft	
	Over	Under	Over	Under
Beams 24 in. and under in nominal depth	3/8	3/8	3/8 plus 1/16 for each additional 5 ft or fraction thereof	3/8
Beams over 24 in. nom. depth; all columns	1/2	1/2	1/2 plus 1/16 for each additional 5 ft or fraction thereof	1/2

OTHER PERMISSIBLE VARIATIONS

Area and Weight Variation: ±2.5% theoretical or specified amount.

Ends Out-of-Square: ¼ in. per in. of depth, or of flange width if it is greater than the depth.

Camber and Sweep:

Sizes	Length	Permissible Variation, In.	
		Camber	Sweep
Sizes with flange width equal to or greater than 6 in.	All	$\frac{1}{8}$ in. $\times \frac{\text{(total length, ft)}}{10}$	
Sizes with flange width less than 6 in.	All	$\frac{1}{8}$ in. $\times \frac{\text{(total length, ft)}}{10}$	$\frac{1}{8}$ in. $\times \frac{\text{(total length, ft)}}{5}$
Certain sections with a flange width approx. equal to depth & specified on order as columns[b]	45 ft and under	$\frac{1}{8}$ in. $\times \frac{\text{(total length, ft)}}{10}$ with $\frac{3}{8}$ in. max.	
	Over 45 ft	$\frac{3}{8}$ in. $+ \left[\frac{1}{8} \text{ in.} \times \frac{\text{(total length, ft} - 45)}{10} \right]$	

[b]Applies only to: W 8 × 31 and heavier, W 10 × 49 and heavier, W 12 × 65 and heavier, W 14 × 90 and heavier. If other sections are specified on the order as columns, the tolerance will be subject to negotiation with the manufacturer.

STANDARD MILL PRACTICE
S shapes, M shapes and channels

PERMISSIBLE VARIATIONS IN CROSS-SECTION

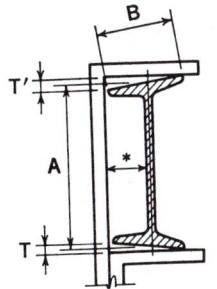 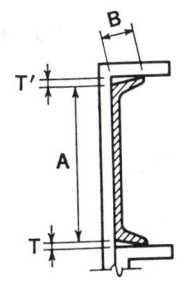

Section	Nominal Size, in.	A, Depth, In.[a]		B, Flange Width, In.		T + T', Out of Square per Inch of B, In.
		Over Theo-retical	Under Theo-retical	Over Theo-retical	Under Theo-retical	
S shapes and M shapes	3 to 7, incl.	3/32	1/16	1/8	1/8	1/32
	Over 7 to 14, incl.	1/8	3/32	5/32	5/32	1/32
	Over 14 to 24, incl.	3/16	1/8	3/16	3/16	1/32
Channels	3 to 7, incl	3/32	1/16	1/8	1/8	1/32
	Over 7 to 14, incl.	1/8	3/32	1/8	5/32	1/32
	Over 14	3/16	1/8	1/8	3/16	1/32

[a] A is measured at centerline of web for beams; and at back of web for channels.
[b] T + T applies when flanges of channels are tied in or out.

PERMISSIBLE VARIATIONS IN LENGTH

Section	Variations from Specified Length for Lengths Given, In.									
	To 30 Ft, incl.		Over 30 to 40 Ft, incl.		Over 40 to 50 Ft, incl.		Over 50 to 65 Ft, incl.		Over 65 Ft	
	Over	Under	Over	Under	Over	Under	Over	Under	Over	Under
S shapes, M shapes and Channels	1/2	1/4	3/4	1/4	1	1/4	1 1/8	1/4	1 1/4	1/4

OTHER PERMISSIBLE VARIATIONS

Area and Weight Variation: ±2.5% theoretical or specified amount.

Ends Out-of-Square: S shapes and channels 1/64 in. per in. of depth.

Camber: 1/8 in. $\times \dfrac{\text{total length, ft}}{5}$

AMERICAN INSTITUTE OF STEEL CONSTRUCTION

STANDARD MILL PRACTICE
Tees split from W, M and S shapes
Angles split from channels

PERMISSIBLE VARIATIONS IN DEPTH

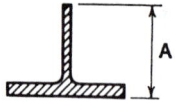 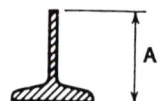 

Dimension A may be approximately ½ beam or channel depth, or any dimension resulting from off-center splitting, or splitting on two lines as specified on the order.

Depth of Beam from which Tees or Angles are Split	Variations in Depth A Over and Under	
	Tees	Angles
To 6 in., excl.	⅛	⅛
6 to 16, excl.	³⁄₁₆	³⁄₁₆
16 to 20, excl.	¼	¼
20 to 24, excl.	⁵⁄₁₆	. . .
24 and over	⅜	. . .

The above variations for depths of tees or angles include the permissible variations in depth for the beams and channels before splitting.

OTHER PERMISSIBLE VARIATIONS

Other permissible variations in cross section, as well as permissible variations in length, area and weight variation and ends out-of-square, will correspond to those of the beam or channel before splitting, except

$$\text{camber} = \frac{1}{8} \text{ in.} \times \frac{\text{total length, ft}}{5}$$

STANDARD MILL PRACTICE
Angles, structural size

PERMISSIBLE VARIATIONS IN CROSS SECTION

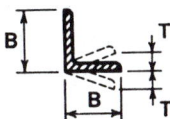

Section	Nominal Size, in.[a]	B Length of Leg, in.		T, Out of Square per In. of B, In.
		Over Theoretical	Under Theoretical	
Angles	3 to 4, incl.	1/8	3/32	3/128[b]
	Over 4 to 6, incl.	1/8	1/8	3/128[b]
	Over 6	3/16	1/8	3/128[b]

[a] For unequal leg angles, longer leg determines classification.
[b] 3/128 in. per in. = 1½ deg.

PERMISSIBLE VARIATIONS IN LENGTH

Section	Variations from Specified Length for Lengths Given, In.									
	To 30 Ft, incl.		Over 30 to 40 Ft, incl.		Over 40 to 50 Ft, incl.		Over 50 to 65 Ft, incl.		Over 65 Ft	
Angles	Over	Under	Over	Under	Over	Under	Over	Under	Over	Under
	1/2	1/4	3/4	1/4	1	1/4	1⅛	1/4	1¼	1/4

OTHER PERMISSIBLE VARIATIONS

Area and weight variation: ±2.5% theoretical or specified amount.

Ends out-of-square: 3/128 in. per in. of leg length, or 1½ deg. Variations based on the longer leg of an unequal angle.

Camber: ⅛ in. $\times \dfrac{\text{total length, ft}}{5}$, applied to either leg.

AMERICAN INSTITUTE OF STEEL CONSTRUCTION

STANDARD MILL PRACTICE
*Angles, bar size

PERMISSIBLE VARIATION IN CROSS SECTION

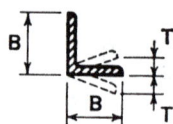

[a]Specified Length of Leg, In.	Variations from Thickness for Thickness Given, Over and Under, In.			B Length of Leg, Over and Under, In.	T, Out of Square per Inch of B, In.
	3/16 and Under	Over 3/16 to 3/8, incl.	Over 3/8		
1 and under	0.008	0.010		1/32	3/128[b]
Over 1 to 2, inc.	0.010	0.010	0.012	3/64	3/128[b]
Over 2 to 3, excl.	0.012	0.015	0.015	1/16	3/128[b]

[a]The longer leg of an unequal angle determines the size for permissible variations.
[b]3/128-in. per in. = 1½ degrees.

PERMISSIBLE VARIATIONS IN LENGTH

Section	Variations Over Specified Length for Lengths Given No Variation Under				
	5 to 10 Ft excl.	10 to 20 Ft excl.	20 to 30 Ft excl.	30 to 40 Ft excl.	40 to 65 Ft incl.
All sizes of bar-size angles	5/8	1	1½	2	2½

OTHER PERMISSIBLE VARIATIONS

Camber: ¼ in. in any 5 ft, or ¼ in. $\times \dfrac{\text{total length, ft}}{5}$

Straightness: Because of warpage, permissible variations for straightness do not apply to bars if any subsequent heating operation has been performed.

Ends Out-of-Square: 3/128-in. per in. of leg length or 1½ degrees. Variation based on longer leg of an unequal angle.

---

*A member is "bar size" when its greatest cross-sectional dimension is less than 3 in.

STANDARD MILL PRACTICE
Steel pipe and tubing

DIMENSIONS AND WEIGHT TOLERANCES
Round Tubing and Pipe

ASTM A53	ASTM A618

Weight—The weight of the pipe as specified in Table X2 and Table X3 (ASTM Specification A53) shall not vary by more than ± 10 percent.

Note that the weight tolerance of ± 10 percent is determined from the weights of the customary lifts of pipe as produced for shipment by the mill, divided by the number of feet of pipe in the lift. On pipe sizes over 4 in. where individual lengths may be weighed, the weight tolerance is applicable to the individual length.

Diameter—For pipe 2 in. and over in nominal diameter, the outside diameter shall not vary more than ± 1 percent from the standard specified.

Thickness—The minimum wall thickness at any point shall be not more than 12.5 percent under the nominal wall thickness specified.

Outside Dimensions—For round hot formed structural tubing 2 in. and over in nominal size, the outside diameter shall not vary more than ± 1 percent from the standard specified.

Mass (A618 only)—The mass of structural tubing shall not be less than the specified value by more than 3.5 percent.

Length—Structural tubing is commonly produced in random mill lengths, in multiple lengths, and in definite cut lengths. When cut lengths are specified for structural tubing, the length tolerances shall be in accordance with the following table:

	22 ft and under		Over 22 to 44 ft, incl.	
	Over	Under	Over	Under
Length tolerance for specified cut lengths, in.	½	¼	¾	¼

Straightness—The permissible variation for straightness of structural tubing shall be ⅛ in. times the number of feet of total length divided by 5.

Square and Rectangular Tubing

ASTM A500 and ASTM A618

Outside Dimensions—The specified dimensions, measured across the flats at positions at least 2 in. from either end of square or rectangular tubing and including an allowance for convexity or concavity, shall not exceed the plus and minus tolerance shown in the following table:

Largest Outside Dimension, Across Flats, in.	Tolerance[a] Plus and Minus, in.
2½ and under	0.020
Over 2½ to 3½, incl.	0.025
Over 3½ to 5½, incl.	0.030
Over 5½	1 percent

[a]The respective outside dimension tolerances include the allowances for convexity and concavity.

Lengths—Structural tubing is commonly produced in random lengths, in multiple lengths, and in definite cut lengths. When cut lengths are specified for structural tubing, the length tolerances shall be in accordance with the following table:

	22 ft and under		Over 22 to 44 ft, incl.	
	Over	Under	Over	Under
Length tolerance for specified cut lengths, in.	½	¼	¾	¼

Mass (A618 only)—The mass of structural tubing shall not be less than the specified value by more than 3.5 percent.

Straightness—The permissible variation for straightness of structural tubing shall be ⅛ in. times the number of feet of total length divided by 5.

Squareness of Sides—For square or rectangular structural tubing, adjacent sides may deviate from 90 degrees by a tolerance of plus or minus 2 degrees max.

Radius of Corners—For square or rectangular structural tubing, the radius of any outside corner of the section shall not exceed three times the specified wall thickness.

Twist—The tolerances for twist or variation with respect to axial alignment of the section, for square and rectangular structural tubing, shall be as shown in the following table:

Specified Dimension of Longest Side, in.	Maximum Twist per 3 ft of Length, in.
1½ and under	0.050
Over 1½ to 2½, incl.	0.062
Over 2½ to 4 incl.	0.075
Over 4 to 6, incl.	0.087
Over 6 to 8, incl.	0.100
Over 8	0.112

Twist is measured by holding down one end of a square or rectangular tube on a flat surface plate with the bottom side of the tube parallel to the surface plate and noting the height that either corner, at the opposite end of the bottom side of the tube, extends above the surface plate.

Wall Thickness (A500 only)—The tolerance for wall thickness exclusive of the weld area shall be plus and minus 10 percent of the nominal wall thickness specified. The wall thickness is to be measured at the center of the flat.

STANDARD MILL PRACTICE
Rectangular sheared plates and Universal mill plates

PERMISSIBLE VARIATIONS IN WIDTH AND LENGTH FOR SHEARED PLATES
(1½ in. and under in thickness)

PERMISSIBLE VARIATIONS IN LENGTH ONLY FOR UNIVERSAL MILL PLATES
(2½ in. and under in thickness)

Specified Dimensions, In.		Variations over Specified Width and Length for Thickness, In., and Equivalent Weights, Lb. Per Sq. Ft; Given							
		To ⅜, excl.		⅜ to ⅝, excl.		⅝ to 1, excl.		1 to 2, incl.*	
Length	Width	To 15.3, excl.		15.3 to 25.5, excl.		25.5 to 40.8, excl.		40.8 to 81.7, incl.	
		Width	Length	Width	Length	Width	Length	Width	Length
To 120, excl.	To 60, excl.	⅜	½	⁷⁄₁₆	⅝	½	¾	⅝	1
	60 to 84, excl.	⁷⁄₁₆	⅝	½	¹¹⁄₁₆	⅝	⅞	¾	1
	84 to 108, excl.	½	¾	⅝	⅞	¾	1	1	1⅛
	108 and over	⅝	⅞	¾	1	⅞	1⅛	1⅛	1¼
120 to 240, excl.	To 60, excl.	⅜	¾	½	⅞	⅝	1	¾	1⅛
	60 to 84, excl.	½	¾	⅝	⅞	¾	1	⅞	1¼
	84 to 108, excl.	⁹⁄₁₆	⅞	¹¹⁄₁₆	¹⁵⁄₁₆	¹³⁄₁₆	1⅛	1	1⅜
	108 and over	⅝	1	¾	1⅛	⅞	1¼	1⅛	1⅜
240 to 360, excl.	To 60, excl.	⅜	1	½	1⅛	⅝	1¼	¾	1½
	60 to 84, excl.	½	1	⅝	1⅛	¾	1¼	⅞	1½
	84 to 108, excl.	⁹⁄₁₆	1	¹¹⁄₁₆	1⅛	⅞	1⅜	1	1½
	108 and over	¹¹⁄₁₆	1⅛	⅞	1¼	1	1⅜	1¼	1⅜
360 to 480, excl.	To 60, excl.	⁷⁄₁₆	1⅛	½	1¼	⅝	1⅜	¾	1⅝
	60 to 84, excl.	½	1¼	⅝	1⅜	¾	1½	⅞	1⅝
	84 to 108, excl.	⁹⁄₁₆	1¼	¾	1⅜	⅞	1½	1	1⅞
	108 and over	¾	1⅜	⅞	1½	1	1⅝	1¼	1⅞
480 to 600, excl.	To 60, excl.	⁷⁄₁₆	1¼	½	1½	⅝	1⅝	¾	1⅞
	60 to 84, excl.	½	1⅜	⅝	1½	¾	1⅝	⅞	1⅞
	84 to 108, excl.	⅝	1⅜	¾	1½	⅞	1⅝	1	1⅞
	108 and over	¾	1½	⅞	1⅝	1	1¾	1¼	1⅞
600 to 720, excl.	To 60, excl.	½	1¾	⅝	1⅞	¾	1⅞	⅞	2¼
	60 to 84, excl.	⅝	1¾	¾	1⅞	⅞	1⅞	1	2¼
	84 to 108, excl.	⅝	1¾	¾	1⅞	⅞	1⅞	1⅛	2¼
	108 and over	⅞	1¾	1	2	1⅛	2¼	1¼	2½
720 and over, excl.	To 60, excl.	⁹⁄₁₆	2	¾	2⅛	⅞	2¼	1	2¾
	60 to 84, excl.	¾	2	⅞	2⅛	1	2¼	1⅛	2¾
	84 to 108, excl.	¾	2	⅞	2⅛	1	2¼	1¼	2¾
	108 and over	1	2	1⅛	2⅜	1¼	2½	1⅜	3

*Permissible variations in length apply also to Universal Mill plates up to 12 in. width for thick-nesses over 2 to 2½ in. incl. except for alloy steels up to 1¾ in. thick.

Notes: Permissible variations under specified width and length, ¼ in.

Table applies to all steels listed in ASTM A6.

STANDARD MILL PRACTICE
Rectangular sheared plates and Universal mill plates

PERMISSIBLE VARIATIONS FROM FLATNESS
(Carbon Steel Only)

Specified Thickness, In.	Variations from Flatness for Specified Widths, In.							
	To 36, excl.	36 to 48, excl.	48 to 60, excl.	60 to 72, excl.	72 to 84, excl.	84 to 96, excl.	96 to 108, excl.	108 to 120, excl.
To 1/4, excl.	9/16	3/4	15/16	1 1/4	1 3/8	1 1/2	1 5/8	1 3/4
1/4 to 3/8, excl.	1/2	5/8	3/4	15/16	1 1/8	1 1/4	1 3/8	1 1/2
3/8 to 1/2, excl.	1/2	9/16	5/8	5/8	3/4	7/8	1	1 1/8
1/2 to 3/4, excl.	7/16	1/2	9/16	5/8	5/8	3/4	1	1
3/4 to 1, excl.	7/16	1/2	9/16	5/8	5/8	5/8	3/4	7/8
1 to 2, excl.	3/8	1/2	1/2	9/16	9/16	5/8	5/8	5/8
2 to 4, excl.	5/16	3/8	7/16	1/2	1/2	1/2	1/2	9/16
4 to 6, excl.	3/8	7/16	1/2	1/2	9/16	9/16	5/8	3/4
6 to 8, excl.	7/16	1/2	1/2	5/8	11/16	3/4	7/8	7/8

General Notes:
1. The longer dimension specified is considered the length, and permissible variations in flatness along the length should not exceed the tabular amount for the specified width in plates up to 12 ft. in length.
2. The flatness variations across the width should not exceed the tabular amount for the specified width.
3. When the longer dimension is under 36 in., the permissible variation should not exceed 1/4 in. When the longer dimension is from 36 to 72 in., incl., the permissible variation should not exceed 75% of the tabular amount for the specified width, but in no case less than 1/4 in.
4. These variations apply to plates which have a specified minimum tensile strength of not more than 60,000 psi or compatible chemistry or hardness. The limits in the table are increased 50% for plates specified to a higher minimum tensile strength or compatible chemistry or hardness.

PERMISSIBLE VARIATIONS IN CAMBER FOR CARBON STEEL SHEARED AND GAS CUT RECTANGULAR PLATES

Maximum permissible camber, in. (all thicknesses) = 1/8 in. × (total length, ft/5)

PERMISSIBLE VARIATIONS IN CAMBER FOR CARBON STEEL UNIVERSAL MILL PLATES, HIGH-STRENGTH AND HIGH-STRENGTH LOW-ALLOY STEEL SHEARED AND GAS CUT RECTANGULAR PLATES, UNIVERSAL MILL PLATES, SPECIAL CUT PLATES

Dimension, In.		Camber for Thicknesses and Widths Given
Thickness	Width	
To 2, incl.	All	1/8 in. × (total length, ft/5)
Over 2 to 15, incl.	To 30, incl.	3/16 in. × (total length, ft/5)
Over 2 to 15, incl.	Over 30 to 60, incl.	1/4 in. × (total length, ft/5)

STANDARD MILL PRACTICE
Rectangular sheared plates and Universal mill plates

PERMISSIBLE VARIATIONS FROM FLATNESS
(High-Strength Low-Alloy and Alloy Steel, Hot Rolled or Thermally Treated)

Specified Thickness, In.	Variations from Flatness for Specified Widths: In.							
	To 36, excl.	36 to 48, excl.	48 to 60, excl.	60 to 72, excl.	72 to 84, excl.	84 to 96, excl.	96 to 108, excl.	108 to 120, excl.
To ¼, excl.	¹³⁄₁₆	1⅛	1⅜	1⅞	2	2¼	2⅜	2⅝
¼ to ⅜, excl.	¾	¹⁵⁄₁₆	1⅛	1⅜	1¾	1⅞	2	2¼
⅜ to ½, excl.	¾	⅞	¹⁵⁄₁₆	¹⁵⁄₁₆	1⅛	1⁵⁄₁₆	1½	1⅝
½ to ¾, excl.	⅝	¾	¹³⁄₁₆	⅞	1	1⅛	1¼	1⅜
¾ to 1, excl.	⅝	¾	⅞	⅞	¹⁵⁄₁₆	1	1⅛	1⁵⁄₁₆
1 to 2, excl.	⁹⁄₁₆	⅝	¾	¹³⁄₁₆	⅞	¹⁵⁄₁₆	1	1
2 to 4, excl.	½	⁹⁄₁₆	¹¹⁄₁₆	¾	¾	¾	¾	⅞
4 to 6, excl.	⁹⁄₁₆	¹¹⁄₁₆	¾	¾	⅞	⅞	¹⁵⁄₁₆	1⅛
6 to 8, excl.	⅝	¾	¾	¹⁵⁄₁₆	1	1⅛	1¼	1⁵⁄₁₆

General Notes:
1. The longer dimension specified is considered the length, and variations from a flat surface along the length should not exceed the tabular amount for the specified width in plates up to 12 ft. in length.
2. The flatness variation across the width should not exceed the tabular amount for the specified width.
3. When the longer dimension is under 36 in., the variation should not exceed ⅜ in. When the longer dimension is from 36 to 72 in., incl. the variation should not exceed 75% of the tabular amount for the specified width.

PERMISSIBLE VARIATIONS IN WIDTH FOR UNIVERSAL MILL PLATES
(15 in. and under in thickness)

Specified Width, In.	Variations Over Specified Width for Thickness, in., and Equivalent Weights, lb. per sq. ft, Given					
	To ⅜, excl.	⅜ to ⅝, excl.	⅝ to 1, excl.	1 to 2, incl.	Over 2 to 10, incl.	Over 10 to 15, incl.
	To 15.3, excl.	15.3 to 25.5, excl.	25.5 to 40.8, excl.	40.8 to 81.7, incl.	81.7 to 409.0, incl.	409.0 to 613.0, incl.
Over 8 to 20, excl.	⅛	⅛	³⁄₁₆	¼	⅜	½
20 to 36, excl.	³⁄₁₆	¼	⁵⁄₁₆	⅜	⁷⁄₁₆	⁹⁄₁₆
36 and over	⁵⁄₁₆	⅜	⁷⁄₁₆	½	⁹⁄₁₆	⅝

Notes: Permissible variation under specified width, ⅛ in.
Table applies to all steels listed in ASTM A6.

AMERICAN INSTITUTE OF STEEL CONSTRUCTION

PART 2
Beam and Girder Design

AMERICAN INSTITUTE OF STEEL CONSTRUCTION

BEAM DIAGRAMS AND FORMULAS

Notes

ALLOWABLE STRESS DESIGN SELECTION TABLE
For shapes used as beams
S_x

This table is provided to facilitate the selection of flexural members designed on the basis of allowable bending stress in accordance with F1 of the AISC ASD Specification. It includes only W and M shapes used as beams. A beam can be selected by entering the table with either the required section modulus, or with the design bending moment, and comparing these with the tabulated values of S_x and M_R, respectively.

The table is applicable to adequately braced beams for which maximum limiting values of allowable stress are permitted by the AISC ASD Specification. For beams not meeting these bracing requirements, the charts of Allowable Moments in Beams with Unbraced Lengths Greater than L_u (Manual Part 2) are recommended.

For most loading conditions, it is convenient to use the selection table. However, for adequately braced simply supported beams with a uniform load over the full length, or equivalent symmetrical loading, the Allowable Uniform Load Tables (Manual Part 2) can also be used.

In this table, the shapes are listed in groups by descending order of section modulus S_X and include corresponding values of F_y' and detailing depth d.

Included also for steels of F_y = 36 ksi and F_y = 50 ksi are values for the maximum resisting moment M_R and the limiting values of unbraced lengths L_c and L_u. The lightest shape is listed at the top of each group, and is shown in boldface type.

The values of M_R are valid for beams with unbraced lengths less than or equal to L_c. When the values of L_c do not appear, the M_R values are valid for unbraced lengths up to L_u.

The symbols used in this table are:

S_x = elastic section modulus, X-X axis, in.^{3}

F_y' = theoretical yield stress at which the shape becomes noncompact, as defined by flange criteria [Sect. B5.1], ksi

L_c = maximum unbraced length, in feet, of the compression flange at which the allowable bending stress may be taken at $0.66F_y$, or from Equation (F1-3) when applicable.

L_u = maximum unbraced length, in feet, of the compression flange for which the allowable bending stress may be taken at $0.60F_y$ when C_b = 1.

M_R = beam resisting moment $F_b S_x/12$, kip-ft, where
$F_b = 0.66F_y$, if shape has compact sections
$F_b = F_y[0.79 - 0.002(b_f/2t_f)\sqrt{F_y}]$, if shape has noncompact flanges

USE OF THE TABLE

Determine the required elastic section modulus S_x from the maximum design moment, using the appropriate F_b for the desired yield strength steel. Enter the column headed S_x and find a value equal to or larger than the section modulus required. Alternately, enter the M_R column and find a value of M_R equal to or greater than the design moment. The beam opposite this value in the shape column, and all beams above it, have sufficient bending capacity. The first beam that appears in boldface type adjacent to or above the required S_x or M_R is the lightest that will serve for the yield strength stated. If the beam must not exceed a certain depth, proceed up the column headed "Shape" until a beam within the required depth is reached; then check to see that no lighter beam of the same depth appears higher in the column.

After a shape has been selected, the following checks should be made: The lateral bracing of the compression flange should be spaced no greater than L_c when an allowable stress of $0.66F_y$ or an allowable stress determined from Equation F1-3 was used in calculating the required S_x, or when M_R value is used as a basis for design. The spacing should be no greater than L_u when an allowable stress of $0.6F_y$ was used in calculating the required S_x. For beams with unbraced lengths greater than these limits, it is recommended that the charts of Allowable Moments in Beams with Unbraced Lengths Greater than L_u be used. A check should be made for web shear capacity of the selected beam by referring to the Allowable Uniform Load Tables or by use of the formula $V = F_v dt$. Also, if a deflection limitation exists, the adequacy of the selected beam should be checked.

Where torsional or other special loading conditions occur, proper provisions must be made in the design. Consult appropriate references for such conditions.

EXAMPLE 1

ASD
Specification
Reference

Given:

Select a beam of $F_y = 36$ ksi steel subjected to a bending moment of 125 kip-ft, having its compression flange braced at 6.0-ft intervals.

Solution (S_x method):

Assume $F_b = 0.66F_y = 23.8$ ksi.

$$S_x \text{ (req'd)} = \frac{M}{F_b} = \frac{125 \times 12}{23.8} = 63.0 \text{ in.}^3$$

Enter the Allowable Stress Design Selection Table and find the nearest tabulated value of S_x is 64.7 in.^{3}, which corresponds to a W16×40 or a W12×50. Try the W16×40 since it is in boldface type.

A check of the F'_y column shows a dash, indicating F'_y is greater than 65 ksi. Therefore, the shape is compact. | **B5.1**

From the table, $L_c = 7.4$ ft > 6.0 ft. ∴ the bracing is adequate and the assumed allowable stress of $0.66F_y$ is correct. | **F1.1**

Use: W16×40

Alternate Solution (M_R method):

Enter the column of M_R values and note the tabulated value nearest the design moment is 128 kip-ft, which corresponds to a W16×40 or a W12×50. A W16×40 is the lightest suitable shape.

Observe that $L_c = 7.4$ ft > 6.0 ft. ∴ M_R is valid.

Use: W16×40

EXAMPLE 2

Given:

Determine the moment capacity of a W16×40 of F_y = 36 ksi steel with the compression flange braced at intervals of 9.0 ft.

Solution:

Enter the Allowable Stress Design Selection Table and note:

L_u = 10.2 ft and L_c = 7.4 ft

L_u > 9.0 ft > L_c ; F_b = 0.60F_y = 21.6 ksi

S_x = 64.7 in.^{3}

$$M = \frac{21.6 \times 64.7}{12} = 116.5 \text{ kip-ft}$$

ASD
Specification
Reference

F1.1
and
F1.3

EXAMPLE 3

Given:

Select a beam of F_y = 50 ksi steel subjected to a bending moment of 20 kip-ft having its compression flange braced at 3.0-ft. intervals.

Solution (S_x method):

Assume F_b = 0.66F_y = 33 ksi

$$S_x \text{ (req'd)} = \frac{M}{F_b} = \frac{20 \times 12}{33} = 7.27 \text{ in.}^3$$

Enter the Allowable Stress Design Selection Table and note the nearest tabulated value of S_x is 7.31 in.^{3} for a W6×12, which is not in boldface type and therefore is not the lightest section.

The lightest shape in the group is an M10×9; however, the L_c and L_u values of 1.9 ft and 2.3 ft, respectively, are less than the required 3.0 ft. The next lightest shape is a W8×10 with L_c = 3.4 ft.

F1.1

A check of the F_y' column shows a value of 45.8 ksi. Since F_y' is less than 50 ksi, the shape is noncompact due to flange criteria. Therefore, the allowable stress is less than 0.66F_y and must be determined from Equation (F1-3). From Properties Tables for W shapes, Part 1, $b_f/2t_f$ for a W8×10 equals 9.6 and the allowable stress is determined to be F_b = 32.7 ksi.

B5.1
F1.2
Equation
(F1-3)

$$f_b = \frac{20 \times 12}{7.81} = 30.7 \text{ ksi} < 32.7 \text{ ksi}$$

Use: W8×10

Solution (M_R method):

Enter the Selection Table in the column of M_R values for F_y = 50 ksi and note the value of M_R = 21 kip-ft. for a W8×10 is greater than the applied bending moment of 20 kip-ft.

Use: W8×10

ALLOWABLE STRESS DESIGN SELECTION TABLE
For shapes used as beams

S_x

F_y = 50 ksi			S_x	Shape	Depth d	F'_y	F_y = 36 ksi		
L_c	L_u	M_R					L_c	L_u	M_R
Ft	Ft	Kip-ft	In.^{3}		In.	Ksi	Ft	Ft	Kip-ft
16.2	64.1	8720	3170	W 36×848	42½	—	19.1	89.0	6280
16.1	61.7	8200	2980	W 36×798	42	—	19.0	85.7	5900
15.9	56.5	7400	2690	W 36×720	41¼	—	18.8	78.5	5330
15.1	45.7	7120	2590	W 40×655	43⅝	—	17.8	63.4	5130
15.7	51.3	6660	2420	W 36×650	40½	—	18.6	71.2	4790
14.9	41.7	6440	2340	W 40×593	43	—	17.6	57.9	4630
15.6	46.9	6000	2180	W 36×588	39⅞	—	18.4	65.2	4320
15.1	52.1	5970	2170	W 33×619	38½	—	17.8	72.3	4300
14.8	37.9	5750	2090	W 40×531	42⅜	—	17.4	52.6	4140
15.0	48.3	5470	1990	W 33×567	37⅞	—	17.7	67.1	3940
15.4	42.7	5360	1950	W 36×527	39¼	—	18.2	59.4	3860
14.7	34.4	5200	1890	W 40×480	41¾	—	17.3	47.7	3740
14.5	53.8	5140	1870	W 30×581	35⅜	—	17.1	74.7	3700
14.9	44.4	4980	1810	W 33×515	37⅜	—	17.5	61.7	3580
15.3	39.2	4920	1790	W 36×485	38¾	—	18.1	54.5	3540
14.5	31.4	4700	1710	W 40×436	41⅜	—	17.1	43.7	3390
14.3	49.8	4620	1680	W 30×526	34¾	—	16.9	69.1	3330
14.7	40.7	4480	1630	W 33×468	36¾	—	17.4	56.5	3230
15.2	36.2	4460	1620	W 36×439	38¼	—	17.9	50.3	3210
13.7	55.6	4320	1570	W 27×539	32½	—	16.1	77.2	3110
14.4	29.0	4290	1560	W 40×397	41	—	17.0	40.3	3090
14.2	45.7	4210	1530	W 30×477	34¼	—	16.7	63.4	3030
14.6	37.0	4070	1480	W 33×424	36⅜	—	17.2	51.4	2930
15.1	32.7	3990	1450	W 36×393	37¾	—	17.8	45.4	2870
13.5	51.3	3960	1440	W 27×494	32	—	15.9	71.2	2850
14.3	26.5	3910	1420	W 40×362	40½	—	16.9	36.7	2810
14.1	41.7	3800	1380	W 30×433	33⅝	—	16.6	57.9	2730
14.5	34.4	3710	1350	W 33×387	36	—	17.1	47.7	2670
16.0	25.8	3690	1340	W 40×328	40	—	18.9	35.9	2650
15.0	30.0	3630	1320	W 36×359	37⅜	—	17.7	41.7	2610
13.4	47.6	3580	1300	W 27×448	31⅜	—	15.8	66.1	2570
12.6	56.5	3550	1290	W 24×492	29⅝	—	14.9	78.5	2550

ALLOWABLE STRESS DESIGN SELECTION TABLE

S_x For shapes used as beams

F_y = 50 ksi			S_x	Shape	Depth d	F'_y	F_y = 36 ksi		
L_c	L_u	M_R					L_c	L_u	M_R
Ft	Ft	Kip-ft	In.^{3}		In.	Ksi	Ft	Ft	Kip-ft
14.2	**23.8**	**3520**	**1280**	**W 40×324**	**40⅛**	—	**16.8**	**33.1**	**2530**
14.0	38.3	3440	1250	W 30×391	33¼	—	16.5	53.2	2480
14.4	31.4	3380	1230	W 33×354	35½	—	17.0	43.7	2440
16.0	**23.6**	**3360**	**1220**	**W 40×298**	**39¾**	—	**18.8**	**32.8**	**2420**
14.9	27.5	3330	1210	W 36×328	37⅛	—	17.6	38.3	2400
14.2	**21.8**	**3220**	**1170**	**W 40×297**	**39⅞**	—	**16.7**	**30.3**	**2320**
13.3	43.3	3220	1170	W 27×407	30⅞	—	15.6	60.1	2320
12.5	52.1	3220	1170	W 24×450	29⅛	—	14.7	72.3	2320
13.9	35.1	3140	1140	W 30×357	32¾	—	16.3	48.7	2260
10.6	**15.9**	**3080**	**1120**	**W 44×285**	**44**	—	**12.5**	**22.0**	**2220**
14.9	25.4	3050	1110	W 36×300	36¾	—	17.6	35.3	2200
14.3	28.7	3050	1110	W 33×318	35⅛	—	16.9	39.9	2200
14.2	**21.0**	**3030**	**1100**	**W 40×277**	**39¾**	—	**16.7**	**29.1**	**2180**
15.9	**21.2**	**3000**	**1090**	**W 40×268**	**39⅜**	—	**18.7**	**29.5**	**2160**
13.1	39.7	2920	1060	W 27×368	30⅜	—	15.5	55.1	2100
12.4	48.3	2920	1060	W 24×408	28½	—	14.6	67.1	2100
14.9	23.8	2830	1030	W 36×280	36½	—	17.5	33.1	2040
13.8	32.4	2830	1030	W 30×326	32⅜	—	16.2	44.9	2040
14.2	26.2	2780	1010	W 33×291	34⅞	—	16.8	36.5	2000
14.1	**18.9**	**2730**	**992**	**W 40×249**	**39⅜**	—	**16.6**	**26.3**	**1960**
15.9	**19.0**	**2700**	**983**	**W 40×244**	**39**	—	**18.7**	**26.5**	**1950**
10.6	14.2	2700	983	W 44×248	43⅝	—	12.5	19.8	1950
13.0	37.0	2670	970	W 27×336	30	—	15.4	51.4	1920
12.2	44.4	2630	957	W 24×370	28	—	14.4	61.7	1890
14.8	21.9	2620	953	W 36×260	36¼	—	17.5	30.5	1890
12.0	53.8	2580	937	W 21×402	26	—	14.1	74.7	1860
13.7	29.5	2550	928	W 30×292	32	—	16.1	41.0	1840
14.2	24.0	2520	917	W 33×263	34½	—	16.7	33.3	1820
14.8	20.6	2460	895	W 36×245	36⅛	—	17.4	28.6	1770
10.9	18.8	2460	895	W 36×256	37⅜	—	12.9	26.2	1770
10.6	**12.9**	**2440**	**889**	**W 44×224**	**43¼**	—	**12.5**	**17.9**	**1760**
12.9	34.0	2430	884	W 27×307	29⅝	—	15.2	47.2	1750
12.1	40.7	2380	864	W 24×335	27½	—	14.3	56.5	1710
14.1	**16.4**	**2360**	**858**	**W 40×215**	**39**	—	**16.6**	**22.8**	**1700**
15.9	17.2	2360	858	W 40×221	38⅝	61.1	18.7	22.6	1700
11.9	49.8	2330	846	W 21×364	25½	—	14.0	69.1	1680
14.8	19.3	2300	837	W 36×230	35⅞	—	17.4	26.8	1660
14.2	21.6	2280	829	W 33×241	34⅛	—	16.7	30.1	1640
13.6	26.5	2270	827	W 30×261	31⅝	—	16.0	36.7	1640
12.9	31.4	2230	811	W 27×281	29¼	—	15.1	43.7	1610
10.9	17.1	2220	809	W 36×232	37⅛	—	12.8	23.7	1600
12.0	37.5	2170	789	W 24×306	27⅛	—	14.1	52.0	1560

ALLOWABLE STRESS DESIGN SELECTION TABLE
For shapes used as beams
S_x

$F_y = 50$ ksi			S_x	Shape	Depth d	F'_y	$F_y = 36$ ksi		
L_c	L_u	M_R					L_c	L_u	M_R
Ft	Ft	Kip-ft	In.³		In.	Ksi	Ft	Ft	Kip-ft
10.6	**11.2**	**2130**	**776**	**W 44×198**	**42⅞**	—	**12.5**	**15.5**	**1540**
14.1	15.2	2110	769	W 40×199	38⅝	—	16.6	20.0	1520
11.8	45.7	2110	769	W 21×333	25	—	13.9	63.4	1520
14.2	19.8	2080	757	W 33×221	33⅞	—	16.7	27.6	1500
13.5	24.0	2050	746	W 30×235	31¼	—	15.9	33.3	1480
12.8	29.0	2040	742	W 27×258	29	—	15.1	40.3	1470
10.9	15.1	1980	719	W 36×210	36¾	—	12.9	20.9	1420
11.9	34.7	1970	718	W 24×279	26¾	—	14.0	48.2	1420
12.8	**16.7**	**1880**	**708**	**W 40×192**	**38¼**	**37.1**	**17.8**	**19.7**	**1400**
11.6	42.7	1900	692	W 21×300	24½	—	13.7	59.4	1370
14.1	17.9	1880	684	W 33×201	33⅝	—	16.6	24.9	1350
10.6	**12.3**	**1880**	**682**	**W 40×183**	**39**	—	**12.5**	**17.1**	**1350**
12.7	26.7	1850	674	W 27×235	28⅝	—	15.0	37.0	1330
10.9	13.9	1830	664	W 36×194	36½	—	12.8	19.4	1310
13.5	21.4	1820	663	W 30×211	31	—	15.9	29.7	1310
11.8	31.4	1770	644	W 24×250	26⅜	—	13.9	43.7	1280
11.5	39.2	1740	632	W 21×275	24⅛	—	13.6	54.5	1250
12.6	24.9	1720	624	W 27×217	28⅜	—	14.9	34.5	1240
10.8	49.0	1720	624	W 18×311	22⅜	—	12.7	68.1	1240
10.8	**13.1**	**1710**	**623**	**W 36×182**	**36⅜**	—	**12.7**	**18.2**	**1230**
10.4	**11.0**	**1650**	**599**	**W 40×167**	**38⅝**	—	**12.5**	**14.5**	**1190**
13.5	19.4	1640	598	W 30×191	30⅝	—	15.9	26.9	1180
11.7	29.0	1620	588	W 24×229	26	—	13.8	40.3	1160
10.8	12.2	1600	580	W 36×170	36⅛	—	12.7	17.0	1150
11.4	35.5	1560	569	W 21×248	23¾	—	13.5	49.3	1130
10.6	45.0	1550	564	W 18×283	21⅞	—	12.6	62.6	1120
12.6	22.4	1530	556	W 27×194	28⅛	—	14.8	31.1	1100
10.3	13.8	1510	549	W 33×169	33⅞	—	12.1	19.2	1090
10.7	**11.4**	**1490**	**542**	**W 36×160**	**36**	—	**12.7**	**15.7**	**1070**
13.4	17.5	1480	539	W 30×173	30½	—	15.8	24.2	1070
11.7	26.5	1460	531	W 24×207	25¾	—	13.7	36.7	1050
10.5	42.2	1410	514	W 18×258	21½	—	12.4	58.6	1020
8.5	**10.7**	**1410**	**512**	**W 40×149**	**38¼**	—	**11.9**	**12.6**	**1010**
11.4	32.7	1400	510	W 21×223	23⅜	—	13.4	45.4	1010
10.5	11.3	1390	504	W 36×150	35⅞	—	12.6	14.6	998
12.6	20.1	1380	502	W 27×178	27¾	—	14.9	27.9	994
11.6	24.7	1350	491	W 24×192	25½	—	13.7	34.3	972
10.4	12.2	1340	487	W 33×152	33½	—	12.2	16.9	964
10.4	38.8	1280	466	W 18×234	21	—	12.3	53.8	923
11.3	29.8	1270	461	W 21×201	23	—	13.3	41.3	913
12.6	18.3	1250	455	W 27×161	27⅝	—	14.8	25.4	901
11.5	22.8	1240	450	W 24×176	25¼	—	13.6	31.7	891

ALLOWABLE STRESS DESIGN SELECTION TABLE
S_x
For shapes used as beams

F_y = 50 ksi			S_x	Shape	Depth d	F_y'	F_y = 36 ksi		
L_c	L_u	M_R					L_c	L_u	M_R
Ft	Ft	Kip-ft	In.³		In.	Ksi	Ft	Ft	Kip-ft
10.3	11.1	1230	448	W 33×141	33¼	—	12.2	15.4	887
8.8	11.0	1210	439	W 36×135	35½	—	12.3	13.0	869
9.4	13.4	1200	436	W 30×148	30⅝	—	11.1	18.7	863
10.3	35.5	1150	419	W 18×211	20⅝	—	12.2	49.3	830
11.2	27.1	1150	417	W 21×182	22¾	—	13.2	37.6	826
11.6	21.1	1140	414	W 24×162	25	—	13.7	29.3	820
12.5	16.6	1130	411	W 27×146	27⅜	—	14.7	23.0	814
9.9	10.8	1120	406	W 33×130	33⅛	—	12.1	13.8	804
9.4	11.6	1050	380	W 30×132	30¼	—	11.1	16.1	752
11.1	25.1	1050	380	W 21×166	22½	—	13.1	34.8	752
10.3	32.7	1050	380	W 18×192	20⅜	—	12.1	45.4	752
11.6	18.9	1020	371	W 24×146	24¾	—	13.6	26.3	735
8.6	10.7	987	359	W 33×118	32⅞	—	12.0	12.6	711
9.4	10.8	976	355	W 30×124	30⅛	—	11.1	15.0	703
9.0	13.3	949	345	W 27×129	27⅝	—	10.6	18.4	683
10.2	30.0	946	344	W 18×175	20	—	12.0	41.7	681
9.4	9.9	905	329	W 30×116	30	—	11.1	13.8	651
11.5	16.8	905	329	W 24×131	24½	—	13.6	23.4	651
11.2	21.8	905	329	W 21×147	22	—	13.2	30.3	651
10.1	27.5	853	310	W 18×158	19¾	—	11.9	38.3	614
8.9	9.8	822	299	W 30×108	29⅞	—	11.1	12.3	592
9.0	11.5	822	299	W 27×114	27¼	—	10.6	15.9	592
11.1	19.6	811	295	W 21×132	21⅞	—	13.1	27.2	584
11.5	14.9	800	291	W 24×117	24¼	—	13.5	20.8	576
10.0	25.3	776	282	W 18×143	19½	—	11.8	35.1	558
11.1	18.3	751	273	W 21×122	21⅝	—	13.1	25.4	541
7.9	9.7	740	269	W 30× 99	29⅝	—	10.9	11.4	533
9.0	10.2	734	267	W 27×102	27⅛	—	10.6	14.2	529
11.4	13.2	710	258	W 24×104	24	58.5	13.5	18.4	511
10.0	23.1	704	256	W 18×130	19¼	—	11.8	32.2	507
11.1	16.8	685	249	W 21×111	21½	—	13.0	23.3	493
7.2	9.6	674	245	W 30× 90	29½	58.1	10.0	11.4	485
8.1	12.0	674	245	W 24×103	24½	—	9.5	16.7	485
8.9	9.5	668	243	W 27× 94	26⅞	—	10.5	12.8	481
10.1	21.0	635	231	W 18×119	19	—	11.9	29.1	457
11.0	15.4	624	227	W 21×101	21⅜	—	13.0	21.3	449
8.1	10.9	611	222	W 24× 94	24¼	—	9.6	15.1	440
8.0	9.4	586	213	W 27× 84	26¾	—	10.5	11.0	422
10.0	18.7	561	204	W 18×106	18¾	—	11.8	26.0	404
8.1	9.6	539	196	W 24× 84	24⅛	—	9.5	13.3	388
7.5	12.1	528	192	W 21× 93	21⅝	—	8.9	16.8	380
13.1	31.7	523	190	W 14×120	14½	—	15.5	44.1	376
10.0	17.4	517	188	W 18× 97	18⅝	—	11.8	24.1	372

ALLOWABLE STRESS DESIGN SELECTION TABLE
For shapes used as beams S_x

F_y = 50 ksi			S_x	Shape	Depth d	F'_y	F_y = 36 ksi		
L_c	L_u	M_R					L_c	L_u	M_R
Ft	Ft	Kip-ft	In.³		In.	Ksi	Ft	Ft	Kip-ft
8.1	**8.6**	**484**	**176**	**W 24× 76**	**23⅞**	**—**	**9.5**	**11.8**	**348**
9.3	20.2	481	175	W 16×100	17	—	11.0	28.1	347
13.1	29.2	476	173	W 14×109	14⅜	58.6	15.4	40.6	343
7.5	10.9	470	171	W 21× 83	21⅜	—	8.8	15.1	339
9.9	15.5	457	166	W 18× 86	18⅜	—	11.7	21.5	329
13.0	26.7	432	157	W 14× 99	14⅛	48.5	15.4	37.0	311
9.3	18.0	426	155	W 16× 89	16¾	—	10.9	25.0	307
7.4	**8.5**	**424**	**154**	**W 24× 68**	**23¾**	**—**	**9.5**	**10.2**	**305**
7.4	9.6	415	151	W 21× 73	21¼	—	8.8	13.4	299
9.9	13.7	402	146	W 18× 76	18¼	64.2	11.6	19.1	289
13.0	24.5	385	143	W 14× 90	14	40.4	15.3	34.0	283
7.4	**8.9**	**385**	**140**	**W 21× 68**	**21⅛**	**—**	**8.7**	**12.4**	**277**
9.2	15.8	369	134	W 16× 77	16½	—	10.9	21.9	265
5.8	**6.4**	**360**	**131**	**W 24× 62**	**23¾**	**—**	**7.4**	**8.1**	**259**
7.4	**8.1**	**349**	**127**	**W 21× 62**	**21**	**—**	**8.7**	**11.2**	**251**
6.8	11.1	349	127	W 18× 71	18½	—	8.1	15.5	251
9.1	20.2	338	123	W 14× 82	14¼	—	10.7	28.1	244
10.9	26.0	325	118	W 12× 87	12½	—	12.8	36.2	234
6.8	10.4	322	117	W 18× 65	18⅜	—	8.0	14.4	232
9.2	13.9	322	117	W 16× 67	16⅜	—	10.8	19.3	232
5.0	**6.3**	**314**	**114**	**W 24× 55**	**23⅝**	**—**	**7.0**	**7.5**	**226**
9.0	18.6	308	112	W 14× 74	14⅛	—	10.6	25.9	222
5.9	6.7	305	111	W 21× 57	21	—	6.9	9.4	220
6.8	9.6	297	108	W 18× 60	18¼	—	8.0	13.3	214
10.8	24.0	294	107	W 12× 79	12⅜	62.6	12.8	33.3	212
9.0	17.2	283	103	W 14× 68	14	—	10.6	23.9	204
6.7	**8.7**	**270**	**98.3**	**W 18× 55**	**18⅛**	**—**	**7.9**	**12.1**	**195**
10.8	21.9	268	97.4	W 12× 72	12¼	52.3	12.7	30.5	193
5.6	**6.0**	**260**	**94.5**	**W 21× 50**	**20⅞**	**—**	**6.9**	**7.8**	**187**
6.4	10.3	254	92.2	W 16× 57	16⅜	—	7.5	14.3	183
9.0	15.5	254	92.2	W 14× 61	13⅞	—	10.6	21.5	183
6.7	**7.9**	**244**	**88.9**	**W 18× 50**	**18**	**—**	**7.9**	**11.0**	**176**
10.7	20.0	238	87.9	W 12× 65	12⅛	43.0	12.7	27.7	174
4.7	**5.9**	**224**	**81.6**	**W 21× 44**	**20⅝**	**—**	**6.6**	**7.0**	**162**
6.3	9.1	223	81.0	W 16× 50	16¼	—	7.5	12.7	160
5.4	6.8	217	78.8	W 18× 46	18	—	6.4	9.4	156
9.0	17.5	215	78.0	W 12× 58	12¼	—	10.6	24.4	154
7.2	12.7	214	77.8	W 14× 53	13⅞	—	8.5	17.7	154
6.3	8.2	200	72.7	W 16× 45	16⅛	—	7.4	11.4	144
9.0	15.9	194	70.6	W 12× 53	12	55.9	10.6	22.0	140
7.2	11.5	193	70.3	W 14× 48	13¾	—	8.5	16.0	139

ALLOWABLE STRESS DESIGN SELECTION TABLE
S_x For shapes used as beams

F_y = 50 ksi			S_x	Shape	Depth d	F_y'	F_y = 36 ksi		
L_c	L_u	M_R					L_c	L_u	M_R
Ft	Ft	Kip-ft	In.³		In.	Ksi	Ft	Ft	Kip-ft
5.4	5.9	188	68.4	W 18×40	17⅞	—	6.3	8.2	135
9.0	22.4	183	66.7	W 10×60	10¼	—	10.6	31.1	132
6.3	7.4	178	64.7	W 16×40	16	—	7.4	10.2	128
7.2	14.1	178	64.7	W 12×50	12¼	—	8.5	19.6	128
7.2	10.4	172	62.7	W 14×43	13⅝	—	8.4	14.4	124
9.0	20.3	165	60.0	W 10×54	10⅛	63.5	10.6	28.2	119
7.2	12.8	160	58.1	W 12×45	12	—	8.5	17.7	115
4.8	5.6	158	57.6	W 18×35	17¾	—	6.3	6.7	114
6.3	6.7	155	56.5	W 16×36	15⅞	64.0	7.4	8.8	112
6.1	8.3	150	54.6	W 14×38	14⅛	—	7.1	11.5	108
9.0	18.7	150	54.6	W 10×49	10	53.0	10.6	26.0	108
7.2	11.5	143	51.9	W 12×40	12	—	8.4	16.0	103
7.2	16.4	135	49.1	W 10×45	10⅛	—	8.5	22.8	97
6.0	7.3	134	48.6	W 14×34	14	—	7.1	10.2	96
4.9	5.2	130	47.2	W 16×31	15⅞	—	5.8	7.1	93
5.9	9.1	125	45.6	W 12×35	12½	—	6.9	12.6	90
7.2	14.2	116	42.1	W 10×39	9⅞	—	8.4	19.8	83
6.0	6.5	116	42.0	W 14×30	13⅞	55.3	7.1	8.7	83
5.8	7.8	106	38.6	W 12×30	12⅜	—	6.9	10.8	76
4.0	5.1	106	38.4	W 16×26	15¾	—	5.6	6.0	76
4.5	5.1	97	35.3	W 14×26	13⅞	—	5.3	7.0	70
7.1	11.9	96	35.0	W 10×33	9¾	50.5	8.4	16.5	69
5.8	6.7	92	33.4	W 12×26	12¼	57.9	6.9	9.4	66
5.2	9.4	89	32.4	W 10×30	10½	—	6.1	13.1	64
7.2	16.3	86	31.2	W 8×35	8⅛	64.4	8.5	22.6	62
4.1	4.7	80	29.0	W 14×22	13¾	—	5.3	5.6	57
5.2	8.2	77	27.9	W 10×26	10⅜	—	6.1	11.4	55
7.2	14.5	76	27.5	W 8×31	8	50.0	8.4	20.1	54
3.6	4.6	70	25.4	W 12×22	12¼	—	4.3	6.4	50
5.9	12.6	67	24.3	W 8×28	8	—	6.9	17.5	48
5.2	6.8	64	23.2	W 10×22	10⅛	—	6.1	9.4	46
3.6	3.8	59	21.3	W 12×19	12⅛	—	4.2	5.3	42
2.6	3.4	58	21.1	M 14×18	14	—	3.6	4.0	42
5.8	10.9	57	20.9	W 8×24	7⅞	64.1	6.9	15.2	41
3.6	5.2	52	18.8	W 10×19	10¼	—	4.2	7.2	37
4.7	8.5	50	18.2	W 8×21	8¼	—	5.6	11.8	36

ALLOWABLE STRESS DESIGN SELECTION TABLE
For shapes used as beams S_x

F_y = 50 ksi			S_x	Shape	Depth d	F_y'	F_y = 36 ksi		
L_c	L_u	M_R					L_c	L_u	M_R
Ft	Ft	Kip-ft	In³		In	Ksi	Ft	Ft	Kip-ft
2.9	**3.6**	**47**	**17.1**	**W 12×16**	**12**	—	**4.1**	**4.3**	**34**
5.4	14.4	46	16.7	W 6×25	6⅜	—	6.4	20.0	33
3.6	4.4	45	16.2	W 10×17	10⅛	—	4.2	6.1	32
4.7	7.1	42	15.2	W 8×18	8⅛	—	5.5	9.9	30
2.5	**3.6**	**41**	**14.9**	**W 12×14**	**11⅞**	**54.3**	**3.5**	**4.2**	**30**
3.6	3.7	38	13.8	W 10×15	10	—	4.2	5.0	27
5.4	11.8	37	13.4	W 6×20	6¼	62.1	6.4	16.4	27
5.3	12.5	36	13.0	M 6×20	6	—	6.3	17.4	26
1.9	**2.6**	**33**	**12.0**	**M 12×11.8**	**12**	—	**2.7**	**3.0**	**24**
3.6	5.2	32	11.8	W 8×15	8⅛	—	4.2	7.2	23
2.8	3.6	30	10.9	W 10×12	9⅞	47.5	3.9	4.3	22
1.8	**2.6**	**30**	**10.9**	**M 12×10.8**	**12**	—	**2.5**	**3.1**	**22**
1.6	**2.8**	**28**	**10.3**	**M 12×10**	**12**	—	**2.3**	**3.3**	**20**
3.6	8.7	28	10.2	W 6×16	6¼	—	4.3	12.0	20
4.5	14.0	28	10.2	W 5×19	5⅛	—	5.3	19.5	20
3.6	4.3	27	9.91	W 8×13	8	—	4.2	5.9	20
5.4	8.7	25	9.72	W 6×15	6	31.8	6.3	12.0	19
4.5	13.9	26	9.63	M 5×18.9	5	—	5.3	19.3	19
4.5	12.0	23	8.51	W 5×16	5	—	5.3	16.7	17
3.4	**3.7**	**21**	**7.81**	**W 8×10**	**7⅞**	**45.8**	**4.2**	**4.7**	**15**
1.9	**2.3**	**21**	**7.76**	**M 10× 9**	**10**	—	**2.6**	**2.7**	**15**
3.6	6.2	20	7.31	W 6×12	6	—	4.2	8.6	14
1.6	**2.3**	**19**	**6.94**	**M 10× 8**	**10**	—	**2.3**	**2.7**	**14**
1.6	**2.3**	**18**	**6.57**	**M 10× 7.5**	**10**	—	**2.2**	**2.7**	**13**
3.5	4.8	15	5.56	W 6× 9	5⅞	50.3	4.2	6.7	11
3.6	11.2	15	5.46	W 4×13	4⅛	—	4.3	15.6	11
1.8	**2.0**	**13**	**4.62**	**M 8× 6.5**	**8**	—	**2.4**	**2.5**	**9**
1.7	**1.8**	**7**	**2.40**	**M 6× 4.4**	**6**	—	**1.9**	**2.4**	**5**

PLASTIC DESIGN SELECTION TABLE
For W and M shapes

$$Z_x$$

When plastic design is used in proportioning continuous beams and structural frames, bending capacity based on ultimate strength is determined by the plastic section modulus of a shape. Fundamentals of plastic design are discussed in various publications, including *Plastic Design of Braced Multistory Frames*, published by the American Iron and Steel Institute in cooperation with AISC.

The AISC ASD Specification permits plastic design with steels of yield strengths up to 65 ksi. Section N2 of the AISC ASD Specification lists the ASTM steels that may be used.

In this table, the plastic section modulus Z_x has been tabulated for hot-rolled shapes which satisfy the requirements of Chapter N of the AISC ASD Specification. Included are W and M shapes of $F_y = 36$ ksi and $F_y = 50$ ksi steel. When no axial load is present, all shapes included in the table can be classified as "plastic design sections" except for those shapes of $F_y = 50$ ksi steel where the values of M_p and P_y do not appear. When axial load is present, shapes marked with an asterisk (*) must be checked for compliance with Equations (N7-1) and (N7-2). Additionally, the tabulated values are valid only for members laterally braced in accordance with AISC ASD Specification Sect. N9.

The use of the Plastic Design Selection Table in determining the lightest shape for the design requirements is similar to the procedure previously outlined for the Allowable Stress Design Selection Table. The boldface type identifies the shapes that are the lightest in weight in each group.

The symbols used in the table are defined below:

Z_x = plastic section modulus, X-X axis, in.[3]

A = area of the shape, in.[2]

d/t_w = depth-thickness ratio of the web. Used to check compliance with Equations (N7-1) and (N7-2).

r_x = radius of gyration with respect to the X-X axis, in. Used in determining the slenderness ratio about the X-X axis.

r_y = radius of gyration with respect to the Y-Y axis, in. Used in determining the slenderness ratio about the Y-Y axis. Also used to determine P_{cr} and M_m, and to determine the lateral bracing requirements in accordance with N9 of the AISC ASD Specification.

M_p = plastic moment, kip-ft, $= (F_y \times Z_x)/12$

P_y = plastic axial load, kips, $= (F_y \times A)$

Z_x

PLASTIC DESIGN SELECTION TABLE
For W and M shapes

$F_y = 50$ ksi		A	Z_x	Shape	$\dfrac{d}{t_w}$	r_x	r_y	$F_y = 36$ ksi	
M_p	P_y							M_p	P_y
Kip-ft	Kip	In.^{2}	In.^{3}			In.	In.	Kip-ft	Kip
16000	12450	249	3830	W 36×848	16.8	16.4	4.27	11490	8960
14900	11700	234	3570	W 36×798	17.6	16.4	4.24	10710	8420
13300	10550	211	3190	W 36×720	19.0	16.2	4.18	9570	7600
12800	9600	192	3060	W 40×655	22.1	17.2	3.86	9180	6910
11800	9500	190	2840	W 36×650	20.5	16.0	4.12	8520	6840
11500	8700	174	2750	W 40×593	24.0	17.0	3.81	8250	6260
10700	9050	181	2560	W 33×619	19.5	15.2	3.98	7680	6520
10600	8600	172	2550	W 36×588	22.3	15.9	4.07	7650	6190
10200	7800	156	2450	W 40×531	26.3	16.9	3.75	7350	5620
9710	8300	166	2330	W 33×567	20.9	15.1	3.94	6990	5980
9460	7700	154	2270	W 36×527	24.4	15.8	4.02	6810	5540
9210	8500	170	2210	W 30×581	18.0	13.9	3.86	6630	6120
9080	7000	140	2180	W 40×480	28.6	16.8	3.72	6540	5040
8790	7550	151	2110	W 33×515	22.6	14.9	3.89	6330	5440
8630	7100	142	2070	W 36×485	25.8	15.6	3.98	6210	5110
8290	7700	154	1990	W 30×526	19.4	13.8	3.80	5970	5540
8250	6400	128	1980	W 40×436	30.9	16.6	3.67	5940	4610
7880	6850	137	1890	W 33×468	24.2	14.8	3.85	5670	4930
7830	7900	158	1880	W 27×539	16.5	12.7	3.66	5640	5690
7750	6400	128	1860	W 36×439	28.1	15.6	3.95	5580	4610
7460	5800	116	1790	W 40×397	33.6	16.6	3.65	5370	4180
7460	7000	140	1790	W 30×477	21.0	13.7	3.75	5370	5040
7130	7250	145	1710	W 27×494	17.7	12.6	3.61	5130	5220
7080	6200	124	1700	W 33×424	26.3	14.7	3.81	5100	4460
6920	5750	115	1660	W 36×393	31.0	15.5	3.90	4980	4140
6920	10750	215	1660	W 14×730	7.3	8.17	4.69	4980	7740
6790	5300	106	1630	W 40×362	36.2	16.5	3.61	4890	3820
6710	6350	127	1610	W 30×433	22.4	13.5	3.71	4830	4570
6460	5650	113	1550	W 33×387	28.5	14.7	3.79	4650	4070
6460	7200	144	1550	W 24×492	15.1	11.5	3.41	4650	5180
6380	6550	131	1530	W 27×448	19.0	12.5	3.57	4590	5220
6290	*4820	96.4	1510	W 40×328	44.0	16.7	4.15	4530	*3470
6290	5250	105	1510	W 36×359	33.4	15.4	3.87	4530	3780
6170	9800	196	1480	W 14×665	7.6	7.98	4.62	4440	7060

* Check shape for compliance with Equations (N7-1) or (N7-2), Sect. N7, AISC ASD Specification, as applicable, when subjected to combined axial force and bending moment at ultimate loading.

PLASTIC DESIGN SELECTION TABLE
For W and M shapes
Z_x

$F_y = 50$ ksi		A	Z_x	Shape	$\dfrac{d}{t_w}$	r_x	r_y	$F_y = 36$ ksi	
M_p	P_y							M_p	P_y
Kip-ft	Kip	In.^{2}	In.^{3}		In.	In.	In.	Kip-ft	Kip
6080	*4770	95.3	1460	W 40×324	40.2	16.4	3.57	4380	3430
5960	5700	114	1430	W 30×391	24.4	13.5	3.68	4290	4100
5920	5200	104	1420	W 33×354	30.6	14.5	3.74	4260	3740
5880	6600	132	1410	W 24×450	16.1	11.4	3.36	4230	4750
5750	*4820	96.4	1380	W 36×328	36.4	15.3	3.84	4140	3470
5750	5950	119	1380	W 27×407	20.3	12.3	3.52	4140	4280
5710	*4380	87.6	1370	W 40×298	47.8	16.6	4.12	4110	*3150
5540	*4370	87.4	1330	W 40×297	42.8	16.3	3.54	3990	3150
5500	8900	178	1320	W 14×605	8.1	7.80	4.55	3960	6410
5460	*4190	83.8	1310	W 44×285	43.0	17.1	2.42	3930	*3020
5420	5200	104	1300	W 30×357	26.5	13.4	3.65	3900	3740
5290	4680	93.5	1270	W 33×318	33.8	14.4	3.71	3810	3370
5250	*4420	88.3	1260	W 36×300	38.9	15.2	3.83	3780	3180
5210	*4070	81.3	1250	W 40×277	47.8	16.4	3.58	3750	*2930
5210	5950	119	1250	W 24×408	17.3	11.3	3.33	3750	4280
5170	5400	108	1240	W 27×368	22.0	12.2	3.48	3720	3890
5080	*3940	78.8	1220	W 40×268	52.5	16.5	4.09	3660	*2840
4960	4790	95.7	1190	W 30×326	28.4	13.2	3.61	3570	3450
4920	8100	162	1180	W 14×550	8.5	7.63	4.49	3540	5830
4880	*4120	82.4	1170	W 36×280	41.3	15.1	3.81	3510	2970
4790	*3640	72.8	1150	W 44×248	50.4	17.2	2.44	3450	*2620
4790	4280	85.6	1150	W 33×291	36.3	14.4	3.69	3450	3080
4710	4940	98.7	1130	W 27×336	23.8	12.1	3.45	3390	3550
4710	5900	118	1130	W 21×402	15.0	10.2	3.27	3390	4250
4670	*3670	73.3	1120	W 40×249	52.5	16.3	3.56	3360	*2640
4670	5400	108	1120	W 24×370	18.4	11.1	3.28	3360	3890
4580	*3590	71.7	1100	W 40×244	55.0	16.4	4.04	3300	*2580
4500	*3830	76.5	1080	W 36×260	43.2	15.0	3.78	3240	*2750
4420	4290	85.7	1060	W 30×292	31.4	13.2	3.58	3180	3090
4380	7350	147	1050	W 14×500	8.9	7.48	4.43	3150	5290
4330	*3770	75.4	1040	W 36×256	39.0	14.9	2.65	3120	2710
4330	*3870	77.4	1040	W 33×263	39.7	14.3	3.66	3120	2790
4290	*3290	65.8	1030	W 44×224	55.0	17.1	2.44	3090	*2370
4250	4510	90.2	1020	W 27×307	25.5	12.0	3.42	3060	3250
4250	4920	98.4	1020	W 24×335	19.9	11.0	3.23	3060	3540
4210	*3610	72.1	1010	W 36×245	45.1	15.0	3.75	3030	*2600
4210	5350	107	1010	W 21×364	16.0	10.0	3.23	3030	3850

*Check shape for compliance with Equation (N7-1) or (N7-2), Sect. N7, AISC ASD Specification, as applicable, when subjected to combined axial force and bending moment at ultimate loading.

Z_x PLASTIC DESIGN SELECTION TABLE
For W and M shapes

F_y = 50 ksi		A	Z_x	Shape	$\dfrac{d}{t_w}$	r_x	r_y	F_y = 36 ksi	
M_p	P_y							M_p	P_y
Kip-ft	Kip	In.^{2}	In.^{3}			In.	In.	Kip-ft	Kip
—	—	64.8	967	W 40×221	54.5	16.0	3.90	2900	*2330
4010	*3170	63.3	963	W 40×215	60.0	16.2	3.54	2890	*2280
3930	*3380	67.6	943	W 36×230	47.2	14.9	3.73	2830	*2430
3920	3840	76.7	941	W 30×261	34.0	13.1	3.54	2820	2760
3910	*3550	70.9	939	W 33×241	41.2	14.1	3.63	2820	2550
3900	*3410	68.1	936	W 36×232	42.7	14.8	2.62	2810	2450
3900	6700	134	936	W 14×455	9.4	7.33	4.38	2810	4820
3890	4130	82.6	933	W 27×281	27.6	12.0	3.40	2800	2970
3840	4490	89.8	922	W 24×306	21.5	10.9	3.20	2770	3230
3810	4900	97.9	915	W 21×333	17.1	9.91	3.19	2750	3520
3760	*2900	58.0	902	W 44×198	60.5	16.9	2.41	2710	*2090
3620	6250	125	869	W 14×426	10.0	7.26	4.34	2610	4500
—	—	58.4	868	W 40×199	59.5	16.0	3.45	2600	*2100
3560	*3250	65.0	855	W 33×221	43.8	14.1	3.59	2570	*2340
3540	3790	75.7	850	W 27×258	29.6	11.9	3.37	2550	2730
3520	*3450	69.0	845	W 30×235	37.7	13.0	3.52	2540	2480
3480	4100	82.0	835	W 24×279	23.0	10.8	3.17	2510	2950
3470	*3090	61.8	833	W 36×210	44.2	14.6	2.58	2500	*2220
3400	4410	88.2	816	W 21×300	18.6	9.81	3.15	2450	3180
3340	5850	117	801	W 14×398	10.3	7.16	4.31	2400	4210
3250	*2690	53.7	781	W 40×183	60.0	15.7	2.50	2340	*1930
3220	*2960	59.1	772	W 33×201	47.1	14.0	3.56	2320	*2130
3200	3460	69.1	769	W 27×235	31.5	11.8	3.33	2310	2490
3200	*2850	57.0	767	W 36×194	47.7	14.6	2.56	2300	*2050
3140	4580	91.5	753	W 18×311	14.7	8.72	2.95	2260	3290
3120	*3100	62.0	749	W 30×211	39.9	12.9	3.49	2250	2230
3100	3680	73.5	744	W 24×250	25.3	10.7	3.14	2230	2650
3090	4040	80.8	741	W 21×275	19.8	9.71	3.12	2220	2910
3070	5450	109	736	W 14×370	10.8	7.07	4.27	2210	3920
2990	*2680	53.6	718	W 36×182	50.1	14.5	2.55	2150	*1930
2950	3190	63.8	708	W 27×217	34.3	11.8	3.32	2120	2300

* Check shape for compliance with Equations (N7-1) or (N7-2), Sect. N7, AISC ASD Specification, as applicable, when subjected to combined axial force and bending moment at ultimate loading.

PLASTIC DESIGN SELECTION TABLE
For W and M shapes

Z_x

F_y = 50 ksi		A	Z_x	Shape	$\dfrac{d}{t_w}$	r_x	r_y	F_y = 36 ksi	
M_p	P_y							M_p	P_y
Kip-ft	Kip	In.²	In.³			In.	In.	Kip-ft	Kip
2880	*2460	49.1	692	W 40×167	59.4	15.3	2.40	2080	*1770
2820	3360	67.2	676	W 24×229	27.1	10.7	3.11	2030	2420
2820	4160	83.2	676	W 18×283	15.6	8.61	2.91	2030	3000
2800	*2805	56.1	673	W 30×191	43.2	12.8	3.46	2020	*2020
2800	5050	101	672	W 14×342	11.4	6.98	4.24	2020	3640
2780	*2500	50.0	668	W 36×170	53.2	14.5	2.53	2000	*1800
2760	3640	72.8	663	W 21×248	21.6	9.63	3.09	1990	2620
2620	*2480	49.5	629	W 33×169	50.5	13.7	2.50	1890	*1780
2620	*2850	57.0	628	W 27×194	37.5	11.7	3.29	1880	2050
2600	*2350	47.0	624	W 36×160	55.4	14.4	2.50	1870	*1690
2550	3800	75.9	611	W 18×258	16.8	8.53	2.88	1830	2730
2530	3040	60.7	606	W 24×207	29.6	10.6	3.08	1820	2190
2520	*2540	50.8	605	W 30×173	46.5	12.7	3.43	1820	*1830
2510	4570	91.4	603	W 14×311	12.1	6.88	4.20	1810	3290
2510	4940	98.8	603	W 12×336	9.5	6.41	3.47	1810	3560
—	—	43.8	597	W 40×149	60.6	14.9	2.29	1790	*1580
2450	3270	65.4	589	W 21×223	23.4	9.54	3.05	1770	2350
2420	*2210	44.2	581	W 36×150	57.4	14.3	2.47	1740	*1590
2360	*2620	52.3	567	W 27×178	38.4	11.6	3.26	1700	1880
2330	*2240	44.7	559	W 33×152	52.7	13.5	2.47	1680	*1610
2330	2820	56.3	559	W 24×192	31.4	10.5	3.07	1680	2030
2290	3440	68.8	549	W 18×234	18.2	8.44	2.85	1650	2480
2260	4170	83.3	542	W 14×283	13.0	6.79	4.17	1630	3000
2240	4480	89.6	537	W 12×305	10.0	6.29	3.42	1610	3230
2210	2960	59.2	530	W 21×201	25.3	9.47	3.02	1590	2130
2140	*2080	41.6	514	W 33×141	55.0	13.4	2.43	1540	*1500
2130	*2370	47.4	512	W 27×161	41.8	11.5	3.24	1540	1710
2130	2590	51.7	511	W 24×176	33.7	10.5	3.04	1530	1860
—	—	39.7	509	W 36×135	59.3	14.0	2.38	1530	*1430
2080	*2180	43.5	500	W 30×148	47.2	12.4	2.28	1500	*1570
2040	3110	62.1	490	W 18×211	19.5	8.35	2.82	1470	2240
2030	3780	75.6	487	W 14×257	13.9	6.71	4.13	1460	2720
2000	4100	81.9	481	W 12×279	10.4	6.16	3.38	1440	2950
1980	2680	53.6	476	W 21×182	27.4	9.40	3.00	1430	1930
1950	2390	47.7	468	W 24×162	35.5	10.4	3.05	1400	1720

* Check shape for compliance with Equations (N7-1) or (N7-2), Sect. N7, AISC ASD Specification, as applicable, when subjected to combined axial force and bending moment at ultimate loading.

AMERICAN INSTITUTE OF STEEL CONSTRUCTION

Z_x PLASTIC DESIGN SELECTION TABLE
For W and M shapes

$F_y = 50$ ksi		A	Z_x	Shape	$\dfrac{d}{t_w}$	r_x	r_y	$F_y = 36$ ksi	
M_p	P_y							M_p	P_y
Kip-ft	Kip	In.²	In.³			In.	In.	Kip-ft	Kip
1950	*1920	38.3	467	W 33×130	57.1	13.2	2.39	1400	*1380
—	—	42.9	461	W 27×146	45.3	11.4	3.21	1380	*1540
1840	2820	56.4	442	W 18×192	21.2	8.28	2.79	1330	2030
1820	*1950	38.9	437	W 30×132	49.3	12.2	2.25	1310	*1400
1820	3430	68.5	436	W 14×233	15.0	6.63	4.10	1310	2470
1800	2440	48.8	432	W 21×166	30.0	9.36	2.98	1300	1760
1780	3710	74.1	428	W 12×252	11.0	6.06	3.34	1280	2670
1740	*2150	43.0	418	W 24×146	38.1	10.3	3.01	1250	1550
—	—	34.7	415	W 33×118	59.7	13.0	2.32	1250	*1250
1700	*1830	36.5	408	W 30×124	51.6	12.1	2.23	1220	*1310
1660	2570	51.3	398	W 18×175	22.5	8.20	2.76	1190	1850
1650	*1890	37.8	395	W 27×129	45.3	11.2	2.21	1190	*1360
1630	3100	62.0	390	W 14×211	16.0	6.55	4.07	1170	2230
1610	3390	67.7	386	W 12×230	11.7	5.97	3.31	1160	2440
1580	*1710	34.2	378	W 30×116	53.1	12.0	2.19	1130	*1230
1550	2160	43.2	373	W 21×147	30.6	9.17	2.95	1120	1560
1540	*1930	38.5	370	W 24×131	40.5	10.2	2.97	1110	1390
1480	2320	46.3	356	W 18×158	24.3	8.12	2.74	1070	1670
1480	2840	56.8	355	W 14×193	17.4	6.50	4.05	1070	2040
1450	3090	61.8	348	W 12×210	12.5	5.89	3.28	1040	2220
1440	*1590	31.7	346	W 30×108	54.7	11.9	2.15	1040	*1140
1430	*1680	33.5	343	W 27×114	47.9	11.0	2.18	1030	*1210
1390	1940	38.8	333	W 21×132	33.6	9.12	2.93	999	1400
—	—	34.4	327	W 24×117	44.1	10.1	2.94	981	*1240
1340	2110	42.1	322	W 18×143	26.7	8.09	2.72	966	1520
1330	2590	51.8	320	W 14×176	18.3	6.43	4.02	960	1860
—	—	29.1	312	W 30× 99	57.0	11.7	2.10	936	*1050
1300	2790	55.8	311	W 12×190	13.6	5.82	3.25	933	2010
1280	1800	35.9	307	W 21×122	36.1	9.09	2.92	921	1290
1270	*1500	30.0	305	W 27×102	52.6	11.0	2.15	915	*1080
1210	1910	38.2	291	W 18×130	28.7	8.03	2.70	873	1380
—	—	30.6	289	W 24×104	48.1	10.1	2.91	867	*1100
1200	2340	46.7	287	W 14×159	20.1	6.38	4.00	861	1680

*Check shape for compliance with Equations (N7-1) or (N7-2), Sect. N7, AISC ASD Specification, as applicable, when subjected to combined axial force and bending moment at ultimate loading.

PLASTIC DESIGN SELECTION TABLE
For W and M shapes

Z_x

F_y = 50 ksi		A	Z_x	Shape	$\dfrac{d}{t_w}$	r_x	r_y	F_y = 36 ksi	
M_p	P_y							M_p	P_y
Kip-ft	Kip	In.^{2}	In.^{3}			In.	In.	Kip-ft	Kip
—	—	26.4	283	W 30× 90	62.8	11.7	2.09	849	* 950
1170	*1520	30.3	280	W 24×103	44.6	9.96	1.99	840	*1090
—	—	32.7	279	W 21×111	39.1	9.05	2.90	837	1180
1160	*1390	27.7	278	W 27× 94	54.9	10.9	2.12	834	* 997
1150	2500	50.0	275	W 12×170	14.6	5.74	3.22	825	1800
1090	1760	35.1	261	W 18×119	29.0	7.90	2.69	783	1260
—	—	42.7	260	W 14×145	21.7	6.33	3.98	780	1540
1060	*1390	27.7	254	W 24× 94	47.2	9.87	1.98	762	* 997
—	—	29.8	253	W 21×101	42.7	9.02	2.89	759	1070
—	—	24.8	244	W 27× 84	58.1	10.7	2.07	732	* 893
1010	2240	44.7	243	W 12×152	15.8	5.66	3.19	729	1610
—	—	38.8	234	W 14×132	22.7	6.28	3.76	702	1400
958	1560	31.1	230	W 18×106	31.7	7.84	2.66	690	1120
933	*1240	24.7	224	W 24× 84	51.3	9.79	1.95	672	* 889
921	*1370	27.3	221	W 21× 93	37.3	8.70	1.84	663	983
892	2000	39.9	214	W 12×136	17.0	5.58	3.16	642	1440
—	—	35.3	212	W 14×120	24.5	6.24	3.74	636	1270
879	1430	28.5	211	W 18× 97	34.7	7.82	2.65	633	1030
833	*1120	22.4	200	W 24× 76	54.4	9.69	1.92	600	* 806
825	1470	29.4	198	W 16×100	29.0	7.10	2.51	594	1060
817	*1220	24.3	196	W 21× 83	41.6	8.67	1.83	588	875
—	—	32.0	192	W 14×109	27.3	6.22	3.73	576	1150
—	—	25.3	186	W 18× 86	38.3	7.77	2.63	558	911
775	1770	35.3	186	W 12×120	18.5	5.51	3.13	558	1270
—	—	20.1	177	W 24× 68	57.2	9.55	1.87	531	* 724
729	1310	26.2	175	W 16× 89	31.9	7.05	2.49	525	943
717	*1080	21.5	172	W 21× 73	46.7	8.64	1.81	516	* 774
683	1560	31.2	164	W 12×106	21.1	5.47	3.11	492	1120
—	—	22.3	163	W 18× 76	42.8	7.73	2.61	489	803
667	*1000	20.0	160	W 21× 68	49.1	8.60	1.80	480	* 720
638	* 910	18.2	153	W 24× 62	55.2	9.23	1.38	459	* 655
625	1130	22.6	150	W 16× 77	36.3	7.00	2.47	450	814
613	1410	28.2	147	W 12× 96	23.1	5.44	3.09	441	1020
613	1650	32.9	147	W 10×112	15.0	4.66	2.68	441	1180
604	*1040	20.8	145	W 18× 71	37.3	7.50	1.70	435	749

* Check shape for compliance with Equations (N7-1) or (N7-2), Sect. N7, AISC ASD Specification, as applicable, when subjected to combined axial force and bending moment at ultimate loading.

AMERICAN INSTITUTE OF STEEL CONSTRUCTION

Z_x
PLASTIC DESIGN SELECTION TABLE
For W and M shapes

$F_y = 50$ ksi		A	Z_x	Shape	$\dfrac{d}{t_w}$	r_x	r_y	$F_y = 36$ ksi	
M_p	P_y							M_p	P_y
Kip-ft	Kip	In.²	In.³			In.	In.	Kip-ft	Kip
600	* 915	18.3	144	W 21× 62	52.5	8.54	1.77	432	* 659
579	1210	24.1	139	W 14×182	28.1	6.05	2.48	417	868
558	* 810	16.2	134	W 24× 55	59.7	9.11	1.34	402	* 583
554	* 955	19.1	133	W 18× 65	40.8	7.49	1.69	399	688
—	—	25.6	132	W 12× 87	24.3	5.38	3.07	396	922
—	—	19.7	130	W 16× 67	41.3	6.96	2.46	390	709
542	1470	29.4	130	W 10×100	16.3	4.60	2.65	390	1060
538	* 835	16.7	129	W 21× 57	52.0	8.36	1.35	387	* 601
525	1090	21.8	126	W 14× 74	31.5	6.04	2.48	378	785
513	* 880	17.6	123	W 18× 60	44.0	7.47	1.69	369	* 634
—	—	23.2	119	W 12× 79	26.3	5.34	3.05	357	835
479	1000	20.0	115	W 14× 68	33.8	6.01	2.46	345	720
471	1300	25.9	113	W 10× 88	17.9	4.54	2.63	339	932
467	* 810	16.2	112	W 18× 55	46.4	7.41	1.67	336	* 583
458	* 735	14.7	110	W 21× 50	54.8	8.18	1.30	330	* 529
438	* 840	16.8	105	W 16× 57	38.2	6.72	1.60	315	605
—	—	17.9	102	W 14× 61	37.0	5.98	2.45	306	644
421	* 735	14.7	101	W 18× 50	50.7	7.38	1.65	303	* 529
407	1130	22.6	97.6	W 10× 77	20.0	4.49	2.60	293	814
—	—	13.0	95.4	W 21× 44	59.0	8.06	1.26	286	* 468
383	* 735	14.7	92.0	W 16× 50	42.8	6.68	1.59	276	529
378	* 675	13.5	90.7	W 18× 46	50.2	7.25	1.29	272	* 486
363	* 780	15.6	87.1	W 14× 53	37.6	5.89	1.92	261	562
—	—	17.0	86.4	W 12× 58	33.9	5.28	2.51	259	612
355	1000	20.0	85.3	W 10× 68	22.1	4.44	2.59	256	720
343	* 665	13.3	82.3	W 16× 45	46.8	6.65	1.57	247	* 479
327	* 590	11.8	78.4	W 18× 40	56.8	7.21	1.27	235	* 425
327	* 705	14.1	78.4	W 14× 48	40.6	5.85	1.91	235	508
—	—	17.6	74.6	W 10× 60	24.3	4.39	2.57	224	634
304	* 590	11.8	72.9	W 16× 40	52.5	6.63	1.57	219	* 425
302	735	14.7	72.4	W 12× 50	32.9	5.18	1.96	217	529
293	985	19.7	70.2	W 8× 67	15.8	3.72	2.12	211	709
—	—	12.6	69.6	W 14× 43	44.8	5.82	1.89	209	* 454
—	—	15.8	66.6	W 10× 54	27.3	4.37	2.56	200	569

* Check shape for compliance with Equations (N7-1) or (N7-2), Sect. N7, AISC ASD Specification, as applicable, when subjected to combined axial force and bending moment at ultimate loading.

AMERICAN INSTITUTE OF STEEL CONSTRUCTION

PLASTIC DESIGN SELECTION TABLE
For W and M shapes Z_x

F_y = 50 ksi		A	Z_x	Shape	$\dfrac{d}{t_w}$	r_x	r_y	F_y = 36 ksi	
M_p	P_y							M_p	P_y
Kip-ft	Kip	In.^{2}	In.^{3}			In.	In.	Kip-ft	Kip
—	—	10.3	66.5	W 18×35	59.0	7.04	1.22	200	*371
270	660	13.2	64.7	W 12×45	36.0	5.15	1.94	194	475
—	—	10.6	64.0	W 16×36	53.8	6.51	1.52	192	*382
256	*560	11.2	61.5	W 14×38	45.5	5.87	1.55	185	*403
249	855	17.1	59.8	W 8×58	17.2	3.65	2.10	179	616
—	—	11.8	57.5	W 12×40	40.5	5.13	1.93	173	425
229	665	13.3	54.9	W 10×45	28.9	4.32	2.01	165	479
—	—	10.0	54.6	W 14×34	49.1	5.83	1.53	164	*360
225	*456	9.12	54.0	W 16×31	57.7	6.41	1.17	162	*328
213	*515	10.3	51.2	W 12×35	41.7	5.25	1.54	154	371
204	705	14.1	49.0	W 8×48	21.3	3.61	2.08	147	508
—	—	11.5	46.8	W 10×39	31.5	4.27	1.98	140	414
—	—	7.68	44.2	W 16×26	62.8	6.26	1.12	133	*276
—	—	8.79	43.1	W 12×30	47.5	5.21	1.52	129	*316
168	*385	7.69	40.2	W 14×26	54.5	5.65	1.08	121	*277
—	—	11.7	39.8	W 8×40	22.9	3.53	2.04	119	421
—	—	7.65	37.2	W 12×26	53.1	5.17	1.51	112	*275
153	442	8.84	36.6	W 10×30	34.9	4.38	1.37	110	318
—	—	10.3	34.7	W 8×35	26.2	3.51	2.03	104	371
—	—	6.49	33.2	W 14×22	59.7	5.54	1.04	100	*234
130	*381	7.61	31.3	W 10×26	39.7	4.35	1.36	94	274
122	*324	6.48	29.3	W 12×22	47.3	4.91	0.847	88	*233
113	413	8.25	27.2	W 8×28	28.3	3.45	1.62	82	297
—	—	6.49	26.0	W 10×22	42.4	4.27	1.33	78	234
—	—	5.10	24.9	M 14×18	65.1	5.38	0.719	75	*184
103	*279	5.57	24.7	W 12×19	51.7	4.82	0.822	74	*201
—	—	7.08	23.2	W 8×24	32.4	3.42	1.61	70	255
90	*281	5.62	21.6	W 10×19	41.0	4.14	0.874	65	202
85	308	6.16	20.4	W 8×21	33.1	3.49	1.26	61	222
—	—	4.71	20.1	W 12×16	54.5	4.67	0.773	60	*170
79	367	7.34	18.9	W 6×25	19.9	2.70	1.52	57	264
78	*250	4.99	18.7	W 10×17	42.1	4.05	0.844	56	180
—	263	5.26	17.0	W 8×18	35.4	3.43	1.23	51	189

* Check shape for compliance with Equations (N7-1) or (N7-2), Sect. N7, AISC ASD Specification, as applicable, when subjected to combined axial force and bending moment at ultimate loading.

Z_x
PLASTIC DESIGN SELECTION TABLE
For W and M shapes

$F_y = 50$ ksi		A	Z_x	Shape	$\dfrac{d}{t_w}$	r_x	r_y	$F_y = 36$ ksi	
M_p	P_y							M_p	P_y
Kip-ft	Kip	In.²	In.³			In.	In.	Kip-ft	Kip
—	—	4.41	16.0	W 10×15	43.4	3.95	0.810	48	*159
—	—	5.87	14.9	W 6×20	23.8	2.66	1.50	45	211
—	—	5.89	14.5	M 6×20	24.0	2.57	1.40	44	212
60	*174	3.47	14.3	M 12×11.8	67.8	4.55	0.532	43	*125
—	—	4.44	13.6	W 8×15	33.1	3.29	0.876	41	160
—	—	3.18	13.2	M 12×10.8	74.8	4.55	0.537	40	*114
—	—	2.94	12.2	M 12×10	80.3	4.57	0.576	—	—
49	237	4.74	11.7	W 6×16	24.2	2.60	0.966	35	171
48	277	5.54	11.6	W 5×19	19.1	2.17	1.28	35	199
—	—	3.84	11.4	W 8×13	34.7	3.21	0.843	34	138
46	278	5.55	11.0	M 5×18.9	15.8	2.08	1.19	33	200
—	—	4.43	10.8	W 6×15	26.0	2.56	1.46	—	—
40	234	4.68	9.59	W 5×16	20.9	2.13	1.27	29	168
38	*133	2.65	9.19	M 10× 9	63.7	3.83	0.480	28	* 95
—	—	3.55	8.30	W 6×12	26.2	2.49	0.918	25	128
—	—	2.35	8.17	M 10× 8	70.6	3.82	0.427	25	* 85
—	—	2.21	7.73	M 10× 7.5	76.8	3.85	0.474	23	* 80
26	192	3.83	6.28	W 4×13	14.9	1.72	1.00	19	138
23	* 96	1.92	5.42	M 8× 6.5	59.3	3.10	0.423	16	* 69
12	* 65	1.29	2.80	M 6× 4.4	52.6	2.36	0.358	8	* 46

* Check shape for compliance with Equations (N7-1) or (N7-2), Sect. N7, AISC ASD Specification, as applicable, when subjected to combined axial force and bending moment at ultimate loading.

MOMENT OF INERTIA SELECTION TABLES
For W and M shapes
$$I_x, I_y$$

These two tables for moment of inertia (I_x and I_y) are provided to facilitate the selection of beams and columns on the basis of their stiffness properties with respect to the X-X axis or Y-Y axis, as applicable, where

I_x = moment of inertia, X-X axis, in.^{4}

I_y = moment of inertia, Y-Y axis, in.^{4}

In each table the shapes are listed in groups by descending order of moment of inertia for all W and M shapes. The boldface type identifies the shapes that are the lightest in weight in each group.

Enter the column headed I_x (or I_y) and find a value of I_x (or I_y) equal to or greater than the moment of inertia required. The shape opposite this value, and all shapes above it, have sufficient stiffness capacity. Note that the member selected must also be checked for compliance with specification provisions governing its specific application.

I_x MOMENT OF INERTIA SELECTION TABLE
For W and M shapes

Shape	I_x In.^{4}	Shape	I_x In.^{4}	Shape	I_x In.^{4}	Shape	I_x In.^{4}
W 36 × 848	**67400**	**W 44 × 248**	**21400**	**W 40 × 167**	**11600**	**W 33 × 118**	**5900**
		W 30×391	20700	W 33×201	11500	W 30×132	5770
W 36 × 798	**62600**	W 27×448	20400	W 36×182	11300	W 24×176	5680
		W 36×300	20300	W 27×258	10800	W 27×146	5630
W 40 × 655	**56500**	W 40×249	19500	W 21×364	10800	W 18×258	5510
W 36×720	55300	W 33×318	19500	W 14×605	10800	W 14×370	5440
				W 24×306	10700	W 30×124	5360
W 40 × 593	**50400**			W 36×170	10500	W 21×201	5310
W 36×650	48900	**W 44 × 224**	**19200**	W 30×211	10300	W 24×162	5170
		W 40×244	19200				
W 40 × 531	**44300**	W 24×492	19100				
W 36×588	43500	W 36×280	18900				
W 33×619	41800	W 30×357	18600	**W 40 × 149**	**9780**	**W 30 × 116**	**4930**
		W 27×407	18100	W 36×160	9750	W 18×234	4900
W 40 × 480	**39500**	W 33×291	17700	W 27×235	9660	W 14×342	4900
W 36×527	38300	W 36×260	17300	W 21×333	9610	W 27×129	4760
W 33×567	37700	W 24×450	17100	W 24×279	9600	W 21×182	4730
		W 36×256	16800	W 14×550	9430	W 24×146	4580
W 40 × 436	**35400**	W 30×326	16800	W 33×169	9290		
W 36×485	34700			W 30×191	9170		
W 33×515	33700			W 36×150	9040		
W 30×581	33000	**W 44 × 198**	**16700**	W 27×217	8870	**W 30 × 108**	**4470**
		W 40×215	16700	W 24×250	8490	W 18×211	4330
W 40 × 397	**32000**	W 40×221	16600	W 21×300	8480	W 14×311	4330
W 36×439	31000	W 36×245	16100	W 14×500	8210	W 21×166	4280
W 33×468	30100	W 27×368	16100	W 30×173	8200	W 27×114	4090
W 30×526	29300	W 33×263	15800	W 33×152	8160	W 12×336	4060
		W 24×408	15100	W 27×194	7820	W 24×131	4020
W 40 × 362	**28900**	W 36×230	15000				
W 36×393	27500	W 36×232	15000				
W 33×424	26900	W 40×199	14900				
		W 30×292	14900	**W 36 × 135**	**7800**	**W 30 × 99**	**3990**
W 40 × 328	**26800**	W 27×336	14500	W 24×229	7650	W 18×192	3870
W 30×477	26100	W 14×730	14300	W 21×275	7620	W 14×283	3840
		W 33×241	14200	W 33×141	7450	W 21×147	3630
W 40 × 324	**25600**			W 14×455	7190		
W 27×539	25500	**W 40 × 192**	**13500**	W 27×178	6990		
W 36×359	24800	W 24×370	13400	W 18×311	6960	**W 30 × 90**	**3620**
				W 24×207	6820	W 27×102	3620
W 44 × 285	**24600**			W 21×248	6760	W 12×305	3550
W 33×387	24300	**W 40 × 183**	**13300**			W 24×117	3540
W 40×298	24200	W 36×210	13200			W 18×175	3450
W 40×297	23200	W 30×261	13100			W 14×257	3400
W 30×433	23200	W 27×307	13100	**W 33 × 130**	**6710**	W 27×94	3270
W 27×494	22900	W 33×221	12800	W 30×148	6680	W 21×132	3220
W 36×328	22500	W 14×665	12400	W 14×426	6600	W 12×279	3110
		W 21×402	12200	W 27×161	6280	W 24×104	3100
W 40 × 277	**21900**	W 36×194	12100	W 24×192	6260	W 18×158	3060
W 33×354	21900	W 27×281	11900	W 18×283	6160	W 14×233	3010
		W 24×335	11900	W 14×398	6000	W 24×103	3000
W 40 × 268	**21500**	W 30×235	11700	W 21×223	5950	W 21×122	2960

MOMENT OF INERTIA SELECTION TABLE
For W and M shapes

I_x

Shape	I_x In.^{4}	Shape	I_x In.^{4}	Shape	I_x In.^{4}	Shape	I_x In.^{4}
W 27× 84	**2850**	**W 21× 50**	**984**	**W 16×26**	**301**	**M 12×10**	**61.6**
W 18×143	2750	W 18× 60	984	W 14×30	291	W 10×12	53.8
W 12×252	2720	W 16× 67	954	W 12×35	285	W 6×25	53.4
W 24× 94	2700	W 12×106	933	W 10×49	272	W 8×15	48.0
W 21×111	2670	W 18× 55	890	W 8×67	272	W 6×20	41.4
W 14×211	2660	W 14× 82	882	W 10×45	248	W 8×13	39.6
W 18×130	2460					M 6×20	29.0
W 21×101	2420						
W 12×230	2420	**W 21× 44**	**843**	**W 14×26**	**245**	**M 10× 9**	**38.8**
W 14×193	2400	W 12× 96	833	W 12×30	238		
		W 18× 50	800	W 8×58	228	**M 10× 8**	**34.5**
W 24× 84	**2370**	W 14× 74	796	W 10×39	209		
W 18×119	2190	W 16× 57	758			**M 10× 7.5**	**32.8**
W 14×176	2140	W 12× 87	740			W 6×16	32.1
W 12×210	2140	W 14× 68	723	**W 12×26**	**204**	W 8×10	30.8
		W 10×112	716			W 6×15	29.1
		W 18× 46	712			W 5×19	26.2
W 24× 76	**2100**	W 12× 79	662	**W 14×22**	**199**	M 5×18.9	24.1
W 21× 93	2070	W 16× 50	659	W 8×48	184	W 6×12	22.1
W 18×106	1910	W 14× 61	640	W 10×30	170	W 5×16	21.3
W 14×159	1900	W 10×100	623	W 10×33	170		
W 12×190	1890						
						M 8× 6.5	**18.5**
W 24× 68	**1830**	**W 18× 40**	**612**	**W 12×22**	**156**	W 6× 9	16.4
W 21× 83	1830	W 12× 72	597			W 4×13	11.3
W 18× 97	1750	W 16× 45	586				
W 14×145	1710	W 14× 53	541	**M 14×18**	**148**	**M 6× 4.4**	**7.20**
W 12×170	1650	W 10× 88	534	W 8×40	146		
W 21× 73	1600	W 12× 65	533	W 10×26	144		
				W 12×19	130		
				W 8×35	127		
W 24× 62	**1550**			W 10×22	118		
W 18× 86	1530	**W 16× 40**	**518**	W 8×31	110		
W 14×132	1530						
W 16×100	1490						
W 21× 68	1480	**W 18× 35**	**510**	**W 12×16**	**103**		
W 12×152	1430	W 14× 48	485	W 8×28	98.0		
W 14×120	1380	W 12× 58	475	W 10×19	96.3		
		W 10× 77	455				
		W 16× 36	448				
W 24× 55	**1350**	W 14× 43	428	**W 12×14**	**88.6**		
W 21× 62	1330	W 12× 53	425	W 8×24	82.8		
W 18× 76	1330	W 12× 50	394	W 10×17	81.9		
W 16× 89	1300	W 10× 68	394	W 8×21	75.3		
W 14×109	1240	W 14× 38	385				
W 12×136	1240						
W 21× 57	1170						
W 18× 71	1170	**W 16× 31**	**375**	**M 12×11.8**	**71.9**		
W 16× 77	1110	W 12× 45	350	W 10×15	68.9		
W 14× 99	1110	W 10× 60	341				
W 18× 65	1070	W 14× 34	340				
W 12×120	1070	W 12× 40	310	**M 12×10.8**	**65.0**		
W 14× 90	999	W 10× 54	303	W 8×18	61.9		

I_y MOMENT OF INERTIA SELECTION TABLE
For W and M shapes

Shape	I_y In.^{4}	Shape	I_y In.^{4}	Shape	I_y In.^{4}	Shape	I_y In.^{4}
W 14×730	**4720**	**W 40×328**	**1660**	**W 14×211**	**1030**	**W 14×145**	**677**
W 36×848	4550	W 33×387	1620	W 33×263	1030	W 30×191	673
W 36×798	4200			W 24×335	1030	W 12×210	664
				W 36×245	1010	W 24×229	651
W 14×665	**4170**	**W 14×311**	**1610**	W 21×333	994	W 18×258	628
		W 36×359	1570	W 40×221	988	W 27×194	618
		W 30×391	1550	W 30×261	959	W 21×223	609
W 14×605	**3680**	W 40×397	1540	W 27×281	953	W 30×173	598
W 36×720	3680			W 36×230	940	W 12×190	589
				W 12×279	937	W 24×207	578
W 14×550	**3250**	**W 40×298**	**1490**	W 33×241	932	W 18×234	558
W 36×650	3230	W 24×450	1490			W 27×178	555
		W 27×407	1480				
W 14×500	**2880**	W 33×354	1460				
W 33×619	2870					**W 14×132**	**548**
W 40×655	2860	**W 14×283**	**1440**			W 21×201	542
W 36×588	2850	W 36×328	1420	**W 14×193**	**931**	W 24×192	530
W 33×567	2580	W 30×357	1390	W 40×249	926	W 36×256	528
		W 40×362	1380	W 24×306	919	W 12×170	517
				W 21×300	873	W 27×161	497
W 14×455	**2560**			W 27×258	859		
W 30×581	2530	**W 40×268**	**1320**	W 30×235	855		
W 40×593	2520	W 24×408	1320	W 33×221	840	**W 14×120**	**495**
W 36×527	2490	W 27×368	1310			W 18×211	493
		W 36×300	1300			W 44×285	490
W 14×426	**2360**					W 21×182	483
W 33×515	2290	**W 14×257**	**1290**			W 24×176	479
W 36×485	2250	W 33×318	1290	**W 14×176**	**838**	W 36×232	468
W 30×526	2230	W 21×402	1270	W 12×252	828	W 12×152	454
W 40×531	2200	W 30×326	1240	W 24×279	823		
		W 40×324	1220	W 40×215	796		
W 14×398	**2170**	W 36×280	1200	W 18×311	795	**W 14×109**	**447**
W 27×539	2110	W 12×336	1190	W 21×275	785	W 27×146	443
W 33×468	2030			W 40×192	770	W 24×162	443
				W 27×235	768	W 18×192	440
W 14×370	**1990**	**W 40×244**	**1170**	W 30×211	757	W 21×166	435
W 36×439	1990	W 27×336	1170	W 33×201	749	W 44×248	435
W 30×477	1970	W 33×291	1160			W 36×210	411
W 40×480	1940	W 24×370	1160				
W 27×494	1890						
		W 14×233	**1150**				
W 14×342	**1810**	W 21×364	1120	**W 14×159**	**748**	**W 14× 99**	**402**
W 33×424	1800	W 30×292	1100	W 12×230	742	W 12×136	398
W 36×393	1750	W 36×260	1090	W 24×250	724	W 24×146	391
W 30×433	1750	W 40×297	1090	W 27×217	704	W 18×175	391
W 40×436	1720	W 12×305	1050	W 18×283	704	W 44×224	391
W 27×448	1670	W 27×307	1050	W 40×199	695	W 21×147	376
W 24×492	1670	W 40×277	1040	W 21×248	694	W 36×194	375

MOMENT OF INERTIA SELECTION TABLE
For W and M shapes

I_y

Shape	I_y In.^{4}	Shape	I_y In.^{4}	Shape	I_y In.^{4}	Shape	I_y In.^{4}
W 14× 90	**362**	**W 12× 65**	**174**	**W 12×45**	**50.0**	**W 6×15**	**9.32**
W 18×158	347	W 30×116	164			W 5×19	9.13
W 36×182	347	W 16× 89	163	**W 8×40**	**49.1**	W 14×26	8.91
W 12×120	345	W 27×114	159	W 14×43	45.2	W 8×18	7.97
W 24×131	340	W 10× 77	154			M 5×18.9	7.86
W 40×183	336	W 18× 76	152			W 5×16	7.51
W 44×198	336	W 14× 82	148	**W 10×39**	**45.0**	W 14×22	7.00
W 21×132	333	W 30×108	146	W 18×55	44.9	W 12×22	4.66
W 36×170	320	W 27×102	139	W 12×40	44.1	W 6×16	4.43
W 18×143	311	W 16× 77	138	W 16×57	43.1	W 10×19	4.29
W 33×169	310	W 10× 68	134				
W 21×122	305	W 14× 74	134	**W 8×35**	**42.6**	**W 4×13**	**3.86**
W 12×106	301	W 30× 99	128	W 18×50	40.1	W 12×19	3.76
W 24×117	297	W 27× 94	124	W 16×50	37.2	W 10×17	3.56
W 36×160	295	W 14× 68	121			W 8×15	3.41
W 40×167	283	W 16× 67	119	**W 8×31**	**37.1**		
W 18×130	278	W 24×103	119	W 10×33	36.6	**W 6×12**	**2.99**
W 21×111	274			W 24×62	34.5	W 10×15	2.89
W 33×152	273	**W 10× 60**	**116**	W 16×45	32.8	W 12×16	2.82
W 12× 96	270	W 30× 90	115	W 21×57	30.6	W 8×13	2.73
W 36×150	270	W 24× 94	109	W 24×55	29.1	M 14×18	2.64
W 24×104	259			W 16×40	28.9	W 12×14	2.36
W 18×119	253	**W 12× 58**	**107**	W 14×38	26.7		
W 21×101	248	W 14× 61	107	W 21×50	24.9	**W 6× 9**	**2.19**
W 33×141	246	W 27× 84	106	W 12×35	24.5	W 10×12	2.18
				W 16×36	24.5	W 8×10	2.09
		W 10× 54	**103**	W 14×34	23.3	M 12×10	0.994
				W 18×46	22.5	M 12×11.8	0.980
W 12× 87	**241**	**W 12× 53**	**95.8**			M 12×10.8	0.905
W 10×112	236	W 24× 84	94.4	**W 8×28**	**21.7**		
W 40×149	229			W 21×44	20.7	**M 10× 9**	**0.609**
W 30×148	227	**W 10× 49**	**93.4**	W 12×30	20.3		
W 36×135	225	W 21× 93	92.9	W 14×30	19.6	**M 10× 8**	**0.537**
W 18×106	220	W 8× 67	88.6	W 18×40	19.1		
W 33×130	218	W 24× 76	82.5			**M 10× 7.5**	**0.498**
		W 21× 83	81.4	**W 8×24**	**18.3**		
		W 8× 58	75.1	W 12×26	17.3	**M 8× 6.5**	**0.343**
W 12× 79	**216**	W 21× 73	70.6	W 6×25	17.1		
W 10×100	207	W 24× 68	70.4	W 10×30	16.7	**M 6× 4.4**	**0.165**
W 18× 97	201	W 21× 68	64.7	W 18×35	15.3		
W 30×132	196			W 10×26	14.1		
		W 8× 48	**60.9**				
		W 18× 71	60.3	**W 6×20**	**13.3**		
		W 14× 53	57.7	W 16×31	12.4		
W 12× 72	**195**	W 21× 62	57.5				
W 33×118	187	W 12× 50	56.3	**M 6×20**	**11.6**		
W 16×100	186	W 18× 65	54.8	W 10×22	11.4		
W 27×129	184			W 8×21	9.77		
W 30×124	181	**W 10× 45**	**53.4**	W 16×26	9.59		
W 10× 88	179	W 14× 48	51.4				
W 18× 86	175	W 18× 60	50.1				

ALLOWABLE LOADS ON BEAMS
General Notes

The tables of allowable loads for W, M and S shapes and channels (C, MC), used as simple laterally supported steel beams, give the total allowable uniformly distributed loads in kips. The tables are based on the allowable stresses specified in F1 of the AISC ASD Specification. Separate tables are presented for $F_y = 36$ ksi and $F_y = 50$ ksi. The tabulated loads include the weight of the beam, which should be deducted to arrive at the net load the beam will support.

The tables are also applicable to laterally supported simple beams for concentrated loading conditions. A method to determine the beam load capacity for several cases is shown in the discussion on "Use of Tables."

It is assumed, in all cases, the loads are applied normal to the X-X axis, shown in the Tables of Properties of shapes in Part 1 of this Manual, and that the beam deflects vertically in the plane of bending. If the conditions of loading involve forces outside of this plane, allowable loads must be determined from the general theory of flexure in accordance with the character of the load and its mode of application.

LATERAL SUPPORT OF BEAMS

The allowable bending stress and resultant allowable load capacity of a beam is dependent upon lateral support of its compression flange in addition to its section properties. In these tables, the notation L_c is used to denote the maximum unbraced length of the compression flange, in feet, for which the allowable loads for compact symmetrical shapes are calculated with an allowable stress of $0.66F_y$. Certain noncompact shapes are calculated with a value of allowable stress between $0.60F_y$ and $0.66F_y$, as permitted by F1.2, i.e., when $65/\sqrt{F_y} < b_f/2t_f < 95/\sqrt{F_y}$. The value of L_c is equal to the smaller value determined from the expressions

$$\frac{76b_f}{12\sqrt{F_y}} \text{ or } \frac{20,000}{12(d/A_f)F_y}$$

in accordance with Sect. F1.1.

The notation L_u is the maximum unbraced length of the compression flange, in feet, beyond which the allowable bending stress is less than $0.60\ F_y$, in accordance with the provisions of F1.3, when $C_b = 1.0$. For most shapes, the value of L_u, in feet, is given as $20,000/[12\ (d/A_f)\ F_y]$ as derived from Equation (F1-8). For a few shapes, L_u is given as $\sqrt{102,000/F_y} \times (r_T/12)$ as derived from Equation (F1-6) where this is more liberal.

These tables are not applicable for beams with unbraced lengths greater than L_u. For such cases, use of the charts of "Allowable Moments in Beams with Unbraced Length Greater than L_u" is recommended.

FLEXURAL STRESS AND TABULATED LOADS

For the symmetrical rolled shapes designated W, M, and S, the allowable bending stress and resultant allowable loads are based on the assumption the compression flanges of the beams are laterally supported at intervals not greater than L_c. When the value of L_c does not appear, L_u is the maximum unbraced length for which the loads are valid.

For compact shapes, the tabulated load is based on an allowable stress of $0.66F_y$ (see F1.1 of the AISC ASD Specification). For noncompact shapes, the tabulated load is based on an allowable stress of $0.60F_y$ or a value between $0.60F_y$ and $0.66F_y$, depending on the flange width-thickness ratio (see F1.2 and F1.3). For noncompact shapes, the allowable stress used to compute the tabulated loads is obtained from AISC ASD Specification Equation (F1-3).

When the unbraced length of a symmetrical member is greater than L_c, but less than L_u, the tabulated load must be reduced by the ratio of $0.60F_y$ over the allowable stress used to compute its capacity.

In the case of channels (C and MC) used as beams, the tabulated loads are based on an allowable stress of $0.60F_y$, in accordance with F1.3, and the assumption that the compression flanges are laterally supported at intervals not greater than L_u.

SHEARING STRESSES

For relatively short spans, the allowable loads for beams and channels may be limited by the shearing stress in the web, instead of by the maximum bending stress in the flanges. This limit is indicated in the tables by solid horizontal lines. Loads shown above these lines will produce the maximum allowable shear in the beam web.

BEAMS WITH CONCENTRATED LOADS

AISC ASD Specification Sect. K1 includes requirements for beam webs under compression due to concentrated loads. When the provisions are exceeded, the webs of the beams should be reinforced or the length of bearing increased.

There are two conditions to be considered:

1. Web yielding - ASD Spec. Sect. K1.3

   Max. end reaction, kips $= 0.66F_{yw}t_w \ (N + 2.5k)$

   Max. interior load, kips $= 0.66F_{yw}t_w \ (N + 5k)$

   where

   t_w = thickness of the web, in.

   k = distance from the outer face of the flange to web toe, in.

   N = length of bearing or length of concentrated load, in.

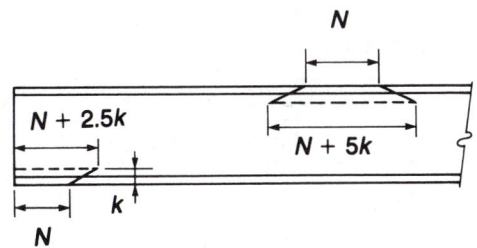

2. Web Crippling - ASD Spec. Sect. K1.4

   When the concentrated load is applied at a distance not less than $d/2$ from the end of the member.

   $$\text{Load} \le 67.5t_w^2 \left[1 + 3\left(\frac{N}{d}\right) \left(\frac{t_w}{t_f}\right)^{1.5} \right] \sqrt{F_{yw} \, t_f/t_w}$$

   When the concentrated load is applied at a distance less than $d/2$ from the end of the member.

   $$\text{Load} \le 34 \, t_w^2 \left[1 + 3\left(\frac{N}{d}\right) \left(\frac{t_w}{t_f}\right)^{1.5} \right] \sqrt{F_{yw} \, t_f/t_w}$$

   where

   d = overall depth of the member, in.

   t_f = flange thickness, in.

VERTICAL DEFLECTION

For rolled shapes designated W, M, S, C and MC, the column at the right of each group of nominal depths gives the deflection for the beams of various spans when supporting the full tabulated allowable loads. These deflections are based on the nominal depth of the beams. The following equation may be used for calculating the maximum deflection of any symmetrical, uniformly loaded beam or girder:

$$\Delta = \frac{5Wl^3}{384EI}$$

where

Δ = deflection, in.
W = total uniform load, including weight of beam, kips
l = span, in.

For $E = 29,000$ ksi and specific values of F_b, this equation reduces to the expressions shown in the table below. In this table, L = span, in feet and d = depth of beam, in inches.

	F_b, ksi	Deflection, in.
$F_y = 36$ ksi	23.8	$0.02458L^2/d$
	21.6	$0.02234L^2/d$
$F_y = 50$ ksi	33	$0.03414L^2/d$
	30	$0.03103L^2/d$

The deflections tabulated for W, M and S shapes are calculated on the basis of $0.66F_y$, regardless of whether the sections are compact or noncompact. Therefore, the tabulated deflections must be reduced to correspond to the lower allowable stresses used to calculate the tabulated loads for noncompact shapes, or compact shapes with unsupported length between L_c and L_u. The table that follows lists the reduction factors.

REDUCTION FACTORS FOR TABULATED DEFLECTION			
	Unbraced length, L_b	Compact	Noncompact
$F_y = 36$ ksi	$L_c \geq L_b$	$1.0 \times \Delta$	$\dfrac{F_b{}^*}{23.8} \times \Delta$
	$L_u \geq L_b > L_c$	$\dfrac{21.6}{23.8} \times \Delta$	
$F_y = 50$ ksi	$L_c \geq L_b$	$1.0 \times \Delta$	$\dfrac{F_b{}^*}{33} \times \Delta$
	$L_u \geq L_b > L_c$	$\dfrac{30}{33} \times \Delta$	

*The value of F_b is computed from AISC ASD Specification Equation (F1-3).

The deflections tabulated for channels are calculated on the basis of $0.60F_y$.

The live load deflection of floor beams supporting plastered ceilings should be limited to not more than 1/360 of the span length. This limit is not reached for the span lengths tabulated when the ratio of live load to dead load is approximately 1.0. For additional guidance on deflection criteria, see AISC ASD Specification Commentary Sect. L3.

ALLOWABLE LOADS ON BEAMS
Use of tables

FOR F_y = 36 KSI STEEL

The loads tabulated for steel of F_y = 36 ksi are based on allowable bending stresses of 23.8 ksi for compact shapes and a reduced stress for noncompact shapes based on Equation (F1-3). The beams must be braced adequately and have an axis of symmetry in the plane of loading. Loads may be read directly from the table when the distance between points of lateral support of the compression flange L_b does not exceed L_c for compact and noncompact W, M and S shapes or L_u for channels.

When $L_u \geq L_b > L_c$, the tabulated loads must be reduced as follows:

1. For a compact shape, multiply load by 21.6/23.8.

2. For a noncompact shape, multiply load by $21.6/F_b$ (calculated from Equation (F1-3)).

When $L_b > L_u$, the allowable bending stress is less than 21.6 ksi and the tables are not applicable. Use of the charts of "Allowable Moments in Beams with Unbraced Length Greater than L_u" is recommended.

FOR F_y = 50 KSI STEEL

The loads tabulated for steel of F_y = 50 ksi are based on allowable bending stresses of 33 ksi for compact shapes and a reduced stress for noncompact shapes based on Equation (F1-3). The beams must be braced adequately and have an axis of symmetry in the plane of loading. Loads may be read directly from the table when the distance between points of lateral support L_b does not exceed L_c for compact and noncompact W, M and S shapes or L_u for channels.

When $L_u \geq L_b > L_c$, the tabulated loads must be reduced as follows:

1. For a compact shape, multiply load by 30/33.

2. For a noncompact shape, multiply load by $30/F_b$ (calculated from Equation (F1-3)).

When $L_b > L_u$, the allowable bending stress is less than 30 ksi and the tables are not applicable. Use of the charts of "Allowable Moments in Beams with Unbraced Length Greater then L_u" is recommended.

CONCENTRATED LOAD CONDITIONS

The load tables are also applicable to laterally supported simple beams with equal concentrated loads spaced as shown in the Table of Concentrated Load Equivalents, p. 2-295. Except for short spans where shear controls the design, the beam load tables may be entered with an equivalent uniform load, equivalent in effect to the sum of the concentrated loads on the beam. Loads which will produce the maximum allowable shear in the beam web are shown in the load tables above the heavy horizontal lines. Deflections listed in the load tables must be multiplied by the proper deflection coefficient to determine the concentrated load deflection.

EXAMPLE 4

Given:

A W16×45 beam of F_y = 36 ksi steel spans 20 ft and is braced at 5-ft intervals. Determine the uniform load capacity, end reaction and required bearing length.

Solution:

Enter the Allowable Uniform Load Table for F_y = 36 ksi and note:

R_1 = 25.6 kips
R_2 = 8.2 kips/in.
R_3 = 31.1 kips
R_4 = 2.76 kips/in.

1. Total allowable uniform load = 58 kips

2. End reaction = 58/2 = 29 kips

3. Bearing length for web yielding
   $N = (29 - 25.6)/8.2 = 0.4$ in.
   Bearing length for web crippling
   $N = (29 - 31.1)/2.76 = -0.8$ in.
   The maximum of N = 0.4 in. governs. From a practical point of view, the bearing length would be longer.

EXAMPLE 5

Given:

A W10×45 beam of F_y = 36 ksi steel spans 6 ft. Determine the uniform load capacity, end reaction and required bearing length.

Solution:

Enter the Allowable Uniform Load Table for F_y = 36 ksi and note:

R_1 = 26.0 kips
R_2 = 8.32 kips/in.
R_3 = 33.3 kips
R_4 = 4.19 kips/in.

1. The beam is above the heavy solid line in the Allowable Uniform Load Table; therefore, span is less than L_v. The total allowable uniform load W is limited by shear in the web.

$$W = 2V = 2 \times 51 = 102 \text{ kips}$$

2. End reaction = V = 51 kips

3. Bearing length for web yielding
   $N = (51 - 26.0)/8.32 = 3.00$ in.
   Bearing length for web crippling
   $N = (51 - 33.3)/4.19 = 4.22$ in.

   Use 4½-in. seat

EXAMPLE 6

Given:

Using F_y = 36 ksi steel, select an 18-in. deep beam to span 30 ft and support three equal concentrated loads of 20 kips located at the quarter points of span. Assume bracing at concentrated load points.

Solution:

Refer to the Table of Concentrated Load Equivalents and note that for a simple span:

Equivalent uniform load = 4.0 P

Deflection coefficient = 0.95

1. Equivalent uniform load = 4.0 × 20 = 80 kips
2. Enter beam load tables for W18 and 30-ft span length.
   Select W18×86 with allowable load = 88 kips
3. Check deflection:
   From load table, uniform load deflection = 1.23 in.
   Concentrated load deflection = 0.95 × 1.23 × 80/88 = 1.06 in.

If the beam depth is not restricted, a shape with less weight can usually be selected by scanning the load tables for deeper sections. For example; W21×73, allowable load = 80 kips; W24×68, allowable load = 81 kips.

EXAMPLE 7

Given:

Using either F_y = 36 ksi steel or F_y = 50 ksi steel, select a 14-in. deep beam to span 25 ft and support a uniform load of 1 kip/ft.

Solution:

1. Required Allowable Uniform Load = wL = 1 × 25 = 25 kips
2. Enter the Allowable Uniform Load Table for F_y = 36 ksi and allowable load = 25 kips

   Select W14 × 30: allowable load = 27 kips
3. Enter the Allowable Uniform Load Table for F_y = 50 ksi and allowable load = 25 kips

   Select W14 × 22: allowable load = 26 kips

ALLOWABLE LOADS ON BEAMS
Reference notes on tables

1. L_c = Maximum unbraced length of compression flange, at which the allowable bending stress may be taken as $0.66F_y$ or as determined by AISC ASD Specification Equation (F1-3), when applicable, ft
2. L_u = Maximum unbraced length of compression flange, at which the allowable bending stress may be taken as $0.60F_y$, ft
3. L_b = Unbraced length of compression flange, ft
4. S = Section modulus, in.^{3}
5. Formulas for reaction values:
   Values of V, R, R_1, R_2, R_3 and R_4 used for connection design and design checks are included at the bottom of the tables for each shape. These symbols and corresponding equations are defined in the table below (see AISC ASD Specification Sect. K1):

Definition of symbols	F_y = 36 ksi	F_y = 50 ksi
V = Max. web shear, kips	14.4 dt	20 dt
R = Max. end reaction for 3½-in. bearing, kips	$R_1 + NR_2$ or $R_3 + NR_4$	$R_1 + NR_2$ or $R_3 + NR_4$
R_1 = Constant for yielding, kips	$59.4kt_w$	$82.5kt_w$
R_2 = Constant for yielding, kips/in.	$23.8t_w$	$33.0t_w$
R_3 = Constant for crippling, kips	$204t_w^{1.5}t_f^{0.5}$	$240t_w^{1.5}t_f^{0.5}$
R_4 = Constant for crippling, kips/in.	$612t_w^3/t_f\,d$	$721t_w^3/t_f\,d$

6. Load above the heavy line in the load column is limited by maximum allowable web shear.
7. Allowable uniform loads are given for span lengths up to the smaller of L/d = 30 or 72 ft.

| W 44 | BEAMS | | | | F_y = 36 ksi |

BEAMS
W Shapes
Allowable uniform loads in kips
for beams laterally supported
For beams laterally unsupported, see page 2-146

Designation		W44				
Wt./ft		285	248	224	198	Deflection In.
Flange Width		11¾	11¾	11¾	11¾	
L_c		12.5	12.5	12.5	12.5	
L_u		22.0	19.8	17.9	15.5	

	Span in Feet		285	248	224	198	Deflection In.
F_y = 36 ksi		13	1300				.09
		14	1270	1090	979	877	.11
		15	1180	1040	939	819	.13
		16	1110	973	880	768	.14
		17	1040	916	828	723	.16
		18	986	865	782	683	.18
		19	934	820	741	647	.20
		20	887	779	704	615	.22
		21	845	741	671	585	.25
		22	806	708	640	559	.27
		23	771	677	612	534	.30
		24	739	649	587	512	.32
		25	710	623	563	492	.35
		26	682	599	542	473	.38
		27	657	577	522	455	.41
		28	634	556	503	439	.44
		29	612	537	486	424	.47
		30	591	519	469	410	.50
		31	572	502	454	397	.54
		32	554	487	440	384	.57
		33	538	472	427	372	.61
		34	522	458	414	362	.65
		35	507	445	402	351	.68
		36	493	433	391	341	.72
		38	467	410	371	323	.81
		40	444	389	352	307	.89
		42	422	371	335	293	.99
		44	403	354	320	279	1.08
		46	386	338	306	267	1.18
		48	370	324	293	256	1.29
		50	355	311	282	246	1.40
		52	341	299	271	236	1.51
		54	329	288	261	228	1.63
		56	317	278	251	219	1.75
		58	306	268	243	212	1.88
		60	296	260	235	205	2.01
		62	286	251	227	198	2.15
		64	277	243	220	192	2.29
		66	269	236	213	186	2.43
		68	261	229	207	181	2.58
		70	253	222	201	176	2.74
		72	246	216	196	171	2.90

Properties and Reaction Values

	285	248	224	198	
S_x in.³	1120	983	889	776	For explanation of deflection, see page 2-32
V kips	650	543	490	439	
R_1 kips	164	128	108	89.6	
R_2 kips/in.	24.4	20.6	18.7	16.9	
R_3 kips	282	206	169	135	
R_4 kips/in.	8.46	5.77	4.83	4.18	
R kips	249	200	173	149	

Load above heavy line is limited by maximum allowable web shear.

AMERICAN INSTITUTE OF STEEL CONSTRUCTION

F_y = 36 ksi	BEAMS W Shapes Allowable uniform loads in kips for beams laterally supported For beams laterally unsupported, see page 2-146						W 40
Designation	W 40						
Wt./ft	328	298	268	244	221	192	Deflection In.
Flange Width	17⅞	17⅞	17¾	17¾	17¾	17¾	
L_c	18.9	18.8	18.7	18.7	18.7	17.8	
L_u	35.9	32.8	29.5	26.5	22.6	19.7	

Span in Feet	W328	W298	W268	W244	W221	W192	Deflection In.
14						781	.12
15						748	.14
16						701	.16
17					791	660	.18
18					755	623	.20
19				799	715	590	.22
20	1050	949	850	779	680	561	.25
21	1010	920	822	741	647	534	.27
22	965	878	785	708	618	510	.30
23	923	840	751	677	591	488	.33
24	884	805	719	649	566	467	.35
25	849	773	691	623	544	449	.38
26	816	743	664	599	523	431	.42
27	786	716	639	577	503	415	.45
28	758	690	617	556	485	401	.48
29	732	666	595	537	469	387	.52
30	708	644	576	519	453	374	.55
31	685	623	557	502	438	362	.59
32	663	604	540	487	425	350	.63
33	643	586	523	472	412	340	.67
34	624	568	508	458	400	330	.71
35	606	552	493	445	388	320	.75
36	590	537	480	433	378	312	.80
37	574	522	467	421	367	303	.84
38	559	509	454	410	358	295	.89
40	531	483	432	389	340	280	.98
42	505	460	411	371	324	267	1.08
44	482	439	392	354	309	255	1.19
46	461	420	375	338	295	244	1.30
48	442	403	360	324	283	234	1.42
50	425	386	345	311	272	224	1.54
52	408	372	332	299	261	216	1.66
54	393	358	320	288	252	208	1.79
56	379	345	308	278	243	200	1.93
58	366	333	298	268	234	193	2.07
60	354	322	288	260	227	187	2.21
62	342	312	278	251	219	181	2.36
64	332	302	270	243	212	175	2.52
66	322	293	262	236	206	170	2.68
68	312	284	254	229	200	165	2.84
70	303	276	247	222	194	160	3.01
72	295	268	240	216	189	156	3.19

Properties and Reaction Values							
S_x in.^{3}	1340	1220	1090	983	858	708	For explanation of deflection, see page 2-32
V kips	524	474	425	399	395	391	
R_1 kips	169	148	125	111	103	94.9	
R_2 kips/in.	21.6	19.7	17.8	16.9	16.9	16.9	
R_3 kips	233	194	158	137	126	111	
R_4 kips/in.	6.66	5.60	4.63	4.45	5.32	6.91	
R kips	245	214	174	153	145	135	

Load above heavy line is limited by maximum allowable web shear.

| W 40 | BEAMS
W Shapes
Allowable uniform loads in kips
for beams laterally supported
For beams laterally unsupported, see page 2-146 | | | | | | | F_y = 36 ksi |

Designation	W 40							
Wt./ft	655	593	531	480	436	397	362	Deflection In.
Flange Width	16⅞	16¾	16½	16⅜	16¼	16⅛	16	
L_c	17.8	17.6	17.4	17.3	17.1	17.0	16.9	
L_u	63.4	57.9	52.6	47.7	43.7	40.3	36.7	

	Span in Feet	655	593	531	480	436	397	362	Deflection In.
	16	2480	2220	1960		1600			.16
	17	2410	2180	1950	1760	1590	1440	1310	.18
	18	2280	2060	1840	1660	1510	1370	1250	.20
	19	2160	1950	1740	1580	1430	1300	1180	.22
	20	2050	1850	1660	1500	1350	1240	1130	.25
	21	1950	1770	1580	1430	1290	1180	1070	.27
	22	1870	1690	1510	1360	1230	1120	1020	.30
	23	1780	1610	1440	1300	1180	1070	978	.33
	24	1710	1540	1380	1250	1130	1030	937	.35
	25	1640	1480	1320	1200	1080	988	900	.38
	26	1580	1430	1270	1150	1040	950	865	.42
	27	1520	1370	1230	1110	1000	915	833	.45
	28	1470	1320	1180	1070	967	883	803	.48
	29	1420	1280	1140	1030	934	852	776	.52
	30	1370	1240	1100	998	903	824	750	.55
	31	1320	1200	1070	966	874	797	726	.59
	32	1280	1160	1040	936	846	772	703	.63
	33	1240	1120	1000	907	821	749	682	.67
	34	1210	1090	974	881	797	727	662	.71
	36	1140	1030	920	832	752	686	625	.80
	38	1080	975	871	788	713	650	592	.89
	40	1030	927	828	748	677	618	562	.98
	42	977	883	788	713	645	588	536	1.08
	44	932	842	752	680	616	562	511	1.19
	46	892	806	720	651	589	537	489	1.30
	48	855	772	690	624	564	515	469	1.42
	50	821	741	662	599	542	494	450	1.54
	52	789	713	637	576	521	475	433	1.66
	54	760	686	613	554	502	458	417	1.79
	56	733	662	591	535	484	441	402	1.93
	58	707	639	571	516	467	426	388	2.07
	60	684	618	552	499	451	412	375	2.21
	62	662	598	534	483	437	399	363	2.36
	64	641	579	517	468	423	386	351	2.52
	66	622	562	502	454	410	374	341	2.68
	68	603	545	487	440	398	363	331	2.84
	70	586	530	473	428	387	353	321	3.01
	72	570	515	460	416	376	343	312	3.19

Properties and Reaction Values								
S_x in.³	2590	2340	2090	1890	1710	1560	1420	For explanation of deflection, see page 2-32
V kips	1240	1110	982	879	798	719	654	
R_1 kips	578	492	412	347	303	263	225	
R_2 kips/in.	46.8	42.5	38.3	34.7	31.8	29.0	26.6	
R_3 kips	1060	878	711	585	490	408	343	
R_4 kips/in.	30.3	25.3	20.7	17.3	14.8	12.3	10.5	
R kips	742	641	546	468	414	365	318	

Load above heavy line is limited by maximum allowable web shear.

$F_y = 36$ ksi		BEAMS							W 40

BEAMS
W Shapes
Allowable uniform loads in kips
for beams laterally supported
For beams laterally unsupported, see page 2-146

Designation		W 40						W 40			
Wt./ft		324	297	277	249	215	199	183	167	149	Deflection In.
Flange Width		15⅞	15⅞	15⅞	15¾	15¾	15¾	11¾	11¾	11¾	
L_c		16.8	16.7	16.7	16.6	16.6	16.6	12.5	12.5	11.9	
L_u		33.1	30.3	29.1	26.3	22.8	20.0	17.1	14.5	12.6	

$F_y = 36$ ksi — Span in Feet		324	297	277	249	215	199	183	167	149	Deflection In.
	11									693	.07
	12									676	.09
	13								722	624	.10
	14							730	678	579	.12
	15							720	633	541	.14
	16						724	675	593	507	.16
	17	1160	1070				717	635	558	477	.18
	18	1130	1030	949	851	730	677	600	527	451	.20
	19	1070	975	917	827	715	641	569	499	427	.22
	20	1010	927	871	786	680	609	540	474	406	.25
	21	965	883	830	748	647	580	514	452	386	.27
	22	922	842	792	714	618	554	491	431	369	.30
	23	882	806	758	683	591	530	470	413	353	.33
	24	845	772	726	655	566	508	450	395	338	.35
	25	811	741	697	629	544	487	432	380	324	.38
	26	780	713	670	604	523	468	415	365	312	.42
	27	751	686	645	582	503	451	400	351	300	.45
	28	724	662	622	561	485	435	386	339	290	.48
	29	699	639	601	542	469	420	373	327	280	.52
	30	676	618	581	524	453	406	360	316	270	.55
	31	654	598	562	507	438	393	348	306	262	.59
	32	634	579	545	491	425	381	338	297	253	.63
	33	614	562	528	476	412	369	327	288	246	.67
	34	596	545	512	462	400	358	318	279	239	.71
	36	563	515	484	436	378	338	300	264	225	.80
	38	534	488	459	414	358	321	284	250	213	.89
	40	507	463	436	393	340	305	270	237	203	.98
	42	483	441	415	374	324	290	257	226	193	1.08
	44	461	421	396	357	309	277	246	216	184	1.19
	46	441	403	379	342	295	265	235	206	176	1.30
	48	422	386	363	327	283	254	225	198	169	1.42
	50	406	371	348	314	272	244	216	190	162	1.54
	52	390	356	335	302	261	234	208	182	156	1.66
	54	375	343	323	291	252	226	200	176	150	1.79
	56	362	331	311	281	243	218	193	169	145	1.93
	58	350	320	300	271	234	210	186	164	140	2.07
	60	338	309	290	262	227	203	180	158	135	2.21
	62	327	299	281	253	219	196	174	153	131	2.36
	64	317	290	272	246	212	190	169	148	127	2.52
	66	307	281	264	238	206	185	164	144	123	2.68
	68	298	273	256	231	200	179	159	140	119	2.84
	70	290	265	249	224	194	174	154	136	116	3.01
	72	282	257	242	218	189	169	150	132	113	3.19

Properties and Reaction Values

	324	297	277	249	215	199	183	167	149	
S_x in.³	1280	1170	1100	992	858	769	682	599	512	
V kips	578	534	474	425	365	362	365	361	347	For explanation of deflection, see page 2-32
R_1 kips	189	169	148	125	101	94.1	101	94.1	84.2	
R_2 kips/in.	23.8	22.1	19.7	17.8	15.4	15.4	15.4	15.4	15.0	
R_3 kips	274	235	194	158	118	110	118	108	92.9	
R_4 kips/in.	8.42	7.49	5.60	4.62	3.53	4.08	3.53	4.25	4.83	
R kips	272	246	214	174	130	124	130	123	110	

Load above heavy line is limited by maximum allowable web shear.

W 36	**BEAMS**	$F_y = 36$ ksi

BEAMS
W Shapes
Allowable uniform loads in kips
for beams laterally supported
For beams laterally unsupported, see page 2-146

Designation		W 36								Deflection In.
Wt./ft		848	798	720	650	588	527	485	439	
Flange Width		18⅛	18	17¾	17⅝	17⅜	17¼	17⅛	17	
L_c		19.1	19.0	18.8	18.6	18.4	18.2	18.1	17.9	
L_u		89.0	85.7	78.5	71.2	65.2	59.4	54.5	50.3	

Span in Feet	848	798	720	650	588	527	485	439	Deflection In.
16	3080	2880	2570	2300	2050	1820	1670		.17
17	2950	2780	2510	2260	2030	1820	1670	1500	.20
18	2790	2620	2370	2130	1920	1720	1580	1430	.22
19	2640	2480	2240	2020	1820	1630	1490	1350	.25
20	2510	2360	2130	1920	1730	1540	1420	1280	.27
21	2390	2250	2030	1830	1640	1470	1350	1220	.30
22	2280	2150	1940	1740	1570	1400	1290	1170	.33
23	2180	2050	1850	1670	1500	1340	1230	1120	.36
24	2090	1970	1780	1600	1440	1290	1180	1070	.39
25	2010	1890	1700	1530	1380	1240	1130	1030	.43
26	1930	1820	1640	1470	1330	1190	1090	987	.46
27	1860	1750	1580	1420	1280	1140	1050	950	.50
28	1790	1690	1520	1370	1230	1100	1010	916	.54
29	1730	1630	1470	1320	1190	1070	978	885	.57
30	1670	1570	1420	1280	1150	1030	945	855	.61
31	1620	1520	1380	1240	1110	996	915	828	.66
32	1570	1480	1330	1200	1080	965	886	802	.70
33	1520	1430	1290	1160	1050	936	859	778	.74
34	1480	1390	1250	1130	1020	908	834	755	.79
36	1400	1310	1180	1070	959	858	788	713	.88
38	1320	1240	1120	1010	909	813	746	675	.99
40	1260	1180	1070	958	863	772	709	642	1.09
42	1200	1120	1020	913	822	735	675	611	1.20
44	1140	1070	968	871	785	702	644	583	1.32
46	1090	1030	926	833	751	671	616	558	1.44
48	1050	983	888	799	719	644	591	535	1.57
50	1000	944	852	767	691	618	567	513	1.71
52	966	908	819	737	664	594	545	493	1.85
54	930	874	789	710	639	572	525	475	1.99
56	897	843	761	685	617	552	506	458	2.14
58	866	814	735	661	595	533	489	442	2.30
60	837	787	710	639	576	515	473	428	2.46
62	810	761	687	618	557	498	457	414	2.62
64	785	738	666	599	540	483	443	401	2.80
66	761	715	646	581	523	468	430	389	2.97
68	738	694	627	564	508	454	417	377	3.16
70	717	674	609	548	493	441	405	367	3.35
72	697	656	592	532	480	429	394	356	3.54

Properties and Reaction Values

	848	798	720	650	588	527	485	439	
S_x in.³	3170	2980	2690	2420	2180	1950	1790	1620	
V kips	1540	1440	1280	1150	1030	909	837	749	For explanation of deflection, see page 2-32
R_1 kips	851	769	651	549	465	389	340	288	
R_2 kips/in.	59.9	56.5	51.4	46.8	42.5	38.3	35.6	32.3	
R_3 kips	1740	1550	1280	1060	878	711	614	505	
R_4 kips/in.	50.9	45.8	38.7	32.7	27.3	22.4	19.9	16.5	
R kips	1060	967	831	713	614	523	465	401	

Load above heavy line is limited by maximum allowable web shear.

$F_y = 36$ ksi

F_y = 36 ksi	BEAMS W Shapes Allowable uniform loads in kips for beams laterally supported For beams laterally unsupported, see page 2-146								W 36

Designation					W 36					
Wt./ft		393	359	328	300	280	260	245	230	Deflection In.
Flange Width		16⅞	16¾	16⅝	16⅝	16⅝	16½	16½	16½	
L_c		17.8	17.7	17.6	17.6	17.5	17.5	17.4	17.4	
L_u		45.4	41.7	38.3	35.3	33.1	30.5	28.6	26.8	

	Span in Feet		393	359	328	300	280	260	245	230	Deflection In.
		16								786	.17
		17	1330	1210	1090	1000	931	877	831	780	.20
		18	1280	1160	1070	977	906	839	788	737	.22
		19	1210	1100	1010	925	859	795	746	698	.25
		20	1150	1050	958	879	816	755	709	663	.27
		21	1090	996	913	837	777	719	675	631	.30
		22	1040	950	871	799	742	686	644	603	.33
		23	999	909	833	764	709	656	616	576	.36
		24	957	871	799	733	680	629	591	552	.39
		25	919	836	767	703	653	604	567	530	.43
		26	883	804	737	676	628	581	545	510	.46
		27	851	774	710	651	604	559	525	491	.50
		28	820	747	685	628	583	539	506	474	.54
		29	792	721	661	606	563	521	489	457	.57
		30	766	697	639	586	544	503	473	442	.61
		31	741	674	618	567	526	487	457	428	.66
		32	718	653	599	549	510	472	443	414	.70
		33	696	634	581	533	494	457	430	402	.74
		34	676	615	564	517	480	444	417	390	.79
		36	638	581	532	488	453	419	394	368	.88
		38	604	550	504	463	429	397	373	349	.99
		40	574	523	479	440	408	377	354	331	1.09
		42	547	498	456	419	388	359	338	316	1.20
		44	522	475	436	400	371	343	322	301	1.32
		46	499	455	417	382	355	328	308	288	1.44
		48	479	436	399	366	340	314	295	276	1.57
		50	459	418	383	352	326	302	284	265	1.71
		52	442	402	369	338	314	290	273	255	1.85
		54	425	387	355	326	302	280	263	246	1.99
		56	410	373	342	314	291	270	253	237	2.14
		58	396	360	330	303	281	260	244	229	2.30
		60	383	348	319	293	272	252	236	221	2.46
		62	370	337	309	284	263	243	229	214	2.62
		64	359	327	299	275	255	236	222	207	2.80
		66	348	317	290	266	247	229	215	201	2.97
		68	338	307	282	259	240	222	208	195	3.16
		70	328	299	274	251	233	216	203	189	3.35
		72	319	290	266	244	227	210	197	184	3.54

Properties and Reaction Values									
S_x in.³	1450	1320	1210	1110	1030	953	895	837	
V kips	664	603	545	500	465	439	416	393	For explanation of deflection, see page 2-32
R_1 kips	240	208	182	158	141	128	119	107	
R_2 kips/in.	29.0	26.6	24.2	22.5	21.0	20.0	19.0	18.1	
R_3 kips	408	343	286	243	213	188	170	152	
R_4 kips/in.	13.4	11.4	9.47	8.37	7.40	6.95	6.43	5.94	
R kips	342	301	267	237	215	198	186	170	

Load above heavy line is limited by maximum allowable web shear.

BEAMS
W Shapes
Allowable uniform loads in kips
for beams laterally supported
For beams laterally unsupported, see page 2-146

W 36

$F_y = 36$ ksi

Designation		W 36									
Wt./ft		256	232	210	194	182	170	160	150	135	Deflection In.
Flange Width		12¼	12⅛	12⅛	12⅛	12⅛	12	12	12	12	
L_c		12.9	12.8	12.9	12.8	12.7	12.7	12.7	12.6	12.3	
L_u		26.2	23.7	20.9	19.4	18.2	17.0	15.7	14.6	13.0	
	11									614	.08
	12			877			708	674	645	579	.10
	13	1040	930	876	804	759	707	660	614	535	.12
	14	1010	915	813	751	705	656	613	570	497	.13
	15	945	854	759	701	658	612	572	532	464	.15
	16	886	801	712	657	617	574	537	499	435	.17
	17	834	754	670	619	580	540	505	470	409	.20
	18	788	712	633	584	548	510	477	444	386	.22
	19	746	674	599	554	519	484	452	420	366	.25
	20	709	641	569	526	493	459	429	399	348	.27
	21	675	610	542	501	470	437	409	380	331	.30
	22	644	582	518	478	449	418	390	363	316	.33
	23	616	557	495	457	429	399	373	347	302	.36
	24	591	534	475	438	411	383	358	333	290	.39
	25	567	513	456	421	395	367	343	319	278	.43
	26	545	493	438	405	380	353	330	307	267	.46
	27	525	475	422	390	365	340	318	296	258	.50
	28	506	458	407	376	352	328	307	285	248	.54
	29	489	442	393	363	340	317	296	275	240	.57
	30	473	427	380	351	329	306	286	266	232	.61
	31	457	413	367	339	318	296	277	258	224	.66
	32	443	400	356	329	308	287	268	249	217	.70
	33	430	388	345	319	299	278	260	242	211	.74
	34	417	377	335	309	290	270	253	235	205	.79
	36	394	356	316	292	274	255	238	222	193	.88
	38	373	337	300	277	260	242	226	210	183	.99
	40	354	320	285	263	247	230	215	200	174	1.09
	42	338	305	271	250	235	219	204	190	166	1.20
	44	322	291	259	239	224	209	195	181	158	1.32
	46	308	279	248	229	215	200	187	174	151	1.44
	48	295	267	237	219	206	191	179	166	145	1.57
	50	284	256	228	210	197	184	172	160	139	1.71
	52	273	246	219	202	190	177	165	154	134	1.85
	54	263	237	211	195	183	170	159	148	129	1.99
	56	253	229	203	188	176	164	153	143	124	2.14
	58	244	221	196	181	170	158	148	138	120	2.30
	60	236	214	190	175	164	153	143	133	116	2.46
	62	229	207	184	170	159	148	138	129	112	2.62
	64	222	200	178	164	154	144	134	125	109	2.80
	66	215	194	173	159	150	139	130	121	105	2.97
	68	208	188	167	155	145	135	126	117	102	3.16
	70	203	183	163	150	141	131	123	114	99	3.35
	72	197	178	158	146	137	128	119	111	97	3.54

Properties and Reaction Values

	256	232	210	194	182	170	160	150	135	
S_x in.³	895	809	719	664	623	580	542	504	439	For explanation of deflection, see page 2-32
V kips	517	465	439	402	379	354	337	323	307	
R_1 kips	150	129	114	99.4	91.5	80.8	74.8	69.6	60.1	
R_2 kips/in.	22.8	20.7	19.7	18.2	17.2	16.2	15.4	14.9	14.3	
R_3 kips	252	207	180	153	137	120	108	97.7	84.3	
R_4 kips/in.	8.36	6.92	7.01	5.96	5.44	4.84	4.58	4.43	4.71	
R kips	230	201	183	163	152	137	124	113	101	

Load above heavy line is limited by maximum allowable web shear.

$F_y = 36$ ksi — Span in Feet

AMERICAN INSTITUTE OF STEEL CONSTRUCTION

| F_y = 36 ksi | BEAMS W Shapes Allowable uniform loads in kips for beams laterally supported | W 33 |

For beams laterally unsupported, see page 2-146

Designation		W 33							
Wt./ft		619	567	515	468	424	387	354	Deflection In.
Flange Width		16⅞	16¾	16⅝	16½	16⅜	16¼	16⅛	
L_c		17.8	17.7	17.5	17.4	17.2	17.1	17.0	
L_u		72.3	67.1	61.7	56.5	51.4	47.7	43.7	

F_y = 36 ksi	Span in Feet	619	567	515	468	424	387	354	Deflection In.
	15	2180	1980						.17
	16	2150	1970	1780	1610	1440	1310	1190	.19
	17	2020	1850	1690	1520	1380	1260	1150	.22
	18	1910	1750	1590	1430	1300	1190	1080	.24
	19	1810	1660	1510	1360	1230	1130	1030	.27
	20	1720	1580	1430	1290	1170	1070	974	.30
	21	1640	1500	1370	1230	1120	1020	928	.33
	22	1560	1430	1300	1170	1070	972	886	.36
	23	1490	1370	1250	1120	1020	930	847	.39
	24	1430	1310	1200	1080	977	891	812	.43
	25	1380	1260	1150	1030	938	855	779	.47
	26	1320	1210	1100	993	902	822	749	.50
	27	1270	1170	1060	956	868	792	722	.54
	28	1230	1130	1020	922	837	764	696	.58
	29	1190	1090	989	890	808	737	672	.63
	30	1150	1050	956	861	781	713	649	.67
	31	1110	1020	925	833	756	690	628	.72
	32	1070	985	896	807	733	668	609	.76
	33	1040	955	869	782	710	648	590	.81
	34	1010	927	843	759	690	629	573	.86
	35	982	901	819	738	670	611	557	.91
	36	955	876	796	717	651	594	541	.97
	37	929	852	775	698	634	578	527	1.02
	38	905	830	754	679	617	563	513	1.08
	40	859	788	717	645	586	535	487	1.19
	42	818	751	683	615	558	509	464	1.31
	44	781	716	652	587	533	486	443	1.44
	46	747	685	623	561	510	465	424	1.58
	48	716	657	597	538	488	446	406	1.72
	50	687	630	573	516	469	428	390	1.86
	52	661	606	551	497	451	411	375	2.01
	54	637	584	531	478	434	396	361	2.17
	56	614	563	512	461	419	382	348	2.34
	58	593	543	494	445	404	369	336	2.51
	60	573	525	478	430	391	356	325	2.68
	62	554	508	462	416	378	345	314	2.86
	64	537	493	448	403	366	334	304	3.05
	66	521	478	434	391	355	324	295	3.24
	68	505	464	422	380	345	314	287	3.44
	70	491	450	410	369	335	305	278	3.65
	72	477	438	398	359	326	297	271	3.86

Properties and Reaction Values									
S_x in.³		2170	1990	1810	1630	1480	1350	1230	
V kips		1090	988	888	806	722	652	594	For explanation of deflection, see page 2-32
R_1 kips		512	437	374	316	272	234	198	
R_2 kips/in.		46.8	43.0	39.2	36.1	32.8	29.9	27.6	
R_3 kips		1060	898	748	630	521	436	368	
R_4 kips/in.		34.4	29.3	24.6	21.5	17.8	14.9	12.9	
R kips		676	588	511	442	387	339	295	

Load above heavy line is limited by maximum allowable web shear.

AMERICAN INSTITUTE OF STEEL CONSTRUCTION

| W 33 | BEAMS | $F_y = 36$ ksi |

W Shapes
Allowable uniform loads in kips
for beams laterally supported
For beams laterally unsupported, see page 2-146

Designation		W 33						
Wt./ft		318	291	263	241	221	201	Deflection In.
Flange Width		16	15⅞	15¾	15⅞	15¾	15¾	
L_c		16.9	16.8	16.7	16.7	16.7	16.6	
L_u		39.9	36.5	33.3	30.1	27.6	24.9	

	Span in Feet								Deflection In.
	15					757	694	.17
	16	1050	963	865	817	749	677	.19
	17	1030	941	854	772	705	637	.22
	18	977	889	807	730	666	602	.24
	19	925	842	764	691	631	570	.27
	20	879	800	726	657	600	542	.30
	21	837	762	692	625	571	516	.33
	22	799	727	660	597	545	492	.36
	23	764	696	632	571	521	471	.39
	24	733	667	605	547	500	451	.43
	25	703	640	581	525	480	433	.47
	26	676	615	559	505	461	417	.50
	27	651	593	538	486	444	401	.54
	28	628	571	519	469	428	387	.58
	29	606	552	501	453	413	374	.63
	30	586	533	484	438	400	361	.67
	31	567	516	469	424	387	350	.72
	32	549	500	454	410	375	339	.76
	33	533	485	440	398	363	328	.81
	34	517	471	427	386	353	319	.86
	35	502	457	415	375	343	310	.91
	36	488	444	403	365	333	301	.97
	37	475	432	393	355	324	293	1.02
	38	463	421	382	346	316	285	1.08
	40	440	400	363	328	300	271	1.19
	42	419	381	346	313	285	258	1.31
	44	400	364	330	298	273	246	1.44
	46	382	348	316	285	261	236	1.58
	48	366	333	303	274	250	226	1.72
	50	352	320	291	263	240	217	1.86
	52	338	308	279	253	231	208	2.01
	54	326	296	269	243	222	201	2.17
	56	314	286	259	234	214	193	2.34
	58	303	276	250	226	207	187	2.51
	60	293	267	242	219	200	181	2.68
	62	284	258	234	212	193	175	2.86
	64	275	250	227	205	187	169	3.05
	66	266	242	220	199	182	164	3.24
	68	259	235	214	193	176	159	3.44
	70	251	229	208	188	171	155	3.65
	72	244	222	202	182	167	150	3.86

Properties and Reaction Values								
S_x in.³		1110	1010	917	829	757	684	
V kips		527	482	433	409	379	347	For
R_1 kips		166	146	123	108	94.9	82.3	explanation
R_2 kips/in.		24.7	22.8	20.7	19.7	18.4	17.0	of deflection,
R_3 kips		297	252	207	183	157	132	see page 2-32
R_4 kips/in.		10.4	8.98	7.43	7.31	6.59	5.78	
R kips		252	226	195	177	159	142	

Load above heavy line is limited by maximum allowable web shear.

$F_y = 36$ ksi

F_y = 36 ksi	BEAMS W Shapes Allowable uniform loads in kips for beams laterally supported	W 33

For beams laterally unsupported, see page 2-146

Designation		W 33					
Wt./ft		169	152	141	130	118	Deflection In.
Flange Width		11½	11⅝	11½	11½	11½	
L_c		12.1	12.2	12.2	12.1	12.0	
L_u		19.2	16.9	15.4	13.8	12.6	

	Span in Feet	169	152	141	130	118	Deflection In.
	10					521	.07
	11				553	517	.09
	12		612	580	536	474	.11
	13	653	593	546	495	437	.13
	14	621	551	507	459	406	.15
	15	580	514	473	429	379	.17
	16	544	482	444	402	355	.19
	17	512	454	417	378	335	.22
	18	483	429	394	357	316	.24
	19	458	406	373	338	299	.27
	20	435	386	355	322	284	.30
	21	414	367	338	306	271	.33
	22	395	351	323	292	258	.36
	23	378	335	309	280	247	.39
	24	362	321	296	268	237	.43
	25	348	309	284	257	227	.47
	26	334	297	273	247	219	.50
	27	322	286	263	238	211	.54
	28	311	276	253	230	203	.58
	29	300	266	245	222	196	.63
	30	290	257	237	214	190	.67
	31	281	249	229	207	183	.72
	32	272	241	222	201	178	.76
	33	264	234	215	195	172	.81
	34	256	227	209	189	167	.86
	36	242	214	197	179	158	.97
	38	229	203	187	169	150	1.08
	40	217	193	177	161	142	1.19
	42	207	184	169	153	135	1.31
	44	198	175	161	146	129	1.44
	46	189	168	154	140	124	1.58
	48	181	161	148	134	118	1.72
	50	174	154	142	129	114	1.86
	52	167	148	136	124	109	2.01
	54	161	143	131	119	105	2.17
	56	155	138	127	115	102	2.34
	58	150	133	122	111	98	2.51
	60	145	129	118	107	95	2.68
	62	140	124	114	104	92	2.86
	64	136	121	111	100	89	3.05
	66	132	117	108	97	86	3.24
	68	128	113	104	95	84	3.44
	70	124	110	101	92	81	3.65
	72	121	107	99	89	79	3.86

F_y = 36 ksi (left margin)

Properties and Reaction Values

	169	152	141	130	118	
S_x in.³	549	487	448	406	359	
V kips	326	306	290	276	260	
R_1 kips	82.1	70.7	62.9	58.1	51	For explanation of deflection, see page 2-32
R_2 kips/in.	15.9	15.1	14.4	13.8	13.1	
R_3 kips	124	106	94.1	83.3	71.6	
R_4 kips/in.	4.46	4.44	4.24	4.22	4.19	
R kips	138	122	109	98	86	

Load above heavy line is limited by maximum allowable web shear.

W 30	BEAMS W Shapes Allowable uniform loads in kips for beams laterally supported For beams laterally unsupported, see page 2-146	$F_y = 36$ ksi

Designation		W 30							
Wt./ft		581	526	477	433	391	357	326	Deflection
Flange Width		16¼	16	15⅞	15¾	15⅝	15½	15⅜	In.
L_c		17.1	16.9	16.7	16.6	16.5	16.3	16.2	
L_u		74.7	69.1	63.4	57.9	53.2	48.7	44.9	
	14	2010	1790						.16
	15	1980	1770	1610	1450	1300	1170	1060	.18
	16	1850	1660	1520	1370	1240	1130	1020	.21
	17	1740	1570	1430	1290	1170	1060	960	.24
	18	1650	1480	1350	1210	1100	1000	906	.27
	19	1560	1400	1280	1150	1040	950	859	.30
	20	1480	1330	1210	1090	990	903	816	.33
	21	1410	1270	1150	1040	943	860	777	.36
	22	1350	1210	1100	994	900	821	742	.40
	23	1290	1160	1050	950	861	785	709	.43
	24	1230	1110	1010	911	825	752	680	.47
	25	1190	1060	969	874	792	722	653	.51
	26	1140	1020	932	841	762	695	628	.55
	27	1100	986	898	810	733	669	604	.60
	28	1060	950	866	781	707	645	583	.64
	29	1020	918	836	754	683	623	563	.69
	30	987	887	808	729	660	602	544	.74
	31	956	858	782	705	639	583	526	.79
	32	926	832	757	683	619	564	510	.84
	33	898	806	734	662	600	547	494	.89
	34	871	783	713	643	582	531	480	.95
	35	846	760	692	625	566	516	466	1.00
	36	823	739	673	607	550	502	453	1.06
	37	801	719	655	591	535	488	441	1.12
	38	779	700	638	575	521	475	429	1.18
	40	741	665	606	546	495	451	408	1.31
	42	705	634	577	520	471	430	388	1.45
	44	673	605	551	497	450	410	371	1.59
	46	644	579	527	475	430	393	355	1.73
	48	617	554	505	455	413	376	340	1.89
	50	592	532	485	437	396	361	326	2.05
	52	570	512	466	420	381	347	314	2.22
	54	549	493	449	405	367	334	302	2.39
	56	529	475	433	390	354	322	291	2.57
	58	511	459	418	377	341	311	281	2.76
	60	494	444	404	364	330	301	272	2.95
	62	478	429	391	353	319	291	263	3.15
	64	463	416	379	342	309	282	255	3.36
	66	449	403	367	331	300	274	247	3.57
	68	436	391	356	321	291	266	240	3.79
	70	423	380	346	312	283	258	233	4.01
	72	411	370	337	304	275	251	227	4.25

The leftmost vertical labels read: $F_y = 36$ ksi and Span in Feet.

Properties and Reaction Values								
S_x in.³	1870	1680	1530	1380	1250	1140	1030	For
V kips	1000	896	803	727	650	586	532	explanation
R_1 kips	505	425	363	306	263	221	190	of deflection,
R_2 kips/in.	46.8	42.5	38.7	35.6	32.3	29.5	27.1	see page 2-32
R_3 kips	1060	878	729	614	505	422	356	
R_4 kips/in.	37.3	31.3	26.3	22.9	19.0	15.9	13.7	
R kips	669	574	498	431	376	324	285	

Load above heavy line is limited by maximum allowable web shear.

| F_y = 36 ksi | BEAMS W Shapes Allowable uniform loads in kips for beams laterally supported For beams laterally unsupported, see page 2-146 | | | | | | W 30 |

Designation				W 30			
Wt./ft	292	261	235	211	191	173	Deflection In.
Flange Width	15¼	15⅛	15	15⅛	15	15	
L_c	16.1	16.0	15.9	15.9	15.9	15.8	
L_u	41.0	36.7	33.3	29.7	26.9	24.2	

Span in Feet	292	261	235	211	191	173	Deflection In.
14						574	.16
15	940	847	748	691	627	569	.18
16	919	819	739	656	592	534	.21
17	865	771	695	618	557	502	.24
18	817	728	656	583	526	474	.27
19	774	689	622	553	499	449	.30
20	735	655	591	525	474	427	.33
21	700	624	563	500	451	407	.36
22	668	595	537	477	431	388	.40
23	639	570	514	457	412	371	.43
24	612	546	492	438	395	356	.47
25	588	524	473	420	379	342	.51
26	565	504	454	404	364	328	.55
27	544	485	438	389	351	316	.60
28	525	468	422	375	338	305	.64
29	507	452	407	362	327	294	.69
30	490	437	394	350	316	285	.74
31	474	423	381	339	306	275	.79
32	459	409	369	328	296	267	.84
34	432	385	348	309	279	251	.95
36	408	364	328	292	263	237	1.06
38	387	345	311	276	249	225	1.18
40	367	327	295	263	237	213	1.31
42	350	312	281	250	226	203	1.45
44	334	298	269	239	215	194	1.59
46	320	285	257	228	206	186	1.73
48	306	273	246	219	197	178	1.89
50	294	262	236	210	189	171	2.05
52	283	252	227	202	182	164	2.22
54	272	243	219	194	175	158	2.39
56	262	234	211	188	169	152	2.57
58	253	226	204	181	163	147	2.76
60	245	218	197	175	158	142	2.95
62	237	211	191	169	153	138	3.15
64	230	205	185	164	148	133	3.36
66	223	198	179	159	144	129	3.57
68	216	193	174	154	139	126	3.79
70	210	187	169	150	135	122	4.01
72	204	182	164	146	132	119	4.25

Properties and Reaction Values							
S_x in.³	928	827	746	663	598	539	
V kips	470	423	374	345	314	287	For explanation of deflection, see page 2-32
R_1 kips	159	135	111	97.8	81.7	73.0	
R_2 kips/in.	24.2	22.1	19.7	18.4	16.9	15.6	
R_3 kips	286	235	189	160	133	112	
R_4 kips/in.	11.0	10.1	7.45	7.00	6.02	5.30	
R kips	244	212	180	162	141	128	

Load above heavy line is limited by maximum allowable web shear.

AMERICAN INSTITUTE OF STEEL CONSTRUCTION

BEAMS
W Shapes
Allowable uniform loads in kips
for beams laterally supported

W 30

F_y = 36 ksi

For beams laterally unsupported, see page 2-146

Designation		W 30							
Wt./ft		148	132	124	116	108	99	90	Deflection In.
Flange Width		10½	10½	10½	10½	10½	10½	10⅜	
L_c		11.1	11.1	11.1	11.1	11.1	10.9	10.0	
L_u		18.7	16.1	15.0	13.8	12.3	11.4	11.4	
	9						444	400	.07
	10				488	468	426	388	.08
	11		537	508	474	431	387	353	.10
	12	574	502	469	434	395	355	323	.12
	13	531	463	433	401	364	328	299	.14
	14	493	430	402	372	338	304	277	.16
	15	460	401	375	347	316	284	259	.18
	16	432	376	351	326	296	266	243	.21
	17	406	354	331	307	279	251	228	.24
	18	384	334	312	290	263	237	216	.27
	19	363	317	296	274	249	224	204	.30
	20	345	301	281	261	237	213	194	.33
	21	329	287	268	248	226	203	185	.36
	22	314	274	256	237	215	194	176	.40
	23	300	262	244	227	206	185	169	.43
	24	288	251	234	217	197	178	162	.47
	25	276	241	225	208	189	170	155	.51
	26	266	232	216	200	182	164	149	.55
Span in Feet	27	256	223	208	193	175	158	144	.60
	28	247	215	201	186	169	152	139	.64
	29	238	208	194	180	163	147	134	.69
	30	230	201	187	174	158	142	129	.74
	31	223	194	181	168	153	137	125	.79
	32	216	188	176	163	148	133	121	.84
	34	203	177	165	153	139	125	114	.95
	36	192	167	156	145	132	118	108	1.06
	38	182	158	148	137	125	112	102	1.18
	40	173	150	141	130	118	107	97	1.31
	42	164	143	134	124	113	101	92	1.45
	44	157	137	128	118	108	97	88	1.59
	46	150	131	122	113	103	93	84	1.73
	48	144	125	117	109	99	89	81	1.89
	50	138	120	112	104	95	85	78	2.05
	52	133	116	108	100	91	82	75	2.22
	54	128	111	104	97	88	79	72	2.39
	56	123	107	100	93	85	76	69	2.57
	58	119	104	97	90	82	73	67	2.76
	60	115	100	94	87	79	71	65	2.95
	62	111	97	91	84	76	69	63	3.15
	64	108	94	88	81	74	67	61	3.36
	66	105	91	85	79	72	65	59	3.57
	68	102	89	83	77	70	63	57	3.79
	70	99	86	80	74	68	61	55	4.01
	72	96	84	78	72	66	59	54	4.25

F_y = 36 ksi

Properties and Reaction Values

	148	132	124	116	108	99	90	
S_x in.³	436	380	355	329	299	269	245	For explanation of deflection, see page 2-32
V kips	287	268	254	244	234	222	200	
R_1 kips	77.2	63.9	58.6	54.5	50.6	44.4	36.6	
R_2 kips/in.	15.4	14.6	13.9	13.4	12.9	12.4	11.2	
R_3 kips	116	98.4	88.0	79.9	71.6	62.6	51.3	
R_4 kips/in.	4.64	4.70	4.37	4.33	4.37	4.33	3.53	
R kips	131	115	103	95	87	78	64	

Load above heavy line is limited by maximum allowable web shear.

AMERICAN INSTITUTE OF STEEL CONSTRUCTION

| Fy = 36 ksi | BEAMS
W Shapes
Allowable uniform loads in kips
for beams laterally supported
For beams laterally unsupported, see page 2-146 | | | | | | | | W 27 |

Designation		W 27								
Wt./ft		539	494	448	407	368	336	307	281	Deflection In.
Flange Width		15¼	15⅛	15	14¾	14⅝	14½	14½	14⅜	
L_c		16.1	15.9	15.8	15.6	15.5	15.4	15.2	15.1	
L_u		77.2	71.2	66.1	60.1	55.1	51.4	47.2	43.7	

	Span in Feet										Deflection
F_y = 36 ksi	13	1850	1670	1490	1350	1210				.15
	14	1780	1630	1470	1320	1200	1090	989	894	.18
	15	1660	1520	1370	1240	1120	1020	934	856	.20
	16	1550	1430	1290	1160	1050	960	875	803	.23
	17	1460	1340	1210	1090	988	904	824	756	.26
	18	1380	1270	1140	1030	933	854	778	714	.29
	19	1310	1200	1080	975	884	809	737	676	.33
	20	1240	1140	1030	927	840	768	700	642	.36
	21	1180	1090	981	883	800	732	667	612	.40
	22	1130	1040	936	842	763	698	636	584	.44
	23	1080	992	895	806	730	668	609	559	.48
	24	1040	950	858	772	700	640	583	535	.52
	25	995	912	824	741	672	615	560	514	.57
	26	956	877	792	713	646	591	539	494	.62
	27	921	845	763	686	622	569	519	476	.66
	28	888	815	735	662	600	549	500	459	.71
	29	858	787	710	639	579	530	483	443	.77
	30	829	760	686	618	560	512	467	428	.82
	31	802	736	664	598	542	496	452	414	.87
	32	777	713	644	579	525	480	438	401	.93
	33	754	691	624	562	509	466	424	389	.99
	34	731	671	606	545	494	452	412	378	1.05
	35	711	652	588	530	480	439	400	367	1.12
	36	691	634	572	515	466	427	389	357	1.18
	38	654	600	542	488	442	404	368	338	1.31
	40	622	570	515	463	420	384	350	321	1.46
	42	592	543	490	441	400	366	333	306	1.61
	44	565	518	468	421	382	349	318	292	1.76
	46	541	496	448	403	365	334	304	279	1.93
	48	518	475	429	386	350	320	292	268	2.10
	50	497	456	412	371	336	307	280	257	2.28
	52	478	439	396	356	323	295	269	247	2.46
	54	461	422	381	343	311	285	259	238	2.65
	56	444	407	368	331	300	274	250	229	2.85
	58	429	393	355	320	289	265	241	221	3.06
	60	414	380	343	309	280	256	233	214	3.28
	62	401	368	332	299	271	248	226	207	3.50
	64	389	356	322	290	262	240	219	201	3.73
	66	377	346	312	281	254	233	212	195	3.97

Properties and Reaction Values										
S_x in.³		1570	1440	1300	1170	1060	970	884	811	For explanation of deflection, see page 2-32
V kips		923	833	747	676	604	544	495	447	
R_1 kips		497	430	361	310	261	225	194	165	
R_2 kips/in.		46.8	43.0	39.2	36.1	32.8	29.9	27.6	25.2	
R_3 kips		1060	898	748	630	521	436	368	309	
R_4 kips/in.		40.6	34.7	29.3	25.6	21.3	17.9	15.4	12.9	
R kips		661	581	498	436	376	330	291	253	

Load above heavy line is limited by maximum allowable web shear.

W 27	BEAMS	F_y = 36 ksi

BEAMS
W Shapes
Allowable uniform loads in kips
for beams laterally supported
For beams laterally unsupported, see page 2-146

Designation		W 27							
Wt./ft		258	235	217	194	178	161	146	Deflection In.
Flange Width		14¼	14¼	14⅛	14	14⅛	14	14	
L_c		15.1	15.0	14.9	14.8	14.9	14.8	14.7	
L_u		40.3	37.0	34.5	31.1	27.9	25.4	23.0	
	13					581	524	477	.15
	14	818	751	680	607	568	515	465	.18
	15	784	712	659	587	530	480	434	.20
	16	735	667	618	550	497	450	407	.23
	17	691	628	581	518	468	424	383	.26
	18	653	593	549	489	442	400	362	.29
	19	619	562	520	464	419	379	343	.33
	20	588	534	494	440	398	360	326	.36
	21	560	508	471	419	379	343	310	.40
	22	534	485	449	400	361	328	296	.44
	23	511	464	430	383	346	313	283	.48
	24	490	445	412	367	331	300	271	.52
	25	470	427	395	352	318	288	260	.57
	26	452	411	380	339	306	277	250	.62
	27	435	395	366	326	295	267	241	.66
	28	420	381	353	315	284	257	233	.71
	29	405	368	341	304	274	249	224	.77
	30	392	356	329	294	265	240	217	.82
	31	379	344	319	284	257	232	210	.87
	32	367	334	309	275	248	225	203	.93
	33	356	324	300	267	241	218	197	.99
	34	346	314	291	259	234	212	191	1.05
	35	336	305	282	252	227	206	186	1.12
	36	326	297	275	245	221	200	181	1.18
	38	309	281	260	232	209	190	171	1.31
	40	294	267	247	220	199	180	163	1.46
	42	280	254	235	210	189	172	155	1.61
	44	267	243	225	200	181	164	148	1.76
	46	256	232	215	191	173	157	142	1.93
	48	245	222	206	183	166	150	136	2.10
	50	235	214	198	176	159	144	130	2.28
	52	226	205	190	169	153	139	125	2.46
	54	218	198	183	163	147	133	121	2.65
	56	210	191	177	157	142	129	116	2.85
	58	203	184	170	152	137	124	112	3.06
	60	196	178	165	147	133	120	109	3.28
	62	190	172	159	142	128	116	105	3.50
	64	184	167	154	138	124	113	102	3.73
	66	178	162	150	133	120	109	99	3.97

(Left margin: F_y = 36 ksi — Span in Feet)

Properties and Reaction Values

	258	235	217	194	178	161	146	
S_x in.³	742	674	624	556	502	455	411	For explanation of deflection, see page 2-32
V kips	409	376	340	304	290	262	239	
R_1 kips	146	125	108	91.9	80.7	71.1	60.6	
R_2 kips/in.	23.3	21.6	19.7	17.8	17.2	15.7	14.4	
R_3 kips	263	225	189	153	137	114	94.8	
R_4 kips/in.	11.2	9.99	8.21	6.85	7.05	5.90	5.08	
R kips	228	201	177	154	141	126	111	

Load above heavy line is limited by maximum allowable web shear.

AMERICAN INSTITUTE OF STEEL CONSTRUCTION

F_y = 36 ksi

BEAMS
W Shapes
Allowable uniform loads in kips for beams laterally supported

For beams laterally unsupported, see page 2-146

W 27

Designation	W 27					Deflection In.
Wt./ft	129	114	102	94	84	
Flange Width	10	10⅛	10	10	10	
L_c	10.6	10.6	10.6	10.5	10.5	
L_u	18.4	15.9	14.2	12.8	11.0	
F_y = 36 ksi — Span in Feet						
9					354	.07
10		448	402	380	337	.09
11	485	431	384	350	307	.11
12	455	395	352	321	281	.13
13	420	364	325	296	260	.15
14	390	338	302	275	241	.18
15	364	316	282	257	225	.20
16	342	296	264	241	211	.23
17	321	279	249	226	198	.26
18	304	263	235	214	187	.29
19	288	249	223	203	178	.33
20	273	237	211	192	169	.36
21	260	226	201	183	161	.40
22	248	215	192	175	153	.44
23	238	206	184	167	147	.48
24	228	197	176	160	141	.52
25	219	189	169	154	135	.57
26	210	182	163	148	130	.62
27	202	175	157	143	125	.66
28	195	169	151	137	120	.71
29	188	163	146	133	116	.77
30	182	158	141	128	112	.82
31	176	153	136	124	109	.87
32	171	148	132	120	105	.93
34	161	139	124	113	99	1.05
36	152	132	117	107	94	1.18
38	144	125	111	101	89	1.31
40	137	118	106	96	84	1.46
42	130	113	101	92	80	1.61
44	124	108	96	87	77	1.76
46	119	103	92	84	73	1.93
48	114	99	88	80	70	2.10
50	109	95	85	77	67	2.28
52	105	91	81	74	65	2.46
54	101	88	78	71	62	2.65
56	98	85	76	69	60	2.85
58	94	82	73	66	58	3.06
60	91	79	70	64	56	3.28
62	88	76	68	62	54	3.50
64	85	74	66	60	53	3.73
66	83	72	64	58	51	3.97

Properties and Reaction Values

	129	114	102	94	84	
S_x in.^{3}	345	299	267	243	213	For explanation of deflection, see page 2-32
V kips	243	224	201	190	177	
R_1 kips	65.7	55.0	47.8	41.8	37.6	
R_2 kips/in.	14.5	13.5	12.2	11.6	10.9	
R_3 kips	102	84.7	68.7	60.4	50.9	
R_4 kips/in.	4.57	4.47	3.72	3.59	3.48	
R kips	116	100	82	73	63	

Load above heavy line is limited by maximum allowable web shear.

W 24	BEAMS W Shapes Allowable uniform loads in kips for beams laterally supported For beams laterally unsupported, see page 2-146									$F_y = 36$ ksi

Designation		W 24									
Wt./ft		492	450	408	370	335	306	279	250	229	
Flange Width		14⅛	14	13¾	13⅝	13½	13⅜	13¼	13⅛	13⅛	Deflection In.
L_c		14.9	14.7	14.6	14.4	14.3	14.1	14.0	13.9	13.8	
L_u		78.5	72.3	67.1	61.7	56.5	52.0	48.2	43.7	40.3	
	12	1680	1520	1360	1230	1090	984	893	789	719	.15
	13	1570	1430	1290	1170	1050	961	875	785	716	.17
	14	1460	1320	1200	1080	978	893	812	729	665	.20
	15	1360	1240	1120	1010	912	833	758	680	621	.23
	16	1280	1160	1050	947	855	781	711	638	582	.26
	17	1200	1090	988	892	805	735	669	600	548	.30
	18	1140	1030	933	842	760	694	632	567	517	.33
	19	1080	975	884	798	720	658	599	537	490	.37
	20	1020	927	840	758	684	625	569	510	466	.41
	21	973	883	800	722	652	595	542	486	444	.45
	22	929	842	763	689	622	568	517	464	423	.50
	23	888	806	730	659	595	543	494	444	405	.54
	24	851	772	700	632	570	521	474	425	388	.59
	25	817	741	672	606	547	500	455	408	373	.64
	26	786	713	646	583	526	481	437	392	358	.69
	27	757	686	622	561	507	463	421	378	345	.75
	28	730	662	600	541	489	446	406	364	333	.80
	29	705	639	579	523	472	431	392	352	321	.86
	30	681	618	560	505	456	417	379	340	310	.92
	31	659	598	542	489	441	403	367	329	300	.98
	32	639	579	525	474	428	391	355	319	291	1.05
	33	619	562	509	459	415	379	345	309	282	1.12
	34	601	545	494	446	403	368	335	300	274	1.18
	35	584	530	480	433	391	357	325	291	266	1.25
	36	568	515	466	421	380	347	316	283	259	1.33
	38	538	488	442	399	360	329	299	268	245	1.48
	40	511	463	420	379	342	312	284	255	233	1.64
	42	487	441	400	361	326	298	271	243	222	1.81
	44	464	421	382	345	311	284	258	232	212	1.98
	46	444	403	365	330	298	272	247	222	202	2.17
	48	426	386	350	316	285	260	237	213	194	2.36
	50	409	371	336	303	274	250	227	204	186	2.56
	52	393	356	323	292	263	240	219	196	179	2.77
	54	378	343	311	281	253	231	211	189	172	2.99
	56	365	331	300	271	244	223	203	182	166	3.21
	58	352	320	289	261	236	215	196	176	161	3.45
	60	341	309	280	253	228	208	190	170	155	3.69

$F_y = 36$ ksi   Span in Feet

Properties and Reaction Values											
S_x in.³		1290	1170	1060	957	864	789	718	644	588	
V kips		841	758	678	613	547	492	446	394	360	For
R_1 kips		505	437	368	316	266	229	198	166	143	explanation
R_2 kips/in.		46.8	43.0	39.2	36.1	32.8	29.9	27.6	24.7	22.8	of deflection,
R_3 kips		1060	898	748	630	521	436	368	297	252	see page 2-32
R_4 kips/in.		44.6	38.2	32.2	28.2	23.6	19.8	17.1	13.8	12.0	
R kips		669	588	505	442	381	334	295	252	223	

Load above heavy line is limited by maximum allowable web shear.

AMERICAN INSTITUTE OF STEEL CONSTRUCTION

| F_y = 36 ksi | BEAMS
W Shapes
Allowable uniform loads in kips
for beams laterally supported
For beams laterally unsupported, see page 2-146 | W 24 |

Designation					W 24					
Wt./ft	207	192	176	162	146	131	117	104	Deflection In.
Flange Width	13	13	12⅞	13	12⅞	12⅞	12¾	12¾	
L_c	13.7	13.7	13.6	13.7	13.6	13.6	13.5	13.5	
L_u	36.7	34.3	31.7	29.3	26.3	23.4	20.8	18.4	

Span in Feet									Deflection In.
11							384	346	.12
12				508	463	427	384	341	.15
13	644	594	545	504	452	401	355	314	.17
14	601	556	509	468	420	372	329	292	.20
15	561	518	475	437	392	347	307	272	.23
16	526	486	446	410	367	326	288	255	.26
17	495	457	419	386	346	307	271	240	.30
18	467	432	396	364	326	290	256	227	.33
19	443	409	375	345	309	274	243	215	.37
20	421	389	356	328	294	261	230	204	.41
21	401	370	339	312	280	248	219	195	.45
22	382	354	324	298	267	237	210	186	.50
23	366	338	310	285	256	227	200	178	.54
24	350	324	297	273	245	217	192	170	.59
25	336	311	285	262	235	208	184	163	.64
26	324	299	274	252	226	200	177	157	.69
27	312	288	264	243	218	193	171	151	.75
28	300	278	255	234	210	186	165	146	.80
29	290	268	246	226	203	180	159	141	.86
30	280	259	238	219	196	174	154	136	.92
31	271	251	230	212	190	168	149	132	.98
32	263	243	223	205	184	163	144	128	1.05
33	255	236	216	199	178	158	140	124	1.12
34	247	229	210	193	173	153	136	120	1.18
36	234	216	198	182	163	145	128	114	1.33
38	221	205	188	173	155	137	121	108	1.48
40	210	194	178	164	147	130	115	102	1.64
42	200	185	170	156	140	124	110	97	1.81
44	191	177	162	149	134	118	105	93	1.98
46	183	169	155	143	128	113	100	89	2.17
48	175	162	149	137	122	109	96	85	2.36
50	168	156	143	131	118	104	92	82	2.56
52	162	150	137	126	113	100	89	79	2.77
54	156	144	132	121	109	97	85	76	2.99
56	150	139	127	117	105	93	82	73	3.21
58	145	134	123	113	101	90	79	70	3.45
60	140	130	119	109	98	87	77	68	3.69

Properties and Reaction Values									
S_x in.³	531	491	450	414	371	329	291	258	
V kips	322	297	273	254	232	213	192	173	For
R_1 kips	123	108	94.7	83.8	72.4	62.9	53.1	44.6	explanation
R_2 kips/in.	20.7	19.2	17.8	16.8	15.4	14.4	13.1	11.9	of deflection,
R_3 kips	207	180	153	133	112	94.1	76.7	62.5	see page 2-32
R_4 kips/in.	9.98	8.75	7.63	7.03	6.23	5.77	4.94	4.24	
R kips	195	175	157	143	126	113	94.0	77.0	

Load above heavy line is limited by maximum allowable web shear.

AMERICAN INSTITUTE OF STEEL CONSTRUCTION

| W 24 | BEAMS
W Shapes
Allowable uniform loads in kips
for beams laterally supported
For beams laterally unsupported, see page 2-146 | | | | | | | F_y = 36 ksi |

Designation		W 24					W 24		
Wt./ft		103	94	84	76	68	62	55	Deflection
Flange Width		9	9⅛	9	9	9	7	7	In.
L_c		9.50	9.60	9.50	9.50	9.50	7.40	7.00	
L_u		16.7	15.1	13.3	11.8	10.2	8.10	7.50	
	6							268	.04
	7						294	258	.05
	8					284	259	226	.07
	9	389	361	326	303	271	231	201	.08
	10	388	352	310	279	244	208	181	.10
	11	353	320	282	253	222	189	164	.12
	12	323	293	259	232	203	173	150	.15
	13	299	270	239	214	188	160	139	.17
	14	277	251	222	199	174	148	129	.20
	15	259	234	207	186	163	138	120	.23
	16	243	220	194	174	152	130	113	.26
	17	228	207	183	164	143	122	106	.30
	18	216	195	172	155	136	115	100	.33
	19	204	185	163	147	128	109	95	.37
	20	194	176	155	139	122	104	90	.41
	21	185	167	148	133	116	99	86	.45
	22	176	160	141	127	111	94	82	.50
	23	169	153	135	121	106	90	79	.54
	24	162	147	129	116	102	86	75	.59
	25	155	141	124	112	98	83	72	.64
	26	149	135	119	107	94	80	69	.69
	27	144	130	115	103	90	77	67	.75
	28	139	126	111	100	87	74	64	.80
	29	134	121	107	96	84	72	62	.86
	30	129	117	103	93	81	69	60	.92
	32	121	110	97	87	76	65	56	1.05
	34	114	103	91	82	72	61	53	1.18
	36	108	98	86	77	68	58	50	1.33
	38	102	93	82	73	64	55	48	1.48
	40	97	88	78	70	61	52	45	1.64
	42	92	84	74	66	58	49	43	1.81
	44	88	80	71	63	55	47	41	1.98
	46	84	76	67	61	53	45	39	2.17
	48	81	73	65	58	51	43	38	2.36
	50	78	70	62	56	49	42	36	2.56
	52	75	68	60	54	47	40	35	2.77
	54	72	65	57	52	45	38	33	2.99
	56	69	63	55	50	44	37	32	3.21
	58	67	61	54	48	42	36	31	3.45
	60	65	59	52	46	41	35	30	3.69
Properties and Reaction Values									
S_x in.³		245	222	196	176	154	131	114	
V kips		194	180	163	152	142	147	134	For
R_1 kips		57.2	49.7	43.6	37.6	33.9	35.1	30.8	explanation
R_2 kips/in.		13.1	12.2	11.2	10.5	9.86	10.2	9.39	of deflection,
R_3 kips		82.4	70.5	57.7	49.1	41.7	44.2	36.0	see page 2-32
R_4 kips/in.		4.24	3.93	3.42	3.21	3.15	3.47	3.17	
R kips		97	84	70	60	53	56	47	

Load above heavy line is limited by maximum allowable web shear.

AMERICAN INSTITUTE OF STEEL CONSTRUCTION

Span in Feet — F_y = 36 ksi

| F_y = 36 ksi | **BEAMS**
W Shapes
Allowable uniform loads in kips
for beams laterally supported
For beams laterally unsupported, see page 2-146 | W 21 |

Designation		W 21								
Wt./ft		402	364	333	300	275	248	223	201	Deflection In.
Flange Width		13⅜	13¼	13⅛	13	12⅞	12¾	12⅝	12⅝	
L_c		14.1	14.0	13.9	13.7	13.6	13.5	13.4	13.3	
L_u		74.7	69.1	63.4	59.4	54.5	49.3	45.4	41.3	

		11	1300	1170	1050	933	848	752			.14
		12	1240	1120	1020	913	834	751	672	604	.17
		13	1140	1030	937	843	770	693	621	562	.20
		14	1060	957	870	783	715	644	577	522	.23
		15	989	893	812	731	667	601	539	487	.26
		16	928	838	761	685	626	563	505	456	.30
		17	873	788	717	645	589	530	475	430	.34
		18	825	744	677	609	556	501	449	406	.38
		19	781	705	641	577	527	474	425	384	.42
		20	742	670	609	548	501	451	404	365	.47
		21	707	638	580	522	477	429	385	348	.52
		22	675	609	554	498	455	410	367	332	.57
		23	645	583	530	477	435	392	351	317	.62
		24	618	558	508	457	417	376	337	304	.67
		25	594	536	487	438	400	361	323	292	.73
		26	571	515	468	422	385	347	311	281	.79
		27	550	496	451	406	371	334	299	270	.85
		28	530	479	435	391	358	322	289	261	.92
F_y = 36 ksi	Span in Feet	29	512	462	420	378	345	311	279	252	.98
		30	495	447	406	365	334	300	269	243	1.05
		31	479	432	393	354	323	291	261	236	1.12
		32	464	419	381	343	313	282	252	228	1.20
		33	450	406	369	332	303	273	245	221	1.27
		34	437	394	358	322	294	265	238	215	1.35
		36	412	372	338	304	278	250	224	203	1.52
		38	391	353	321	288	263	237	213	192	1.69
		40	371	335	305	274	250	225	202	183	1.87
		42	353	319	290	261	238	215	192	174	2.06
		44	337	305	277	249	228	205	184	166	2.27
		46	323	291	265	238	218	196	176	159	2.48
		48	309	279	254	228	209	188	168	152	2.70
		50	297	268	244	219	200	180	162	146	2.93
		52	285	258	234	211	193	173	155	140	3.16

Properties and Reaction Values										
S_x in.³		937	846	769	692	632	569	510	461	
V kips		648	583	526	466	424	376	336	302	For
R_1 kips		398	342	293	245	217	180	152	128	explanation
R_2 kips/in.		41.1	37.8	34.7	31.4	29.0	26.1	23.8	21.6	of deflection,
R_3 kips		821	690	583	477	407	332	273	226	see page 2-32
R_4 kips/in.		38.9	33.9	29.1	24.1	21.0	17.2	14.6	12.3	
R kips		542	474	414	355	319	271	235	204	

Load above heavy line is limited by maximum allowable web shear.

| W 21 | **BEAMS** | | | | | | | F_y = 36 ksi |

BEAMS
W Shapes
Allowable uniform loads in kips
for beams laterally supported
For beams laterally unsupported, see page 2-146

Designation		W 21							
Wt./ft		182	166	147	132	122	111	101	Deflection In.
Flange Width		12½	12⅜	12½	12½	12⅜	12⅜	12¼	
L_c		13.2	13.1	13.2	13.1	13.1	13.0	13.0	
L_u		37.6	34.8	30.3	27.2	25.4	23.3	21.3	

Span in Feet		182	166	147	132	122	111	101	Deflection
	11			457	409	375	341	308	.14
	12	543	486	434	389	360	329	300	.17
	13	508	463	401	359	333	303	277	.20
	14	472	430	372	334	309	282	257	.23
	15	440	401	347	312	288	263	240	.26
	16	413	376	326	292	270	247	225	.30
	17	389	354	307	275	254	232	212	.34
	18	367	334	290	260	240	219	200	.38
	19	348	317	274	246	228	208	189	.42
	20	330	301	261	234	216	197	180	.47
	21	315	287	248	223	206	188	171	.52
	22	300	274	237	212	197	179	163	.57
	23	287	262	227	203	188	171	156	.62
	24	275	251	217	195	180	164	150	.67
	25	264	241	208	187	173	158	144	.73
	26	254	232	200	180	166	152	138	.79
	27	245	223	193	173	160	146	133	.85
	28	236	215	186	167	154	141	128	.92
	29	228	208	180	161	149	136	124	.98
	30	220	201	174	156	144	131	120	1.05
	31	213	194	168	151	139	127	116	1.12
	32	206	188	163	146	135	123	112	1.20
	33	200	182	158	142	131	120	109	1.27
	34	194	177	153	137	127	116	106	1.35
	36	183	167	145	130	120	110	100	1.52
	38	174	158	137	123	114	104	95	1.69
	40	165	150	130	117	108	99	90	1.87
	42	157	143	124	111	103	94	86	2.06
	44	150	137	118	106	98	90	82	2.27
	46	144	131	113	102	94	86	78	2.48
	48	138	125	109	97	90	82	75	2.70
	50	132	120	104	93	86	79	72	2.93
	52	127	116	100	90	83	76	69	3.16

Properties and Reaction Values									
S_x in.³		417	380	329	295	273	249	227	
V kips		272	243	229	204	187	170	154	For
R_1 kips		111	94.7	80.2	70.0	60.1	53.1	46.4	explanation
R_2 kips/in.		19.7	17.8	17.1	15.4	14.3	13.1	11.9	of deflection,
R_3 kips		188	155	134	109	92.9	77.8	64.5	see page 2-32
R_4 kips/in.		10.4	8.45	9.00	7.44	6.35	5.41	4.48	
R kips		180	157	140	124	110	97	80	

Load above heavy line is limited by maximum allowable web shear.

| F_y = 36 ksi | BEAMS W Shapes Allowable uniform loads in kips for beams laterally supported | W 21 |

For beams laterally unsupported, see page 2-146

Designation	W 21					W 21			Deflection In.
Wt./ft	93	83	73	68	62	57	50	44	
Flange Width	8⅜	8⅜	8¼	8¼	8¼	6½	6½	6½	
L_c	8.90	8.80	8.80	8.70	8.70	6.90	6.90	6.60	
L_u	16.8	15.1	13.4	12.4	11.2	9.40	7.80	7.00	

Span in Feet (F_y = 36 ksi)

Span	93	83	73	68	62	57	50	44	Defl.
6							228	208	.04
7						246	214	185	.06
8	361	318	278	262	242	220	187	162	.07
9	338	301	266	246	224	195	166	144	.09
10	304	271	239	222	201	176	150	129	.12
11	276	246	217	202	183	160	136	118	.14
12	253	226	199	185	168	147	125	108	.17
13	234	208	184	171	155	135	115	99	.20
14	217	193	171	158	144	126	107	92	.23
15	203	181	159	148	134	117	100	86	.26
16	190	169	149	139	126	110	94	81	.30
17	179	159	141	130	118	103	88	76	.34
18	169	150	133	123	112	98	83	72	.38
19	160	143	126	117	106	93	79	68	.42
20	152	135	120	111	101	88	75	65	.47
21	145	129	114	106	96	84	71	62	.52
22	138	123	109	101	91	80	68	59	.57
23	132	118	104	96	87	76	65	56	.62
24	127	113	100	92	84	73	62	54	.67
25	122	108	96	89	80	70	60	52	.73
26	117	104	92	85	77	68	58	50	.79
27	113	100	89	82	75	65	55	48	.85
28	109	97	85	79	72	63	53	46	.92
29	105	93	82	76	69	61	52	45	.98
30	101	90	80	74	67	59	50	43	1.05
31	98	87	77	72	65	57	48	42	1.12
32	95	85	75	69	63	55	47	40	1.20
33	92	82	72	67	61	53	45	39	1.27
34	89	80	70	65	59	52	44	38	1.35
35	87	77	68	63	57	50	43	37	1.43
36	84	75	66	62	56	49	42	36	1.52
38	80	71	63	58	53	46	39	34	1.69
40	76	68	60	55	50	44	37	32	1.87
42	72	64	57	53	48	42	36	31	2.06
44	69	62	54	50	46	40	34	29	2.27
46	66	59	52	48	44	38	33	28	2.48
48	63	56	50	46	42	37	31	27	2.70
50	61	54	48	44	40	35	30	26	2.93
52	58	52	46	43	39	34	29	25	3.16

Properties and Reaction Values

	93	83	73	68	62	57	50	44	
S_x in.³	192	171	151	140	127	111	94.5	81.6	
V kips	181	159	139	131	121	123	114	104	For explanation of deflection, see page 2-32
R_1 kips	58.1	47.8	40.5	36.7	32.7	33.1	29.6	24.7	
R_2 kips/in.	13.8	12.2	10.8	10.2	9.50	9.62	9.03	8.32	
R_3 kips	86.9	68.9	53.9	47.6	40.5	42.4	35.0	28.3	
R_4 kips/in.	5.94	4.67	3.67	3.36	3.03	2.97	3.01	2.82	
R kips	106	85	67	59	51	53	46	38	

Load above heavy line is limited by maximum allowable web shear.

AMERICAN INSTITUTE OF STEEL CONSTRUCTION

BEAMS
W Shapes
Allowable uniform loads in kips
for beams laterally supported
For beams laterally unsupported, see page 2-146

W 18			F_y = 36 ksi

Designation		W 18								
Wt./ft		311	283	258	234	211	192	175	158	Deflection
Flange Width		12	11⅞	11¾	11⅝	11½	11½	11⅜	11¼	In.
L_c		12.7	12.6	12.4	12.3	12.2	12.1	12.0	11.9	
L_u		68.1	62.6	58.6	53.8	49.3	45.4	41.7	38.3	

Span in Feet / F_y = 36 ksi

Span	311	283	258	234	211	192	175	158	Deflection
10	977	881	791	704	631	563	514	460	.14
11	899	812	740	671	603	547	495	446	.17
12	824	744	678	615	553	502	454	409	.20
13	760	687	626	568	511	463	419	378	.23
14	706	638	582	527	474	430	389	351	.27
15	659	596	543	492	442	401	363	327	.31
16	618	558	509	461	415	376	341	307	.35
17	581	526	479	434	390	354	321	289	.39
18	549	496	452	410	369	334	303	273	.44
19	520	470	429	388	349	317	287	258	.49
20	494	447	407	369	332	301	272	246	.55
21	471	425	388	351	316	287	259	234	.60
22	449	406	370	336	302	274	248	223	.66
23	430	388	354	321	289	262	237	213	.72
24	412	372	339	308	277	251	227	205	.79
25	395	357	326	295	265	241	218	196	.85
26	380	344	313	284	255	232	210	189	.92
27	366	331	302	273	246	223	202	182	1.00
28	353	319	291	264	237	215	195	175	1.07
29	341	308	281	255	229	208	188	169	1.15
30	329	298	271	246	221	201	182	164	1.23
31	319	288	263	238	214	194	176	158	1.31
32	309	279	254	231	207	188	170	153	1.40
33	300	271	247	224	201	182	165	149	1.49
34	291	263	239	217	195	177	160	144	1.58
35	282	255	233	211	190	172	156	140	1.67
36	275	248	226	205	184	167	151	136	1.77
37	267	241	220	199	179	163	147	133	1.87
38	260	235	214	194	175	158	143	129	1.97
39	253	229	209	189	170	154	140	126	2.08
40	247	223	204	185	166	150	136	123	2.18
42	235	213	194	176	158	143	130	117	2.41
44	225	203	185	168	151	137	124	112	2.64

Properties and Reaction Values

	311	283	258	234	211	192	175	158	
S_x in.³	624	564	514	466	419	380	344	310	
V kips	489	440	396	352	316	281	257	230	For
R_1 kips	310	265	228	189	161	139	119	102	explanation
R_2 kips/in.	36.1	33.3	30.4	27.6	25.2	22.8	21.1	19.2	of deflection,
R_3 kips	633	534	448	370	308	254	216	178	see page 2-32
R_4 kips/in.	35.1	30.7	26.0	21.5	18.5	15.2	13.5	11.5	
R kips	436	382	334	286	249	219	193	169	

Load above heavy line is limited by maximum allowable web shear.

AMERICAN INSTITUTE OF STEEL CONSTRUCTION

| F_y = 36 ksi | BEAMS
W Shapes
Allowable uniform loads in kips
for beams laterally supported
For beams laterally unsupported, see page 2-146 | | | | | | | W 18 |

Designation		W 18							
Wt./ft		143	130	119	106	97	86	76	Deflection In.
Flange Width		11¼	11⅛	11¼	11¼	11⅛	11⅛	11	
L_c		11.8	11.8	11.9	11.8	11.8	11.7	11.6	
L_u		35.1	32.2	29.1	26.0	24.1	21.5	19.1	
	10	410	371	358	318	286	254	223	.14
	11	406	369	333	294	271	239	210	.17
	12	372	338	305	269	248	219	193	.20
	13	344	312	281	249	229	202	178	.23
	14	319	290	261	231	213	188	165	.27
	15	298	270	244	215	199	175	154	.31
	16	279	253	229	202	186	164	145	.35
	17	263	239	215	190	175	155	136	.39
	18	248	225	203	180	165	146	128	.44
	19	235	213	193	170	157	138	122	.49
	20	223	203	183	162	149	131	116	.55
	21	213	193	174	154	142	125	110	.60
	22	203	184	166	147	135	120	105	.66
	23	194	176	159	140	129	114	101	.72
	24	186	169	152	135	124	110	96	.79
	25	179	162	146	129	119	105	93	.85
	26	172	156	141	124	115	101	89	.92
	27	165	150	136	120	110	97	86	1.00
	28	160	145	131	115	106	94	83	1.07
	29	154	140	126	111	103	91	80	1.15
	30	149	135	122	108	99	88	77	1.23
	31	144	131	118	104	96	85	75	1.31
	32	140	127	114	101	93	82	72	1.40
	33	135	123	111	98	90	80	70	1.49
	34	131	119	108	95	88	77	68	1.58
	36	124	113	102	90	83	73	64	1.77
	38	118	107	96	85	78	69	61	1.97
	40	112	101	91	81	74	66	58	2.18
	42	106	97	87	77	71	63	55	2.41
	44	102	92	83	73	68	60	53	2.64

F_y = 36 ksi — *Span in Feet*

Properties and Reaction Values								
S_x in.³	282	256	231	204	188	166	146	
V kips	205	186	179	159	143	127	111	For
R_1 kips	86.7	74.6	68.1	56.9	49.7	41.0	34.7	explanation
R_2 kips/in.	17.3	15.9	15.6	14.0	12.7	11.4	10.1	of deflection,
R_3 kips	146	123	111	89.6	74.5	59.5	46.6	see page 2-32
R_4 kips/in.	9.25	7.97	8.55	7.14	5.79	4.78	3.79	
R kips	147	130	123	106	94	76	60	

Load above heavy line is limited by maximum allowable web shear.

AMERICAN INSTITUTE OF STEEL CONSTRUCTION

| W 18 | BEANS W Shapes Allowable uniform loads in kips for beams laterally supported For beams laterally unsupported, see page 2-146 | | | | | | | | $F_y = 36$ ksi |

Designation		W 18					W 18			Deflection In.
Wt./ft	71	65	60	55	50	46	40	35	
Flange Width	7⅝	7⅝	7½	7½	7½	6	6	6	
L_c	8.10	8.00	8.00	7.90	7.90	6.40	6.30	6.30	
L_u	15.5	14.4	13.3	12.1	11.0	9.40	8.20	6.70	

Span in Feet	71	65	60	55	50	46	40	35	Deflection In.
5								153	.03
6						187	162	152	.05
7	263	238	218	203	184	178	155	130	.07
8	251	232	214	195	176	156	135	114	.09
9	224	206	190	173	156	139	120	101	.11
10	201	185	171	156	141	125	108	91	.14
11	183	168	156	142	128	113	98	83	.17
12	168	154	143	130	117	104	90	76	.20
13	155	143	132	120	108	96	83	70	.23
14	144	132	122	111	101	89	77	65	.27
15	134	124	114	104	94	83	72	61	.31
16	126	116	107	97	88	78	68	57	.35
17	118	109	101	92	83	73	64	54	.39
18	112	103	95	87	78	69	60	51	.44
19	106	98	90	82	74	66	57	48	.49
20	101	93	86	78	70	62	54	46	.55
21	96	88	81	74	67	59	52	43	.60
22	91	84	78	71	64	57	49	41	.66
24	84	77	71	65	59	52	45	38	.79
26	77	71	66	60	54	48	42	35	.92
28	72	66	61	56	50	45	39	33	1.07
30	67	62	57	52	47	42	36	30	1.23
32	63	58	53	49	44	39	34	29	1.40
34	59	55	50	46	41	37	32	27	1.58
36	56	51	48	43	39	35	30	25	1.77
38	53	49	45	41	37	33	29	24	1.97
40	50	46	43	39	35	31	27	23	2.18
42	48	44	41	37	34	30	26	22	2.41
44	46	42	39	35	32	28	25	21	2.64

Properties and Reaction Values

	71	65	60	55	50	46	40	35	
S_x in.³	127	117	108	98.3	88.9	78.8	68.4	57.6	For explanation of deflection, see page 2-32
V kips	132	119	109	102	92	94	81	76	
R_1 kips	44.1	38.4	33.9	30.4	26.4	26.7	22.2	20.0	
R_2 kips/in.	11.8	10.7	9.86	9.27	8.43	8.55	7.48	7.13	
R_3 kips	63.9	53.3	45.5	39.4	32.6	34.3	26.1	21.9	
R_4 kips/in.	4.96	4.05	3.45	3.18	2.67	2.61	2.04	2.20	
R kips	81	67	58	51	42	43	33	30	

Load above heavy line is limited by maximum allowable web shear.

F_y = 36 ksi		BEANS W Shapes Allowable uniform loads in kips for beams laterally supported For beams laterally unsupported, see page 2-146				W 16

Designation		W 16				
Wt./ft		100	89	77	67	Deflection In.
Flange Width		10⅜	10⅜	10¼	10¼	
L_c		11.0	10.9	10.9	10.8	
L_u		28.1	25.0	21.9	19.3	
	9	286	253	216	186	.12
	10	277	246	212	185	.15
	11	252	223	193	168	.19
	12	231	205	177	154	.22
	13	213	189	163	143	.26
	14	198	175	152	132	.30
	15	185	164	142	124	.35
	16	173	153	133	116	.39
	17	163	144	125	109	.44
	18	154	136	118	103	.50
	19	146	129	112	98	.55
	20	139	123	106	93	.61
	21	132	117	101	88	.68
	22	126	112	96	84	.74
	23	121	107	92	81	.81
	24	116	102	88	77	.88
	25	111	98	85	74	.96
	26	107	94	82	71	1.04
	28	99	88	76	66	1.20
	30	92	82	71	62	1.38
	32	87	77	66	58	1.57
	34	82	72	62	55	1.78
	36	77	68	59	51	1.99
	38	73	65	56	49	2.22
	40	69	61	53	46	2.46

F_y = 36 ksi — Span in Feet

Properties and Reaction Values						
S_x in.³		175	155	134	117	
V kips		143	127	108	93	For explanation of deflection, see page 2-32
R_1 kips		58.6	48.7	38.9	32.3	
R_2 kips/in.		13.9	12.5	10.8	9.39	
R_3 kips		90.6	72.6	54.6	41.3	
R_4 kips/in.		7.33	6.04	4.59	3.47	
R kips		107	92	71	53	

Load above heavy line is limited by maximum allowable web shear.

AMERICAN INSTITUTE OF STEEL CONSTRUCTION

W 16	BEAMS	F_y = 36 ksi

W Shapes
Allowable uniform loads in kips
for beams laterally supported
For beams laterally unsupported, see page 2-146

Designation		W 16					W 16		
Wt./ft		57	50	45	40	36	31	26	Deflection
Flange Width		7⅛	7⅛	7	7	7	5½	5½	In.
L_c		7.50	7.50	7.40	7.40	7.40	5.80	5.60	
L_u		14.3	12.7	11.4	10.2	8.80	7.10	6.00	
	5						126	113	.04
	6					135	125	101	.06
	7	203	178	160	141	128	107	87	.08
	8	183	160	144	128	112	93	76	.10
	9	162	143	128	114	99	83	68	.12
	10	146	128	115	102	89	75	61	.15
	11	133	117	105	93	81	68	55	.19
	12	122	107	96	85	75	62	51	.22
	13	112	99	89	79	69	58	47	.26
	14	104	92	82	73	64	53	43	.30
	15	97	86	77	68	60	50	41	.35
	16	91	80	72	64	56	47	38	.39
	17	86	75	68	60	53	44	36	.44
	18	81	71	64	57	50	42	34	.50
	19	77	68	61	54	47	39	32	.55
	20	73	64	58	51	45	37	30	.61
	21	70	61	55	49	43	36	29	.68
	22	66	58	52	47	41	34	28	.74
	23	63	56	50	45	39	33	26	.81
	24	61	53	48	43	37	31	25	.88
	25	58	51	46	41	36	30	24	.96
	26	56	49	44	39	34	29	23	1.04
	27	54	48	43	38	33	28	23	1.12
	28	52	46	41	37	32	27	22	1.20
	29	50	44	40	35	31	26	21	1.29
	30	49	43	38	34	30	25	20	1.38
	31	47	41	37	33	29	24	20	1.48
	32	46	40	36	32	28	23	19	1.57
	33	44	39	35	31	27	23	18	1.67
	34	43	38	34	30	26	22	18	1.78
	35	42	37	33	29	26	21	17	1.88
	36	41	36	32	28	25	21	17	1.99
	37	39	35	31	28	24	20	16	2.10
	38	38	34	30	27	24	20	16	2.22
	39	37	33	30	26	23	19	16	2.34
	40	37	32	29	26	22	19	15	2.46

(Left margin: F_y = 36 ksi · Span in Feet)

Properties and Reaction Values								
S_x in.³	92.2	81.0	72.7	64.7	56.5	47.2	38.4	
V kips	102	89	80	70	67	63	56	For
R_1 kips	35.1	29.6	25.6	21.5	19.7	18.4	15.8	explanation
R_2 kips/in.	10.2	9.03	8.20	7.25	7.01	6.53	5.94	of deflection,
R_3 kips	48.6	37.9	31.1	24.4	21.4	19.5	15.0	see page 2-32
R_4 kips/in.	4.14	3.28	2.76	2.15	2.30	1.82	1.77	
R kips	63	49	41	32	29	26	21	

Load above heavy line is limited by maximum allowable web shear.

AMERICAN INSTITUTE OF STEEL CONSTRUCTION

| F_y = 36 ksi | BEABeams W Shapes Allowable uniform loads in kips for beams laterally supported For beams laterally unsupported, see page 2-146 | W 14 |

<table>
<tr><td colspan="2">Designation</td><td colspan="6" align="center">W 14</td><td rowspan="5">Deflection In.</td></tr>
<tr><td colspan="2">Wt./ft</td><td>233</td><td>211</td><td>193</td><td>176</td><td>159</td><td>145</td></tr>
<tr><td colspan="2">Flange Width</td><td>15⅞</td><td>15¾</td><td>15¾</td><td>15⅝</td><td>15⅝</td><td>15½</td></tr>
<tr><td colspan="2">L_c</td><td>16.8</td><td>16.7</td><td>16.6</td><td>16.5</td><td>16.4</td><td>16.4</td></tr>
<tr><td colspan="2">L_u</td><td>78.5</td><td>72.3</td><td>68.1</td><td>62.6</td><td>57.2</td><td>53.2</td></tr>
<tr><td rowspan="23">F_y = 36 ksi</td><td rowspan="23">Span in Feet</td></tr>
<tr><td>12</td><td>494</td><td>444</td><td>397</td><td>364</td><td>321</td><td>289</td><td>.25</td></tr>
<tr><td>13</td><td>457</td><td>412</td><td>378</td><td>342</td><td>309</td><td>283</td><td>.30</td></tr>
<tr><td>14</td><td>424</td><td>382</td><td>351</td><td>318</td><td>287</td><td>262</td><td>.34</td></tr>
<tr><td>15</td><td>396</td><td>357</td><td>327</td><td>297</td><td>268</td><td>245</td><td>.40</td></tr>
<tr><td>16</td><td>371</td><td>335</td><td>307</td><td>278</td><td>251</td><td>230</td><td>.45</td></tr>
<tr><td>17</td><td>349</td><td>315</td><td>289</td><td>262</td><td>237</td><td>216</td><td>.51</td></tr>
<tr><td>18</td><td>330</td><td>297</td><td>273</td><td>247</td><td>224</td><td>204</td><td>.57</td></tr>
<tr><td>19</td><td>313</td><td>282</td><td>258</td><td>234</td><td>212</td><td>193</td><td>.63</td></tr>
<tr><td>20</td><td>297</td><td>268</td><td>246</td><td>223</td><td>201</td><td>184</td><td>.70</td></tr>
<tr><td>21</td><td>283</td><td>255</td><td>234</td><td>212</td><td>192</td><td>175</td><td>.77</td></tr>
<tr><td>22</td><td>270</td><td>243</td><td>223</td><td>202</td><td>183</td><td>167</td><td>.85</td></tr>
<tr><td>23</td><td>258</td><td>233</td><td>213</td><td>194</td><td>175</td><td>160</td><td>.93</td></tr>
<tr><td>24</td><td>248</td><td>223</td><td>205</td><td>185</td><td>168</td><td>153</td><td>1.01</td></tr>
<tr><td>25</td><td>238</td><td>214</td><td>196</td><td>178</td><td>161</td><td>147</td><td>1.10</td></tr>
<tr><td>26</td><td>228</td><td>206</td><td>189</td><td>171</td><td>155</td><td>141</td><td>1.19</td></tr>
<tr><td>27</td><td>220</td><td>198</td><td>182</td><td>165</td><td>149</td><td>136</td><td>1.28</td></tr>
<tr><td>28</td><td>212</td><td>191</td><td>175</td><td>159</td><td>144</td><td>131</td><td>1.38</td></tr>
<tr><td>29</td><td>205</td><td>185</td><td>169</td><td>153</td><td>139</td><td>127</td><td>1.48</td></tr>
<tr><td>30</td><td>198</td><td>178</td><td>164</td><td>148</td><td>134</td><td>122</td><td>1.58</td></tr>
<tr><td>31</td><td>192</td><td>173</td><td>158</td><td>144</td><td>130</td><td>119</td><td>1.69</td></tr>
<tr><td>32</td><td>186</td><td>167</td><td>153</td><td>139</td><td>126</td><td>115</td><td>1.80</td></tr>
<tr><td>33</td><td>180</td><td>162</td><td>149</td><td>135</td><td>122</td><td>111</td><td>1.91</td></tr>
<tr><td>34</td><td>175</td><td>157</td><td>144</td><td>131</td><td>118</td><td>108</td><td>2.03</td></tr>
</table>

	Properties and Reaction Values						
S_x in.³	375	338	310	281	254	232	For explanation of deflection, see page 2-32
V kips	247	222	198	182	161	145	
R_1 kips	151	131	112	98.6	83.0	70.7	
R_2 kips/in.	25.4	23.3	21.1	19.7	17.7	16.2	
R_3 kips	296	247	206	177	143	119	
R_4 kips/in.	27.2	23.5	19.4	17.6	14.2	11.9	
R kips	240	213	186	168	145	127	

Load above heavy line is limited by maximum allowable web shear.

					BEAMS				$F_y = 36$ ksi	

W 14 — BEAMS / W Shapes / Allowable uniform loads in kips / for beams laterally supported
For beams laterally unsupported, see page 2-146

Designation		W 14					W 14				
Wt./ft		132	120	109	99	90	82	74	68	61	Deflection
Flange Width		14¾	14⅝	14⅝	14⅝	14½	10⅛	10⅛	10	10	In.
L_c		15.5	15.5	15.4	15.4	15.3	10.7	10.6	10.6	10.6	
L_u		47.7	44.1	40.6	37.0	34.0	28.1	25.9	23.9	21.5	
	9						210	184	168	150	.14
	10						195	177	163	146	.18
	11						177	161	148	133	.21
	12	272	246	217	198	178	162	148	136	122	.25
	13	255	232	211	191	174	150	136	126	112	.30
	14	236	215	196	178	162	139	127	117	104	.34
	15	221	201	183	166	151	130	118	109	97	.40
	16	207	188	171	155	142	122	111	102	91	.45
	17	195	177	161	146	133	115	104	96	86	.51
	18	184	167	152	138	126	108	99	91	81	.57
	19	174	158	144	131	119	103	93	86	77	.63
	20	166	150	137	124	113	97	89	82	73	.70
	21	158	143	130	118	108	93	84	78	70	.77
	22	150	137	125	113	103	89	81	74	66	.85
	23	144	131	119	108	98	85	77	71	63	.93
	24	138	125	114	104	94	81	74	68	61	1.01
	25	132	120	110	99	91	78	71	65	58	1.10
	26	127	116	105	96	87	75	68	63	56	1.19
	27	123	111	101	92	84	72	66	60	54	1.28
	28	118	107	98	89	81	70	63	58	52	1.38
	29	114	104	94	86	78	67	61	56	50	1.48
	30	110	100	91	83	76	65	59	54	49	1.58
	31	107	97	88	80	73	63	57	53	47	1.69
	32	103	94	86	78	71	61	55	51	46	1.80
	33	100	91	83	75	69	59	54	49	44	1.91
	34	97	89	81	73	67	57	52	48	43	2.03

Left margin: $F_y = 36$ ksi — Span in Feet

Properties and Reaction Values

	132	120	109	99	90	82	74	68	61	
S_x in.³	209	190	173	157	143	123	112	103	92.2	
V kips	136	123	108	99	89	105	92	84	75	For
R_1 kips	64.7	56.9	48.7	41.4	35.9	49.2	41.8	37.0	32.0	explanation
R_2 kips/in.	15.3	14.0	12.5	11.5	10.5	12.1	10.7	9.86	8.91	of deflection,
R_3 kips	107	89.6	72.0	60.9	50.2	68.7	54.6	46.3	37.6	see page 2-32
R_4 kips/in.	10.9	9.23	7.19	6.32	5.24	6.64	5.01	4.33	3.60	
R kips	118	106	92	82	69	92	72	61	50	

Load above heavy line is limited by maximum allowable web shear.

AMERICAN INSTITUTE OF STEEL CONSTRUCTION

| F_y = 36 ksi | BEAMS
W Shapes
Allowable uniform loads in kips
for beams laterally supported
For beams laterally unsupported, see page 2-146 | | | | | | | | W 14 |

Designation		W 14			W 14			W 14		
Wt./ft		53	48	43	38	34	30	26	22	Deflection In.
Flange Width		8	8	8	6¾	6¾	6¾	5	5	
L_c		8.50	8.50	8.40	7.10	7.10	7.10	5.30	5.30	
L_u		17.7	16.0	14.4	11.5	10.2	8.70	7.00	5.60	
	5							102	91	.04
	6				126	115	108	93	77	.06
	7				124	110	95	80	66	.09
	8	148	135	120	108	96	83	70	57	.11
	9	137	124	110	96	86	74	62	51	.14
	10	123	111	99	86	77	67	56	46	.18
	11	112	101	90	79	70	60	51	42	.21
	12	103	93	83	72	64	55	47	38	.25
	13	95	86	76	67	59	51	43	35	.30
	14	88	80	71	62	55	48	40	33	.34
	15	82	74	66	58	51	44	37	31	.40
	16	77	70	62	54	48	42	35	29	.45
	17	72	66	58	51	45	39	33	27	.51
	18	68	62	55	48	43	37	31	26	.57
	19	65	59	52	46	41	35	29	24	.63
	20	62	56	50	43	38	33	28	23	.70
	21	59	53	47	41	37	32	27	22	.77
	22	56	51	45	39	35	30	25	21	.85
	23	54	48	43	38	33	29	24	20	.93
	24	51	46	41	36	32	28	23	19	1.01
	25	49	45	40	35	31	27	22	18	1.10
	26	47	43	38	33	30	26	22	18	1.19
	27	46	41	37	32	29	25	21	17	1.28
	28	44	40	35	31	27	24	20	16	1.38
	30	41	37	33	29	26	22	19	15	1.58
	32	39	35	31	27	24	21	17	14	1.80
	34	36	33	29	25	23	20	16	14	2.03

Span in Feet — F_y = 36 ksi

Properties and Reaction Values									
S_x in.³	77.8	70.3	62.7	54.6	48.6	42.0	35.3	29.0	
V kips	74	68	60	63	57	54	51	46	For
R_1 kips	31.6	27.8	23.8	19.6	16.9	15.0	14.2	12.0	explanation
R_2 kips/in.	8.79	8.08	7.25	7.37	6.77	6.42	6.06	5.46	of deflection,
R_3 kips	37.3	31.2	25.0	25.3	20.9	17.8	17.0	13.0	see page 2-32
R_4 kips/in.	3.37	2.93	2.40	2.51	2.23	2.26	1.74	1.62	
R kips	49	41	33	34	29	26	23	19	

Load above heavy line is limited by maximum allowable web shear.

AMERICAN INSTITUTE OF STEEL CONSTRUCTION

W 12	BEAMS										F_y = 36 ksi

W Shapes
Allowable uniform loads in kips
for beams laterally supported
For beams laterally unsupported, see page 2-146

Designation		W 12								W 12		
Wt./ft		136	120	106	96	87	79	72	65	58	53	Deflection In.
Flange Width		12⅜	23⅜	12¼	12⅛	12⅛	12⅛	12	12	10	10	
L_c		13.1	13.0	12.9	12.8	12.8	12.8	12.7	12.7	10.6	10.6	
L_u		53.2	48.2	43.3	39.9	36.2	33.3	30.5	27.7	24.4	22.0	

	Span in Feet												Deflection In.
	9	305	268							126	120	.17
	10	295	258	226	201	186	168	152	136	124	112	.20
	11	268	235	209	189	170	154	140	127	112	102	.25
	12	246	215	191	173	156	141	129	116	103	93	.29
	13	227	199	177	160	144	130	119	107	95	86	.35
	14	210	184	164	148	134	121	110	99	88	80	.40
	15	196	172	153	138	125	113	103	93	82	75	.46
	16	184	161	144	130	117	106	96	87	77	70	.52
	17	173	152	135	122	110	100	91	82	73	66	.59
	18	164	143	128	115	104	94	86	77	69	62	.66
	19	155	136	121	109	98	89	81	73	65	59	.74
	20	147	129	115	104	93	85	77	70	62	56	.82
	21	140	123	109	99	89	81	73	66	59	53	.90
	22	134	117	104	94	85	77	70	63	56	51	.99
	23	128	112	100	90	81	74	67	61	54	49	1.08
	24	123	108	96	86	78	71	64	58	51	47	1.18
	25	118	103	92	83	75	68	62	56	49	45	1.28
	26	113	99	88	80	72	65	59	54	48	43	1.38
	27	109	96	85	77	69	63	57	52	46	41	1.49
	28	105	92	82	74	67	61	55	50	44	40	1.61
	29	102	89	79	72	64	58	53	48	43	39	1.72
	30	98	86	77	69	62	56	51	46	41	37	1.84

F_y = 36 ksi

Properties and Reaction Values

	136	120	106	96	87	79	72	65	58	53	
S_x in.³	186	163	145	131	118	107	97.4	87.9	78.0	70.6	
V kips	153	134	113	101	93	84	76	68	63	60	For explanation of deflection, see page 2-32
R_1 kips	90.9	76.4	61.1	53.1	45.9	40.1	35.1	30.4	29.4	25.6	
R_2 kips/in.	18.8	16.9	14.5	13.1	12.2	11.2	10.2	9.27	8.55	8.20	
R_3 kips	160	128	96.7	78.9	67.9	56.4	47.1	38.6	35.3	31.3	
R_4 kips/in.	18.0	15.1	10.9	8.90	8.24	6.98	5.93	4.95	3.66	3.62	
R kips	**157**	**136**	112	99	89	79	68	56	48	44	

Load above heavy line is limited by maximum allowable web shear.
Values of R in **bold face** exceed maximum web shear V.

AMERICAN INSTITUTE OF STEEL CONSTRUCTION

F_y = 36 ksi											**BEAMS** **W Shapes** **Allowable uniform loads in kips** **for beams laterally supported** For beams laterally unsupported, see page 2-146	W 12

Designation		W 12			W 12			W 12				
Wt./ft		50	45	40	35	30	26	22	19	16	14	Deflection In.
Flange Width		8⅛	8	8	6½	6½	6½	4	4	4	4	
L_c		8.50	8.50	8.40	6.90	6.90	6.90	4.30	4.20	4.10	3.50	
L_u		19.6	17.7	16.0	12.6	10.8	9.40	6.40	5.30	4.30	4.20	
	3									76	69	.02
	4							92	82	68	59	.03
	5							80	67	54	47	.05
	6				108	92	81	67	56	45	39	.07
	7	130	116		103	87	76	57	48	39	34	.10
	8	128	115	101	90	76	66	50	42	34	30	.13
	9	114	102	91	80	68	59	45	37	30	26	.17
	10	102	92	82	72	61	53	40	34	27	24	.20
	11	93	84	75	66	56	48	37	31	25	21	.25
	12	85	77	69	60	51	44	34	28	23	20	.29
	13	79	71	63	56	47	41	31	26	21	18	.35
	14	73	66	59	52	44	38	29	24	19	17	.40
	15	68	61	55	48	41	35	27	22	18	16	.46
	16	64	58	51	45	38	33	25	21	17	15	.52
	17	60	54	48	42	36	31	24	20	16	14	.59
	18	57	51	46	40	34	29	22	19	15	13	.66
	19	54	48	43	38	32	28	21	18	14	12	.74
	20	51	46	41	36	31	26	20	17	14	12	.82
	21	49	44	39	34	29	25	19	16	13	11	.90
	22	47	42	37	33	28	24	18	15	12	11	.99
	23	45	40	36	31	27	23	17	15	12	10	1.08
	24	43	38	34	30	25	22	17	14	11	10	1.18
	25	41	37	33	29	24	21	16	13	11	9	1.28
	26	39	35	32	28	24	20	15	13	10	9	1.38
	28	37	33	29	26	22	19	14	12	10	8	1.61
	30	34	31	27	24	20	18	13	11	9	8	1.84

(Left margin, rotated: F_y = 36 ksi · Span in Feet)

Properties and Reaction Values											
S_x in.³	64.7	58.1	51.9	45.6	38.6	33.4	25.4	21.3	17.1	14.9	
V kips	65	58	51	54	46	40	46	41	38	34	For explanation of deflection, see page 2-32
R_1 kips	30.2	24.9	21.9	17.8	14.5	12.0	13.5	11.3	9.80	8.17	
R_2 kips/in.	8.79	7.96	7.01	7.13	6.18	5.46	6.18	5.58	5.23	4.75	
R_3 kips	36.7	30.0	23.5	24.2	17.9	13.9	17.6	13.7	10.8	8.65	
R_4 kips/in.	3.97	3.32	2.56	2.54	1.98	1.60	2.06	1.87	2.05	1.83	
R kips	51	42	32	33	25	20	25	20	18	15	

Load above heavy line is limited by maximum allowable web shear.

AMERICAN INSTITUTE OF STEEL CONSTRUCTION

BEAMS
W Shapes
Allowable uniform loads in kips
for beams laterally supported
For beams laterally unsupported, see page 2-146

W 10

F_y = 36 ksi

Designation			W 10								
Wt./ft			112	100	88	77	68	60	54	49	Deflection In.
Flange Width			10⅜	10⅜	10¼	10¼	10⅛	10⅛	10	10	
L_c			11.0	10.9	10.8	10.8	10.7	10.6	10.6	10.6	
L_u			53.2	48.2	43.3	38.6	34.8	31.1	28.2	26.0	
		8	247	217	189	162	141	124	108	98	.16
		9	222	197	173	151	133	117	106	96	.20
		10	200	177	156	136	120	106	95	86	.25
		11	181	161	142	124	109	96	86	79	.30
		12	166	148	130	113	100	88	79	72	.35
		13	154	136	120	105	92	81	73	67	.42
		14	143	127	111	97	86	75	68	62	.48
		15	133	118	104	91	80	70	63	58	.55
		16	125	111	98	85	75	66	59	54	.63
		17	117	104	92	80	71	62	56	51	.71
		18	111	99	87	76	67	59	53	48	.80
		19	105	93	82	72	63	56	50	46	.89
		20	100	89	78	68	60	53	48	43	.98
		21	95	84	74	65	57	50	45	41	1.08
		22	91	81	71	62	55	48	43	39	1.19
		23	87	77	68	59	52	46	41	38	1.30
		24	83	74	65	57	50	44	40	36	1.42

F_y = 36 ksi

Span in Feet

Properties and Reaction Values									
S_x in.³	126	112	98.5	85.9	75.7	66.7	60.0	54.6	
V kips	124	109	94	81	70	62	54	49	For
R_1 kips	84.1	70.7	58.4	47.2	38.4	32.7	27.5	24.0	explanation
R_2 kips/in.	17.9	16.2	14.4	12.6	11.2	9.98	8.79	8.08	of deflection,
R_3 kips	150	121	95.5	73.4	57.7	45.8	36.0	30.3	see page 2-32
R_4 kips/in.	18.5	15.5	12.6	9.88	7.93	6.52	5.00	4.30	
R kips	**147**	**127**	**109**	**91**	**78**	**68**	54	45	

Load above heavy line is limited by maximum allowable web shear.
Values of R in **bold face** exceed maximum web shear V.

AMERICAN INSTITUTE OF STEEL CONSTRUCTION

| F_y = 36 ksi | | BEAMS W Shapes — Allowable uniform loads in kips for beams laterally supported. For beams laterally unsupported, see page 2-146 | | | | | | | | | W 10 |

BEAMS — W Shapes — Allowable uniform loads in kips for beams laterally supported

For beams laterally unsupported, see page 2-146

Designation	W 10			W 10			W 10				Deflection In.
Wt./ft	45	39	33	30	26	22	19	17	15	12	
Flange Width	8	8	8	5¾	5¾	5¾	4	4	4	4	
L_c	8.50	8.40	8.40	6.10	6.10	6.10	4.20	4.20	4.20	3.90	
L_u	22.8	19.8	16.5	13.1	11.4	9.40	7.20	6.10	5.00	4.30	
3								70	66	54	.02
4							74	64	55	43	.04
5				90	77	70	60	51	44	35	.06
6			81	86	74	61	50	43	36	29	.09
7	102	90	79	73	63	52	43	37	31	25	.12
8	97	83	69	64	55	46	37	32	27	22	.16
9	86	74	62	57	49	41	33	29	24	19	.20
10	78	67	55	51	44	37	30	26	22	17	.25
11	71	61	50	47	40	33	27	23	20	16	.30
12	65	56	46	43	37	31	25	21	18	14	.35
13	60	51	43	39	34	28	23	20	17	13	.42
14	56	48	40	37	32	26	21	18	16	12	.48
16	49	42	35	32	28	23	19	16	14	11	.63
18	43	37	31	29	25	20	17	14	12	10	.80
20	39	33	28	26	22	18	15	13	11	8.6	.98
22	35	30	25	23	20	17	14	12	10	7.8	1.19
24	32	28	23	21	18	15	12	11	9.1	7.2	1.42

F_y = 36 ksi — Span in Feet

Properties and Reaction Values

S_x in.³	49.1	42.1	35.0	32.4	27.9	23.2	18.8	16.2	13.8	10.9	For explanation of deflection, see page 2-32
V kips	51	45	41	45	39	35	37	35	33	27	
R_1 kips	26.0	21.0	18.3	16.7	13.5	10.7	12.1	10.7	9.39	7.05	
R_2 kips/in.	8.32	7.48	6.89	7.13	6.18	5.70	5.94	5.70	5.46	4.51	
R_3 kips	33.3	26.3	21.0	23.9	17.9	14.4	16.0	13.8	11.7	7.74	
R_4 kips/in.	4.19	3.64	3.53	3.09	2.37	2.31	2.36	2.54	2.76	2.03	
R kips	48	39	33	35	26	22	24	23	21	15	

Load above heavy line is limited by maximum allowable web shear.

AMERICAN INSTITUTE OF STEEL CONSTRUCTION

| W 8 | BEAMS W Shapes Allowable uniform loads in kips for beams laterally supported For beams laterally unsupported, see page 2-146 | | | | | | $F_y = 36$ ksi |

Designation		W 8						
Wt./ft		67	58	48	40	35	31	Deflection In.
Flange Width		8¼	8¼	8⅛	8⅛	8	8	
L_c		8.70	8.70	8.60	8.50	8.50	8.40	
L_u		39.9	35.3	30.3	25.3	22.6	20.1	
	6	148	129		86	72	66	.11
	7	137	118	98	80	71	62	.15
	8	120	103	86	70	62	54	.20
	9	106	92	76	62	55	48	.25
	10	96	82	69	56	49	44	.31
	11	87	75	62	51	45	40	.37
	12	80	69	57	47	41	36	.44
	13	74	63	53	43	38	34	.52
	14	68	59	49	40	35	31	.60
	15	64	55	46	37	33	29	.69
	16	60	51	43	35	31	27	.79
	17	56	48	40	33	29	26	.89
	18	53	46	38	31	27	24	1.00
	19	50	43	36	30	26	23	1.11
	20	48	41	34	28	25	22	1.23

$F_y = 36$ ksi Span in Feet

Properties and Reaction Values								
S_x in.³		60.4	52.0	43.3	35.5	31.2	27.5	For explanation of deflection, see page 2-32
V kips		74	64	49	43	36	33	
R_1 kips		48.7	39.8	28.2	22.7	18.4	15.9	
R_2 kips/in.		13.5	12.1	9.50	8.55	7.37	6.77	
R_3 kips		84.9	66.9	42.7	33.0	24.8	20.5	
R_4 kips/in.		13.5	11.5	6.73	6.18	4.54	4.07	
R kips		**96**	**82**	**61**	**53**	**41**	**35**	

Load above heavy line is limited by maximum allowable web shear.
Values of R in **bold face** exceed maximum web shear V.

AMERICAN INSTITUTE OF STEEL CONSTRUCTION

| F_y = 36 ksi | | BEAMS W Shapes Allowable uniform loads in kips for beams laterally supported | | | | | | | W 8 |

For beams laterally unsupported, see page 2-146

Designation		W 8		W 8		W 8			
Wt./ft		28	24	21	18	15	13	10	Deflection In.
Flange Width		6½	6½	5¼	5¼	4	4	4	
L_c		6.90	6.90	5.60	5.50	4.20	4.20	4.20	
L_u		17.5	15.2	11.8	9.90	7.20	5.90	4.70	

F_y = 36 ksi	Span in Feet		W 28	W 24	W 21	W 18	W 15	W 13	W 10	Deflection In.
		2						53		.01
		3					57	52	39	.03
		4			60	54	47	39	31	.05
		5	66	56	58	48	37	31	25	.08
		6	64	55	48	40	31	26	21	.11
		7	55	47	41	34	27	22	18	.15
		8	48	41	36	30	23	20	15	.20
		9	43	37	32	27	21	17	14	.25
		10	38	33	29	24	19	16	12	.31
		11	35	30	26	22	17	14	11	.37
		12	32	28	24	20	16	13	10	.44
		13	30	25	22	19	14	12	10	.52
		14	27	24	21	17	13	11	8.8	.60
		15	26	22	19	16	12	10	8.2	.69
		16	24	21	18	15	12	10	7.7	.79
		17	23	19	17	14	11	9.2	7.3	.89
		18	21	18	16	13	10	8.7	6.9	1.00
		19	20	17	15	13	10	8.3	6.5	1.11
		20	19	17	14	12	9.3	7.8	6.2	1.23

Properties and Reaction Values

S_x in.³		24.3	20.9	18.2	15.2	11.8	9.91	7.81	
V kips		33	28	30	27	29	26	19	For explanation of deflection, see page 2-32
R_1 kips		15.9	12.7	12.1	10.2	10.9	9.39	6.31	
R_2 kips/in.		6.77	5.82	5.94	5.46	5.82	5.46	4.04	
R_3 kips		21.2	15.6	16.1	12.9	13.9	11.4	6.47	
R_4 kips/in.		3.78	2.84	2.89	2.77	3.52	3.65	1.86	
R kips		**34**	26	26	23	26	24	13	

Load above heavy line is limited by maximum allowable web shear.
Values of R in **bold face** exceed maximum web shear V.

| W 6–5–4 | | | | **BEAMS**
W Shapes
Allowable uniform loads in kips
for beams laterally supported
For beams laterally unsupported, see page 2-146 | | | | | | | F_y = 36 ksi |

Designation	W 6			W 6				W 5			W 4	
Wt./ft	25	20	15*	16	12	9	Deflection	19	16	Deflection	13	Deflection
Flange Width	6⅛	6	6	4	4	4	In.	5	5	In.	4	In.
L_c	6.40	6.40	6.30	4.30	4.20	4.20		5.30	5.30		4.30	
L_u	20.0	16.4	12.0	12.0	8.60	6.70		19.5	16.7		15.6	
2				40			.02				34	.02
3			40	47	39	29	.04		35	.04	29	.06
4	59	46	38	40	29	22	.07	40	34	.08	22	.10
5	53	42	30	32	23	18	.10	32	27	.12	17	.15
6	44	35	25	27	19	15	.15	27	22	.18	14	.22
7	38	30	22	23	17	13	.20	23	19	.24	12	.30
8	33	27	19	20	14	11	.26	20	17	.31	11	.39
9	29	24	17	18	13	10	.33	18	15	.40	10	.50
10	26	21	15	16	12	8.8	.41	16	13	.49	8.6	.61
11	24	19	14	15	11	8.0	.50	15	12	.59		
12	22	18	13	13	10	7.3	.59	13	11	.71		
13	20	16	12	12	8.9	6.8	.69					
14	19	15	11	12	8.3	6.3	.80					

Span in Feet — F_y = 36 ksi

Properties and Reaction Values

S_x in.³	16.7	13.4	9.72	10.2	7.31	5.56		10.2	8.51		5.46	
V kips	29	23	20	24	20	14	For	20	17	For	17	For
R_1 kips	15.4	11.6	8.54	11.6	8.54	5.68	explanation	13.0	10.7	explanation	11.4	explanation
R_2 kips/in.	7.60	6.18	5.46	6.18	5.46	4.04	of deflection,	6.42	5.70	of deflection,	6.65	of deflection,
R_3 kips	24.9	16.3	11.5	17.2	11.9	6.63	see	18.8	14.4	see	17.8	see
R_4 kips/in.	6.91	4.75	4.78	4.23	4.41	2.37	page 2-32	5.44	4.69	page 2-32	9.36	page 2-32
R kips	**42**	**33**	**28**	**32**	**27**	**15**		**35**	**31**		**35**	

*Indicates noncompact shape.
Load above heavy line is limited by maximum allowable web shear.
Values of R in **bold face** exceed maximum web shear V.

AMERICAN INSTITUTE OF STEEL CONSTRUCTION

F_y = 36 ksi							

BEAMS
M Shapes
Allowable uniform loads in kips
for beams laterally supported
For beams laterally unsupported, see page 2-146

M 14–12

Designation		M 14			M 12			
Wt./ft		18	Deflection In.	11.8	10.8	10	Deflection In.
Flange Width		4		3⅛	3⅛	3¼	
L_c		3.60		2.70	2.50	2.30	
L_u		4.00		3.00	3.10	3.30	
	3	87	.02	61	55	51	.02
	4	84	.03	48	43	41	.03
	5	67	.04	38	35	33	.05
	6	56	.06	32	29	27	.07
	7	48	.09	27	25	23	.10
	8	42	.11	24	22	20	.13
	9	37	.14	21	19	18	.17
	10	33	.18	19	17	16	.20
	11	30	.21	17	16	15	.25
	12	28	.25	16	14	14	.29
	13	26	.30	15	13	13	.35
	14	24	.34	14	12	12	.40
	15	22	.40	13	12	11	.46
	16	21	.45	12	11	10	.52
	17	20	.51	11	10	10	.59
	18	19	.57	11	10	9.1	.66
	19	18	.63	10	9.1	8.6	.74
	20	17	.70	10	8.6	8.2	.82
	22	15	.85	8.6	7.8	7.4	.99
	24	14	1.01	7.9	7.2	6.8	1.18
	26	13	1.19	7.3	6.6	6.3	1.38
	28	12	1.38	6.8	6.2	5.8	1.61
	30	11	1.58	6.3	5.8	5.4	1.84
	32	10	1.80				
	34	10	2.03				

F_y = 36 ksi — Span in Feet

Properties and Reaction Values							
S_x in.³		21.1		12.0	10.9	10.3	
V kips		43	For	31	28	26	For
R_1 kips		7.98	explanation	5.91	4.75	4.43	explanation
R_2 kips/in.		5.11	of deflection,	4.21	3.80	3.54	of deflection,
R_3 kips		10.6	see page 2-32	7.21	5.98	4.98	see page 2-32
R_4 kips/in.		1.61		1.26	1.00	.940	
R kips		16		12	9	8	

Load above heavy line is limited by maximum allowable web shear.

AMERICAN INSTITUTE OF STEEL CONSTRUCTION

BEAMS
M Shapes
Allowable uniform loads in kips
for beams laterally supported

M 10–8–6–5

$F_y = 36$ ksi

For beams laterally unsupported, see page 2-146

Designation		M 10				M 8		M 6		M 5	
Wt./ft		9	8	7.5	Deflection In.	6.5	Deflection In.	4.4	Deflection In.	18.9	Deflection In.
Flange Width		2¾	2¾	2¾		2¼		1⅞		5	
L_c		2.60	2.30	2.20		2.40		1.90		5.30	
L_u		2.70	2.70	2.70		2.50		2.40		19.3	
	1							20	.004		
	2	45	40	37	.01	31	.01	19	.02		
	3	41	37	35	.02	24	.03	13	.04	46	.04
	4	31	27	26	.04	18	.05	10	.07	38	.08
	5	25	22	21	.06	15	.08	7.6	.10	31	.12
	6	20	18	17	.09	12	.11	6.3	.15	25	.18
	7	18	16	15	.12	10	.15	5.4	.20	22	.24
	8	15	14	13	.16	9.1	.20	4.8	.26	19	.31
	9	14	12	12	.20	8.1	.25	4.2	.33	17	.40
	10	12	11	10	.25	7.3	.31	3.8	.41	15	.49
	11	11	10	9.5	.30	6.7	.37	3.5	.50	14	.59
	12	10	9.2	8.7	.35	6.1	.44	3.2	.59	13	.71
	13	9.5	8.5	8.0	.42	5.6	.52	2.9	.69		
	14	8.8	7.9	7.4	.48	5.2	.60	2.7	.80		
	15	8.2	7.3	6.9	.55	4.9	.69				
	16	7.7	6.9	6.5	.63	4.6	.79				
	17	7.2	6.5	6.1	.71	4.3	.89				
	18	6.8	6.1	5.8	.80	4.1	1.00				
	20	6.1	5.5	5.2	.98	3.7	1.23				
	22	5.6	5.0	4.7	1.19						
	24	5.1	4.6	4.3	1.42						

($F_y = 36$ ksi, Span in Feet)

Properties and Reaction Values

	9	8	7.5		6.5		4.4		18.9	
S_x in.³	7.76	6.94	6.57		4.62		2.40		9.63	
V kips	23	20	19	For explanation of deflection, see page 2-32	16	For explanation of deflection, see page 2-32	10	For explanation of deflection, see page 2-32	23	For explanation of deflection, see page 2-32
R_1 kips	5.25	3.66	3.38		4.01		2.96		16.4	
R_2 kips/in.	3.73	3.35	3.09		3.21		2.71		7.51	
R_3 kips	5.76	4.61	3.98		4.40		3.25		23.4	
R_4 kips/in.	1.15	.947	.778		1.00		.884		9.28	
R kips	10	8	7		8		6		**43**	

Load above heavy line is limited by maximum allowable web shear.
Values of R in **bold face** exceed maximum web shear V.

AMERICAN INSTITUTE OF STEEL CONSTRUCTION

| F_y = 36 ksi | BEAMS S Shapes — Allowable uniform loads in kips for beams laterally supported. For beams laterally unsupported, see page 2-146 | S 24–20 |

Designation	S 24		S 24			Deflection In.	S 20		S 20		Deflection In.
Wt./ft	121	106	100	90	80		96	86	75	66	
Flange Width	8	7⅞	7¼	7⅛	7		7¼	7	6⅜	6¼	
L_c	8.50	8.30	7.60	7.50	7.40		7.60	7.50	6.70	6.60	
L_u	16.6	16.2	12.2	12.0	11.8		15.1	14.8	11.8	11.5	

Span in Feet (F_y = 36 ksi)

Span	121	106	100	90	80	Defl.	96	86	75	66	Defl.
5							468		366		.03
6			515	432		.04	436	386	338	291	.04
7	564		450	423		.05	373	351	290	269	.06
8	511	437	394	370	346	.07	327	307	253	236	.08
9	454	422	350	329	308	.08	290	273	225	209	.10
10	409	380	315	296	277	.10	261	246	203	188	.12
11	372	346	287	269	252	.12	238	223	184	171	.15
12	341	317	263	247	231	.15	218	205	169	157	.18
13	314	292	242	228	213	.17	201	189	156	145	.21
14	292	272	225	212	198	.20	187	175	145	135	.24
15	272	253	210	197	185	.23	174	164	135	126	.28
16	255	238	197	185	173	.26	163	153	127	118	.31
17	240	224	185	174	163	.30	154	144	119	111	.36
18	227	211	175	165	154	.33	145	136	113	105	.40
19	215	200	166	156	146	.37	138	129	107	99	.44
20	204	190	158	148	139	.41	131	123	101	94	.49
21	195	181	150	141	132	.45	124	117	97	90	.54
22	186	173	143	135	126	.50	119	112	92	86	.59
23	178	165	137	129	121	.54	114	107	88	82	.65
24	170	158	131	123	116	.59	109	102	84	79	.71
25	163	152	126	118	111	.64	105	98	81	75	.77
26	157	146	121	114	107	.69	101	94	78	72	.83
27	151	141	117	110	103	.75	97	91	75	70	.90
28	146	136	113	106	99	.80	93	88	72	67	.96
30	136	127	105	99	92	.92	87	82	68	63	1.11
32	128	119	99	93	87	1.05	82	77	63	59	1.26
34	120	112	93	87	82	1.18	77	72	60	55	1.42
36	114	106	88	82	77	1.33	73	68	56	52	1.59
38	108	100	83	78	73	1.48	69	65	53	50	1.77
40	102	95	79	74	69	1.64	65	61	51	47	1.97
42	97	91	75	71	66	1.81	62	58	48	45	2.17
44	93	86	72	67	63	1.98	59	56	46	43	2.38
46	89	83	69	64	60	2.17	57	53	44	41	2.60
48	85	79	66	62	58	2.36	54	51	42	39	2.83
50	82	76	63	59	55	2.56	52	49	41	38	3.07
52	79	73	61	57	53	2.77					
54	76	70	58	55	51	2.99					
56	73	68	56	53	50	3.21					
58	70	66	54	51	48	3.45					
60	68	63	53	49	46	3.69					

Properties and Reaction Values

	121	106	100	90	80		96	86	75	66	
S_x in.³	258	240	199	187	175		165	155	128	119	
V kips	282	219	257	216	173	For explanation of deflection, see page 2-32	234	193	183	145	For explanation of deflection, see page 2-32
R_1 kips	95.0	73.7	77.4	65.0	52.0		83.2	68.6	61.3	48.7	
R_2 kips/in.	19.0	14.7	17.7	14.9	11.9		19.0	15.7	15.1	12.0	
R_3 kips	152	104	122	94.0	67.3		140	105	92.0	65.3	
R_4 kips/in.	11.7	5.46	12.1	7.16	3.66		16.8	9.42	9.86	4.96	
R kips	162	123	139	117	80		150	124	114	83	

Load above heavy line is limited by maximum allowable web shear.

AMERICAN INSTITUTE OF STEEL CONSTRUCTION

BEAMS
S Shapes
Allowable uniform loads in kips
for beams laterally supported
For beams laterally unsupported, see page 2-146

S 18–15–12 F_y = 36 ksi

Designation	S 18			S 15			S 12		S 12		
Wt./ft	70	54.7	Deflection In.	50	42.9	Deflection In.	50	40.8	35	31.8	Deflection In.
Flange Width	6¼	6		5⅝	5½		5½	5¼	5⅛	5	
L_c	6.60	6.30		6.00	5.80		5.80	5.50	5.40	5.30	
L_u	11.1	10.7		10.8	10.6		13.9	13.4	10.7	10.5	
Span in Feet											
3							237				.02
4	369		.02	238		.03	201	160	148	121	.03
5	326	239	.03	205	178	.04	161	144	121	115	.05
6	272	236	.05	171	157	.06	134	120	101	96	.07
7	233	202	.07	147	135	.08	115	103	86	82	.10
8	204	177	.09	128	118	.10	101	90	76	72	.13
9	181	157	.11	114	105	.13	89	80	67	64	.17
10	163	142	.14	103	94	.16	80	72	61	58	.20
11	148	129	.17	93	86	.20	73	65	55	52	.25
12	136	118	.20	86	79	.24	67	60	50	48	.29
13	126	109	.23	79	73	.28	62	55	47	44	.35
14	117	101	.27	73	67	.32	57	51	43	41	.40
15	109	94	.31	68	63	.37	54	48	40	38	.46
16	102	89	.35	64	59	.42	50	45	38	36	.52
17	96	83	.39	60	56	.47	47	42	36	34	.59
18	91	79	.44	57	52	.53	45	40	34	32	.66
19	86	75	.49	54	50	.59	42	38	32	30	.74
20	82	71	.55	51	47	.66	40	36	30	29	.82
21	78	67	.60	49	45	.72	38	34	29	27	.90
22	74	64	.66	47	43	.79	37	33	28	26	.99
23	71	62	.72	45	41	.87	35	31	26	25	1.08
24	68	59	.79	43	39	.94	34	30	25	24	1.18
25	65	57	.85	41	38	1.02	32	29	24	23	1.28
26	63	54	.92	39	36	1.11	31	28	23	22	1.38
27	60	52	1.00	38	35	1.19	30	27	22	21	1.49
28	58	51	1.07	37	34	1.28	29	26	22	21	1.61
29	56	49	1.15	35	33	1.38	28	25	21	20	1.72
30	54	47	1.23	34	31	1.47	27	24	20	19	1.84
31	53	46	1.31	33	30	1.57					
32	51	44	1.40	32	30	1.68					
33	49	43	1.49	31	29	1.78					
34	48	42	1.58	30	28	1.89					
35	47	40	1.67	29	27	2.01					
36	45	39	1.77	29	26	2.12					
37	44	38	1.87								
38	43	37	1.97								
40	41	35	2.18								
42	39	34	2.41								
44	37	32	2.64								

F_y = 36 ksi

Properties and Reaction Values

	70	54.7		50	42.9		50	40.8	35	31.8	
S_x in.³	103	89.4		64.8	59.6		50.8	45.4	38.2	36.4	
V kips	184	119		119	89		119	80	74	60	
R_1 kips	63.4	41.1	For explanation of deflection, see page 2-32	44.9	33.6	For explanation of deflection, see page 2-32	58.7	39.4	30.2	24.7	For explanation of deflection, see page 2-32
R_2 kips/in.	16.9	11.0		13.1	9.77		16.3	11.0	10.2	8.32	
R_3 kips	102	53.1		65.6	42.4		94.3	52.0	42.1	31.2	
R_4 kips/in.	17.7	4.82		10.9	4.55		25.1	7.63	7.35	4.02	
R kips	123	70		91	58		116	78	66	45	

Load above heavy line is limited by maximum allowable web shear.

AMERICAN INSTITUTE OF STEEL CONSTRUCTION

Fy = 36 ksi

BEAMS
S Shapes
Allowable uniform loads in kips
for beams laterally supported
For beams laterally unsupported, see page 2-146

S 10–8–7–6

Designation	S 10		Deflection In.	S 8		Deflection In.	S 7		Deflection In.	S 6		Deflection In.
Wt./ft	35	25.4		23	18.4		20	15.3		17.25	12.5	
Flange Width	5	4⅝		4⅛	4		3⅞	3⅝		3⅜	3⅜	
L_c	5.20	4.90		4.40	4.20		4.10	3.90		3.80	3.50	
L_u	11.2	10.6		10.3	9.90		10.0	9.50		9.90	9.20	
Span in Feet												
1										80		.004
2	171		.01	102		.01	91		.01	69	40	.02
3	155		.02	86	62	.03	64	51	.03	46	39	.04
4	116	90	.04	64	57	.05	48	42	.06	35	29	.07
5	93	78	.06	51	46	.08	38	33	.09	28	23	.10
6	78	65	.09	43	38	.11	32	28	.13	23	19	.15
7	67	56	.12	37	33	.15	27	24	.17	20	17	.20
8	58	49	.16	32	29	.20	24	21	.22	17	15	.26
9	52	43	.20	29	25	.25	21	18	.28	15	13	.33
10	47	39	.25	26	23	.31	19	17	.35	14	12	.41
11	42	36	.30	23	21	.37	17	15	.42	13	11	.50
12	39	33	.35	21	19	.44	16	14	.51	12	10	.59
13	36	30	.42	20	18	.52	15	13	.59	11	9.0	.69
14	33	28	.48	18	16	.60	14	12	.69	10	8.3	.80
15	31	26	.55	17	15	.69	13	11	.79			
16	29	24	.63	16	14	.79	12	10	.90			
17	27	23	.71	15	13	.89						
18	26	22	.80	14	13	1.00						
20	23	20	.98	13	11	1.23						
22	21	18	1.19									
24	19	16	1.42									

Fy = 36 ksi

Properties and Reaction Values

	S 10			S 8			S 7			S 6		
S_x in.³	29.4	24.7	For explanation of deflection, see page 2-32	16.2	14.4	For explanation of deflection, see page 2-32	12.1	10.5	For explanation of deflection, see page 2-32	8.77	7.37	For explanation of deflection, see page 2-32
V kips	86	45		51	31		45	25		40	20	
R_1 kips	39.7	20.8		26.2	16.1		25.1	14.0		24.2	12.1	
R_2 kips/in.	14.1	7.39		10.5	6.44		10.7	5.99		11.0	5.51	
R_3 kips	65.4	24.8		39.0	18.8		38.6	16.2		38.8	13.7	
R_4 kips/in.	26.1	3.75		15.4	3.57		20.3	3.57		28.6	3.55	
R kips	**89**	38		**63**	31		**63**	**29**		**63**	**26**	

Load above heavy line is limited by maximum allowable web shear.
Values of R in **bold face** exceed maximum web shear V.

AMERICAN INSTITUTE OF STEEL CONSTRUCTION

S 5–4–3	**BEAMS**	F_y = 36 ksi

BEAMS
S Shapes
Allowable uniform loads in kips
for beams laterally supported
For beams laterally unsupported, see page 2-146

Designation	S 5			S 4			S 3		
Wt./ft	14.75	10	Deflection In.	9.5	7.7	Deflection In.	7.5	5.7	Deflection In.
Flange Width	3¼	3		2¾	2⅝		2½	2⅜	
L_c	3.50	3.20		3.00	2.80		2.60	2.50	
L_u	9.90	9.10		9.50	9.00		10.1	9.40	

Span in Feet | F_y = 36 ksi

Span	S5 14.75	S5 10	Defl	S4 9.5	S4 7.7	Defl	S3 7.5	S3 5.7	Defl
1	71		.005	38		.01	30	15	.01
2	48	31	.02	27	22	.02	15	13	.03
3	32	26	.04	18	16	.06	10	8.9	.07
4	24	19	.08	13	12	.10	7.7	6.7	.13
5	19	16	.12	11	9.6	.15	6.2	5.3	.20
6	16	13	.18	8.9	8.0	.22	5.1	4.4	.29
7	14	11	.24	7.7	6.9	.30			
8	12	9.7	.31	6.7	6.0	.39			
9	11	8.7	.40	6.0	5.4	.50			
10	9.6	7.8	.49	5.4	4.8	.61			
11	8.8	7.1	.59						
12	8.0	6.5	.71						

Properties and Reaction Values

	S5 14.75	S5 10		S4 9.5	S4 7.7		S3 7.5	S3 5.7	
S_x in.³	6.09	4.92		3.39	3.04		1.95	1.68	
V kips	36	15	For explanation of deflection, see page 2-32	19	11	For explanation of deflection, see page 2-32	15	7	For explanation of deflection, see page 2-32
R_1 kips	23.8	10.3		14.5	8.60		14.3	6.94	
R_2 kips/in.	11.7	5.08		7.75	4.59		8.29	4.04	
R_3 kips	40.4	11.5		20.6	9.36		21.4	7.29	
R_4 kips/in.	45.3	3.68		18.1	3.75		33.4	3.85	
R kips	**65**	24		**42**	**22**		**43**	21	

Load above heavy line is limited by maximum allowable web shear.
Values of R in **bold face** exceed maximum web shear V.

AMERICAN INSTITUTE OF STEEL CONSTRUCTION

| **F_y = 36 ksi** | | BEAMS | | | | | | | 18–15 |

BEAMS
Channels
Allowable uniform loads in kips
for beams laterally supported
For beams laterally unsupported, see page 2-146

Designation		MC 18					C 15			
Wt./ft		58	51.9	45.8	42.7	Deflection In.	50	40	33.9	Deflection In.
Flange Width		4¼	4⅛	4	4		3¾	3½	3⅜	
L_u		6.70	6.60	6.40	6.40		7.50	7.10	6.80	
	2	363				.005	309	225		.01
	3	360	311	259	233	.01	258	223	173	.01
	4	270	251	231	222	.02	194	167	151	.02
	5	216	201	185	177	.03	155	134	121	.04
	6	180	167	154	148	.04	129	112	101	.05
	7	154	143	132	127	.06	111	96	86	.07
	8	135	125	116	111	.08	97	84	76	.10
	9	120	112	103	99	.10	86	74	67	.12
	10	108	100	93	89	.12	77	67	60	.15
	11	98	91	84	81	.15	70	61	55	.18
	12	90	84	77	74	.18	65	56	50	.21
	13	83	77	71	68	.21	60	52	47	.25
	14	77	72	66	63	.24	55	48	43	.29
	15	72	67	62	59	.28	52	45	40	.33
	16	68	63	58	55	.32	48	42	38	.38
	17	64	59	54	52	.36	46	39	36	.43
	18	60	56	51	49	.40	43	37	34	.48
	19	57	53	49	47	.45	41	35	32	.54
	20	54	50	46	44	.50	39	33	30	.60
	21	51	48	44	42	.55	37	32	29	.66
	22	49	46	42	40	.60	35	30	27	.72
	23	47	44	40	39	.66	34	29	26	.79
	24	45	42	39	37	.71	32	28	25	.86
	25	43	40	37	35	.78	31	27	24	.93
	26	42	39	36	34	.84	30	26	23	1.01
	28	39	36	33	32	.97	28	24	22	1.17
	30	36	33	31	30	1.12	26	22	20	1.34
	32	34	31	29	28	1.27	24	21	19	1.52
	34	32	30	27	26	1.43	23	20	18	1.72
	36	30	28	26	25	1.61	22	19	17	1.93
	38	28	26	24	23	1.79				
	40	27	25	23	22	1.98				
	42	26	24	22	21	2.19				
	44	25	23	21	20	2.40				

F_y = 36 ksi · *Span in Feet*

Properties and Reaction Values

S_x in.³	75.1	69.7	64.3	61.6		53.8	46.5	42.0	
V kips	181	156	130	117	For	155	112	86	For
R_1 kips	57.2	49.0	40.8	36.8	explanation	61.1	44.4	34.2	explanation
R_2 kips/in.	16.6	14.3	11.9	10.7	of deflection,	17.0	12.4	9.50	of deflection,
R_3 kips	94.5	75.0	57.0	48.7	see	99.6	61.7	41.6	see
R_4 kips/in.	18.7	11.8	6.80	4.96	page 2-32	23.0	8.83	4.02	page 2-32
R kips	115	99	81	66		121	88	56	

Load above heavy line is limited by maximum allowable web shear.

| 13 | **BEAMS** | | | | F_y = 36 ksi |

BEAMS
Channels
Allowable uniform loads in kips
for beams laterally supported
For beams laterally unsupported, see page 2-146

Designation		MC 13				
Wt./ft		50	40	35	31.8	Deflection
Flange Width		4⅜	4⅛	4⅛	4	In.
L_u		9.60	9.10	8.90	8.70	
	2	295	210			.01
	3	232	202	167	140	.02
	4	174	151	140	132	.03
	5	139	121	112	106	.04
	6	116	101	93	88	.06
	7	100	86	80	76	.08
	8	87	76	70	66	.11
	9	77	67	62	59	.14
	10	70	60	56	53	.17
	11	63	55	51	48	.21
	12	58	50	47	44	.25
	13	54	47	43	41	.29
	14	50	43	40	38	.34
	15	46	40	37	35	.39
	16	44	38	35	33	.44
	17	41	36	33	31	.50
	18	39	34	31	29	.56
	19	37	32	29	28	.62
	20	35	30	28	26	.69
	21	33	29	27	25	.76
	22	32	27	25	24	.83
	23	30	26	24	23	.91
	24	29	25	23	22	.99
	25	28	24	22	21	1.07
	26	27	23	21	20	1.16
	27	26	22	21	20	1.25
	28	25	22	20	19	1.35
	29	24	21	19	18	1.44
	30	23	20	19	18	1.55
	31	22	20	18	17	1.65
	32	22	19	17	17	1.76

F_y = 36 ksi — Span in Feet

Properties and Reaction Values

S_x in.³	48.4	42.0	38.8	36.8	
V kips	147	105	84	70	For
R_1 kips	64.3	45.7	36.5	30.6	explanation
R_2 kips/in.	18.7	13.3	10.6	8.91	of deflection,
R_3 kips	111	66.8	47.6	36.6	see page 2-32
R_4 kips/in.	37.6	13.6	6.89	4.07	
R kips	130	92	72	51	

Load above heavy line is limited by maximum allowable web shear.

AMERICAN INSTITUTE OF STEEL CONSTRUCTION

BEAMS
Channels
Allowable uniform loads in kips for beams laterally supported

F_y = 36 ksi

For beams laterally unsupported, see page 2-146

12

Designation		C 12			MC 12					MC 12	Deflection In.
Wt./ft		30	25	20.7	50	45	40	35	31	10.6	
Flange Width		3⅛	3	3	4⅛	4	3⅞	3¾	3⅝	1½	
L_u		6.10	5.90	5.70	11.2	10.8	10.5	10.2	9.90	1.80	
	2	176	134		289	246	204			66	.01
	3	130	116	97	216	202	187	161	128	44	.02
	4	97	87	77	162	151	140	130	122	33	.03
	5	78	69	62	129	121	112	104	97	27	.05
	6	65	58	52	108	101	94	87	81	22	.07
	7	56	50	44	92	86	80	74	70	19	.09
	8	49	43	39	81	76	70	65	61	17	.12
	9	43	39	34	72	67	62	58	54	15	.15
	10	39	35	31	65	60	56	52	49	13	.19
	11	35	32	28	59	55	51	47	44	12	.23
	12	32	29	26	54	50	47	43	41	11	.27
	13	30	27	24	50	47	43	40	37	10	.31
	14	28	25	22	46	43	40	37	35	9.5	.36
	15	26	23	21	43	40	37	35	32	8.9	.42
	16	24	22	19	40	38	35	32	30	8.3	.48
	17	23	20	18	38	36	33	31	29	7.8	.54
	18	22	19	17	36	34	31	29	27	7.4	.60
	19	20	18	16	34	32	30	27	26	7.0	.67
	20	19	17	15	32	30	28	26	24	6.6	.74
	21	19	17	15	31	29	27	25	23	6.3	.82
	22	18	16	14	29	27	26	24	22	6.0	.90
	23	17	15	13	28	26	24	23	21	5.8	.98
	24	16	14	13	27	25	23	22	20	5.5	1.07
	25	16	14	12	26	24	22	21	19	5.3	1.16
	26	15	13	12	25	23	22	20	19	5.1	1.26
	28	14	12	11	23	22	20	19	17	4.7	1.46
	30	13	12	10	22	20	19	17	16	4.4	1.67

F_y = 36 ksi — Span in Feet

For explanation of deflection, see page 2-32

Properties and Reaction Values

	C 30	C 25	C 20.7	MC 50	MC 45	MC 40	MC 35	MC 31	MC 10.6
S_x in.³	27.0	24.1	21.5	44.9	42.0	39.0	36.1	33.8	9.23
V kips	88	67	49	144	123	102	81	64	33
R_1 kips	34.1	25.9	18.8	65.1	55.5	46.0	36.4	28.8	7.76
R_2 kips/in.	12.1	9.20	6.70	19.8	16.9	14.0	11.1	8.79	4.51
R_3 kips	52.6	34.8	21.6	130	103	77.3	54.5	38.4	9.39
R_4 kips/in.	13.5	5.90	2.28	42.4	26.3	15.0	7.42	3.69	1.13
R kips	76	55	30	134	115	95	75	51	13

Load above heavy line is limited by maximum allowable web shear.

AMERICAN INSTITUTE OF STEEL CONSTRUCTION

BEAMS
Channels
Allowable uniform loads in kips
for beams laterally supported
For beams laterally unsupported, see page 2-146

$F_y = 36$ ksi

10

Designation		C 10				MC 10			MC 10		MC 10	MC 10	
Wt./ft		30	25	20	15.3	41.1	33.6	28.5	25	22	8.4	6.5	Deflection In.
Flange Width		3	2⅞	2¾	2⅝	4⅜	4⅛	4	3⅜	3⅜	1½	1⅛	
L_u		6.10	5.80	5.50	5.30	11.5	10.9	10.5	9.10	8.80	1.90	1.10	
	1	194	151			229					49	44	.002
	2	149	131	109	69	227	166	122	109		46	32	.01
	3	99	87	76	65	151	133	121	106	84	31	21	.02
	4	75	66	57	49	113	100	91	79	74	23	16	.04
	5	60	52	46	39	91	80	73	63	59	18	13	.06
	6	50	44	38	32	76	67	61	53	49	15	11	.08
	7	43	37	33	28	65	57	52	45	42	13	9.1	.11
	8	37	33	28	24	57	50	46	40	37	12	8.0	.14
	9	33	29	25	22	50	44	40	35	33	10	7.1	.18
	10	30	26	23	19	45	40	36	32	30	9.2	6.4	.22
	11	27	24	21	18	41	36	33	29	27	8.4	5.8	.27
	12	25	22	19	16	38	33	30	26	25	7.7	5.3	.32
	13	23	20	18	15	35	31	28	24	23	7.1	4.9	.38
	14	21	19	16	14	32	29	26	23	21	6.6	4.5	.44
	15	20	17	15	13	30	27	24	21	20	6.1	4.2	.50
	16	19	16	14	12	28	25	23	20	18	5.8	4.0	.57
	17	18	15	13	11	27	24	21	19	17	5.4	3.7	.65
	18	17	15	13	11	25	22	20	18	16	5.1	3.5	.72
	19	16	14	12	10	24	21	19	17	16	4.9	3.3	.81
	20	15	13	11	9.7	23	20	18	16	15	4.6	3.2	.89
	21	14	12	11	9.3	22	19	17	15	14	4.4	3.0	.98
	22	14	12	10	8.8	21	18	17	14	13	4.2	2.9	1.08
	23	13	11	9.9	8.5	20	17	16	14	13	4.0	2.8	1.18
	24	12	11	9.5	8.1	19	17	15	13	12	3.8	2.7	1.29

$F_y = 36$ ksi — Span in Feet

Properties and Reaction Values

	30	25	20	15.3	41.1	33.6	28.5	25	22	8.4	6.5	
S_x in.³	20.7	18.2	15.8	13.5	31.5	27.8	25.3	22.0	20.5	6.40	4.42	For explanation of deflection, see page 2-32
V kips	97	76	55	35	115	83	61	55	42	24	22	
R_1 kips	40.0	31.2	22.5	14.3	59.1	42.7	31.6	28.2	21.5	6.94	3.95	
R_2 kips/in.	16.0	12.5	9.01	5.70	18.9	13.7	10.1	9.03	6.89	4.04	3.61	
R_3 kips	74.4	51.4	31.4	15.8	110	67.4	42.9	36.2	24.2	7.57	5.43	
R_4 kips/in.	42.8	20.4	7.64	1.94	53.7	20.2	8.17	5.84	2.60	1.07	1.06	
R kips	96	75	54	23	**125**	**91**	**67**	**57**	33	11	9	

Load above heavy line is limited by maximum allowable web shear.
Values of R in **bold face** exceed maximum web shear.

| F_y = 36 ksi | BEAMS
Channels
Allowable uniform loads in kips
for beams laterally supported
For beams laterally unsupported, see page 2-146 | | | | | 9 [|

Designation		C 9			MC 9		
Wt./ft		20	15	13.4	25.4	23.9	Deflection In.
Flange Width		2⅝	2½	2⅜	3½	3½	
L_u		5.60	5.30	5.20	9.90	9.80	

Span in Feet							Deflection
1	116						.002
2	97	74	60	117	104		.01
3	65	54	51	94	91		.02
4	49	41	38	71	68		.04
5	39	33	31	56	54		.06
6	32	27	25	47	45		.09
7	28	23	22	40	39		.12
8	24	20	19	35	34		.16
9	22	18	17	31	30		.20
10	19	16	15	28	27		.25
11	18	15	14	26	25		.30
12	16	14	13	24	23		.36
13	15	13	12	22	21		.42
14	14	12	11	20	19		.49
15	13	11	10	19	18		.56
16	12	10	9.5	18	17		.63
17	11	9.6	9.0	17	16		.72
18	11	9.0	8.5	16	15		.80
19	10	8.6	8.0	15	14		.90
20	9.7	8.1	7.6	14	14		.99
21	9.3	7.7	7.3	13	13		1.09
22	8.8	7.4	6.9	13	12		1.20

Properties and Reaction Values						
S_x in.³	13.5	11.3	10.6	19.6	18.9	
V kips	58	37	30	58	52	For
R_1 kips	24.9	15.9	13.0	31.7	28.2	explanation
R_2 kips/in.	10.6	6.77	5.54	10.7	9.50	of deflection
R_3 kips	39.3	19.9	14.7	45.7	38.3	see page 2-32
R_4 kips/in.	14.8	3.81	2.08	11.3	7.91	
R kips	**62**	33	22	**69**	61	

Load above heavy line is limited by maximum allowable web shear.
Values of R in **bold face** exceed maximum web shear.

BEEAMS
Channels
Allowable uniform loads in kips
for beams laterally supported
For beams laterally unsupported, see page 2-146

F_y = 36 ksi

8

Designation		C 8			MC 8		MC 8		MC 8	Deflection In.
Wt./ft		18.75	13.75	11.5	22.8	21.4	20	18.7	8.5	
Flange Width		2½	2⅜	2¼	3½	3½	3	3	1⅞	
L_u		5.70	5.30	5.10	10.6	10.5	8.80	8.60	3.40	
Span in Feet	1	112	70							.003
	2	79	65	51	98	86	92	81	41	.01
	3	53	43	39	77	74	65	63	28	.03
	4	40	33	29	58	55	49	47	21	.04
	5	32	26	23	46	44	39	38	17	.07
	6	26	22	20	38	37	33	31	14	.10
	7	23	19	17	33	32	28	27	12	.14
	8	20	16	15	29	28	24	24	10	.18
	9	18	14	13	26	25	22	21	9.3	.23
	10	16	13	12	23	22	20	19	8.4	.28
	11	14	12	11	21	20	18	17	7.6	.34
	12	13	11	9.8	19	18	16	16	7.0	.40
	13	12	10	9.0	18	17	15	15	6.5	.47
	14	11	9.3	8.4	16	16	14	13	6.0	.55
	15	11	8.7	7.8	15	15	13	13	5.6	.63
	16	9.9	8.1	7.3	14	14	12	12	5.2	.71
	17	9.3	7.6	6.9	14	13	12	11	4.9	.81
	18	8.8	7.2	6.5	13	12	11	10	4.7	.90
	19	8.3	6.8	6.2	12	12	10	9.9	4.4	1.01
	20	7.9	6.5	5.9	12	11	9.8	9.4	4.2	1.12

F_y = 36 ksi

Properties and Reaction Values										
S_x in.³		11.0	9.03	8.14	16.0	15.4	13.6	13.1	5.83	For
V kips		56	35	25	49	43	46	41	21	explanation
R_1 kips		27.1	16.9	12.3	30.1	26.5	26.7	23.6	7.97	of deflection,
R_2 kips/in.		11.6	7.20	5.23	10.1	8.91	9.50	8.39	4.25	see page 2-32
R_3 kips		43.3	21.2	13.1	41.2	33.9	36.5	30.3	8.62	
R_4 kips/in.		22.7	5.46	2.09	11.3	7.68	9.79	6.73	1.41	
R kips		**68**	**40**	20	**65**	**58**	**60**	**53**	14	

Load above heavy line is limited by maximum allowable web shear.
Values of R in **bold face** exceed maximum web shear.

AMERICAN INSTITUTE OF STEEL CONSTRUCTION

F_y = 36 ksi	BEAMS Channels Allowable uniform loads in kips for beams laterally supported For beams laterally unsupported, see page 2-146					7 [

Designation		C 7			MC 7		
Wt./ft		14.75	12.25	9.8	22.7	19.1	Deflection In.
Flange Width		2¼	2¼	2⅛	3⅝	3½	
L_u		5.60	5.30	5.10	11.90	11.40	

Span in Feet		C 7 14.75	C 7 12.25	C 7 9.8	MC 7 22.7	MC 7 19.1	Deflection In.
	1	84	63		101		.003
	2	56	50	42	98	71	.01
	3	37	33	29	65	59	.03
	4	28	25	22	49	44	.05
	5	22	20	18	39	35	.08
	6	19	17	15	33	30	.11
	7	16	14	13	28	25	.16
	8	14	12	11	24	22	.20
	9	12	11	9.7	22	20	.26
	10	11	10	8.8	20	18	.32
	11	10	9.1	8.0	18	16	.39
	12	9.3	8.3	7.3	16	15	.46
	13	8.6	7.7	6.7	15	14	.54
	14	8.0	7.1	6.3	14	13	.62
	15	7.5	6.7	5.8	13	12	.72
	16	7.0	6.2	5.5	12	11	.82

(F_y = 36 ksi)

Properties and Reaction Values							
S_x in.³		7.78	6.93	6.08	13.6	12.3	
V kips		42	32	21	51	35	For explanation of deflection, see page 2-32
R_1 kips		21.8	16.3	10.9	33.6	23.5	
R_2 kips/in.		9.96	7.46	4.99	12.0	8.36	
R_3 kips		33.5	21.7	11.9	51.5	30.1	
R_4 kips/in.		17.6	7.40	2.21	22.3	7.63	
R kips		**57**	**42**	20	**76**	**53**	

Load above heavy line is limited by maximum allowable web shear.
Values of R in **bold face** exceed maximum web shear.

| 6 | | | | BEAMS | | | | | F_y = 36 ksi |

BEAMS
Channels
Allowable uniform loads in kips
for beams laterally supported
For beams laterally unsupported, see page 2-146

Designation		C 6			MC 6		MC 6		MC 6	Deflection In.
Wt./ft		13	10.5	8.2	18	15.3	16.3	15.1	12	
Flange Width		2⅛	2	1⅞	3½	3½	3	3	2½	
L_u		5.70	5.40	5.10	12.9	10.4	11.0	10.8	7.20	
	1	76	54	35			65		54	.004
	2	42	36	32	65	59	62	55	45	.01
	3	28	24	21	48	41	42	40	30	.03
	4	21	18	16	36	30	31	30	22	.06
	5	17	15	13	29	24	25	24	18	.09
	6	14	12	11	24	20	21	20	15	.13
	7	12	10	9.0	20	17	18	17	13	.18
	8	10	9.1	7.9	18	15	16	15	11	.24
	9	9.3	8.1	7.0	16	14	14	13	10	.30
	10	8.4	7.3	6.3	14	12	12	12	9.0	.37
	11	7.6	6.6	5.7	13	11	11	11	8.2	.45
	12	7.0	6.1	5.3	12	10	10	10	7.5	.54
	13	6.4	5.6	4.9	11	9.4	9.6	9.2	6.9	.63
	14	6.0	5.2	4.5	10	8.7	8.9	8.6	6.4	.73

F_y = 36 ksi — Span in Feet

Properties and Reaction Values

									For explanation of deflection, see page 2-32
S_x in.³	5.80	5.06	4.38	9.91	8.47	8.68	8.32	6.24	
V kips	38	27	17	33	29	32	27	27	
R_1 kips	21.1	15.2	9.65	23.9	17.7	23.7	19.9	15.0	
R_2 kips/in.	10.4	7.46	4.75	9.01	8.08	8.91	7.51	7.37	
R_3 kips	34.5	21.0	10.7	32.8	25.1	32.3	25.0	21.6	
R_4 kips/in.	24.8	9.21	2.38	11.7	10.4	11.3	6.78	8.10	
R kips	**58**	**41**	**19**	**55**	**46**	**55**	**46**	**41**	

Load above heavy line is limited by maximum allowable web shear.
Values of R in **bold face** exceed maximum web shear.

AMERICAN INSTITUTE OF STEEL CONSTRUCTION

| F_y = 36 ksi | | | BEAMS
Channels
Allowable uniform loads in kips
for beams laterally supported
For beams laterally unsupported, see page 2-146 | | | | | | | 5-4-3 [|

Designation		C 5				C 4				C 3			
Wt./ft		9	6.7	Deflection In.		7.25	5.4	Deflection In.		6	5	4.1	Deflection In.
Flange Width		1⅞	1¾			1¾	1⅝			1⅝	1½	1⅜	
L_u		5.60	5.20			5.90	5.40			6.70	6.30	6.00	
	1	47	27	.004		33	21	.01		20	18	15	.01
	2	26	22	.02		16	14	.02		9.9	8.9	7.9	.03
	3	17	14	.04		11	9.3	.05		6.6	6.0	5.3	.07
	4	13	11	.07		8.2	6.9	.09		5.0	4.5	4.0	.12
	5	10	8.6	.11		6.6	5.6	.14		4.0	3.6	3.2	.19
	6	8.5	7.2	.16		5.5	4.6	.20		3.3	3.0	2.6	.27
	7	7.3	6.2	.22		4.7	4.0	.27					
	8	6.4	5.4	.29		4.1	3.5	.36					
	9	5.7	4.8	.36		3.7	3.1	.45					
	10	5.1	4.3	.45		3.3	2.8	.56					
	11	4.7	3.9	.54									
	12	4.3	3.6	.64									

F_y = 36 ksi — Span in Feet

Properties and Reaction Values

	C 5				C 4				C 3			
S_x in.³	3.56	3.00			2.29	1.93			1.38	1.24	1.10	
V kips	23	14	For explanation of deflection, see page 2-32		18	11	For explanation of deflection, see page 2-32		15	11	7	For explanation of deflection, see page 2-32
R_1 kips	14.5	8.46			13.1	7.51			14.5	10.5	6.94	
R_2 kips/in.	7.72	4.51			7.63	4.37			8.46	6.13	4.04	
R_3 kips	21.4	9.56			20.2	8.76			22.6	14.0	7.47	
R_4 kips/in.	13.1	2.62			17.1	3.22			33.7	12.8	3.67	
R kips	**42**	**19**			**40**	**20**			**44**	**32**	**20**	

Load above heavy line is limited by maximum allowable web shear.
Values of R in **bold face** exceed maximum web shear.

W 44	BEAMS				F_y = 50 ksi

BEAMS
W Shapes
Allowable uniform loads in kips
for beams laterally supported
For beams laterally unsupported, see page 2-146

Designation		W 44				
Wt./ft		285	248	224	198	Deflection In.
Flange Width		11¾	11¾	11¾	11¾	
L_c		10.6	10.6	10.6	10.6	
L_u		15.9	14.2	12.9	11.2	

	Span in Feet	285	248	224	198	Deflection In.
	13	1810				.13
	14	1760	1510	1360	1220	.15
	15	1640	1440	1300	1140	.17
	16	1540	1350	1220	1070	.20
	17	1450	1270	1150	1000	.22
	18	1370	1200	1090	948	.25
	19	1300	1140	1030	899	.28
	20	1230	1080	978	854	.31
	21	1170	1030	931	813	.34
	22	1120	983	889	776	.38
	23	1070	940	850	742	.41
	24	1030	901	815	711	.45
	25	986	865	782	683	.48
	26	948	832	752	657	.52
	27	913	801	724	632	.57
	28	880	772	699	610	.61
	29	850	746	674	589	.65
	30	821	721	652	569	.70
	31	795	698	631	551	.75
	32	770	676	611	534	.79
	33	747	655	593	517	.84
	34	725	636	575	502	.90
	35	704	618	559	488	.95
	36	684	601	543	474	1.01
	38	648	569	515	449	1.12
	40	616	541	489	427	1.24
	42	587	515	466	406	1.37
	44	560	492	445	388	1.50
	46	536	470	425	371	1.64
	48	513	451	407	356	1.79
	50	493	433	391	341	1.94
	52	474	416	376	328	2.10
	54	456	400	362	316	2.26
	56	440	386	349	305	2.43
	58	425	373	337	294	2.61
	60	411	360	326	285	2.79
	62	397	349	315	275	2.98
	64	385	338	306	267	3.18
	66	373	328	296	259	3.38
	68	362	318	288	251	3.59
	70	352	309	279	244	3.80
	72	342	300	272	237	4.02

Properties and Reaction Values						
S_x in.³		1120	983	889	776	
V kips		902	755	680	609	For explanation of deflection, see page 2-32
R_1 kips		227	178	150	124	
R_2 kips/in.		33.8	28.5	25.9	23.4	
R_3 kips		332	243	199	159	
R_4 kips/in.		9.97	6.79	5.69	4.93	
R kips		345	267	219	176	

Load above heavy line is limited by maximum allowable web shear.

F_y = 50 ksi

AMERICAN INSTITUTE OF STEEL CONSTRUCTION

| F_y = 50 ksi | BEAMS W Shapes Allowable uniform loads in kips for beams laterally supported For beams laterally unsupported, see page 2-146 | | | | | | W 40 |

Designation	W 40						
Wt./ft	328	298	268	244	221	192*	Deflection In.
Flange Width	17⅞	17⅞	17¾	17¾	17¾	17¾	
L_c	16.0	16.0	15.9	15.9	15.9	12.8	
L_u	25.8	23.6	21.2	19.0	17.2	16.7	
13						1090	.14
14						1080	.17
15						1010	.19
16						942	.22
17					1100	887	.25
18					1050	837	.28
19				1110	993	793	.31
20	1460	1320	1180	1080	944	754	.34
21	1400	1280	1140	1030	899	718	.38
22	1340	1220	1090	983	858	685	.41
23	1280	1170	1040	940	821	655	.45
24	1230	1120	999	901	787	628	.49
25	1180	1070	959	865	755	603	.53
26	1130	1030	922	832	726	580	.58
27	1090	994	888	801	699	558	.62
28	1050	959	856	772	674	538	.67
29	1020	926	827	746	651	520	.72
30	983	895	799	721	629	502	.77
31	951	866	774	698	609	486	.82
32	921	839	749	676	590	471	.87
33	893	813	727	655	572	457	.93
34	867	789	705	636	555	443	.99
35	842	767	685	618	539	431	1.05
36	819	746	666	601	524	419	1.11
38	776	706	631	569	497	397	1.23
40	737	671	600	541	472	377	1.37
42	702	639	571	515	449	359	1.51
44	670	610	545	492	429	343	1.65
46	641	583	521	470	410	328	1.81
48	614	559	500	451	393	314	1.97
50	590	537	480	433	378	301	2.13
52	567	516	461	416	363	290	2.31
54	546	497	444	400	350	279	2.49
56	526	479	428	386	337	269	2.68
58	508	463	413	373	325	260	2.87
60	491	447	400	360	315	251	3.07
62	475	433	387	349	304	243	3.28
64	461	419	375	338	295	236	3.50
66	447	407	363	328	286	228	3.72
68	434	395	353	318	278	222	3.95
70	421	383	343	309	270	215	4.18
72	409	373	333	300	262	209	4.42

Properties and Reaction Values							
S_x in.³	1340	1220	1090	983	858	708	
V kips	728	659	591	555	549	542	For explanation of deflection, see page 2-32
R_1 kips	235	205	174	154	143	132	
R_2 kips/in.	30.0	27.4	24.8	23.4	23.4	23.4	
R_3 kips	275	228	186	161	148	131	
R_4 kips/in.	7.85	6.60	5.46	5.25	6.27	8.14	
R kips	302	251	205	179	170	159	

*Indicates noncompact shape.
Load above heavy line is limited by maximum allowable web shear.

AMERICAN INSTITUTE OF STEEL CONSTRUCTION

W 40		BEAMS						F_y = 50 ksi	

BEAMS
W Shapes
Allowable uniform loads in kips
for beams laterally supported
For beams laterally unsupported, see page 2-146

Designation		W 40							Deflection In.
Wt./ft		655	593	531	480	436	397	362	
Flange Width		16⅞	16¾	16½	16⅜	16¼	16⅛	16	
L_c		15.1	14.9	14.8	14.7	14.5	14.4	14.3	
L_u		45.7	41.7	37.9	34.4	31.4	29.0	26.5	

		Span in Feet	655	593	531	480	436	397	362	Deflection
F_y = 50 ksi		16	3440	3080	2730	2440	2220			.22
		17	3350	3030	2710	2440	2210	2000	1820	.25
		18	3170	2860	2550	2310	2090	1910	1740	.28
		19	3000	2710	2420	2190	1980	1810	1640	.31
		20	2850	2570	2300	2080	1880	1720	1560	.34
		21	2710	2450	2190	1980	1790	1630	1490	.38
		22	2590	2340	2090	1890	1710	1560	1420	.41
		23	2480	2240	2000	1810	1640	1490	1360	.45
		24	2370	2150	1920	1730	1570	1430	1300	.49
		25	2280	2060	1840	1660	1510	1370	1250	.53
		26	2190	1980	1770	1600	1450	1320	1200	.58
		27	2110	1910	1700	1540	1390	1270	1160	.62
		28	2040	1840	1640	1490	1340	1230	1120	.67
		29	1970	1780	1590	1430	1300	1180	1080	.72
		30	1900	1720	1530	1390	1250	1140	1040	.77
		31	1840	1660	1480	1340	1210	1110	1010	.82
		32	1780	1610	1440	1300	1180	1070	976	.87
		33	1730	1560	1390	1260	1140	1040	947	.93
		34	1680	1510	1350	1220	1110	1010	919	.99
		36	1580	1430	1280	1160	1050	953	868	1.11
		38	1500	1360	1210	1090	990	903	822	1.23
		40	1430	1290	1150	1040	941	858	781	1.37
		42	1360	1230	1100	990	896	817	744	1.51
		44	1300	1170	1050	945	855	780	710	1.65
		46	1240	1120	1000	904	818	746	679	1.81
		48	1190	1070	958	866	784	715	651	1.97
		50	1140	1030	920	832	752	686	625	2.13
		52	1100	990	884	800	723	660	601	2.31
		54	1060	953	851	770	697	636	579	2.49
		56	1020	919	821	743	672	613	558	2.68
		58	982	888	793	717	649	592	539	2.87
		60	950	858	766	693	627	572	521	3.07
		62	919	830	742	671	607	554	504	3.28
		64	890	804	718	650	588	536	488	3.50
		66	863	780	697	630	570	520	473	3.72
		68	838	757	676	611	553	505	459	3.95
		70	814	735	657	594	537	490	446	4.18
		72	791	715	639	578	523	477	434	4.42

Properties and Reaction Values

S_x in.³		2590	2340	2090	1890	1710	1560	1420	For explanation of deflection, see page 2-32
V kips		1720	1540	1360	1220	1110	999	908	
R_1 kips		802	683	573	482	421	365	312	
R_2 kips/in.		65.0	59.1	53.1	48.2	44.2	40.3	37.0	
R_3 kips		1250	1030	838	689	578	481	404	
R_4 kips/in.		35.7	29.8	24.4	20.3	17.5	14.5	12.4	
R kips		1030	890	759	651	576	506	442	

Load above heavy line is limited by maximum allowable web shear.

| F_y = 50 ksi | BEAMS W Shapes Allowable uniform loads in kips for beams laterally supported For beams laterally unsupported, see page 2-146 | W 40 |

Designation	W 40						W 40			
Wt./ft	324	297	277	249	215	199	183	167	149	Deflection In.
Flange Width	15⅞	15⅞	15⅞	15¾	15¾	15¾	11¾	11¾	11¾	
L_c	14.2	14.2	14.2	14.1	14.1	14.1	10.6	10.4	8.5	
L_u	23.8	21.8	21.0	18.9	16.4	15.2	12.3	11.0	10.7	

Span in Feet (F_y = 50 ksi):

Span	324	297	277	249	215	199	183	167	149	Deflection
11									963	.10
12									939	.12
13								1000	866	.14
14							1010	941	805	.17
15							1000	879	751	.19
16						1010	938	824	704	.22
17	1610	1480				995	883	775	663	.25
18	1560	1430	1320	1180	1010	940	834	732	626	.28
19	1480	1360	1270	1150	993	890	790	694	593	.31
20	1410	1290	1210	1090	944	846	750	659	563	.34
21	1340	1230	1150	1040	899	806	714	628	536	.38
22	1280	1170	1100	992	858	769	682	599	512	.41
23	1220	1120	1050	949	821	736	652	573	490	.45
24	1170	1070	1010	909	787	705	625	549	469	.49
25	1130	1030	968	873	755	677	600	527	451	.53
26	1080	990	931	839	726	651	577	507	433	.58
27	1040	953	896	808	699	627	556	488	417	.62
28	1010	919	864	779	674	604	536	471	402	.67
29	971	888	834	753	651	583	517	454	388	.72
30	939	858	807	727	629	564	500	439	375	.77
31	908	830	781	704	609	546	484	425	363	.82
32	880	804	756	682	590	529	469	412	352	.87
33	853	780	733	661	572	513	455	399	341	.93
34	828	757	712	642	555	498	441	388	331	.99
36	782	715	672	606	524	470	417	366	313	1.11
38	741	677	637	574	497	445	395	347	296	1.23
40	704	644	605	546	472	423	375	329	282	1.37
42	670	613	576	520	449	403	357	314	268	1.51
44	640	585	550	496	429	385	341	300	256	1.65
46	612	560	526	474	410	368	326	286	245	1.81
48	587	536	504	455	393	352	313	275	235	1.97
50	563	515	484	436	378	338	300	264	225	2.13
52	542	495	465	420	363	325	289	253	217	2.31
54	521	477	448	404	350	313	278	244	209	2.49
56	503	460	432	390	337	302	268	235	201	2.68
58	486	444	417	376	325	292	259	227	194	2.87
60	469	429	403	364	315	282	250	220	188	3.07
62	454	415	390	352	304	273	242	213	182	3.28
64	440	402	378	341	295	264	234	206	176	3.50
66	427	390	367	331	286	256	227	200	171	3.72
68	414	379	356	321	278	249	221	194	166	3.95
70	402	368	346	312	270	242	214	188	161	4.18
72	391	358	336	303	262	235	208	183	156	4.42

Properties and Reaction Values

	324	297	277	249	215	199	183	167	149	
S_x in.^{3}	1280	1170	1100	992	858	769	682	599	512	
V kips	803	741	659	591	507	503	507	502	481	For explanation of deflection, see page 2-32
R_1 kips	263	235	205	174	141	131	141	131	117	
R_2 kips/in.	33.0	30.7	27.4	24.8	21.4	21.4	21.4	21.4	20.8	
R_3 kips	323	277	228	186	139	130	139	128	110	
R_4 kips/in.	9.92	8.83	6.60	5.44	4.17	4.81	4.17	5.01	5.69	
R kips	358	308	251	205	154	147	154	146	130	

Load above heavy line is limited by maximum allowable web shear.

W 36		BEAMS W Shapes Allowable uniform loads in kips for beams laterally supported For beams laterally unsupported, see page 2-146							F_y = 50 ksi

Designation		W 36								
Wt./ft		848	798	720	650	588	527	485	439	Deflection
Flange Width		18⅛	18	17¾	17⅝	17⅜	17¼	17⅛	17	In.
L_c		16.2	16.1	15.9	15.7	15.6	15.4	15.3	15.2	
L_u		64.1	61.7	56.5	51.3	46.9	42.7	39.2	36.2	

F_y = 50 ksi	Span in Feet										Deflection In.
		16	4280	4000	3570	3190	2850	2530	2320		.24
		17	4100	3860	3480	3130	2820	2520	2320	2080	.27
		18	3870	3640	3290	2960	2660	2380	2190	1980	.31
		19	3670	3450	3120	2800	2520	2260	2070	1880	.34
		20	3490	3280	2960	2660	2400	2150	1970	1780	.38
		21	3320	3120	2820	2540	2280	2040	1880	1700	.42
		22	3170	2980	2690	2420	2180	1950	1790	1620	.46
		23	3030	2850	2570	2320	2090	1870	1710	1550	.50
		24	2910	2730	2470	2220	2000	1790	1640	1490	.55
		25	2790	2620	2370	2130	1920	1720	1580	1430	.59
		26	2680	2520	2280	2050	1850	1650	1520	1370	.64
		27	2580	2430	2190	1970	1780	1590	1460	1320	.69
		28	2490	2340	2110	1900	1710	1530	1410	1270	.74
		29	2410	2260	2040	1840	1650	1480	1360	1230	.80
		30	2330	2190	1970	1780	1600	1430	1310	1190	.85
		31	2250	2120	1910	1720	1550	1380	1270	1150	.91
		32	2180	2050	1850	1660	1500	1340	1230	1110	.97
		33	2110	1990	1790	1610	1450	1300	1190	1080	1.03
		34	2050	1930	1740	1570	1410	1260	1160	1050	1.10
		36	1940	1820	1640	1480	1330	1190	1090	990	1.23
		38	1840	1730	1560	1400	1260	1130	1040	938	1.37
		40	1740	1640	1480	1330	1200	1070	985	891	1.52
		42	1660	1560	1410	1270	1140	1020	938	849	1.67
		44	1590	1490	1350	1210	1090	975	895	810	1.84
		46	1520	1430	1290	1160	1040	933	856	775	2.01
		48	1450	1370	1230	1110	999	894	820	743	2.18
		50	1400	1310	1180	1070	959	858	788	713	2.37
		52	1340	1260	1140	1020	922	825	757	685	2.56
		54	1290	1210	1100	986	888	794	729	660	2.77
		56	1250	1170	1060	951	856	766	703	636	2.97
		58	1200	1130	1020	918	827	740	679	614	3.19
		60	1160	1090	986	887	799	715	656	594	3.41
		62	1130	1060	955	859	774	692	635	575	3.65
		64	1090	1020	925	832	749	670	615	557	3.88
		66	1060	993	897	807	727	650	597	540	4.13
		68	1030	964	870	783	705	631	579	524	4.39
		70	996	937	845	761	685	613	563	509	4.65
		72	969	911	822	739	666	596	547	495	4.92

Properties and Reaction Values										
S_x in.³		3170	2980	2690	2420	2180	1950	1790	1620	For
V kips		2140	2000	1780	1600	1430	1260	1160	1040	explanation
R_1 kips		1180	1070	904	762	646	540	472	400	of deflection,
R_2 kips/in.		83.2	78.5	71.4	65.0	59.1	53.1	49.5	44.9	see page 2-32
R_3 kips		2050	1830	1510	1250	1030	838	723	596	
R_4 kips/in.		60.0	54.0	45.6	38.5	32.1	26.4	23.4	19.4	
R kips		1470	1340	1150	990	853	726	645	557	

Load above heavy line is limited by maximum allowable web shear.

AMERICAN INSTITUTE OF STEEL CONSTRUCTION

| F_y = 50 ksi | BEAMS
W Shapes
Allowable uniform loads in kips
for beams laterally supported
For beams laterally unsupported, see page 2-146 | | | | | | | | W 36 |

Designation		W 36								
Wt./ft		393	359	328	300	280	260	245	230	Deflection In.
Flange Width		16⅞	16¾	16⅝	16⅝	16⅝	16½	16½	16½	
L_c		15.1	15.0	14.9	14.9	14.9	14.8	14.8	14.8	
L_u		32.7	30.0	27.5	25.4	23.8	21.9	20.6	19.3	
	16								1090	.24
	17	1850	1680	1510	1390	1290	1220	1160	1080	.27
	18	1770	1610	1480	1360	1260	1170	1090	1020	.31
	19	1680	1530	1400	1290	1190	1100	1040	969	.34
	20	1600	1450	1330	1220	1130	1050	985	921	.38
	21	1520	1380	1270	1160	1080	998	938	877	.42
	22	1450	1320	1210	1110	1030	953	895	837	.46
	23	1390	1260	1160	1060	985	912	856	801	.50
	24	1330	1210	1110	1020	944	874	820	767	.55
	25	1280	1160	1070	977	906	839	788	737	.59
	26	1230	1120	1020	939	872	806	757	708	.64
	27	1180	1080	986	904	839	777	729	682	.69
	28	1140	1040	951	872	809	749	703	658	.74
	29	1100	1000	918	842	781	723	679	635	.80
	30	1060	968	887	814	755	699	656	614	.85
	31	1030	937	859	788	731	676	635	594	.91
	32	997	908	832	763	708	655	615	575	.97
	33	967	880	807	740	687	635	597	558	1.03
	34	938	854	783	718	666	617	579	542	1.10
	36	886	807	739	678	629	582	547	512	1.23
	38	839	764	701	643	596	552	518	485	1.37
	40	798	726	666	611	567	524	492	460	1.52
	42	760	691	634	581	540	499	469	438	1.67
	44	725	660	605	555	515	477	448	419	1.84
	46	693	631	579	531	493	456	428	400	2.01
	48	665	605	555	509	472	437	410	384	2.18
	50	638	581	532	488	453	419	394	368	2.37
	52	613	558	512	470	436	403	379	354	2.56
	54	591	538	493	452	420	388	365	341	2.77
	56	570	519	475	436	405	374	352	329	2.97
	58	550	501	459	421	391	361	339	317	3.19
	60	532	484	444	407	378	349	328	307	3.41
	62	515	468	429	394	365	338	318	297	3.65
	64	498	454	416	382	354	328	308	288	3.88
	66	483	440	403	370	343	318	298	279	4.13
	68	469	427	391	359	333	308	290	271	4.39
	70	456	415	380	349	324	300	281	263	4.65
	72	443	403	370	339	315	291	273	256	4.92

Span in Feet — F_y = 50 ksi

Properties and Reaction Values									
S_x in.³	1450	1320	1210	1110	1030	953	895	837	
V kips	922	838	757	694	646	609	577	546	
R_1 kips	333	289	252	219	196	178	165	149	For explanation of deflection, see page 2-32
R_2 kips/in.	40.3	37.0	33.7	31.2	29.2	27.7	26.4	25.1	
R_3 kips	481	404	337	286	251	222	200	179	
R_4 kips/in.	15.7	13.5	11.2	9.86	8.72	8.19	7.58	7.00	
R kips	474	419	370	321	282	251	227	204	

Load above heavy line is limited by maximum allowable web shear.

BEAMS
W Shapes
Allowable uniform loads in kips
for beams laterally supported

W 36

For beams laterally unsupported, see page 2-146

F_y = 50 ksi

Designation					W 36					
Wt./ft	256	232	210	194	182	170	160	150	135	Deflection In.
Flange Width	12¼	12⅛	12⅛	12⅛	12⅛	12	12	12	12	
L_c	10.9	10.9	10.9	10.9	10.8	10.8	10.7	10.5	8.80	
L_u	18.8	17.1	15.1	13.9	13.1	12.2	11.4	11.3	11.0	

Span in Feet (F_y = 50 ksi)

Span	256	232	210	194	182	170	160	150	135	Deflection
11									853	.11
12			1220			984	936	896	805	.14
13	1440	1290	1220	1120	1050	982	917	853	743	.16
14	1410	1270	1130	1040	979	911	852	792	690	.19
15	1310	1190	1060	974	914	851	795	739	644	.21
16	1230	1110	989	913	857	798	745	693	604	.24
17	1160	1050	930	859	806	751	701	652	568	.27
18	1090	989	879	812	761	709	662	616	537	.31
19	1040	937	833	769	721	672	628	584	508	.34
20	985	890	791	730	685	638	596	554	483	.38
21	938	848	753	696	653	608	568	528	460	.42
22	895	809	719	664	623	580	542	504	439	.46
23	856	774	688	635	596	555	518	482	420	.50
24	820	742	659	609	571	532	497	462	402	.55
25	788	712	633	584	548	510	477	444	386	.59
26	757	685	608	562	527	491	459	426	371	.64
27	729	659	586	541	508	473	442	411	358	.69
28	703	636	565	522	490	456	426	396	345	.74
29	679	614	545	504	473	440	411	382	333	.80
30	656	593	527	487	457	425	397	370	322	.85
31	635	574	510	471	442	412	385	358	312	.91
32	615	556	494	457	428	399	373	347	302	.97
33	597	539	479	443	415	387	361	336	293	1.03
34	579	523	465	430	403	375	351	326	284	1.10
36	547	494	439	406	381	354	331	308	268	1.23
38	518	468	416	384	361	336	314	292	254	1.37
40	492	445	395	365	343	319	298	277	241	1.52
42	469	424	377	348	326	304	284	264	230	1.67
44	448	405	360	332	312	290	271	252	220	1.84
46	428	387	344	318	298	277	259	241	210	2.01
48	410	371	330	304	286	266	248	231	201	2.18
50	394	356	316	292	274	255	238	222	193	2.37
52	379	342	304	281	264	245	229	213	186	2.56
54	365	330	293	271	254	236	221	205	179	2.77
56	352	318	282	261	245	228	213	198	172	2.97
58	339	307	273	252	236	220	206	191	167	3.19
60	328	297	264	243	228	213	199	185	161	3.41
62	318	287	255	236	221	206	192	179	156	3.65
64	308	278	247	228	214	199	186	173	151	3.88
66	298	270	240	221	208	193	181	168	146	4.13
68	290	262	233	215	202	188	175	163	142	4.39
70	281	254	226	209	196	182	170	158	138	4.65
72	273	247	220	203	190	177	166	154	134	4.92

Properties and Reaction Values

	256	232	210	194	182	170	160	150	135	
S_x in.³	895	809	719	664	623	580	542	504	439	For explanation of deflection, see page 2-32
V kips	719	646	609	558	527	492	468	448	427	
R_1 kips	208	179	158	138	127	112	104	96.7	83.5	
R_2 kips/in.	31.7	28.7	27.4	25.2	23.9	22.4	21.4	20.6	19.8	
R_3 kips	297	244	212	181	161	141	127	115	99.3	
R_4 kips/in.	9.85	8.15	8.26	7.02	6.41	5.70	5.39	5.23	5.55	
R kips	319	273	241	206	183	161	146	133	119	

Load above heavy line is limited by maximum allowable web shear.

| F_y = 50 ksi | | **BEAMS** W Shapes Allowable uniform loads in kips for beams laterally supported For beams laterally unsupported, see page 2-146 | | | | | | | W 33 |

Designation						W 33				
Wt./ft		619	567	515	468	424	387	354	Deflection In.
Flange Width		16⅞	16¾	16⅝	16½	16⅜	16¼	16⅛	
L_c		15.1	15.0	14.9	14.7	14.6	14.5	14.4	
L_u		52.1	48.3	44.4	40.7	37.0	34.4	31.4	

	Span in Feet									Deflection In.
	15	3030	2750						.23
	16	2980	2740	2470	2240	2010	1810	1650	.26
	17	2810	2580	2340	2110	1920	1750	1590	.30
	18	2650	2430	2210	1990	1810	1650	1500	.34
	19	2510	2300	2100	1890	1710	1560	1420	.37
	20	2390	2190	1990	1790	1630	1490	1350	.41
	21	2270	2090	1900	1710	1550	1410	1290	.46
	22	2170	1990	1810	1630	1480	1350	1230	.50
	23	2080	1900	1730	1560	1420	1290	1180	.55
	24	1990	1820	1660	1490	1360	1240	1130	.60
	25	1910	1750	1590	1430	1300	1190	1080	.65
	26	1840	1680	1530	1380	1250	1140	1040	.70
	27	1770	1620	1480	1330	1210	1100	1000	.75
	28	1710	1560	1420	1280	1160	1060	966	.81
	29	1650	1510	1370	1240	1120	1020	933	.87
	30	1590	1460	1330	1200	1090	990	902	.93
	31	1540	1410	1290	1160	1050	958	873	.99
	32	1490	1370	1240	1120	1020	928	846	1.06
	33	1450	1330	1210	1090	987	900	820	1.13
	34	1400	1290	1170	1060	958	874	796	1.20
	35	1360	1250	1140	1030	930	849	773	1.27
	36	1330	1220	1110	996	904	825	752	1.34
	37	1290	1180	1080	969	880	803	731	1.42
	38	1260	1150	1050	944	857	782	712	1.49
	40	1190	1100	996	897	814	743	677	1.66
	42	1140	1040	948	854	775	707	644	1.82
	44	1090	995	905	815	740	675	615	2.00
	46	1040	952	866	780	708	646	588	2.19
	48	995	912	830	747	678	619	564	2.38
	50	955	876	796	717	651	594	541	2.59
	52	918	842	766	690	626	571	520	2.80
	54	884	811	737	664	603	550	501	3.02
	56	853	782	711	640	581	530	483	3.24
	58	823	755	687	618	561	512	467	3.48
	60	796	730	664	598	543	495	451	3.72
	62	770	706	642	578	525	479	436	3.98
	64	746	684	622	560	509	464	423	4.24
	66	723	663	603	543	493	450	410	4.51
	68	702	644	586	527	479	437	398	4.78
	70	682	625	569	512	465	424	387	5.07
	72	663	608	553	498	452	413	376	5.36

F_y = 50 ksi

Properties and Reaction Values								
S_x in.³	2170	1990	1810	1630	1480	1350	1230	
V kips	1520	1370	1230	1120	1000	906	825	For explanation of deflection, see page 2-32
R_1 kips	711	607	519	439	377	325	275	
R_2 kips/in.	65.0	59.7	54.5	50.2	45.5	41.6	38.3	
R_3 kips	1250	1060	881	743	614	513	434	
R_4 kips/in.	40.5	34.5	29.0	25.3	21.0	17.6	15.2	
R kips	939	816	710	615	536	471	409	

Load above heavy line is limited by maximum allowable web shear.

W 33	**BEAMS** **W Shapes** **Allowable uniform loads in kips** **for beams laterally supported** For beams laterally unsupported, see page 2-146	F_y = 50 ksi

I

Designation		W 33						
Wt./ft		318	291	263	241	221	201	Deflection
Flange Width		16	15⅞	15¾	15⅞	15¾	15¾	In.
L_c		14.3	14.2	14.2	14.2	14.2	14.1	
L_u		28.7	26.2	24.0	21.6	19.8	17.9	

	Span in Feet	318	291	263	241	221	201	Deflection In.
	15					1050	963	.23
	16	1460	1340	1200	1140	1040	941	.26
	17	1440	1310	1190	1070	980	885	.30
	18	1360	1230	1120	1010	925	836	.34
	19	1290	1170	1060	960	877	792	.37
	20	1220	1110	1010	912	833	752	.41
	21	1160	1060	961	868	793	717	.46
	22	1110	1010	917	829	757	684	.50
	23	1060	966	877	793	724	654	.55
	24	1020	926	841	760	694	627	.60
	25	977	889	807	730	666	602	.65
	26	939	855	776	701	641	579	.70
	27	904	823	747	675	617	557	.75
	28	872	794	721	651	595	537	.81
	29	842	766	696	629	574	519	.87
	30	814	741	672	608	555	502	.93
	31	788	717	651	588	537	485	.99
	32	763	694	630	570	520	470	1.06
	34	718	654	593	536	490	443	1.20
	36	678	617	560	507	463	418	1.34
	38	643	585	531	480	438	396	1.49
	40	611	556	504	456	416	376	1.66
	42	581	529	480	434	397	358	1.82
	44	555	505	459	415	379	342	2.00
	46	531	483	439	396	362	327	2.19
	48	509	463	420	380	347	314	2.38
	50	488	444	403	365	333	301	2.59
	52	470	427	388	351	320	289	2.80
	54	452	411	374	338	308	279	3.02
	56	436	397	360	326	297	269	3.24
	58	421	383	348	314	287	259	3.48
	60	407	370	336	304	278	251	3.72
	62	394	358	325	294	269	243	3.98
	64	382	347	315	285	260	235	4.24
	66	370	337	306	276	252	228	4.51
	68	359	327	297	268	245	221	4.78
	70	349	317	288	261	238	215	5.07
	72	339	309	280	253	231	209	5.36

F_y = 50 ksi

Properties and Reaction Values								
S_x in.³		1110	1010	917	829	757	684	
V kips		731	669	601	567	526	482	For
R_1 kips		231	203	170	150	132	114	explanation
R_2 kips/in.		34.3	31.7	28.7	27.4	25.6	23.6	of deflection,
R_3 kips		351	297	244	215	185	156	see page 2-32
R_4 kips/in.		12.2	10.6	8.76	8.62	7.76	6.81	
R kips		351	314	270	245	212	180	

Load above heavy line is limited by maximum allowable web shear.

AMERICAN INSTITUTE OF STEEL CONSTRUCTION

| F_y = 50 ksi | BEAMS
W Shapes
Allowable uniform loads in kips
for beams laterally supported
For beams laterally unsupported, see page 2-146 | W 33 |

Designation						
Wt./ft		169	152	141	130	118
Flange Width		11½	11⅝	11½	11½	11½
L_c		10.3	10.4	10.3	9.90	8.60
L_u		13.8	12.2	11.1	10.8	10.7

Deflection In.

Span in Feet	Designation →	169	152	141	130	118	Deflection In.
10						723	.10
11					768	718	.13
12			851	806	744	658	.15
13		906	824	758	687	608	.17
14		863	765	704	638	564	.20
15		805	714	657	595	527	.23
16		755	670	616	558	494	.26
17		710	630	580	525	465	.30
18		671	595	548	496	439	.34
19		636	564	519	470	416	.37
20		604	536	493	447	395	.41
21		575	510	469	425	376	.46
22		549	487	448	406	359	.50
23		525	466	429	388	343	.55
24		503	446	411	372	329	.60
25		483	429	394	357	316	.65
26		465	412	379	344	304	.70
27		447	397	365	331	293	.75
28		431	383	352	319	282	.81
29		416	369	340	308	272	.87
30		403	357	329	298	263	.93
31		390	346	318	288	255	.99
32		377	335	308	279	247	1.06
33		366	325	299	271	239	1.13
34		355	315	290	263	232	1.20
36		336	298	274	248	219	1.34
38		318	282	259	235	208	1.49
40		302	268	246	223	197	1.66
42		288	255	235	213	188	1.82
44		275	244	224	203	180	2.00
46		263	233	214	194	172	2.19
48		252	223	205	186	165	2.38
50		242	214	197	179	158	2.59
52		232	206	190	172	152	2.80
54		224	198	183	165	146	3.02
56		216	191	176	160	141	3.24
58		208	185	170	154	136	3.48
60		201	179	164	149	132	3.72
62		195	173	159	144	127	3.98
64		189	167	154	140	123	4.24
66		183	162	149	135	120	4.51
68		178	158	145	131	116	4.78
70		173	153	141	128	113	5.07
72		168	149	137	124	110	5.36

F_y = 50 ksi

Properties and Reaction Values

	169	152	141	130	118	
S_x in.³	549	487	448	406	359	
V kips	453	425	403	384	361	For
R_1 kips	114	98.2	87.3	80.7	70.9	explanation
R_2 kips/in.	22.1	21.0	20.0	19.1	18.2	of deflection,
R_3 kips	146	125	111	98.2	84.4	see page 2-32
R_4 kips/in.	5.26	5.23	5.00	4.97	4.93	
R kips	164	143	129	116	102	

Load above heavy line is limited by maximum allowable web shear.

AMERICAN INSTITUTE OF STEEL CONSTRUCTION

W 30		BEAMS W Shapes Allowable uniform loads in kips for beams laterally supported For beams laterally unsupported, see page 2-146							F_y = 50 ksi

Designation		W 30							
Wt./ft		581	526	477	433	391	357	326	Deflection In.
Flange Width		16¼	16	15⅞	15¾	15⅝	15½	15⅜	
L_c		14.5	14.3	14.2	14.1	14.0	13.9	13.8	
L_u		53.8	49.8	45.7	41.7	38.3	35.1	32.4	

Span in Feet		581	526	477	433	391	357	326	Deflection In.
	14	2790	2490						.22
	15	2740	2460	2230	2020	1810	1630	1480	.26
	16	2570	2310	2100	1900	1720	1570	1420	.29
	17	2420	2170	1980	1790	1620	1480	1330	.33
	18	2290	2050	1870	1690	1530	1390	1260	.37
	19	2170	1950	1770	1600	1450	1320	1190	.41
	20	2060	1850	1680	1520	1380	1250	1130	.46
	21	1960	1760	1600	1450	1310	1190	1080	.50
	22	1870	1680	1530	1380	1250	1140	1030	.55
	23	1790	1610	1460	1320	1200	1090	985	.60
	24	1710	1540	1400	1270	1150	1050	944	.66
	25	1650	1480	1350	1210	1100	1000	906	.71
	26	1580	1420	1300	1170	1060	965	872	.77
	27	1520	1370	1250	1120	1020	929	839	.83
	28	1470	1320	1200	1080	982	896	809	.89
	29	1420	1270	1160	1050	948	865	781	.96
	30	1370	1230	1120	1010	917	836	755	1.02
	31	1330	1190	1090	979	887	809	731	1.09
	32	1290	1160	1050	949	859	784	708	1.17
	33	1250	1120	1020	920	833	760	687	1.24
	34	1210	1090	990	893	809	738	666	1.32
	35	1180	1060	962	867	786	717	647	1.39
	36	1140	1030	935	843	764	697	629	1.47
	37	1110	999	910	821	743	678	612	1.56
	38	1080	973	886	799	724	660	596	1.64
	40	1030	924	842	759	688	627	567	1.82
	42	980	880	801	723	655	597	540	2.01
	44	935	840	765	690	625	570	515	2.20
	46	894	803	732	660	598	545	493	2.41
	48	857	770	701	633	573	523	472	2.62
	50	823	739	673	607	550	502	453	2.85
	52	791	711	647	584	529	482	436	3.08
	54	762	684	623	562	509	464	420	3.32
	56	735	660	601	542	491	448	405	3.57
	58	709	637	580	523	474	432	391	3.83
	60	686	616	561	506	458	418	378	4.10
	62	664	596	543	490	444	405	365	4.37
	64	643	578	526	474	430	392	354	4.66
	66	623	560	510	460	417	380	343	4.96
	68	605	544	495	446	404	369	333	5.26
	70	588	528	481	434	393	358	324	5.58
	72	571	513	468	422	382	348	315	5.90

Properties and Reaction Values									
S_x in.³		1870	1680	1530	1380	1250	1140	1030	For explanation of deflection, see page 2-32
V kips		1390	1240	1120	1010	903	813	739	
R_1 kips		701	591	504	425	365	307	265	
R_2 kips/in.		65.0	59.1	53.8	49.5	44.9	40.9	37.6	
R_3 kips		1250	1030	859	723	596	497	419	
R_4 kips/in.		44.0	36.8	31.0	27.0	22.4	18.7	16.1	
R kips		929	798	692	598	522	450	397	

Load above heavy line is limited by maximum allowable web shear.

AMERICAN INSTITUTE OF STEEL CONSTRUCTION

| F_y = 50 ksi | BEANS
W Shapes
Allowable uniform loads in kips
for beams laterally supported
For beams laterally unsupported, see page 2-146 | | | | | | W 30 |

Designation		W 30						
Wt./ft		292	261	235	211	191	173	Deflection In.
Flange Width		15¼	15⅛	15	15⅛	15	15	
L_c		13.7	13.6	13.5	13.5	13.5	13.4	
L_u		29.5	26.5	24.0	21.4	19.4	17.5	

	Span in Feet							
	14						798	.22
	15	1310	1180	1040	959	871	791	.26
	16	1280	1140	1030	912	822	741	.29
	17	1200	1070	965	858	774	698	.33
	18	1130	1010	912	810	731	659	.37
	19	1080	958	864	768	692	624	.41
	20	1020	910	821	729	658	593	.46
	21	972	866	782	695	626	565	.50
	22	928	827	746	663	598	539	.55
	23	888	791	714	634	572	516	.60
	24	851	758	684	608	548	494	.66
	25	817	728	656	583	526	474	.71
	26	785	700	631	561	506	456	.77
	27	756	674	608	540	487	439	.83
	28	729	650	586	521	470	424	.89
	29	704	627	566	503	454	409	.96
	30	681	606	547	486	439	395	1.02
	31	659	587	529	471	424	383	1.09
	32	638	569	513	456	411	371	1.17
	33	619	551	497	442	399	359	1.24
	34	600	535	483	429	387	349	1.32
	35	583	520	469	417	376	339	1.39
	36	567	505	456	405	365	329	1.47
	37	552	492	444	394	356	320	1.56
	38	537	479	432	384	346	312	1.64
	40	510	455	410	365	329	296	1.82
	42	486	433	391	347	313	282	2.01
	44	464	414	373	332	299	270	2.20
	46	444	396	357	317	286	258	2.41
	48	425	379	342	304	274	247	2.62
	50	408	364	328	292	263	237	2.85
	52	393	350	316	281	253	228	3.08
	54	378	337	304	270	244	220	3.32
	56	365	325	293	260	235	212	3.57
	58	352	314	283	251	227	204	3.83
	60	340	303	274	243	219	198	4.10
	62	329	293	265	235	212	191	4.37
	64	319	284	256	228	206	185	4.66
	66	309	276	249	221	199	180	4.96
	68	300	268	241	215	193	174	5.26
	70	292	260	234	208	188	169	5.58
	72	284	253	228	203	183	165	5.90

Properties and Reaction Values								
S_x in.³		928	827	746	663	598	539	
V kips		653	588	520	480	436	399	For
R_1 kips		221	187	154	136	113	101	explanation
R_2 kips/in.		33.7	30.7	27.4	25.6	23.4	21.6	of deflection,
R_3 kips		337	277	223	188	157	132	see page 2-32
R_4 kips/in.		12.9	11.1	8.78	8.25	7.10	6.25	
R kips		339	294	250	217	182	154	

Load above heavy line is limited by maximum allowable web shear.

BEAMS
W Shapes
Allowable uniform loads in kips
for beams laterally supported
For beams laterally unsupported, see page 2-146

W 30

$F_y = 50$ ksi

Designation					W 30					
Wt./ft			148	132	124	116	108	99	90	Deflection In.
Flange Width			10½	10½	10½	10½	10½	10½	10⅜	
L_c			9.40	9.40	9.40	9.40	8.90	7.90	7.20	
L_u			13.4	11.6	10.8	9.90	9.80	9.70	9.60	
		9						617	555	.09
		10				678	650	592	539	.11
		11		746	706	658	598	538	490	.14
		12	797	697	651	603	548	493	449	.16
		13	738	643	601	557	506	455	415	.19
		14	685	597	558	517	470	423	385	.22
		15	639	557	521	483	439	395	359	.26
		16	600	523	488	452	411	370	337	.29
		17	564	492	459	426	387	348	317	.33
		18	533	464	434	402	365	329	299	.37
		19	505	440	411	381	346	311	284	.41
		20	480	418	391	362	329	296	270	.46
		21	457	398	372	345	313	282	257	.50
		22	436	380	355	329	299	269	245	.55
		23	417	363	340	315	286	257	234	.60
		24	400	348	325	302	274	247	225	.66
		25	384	334	312	290	263	237	216	.71
		26	369	322	300	278	253	228	207	.77
		27	355	310	289	268	244	219	200	.83
		28	343	299	279	259	235	211	193	.89
		29	331	288	269	250	227	204	186	.96
		30	320	279	260	241	219	197	180	1.02
		31	309	270	252	233	212	191	174	1.09
		32	300	261	244	226	206	185	168	1.17
		34	282	246	230	213	193	174	159	1.32
		36	266	232	217	201	183	164	150	1.47
		38	252	220	206	190	173	156	142	1.64
		40	240	209	195	181	164	148	135	1.82
		42	228	199	186	172	157	141	128	2.01
		44	218	190	178	165	150	135	123	2.20
		46	209	182	170	157	143	129	117	2.41
		48	200	174	163	151	137	123	112	2.62
		50	192	167	156	145	132	118	108	2.85
		52	184	161	150	139	127	114	104	3.08
		54	178	155	145	134	122	110	100	3.32
		56	171	149	139	129	117	106	96	3.57
		58	165	144	135	125	113	102	93	3.83
		60	160	139	130	121	110	99	90	4.10
		62	155	135	126	117	106	95	87	4.37
		64	150	131	122	113	103	92	84	4.66
		66	145	127	118	110	100	90	82	4.96
		68	141	123	115	106	97	87	79	5.26
		70	137	119	112	103	94	85	77	5.58
		72	133	116	108	101	91	82	75	5.90

$F_y = 50$ ksi · *Span in Feet* (left margin labels)

Properties and Reaction Values										
S_x in.³			436	380	355	329	299	269	245	
V kips			399	373	353	339	339	325	308	278	For
R_1 kips			107	88.8	81.4	75.7	70.3	61.7	50.9	explanation
R_2 kips/in.			21.4	20.3	19.3	18.6	18.0	17.2	15.5	of deflection,
R_3 kips			137	116	104	94.1	84.3	73.8	60.5	see page 2-32
R_4 kips/in.			5.47	5.54	5.15	5.10	5.15	5.10	4.16	
R kips			156	135	122	112	102	92	75	

Load above heavy line is limited by maximum allowable web shear.

AMERICAN INSTITUTE OF STEEL CONSTRUCTION

F_y = 50 ksi	BEAMS W Shapes Allowable uniform loads in kips for beams laterally supported For beams laterally unsupported, see page 2-146								W 27

Designation		W 27								
Wt./ft		539	494	448	408	368	336	307	281	Deflection In.
Flange Width		15¼	15⅛	15	14¾	14⅝	14½	14½	14⅜	
L_c		13.7	13.5	13.4	13.3	13.1	13.0	12.9	12.9	
L_u		55.6	51.3	47.6	43.3	39.7	37.0	34.0	31.4	
	13	2560	2320	2070	1880	1680				.21
	14	2470	2260	2040	1840	1670	1510	1370	1240	.25
	15	2300	2110	1910	1720	1560	1420	1300	1190	.28
	16	2160	1980	1790	1610	1460	1330	1220	1120	.32
	17	2030	1860	1680	1510	1370	1260	1140	1050	.37
	18	1920	1760	1590	1430	1300	1190	1080	991	.41
	19	1820	1670	1510	1360	1230	1120	1020	939	.46
	20	1730	1580	1430	1290	1170	1070	972	892	.51
	21	1650	1510	1360	1230	1110	1020	926	850	.56
	22	1570	1440	1300	1170	1060	970	884	811	.61
	23	1500	1380	1240	1120	1010	928	846	776	.67
	24	1440	1320	1190	1070	972	889	810	743	.73
	25	1380	1270	1140	1030	933	854	778	714	.79
	26	1330	1220	1100	990	897	821	748	686	.85
	27	1280	1170	1060	953	864	790	720	661	.92
	28	1230	1130	1020	919	833	762	695	637	.99
	29	1190	1090	986	888	804	736	671	615	1.06
	30	1150	1060	953	858	777	711	648	595	1.14
	31	1110	1020	923	830	752	688	627	576	1.22
	32	1080	990	894	804	729	667	608	558	1.29
	33	1050	960	867	780	707	647	589	541	1.38
	34	1020	932	841	757	686	628	572	525	1.46
	35	987	905	817	735	666	610	556	510	1.55
	36	959	880	794	715	648	593	540	496	1.64
	38	909	834	753	677	614	562	512	470	1.83
	40	864	792	715	644	583	534	486	446	2.02
	42	822	754	681	613	555	508	463	425	2.23
	44	785	720	650	585	530	485	442	406	2.45
	46	751	689	622	560	507	464	423	388	2.68
	48	720	660	596	536	486	445	405	372	2.91
	50	691	634	572	515	466	427	389	357	3.16
	52	664	609	550	495	448	410	374	343	3.42
	54	640	587	530	477	432	395	360	330	3.69
	56	617	566	511	460	416	381	347	319	3.97
	58	596	546	493	444	402	368	335	308	4.25
	60	576	528	477	429	389	356	324	297	4.55
	62	557	511	461	415	376	344	314	288	4.86
	64	540	495	447	402	364	333	304	279	5.18
	66	523	480	433	390	353	323	295	270	5.51

Left vertical labels: F_y = 50 ksi, Span in Feet

Properties and Reaction Values									
S_x in.³	1570	1440	1300	1170	1060	970	884	811	
V kips	1280	1160	1040	938	839	756	687	621	For explanation of deflection, see page 2-32
R_1 kips	691	597	502	431	363	312	269	230	
R_2 kips/in.	65.0	59.7	54.5	50.2	45.5	41.6	38.3	35.0	
R_3 kips	1250	1060	881	743	614	513	434	365	
R_4 kips/in.	47.9	40.9	34.5	30.2	25.2	21.1	18.2	15.2	
R kips	919	806	693	607	522	458	403	353	

Load above heavy line is limited by maximum allowable web shear.

| W 27 | BEAMS | | | | | | | F_y = 50 ksi |

I

BEAMS
W Shapes
Allowable uniform loads in kips
for beams laterally supported
For beams laterally unsupported, see page 2-146

Designation		W 27							
Wt./ft		258	235	217	194	178	161	146	Deflection
Flange Width		14¼	14¼	14⅛	14	14⅛	14	14	In.
L_c		12.8	12.7	12.6	12.6	12.6	12.6	12.5	
L_u		29.0	26.7	24.9	22.4	20.1	18.3	16.6	
	13					806	728	663	.21
	14	1140	1040	944	843	789	715	646	.25
	15	1090	989	915	815	736	667	603	.28
	16	1020	927	858	765	690	626	565	.32
	17	960	872	808	720	650	589	532	.37
	18	907	824	763	680	614	556	502	.41
	19	859	780	723	644	581	527	476	.46
	20	816	741	686	612	552	501	452	.51
	21	777	706	654	582	526	477	431	.56
	22	742	674	624	556	502	455	411	.61
	23	710	645	597	532	480	435	393	.67
	24	680	618	572	510	460	417	377	.73
	25	653	593	549	489	442	400	362	.79
	26	628	570	528	470	425	385	348	.85
	27	605	549	508	453	409	371	335	.92
	28	583	530	490	437	394	358	323	.99
	29	563	511	473	422	381	345	312	1.06
	30	544	494	458	408	368	334	301	1.14
	32	510	463	429	382	345	313	283	1.29
	34	480	436	404	360	325	294	266	1.46
	36	453	412	381	340	307	278	251	1.64
	38	430	390	361	322	291	263	238	1.83
	40	408	371	343	306	276	250	226	2.02
	42	389	353	327	291	263	238	215	2.23
	44	371	337	312	278	251	228	206	2.45
	46	355	322	298	266	240	218	197	2.68
	48	340	309	286	255	230	209	188	2.91
	50	326	297	275	245	221	200	181	3.16
	52	314	285	264	235	212	193	174	3.42
	54	302	275	254	227	205	185	167	3.69
	56	292	265	245	218	197	179	161	3.97
	58	281	256	237	211	190	173	156	4.25
	60	272	247	229	204	184	167	151	4.55
	62	263	239	221	197	178	161	146	4.86
	64	255	232	215	191	173	156	141	5.18
	66	247	225	208	185	167	152	137	5.51

F_y = 50 ksi — Span in Feet

Properties and Reaction Values									
S_x in.³		742	674	624	556	502	455	411	
V kips		568	522	472	422	403	364	331	For
R_1 kips		202	174	150	128	112	98.7	84.2	explanation
R_2 kips/in.		32.3	30.0	27.4	24.8	23.9	21.8	20.0	of deflection,
R_3 kips		310	265	223	181	162	134	112	see page 2-32
R_4 kips/in.		13.2	11.8	9.67	8.08	8.31	6.96	5.98	
R kips		315	279	246	209	191	158	133	

Load above heavy line is limited by maximum allowable web shear.

F_y = 50 ksi	BEAMS W Shapes Allowable uniform loads in kips for beams laterally supported For beams laterally unsupported, see page 2-146					W 27

Designation		W 27					Deflection In.
Wt./ft		129	114	102	94	84	
Flange Width		10	10⅛	10	10	10	
L_c		9.00	9.00	9.00	8.90	8.00	
L_u		13.3	11.5	10.2	9.50	9.40	

Span in Feet		129	114	102	94	84	Deflection In.
	9					491	.10
	10		622	558	528	469	.13
	11	674	598	534	486	426	.15
	12	633	548	490	446	391	.18
	13	584	506	452	411	360	.21
	14	542	470	420	382	335	.25
	15	506	439	392	356	312	.28
	16	474	411	367	334	293	.32
	17	446	387	346	314	276	.37
	18	422	365	326	297	260	.41
	19	399	346	309	281	247	.46
	20	380	329	294	267	234	.51
	21	361	313	280	255	223	.56
	22	345	299	267	243	213	.61
	23	330	286	255	232	204	.67
	24	316	274	245	223	195	.73
	25	304	263	235	214	187	.79
	26	292	253	226	206	180	.85
	27	281	244	218	198	174	.92
	28	271	235	210	191	167	.99
	29	262	227	203	184	162	1.06
	30	253	219	196	178	156	1.14
	31	245	212	189	172	151	1.22
	32	237	206	184	167	146	1.29
	34	223	193	173	157	138	1.46
	36	211	183	163	149	130	1.64
	38	200	173	155	141	123	1.83
	40	190	164	147	134	117	2.02
	42	181	157	140	127	112	2.23
	44	173	150	134	122	107	2.45
	46	165	143	128	116	102	2.68
	48	158	137	122	111	98	2.91
	50	152	132	117	107	94	3.16
	52	146	127	113	103	90	3.42
	54	141	122	109	99	87	3.69
	56	136	117	105	95	84	3.97
	58	131	113	101	92	81	4.25
	60	127	110	98	89	78	4.55
	62	122	106	95	86	76	4.86
	64	119	103	92	84	73	5.18
	66	115	100	89	81	71	5.51

F_y = 50 ksi

Properties and Reaction Values							
S_x in.³		345	299	267	243	213	For explanation of deflection, see page 2-32
V kips		337	311	279	264	246	
R_1 kips		91.2	76.4	66.4	58.1	52.2	
R_2 kips/in.		20.1	18.8	17.0	16.2	15.2	
R_3 kips		120	99.8	80.9	71.2	60.0	
R_4 kips/in.		5.39	5.26	4.38	4.23	4.11	
R kips		139	118	96	86	74	

Load above heavy line is limited by maximum allowable web shear.

| W 24 | BEAMS | | | | | | | | | F_y = 50 ksi |

W 24

BEAMS
W Shapes
Allowable uniform loads in kips
for beams laterally supported
For beams laterally unsupported, see page 2-146

Designation					W 24						
Wt./ft	492	450	408	370	335	306	279	250	229		Deflection In.
Flange Width	14⅛	14	13¾	13⅝	13½	13⅜	13¼	13⅛	13⅛		
L_c	12.6	12.5	12.4	12.2	12.1	12.0	11.9	11.8	11.7		
L_u	56.5	52.1	48.3	44.4	40.7	37.5	34.7	31.4	29.0		

Span in Feet / F_y = 50 ksi

Span	492	450	408	370	335	306	279	250	229	Deflection
12	2340	2110	1880	1700	1520	1370	1240	1100	999	.20
13	2180	1980	1790	1620	1460	1340	1220	1090	995	.24
14	2030	1840	1670	1500	1360	1240	1130	1010	924	.28
15	1890	1720	1560	1400	1270	1160	1050	945	862	.32
16	1770	1610	1460	1320	1190	1090	987	886	809	.36
17	1670	1510	1370	1240	1120	1020	929	833	761	.41
18	1580	1430	1300	1170	1060	964	878	787	719	.46
19	1490	1360	1230	1110	1000	914	831	746	681	.51
20	1420	1290	1170	1050	950	868	790	708	647	.57
21	1350	1230	1110	1000	905	827	752	675	616	.63
22	1290	1170	1060	957	864	789	718	644	588	.69
23	1230	1120	1010	915	826	755	687	616	562	.75
24	1180	1070	972	877	792	723	658	590	539	.82
25	1140	1030	933	842	760	694	632	567	517	.89
26	1090	990	897	810	731	668	608	545	498	.96
27	1050	953	864	780	704	643	585	525	479	1.04
28	1010	919	833	752	679	620	564	506	462	1.12
29	979	888	804	726	655	599	545	489	446	1.20
30	946	858	777	702	634	579	527	472	431	1.28
31	915	830	752	679	613	560	510	457	417	1.37
32	887	804	729	658	594	542	494	443	404	1.46
33	860	780	707	638	576	526	479	429	392	1.55
34	835	757	686	619	559	511	465	417	380	1.64
35	811	735	666	602	543	496	451	405	370	1.74
36	788	715	648	585	528	482	439	394	359	1.84
38	747	677	614	554	500	457	416	373	340	2.05
40	710	644	583	526	475	434	395	354	323	2.28
42	676	613	555	501	453	413	376	337	308	2.51
44	645	585	530	479	432	395	359	322	294	2.75
46	617	560	507	458	413	377	343	308	281	3.01
48	591	536	486	439	396	362	329	295	270	3.28
50	568	515	466	421	380	347	316	283	259	3.56
52	546	495	448	405	366	334	304	272	249	3.85
54	526	477	432	390	352	321	293	262	240	4.15
56	507	460	416	376	339	310	282	253	231	4.46
58	489	444	402	363	328	299	272	244	223	4.79
60	473	429	389	351	317	289	263	236	216	5.12

Properties and Reaction Values

	492	450	408	370	335	306	279	250	229	
S_x in.³	1290	1170	1060	957	864	789	718	644	588	For explanation of deflection, see page 2-32
V kips	1170	1050	942	851	760	684	620	548	500	
R_1 kips	701	607	510	439	370	318	275	231	198	
R_2 kips/in.	65.0	59.7	54.5	50.2	45.5	41.6	38.3	34.3	31.7	
R_3 kips	1250	1060	881	743	614	513	434	351	297	
R_4 kips/in.	52.5	45.0	38.0	33.3	27.8	23.3	20.2	16.3	14.2	
R kips	929	816	701	615	529	464	409	351	309	

Load above heavy line is limited by maximum allowable web shear.

AMERICAN INSTITUTE OF STEEL CONSTRUCTION

F_y = 50 ksi	BEAMS W Shapes Allowable uniform loads in kips for beams laterally supported For beams laterally unsupported, see page 2-146								W 24

Designation		W 24							
Wt./ft		207	192	176	162	146	131	117	104	Deflection In.
Flange Width		13	13	12⅞	13	12⅞	12⅞	12¾	12¾	
L_c		11.7	11.6	11.5	11.6	11.6	11.5	11.5	11.4	
L_u		26.5	24.7	22.8	21.1	18.9	16.8	14.9	13.2	

	Span in Feet		207	192	176	162	146	131	117	104	Deflection In.
F_y = 50 ksi		11							534	481	.17
		12				705	643	592	534	473	.20
		13	895	825	757	701	628	557	492	437	.24
		14	834	772	707	651	583	517	457	405	.28
		15	779	720	660	607	544	483	427	378	.32
		16	730	675	619	569	510	452	400	355	.36
		17	687	635	582	536	480	426	377	334	.41
		18	649	600	550	506	453	402	356	315	.46
		19	615	569	521	479	430	381	337	299	.51
		20	584	540	495	455	408	362	320	284	.57
		21	556	514	471	434	389	345	305	270	.63
		22	531	491	450	414	371	329	291	258	.69
		23	508	470	430	396	355	315	278	247	.75
		24	487	450	413	380	340	302	267	237	.82
		25	467	432	396	364	326	290	256	227	.89
		26	449	415	381	350	314	278	246	218	.96
		27	433	400	367	337	302	268	237	210	1.04
		28	417	386	354	325	292	259	229	203	1.12
		29	403	372	341	314	281	250	221	196	1.20
		30	389	360	330	304	272	241	213	189	1.28
		31	377	348	319	294	263	233	207	183	1.37
		32	365	338	309	285	255	226	200	177	1.46
		33	354	327	300	276	247	219	194	172	1.55
		34	344	318	291	268	240	213	188	167	1.64
		36	325	300	275	253	227	201	178	158	1.84
		38	307	284	261	240	215	190	168	149	2.05
		40	292	270	248	228	204	181	160	142	2.28
		42	278	257	236	217	194	172	152	135	2.51
		44	266	246	225	207	186	165	146	129	2.75
		46	254	235	215	198	177	157	139	123	3.01
		48	243	225	206	190	170	151	133	118	3.28
		50	234	216	198	182	163	145	128	114	3.56
		52	225	208	190	175	157	139	123	109	3.85
		54	216	200	183	169	151	134	119	105	4.15
		56	209	193	177	163	146	129	114	101	4.46
		58	201	186	171	157	141	125	110	98	4.79
		60	195	180	165	152	136	121	107	95	5.12

Properties and Reaction Values									
S_x in.³	531	491	450	414	371	329	291	258	
V kips	447	413	379	353	322	296	267	241	For explanation of deflection, see page 2-32
R_1 kips	170	150	131	116	101	87.3	73.7	61.9	
R_2 kips/in.	28.7	26.7	24.8	23.3	21.4	20.0	18.2	16.5	
R_3 kips	244	212	181	157	132	111	90.4	73.6	
R_4 kips/in.	11.8	10.3	9.00	8.29	7.35	6.80	5.82	5.00	
R kips	270	243	213	186	158	135	111	91	

Load above heavy line is limited by maximum allowable web shear.

BEAMS
W Shapes
Allowable uniform loads in kips
for beams laterally supported
For beams laterally unsupported, see page 2-146

W 24

$F_y = 50$ ksi

Designation		W 24					W 24		Deflection In.
Wt./ft		103	94	84	76	68	62	55	
Flange Width		9	9⅛	9	9	9	7	7	
L_c		8.10	8.10	8.10	8.10	7.40	5.80	5.00	
L_u		12.0	10.9	9.60	8.60	8.50	6.40	6.30	
	6							372	.05
	7						408	358	.07
	8					394	360	314	.09
	9	540	501	453	421	376	320	279	.12
	10	539	488	431	387	339	288	251	.14
	11	490	444	392	352	308	262	228	.17
	12	449	407	359	323	282	240	209	.20
	13	415	376	332	298	261	222	193	.24
	14	385	349	308	277	242	206	179	.28
	15	359	326	287	258	226	192	167	.32
	16	337	305	270	242	212	180	157	.36
	17	317	287	254	228	199	170	148	.41
	18	299	271	240	215	188	160	139	.46
	19	284	257	227	204	178	152	132	.51
	20	270	244	216	194	169	144	125	.57
	21	257	233	205	184	161	137	119	.63
	22	245	222	196	176	154	131	114	.69
	23	234	212	187	168	147	125	109	.75
	24	225	204	180	161	141	120	105	.82
	25	216	195	172	155	136	115	100	.89
	26	207	188	166	149	130	111	96	.96
	27	200	181	160	143	125	107	93	1.04
	28	193	174	154	138	121	103	90	1.12
	29	186	168	149	134	117	99	86	1.20
	30	180	163	144	129	113	96	84	1.28
	32	168	153	135	121	106	90	78	1.46
	34	159	144	127	114	100	85	74	1.64
	36	150	136	120	108	94	80	70	1.84
	38	142	129	113	102	89	76	66	2.05
	40	135	122	108	97	85	72	63	2.28
	42	128	116	103	92	81	69	60	2.51
	44	123	111	98	88	77	66	57	2.75
	46	117	106	94	84	74	63	55	3.01
	48	112	102	90	81	71	60	52	3.28
	50	108	98	86	77	68	58	50	3.56
	52	104	94	83	74	65	55	48	3.85
	54	100	90	80	72	63	53	46	4.15
	56	96	87	77	69	61	51	45	4.46
	58	93	84	74	67	58	50	43	4.79
	60	90	81	72	65	56	48	42	5.12

($F_y = 50$ ksi — Span in Feet)

Properties and Reaction Values

	103	94	84	76	68	62	55	
S_x in.³	245	222	196	176	154	131	114	For
V kips	270	250	227	210	197	204	186	explanation
R_1 kips	79.4	69.0	60.6	52.2	47.1	48.8	42.8	of deflection,
R_2 kips/in.	18.2	17.0	15.5	14.5	13.7	14.2	13.0	see page 2-32
R_3 kips	97.1	83.1	68.0	57.9	49.2	52.1	42.4	
R_4 kips/in.	4.99	4.63	4.04	3.78	3.71	4.09	3.73	
R kips	115	99	82	71	62	66	55	

Load above heavy line is limited by maximum allowable web shear.

AMERICAN INSTITUTE OF STEEL CONSTRUCTION

F_y = 50 ksi	BEAMS W Shapes — Allowable uniform loads in kips for beams laterally supported	W 21

For beams laterally unsupported, see page 2-146

Designation	W 21								Deflection In.
Wt./ft	402	364	333	300	275	248	223	201	
Flange Width	13⅜	13¼	13⅛	13	12⅞	12¾	12⅝	12⅝	
L_c	12.0	11.9	11.8	11.6	11.5	11.4	11.4	11.3	
L_u	53.8	49.8	45.7	42.7	39.2	35.5	32.7	29.8	

Span in Feet | F_y = 50 ksi

Span	402	364	333	300	275	248	223	201	Defl.
11	1800	1620	1460	1300	1180	1050			.20
12	1720	1550	1410	1270	1160	1040	934	838	.23
13	1590	1430	1300	1170	1070	963	863	780	.27
14	1470	1330	1210	1090	993	894	801	724	.32
15	1370	1240	1130	1020	927	835	748	676	.37
16	1290	1160	1060	952	869	782	701	634	.42
17	1210	1100	995	896	818	736	660	597	.47
18	1150	1030	940	846	772	695	623	563	.53
19	1090	980	890	801	732	659	591	534	.59
20	1030	931	846	761	695	626	561	507	.65
21	982	886	806	725	662	596	534	483	.72
22	937	846	769	692	632	569	510	461	.79
23	896	809	736	662	605	544	488	441	.86
24	859	776	705	634	579	522	468	423	.94
25	825	744	677	609	556	501	449	406	1.02
26	793	716	651	586	535	481	432	390	1.10
27	763	689	627	564	515	464	416	376	1.19
28	736	665	604	544	497	447	401	362	1.27
29	711	642	583	525	479	432	387	350	1.37
30	687	620	564	507	463	417	374	338	1.46
31	665	600	546	491	449	404	362	327	1.56
32	644	582	529	476	435	391	351	317	1.66
33	625	564	513	461	421	379	340	307	1.77
34	606	547	498	448	409	368	330	298	1.88
36	573	517	470	423	386	348	312	282	2.11
38	542	490	445	401	366	329	295	267	2.35
40	515	465	423	381	348	313	281	254	2.60
42	491	443	403	362	331	298	267	241	2.87
44	469	423	385	346	316	285	255	231	3.15
46	448	405	368	331	302	272	244	220	3.44
48	429	388	352	317	290	261	234	211	3.75
50	412	372	338	304	278	250	224	203	4.06
52	396	358	325	293	267	241	216	195	4.40

Properties and Reaction Values

	402	364	333	300	275	248	223	201	
S_x in.³	937	846	769	692	632	569	510	461	
V kips	900	810	730	648	589	522	467	419	
R_1 kips	553	476	407	340	302	250	211	178	For explanation of deflection, see page 2-32
R_2 kips/in.	57.1	52.5	48.2	43.6	40.3	36.3	33.0	30.0	
R_3 kips	968	814	687	562	479	391	322	266	
R_4 kips/in.	45.9	39.9	34.3	28.4	24.8	20.3	17.3	14.5	
R kips	753	660	576	493	443	377	327	283	

Load above heavy line is limited by maximum allowable web shear.

AMERICAN INSTITUTE OF STEEL CONSTRUCTION

		BEAMS		F_y = 50 ksi

W 21

I

BEAMS
W Shapes
Allowable uniform loads in kips
for beams laterally supported
For beams laterally unsupported, see page 2-146

Designation		W 21							
Wt./ft		182	166	147	132	122	111	101	Deflection In.
Flange Width		12½	12⅜	12½	12½	12⅜	12⅜	12¼	
L_c		11.2	11.1	11.2	11.1	11.1	11.1	11.0	
L_u		27.1	25.1	21.8	19.6	18.3	16.8	15.4	

F_y = 50 ksi	Span in Feet		182	166	147	132	122	111	101	Deflection
		11			635	568	520	473	427	.20
		12	754	674	603	541	501	457	416	.23
		13	706	643	557	499	462	421	384	.27
		14	655	597	517	464	429	391	357	.32
		15	612	557	483	433	400	365	333	.37
		16	573	523	452	406	375	342	312	.42
		17	540	492	426	382	353	322	294	.47
		18	510	464	402	361	334	304	277	.53
		19	483	440	381	342	316	288	263	.59
		20	459	418	362	325	300	274	250	.65
		21	437	398	345	309	286	261	238	.72
		22	417	380	329	295	273	249	227	.79
		23	399	363	315	282	261	238	217	.86
		24	382	348	302	270	250	228	208	.94
		25	367	334	290	260	240	219	200	1.02
		26	353	322	278	250	231	211	192	1.10
		27	340	310	268	240	222	203	185	1.19
		28	328	299	259	232	215	196	178	1.27
		29	316	288	250	224	207	189	172	1.37
		30	306	279	241	216	200	183	166	1.46
		31	296	270	233	209	194	177	161	1.56
		32	287	261	226	203	188	171	156	1.66
		33	278	253	219	197	182	166	151	1.77
		34	270	246	213	191	177	161	147	1.88
		36	255	232	201	180	167	152	139	2.11
		38	241	220	190	171	158	144	131	2.35
		40	229	209	181	162	150	137	125	2.60
		42	218	199	172	155	143	130	119	2.87
		44	209	190	165	148	137	125	114	3.15
		46	199	182	157	141	131	119	109	3.44
		48	191	174	151	135	125	114	104	3.75
		50	183	167	145	130	120	110	100	4.06
		52	176	161	139	125	116	105	96	4.40

Properties and Reaction Values

	182	166	147	132	122	111	101	
S_x in.³	417	380	329	295	273	249	227	
V kips	377	337	318	284	260	237	214	For
R_1 kips	154	131	111	97.2	83.5	73.7	64.5	explanation
R_2 kips/in.	27.4	24.8	23.8	21.4	19.8	18.2	16.5	of deflection,
R_3 kips	221	182	158	128	109	91.7	76.0	see page 2-32
R_4 kips/in.	12.3	9.95	10.6	8.77	7.49	6.38	5.28	
R kips	250	217	194	159	135	114	94	

Load above heavy line is limited by maximum allowable web shear.

AMERICAN INSTITUTE OF STEEL CONSTRUCTION

F_y = 50 ksi		BEAMS					W 21		

BEAMS
W Shapes
Allowable uniform loads in kips
for beams laterally supported
For beams laterally unsupported, see page 2-146

Designation		W 21					W 21			
Wt./ft	93	83	73	68	62	57	50	44	Deflection In.
Flange Width	8⅜	8⅜	8¼	8¼	8¼	6½	6½	6½	
L_c	7.50	7.50	7.40	7.40	7.40	5.90	5.60	4.70	
L_u	12.1	10.9	9.60	8.90	8.10	6.70	6.00	5.90	

Span in Feet — F_y = 50 ksi

Span	93	83	73	68	62	57	50	44	Deflection In.
6							317	289	.06
7						341	297	256	.08
8	502	441	387	363	336	305	260	224	.10
9	469	418	369	342	310	271	231	199	.13
10	422	376	332	308	279	244	208	180	.16
11	384	342	302	280	254	222	189	163	.20
12	352	314	277	257	233	204	173	150	.23
13	325	289	256	237	215	188	160	138	.27
14	302	269	237	220	200	174	149	128	.32
15	282	251	221	205	186	163	139	120	.37
16	264	235	208	193	175	153	130	112	.42
17	248	221	195	181	164	144	122	106	.47
18	235	209	185	171	155	136	116	100	.53
19	222	198	175	162	147	129	109	94	.59
20	211	188	166	154	140	122	104	90	.65
21	201	179	158	147	133	116	99	85	.72
22	192	171	151	140	127	111	95	82	.79
23	184	164	144	134	121	106	90	78	.86
24	176	157	138	128	116	102	87	75	.94
25	169	150	133	123	112	98	83	72	1.02
26	162	145	128	118	107	94	80	69	1.10
27	156	139	123	114	103	90	77	66	1.19
28	151	134	119	110	100	87	74	64	1.27
29	146	130	115	106	96	84	72	62	1.37
30	141	125	111	103	93	81	69	60	1.46
31	136	121	107	99	90	79	67	58	1.56
32	132	118	104	96	87	76	65	56	1.66
33	128	114	101	93	85	74	63	54	1.77
34	124	111	98	91	82	72	61	53	1.88
35	121	107	95	88	80	70	59	51	1.99
36	117	105	92	86	78	68	58	50	2.11
38	111	99	87	81	74	64	55	47	2.35
40	106	94	83	77	70	61	52	45	2.60
42	101	90	79	73	67	58	50	43	2.87
44	96	86	76	70	64	56	47	41	3.15
46	92	82	72	67	61	53	45	39	3.44
48	88	78	69	64	58	51	43	37	3.75
50	84	75	66	62	56	49	42	36	4.06
52	81	72	64	59	54	47	40	35	4.40

Properties and Reaction Values

	93	83	73	68	62	57	50	44	
S_x in.³	192	171	151	140	127	111	94.5	81.6	
V kips	251	221	193	182	168	171	158	145	For explanation of deflection, see page 2-32
R_1 kips	80.7	66.4	56.3	51.0	45.4	45.9	41.1	34.3	
R_2 kips/in.	19.1	17.0	15.0	14.2	13.2	13.4	12.5	11.5	
R_3 kips	102	81.2	63.5	56.1	47.7	50.0	41.2	33.4	
R_4 kips/in.	7.00	5.51	4.32	3.96	3.58	3.50	3.55	3.33	
R kips	127	100	79	70	60	62	54	45	

Load above heavy line is limited by maximum allowable web shear.

BEAMS
W Shapes
Allowable uniform loads in kips
for beams laterally supported
For beams laterally unsupported, see page 2-146

W 18

I

F_y = 50 ksi

Designation		W 18								Deflection In.
Wt./ft		311	283	258	234	211	192	175	158	
Flange Width		12	11⅞	11¾	11⅝	11½	11½	11⅜	11¼	
L_c		10.8	10.6	10.5	10.4	10.3	10.3	10.2	10.1	
L_u		49.0	45.0	42.2	38.8	35.5	32.7	30.0	27.5	
	10	1360	1220	1100	977	876	781	713	639	.19
	11	1250	1130	1030	932	838	760	688	620	.23
	12	1140	1030	942	854	768	697	631	568	.27
	13	1060	954	870	789	709	643	582	525	.32
	14	981	886	808	732	658	597	541	487	.37
	15	915	827	754	683	615	557	505	455	.43
	16	858	776	707	641	576	523	473	426	.49
	17	808	730	665	603	542	492	445	401	.55
	18	763	689	628	570	512	464	420	379	.61
	19	723	653	595	540	485	440	398	359	.68
	20	686	620	565	513	461	418	378	341	.76
	21	654	591	538	488	439	398	360	325	.84
	22	624	564	514	466	419	380	344	310	.92
	23	597	539	492	446	401	363	329	297	1.00
	24	572	517	471	427	384	348	315	284	1.09
	25	549	496	452	410	369	334	303	273	1.19
	26	528	477	435	394	355	322	291	262	1.28
	27	508	460	419	380	341	310	280	253	1.38
	28	490	443	404	366	329	299	270	244	1.49
	29	473	428	390	354	318	288	261	235	1.60
	30	458	414	377	342	307	279	252	227	1.71
	31	443	400	365	331	297	270	244	220	1.82
	32	429	388	353	320	288	261	237	213	1.94
	33	416	376	343	311	279	253	229	207	2.07
	34	404	365	333	302	271	246	223	201	2.19
	36	381	345	314	285	256	232	210	189	2.46
	38	361	327	298	270	243	220	199	179	2.74
	40	343	310	283	256	230	209	189	171	3.03
	42	327	295	269	244	219	199	180	162	3.35
	44	312	282	257	233	210	190	172	155	3.67

F_y = 50 ksi — Span in Feet

Properties and Reaction Values

	311	283	258	234	211	192	175	158	
S_x in.³	624	564	514	466	419	380	344	310	
V kips	679	612	549	489	438	391	357	319	For
R_1 kips	431	368	317	263	224	193	165	142	explanation
R_2 kips/in.	50.2	46.2	42.2	38.3	35.0	31.7	29.4	26.7	of deflection,
R_3 kips	746	630	528	436	363	299	255	210	see page 2-32
R_4 kips/in.	41.4	36.2	30.6	25.3	21.8	17.9	16.0	13.5	
R kips	607	530	465	397	347	304	268	235	

Load above heavy line is limited by maximum allowable web shear.

F_y = 50 ksi

BEAMS
W Shapes
Allowable uniform loads in kips for beams laterally supported

For beams laterally unsupported, see page 2-146

W 18

Designation	W 18							Deflection In.
Wt./ft	143	130	119	106	97	86	76	
Flange Width	11¼	11⅛	11¼	11¼	11⅛	11⅛	11	
L_c	10.0	10.0	10.1	10.0	10.0	9.90	9.90	
L_u	25.3	23.1	21.0	18.7	17.4	15.5	13.7	
F_y = 50 ksi — Span in Feet 10	569	516	497	442	398	353	310	.19
11	564	512	462	408	376	332	292	.23
12	517	469	424	374	345	304	268	.27
13	477	433	391	345	318	281	247	.32
14	443	402	363	321	295	261	229	.37
15	414	375	339	299	276	243	214	.43
16	388	352	318	281	259	228	201	.49
17	365	331	299	264	243	215	189	.55
18	345	313	282	249	230	203	178	.61
19	327	296	267	236	218	192	169	.68
20	310	282	254	224	207	183	161	.76
21	295	268	242	214	197	174	153	.84
22	282	256	231	204	188	166	146	.92
23	270	245	221	195	180	159	140	1.00
24	259	235	212	187	172	152	134	1.09
25	248	225	203	180	165	146	128	1.19
26	239	217	195	173	159	140	124	1.28
27	230	209	188	166	153	135	119	1.38
28	222	201	182	160	148	130	115	1.49
29	214	194	175	155	143	126	111	1.60
30	207	188	169	150	138	122	107	1.71
31	200	182	164	145	133	118	104	1.82
32	194	176	159	140	129	114	100	1.94
33	188	171	154	136	125	111	97	2.07
34	182	166	149	132	122	107	94	2.19
35	177	161	145	128	118	104	92	2.32
36	172	156	141	125	115	101	89	2.46
37	168	152	137	121	112	99	87	2.60
38	163	148	134	118	109	96	85	2.74
39	159	144	130	115	106	94	82	2.88
40	155	141	127	112	103	91	80	3.03
42	148	134	121	107	98	87	76	3.35
44	141	128	116	102	94	83	73	3.67

Properties and Reaction Values

	143	130	119	106	97	86	76	
S_x in.^{3}	282	256	231	204	188	166	146	For explanation of deflection, see page 2-32
V kips	285	258	249	221	199	177	155	
R_1 kips	120	104	94.6	79.1	69.0	56.9	48.2	
R_2 kips/in.	24.1	22.1	21.6	19.5	17.7	15.8	14.0	
R_3 kips	172	144	131	106	87.8	70.2	54.9	
R_4 kips/in.	10.9	9.39	10.1	8.41	6.83	5.63	4.47	
R kips	204	177	166	135	112	90	71	

Load above heavy line is limited by maximum allowable web shear.

| W 18 | BEAMS W Shapes Allowable uniform loads in kips for beams laterally supported For beams laterally unsupported, see page 2-146 | | | | | | | | F_y = 50 ksi |

F_y = 50 ksi

W 18

BEAMS
W Shapes
Allowable uniform loads in kips
for beams laterally supported
For beams laterally unsupported, see page 2-146

Designation			W 18				W 18			
Wt./ft	71	65	60	55	50	46	40	35	Deflection In.
Flange Width	7⅝	7⅝	7½	7½	7½	6	6	6	
L_c	6.80	6.80	6.80	6.70	6.70	5.40	5.40	4.80	
L_u	11.1	10.4	9.60	8.70	7.90	6.80	5.90	5.60	

Span in Feet — F_y = 50 ksi

Span	71	65	60	55	50	46	40	35	Deflection In.
5								212	.05
6						260	226	211	.07
7	366	330	303	283	255	248	215	181	.09
8	349	322	297	270	244	217	188	158	.12
9	310	286	264	240	217	193	167	141	.15
10	279	257	238	216	196	173	150	127	.19
11	254	234	216	197	178	158	137	115	.23
12	233	215	198	180	163	144	125	106	.27
13	215	198	183	166	150	133	116	97	.32
14	200	184	170	154	140	124	107	91	.37
15	186	172	158	144	130	116	100	84	.43
16	175	161	149	135	122	108	94	79	.49
17	164	151	140	127	115	102	89	75	.55
18	155	143	132	120	109	96	84	70	.61
19	147	135	125	114	103	91	79	67	.68
20	140	129	119	108	98	87	75	63	.76
21	133	123	113	103	93	83	72	60	.84
22	127	117	108	98	89	79	68	58	.92
23	121	112	103	94	85	75	65	55	1.00
24	116	107	99	90	81	72	63	53	1.09
25	112	103	95	87	78	69	60	51	1.19
26	107	99	91	83	75	67	58	49	1.28
27	103	95	88	80	72	64	56	47	1.38
28	100	92	85	77	70	62	54	45	1.49
30	93	86	79	72	65	58	50	42	1.71
32	87	80	74	68	61	54	47	40	1.94
34	82	76	70	64	58	51	44	37	2.19
36	78	72	66	60	54	48	42	35	2.46
38	74	68	63	57	51	46	40	33	2.74
40	70	64	59	54	49	43	38	32	3.03
42	67	61	57	51	47	41	36	30	3.35
44	64	59	54	49	44	39	34	29	3.67

Properties and Reaction Values

	71	65	60	55	50	46	40	35	
S_x in.³	127	117	108	98.3	88.9	78.8	68.4	57.6	
V kips	183	165	151	141	128	130	113	106	For
R_1 kips	61.3	53.4	47.1	42.2	36.6	37.1	30.9	27.8	explanation
R_2 kips/in.	16.3	14.9	13.7	12.9	11.7	11.9	10.4	9.90	of deflection,
R_3 kips	75.4	62.9	53.6	46.5	38.4	40.4	30.8	25.8	see page 2-32
R_4 kips/in.	5.85	4.78	4.07	3.75	3.15	3.08	2.40	2.59	
R kips	96	80	68	60	49	51	39	35	

Load above heavy line is limited by maximum allowable web shear.

F_y = 50 ksi

BEAMS
W Shapes
Allowable uniform loads in kips for beams laterally supported

For beams laterally unsupported, see page 2-146

W 16

Designation		W 16				Deflection In.
Wt./ft		100	89	77	67	
Flange Width		10⅜	10⅜	10¼	10¼	
L_c		9.30	9.30	9.20	9.20	
L_u		20.2	18.0	15.8	13.9	
F_y = 50 ksi, Span in Feet	9	397	352	301	258	.17
	10	385	341	295	257	.21
	11	350	310	268	234	.26
	12	321	284	246	215	.31
	13	296	262	227	198	.36
	14	275	244	211	184	.42
	15	257	227	197	172	.48
	16	241	213	184	161	.55
	17	226	201	173	151	.62
	18	214	189	164	143	.69
	19	203	179	155	135	.77
	20	193	171	147	129	.85
	21	183	162	140	123	.94
	22	175	155	134	117	1.03
	23	167	148	128	112	1.13
	24	160	142	123	107	1.23
	25	154	136	118	103	1.33
	26	148	131	113	99	1.44
	27	143	126	109	95	1.56
	28	138	122	105	92	1.67
	29	133	118	102	89	1.79
	30	128	114	98	86	1.92
	31	124	110	95	83	2.05
	32	120	107	92	80	2.18
	34	113	100	87	76	2.47
	36	107	95	82	72	2.77
	38	101	90	78	68	3.08
	40	96	85	74	64	3.41

Properties and Reaction Values

	100	89	77	67	
S_x in.^{3}	175	155	134	117	For explanation of deflection, see page 2-32
V kips	199	176	150	129	
R_1 kips	81.4	67.7	54.0	44.8	
R_2 kips/in.	19.3	17.3	15.0	13.0	
R_3 kips	107	85.5	64.3	48.7	
R_4 kips/in.	8.64	7.12	5.41	4.09	
R kips	137	110	83	63	

Load above heavy line is limited by maximum allowable web shear.

		BEAMS W Shapes Allowable uniform loads in kips for beams laterally supported For beams laterally unsupported, see page 2-146							F_y = 50 ksi

W 16

Designation		W 16					W 16		
Wt./ft		57	50	45	40	36	31	26	Deflection
Flange Width		7⅛	7⅛	7	7	7	5½	5½	In.
L_c		6.40	6.30	6.30	6.30	6.30	4.90	4.00	
L_u		10.3	9.10	8.20	7.40	6.70	5.20	5.10	
	5						175	157	.05
	6					187	173	141	.08
	7	283	247	223	195	178	148	121	.10
	8	254	223	200	178	155	130	106	.14
	9	225	198	178	158	138	115	94	.17
	10	203	178	160	142	124	104	84	.21
	11	184	162	145	129	113	94	77	.26
	12	169	149	133	119	104	87	70	.31
	13	156	137	123	109	96	80	65	.36
	14	145	127	114	102	89	74	60	.42
	15	135	119	107	95	83	69	56	.48
	16	127	111	100	89	78	65	53	.55
	17	119	105	94	84	73	61	50	.62
	18	113	99	89	79	69	58	47	.69
	19	107	94	84	75	65	55	44	.77
	20	101	89	80	71	62	52	42	.85
	21	97	85	76	68	59	49	40	.94
	22	92	81	73	65	57	47	38	1.03
	24	85	74	67	59	52	43	35	1.23
	26	78	69	62	55	48	40	32	1.44
	28	72	64	57	51	44	37	30	1.67
	30	68	59	53	47	41	35	28	1.92
	32	63	56	50	44	39	32	26	2.18
	34	60	52	47	42	37	31	25	2.47
	36	56	50	44	40	35	29	23	2.77
	38	53	47	42	37	33	27	22	3.08
	40	51	45	40	36	31	26	21	3.41

F_y = 50 ksi — Span in Feet

Properties and Reaction Values									
S_x in.^{3}		92.2	81.0	72.7	64.7	56.5	47.2	38.4	For
V kips		141	124	111	98	94	87	78	explanation
R_1 kips		48.8	41.1	35.6	29.9	27.4	25.5	21.9	of deflection,
R_2 kips/in.		14.2	12.5	11.4	10.1	9.73	9.08	8.25	see page 2-32
R_3 kips		57.3	44.7	36.6	28.8	25.3	23.0	17.7	
R_4 kips/in.		4.88	3.86	3.25	2.53	2.72	2.15	2.08	
R kips		74	58	48	38	35	31	25	

Load above heavy line is limited by maximum allowable web shear.

F_y = 50 ksi	BEAMS W Shapes Allowable uniform loads in kips for beams laterally supported For beams laterally unsupported, see page 2-146						W 14

Designation		W 14						
Wt./ft		233	211	193	176	159	145	Deflection In.
Flange Width		15⅞	15¾	15¾	15⅝	15⅝	15½	
L_c		14.2	14.2	14.1	14.0	13.9	13.9	
L_u		56.5	52.1	49.0	45.0	41.2	38.3	
	12	687	616	551	505	446	402	.35
	13	635	572	525	476	430	393	.41
	14	589	531	487	442	399	365	.48
	15	550	496	455	412	373	340	.55
	16	516	465	426	386	349	319	.62
	17	485	437	401	364	329	300	.70
	18	458	413	379	343	310	284	.79
	19	434	391	359	325	294	269	.88
	20	413	372	341	309	279	255	.98
	21	393	354	325	294	266	243	1.08
	22	375	338	310	281	254	232	1.18
	23	359	323	297	269	243	222	1.29
	24	344	310	284	258	233	213	1.40
	25	330	297	273	247	224	204	1.52
	26	317	286	262	238	215	196	1.65
	27	306	275	253	229	207	189	1.78
	28	295	266	244	221	200	182	1.91
	29	284	256	235	213	193	176	2.05
	30	275	248	227	206	186	170	2.19
	31	266	240	220	199	180	165	2.34
	32	258	232	213	193	175	160	2.50
	33	250	225	207	187	169	155	2.66
	34	243	219	201	182	164	150	2.82

F_y = 50 ksi — Span in Feet

Properties and Reaction Values

	233	211	193	176	159	145	
S_x in.³	375	338	310	281	254	232	
V kips	343	308	276	253	223	201	
R_1 kips	210	182	156	137	115	98.2	For explanation of deflection, see page 2-32
R_2 kips/in.	35.3	32.3	29.4	27.4	24.6	22.4	
R_3 kips	349	291	242	208	169	141	
R_4 kips/in.	32.0	27.7	22.8	20.7	16.7	14.1	
R kips	334	295	259	233	201	177	

Load above heavy line is limited by maximum allowable web shear.

W 14	BEAMS	F_y = 50 ksi

BEAMS
W Shapes
Allowable uniform loads in kips
for beams laterally supported
For beams laterally unsupported, see page 2-146

Designation		W 14					W 14				
Wt./ft		132	120	109	99*	90*	82	74	68	61	Deflection
Flange Width		14¾	14⅝	14⅝	14⅝	14½	10⅛	10⅛	10	10	In.
L_c		13.2	13.1	13.1	13.0	13.0	9.10	9.00	9.00	9.00	
L_u		34.4	31.7	29.2	26.7	24.5	20.2	18.6	17.2	15.5	

	Span in Feet	132	120	109	99*	90*	82	74	68	61	Deflection In.
	9						292	255	233	208	.20
	10						271	246	227	203	.24
	11						246	224	206	184	.30
	12	378	342	301	275	247	226	205	189	169	.35
	13	354	322	293	265	237	208	190	174	156	.41
	14	328	299	272	246	220	193	176	162	145	.48
	15	307	279	254	230	205	180	164	151	135	.55
	16	287	261	238	215	192	169	154	142	127	.62
	17	270	246	224	203	181	159	145	133	119	.70
	18	255	232	211	191	171	150	137	126	113	.79
	19	242	220	200	181	162	142	130	119	107	.88
	20	230	209	190	172	154	135	123	113	101	.98
	21	219	199	181	164	147	129	117	108	97	1.08
	22	209	190	173	157	140	123	112	103	92	1.18
	23	200	182	165	150	134	118	107	99	88	1.29
	24	192	174	159	144	128	113	103	94	85	1.40
	25	184	167	152	138	123	108	99	91	81	1.52
	26	177	161	146	133	118	104	95	87	78	1.65
	27	170	155	141	128	114	100	91	84	75	1.78
	28	164	149	136	123	110	97	88	81	72	1.91
	29	159	144	131	119	106	93	85	78	70	2.05
	30	153	139	127	115	103	90	82	76	68	2.19
	31	148	135	123	111	99	87	79	73	65	2.34
	32	144	131	119	108	96	85	77	71	63	2.50
	33	139	127	115	104	93	82	75	69	61	2.66
	34	135	123	112	101	91	80	72	67	60	2.82

F_y = 50 ksi

Properties and Reaction Values

	132	120	109	99*	90*	82	74	68	61	
S_x in.³	209	190	173	157	143	123	112	103	92.2	For explanation of deflection, see page 2-32
V kips	189	171	150	137	123	146	128	117	104	
R_1 kips	89.8	79.1	67.7	57.5	49.9	68.4	58.0	51.4	44.5	
R_2 kips/in.	21.3	19.5	17.3	16.0	14.5	16.8	14.9	13.7	12.4	
R_3 kips	126	106	84.8	71.7	59.1	81.0	64.3	54.5	44.3	
R_4 kips/in.	12.8	10.9	8.47	7.45	6.17	7.82	5.91	5.10	4.25	
R kips	164	144	114	98	81	108	85	72	59	

*Indicates noncompact shape.
Load above heavy line is limited by maximum allowable web shear.

F_y = 50 ksi									BEAMS W Shapes — W 14

BEAMS
W Shapes
Allowable uniform loads in kips
for beams laterally supported
For beams laterally unsupported, see page 2-146

W 14

Designation		W 14			W 14			W14		
Wt./ft		53	48	43	38	34	30	26	22	Deflection In.
Flange Width		8	8	8	6¾	6¾	6¾	5	5	
L_c		7.20	7.20	7.20	6.10	6.00	6.00	4.50	4.10	
L_u		12.7	11.5	10.4	8.30	7.30	6.50	5.10	4.70	
	5							142	126	.06
	6				175	159	149	129	106	.09
	7				172	153	132	111	91	.12
	8	206	188	167	150	134	116	97	80	.16
	9	190	172	153	133	119	103	86	71	.20
	10	171	155	138	120	107	92	78	64	.24
	11	156	141	125	109	97	84	71	58	.30
	12	143	129	115	100	89	77	65	53	.35
	13	132	119	106	92	82	71	60	49	.41
	14	122	110	99	86	76	66	55	46	.48
	15	114	103	92	80	71	62	52	43	.55
	16	107	97	86	75	67	58	49	40	.62
	17	101	91	81	71	63	54	46	38	.70
	18	95	86	77	67	59	51	43	35	.79
	19	90	81	73	63	56	49	41	34	.88
	20	86	77	69	60	53	46	39	32	.98
	21	82	74	66	57	51	44	37	30	1.08
	22	78	70	63	55	49	42	35	29	1.18
	23	74	67	60	52	46	40	34	28	1.29
	24	71	64	57	50	45	39	32	27	1.40
	25	68	62	55	48	43	37	31	26	1.52
	26	66	59	53	46	41	36	30	25	1.65
	27	63	57	51	44	40	34	29	24	1.78
	28	61	55	49	43	38	33	28	23	1.91
	29	59	53	48	41	37	32	27	22	2.05
	30	57	52	46	40	36	31	26	21	2.19
	31	55	50	44	39	34	30	25	21	2.34
	32	53	48	43	38	33	29	24	20	2.50
	33	52	47	42	36	32	28	24	19	2.66
	34	50	45	41	35	31	27	23	19	2.82

F_y = 50 ksi — Span in Feet

Properties and Reaction Values									
S_x in.³	77.8	70.3	62.7	54.6	48.6	42.0	35.3	29.0	
V kips	103	94	83	87	80	75	71	63	For explanation of deflection, see page 2-32
R_1 kips	43.9	38.6	33.0	27.2	23.5	20.9	19.7	16.6	
R_2 kips/in.	12.2	11.2	10.1	10.2	9.41	8.91	8.42	7.59	
R_3 kips	44.0	36.8	29.5	29.8	24.7	20.9	20.1	15.3	
R_4 kips/in.	3.98	3.45	2.83	2.96	2.62	2.66	2.05	1.91	
R kips	58	49	39	40	34	30	27	22	

Load above heavy line is limited by maximum allowable web shear.

BEAMS
W Shapes
Allowable uniform loads in kips
for beams laterally supported
For beams laterally unsupported, see page 2-146

W 12

I

F_y = 50 ksi

F_y = 50 ksi   Span in Feet

Designation		W 12								W 12		Deflection In.
Wt./ft		136	120	106	96	87	79	72	65*	58	53	
Flange Width		12⅜	12⅜	12¼	12⅛	12⅛	12⅛	12	12	10	10	
L_c		11.1	11.0	10.9	10.9	10.9	10.8	10.8	10.7	9.00	9.00	
L_u		38.3	34.7	31.2	28.7	26.0	24.0	21.9	20.0	17.5	15.9	
	9	424	373							176	166	.23
	10	409	359	315	280	258	233	211	189	172	155	.28
	11	372	326	290	262	236	214	195	173	156	141	.34
	12	341	299	266	240	216	196	179	159	143	129	.41
	13	315	276	245	222	200	181	165	146	132	119	.48
	14	292	256	228	206	185	168	153	136	123	111	.56
	15	273	239	213	192	173	157	143	127	114	104	.64
	16	256	224	199	180	162	147	134	119	107	97	.73
	17	241	211	188	170	153	138	126	112	101	91	.82
	18	227	199	177	160	144	131	119	106	95	86	.92
	19	215	189	168	152	137	124	113	100	90	82	1.03
	20	205	179	160	144	130	118	107	95	86	78	1.14
	21	195	171	152	137	124	112	102	91	82	74	1.25
	22	186	163	145	131	118	107	97	87	78	71	1.38
	23	178	156	139	125	113	102	93	83	75	68	1.51
	24	171	149	133	120	108	98	89	79	72	65	1.64
	25	164	143	128	115	104	94	86	76	69	62	1.78
	26	157	138	123	111	100	91	82	73	66	60	1.92
	27	152	133	118	107	96	87	79	71	64	58	2.07
	28	146	128	114	103	93	84	77	68	61	55	2.23
	29	141	124	110	99	90	81	74	66	59	54	2.39
	30	136	120	106	96	87	78	71	63	57	52	2.56

Properties and Reaction Values

	136	120	106	96	87	79	72	65*	58	53	
S_x in.³	186	163	145	131	118	107	97.4	87.9	78.0	70.6	
V kips	212	186	157	140	129	116	105	95	88	83	For
R_1 kips	126	106	84.9	73.7	63.7	55.7	48.8	42.2	40.8	35.6	explanation
R_2 kips/in.	26.1	23.4	20.1	18.2	17.0	15.5	14.2	12.9	11.9	11.4	of deflection,
R_3 kips	189	151	114	93.0	80.0	66.4	55.5	45.5	41.5	36.9	see page 2-32
R_4 kips/in.	21.2	17.8	12.8	10.5	9.71	8.23	6.99	5.83	4.31	4.27	
R kips	**217**	**188**	155	130	114	95	80	66	57	52	

*Indicates noncompact shape.
Load above heavy line is limited by maximum allowable web shear.
Values of R in **bold face** exceed maximum web shear V.

AMERICAN INSTITUTE OF STEEL CONSTRUCTION

F_y = 50 ksi							BEAMS W Shapes Allowable uniform loads in kips for beams laterally supported For beams laterally unsupported, see page 2-146				W 12

Designation		W 12			W 12			W 12				
Wt./ft		50	45	40	35	30	26	22	19	16	14	Deflection In.
Flange Width		8⅛	8	8	6½	6½	6½	4	4	4	4	
L_c		7.20	7.20	7.20	5.90	5.80	5.80	3.60	3.60	2.90	2.50	
L_u		14.1	12.8	11.5	9.10	7.80	6.70	4.60	3.80	3.60	3.60	
	3									106	95	.03
	4							128	114	94	82	.05
	5							112	94	75	66	.07
	6				150	128	112	93	78	63	55	.10
	7	180	162		143	121	105	80	67	54	47	.14
	8	178	160	141	125	106	92	70	59	47	41	.18
	9	158	142	127	111	94	82	62	52	42	36	.23
	10	142	128	114	100	85	73	56	47	38	33	.28
	11	129	116	104	91	77	67	51	43	34	30	.34
	12	119	107	95	84	71	61	47	39	31	27	.41
	13	109	98	88	77	65	57	43	36	29	25	.48
	14	102	91	82	72	61	52	40	33	27	23	.56
	15	95	85	76	67	57	49	37	31	25	22	.64
	16	89	80	71	63	53	46	35	29	24	20	.73
	17	84	75	67	59	50	43	33	28	22	19	.82
	18	79	71	63	56	47	41	31	26	21	18	.92
	19	75	67	60	53	45	39	29	25	20	17	1.03
	20	71	64	57	50	42	37	28	23	19	16	1.14
	22	65	58	52	46	39	33	25	21	17	15	1.38
	24	59	53	48	42	35	31	23	20	16	14	1.64
	26	55	49	44	39	33	28	21	18	14	13	1.92
	28	51	46	41	36	30	26	20	17	13	12	2.23
	30	47	43	38	33	28	24	19	16	13	11	2.56

_F_y = 50 ksi_ — _Span in Feet_

Properties and Reaction Values												
S_x in.³		64.7	58.1	51.9	45.6	38.6	33.4	25.4	21.3	17.1	14.9	For explanation of deflection, see page 2-32
V kips		90	81	70	75	64	56	64	57	53	48	
R_1 kips		42.0	34.5	30.4	24.8	20.1	16.6	18.8	15.8	13.6	11.3	
R_2 kips/in.		12.2	11.1	9.73	9.90	8.58	7.59	8.58	7.76	7.26	6.60	
R_3 kips		43.3	35.3	27.6	28.5	21.1	16.3	20.8	16.2	12.8	10.2	
R_4 kips/in.		4.68	3.91	3.01	3.00	2.33	1.89	2.42	2.20	2.42	2.15	
R kips		60	49	38	39	29	23	29	24	21	18	

Load above heavy line is limited by maximum allowable web shear.

| W 10 | BEAMS
W Shapes
Allowable uniform loads in kips
for beams laterally supported
For beams laterally unsupported, see page 2-146 | | | | | | | | F_y = 50 ksi |

Designation	W 10								
Wt./ft	112	100	88	77	68	60	54	49	Deflection In.
Flange Width	10⅜	10⅜	10¼	10¼	10⅛	10⅛	10	10	
L_c	9.30	9.30	9.20	9.10	9.10	9.00	9.00	9.00	
L_u	38.3	34.7	31.2	27.8	25.1	22.4	20.3	18.7	

Span in Feet	112	100	88	77	68	60	54	49	Deflection
8	343	302	262	225	196	172	149	136	.22
9	308	274	241	210	185	163	147	133	.28
10	277	246	217	189	167	147	132	120	.34
11	252	224	197	172	151	133	120	109	.41
12	231	205	181	157	139	122	110	100	.49
13	213	190	167	145	128	113	102	92	.58
14	198	176	155	135	119	105	94	86	.67
15	185	164	144	126	111	98	88	80	.77
16	173	154	135	118	104	92	83	75	.87
17	163	145	127	111	98	86	78	71	.99
18	154	137	120	105	93	82	73	67	1.11
19	146	130	114	99	88	77	69	63	1.23
20	139	123	108	94	83	73	66	60	1.37
21	132	117	103	90	79	70	63	57	1.51
22	126	112	99	86	76	67	60	55	1.65
23	121	107	94	82	72	64	57	52	1.81
24	116	103	90	79	69	61	55	50	1.97

F_y = 50 ksi

Properties and Reaction Values									
S_x in.³	126	112	98.5	85.9	75.7	66.7	60.0	54.6	For explanation of deflection, see page 2-32
V kips	172	151	131	112	98	86	75	68	
R_1 kips	117	98.2	81.1	65.6	53.3	45.5	38.2	33.3	
R_2 kips/in.	24.9	22.4	20.0	17.5	15.5	13.9	12.2	11.2	
R_3 kips	176	143	113	86.5	68.0	54.0	42.4	35.7	
R_4 kips/in.	21.9	18.2	14.9	11.6	9.35	7.69	5.89	5.07	
R kips	**204**	**177**	**151**	**127**	**101**	81	63	53	

Load above heavy line is limited by maximum allowable web shear.
Values of R in **bold face** exceed maximum web shear V.

$F_y = 50$ ksi	BEAMS W Shapes Allowable uniform loads in kips for beams laterally supported	W 10

BEAMS
W Shapes
Allowable uniform loads in kips
for beams laterally supported
For beams laterally unsupported, see page 2-146

Designation		W 10			W 10			W 10				Deflection In.
Wt./ft		45	39	33	30	26	22	19	17	15	12*	
Flange Width		8	8	8	5¾	5¾	5¾	4	4	4	4	
L_c		7.20	7.20	7.10	5.20	5.20	5.20	3.60	3.60	3.60	2.80	
L_u		16.4	14.2	11.9	9.40	8.20	6.80	5.20	4.40	3.70	3.60	
	3								97	92	75	.03
	4							102	89	76	60	.05
	5				126	107	98	83	71	61	48	.09
	6			113	119	102	85	69	59	51	40	.12
	7	141	125	110	102	88	73	59	51	43	34	.17
	8	135	116	96	89	77	64	52	45	38	30	.22
	9	120	103	86	79	68	57	46	40	34	27	.28
	10	108	93	77	71	61	51	41	36	30	24	.34
	11	98	84	70	65	56	46	38	32	28	22	.41
	12	90	77	64	59	51	43	34	30	25	20	.49
	13	83	71	59	55	47	39	32	27	23	18	.58
	14	77	66	55	51	44	36	30	25	22	17	.67
	15	72	62	51	48	41	34	28	24	20	16	.77
	16	68	58	48	45	38	32	26	22	19	15	.87
	17	64	54	45	42	36	30	24	21	18	14	.99
	18	60	51	43	40	34	28	23	20	17	13	1.11
	19	57	49	41	38	32	27	22	19	16	13	1.23
	20	54	46	39	36	31	26	21	18	15	12	1.37
	21	51	44	37	34	29	24	20	17	14	11	1.51
	22	49	42	35	32	28	23	19	16	14	11	1.65
	23	47	40	33	31	27	22	18	15	13	10	1.81
	24	45	39	32	30	26	21	17	15	13	10	1.97

($F_y = 50$ ksi — Span in Feet)

Properties and Reaction Values												
S_x in.³		49.1	42.1	35.0	32.4	27.9	23.2	18.8	16.2	13.8	10.9	
V kips		71	62	56	63	54	49	51	49	46	38	For explanation of deflection, see page 2-32
R_1 kips		36.1	29.2	25.4	23.2	18.8	14.9	16.8	14.9	13.0	9.80	
R_2 kips/in.		11.5	10.4	9.57	9.90	8.58	7.92	8.25	7.92	7.59	6.27	
R_3 kips		39.2	30.9	24.8	28.2	21.1	17.0	18.9	16.2	13.8	9.12	
R_4 kips/in.		4.94	4.29	4.16	3.65	2.79	2.72	2.79	2.99	3.25	2.39	
R kips		56	46	39	41	31	27	29	27	25	17	

* Indicates noncompact shape.
Load above heavy line is limited by maximum allowable web shear.

AMERICAN INSTITUTE OF STEEL CONSTRUCTION

					BEAMS				F_y = 50 ksi

W 8

BEAMS
W Shapes
Allowable uniform loads in kips
for beams laterally supported
For beams laterally unsupported, see page 2-146

Designation					W 8				
Wt./ft		67	58	48	40	35	31	Deflection In.
Flange Width		8¼	8¼	8⅛	8⅛	8	8	
L_c		7.40	7.40	7.30	7.20	7.20	7.20	
L_u		28.7	25.4	21.8	18.2	16.3	14.5	
	6	205	179		119	101	91	.15
	7	190	163	136	112	98	86	.21
	8	166	143	119	98	86	76	.27
	9	148	127	106	87	76	67	.35
	10	133	114	95	78	69	60	.43
	11	121	104	87	71	62	55	.52
	12	111	95	79	65	57	50	.61
	13	102	88	73	60	53	47	.72
	14	95	82	68	56	49	43	.84
	15	89	76	64	52	46	40	.96
	16	83	72	60	49	43	38	1.09
	17	78	67	56	46	40	36	1.23
	18	74	64	53	43	38	34	1.38
	19	70	60	50	41	36	32	1.54
	20	66	57	48	39	34	30	1.71

F_y = 50 ksi — Span in Feet

Properties and Reaction Values

	67	58	48	40	35	31	
S_x in.³	60.4	52.0	43.3	35.5	31.2	27.5	For
V kips	103	89	68	59	50	46	explanation
R_1 kips	67.6	55.2	39.2	31.6	25.6	22.0	of deflection,
R_2 kips/in.	18.8	16.8	13.2	11.9	10.2	9.41	see page 2-32
R_3 kips	100	78.8	50.3	38.9	29.2	24.1	
R_4 kips/in.	15.9	13.5	7.93	7.28	5.35	4.80	
R kips	**133**	**114**	**78**	**64**	48	41	

* Indicates noncompact shape.
Load above heavy line is limited by maximum allowable web shear.
Values of R in **bold face** exceed maximum web shear V.

F_y = 50 ksi								BEAMS W Shapes Allowable uniform loads in kips for beams laterally supported For beams laterally unsupported, see page 2-146	W 8

Designation		W 8		W 8		W 8			
Wt./ft		28	24	21	18	15	13	10*	Deflection In.
Flange Width		6½	6½	5¼	5¼	4	4	4	
L_c		5.90	5.80	4.70	4.70	3.60	3.60	3.40	
L_u		12.6	10.9	8.50	7.10	5.20	4.30	3.70	

Span in Feet		W 8 (28)	W 8 (24)	W 8 (21)	W 8 (18)	W 8 (15)	W 8 (13)	W 8 (10*)	Deflection In.
	2							74	.02
	3					79	73		.04
	4			83	75	65	55	54	.07
	5	92	78	80	67	52	44	43	.11
	6	89	77	67	56	43	36	34	.15
	7	76	66	57	48	37	31	28	.21
	8	67	57	50	42	32	27	24	.27
	9	59	51	44	37	29	24	21	.35
	10	53	46	40	33	26	22	19	.43
	11	49	42	36	30	24	20	17	.52
	12	45	38	33	28	22	18	15	.61
	13	41	35	31	26	20	17	14	.72
	14	38	33	29	24	19	16	13	.84
	15	36	31	27	22	17	15	12	.96
	16	33	29	25	21	16	14	11	1.09
	17	31	27	24	20	15	13	11	1.23
	18	30	26	22	19	14	12	10	1.38
	19	28	24	21	18	14	11	9	1.54
	20	27	23	20	17	13	11	9	1.71

Properties and Reaction Values									
S_x in.³		24.3	20.9	18.2	15.2	11.8	9.91	7.81	
V kips		46	39	41	37	40	37	27	For
R_1 kips		22.0	17.7	16.8	14.2	15.2	13.0	8.77	explanation
R_2 kips/in.		9.41	8.09	8.25	7.59	8.09	7.59	5.61	of deflection,
R_3 kips		24.9	18.4	19.0	15.2	16.4	13.4	7.63	see page
R_4 kips/in.		4.45	3.34	3.40	3.27	4.15	4.31	2.19	2-32
R kips		40	30	31	27	31	28	15	

* Indicates noncompact shape.
Load above heavy line is limited by maximum allowable web shear.

	BEAMS	F_y = 50 ksi

W 6–5–4

BEAMS
W Shapes
Allowable uniform loads in kips
for beams laterally supported
For beams laterally unsupported, see page 2-146

Designation	W 6			W 6			Deflection In.	W 5		Deflection In.	W 4	Deflection In.
Wt./ft	25	20	15*	16	12	9		19	16		13	
Flange Width	6⅛	6	6	4	4	4		5	5		4	
L_c	5.40	5.40	5.40	3.60	3.60	3.50		4.50	4.50		3.60	
L_u	14.4	11.18	8.70	8.70	6.20	4.80		14.0	12.0		11.2	

Span in Feet	W6-25	W6-20	W6-15*	W6-16	W6-12	W6-9	Deflection In.	W5-19	W5-16	Deflection In.	W4-13	Deflection In.
2					55		.02				47	.03
3			55	65	54	40	.05		48	.06	40	.08
4	82	64	51	56	40	31	.09	56	47	.11	30	.14
5	73	59	41	45	32	24	.14	45	37	.17	24	.21
6	61	49	34	37	27	20	.20	37	31	.25	20	.31
7	52	42	29	32	23	17	.28	32	27	.33	17	.42
8	46	37	25	28	20	15	.36	28	23	.44	15	.55
9	41	33	23	25	18	14	.46	25	21	.55	13	.69
10	37	29	20	22	16	12	.57	22	19	.68	12	.85
11	33	27	18	20	15	11	.69	20	17	.83		
12	31	25	17	19	13	10	.82	19	16	.98		
13	28	23	16	17	12	9.4	.96					
14	26	21	15	16	11	8.7	1.12					

F_y = 50 ksi

Properties and Reaction Values

	W6-25	W6-20	W6-15*	W6-16	W6-12	W6-9		W5-19	W5-16		W4-13	
S_x in.³	16.7	13.4	9.72	10.2	7.31	5.56		10.2	8.51		5.46	
V kips	41	32	28	33	28	20	For	28	24	For	23	For
R_1 kips	21.4	16.1	11.9	16.1	11.9	7.89	explanation	18.1	14.9	explanation	15.9	explanation
R_2 kips/in.	10.6	8.58	7.59	8.58	7.59	5.61	of deflection,	8.91	7.92	of deflection,	9.24	of deflection,
R_3 kips	29.4	19.3	13.5	20.3	14.0	7.81	see page	22.1	17.0	see page	20.9	see page
R_4 kips/in.	8.14	5.60	5.63	4.98	5.20	2.79	2-32	6.41	5.53	2-32	11.0	2-32
R kips	**58**	**39**	**33**	**38**	**32**	18		**45**	**36**		**48**	

* Indicates noncompact shape.
Load above heavy line is limited by maximum allowable web shear.
Values of R in **bold face** exceed maximum web shear V.

AMERICAN INSTITUTE OF STEEL CONSTRUCTION

$F_y = 50$ ksi							

BEEAMS
M Shapes
Allowable uniform loads in kips
for beams laterally supported
For beams laterally unsupported, see page 2-146

M 14–12

Designation	M 14	Deflection In.	M 12			Deflection In.
Wt./ft	18		11.8	10.8	10	
Flange Width	4		3⅛	3⅛	3¼	
L_c	2.60		1.90	1.80	1.60	
L_u	3.40		2.60	2.60	2.80	

Span in Feet	M 14	Deflection In.	11.8	10.8	10	Deflection In.
3	120	.02	85	77	71	.03
4	116	.04	66	60	57	.05
5	93	.06	53	48	45	.07
6	77	.09	44	40	38	.10
7	66	.12	38	34	32	.14
8	58	.16	33	30	28	.18
9	52	.20	29	27	25	.23
10	46	.24	26	24	23	.28
11	42	.30	24	22	21	.34
12	39	.35	22	20	19	.41
13	36	.41	20	18	17	.48
14	33	.48	19	17	16	.56
15	31	.55	18	16	15	.64
16	29	.62	17	15	14	.73
17	27	.70	16	14	13	.82
18	26	.79	15	13	13	.92
19	24	.88	14	13	12	1.03
20	23	.98	13	12	11	1.14
22	21	1.18	12	11	10	1.38
24	19	1.40	11	10	9.4	1.64
26	18	1.65	10	9.2	8.7	1.92
28	17	1.91	9.4	8.6	8.1	2.23
30	15	2.19	8.8	8.0	7.6	2.56
32	15	2.50				
34	14	2.82				

$F_y = 50$ ksi

Properties and Reaction Values

	M 14					
S_x in.^{3}	21.1		12.0	10.9	10.3	
V kips	60		42	38	36	
R_1 kips	11.1	For explanation of deflection, see page 2-32	8.21	6.60	6.15	For explanation of deflection, see page 2-32
R_2 kips/in.	7.10		5.84	5.28	4.92	
R_3 kips	12.5		8.49	7.05	5.87	
R_4 kips/in.	1.90		1.48	1.18	1.11	
R kips	19		14	11	10	

Load above heavy line is limited by maximum allowable web shear.

AMERICAN INSTITUTE OF STEEL CONSTRUCTION

M 10–8–6–5	BEAMS									F_y = 50 ksi

BEAMS
M Shapes
Allowable uniform loads in kips
for beams laterally supported
For beams laterally unsupported, see page 2-146

Designation	M 10				M 8		M 6		M 5	
Wt./ft	9	8	7.5	Deflection In.	6.5	Deflection In.	4.4	Deflection In.	18.9	Deflection In.
Flange Width	2¾	2¾	2¾		2¼		1⅞		5	
L_c	1.90	1.60	1.60		1.80		1.70		4.50	
L_u	2.30	2.30	2.30		2.00		1.80		13.9	

Span in Feet / F_y = 50 ksi

Span	M10-9	M10-8	M10-7.5	Defl.	M8-6.5	Defl.	M6-4.4	Defl.	M5-18.9	Defl.
1							27	.01		
2	63	56	52	.01	43	.02	26	.02		
3	57	51	48	.03	34	.04	18	.05	63	.06
4	43	38	36	.05	25	.07	13	.09	53	.11
5	34	31	29	.09	20	.11	11	.14	42	.17
6	28	25	24	.12	17	.15	8.8	.20	35	.25
7	24	22	21	.17	15	.21	7.5	.28	30	.33
8	21	19	18	.22	13	.27	8.6	.36	26	.44
9	19	17	16	.28	11	.35	7.7	.46	24	.55
10	17	15	14	.34	10	.43	6.9	.57	21	.68
11	16	14	13	.41	9.2	.52	4.8	.69	19	.83
12	14	13	12	.49	8.5	.61	4.4	.82	18	.98
13	13	12	11	.58	7.8	.72	4.1	.96		
14	12	11	10	.67	7.3	.84	3.8	1.12		
15	11	10	10	.77	6.8	.96				
16	11	10	9.0	.87	6.4	1.09				
17	10	9.0	8.5	.99	6.0	1.23				
18	9.5	8.5	8.0	1.11	5.6	1.38				
19	9.0	8.0	7.6	1.23	5.3	1.54				
20	8.5	7.6	7.2	1.37	5.1	1.71				
21	8.1	7.3	6.9	1.51						
22	7.8	6.9	6.6	1.65						
23	7.4	6.6	6.3	1.81						
24	7.1	6.4	6.0	1.97						

Properties and Reaction Values

	M10-9	M10-8	M10-7.5		M8-6.5		M6-4.4		M5-18.9	
S_x in.³	7.76	6.94	6.57	For explanation of deflection, see page 2-32	4.62	For explanation of deflection, see page 2-32	2.40	For explanation of deflection, see page 2-32	9.63	For explanation of deflection, see page 2-32
V kips	31	28	26		22		14		32	
R_1 kips	7.29	5.09	4.69		5.57		4.11		22.8	
R_2 kips/in.	5.18	4.65	4.29		4.46		3.76		10.4	
R_3 kips	6.79	5.43	4.69		5.18		3.83		27.5	
R_4 kips/in.	1.35	1.12	.917		1.17		1.04		10.9	
R kips	12	9	8		9		7		**59**	

Load above heavy line is limited by maximum allowable web shear.
Values of R in **bold face** exceed maximum web shear V.

F_y = 50 ksi

BEAMS
S Shapes
Allowable uniform loads in kips
for beams laterally supported
For beams laterally unsupported, see page 2-146

S 24–20

Designation	S 24		S 24			Deflection In.	S 20		S 20		Deflection In.
Wt./ft	121	106	100	90	80		96	86	75	66	
Flange Width	8	7⅞	7¼	7⅛	7		7¼	7	6⅜	6¼	
L_c	7.20	7.00	6.50	6.40	6.30		6.40	6.30	5.70	5.60	
L_u	11.9	11.7	8.70	8.60	8.50		10.9	10.6	8.50	8.30	
Span in Feet											
5							650		508		.04
6			715	600		.05	605	536	469	404	.06
7	784		625	588		.07	519	487	402	374	.08
8	710	608	547	514	480	.09	454	426	352	327	.11
9	631	587	486	457	428	.12	403	379	313	291	.14
10	568	528	438	411	385	.14	363	341	282	262	.17
11	516	480	398	374	350	.17	330	310	256	238	.21
12	473	440	365	343	321	.20	303	284	235	218	.25
13	437	406	337	316	296	.24	279	262	217	201	.29
14	405	377	313	294	275	.28	259	244	201	187	.33
15	378	352	292	274	257	.32	242	227	188	175	.38
16	355	330	274	257	241	.36	227	213	176	164	.44
17	334	311	258	242	226	.41	214	201	166	154	.49
18	315	293	243	229	214	.46	202	189	156	145	.55
19	299	278	230	217	203	.51	191	179	148	138	.62
20	284	264	219	206	193	.57	182	171	141	131	.68
21	270	251	208	196	183	.63	173	162	134	125	.75
22	258	240	199	187	175	.69	165	155	128	119	.83
23	247	230	190	179	167	.75	158	148	122	114	.90
24	237	220	182	171	160	.82	151	142	117	109	.98
25	227	211	175	165	154	.89	145	136	113	105	1.07
26	218	203	168	158	148	.96	140	131	108	101	1.15
27	210	196	162	152	143	1.04	134	126	104	97	1.24
28	203	189	156	147	138	1.12	130	122	101	94	1.34
29	196	182	151	142	133	1.20	125	118	97	90	1.44
30	189	176	146	137	128	1.28	121	114	94	87	1.54
31	183	170	141	133	124	1.37	117	110	91	84	1.64
32	177	165	137	129	120	1.46	113	107	88	82	1.75
33	172	160	133	125	117	1.55	110	103	85	79	1.86
34	167	155	129	121	113	1.64	107	100	83	77	1.97
36	158	147	122	114	107	1.84	101	95	78	73	2.21
38	149	139	115	108	101	2.05	96	90	74	69	2.46
40	142	132	109	103	96	2.28	91	85	70	65	2.73
42	135	126	104	98	92	2.51	86	81	67	62	3.01
44	129	120	100	94	88	2.75	83	78	64	60	3.30
46	123	115	95	89	84	3.01	79	74	61	57	3.61
48	118	110	91	86	80	3.28	76	71	59	55	3.93
50	114	106	88	82	77	3.56	73	68	56	52	4.27
52	109	102	84	79	74	3.85					
54	105	98	81	76	71	4.15					
56	101	94	78	73	69	4.46					
58	98	91	75	71	66	4.79					
60	95	88	73	69	64	5.12					

F_y = 50 ksi (left margin, vertical)

Properties and Reaction Values

S_x in.³	258	240	199	187	175		165	155	128	119	
V kips	392	304	358	300	240	For explanation of deflection, see page 2-32	325	268	254	202	For explanation of deflection, see page 2-32
R_1 kips	132	102	108	90.2	72.2		116	95.3	85.1	67.7	
R_2 kips/in.	26.4	20.5	24.6	20.6	16.5		26.4	21.8	21.0	16.7	
R_3 kips	180	123	144	111	79.3		165	124	108	76.9	
R_4 kips/in.	13.8	6.44	14.3	8.43	4.32		19.8	11.1	11.6	5.84	
R kips	224	146	194	141	94		208	163	149	97	

Load above heavy line is limited by maximum allowable web shear.

AMERICAN INSTITUTE OF STEEL CONSTRUCTION

BEAMS
S Shapes
Allowable uniform loads in kips
for beams laterally supported

S 18–15–12

$F_y = 50$ ksi

For beams laterally unsupported, see page 2-146

Designation	S 18		Deflection In.	S 15		Deflection In.	S 12		S 12		Deflection In.
Wt./ft	70	54.7		50	42.9		50	40.8	35	31.8	
Flange Width	6¼	6		5⅝	5½		5½	5¼	5⅛	5	
L_c	5.60	5.40		5.10	4.90		4.90	4.70	4.50	4.50	
L_u	8.00	7.70		7.80	7.60		10.0	9.60	7.70	7.60	

Span in Feet — $F_y = 50$ ksi

Span	S18 70	S18 54.7	Defl.	S15 50	S15 42.9	Defl.	S12 50	S12 40.8	S12 35	S12 31.8	Defl.
3							330				.03
4	512		.03	330		.04	279	222	205	168	.05
5	453	332	.05	285	247	.06	224	200	168	160	.07
6	378	328	.07	238	219	.08	186	166	140	133	.10
7	324	281	.09	204	187	.11	160	143	120	114	.14
8	283	246	.12	178	164	.15	140	125	105	100	.18
9	252	219	.15	158	146	.18	124	111	93	89	.23
10	227	197	.19	143	131	.23	112	100	84	80	.28
11	206	179	.23	130	119	.28	102	91	76	73	.34
12	189	164	.27	119	109	.33	93	83	70	67	.41
13	174	151	.32	110	101	.38	86	77	65	62	.48
14	162	140	.37	102	94	.45	80	71	60	57	.56
15	151	131	.43	95	87	.51	75	67	56	53	.64
16	142	123	.49	89	82	.58	70	62	53	50	.73
17	133	116	.55	84	77	.66	66	59	49	47	.82
18	126	109	.61	79	73	.74	62	55	47	44	.92
19	119	104	.68	75	69	.82	59	53	44	42	1.03
20	113	98	.76	71	66	.91	56	50	42	40	1.14
21	108	94	.84	68	62	1.00	53	48	40	38	1.25
22	103	89	.92	65	60	1.10	51	45	38	36	1.38
23	99	86	1.00	62	57	1.20	49	43	37	35	1.51
24	94	82	1.09	59	55	1.31	47	42	35	33	1.64
25	91	79	1.19	57	52	1.42	45	40	34	32	1.78
26	87	76	1.28	55	50	1.54	43	38	32	31	1.92
27	84	73	1.38	53	49	1.66	41	37	31	30	2.07
28	81	70	1.49	51	47	1.78	40	36	30	29	2.23
29	78	68	1.60	49	45	1.91	39	34	29	28	2.39
30	76	66	1.71	48	44	2.05	37	33	28	27	2.56
31	73	63	1.82	46	42	2.19					
32	71	61	1.94	45	41	2.33					
33	69	60	2.07	43	40	2.48					
34	67	58	2.19	42	39	2.63					
35	65	56	2.32	41	37	2.79					
36	63	55	2.46	40	36	2.95					
37	61	53	2.60								
38	60	52	2.74								
40	57	49	3.03								
42	54	47	3.35								
44	52	45	3.67								

Properties and Reaction Values

	S18 70	S18 54.7		S15 50	S15 42.9		S12 50	S12 40.8	S12 35	S12 31.8	
S_x in.³	103	89.4		64.8	59.6		50.8	45.4	38.2	36.4	
V kips	256	166		165	123		165	111	103	84	
R_1 kips	88.0	57.0	For explanation of deflection, see page 2-32	62.4	46.6	For explanation of deflection, see page 2-32	81.5	54.8	41.9	34.3	For explanation of deflection, see page 2-32
R_2 kips/in.	23.5	15.2		18.2	13.6		22.7	15.2	14.1	11.5	
R_3 kips	120	62.6		77.3	50.0		111	61.3	49.7	36.7	
R_4 kips/in.	20.8	5.68		12.9	5.37		29.6	8.99	8.66	4.74	
R kips	170	82		122	69		161	93	80	53	

Load above heavy line is limited by maximum allowable web shear.

AMERICAN INSTITUTE OF STEEL CONSTRUCTION

F_y = 50 ksi

BEAMS
S Shapes
Allowable uniform loads in kips for beams laterally supported

S 10–8–7–6

For beams laterally unsupported, see page 2-146

Designation	S 10		Deflection In.	S 8		Deflection In.	S 7		Deflection In.	S 6		Deflection In.
Wt./ft	35	25.4		23	18.4		20	15.3		17.25	12.5	
Flange Width	5	4⅝		4⅛	4		3⅞	3⅝		3⅝	3⅜	
L_c	4.40	4.20		3.70	3.60		3.50	3.30		3.20	3.00	
L_u	8.10	7.60		7.40	7.10		7.20	6.80		7.10	6.60	
F_y = 50 ksi — Span in Feet												
1										112		.01
2	238		.01	141		.02	126		.02	96	56	.02
3	216		.03	119	87	.04	89	71	.04	64	54	.05
4	162	124	.05	89	79	.07	67	58	.08	48	41	.09
5	129	109	.09	71	63	.11	53	46	.12	39	32	.14
6	108	91	.12	59	53	.15	44	39	.18	32	27	.20
7	92	78	.17	51	45	.21	38	33	.24	28	23	.28
8	81	68	.22	45	40	.27	33	29	.31	24	20	.36
9	72	60	.28	40	35	.35	30	26	.40	21	18	.46
10	65	54	.34	36	32	.43	27	23	.49	19	16	.57
11	59	49	.41	32	29	.52	24	21	.59	18	15	.69
12	54	45	.49	30	26	.61	22	19	.70	16	14	.82
13	50	42	.58	27	24	.72	20	18	.82	15	12	.96
14	46	39	.67	25	23	.84	19	17	.96	14	12	1.12
15	43	36	.77	24	21	.96	18	15	1.10			
16	40	34	.87	22	20	1.09	17	14	1.25			
17	38	32	.99	21	19	1.23						
18	36	30	1.11	20	18	1.38						
20	32	27	1.37	18	16	1.71						
22	29	25	1.65									
24	27	23	1.97									
Properties and Reaction Values												
S_x in.^{3}	29.4	24.7	For explanation of deflection, see page 2-32	16.2	14.4	For explanation of deflection, see page 2-32	12.1	10.5	For explanation of deflection, see page 2-32	8.77	7.37	For explanation of deflection, see page 2-32
V kips	119	62		71	43		63	35		56	28	
R_1 kips	55.1	28.9		36.4	22.4		34.8	19.5		33.6	16.7	
R_2 kips/in.	19.6	10.3		14.6	8.94		14.9	8.32		15.3	7.66	
R_3 kips	77.1	29.2		46.0	22.1		45.4	19.0		45.7	16.1	
R_4 kips/in.	30.8	4.42		18.2	4.21		24.0	4.21		33.7	4.18	
R kips	**124**	45		**88**	37		**87**	34		**87**	**31**	

Load above heavy line is limited by maximum allowable web shear.
Values of R in **bold face** exceed maximum web shear V.

| S 5–4–3 | | | BEAMS
S Shapes
Allowable uniform loads in kips
for beams laterally supported
For beams laterally unsupported, see page 2-146 | | | | | | $F_y = 50$ ksi |

Designation	S 5			S 4			S 3		
Wt./ft	14.75	10	Deflection In.	9.5	7.7	Deflection In.	7.5	5.7	Deflection In.
Flange Width	3¼	3		2¾	2⅝		2½	2⅜	
L_c	2.90	2.70		2.50	2.40		2.20	2.10	
L_u	7.20	6.50		6.80	6.50		7.20	6.70	

Span in Feet ($F_y = 50$ ksi)

Span	S5 (14.75)	S5 (10)	Defl.	S4 (9.5)	S4 (7.7)	Defl.	S3 (7.5)	S3 (5.7)	Defl.
1	99		.01	52		.01	42	20	.01
2	67	43	.03	37	31	.03	21	18	.05
3	45	36	.06	25	22	.08	14	12	.10
4	33	27	.11	19	17	.14	11	9.2	.18
5	27	22	.17	15	13	.21	8.6	7.4	.28
6	22	18	.25	12	11	.31	7.2	6.2	.41
7	19	15	.33	11	10	.42			
8	17	14	.44	9.3	8.4	.55			
9	15	12	.55	8.3	7.4	.69			
10	13	11	.68	7.5	6.7	.85			
11	12	10	.83						
12	11	9.0	.98						

Properties and Reaction Values

	S5 (14.75)	S5 (10)		S4 (9.5)	S4 (7.7)		S3 (7.5)	S3 (5.7)	
S_x in.^{3}	6.09	4.92		3.39	3.04		1.95	1.68	
V kips	49	21	For explanation of deflection, see page 2-32	26	15	For explanation of deflection, see page 2-32	21	10	For explanation of deflection, see page 2-32
R_1 kips	33.1	14.3		20.2	11.9		19.8	9.64	
R_2 kips/in.	16.3	7.06		10.8	6.37		11.5	5.61	
R_3 kips	47.7	13.6		24.2	11.0		25.3	8.59	
R_4 kips/in.	53.3	4.34		21.3	4.42		39.3	4.54	
R kips	**90**	**29**		**58**	**26**		**60**	**24**	

Load above heavy line is limited by maximum allowable web shear.
Values of R in **bold face** exceed maximum web shear V.

AMERICAN INSTITUTE OF STEEL CONSTRUCTION

F_y = 50 ksi	BEAMS	18–15

BEAMS
Channels
Allowable uniform loads in kips
for beams laterally supported
For beams laterally unsupported, see page 2-146

Designation		MC 18				Deflection In.	C 15			Deflection In.
Wt./ft		58	51.9	45.8	42.7		50	40	33.9	
Flange Width		4¼	4⅛	4	4		3¾	3½	3⅜	
L_u		4.90	4.70	4.60	4.60		5.40	5.10	4.90	

Span in Feet	MC 18 58	51.9	45.8	42.7	Deflection In.	C 15 50	40	33.9	Deflection In.
2	504				.01	430	312		.01
3	501	432	360	324	.02	359	310	240	.02
4	376	349	322	308	.03	269	233	210	.03
5	300	279	257	246	.04	215	186	168	.05
6	250	232	214	205	.06	179	155	140	.07
7	215	199	184	176	.08	154	133	120	.10
8	188	174	161	154	.11	135	116	105	.13
9	167	155	143	137	.14	120	103	93	.17
10	150	139	129	123	.17	108	93	84	.21
11	137	127	117	112	.21	98	85	76	.25
12	125	116	107	103	.25	90	78	70	.30
13	116	107	99	95	.29	83	72	65	.35
14	107	100	92	88	.34	77	66	60	.41
15	100	93	86	82	.39	72	62	56	.47
16	94	87	80	77	.44	67	58	53	.53
17	88	82	76	72	.50	63	55	49	.60
18	83	77	71	68	.56	60	52	47	.67
19	79	73	68	65	.62	57	49	44	.75
20	75	70	64	62	.69	54	47	42	.83
21	72	66	61	59	.76	51	44	40	.91
22	68	63	58	56	.83	49	42	38	1.00
23	65	61	56	54	.91	47	40	37	1.09
24	63	58	54	51	.99	45	39	35	1.19
25	60	56	51	49	1.08	43	37	34	1.29
26	58	54	49	47	1.16	41	36	32	1.40
27	56	52	48	46	1.26	40	34	31	1.51
28	54	50	46	44	1.35	38	33	30	1.62
29	52	48	44	42	1.45	37	32	29	1.74
30	50	46	43	41	1.55	36	31	28	1.86
31	48	45	41	40	1.66	35	30	27	1.99
32	47	44	40	39	1.76	34	29	26	2.12
34	44	41	38	36	1.99	32	27	25	2.39
36	42	39	36	34	2.23	30	26	23	2.68
38	40	37	34	32	2.49				
40	38	35	32	31	2.76				
42	36	33	31	29	3.04				
44	34	32	29	28	3.33				

Properties and Reaction Values

	MC 18					C 15			
S_x in.³	75.1	69.7	64.3	61.6		53.8	46.5	42.0	
V kips	252	216	180	162	For explanation of deflection, see page 2-32	215	156	120	For explanation of deflection, see page 2-32
R_1 kips	79.4	68.1	56.7	51		84.9	61.7	47.4	
R_2 kips/in.	23.1	19.8	16.5	14.9		23.6	17.2	13.2	
R_3 kips	111	88.3	67.2	57.4		117	72.7	49.0	
R_4 kips/in.	22.0	13.8	8.01	5.84		27.2	10.4	4.73	
R kips	160	137	95	78		168	109	66	

Load above heavy line is limited by maximum allowable web shear.

AMERICAN INSTITUTE OF STEEL CONSTRUCTION

13	BEAMS Channels Allowable uniform loads in kips for beams laterally supported For beams laterally unsupported, see page 2-146				$F_y = 50$ ksi

Designation	MC 13				Deflection In.
Wt./ft	50	40	35	31.8	
Flange Width	4⅜	4⅛	4⅛	4	
L_u	6.90	6.50	6.40	6.30	

Span in Feet	50	40	35	31.8	Deflection In.
2	409	291			.01
3	323	280	232	195	.02
4	242	210	194	184	.04
5	194	168	155	147	.06
6	161	140	129	123	.09
7	138	120	111	105	.12
8	121	105	97	92	.15
9	108	93	86	82	.19
10	97	84	78	74	.24
11	88	76	71	67	.29
12	81	70	65	61	.34
13	74	65	60	57	.40
14	69	60	55	53	.47
15	65	56	52	49	.54
16	61	53	49	46	.61
17	57	49	46	43	.69
18	54	47	43	41	.77
19	51	44	41	39	.86
20	48	42	39	37	.95
21	46	40	37	35	1.05
22	44	38	35	33	1.15
23	42	37	34	32	1.26
24	40	35	32	31	1.37
25	39	34	31	29	1.49
26	37	32	30	28	1.61
27	36	31	29	27	1.74
28	35	30	28	26	1.87
29	33	29	27	25	2.01
30	32	28	26	25	2.15
31	31	27	25	24	2.29
32	30	26	24	23	2.44

Properties and Reaction Values					
S_x in.³	48.4	42.0	38.8	36.8	
V kips	205	146	116	98	For
R_1 kips	89.3	63.5	50.7	42.5	explanation
R_2 kips/in.	26.0	18.5	14.8	12.4	of deflection,
R_3 kips	131	78.7	56.1	43.1	see page 2-32
R_4 kips/in.	44.3	16.0	8.12	4.80	
R kips	180	128	85	60	

Load above heavy line is limited by maximum allowable web shear.

$F_y = 50$ ksi

| F_y = 50 ksi | BEAMS | | | | | | | | | 12 |

BEAMS
Channels
Allowable uniform loads in kips
for beams laterally supported
For beams laterally unsupported, see page 2-146

Designation		C 12			MC 12					MC 12	Deflection In.
Wt./ft		30	25	20.7	50	45	40	35	31	10.6	
Flange Width		3⅛	3	3	4⅛	4	3⅞	3¾	3⅝	1½	
L_u		4.40	4.20	4.10	8.00	7.80	7.60	7.30	7.10	1.30	
	2	245	186		401	342	283			91	.01
	3	180	161	135	299	280	260	224	178	62	.02
	4	135	121	108	225	210	195	181	169	46	.04
	5	108	96	86	180	168	156	144	135	37	.06
	6	90	80	72	150	140	130	120	113	31	.09
	7	77	69	61	128	120	111	103	97	26	.13
	8	68	60	54	112	105	98	90	85	23	.17
	9	60	54	48	100	93	87	80	75	21	.21
	10	54	48	43	90	84	78	72	68	18	.26
	11	49	44	39	82	76	71	66	61	17	.31
	12	45	40	36	75	70	65	60	56	15	.37
	13	42	37	33	69	65	60	56	52	14	.44
	14	39	34	31	64	60	56	52	48	13	.51
	15	36	32	29	60	56	52	48	45	12	.58
	16	34	30	27	56	53	49	45	42	12	.66
	17	32	28	25	53	49	46	42	40	11	.75
	18	30	27	24	50	47	43	40	38	10	.84
	19	28	25	23	47	44	41	38	36	9.7	.93
	20	27	24	22	45	42	39	36	34	9.2	1.03
	21	26	23	20	43	40	37	34	32	8.8	1.14
	22	25	22	20	41	38	35	33	31	8.4	1.25
	23	23	21	19	39	37	34	31	29	8.0	1.37
	24	23	20	18	37	35	33	30	28	7.7	1.48
	25	22	19	17	36	34	31	29	27	7.4	1.61
	26	21	19	17	35	32	30	28	26	7.1	1.74
	28	19	17	15	32	30	28	26	24	6.6	2.02
	30	18	16	14	30	28	26	24	23	6.2	2.32

(F_y = 50 ksi — Span in Feet)

Properties and Reaction Values

	C 12			MC 12					MC 12	
S_x in.³	27.0	24.1	21.5	44.9	42.0	39.0	36.1	33.8	9.23	
V kips	122	93	68	200	171	142	112	89	46	For explanation of deflection, see page 2-32
R_1 kips	47.3	35.9	26.2	90.4	77.1	63.9	50.6	40.1	10.8	
R_2 kips/in.	16.8	12.8	9.31	27.6	23.5	19.5	15.4	12.2	6.27	
R_3 kips	62.0	41.0	25.5	153	121	91.2	64.2	45.3	11.1	
R_4 kips/in.	15.9	6.95	2.69	50.0	31.0	17.6	8.74	4.35	1.33	
R kips	106	65	35	187	159	132	95	61	16	

Load above heavy line is limited by maximum allowable web shear.

| | BEAMS | | | | | | | | | | F_y = 50 ksi | |

10

BEAMS
Channels
Allowable uniform loads in kips
for beams laterally supported
For beams laterally unsupported, see page 2-146

Designation		C 10				MC 10			MC 10		MC 10	MC 10	Deflection In.
Wt./ft		30	25	20	15.3	41.1	33.6	28.5	25	22	8.4	6.5	
Flange Width		3	2⅞	2¾	2⅝	4⅜	4⅛	4	3⅜	3⅜	1½	1⅛	
L_u		4.40	4.20	4.00	3.80	8.30	7.90	7.60	6.50	6.30	1.40	.759	
	1	269	210			318					68	61	.003
	2	207	182	152	96	315	230	170	152		64	44	.01
	3	138	121	105	90	210	185	169	147	116	43	29	.03
	4	104	91	79	68	158	139	127	110	103	32	22	.05
	5	83	73	63	54	126	111	101	88	82	26	18	.08
	6	69	61	53	45	105	93	84	73	68	21	15	.11
	7	59	52	45	39	90	79	72	63	59	18	13	.15
	8	52	46	40	34	79	70	63	55	51	16	11	.20
	9	46	40	35	30	70	62	56	49	46	14	9.8	.25
	10	41	36	32	27	63	56	51	44	41	13	8.8	.31
	11	38	33	29	25	57	51	46	40	37	12	8.0	.38
	12	35	30	26	23	53	46	42	37	34	11	7.4	.45
	13	32	28	24	21	48	43	39	34	32	9.8	6.8	.52
	14	30	26	23	19	45	40	36	31	29	9.1	6.3	.61
	15	28	24	21	18	42	37	34	29	27	8.5	5.9	.70
	16	26	23	20	17	39	35	32	28	26	8.0	5.5	.79
	17	24	21	19	16	37	33	30	26	24	7.5	5.2	.90
	18	23	20	18	15	35	31	28	24	23	7.1	4.9	1.00
	19	22	19	17	14	33	29	27	23	22	6.7	4.7	1.12
	20	21	18	16	14	32	28	25	22	21	6.4	4.4	1.24
	21	20	17	15	13	30	26	24	21	20	6.1	4.2	1.37
	22	19	17	14	12	29	25	23	20	19	5.8	4.0	1.50
	23	18	16	14	12	27	24	22	19	18	5.6	3.8	1.64
	24	17	15	13	11	26	23	21	18	17	5.3	3.7	1.79

F_y = 50 ksi — Span in Feet

Properties and Reaction Values

S_x in.³	20.7	18.2	15.8	13.5	31.5	27.8	25.3	22.0	20.5	6.40	4.42	For explanation of deflection, see page 2-32
V kips	135	105	76	48	159	115	85	76	58	34	30	
R_1 kips	55.5	43.4	31.3	19.8	82.1	59.3	43.8	39.2	29.9	9.64	5.49	
R_2 kips/in.	22.2	17.4	12.5	7.92	26.3	19.0	14.0	12.5	9.57	5.61	5.02	
R_3 kips	87.6	60.6	37.0	18.7	129	79.5	50.5	42.7	28.5	8.92	6.40	
R_4 kips/in.	50.4	24.1	9.01	2.29	63.3	23.8	9.63	6.88	3.06	1.27	1.25	
R kips	133	104	69	27	**174**	**126**	84	67	39	13	11	

Load above heavy line is limited by maximum allowable web shear.
Values of R in **bold face** exceed maximum web shear.

| F_y = 50 ksi | | BEAMS
Channels
Allowable uniform loads in kips
for beams laterally supported
For beams laterally unsupported, see page 2-146 | | | | 9 $\lceil$ |

Designation		C 9			MC 9		
Wt./ft		20	15	13.4	25.4	23.9	Deflection In.
Flange Width		2⅝	2½	2⅜	3½	3½	
L_u		4.10	3.80	3.70	7.10	7.00	

	Span in Feet	20	15	13.4	25.4	23.9	Deflection In.
	1	161					.003
	2	135	103	84	162	144	.01
	3	90	75	71	131	126	.03
	4	68	57	53	98	95	.06
	5	54	45	42	78	76	.09
	6	45	38	35	65	63	.12
	7	39	32	30	56	54	.17
	8	34	28	27	49	47	.22
	9	30	25	24	44	42	.28
	10	27	23	21	39	38	.34
	11	25	21	19	36	34	.42
	12	23	19	18	33	32	.50
	13	21	17	16	30	29	.58
	14	19	16	15	28	27	.68
	15	18	15	14	26	25	.78
	16	17	14	13	25	24	.88
	17	16	13	12	23	22	1.00
	18	15	13	12	22	21	1.12
	19	14	12	11	21	20	1.24
	20	14	11	11	20	19	1.38
	21	13	11	10	19	18	1.52
	22	12	10	9.6	18	17	1.67

F_y = 50 ksi

Properties and Reaction Values							
S_x in.³		13.5	11.3	10.6	19.6	18.9	
V kips		81	51	42	81	72	For explanation of deflection, see page 2-32
R_1 kips		34.6	22.0	18.0	44.1	39.2	
R_2 kips/in.		14.8	9.41	7.69	14.9	13.2	
R_3 kips		46.3	23.5	17.4	53.8	45.1	
R_4 kips/in.		17.4	4.49	2.45	13.3	9.33	
R kips		**86**	39	26	**96**	**78**	

Load above heavy line is limited by maximum allowable web shear.
Values of R in **bold face** exceed maximum web shear.

BEAMS
Channels
Allowable uniform loads in kips
for beams laterally supported
For beams laterally unsupported, see page 2-146

8

F_y = 50 ksi

Designation		C 8			MC 8		MC 8		MC 8	
Wt./ft		18.75	13.75	11.5	22.8	21.4	20	18.7	8.5	Deflection
Flange Width		2½	2⅜	2¼	3½	3½	3	3	1⅞	In.
L_u		4.10	3.80	3.70	7.70	7.50	6.30	6.20	2.40	
	1	156	97							.004
	2	110	90	70	137	120	128	113	57	.02
	3	73	60	54	107	103	91	87	39	.03
	4	55	45	41	80	77	68	66	29	.06
	5	44	36	33	64	62	54	52	23	.10
	6	37	30	27	53	51	45	44	19	.14
	7	31	26	23	46	44	39	37	17	.19
	8	28	23	20	40	39	34	33	15	.25
	9	24	20	18	36	34	30	29	13	.31
	10	22	18	16	32	31	27	26	12	.39
	11	20	16	15	29	28	25	24	11	.47
	12	18	15	14	27	26	23	22	9.7	.56
	13	17	14	13	25	24	21	20	9.0	.65
	14	16	13	12	23	22	19	19	8.3	.76
	15	15	12	11	21	21	18	17	7.8	.87
	16	14	11	10	20	19	17	16	7.3	.99
	17	13	11	9.6	19	18	16	15	6.9	1.12
	18	12	10	9.0	18	17	15	15	6.5	1.26
	19	12	9.5	8.6	17	16	14	14	6.1	1.40
	20	11	9.0	8.1	16	15	14	13	5.8	1.55

F_y = 50 ksi — Span in Feet

Properties and Reaction Values

S_x in.³	11.0	9.03	8.14	16.0	15.4	13.6	13.1	5.83	For
V kips	78	48	35	68	60	64	56	29	explanation
R_1 kips	37.7	23.4	17.0	41.8	36.7	37.1	32.8	11.1	of deflection,
R_2 kips/in.	16.1	10.0	7.26	14.1	12.4	13.2	11.6	5.91	see page 2-32
R_3 kips	51.0	25.0	15.5	48.6	40.0	43.0	35.7	10.2	
R_4 kips/in.	26.7	6.43	2.46	13.4	9.06	11.5	7.93	1.66	
R kips	**94**	48	24	**91**	**72**	**83**	**63**	16	

Load above heavy line is limited by maximum allowable web shear.
Values of R in **bold face** exceed maximum web shear.

BEAMS
Channels
Allowable uniform loads in kips
for beams laterally supported
For beams laterally unsupported, see page 2-146

F_y = 50 ksi

7

Designation		C 7			MC 7		
Wt./ft	14.75	12.25	9.8	22.7	19.1	Deflection In.
Flange Width	2¼	2¼	2⅛	3⅝	3½	
L_u	4.00	3.80	3.60	8.60	8.20	
1	117	88		141		.004
2	78	69	59	136	99	.02
3	52	46	41	91	82	.04
4	39	35	30	68	62	.07
5	31	28	24	54	49	.11
6	26	23	20	45	41	.16
7	22	20	17	39	35	.22
8	19	17	15	34	31	.28
9	17	15	14	30	27	.36
10	16	14	12	27	25	.44
11	14	13	11	25	22	.54
12	13	12	10	23	21	.64
13	12	11	9.4	21	19	.75
14	11	9.9	8.7	19	18	.87
15	10	9.2	8.1	18	16	1.00
16	9.7	8.7	7.6	17	15	1.13

F_y = 50 ksi — Span in Feet

Properties and Reaction Values

S_x in.³	7.78	6.93	6.08	13.6	12.3	
V kips	59	44	29	70	49	For
R_1 kips	30.2	22.7	15.2	46.7	32.7	explanation
R_2 kips/in.	13.8	10.4	6.93	16.6	11.6	of deflection,
R_3 kips	39.4	25.6	14.0	60.6	35.5	see page 2-32
R_4 kips/in.	20.7	8.72	2.61	26.2	8.99	
R kips	**79**	**56**	23	**105**	**67**	

Load above heavy line is limited by maximum allowable web shear.
Values of R in **bold face** exceed maximum web shear.

AMERICAN INSTITUTE OF STEEL CONSTRUCTION

BEAMS
Channels
Allowable uniform loads in kips
for beams laterally supported

$F_y = 50$ ksi

6

For beams laterally unsupported, see page 2-146

Designation		C 6			MC 6		MC 6		MC 6	Deflection In.
Wt./ft	13	10.5	8.2	18	15.3	16.3	15.1	12	
Flange Width	2⅛	2	1⅞	3½	3½	3	3	2½	
L_u	4.10	3.90	3.70	9.30	7.50	7.90	7.80	5.20	
1	105	75	48			90		74	.01
2	58	51	44	91	82	87	76	62	.02
3	39	34	29	66	56	58	55	42	.05
4	29	25	22	50	42	43	42	31	.08
5	23	20	18	40	34	35	33	25	.13
6	19	17	15	33	28	29	28	21	.19
7	17	14	13	28	24	25	24	18	.25
8	15	13	11	25	21	22	21	16	.33
9	13	11	9.7	22	19	19	18	14	.42
10	12	10	8.8	20	17	17	17	12	.52
11	11	9.2	8.0	18	15	16	15	11	.63
12	9.7	8.4	7.3	17	14	14	14	10	.74
13	8.9	7.8	6.7	15	13	13	13	9.6	.87
14	8.3	7.2	6.3	14	12	12	12	8.9	1.01

$F_y = 50$ ksi · Span in Feet

Properties and Reaction Values

S_x in.³	5.80	5.06	4.38	9.91	8.47	8.68	8.32	6.24	
V kips	52	38	24	45	41	45	38	37	For
R_1 kips	29.3	21.0	13.4	33.2	24.5	32.9	27.7	20.8	explanation
R_2 kips/in.	14.4	10.4	6.6	12.5	11.2	12.4	10.4	10.2	of deflection,
R_3 kips	40.7	24.8	12.6	38.7	29.6	38.1	29.4	25.4	see page 2-32
R_4 kips/in.	29.2	10.8	2.8	13.8	12.3	13.3	7.99	9.55	
R kips	**80**	**57**	22	**77**	**64**	**76**	**57**	**57**	

Load above heavy line is limited by maximum allowable web shear.
Values of R in **bold face** exceed maximum web shear.

| F_y = 50 ksi | | | BEAMS | | | | | | 5–4–3 |

BEAMS
Channels
Allowable uniform loads in kips
for beams laterally supported
For beams laterally unsupported, see page 2-146

Designation		C 5		Deflection In.	C 4		Deflection In.	C 3			Deflection In.
Wt./ft		9	6.7		7.25	5.4		6	5	4.1	
Flange Width		1⅞	1¾		1¾	1⅝		1⅝	1½	1⅜	
L_u		4.00	3.70		4.30	3.90		4.90	4.60	4.30	
Span in Feet	1	65	38	.01	46	29	.01	28	25	20	.01
	2	36	30	.02	23	19	.03	14	12	11	.04
	3	24	20	.06	15	13	.07	9.2	8.3	7.3	.09
	4	18	15	.10	11	9.7	.12	6.9	6.2	5.5	.17
	5	14	12	.16	9.2	7.7	.19	5.5	5.0	4.4	.26
	6	12	10	.22	7.6	6.4	.28	4.6	4.1	3.7	.37
	7	10	8.6	.30	6.5	5.5	.38				
	8	8.9	7.5	.40	5.7	4.8	.50				
	9	7.9	6.7	.50	5.1	4.3	.63				
	10	7.1	6.0	.62	4.6	3.9	.78				
	11	6.5	5.5	.75							
	12	5.9	5.0	.89							

Properties and Reaction Values											
S_x in.³		3.56	3.00		2.29	1.93		1.38	1.24	1.10	
V kips		33	19		26	15		21	15	10	
R_1 kips		20.1	11.8	For explanation of deflection, see page 2-32	18.2	10.4	For explanation of deflection, see page 2-32	20.2	14.6	9.64	For explanation of deflection, see page 2-32
R_2 kips/in.		10.7	6.27		10.6	6.07		11.7	8.51	5.61	
R_3 kips		25.2	11.3		23.8	10.3		26.7	16.5	8.80	
R_4 kips/in.		15.5	3.09		20.1	3.79		39.7	15.1	4.33	
R kips		**58**	**22**		**55**	**24**		**61**	**44**	**24**	

Load above heavy line is limited by maximum allowable web shear.
Values of R in **bold face** exceed maximum web shear.

AMERICAN INSTITUTE OF STEEL CONSTRUCTION

BEAMS
Design of Bearing Plates

When a beam is supported by a masonry wall or pilaster, it is essential that the beam reaction be distributed over an area sufficient to keep the average pressure on the masonry within allowable limits. In the absence of code provisions, an allowable F_p, depending on the type of construction, may be selected from AISC ASD Specification Sect. J9.

The following method of design makes use of the Allowable Uniform Load tables and is recommended for bearing plates on concrete supports.*

R = Reaction of beam, kips

A = $B \times N$ = Area of plate, in.²

F_b = Allowable bending stress of plate, ksi

F_p = Allowable bearing pressure on support, ksi

f_p = Actual bearing pressure on support, ksi

R_1 = $1.65kF_{yw}t_w$ = first part of ASD Spec. Equation (K1-3), kips

R_2 = $0.66F_{yw}t_w$ = second part of ASD Spec. Equation (K1-3), kips/in.

R_3 = $34t_w^2 \sqrt{\dfrac{F_{yw}t_f}{t_w}}$ = first part of ASD Spec. Equation (K1-5) kips

R_4 = $34t_w^2 \left[3\left(\dfrac{1}{d}\right)\left(\dfrac{t_w}{t_f}\right)^{1.5} \right] \sqrt{\dfrac{F_{yw}t_f}{t_w}}$ = second part of Spec. Equation (K1-5), based on $N = 1.0$, kips/in.

k = Distance from bottom of beam to web toe of fillet, in. (from Manual Part 1)

t = Thickness of plate, in.

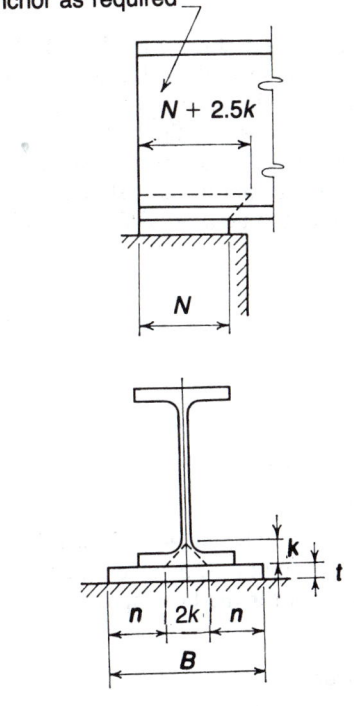

Anchor as required

$N + 2.5k$

N

k

t

n $2k$ n

B

Calculate the minimum bearing length N based on local web yielding AISC ASD Specification Equation (K1-3) or web crippling Specification Equation (K1-5).

---

*For concrete supports where the bearing plate does not cover the full concrete area, see Example 9.

The equation yielding the larger N value controls. By replacing each portion of the above Equation with a variable, as defined previously, and solving for N, the following equations result:

$$\text{local web yielding} \quad N = \frac{R - R_1}{R_2}, \text{ in.}$$

$$\text{web crippling} \quad N = \frac{R - R_3}{R_2}, \text{ in.}$$

The values for R_1, R_2, R_3 and R_4 are tabulated in the Allowable Uniform Load Tables for each shape.

Determine the required bearing plate area, A_1, in.^{2}, by rearranging the formulas given in J9 of the AISC ASD Specification and solving for A_1.

On full area of concrete support

$$A_1 = \frac{R}{0.35 f'_c}$$

On less than full area of concrete support

$$A_1 = \frac{1}{A_2} \left(\frac{R}{0.35 f'_c} \right)^2$$

Establish N and solve for $B = A_1/N$. The length of bearing N is usually governed by the available wall thickness or some other structural consideration. Preferably, B and N should be in full inches, and B rounded off so that $B \times N \geq A_1$ (req'd).

Solve for t in the following formula based on cantilever bending of the plate under uniform concrete pressure.

Determine the actual bearing pressure, $f_p = R/(B \times N)$.

Determine $n = (B/2) - k$ and, using the actual f_p, solve for t in the formula:

$$t = \sqrt{\frac{3 f_p n^2}{F_b}}$$

EXAMPLE 8

Given:

A W18 × 50 beam, $F_y = 36$ ksi, for which $k = 1\frac{1}{4}$ in., has a reaction of 49 kips and is to be supported by a 10 in. concrete wall. Using the entire width of the 3 ksi concrete wall for the bearing length N, design a bearing plate for the beam.

Solution:

From the Allowable Uniform Load Tables:

$R_1 = 26.4$ kips
$R_2 = 8.43$ kips/in.
$R_3 = 32.6$ kips
$R_4 = 2.67$ kips/in.

$$N = \frac{R - R_1}{R_2} = \frac{49 - 26.4}{8.43} = 2.68 \text{ in.} < 10 \text{ in.} \quad \textbf{o.k.}$$

$$N = \frac{R - R_3}{R_4} = \frac{49 - 32.6}{2.67} = 6.14 \text{ in.} < 10 \text{ in.} \quad \textbf{o.k.}$$

$$A_1 \text{ (Req'd)} = \frac{R}{0.35f'_c} = \frac{49}{(0.35) \, 3} = 46.7 \text{ in.}^2$$

$B = A_1/N = 46.7/10 = 4.7 \text{ in.; use 8 in. (flange width controls)}$

$A_1 = B \times N = 8 \times 10 = 80.0 \text{ in.}^2 \geq 46.7 \text{ in.}^2$

$n = 8/2 - 1.25 = 2.75 \text{ in.}$

$f_p = 49/80 = 0.613 \text{ ksi}$

$$t = \sqrt{\frac{3f_p n^2}{F_b}} = \sqrt{\frac{3 \, (0.613) \, (2.75)^2}{27}} = 0.718 \text{ in.}$$

Use: Bearing plate ¾ × 10 × 0' − 8

EXAMPLE 9

Given:

Investigate a 1 × 6½ × 0' − 8 bearing plate for the beam in Example 8 supported on a 10½″ concrete wall. The least distance from the edge of bearing plate to the edge of concrete support, b_1, is 2 in. $f'_c = 3.0$ ksi

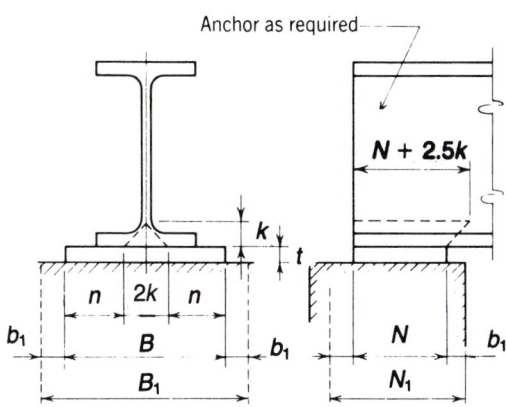

Solution:

$A_1 = B \times N = 8 \times 6.5 = 52.0 \text{ in.}^2$

Assumed area of concrete support:

$$A_2 = B_1 \times N_1 = [8 + (2 + 2)] \times [6.5 + (2 + 2)] = 126.0 \text{ in.}^2$$

$f_p = 49/52 = 0.942$ ksi

$F_p = 0.35f'_c \sqrt{A_2/A_1} = 0.35 \times 3 \times \sqrt{126.0/52.0}$
$\qquad\qquad = 1.63$ ksi > 0.942 ksi **o.k.**

and $1.63 \leq 0.7f'_c = 0.7 \times 3 = 2.1$ **o.k.** (AISC ASD Specification Sect. J9)

$N = 6.5$ in. > 6.14 in. **o.k.** (see Ex. 8)

$n = B/2 - k = 8/2 - 1.25 = 2.75$ in.

Min. $t = \sqrt{\dfrac{3\ (0.942)\ (2.75)^2}{27}} = 0.89$ in. < 1 in. **o.k.**

Investigate beam without bearing plate ($t_f = 0.57$ in.):

$f_p = \dfrac{49}{7.5 \times 6.5} = 1.01$ ksi

Min. $t = \sqrt{\dfrac{3f_p\ [(b_f/2) - k]^2}{F_b}}$

$\qquad = \sqrt{\dfrac{(3)\ 1.01\ [(7.5/2) - 1.25]^2}{27.0}}$

$\qquad = 0.84$ in. > 0.57 in. **n.g.**

Use bearing plate.

FLOOR PLATE
BENDING CAPACITY
Pounds per Sq. Ft

Theoretical Weight per sq. ft in lbs.	Plate Thickness inches	SPAN										Section Modulus per ft of width
		1'−6"	2'−0"	2'−6"	3'−0"	3'−6"	4'−0"	4'−6"	5'−0"	6'−0"	7'−0"	
6.15	⅛	148	83	53	37	—	—	—	—	—	—	.031
8.70	3⁄16	333	188	120	83	61	47	—	—	—	—	.070
11.25	¼	593	333	213	148	109	83	66	53	—	—	.125
13.80	5⁄16	927	521	333	232	170	130	103	83	58	—	.195
16.35	⅜	1333	750	480	333	245	188	148	120	83	61	.281
18.90	7⁄16	1814	1021	653	453	333	255	201	163	113	83	.383
21.45	½	2370	1333	853	593	435	333	263	213	148	109	.500
24.00	9⁄16	3000	1688	1080	750	551	422	333	270	188	138	.633
26.55	⅝	3703	2084	1333	926	680	521	412	333	232	170	.781
29.10	11⁄16	4482	2521	1613	1121	823	630	498	403	280	206	.945
31.65	¾	5333	3000	1920	1333	980	750	593	480	333	245	1.125
34.20	13⁄16	6260	3521	2253	1565	1150	880	696	563	391	287	1.320
36.75	⅞	7258	4088	2613	1815	1333	1022	807	653	454	333	1.531
39.30	15⁄16	8333	4688	3000	2083	1531	1172	926	750	521	383	1.758
41.85	1	9481	5333	3413	2370	1741	1333	1053	853	593	435	2.000
44.40	1-1⁄16	10705	6021	3853	2676	1966	1505	1190	963	669	492	2.258
46.95	1-⅛	11999	6749	4319	3000	2204	1687	1333	1080	750	551	2.531
49.50	1-3⁄16	13369	7520	4813	3343	2456	1880	1486	1203	836	614	2.820
52.05	1-¼	14815	8333	5333	3704	2721	2083	1646	1333	926	680	3.125
54.60	1-5⁄16	16332	9186	5879	4083	3000	2297	1815	1470	1021	750	3.445
57.15	1-⅜	17925	10082	6453	4482	3292	2521	1992	1613	1120	823	3.781
59.70	1-7⁄16	19593	11021	7053	4898	3599	2755	2177	1764	1225	900	4.133
62.25	1-½	21333	12000	7680	5333	3919	3000	2371	1920	1333	980	4.500
64.80	1-9⁄16	23149	13021	8333	5787	4252	3255	2572	2084	1447	1063	4.883
67.35	1-⅝	25036	14082	9013	6260	4599	3521	2782	2253	1565	1150	5.281
69.90	1-11⁄16	26998	15186	9719	6750	4959	3797	3000	2430	1687	1240	5.695
72.45	1-¾	29037	16333	10453	7259	5334	4084	3227	2614	1815	1333	6.125
75.00	1-13⁄16	31146	17520	11212	7787	5721	4380	3461	2803	1947	1430	6.570
77.55	1-⅞	33332	18749	11999	8333	6123	4688	3704	3000	2083	1531	7.031
80.10	1-15⁄16	35593	20021	12813	8898	6538	5006	3955	3204	2225	1634	7.508
82.65	2	37926	21333	13653	9482	6966	5334	4214	3414	2370	1742	8.000
Deflection Coefficient E = 29,000 ksi		.037	.066	.104	.149	.203	.265	.335	.414	.596	.700	.810

Loads above and to the right of the heavy black lines will cause deflections of more than 1/164 of the span.

To find the actual deflections for the loads given above, divide the coefficient of deflection for the span by the thickness of the plate in inches.

To find the deflection caused by loads less than shown above, first find the deflection caused by the loads given above. Multiply this by the actual load and divide by the load given above. For safety, loads greater than those given in the above table should not be used.

Loads are based on an extreme fiber stress of 16 ksi and simple span bending.

ALLOWABLE MOMENTS IN BEAMS
With unbraced length greater than L_u
General Notes

Spacing of lateral bracing at distances greater than L_u creates a problem in which the designer is confronted with a given laterally unbraced length (usually less than the total span) along the compression flange, and a calculated required bending moment. The beam cannot be selected from its section modulus alone since depth and flange proportions have an influence on its bending strength.

The following charts show the total allowable bending moment for W and M shapes of $F_y = 36$ ksi and $F_y = 50$ ksi steels, used as beams, with respect to the maximum unbraced length for which this moment is permissible. The charts extend over varying unbraced lengths, depending upon the size of beams represented. In general, they cover most lengths frequently encountered in design practice.

The total allowable bending moment, in kip-ft, is plotted with respect to unbraced length with no consideration of the moment due to weight of the beam. Total allowable moments are shown for unbraced lengths in feet, starting at spans less than L_c, of spans between L_c and L_u, and of spans beyond L_u.

The unbraced length L_c, in feet, with the limit indicated by a solid symbol (●), is the maximum unbraced length of the compression flange for which the allowable bending stress F_b may be taken at $0.66F_y$ for compact sections by AISC ASD Specification Sect. F1.1, and for noncompact shapes that are permitted an allowable stress higher than $0.60F_y$ by Sect. F1.2. For these noncompact shapes, which meet the requirements of compact sections except that $b_f/2t_f$ exceeds $65/\sqrt{F_y}$, but is less than $95\sqrt{F_y}$, the allowable bending stress is obtained from Equation (F1-3). L_c is equal to the smaller value obtained from the expressions $76b_f/\sqrt{F_y}$ and $20000/[(d/A_f)F_y]$. This criterion applies to one beam when F_y is equal to 36 ksi and applies to eight beams when F_y is equal to 50 ksi. L_c for these beams are indicated by a half-filled circle ◐.

The unbraced length L_u, in feet, with the limit indicated by an open symbol (○), is the maximum unbraced length of the compression flange beyond which the allowable bending stress F_b is less than $0.60F_y$. L_u is equal to the greater value obtained from Equations (F1-6) and (F1-8) when F_b is $0.60F_y$ and C_b equals unity. For lengths greater than L_c, but not greater than L_u, F_b may be taken at $0.60F_y$. In no case is L_c taken greater than L_u.

The unbraced length is the maximum laterally unbraced length of the compression flange corresponding to the total allowable moment. It may be either the total span or any part of the total span between braced points. The curves shown in these charts were computed for beams subjected to loading conditions which produce bending moments within the unbraced length greater than that at both ends of this length. In these cases, C_b is taken as unity in accordance with Sect. F1.3. When the unbraced length is greater than L_u and the bending moment within the unbraced length is smaller than that at either end of this length, C_b is larger than unity and the section may provide a more liberal moment capacity. In these cases the allowable moment can be determined using the provisions of Sect. F1.3 of the AISC ASD Specification.

In all cases where the unbraced length of the compression flange exceeds L_u, F_b must be calculated according to the provisions of Sect. F1.3, and may neither exceed the larger value given by the following formulas, nor $0.60F_y$:

When $l/r_T \leq \sqrt{\dfrac{102 \times 10^3 \times C_b}{F_y}}$: $F_b = 0.60F_y$

When $\sqrt{\dfrac{102 \times 10^3 C_b}{F_y}} \leq \dfrac{l}{r_T} \leq \sqrt{\dfrac{510 \times 10^3 C_b}{F_y}}$:

$$F_b = \left[\frac{2}{3} - \frac{F_y(l/r_T)^2}{1530 \times 10^3 C_b} \right] F_y \qquad \text{(F1-6)}$$

When $l/r_T \geq \sqrt{\dfrac{510 \times 10^3 C_b}{F_y}}$: $F_b = \dfrac{170 \times 10^3 C_b}{(l/r_T)^2}$ \qquad (F1-7)

For any value of l/r_T :

$$F_b = \frac{12 \times 10^3 C_b}{ld/A_f} \qquad \text{(F1-8)}$$

In computing the points for the curves, C_b in the above formulas was taken as unity; the radius of gyration r_T about an axis in the plane of the web and the depth-flange area ratio d/A_f are taken from the Tables of Dimensions and Properties in Part 1 of this Manual.

Over a limited range of length, a given beam is the lightest available for various combinations of unbraced length and total moment. The charts are designed to assist in selection of the lightest available beam for the given combination.

The solid portion of each curve indicates the most economical section by weight. The dashed portion of each curve indicates ranges in which a lighter weight beam will satisfy the loading conditions. For beams of equal weight, where both would satisfy the loading conditions, the deeper beam, when having a lesser moment capacity than the shallower beam, is indicated as a dashed curve to assist in making a selection for reduced deflection or a limited depth condition.

In the case of W and M shapes of equal weight and the same nominal depth, the M shape is shown dashed when its design moment capacity is less than the W shape, to indicate that the W shape is usually more readily available.

The curves are plotted without regard to deflection, therefore due care must be exercised in their use. The curves do not extend beyond an arbitrary span/depth limit of 30. In no case is the dashed line or the solid line extended beyond the point where the calculated bending stress is less than 11 ksi for $F_y = 36$ ksi steel, or 15 ksi for $F_y = 50$ ksi steel.

The following example illustrates the use of the charts for selection of a proper size beam with an unbraced length greater than L_u.

EXAMPLE 10

Given:

Using $F_y = 36$ ksi steel, determine the size of a "simple" framed girder with a span of 35 ft, which supports two equal concentrated loads located 10 ft from its left and right reaction points. The compression flange is laterally supported at the concentrated load points only. The loads produce a maximum calculated moment of 220 kip-ft in the center 15-ft section between the loads.

Solution:

For this loading condition, $C_b = 1.0$.

Center section of 15 ft is longest unbraced length.

With total span equal to 35 ft and $M = 220$ kip-ft, assume approximate weight of beam at 70 lbs./ft (equal to 0.07 kips/ft).

$$\text{Total } M = 220 + \frac{0.07 \times (35)^2}{8} = 231 \text{ kip-ft}$$

Entering chart, with unbraced length equal to 15 ft on the bottom scale (abscissa), proceed upward to meet the horizontal line corresponding to a moment equal to 231 kip-ft on the left hand scale (ordinate). Any beam listed above and to the right of the point so located satisfies the allowable bending stress requirement. In this case, the lightest section satisfying this criterion is a W 24 × 68, for which the total allowable moment with an unbraced length of 15 ft is 239 kip-ft.

Use: **W24×68**

Note: If depth is limited, a W18×71 could be selected, provided deflection conditions are satisfied.

ALLOWABLE MOMENTS IN BEAMS ($C_b = 1$, $F_y = 36$ ksi)

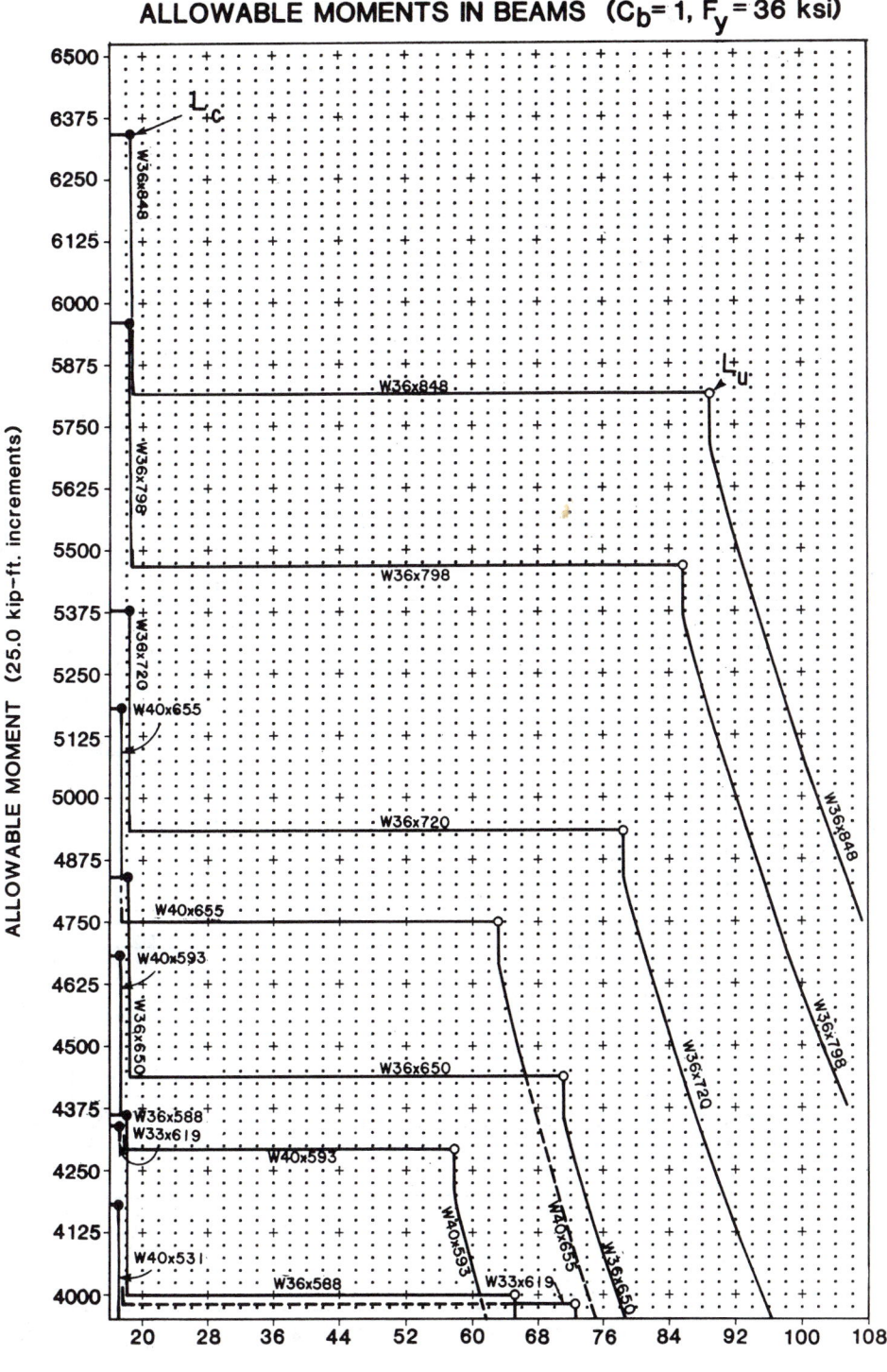

ALLOWABLE MOMENT (25.0 kip-ft. increments)

UNBRACED LENGTH (2.0 ft. increments)

ALLOWABLE MOMENTS IN BEAMS (C_b=1, F_y=36 ksi)

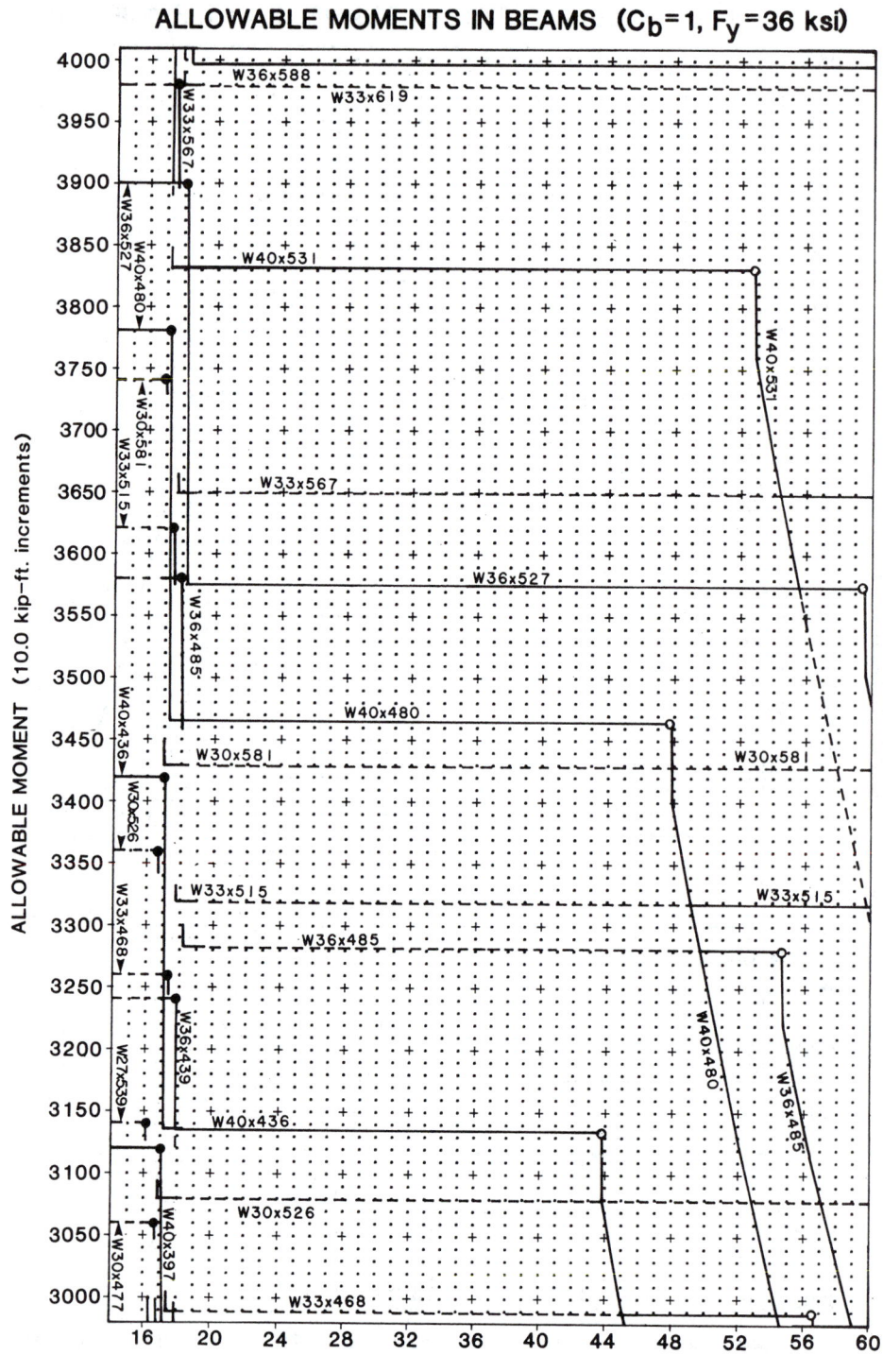

ALLOWABLE MOMENT (10.0 kip–ft. increments)

UNBRACED LENGTH (1.0 ft. increments)

AMERICAN INSTITUTE OF STEEL CONSTRUCTION

ALLOWABLE MOMENTS IN BEAMS ($C_b = 1$, $F_y = 36$ ksi)

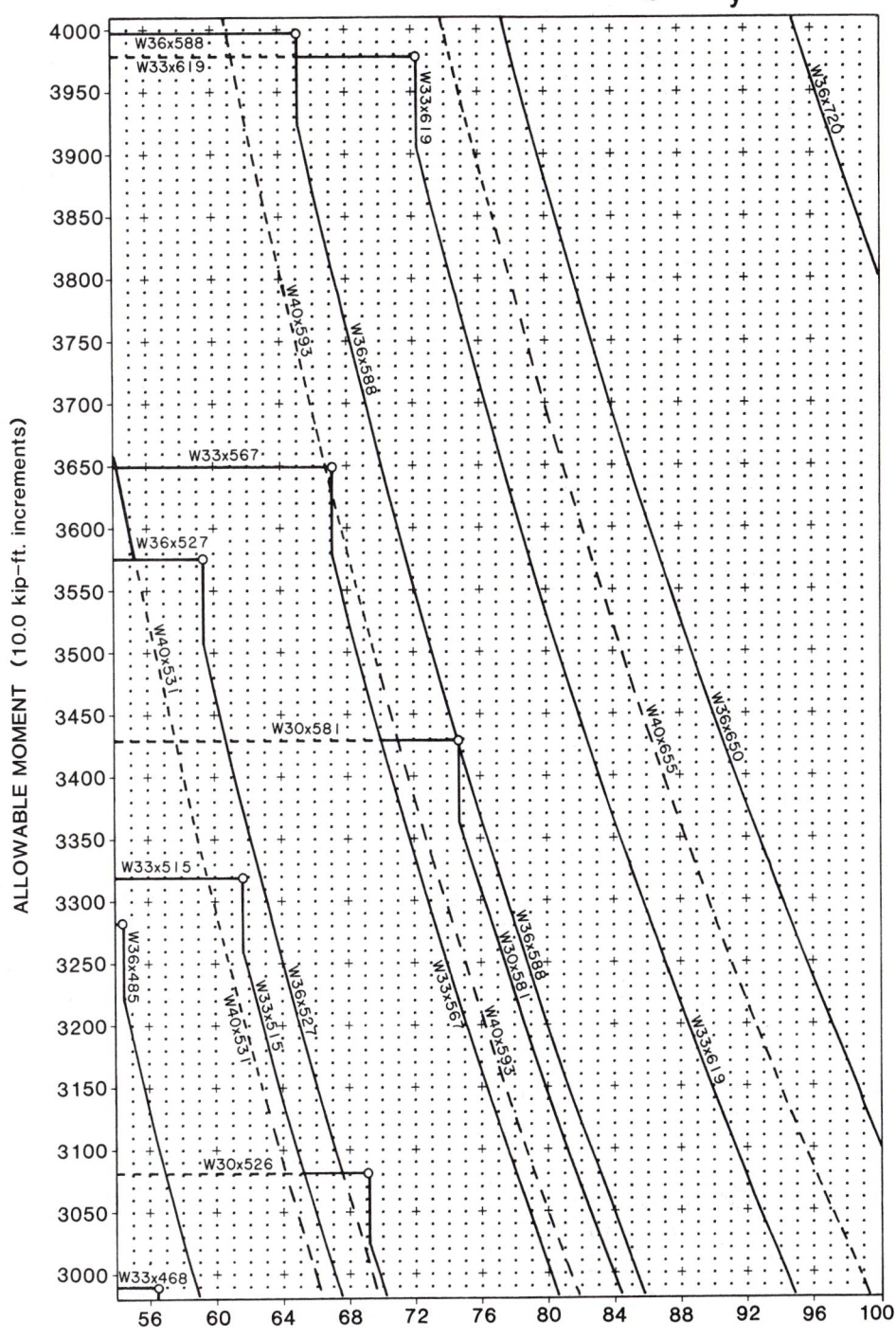

ALLOWABLE MOMENT (10.0 kip-ft. increments)

UNBRACED LENGTH (1.0 ft. increments)

ALLOWABLE MOMENTS IN BEAMS (C_b=1, F_y=36 ksi)

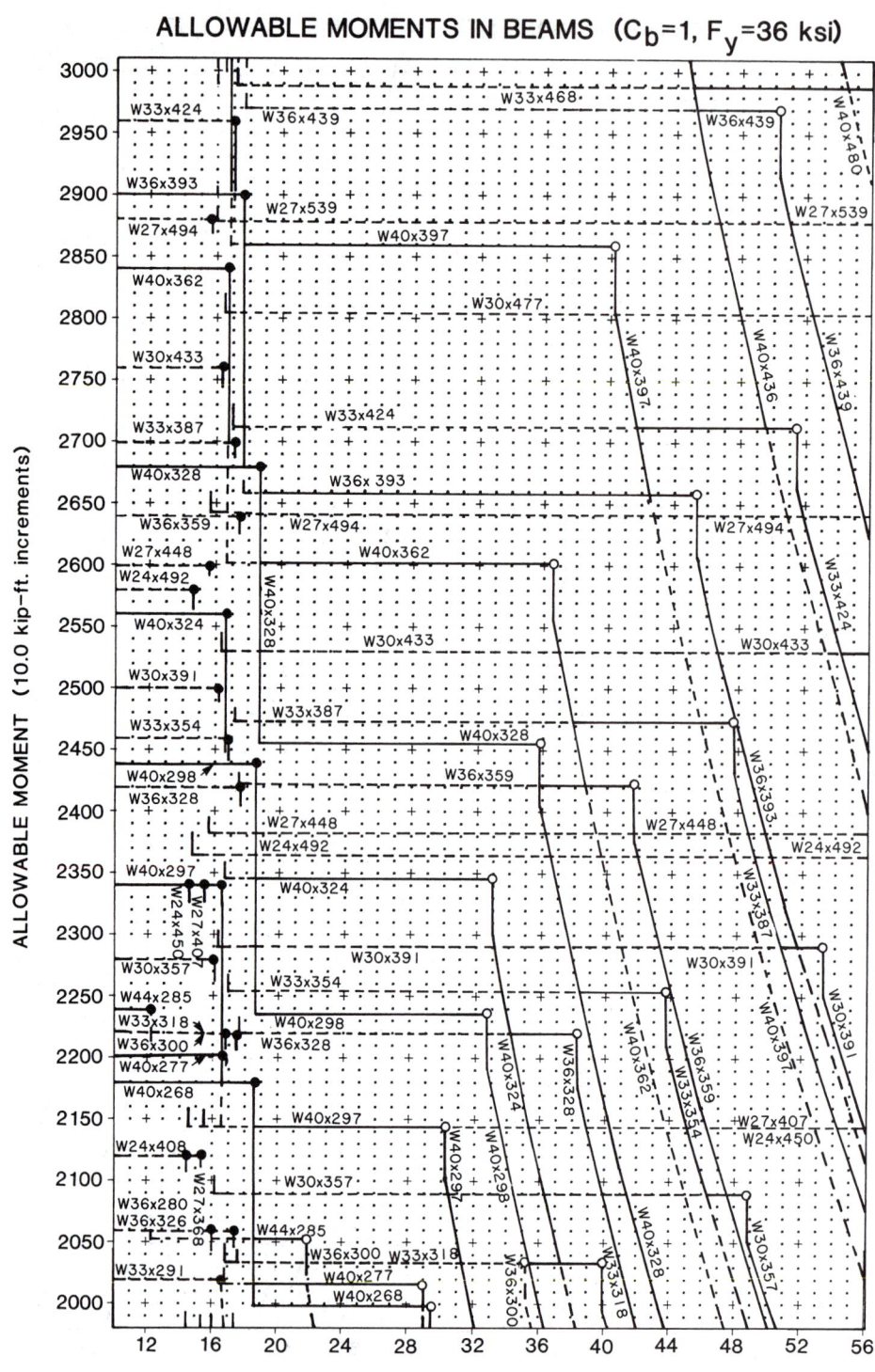

UNBRACED LENGTH (1.0 ft. increments)

AMERICAN INSTITUTE OF STEEL CONSTRUCTION

ALLOWABLE MOMENTS IN BEAMS (C_b=1, F_y=36 ksi)

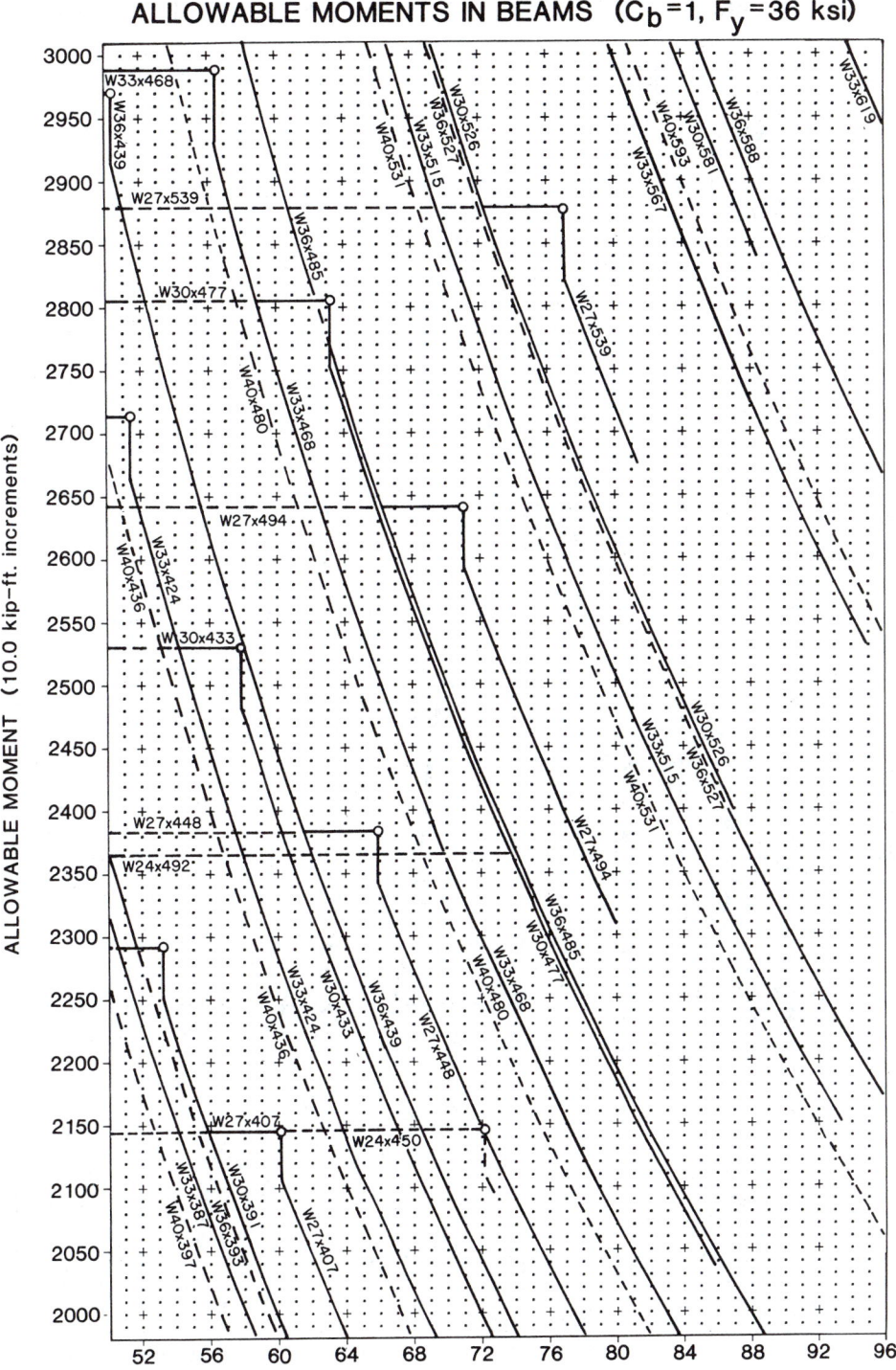

ALLOWABLE MOMENT (10.0 kip-ft. increments)

UNBRACED LENGTH (1.0 ft. increments)

ALLOWABLE MOMENTS IN BEAMS (C_b=1, F_y =36 ksi)

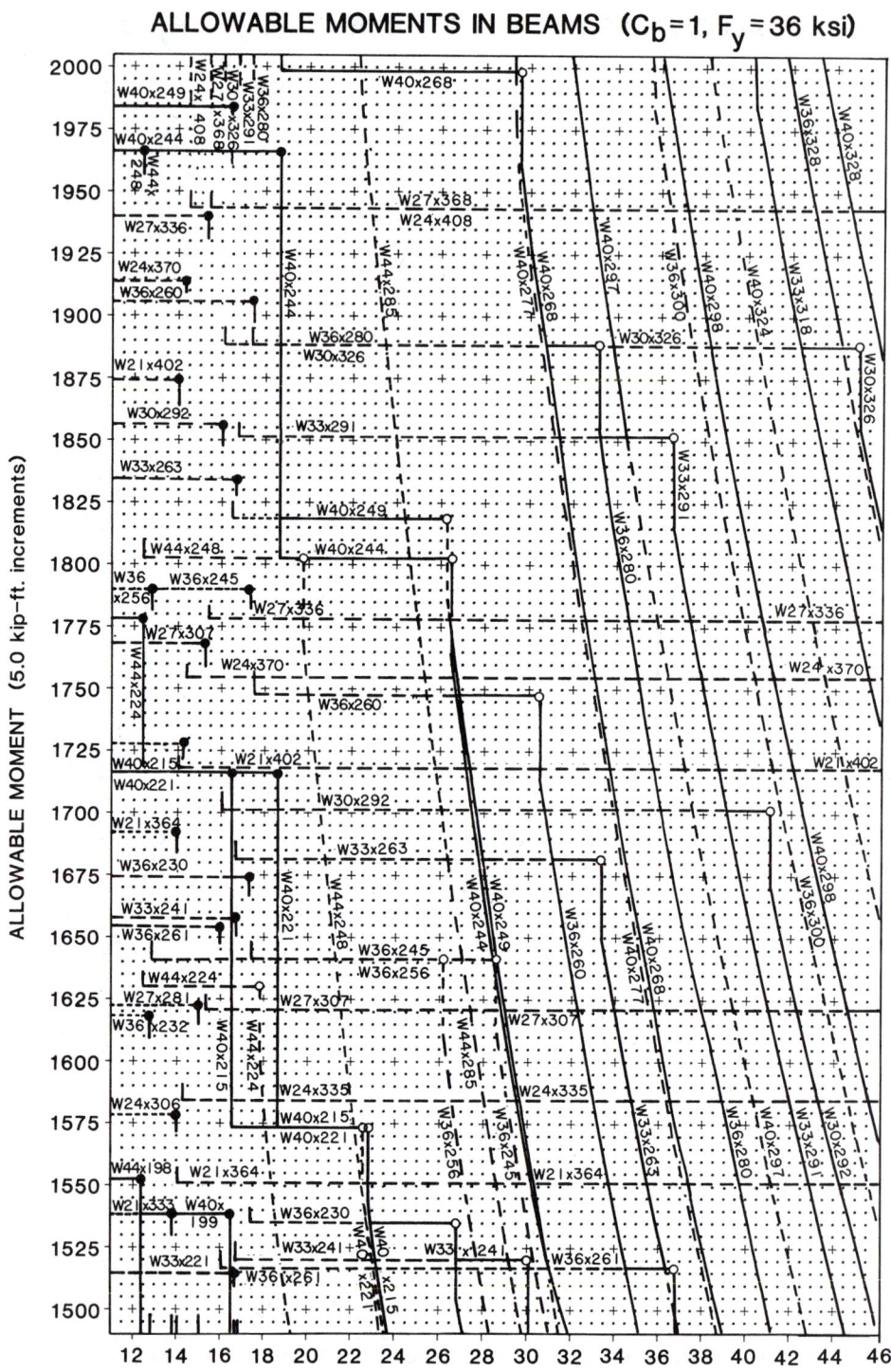

ALLOWABLE MOMENT (5.0 kip-ft. increments)

UNBRACED LENGTH (0.5 ft. increments)

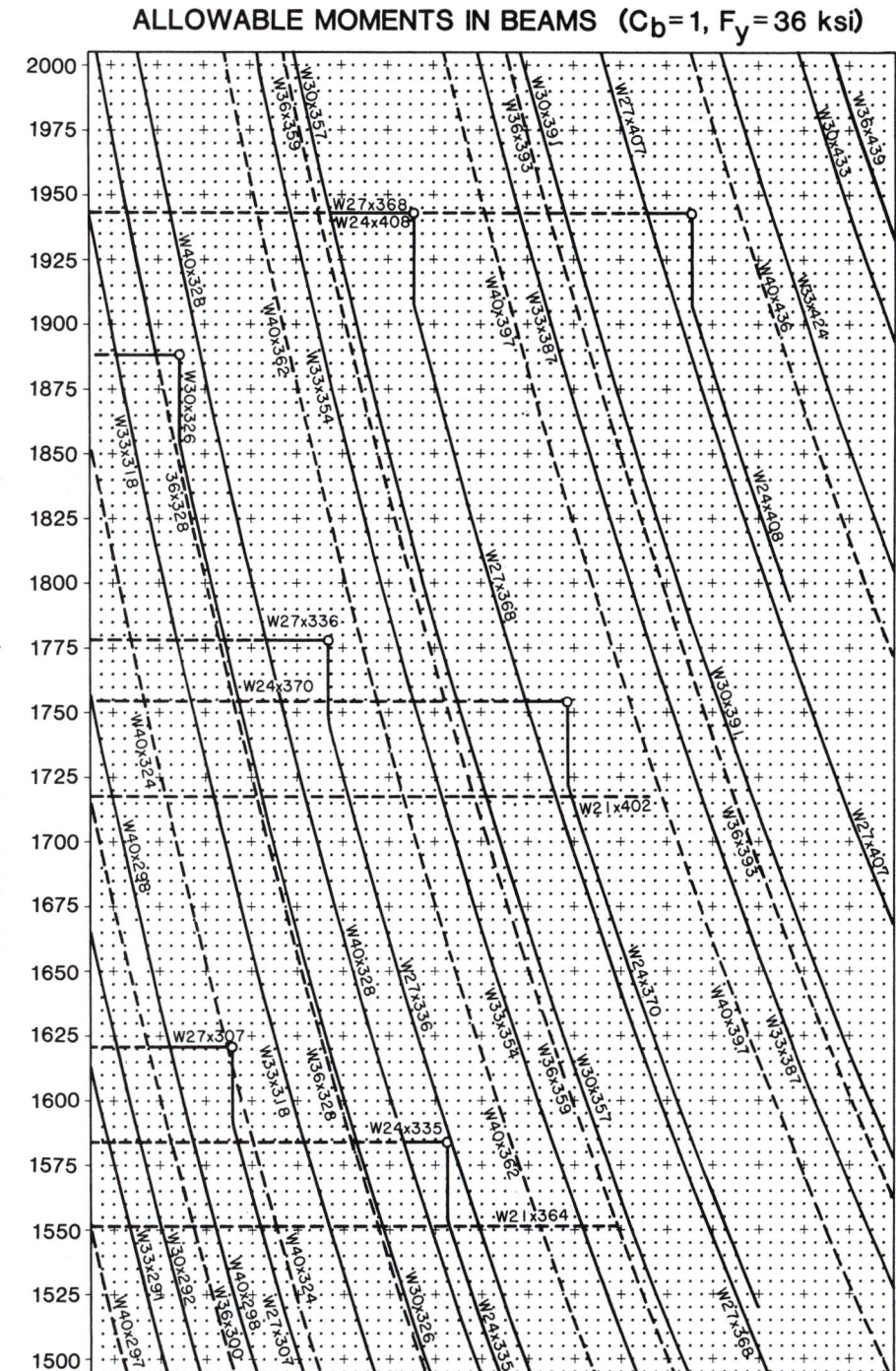

ALLOWABLE MOMENTS IN BEAMS ($C_b = 1$, $F_y = 36$ ksi)

ALLOWABLE MOMENT (5.0 kip-ft. increments)

UNBRACED LENGTH (0.5 ft. increments)

AMERICAN INSTITUTE OF STEEL CONSTRUCTION

ALLOWABLE MOMENTS IN BEAMS (C_b=1, F_y=36 ksi)

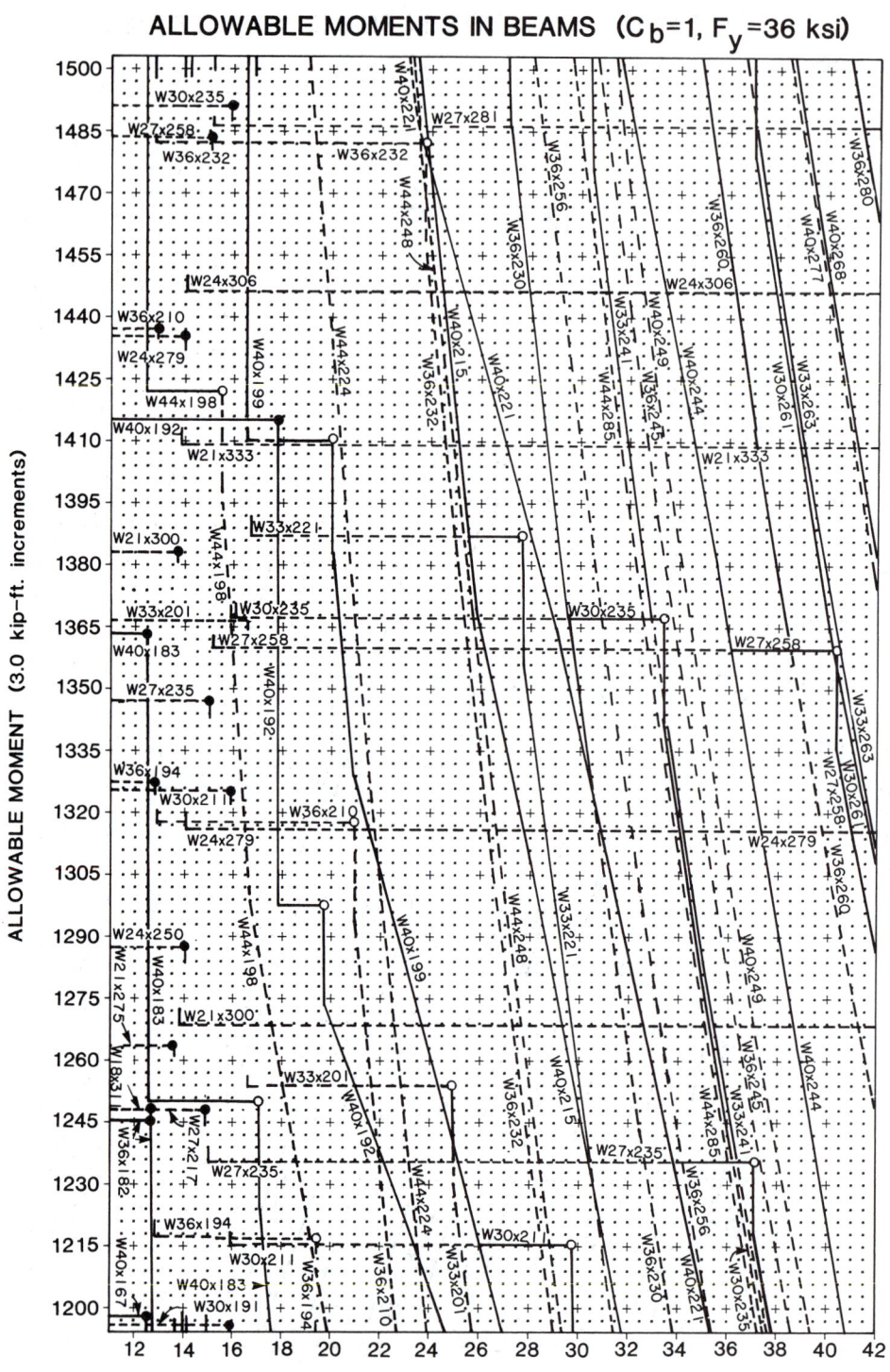

ALLOWABLE MOMENT (3.0 kip-ft. increments)

UNBRACED LENGTH (0.5 ft. increments)

ALLOWABLE MOMENTS IN BEAMS (C$_b$=1, F$_y$=36 ksi)

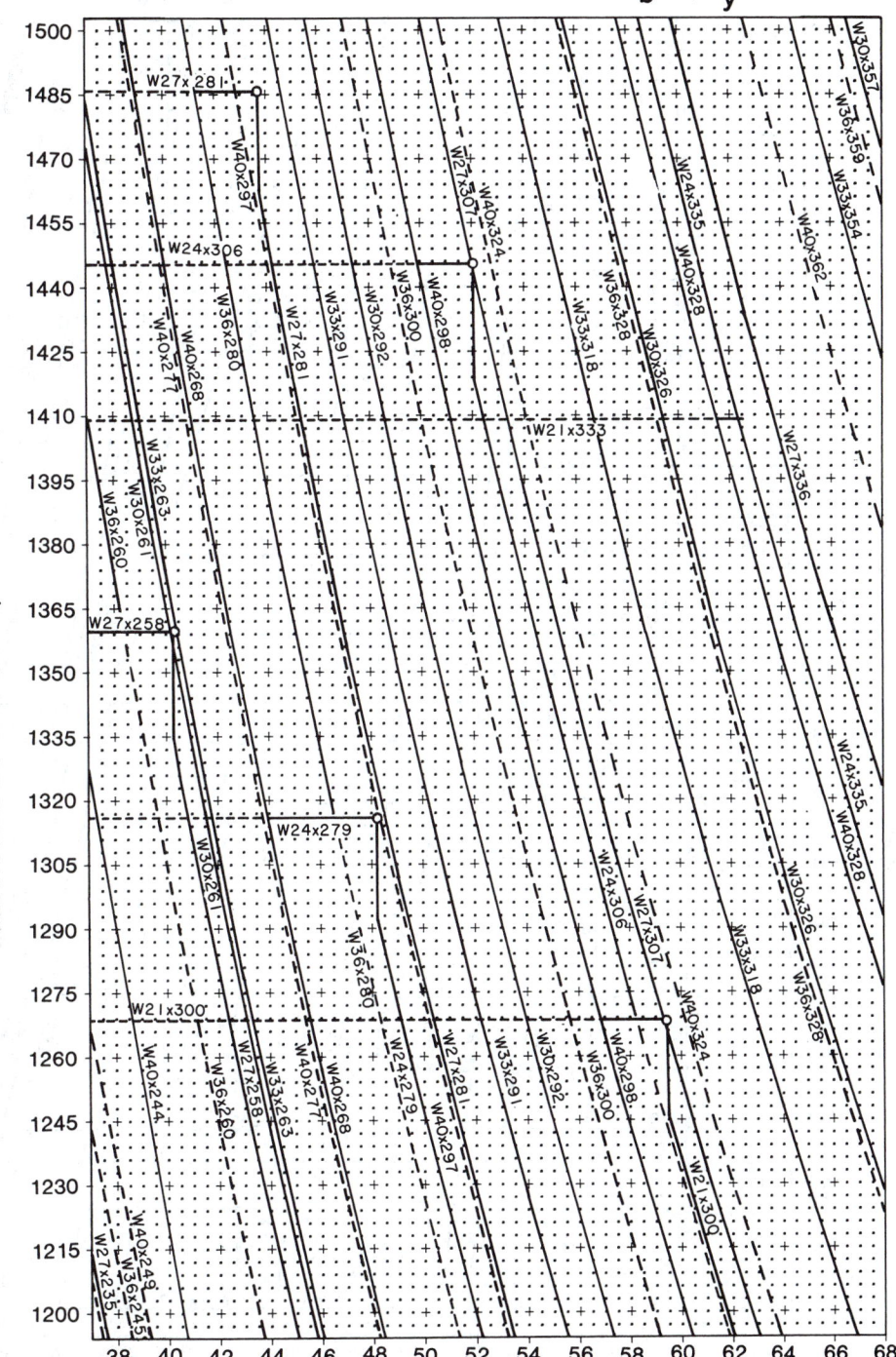

ALLOWABLE MOMENT (3.0 kip-ft. increments)

UNBRACED LENGTH (0.5 ft. increments)

AMERICAN INSTITUTE OF STEEL CONSTRUCTION

ALLOWABLE MOMENTS IN BEAMS (C_b=1, F_y=36 ksi)

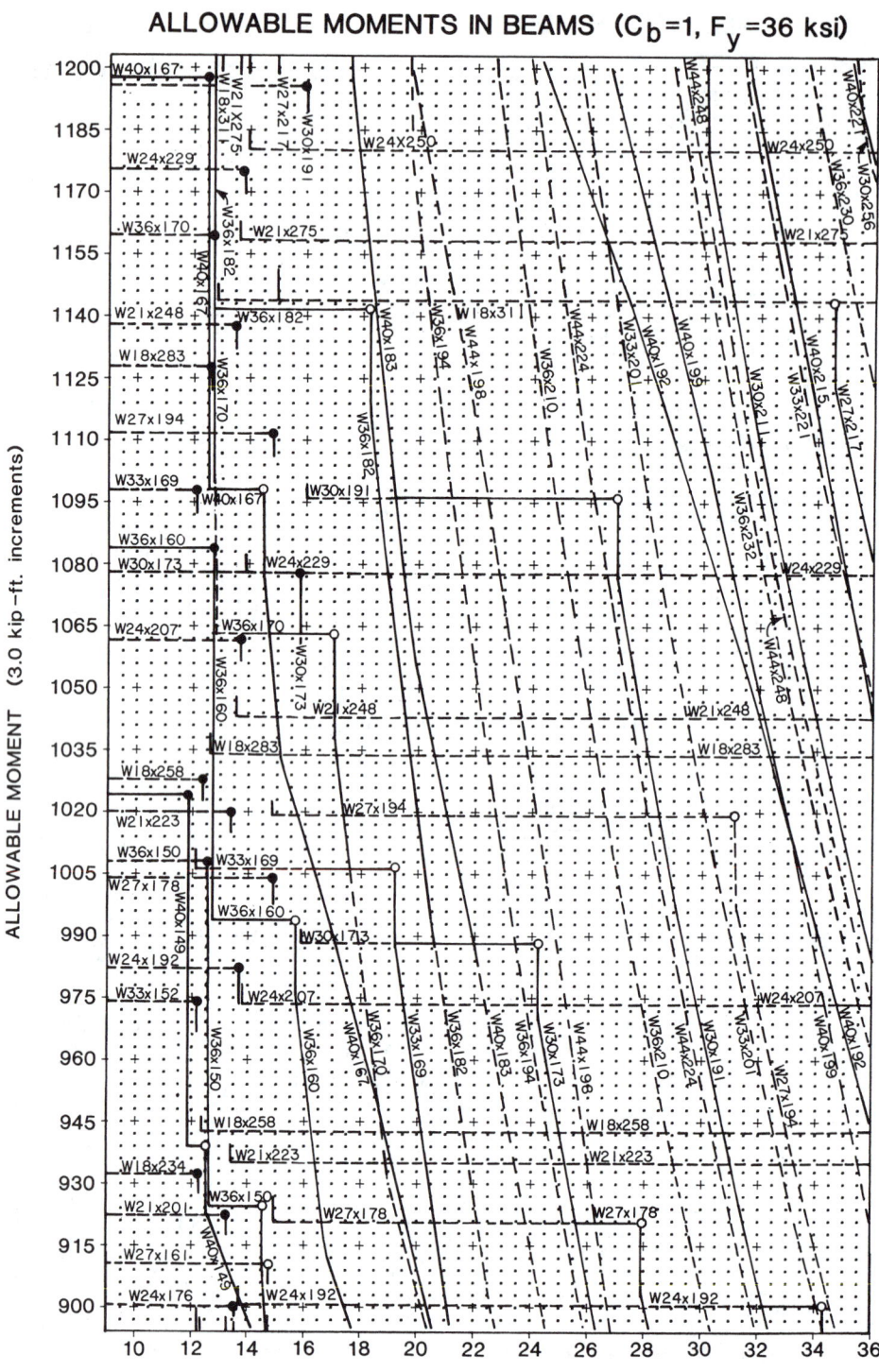

ALLOWABLE MOMENT (3.0 kip-ft. increments)

UNBRACED LENGTH (0.5 ft. increments)

ALLOWABLE MOMENTS IN BEAMS (C_b = 1, F_y = 36 ksi)

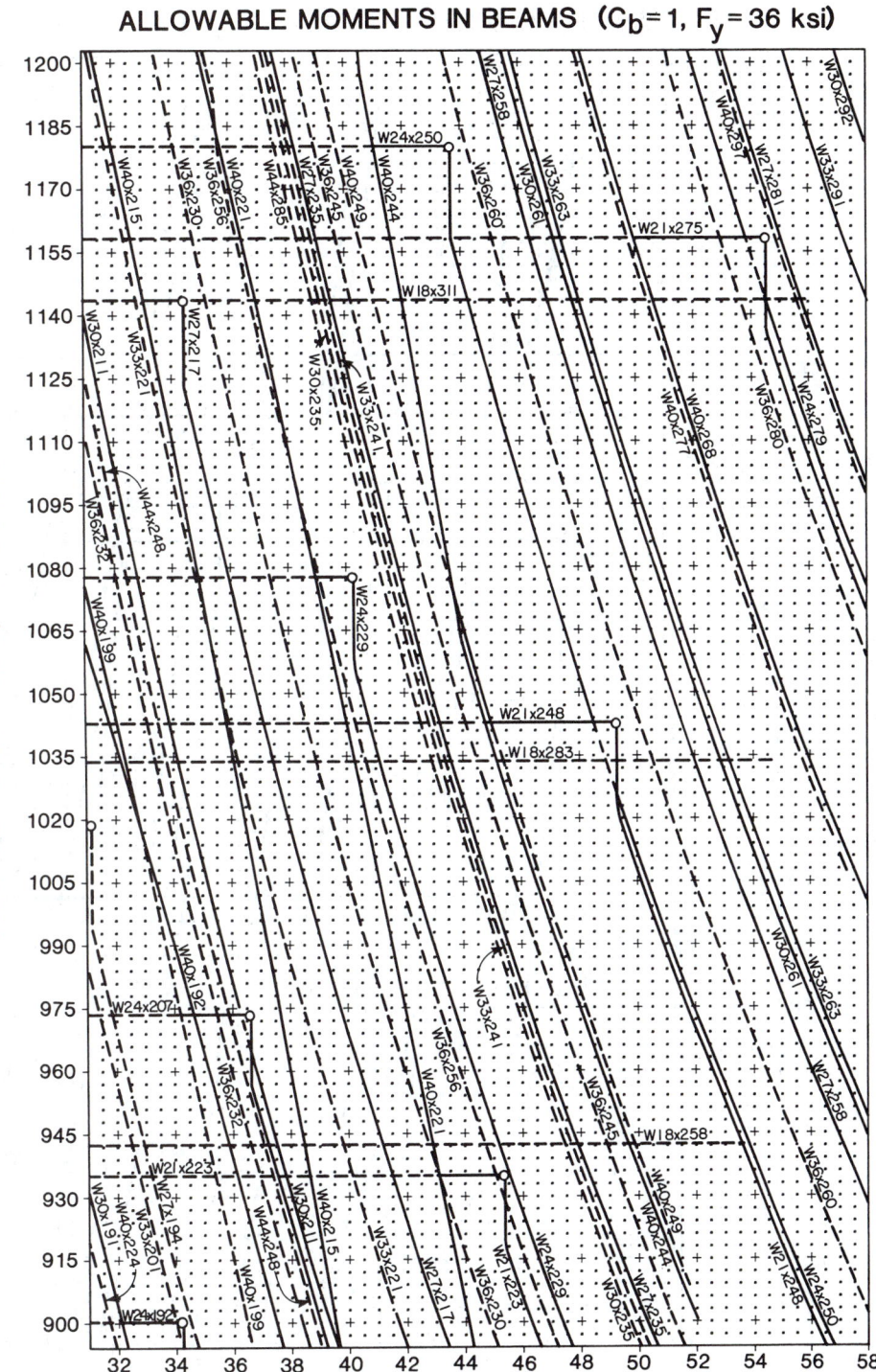

ALLOWABLE MOMENT (3.0 kip-ft. increments)

UNBRACED LENGTH (0.5 ft. increments)

ALLOWABLE MOMENTS IN BEAMS (C_b =1, F_y = 36 ksi)

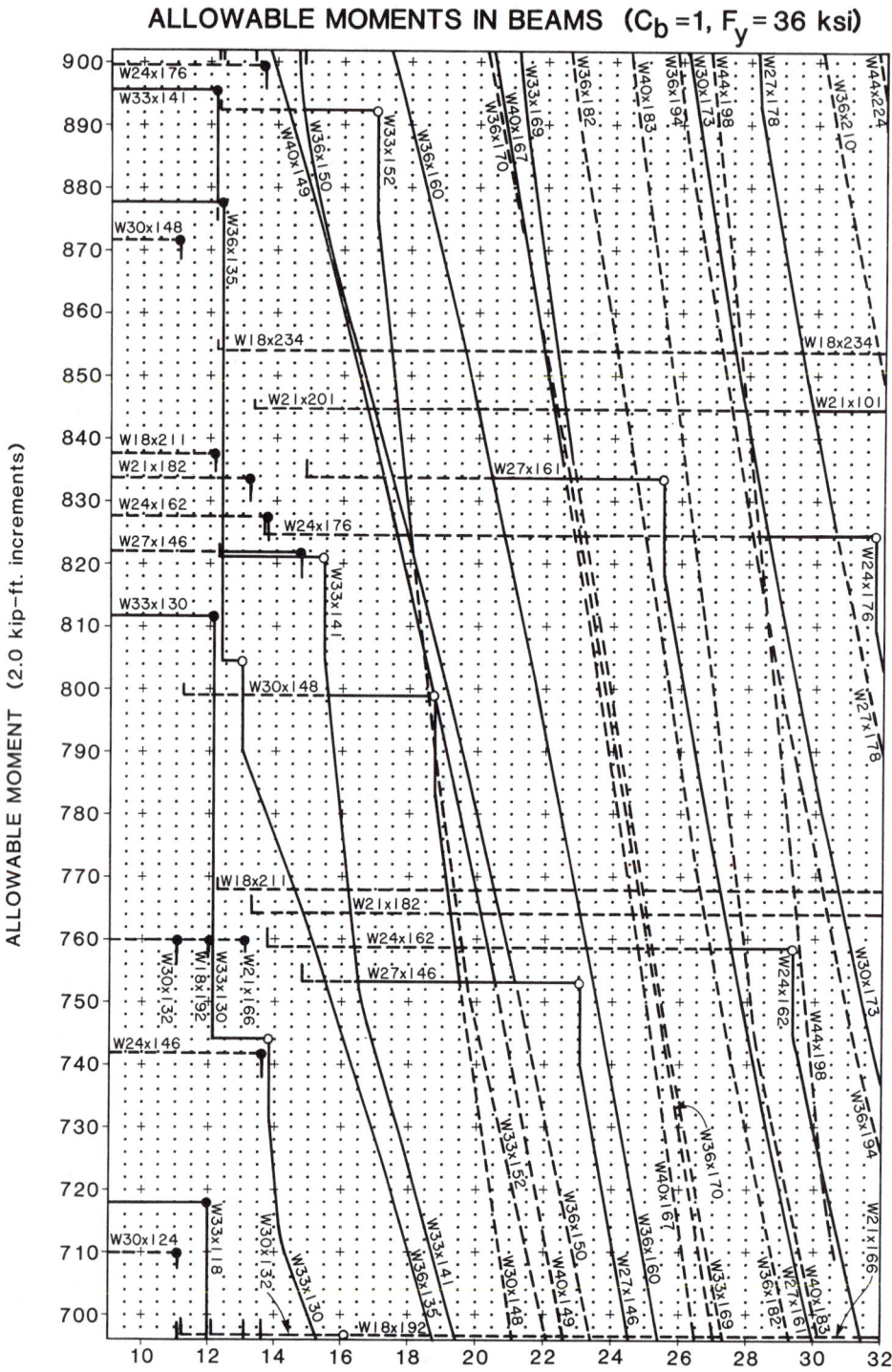

UNBRACED LENGTH (0.5 ft. increments)

ALLOWABLE MOMENTS IN BEAMS ($C_b = 1$, $F_y = 36$ ksi)

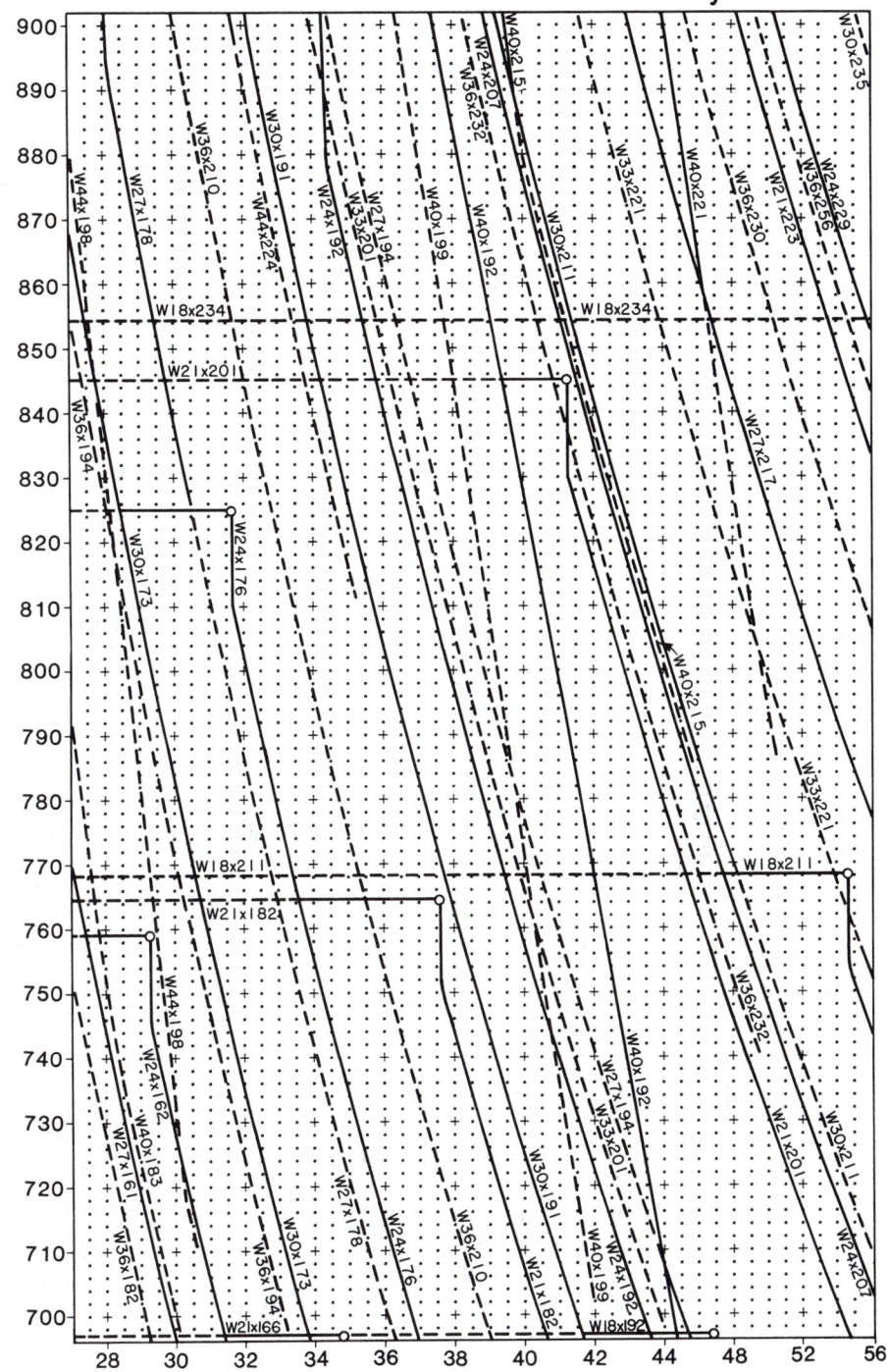

UNBRACED LENGTH (0.5 ft. increments)

ALLOWABLE MOMENT (2.0 kip-ft. increments)

AMERICAN INSTITUTE OF STEEL CONSTRUCTION

ALLOWABLE MOMENTS IN BEAMS (C_b =1, F_y =36 ksi)

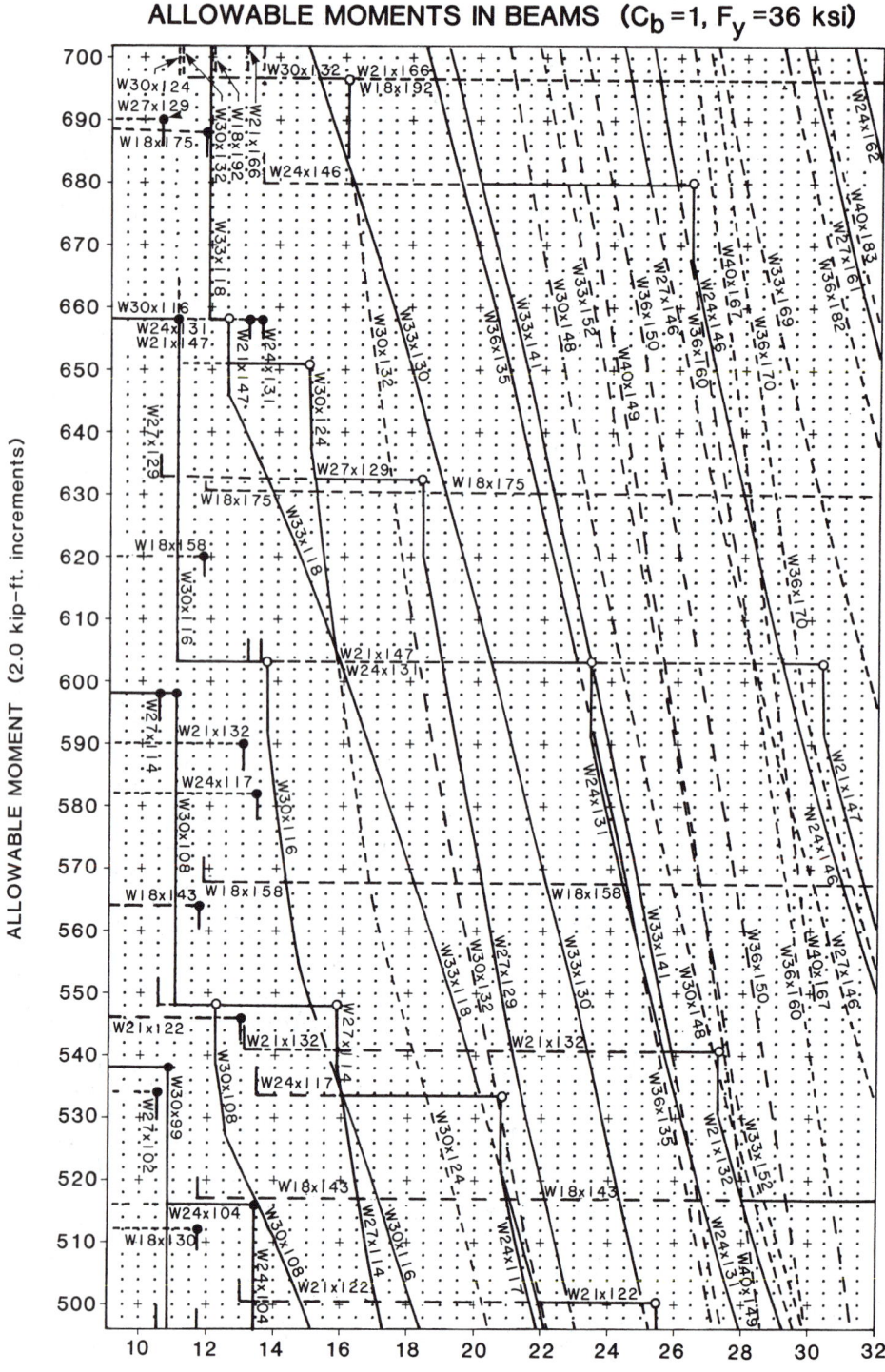

ALLOWABLE MOMENT (2.0 kip-ft. increments)

UNBRACED LENGTH (0.5 ft. increments)

ALLOWABLE MOMENTS IN BEAMS (C_b =1, F_y =36 ksi)

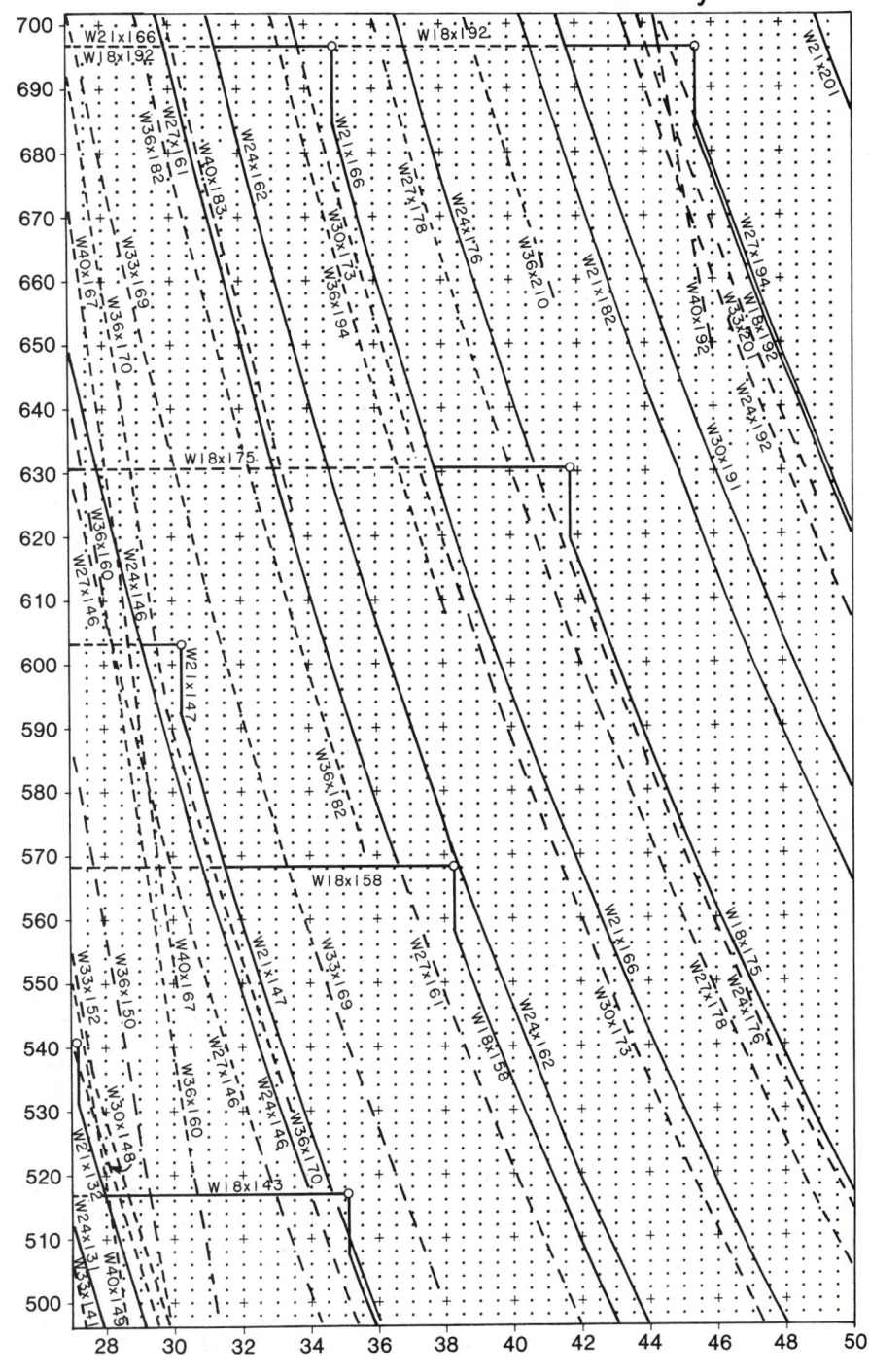

ALLOWABLE MOMENT (2.0 kip–ft. increments)

UNBRACED LENGTH (0.5 ft. increments)

ALLOWABLE MOMENTS IN BEAMS (C_b=1, F_y = 36 ksi)

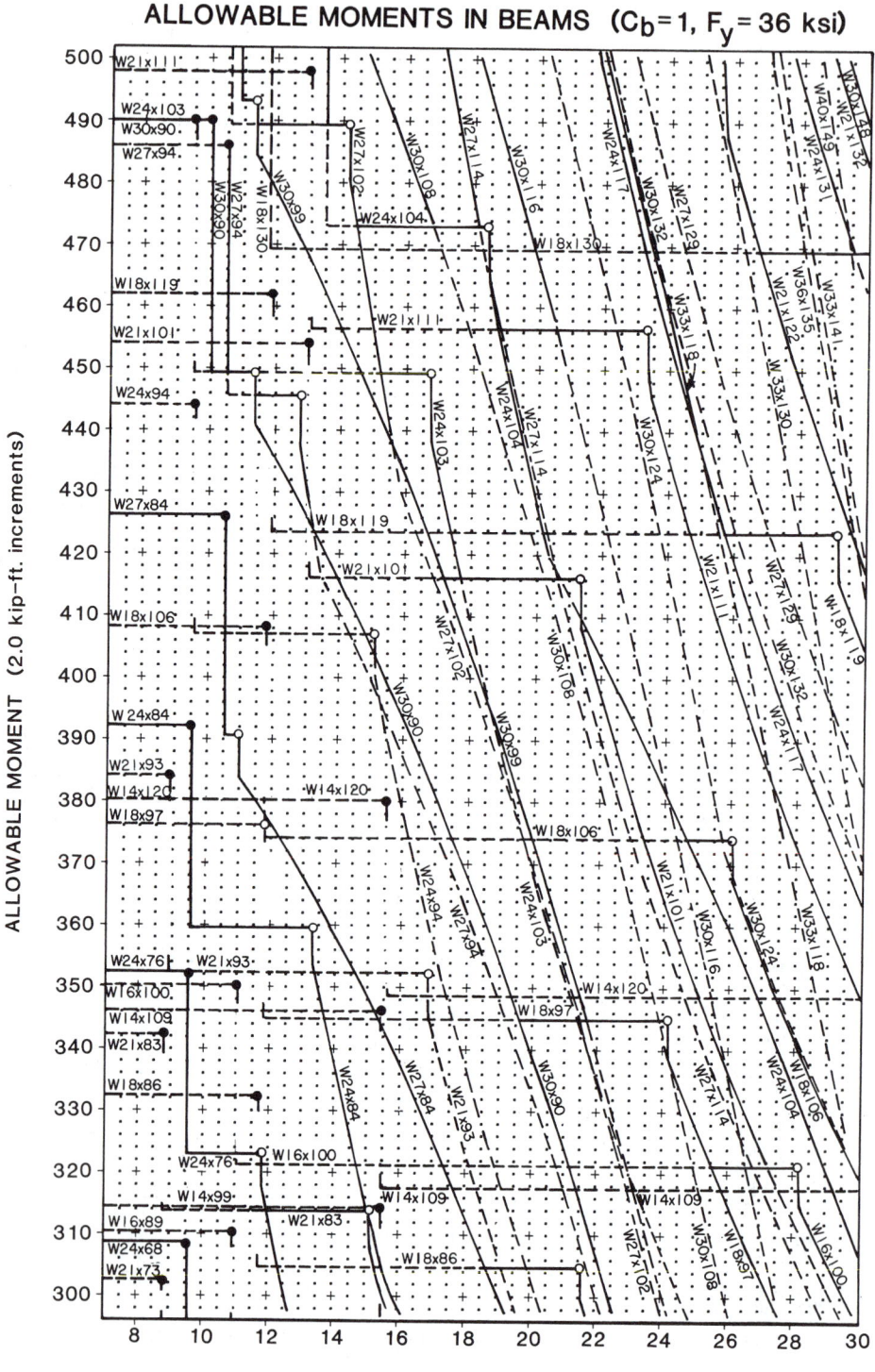

ALLOWABLE MOMENT (2.0 kip-ft. increments)

UNBRACED LENGTH (0.5 ft. increments)

ALLOWABLE MOMENTS IN BEAMS (C_b =1, F_y =36 ksi)

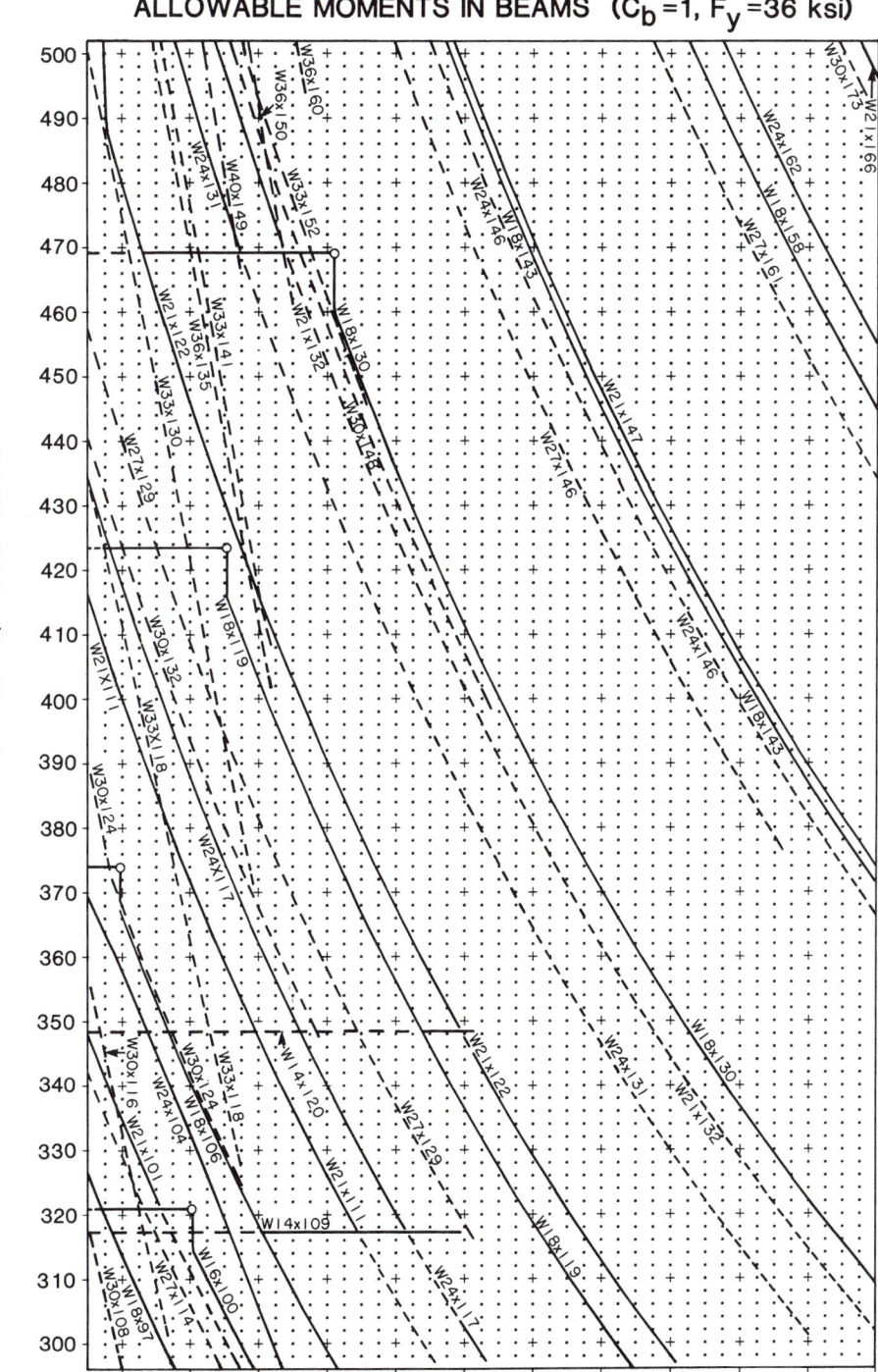

ALLOWABLE MOMENT (2.0 kip-ft. increments)

UNBRACED LENGTH (0.5 ft. increments)

ALLOWABLE MOMENTS IN BEAMS (C_b=1, F_y=36 ksi)

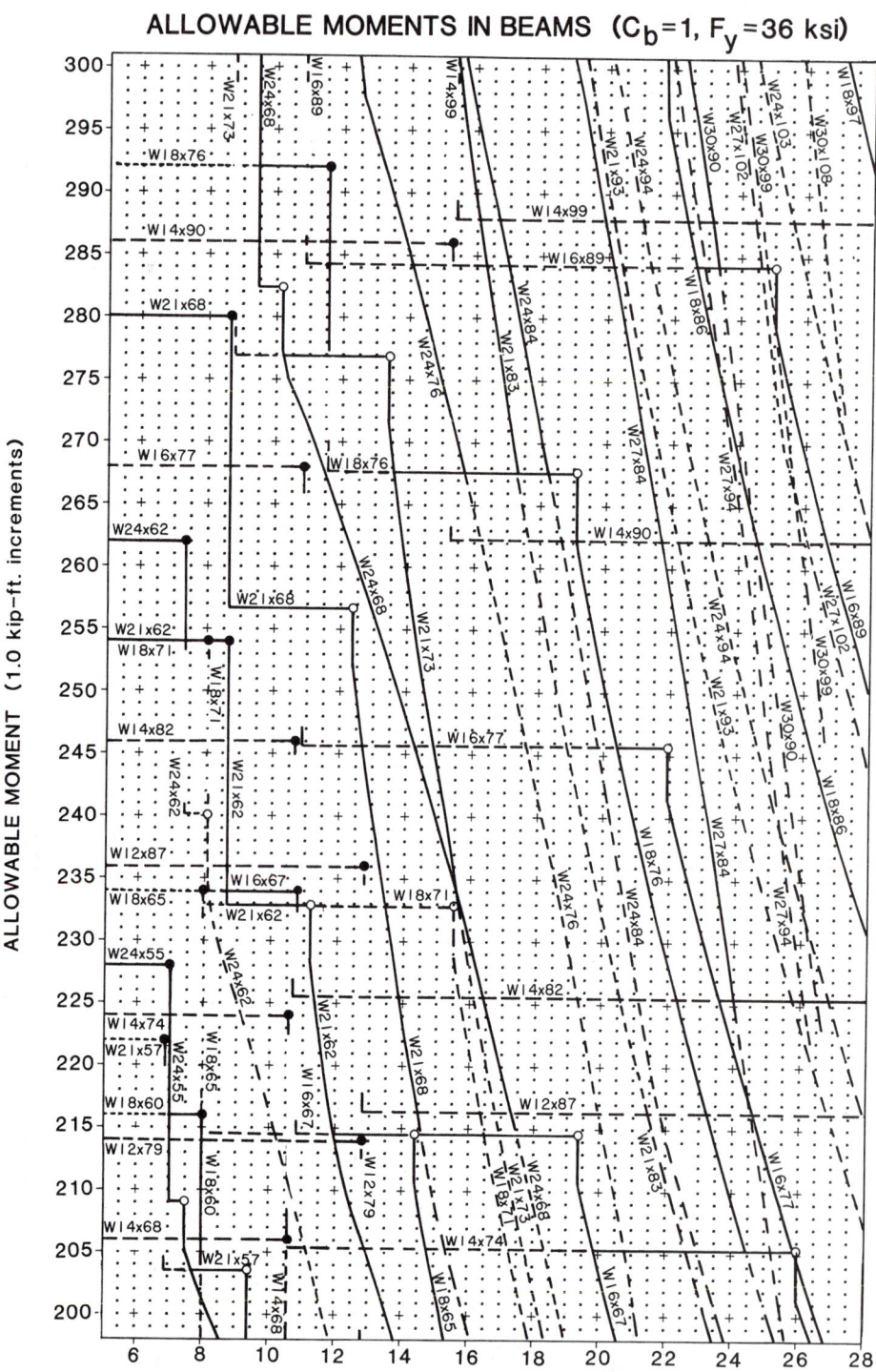

ALLOWABLE MOMENT (1.0 kip-ft. increments)

UNBRACED LENGTH (0.5 ft. increments)

ALLOWABLE MOMENTS IN BEAMS ($C_b = 1$, $F_y = 36$ ksi)

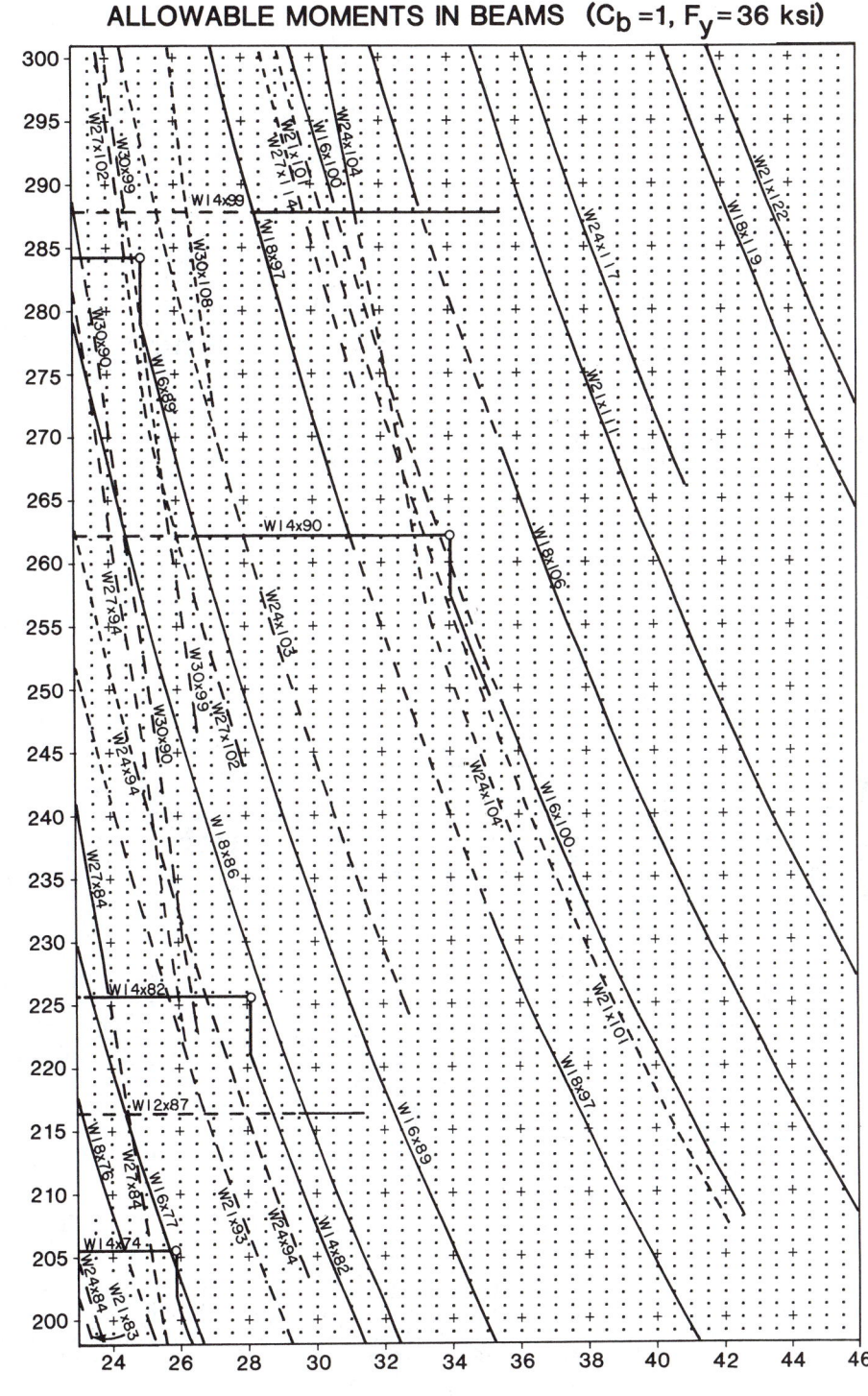

ALLOWABLE MOMENT (1.0 kip-ft. increments)

UNBRACED LENGTH (0.5 ft. increments)

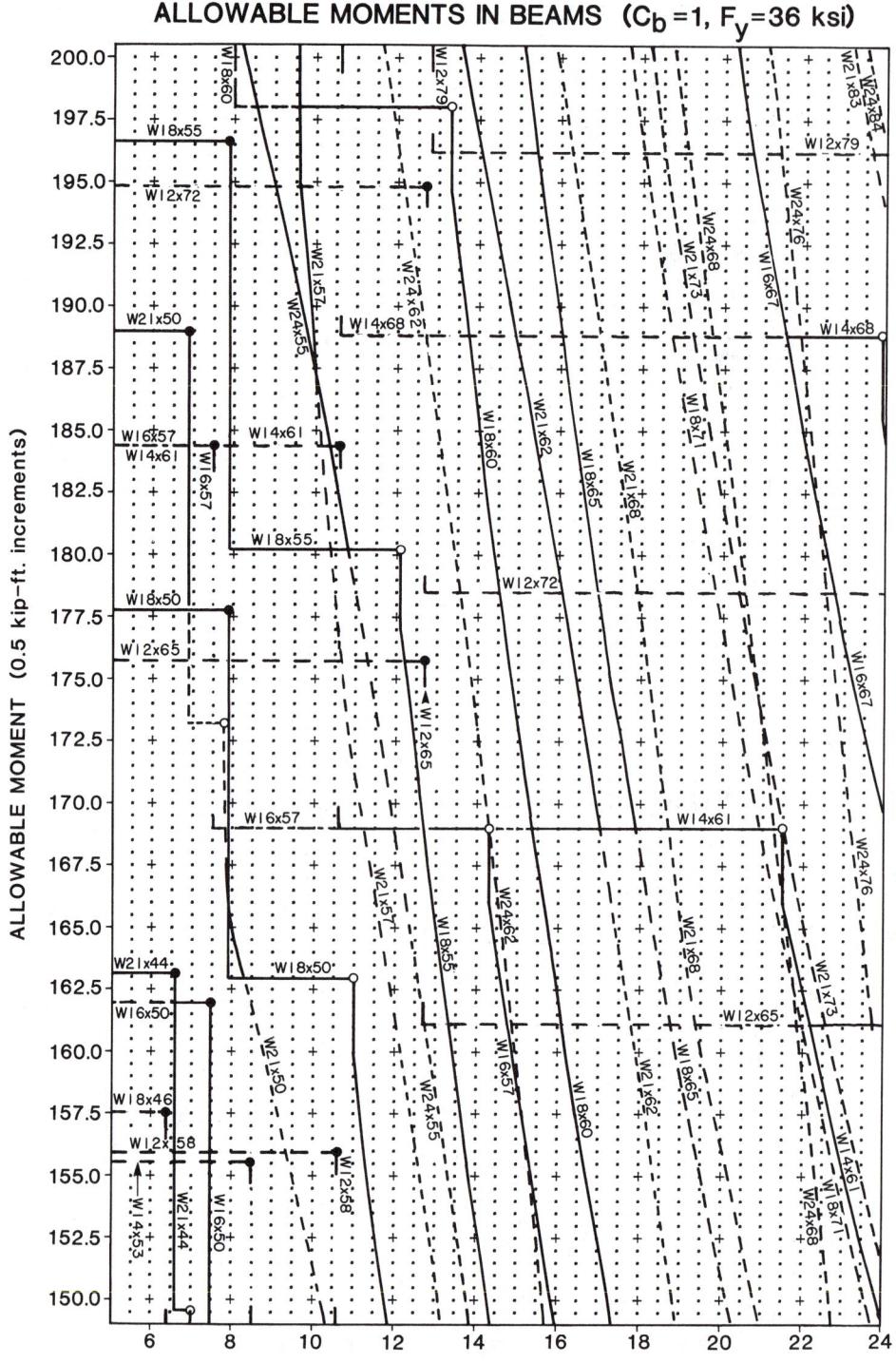

ALLOWABLE MOMENTS IN BEAMS $(C_b = 1, F_y = 36 \text{ ksi})$

ALLOWABLE MOMENT (0.5 kip–ft. increments)

UNBRACED LENGTH (0.5 ft. increments)

ALLOWABLE MOMENTS IN BEAMS (C_b=1, F_y=36 ksi)

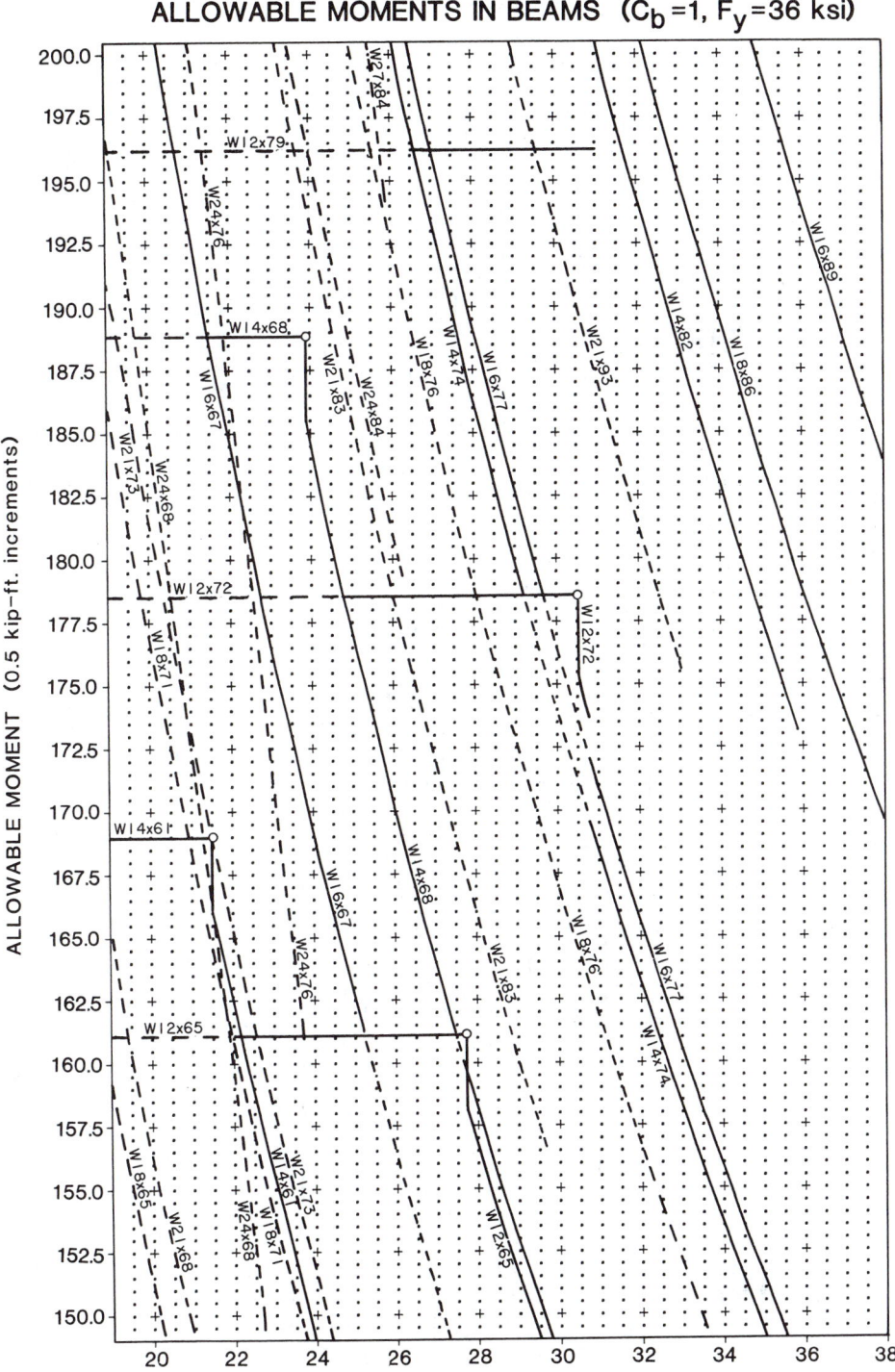

ALLOWABLE MOMENT (0.5 kip-ft. increments)

UNBRACED LENGTH (0.5 ft. increments)

ALLOWABLE MOMENTS IN BEAMS (C_b =1, F_y =36 ksi)

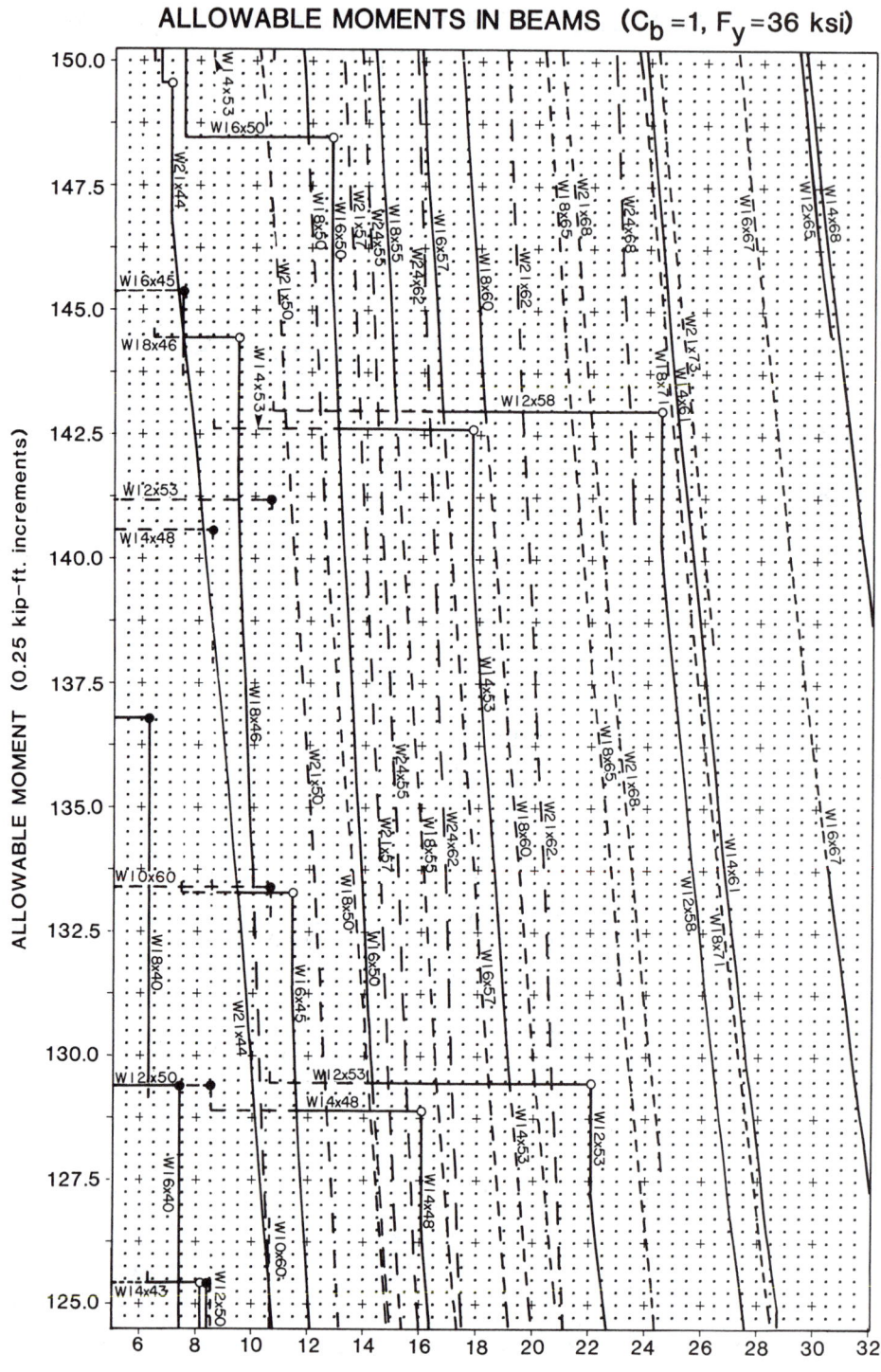

ALLOWABLE MOMENT (0.25 kip-ft. increments)

UNBRACED LENGTH (0.5 ft. increments)

ALLOWABLE MOMENTS IN BEAMS ($C_b = 1$, $F_y = 36$ ksi)

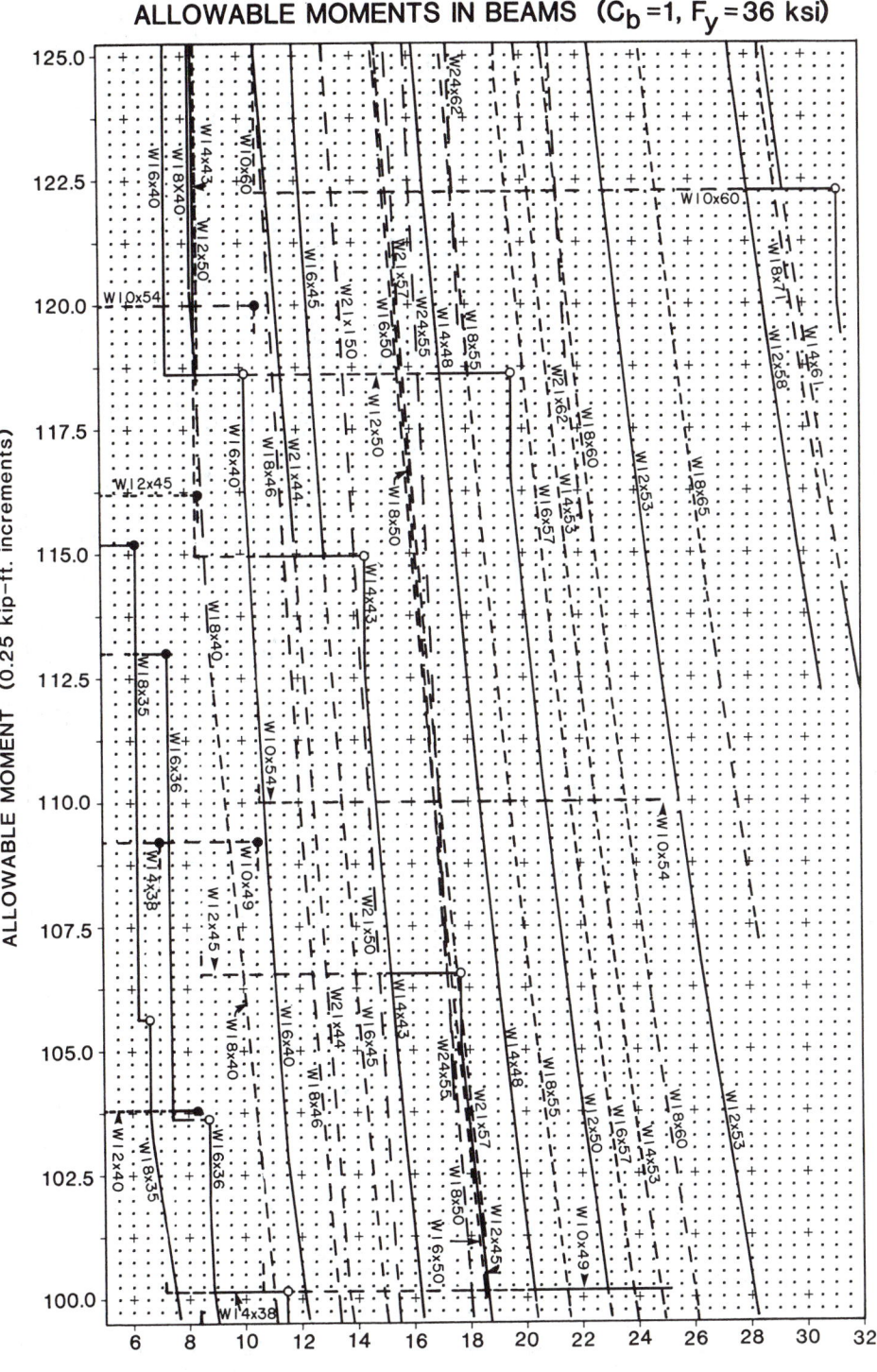

ALLOWABLE MOMENT (0.25 kip-ft. increments)

UNBRACED LENGTH (0.5 ft. increments)

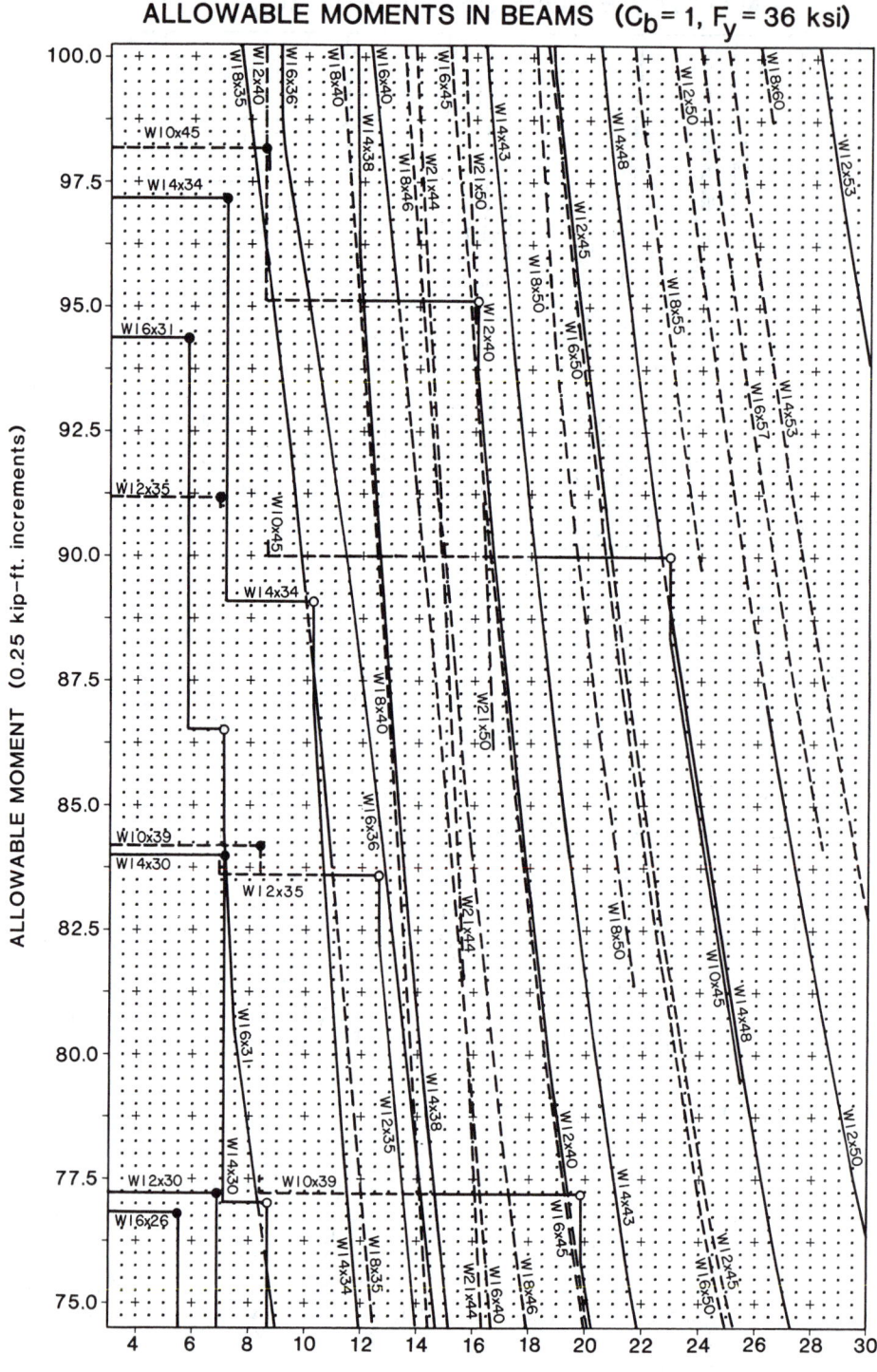

ALLOWABLE MOMENTS IN BEAMS (C_b = 1, F_y = 36 ksi)

ALLOWABLE MOMENT (0.25 kip-ft. increments)

UNBRACED LENGTH (0.5 ft. increments)

AMERICAN INSTITUTE OF STEEL CONSTRUCTION

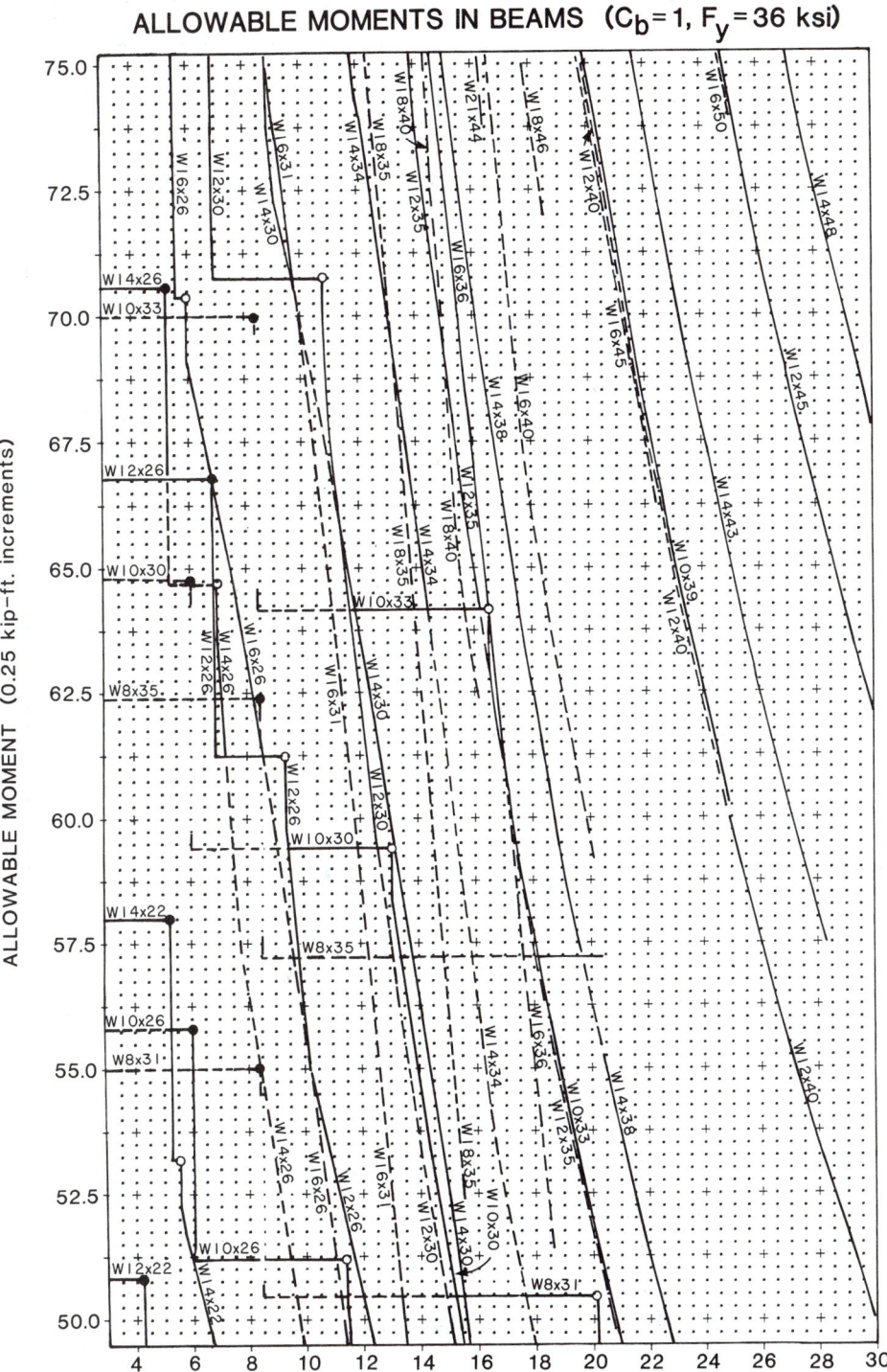

ALLOWABLE MOMENTS IN BEAMS ($C_b = 1$, $F_y = 36$ ksi)

ALLOWABLE MOMENT (0.25 kip-ft. increments)

UNBRACED LENGTH (0.5 ft. increments)

AMERICAN INSTITUTE OF STEEL CONSTRUCTION

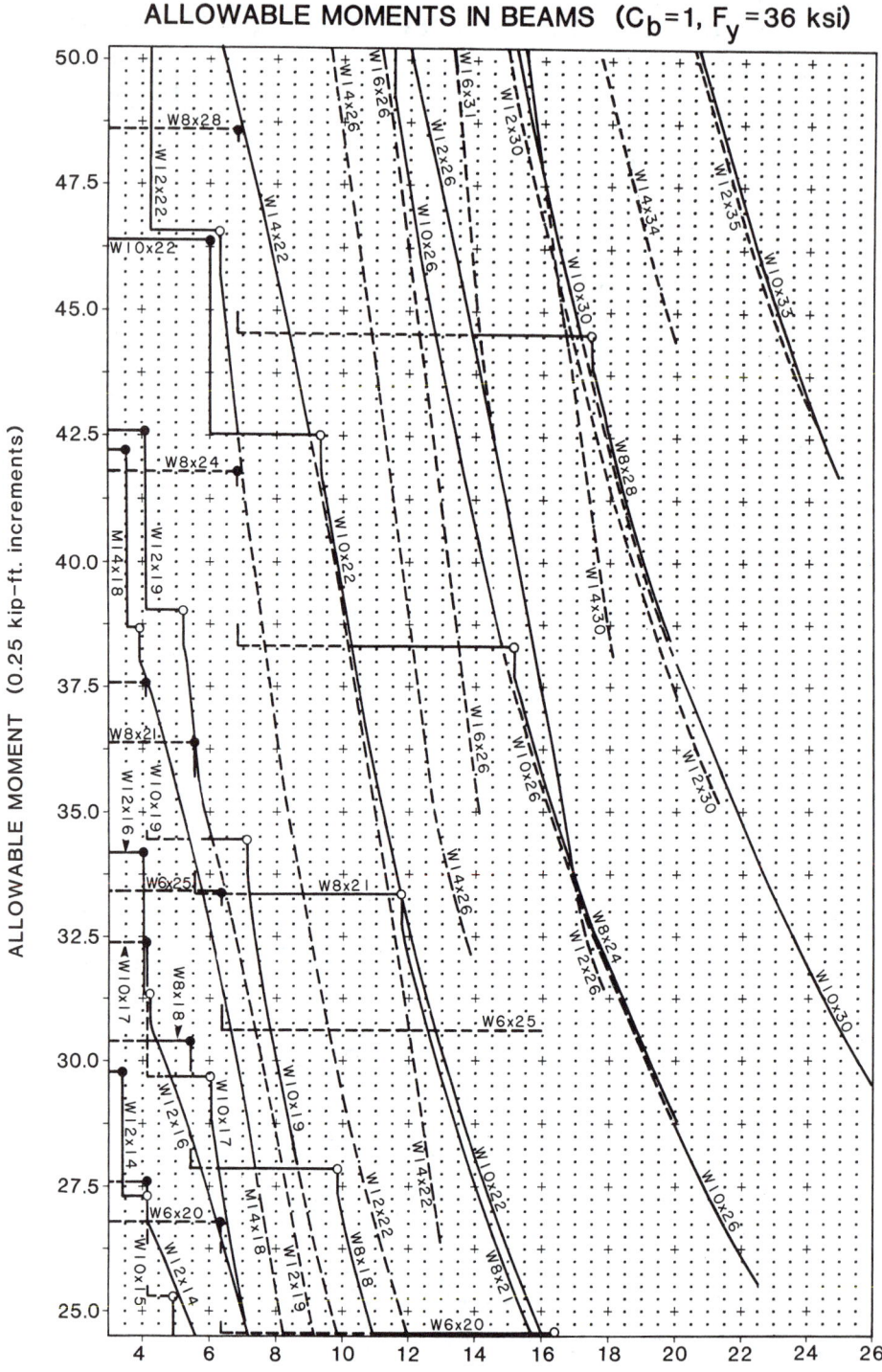

ALLOWABLE MOMENTS IN BEAMS (C_b=1, F_y=36 ksi)

ALLOWABLE MOMENT (0.25 kip-ft. increments)

UNBRACED LENGTH (0.5 ft. increments)

AMERICAN INSTITUTE OF STEEL CONSTRUCTION

ALLOWABLE MOMENTS IN BEAMS $(C_b = 1, F_y = 36 \text{ ksi})$

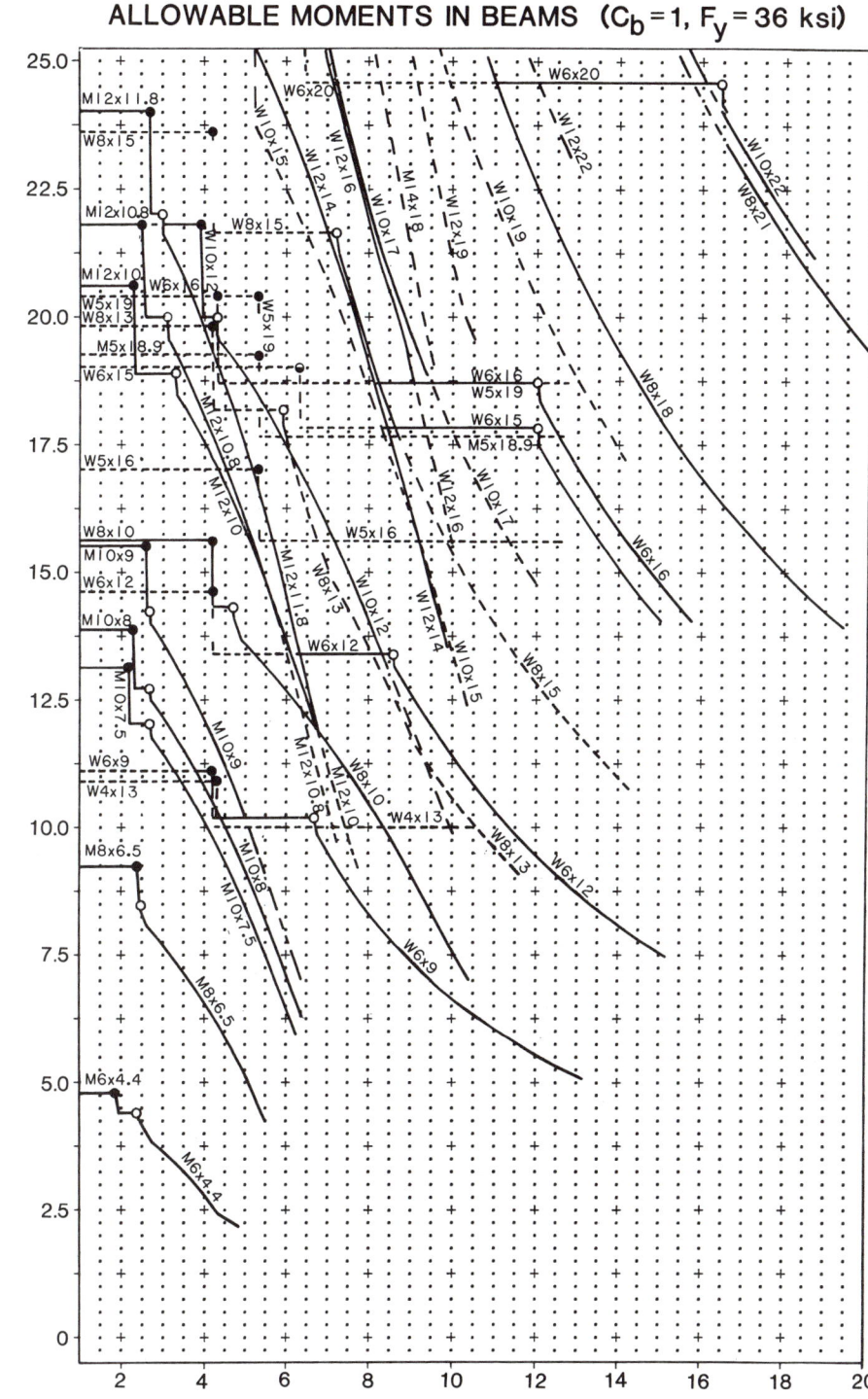

ALLOWABLE MOMENT (0.25 kip-ft. increments)

UNBRACED LENGTH (0.5 ft. increments)

AMERICAN INSTITUTE OF STEEL CONSTRUCTION

ALLOWABLE MOMENTS IN BEAMS (C_b =1, F_y =50 ksi)

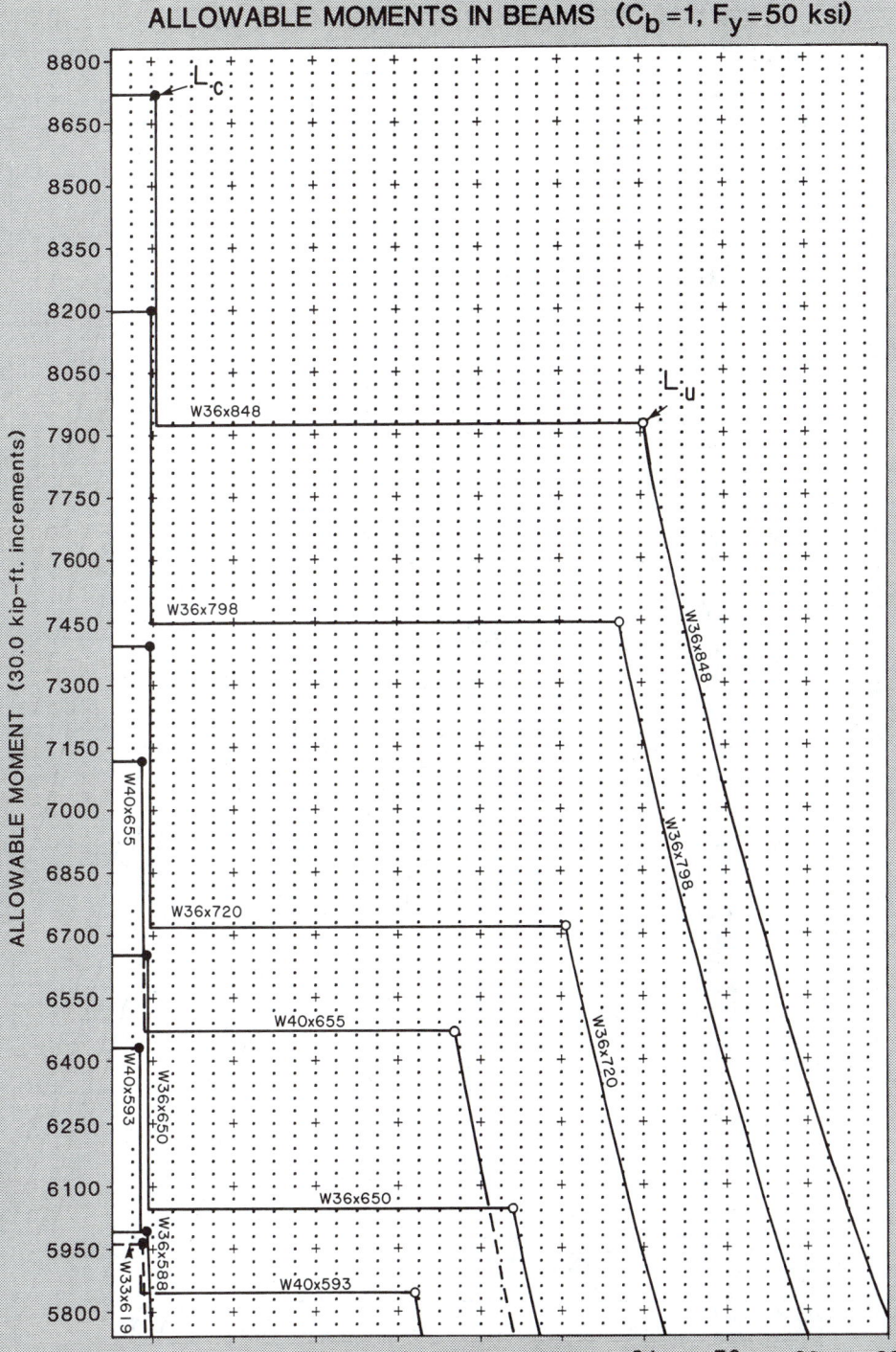

ALLOWABLE MOMENT (30.0 kip-ft. increments)

UNBRACED LENGTH (2.0 ft. increments)

ALLOWABLE MOMENTS IN BEAMS (C_b=1, F_y=50 ksi)

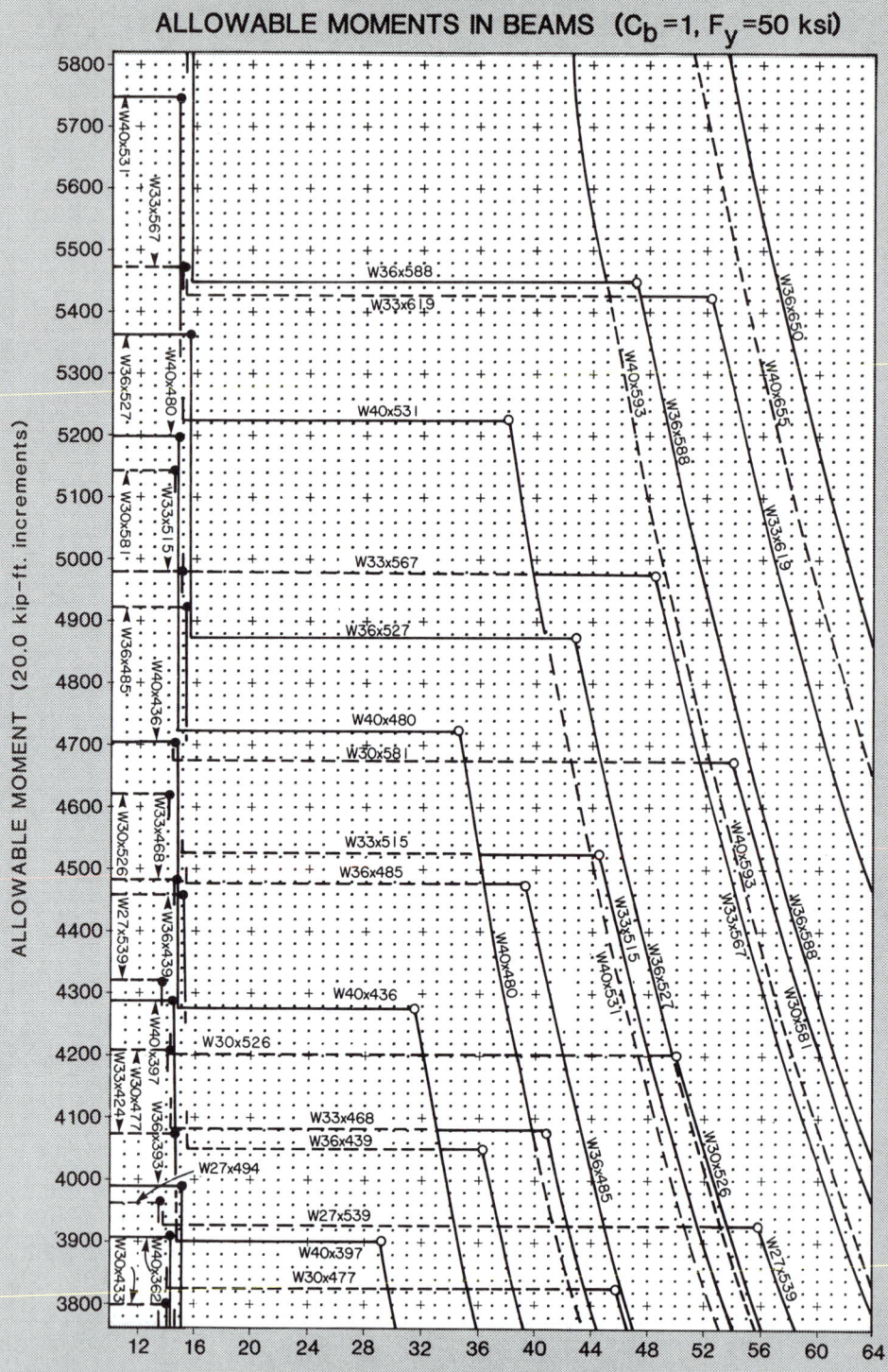

ALLOWABLE MOMENT (20.0 kip-ft. increments)

UNBRACED LENGTH (1.0 ft. increments)

ALLOWABLE MOMENTS IN BEAMS ($C_b = 1$, $F_y = 50$ ksi)

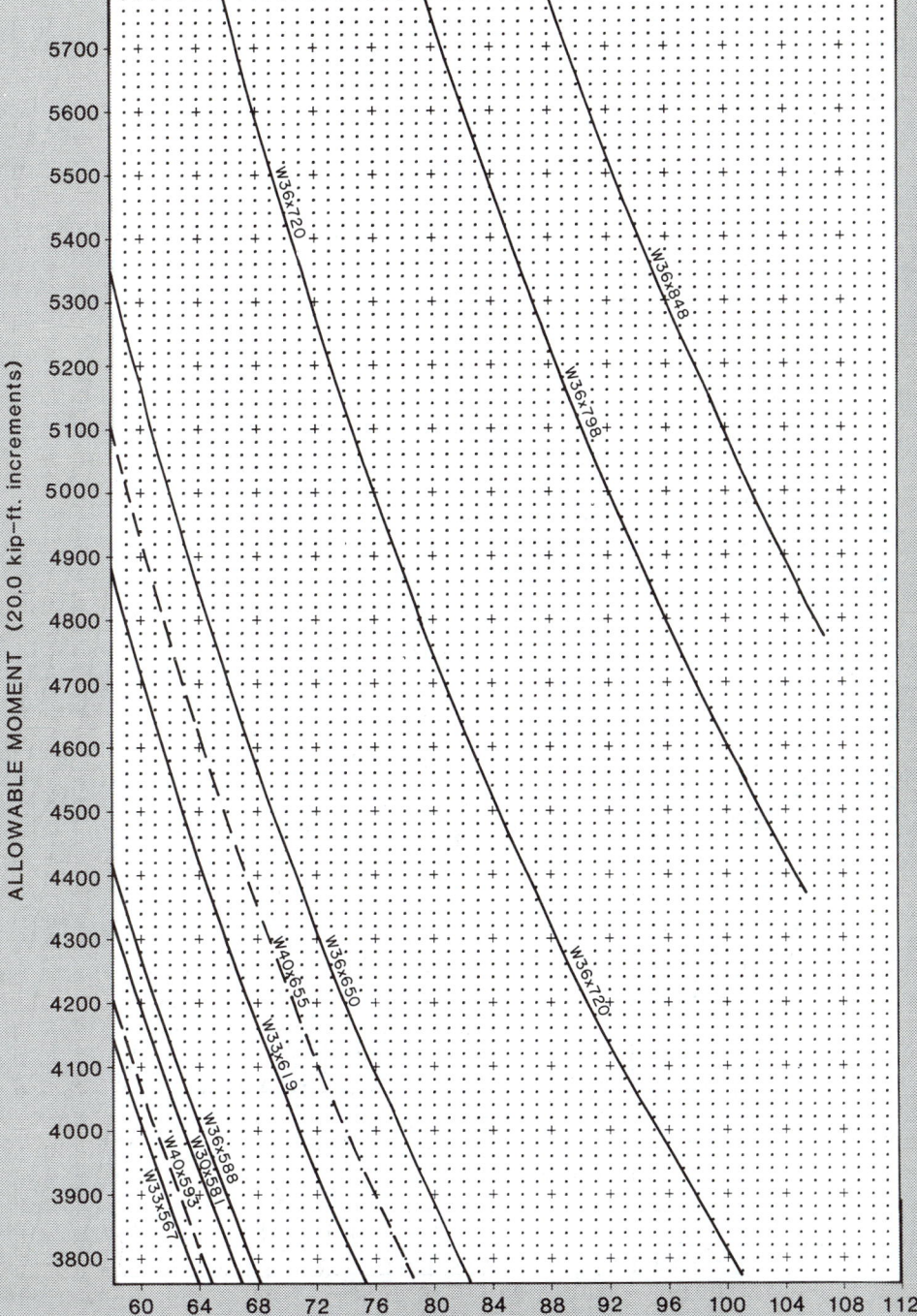

ALLOWABLE MOMENT (20.0 kip-ft. increments)

UNBRACED LENGTH (1.0 ft. increments)

AMERICAN INSTITUTE OF STEEL CONSTRUCTION

ALLOWABLE MOMENTS IN BEAMS ($C_b = 1$, $F_y = 50$ ksi)

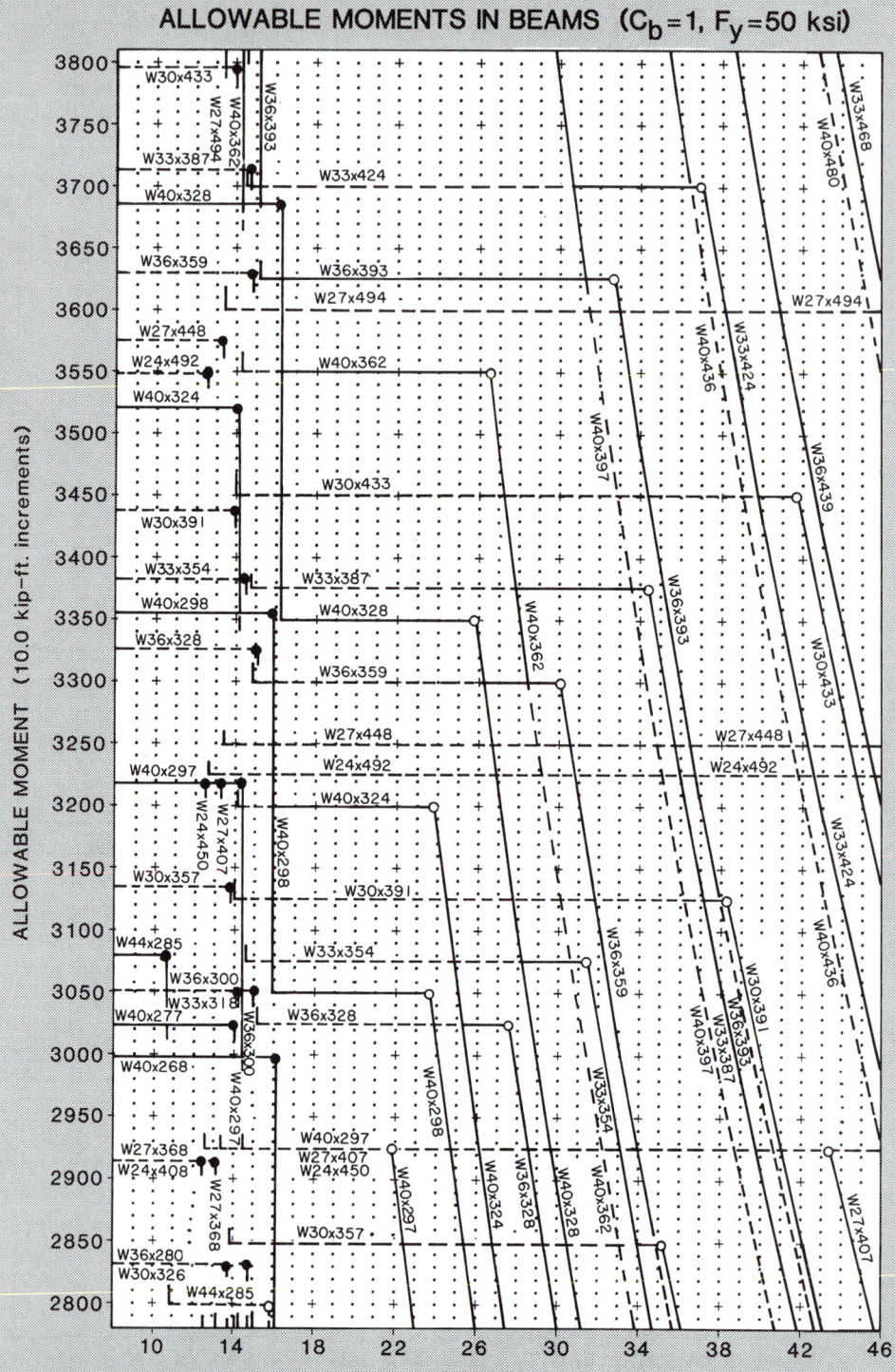

ALLOWABLE MOMENT (10.0 kip–ft. increments)

UNBRACED LENGTH (1.0 ft. increments)

AMERICAN INSTITUTE OF STEEL CONSTRUCTION

ALLOWABLE MOMENTS IN BEAMS (C_b=1, F_y=50 ksi)

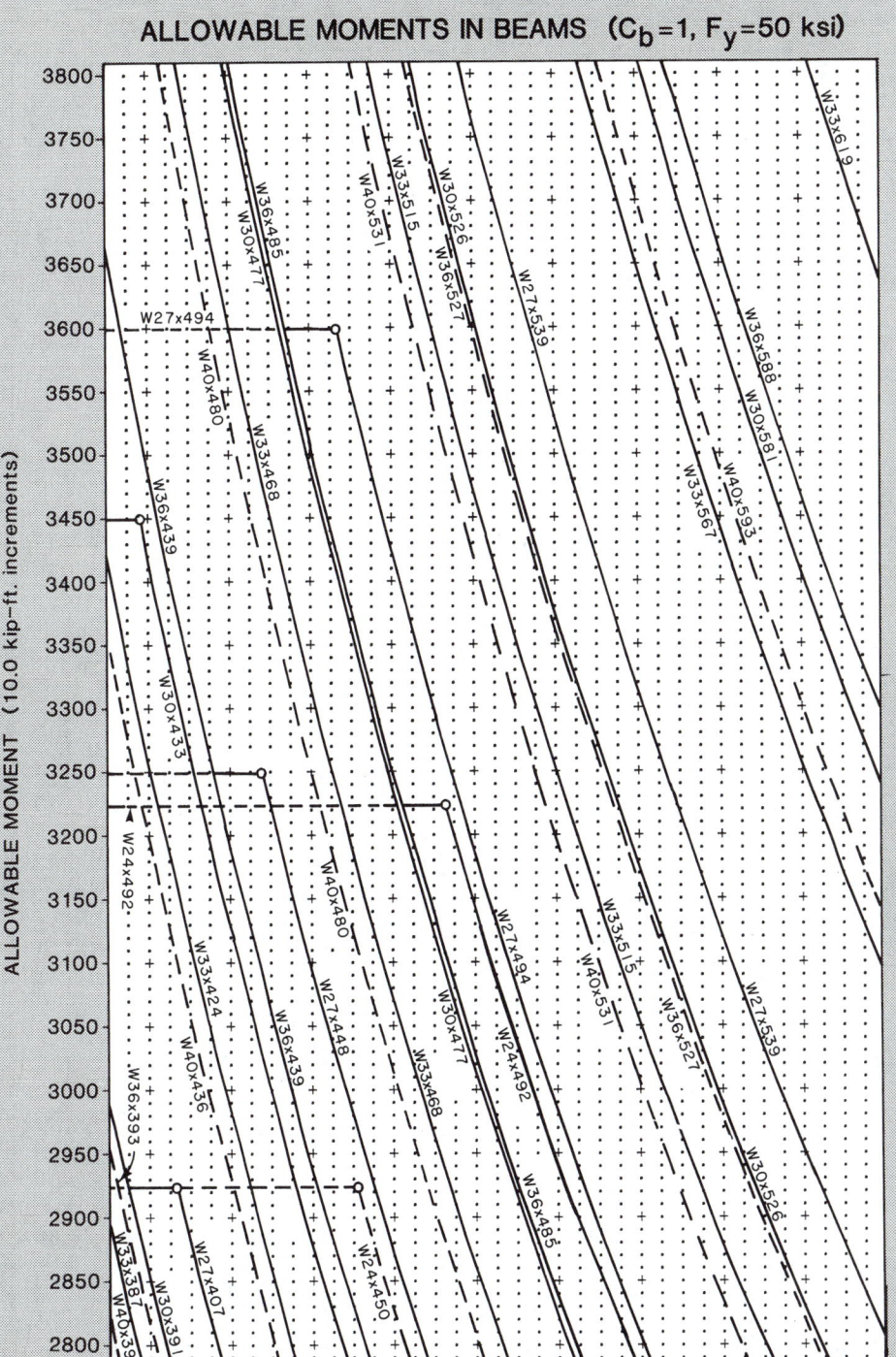

ALLOWABLE MOMENT (10.0 kip-ft. increments)

UNBRACED LENGTH (1.0 ft. increments)

ALLOWABLE MOMENTS IN BEAMS (C_b=1, F_y=50 ksi)

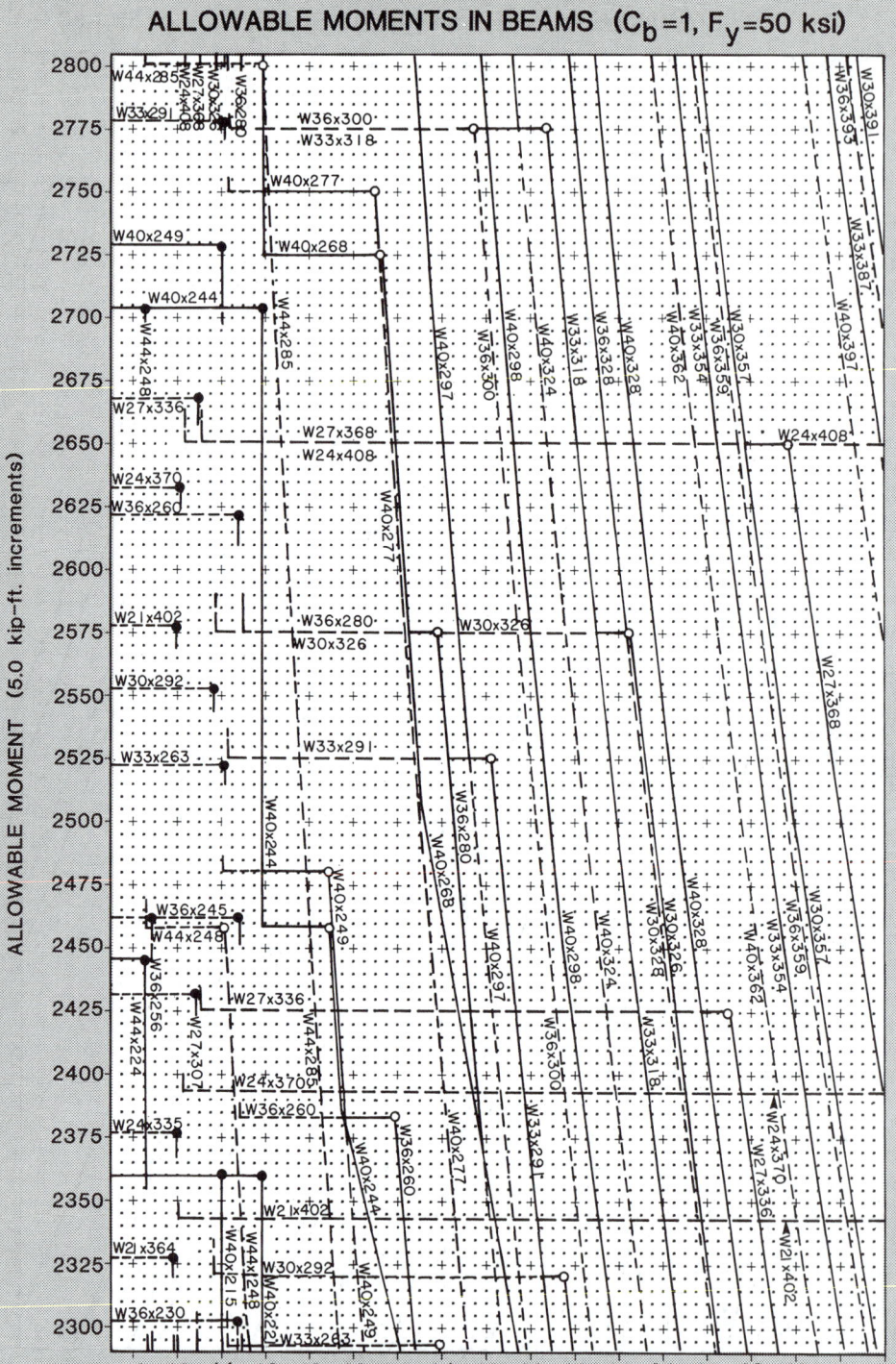

ALLOWABLE MOMENT (5.0 kip-ft. increments)

UNBRACED LENGTH (0.5 ft. increments)

AMERICAN INSTITUTE OF STEEL CONSTRUCTION

ALLOWABLE MOMENTS IN BEAMS (C_b=1, F_y=50 ksi)

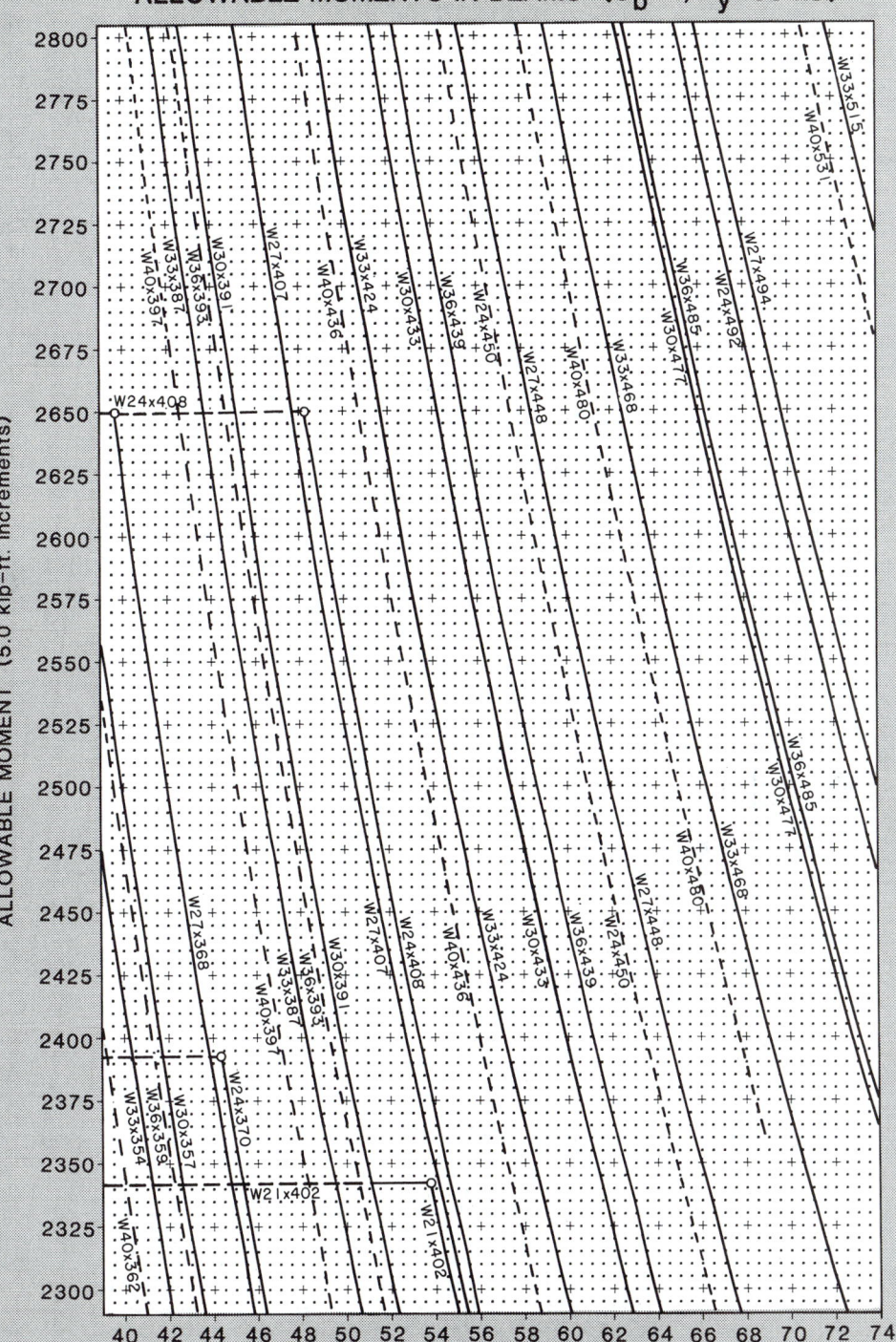

ALLOWABLE MOMENT (5.0 kip-ft. increments)

UNBRACED LENGTH (0.5 ft. increments)

AMERICAN INSTITUTE OF STEEL CONSTRUCTION

ALLOWABLE MOMENTS IN BEAMS (C_b =1, F_y =50 ksi)

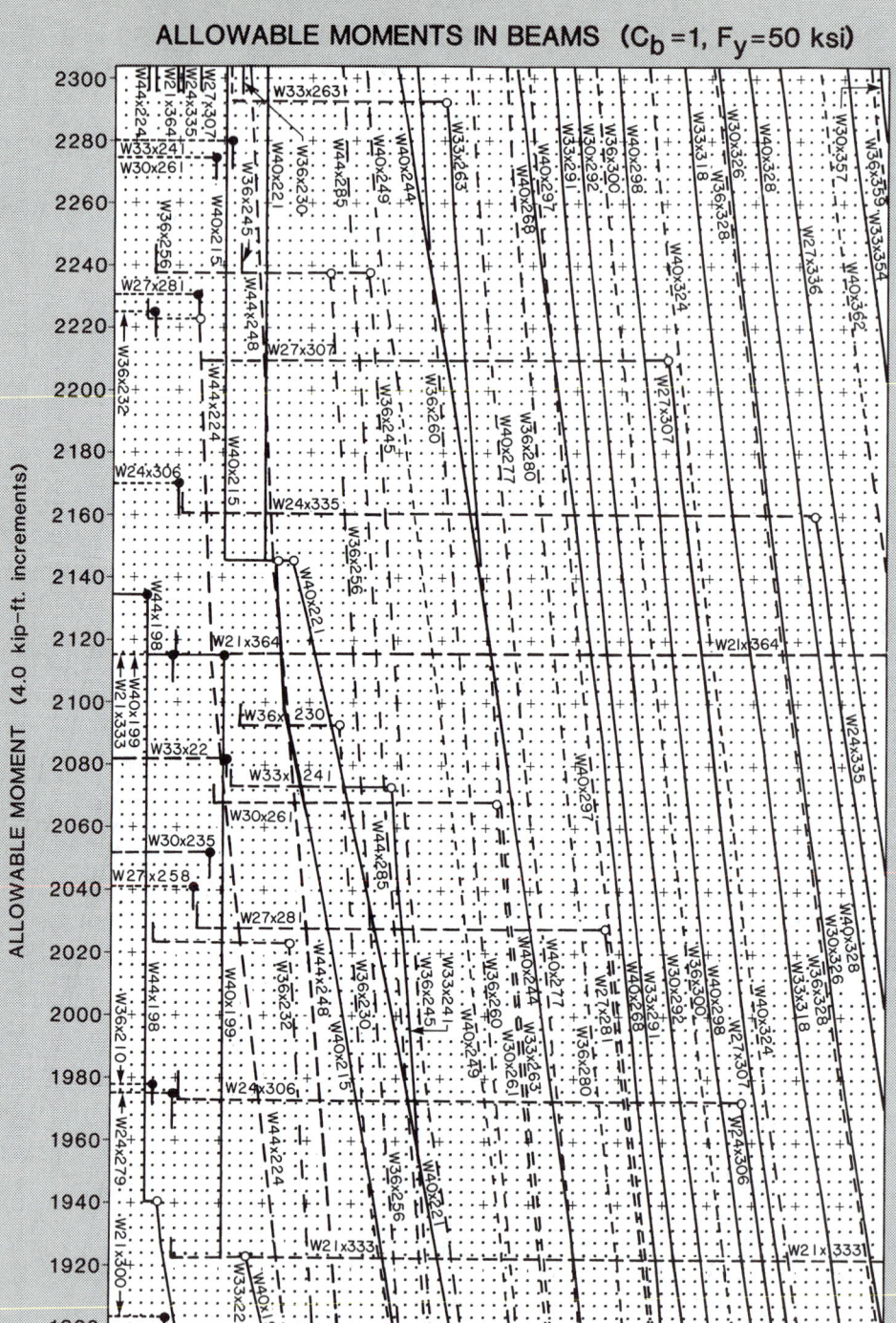

ALLOWABLE MOMENT (4.0 kip-ft. increments)

UNBRACED LENGTH (0.5 ft. increments)

AMERICAN INSTITUTE OF STEEL CONSTRUCTION

ALLOWABLE MOMENTS IN BEAMS (C_b =1, F_y =50 ksi)

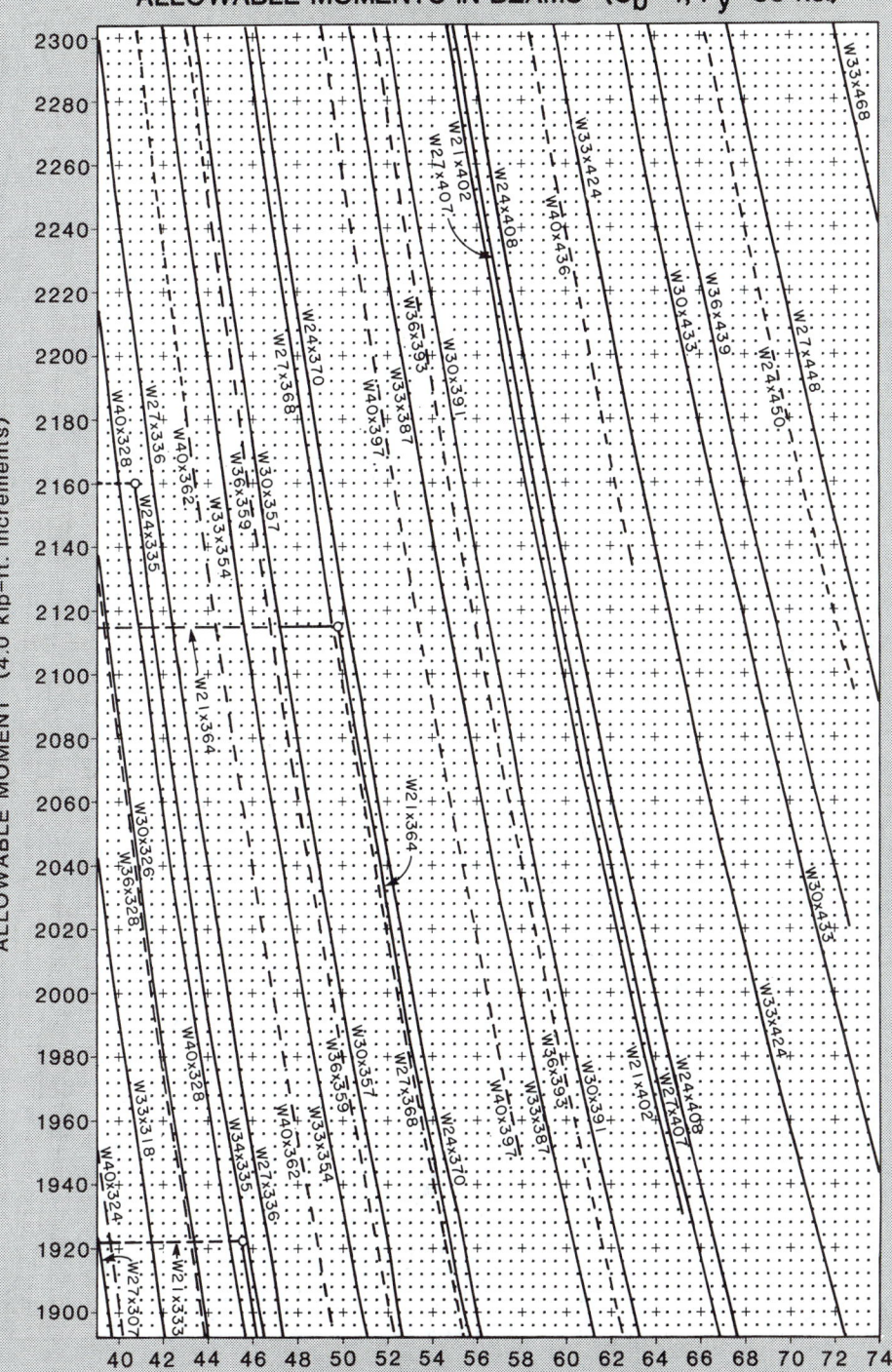

ALLOWABLE MOMENT (4.0 kip-ft. increments)

UNBRACED LENGTH (0.5 ft. increments)

ALLOWABLE MOMENTS IN BEAMS (C_b=1, F_y=50 ksi)

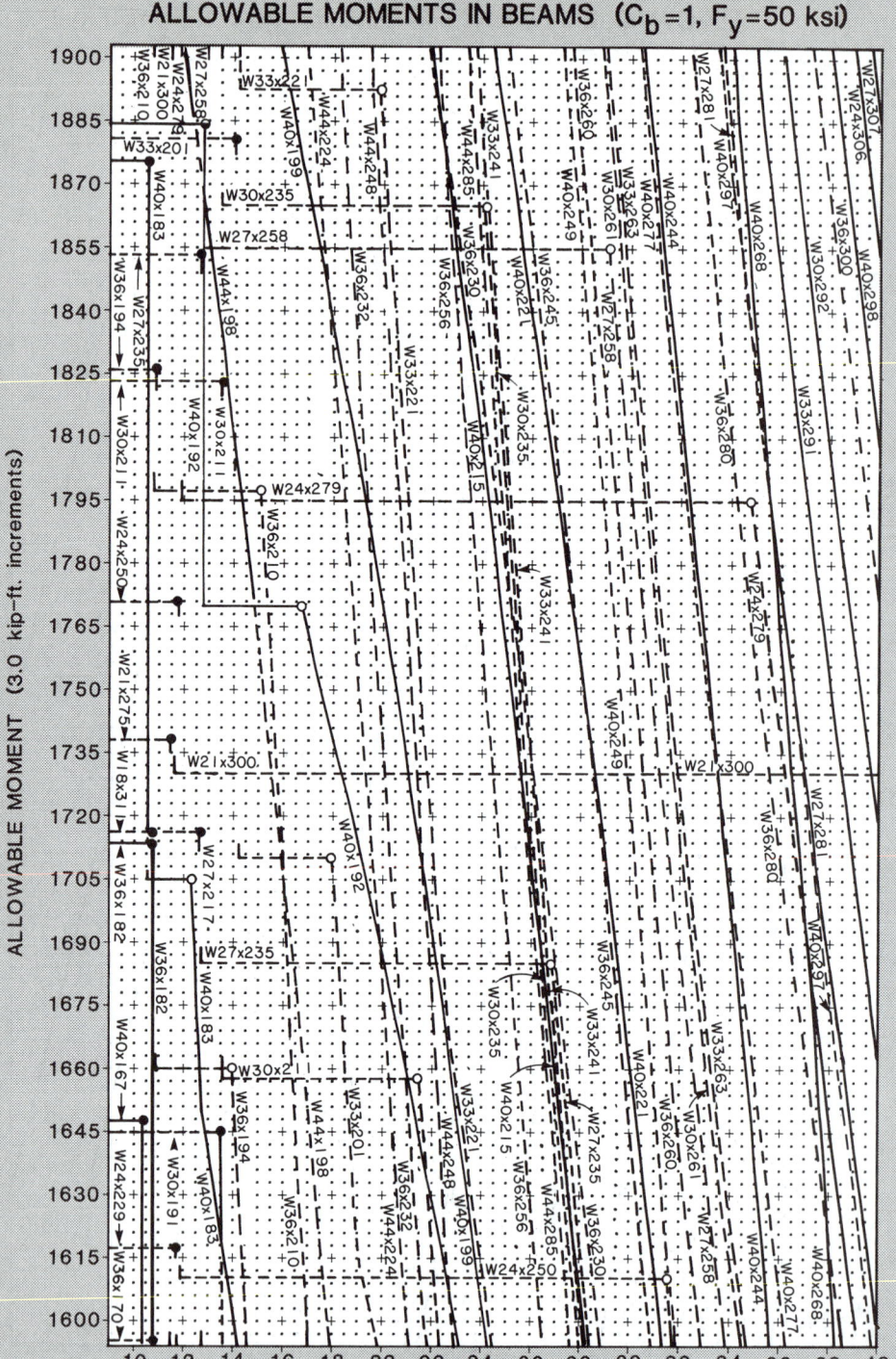

ALLOWABLE MOMENT (3.0 kip-ft. increments)

UNBRACED LENGTH (0.5 ft. increments)

AMERICAN INSTITUTE OF STEEL CONSTRUCTION

ALLOWABLE MOMENTS IN BEAMS (C_b =1, F_y=50 ksi)

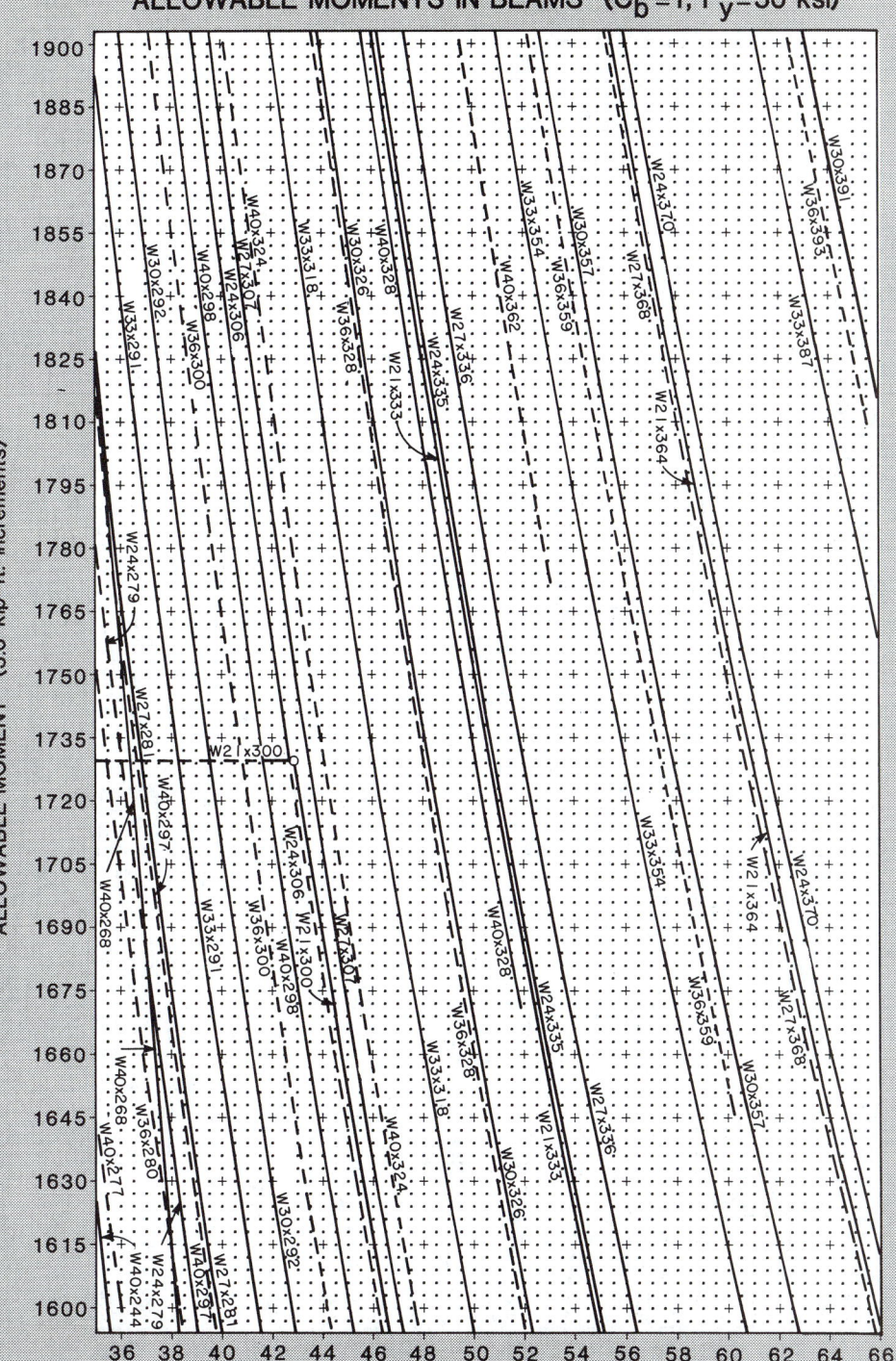

ALLOWABLE MOMENT (3.0 kip-ft. increments)

UNBRACED LENGTH (0.5 ft. increments)

AMERICAN INSTITUTE OF STEEL CONSTRUCTION

ALLOWABLE MOMENTS IN BEAMS ($C_b = 1$, $F_y = 50$ ksi)

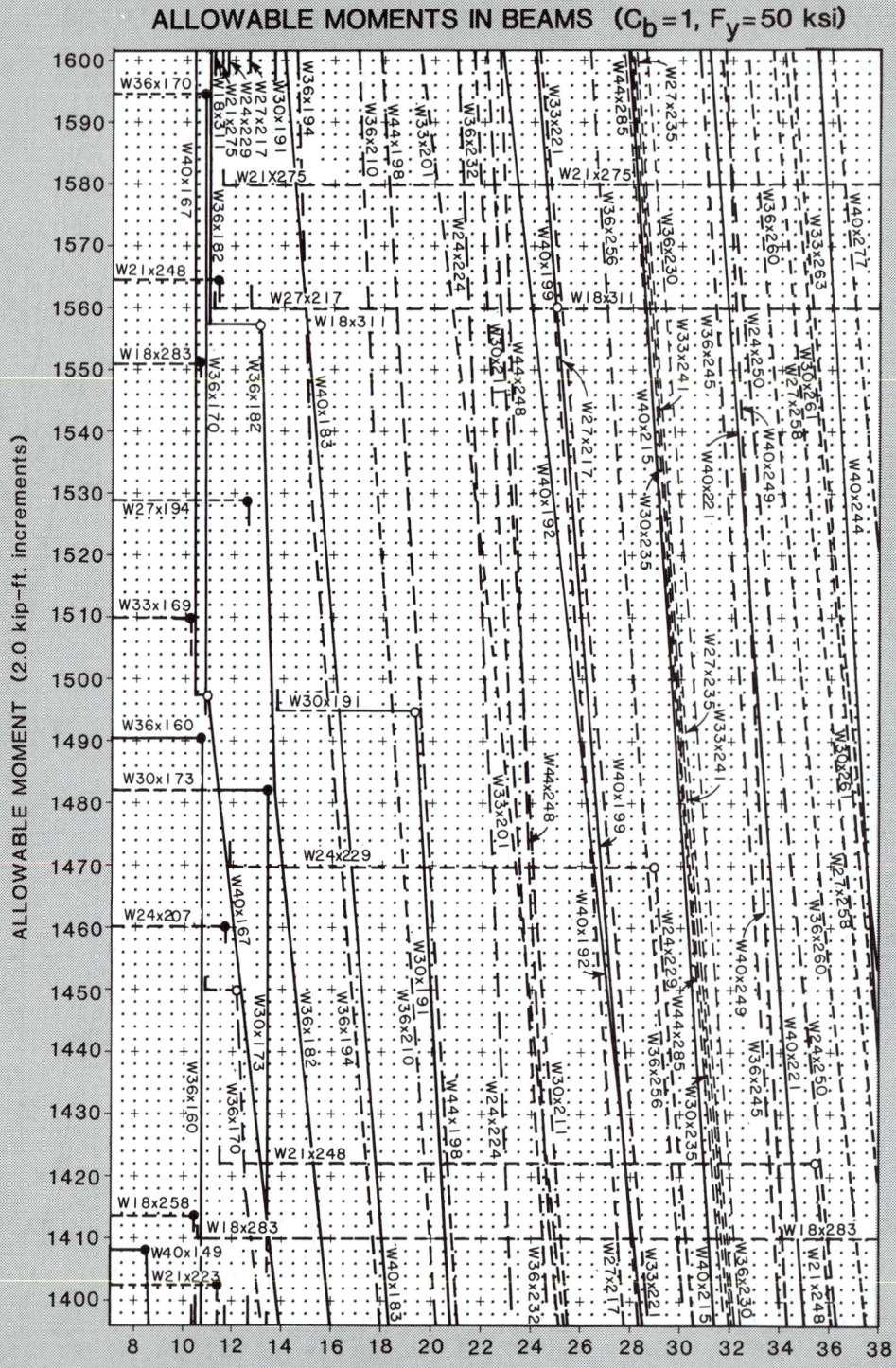

ALLOWABLE MOMENT (2.0 kip-ft. increments)

UNBRACED LENGTH (0.5 ft. increments)

ALLOWABLE MOMENTS IN BEAMS (C_b=1, F_y=50 ksi)

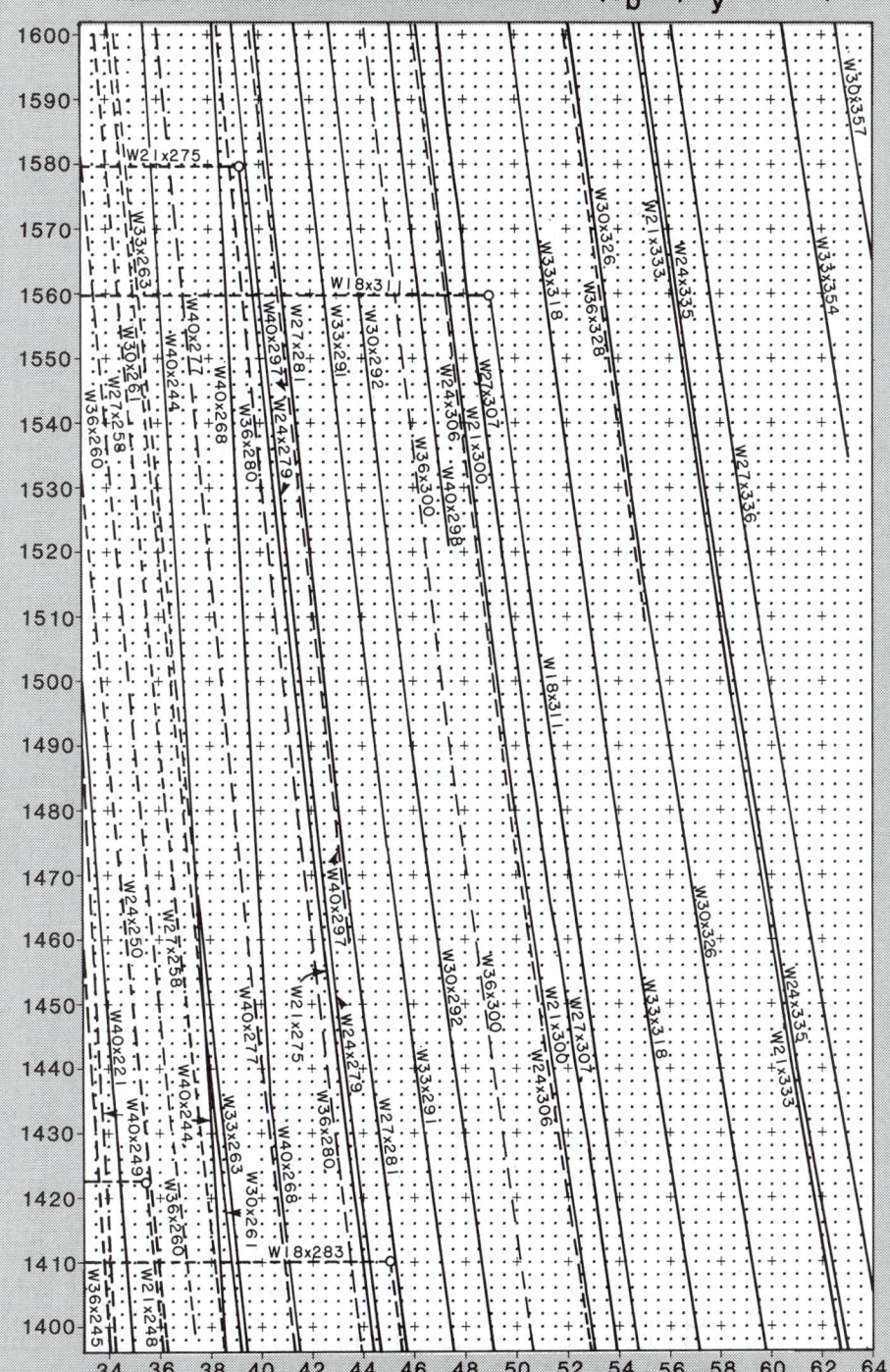

ALLOWABLE MOMENT (2.0 kip–ft. increments)

UNBRACED LENGTH (0.5 ft. increments)

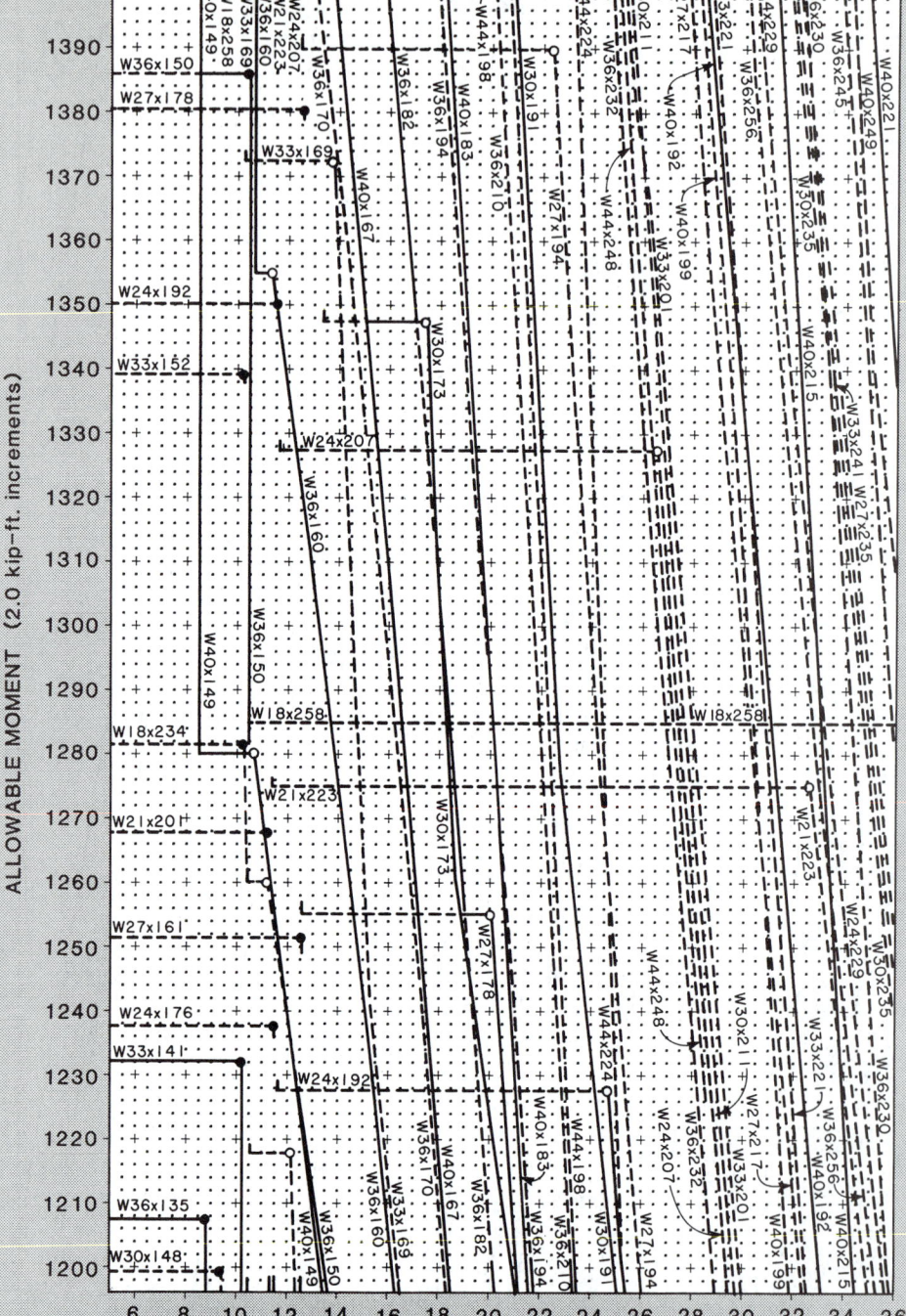

ALLOWABLE MOMENTS IN BEAMS ($C_b=1$, $F_y=50$ ksi)

ALLOWABLE MOMENT (2.0 kip-ft. increments)

UNBRACED LENGTH (0.5 ft. increments)

ALLOWABLE MOMENTS IN BEAMS (C_b=1, F_y=50 ksi)

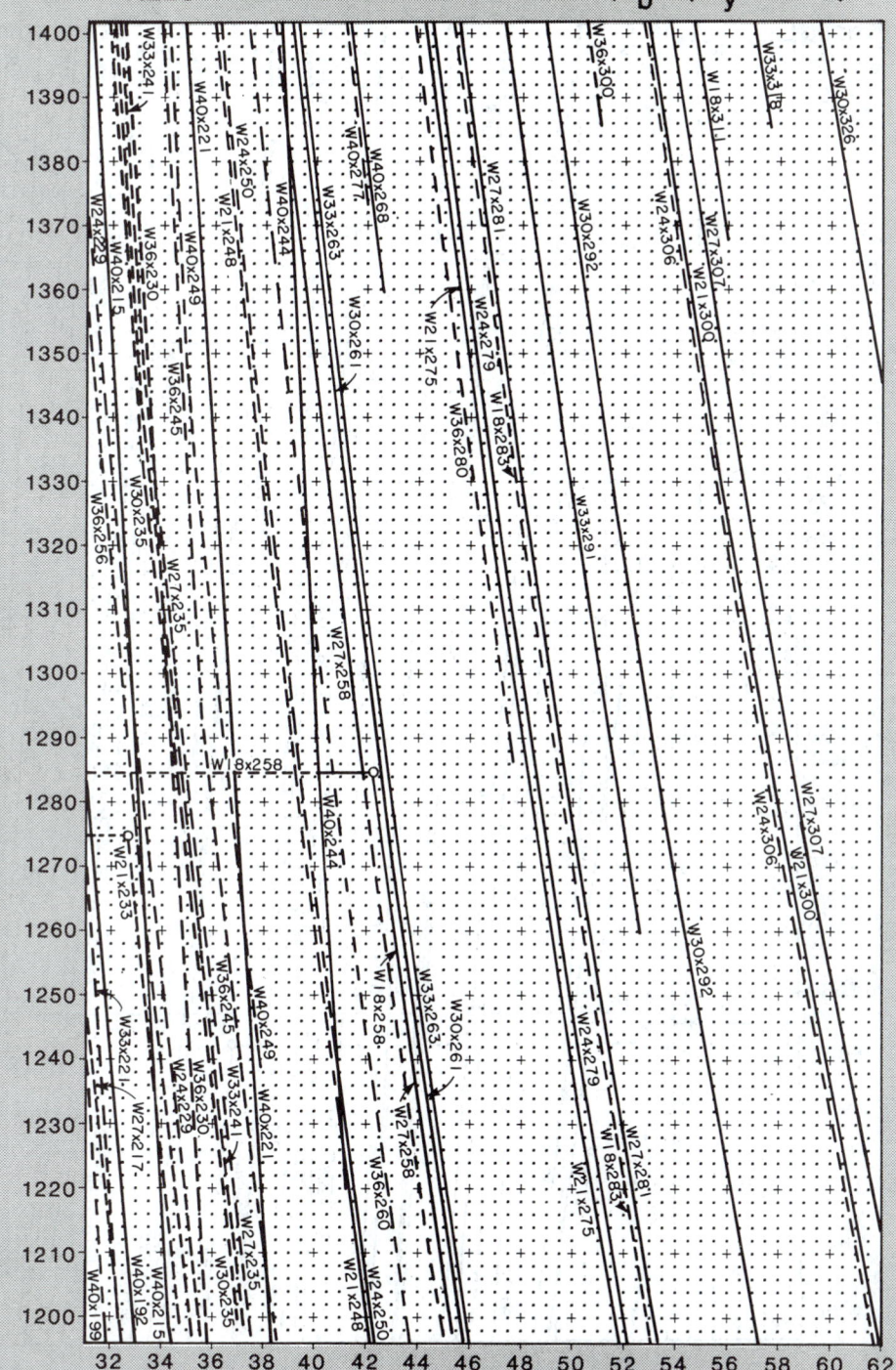

AMERICAN INSTITUTE OF STEEL CONSTRUCTION

ALLOWABLE MOMENTS IN BEAMS (C$_b$=1, F$_y$=50 ksi)

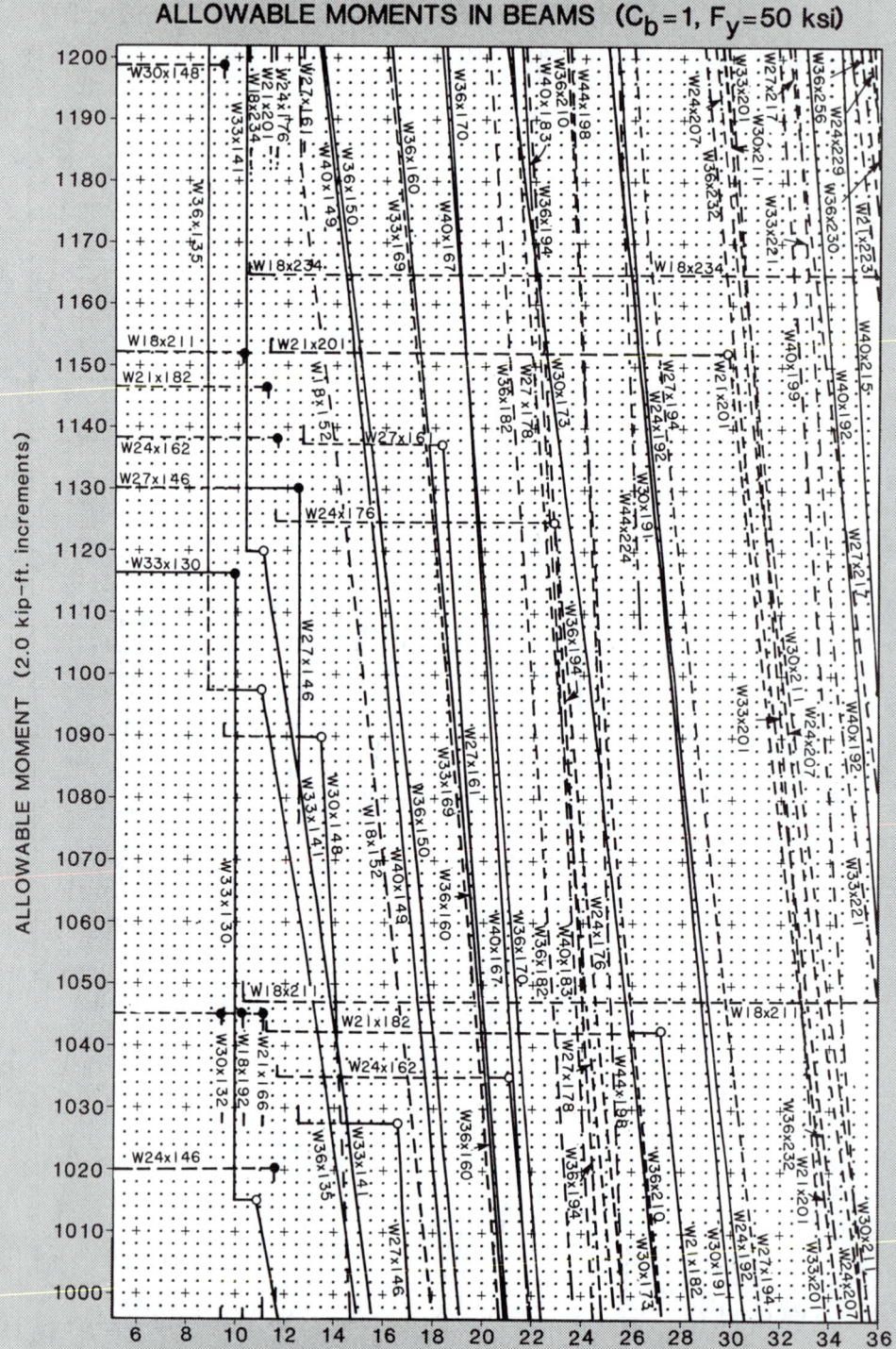

ALLOWABLE MOMENT (2.0 kip-ft. increments)

UNBRACED LENGTH (0.5 ft. increments)

ALLOWABLE MOMENTS IN BEAMS ($C_b = 1$, $F_y = 50$ ksi)

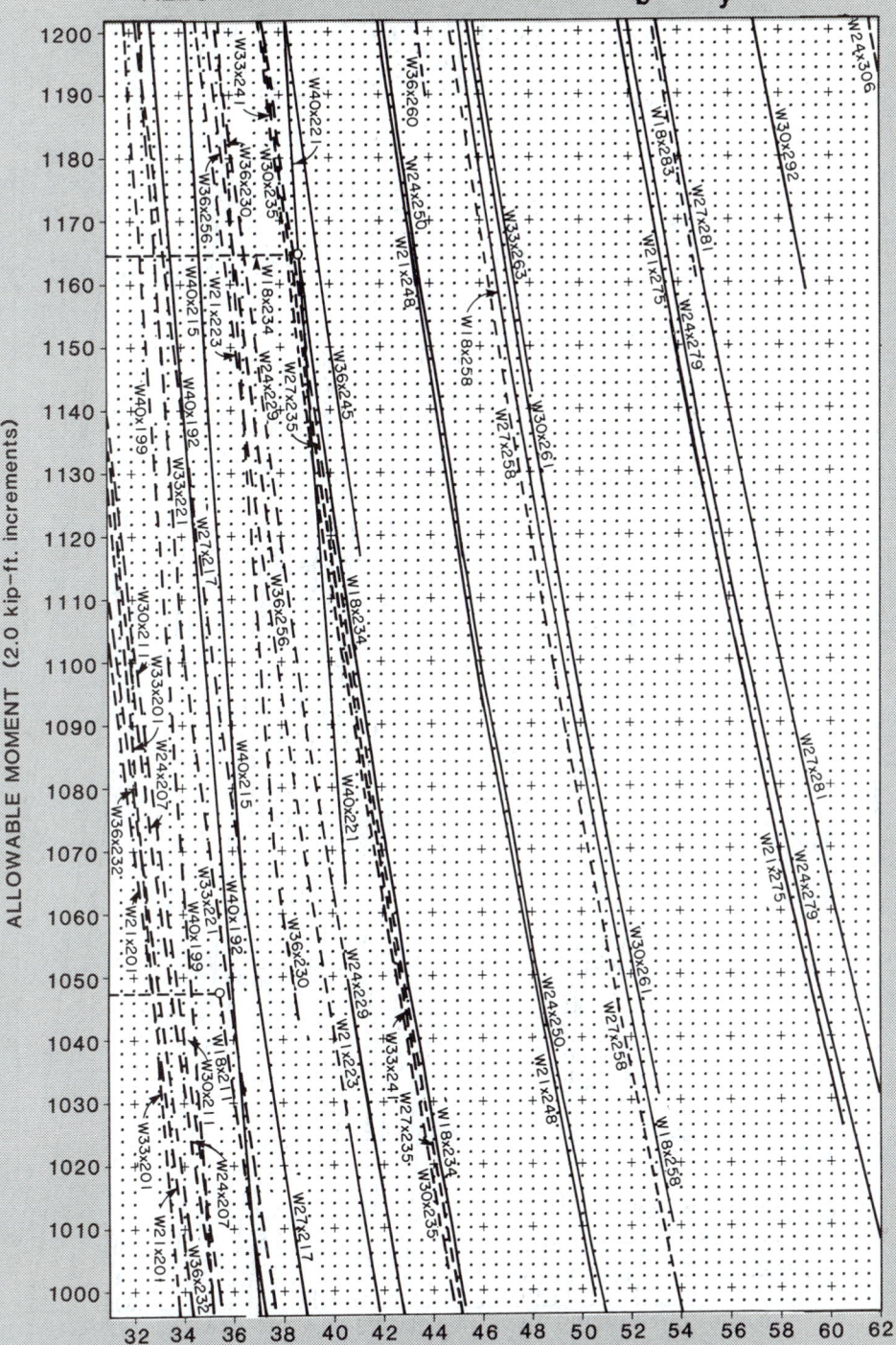

ALLOWABLE MOMENT (2.0 kip–ft. increments)

UNBRACED LENGTH (0.5 ft. increments)

ALLOWABLE MOMENTS IN BEAMS (C_b=1, F_y=50 ksi)

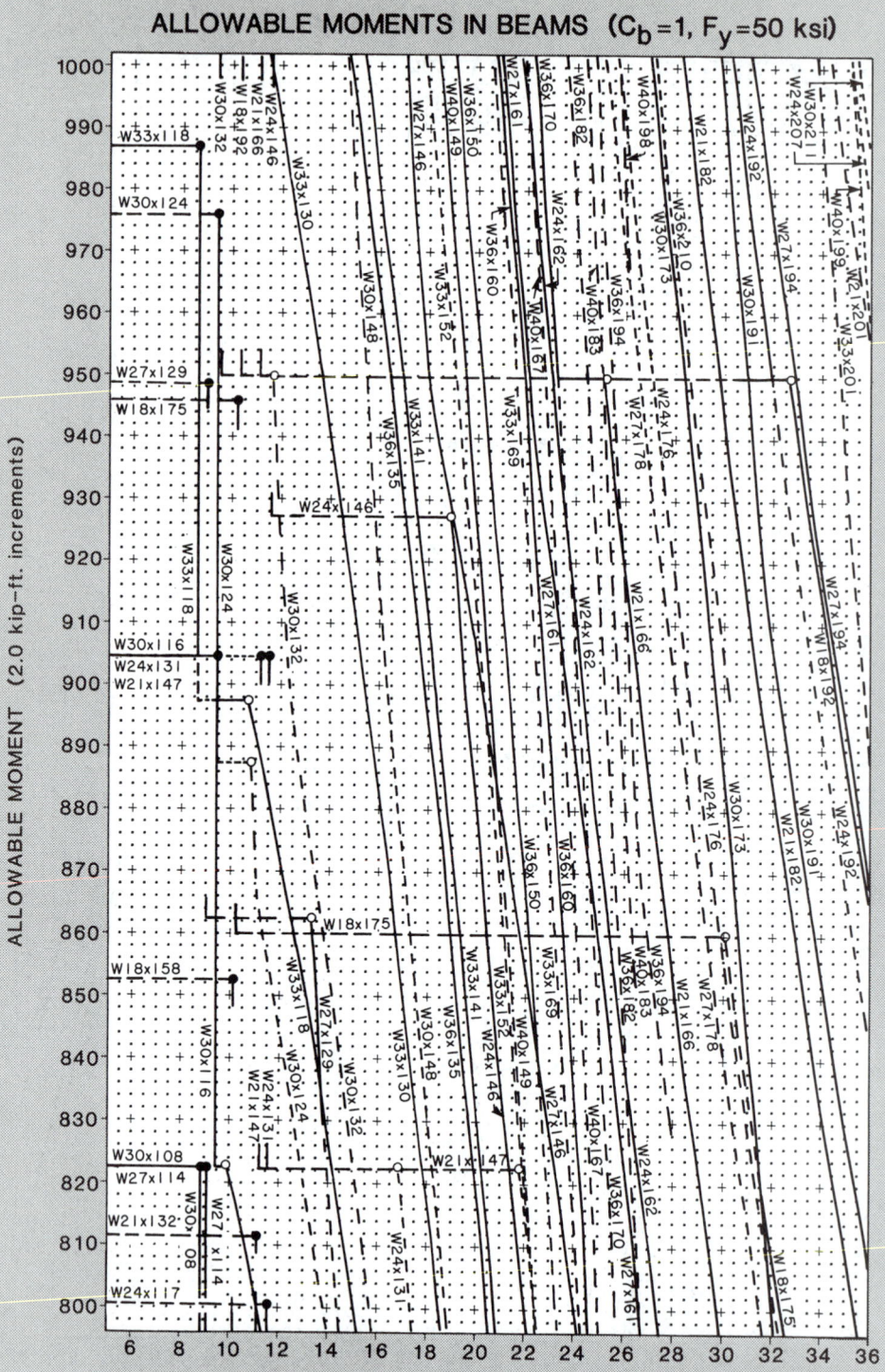

ALLOWABLE MOMENT (2.0 kip–ft. increments)

UNBRACED LENGTH (0.5 ft. increments)

ALLOWABLE MOMENTS IN BEAMS (C_b=1, F_y=50 ksi)

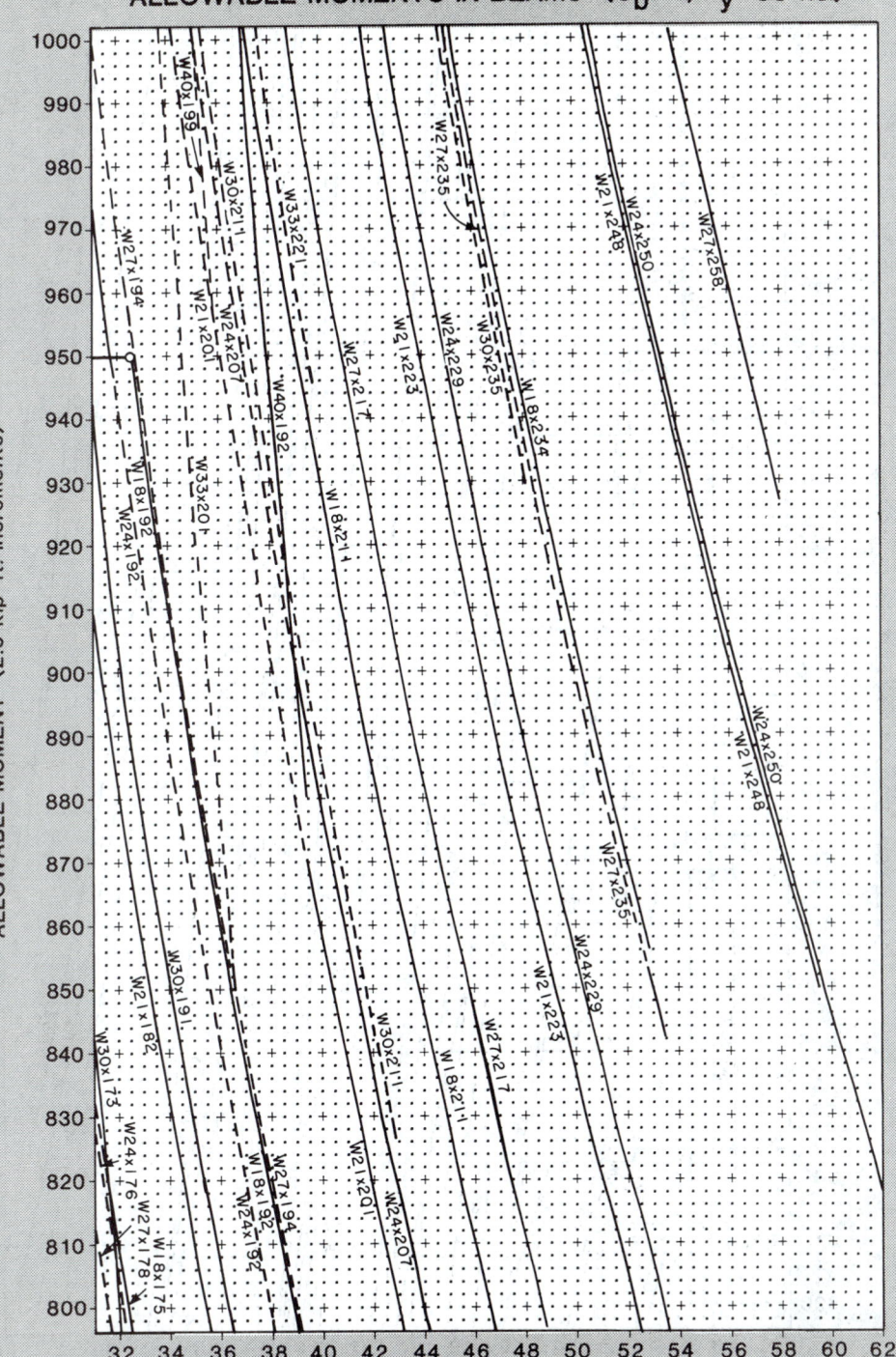

ALLOWABLE MOMENT (2.0 kip-ft. increments)

UNBRACED LENGTH (0.5 ft. increments)

ALLOWABLE MOMENTS IN BEAMS (C_b=1, F_y=50 ksi)

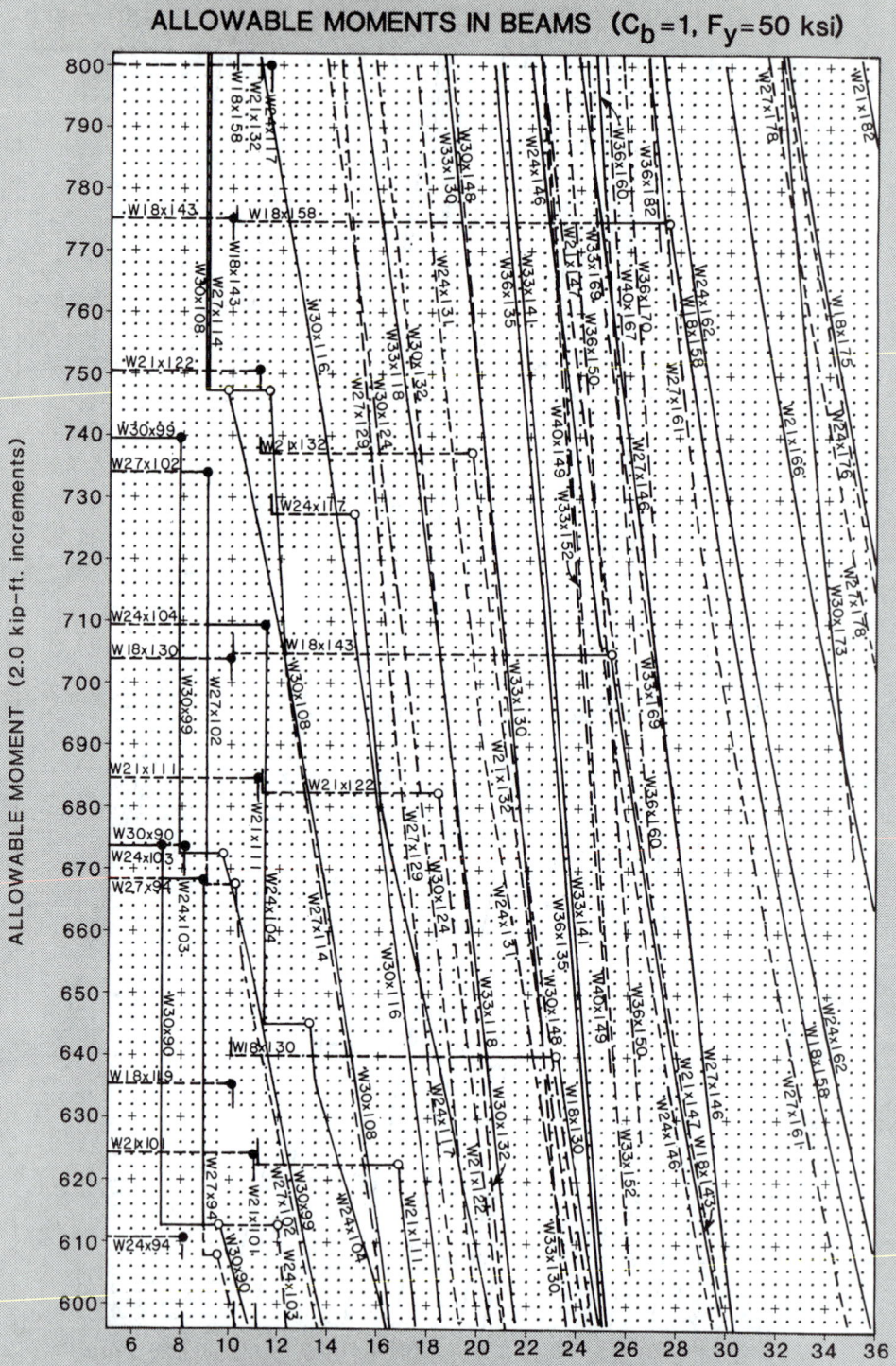

ALLOWABLE MOMENT (2.0 kip-ft. increments)

UNBRACED LENGTH (0.5 ft. increments)

AMERICAN INSTITUTE OF STEEL CONSTRUCTION

ALLOWABLE MOMENTS IN BEAMS (C_b=1, F_y=50 ksi)

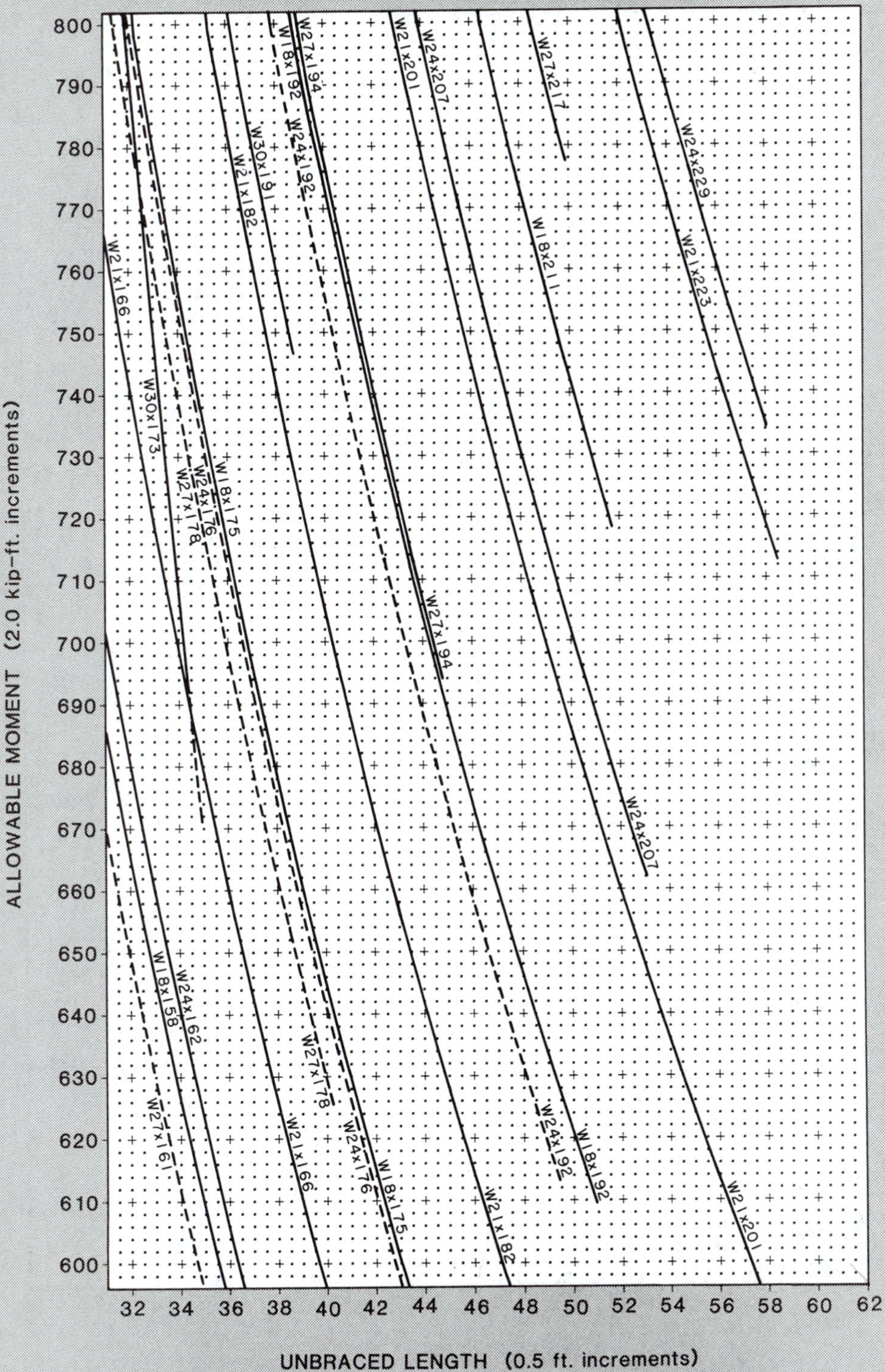
ALLOWABLE MOMENT (2.0 kip-ft. increments)
UNBRACED LENGTH (0.5 ft. increments)
800
790
780
770
760
750
740
730
720
710
700
690
680
670
660
650
640
630
620
610
600
32
34
36
38
40
42
44
46
48
50
52
54
56
58
60
62
W18x192
W27x194
W21x201
W24x207
W27x217
W24x229
W21x223
W18x211
W30x191
W24x192
W21x182
W21x166
W30x173
W27x178
W24x176
W18x175
W27x194
W24x207
W18x158
W24x162
W27x161
W27x178
W21x166
W24x176
W18x175
W21x182
W24x192
W18x192
W21x201

ALLOWABLE MOMENTS IN BEAMS ($C_b = 1$, $F_y = 50$ ksi)

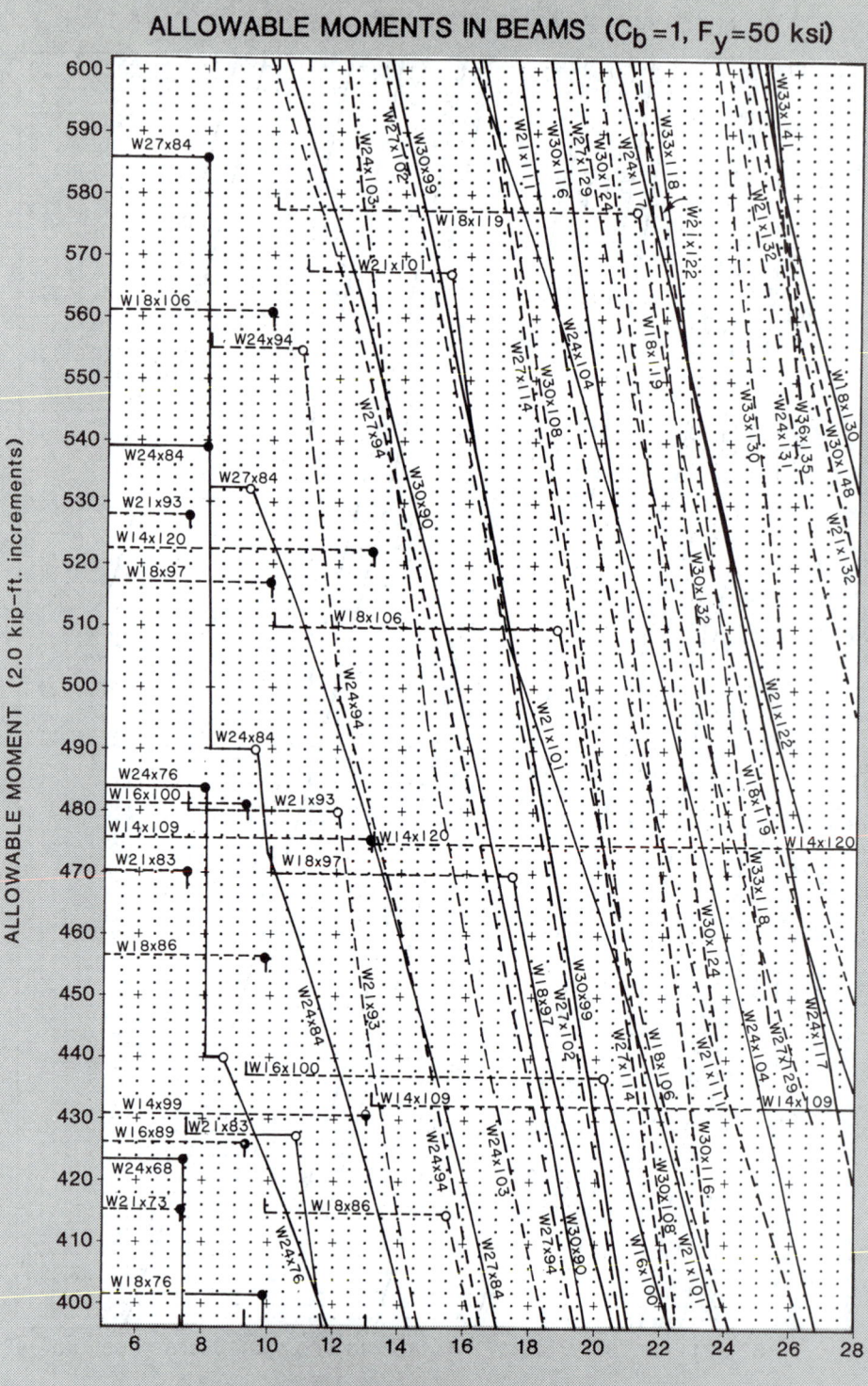

ALLOWABLE MOMENT (2.0 kip-ft. increments)

UNBRACED LENGTH (0.5 ft. increments)

ALLOWABLE MOMENTS IN BEAMS (C_b=1, F_y=50 ksi)

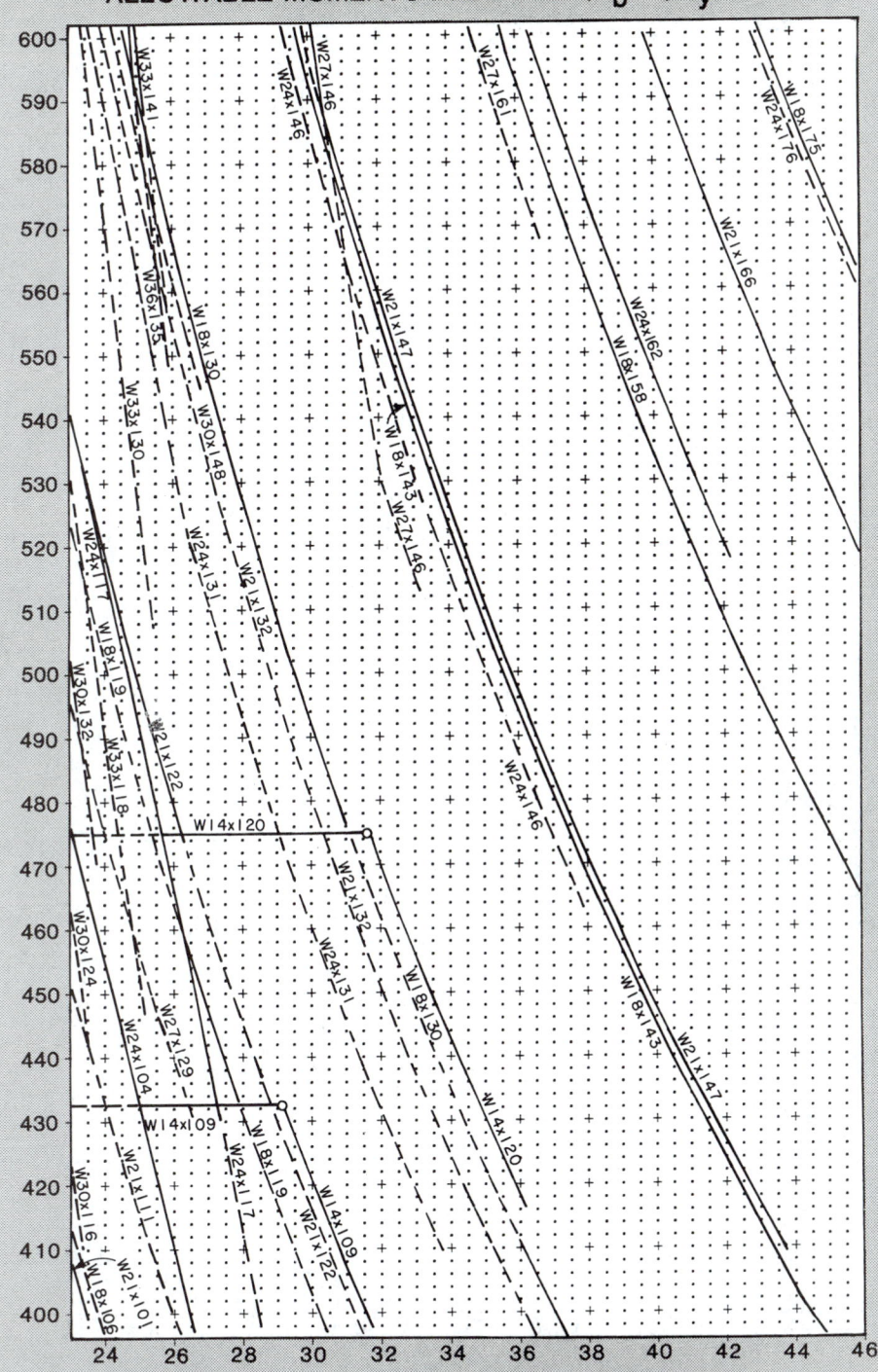

ALLOWABLE MOMENT (2.0 kip-ft. increments)

UNBRACED LENGTH (0.5 ft. increments)

AMERICAN INSTITUTE OF STEEL CONSTRUCTION

ALLOWABLE MOMENTS IN BEAMS (C$_b$=1, F$_y$=50 ksi)

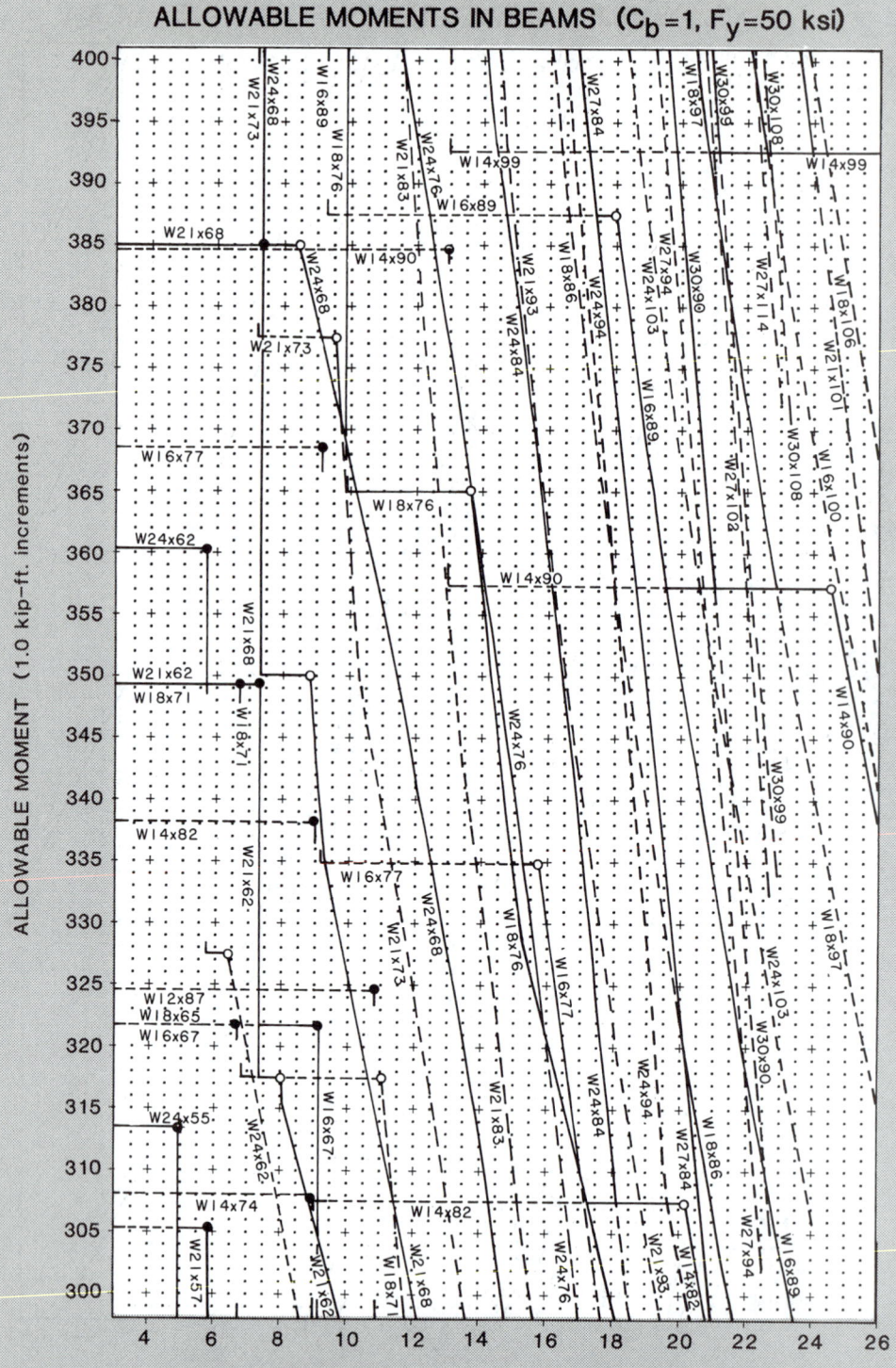

ALLOWABLE MOMENT (1.0 kip-ft. increments)

UNBRACED LENGTH (0.5 ft. increments)

AMERICAN INSTITUTE OF STEEL CONSTRUCTION

ALLOWABLE MOMENTS IN BEAMS ($C_b=1$, $F_y=50$ ksi)

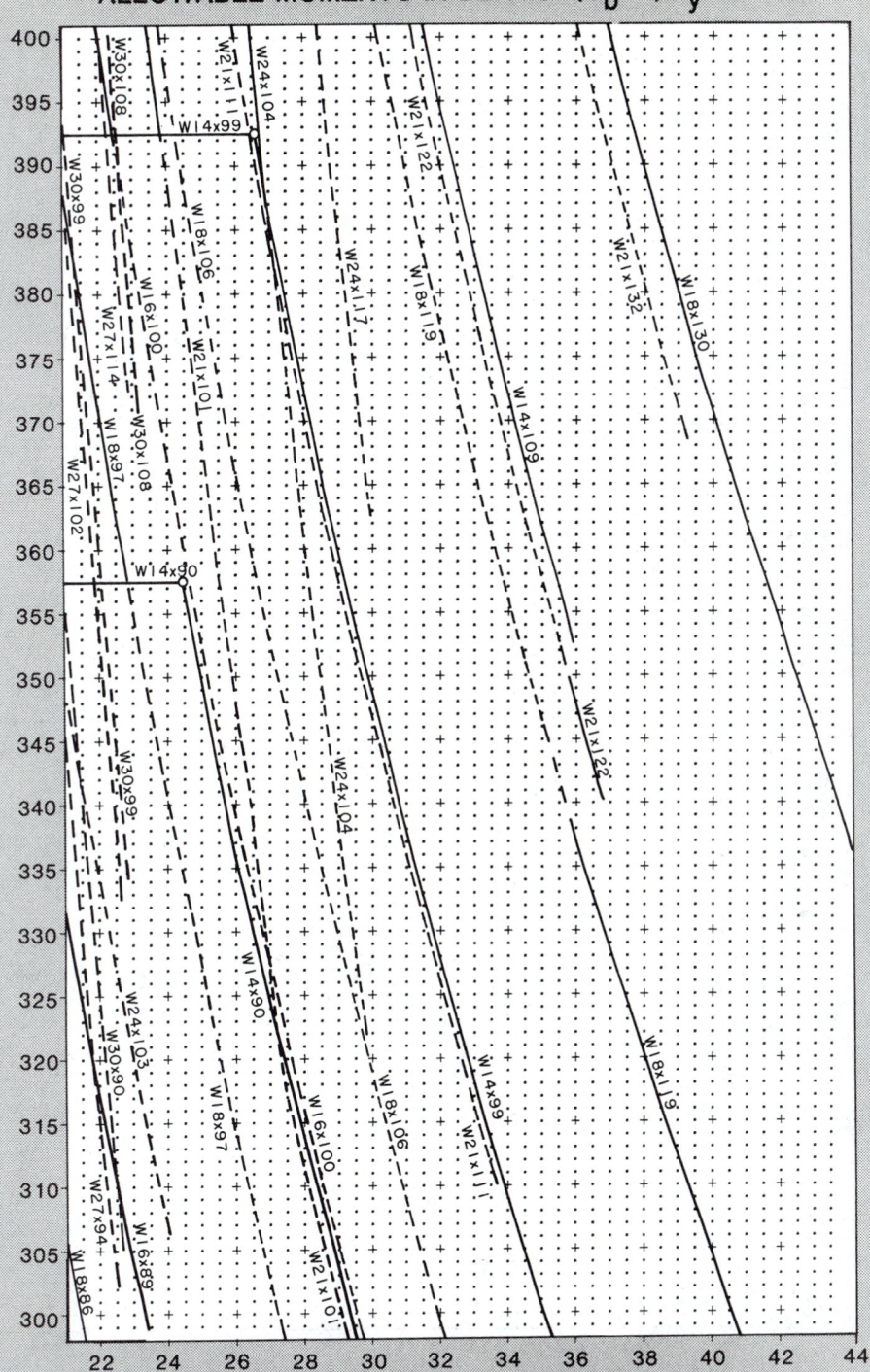

ALLOWABLE MOMENT (1.0 kip–ft. increments)

UNBRACED LENGTH (0.5 ft. increments)

AMERICAN INSTITUTE OF STEEL CONSTRUCTION

ALLOWABLE MOMENTS IN BEAMS (C_b=1, F_y=50 ksi)

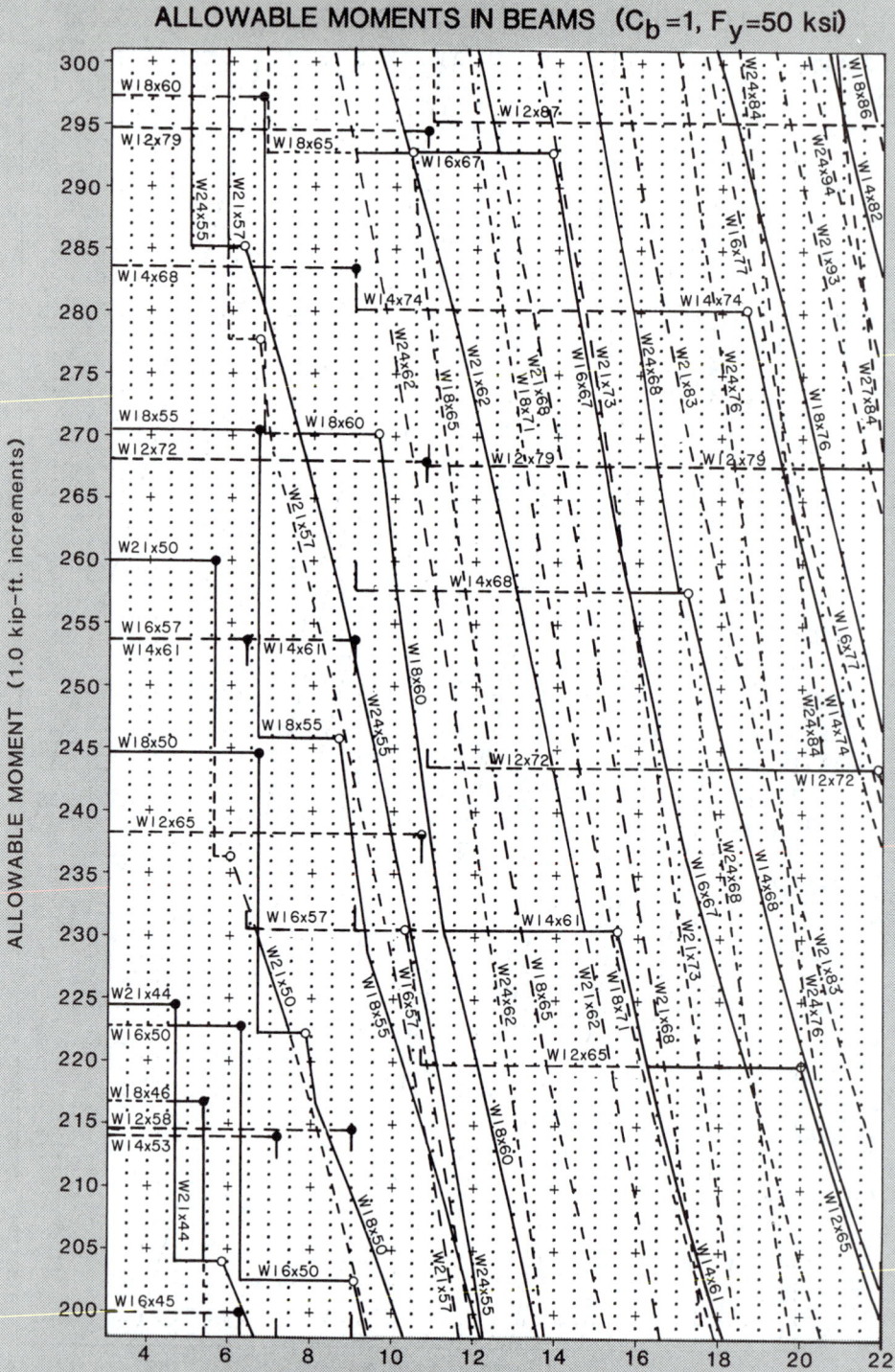

ALLOWABLE MOMENT (1.0 kip-ft. increments)

UNBRACED LENGTH (0.5 ft. increments)

ALLOWABLE MOMENTS IN BEAMS ($C_b=1$, $F_y=50$ ksi)

ALLOWABLE MOMENT (1.0 kip-ft. increments)

UNBRACED LENGTH (0.5 ft. increments)

AMERICAN INSTITUTE OF STEEL CONSTRUCTION

ALLOWABLE MOMENTS IN BEAMS (C_b =1, F_y=50 ksi)

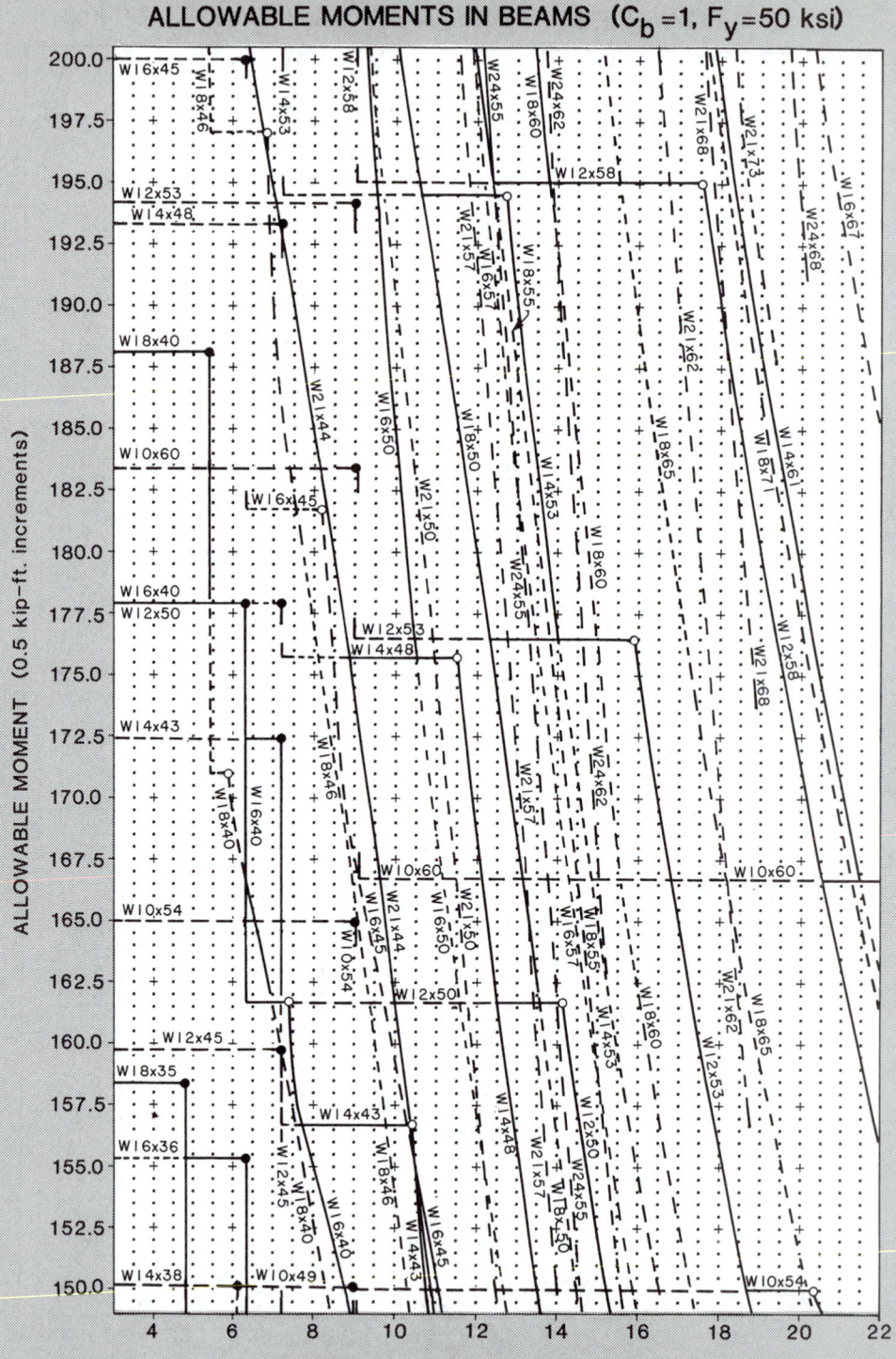

ALLOWABLE MOMENT (0.5 kip-ft. increments)

UNBRACED LENGTH (0.5 ft. increments)

AMERICAN INSTITUTE OF STEEL CONSTRUCTION

ALLOWABLE MOMENTS IN BEAMS ($C_b=1$, $F_y=50$ ksi)

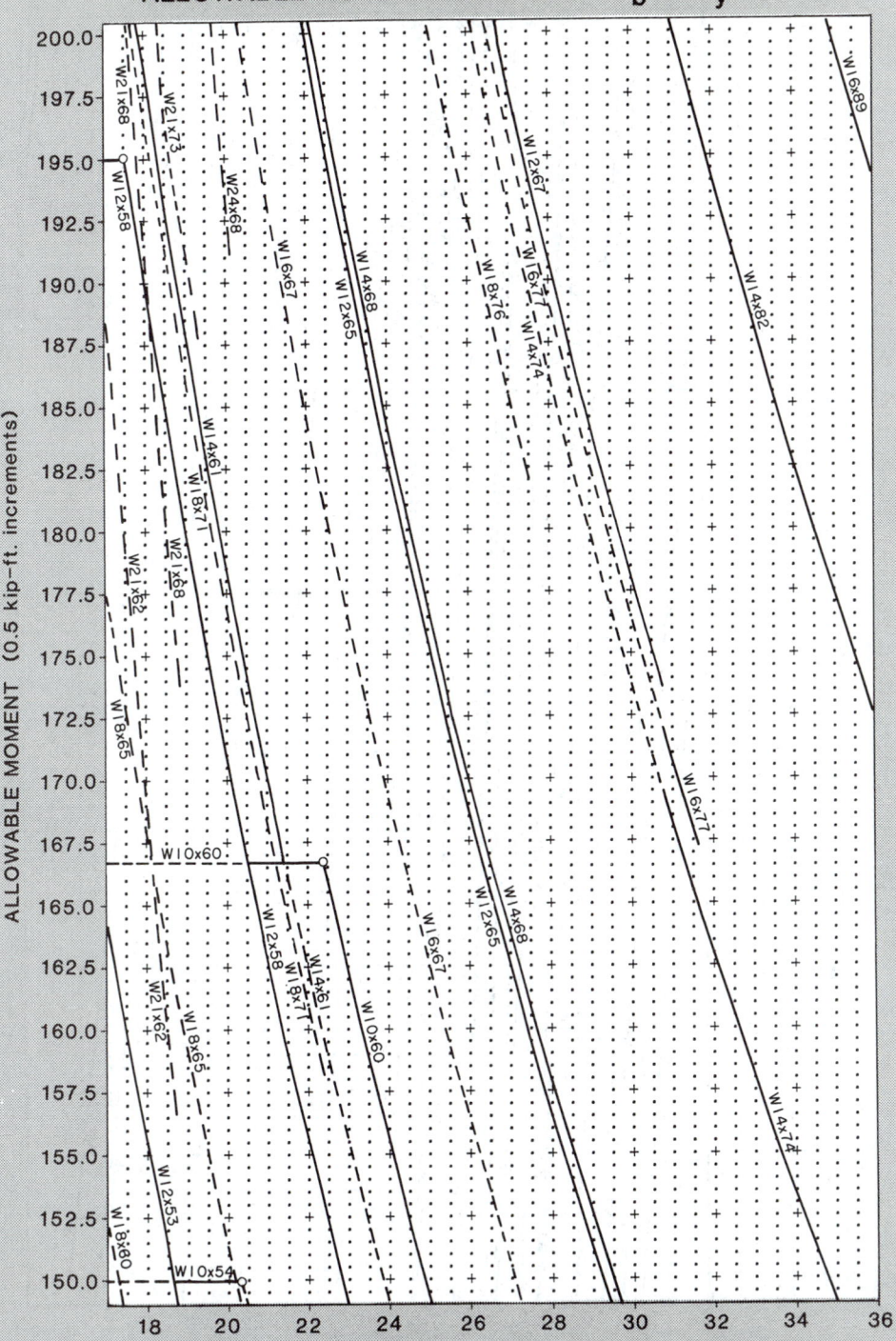

ALLOWABLE MOMENT   (0.5 kip–ft. increments)

UNBRACED LENGTH   (0.5 ft. increments)

ALLOWABLE MOMENTS IN BEAMS $(C_b = 1, F_y = 50$ ksi)

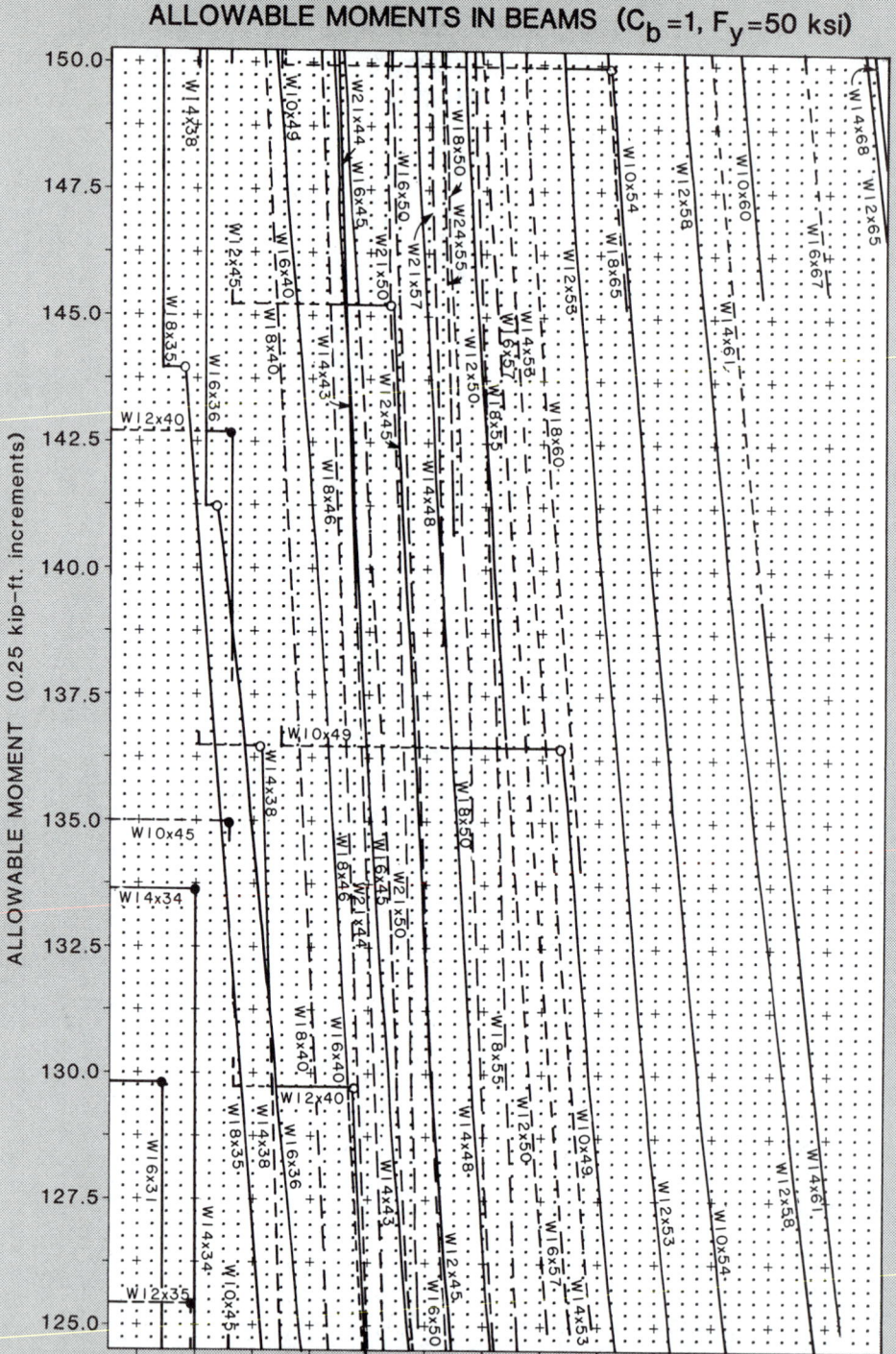

ALLOWABLE MOMENT (0.25 kip-ft. increments)

UNBRACED LENGTH (0.5 ft. increments)

ALLOWABLE MOMENTS IN BEAMS (C_b =1, F_y =50 ksi)

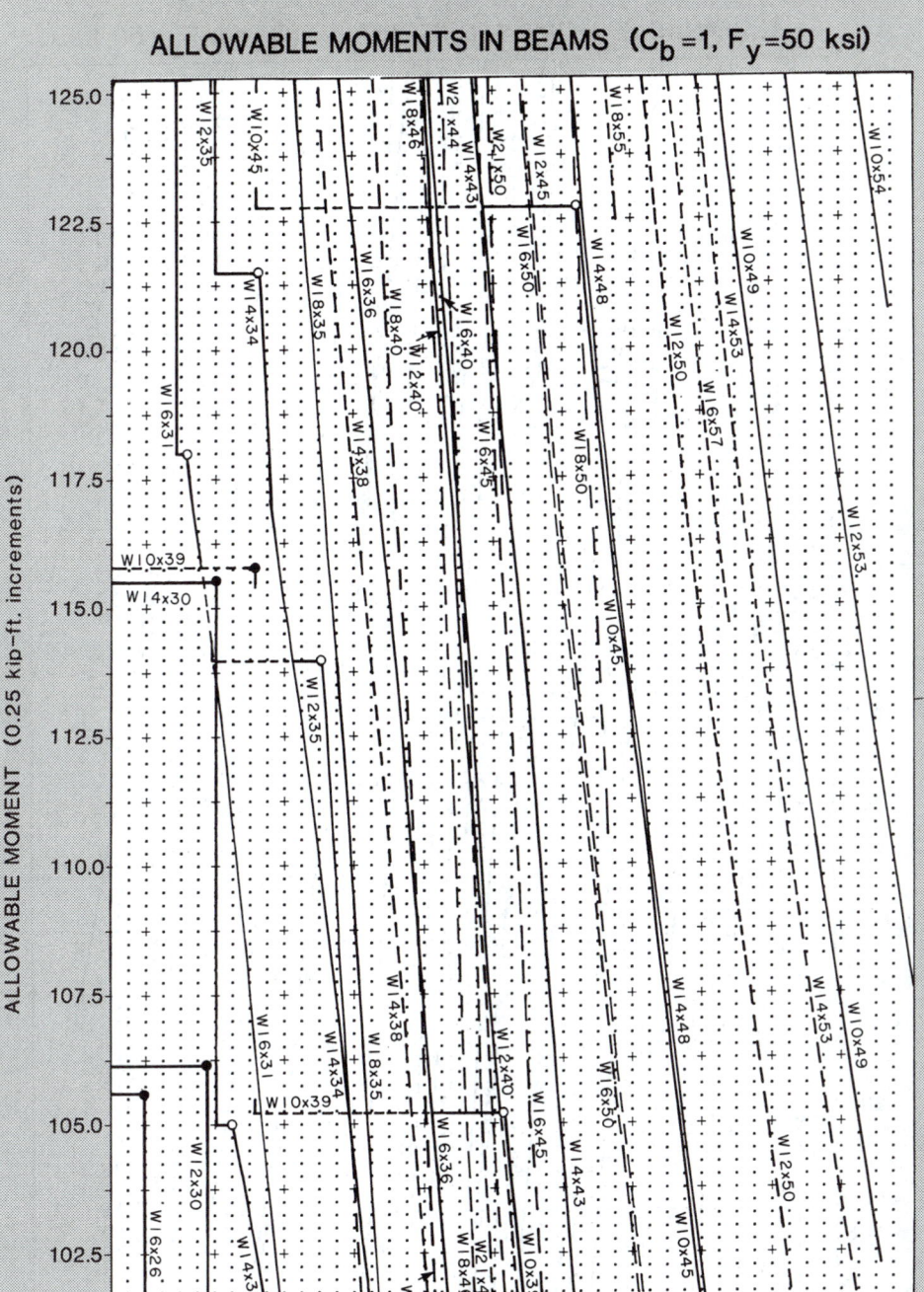

ALLOWABLE MOMENT (0.25 kip-ft. increments)

UNBRACED LENGTH (0.5 ft. increments)

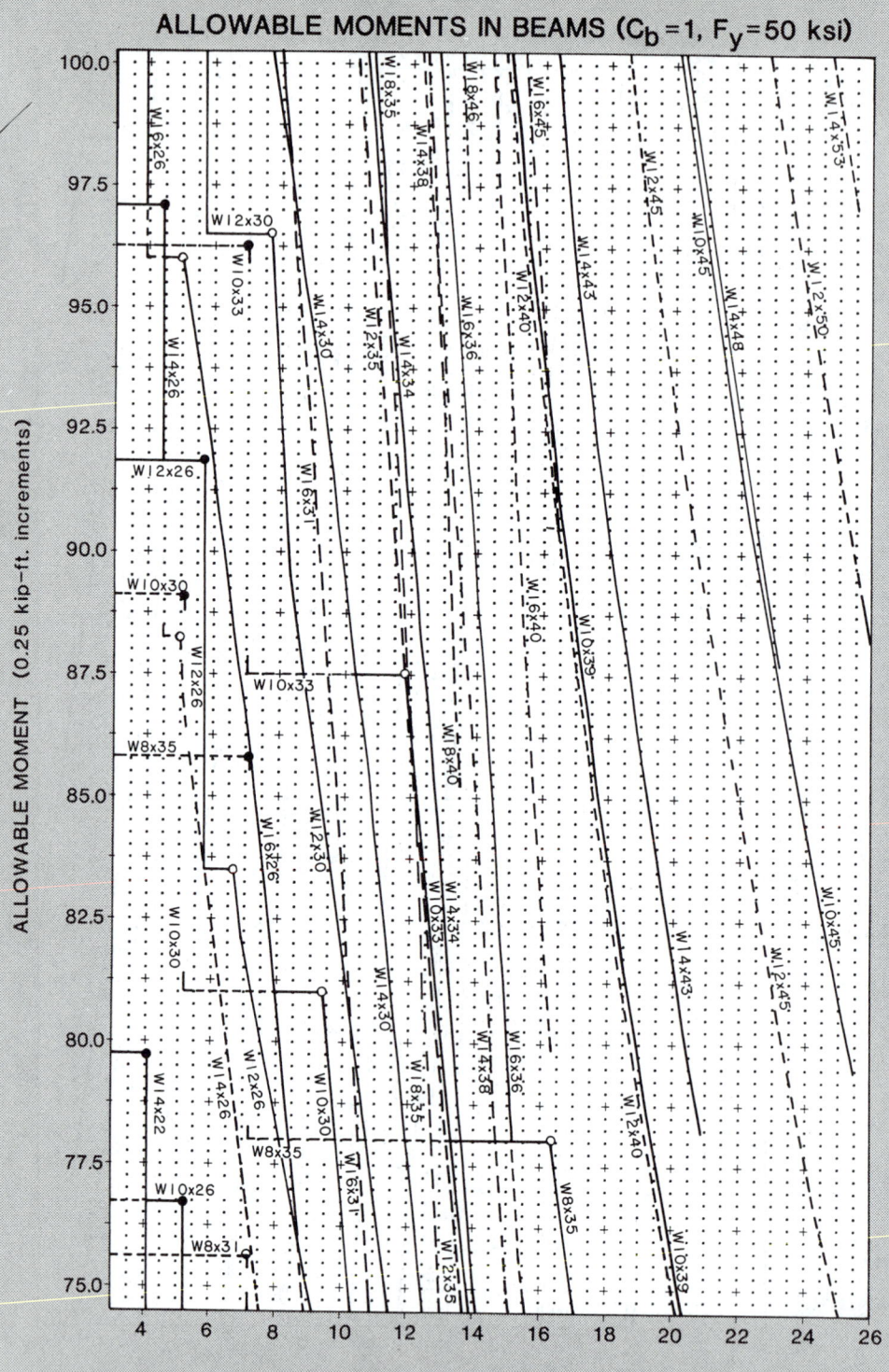

ALLOWABLE MOMENTS IN BEAMS (C_b=1, F_y=50 ksi)

ALLOWABLE MOMENT (0.25 kip-ft. increments)

UNBRACED LENGTH (0.5 ft. increments)

ALLOWABLE MOMENTS IN BEAMS (C_b =1, F_y=50 ksi)

ALLOWABLE MOMENT (0.25 kip-ft. increments)

UNBRACED LENGTH (0.5 ft. increments)

AMERICAN INSTITUTE OF STEEL CONSTRUCTION

ALLOWABLE MOMENTS IN BEAMS (C_b=1, F_y=50 ksi)

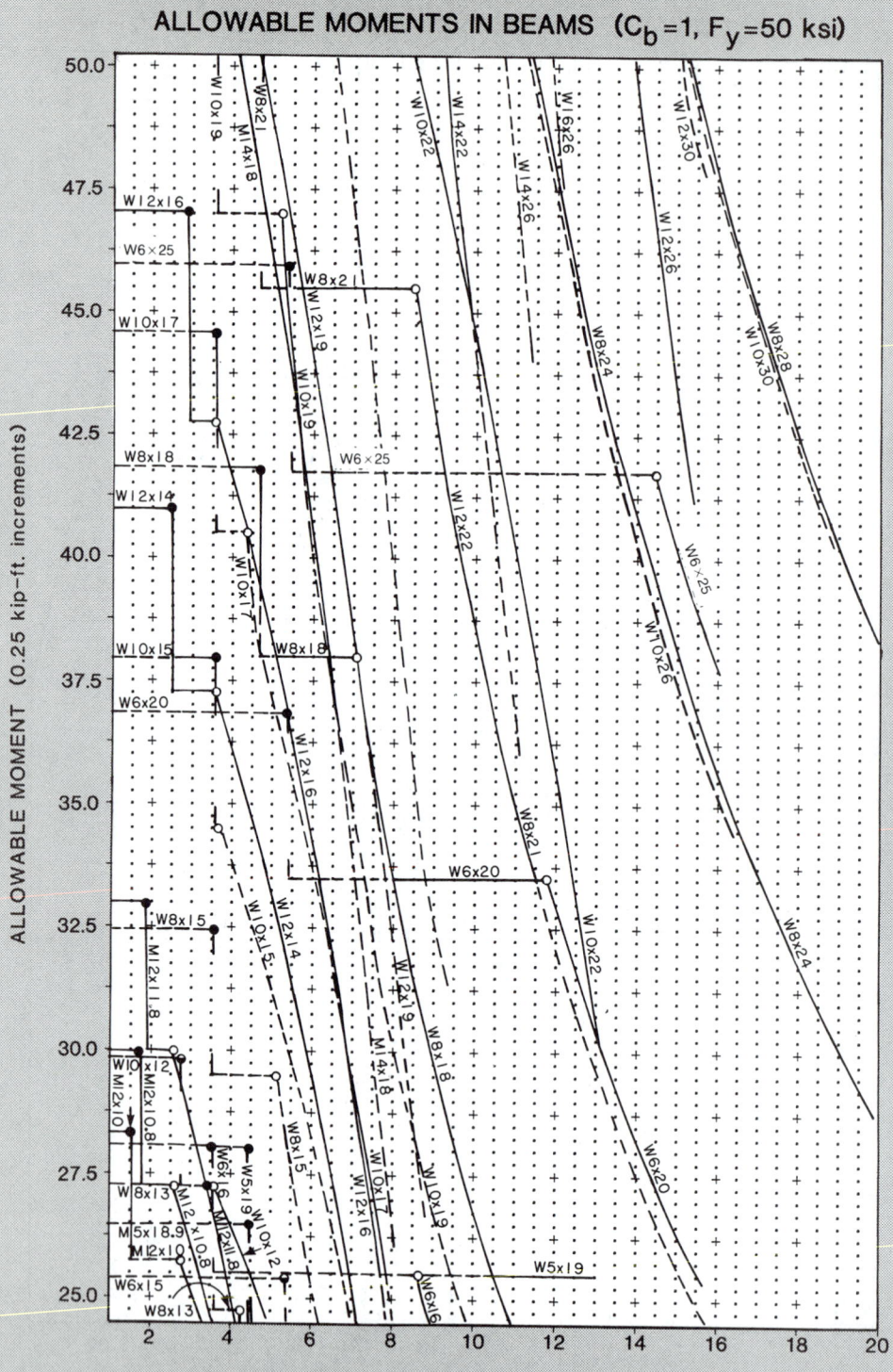

ALLOWABLE MOMENT (0.25 kip-ft. increments)

UNBRACED LENGTH (0.5 ft. increments)

AMERICAN INSTITUTE OF STEEL CONSTRUCTION

ALLOWABLE MOMENTS IN BEAMS $(C_b=1, F_y=50$ ksi)

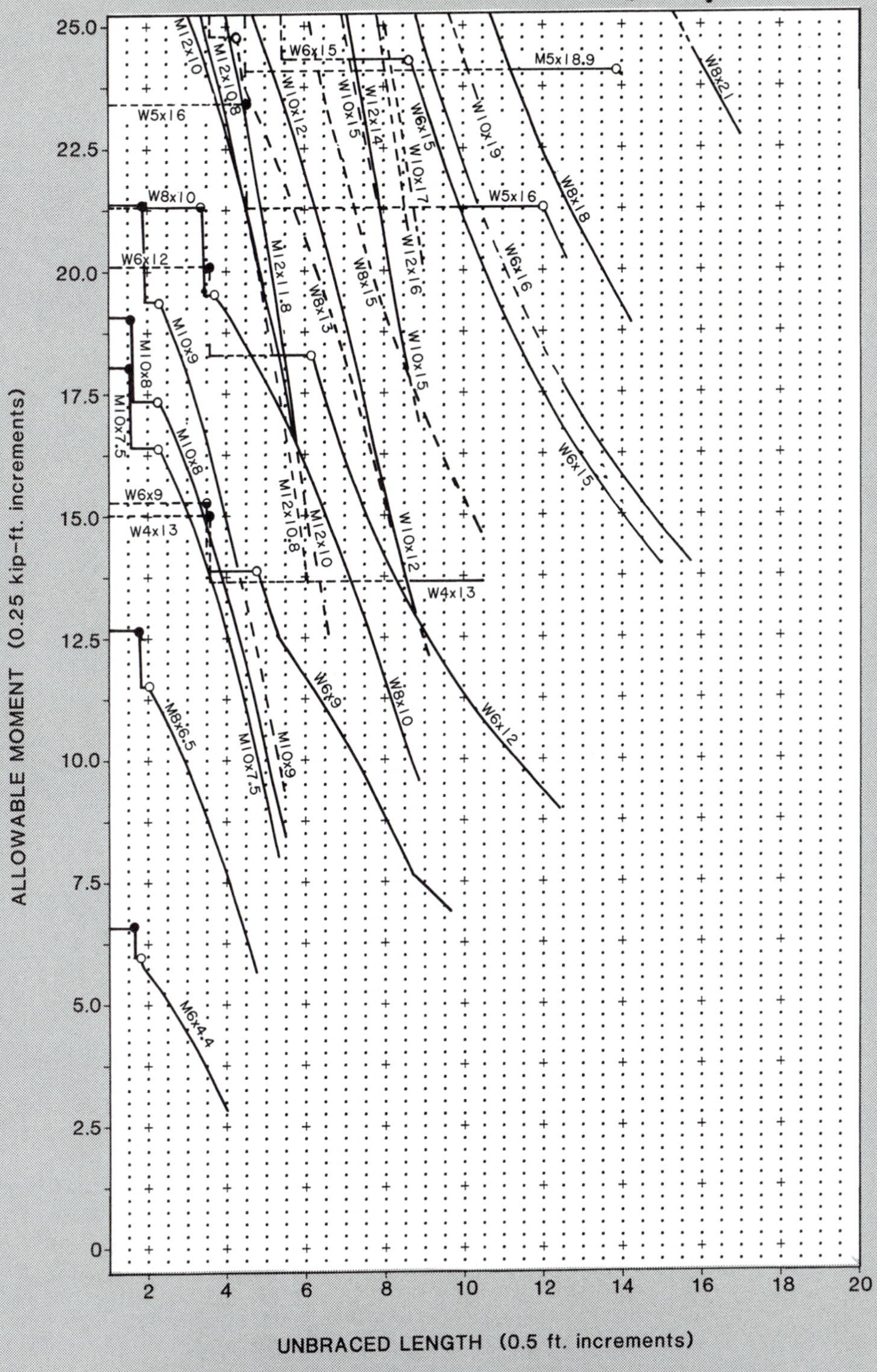

ALLOWABLE MOMENT (0.25 kip–ft. increments)

UNBRACED LENGTH (0.5 ft. increments)

PLATE GIRDERS
General Notes

The *Specification for Structural Steel Buildings for Allowable Stress Design (ASD)*, adopted by the American Institute of Steel Construction effective June 1, 1989, is the basis for the material presented in this section on the design of plate girders.

TABLE OF DIMENSIONS AND PROPERTIES OF WELDED PLATE GIRDERS

This table serves as a guide for selecting welded plate girders of economical proportions. It provides dimensions and properties for a wide range of sections with nominal depths from 45 to 92 in.

No preference is intended for the tabulated flange plate dimensions, as compared to other flange plates having the same area. Substitution of wider but thinner flange plates, without a change in flange area, will result in a slight reduction in section modulus.

All flange plates listed have width-thickness ratios that are within the maximum limitations of Sect. B5.1 of the AISC ASD Specification for $F_y = 36$ ksi steel. If thinner compression flange plates are used, or if steels of higher yield stresses are used, the proportions of the girder flange should be checked for compliance with Sect. B5.1 or Appendix B5, as applicable.

In analyzing overall economy, weight savings must be balanced against higher fabrication costs incurred in splicing the flanges. In some cases, it may prove economical to reduce the size of flange plates at one or more points near the girder ends, where the bending moment is substantially less. Economy through reduction of flange plate sizes is most likely to be realized with long girders where flanges must be spliced in any case.

Only one thickness of web plate is given for each depth of girder. When the design is primarily dominated by shear in the web, rather than moment capacity, overall economy may dictate selection of a thicker web plate. The resulting increase in section modulus can be obtained by multiplying the value S', given in the table, by the number of sixteenths of an inch increase in web thickness, and adding the value obtained to the section modulus value S for the girder profile shown in the table.

Overall economy may often be obtained by using a web plate of such thickness that intermediate stiffeners are not required. However, this is not always the case. The girder sections listed in the table will provide a "balanced" design with respect to bending moment and web shear without excessive use of intermediate stiffeners. When stiffeners are required, their proper spacing can be determined by tables of "Allowable Shear Stress (ksi) in Webs of Plate Girders." Tables for the case of tension field action not included and tension field action included are shown starting on page 2-232 with headings 1-36, 1-50, 2-36 and 2-50. Tension field action is not applicable to hybrid girders since tension field action is not allowed in this case.

The maximum end reaction permissible without intermediate stiffeners for the tabulated web plate thicknesses for $F_y = 36$ ksi steel is listed in the table column headed R. If a thicker web plate is used, the value R will be increased in proportion to the increase in web plate area. Use of a thicker web plate will also result in an increase in the allowable shear stress, through reduction of web depth-thickness ratio h/t. In Tables 1 and 2, "Allowable Shear Stress (ksi) in Webs of Plate Girders," allowable values for shear stress in the case where intermediate stiffeners are not required are given in the right hand column headed "Over 3."

It should be noted the table does not include local effects on the web due to concentrated loads and reactions. See AISC ASD Specification Sect. K1.

DESIGN EXAMPLES

Design of a plate girder by the moment of inertia method recommended in the AISC ASD Specification should start with the preliminary design or selection of a trial section. The initial choice may require one or more adjustments before a final cross section is obtained that satisfies all the provisions of the AISC ASD Specification with maximum economy. In the following design examples, all applicable provisions of the AISC ASD Specification are listed at the right of each page.

Example 11 illustrates a recommended procedure for designing a welded plate girder of constant depth. The selection of a suitable trial cross section is obtained by the "flange area method" and then checked by the "moment of inertia" method.

Example 12 shows a recommended procedure for designing a welded hybrid girder of constant depth.

Example 13 illustrates use of the table of "Welded Plate Girders Dimensions and Properties," to obtain an efficient trial profile. The 52-in. depth specified for this example demonstrates how the tabular data may be used for girder depths intermediate to those listed. Another design requirement in this example is the omission of intermediate web stiffeners. The final girder cross section is checked using the "moment of inertia" method.

Example 14 is similar to Ex. 13, except it illustrates the selection of a girder section whose web requires intermediate stiffeners.

EXAMPLE 11

Design a welded plate girder to support a uniform load of 3 kips per ft and two concentrated loads of 70 kips located 17 ft from each end. The compression flange of the girder will be supported laterally only at points of concentrated load.

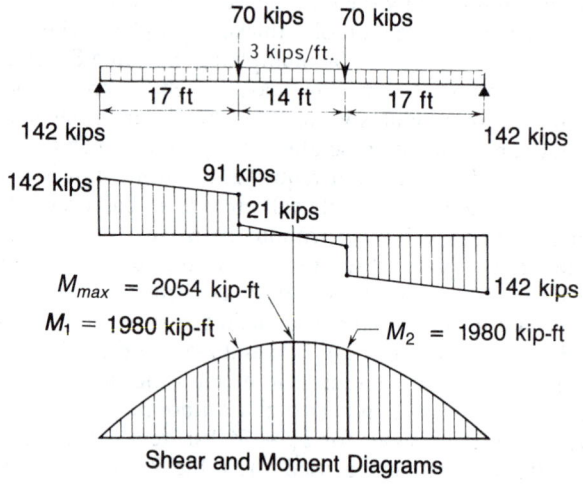

Shear and Moment Diagrams

Given:

Maximum bending moment: 2054 kip-ft
Maximum vertical shear: 142 kips
Span: 48 ft
Maximum depth: 72 in.
Steel: $F_y = 36$ ksi

Solution:

A. Preliminary web design:

1. Assume web depth, $h = 70$ in.

   For no reduction in flange stress, $h/t \leq 970/\sqrt{36} = 162$ **G2**

   Corresponding thickness of web $= 70/162 = 0.43$ in.

2. Minimum thickness of web $= 70/322 = 0.22$ in.   **G1**
   and
   Try web plate $\frac{5}{16} \times 70$: $A_w = 21.9$ in.^{2};   Numerical
   $$h/t = 70/0.313 = 224$$ Values
   Table 5

B. Preliminary flange design:

1. Required flange area:

   An approximate formula for the area of one flange is:

   $$A_f \approx \frac{M}{F_b\,h} = \frac{2054 \times 12}{22 \times 70} = 16.0 \text{ in.}^2$$

   Try 1×18 plate: $A_f = 18$ in.^{2}

2. Check for adequacy against local buckling: Table B5.1
   and
   $$k_c = \frac{4.05}{(h/t)^{0.46}} = \frac{4.05}{(224)^{0.46}} = 0.34$$ Numerical
   Values
   $$\frac{b_f}{2t_f} = \frac{18}{2 \times 1.0} = 9.0 < 95/\sqrt{36/0.34} = 9.23 \quad \textbf{o.k.}$$ Table 5

C. Trial girder section:

   Web $\frac{5}{16} \times 70$; 2 flange plates 1×18

1. Check by "moment of inertia" method: **B10**

Section	A In.^{2}	y In.	$\Sigma A y^2$ In.^{4}	I_o In.^{4}	I_{gr} In.^{4}
1 web $\frac{5}{16} \times 70$	21.9			8932	8932
1 flange 1 x 18	18.0	35.5	45,369	3	45,372
1 flange 1 x 18	18.0				
Moment of inertia					54,304

Section modulus furnished: $54,304/36 = 1508$

2. Check flange stresses:

   a. Check bending stress in 14 ft. panel:

Maximum bending stress at midspan:

$$f_b = \frac{2054 \times 12}{1508} = 16.3 \text{ ksi}$$

Moment of inertia of flange plus ⅙ web about Y-Y axis:

$I_{oy} = 1 \times (18^3/12) = 486 \text{ in.}^4$

$A_f + \frac{1}{6} A_w = 18 + \frac{1}{6} (21.9) = 21.65 \text{ in.}^2$

$r_T = \sqrt{486/21.65} = 4.74 \text{ in.}$

$M_{max} > M_1$ and $M_2 \therefore C_b = 1$

$$\frac{l}{r_T} = \frac{14 \times 12}{4.74} = 35.4 < 53\sqrt{C_b} = 53$$

Allowable stress based upon lateral buckling criteria:

$F_b = 0.60F_y = 21.6 \text{ ksi}$

Reduced allowable bending stress in compression flange:

$$R_{PG} = 1 - 0.0005 \left(\frac{A_w}{A_f}\right) \left(\frac{h}{t} - \frac{760}{\sqrt{F_b}}\right)$$

$$R_{PG} = 1 - 0.0005 \left(\frac{21.9}{18.0}\right) \left(224 - \frac{760}{\sqrt{21.6}}\right) = 0.963$$

$R_e = 1.0$

$F_b' = 21.6 (0.963) 1.0$

$\quad = 20.8 \text{ ksi} > 16.3 \text{ ksi}$ **o.k.**

b. Bending stress in 17-ft panel:

Maximum bending stress:

$$f_b = \frac{1980 \times 12}{15.08} = 15.76 \text{ ksi}$$

$$C_b = 1.75 + 1.05 \frac{M_1}{M_2} + 0.3 \left(\frac{M_1}{M_2}\right)^2$$

where $M_1 = 0$; then $\dfrac{M_1}{M_2} = 0$ $\therefore$ $C_b = 1.75$

$$\frac{l}{r_T} = \frac{17 \times 12}{4.74} = 43.0 < 53\sqrt{C_b} = 70.1$$

Allowable stress in 17-ft panel based upon lateral buckling criteria:

$F_b = 0.60F_y = 21.6 \text{ ksi}$

Reduced allowable bending stress in compression flange:

$F_b' = 20.8 \text{ ksi}$ (see Step C2a)

20.8 ksi > 15.76 ksi   **o.k.**

Use: Web: One plate 5⁄16 × 70
Flanges: Two plates 1 × 18

AMERICAN INSTITUTE OF STEEL CONSTRUCTION

ASD Specification Reference

F1.3
and
Numerical
Values
Table 5

G2

Equation
(G2-1)
F1.2

Numerical
Values
Table 5

G2

D. Stiffener requirements:

   1. Bearing stiffeners:

      a. Bearing stiffeners are required at unframed girder ends.   **K1.8**

      b. Check bearing under concentrated loads:

Assume point bearing and ¼ in. web-to-flange welds.

Local web yielding:

$$\frac{R}{t_w\,(N + 5k)} \leq 0.66\,F_y: \qquad k = 1 + \tfrac{1}{4} = 1\tfrac{1}{4}\ \text{in.}$$

   Equation
   (K1-2)

$$\frac{70}{\tfrac{5}{16}[0 + (5 \times 1)]} > 0.66 \times 36 = 23.8\ \text{ksi} \quad \textbf{n.g.}$$

Note: If local web yielding criterion is satisfied, criteria for web crippling in Sect. K1.4 and Sect. K1.5 would have to be checked.

∴ Provide bearing stiffeners under concentrated loads.

   2. Intermediate stiffeners:

      a. Check shear stress in unstiffened end panel:

F4
Table 1-36
or
Equation
(F4-2)

$h/t = 224;\quad a/h = (17 \times 12)/70 = 2.9$

$F_v = 1.8\ \text{ksi}$

$f_v = 142/21.9 = 6.48\ \text{ksi} > 1.8\ \text{ksi}$

∴ Provide intermediate stiffeners.

      b. End panel stiffener spacing (tension field action not permitted):

F5

Equation
(F4-2)
or
Table 1-36

$F_v = 6.48\ \text{ksi} \therefore a/h = 0.57$

$a \leq 0.57 \times 70 \leq 39.9\ \text{in.}$

Use: 36 in.

      c. Check for additional stiffeners:

Shear at first intermediate stiffener:

$$V = 142 - \left(3 \times \frac{36}{12}\right) = 133\ \text{kips}$$

$f_v = 133/21.9 = 6.07\ \text{ksi}$

Distance between first intermediate stiffener and concentrated load:

$a = (17 \times 12) - 36 = 168\ \text{in.}$

$a/h = 168/70 = 2.4$

Equation
(F4-2)
or
Table 1-36

$F_v = 1.7\ \text{ksi} < 6.07\ \text{ksi}$

∴ Provide intermediate stiffener spaced at $168/2 = 84$ in.

$a/h = 84/70 = 1.2$

$$\text{Maximum } a/h = \left(\frac{260}{h/t}\right)^2 = \left(\frac{260}{224}\right)^2 = 1.35 > 1.2 \quad \textbf{o.k.}$$

$F_v = 8.1 \text{ ksi} > 6.07 \text{ ksi} \quad \textbf{o.k.}$

d. Check center 14-ft panel:

$h/t = 224; \ a/h = (14 \times 12)/70 = 2.4$

$F_v = 2.0 \text{ ksi}$

$f_v = 21/21.9 = 0.96 \text{ ksi} < 1.86 \text{ ksi} \quad \textbf{o.k.}$

3. Combined shear and tension stress:

Check interaction at concentrated load in tension field panel:

$f_v = 91/21.9 = 4.16 \text{ ksi}$

Allowable bending tensile stress:

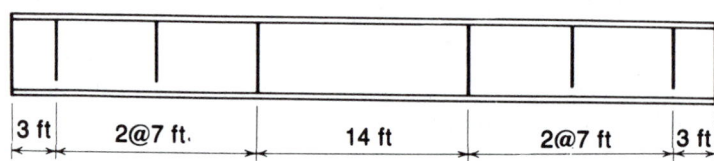

$$F_b = \left(0.825 - 0.375 \times \frac{4.16}{8.1}\right) 36 = 22.8 \text{ ksi}$$

$\therefore F_b = 0.60F_y = 21.6 \text{ ksi} > f_b \quad \textbf{o.k.}$

Summary: Space stiffeners as shown:

| 3 ft | 2@7 ft. | 14 ft | 2@7 ft | 3 ft |

E. Stiffener size:

1. For intermediate stiffeners:

a. Area required (single plate stiffener):

$A_{st} = \%$ web area $\times D(f_v/F_v)$

$D = 2.4$ for single plate stiffeners

$h/t = 224$

$a/h = 84/70 = 1.2$

$$A_{st} = 0.111 \times 21.9 \times 2.4 \left(\frac{6.04}{8.10}\right) = 4.35 \text{ in.}^2$$

Table 2-36

Try one bar 9/16 × 8:

$A_{st} = 4.5 \text{ in.}^2 > 4.35 \text{ in.}^2 \quad \textbf{o.k.}$

b. Check width-thickness ratio:

$8/0.5625 = 14.2 < 15.8 \quad \textbf{o.k.}$

c. Check moment of inertia:

$I_{req'd} = (70/50)^4 = 3.84 \text{ in.}^4$

$I_{furn.} \approx \frac{1}{3} (0.5625)(8.15)^3 = 102$ in.^{4} > 3.84 in.^{4} **o.k.**

d. Min. length required:

$70 - \frac{5}{16} - (6 \times \frac{5}{16}) = 67\frac{13}{16}$ in.

Use for intermediate stiffeners: One plate $\frac{9}{16} \times 8 \times 5$ ft-9 in., fillet-welded to the compression flange and web.

2. Design bearing stiffeners:

At end of girder, design for end reaction.

Try two $\frac{9}{16} \times 8$ in. bars.

a. Check width-thickness ratio:

$$\frac{8}{0.5625} = 14.2 < 15.8 \quad \textbf{o.k.}$$

b. Check compressive stress:

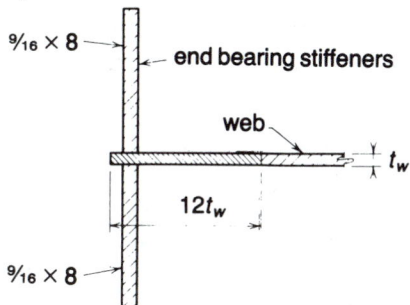

$I = \frac{9}{16} \times \dfrac{(16.31)^3}{12} = 203$ in.^{4}

$A_{eff} = (2 \times 8 \times \frac{9}{16}) + [12 \times (\frac{5}{16})^2] = 10.17$ in.^{2}

$r = \sqrt{\dfrac{203}{10.17}} = 4.47$ in.

$Kl = \frac{3}{4} \times 70 = 52.5$ in.

$\dfrac{Kl}{r} = \dfrac{52.5}{4.47} = 11.7$

Allowable stress: $F_a = 21.06$ ksi

$f_a = \dfrac{142}{10.17} = 13.96$ ksi < 21.06 ksi **o.k.**

Use for bearing stiffeners: Two plates $\frac{9}{16} \times 8 \times 5$ ft-9$\frac{3}{4}$ in. with close bearing on flange receiving reaction or concentrated loads.

Use same size stiffeners for bearing under concentrated loads.*

K1.8

B5.1

K1.8

Table C-36,
part 3

---

* In this example, bearing stiffeners were designed for end bearing; however, $25t$ may be used in determining effective area of web for bearing stiffeners under concentrated loads at interior panels (Sect. K1.8).

EXAMPLE 12

Design a hybrid girder to support a uniform load of 2 kips per ft and three concentrated loads of 200 kips located at the quarter points. The girder depth must be limited to 5 ft. The compression flange will be laterally supported throughout its length.

Given:

Maximum bending moment: 9600 kip-ft
Maximum vertical shear: 380 kips
Span: 80 ft
Maximum depth: 60 in.
Steel: Flanges: $F_y = 50$ ksi
$\quad\quad$ Web: $\quad F_y = 36$ ksi

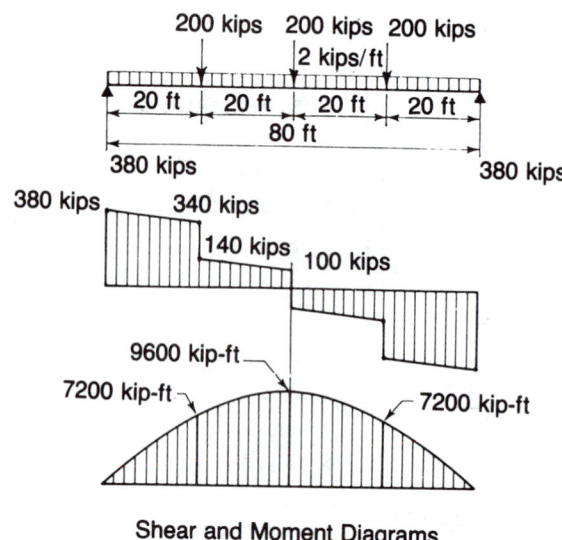

Shear and Moment Diagrams

	ASD Specification Reference

Solution:

A. Preliminary web design:

$\quad$ Assume web depth, $h = 54$ in.

$\quad$ Minimum thickness of web: $54/243 = 0.22$ in.
$\quad$ For no reduction in flange compression stress due to web slenderness:

$$h/t_w \leq 970/\sqrt{50} = 137$$

$\quad$ Corresponding web thickness $= 54/137 = 0.394$

$\quad$ Minimum t_w required for maximum allowable shear stress of 14.5 ksi:

$$t_w = \frac{V}{F_v h} = \frac{380}{14.5 \times 54} = 0.486 \text{ in.}$$

G1 and Numerical Values Table 5

G2

Numerical Values Table 1

Try web plate $\%_{16} \times 54$; $A_w = 30.38$ in.^{2}

$f_v = 380/30.38 = 12.5$ ksi < 14.5 ksi   **o.k.**

$h/t_w = 54/0.563 = 96$

B. Preliminary flange design:

  1. An approximate formula for the area of one flange plate for a hybrid girder is:

$$A_f \approx \frac{M}{F_b h} = \frac{9600(12)}{30 \times 54} = 71.1 \text{ in.}^2$$

Try flange plate $2\%_8 \times 24$:

$A_f = 69$ in.^{2}

  2. Check adequacy against local buckling:

B5.1

$$k_c = \frac{4.05}{(h/t)^{0.46}} = \frac{4.05}{(96)^{0.46}} = 0.50$$

$$\frac{b_f}{2t_f} = \frac{24}{2 \times 2.875} = 4.17 < 95/\sqrt{50/0.50} = 9.5 \quad \textbf{o.k.}$$

C. Trial girder section:

  1 web: $\%_{16} \times 54$

  2 flange plates: $2\%_8 \times 24$

  1. Check by "moment of inertia" method:

B10

Section	A In.^{2}	y In.	Ay^2 In.^{4}	I_o In.^{4}	I_{gr} In.^{4}
1 web $\%_{16}$ x 54	30.38			7380	7380
1 flange $2\%_8$ x 24	69 }	28.44	111599	95	111694
1 flange $2\%_8$ x 24	69 }				
Moment of inertia					119074

Section modulus furnished $= \dfrac{119074}{29.875} = 3986$ in.^{3}

  2. Check allowable flange stresses:

G2

  a. Compression flange is supported laterally for full length.

  $F_b = 30$ ksi

  b. Allowable flange stress (applies to either flange) from Sect. G2:

$$R_{PG} = 1 \text{ since } \frac{h}{t} < \frac{970}{\sqrt{F_y}}$$

$$\frac{A_w}{A_f} = \frac{30.38}{69} = 0.440$$

$$\alpha = 0.72$$

$$R_e = \frac{12 + (A_w/A_f)\,(3\alpha - \alpha^3)}{12 + 2\,(A_w/A_f)}$$

$$R_e = \frac{12 + 0.44\,[3\,(0.72) - (0.72)^3]}{12 + 2\,(0.44)}$$

$$R_e = 0.99$$

$$F_b' = 0.99F_b = 0.99\,(30) = 29.7 \text{ ksi}$$

Use allowable flange stress of $F_b = 29.7$ ksi.

$$\text{Section modulus required} = \frac{9600 \times 12}{29.7} = 3880 \text{ in.}^3$$

3986 in.^{3} > 3880 in.^{3} **o.k.**

Use: Web: One plate $9/16 \times 54$ ($F_y = 36$ ksi)
    Flanges: Two plates $2\frac{7}{8} \times 24$ ($F_y = 50$ ksi)

D. Stiffener requirements:

1. Bearing stiffeners at ends of girder:

   For design of end bearing stiffener, see step E-2, Ex. 11.   K1.8

   Use: Two plates $3/4 \times 11 \times 4$ ft-$5\frac{3}{4}$ in. with close bearing on flange receiving reaction.

2. Bearing stiffener at concentrated loads:

   Check web yielding by Equation (K1-2):   K1.3

   $R = 200$ kips
   Assume $N = 10$ in., $k = 2\frac{7}{8} + 5/16 = 3\frac{3}{16}$ in.

   Allowable compressive stress $= 0.66\,F_y = 23.8$ ksi.

   $$\text{Computed compressive stress} = \frac{200}{9/16[10 + (5 \times 3\frac{3}{16})]}$$

   $$\approx 5.8 \text{ ksi} < 23.8 \text{ ksi} \textbf{o.k.}$$

   Check web crippling by Equation (K1-4).

   $$P_{all.} = 67.5t_w^2\left[1 + 3\left(\frac{N}{d}\right)\left(\frac{t_w}{t_f}\right)^{1.5}\right]\sqrt{\frac{F_y\,t_f}{t_w}}$$

   $$P_{all.} = 67.5\left(\frac{9}{16}\right)^2\left[1 + 3\left(\frac{10}{54}\right)\left(\frac{0.5625}{2.875}\right)^{1.5}\right]\sqrt{\frac{(36)2.875}{0.5625}}$$

   $P_{all.} = 304$ kips < 380 kips **n.g.**

   Check sidesway buckling by Sect. K1.5.

   Assume flange continuously restrained against rotation.

   $$\frac{d_c/t_w}{l/b_f} = \frac{96}{960/24} = 2.4 > 2.3 \textbf{o.k.}$$

   ∴Bearing stiffeners at points of concentrated loads are required.

3. The AISC ASD Specification does not permit design of hybrid girders on the basis of tension field action. Therefore, determine need for intermediate stiffeners by use of Equation (F4-2).

ASD
Specification
Reference

G3

$h/t = 96$

a/h is over 3.

Allowable shear stress:

$F_v = 9.0$ ksi (by interpolation)

Table 1-36

Vertical shear at end of girder:

$V = 380$ kips

Calculated shear stress:

$f_v = 380/30.38 = 12.5$ ksi > 9.0 ksi

$\therefore$ Intermediate stiffeners required

4. Intermediate stiffener spacing:

Table 1-36

End panel:

$f_v = 12.5$ ksi; $a/h = 1.0$ (by interpolation)

Max. $a_1 = 54$ in.

Next panel, shear at 54 in. from centerline bearing:

$$V = 380 - \left(2 \times \frac{54}{12}\right) = 371 \text{ kips}$$

$f_v = 371/30.38 = 12.2$ ksi

Table 1-36

$a/h = 1.08$

Max. $a_2 = 1.08 \times 54 = 58.3$ in. (use 58 in.)

Next panel, shear at $54 + 58 = 112$ in. from centerline bearing:

$$V = 380 - \left(2 \times \frac{112}{12}\right) = 361 \text{ kips}$$

$f_v = 361/30.38 = 11.88$ ksi

Table 1-36

$a/h = 1.16$

Max. $a_3 = 1.16 \times 54 = 62.7$ in. (use 62 in.)

Next panel, shear at $54 + 58 + 62 = 174$ in. from centerline bearing:

$$V = 380 - \left(2 \times \frac{174}{12}\right) = 351 \text{ kips}$$

$f_v = 351/30.38 = 11.55$ ksi

$a/h = 1.26$

Table 1-36

Max. $a_4 = 1.26 \times 54 = 68$ in.

$[240 - (54 + 58 + 62)] = 66$ in. (use 66 in.)

5. Check need for stiffeners between concentrated loads:

$V \approx 140$ kips (from shear diagram)

$f_v = 140/30.38 = 4.6$ ksi

For $a/h = 3.0$, $F_v = 9.0$ ksi > 4.6 ksi **o.k.**

$h/t = 96 < 260$ **o.k.**

∴ No intermediate stiffeners are required between the concentrated loads.

ASD
Specification
Reference

Table 1-36

F5

Summary:

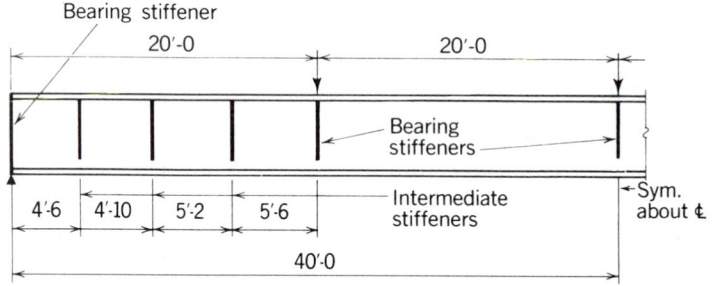

E. Stiffener size:

1. Bearing stiffeners:

See Step E2, Ex. 11, for design procedure.

Use two plates ¾ × 11 × 4 ft-5¾ in. with close bearing on flange receiving reaction.

2. For intermediate stiffeners:

a. Check width-thickness ratio:

Assume ⁵⁄₁₆ × 4 in., $F_y = 36$ ksi, one side only.

$4/0.313 = 12.8 < 15.8$ **o.k.**

b. Check moment of inertia:

$I_{req'd} = (54/50)^4 = 1.36$ in.⁴

$I_{furn} = ⅓ (0.313) (4.28)^3 = 8.18$ in.⁴ > 1.36 in.⁴ **o.k.**

c. Length required $= 54 - ⁵⁄₁₆ - (6 \times 0.5625)$

$= 50.3125$ in. (use 51 in.)

Use for intermediate stiffeners: One plate ¼ × 4 × 4 ft-3 in., one side of web only.

B5.1
and
Numerical
Values
Table 5

G4

G4
Table 1-36

EXAMPLE 13

Given:

Using $F_y = 36$ ksi, design the section of a nominal 52-in. deep-welded plate girder with no intermediate stiffeners to support a uniform load of 2.4 kips per linear foot on an 85-ft span. The girder will be framed between columns and its compression flange will be laterally supported for its entire length.

Solution:

Maximum bending moment: 2168 kip-ft

Maximum vertical shear: 102 kips

Required section modulus: 1182 in.^{3}

Enter table of "Welded Plate Girders, Dimensions and Properties":

For girder having $\frac{3}{8} \times 48$ web with $1\frac{1}{4} \times 16$ flange plates:

$S = 1100$ in.$^3 < 1182$ in.^{3}

For girder having $\frac{3}{8} \times 52$ web with $1\frac{1}{4} \times 18$ flange plates:

$S = 1330$ in.$^3 > 1182$ in.^{3}

A. Determine web required:

Try: Web = $\frac{3}{8} \times 50$; $A_w = 18.75$ in.^{2}

Check web:

For $h/t = 50/0.375 = 133$; from Table 1-36 under column headed "Over 3," allowable shear stress without intermediate stiffeners = 4.7 ksi (by interpolation).

Allowable vertical shear = 18.75×4.7
= 88 kips < 102 kips **n.g.**

Try: Web = $\frac{7}{16} \times 50$; $A_w = 21.8$ in.^{2}

For $h/t = 50/0.4375 = 114$, re-enter table; allowable shear stress without intermediate stiffeners = 6.5 ksi (by interpolation).

Allowable vertical shear = 21.8×6.5
= 141.7 kips > 102 kips **o.k.**

B. Determine flange required:

$$A_f \approx \frac{M}{F_b h} \text{ (see Ex. 11, Step B1)}$$

$$\approx \frac{2168 \times 12}{22 \times 50} = 23.7 \text{ in.}^2$$

Try $1\frac{1}{8} \times 18$ plate: $A_f = 20.25$ in.^{2} **o.k.**

C. Check by "moment of inertia" method:

Table 1-36

Section	A In.^{2}	y In.	Ay^2 In.^{4}	I_o In.^{4}	I_{gr} In.^{4}
1 web $\frac{7}{16} \times 50$			0	4557	4557
1 flange $1\frac{1}{8} \times 18$	20.25	25.56	26,464	4	26,468
1 flange $1\frac{1}{8} \times 18$	20.25				
Moment of inertia					31,025

$S_{furn.} = 31,025/26.125 = 1188$ in.$^3 > 1182$ in.^{3} **o.k.**

AMERICAN INSTITUTE OF STEEL CONSTRUCTION

D. Check web proportions:

$$h/t = 114 \begin{cases} < 322 & \textbf{o.k.} \\ < 260 & \textbf{o.k.} \\ < 106.7 & \text{not compact } (F_b = 22 \text{ ksi}) \end{cases}$$

G1

E. Check bearing stiffener requirements. Since there are no concentrated loads, intermediate bearing stiffeners are not required. Because the girder will be framed between columns, the usual end-bearing stiffeners are not required.

F5
B5.1
and
F1

Use: Web: One plate $\frac{7}{16} \times 50$
  * Flanges: Two plates $1\frac{1}{8} \times 18$

EXAMPLE 14

Given:

Design conditions are the same as given in Ex. 13, except intermediate stiffeners are to be used.

Solution:

A. Preliminary web design:

G1
and
Numerical
Values
Table 5

Minimum web thickness:

$$t_w = h/322 = 50/322 = 0.16 \text{ in.}$$

or

$$h/333 = 0.15 \text{ in. (if } a \le 1.5d)$$

Try web $\frac{5}{16} \times 50$:

$$A_w = 15.63 \text{ in.}^2, \ h/t = 160$$

B. Flange design:

$$A_f \approx \frac{M}{F_b h} \quad \text{(see Ex. 11, Step B1)}$$

$$\approx \frac{2168 \times 12}{22 \times 50} = 23.7 \text{ in.}^2$$

Try flange $1\frac{1}{8} \times 22$:

$$A_f = 25.0 \text{ in.}^2$$

$$k_c = \frac{4.05}{(160)^{0.46}} = 0.39$$

$$b_f/2t_f = 22/(2 \times 1.125) = 9.8 < 95/\sqrt{36/0.39} = 9.89$$

B5.1

---

*Because this girder is no longer than 60 ft, some economy may be gained by decreasing the flange size in areas of smaller moment near ends of girder.

Check by "moment of inertia" method:

Section	A In.²	y In.	Ay² In.⁴	I_o In.⁴	I_{gr} In.⁴
1 web 5⁄16 x 50			0	3255	3255
1 flange 1⅛ x 22	24.75⎱	25.56	32,339	5	32,344
1 flange 1⅛ x 22	24.75⎰				
Moment of inertia					35,599

$$S_{furn} = \frac{35,599}{25.875} = 1376 \text{ in.}^3 > 1182 \text{ in.}^3 \quad \textbf{o.k.}$$

Use: Two flange plates 7⁄8 × 25

C. Stiffener spacing:

F5
and
Table 1-36

1. Because established $h/t < 260$, stiffeners are not required when $f_v \leq 3.2$. Corresponding $V = 3.2 \times 15.63 = 50$ kips.

   Corresponding distance each side of point of zero shear at centerline of girder:

   $$a = (50/2.4)12 = 250 \text{ in.}$$

   Stiffeners are required from each end to 260 in. from each end of the girder.

2. End panel (Equation (F4-2) governs):

   G4

   $V = 102$ kips

   $f_v = 102/15.63 = 6.53$ ksi

   $h/t = 160; \quad \therefore a/h = 0.9$

   Table 1-36

   Max. $a_1 = 0.9 \times 50 = 45$ in.

3. Next panel (tension field action is allowed):

   $$V = 102 - \left(2.4 \times \frac{45}{12}\right) = 93 \text{ kips}$$

   $f_v = 93/15.63 = 5.95$ ksi

   $h/t = 160; \quad \therefore a/h = 2.5$

   Max. $a_2 = 2.5 \times 50 = 125$ in.

   $260 - 45 - 125 = 90$ in. < 125 in.

   Therefore a total of six stiffeners is required. Space at convenient dimension so that a is not greater than 45 in. for end panels and 500 in. for middle panel. See sketch.

   Table 2-36

   $$a/h < \left(\frac{260}{h/t}\right)^2 = 2.64$$

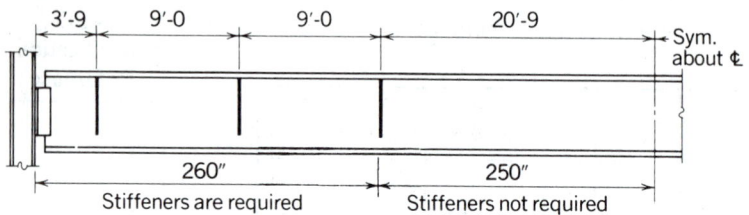

Stiffeners are required | Stiffeners not required

Use: Web: One plate $\frac{5}{16} \times 50$ stiffened as shown in sketch.

Flanges: Two plates $\frac{7}{8} \times 25$

D. Stiffener size:

Intermediate stiffeners:

For stiffener size calculation, see Ex. 11, Step E1.

Use: One plate $\frac{1}{2} \times 5\frac{1}{2} \times$ 4-ft welded to the compression flange and web.

Notes

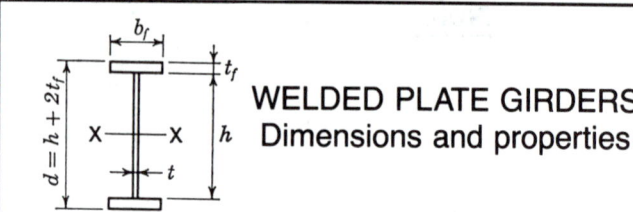

WELDED PLATE GIRDERS
Dimensions and properties

Nominal Size	Wt. per Ft	Area	Depth d	Flange Width b_f	Flange Thick t_f	Web Depth h	Web Thick t	I	S	S'^a	$r_T{}^b$	R^c	$\dfrac{d}{A_f}$
In.	Lb.	In.²	In.	In.	In.	In.	In.	In.⁴	In.³	In.³	In.	Kips	In.⁻¹
92 × 30	823	242	96.00	30	3	90	11/16	431000	8980	79.1	8.20	300.2	1.07
h/t = 131	721	212	95.00	30	2½	90	11/16	363000	7640	79.9	8.12	300.2	1.27
	619	182	94.00	30	2	90	11/16	296000	6290	80.8	8.00	300.2	1.57
	568	167	93.50	30	1¾	90	11/16	263000	5620	81.2	7.92	300.2	1.78
	517	152	93.00	30	1½	90	11/16	230000	4950	81.7	7.81	300.2	2.07
	466	137	92.50	30	1¼	90	11/16	198000	4280	82.1	7.67	300.2	2.47
	415	122	92.00	30	1	90	11/16	166000	3610	82.5	7.47	300.2	3.07
86 × 28	750	220	90.00	28	3	84	5/8	349000	7750	68.6	7.69	241.6	1.07
h/t = 134	654	192	89.00	28	2½	84	5/8	293000	6580	69.4	7.62	241.6	1.27
	559	164	88.00	28	2	84	5/8	238000	5410	70.2	7.52	241.6	1.57
	512	150	87.50	28	1¾	84	5/8	211000	4820	70.6	7.45	241.6	1.79
	464	136	87.00	28	1½	84	5/8	184000	4240	71.0	7.35	241.6	2.07
	416	122	86.50	28	1¼	84	5/8	158000	3650	71.4	7.23	241.6	2.47
	369	108	86.00	28	1	84	5/8	132000	3070	71.8	7.06	241.6	3.07
80 × 26	696	205	84.00	26	3	78	5/8	281000	6680	58.8	7.14	260.2	1.08
h/t = 125	608	179	83.00	26	2½	78	5/8	235000	5670	59.6	7.08	260.2	1.28
	519	153	82.00	26	2	78	5/8	191000	4660	60.3	6.98	260.2	1.58
	475	140	81.50	26	1¾	78	5/8	169000	4160	60.7	6.91	260.2	1.79
	431	127	81.00	26	1½	78	5/8	148000	3650	61.0	6.83	260.2	2.08
	387	114	80.50	26	1¼	78	5/8	127000	3150	61.4	6.71	260.2	2.48
	343	101	80.00	26	1	78	5/8	106000	2690	61.8	6.55	260.2	3.08
	320	94.2	79.75	26	7/8	78	5/8	95500	2390	62.0	6.44	260.2	3.51
74 × 24	627	184	78.00	24	3	72	9/16	220000	5640	49.8	6.62	205.5	1.08
h/t = 128	546	160	77.00	24	2½	72	9/16	184000	4780	50.5	6.57	205.5	1.28
	464	136	76.00	24	2	72	9/16	149000	3920	51.2	6.49	205.5	1.58
	423	124	75.50	24	1¾	72	9/16	132000	3490	51.5	6.43	205.5	1.80
	382	112	75.00	24	1½	72	9/16	115000	3060	51.8	6.36	205.5	2.08
	342	100	74.50	24	1¼	72	9/16	98000	2630	52.2	6.26	205.5	2.48
	301	88.5	74.00	24	1	72	9/16	81400	2200	52.5	6.12	205.5	3.08
	280	82.5	73.75	24	7/8	72	9/16	73300	1990	52.7	6.03	205.5	3.51
68 × 22	561	165	72.00	22	3	66	½	169000	4700	41.6	6.10	157.5	1.09
h/t = 132	486	143	71.00	22	2½	66	½	141000	3970	42.2	6.06	157.5	1.29
	411	121	70.00	22	2	66	½	114000	3250	42.8	5.99	157.5	1.59
	374	110	69.50	22	1¾	66	½	100000	2890	43.1	5.94	157.5	1.81
	337	99.0	69.00	22	1½	66	½	87200	2530	43.4	5.88	157.5	2.09
	299	88.0	68.50	22	1¼	66	½	74200	2170	43.7	5.80	157.5	2.49
	262	77.0	68.00	22	1	66	½	61400	1800	44.0	5.68	157.5	3.09
	243	71.5	67.75	22	7/8	66	½	55000	1620	44.2	5.60	157.5	3.52
	224	66.0	67.50	22	¾	66	½	48700	1440	44.4	5.50	157.5	4.09
61 × 20	429	126	65.00	20	2½	60	7/16	106000	3250	34.6	5.54	116.0	1.30
h/t = 137	361	106	64.00	20	2	60	7/16	84800	2650	35.2	5.48	116.0	1.60
	327	96.2	63.50	20	1¾	60	7/16	74600	2350	35.4	5.44	116.0	1.81
	293	86.2	63.00	20	1½	60	7/16	64600	2090	35.7	5.39	116.0	2.10
	259	76.2	62.50	20	1¼	60	7/16	54800	1750	36.0	5.33	116.0	2.50
	225	66.2	62.00	20	1	60	7/16	45100	1450	36.3	5.23	116.0	3.10
	208	61.2	61.75	20	7/8	60	7/16	40300	1310	36.4	5.16	116.0	3.53
	191	56.2	61.50	20	¾	60	7/16	35600	1160	36.6	5.08	116.0	4.10

WELDED PLATE GIRDERS
Dimensions and properties

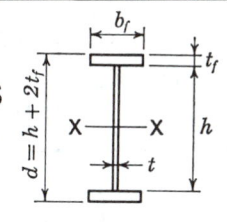

Nominal Size	Wt. per Ft	Area	Depth d	Flange Width b_f	Flange Thick t_f	Web Depth h	Web Thick t	Axis X-X I	Axis X-X S	Axis X-X S'^a	$r_T{}^b$	R^c	$\dfrac{d}{A_f}$
In.	Lb.	In.²	In.	In.	In.	In.	In.	In.⁴	In.³	In.³	In.	Kips.	In.⁻¹
57 × 18	389	115	61.00	18	2½	56	7/16	83500	2740	30.0	4.98	124.3	1.36
h/t = 128	328	96.5	60.00	18	2	56	7/16	67000	2230	30.5	4.92	124.3	1.67
	298	87.5	59.50	18	1¾	56	7/16	58900	1980	30.7	4.89	124.3	1.89
	267	78.5	59.00	18	1½	56	7/16	51000	1730	31.0	4.84	124.3	2.19
	236	69.5	58.50	18	1¼	56	7/16	43300	1480	31.3	4.78	124.3	2.60
	206	60.5	58.00	18	1	56	7/16	35600	1230	31.5	4.69	124.3	3.22
	190	56.0	57.75	18	7/8	56	7/16	31900	1100	31.7	4.63	124.3	3.67
	175	51.5	57.50	18	¾	56	7/16	28100	979	31.8	4.55	124.3	4.26
	160	47.0	57.25	18	5/8	56	7/16	24400	854	32.0	4.45	124.3	5.09
53 × 18	342	100	56.50	18	2¼	52	3/8	64000	2270	25.9	5.00	84.3	1.40
h/t = 138	311	91.5	56.00	18	2	52	3/8	56900	2030	26.2	4.98	84.3	1.56
	280	82.5	55.50	18	1¾	52	3/8	49900	1800	26.4	4.95	84.3	1.76
	250	73.5	55.00	18	1½	52	3/8	43000	1566	26.6	4.91	84.3	2.04
	219	64.5	54.50	18	1¼	52	3/8	36300	1330	26.9	4.86	84.3	2.42
	189	55.5	54.00	18	1	52	3/8	29700	1100	27.1	4.78	84.3	3.00
	173	51.0	53.75	18	7/8	52	3/8	26400	983	27.2	4.73	84.3	3.41
	158	46.5	53.50	18	¾	52	3/8	23200	866	27.4	4.67	84.3	3.96
	143	42.0	53.25	18	5/8	52	3/8	20000	750	27.5	4.58	84.3	4.73
49 × 16	306	90.0	52.50	16	2¼	48	3/8	48900	1860	21.9	4.44	91.3	1.46
h/t = 128	279	82.0	52.00	16	2	48	3/8	43500	1670	22.2	4.42	91.3	1.63
	252	74.0	51.50	16	1¾	48	3/8	38100	1480	22.4	4.39	91.3	1.84
	224	66.0	51.00	16	1½	48	3/8	32900	1290	22.6	4.35	91.3	2.13
	197	58.0	50.50	16	1¼	48	3/8	27700	1100	22.8	4.31	91.3	2.53
	170	50.0	50.00	16	1	48	3/8	22700	907	23.0	4.24	91.3	3.13
	156	46.0	49.75	16	7/8	48	3/8	20200	811	23.2	4.19	91.3	3.55
	143	42.0	49.50	16	¾	48	3/8	17700	716	23.3	4.13	91.3	4.13
	129	38.0	49.25	16	5/8	48	3/8	15300	620	23.4	4.05	91.3	4.93
45 × 16	237	69.8	47.50	16	1¾	44	5/16	31500	1330	18.7	4.44	57.7	1.70
h/t = 141	210	61.8	47.00	16	1½	44	5/16	27100	1150	18.9	4.41	57.7	1.96
	183	53.8	46.50	16	1¼	44	5/16	22700	976	19.1	4.38	57.7	2.33
	156	45.8	46.00	16	1	44	5/16	18400	801	19.3	4.32	57.7	2.88
	142	41.8	45.75	16	7/8	44	5/16	16300	713	19.4	4.28	57.7	3.27
	128	37.8	45.50	16	¾	44	5/16	14200	626	19.5	4.23	57.7	3.79
	115	33.8	45.25	16	5/8	44	5/16	12200	538	19.6	4.17	57.7	4.53

$^a S'$ = Additional section modulus corresponding to 1/16-in. increase in web thickness.

$^b r_T$ = Radius of gyration of the "T" section comprising the compression flange plus 1/3 of the compression web area, about an axis in the plane of the web.

$^c R$ = Maximum end reaction permissible without intermediate stiffeners for tabulated web plate.

Notes:

The width-thickness ratios for girders in this table comply with AISC ASD Specification Sect. B5.1 for F_y = 36 ksi steel. For steels of higher yield strengths, check flanges for compliance with this section.

See Sects. F5, G4 and K1.8 for design of stiffeners.
Welds not included in tabulated weight per foot.

This table does not include local effects on web due to concentrated loads and reactions. See K1.

TABLE 1-36
ALLOWABLE SHEAR STRESS (KSI)
IN WEBS OF PLATE GIRDERS BY EQUATION (F4-2)
For 36 ksi Yield Stress Steel, Tension Field Action Not Included

F_y = 36 ksi

Slenderness Ratios h/t: Web Depth to Web Thickness

h/t	Aspect Ratios a/h: Stiffener Spacing to Web Depth													
	0.5	0.6	0.7	0.8	0.9	1.0	1.2	1.4	1.6	1.8	2.0	2.5	3.0	over 3
60											14.5	14.5	14.5	14.5
70									14.5	14.5	14.2	13.8	13.6	13.0
80						14.5	14.1	13.4	13.0	12.6	12.4	12.0	11.9	11.4
90				14.5	14.3	13.4	12.5	11.9	11.5	11.2	11.0	10.7	10.5	10.1
100			14.5	13.9	12.8	12.0	11.2	10.7	10.4	10.1	9.9	9.3	9.0	8.3
110		14.5	13.8	12.6	11.7	10.9	10.2	9.5	8.9	8.5	8.2	7.7	7.4	6.9
120		14.3	12.7	11.6	10.7	10.0	8.8	8.0	7.5	7.1	6.9	6.5	6.3	5.8
130	14.5	13.2	11.7	10.7	9.8	8.6	7.5	6.8	6.4	6.1	5.8	5.5	5.3	4.9
140	14.2	12.2	10.9	9.8	8.4	7.4	6.5	5.9	5.5	5.2	5.0	4.8	4.6	4.2
150	13.2	11.4	10.1	8.5	7.3	6.5	5.6	5.1	4.8	4.5	4.4	4.1	4.0	3.7
160	12.4	10.7	9.1	7.5	6.5	5.7	4.9	4.5	4.2	4.0	3.9	3.6		3.2
170	11.7	10.1	8.0	6.7	5.7	5.0	4.4	4.0	3.7	3.5	3.4			2.9
180	11.0	9.1	7.2	5.9	5.1	4.5	3.9	3.5	3.3	3.2	3.0			2.6
200	9.9	7.3	5.8	4.8	4.1	3.6	3.2	2.9	2.7					2.1
220	8.2	6.1	4.8	4.0	3.4	3.0	2.6							1.7
240	6.9	5.1	4.0	3.3	2.9	2.5								1.4
260	5.8	4.3	3.4	2.8	2.4	2.2								1.2
280	5.0	3.7	3.0	2.5										
300	4.4	3.3	2.6											
320	3.9	2.9												

TABLE 1-50
ALLOWABLE SHEAR STRESS (KSI)
IN WEBS OF PLATE GIRDERS BY EQUATION (F4-2)

For 50 ksi Yield Stress Steel, Tension Field Action Not Included

$F_y = 50$ ksi

| h/t | Aspect Ratios a/h: Stiffener Spacing to Web Depth | | | | | | | | | | | | | |
	0.5	0.6	0.7	0.8	0.9	1.0	1.2	1.4	1.6	1.8	2.0	2.5	3.0	over 3.0
60							20.0	20.0	20.0	19.9	19.5	18.9	18.6	17.9
70					20.0	20.0	18.9	18.0	17.4	17.0	16.7	16.2	16.0	15.3
80			20.0	20.0	18.9	17.8	16.6	15.8	15.3	14.9	14.6	14.2	14.0	13.0
90			19.9	18.1	16.8	15.8	14.7	14.0	13.3	12.6	12.2	11.5	11.1	10.3
100		20.0	17.9	16.3	15.1	14.2	12.6	11.5	10.7	10.2	9.9	9.3	9.0	8.3
110	20.0	18.3	16.3	14.8	13.6	12.0	10.4	9.5	8.9	8.5	8.2	7.7	7.4	6.9
120	19.5	16.8	15.0	13.3	11.5	10.1	8.8	8.0	7.5	7.1	6.9	6.5	6.3	5.8
130	18.0	15.5	13.7	11.4	9.8	8.6	7.5	6.8	6.4	6.1	5.8	5.5	5.3	4.9
140	16.7	14.4	11.8	9.8	8.4	7.4	6.4	5.9	5.5	5.2	5.0	4.8	4.6	4.2
150	15.6	13.0	10.3	8.5	7.3	6.5	5.6	5.1	4.8	4.5	4.4	4.1	4.0	3.7
160	14.6	11.5	9.1	7.5	6.4	5.7	4.9	4.5	4.2	4.0	3.9	3.6		3.2
170	13.7	10.1	8.0	6.7	5.7	5.0	4.4	4.0	3.7	3.5	3.4			2.9
180	12.2	9.1	7.2	5.9	5.1	4.5	3.9	3.5	3.3	3.2	3.0			2.6
200	9.9	7.3	5.8	4.8	4.1	3.6	3.2	2.9	2.7					2.1
220	8.2	6.1	4.8	4.0	3.4	3.0	2.6							1.7
240	6.9	5.1	4.0	3.3	2.9	2.5								1.4
260	5.8	4.3	3.4	2.8	2.4	2.2								
280	5.0	3.7	3.0	2.5										

Slenderness Ratios h/t: Web Depth to Web Thickness

2-36
ALLOWABLE SHEAR STRESS (KSI)
IN WEBS OF PLATE GIRDERS

For 36 ksi Yield Stress Steel, Tension Field Action Included

(*Italic* values indicate gross area, as percent of web area, required for pairs of intermediate stiffeners of 36 ksi yield stress steel.)[a]

$F_y = 36$ ksi

Slenderness Ratios h/t: Web Depth to Web Thickness

h/t	\multicolumn{13}{c}{Aspect Ratios a/h: Stiffener Spacing to Web Depth}	
	0.5	0.6	0.7	0.8	0.9	1.0	1.2	1.4	1.6	1.8	2.0	2.5	3.0	over 3
60										14.5	14.5	14.5	14.5	14.5
70							14.5	14.5	14.5	14.4	14.2	13.8	13.6	13.0
80					14.5	14.5	14.0	13.4	13.0	12.6	12.4	12.2	12.0	11.4
												0.3	*0.4*	
90				14.5	14.3	13.4	12.5	12.2	11.9	11.8	11.6	11.3	11.1	10.1
								0.6	*0.9*	*1.1*	*1.2*	*1.2*	*1.2*	
100			14.5	13.9	12.8	12.3	11.9	11.6	11.3	11.1	10.9	10.3	10.0	8.3
						0.5	*1.4*	*1.8*	*2.0*	*2.1*	*2.2*	*2.3*	*2.1*	
110		14.5	13.8	12.6	12.2	11.9	11.5	11.0	10.5	10.1	9.8	9.2	8.8	6.9
					0.9	*1.8*	*2.5*	*3.1*	*3.5*	*3.6*	*3.6*	*3.4*	*3.1*	
120		14.3	12.7	12.2	11.8	11.5	10.8	10.2	9.8	9.4	9.0	8.4	8.0	5.8
				1.1	*2.1*	*2.8*	*4.1*	*4.7*	*4.9*	*4.9*	*4.7*	*4.3*	*3.8*	
130	14.5	13.2	12.2	11.9	11.5	11.0	10.3	9.7	9.2	8.8	8.4	7.8	7.3	4.9
			0.9	*2.2*	*3.2*	*4.5*	*5.6*	*5.9*	*6.0*	*5.8*	*5.6*	*5.0*	*4.4*	
140	14.2	12.4	12.0	11.6	11.0	10.5	9.8	9.2	8.7	8.3	7.9	7.2	6.8	4.2
		0.3	*1.9*	*3.2*	*4.8*	*5.9*	*6.7*	*6.9*	*6.8*	*6.6*	*6.3*	*5.5*	*4.9*	
150	13.2	12.2	11.8	11.2	10.6	10.1	9.4	8.8	8.3	7.9	7.5	6.8	6.3	3.7
		1.2	*2.8*	*4.7*	*6.1*	*7.0*	*7.6*	*7.7*	*7.5*	*7.2*	*6.8*	*6.0*	*5.2*	
160	12.4	12.0	11.5	10.9	10.3	9.8	9.1	8.5	8.0	7.6	7.2	6.5		3.2
		2.1	*4.1*	*6.0*	*7.2*	*8.0*	*8.4*	*8.3*	*8.1*	*7.7*	*7.3*	*6.3*		
170	12.3	11.8	11.2	10.6	10.1	9.6	8.9	8.3	7.7	7.3	6.9			2.9
	0.9	*2.8*	*5.3*	*7.0*	*8.1*	*8.7*	*9.0*	*8.9*	*8.5*	*8.1*	*7.7*			
180	12.1	11.6	10.9	10.4	9.9	9.4	8.7	8.1	7.5	7.1	6.7			2.6
	1.6	*4.0*	*6.3*	*7.9*	*8.8*	*9.4*	*9.6*	*9.3*	*8.9*	*8.5*	*8.0*			
200	11.9	11.2	10.5	10.0	9.5	9.1	8.3	7.7	7.2					2.1
	2.9	*6.0*	*8.0*	*9.2*	*10.0*	*10.4*	*10.4*	*10.0*	*9.5*					
220	11.5	10.8	10.3	9.7	9.3	8.8	8.1	7.5						1.7
	4.8	*7.5*	*9.2*	*10.2*	*10.8*	*11.1*	*11.0*	*10.6*						
240	11.2	10.6	10.0	9.5	9.1	8.6								1.4
	6.2	*8.6*	*10.1*	*11.0*	*11.5*	*11.7*								
260	11.0	10.4	9.9	9.4	8.9	8.5								1.2
	7.3	*9.5*	*10.8*	*11.6*	*12.0*	*12.1*								
280	10.8	10.2	9.7	9.2										
	8.2	*10.2*	*11.4*	*12.1*										
300	10.7	10.1	9.6											
	9.0	*10.8*	*11.8*											
320	10.5	10.0												
	9.5	*11.2*												

[a]For single angle stiffeners, multiply by 1.8; for single plate stiffeners, multiply by 2.4.

Note: Girders so proportioned that the computed shear is less than that given in right-hand column do not require intermediate stiffeners.

ALLOWABLE SHEAR STRESS (KSI) IN WEBS OF PLATE GIRDERS

For 50 ksi Yield Stress Steel, Tension Field Action Included

(*Italic* values indicate gross area, as percent of web area, required for pairs of intermediate stiffeners of 50 ksi yield stress steel.)[a]

$F_y = 50$ ksi

h/t		Aspect Ratios a/h: Stiffener Spacing to Web Depth												
Slenderness Ratios h/t: Web Depth to Web Thickness	0.5	0.6	0.7	0.8	0.9	1.0	1.2	1.4	1.6	1.8	2.0	2.5	3.0	over 3
50										20.0	20.0	20.0	20.0	20.0
60							20.0	20.0	20.0	19.9	19.5	18.9	18.6	17.9
70				20.0	20.0	18.9	18.0	17.4	17.1	16.9	16.6	16.3		15.3
										0.2	*0.4*	*0.5*	*0.6*	
80			20.0	20.0	18.9	17.8	17.0	16.6	16.2	15.9	15.7	15.2	14.9	13.0
						0.6	*1.1*	*1.4*	*1.6*	*1.6*	*1.6*	*1.5*		
90			19.9	18.1	17.1	16.7	16.2	15.7	15.1	14.6	14.2	13.4	12.8	10.3
					0.4	*1.3*	*2.1*	*2.5*	*2.8*	*3.1*	*3.1*	*3.0*	*2.8*	
100		20.0	17.9	17.0	16.5	16.1	15.2	14.4	13.8	13.2	12.8	11.9	11.3	8.3
				0.8	*1.9*	*2.6*	*3.7*	*4.4*	*4.6*	*4.6*	*4.5*	*4.1*	*3.7*	
110	20.0	18.3	17.0	16.5	16.0	15.3	14.3	13.4	12.8	12.2	11.7	10.8	10.2	6.9
			0.9	*2.1*	*3.2*	*4.5*	*5.5*	*5.9*	*5.9*	*5.8*	*5.6*	*5.0*	*4.4*	
120	19.5	17.2	16.6	16.0	15.2	14.5	13.5	12.7	12.0	11.4	10.9	10.0	9.3	5.8
		0.4	*2.0*	*3.4*	*5.0*	*6.1*	*6.9*	*7.0*	*6.9*	*6.7*	*6.4*	*5.6*	*4.9*	
130	18.0	16.8	16.3	15.4	14.6	14.0	12.9	12.1	11.4	10.8	10.3	9.3	8.6	4.9
		1.5	*3.1*	*5.1*	*6.5*	*7.4*	*7.9*	*7.9*	*7.7*	*7.4*	*7.0*	*6.1*	*5.3*	
140	17.2	16.6	15.7	14.9	14.2	13.5	12.5	11.6	10.9	10.3	9.8	8.8	8.1	4.2
	0.5	*2.4*	*4.7*	*6.5*	*7.7*	*8.4*	*8.7*	*8.6*	*8.3*	*7.9*	*7.5*	*6.5*	*5.7*	
150	16.9	16.2	15.3	14.5	13.8	13.1	12.1	11.3	10.5	9.9	9.4	8.4	7.7	3.7
	1.4	*3.6*	*6.0*	*7.6*	*8.6*	*9.2*	*9.4*	*9.2*	*8.8*	*8.3*	*7.9*	*6.8*	*5.9*	
160	16.7	15.8	14.9	14.2	13.5	12.8	11.8	11.0	10.2	9.6	9.1	8.0		3.2
	2.1	*4.9*	*7.1*	*8.5*	*9.3*	*9.8*	*9.9*	*9.7*	*9.2*	*8.7*	*8.2*	*7.1*		
170	16.5	15.5	14.6	13.9	13.2	12.6	11.6	10.7	10.0	9.4	8.8			2.9
	2.9	*6.0*	*8.0*	*9.2*	*10.0*	*10.4*	*10.4*	*10.0*	*9.5*	*9.0*	*8.5*			
180	16.2	15.2	14.4	13.7	13.0	12.4	11.4	10.5	9.8	9.1	8.6			2.6
	4.1	*6.9*	*8.8*	*9.9*	*10.5*	*10.8*	*10.8*	*10.4*	*9.8*	*9.3*	*8.7*			
200	15.7	14.8	14.0	13.3	12.6	12.0	11.0	10.2	9.4					2.1
	5.9	*8.4*	*9.9*	*10.8*	*11.3*	*11.6*	*11.4*	*10.9*	*10.3*					
220	15.3	14.4	13.7	13.0	12.4	11.8	10.8	9.9						1.7
	7.3	*9.5*	*10.8*	*11.6*	*12.0*	*12.1*	*11.8*	*11.2*						
240	15.0	14.2	13.5	12.8	12.2	11.6								1.4
	8.3	*10.3*	*11.5*	*12.1*	*12.4*	*12.5*								
260	14.8	14.0	13.3	12.7	12.0	11.5								
	9.2	*10.9*	*12.0*	*12.5*	*12.8*	*12.8*								
280	14.6	13.9	13.2	12.5										
	9.8	*11.4*	*12.4*	*12.9*										

[a]For areas of other intermediate stiffeners, multiply *italic* values by appropriate factor:

Stiffener Steel Grade	Pairs of Stiffeners	Single Angle Stiffeners	Single Plate Stiffeners
$F_y = 50$ ksi	1.0	1.8	2.4
$F_y = 36$ ksi	1.4	2.5	3.3

Note: Girders so proportioned that the computed shear is less than that given in right-hand column do not require intermediate stiffeners.

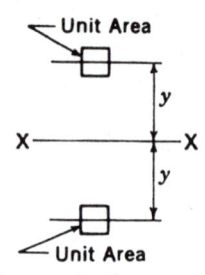

PLATE GIRDERS
Values of $2y^2$ for computing
Moment of Inertia
of areas about axis X-X

$(\Sigma Ay^2 = $ Area of one flange $\times 2y^2)$

2y	.0	.1	.2	.3	.4	.5	.6	.7	.8	.9
10	50	51	52	53	54	55	56	57	58	59
11	61	62	63	64	65	66	67	68	70	71
12	72	73	74	76	77	78	79	81	82	83
13	85	86	87	88	90	91	92	94	95	97
14	98	99	101	102	104	105	107	108	110	111
15	113	114	116	117	119	120	122	123	125	126
16	128	130	131	133	134	136	138	139	141	143
17	145	146	148	150	151	153	155	157	158	160
18	162	164	166	167	169	171	173	175	177	179
19	181	182	184	186	188	190	192	194	196	198
20	200	202	204	206	208	210	212	214	216	218
21	221	223	225	227	229	231	233	235	238	240
22	242	244	246	249	251	253	255	258	260	262
23	265	267	269	271	274	276	278	281	283	286
24	288	290	293	295	298	300	303	305	308	310
25	313	315	318	320	323	325	328	330	333	335
26	338	341	343	346	348	351	354	356	359	362
27	365	367	370	373	375	378	381	384	386	389
28	392	395	398	400	403	406	409	412	415	418
29	421	423	426	429	432	435	438	441	444	447
30	450	453	456	459	462	465	468	471	474	477
31	481	484	487	490	493	496	499	502	506	509
32	512	515	518	522	525	528	531	535	538	541
33	545	548	551	554	558	561	564	568	571	575
34	578	581	585	588	592	595	599	602	606	609
35	613	616	620	623	627	630	634	637	641	644
36	648	652	655	659	662	666	670	673	677	681
37	685	688	692	696	699	703	707	711	714	718
38	722	726	730	733	737	741	745	749	753	757
39	761	764	768	772	776	780	784	788	792	796
40	800	804	808	812	816	820	824	828	832	836
41	841	845	849	853	857	861	865	869	874	878
42	882	886	890	895	899	903	907	912	916	920
43	925	929	933	937	942	946	950	955	959	964
44	968	972	977	981	986	990	995	999	1004	1008
45	1013	1017	1022	1026	1031	1035	1040	1044	1049	1053
46	1058	1063	1067	1072	1076	1081	1086	1090	1095	1100
47	1105	1109	1114	1119	1123	1128	1133	1138	1142	1147
48	1152	1157	1162	1166	1171	1176	1181	1186	1191	1196
49	1201	1205	1210	1215	1220	1225	1230	1235	1240	1245

PLATE GIRDERS
Values of $2y^2$ for computing Moment of Inertia of areas about axis X-X

$(\Sigma Ay^2 = \text{Area of one flange} \times 2y^2)$

Unit Area

2y	.0	.1	.2	.3	.4	.5	.6	.7	.8	.9
50	1250	1255	1260	1265	1270	1275	1280	1285	1290	1295
51	1301	1306	1311	1316	1321	1326	1331	1336	1342	1347
52	1352	1357	1362	1368	1373	1378	1383	1389	1394	1399
53	1405	1410	1415	1420	1426	1431	1436	1442	1447	1453
54	1458	1463	1469	1474	1480	1485	1491	1496	1502	1507
55	1513	1518	1524	1529	1535	1540	1546	1551	1557	1562
56	1568	1574	1579	1585	1590	1596	1602	1607	1613	1619
57	1625	1630	1636	1642	1647	1653	1659	1665	1670	1676
58	1682	1688	1694	1699	1705	1711	1717	1723	1729	1735
59	1741	1746	1752	1758	1764	1770	1776	1782	1788	1794
60	1800	1806	1812	1818	1824	1830	1836	1842	1848	1854
61	1861	1867	1873	1879	1885	1891	1897	1903	1910	1916
62	1922	1928	1934	1941	1947	1953	1959	1966	1972	1978
63	1985	1991	1997	2003	2010	2016	2022	2029	2035	2042
64	2048	2054	2061	2067	2074	2080	2087	2093	2100	2106
65	2113	2119	2126	2132	2139	2145	2152	2158	2165	2171
66	2178	2185	2191	2198	2204	2211	2218	2224	2231	2238
67	2245	2251	2258	2265	2271	2278	2285	2292	2298	2305
68	2312	2319	2326	2332	2339	2346	2353	2360	2367	2374
69	2381	2387	2394	2401	2408	2415	2422	2429	2436	2443
70	2450	2457	2464	2471	2478	2485	2492	2499	2506	2513
71	2521	2528	2535	2542	2549	2556	2563	2570	2578	2585
72	2592	2599	2606	2614	2621	2628	2635	2643	2650	2657
73	2665	2672	2679	2686	2694	2701	2708	2716	2723	2731
74	2738	2745	2753	2760	2768	2775	2783	2790	2798	2805
75	2813	2820	2828	2835	2843	2850	2858	2865	2873	2880
76	2888	2896	2903	2911	2918	2926	2934	2941	2949	2957
77	2965	2972	2980	2988	2995	3003	3011	3019	3026	3034
78	3042	3050	3058	3065	3073	3081	3089	3097	3105	3113
79	3121	3128	3136	3144	3152	3160	3168	3176	3184	3192
80	3200	3208	3216	3224	3232	3240	3248	3256	3264	3272
81	3281	3289	3297	3305	3313	3321	3329	3337	3346	3354
82	3362	3370	3378	3387	3395	3403	3411	3420	3428	3436
83	3445	3453	3461	3469	3478	3486	3494	3503	3511	3520
84	3528	3536	3545	3553	3562	3570	3579	3587	3596	3604
85	3613	3621	3630	3638	3647	3655	3664	3672	3681	3689
86	3698	3707	3715	3724	3732	3741	3750	3758	3767	3776
87	3785	3793	3802	3811	3819	3828	3837	3846	3854	3863
88	3872	3881	3890	3898	3907	3916	3925	3934	3943	3952
89	3961	3969	3978	3987	3996	4005	4014	4023	4032	4041

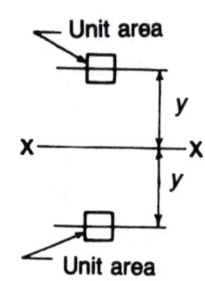

Unit area

PLATE GIRDERS
Values of $2y^2$ for computing
Moment of Inertia
of areas about axis X-X

$(\Sigma Ay^2 = \text{Area of one flange} \times 2y^2)$

Unit area

2y	.0	.1	.2	.3	.4	.5	.6	.7	.8	.9
90	4050	4059	4068	4077	4086	4095	4104	4113	4122	4131
91	4141	4150	4159	4168	4177	4186	4195	4204	4214	4223
92	4232	4241	4250	4260	4269	4278	4287	4297	4306	4315
93	4325	4334	4343	4352	4362	4371	4380	4390	4399	4409
94	4418	4427	4437	4446	4456	4465	4475	4484	4494	4503
95	4513	4522	4532	4541	4551	4560	4570	4579	4589	4598
96	4608	4618	4627	4637	4646	4656	4666	4675	4685	4695
97	4705	4714	4724	4734	4743	4753	4763	4773	4782	4792
98	4802	4812	4822	4831	4841	4851	4861	4871	4881	4891
99	4901	4910	4920	4930	4940	4950	4960	4970	4980	4990
100	5000	5010	5020	5030	5040	5050	5060	5070	5080	5090
101	5101	5111	5121	5131	5141	5151	5161	5171	5182	5192
102	5202	5212	5222	5233	5243	5253	5263	5274	5284	5294
103	5305	5315	5325	5335	5346	5356	5366	5377	5387	5398
104	5408	5418	5429	5439	5450	5460	5471	5481	5492	5502
105	5513	5523	5534	5544	5555	5565	5576	5586	5597	5607
106	5618	5629	5639	5650	5660	5671	5682	5692	5703	5714
107	5725	5735	5746	5757	5767	5778	5789	5800	5810	5821
108	5832	5843	5854	5864	5875	5886	5897	5908	5919	5930
109	5941	5951	5962	5973	5984	5995	6006	6017	6028	6039
110	6050	6061	6072	6083	6094	6105	6116	6127	6138	6149
111	6161	6172	6183	6194	6205	6216	6227	6238	6250	6261
112	6272	6283	6294	6306	6317	6328	6339	6351	6362	6373
113	6385	6396	6407	6418	6430	6441	6452	6464	6475	6487
114	6498	6509	6521	6532	6544	6555	6567	6578	6590	6601
115	6613	6624	6636	6647	6659	6670	6682	6693	6705	6716
116	6728	6740	6751	6763	6774	6786	6798	6809	6821	6833
117	6845	6856	6868	6880	6891	6903	6915	6927	6938	6950
118	6962	6974	6986	6997	7009	7021	7033	7045	7057	7069
119	7081	7092	7104	7116	7128	7140	7152	7164	7176	7188
120	7200	7212	7224	7236	7248	7260	7272	7284	7296	7308
121	7321	7333	7345	7357	7369	7381	7393	7405	7418	7430
122	7442	7454	7466	7479	7491	7503	7515	7528	7540	7552
123	7565	7577	7589	7601	7614	7626	7638	7651	7663	7676
124	7688	7700	7713	7725	7738	7750	7763	7775	7788	7800
125	7813	7825	7838	7850	7863	7875	7888	7900	7913	7925
126	7938	7951	7963	7976	7988	8001	8014	8026	8039	8052
127	8065	8077	8090	8103	8115	8123	8141	8154	8166	8179
128	8192	8205	8218	8230	8243	8256	8269	8282	8295	8308
129	8321	8333	8346	8359	8372	8385	8398	8411	8424	8437

PLATE GIRDERS
Values of $2y^2$ for computing Moment of Inertia of areas about axis X-X

$(\Sigma Ay^2 = \text{Area of one flange} \times 2y^2)$

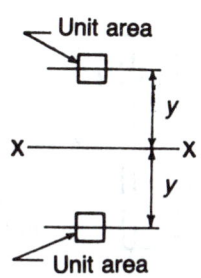

2y	.0	.1	.2	.3	.4	.5	.6	.7	.8	.9
130	8450	8463	8476	8489	8502	8515	8528	8541	8554	8567
131	8581	8594	8607	8620	8633	8646	8659	8672	8686	8699
132	8712	8725	8738	8752	8765	8778	8791	8805	8818	8831
133	8845	8858	8871	8884	8898	8911	8924	8938	8951	8965
134	8978	8991	9005	9018	9032	9045	9059	9072	9086	9099
135	9113	9126	9140	9153	9167	9180	9194	9207	9221	9234
136	9248	9262	9275	9289	9302	9316	9330	9343	9357	9371
137	9385	9398	9412	9426	9439	9453	9467	9481	9494	9508
138	9522	9536	9550	9563	9577	9591	9605	9619	9633	9647
139	9661	9674	9688	9702	9716	9730	9744	9758	9772	9786
140	9800	9814	9828	9842	9856	9870	9884	9898	9912	9926
141	9941	9955	9969	9983	9997	10011	10025	10039	10054	10068
142	10082	10096	10110	10125	10139	10153	10167	10182	10196	10210
143	10225	10239	10253	10267	10282	10296	10310	10325	10339	10354
144	10368	10382	10397	10411	10426	10440	10455	10469	10484	10498
145	10513	10527	10542	10556	10571	10585	10600	10614	10629	10643
146	10658	10673	10687	10702	10716	10731	10746	10760	10775	10790
147	10805	10819	10834	10849	10863	10878	10893	10908	10922	10937
148	10952	10967	10982	10996	11011	11026	11041	11056	11071	11086
149	11101	11115	11130	11145	11160	11175	11190	11205	11220	11235
150	11250	11265	11280	11295	11310	11325	11340	11355	11370	11385
151	11401	11416	11431	11446	11461	11476	11491	11506	11522	11537
152	11552	11567	11582	11598	11613	11628	11643	11659	11674	11689
153	11705	11720	11735	11750	11766	11781	11796	11812	11827	11843
154	11858	11873	11889	11904	11920	11935	11951	11966	11982	11997
155	12013	12028	12044	12059	12075	12090	12106	12121	12137	12152
156	12168	12184	12199	12215	12230	12246	12262	12277	12293	12309
157	12325	12340	12356	12372	12387	12403	12419	12435	12450	12466
158	12482	12498	12514	12529	12545	12561	12577	12593	12609	12625
159	12641	12656	12672	12688	12704	12720	12736	12752	12768	12784
160	12800	12816	12832	12848	12864	12880	12896	12912	12928	12944
161	12961	12977	12993	13009	13025	13041	13057	13073	13090	13106
162	13122	13138	13154	13171	13187	13203	13219	13236	13252	13268
163	13285	13301	13317	13333	13350	13366	13382	13399	13415	13432
164	13448	13464	13481	13497	13514	13530	13547	13563	13580	13596
165	13613	13629	13646	13662	13679	13695	13712	13728	13745	13761
166	13778	13795	13811	13828	13844	13861	13878	13894	13911	13928
167	13945	13961	13978	13995	14011	14028	14045	14062	14078	14095
168	14112	14129	14146	14162	14179	14196	14213	14230	14247	14264
169	14281	14297	14314	14331	14348	14365	14382	14399	14416	14433

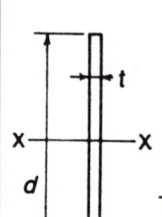

PLATE GIRDERS
Moment of Inertia
of one plate about axis X-X

To obtain the moment of inertia for any thickness of plate not listed below, multiply the value for a plate one inch thick by the desired thickness.

Depth d, In.	Thickness t, In.							
	3/8	7/16	1/2	9/16	5/8	3/4	7/8	1
10	31.3	36.5	41.7	46.9	52.1	62.5	72.9	83.3
11	41.6	48.5	55.5	62.4	69.3	83.2	97.1	110.9
12	54.0	63.0	72.0	81.0	90.0	108.0	126.0	144.0
13	68.7	80.1	91.5	103.0	114.4	137.3	160.2	183.1
14	85.8	100.0	114.3	128.6	142.9	171.5	200.1	228.7
15	105.5	123.0	140.6	158.2	175.8	210.9	246.1	281.3
16	128.0	149.3	170.7	192.0	213.3	256.0	298.7	341.3
17	153.5	179.1	204.7	230.3	255.9	307.1	358.2	409.4
18	182.3	212.6	243.0	273.4	303.8	364.5	425.3	486.0
19	214.3	250.1	285.8	321.5	357.2	428.7	500.1	571.6
20	250.0	291.7	333.3	375.0	416.7	500.0	583.3	666.7
21	289.4	337.6	385.9	434.1	482.3	578.8	675.3	771.8
22	332.8	388.2	443.7	499.1	554.6	665.5	776.4	887.3
23	380.2	443.6	507.0	570.3	633.7	760.4	887.2	1013.9
24	432.0	504.0	576.0	648.0	720.0	864.0	1008.0	1152.0
25	488.3	569.7	651.0	732.4	813.8	976.6	1139.3	1302.1
26	549.3	640.8	732.3	823.9	915.4	1098.5	1281.6	1464.7
27	615.1	717.6	820.1	922.6	1025.2	1230.2	1435.2	1640.3
28	686.0	800.3	914.7	1029.0	1143.3	1372.0	1600.7	1829.3
29	762.2	889.2	1016.2	1143.2	1270.3	1524.3	1778.4	2032.4
30	843.8	984.4	1125.0	1265.6	1406.3	1687.5	1968.8	2250.0
31	931.0	1086.1	1241.3	1396.5	1551.6	1861.9	2172.3	2482.6
32	1024.0	1194.7	1365.3	1536.0	1706.7	2048.0	2389.3	2730.7
33	1123.0	1310.2	1497.4	1684.5	1871.7	2246.1	2620.4	2994.8
34	1228.3	1433.0	1637.7	1842.4	2047.1	2456.5	2865.9	3275.3
35	1339.8	1563.2	1786.5	2009.8	2233.1	2679.7	3126.3	3572.9
36	1458.0	1701.0	1944.0	2187.0	2430.0	2916.0	3402.0	3888.0
37	1582.9	1846.7	2110.5	2374.4	2638.2	3165.8	3693.4	4221.1
38	1714.8	2000.5	2286.3	2572.1	2857.9	3429.5	4001.1	4572.7
39	1853.7	2162.7	2471.6	2780.6	3089.5	3707.4	4325.3	4943.3
40	2000.0	2333.3	2666.7	3000.0	3333.3	4000.0	4666.7	5333.3
41	2153.8	2512.7	2871.7	3230.7	3589.6	4307.6	5025.5	5743.4
42	2315.3	2701.1	3087.0	3472.9	3858.8	4630.5	5402.3	6174.0
43	2484.6	2898.7	3312.8	3726.9	4141.0	4969.2	5797.4	6625.6
44	2662.0	3105.7	3549.3	3993.0	4436.7	5324.0	6211.3	7098.7
45	2847.7	3322.3	3796.9	4271.5	4746.1	5695.3	6644.5	7593.8
46	3041.8	3548.7	4055.7	4562.6	5069.6	6083.5	7097.4	8111.3
47	3244.5	3785.2	4326.0	4866.7	5407.4	6488.9	7570.4	8651.9
48	3456.0	4032.0	4608.0	5184.0	5760.0	6912.0	8064.0	9216.0
49	3676.5	4289.3	4902.0	5514.8	6127.6	7353.1	8578.6	9804.1

PLATE GIRDERS
Moment of Inertia
of one plate about axis X-X

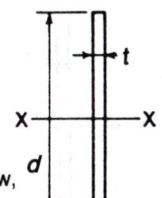

To obtain the moment of inertia for any thickness of plate not listed below, multiply the value for a plate one inch thick by the desired thickness.

Depth d, In.	Thickness t, In.							
	3/8	7/16	1/2	9/16	5/8	3/4	7/8	1
50	3906.3	4557.3	5208.3	5859.4	6510.4	7812.5	9114.6	10417
51	4145.3	4836.2	5527.1	6218.0	6908.9	8290.7	9672.5	11054
52	4394.0	5126.3	5858.7	6591.0	7323.3	8788.0	10253	11717
53	4652.4	5427.8	6203.2	6978.6	7754.0	9304.8	10856	12406
54	4920.8	5740.9	6561.0	7381.1	8201.3	9841.5	11482	13122
55	5199.2	6065.8	6932.3	7798.8	8665.4	10398	12132	13865
56	5488.0	6402.7	7317.3	8232.0	9146.7	10976	12805	14635
57	5787.3	6751.8	7716.4	8680.9	9645.5	11575	13504	15433
58	6097.3	7113.5	8129.7	9145.9	10162	12195	14227	16259
59	6418.1	7487.8	8557.5	9627.1	10697	12836	14976	17115
60	6750.0	7875.0	9000.0	10125	11250	13500	15750	18000
61	7093.2	8275.3	9457.5	10640	11822	14186	16551	18915
62	7447.8	8689.0	9930.3	11172	12413	14896	17378	19861
63	7814.0	9116.3	10419	11721	13023	15628	18232	20837
64	8192.0	9557.3	10923	12288	13653	16384	19115	21845
65	8582.0	10012	11443	12873	14303	17164	20025	22885
66	8984.3	10482	11979	13476	14974	17969	20963	23958
67	9398.8	10965	12532	14098	15665	18798	21931	25064
68	9826.0	11464	13101	14739	16377	19652	22927	26203
69	10266	11977	13688	15399	17110	20532	23954	27376
70	10719	12505	14292	16078	17865	21438	25010	28583
72	11664	13608	15552	17496	19440	23328	27216	31104
74	12663	14774	16884	18995	21105	25327	29548	33769
76	13718	16004	18291	20577	22863	27436	32009	36581
78	14830	17301	19773	22245	24716	29660	34603	39546
80	16000	18667	21333	24000	26667	32000	37333	42667
82	17230	20102	22974	25845	28717	34461	40204	45947
84	18522	21609	24696	27783	30870	37044	43218	49392
86	19877	23190	26502	29815	33128	39754	46379	53005
88	21296	24845	28395	31944	35493	42592	49691	56789
90	22781	26578	30375	34172	37969	45563	53156	60750
92	24334	28390	32445	36501	40557	48668	56779	64891
94	25956	30282	34608	38934	43260	51912	60563	69215
96	27648	32256	36864	41472	46080	55296	64512	73728
98	29412	34314	39216	44118	49020	58825	68629	78433
100	31250	36458	41667	46875	52083	62500	72917	83333
102	33163	38690	44217	49744	55271	66326	77380	88434
104	35152	41011	46869	52728	58587	70304	82021	93739
106	37219	43422	49626	55829	62032	74439	86845	99251
108	39366	45927	52488	59049	65610	78732	91854	104976

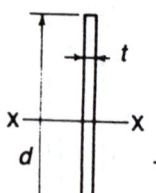

PLATE GIRDERS
Moment of Inertia
of one plate about axis X-X

To obtain the moment of inertia for any thickness of plate not listed below, multiply the value for a plate one inch thick by the desired thickness.

Depth d, In.	Thickness t, In.							
	3/8	7/16	1/2	9/16	5/8	3/4	7/8	1
110	41594	48526	55458	62391	69323	83188	97052	110917
112	43904	51221	58539	65856	73173	87808	102443	117077
114	46298	54015	61731	69447	77164	92597	108029	123462
116	48778	56908	65037	73167	81297	97556	113815	130075
118	51345	59902	68460	77017	85575	102690	119804	136919
120	54000	63000	72000	81000	90000	108000	126000	144000
122	56745	66203	75660	85118	94575	113491	132406	151321
124	59582	69512	79443	89373	99303	119164	139025	158885
126	62512	72930	83349	93768	104186	125024	145861	166698
128	65536	76459	87381	98304	109227	131072	152917	174763
130	68656	80099	91542	102984	114427	137312	160198	183083
132	71874	83853	95832	107811	119790	143748	167706	191664
134	75191	87723	100254	112786	125318	150381	175445	200509
136	78608	91709	104811	117912	131013	157216	183418	209621
138	82127	95815	109503	123191	136879	164255	191630	219006
140	85750	100042	114333	128625	142917	171500	200083	228667
142	89478	104391	119304	134216	149129	178955	208781	238607
144	93312	108864	124416	139968	155520	186624	217728	248832
146	97254	113463	129672	145881	162090	194508	226927	259345
148	101306	118190	135075	151959	168843	202612	236380	270149
150	105469	123047	140625	158203	175781	210938	246094	281250
152	109744	128035	146325	164616	182907	219488	256069	292651
154	114133	133155	152178	171200	190222	228266	266311	304355
156	118638	138411	158184	177957	197730	237276	276822	316368
158	123260	143803	164346	184890	205433	246519	287606	328693
160	128000	149333	170667	192000	213333	256000	298666	341333
162	132860	155004	177147	199290	221434	265721	310007	354294
164	137842	160815	183789	206763	229737	275684	321631	367579
166	142947	166771	190596	214420	238244	285893	333542	381191
168	148176	172872	197568	222264	246960	296352	345744	395136
170	153531	179120	204708	230297	255885	307062	358240	409417

COMPOSITE DESIGN
For building construction
General Notes

The AISC ASD Specification contains provisions for designing composite steel-concrete beams as follows:

1. For totally encased unshored steel beams not requiring mechanical anchorage (shear connectors), see Sects. I1 and I2.1.

2. For both shored and unshored beams with mechanically anchored slabs, design of the steel beam is based on the assumption composite action resists the total design moment (I2.2). In *shored* construction, flexural stress in the concrete slab due to composite action is determined from the total moment.

   In *unshored* construction, flexural stress in the concrete slab due to composite action is determined from moment M_L, produced by loads imposed after the concrete has achieved 75% of its required strength. Shored construction may be used to reduce dead load deflection and *must* be used if the combined bending stress in the steel exceeds $0.9 F_y$.

3. For partial composite action, see I2.2.

4. For negative moment zones, see I2.2.

5. For composite beams with formed steel deck (FSD), see I5.

GENERAL CONSIDERATIONS

1. Composite construction is appropriate for any loading. It is most efficient with heavy loading, relatively long spans, and beams spaced as far apart as permissible. The decision to use fully or partially composite beams is usually economic and will generally be based on a comparison of the installed cost of the shear connectors and the savings in beam weight.

2. For unshored construction, concrete compressive stress will seldom be critical for the beams listed in the Composite Beam Selection Tables if a full width slab and $F_y = 36$ ksi steel are used. It is more likely to be critical when a fully composite narrow concrete flange or $F_y = 50$ ksi is used, and is frequently critical if both $F_y = 50$ ksi steel and a narrow concrete flange are used. Shored construction also results in a higher concrete stress while partial composite construction reduces this stress.

   The following rational procedure for a standard transformed section may be used to calculate the maximum concrete compressive stress for any degree of partially composite construction with solid slabs.* (see Fig. 1)

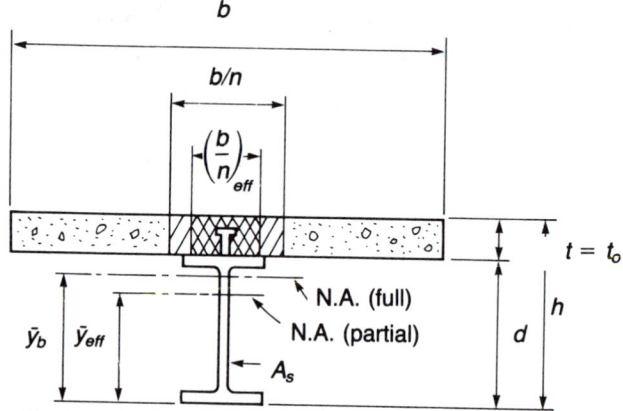

Fig. 1. *Composite beam with solid slab*

A. The location of the elastic neutral axis from the bottom of the steel beam is equal to

$$\bar{y}_{eff} = \frac{I_{eff}}{S_{eff}} \tag{1}$$

where

I_{eff} is determined from AISC ASD Equation (I4-4)

S_{eff} is determined from AISC ASD Equation (I2-1)

B. The effective width of the concrete slab $\left(\dfrac{b}{n}\right)_{eff}$, where $n = E/E_c$, varies from the full composite value, (b/n), to zero for the non-composite condition, as described by the relationship from the definition of neutral axis:

$$\left(\frac{b}{n}\right)_{eff} = \frac{A_s}{t}\left[\frac{\bar{y}_{eff} - d/2}{d + t/2 - \bar{y}_{eff}}\right] \leq \left(\frac{b}{n}\right) \tag{2}$$

where the quantities are as shown in Fig. 1.

C. The section modulus relative to the top of the equivalent transformed steel section is

$$S_{t\text{-}eff} = \frac{I_{eff}}{(h - \bar{y}_{eff})} \tag{3}$$

and the concrete stress at that point is

$$f_c = \frac{M}{S_{t\text{-}eff}}\frac{\left(\dfrac{b}{n}\right)_{eff}}{b} \tag{4}$$

---

*Lorenz, R. F. and F. W. Stockwell, "Concrete Slab Stresses in Partial Composite Beams and Girders," AISC *Engineering Journal*, 3rd Qtr., 1984.

As I_{eff} and S_{eff} approach I_s and S_s, $\bar{y}_{eff}$ approaches $d/2$, $\left(\dfrac{b}{n}\right)_{eff}$ approaches zero, $S_{t\text{-}eff}$ approaches $S_t = I_s/(t + d/2)$, and f_c' approaches zero — the non-composite case. As I_{eff} and S_{eff} approach I_{tr} and S_{tr} (full composite), $\bar{y}_{eff} = I_{tr}/S_{tr} = \bar{y}_b$, $\left(\dfrac{b}{n}\right)_{eff} = b/n$, $S_{t\text{-}eff} = I_{tr}\,(h-\bar{y}_b)$ and $f_c = M/(nS_t)$. The concrete stress moment M is determined in accordance with General Note 2.

To extend this procedure more generally to composite beams with steel deck or with the neutral axis in the concrete, only replace $(t/2)$ in Eq. 2 by $Y2$, the distance from steel beam top flange to centroid of concrete compressive area, (Fig. 2), so that

$$\left(\frac{b}{n}\right)_{eff} = \frac{A_s}{t}\left[\frac{\bar{y}_{eff} - d/2}{d + Y2 - \bar{y}_{eff}}\right] \le \left(\frac{b}{n}\right) \tag{2a}$$

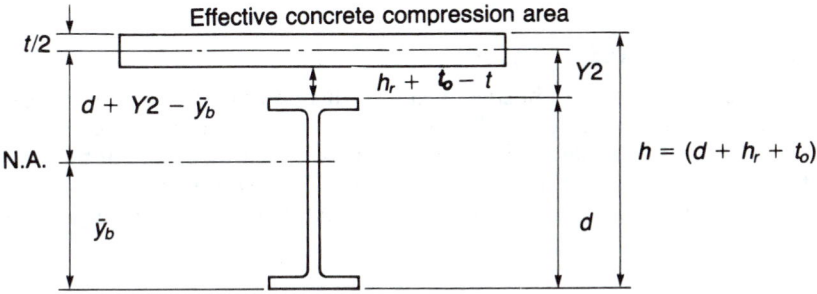

Fig. 2.   *Fully composite beam with steel deck*

Note that with deck, t represents the thickness of the concrete in compression, which is often conservatively assumed to be only the full concrete topping t_o above the deck (both for deck perpendicular and parallel to beam) such that $Y2 = \left(h_r + \dfrac{t}{2}\right)$, where h_r is the nominal deck rib height. The effective concrete thickness needs to be appropriately reduced if the elastic neutral axis falls within this topping area, as discussed later.

3. The previous dependence of concrete effective width on slab thickness has been removed to be consistent with the most current research findings.

4. Steel and concrete materials of various strengths may be used.

END REACTIONS

End reactions for composite beams are likely to be higher than for non-composite beams of the same size. They must be calculated by the engineer and shown on the contract documents.

DEFLECTION

A composite beam has much greater stiffness than a non-composite beam of equal depth, size, loads and span length. Deflection of composite beams will usually be about ⅓ to ½ less than deflection of non-composite beams. In practice, shallower beams are used and deflections, particularly of the steel section alone under construction loads, should be calculated and listed on the contract documents as a guide for cambering or estimating slab quantities.

If the desire is to minimize the transient vibration because of pedestrian traffic or other moving loads when composite beams support large open floor areas free of partitions or other damping sources, a suitable dynamic analysis should be made (see AISC *Engineering Journal*, Vol. 12, No. 3, 3rd Qtr., 1975).

Long-term creep deflections are usually not significant for composite beams. However, since a part of the section is concrete, which can be susceptible to creep, creep deflection should be investigated if considered undesirable by the design professional. If the desire is to investigate long-term creep deflections, I_{tr} should be based on a modular ratio n, double that used for stress calculations.

When lightweight concrete is used in composite construction, deflection should be calculated using I_{tr} based on the actual modulus of elasticity of the concrete E_c, even though stress calculations are based on the E_c of normal weight concrete.

USE OF COVER PLATES

Bottom cover plates are an effective means to increase the strength or reduce the depth of composite beams when deflections are not critical, but they should be used with overall economy in mind. High labor cost makes their use rare. For this reason, they have not been retained in the selection table. However, their use is permitted by the Specification. Section properties may be calculated or obtained from other sources.

USE OF FORMED STEEL DECK (FSD)

Although use of FSD in composite construction has been permitted by the AISC Specification for many years, specific provisions for its use were not included in the Specification until 1978. The following limitation on parameters were established to keep composite construction with FSD within the currently available research data (see AISC ASD Specification Sect. I5.1):

1. Deck rib height (h_r): Max. 3 in.

2. Average width of concrete rib or haunch (w_r): Min. 2 in.

3. Shear connectors: Welded studs only, maximum ¾-in. dia.

4. Stud length: Min. = rib height + 1½ in.

5. Slab thickness above deck: Min. 2 in.

The composite beam selection tables may be used with FSD as well as with solid slabs. There are numerous proprietary FSDs and most manufacturers can furnish tables of properties for beams using their own particular one. The FSD used in the examples is arbitrary and is not intended to agree dimensionally with any proprietary deck.

The FSD itself may be either composite or non-composite. Under this AISC ASD Specification both will produce the same composite beam properties.

OTHER CONSIDERATIONS

The AISC ASD Specification provisions for the design of composite beams are based on ultimate load considerations, even though they are presented in terms of working stresses. Because of this, for unshored construction, actual stresses in the steel beam under working load are higher than calculated stresses. Section I2.2 limits this steel stress to 90% of the minimum yield stress. Section I2.2 also provides requirements for limiting the steel beam compression flange stress under construction loading.

Adequate lateral support for the compression flange of the steel section will be provided by the concrete slab after hardening. During construction, however, lateral support must be provided, or working stresses must be reduced in accordance with F1 of the Specification. Steel deck with adequate attachment to the compression flange, or properly constructed concrete forms, will usually provide the necessary lateral support. For construction using fully encased beams, particular attention should be given to lateral support during construction.

The design of the concrete slab should conform to the current ACI Building Code.

COMPOSITE BEAM SELECTION TABLES

New composite beam selection tables for I_{tr} and S_{tr} based on an elastic transformed section have been prepared to assist engineers in designing these members or in checking steel and concrete allowable stresses. They are applicable to both solid slab and FSD floors, fully composite construction, for all steel beam sizes and without any restrictions on concrete strength f'_c, modular ratio n or effective width b.

By definition of elastic neutral axis for a fully composite beam (see Fig. 2)

$$\bar{y}_b = \frac{A_s\left(\dfrac{d}{2}\right) + \left(\dfrac{b}{n}\right) t\,(d + Y2)}{A_s + \left(\dfrac{b}{n}\right)t} \tag{5}$$

Let

$$A_{ctr} = \left(\frac{b}{n}\right)t \tag{6}$$

then Eq. (5) reduces to

$$\bar{y}_b = \frac{A_s\left(\dfrac{d}{2}\right) + A_{ctr}\,(d + Y2)}{(A_s + A_{ctr})}$$

or

$$\bar{y}_b = \frac{d}{2} + \left[\frac{\left(\dfrac{A_{ctr}}{A_s}\right)}{\left(1 + \dfrac{A_{ctr}}{A_s}\right)}\right]\left(Y2 + \frac{d}{2}\right)$$

Inclusion of the effective concrete width, strength, and modular ratio in the A_{ctr} variable permits direct computation of the $\bar{y}_b$ given a steel beam size and $Y2$. Once the neutral axis location has been identified, it follows,

$$I_{tr} = I_s + A_s\left(\bar{y}_b - \frac{d}{2}\right)^2 + \frac{1}{12}\left(\frac{b}{n}\right)t^3 + A_{ctr}\,(d + Y2 - \bar{y}_b)^2 \tag{7}$$

or

$$I_{tr} = I_s + A_{ctr} \left[\frac{t^2}{12} + \left(Y2 + \frac{d}{2} \right)^2 \frac{1}{\left(\frac{A_{ctr}}{A_s} + 1 \right)} \right]$$

To avoid introduction of another variable (concrete compression thickness) into the tables, the concrete moment of inertia about its own axis is conservatively simplified to $\frac{1}{12} A_{ctr} (1)^2$ (assumes $t = 1$ in.) in the tables, which lowers the I_{tr} value by usually a small percentage. Finally, referring to $\bar{I}_{tr}$ as the tabulated value,

$$\bar{S}_{tr} = \frac{\bar{I}_{tr}}{\bar{y}_b} \tag{8}$$

For many situations, assume $I_{tr} = \bar{I}_{tr}$ and $S_{tr} = \bar{S}_{tr}$ and it will be sufficiently accurate.

If it is desirable to determine the slightly higher and more exact elastic transformed section properties I_{tr} and S_{tr}, a quick correction can be made to the tabulated values, using the actual concrete thickness t in compression, as follows:

$$I_{tr} = \bar{I}_{tr} \text{ (tabulated)} + \frac{1}{12} A_{ctr} (t^2 - 1) \tag{9}$$

$$S_{tr} = (I_{tr}) \left[\frac{\bar{S}_{tr} \text{ (tabulated)}}{\bar{I}_{tr} \text{ (tabulated)}} \right] \tag{10}$$

When the neutral axis is located within the concrete flange (t_o), only the concrete thickness in compression (above the neutral axis) may be used for t and A_{ctr}.

When the neutral axis is in the steel section or deck, $A_{ctr} = \left(\frac{b}{n} \right) t_o$ and $Y2 = \left(h_r + \frac{t_o}{2} \right)$. The maximum depth of the transformed section neutral axis for $t = t_o$ (all concrete above deck in compression) is $(d + h_r)$ to the top of the steel deck. Applying this limit to $\bar{y}_b$ in (5) and solving for A_{ctr}^{max} results in

$$A_{ctr}^{max} = max \left(\frac{b}{n} \right) (t_o) = \frac{A_s \left(\frac{d}{2} + h_r \right)}{(Y2 - h_r)} \tag{11}$$

Note that the distance $Y2$ must be greater than h_r. If $\left(\frac{b}{n} \right) t_o$ is larger than A_{ctr}^{max} (indicated by shaded area in tables for $h_r = 0$) for a particular $Y2$ and h_r, a reduced A_{ctr} using $t < t_o$ and the corresponding revised $Y2$ need to be used for the actual section properties (see Ex. 18).

In such cases, with the elastic neutral axis within the concrete, the reduced concrete thickness in compression may be calculated as

$$t = A_s \left[\frac{-1 + \sqrt{1 + \frac{2 \left(\frac{b}{n} \right)}{A_s} \left(\frac{d}{2} + h_r + t_o \right)}}{\left(\frac{b}{n} \right)} \right] \leq t_o \tag{12}$$

and

$$Y2 = \left(h_r + t_o - \frac{t}{2} \right) \tag{13}$$

Linear interpolation may be used to estimate properties between the listed A_{ctr} and $Y2$ values as an alternative to direct computation. The following steps are suggested for general application of the Composite Beam Selection Tables:

Design

1. With a given $S_{req} = \dfrac{M_{max}}{0.66F_y}$, approximate A_{ctr} $\left(\text{for full composite } b_{eff} = \dfrac{b}{n}\right)$ and $Y2$, select appropriate steel beam size with $\bar{S}_{tr} > S_{req}$; check A_{ctr}^{max} and neutral axis location and make necessary corrections to tabulated values or calculate exact I_{tr}, S_{tr} directly with selected beam size.

2. Compute V_h for shear stud design using (I4-1) or (I4-2). The latter equation representing max V_h in steel is tabulated.

3. (Optional) Economize with partial composite,

$$V_h' = V_h \left[\frac{(S_{req} - S_s)}{(S_{tr} - S_s)} \right]^2 \geq 0.25$$

4. (Optional) Compute corresponding $\bar{I}_{eff}$ by AISC ASD Specification Equation (I4-4) to check deflections.

Analysis

With a given A_{ctr}, $Y2$ and steel beam size, read $\bar{I}_{tr}$ and $\bar{S}_{tr}$ $(\bar{S}_{tr} \geq S_{req})$, check A_{ctr}^{max}, as previously, and if necessary, compute corrected I_{tr} and S_{tr} by Eqs. 9 & 10.

EXAMPLE 15

Design a non-coverplated fully composite interior floor beam of an office building. There is no depth restriction. Do not use temporary shores. Limit dead-load deflection to 1½ in. and live-load deflection to $L/360$.

Given:

Span length, $L = 36$ ft
Beam spacing, $s = 8$ ft
Slab thickness, $t_o = 4$ in.
Concrete: $f_c' = 3.0$ ksi $(n = 9)$
Weight $= 145$ pcf
Steel: $F_y = 36$ ksi

Live load $= 100$ lbs./ft^2
Partition load $= 20$ lbs./ft^2
Ceiling load $= 8$ lbs./ft^2

Solution:

A. Bending moments:

1. Construction loads:

$$\begin{array}{ll} \text{4 in. slab} & = 0.048 \text{ kip/ft}^2 \\ \text{Steel (assumed)} & = \underline{0.007} \\ & 0.055 \text{ kip/ft}^2 \times 8 \text{ ft} = 0.44 \text{ kip/ft} \end{array}$$

$M_D = 71.3$ kip-ft

2. Loads applied after concrete has hardened:

Live load   = 0.100 kip/ft²*
Partition load = 0.020
Ceiling load   = 0.008
         0.128 kip/ft² × 8 ft = 1.024 kip/ft

M_L = 166 kip-ft

3. Maximum moment:

$M_{max} = M_D + M_L$
    = 71.3 + 166 = 237.3 kip-ft

4. Maximum shear:

$V = (0.44 + 1.024)\dfrac{36}{2} = 26.4$ kips

B. Check effective width of concrete slab:

I1

$b = \frac{1}{4}L = \frac{1}{4} \times 36 \times 12 = 108$ in.

$b = s = 8 \times 12 = 96$ in.   **(governs)**

C. Required section moduli (for F_y = 36 ksi):

For M_{D+L}:

$S_{tr} = \dfrac{12 \times 237}{24} = 119$ in.³

F1.1

For M_D (assume compression flange is adequately braced):

$S_s = \dfrac{12 \times 71.3}{24} = 35.6$ in.³

D. Select section and determine properties:

$Y2$ = 2 in.
A_{ctr} = (96/9) (4) = 42.67 in.²

Enter selection table for
S_{tr} (required) = 119 in.³

Select W21×44 trial section, since
S_{tr} = 119 with $Y2$ = 2 and A_{ctr} = 40 < A_{ctr}^{max} = 67

From Tables of Properties for Designing, W shapes (Manual, Part 1):

S_s = 81.6 in.³, A_s = 13.0 in.², t_f = 0.450 in.
I_s = 843 in.⁴, d = 20.66 in., t_w = 0.350 in.

Calculate section properties:

1. From table (at $Y2$ = 2 and A_{ctr} = 42.67 by interpolation)

$\bar{S}_{tr}$ = 119.3 in.³
$\bar{I}_{tr}$ = 2356 in.⁴

---

*Live load reduction is allowed by most building codes. It is omitted here for simplicity.

2. Add correction by Eqs. 9, 10 (optional)

$t = 4$ in.

$I_{tr} = 2409$ in.^{4}

$S_{tr} = 122$ in.^{3}

$\bar{y}_b = 2409/122 = 19.8$ in.

3. Compute directly (optional)

$I_{tr} = 2420$ in.^{4}

$S_{tr} = 122$ in.^{3}

$\bar{y}_b = 19.8$ in.

E. Check concrete stress (unshored):

$S_t = 2420/(20.66 + 4 - 19.8) = 498$ in.^{3}

$f_c = (166 \times 12)/(498 \times 9)$

$= 0.44$ ksi $< 0.45 \times 3 = 1.35$ ksi   **o.k.**

I2

F. Check steel stresses:

Total load: $S_{str} = 122$ in.$^3 > 119$ in.^{3} **o.k.**

Dead load: $S_s = 81.6$ in.$^3 > 35.6$ in.^{3} **o.k.**

f_b is   **o.k.**

$f_v = 26.4/(20.66 \times .350) = 3.65$ ksi < 14.5 ksi   **o.k.**

Note: Beam is not coped; block shear does not control.

G. Check deflection:

$\triangle_{DL} = M_D L^2/161I_s = 71.3 \times 36^2/(161 \times 843)$

$= 0.68$ in. < 1.5 **o.k.**

$\triangle_{LL} = M_L L^2/161I_{tr} = 166 \times 36^2/(161 \times 2420)$

$= 0.55$ in. $< L/360 = (36 \times 12)/360 = 1.20$ in.   **o.k.**

(Long-term creep deflection is not considered significant.)

H. Check bottom flange tension stress:

$f_b = (71.3 \times 12/81.6) + (166 \times 12/122)$

$= 26.8$ ksi $< 0.9 \times 36 = 32.4$ ksi

I. Shear connectors (for full composite action):

Try ¾-in. dia. × 3½-in. studs:

Max. stud dia. (unless located directly over the web):

$2.5t_f = 2.5 \times 0.450$

$= 1.13$ in. > 0.75 in.   **o.k.**

I4

Total horizontal shear:

Equation (I4-1):

$$V_h = 0.85 \times f_c' A_c/2 = 0.85 \times 3 \times \frac{(4 \times 96)}{2} = 490 \text{ kips}$$

ASD
Specification
Reference

Equation (I4-2):

Tabulated max V_h (steel) = 234 kips

$V_h = A_s F_y/2 = 13.0 \times (36/2) = 234$ kips **governs**

$N = V_h/q = 234/11.5 = 20.3$

I4

Use: 42 – ¾-in. dia. × 3½-in. studs, equally spaced (21 each side of the point of maximum moment)*, W21×44, A36.

EXAMPLE 16

Design the composite beam in Ex. 15 using 2-in. formed steel deck with ribs running perpendicular to the beam.

Given:

Same data as Ex. 15, except $t_o = 2$ in.

Solution:

A. Bending moments:

1. Construction loads:

4-in. slab + deck + mesh = 0.041 kip/ft^2
Steel (assumed)   = 0.007
                           0.048 kip/ft^2 × 8 ft = .384 kips/ft

$M_D = 62.2$ kip-ft

2. Loads applied after concrete has hardened (from Ex. 15):

0.128 kip/ft^2

$M_L = 166$ kip-ft

3. Maximum moment:

$M_{max} = M_D + M_L$

$= 62.2 + 166 = 228$ kip-ft

4. Maximum shear:

$V = 8 (0.048 + 0.128)36/2 = 25.3$ kips

B. Check effective width of concrete slab:

$b = (1/4)L = ¼ \times 36 \times 12 = 108$ in.
$b = s = 8 \times 12 = 96$ in. **governs**

C. Required section moduli: (for $F_y = 36$ ksi)

For M_{D+L}: $S_{tr} = (12 \times 228)/24 = 114$ in.^{3}

For M_D: $S_s = (12 \times 62.2)/24 = 31.1$ in.^{3}

---

*Partial composite action could be used to reduce the number of studs (see Ex. 16, step J).

D. Select section and determine properties:

$Y2 = 2 + 1 = 3$ in.

$A_{ctr} = (96/9)\,(2) = 21.33$ in.^{2}

Enter selection table for

S_t (required) $= 114$ in.^{3}

Select W21×44 trial section, since

$\tilde{S}_{tr} = 122$ with $Y2 = 3$ and $A_{ctr} = 20 < A_{ctr}^{max} = 160$

From Tables of Properties for Designing W shapes (Part 1):

$S_s = 81.6$ in.^{3}, $A_s = 13.0$ in.^{2}, $t_f = 0.450$ in.

$I_s = 843$ in.^{4}, $d = 20.66$ in., $t_w = 0.350$ in.

Calculate composite section properties:

1. From table ($Y2 = 3$ and $A_{ctr} = 21.33$) by interpolation.

$\tilde{S}_{tr} = 122.7$ in.^{3}

$\tilde{I}_{tr} = 2273$ in.^{4}

2. Add correction by Eqs. 9,10 (optional)

$t = 2$ in.

$I_{tr} = 2278$ in.^{4}

$S_{tr} = 123$ in.^{3}

$\bar{y}_b = 18.52$ in.

3. Compute directly (optional)

$I_{tr} = 2290$ in.^{4}

$S_{tr} = 123$ in.^{3}

$\bar{y}_b = 18.6$ in.

E. Check concrete stress:

$S_t = 2290/(20.66 + 4 - 18.6) = 378$ in.^{3}

$f_c = (166 \times 12)/(378 \times 9) = 0.59$ ksi < 1.35 ksi **o.k.**

F. Check steel stresses:

Total load: $S_{tr} = 123$ in.$^3 > 114$ in.^{3} **o.k.**

Dead load: $S_s = 81.6$ in.$^3 > 31.1$ in.^{3} **o.k.**

$\therefore f_b$ is **o.k.**

Web shear $f_v = 25.3/(20.66 \times 0.350)$

$= 3.50$ ksi < 14.5 ksi **o.k.**

Note: Beam is not coped; block shear does not control.

AMERICAN INSTITUTE OF STEEL CONSTRUCTION

G. Check deflection:

$$\Delta_{DL} = M_D L^2/161 I_s = (62.2 \times 36^2)/(161 \times 843)$$
$$= 0.59 \text{ in.} < 1\frac{1}{2} \text{ in.}\quad \textbf{o.k.}$$

$$\Delta_{LL} = M_L L^2/161 I_{tr} = (166 \times 36^2)/(161 \times 2290)$$
$$= 0.58 \text{ in.} < L/360 = (36 \times 12)/360 = 1.20 \text{ in.}\quad \textbf{o.k.}$$

(Long-term creep deflection is not considered significant.)

H. Check bottom flange tension stress:

$$f_b = \frac{62.2 \times 12}{81.6} + \frac{166 \times 12}{123}$$
$$= 25.3 \text{ ksi} < 0.9\, F_y = 32.4 \text{ ksi}\quad \textbf{o.k.}$$

I. Shear connectors (for full composite action):

Try ¾-in. dia. × 3½-in. studs:

Max. stud dia. (unless located directly over the web):

$$2.5t_f = 2.5 \times 0.450 = 1.13 \text{ in.} > 0.75 \text{ in.}\quad \textbf{o.k.}$$

Total horizontal shear:

Equation (I4-1):

$$V_h = 0.85 \times f_c' A_c/2$$
$$= 0.85 \times 3 \times \frac{(2 \times 96)}{2} = 245 \text{ kips}$$

Equation (I4-2):

Tabulated max. V_h (steel) = 234 kips **governs**

$$V_h = A_s F_y/2 = 13.0 \times \frac{36}{2} = 234 \text{ kips}$$

Calculated stud reduction factor:

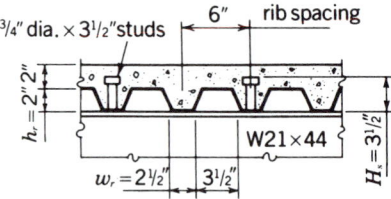

Assume $N_r = 1$ and $H_s = 3\frac{1}{2}$ in.

Given: $h_r = 2$ in., $w_r = 2\frac{1}{2}$ in.

Reduction Factor $= \left(\frac{0.85}{\sqrt{N_r}}\right)\left(\frac{w_r}{h_r}\right)\left(\frac{H_s}{h_r} - 1.0\right) \le 1.0$

$$= \left(\frac{0.85}{\sqrt{1}}\right)\left(\frac{2.5}{2}\right)\left(\frac{3\frac{1}{2}}{2} - 1.0\right) = 0.80$$

$q = (11.5 \times 0.80) = 9.2$ kips

$$N_1 = \frac{V_h}{q} = \frac{234}{9.2} = 25.4$$

Use: 52 − ¾-in. dia. × 3½-in. studs (26 each side of mid span)

J. Shear connectors (for partial composite action):*

Solve Equation (I2-1) for V_h':

$$V_h' = V_h \left[\frac{(S_{eff} - S_s)}{(S_{tr} - S_s)} \right]^2$$

$$= 234 \left[\frac{114 - 81.6}{123 - 81.6} \right]^2$$

$$= 143 \text{ kips} > \frac{1}{4} \times 234 = 58.5 \text{ kips} \textbf{o.k.}$$

$$N = \frac{V_h'}{q} = \frac{143}{9.2} = 15.5$$

Use: 30 − ¾-in. dia. 3½-in. studs (15 each side of mid-span), W21×44, A36

Spacing:

Studs must be placed in the deck ribs; therefore, stud spacing must be in 6 in. multiples. Additional studs or "puddle welds" must be used, as required, so that the space between deck attachments (welds or studs) does not exceed 16 in. Space studs as shown below.

℄ Beam

1'-0"   Puddle welds | 2 spa.@2'-0"

¾" diam.×3½" studs   12 spa.@1'-0"   2 spa.@2'-0"   1'-0"

K. Check deflection with partial composite action:

Effective moment of inertia:

$$V_h' = 9.2 \times 15 = 138 \text{ kips}$$

$$I_{eff} = I_s + \sqrt{\frac{V_h'}{V_h}} (I_{tr} - I_s)$$

$$= 843 + \sqrt{\frac{138}{234}} (2290 - 843)$$

$$= 1954 \text{ in.}^4$$

$$\Delta_{LL} \text{ (partial)} = \frac{2290}{1954} \times 0.58$$

$$= 0.68 \text{ in.} < L/360 = \frac{36 \times 12}{360}$$

$$= 1.20 \text{ in.} \textbf{o.k.}$$

---

*When the beam selected by the designer has a section modulus S_{tr} greater than the required section modulus, even by a small amount, the number of shear connectors can be reduced substantially by using the partial composite action provisions of the Specification. Reducing the number of shear connectors by 10 to 30% will usually reduce S_{tr} by only 3 to 10%. Live-load deflection will be increased slightly because of the corresponding reduction in the moment of inertia.

EXAMPLE 17 (Shear Connector Spacing for Concentrated Loads)

Design shear connectors for the composite beam shown below.

Given:

Slab thickness, t = 4 in.
F_y = 36 ksi
f'_c = 3.0 ksi
n = 9
b = 70 in.
Stud diameter = ⅝ in.

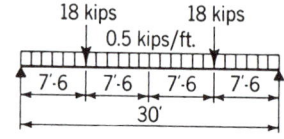

Solution:

A. Calculate moments and design the beam and shear connectors as in previous examples:

M_{max} (at mid-span) = 191 kip-ft

M (at concentrated load point) = 177 kip-ft

Beam = W18×40, F_y = 36 ksi

$$S_{tr} = 99.4; \quad S_s = 68.4; \quad \beta = \frac{S_{tr}}{S_s} = 1.45$$

N_1 = 27 − ⅝-in. dia. × 2½-in. studs required between the maximum and zero moment points.

B. Solve for N_2 (number of studs required between concentrated load and zero moment point):

$$N_2 = \frac{N_1 \left[\dfrac{M\beta}{M_{max}} - 1 \right]}{\beta - 1} = \frac{27 \left[\dfrac{177 \times 1.45}{191} - 1 \right]}{1.45 - 1}$$

= 21 − ⅝-in. dia. × 2½-in. studs

The stud spacing must be adjusted to place 21 studs between the concentrated load and the end of the beam (zero moment point). The balance ($N_1 - N_2$ = 6 studs) are placed between the concentrated load and the beam centerline (maximum moment point), as shown below.

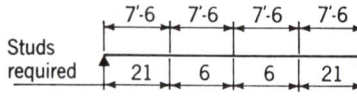

If the term $M\beta/M_{max}$ is less than unity, or if calculation of N_2 results in fewer shear connectors required between the concentrated load and the zero moment point than would be required by spacing N_1 uniformly between the maximum and zero moment points, then Equation (I4-5) does not apply.

EXAMPLE 18:

Given:

Neutral axis in concrete

Determine transformed section properties for a 70% composite beam: W14×22, $F_y = 36$ ksi. Use formed steel deck with ribs running perpendicular to beam $h_r = 2$; concrete cover $t_o = 4.5$ in.; normal weight concrete $(n = 9)$ $f_c' = 3.5$ ksi; effective width $b = 72$ in.

W14×22: $d = 13.74$ in.
$\quad\quad\quad A_s = 6.49$ in.
$\quad\quad\quad I_s = 199$ in.^{4}
$\quad\quad\quad S_s = 29$ in.^{3}

Solution:

A. Fully composite

$$A_{ctr} = \frac{(72)\,(4.5)}{9} = 36$$

$\quad$ (assuming $t = t_o = 4.5$ in.)

$\quad Y2 = 2 + (4.5/2) = 4.25$

Enter Composite Beam Selection Tables and interpolate:

$\quad \bar{I}_{tr} = 881$ in.^{4}

$\quad \bar{S}_{tr} = 54.1$ in.^{3}

$\quad \bar{y}_b = 881/54.1 = 16.3$ in.

since $\bar{y}_b > (d + h_r) = 15.74$ (or equivalently, $(b/n)\,t_o = 36 > A_{ctr}^{max} = 25.6$ for $h_r = 2$) a part of the concrete is in tension and must be disregarded:

$\quad t = h - \bar{y}_b = (13.74 + 2 + 4.5) - 16.3 = 3.94$ in.

alternately solve directly for t:

$$t = 6.49\left[-1 + \sqrt{1 + \frac{2\,(8)}{6.49}\,(13.74 + 2 + 4.5)}\,\right] = 3.92 \text{ in.}$$

$\quad Y2 = 2 + 4.5 - (3.92/2) = 4.54$ in.

$\quad A_{ctr} = (72/9)\,(3.92) = 31.3$ in.^{2}

From tables interpolate:

$\quad \bar{I}_{tr} = 900$ in.^{4}

$\quad \bar{S}_{tr} = 55$ in.^{3}

Optional:

$\quad$ add thickness correction terms

$\quad I_{tr} = 900 + (1/12)\,(31.3)\,(3.92^2 - 1) = 937$ in.^{4}

$\quad S_{tr} = (937/900)\,55 = 57.3$ in.^{3}

$\quad \bar{y}_b = (937/57.3) = 16.35$ in.

$\quad y_t = h - \bar{y}_b = 13.74 + 2 + 4.5 - 16.35 = 3.9$ in.

$\quad S_t = 937/3.9 = 240$ in.^{3}

B.   70% composite

$$V_h = \frac{0.85\ (3.5)\ (3.92)\ (72)}{2} = 422 \text{ kips}$$

or

$$V_h = 6.49 \times 36/2 = 116.8 \text{ kips} \quad \textbf{steel governs}$$

$$V_h' = 0.7\ V_h = (0.7)\ 116.8 = 81.8 \text{ kips}$$

$$I_{eff} = 199 + \sqrt{0.7}\ (937 - 199) = 816 \text{ in.}^4$$

$$S_{eff} = 29 + \sqrt{0.7}\ (57.3 - 29) = 52.7 \text{ in.}^3$$

$$\bar{y}_{eff} = 816/52.7 = 15.5 \text{ in.}$$

$$\left(\frac{b}{n}\right)_{eff} = \frac{6.49}{3.92}\left[\frac{15.5 - (13.74/2)}{13.74 + 4.54 - 15.5}\right] = 5.1 \text{ in.}$$

$$S_{t-eff} = \frac{816}{(13.74 + 2 + 4.5) - 15.5} = 172 \text{ in.}^3$$

Equation
(14-1)

Equation
(14-2)

COMPOSITE BEAM SELECTION TABLE
Transformed Section Properties

$$\overline{I}_{tr}, \text{ In.}^4$$
$$\overline{S}_{tr}, \text{ In.}^3$$

Shape											
W 36×300	$A_s = 88.3 \text{ In.}^2$ $\quad I_s = 20300 \text{ In.}^4$ $\quad$ max V_h (steel) = 1589 kips for F_y = 36 ksi										
	$d = 36.7 \text{ In.}$ $\quad S_s = 1110 \text{ In.}^3$ $\quad$ max V_h (steel) = 2208 kips for F_y = 50 ksi										

A_{ctr}, In.[2] [b,e]	$Y2$, In.[a]										
	2.00	3.00	3.25	3.50	3.75	4.00	4.25	4.50	4.75	5.00	6.00
200	45734	48291	48949	49615	50289	50970	51659	52356	53060	53772	56696
	1407	*1455*	*1467*	*1479*	*1492*	*1504*	*1517*	*1529*	*1542*	*1555*	*1607*
100	39766	41723	42227	42737	43253	43775	44302	44835	45374	45919	48158
	1362	*1404*	*1415*	*1425*	*1436*	*1447*	*1458*	*1469*	*1480*	*1492*	*1538*
60	35129	36620	37004	37392	37785	38182	38584	38990	39401	39816	41522
	1320	*1355*	*1365*	*1374*	*1383*	*1392*	*1402*	*1412*	*1421*	*1431*	*1471*
50	33550	34883	35226	35573	35924	36279	36638	37001	37368	37739	39263
	1304	*1337*	*1345*	*1354*	*1362*	*1371*	*1380*	*1389*	*1398*	*1407*	*1445*
40	31726	32875	33171	33470	33773	34079	34389	34702	35019	35339	36653
	1283	*1313*	*1321*	*1329*	*1337*	*1345*	*1353*	*1361*	*1369*	*1377*	*1411*
30	29594	30529	30769	31013	31259	31508	31760	32014	32272	32532	33601
	1257	*1283*	*1290*	*1297*	*1304*	*1310*	*1317*	*1325*	*1332*	*1339*	*1369*
20	27068	27748	27924	28101	28280	28462	28645	28831	29018	29208	29986
	1223	*1243*	*1249*	*1254*	*1259*	*1265*	*1270*	*1276*	*1282*	*1287*	*1311*
10	24028	24403	24500	24597	24696	24796	24897	24999	25102	25207	25636
	1175	*1188*	*1191*	*1194*	*1198*	*1201*	*1204*	*1208*	*1211*	*1215*	*1230*

h_r, In.[c]	max A_{ctr}, In.[2] [d]										
0	811	541	499	463	433	406	382	360	341	324	270
2		1799	1439	1199	1028	899	799	719	654	600	450
3			7548	3774	2516	1887	1510	1258	1078	943	629

Shape											
W 36×280	$A_s = 82.4 \text{ In.}^2$ $\quad I_s = 18900 \text{ In.}^4$ $\quad$ max V_h (steel) = 1483 kips for F_y = 36 ksi										
	$d = 36.5 \text{ In.}$ $\quad S_s = 1030 \text{ In.}^3$ $\quad$ max V_h (steel) = 2060 kips for F_y = 50 ksi										

A_{ctr}, In.[2] [b,e]	$Y2$, In.[a]										
	2.00	3.00	3.25	3.50	3.75	4.00	4.25	4.50	4.75	5.00	6.00
200	42870	45293	45917	46549	47187	47833	48486	49147	49814	50489	53263
	1315	*1359*	*1371*	*1382*	*1394*	*1406*	*1418*	*1430*	*1442*	*1454*	*1503*
100	37451	39327	39810	40299	40793	41293	41799	42310	42827	43349	45496
	1275	*1315*	*1325*	*1335*	*1345*	*1355*	*1366*	*1376*	*1387*	*1398*	*1442*
60	33156	34598	34969	35344	35724	36109	36497	36890	37287	37689	39339
	1237	*1271*	*1280*	*1289*	*1297*	*1306*	*1315*	*1325*	*1334*	*1343*	*1381*
50	31677	32969	33302	33638	33979	34323	34672	35024	35380	35740	37218
	1223	*1254*	*1262*	*1270*	*1279*	*1287*	*1296*	*1304*	*1313*	*1322*	*1357*
40	29956	31075	31362	31654	31948	32246	32548	32853	33161	33472	34752
	1204	*1233*	*1240*	*1248*	*1255*	*1263*	*1271*	*1278*	*1286*	*1294*	*1327*
30	27930	28843	29078	29316	29557	29800	30046	30295	30547	30801	31846
	1180	*1205*	*1212*	*1218*	*1225*	*1231*	*1238*	*1245*	*1252*	*1259*	*1287*
20	25508	26176	26348	26522	26698	26876	27056	27239	27423	27609	28374
	1148	*1168*	*1173*	*1178*	*1183*	*1189*	*1194*	*1200*	*1205*	*1211*	*1234*
10	22561	22932	23027	23123	23221	23320	23419	23520	23622	23726	24149
	1103	*1115*	*1118*	*1122*	*1125*	*1128*	*1132*	*1135*	*1138*	*1142*	*1156*

h_r, In.[c]	max A_{ctr}, In.[2] [d]										
0	752	502	463	430	401	376	354	334	317	301	251
2		1669	1336	1113	954	835	742	668	607	556	417
3			7007	3504	2336	1752	1401	1168	1001	876	584

[a] $Y2$ = distance from top of steel beam to concrete centroid
[b] A_{ctr} = transformed effective concrete area = $(b/n)\,t$ for full composite
[c] h_r = height of steel deck ribs
[d] max A_{ctr} = maximum transformed effective concrete area based on full concrete thickness $t = t_o$ for elastic neutral axis to be below the top of the steel deck
[e] Shaded area represents A_{ctr} range wherein elastic neutral axis falls within the concrete slab ($h_r = 0$)

COMPOSITE BEAM SELECTION TABLE
Transformed Section Properties

$$\bar{I}_{tr}, \text{ In.}^4$$
$$\bar{S}_{tr}, \text{ In.}^3$$

Shape

W 36×260

$A_s = 76.5$ In.² $I_s = 17300$ In.⁴   max V_h (steel) = 1377 kips for $F_y = 36$ ksi
$d = 36.3$ In. $S_s = 953$ In.³   max V_h (steel) = 1913 kips for $F_y = 50$ ksi

A_{ctr}, In.² [b,e]	Y2, In. [a]										
	2.00	3.00	3.25	3.50	3.75	4.00	4.25	4.50	4.75	5.00	6.00
200	39739	42022	42610	43205	43807	44416	45032	45654	46284	46920	49536
	1216	*1258*	*1268*	*1279*	*1290*	*1301*	*1312*	*1323*	*1335*	*1346*	*1392*
100	34872	36660	37121	37587	38058	38535	39017	39505	39998	40497	42545
	1181	*1218*	*1227*	*1237*	*1247*	*1257*	*1266*	*1276*	*1286*	*1297*	*1338*
60	30931	32318	32676	33037	33403	33773	34147	34526	34908	35295	36884
	1147	*1179*	*1187*	*1195*	*1204*	*1212*	*1221*	*1230*	*1238*	*1247*	*1284*
50	29557	30804	31126	31451	31780	32112	32449	32789	33133	33481	34910
	1133	*1163*	*1171*	*1179*	*1187*	*1195*	*1203*	*1211*	*1219*	*1228*	*1262*
40	27947	29031	29310	29592	29878	30167	30459	30755	31053	31356	32597
	1116	*1144*	*1151*	*1158*	*1165*	*1173*	*1180*	*1187*	*1195*	*1203*	*1234*
30	26035	26924	27153	27384	27619	27856	28096	28338	28583	28831	29850
	1094	*1118*	*1124*	*1131*	*1137*	*1143*	*1150*	*1156*	*1163*	*1170*	*1197*
20	23726	24381	24549	24720	24892	25066	25243	25421	25602	25784	26533
	1064	*1083*	*1088*	*1093*	*1098*	*1103*	*1109*	*1114*	*1119*	*1125*	*1147*
10	20885	21249	21343	21439	21535	21632	21730	21830	21931	22032	22450
	1021	*1033*	*1036*	*1039*	*1042*	*1046*	*1049*	*1052*	*1056*	*1059*	*1073*

h_r, In. [c]	max A_{ctr}, In.² [d]										
0	693	462	427	396	370	347	326	308	292	277	231
2		1540	1232	1027	880	770	684	616	560	513	385
3			6466	3233	2155	1616	1293	1078	924	808	539

W 36×245

$A_s = 72.1$ In.² $I_s = 16100$ In.⁴   max V_h (steel) = 1298 kips for $F_y = 36$ ksi
$d = 36.1$ In. $S_s = 895$ In.³   max V_h (steel) = 1803 kips for $F_y = 50$ ksi

A_{ctr}, In.² [b,e]	Y2, In. [a]										
	2.00	3.00	3.25	3.50	3.75	4.00	4.25	4.50	4.75	5.00	6.00
200	37400	39577	40137	40705	41279	41860	42447	43041	43642	44249	46744
	1141	*1181*	*1191*	*1202*	*1212*	*1223*	*1233*	*1244*	*1254*	*1265*	*1309*
100	32933	34654	35097	35546	36000	36459	36923	37393	37868	38348	40320
	1109	*1145*	*1154*	*1163*	*1173*	*1182*	*1191*	*1201*	*1211*	*1220*	*1260*
60	29257	30602	30948	31299	31654	32013	32376	32743	33114	33489	35031
	1078	*1109*	*1117*	*1125*	*1133*	*1141*	*1150*	*1158*	*1166*	*1175*	*1210*
50	27961	29174	29487	29803	30123	30446	30773	31104	31439	31777	33167
	1065	*1094*	*1102*	*1110*	*1117*	*1125*	*1133*	*1141*	*1149*	*1157*	*1189*
40	26435	27492	27764	28040	28319	28601	28886	29174	29466	29760	30972
	1049	*1076*	*1083*	*1090*	*1097*	*1104*	*1111*	*1119*	*1126*	*1133*	*1164*
30	24610	25481	25705	25932	26161	26393	26628	26866	27106	27348	28346
	1029	*1052*	*1058*	*1064*	*1070*	*1077*	*1083*	*1089*	*1096*	*1102*	*1129*
20	22390	23033	23198	23366	23536	23707	23881	24056	24234	24413	25150
	1000	*1019*	*1024*	*1029*	*1034*	*1039*	*1044*	*1049*	*1054*	*1059*	*1081*
10	19628	19988	20081	20175	20271	20367	20464	20563	20662	20763	21176
	958	*970*	*973*	*976*	*980*	*983*	*986*	*989*	*993*	*996*	*1010*

h_r, In. [c]	max A_{ctr}, In.² [d]										
0	650	434	400	372	347	325	306	289	274	260	217
2		1445	1156	963	826	722	642	578	525	482	361
3			6068	3034	2023	1517	1214	1011	867	758	506

[a] Y2 = distance from top of steel beam to concrete centroid
[b] A_{ctr} = transformed effective concrete area = (b/n) t for full composite
[c] h_r = height of steel deck ribs
[d] max A_{ctr} = maximum transformed effective concrete area based on full concrete thickness $t = t_o$ for elastic neutral axis to be below the top of the steel deck
[e] Shaded area represents A_{ctr} range wherein elastic neutral axis falls within the concrete slab ($h_r = 0$)

COMPOSITE BEAM SELECTION TABLE
Transformed Section Properties

$$\overline{I}_{tr}, \text{ In.}^4$$
$$\overline{S}_{tr}, \text{ In.}^3$$

Shape												
W 36×230	A_s = 67.6 In.^{2} d = 35.9 In.	I_s = 15000 In.^{4} S_s = 837 In.^{3}			max V_h (steel) = 1217 kips for F_y = 36 ksi max V_h (steel) = 1690 kips for F_y = 50 ksi							

W 36×230

A_{ctr}, In.$^{2\ b,e}$	Y2, In.^{a}										
	2.00	3.00	3.25	3.50	3.75	4.00	4.25	4.50	4.75	5.00	6.00
200	35125	37191	37724	38263	38808	39359	39917	40480	41051	41627	43997
	1069	1107	1116	1126	1136	1146	1156	1166	1176	1186	1227
100	31061	32711	33136	33566	34001	34441	34887	35337	35792	36252	38144
	1040	1074	1083	1092	1100	1109	1118	1127	1136	1146	1183
60	27656	28956	29291	29630	29973	30320	30671	31026	31384	31747	33238
	1012	1042	1049	1057	1065	1072	1080	1088	1096	1105	1138
50	26443	27619	27922	28228	28538	28852	29169	29490	29814	30142	31490
	1000	1028	1036	1043	1050	1058	1065	1073	1080	1088	1119
40	25005	26033	26298	26566	26837	27111	27388	27669	27953	28239	29418
	986	1011	1018	1025	1032	1038	1045	1052	1059	1066	1096
30	23272	24122	24341	24563	24787	25014	25243	25475	25710	25947	26921
	966	989	995	1001	1007	1013	1019	1025	1031	1038	1064
20	21144	21776	21938	22103	22269	22438	22608	22780	22955	23131	23855
	940	958	963	967	972	977	982	987	992	997	1019
10	18468	18824	18916	19009	19103	19198	19294	19391	19490	19589	19998
	900	912	915	918	921	924	927	930	934	937	951

h_r, In.^{c}	max A_{ctr}, In.$^{2\ d}$										
0	607	404	373	347	324	303	286	270	255	243	202
2		1349	1079	899	771	674	599	539	490	450	337
3			5665	2832	1888	1416	1133	944	809	708	472

Shape												
W 36×210	A_s = 61.8 In.^{2} d = 36.7 In.	I_s = 13200 In.^{4} S_s = 719 In.^{3}			max V_h (steel) = 1112 kips for F_y = 36 ksi max V_h (steel) = 1545 kips for F_y = 50 ksi							

A_{ctr}, In.$^{2\ b,e}$	Y2, In.^{a}										
	2.00	3.00	3.25	3.50	3.75	4.00	4.25	4.50	4.75	5.00	6.00
200	32758	34727	35234	35746	36265	36789	37320	37856	38398	38946	41198
	967	1002	1011	1020	1030	1039	1048	1058	1067	1076	1115
100	29018	30610	31020	31435	31855	32279	32708	33142	33581	34024	35846
	939	971	979	987	995	1004	1012	1021	1029	1038	1074
60	25806	27075	27402	27733	28067	28405	28747	29093	29443	29796	31248
	910	938	945	953	960	968	975	983	991	998	1030
50	24644	25797	26093	26393	26697	27004	27315	27629	27946	28267	29585
	898	925	932	939	946	953	960	967	975	982	1012
40	23254	24267	24528	24791	25058	25328	25601	25876	26155	26437	27595
	883	908	914	921	927	934	940	947	954	961	989
30	21562	22404	22621	22840	23062	23286	23513	23743	23975	24209	25172
	863	885	891	896	902	908	914	920	926	932	957
20	19456	20086	20248	20412	20578	20746	20916	21087	21261	21436	22157
	834	852	857	862	867	871	876	881	886	891	912
10	16764	17122	17215	17308	17403	17498	17595	17693	17792	17892	18302
	792	803	806	809	812	816	819	822	825	828	842

h_r, In.^{c}	max A_{ctr}, In.$^{2\ d}$										
0	567	378	349	324	302	283	267	252	239	227	189
2		1257	1006	838	718	629	559	503	457	419	314
3			5276	2638	1759	1319	1055	879	754	660	440

^{a}Y2 = distance from top of steel beam to concrete centroid
$^b A_{ctr}$ = transformed effective concrete area
$^c h_r$ = $(b/n)\,t$ for full composite
= height of steel deck ribs
^{d}max A_{ctr} = maximum transformed effective concrete area based on full concrete thickness $t = t_o$ for elastic neutral axis to be below the top of the steel deck
^{e}Shaded area represents A_{ctr} range wherein elastic neutral axis falls within the concrete slab (h_r = 0)

COMPOSITE BEAM SELECTION TABLE
Transformed Section Properties

$$\overline{I}_{tr}, \text{ In.}^4$$
$$\overline{S}_{tr}, \text{ In.}^3$$

Shape												
W 36×194	$A_s = 57.0 \text{ In.}^2$ $I_s = 12100 \text{ In.}^4$ max V_h (steel) = 1026 kips for F_y = 36 ksi											
	$d = 36.5 \text{ In.}$ $S_s = 664 \text{ In.}^3$ max V_h (steel) = 1425 kips for F_y = 50 ksi											

W 36×194 — $Y2$, In.[a]

A_{ctr}, In.[2 b,e]	2.00	3.00	3.25	3.50	3.75	4.00	4.25	4.50	4.75	5.00	6.00
200	30297	32138	32612	33091	33576	34067	34563	35065	35572	36085	38191
	891	924	932	941	950	958	967	975	984	993	1029
100	26989	28495	28883	29275	29672	30074	30480	30891	31306	31725	33450
	867	897	904	912	920	928	936	944	952	960	993
60	24086	25298	25611	25927	26246	26570	26897	27227	27561	27899	29287
	841	868	875	882	889	896	903	910	918	925	955
50	23021	24126	24411	24699	24990	25284	25582	25884	26188	26496	27761
	831	856	863	869	876	883	890	896	903	910	939
40	21737	22712	22964	23218	23475	23735	23998	24263	24532	24804	25920
	817	841	847	853	859	866	872	878	885	891	918
30	20158	20974	21184	21396	21611	21829	22049	22271	22496	22723	23656
	799	820	826	831	837	842	848	854	859	865	889
20	18170	18784	18942	19102	19264	19428	19593	19761	19930	20101	20804
	773	790	795	799	804	809	813	818	823	828	848
10	15588	15941	16032	16124	16217	16311	16406	16502	16599	16698	17102
	733	744	747	750	753	756	759	763	766	769	782

h_r, In.[c]		max A_{ctr}, In.[2 d]									
0	520	347	320	297	277	260	245	231	219	208	173
2		1154	923	769	659	577	513	462	420	385	288
3			4844	2422	1615	1211	969	807	692	605	404

Shape												
W 36×182	$A_s = 53.6 \text{ In.}^2$ $I_s = 11300 \text{ In.}^4$ max V_h (steel) = 965 kips for F_y = 36 ksi											
	$d = 36.3 \text{ In.}$ $S_s = 623 \text{ In.}^3$ max V_h (steel) = 1340 kips for F_y = 50 ksi											

W 36×182 — $Y2$, In.[a]

A_{ctr}, In.[2 b,e]	2.00	3.00	3.25	3.50	3.75	4.00	4.25	4.50	4.75	5.00	6.00
200	28505	30252	30702	31158	31618	32084	32555	33032	33513	34000	36001
	837	868	876	884	892	900	908	917	925	933	967
100	25498	26940	27312	27687	28068	28452	28841	29234	29632	30034	31686
	815	843	851	858	865	873	880	888	896	903	935
60	22817	23987	24288	24593	24901	25213	25529	25848	26170	26497	27836
	792	817	824	831	837	844	851	858	865	872	900
50	21823	22892	23168	23446	23728	24013	24301	24593	24888	25186	26410
	782	807	813	819	826	832	838	845	852	858	885
40	20618	21564	21808	22055	22304	22557	22812	23070	23331	23595	24679
	770	793	798	804	810	816	822	828	835	841	866
30	19124	19919	20123	20331	20540	20752	20967	21183	21402	21624	22534
	753	773	778	784	789	795	800	805	811	817	840
20	17224	17826	17981	18138	18297	18457	18620	18784	18950	19118	19807
	728	745	750	754	759	763	768	772	777	782	801
10	14728	15076	15166	15257	15348	15441	15535	15630	15726	15823	16222
	690	701	704	707	710	713	716	719	722	726	739

h_r, In.[c]		max A_{ctr}, In.[2 d]									
0	487	325	300	278	260	243	229	216	205	195	162
2		1081	865	721	618	540	480	432	393	360	270
3			4538	2269	1513	1134	908	756	648	567	378

[a] $Y2$ = distance from top of steel beam to concrete centroid
[b] A_{ctr} = transformed effective concrete area
        = $(b/n)\,t$ for full composite
[c] h_r = height of steel deck ribs
[d] max A_{ctr} = maximum transformed effective concrete area based on full concrete thickness $t = t_o$ for elastic neutral axis to be below the top of the steel deck
[e] Shaded area represents A_{ctr} range wherein elastic neutral axis falls within the concrete slab (h_r = 0)

COMPOSITE BEAM SELECTION TABLE
Transformed Section Properties

$$\overline{I}_{tr}, \text{ In.}^4$$
$$\overline{S}_{tr}, \text{ In.}^3$$

Shape												
W 36×170	$A_s = 50.0 \text{ In.}^2$ $I_s = 10500 \text{ In.}^4$ max V_h (steel) = 900 kips for F_y = 36 ksi											
	$d = 36.2 \text{ In.}$ $S_s = 580 \text{ In.}^3$ max V_h (steel) = 1250 kips for F_y = 50 ksi											

A_{ctr}, In.$^{2\ b,e}$	Y2, In.^{a}										
	2.00	3.00	3.25	3.50	3.75	4.00	4.25	4.50	4.75	5.00	6.00
200	26653	28300	28724	29153	29587	30027	30471	30920	31374	31833	33720
	780	810	817	825	832	840	848	855	863	871	903
100	23955	25328	25681	26039	26401	26767	27137	27511	27890	28272	29845
	761	788	795	802	809	816	823	830	837	845	874
60	21507	22630	22919	23212	23508	23807	24110	24416	24726	25039	26326
	741	765	771	777	784	790	797	803	810	816	843
50	20589	21619	21884	22152	22423	22698	22975	23256	23540	23827	25006
	732	755	761	767	773	779	785	792	798	804	830
40	19468	20383	20618	20857	21098	21342	21589	21838	22091	22346	23394
	721	742	748	754	759	765	771	777	782	788	813
30	18066	18838	19037	19238	19442	19648	19856	20067	20279	20495	21379
	705	725	730	735	740	745	750	756	761	766	788
20	16265	16853	17004	17158	17313	17469	17628	17789	17951	18115	18789
	683	699	703	707	712	716	721	725	729	734	753
10	13863	14206	14294	14383	14474	14565	14658	14752	14846	14942	15335
	647	658	661	663	666	669	672	675	678	681	694

h_r, In.^{c}	max A_{ctr}, In.$^{2\ d}$										
0	452	301	278	258	241	226	213	201	190	181	151
2		1004	803	670	574	502	446	402	365	335	251
3			4217	2109	1406	1054	843	703	602	527	351

Shape												
W 36×160	$A_s = 47.0 \text{ In.}^2$ $I_s = 9750 \text{ In.}^4$ max V_h (steel) = 846 kips for F_y = 36 ksi											
	$d = 36.0 \text{ In.}$ $S_s = 542 \text{ In.}^3$ max V_h (steel) = 1175 kips for F_y = 50 ksi											

A_{ctr}, In.$^{2\ b,e}$	Y2, In.^{a}										
	2.00	3.00	3.25	3.50	3.75	4.00	4.25	4.50	4.75	5.00	6.00
200	24997	26558	26960	27367	27778	28194	28616	29041	29472	29907	31696
	731	759	766	773	780	787	794	802	809	816	847
100	22554	23865	24203	24545	24890	25240	25594	25952	26314	26679	28182
	713	739	746	752	759	765	772	779	786	793	821
60	20302	21383	21662	21943	22228	22517	22808	23103	23401	23703	24942
	695	718	724	730	736	742	748	754	761	767	793
50	19450	20443	20699	20958	21220	21485	21753	22024	22299	22576	23715
	687	709	715	720	726	732	738	744	750	756	781
40	18401	19288	19516	19747	19981	20217	20456	20698	20942	21190	22205
	676	697	703	708	713	719	724	730	736	741	765
30	17081	17832	18025	18221	18419	18619	18822	19027	19234	19444	20304
	662	681	686	691	696	701	706	711	716	721	742
20	15366	15942	16090	16240	16392	16545	16700	16857	17016	17177	17836
	641	657	661	665	669	673	678	682	686	691	709
10	13051	13389	13476	13564	13653	13744	13835	13927	14020	14115	14502
	607	617	620	623	626	629	631	634	637	640	653

h_r, In.^{c}	max A_{ctr}, In.$^{2\ d}$										
0	423	282	260	242	226	212	199	188	178	169	141
2		940	752	627	537	470	418	376	342	313	235
3			3949	1974	1316	987	790	658	564	494	329

^{a}Y2 = distance from top of steel beam to concrete centroid
^{b}A_{ctr} = transformed effective concrete area
     = $(b/n)\,t$ for full composite
^{c}h_r = height of steel deck ribs
^{d}max A_{ctr} = maximum transformed effective concrete area based on full concrete thickness $t = t_o$ for elastic neutral axis
     to be below the top of the steel deck
^{e}Shaded area represents A_{ctr} range wherein elastic neutral axis falls within the concrete slab ($h_r = 0$)

COMPOSITE BEAM SELECTION TABLE
Transformed Section Properties

$$\overline{I}_{tr},\ \text{In.}^4$$
$$\overline{S}_{tr},\ \text{In.}^3$$

Shape

W 36×150

$A_s = 44.2$ In.^{2} $I_s = 9040$ In.^{4} max V_h (steel) = 796 kips for F_y = 36 ksi
$d = 35.9$ In. $S_s = 504$ In.^{3} max V_h (steel) = 1105 kips for F_y = 50 ksi

A_{ctr}, In.$^{2\ b,e}$	Y2, In.^{a}										
	2.00	3.00	3.25	3.50	3.75	4.00	4.25	4.50	4.75	5.00	6.00
200	23428	24907	25288	25674	26064	26458	26857	27261	27669	28082	29778
	684	710	717	724	731	737	744	751	758	765	794
100	21217	22469	22792	23118	23449	23783	24121	24463	24808	25158	26594
	668	693	699	705	712	718	724	731	737	744	770
60	19149	20189	20457	20728	21002	21279	21560	21844	22131	22421	23613
	651	674	679	685	691	697	702	708	714	720	745
50	18358	19317	19563	19813	20066	20322	20581	20842	21107	21374	22473
	644	665	671	676	682	687	693	699	704	710	734
40	17380	18237	18458	18682	18908	19137	19369	19603	19839	20079	21062
	635	654	660	665	670	675	681	686	691	697	719
30	16137	16867	17055	17246	17438	17633	17830	18029	18231	18435	19272
	621	639	644	649	653	658	663	668	673	678	698
20	14508	15071	15216	15362	15511	15661	15813	15966	16121	16278	16923
	601	617	620	624	629	633	637	641	645	649	667
10	12278	12612	12697	12784	12872	12961	13051	13142	13234	13327	13709
	568	579	582	584	587	590	593	596	599	602	614

h_r, In.^{c}	max A_{ctr}, In.$^{2\ d}$										
	2.00	3.00	3.25	3.50	3.75	4.00	4.25	4.50	4.75	5.00	6.00
0	396	264	244	226	211	198	186	176	167	158	132
2		881	705	587	503	440	391	352	320	294	220
3			3700	1850	1233	925	740	617	529	462	308

W 36×135

$A_s = 39.7$ In.^{2} $I_s = 7800$ In.^{4} max V_h (steel) = 715 kips for F_y = 36 ksi
$d = 35.6$ In. $S_s = 439$ In.^{3} max V_h (steel) = 993 kips for F_y = 50 ksi

A_{ctr}, In.$^{2\ b,e}$	Y2, In.^{a}										
	2.00	3.00	3.25	3.50	3.75	4.00	4.25	4.50	4.75	5.00	6.00
200	20770	22113	22459	22810	23164	23523	23885	24252	24623	24998	26540
	606	630	636	642	648	654	661	667	673	680	706
100	18921	20074	20371	20671	20975	21283	21594	21909	22227	22549	23872
	593	615	621	626	632	638	644	650	656	662	686
60	17148	18117	18366	18619	18875	19133	19395	19659	19927	20198	21310
	578	598	604	609	614	620	625	631	636	642	664
50	16458	17355	17586	17820	18057	18297	18539	18784	19032	19283	20313
	571	591	596	601	606	612	617	622	627	633	655
40	15595	16403	16611	16822	17035	17251	17469	17689	17913	18138	19066
	563	582	586	591	596	601	606	611	616	621	642
30	14485	15177	15356	15537	15720	15905	16092	16281	16472	16666	17461
	551	568	572	577	581	586	590	595	600	604	623
20	13003	13542	13681	13822	13964	14108	14253	14401	14550	14700	15319
	533	547	551	555	559	563	567	571	575	579	595
10	10925	11248	11332	11416	11502	11588	11676	11764	11854	11944	12316
	502	512	515	518	520	523	526	529	531	534	546

h_r, In.^{c}	max A_{ctr}, In.$^{2\ d}$										
	2.00	3.00	3.25	3.50	3.75	4.00	4.25	4.50	4.75	5.00	6.00
0	353	235	217	202	188	176	166	157	149	141	118
2		785	628	523	449	393	349	314	285	262	196
3			3299	1650	1100	825	660	550	471	412	275

[a] $Y2$ = distance from top of steel beam to concrete centroid
[b] A_{ctr} = transformed effective concrete area = $(b/n)\,t$ for full composite
[c] h_r = height of steel deck ribs
[d] max A_{ctr} = maximum transformed effective concrete area based on full concrete thickness $t = t_o$ for elastic neutral axis to be below the top of the steel deck
[e] Shaded area represents A_{ctr} range wherein elastic neutral axis falls within the concrete slab ($h_r = 0$)

AMERICAN INSTITUTE OF STEEL CONSTRUCTION

COMPOSITE BEAM SELECTION TABLE
Transformed Section Properties

$$\bar{I}_{tr}, \text{ In.}^4$$
$$\bar{S}_{tr}, \text{ In.}^3$$

Shape												

W 33×221 $A_s = 65.0$ In.^{2} $I_s = 12800$ In.^{4} max V_h (steel) = 1170 kips for $F_y = 36$ ksi
$d = 33.9$ In. $S_s = 757$ In.^{3} max V_h (steel) = 1625 kips for $F_y = 50$ ksi

A_{ctr}, In.$^{2\ b,e}$	Y2, In.^{a}										
	2.00	3.00	3.25	3.50	3.75	4.00	4.25	4.50	4.75	5.00	6.00
200	30461	32371	32863	33362	33867	34379	34896	35419	35949	36485	38689
	974	1011	1020	1029	1039	1049	1058	1068	1078	1088	1128
100	26977	28511	28907	29307	29713	30123	30539	30959	31384	31814	33584
	948	981	989	998	1007	1015	1024	1033	1042	1051	1087
60	24027	25241	25555	25872	26193	26518	26847	27180	27517	27858	29260
	922	951	958	966	973	981	989	997	1005	1013	1045
50	22969	24069	24353	24640	24931	25226	25524	25825	26130	26439	27709
	911	939	946	953	960	967	975	982	990	997	1028
40	21709	22673	22922	23174	23429	23687	23948	24212	24480	24750	25863
	897	923	929	936	943	949	956	963	970	977	1006
30	20185	20984	21191	21399	21611	21824	22041	22260	22482	22706	23628
	879	902	908	913	919	925	931	938	944	950	976
20	18303	18898	19052	19207	19365	19524	19685	19848	20013	20180	20868
	854	872	877	882	887	892	897	902	907	912	933
10	15918	16255	16342	16431	16520	16610	16701	16794	16888	16982	17372
	817	828	831	834	837	841	844	847	850	854	867

h_r, In.^{c}	max A_{ctr}, In.$^{2\ d}$										
0	551	368	339	315	294	276	259	245	232	221	184
2		1233	986	822	704	616	548	493	448	411	308
3			5191	2595	1730	1298	1038	865	742	649	433

W 33×201 $A_s = 59.1$ In.^{2} $I_s = 11500$ In.^{4} max V_h (steel) = 1064 kips for $F_y = 36$ ksi
$d = 33.7$ In. $S_s = 684$ In.^{3} max V_h (steel) = 1478 kips for $F_y = 50$ ksi

A_{ctr}, In.$^{2\ b,e}$	Y2, In.^{a}										
	2.00	3.00	3.25	3.50	3.75	4.00	4.25	4.50	4.75	5.00	6.00
200	27709	29474	29929	30390	30857	31329	31808	32292	32781	33276	35315
	883	917	925	934	943	951	960	969	978	987	1024
100	24693	26130	26501	26876	27256	27641	28031	28425	28823	29227	30886
	861	892	899	907	915	923	931	940	948	956	990
60	22073	23225	23522	23823	24127	24436	24748	25064	25383	25706	27037
	838	865	872	879	887	894	901	908	916	923	954
50	21118	22166	22436	22710	22987	23267	23551	23839	24129	24423	25634
	829	855	861	868	875	882	889	896	903	910	939
40	19970	20893	21131	21372	21616	21864	22114	22367	22623	22882	23948
	817	841	847	853	859	866	872	879	885	892	919
30	18566	19335	19534	19735	19939	20145	20353	20564	20778	20994	21883
	801	822	828	833	839	844	850	856	862	868	892
20	16806	17384	17533	17684	17837	17992	18148	18307	18467	18629	19297
	778	795	800	804	809	814	819	823	828	833	853
10	14537	14867	14953	15039	15127	15215	15305	15396	15488	15580	15963
	743	754	757	760	763	766	769	773	776	779	792

h_r, In.^{c}	max A_{ctr}, In.$^{2\ d}$										
0	498	332	306	284	265	249	234	221	210	199	166
2		1113	891	742	636	557	495	445	405	371	278
3			4690	2345	1563	1173	938	782	670	586	391

[a] $Y2$ = distance from top of steel beam to concrete centroid
[b] A_{ctr} = transformed effective concrete area
    = $(b/n)\, t$ for full composite
[c] h_r = height of steel deck ribs
[d] max A_{ctr} = maximum transformed effective concrete area based on full concrete thickness $t = t_o$ for elastic neutral axis to be below the top of the steel deck
[e] Shaded area represents A_{ctr} range wherein elastic neutral axis falls within the concrete slab ($h_r = 0$)

AMERICAN INSTITUTE OF STEEL CONSTRUCTION

COMPOSITE BEAM SELECTION TABLE
Transformed Section Properties

$$\overline{I}_{tr}, \text{ In.}^4$$
$$\overline{S}_{tr}, \text{ In.}^3$$

Shape											
W 33×141	$A_s = 41.6$ In.^{2} $I_s = 7450$ In.^{4} max V_h (steel) = 749 kips for $F_y = 36$ ksi										
	$d = 33.3$ In. $S_s = 448$ In.^{3} max V_h (steel) = 1040 kips for $F_y = 50$ ksi										

A_{ctr}, In.$^{2\ b,e}$	Y2, In.^{a}										
	2.00	3.00	3.25	3.50	3.75	4.00	4.25	4.50	4.75	5.00	6.00
200	19445	20764	21104	21449	21798	22151	22509	22871	23237	23608	25134
	606	*631*	*637*	*644*	*650*	*656*	*663*	*670*	*676*	*683*	*710*
100	17677	18802	19093	19387	19685	19986	20291	20600	20913	21229	22530
	593	*616*	*622*	*628*	*634*	*640*	*646*	*652*	*658*	*665*	*690*
60	16000	16941	17184	17430	17679	17931	18186	18444	18706	18970	20058
	578	*600*	*605*	*611*	*616*	*622*	*627*	*633*	*639*	*644*	*668*
50	15352	16222	16447	16674	16904	17137	17373	17612	17853	18098	19104
	572	*593*	*598*	*603*	*608*	*614*	*619*	*625*	*630*	*636*	*658*
40	14546	15327	15529	15733	15940	16149	16361	16575	16792	17012	17915
	564	*583*	*588*	*593*	*598*	*603*	*608*	*613*	*619*	*624*	*646*
30	13515	14183	14355	14530	14706	14885	15066	15249	15435	15622	16395
	552	*570*	*574*	*579*	*584*	*588*	*593*	*598*	*603*	*607*	*627*
20	12150	12667	12800	12936	13073	13211	13351	13493	13637	13782	14381
	535	*550*	*554*	*558*	*562*	*566*	*570*	*574*	*578*	*582*	*599*
10	10255	10564	10643	10724	10806	10889	10972	11057	11143	11230	11587
	506	*516*	*519*	*522*	*524*	*527*	*530*	*533*	*536*	*539*	*551*

h_r, In.^{c}	max A_{ctr}, In.$^{2\ d}$										
0	346	231	213	198	185	173	163	154	146	139	115
2		776	621	517	443	388	345	310	282	259	194
3			3270	1635	1090	817	654	545	467	409	272

Shape											
W 33×130	$A_s = 38.3$ In.^{2} $I_s = 6710$ In.^{4} max V_h (steel) = 689 kips for $F_y = 36$ ksi										
	$d = 33.1$ In. $S_s = 406$ In.^{3} max V_h (steel) = 958 kips for $F_y = 50$ ksi										

A_{ctr}, In.$^{2\ b,e}$	Y2, In.^{a}										
	2.00	3.00	3.25	3.50	3.75	4.00	4.25	4.50	4.75	5.00	6.00
200	17782	19006	19322	19642	19967	20295	20627	20963	21303	21648	23065
	554	*577*	*583*	*589*	*595*	*601*	*607*	*613*	*619*	*625*	*650*
100	16243	17297	17570	17846	18125	18408	18694	18984	19277	19573	20794
	542	*564*	*569*	*575*	*581*	*586*	*592*	*598*	*603*	*609*	*633*
60	14755	15645	15875	16108	16344	16583	16824	17069	17316	17566	18597
	530	*549*	*555*	*560*	*565*	*570*	*575*	*581*	*586*	*592*	*614*
50	14173	14999	15212	15428	15647	15868	16093	16319	16549	16781	17737
	524	*543*	*548*	*553*	*558*	*563*	*568*	*573*	*579*	*584*	*605*
40	13442	14188	14380	14575	14772	14972	15174	15379	15586	15796	16658
	517	*535*	*539*	*544*	*549*	*554*	*559*	*563*	*568*	*573*	*594*
30	12498	13139	13304	13472	13642	13813	13987	14163	14341	14521	15263
	506	*523*	*527*	*531*	*536*	*540*	*545*	*549*	*554*	*558*	*577*
20	11230	11731	11860	11991	12123	12258	12393	12531	12670	12811	13390
	490	*505*	*508*	*512*	*516*	*520*	*523*	*527*	*531*	*535*	*551*
10	9438	9740	9818	9897	9977	10058	10140	10223	10307	10392	10741
	463	*473*	*476*	*478*	*481*	*484*	*486*	*489*	*492*	*495*	*506*

h_r, In.^{c}	max A_{ctr}, In.$^{2\ d}$										
0	317	211	195	181	169	158	149	141	133	127	106
2		710	568	474	406	355	316	284	258	237	178
3			2994	1497	998	749	599	499	428	374	250

^{a}Y2   = distance from top of steel beam to concrete centroid
^{b}A_{ctr} = transformed effective concrete area
      = $(b/n)\, t$ for full composite
^{c}h_r = height of steel deck ribs
^{d}max A_{ctr} = maximum transformed effective concrete area based on full concrete thickness $t = t_o$ for elastic neutral axis
      to be below the top of the steel deck
^{e}Shaded area represents A_{ctr} range wherein elastic neutral axis falls within the concrete slab ($h_r = 0$)

COMPOSITE BEAM SELECTION TABLE
Transformed Section Properties

$$\overline{I}_{tr}, \text{In.}^4$$
$$\overline{S}_{tr}, \text{In.}^3$$

Shape		
W 33×118	A_s = 34.7 In.² I_s = 5900 In.⁴	max V_h (steel) = 625 kips for F_y = 36 ksi
	d = 32.9 In. S_s = 359 In.³	max V_h (steel) = 868 kips for F_y = 50 ksi

A_{ctr}, In.² [b,e]	Y2, In. [a]										
	2.00	3.00	3.25	3.50	3.75	4.00	4.25	4.50	4.75	5.00	6.00
200	15960	17080	17369	17662	17958	18259	18562	18870	19181	19496	20793
	497	518	523	529	534	540	545	551	556	562	585
100	14658	15634	15886	16141	16399	16661	16925	17193	17464	17739	18869
	487	507	512	517	522	527	533	538	543	549	570
60	13373	14205	14420	14638	14858	15081	15307	15536	15767	16002	16966
	476	494	499	504	509	513	518	523	528	533	554
50	12862	13637	13838	14041	14246	14454	14664	14878	15093	15311	16210
	471	489	493	498	503	507	512	517	522	527	546
40	12215	12918	13100	13284	13470	13649	13850	14043	14239	14437	15252
	464	481	486	490	495	499	504	508	513	517	536
30	11368	11977	12134	12293	12455	12618	12783	12951	13120	13292	13997
	455	471	475	479	483	487	491	496	500	504	522
20	10211	10691	10816	10941	11068	11197	11328	11460	11593	11728	12285
	441	454	458	461	465	469	472	476	480	483	499
10	8538	8832	8907	8984	9062	9141	9221	9301	9383	9466	9806
	415	425	428	430	433	435	438	441	443	446	457

h_r, In. [c]	max A_{ctr}, In.² [d]										
0	285	190	175	163	152	143	134	127	120	114	95
2		640	512	426	365	320	284	256	233	213	160
3			2697	1348	899	674	539	449	385	337	225

Shape		
W 30×116	A_s = 34.2 In.² I_s = 4930 In.⁴	max V_h (steel) = 616 kips for F_y = 36 ksi
	d = 30.0 In. S_s = 329 In.³	max V_h (steel) = 855 kips for F_y = 50 ksi

A_{ctr}, In.² [b,e]	Y2, In. [a]										
	2.00	3.00	3.25	3.50	3.75	4.00	4.25	4.50	4.75	5.00	6.00
160	13091	14078	14333	14592	14855	15121	15390	15663	15940	16220	17375
	451	472	477	482	488	493	499	504	510	515	538
100	12308	13200	13431	13665	13902	14143	14387	14634	14884	15137	16182
	445	464	469	475	480	485	490	495	501	506	528
60	11234	11997	12194	12394	12597	12803	13011	13222	13436	13653	14546
	435	453	458	463	467	472	477	482	487	492	512
40	10265	10910	11077	11247	11418	11592	11769	11947	12128	12312	13068
	425	442	446	450	455	459	464	468	473	477	496
30	9554	10113	10258	10405	10554	10705	10858	11013	11169	11328	11984
	416	432	436	440	444	448	452	457	461	465	483
20	8581	9023	9137	9253	9371	9490	9611	9733	9857	9982	10500
	403	417	420	424	427	431	435	438	442	446	461
15	7946	8311	8406	8502	8599	8697	8797	8898	9000	9104	9532
	394	406	409	412	415	418	421	425	428	431	445
10	7168	7439	7509	7580	7653	7726	7800	7875	7950	8027	8345
	380	390	392	395	398	400	403	406	408	411	422

h_r, In. [c]	max A_{ctr}, In.² [d]										
0	257	171	158	147	137	128	121	114	108	103	86
2		582	465	388	332	291	258	233	211	194	145
3			2463	1232	821	616	493	411	352	308	205

[a] Y2 = distance from top of steel beam to concrete centroid
[b] A_{ctr} = transformed effective concrete area
= $(b/n) t$ for full composite
[c] h_r = height of steel deck ribs
[d] max A_{ctr} = maximum transformed effective concrete area based on full concrete thickness $t = t_o$ for elastic neutral axis to be below the top of the steel deck
[e] Shaded area represents A_{ctr} range wherein elastic neutral axis falls within the concrete slab ($h_r = 0$)

COMPOSITE BEAM SELECTION TABLE
Transformed Section Properties

$$\bar{I}_{tr}, \text{ In.}^4$$
$$\bar{S}_{tr}, \text{ In.}^3$$

Shape												
W 30×108	$A_s = 31.7$ In.^{2} $d = 29.8$ In.			$I_s = 4470$ In.^{4} $S_s = 299$ In.^{3}			max V_h (steel) = 571 kips for F_y = 36 ksi max V_h (steel) = 793 kips for F_y = 50 ksi					

A_{ctr}, In.$^{2\ b,e}$	Y2, In.[a]										
	2.00	3.00	3.25	3.50	3.75	4.00	4.25	4.50	4.75	5.00	6.00
160	12053 / 415	12975 / 434	13214 / 439	13456 / 444	13701 / 449	13949 / 454	14201 / 459	14456 / 465	14715 / 470	14977 / 475	16057 / 496
100	11365 / 409	12203 / 428	12421 / 433	12641 / 437	12864 / 442	13090 / 447	13319 / 452	13551 / 457	13786 / 462	14025 / 467	15007 / 487
60	10410 / 401	11132 / 418	11319 / 422	11509 / 427	11701 / 431	11896 / 436	12093 / 440	12293 / 445	12496 / 450	12701 / 455	13548 / 474
40	9533 / 391	10149 / 407	10309 / 412	10470 / 416	10634 / 420	10801 / 424	10969 / 428	11139 / 433	11312 / 437	11487 / 441	12209 / 459
30	8883 / 384	9419 / 399	9558 / 403	9699 / 406	9842 / 410	9987 / 414	10134 / 418	10282 / 422	10433 / 426	10586 / 430	11215 / 447
20	7980 / 372	8407 / 385	8518 / 388	8630 / 392	8744 / 395	8859 / 398	8976 / 402	9094 / 406	9214 / 409	9335 / 413	9836 / 428
15	7384 / 363	7739 / 374	7831 / 377	7924 / 380	8018 / 383	8114 / 387	8211 / 390	8309 / 393	8409 / 396	8510 / 399	8925 / 413
10	6646 / 350	6911 / 360	6979 / 362	7049 / 365	7119 / 367	7191 / 370	7263 / 372	7336 / 375	7411 / 377	7486 / 380	7796 / 391

h_r, In.[c]	max A_{ctr}, In.$^{2\ d}$										
0	236										
2		158	145	135	126	118	111	105	100	95	79
3		536	429 / 2272	357 / 1136	306 / 757	268 / 568	238 / 454	214 / 379	195 / 325	179 / 284	134 / 189

W 30×99	$A_s = 29.1$ In.^{2} $d = 29.7$ In.			$I_s = 3990$ In.^{4} $S_s = 269$ In.^{3}			max V_h (steel) = 524 kips for F_y = 36 ksi max V_h (steel) = 728 kips for F_y = 50 ksi					

A_{ctr}, In.$^{2\ b,e}$	Y2, In.[a]										
	2.00	3.00	3.25	3.50	3.75	4.00	4.25	4.50	4.75	5.00	6.00
160	10973 / 378	11826 / 395	12047 / 400	12272 / 405	12499 / 409	12729 / 414	12962 / 419	13199 / 423	13438 / 428	13680 / 433	14681 / 452
100	10379 / 373	11160 / 390	11362 / 394	11568 / 399	11776 / 403	11986 / 408	12200 / 412	12416 / 417	12635 / 421	12858 / 426	13774 / 445
60	9542 / 365	10221 / 381	10397 / 385	10575 / 389	10756 / 394	10939 / 398	11125 / 402	11313 / 406	11504 / 411	11697 / 415	12493 / 433
40	8762 / 357	9346 / 372	9497 / 376	9650 / 379	9805 / 383	9963 / 387	10123 / 391	10284 / 395	10448 / 399	10614 / 404	11299 / 420
30	8174 / 350	8686 / 364	8818 / 367	8953 / 371	9089 / 375	9227 / 378	9367 / 382	9509 / 386	9653 / 390	9798 / 394	10399 / 409
20	7347 / 339	7758 / 351	7864 / 354	7972 / 358	8081 / 361	8192 / 364	8305 / 368	8418 / 371	8534 / 374	8650 / 378	9132 / 392
15	6793 / 331	7136 / 342	7225 / 344	7315 / 347	7406 / 350	7499 / 353	7593 / 356	7688 / 359	7784 / 362	7881 / 365	8284 / 378
10	6098 / 319	6356 / 328	6422 / 330	6490 / 333	6559 / 335	6628 / 337	6699 / 340	6770 / 342	6843 / 345	6916 / 348	7218 / 358

h_r, In.[c]	max A_{ctr}, In.$^{2\ d}$										
0	216										
2		144	133	123	115	108	102	96	91	86	72
3		490	392 / 2075	326 / 1037	280 / 692	245 / 519	218 / 415	196 / 346	178 / 296	163 / 259	122 / 173

[a] $Y2$ = distance from top of steel beam to concrete centroid
[b] A_{ctr} = transformed effective concrete area = $(b/n)\,t$ for full composite
[c] h_r = height of steel deck ribs
[d] max A_{ctr} = maximum transformed effective concrete area based on full concrete thickness $t = t_o$ for elastic neutral axis to be below the top of the steel deck
[e] Shaded area represents A_{ctr} range wherein elastic neutral axis falls within the concrete slab ($h_r = 0$)

COMPOSITE BEAM SELECTION TABLE
Transformed Section Properties

$$\overline{I}_{tr}, \text{ In.}^4$$
$$\overline{S}_{tr}, \text{ In.}^3$$

Shape												
W 30×90	A_s = 26.4 In.^{2} d = 29.5 In.	I_s = 3620 In.^{4} S_s = 245 In.^{3}		max V_h (steel) = 475 kips for F_y = 36 ksi max V_h (steel) = 660 kips for F_y = 50 ksi								

A_{ctr}, In.^{2} b,e	Y2, In.^{a}										
	2.00	3.00	3.25	3.50	3.75	4.00	4.25	4.50	4.75	5.00	6.00
160	10003	10785	10988	11193	11402	11613	11827	12044	12263	12486	13404
	343	*359*	*363*	*368*	*372*	*376*	*380*	*385*	*389*	*393*	*411*
100	9499	10220	10407	10596	10788	10983	11180	11380	11582	11788	12634
	339	*355*	*359*	*363*	*367*	*371*	*375*	*379*	*383*	*388*	*405*
60	8778	9411	9575	9741	9910	10081	10254	10429	10607	10787	11530
	332	*347*	*351*	*355*	*359*	*363*	*367*	*371*	*375*	*379*	*395*
40	8093	8642	8785	8929	9075	9223	9374	9526	9680	9836	10481
	325	*339*	*343*	*347*	*350*	*354*	*358*	*361*	*365*	*369*	*384*
30	7569	8054	8180	8307	8436	8567	8700	8834	8970	9108	9677
	320	*333*	*336*	*339*	*343*	*346*	*350*	*353*	*357*	*360*	*375*
20	6820	7213	7315	7418	7523	7629	7736	7845	7955	8067	8528
	310	*322*	*325*	*328*	*331*	*334*	*337*	*340*	*343*	*346*	*360*
15	6310	6640	6726	6812	6900	6989	7080	7171	7264	7358	7746
	303	*313*	*316*	*319*	*321*	*324*	*327*	*330*	*333*	*336*	*348*
10	5659	5910	5975	6040	6107	6175	6243	6313	6383	6454	6748
	292	*301*	*303*	*305*	*308*	*310*	*312*	*315*	*317*	*320*	*330*

h_r, In.^{c}	max A_{ctr}, In.$^{2\ d}$										
0	195	130	120	111	104	97	92	87	82	78	65
2		443	354	295	253	221	197	177	161	148	111
3			1876	938	625	469	375	313	268	234	155

Shape												
W 27×102	A_s = 30.0 In.^{2} d = 27.1 In.	I_s = 3620 In.^{4} S_s = 267 In.^{3}		max V_h (steel) = 540 kips for F_y = 36 ksi max V_h (steel) = 750 kips for F_y = 50 ksi								

A_{ctr}, In.^{2} b,e	Y2, In.^{a}										
	2.00	3.00	3.25	3.50	3.75	4.00	4.25	4.50	4.75	5.00	6.00
160	9738	10549	10759	10973	11190	11410	11633	11860	12089	12322	13284
	366	*384*	*389*	*393*	*398*	*403*	*408*	*413*	*418*	*423*	*443*
100	9205	9945	10138	10333	10531	10732	10936	11143	11352	11565	12444
	361	*379*	*383*	*388*	*392*	*397*	*402*	*406*	*411*	*416*	*435*
60	8458	9100	9266	9436	9607	9782	9958	10137	10319	10503	11265
	354	*370*	*375*	*379*	*383*	*388*	*392*	*396*	*401*	*405*	*424*
40	7766	8316	8459	8604	8751	8900	9052	9205	9361	9519	10172
	346	*362*	*366*	*370*	*374*	*378*	*382*	*386*	*390*	*394*	*412*
30	7247	7729	7854	7980	8109	8240	8372	8507	8643	8781	9353
	340	*354*	*358*	*362*	*365*	*369*	*373*	*377*	*381*	*385*	*401*
20	6521	6907	7007	7108	7211	7316	7422	7529	7638	7749	8206
	330	*343*	*346*	*349*	*352*	*356*	*359*	*363*	*366*	*370*	*384*
15	6038	6359	6442	6527	6612	6700	6788	6877	6968	7060	7441
	322	*334*	*337*	*339*	*342*	*345*	*349*	*352*	*355*	*358*	*371*
10	5433	5674	5736	5800	5864	5930	5996	6063	6131	6200	6486
	312	*321*	*323*	*326*	*328*	*331*	*333*	*336*	*338*	*341*	*352*

h_r, In.^{c}	max A_{ctr}, In.$^{2\ d}$										
0	203	135	125	116	108	102	96	90	86	81	68
2		466	373	311	266	233	207	187	170	155	117
3			1985	993	662	496	397	331	284	248	165

$^a Y2$ = distance from top of steel beam to concrete centroid
$^b A_{ctr}$ = transformed effective concrete area
   = $(b/n)\ t$ for full composite
$^c h_r$ = height of steel deck ribs
^{d}max A_{ctr} = maximum transformed effective concrete area based on full concrete thickness $t = t_o$ for elastic neutral axis to be below the top of the steel deck
^{e}Shaded area represents A_{ctr} range wherein elastic neutral axis falls within the concrete slab (h_r = 0)

COMPOSITE BEAM SELECTION TABLE
Transformed Section Properties

$$\overline{I}_{tr}, \ \text{In.}^4$$
$$\overline{S}_{tr}, \ \text{In.}^3$$

Shape												

W 27×94

$A_s = 27.7 \ \text{In.}^2$ $\qquad I_s = 3270 \ \text{In.}^4$ $\qquad$ max V_h (steel) = 499 kips for F_y = 36 ksi
$d = 26.9 \ \text{In.}$ $\qquad S_s = 243 \ \text{In.}^3$ $\qquad$ max V_h (steel) = 693 kips for F_y = 50 ksi

A_{ctr}, In.$^{2\ b,e}$	Y2, In.^{a}										
	2.00	3.00	3.25	3.50	3.75	4.00	4.25	4.50	4.75	5.00	6.00
160	8927	9681	9876	10075	10277	10482	10689	10900	11113	11330	12225
	335	352	356	361	365	370	374	379	383	388	407
100	8463	9155	9335	9518	9703	9891	10082	10275	10471	10670	11493
	331	347	352	356	360	365	369	373	378	382	400
60	7805	8409	8567	8726	8888	9052	9219	9388	9559	9733	10452
	325	340	344	348	352	356	360	365	369	373	390
40	7185	7707	7843	7981	8121	8263	8407	8552	8700	8851	9471
	318	332	336	340	344	348	351	355	359	363	379
30	6715	7174	7294	7415	7538	7663	7790	7918	8048	8180	8726
	312	326	329	333	336	340	344	347	351	355	370
20	6048	6418	6515	6612	6712	6812	6914	7018	7123	7229	7670
	303	315	318	321	325	328	331	334	338	341	355
15	5597	5908	5988	6070	6153	6238	6323	6410	6498	6587	6956
	296	307	310	313	315	318	321	324	327	330	343
10	5027	5261	5322	5384	5447	5511	5575	5641	5707	5775	6053
	286	295	297	300	302	305	307	310	312	315	325

h_r, In.^{c}	max A_{ctr}, In.$^{2\ d}$										
0	186	124	115	107	99	93	88	83	78	75	62
2		428	343	285	245	214	190	171	156	143	107
3			1824	912	608	456	365	304	261	228	152

W 27×84

$A_s = 24.8 \ \text{In.}^2$ $\qquad I_s = 2850 \ \text{In.}^4$ $\qquad$ max V_h (steel) = 446 kips for F_y = 36 ksi
$d = 26.7 \ \text{In.}$ $\qquad S_s = 213 \ \text{In.}^3$ $\qquad$ max V_h (steel) = 620 kips for F_y = 50 ksi

A_{ctr}, In.$^{2\ b,e}$	Y2, In.^{a}										
	2.00	3.00	3.25	3.50	3.75	4.00	4.25	4.50	4.75	5.00	6.00
160	7926	8607	8784	8963	9146	9331	9518	9709	9902	10097	10907
	297	313	317	321	325	329	333	337	341	345	362
100	7544	8174	8338	8504	8672	8844	9017	9193	9372	9553	10303
	294	309	313	317	320	324	328	332	336	340	357
60	6992	7549	7693	7840	7989	8140	8294	8449	8607	8767	9428
	289	303	306	310	314	318	321	325	329	333	349
40	6463	6948	7074	7202	7332	7464	7598	7734	7871	8011	8588
	283	296	300	303	307	310	314	317	321	325	339
30	6054	6484	6596	6710	6825	6942	7060	7181	7303	7427	7939
	278	291	294	297	300	304	307	310	314	317	331
20	5462	5813	5904	5997	6091	6186	6283	6381	6481	6582	6999
	270	281	284	287	290	293	296	299	302	305	318
15	5055	5351	5428	5507	5586	5666	5748	5831	5915	6000	6353
	264	274	277	279	282	285	288	290	293	296	308
10	4531	4757	4816	4875	4936	4997	5060	5123	5187	5252	5521
	255	263	266	268	270	272	275	277	279	282	292

h_r, In.^{c}	max A_{ctr}, In.$^{2\ d}$										
0	166	110	102	95	88	83	78	74	70	66	55
2		381	305	254	218	190	169	152	138	127	95
3			1622	811	541	406	324	270	232	203	135

^{a}Y2 = distance from top of steel beam to concrete centroid
^{b}A_{ctr} = transformed effective concrete area
= $(b/n)\ t$ for full composite
^{c}h_r = height of steel deck ribs
^{d}max A_{ctr} = maximum transformed effective concrete area based on full concrete thickness $t = t_o$ for elastic neutral axis to be below the top of the steel deck
^{e}Shaded area represents A_{ctr} range wherein elastic neutral axis falls within the concrete slab (h_r = 0)

COMPOSITE BEAM SELECTION TABLE
Transformed Section Properties

$$\overline{I}_{tr}, \text{ In.}^4$$
$$\overline{S}_{tr}, \text{ In.}^3$$

Shape												
W 24×76	$A_s = 22.4$ In.^{2} $d = 23.9$ In.		$I_s = 2100$ In.^{4} $S_s = 176$ In.^{3}			max V_h (steel) = 403 kips for F_y = 36 ksi max V_h (steel) = 560 kips for F_y = 50 ksi						

A_{ctr}, In.^{2} [b,e]	Y2, In.[a]										
	2.00	3.00	3.25	3.50	3.75	4.00	4.25	4.50	4.75	5.00	6.00
100	5675 243	6204 257	6342 260	6482 264	6625 267	6770 271	6917 274	7067 278	7218 282	7372 286	8011 301
60	5284 239	5755 252	5878 255	6003 259	6131 262	6260 265	6391 269	6524 272	6659 276	6797 280	7366 294
40	4902 234	5317 247	5425 250	5535 253	5647 256	5761 260	5876 263	5994 266	6113 270	6234 273	6735 287
30	4602 231	4973 242	5069 245	5168 248	5268 251	5369 254	5472 258	5577 261	5683 264	5791 267	6239 281
20	4161 224	4466 235	4546 238	4627 240	4709 243	4793 246	4878 249	4964 252	5052 255	5141 258	5510 270
15	3852 219	4112 229	4180 231	4249 234	4319 236	4390 239	4462 242	4535 244	4610 247	4685 250	4999 261
10	3448 212	3648 220	3700 222	3753 224	3807 226	3862 229	3917 231	3974 233	4031 236	4089 238	4331 247
5	2897 200	3015 205	3046 207	3077 208	3109 210	3142 211	3174 213	3208 214	3242 216	3276 218	3419 224

h_r, In.[c]	max A_{ctr}, In.^{2} [d]										
0	134	89	82	77	71	67	63	60	56	54	45
2		313	250	208	179	156	139	125	114	104	78
3			1340	670	447	335	268	223	191	168	112

Shape												
W 24×68	$A_s = 20.1$ In.^{2} $d = 23.7$ In.		$I_s = 1830$ In.^{4} $S_s = 154$ In.^{3}			max V_h (steel) = 362 kips for F_y = 36 ksi max V_h (steel) = 503 kips for F_y = 50 ksi						

A_{ctr}, In.^{2} [b,e]	Y2, In.[a]										
	2.00	3.00	3.25	3.50	3.75	4.00	4.25	4.50	4.75	5.00	6.00
100	5056 216	5536 228	5662 232	5789 235	5919 238	6051 241	6185 245	6320 248	6458 251	6599 255	7180 269
60	4729 213	5162 224	5275 227	5390 231	5506 234	5625 237	5745 240	5867 243	5991 246	6117 250	6640 263
40	4405 209	4789 220	4890 223	4992 226	5095 229	5200 232	5307 235	5416 238	5526 241	5638 244	6103 257
30	4146 206	4492 216	4582 219	4674 222	4767 225	4862 228	4958 230	5056 233	5155 236	5256 239	5674 251
20	3759 200	4047 210	4122 212	4198 215	4276 218	4355 220	4435 223	4516 226	4599 228	4683 231	5031 242
15	3483 196	3729 205	3794 207	3859 209	3926 212	3993 214	4062 217	4132 219	4203 222	4274 224	4573 235
10	3115 189	3306 197	3356 199	3407 201	3459 203	3512 205	3565 207	3619 209	3674 211	3730 214	3962 223
5	2600 178	2715 183	2745 185	2776 186	2807 187	2838 189	2870 190	2903 192	2936 193	2969 195	3108 202

h_r, In.[c]	max A_{ctr}, In.^{2} [d]										
0	119	79	73	68	64	60	56	53	50	48	40
2		279	223	186	159	139	124	111	101	93	70
3			1195	598	398	299	239	199	171	149	100

[a] Y2 = distance from top of steel beam to concrete centroid
[b] A_{ctr} = transformed effective concrete area = $(b/n)\,t$ for full composite
[c] h_r = height of steel deck ribs
[d] max A_{ctr} = maximum transformed effective concrete area based on full concrete thickness $t = t_o$ for elastic neutral axis to be below the top of the steel deck
[e] Shaded area represents A_{ctr} range wherein elastic neutral axis falls within the concrete slab ($h_r = 0$)

COMPOSITE BEAM SELECTION TABLE
Transformed Section Properties

$$\overline{I}_{tr}, \text{ In.}^4$$
$$\overline{S}_{tr}, \text{ In.}^3$$

Shape												
W 24×62	$A_s = 18.2 \text{ In.}^2$ $I_s = 1550 \text{ In.}^4$ max V_h (steel) = 328 kips for F_y = 36 ksi											
	$d = 23.7 \text{ In.}$ $S_s = 131 \text{ In.}^3$ max V_h (steel) = 455 kips for F_y = 50 ksi											

A_{ctr}, In.$^{2\ b,e}$

	Y2, In.^{a}										
	2.00	3.00	3.25	3.50	3.75	4.00	4.25	4.50	4.75	5.00	6.00
100	4520	4963	5078	5196	5315	5436	5559	5685	5812	5940	6475
	192	203	206	209	212	215	218	221	224	227	240
60	4241	4643	4747	4854	4962	5072	5184	5297	5412	5529	6014
	188	199	202	205	208	211	214	217	220	223	235
40	3960	4319	4413	4508	4605	4704	4804	4905	5009	5113	5548
	185	196	198	201	204	207	209	212	215	218	230
30	3732	4057	4142	4229	4316	4405	4496	4588	4682	4776	5170
	182	192	195	197	200	203	205	208	211	214	225
20	3385	3659	3730	3803	3877	3952	4028	4105	4184	4264	4595
	177	186	189	191	193	196	198	201	203	206	216
15	3133	3369	3431	3494	3558	3622	3688	3755	3823	3891	4177
	173	181	183	186	188	190	193	195	197	200	209
10	2792	2978	3026	3075	3125	3176	3228	3280	3334	3388	3612
	166	174	176	178	180	182	184	186	188	190	198
5	2305	2418	2447	2477	2507	2538	2570	2602	2634	2667	2803
	155	160	162	163	165	166	167	169	170	172	178

h_r, In.^{c}	max A_{ctr}, In.$^{2\ d}$										
0	108	72	66	62	58	54	51	48	45	43	36
2		252	202	168	144	126	112	101	92	84	63
3			1083	541	361	271	217	180	155	135	90

Shape												
W 24×55	$A_s = 16.2 \text{ In.}^2$ $I_s = 1350 \text{ In.}^4$ max V_h (steel) = 292 kips for F_y = 36 ksi											
	$d = 23.6 \text{ In.}$ $S_s = 114 \text{ In.}^3$ max V_h (steel) = 405 kips for F_y = 50 ksi											

A_{ctr}, In.$^{2\ b,e}$

	Y2, In.^{a}										
	2.00	3.00	3.25	3.50	3.75	4.00	4.25	4.50	4.75	5.00	6.00
100	4008	4406	4510	4615	4723	4832	4943	5056	5170	5286	5768
	169	180	182	185	188	190	193	196	199	202	213
60	3779	4143	4238	4335	4433	4533	4635	4738	4843	4949	5390
	167	177	179	182	185	187	190	193	195	198	209
40	3544	3874	3960	4047	4136	4226	4318	4411	4506	4602	5000
	164	174	176	179	181	184	186	189	191	194	205
30	3351	3652	3730	3810	3891	3974	4057	4142	4229	4316	4680
	162	171	173	176	178	180	183	185	188	190	201
20	3052	3308	3375	3443	3512	3582	3653	3725	3799	3873	4183
	157	166	168	170	172	175	177	179	182	184	194
15	2831	3054	3112	3171	3231	3292	3354	3417	3481	3546	3815
	154	162	164	166	168	170	172	174	176	179	188
10	2526	2702	2749	2795	2843	2891	2941	2991	3041	3093	3307
	148	155	157	159	160	162	164	166	168	170	178
5	2076	2186	2214	2243	2273	2302	2333	2364	2395	2427	2559
	138	143	144	146	147	148	150	151	153	154	160

h_r, In.^{c}	max A_{ctr}, In.$^{2\ d}$										
0	95	64	59	55	51	48	45	42	40	38	32
2		223	179	149	128	112	99	89	81	74	56
3			958	479	319	240	192	160	137	120	80

^{a}Y2 = distance from top of steel beam to concrete centroid
^{b}A_{ctr} = transformed effective concrete area
     = $(b/n)\ t$ for full composite
^{c}h_r = height of steel deck ribs
^{d}max A_{ctr} = maximum transformed effective concrete area based on full concrete thickness $t = t_o$ for elastic neutral axis to be below the top of the steel deck
^{e}Shaded area represents A_{ctr} range wherein elastic neutral axis falls within the concrete slab (h_r = 0)

COMPOSITE BEAM SELECTION TABLE
Transformed Section Properties

$$\overline{I}_{tr}, \text{ In.}^4$$
$$\overline{S}_{tr}, \text{ In.}^3$$

Shape												

W 21×62

$A_s = 18.3 \text{ In.}^2$ $I_s = 1330 \text{ In.}^4$ max V_h (steel) = 329 kips for $F_y = 36$ ksi
$d = 21.0 \text{ In.}$ $S_s = 127 \text{ In.}^3$ max V_h (steel) = 458 kips for $F_y = 50$ ksi

$A_{ctr}, \text{ In.}^{2 \text{ b,e}}$	Y2, In.[a]										
	2.00	3.00	3.25	3.50	3.75	4.00	4.25	4.50	4.75	5.00	6.00
100	3753	4155	4261	4368	4477	4588	4702	4817	4934	5052	5547
	178	190	193	196	199	202	205	208	211	214	227
60	3524	3889	3984	4082	4181	4281	4384	4488	4594	4702	5150
	176	187	189	192	195	198	201	204	207	210	223
40	3294	3620	3705	3793	3881	3971	4063	4156	4251	4348	4750
	173	183	186	189	191	194	197	200	203	206	218
30	3107	3403	3480	3559	3639	3721	3804	3888	3974	4062	4425
	170	180	183	185	188	191	194	196	199	202	213
20	2824	3072	3137	3203	3271	3339	3409	3480	3553	3626	3932
	166	175	178	180	182	185	187	190	192	195	206
15	2618	2832	2889	2946	3004	3063	3123	3185	3247	3310	3574
	162	171	173	175	178	180	182	185	187	189	199
10	2340	2508	2553	2597	2643	2689	2737	2785	2834	2883	3090
	157	164	166	168	170	172	174	176	178	181	189
5	1944	2046	2072	2100	2127	2156	2184	2213	2243	2273	2399
	148	153	154	156	157	158	160	161	163	164	171

$h_r, \text{ In.}^{\text{c}}$	max $A_{ctr}, \text{ In.}^{2 \text{ d}}$										
0	96	64	59	55	51	48	45	43	40	38	32
2		229	183	152	131	114	102	91	83	76	57
3			988	494	329	247	198	165	141	123	82

W 21×57

$A_s = 16.7 \text{ In.}^2$ $I_s = 1170 \text{ In.}^4$ max V_h (steel) = 301 kips for $F_y = 36$ ksi
$d = 21.1 \text{ In.}$ $S_s = 111 \text{ In.}^3$ max V_h (steel) = 418 kips for $F_y = 50$ ksi

$A_{ctr}, \text{ In.}^{2 \text{ b,e}}$	Y2, In.[a]										
	2.00	3.00	3.25	3.50	3.75	4.00	4.25	4.50	4.75	5.00	6.00
100	3425	3798	3896	3995	4096	4200	4304	4411	4519	4630	5088
	161	172	174	177	180	183	186	188	191	194	206
60	3226	3566	3656	3747	3839	3933	4029	4126	4225	4326	4745
	159	169	172	174	177	180	182	185	188	191	202
40	3023	3330	3410	3492	3576	3661	3747	3835	3924	4015	4392
	156	166	168	171	174	176	179	181	184	187	198
30	2857	3136	3210	3284	3360	3437	3516	3596	3677	3760	4104
	154	163	166	168	171	173	176	178	181	183	194
20	2601	2838	2900	2963	3027	3093	3160	3228	3297	3367	3658
	150	158	161	163	165	168	170	172	175	177	187
15	2412	2618	2672	2727	2783	2840	2897	2956	3016	3077	3330
	147	155	157	159	161	163	165	168	170	172	181
10	2153	2316	2359	2402	2446	2491	2537	2584	2631	2679	2880
	141	148	150	152	154	156	158	160	162	164	172
5	1775	1875	1901	1928	1955	1983	2011	2040	2069	2098	2222
	132	137	139	140	141	143	144	146	147	149	155

$h_r, \text{ In.}^{\text{c}}$	max $A_{ctr}, \text{ In.}^{2 \text{ d}}$										
0	88	59	54	50	47	44	41	39	37	35	29
2		209	167	140	120	105	93	84	76	70	52
3			904	452	301	226	181	151	129	113	75

[a] Y2 = distance from top of steel beam to concrete centroid
[b] A_{ctr} = transformed effective concrete area
= (b/n) t for full composite
[c] h_r = height of steel deck ribs
[d] max A_{ctr} = maximum transformed effective concrete area based on full concrete thickness $t = t_o$ for elastic neutral axis to be below the top of the steel deck
[e] Shaded area represents A_{ctr} range wherein elastic neutral axis falls within the concrete slab ($h_r = 0$)

AMERICAN INSTITUTE OF STEEL CONSTRUCTION

COMPOSITE BEAM SELECTION TABLE
Transformed Section Properties

$$\overline{I}_{tr}, \text{In.}^4$$
$$\overline{S}_{tr}, \text{In.}^3$$

Shape												
W 21×50	$A_s = 14.7$ In.^{2} $d = 20.8$ In.	$I_s = 984$ In.^{4} $S_s = 95$ In.^{3}			max V_h (steel) = 265 kips for F_y = 36 ksi max V_h (steel) = 368 kips for F_y = 50 ksi							

A_{ctr}, In.$^{2\ b,e}$	— Y2, In.^{a}										
	2.00	3.00	3.25	3.50	3.75	4.00	4.25	4.50	4.75	5.00	6.00
100	2968 / 140	3299 / 149	3386 / 152	3474 / 154	3564 / 157	3655 / 159	3749 / 162	3843 / 164	3940 / 167	4038 / 169	4446 / 180
60	2809 / 138	3114 / 147	3194 / 149	3275 / 152	3358 / 154	3442 / 157	3528 / 159	3616 / 161	3704 / 164	3795 / 166	4170 / 177
40	2644 / 136	2922 / 144	2995 / 147	3069 / 149	3144 / 151	3221 / 154	3299 / 156	3379 / 158	3459 / 161	3542 / 163	3884 / 173
30	2507 / 134	2762 / 142	2829 / 144	2897 / 147	2966 / 149	3037 / 151	3108 / 153	3181 / 156	3255 / 158	3331 / 160	3645 / 170
20	2292 / 130	2510 / 138	2568 / 140	2626 / 142	2686 / 145	2746 / 147	2808 / 149	2870 / 151	2934 / 153	2999 / 155	3269 / 164
15	2130 / 128	2321 / 135	2372 / 137	2423 / 139	2475 / 141	2528 / 143	2582 / 145	2637 / 147	2693 / 149	2749 / 151	2986 / 160
10	1902 / 123	2056 / 130	2096 / 131	2137 / 133	2179 / 135	2221 / 137	2265 / 138	2309 / 140	2354 / 142	2399 / 144	2588 / 152
5	1559 / 115	1656 / 120	1681 / 121	1707 / 122	1733 / 124	1760 / 125	1787 / 126	1814 / 128	1842 / 129	1871 / 131	1990 / 136

h_r, In.^{c}	— max A_{ctr}, In.$^{2\ d}$										
0	77	51	47	44	41	38	36	34	32	31	26
2		183	146	122	104	91	81	73	66	61	46
3			789	394	263	197	158	131	113	99	66

Shape												
W 21×44	$A_s = 13.0$ In.^{2} $d = 20.7$ In.	$I_s = 843$ In.^{4} $S_s = 82$ In.^{3}			max V_h (steel) = 234 kips for F_y = 36 ksi max V_h (steel) = 325 kips for F_y = 50 ksi							

A_{ctr}, In.$^{2\ b,e}$	— Y2, In.^{a}										
	2.00	3.00	3.25	3.50	3.75	4.00	4.25	4.50	4.75	5.00	6.00
100	2600 / 122	2896 / 131	2973 / 133	3052 / 135	3132 / 137	3214 / 140	3297 / 142	3381 / 144	3468 / 146	3555 / 149	3919 / 158
60	2472 / 121	2747 / 129	2818 / 131	2892 / 133	2966 / 135	3042 / 138	3119 / 140	3198 / 142	3278 / 144	3359 / 146	3697 / 156
40	2338 / 119	2590 / 127	2656 / 129	2723 / 131	2791 / 133	2861 / 135	2932 / 137	3004 / 140	3077 / 142	3152 / 144	3463 / 153
30	2224 / 117	2457 / 125	2518 / 127	2580 / 129	2644 / 131	2708 / 133	2774 / 135	2840 / 137	2908 / 139	2977 / 142	3264 / 150
20	2042 / 115	2245 / 122	2298 / 124	2352 / 126	2407 / 128	2463 / 130	2520 / 131	2577 / 133	2636 / 135	2696 / 137	2946 / 146
15	1903 / 112	2082 / 119	2129 / 121	2176 / 123	2225 / 124	2274 / 126	2325 / 128	2376 / 130	2428 / 132	2481 / 134	2701 / 142
10	1703 / 109	1848 / 115	1886 / 116	1925 / 118	1964 / 119	2005 / 121	2045 / 123	2087 / 124	2129 / 126	2172 / 128	2351 / 135
5	1392 / 101	1485 / 106	1509 / 107	1534 / 108	1559 / 109	1585 / 111	1611 / 112	1638 / 113	1665 / 115	1692 / 116	1806 / 122

h_r, In.^{c}	— max A_{ctr}, In.$^{2\ d}$										
0	67	45	41	38	36	34	32	30	28	27	22
2		160	128	107	92	80	71	64	58	53	40
3			693	347	231	173	139	116	99	87	58

^{a}Y2 = distance from top of steel beam to concrete centroid
^{b}A_{ctr} = transformed effective concrete area
  = $(b/n)\ t$ for full composite
^{c}h_r = height of steel deck ribs
^{d}max A_{ctr} = maximum transformed effective concrete area based on full concrete thickness $t = t_o$ for elastic neutral axis to be below the top of the steel deck
^{e}Shaded area represents A_{ctr} range wherein elastic neutral axis falls within the concrete slab ($h_r = 0$)

COMPOSITE BEAM SELECTION TABLE
Transformed Section Properties

$$\overline{I}_{tr}, \text{ In.}^4$$
$$\overline{S}_{tr}, \text{ In.}^3$$

Shape												
W 18×60	A_s = 17.6 In.^{2} d = 18.2 In.	I_s = 984 In.^{4} S_s = 108 In.^{3}		max V_h (steel) = 317 kips for F_y = 36 ksi max V_h (steel) = 440 kips for F_y = 50 ksi								

A_{ctr}, In.$^{2 \ b,e}$	Y2, In.^{a}										
	2.00	3.00	3.25	3.50	3.75	4.00	4.25	4.50	4.75	5.00	6.00
100	2843 *153*	3191 *164*	3282 *167*	3376 *170*	3471 *173*	3568 *176*	3668 *179*	3769 *182*	3871 *185*	3976 *188*	4414 *201*
60	2672 *151*	2988 *162*	3071 *164*	3156 *167*	3243 *170*	3331 *173*	3422 *176*	3513 *179*	3607 *182*	3702 *185*	4100 *197*
40	2499 *148*	2783 *159*	2858 *161*	2934 *164*	3012 *167*	3091 *170*	3172 *172*	3255 *175*	3339 *178*	3424 *181*	3782 *193*
30	2358 *146*	2616 *156*	2684 *159*	2753 *161*	2824 *164*	2896 *167*	2969 *169*	3044 *172*	3120 *175*	3198 *177*	3522 *189*
20	2143 *143*	2361 *152*	2418 *154*	2477 *156*	2536 *159*	2597 *161*	2659 *164*	2722 *166*	2787 *169*	2852 *171*	3126 *182*
15	1987 *140*	2175 *148*	2224 *150*	2275 *152*	2327 *155*	2379 *157*	2433 *159*	2487 *162*	2543 *164*	2600 *166*	2837 *176*
10	1773 *135*	1922 *142*	1961 *144*	2000 *146*	2041 *148*	2083 *150*	2125 *152*	2168 *154*	2212 *156*	2256 *158*	2443 *167*
5	1466 *127*	1556 *132*	1580 *133*	1605 *135*	1629 *136*	1655 *138*	1680 *139*	1707 *141*	1733 *142*	1761 *144*	1875 *150*

h_r, In.^{c}	max A_{ctr}, In.$^{2 \ d}$										
0 2 3	80	54 196	49 157 853	46 130 427	43 112 284	40 98 213	38 87 171	36 78 142	34 71 122	32 65 107	27 49 71

Shape												
W 18×55	A_s = 16.2 In.^{2} d = 18.1 In.	I_s = 890 In.^{4} S_s = 98 In.^{3}		max V_h (steel) = 292 kips for F_y = 36 ksi max V_h (steel) = 405 kips for F_y = 50 ksi								

A_{ctr}, In.$^{2 \ b,e}$	Y2, In.^{a}										
	2.00	3.00	3.25	3.50	3.75	4.00	4.25	4.50	4.75	5.00	6.00
100	2602 *140*	2924 *151*	3009 *153*	3096 *156*	3184 *159*	3274 *161*	3366 *164*	3460 *167*	3555 *170*	3652 *173*	4058 *184*
60	2454 *138*	2749 *148*	2826 *151*	2906 *153*	2987 *156*	3069 *159*	3153 *161*	3239 *164*	3326 *167*	3415 *170*	3786 *181*
40	2302 *136*	2569 *146*	2639 *148*	2711 *151*	2784 *153*	2858 *156*	2934 *158*	3012 *161*	3091 *164*	3171 *166*	3507 *177*
30	2178 *134*	2421 *143*	2485 *146*	2551 *148*	2617 *151*	2685 *153*	2755 *156*	2825 *158*	2897 *161*	2971 *163*	3277 *174*
20	1986 *131*	2192 *140*	2247 *142*	2302 *144*	2359 *146*	2417 *149*	2476 *151*	2536 *153*	2597 *156*	2660 *158*	2920 *168*
15	1843 *128*	2023 *136*	2071 *138*	2119 *140*	2168 *143*	2219 *145*	2270 *147*	2322 *149*	2376 *151*	2430 *154*	2657 *163*
10	1647 *124*	1789 *131*	1827 *133*	1865 *135*	1905 *137*	1945 *139*	1985 *140*	2027 *142*	2069 *144*	2112 *146*	2292 *155*
5	1357 *116*	1446 *122*	1469 *123*	1493 *124*	1517 *126*	1542 *127*	1567 *128*	1592 *130*	1619 *131*	1645 *133*	1756 *139*

h_r, In.^{c}	max A_{ctr}, In.$^{2 \ d}$										
0 2 3	73	49 179	45 143 781	42 119 391	39 102 260	37 90 195	35 80 156	33 72 130	31 65 112	29 60 98	24 45 65

^{a}Y2 = distance from top of steel beam to concrete centroid
$^b A_{ctr}$ = transformed effective concrete area = $(b/n) \, t$ for full composite
$^c h_r$ = height of steel deck ribs
^{d}max A_{ctr} = maximum transformed effective concrete area based on full concrete thickness $t = t_o$ for elastic neutral axis to be below the top of the steel deck
^{e}Shaded area represents A_{ctr} range wherein elastic neutral axis falls within the concrete slab ($h_r = 0$)

AMERICAN INSTITUTE OF STEEL CONSTRUCTION

COMPOSITE BEAM SELECTION TABLE
Transformed Section Properties

$$\overline{I}_{tr}, \text{ In.}^4$$
$$\overline{S}_{tr}, \text{ In.}^3$$

Shape												

W 18×50

$A_s = 14.7$ In.^{2} $\quad I_s = 800$ In.^{4} $\quad$ max V_h (steel) = 265 kips for $F_y = 36$ ksi
$d = 18.0$ In. $\quad S_s = 89$ In.^{3} $\quad$ max V_h (steel) = 368 kips for $F_y = 50$ ksi

A_{ctr}, In.$^{2\ b,e}$	Y2, In.^{a}										
	2.00	3.00	3.25	3.50	3.75	4.00	4.25	4.50	4.75	5.00	6.00
100	2358	2652	2730	2809	2890	2973	3057	3142	3230	3318	3690
	127	136	139	141	144	146	149	151	154	157	167
60	2232	2504	2575	2648	2723	2799	2876	2955	3036	3118	3460
	125	134	137	139	142	144	147	149	152	154	164
40	2103	2350	2415	2482	2549	2619	2689	2761	2834	2909	3220
	123	132	135	137	139	142	144	146	149	151	161
30	1995	2222	2282	2343	2405	2469	2533	2599	2666	2735	3021
	122	130	133	135	137	139	142	144	146	149	159
20	1826	2021	2072	2124	2178	2232	2288	2345	2402	2461	2707
	119	127	129	131	133	135	138	140	142	144	153
15	1699	1869	1914	1960	2007	2055	2104	2153	2204	2255	2471
	117	124	126	128	130	132	134	136	138	140	149
10	1520	1657	1693	1730	1768	1806	1845	1885	1925	1966	2139
	113	120	121	123	125	127	128	130	132	134	142
5	1251	1337	1360	1383	1406	1430	1455	1480	1505	1531	1639
	106	111	112	114	115	116	118	119	121	122	128

h_r, In.^{c}	max A_{ctr}, In.$^{2\ d}$										
0	66	44	41	38	35	33	31	29	28	26	22
2		162	129	108	92	81	72	65	59	54	40
3			705	353	235	176	141	118	101	88	59

W 18×46

$A_s = 13.5$ In.^{2} $\quad I_s = 712$ In.^{4} $\quad$ max V_h (steel) = 243 kips for $F_y = 36$ ksi
$d = 18.1$ In. $\quad S_s = 78.8$ In.^{3} $\quad$ max V_h (steel) = 338 kips for $F_y = 50$ ksi

A_{ctr}, In.$^{2\ b,e}$	Y2, In.^{a}										
	2.00	3.00	3.25	3.50	3.75	4.00	4.25	4.50	4.75	5.00	6.00
100	2167	2442	2514	2588	2663	2740	2818	2898	2979	3062	3407
	116	124	127	129	131	134	136	138	141	143	153
60	2058	2312	2379	2447	2517	2588	2661	2734	2810	2886	3207
	114	123	125	127	129	132	134	136	139	141	151
40	1943	2176	2237	2300	2364	2429	2495	2563	2632	2702	2995
	112	121	123	125	127	129	132	134	136	138	148
30	1847	2062	2118	2176	2235	2295	2356	2419	2482	2547	2818
	111	119	121	123	125	127	130	132	134	136	145
20	1694	1880	1929	1979	2030	2082	2135	2189	2244	2300	2534
	108	116	118	120	122	124	126	128	130	132	141
15	1578	1742	1785	1829	1874	1920	1966	2014	2062	2112	2318
	106	113	115	117	119	121	123	125	127	129	137
10	1412	1544	1579	1615	1651	1688	1726	1764	1804	1844	2011
	103	109	111	112	114	116	118	119	121	123	130
5	1156	1240	1263	1285	1308	1332	1356	1380	1405	1431	1537
	96	101	102	104	105	106	107	109	110	112	117

h_r, In.^{c}	max A_{ctr}, In.$^{2\ d}$										
0	61	41	38	35	33	30	29	27	26	24	20
2		149	119	99	85	74	66	60	54	50	37
3			650	325	217	162	130	108	93	81	54

[a] $Y2$ = distance from top of steel beam to concrete centroid
[b] A_{ctr} = transformed effective concrete area
= (b/n) t for full composite
[c] h_r = height of steel deck ribs
[d] max A_{ctr} = maximum transformed effective concrete area based on full concrete thickness $t = t_o$ for elastic neutral axis to be below the top of the steel deck
[e] Shaded area represents A_{ctr} range wherein elastic neutral axis falls within the concrete slab ($h_r = 0$)

COMPOSITE BEAM SELECTION TABLE
Transformed Section Properties

$$\overline{I}_{tr}, \text{ In.}^4$$
$$\overline{S}_{tr}, \text{ In.}^3$$

Shape												

W 18×40

$A_s = 11.8$ In.² $\quad I_s = 612$ In.⁴ $\quad$ max V_h (steel) = 212 kips for $F_y = 36$ ksi
$d = 17.9$ In. $\quad S_s = 68.4$ In.³ $\quad$ max V_h (steel) = 295 kips for $F_y = 50$ ksi

A_{ctr}, In.² [b,e]	Y2, In.[a]										
	2.00	3.00	3.25	3.50	3.75	4.00	4.25	4.50	4.75	5.00	6.00
100	1886 / 101	2128 / 108	2191 / 110	2256 / 112	2323 / 114	2390 / 116	2459 / 118	2530 / 121	2601 / 123	2674 / 125	2979 / 133
60	1799 / 99	2025 / 107	2085 / 109	2145 / 111	2207 / 113	2271 / 115	2335 / 117	2401 / 119	2468 / 121	2536 / 123	2821 / 132
40	1708 / 98	1917 / 105	1972 / 107	2028 / 109	2085 / 111	2143 / 113	2203 / 115	2264 / 117	2326 / 119	2389 / 121	2652 / 129
30	1630 / 97	1824 / 104	1875 / 106	1927 / 108	1980 / 110	2035 / 112	2090 / 113	2147 / 115	2204 / 117	2263 / 119	2507 / 127
20	1504 / 95	1673 / 102	1718 / 103	1764 / 105	1811 / 107	1858 / 109	1907 / 111	1956 / 112	2007 / 114	2058 / 116	2272 / 124
15	1405 / 93	1556 / 100	1596 / 101	1637 / 103	1678 / 105	1721 / 106	1764 / 108	1808 / 110	1853 / 111	1898 / 113	2089 / 121
10	1262 / 90	1386 / 96	1418 / 98	1452 / 99	1486 / 101	1521 / 102	1556 / 104	1592 / 105	1629 / 107	1666 / 109	1823 / 115
5	1034 / 85	1114 / 89	1135 / 90	1157 / 91	1179 / 93	1201 / 94	1224 / 95	1248 / 96	1272 / 98	1296 / 99	1397 / 104

h_r, In.[c]	max A_{ctr}, In.² [d]										
0	53	35	32	30	28	26	25	23	22	21	18
2		129	103	86	74	65	57	52	47	43	32
3			564	282	188	141	113	94	81	71	47

W 18×35

$A_s = 10.3$ In.² $\quad I_s = 510$ In.⁴ $\quad$ max V_h (steel) = 185 kips for $F_y = 36$ ksi
$d = 17.7$ In. $\quad S_s = 57.6$ In.³ $\quad$ max V_h (steel) = 258 kips for $F_y = 50$ ksi

A_{ctr}, In.² [b,e]	Y2, In.[a]										
	2.00	3.00	3.25	3.50	3.75	4.00	4.25	4.50	4.75	5.00	6.00
100	1618 / 87	1830 / 93	1886 / 95	1943 / 97	2001 / 99	2060 / 101	2121 / 102	2183 / 104	2246 / 106	2310 / 108	2578 / 116
60	1550 / 86	1749 / 92	1802 / 94	1856 / 96	1911 / 97	1967 / 99	2024 / 101	2082 / 103	2141 / 105	2201 / 106	2454 / 114
40	1478 / 85	1664 / 91	1713 / 93	1763 / 94	1814 / 96	1866 / 98	1919 / 100	1973 / 101	2028 / 103	2085 / 105	2320 / 112
30	1415 / 84	1589 / 90	1635 / 92	1682 / 93	1730 / 95	1779 / 97	1828 / 98	1879 / 100	1931 / 102	1983 / 104	2203 / 111
20	1312 / 82	1466 / 88	1507 / 90	1549 / 91	1591 / 93	1634 / 94	1678 / 96	1723 / 98	1769 / 99	1816 / 101	2011 / 108
15	1230 / 80	1369 / 86	1405 / 88	1443 / 89	1481 / 91	1520 / 92	1559 / 94	1600 / 95	1641 / 97	1683 / 99	1858 / 105
10	1108 / 78	1223 / 83	1254 / 85	1285 / 86	1316 / 87	1349 / 89	1382 / 90	1415 / 92	1449 / 93	1484 / 95	1630 / 101
5	907 / 73	983 / 77	1003 / 78	1024 / 79	1045 / 81	1066 / 82	1088 / 83	1110 / 84	1133 / 85	1156 / 86	1253 / 91

h_r, In.[c]	max A_{ctr}, In.² [d]										
0	46	30	28	26	24	23	21	20	19	18	15
2		112	89	75	64	56	50	45	41	37	28
3			488	244	163	122	98	81	70	61	41

[a] $Y2$ = distance from top of steel beam to concrete centroid
[b] A_{ctr} = transformed effective concrete area
= $(b/n)\,t$ for full composite
[c] h_r = height of steel deck ribs
[d] max A_{ctr} = maximum transformed effective concrete area based on full concrete thickness $t = t_o$ for elastic neutral axis to be below the top of the steel deck
[e] Shaded area represents A_{ctr} range wherein elastic neutral axis falls within the concrete slab ($h_r = 0$)

AMERICAN INSTITUTE OF STEEL CONSTRUCTION

COMPOSITE BEAM SELECTION TABLE
Transformed Section Properties

$$\bar{I}_{tr}, \ \text{In.}^4$$
$$\bar{S}_{tr}, \ \text{In.}^3$$

Shape												

W 16×36

$A_s = 10.6 \ \text{In.}^2$ $I_s = 448 \ \text{In.}^4$ max V_h (steel) = 191 kips for $F_y = 36$ ksi
$d = 15.9 \ \text{In.}$ $S_s = 56.5 \ \text{In.}^3$ max V_h (steel) = 265 kips for $F_y = 50$ ksi

A_{ctr}, In.$^{2 \ b,e}$	Y2, In.^{a}										
	2.00	3.00	3.25	3.50	3.75	4.00	4.25	4.50	4.75	5.00	6.00
80	1378 *82.5*	1573 *89.5*	1625 *91.3*	1677 *93.1*	1732 *94.9*	1787 *96.8*	1843 *98.6*	1901 *100.5*	1960 *102.5*	2019 *104.4*	2271 *112.3*
50	1315 *81.5*	1497 *88.3*	1545 *90.1*	1595 *91.9*	1645 *93.7*	1697 *95.5*	1750 *97.3*	1803 *99.2*	1858 *101.0*	1914 *102.9*	2149 *110.7*
35	1253 *80.6*	1423 *87.2*	1468 *88.9*	1514 *90.6*	1561 *92.4*	1609 *94.2*	1658 *96.0*	1708 *97.8*	1759 *99.6*	1811 *101.4*	2030 *109.0*
25	1184 *79.5*	1339 *85.8*	1381 *87.5*	1423 *89.2*	1466 *90.8*	1510 *92.6*	1554 *94.3*	1600 *96.1*	1647 *97.8*	1695 *99.6*	1895 *107.0*
20	1133 *78.6*	1277 *84.7*	1316 *86.3*	1355 *88.0*	1395 *89.6*	1436 *91.3*	1477 *93.0*	1520 *94.7*	1564 *96.4*	1608 *98.2*	1794 *105.3*
15	1062 *77.2*	1191 *83.1*	1226 *84.6*	1261 *86.2*	1297 *87.8*	1333 *89.4*	1371 *91.0*	1409 *92.6*	1448 *94.3*	1488 *95.9*	1654 *102.8*
10	956 *75.0*	1064 *80.4*	1092 *81.8*	1121 *83.2*	1151 *84.6*	1181 *86.1*	1212 *87.6*	1244 *89.1*	1276 *90.6*	1309 *92.1*	1447 *98.5*
5	783 *70.5*	854 *74.7*	873 *75.8*	892 *77.0*	912 *78.1*	932 *79.3*	952 *80.5*	973 *81.7*	995 *82.9*	1016 *84.2*	1108 *89.4*

h_r, In.^{c}	max A_{ctr}, In.$^{2 \ d}$										
0	42.0	28.0	25.9	24.0	22.4	21.0	19.8	18.7	17.7	16.8	14.0
2		105.3	84.2	70.2	60.1	52.6	46.8	42.1	38.3	35.1	26.3
3			463.4	231.7	154.5	115.9	92.7	77.2	66.2	57.9	38.6

W 16×31

$A_s = 9.1 \ \text{In.}^2$ $I_s = 375 \ \text{In.}^4$ max V_h (steel) = 164 kips for $F_y = 36$ ksi
$d = 15.9 \ \text{In.}$ $S_s = 47.2 \ \text{In.}^3$ max V_h (steel) = 228 kips for $F_y = 50$ ksi

A_{ctr}, In.$^{2 \ b,e}$	Y2, In.^{a}										
	2.00	3.00	3.25	3.50	3.75	4.00	4.25	4.50	4.75	5.00	6.00
80	1191 *70.6*	1361 *76.7*	1407 *78.2*	1453 *79.8*	1500 *81.4*	1549 *83.0*	1598 *84.6*	1649 *86.3*	1700 *87.9*	1752 *89.6*	1973 *96.4*
50	1141 *69.8*	1302 *75.7*	1345 *77.3*	1389 *78.8*	1433 *80.4*	1479 *82.0*	1525 *83.6*	1573 *85.2*	1621 *86.8*	1671 *88.5*	1878 *95.2*
35	1093 *69.1*	1244 *74.8*	1284 *76.3*	1325 *77.9*	1367 *79.4*	1409 *80.9*	1453 *82.5*	1498 *84.1*	1543 *85.7*	1589 *87.3*	1784 *93.9*
25	1037 *68.1*	1177 *73.8*	1214 *75.2*	1252 *76.7*	1290 *78.2*	1330 *79.7*	1370 *81.2*	1411 *82.7*	1453 *84.3*	1496 *85.9*	1676 *92.3*
20	996 *67.4*	1126 *72.9*	1161 *74.3*	1196 *75.7*	1233 *77.2*	1270 *78.7*	1307 *80.2*	1346 *81.7*	1385 *83.2*	1425 *84.7*	1594 *91.0*
15	937 *66.3*	1055 *71.6*	1086 *72.9*	1119 *74.3*	1151 *75.7*	1185 *77.1*	1219 *78.5*	1254 *80.0*	1290 *81.5*	1326 *82.9*	1478 *89.0*
10	847 *64.5*	947 *69.3*	973 *70.6*	1000 *71.8*	1028 *73.1*	1056 *74.4*	1085 *75.8*	1114 *77.1*	1144 *78.5*	1175 *79.9*	1303 *85.5*
5	694 *60.6*	762 *64.5*	780 *65.5*	798 *66.6*	817 *67.6*	836 *68.7*	855 *69.8*	875 *70.9*	895 *72.0*	916 *73.2*	1003 *77.9*

h_r, In.^{c}	max A_{ctr}, In.$^{2 \ d}$										
0	36.2	24.1	22.3	20.7	19.3	18.1	17.0	16.1	15.2	14.5	12.1
2		90.7	72.5	60.4	51.8	45.3	40.3	36.3	33.0	30.2	22.7
3			399.1	199.5	133.0	99.8	79.8	66.5	57.0	49.9	33.3

[a] $Y2$ = distance from top of steel beam to concrete centroid
[b] A_{ctr} = transformed effective concrete area
  = $(b/n) \ t$ for full composite
[c] h_r = height of steel deck ribs
[d] max A_{ctr} = maximum transformed effective concrete area based on full concrete thickness $t = t_o$ for elastic neutral axis to be below the top of the steel deck
[e] Shaded area represents A_{ctr} range wherein elastic neutral axis falls within the concrete slab ($h_r = 0$)

COMPOSITE BEAM SELECTION TABLE
Transformed Section Properties

$$\overline{I}_{tr}, \text{ In.}^4$$
$$\overline{S}_{tr}, \text{ In.}^3$$

Shape												
W 16×26	$A_s = 7.68$ In.^{2} $\quad$ $I_s = 301$ In.^{4} $\quad$ max V_h (steel) = 138 kips for F_y = 36 ksi											
	$d = 15.69$ In. $\quad$ $S_s = 38.4$ In.^{3} $\quad$ max V_h (steel) = 192 kips for F_y = 50 ksi											

A_{ctr}, In.$^{2\ b,e}$	Y2, In.^{a}										
	2.00	3.00	3.25	3.50	3.75	4.00	4.25	4.50	4.75	5.00	6.00
80	987 *58.6*	1132 *63.8*	1170 *65.1*	1210 *66.5*	1250 *67.8*	1291 *69.2*	1333 *70.6*	1376 *72.0*	1419 *73.4*	1464 *74.8*	1651 *80.6*
50	950 *58.0*	1088 *63.1*	1125 *64.4*	1162 *65.7*	1200 *67.1*	1239 *68.4*	1279 *69.8*	1320 *71.2*	1361 *72.6*	1404 *74.0*	1581 *79.7*
35	914 *57.4*	1045 *62.4*	1079 *63.7*	1115 *65.0*	1151 *66.3*	1188 *67.6*	1225 *69.0*	1264 *70.3*	1303 *71.7*	1343 *73.1*	1511 *78.7*
25	873 *56.7*	994 *61.6*	1026 *62.8*	1059 *64.1*	1093 *65.4*	1127 *66.7*	1163 *68.0*	1198 *69.3*	1235 *70.7*	1272 *72.0*	1429 *77.5*
20	841 *56.2*	955 *60.9*	986 *62.1*	1017 *63.4*	1049 *64.6*	1081 *65.9*	1114 *67.2*	1148 *68.5*	1183 *69.8*	1218 *71.1*	1366 *76.6*
15	795 *55.3*	900 *59.9*	928 *61.1*	956 *62.3*	985 *63.5*	1015 *64.7*	1045 *66.0*	1076 *67.2*	1108 *68.5*	1140 *69.8*	1276 *75.0*
10	723 *53.9*	813 *58.1*	837 *59.2*	861 *60.4*	886 *61.5*	911 *62.7*	937 *63.8*	964 *65.0*	991 *66.2*	1019 *67.4*	1134 *72.4*
5	595 *50.7*	658 *54.3*	674 *55.2*	691 *56.1*	709 *57.1*	726 *58.0*	744 *59.0*	763 *60.0*	782 *61.0*	801 *62.1*	882 *66.3*

h_r, In.^{c}	max A_{ctr}, In.$^{2\ d}$										
0	30.1	20.1	18.5	17.2	16.1	15.1	14.2	13.4	12.7	12.0	10.0
2		75.6	60.5	50.4	43.2	37.8	33.6	30.2	27.5	25.2	18.9
3			333.2	166.6	111.1	83.3	66.6	55.5	47.6	41.6	27.8

Shape												
W 14×38	$A_s = 11.20$ In.^{2} $\quad$ $I_s = 385$ In.^{4} $\quad$ max V_h (steel) = 202 kips for F_y = 36 ksi											
	$d = 14.10$ In. $\quad$ $S_s = 54.6$ In.^{3} $\quad$ max V_h (steel) = 280 kips for F_y = 50 ksi											

A_{ctr}, In.$^{2\ b,e}$	Y2, In.^{a}										
	2.00	3.00	3.25	3.50	3.75	4.00	4.25	4.50	4.75	5.00	6.00
80	1196 *79.8*	1384 *87.2*	1434 *89.1*	1485 *91.1*	1538 *93.1*	1591 *95.0*	1646 *97.0*	1702 *99.1*	1760 *101.1*	1818 *103.2*	2065 *111.6*
50	1139 *78.8*	1313 *86.1*	1360 *87.9*	1408 *89.8*	1456 *91.8*	1506 *93.7*	1558 *95.7*	1610 *97.6*	1663 *99.7*	1718 *101.7*	1947 *110.0*
35	1083 *77.9*	1245 *84.9*	1288 *86.7*	1332 *88.6*	1378 *90.4*	1424 *92.3*	1471 *94.3*	1520 *96.2*	1569 *98.1*	1620 *100.1*	1833 *108.2*
25	1021 *76.7*	1168 *83.5*	1208 *85.3*	1248 *87.1*	1289 *88.9*	1332 *90.7*	1375 *92.6*	1419 *94.4*	1464 *96.3*	1510 *98.2*	1704 *106.1*
20	975 *75.8*	1112 *82.4*	1148 *84.1*	1186 *85.8*	1224 *87.6*	1263 *89.4*	1303 *91.2*	1344 *93.0*	1386 *94.9*	1429 *96.7*	1609 *104.4*
15	911 *74.5*	1034 *80.7*	1067 *82.4*	1100 *84.0*	1134 *85.7*	1169 *87.4*	1205 *89.1*	1242 *90.9*	1279 *92.6*	1317 *94.4*	1478 *101.8*
10	819 *72.3*	919 *78.0*	946 *79.5*	974 *81.0*	1002 *82.5*	1031 *84.1*	1060 *85.7*	1091 *87.3*	1121 *88.9*	1153 *90.5*	1286 *97.3*
5	669 *67.9*	735 *72.4*	752 *73.5*	770 *74.7*	789 *76.0*	807 *77.2*	827 *78.5*	847 *79.8*	867 *81.1*	887 *82.4*	974 *87.9*

h_r, In.^{c}	max A_{ctr}, In.$^{2\ d}$										
0	39.5	26.3	24.3	22.6	21.1	19.7	18.6	17.5	16.6	15.8	13.2
2		101.4	81.1	67.6	57.9	50.7	45.0	40.5	36.9	33.8	25.3
3			450.2	225.1	150.1	112.6	90.0	75.0	64.3	56.3	37.5

^{a}Y2 = distance from top of steel beam to concrete centroid
$^b A_{ctr}$ = transformed effective concrete area
$\quad$ = $(b/n)\,t$ for full composite
$^c h_r$ = height of steel deck ribs
^{d}max A_{ctr} = maximum transformed effective concrete area based on full concrete thickness $t = t_o$ for elastic neutral axis to be below the top of the steel deck
^{e}Shaded area represents A_{ctr} range wherein elastic neutral axis falls within the concrete slab ($h_r = 0$)

AMERICAN INSTITUTE OF STEEL CONSTRUCTION

COMPOSITE BEAM SELECTION TABLE
Transformed Section Properties

$$\overline{I}_{tr},\ \text{In.}^4$$
$$\overline{S}_{tr},\ \text{In.}^3$$

Shape													

W 14×34

$A_s = 10.00$ In.^{2} $I_s = 340$ In.^{4} max V_h (steel) = 180 kips for F_y = 36 ksi
$d = 13.98$ In. $S_s = 48.6$ In.^{3} max V_h (steel) = 250 kips for F_y = 50 ksi

A_{ctr}, In.$^{2\ b,e}$	Y2, In.^{a}										
	2.00	3.00	3.25	3.50	3.75	4.00	4.25	4.50	4.75	5.00	6.00
80	1065	1234	1279	1325	1372	1420	1470	1520	1572	1625	1847
	71.1	*77.7*	*79.5*	*81.2*	*83.0*	*84.7*	*86.5*	*88.4*	*90.2*	*92.1*	*99.6*
50	1018	1176	1218	1261	1305	1351	1397	1444	1493	1542	1750
	70.3	*76.8*	*78.5*	*80.2*	*81.9*	*83.6*	*85.4*	*87.2*	*89.0*	*90.8*	*98.3*
35	972	1119	1158	1199	1240	1282	1326	1370	1415	1461	1655
	69.5	*75.8*	*77.5*	*79.1*	*80.8*	*82.5*	*84.3*	*86.0*	*87.8*	*89.5*	*96.8*
25	919	1055	1091	1128	1166	1205	1244	1285	1327	1369	1547
	68.6	*74.7*	*76.3*	*77.9*	*79.5*	*81.2*	*82.9*	*84.6*	*86.3*	*88.0*	*95.1*
20	880	1007	1041	1075	1111	1147	1184	1222	1261	1300	1467
	67.8	*73.8*	*75.3*	*76.9*	*78.5*	*80.1*	*81.7*	*83.4*	*85.1*	*86.8*	*93.7*
15	826	940	970	1001	1033	1066	1099	1133	1168	1204	1354
	66.7	*72.4*	*73.9*	*75.4*	*76.9*	*78.5*	*80.0*	*81.6*	*83.2*	*84.9*	*91.6*
10	745	840	865	891	918	945	973	1001	1030	1060	1185
	64.9	*70.1*	*71.4*	*72.8*	*74.2*	*75.7*	*77.1*	*78.6*	*80.1*	*81.6*	*87.8*
5	610	673	690	707	725	743	762	780	800	820	903
	61.1	*65.2*	*66.3*	*67.4*	*68.6*	*69.7*	*70.9*	*72.1*	*73.4*	*74.6*	*79.8*

h_r, In.^{c}	max A_{ctr}, In.$^{2\ d}$										
0	35.0	23.3	21.5	20.0	18.6	17.5	16.4	15.5	14.7	14.0	11.7
2		89.9	71.9	59.9	51.4	45.0	40.0	36.0	32.7	30.0	22.5
3			399.6	199.8	133.2	99.9	79.9	66.6	57.1	50.0	33.3

W 14×30

$A_s = 8.85$ In.^{2} $I_s = 291$ In.^{4} max V_h (steel) = 159 kips for F_y = 36 ksi
$d = 13.84$ In. $S_s = 42.0$ In.^{3} max V_h (steel) = 221 kips for F_y = 50 ksi

A_{ctr}, In.$^{2\ b,e}$	Y2, In.^{a}										
	2.00	3.00	3.25	3.50	3.75	4.00	4.25	4.50	4.75	5.00	6.00
80	932	1082	1122	1163	1205	1248	1292	1337	1383	1430	1628
	62.3	*68.2*	*69.8*	*71.3*	*72.9*	*74.5*	*76.1*	*77.7*	*79.4*	*81.0*	*87.7*
50	893	1035	1073	1112	1151	1192	1233	1276	1319	1364	1550
	61.6	*67.4*	*68.9*	*70.5*	*72.0*	*73.6*	*75.2*	*76.7*	*78.4*	*80.0*	*86.6*
35	856	989	1025	1061	1098	1136	1175	1215	1256	1298	1473
	61.0	*66.7*	*68.1*	*69.6*	*71.1*	*72.7*	*74.2*	*75.8*	*77.4*	*79.0*	*85.5*
25	813	936	969	1003	1037	1073	1109	1146	1183	1222	1384
	60.2	*65.7*	*67.2*	*68.6*	*70.1*	*71.6*	*73.1*	*74.6*	*76.1*	*77.7*	*84.1*
20	781	896	927	959	991	1024	1058	1093	1128	1164	1317
	59.6	*65.0*	*66.4*	*67.8*	*69.2*	*70.7*	*72.2*	*73.7*	*75.2*	*76.7*	*82.9*
15	735	840	868	897	926	956	987	1018	1050	1083	1221
	58.7	*63.8*	*65.2*	*66.5*	*67.9*	*69.3*	*70.8*	*72.2*	*73.7*	*75.1*	*81.2*
10	665	754	777	802	826	852	878	904	931	959	1076
	57.1	*61.9*	*63.1*	*64.4*	*65.7*	*67.0*	*68.3*	*69.7*	*71.0*	*72.4*	*78.1*
5	546	606	622	638	655	672	690	708	727	745	825
	53.8	*57.7*	*58.7*	*59.8*	*60.8*	*61.9*	*63.0*	*64.1*	*65.3*	*66.4*	*71.2*

h_r, In.^{c}	max A_{ctr}, In.$^{2\ d}$										
0	30.6	20.4	18.8	17.5	16.3	15.3	14.4	13.6	12.9	12.2	10.2
2		78.9	63.2	52.6	45.1	39.5	35.1	31.6	28.7	26.3	19.7
3			351.2	175.6	117.1	87.8	70.2	58.5	50.2	43.9	29.3

^{a}Y2 = distance from top of steel beam to concrete centroid
^{b}A_{ctr} = transformed effective concrete area
= (b/n) t for full composite
^{c}h_r = height of steel deck ribs
^{d}max A_{ctr} = maximum transformed effective concrete area based on full concrete thickness $t = t_o$ for elastic neutral axis to be below the top of the steel deck
^{e}Shaded area represents A_{ctr} range wherein elastic neutral axis falls within the concrete slab (h_r = 0)

COMPOSITE BEAM SELECTION TABLE
Transformed Section Properties

$$\overline{I}_{tr}, \text{ In.}^4$$
$$\overline{S}_{tr}, \text{ In.}^3$$

Shape

W 14×26

$A_s = 7.69$ In.² $I_s = 241$ In.⁴   max V_h (steel) = 138 kips for F_y = 36 ksi
$d = 13.91$ In. $S_s = 35.3$ In.³   max V_h (steel) = 192 kips for F_y = 50 ksi

A_{ctr}, In.² [b,e]	Y2, In.[a]										
	2.00	3.00	3.25	3.50	3.75	4.00	4.25	4.50	4.75	5.00	6.00
80	814 / *53.8*	947 / *59.0*	982 / *60.4*	1019 / *61.8*	1056 / *63.1*	1094 / *64.5*	1132 / *65.9*	1172 / *67.3*	1213 / *68.8*	1254 / *70.2*	1429 / *76.1*
50	784 / *53.2*	910 / *58.4*	943 / *59.7*	978 / *61.0*	1013 / *62.4*	1049 / *63.8*	1086 / *65.2*	1124 / *66.6*	1162 / *68.0*	1202 / *69.4*	1368 / *75.2*
35	754 / *52.7*	873 / *57.7*	905 / *59.0*	937 / *60.4*	970 / *61.7*	1005 / *63.0*	1039 / *64.4*	1075 / *65.8*	1112 / *67.2*	1149 / *68.6*	1306 / *74.3*
25	719 / *52.1*	830 / *57.0*	860 / *58.2*	890 / *59.5*	921 / *60.8*	953 / *62.1*	985 / *63.5*	1019 / *64.8*	1053 / *66.2*	1088 / *67.6*	1234 / *73.2*
20	692 / *51.6*	797 / *56.4*	825 / *57.6*	854 / *58.9*	883 / *60.1*	913 / *61.4*	944 / *62.7*	975 / *64.1*	1008 / *65.4*	1041 / *66.7*	1179 / *72.3*
15	654 / *50.8*	750 / *55.4*	776 / *56.6*	802 / *57.8*	829 / *59.1*	856 / *60.3*	885 / *61.6*	913 / *62.9*	943 / *64.2*	973 / *65.5*	1099 / *70.8*
10	594 / *49.5*	677 / *53.8*	699 / *54.9*	721 / *56.0*	744 / *57.2*	768 / *58.4*	792 / *59.6*	816 / *60.8*	841 / *62.0*	867 / *63.2*	975 / *68.3*
5	488 / *46.6*	546 / *50.2*	561 / *51.5*	577 / *52.1*	593 / *53.0*	609 / *54.0*	626 / *55.0*	643 / *56.1*	661 / *57.1*	678 / *58.2*	754 / *62.5*

h_r, In.[c]	max A_{ctr}, In.² [d]										
0	26.7		16.5	15.3	14.3	13.4	12.6	11.9	11.3	10.7	8.9
2		68.9	55.1	45.9	39.4	34.4	30.6	27.5	25.0	23.0	17.2
3			306.2	153.1	102.1	76.6	61.2	51.0	43.7	38.3	25.5

W 14×22

$A_s = 6.49$ In.² $I_s = 199$ In.⁴   max V_h (steel) = 117 kips for F_y = 36 ksi
$d = 13.74$ In. $S_s = 29.0$ In.³   max V_h (steel) = 162 kips for F_y = 50 ksi

A_{ctr}, In.² [b,e]	Y2, In.[a]										
	2.00	3.00	3.25	3.50	3.75	4.00	4.25	4.50	4.75	5.00	6.00
80	678 / *45.0*	790 / *49.4*	820 / *50.6*	851 / *51.7*	883 / *52.9*	915 / *54.1*	948 / *55.3*	982 / *56.5*	1016 / *57.7*	1051 / *58.9*	1200 / *63.9*
50	655 / *44.5*	763 / *48.9*	791 / *50.0*	821 / *51.2*	851 / *52.3*	882 / *53.5*	913 / *54.7*	946 / *55.9*	979 / *57.1*	1013 / *58.3*	1155 / *63.2*
35	633 / *44.1*	735 / *48.4*	763 / *49.5*	791 / *50.6*	819 / *51.8*	849 / *52.9*	879 / *54.1*	910 / *55.3*	941 / *56.4*	973 / *57.6*	1109 / *62.5*
25	606 / *43.6*	703 / *47.8*	729 / *48.9*	755 / *50.0*	782 / *51.1*	810 / *52.3*	838 / *53.4*	867 / *54.6*	897 / *55.7*	927 / *56.9*	1055 / *61.7*
20	586 / *43.2*	678 / *47.3*	702 / *48.4*	728 / *49.5*	753 / *50.6*	780 / *51.7*	807 / *52.8*	834 / *54.0*	862 / *55.1*	891 / *56.3*	1012 / *61.0*
15	557 / *42.6*	642 / *46.6*	664 / *47.7*	687 / *48.7*	711 / *49.8*	736 / *50.9*	760 / *52.0*	786 / *53.1*	812 / *54.2*	839 / *55.3*	951 / *60.0*
10	509 / *41.6*	583 / *45.4*	603 / *46.4*	623 / *47.4*	644 / *48.4*	665 / *49.4*	687 / *50.4*	709 / *51.5*	731 / *52.5*	754 / *53.6*	852 / *58.0*
5	422 / *39.3*	475 / *42.5*	489 / *43.3*	503 / *44.2*	518 / *45.1*	533 / *46.0*	549 / *46.9*	565 / *47.8*	581 / *48.7*	597 / *49.6*	667 / *53.5*

h_r, In.[c]	max A_{ctr}, In.² [d]										
0	22.3		13.7	12.7	11.9	11.1	10.5	9.9	9.4	8.9	7.4
2		57.6	46.1	38.4	32.9	28.8	25.6	23.0	20.9	19.2	14.4
3			256.2	128.1	85.4	64.1	51.2	42.7	36.6	32.0	21.4

[a] $Y2$ = distance from top of steel beam to concrete centroid
[b] A_{ctr} = transformed effective concrete area = $(b/n)\,t$ for full composite
[c] h_r = height of steel deck ribs
[d] max A_{ctr} = maximum transformed effective concrete area based on full concrete thickness $t = t_o$ for elastic neutral axis to be below the top of the steel deck
[e] Shaded area represents A_{ctr} range wherein elastic neutral axis falls within the concrete slab ($h_r = 0$)

COMPOSITE BEAM SELECTION TABLE
Transformed Section Properties

$$\overline{I}_{tr},\ \text{In.}^4$$
$$\overline{S}_{tr},\ \text{In.}^3$$

Shape												
W 12×30	$A_s = 8.79$ In.² $d = 12.34$ In.				$I_s = 238$ In.⁴ $S_s = 38.6$ In.³			max V_h (steel) = 158 kips for $F_y = 36$ ksi max V_h (steel) = 220 kips for $F_y = 50$ ksi				

A_{ctr}, In.² [b,e]	Y2, In. [a]										
	2.00	3.00	3.25	3.50	3.75	4.00	4.25	4.50	4.75	5.00	6.00
80	773 / 57.1	911 / 63.1	947 / 64.6	985 / 66.2	1024 / 67.8	1064 / 69.4	1105 / 71.0	1146 / 72.6	1189 / 74.3	1233 / 75.9	1418 / 82.7
50	741 / 56.5	871 / 62.3	906 / 63.9	941 / 65.4	978 / 66.9	1015 / 68.5	1054 / 70.1	1093 / 71.7	1134 / 73.3	1175 / 75.0	1349 / 81.7
35	710 / 55.9	832 / 61.6	864 / 63.1	898 / 64.6	932 / 66.1	968 / 67.7	1004 / 69.2	1041 / 70.8	1079 / 72.4	1117 / 74.0	1281 / 80.6
25	674 / 55.2	787 / 60.7	817 / 62.2	848 / 63.7	880 / 65.1	913 / 66.6	946 / 68.2	980 / 69.7	1016 / 71.3	1052 / 72.8	1203 / 79.3
20	647 / 54.6	753 / 60.1	782 / 61.5	811 / 62.9	841 / 64.4	871 / 65.8	903 / 67.3	935 / 68.8	968 / 70.4	1002 / 71.9	1144 / 78.2
15	609 / 53.8	705 / 59.0	731 / 60.4	757 / 61.8	785 / 63.2	812 / 64.6	841 / 66.0	870 / 67.5	900 / 68.9	931 / 70.4	1060 / 76.6
10	551 / 52.4	632 / 57.2	654 / 58.5	676 / 59.8	699 / 61.1	723 / 62.4	747 / 63.7	771 / 65.1	797 / 66.5	823 / 67.9	932 / 73.7
5	451 / 49.4	506 / 53.3	521 / 54.4	536 / 55.4	552 / 56.5	568 / 57.6	584 / 58.8	601 / 59.9	618 / 61.1	636 / 62.2	710 / 67.1

h_r, In. [c]	max A_{ctr}, In.² [d]										
0	27.1	18.1	16.7	15.5	14.5	13.6	12.8	12.1	11.4	10.8	9.0
2		71.8	57.5	47.9	41.0	35.9	31.9	28.7	26.1	23.9	18.0
3			322.4	161.2	107.5	80.6	64.5	53.7	46.1	40.3	26.9

Shape												
W 12×26	$A_s = 7.65$ In.² $d = 12.22$ In.				$I_s = 204$ In.⁴ $S_s = 33.4$ In.³			max V_h (steel) = 138 kips for $F_y = 36$ ksi max V_h (steel) = 191 kips for $F_y = 50$ ksi				

A_{ctr}, In.² [b,e]	Y2, In. [a]										
	2.00	3.00	3.25	3.50	3.75	4.00	4.25	4.50	4.75	5.00	6.00
80	670 / 49.6	790 / 54.8	822 / 56.1	855 / 57.5	889 / 58.9	924 / 60.3	960 / 61.7	997 / 63.1	1034 / 64.5	1073 / 66.0	1235 / 71.9
50	645 / 49.0	759 / 54.2	789 / 55.5	821 / 56.8	853 / 58.2	886 / 59.6	920 / 61.0	955 / 62.4	991 / 63.8	1027 / 65.2	1181 / 71.1
35	620 / 48.6	728 / 53.6	757 / 54.9	787 / 56.2	817 / 57.5	849 / 58.9	881 / 60.3	914 / 61.7	947 / 63.1	982 / 64.5	1128 / 70.3
25	591 / 48.0	692 / 52.9	719 / 54.2	747 / 55.5	776 / 56.8	805 / 58.1	835 / 59.4	865 / 60.8	897 / 62.2	929 / 63.6	1065 / 69.2
20	570 / 47.6	665 / 52.4	690 / 53.6	717 / 54.9	744 / 56.2	771 / 57.5	800 / 58.8	829 / 60.1	858 / 61.5	889 / 62.8	1017 / 68.4
15	538 / 46.9	626 / 51.5	649 / 52.7	673 / 54.0	698 / 55.2	723 / 56.5	749 / 57.7	776 / 59.0	803 / 60.3	831 / 61.7	948 / 67.1
10	490 / 45.8	565 / 50.1	585 / 51.2	605 / 52.4	626 / 53.5	648 / 54.7	670 / 55.9	693 / 57.2	716 / 58.4	740 / 59.6	840 / 64.8
5	403 / 43.3	455 / 46.9	469 / 47.8	484 / 48.8	498 / 49.8	513 / 50.8	529 / 51.8	545 / 52.9	561 / 53.9	578 / 55.0	648 / 59.5

h_r, In. [c]	max A_{ctr}, In.² [d]										
0	23.4	15.6	14.4	13.4	12.5	11.7	11.0	10.4	9.8	9.3	7.8
2		62.0	49.6	41.4	35.5	31.0	27.6	24.8	22.6	20.7	15.5
3			278.8	139.4	92.9	69.7	55.8	46.5	39.8	34.8	23.2

[a] Y2 = distance from top of steel beam to concrete centroid
[b] A_{ctr} = transformed effective concrete area = $(b/n)\,t$ for full composite
[c] h_r = height of steel deck ribs
[d] max A_{ctr} = maximum transformed effective concrete area based on full concrete thickness $t = t_o$ for elastic neutral axis to be below the top of the steel deck
[e] Shaded area represents A_{ctr} range wherein elastic neutral axis falls within the concrete slab ($h_r = 0$)

COMPOSITE BEAM SELECTION TABLE
Transformed Section Properties

$$\overline{I}_{tr}, \text{ In.}^4$$
$$\overline{S}_{tr}, \text{ In.}^3$$

Shape												
W 12×22	$A_s = 6.48$ In.^{2} $I_s = 156$ In.^{4} max V_h (steel) = 117 kips for F_y = 36 ksi										
	$d = 12.31$ In. $S_s = 25.4$ In.^{3} max V_h (steel) = 162 kips for F_y = 50 ksi										

A_{ctr}, In.$^{2\ b,e}$	Y2, In.^{a}										
	2.00	3.00	3.25	3.50	3.75	4.00	4.25	4.50	4.75	5.00	6.00
80	561	665	693	721	751	781	812	843	876	909	1048
	41.0	*45.5*	*46.6*	*47.8*	*49.0*	*50.2*	*51.4*	*52.7*	*53.9*	*55.2*	*60.3*
50	542	641	668	695	723	752	781	811	842	874	1008
	40.5	*44.9*	*46.1*	*47.3*	*48.4*	*49.6*	*50.8*	*52.1*	*53.3*	*54.5*	*59.6*
35	523	617	643	669	695	723	751	780	809	839	967
	40.1	*44.5*	*45.6*	*46.8*	*47.9*	*49.1*	*50.3*	*51.5*	*52.7*	*53.9*	*58.9*
25	500	589	613	638	663	689	715	742	770	798	918
	39.6	*43.9*	*45.0*	*46.1*	*47.3*	*48.4*	*49.6*	*50.8*	*52.0*	*53.2*	*58.1*
20	483	568	591	614	638	662	688	713	740	767	881
	39.2	*43.4*	*44.5*	*45.7*	*46.8*	*47.9*	*49.1*	*50.2*	*51.4*	*52.6*	*57.4*
15	458	537	558	579	601	624	647	671	695	720	826
	38.7	*42.8*	*43.8*	*44.9*	*46.0*	*47.1*	*48.2*	*49.4*	*50.5*	*51.7*	*56.4*
10	418	486	505	523	543	562	583	603	624	646	738
	37.7	*41.5*	*42.5*	*43.6*	*44.6*	*45.7*	*46.7*	*47.8*	*48.9*	*50.0*	*54.5*
5	344	393	406	420	433	447	462	477	492	508	573
	35.5	*38.7*	*39.6*	*40.5*	*41.4*	*42.3*	*43.2*	*44.2*	*45.1*	*46.1*	*50.1*

h_r, In.^{c}	max A_{ctr}, In.$^{2\ d}$										
0	19.9	13.3	12.3	11.4	10.6	10.0	9.4	8.9	8.4	8.0	6.6
2		52.8	42.3	35.2	30.2	26.4	23.5	21.1	19.2	17.6	13.2
3			237.3	118.6	79.1	59.3	47.5	39.5	33.9	29.7	19.8

Shape												
W 12×19	$A_s = 5.57$ In.^{2} $I_s = 130$ In.^{4} max V_h (steel) = 100 kips for F_y = 36 ksi										
	$d = 12.16$ In. $S_s = 21.3$ In.^{3} max V_h (steel) = 139 kips for F_y = 50 ksi										

A_{ctr}, In.$^{2\ b,e}$	Y2, In.^{a}										
	2.00	3.00	3.25	3.50	3.75	4.00	4.25	4.50	4.75	5.00	6.00
80	477	566	590	615	640	666	692	720	747	776	897
	35.0	*38.8*	*39.9*	*40.9*	*41.9*	*42.9*	*44.0*	*45.1*	*46.1*	*47.2*	*51.6*
50	461	547	570	594	618	643	669	695	722	749	866
	34.6	*38.4*	*39.4*	*40.4*	*41.4*	*42.5*	*43.5*	*44.6*	*45.6*	*46.7*	*51.1*
35	447	529	551	574	597	621	646	671	697	723	834
	34.2	*38.0*	*39.0*	*40.0*	*41.0*	*42.0*	*43.1*	*44.1*	*45.2*	*46.2*	*50.5*
25	429	508	529	550	572	595	618	642	666	691	797
	33.8	*37.6*	*38.6*	*39.5*	*40.5*	*41.5*	*42.5*	*43.6*	*44.6*	*45.7*	*49.9*
20	416	491	511	532	553	574	597	619	643	667	767
	33.6	*37.2*	*38.2*	*39.2*	*40.1*	*41.1*	*42.1*	*43.1*	*44.2*	*45.2*	*49.4*
15	396	466	485	504	524	544	565	586	608	630	724
	33.1	*36.7*	*37.6*	*38.6*	*39.5*	*40.5*	*41.5*	*42.5*	*43.5*	*44.5*	*48.6*
10	364	426	442	459	477	494	513	531	550	570	653
	32.3	*35.7*	*36.6*	*37.5*	*38.4*	*39.4*	*40.3*	*41.3*	*42.2*	*43.2*	*47.2*
5	302	348	360	372	385	398	412	425	439	454	515
	30.5	*33.5*	*34.3*	*35.1*	*35.9*	*36.7*	*37.5*	*38.4*	*39.2*	*40.1*	*43.7*

h_r, In.^{c}	max A_{ctr}, In.$^{2\ d}$										
0	16.9	11.3	10.4	9.7	9.0	8.5	8.0	7.5	7.1	6.8	5.6
2		45.0	36.0	30.0	25.7	22.5	20.0	18.0	16.4	15.0	11.3
3			202.3	101.2	67.4	50.6	40.5	33.7	28.9	25.3	16.9

[a] $Y2$ = distance from top of steel beam to concrete centroid
[b] A_{ctr} = transformed effective concrete area
= $(b/n)\, t$ for full composite
[c] h_r = height of steel deck ribs
[d] max A_{ctr} = maximum transformed effective concrete area based on full concrete thickness $t = t_o$ for elastic neutral axis to be below the top of the steel deck
[e] Shaded area represents A_{ctr} range wherein elastic neutral axis falls within the concrete slab ($h_r = 0$)

COMPOSITE BEAM SELECTION TABLE
Transformed Section Properties

$$\overline{I}_{tr}, \text{In.}^4$$
$$\overline{S}_{tr}, \text{In.}^3$$

Shape												

W 12×16

$A_s = 4.71$ In.^{2} $I_s = 103.0$ In.^{4} max V_h (steel) = 85 kips for F_y = 36 ksi
$d = 11.99$ In. $S_s = 17.1$ In.^{3} max V_h (steel) = 118 kips for F_y = 50 ksi

A_{ctr}, In.$^{2\ b,e}$	Y2, In.^{a}										
	2.00	3.00	3.25	3.50	3.75	4.00	4.25	4.50	4.75	5.00	6.00
80	394	470	490	511	532	554	577	600	623	647	750
	29.1	32.4	33.3	34.1	35.0	35.9	36.8	37.7	38.6	39.5	43.3
50	382	455	475	495	516	537	559	581	604	628	727
	28.7	32.0	32.9	33.8	34.6	35.5	36.4	37.3	38.2	39.1	42.8
35	371	442	461	480	500	521	542	563	585	608	703
	28.5	31.7	32.6	33.4	34.3	35.2	36.0	36.9	37.8	38.7	42.4
25	358	426	444	462	481	501	521	542	563	584	675
	28.2	31.4	32.2	33.1	33.9	34.8	35.7	36.5	37.4	38.3	42.0
20	348	413	430	448	467	486	505	525	545	566	653
	27.9	31.1	31.9	32.8	33.6	34.5	35.3	36.2	37.1	38.0	41.6
15	333	394	411	427	445	462	480	499	518	538	620
	27.6	30.7	31.5	32.3	33.2	34.0	34.8	35.7	36.6	37.4	41.0
10	308	363	378	393	408	424	440	457	474	491	565
	27.0	30.0	30.7	31.5	32.3	33.1	33.9	34.8	35.6	36.4	39.9
5	258	300	311	322	334	346	358	371	383	397	452
	25.6	28.2	28.9	29.6	30.3	31.0	31.8	32.5	33.3	34.0	37.2

h_r, In.^{c}	max A_{ctr}, In.$^{2\ d}$										
0	14.1	9.4	8.7	8.1	7.5	7.1	6.6	6.3	5.9	5.6	4.7
2		37.7	30.1	25.1	21.5	18.8	16.7	15.1	13.7	12.6	9.4
3			169.5	84.7	56.5	42.4	33.9	28.2	24.2	21.2	14.1

W 12×14

$A_s = 4.16$ In.^{2} $I_s = 88.6$ In.^{4} max V_h (steel) = 75 kips for F_y = 36 ksi
$d = 11.91$ In. $S_s = 14.9$ In.^{3} max V_h (steel) = 104 kips for F_y = 50 ksi

A_{ctr}, In.$^{2\ b,e}$	Y2, In.^{a}										
	2.00	3.00	3.25	3.50	3.75	4.00	4.25	4.50	4.75	5.00	6.00
80	346	412	430	449	468	487	507	528	548	570	660
	25.6	28.5	29.3	30.0	30.8	31.6	32.4	33.2	34.0	34.8	38.1
50	336	401	418	436	454	473	493	513	533	554	642
	25.3	28.2	28.9	29.7	30.5	31.3	32.0	32.8	33.6	34.5	37.8
35	327	390	407	424	442	460	479	498	518	538	623
	25.0	27.9	28.7	29.4	30.2	31.0	31.8	32.5	33.3	34.1	37.4
25	316	377	393	410	427	444	462	481	499	519	600
	24.8	27.6	28.4	29.1	29.9	30.7	31.4	32.2	33.0	33.8	37.1
20	308	366	382	398	415	432	449	467	485	504	582
	24.6	27.4	28.1	28.9	29.6	30.4	31.2	31.9	32.7	33.5	36.7
15	296	351	366	381	397	413	429	446	463	481	555
	24.3	27.1	27.8	28.5	29.3	30.0	30.8	31.5	32.3	33.1	36.3
10	275	325	338	352	366	381	395	411	426	442	509
	23.8	26.5	27.2	27.9	28.6	29.3	30.0	30.8	31.5	32.3	35.4
5	233	271	281	292	303	314	325	337	349	362	414
	22.6	25.0	25.6	26.3	26.9	27.6	28.2	28.9	29.6	30.3	33.1

h_r, In.^{c}	max A_{ctr}, In.$^{2\ d}$										
0	12.4	8.3	7.6	7.1	6.6	6.2	5.8	5.5	5.2	5.0	4.1
2		33.1	26.5	22.1	18.9	16.5	14.7	13.2	12.0	11.0	8.3
3			149.0	74.5	49.7	37.3	29.8	24.8	21.3	18.6	12.4

^{a}Y2 = distance from top of steel beam to concrete centroid
^{b}A_{ctr} = transformed effective concrete area
= $(b/n)\ t$ for full composite
^{c}h_r = height of steel deck ribs
^{d}max A_{ctr} = maximum transformed effective concrete area based on full concrete thickness $t = t_o$ for elastic neutral axis to be below the top of the steel deck
^{e}Shaded area represents A_{ctr} range wherein elastic neutral axis falls within the concrete slab ($h_r = 0$)

COMPOSITE BEAM SELECTION TABLE
Transformed Section Properties

$$\overline{I}_{tr}, \text{ In.}^4$$
$$\overline{S}_{tr}, \text{ In.}^3$$

W 10×26

$A_s = 7.61$ In.² $I_s = 144.0$ In.⁴   max V_h (steel) = 137 kips for F_y = 36 ksi
$d = 10.33$ In. $S_s = 27.9$ In.³   max V_h (steel) = 190 kips for F_y = 50 ksi

A_{ctr}, In.² [b,e]	\multicolumn{11}{c}{Y2, In. [a]}
	2.00	3.00	3.25	3.50	3.75	4.00	4.25	4.50	4.75	5.00	6.00
80	507	614	643	672	703	734	767	800	834	869	1017
	43.3	48.6	50.0	51.4	52.8	54.3	55.7	57.2	58.6	60.1	66.2
50	487	588	616	644	673	703	734	765	797	831	971
	42.8	48.0	49.4	50.8	52.2	53.6	55.0	56.5	57.9	59.4	65.4
35	468	564	590	616	644	672	701	731	761	793	926
	42.3	47.5	48.8	50.2	51.5	52.9	54.3	55.8	57.2	58.7	64.6
25	446	535	559	584	610	636	663	691	720	749	873
	41.8	46.8	48.1	49.5	50.8	52.2	53.6	55.0	56.4	57.8	63.6
20	429	513	536	560	584	609	634	661	688	715	833
	41.4	46.3	47.6	48.9	50.2	51.6	52.9	54.3	55.7	57.1	62.8
15	404	482	503	524	547	569	593	617	642	667	775
	40.8	45.5	46.8	48.0	49.3	50.6	51.9	53.3	54.6	56.0	61.6
10	367	433	451	469	488	508	528	549	570	591	684
	39.7	44.2	45.3	46.5	47.7	49.0	50.2	51.5	52.8	54.1	59.4
5	299	346	358	371	384	398	412	426	441	456	521
	37.4	41.1	42.1	43.1	44.2	45.2	46.3	47.4	48.5	49.6	54.3
h_r, In. [c]	\multicolumn{11}{c}{max A_{ctr}, In.² [d]}
0	19.7	13.1	12.1	11.2	10.5	9.8	9.2	8.7	8.3	7.9	6.6
2		54.5	43.6	36.4	31.2	27.3	24.2	21.8	19.8	18.2	13.6
3			248.5	124.3	82.8	62.1	49.7	41.4	35.5	31.1	20.7

W 10×22

$A_s = 6.49$ In.² $I_s = 118.0$ In.⁴   max V_h (steel) = 117 kips for F_y = 36 ksi
$d = 10.17$ In. $S_s = 23.2$ In.³   max V_h (steel) = 162 kips for F_y = 50 ksi

A_{ctr}, In.² [b,e]	\multicolumn{11}{c}{Y2, In. [a]}
	2.00	3.00	3.25	3.50	3.75	4.00	4.25	4.50	4.75	5.00	6.00
80	426	517	542	567	593	620	648	676	705	735	862
	36.6	41.2	42.3	43.5	44.7	46.0	47.2	48.5	49.7	51.0	56.2
50	411	498	521	546	571	596	623	650	678	706	828
	36.1	40.7	41.8	43.0	44.2	45.4	46.7	47.9	49.2	50.4	55.6
35	396	479	501	524	548	573	598	624	650	678	794
	35.8	40.2	41.4	42.5	43.7	44.9	46.1	47.4	48.6	49.9	55.0
25	379	457	478	500	522	545	569	593	618	644	753
	35.4	39.7	40.9	42.0	43.2	44.3	45.5	46.7	48.0	49.2	54.2
20	366	440	460	481	502	524	547	570	594	618	722
	35.0	39.3	40.4	41.6	42.7	43.9	45.1	46.2	47.5	48.7	53.6
15	347	415	434	453	473	493	514	535	557	580	676
	34.6	38.7	39.8	40.9	42.0	43.2	44.3	45.5	46.6	47.8	52.7
10	316	376	392	409	426	444	462	480	500	519	602
	33.7	37.7	38.7	39.7	40.8	41.9	43.0	44.1	45.2	46.3	51.0
5	260	303	315	327	339	352	365	378	392	406	465
	31.9	35.2	36.1	37.0	37.9	38.9	39.9	40.8	41.8	42.8	47.0
h_r, In. [c]	\multicolumn{11}{c}{max A_{ctr}, In.² [d]}
0	16.5	11.0	10.2	9.4	8.8	8.3	7.8	7.3	6.9	6.6	5.5
2		46.0	36.8	30.7	26.3	23.0	20.4	18.4	16.7	15.3	11.5
3			209.9	104.9	70.0	52.5	42.0	35.0	30.0	26.2	17.5

[a] $Y2$ = distance from top of steel beam to concrete centroid
[b] A_{ctr} = transformed effective concrete area = $(b/n) t$ for full composite
[c] h_r = height of steel deck ribs
[d] Max A_{ctr} = maximum transformed effective concrete area based on full concrete thickness $t = t_o$ for elastic neutral axis to be below the top of the steel deck
[e] Shaded area represents A_{ctr} range wherein elastic neutral axis falls within the concrete slab ($h_r = 0$)

AMERICAN INSTITUTE OF STEEL CONSTRUCTION

COMPOSITE BEAM SELECTION TABLE
Transformed Section Properties

$$\overline{I}_{tr},\ \text{In.}^4$$
$$\overline{S}_{tr},\ \text{In.}^3$$

Shape												
W 10×19	$A_s = 5.62\ \text{In.}^2$		$I_s = 96.3\ \text{In.}^4$		max V_h (steel) = 101 kips for F_y = 36 ksi							
	$d = 10.24\ \text{In.}$		$S_s = 18.8\ \text{In.}^3$		max V_h (steel) = 141 kips for F_y = 50 ksi							

A_{ctr}, In.$^{2\ b,e}$	Y2, In.^{a}										
	2.00	3.00	3.25	3.50	3.75	4.00	4.25	4.50	4.75	5.00	6.00
80	369	449	471	493	516	540	564	589	615	641	752
	31.4	35.4	36.4	37.4	38.5	39.6	40.6	41.7	42.8	44.0	48.5
50	357	434	454	476	498	521	544	568	593	618	725
	31.0	34.9	35.9	37.0	38.0	39.1	40.2	41.3	42.4	43.5	48.0
35	345	419	438	459	480	502	524	547	571	595	698
	30.6	34.5	35.6	36.6	37.6	38.7	39.7	40.8	41.9	43.0	47.5
25	331	401	420	439	459	480	501	523	545	568	666
	30.3	34.1	35.1	36.1	37.2	38.2	39.3	40.3	41.4	42.5	46.9
20	320	387	405	424	443	463	483	504	525	547	640
	30.0	33.8	34.8	35.8	36.8	37.8	38.9	39.9	41.0	42.0	46.4
15	305	367	384	401	419	438	456	476	496	516	603
	29.6	33.3	34.3	35.2	36.2	37.2	38.2	39.3	40.3	41.4	45.7
10	280	334	349	364	380	396	413	430	448	466	542
	28.9	32.4	33.3	34.3	35.2	36.2	37.1	38.1	39.1	40.1	44.3
5	231	271	282	293	305	317	329	342	354	368	424
	27.2	30.3	31.1	32.0	32.8	33.7	34.5	35.4	36.3	37.2	40.9

h_r, In.^{c}	max A_{ctr}, In.$^{2\ d}$										
0	14.4	9.6	8.9	8.2	7.7	7.2	6.8	6.4	6.1	5.8	4.8
2		40.0	32.0	26.7	22.9	20.0	17.8	16.0	14.6	13.3	10.0
3			182.5	91.3	60.8	45.6	36.5	30.4	26.1	22.8	15.2

W 10×17	$A_s = 4.99\ \text{In.}^2$		$I_s = 81.9\ \text{In.}^4$		max V_h (steel) = 90 kips for F_y = 36 ksi							
	$d = 10.11\ \text{In.}$		$S_s = 16.2\ \text{In.}^3$		max V_h (steel) = 125 kips for F_y = 50 ksi							

A_{ctr}, In.$^{2\ b,e}$	Y2, In.^{a}										
	2.00	3.00	3.25	3.50	3.75	4.00	4.25	4.50	4.75	5.00	6.00
80	322	393	413	432	453	474	495	517	540	563	663
	27.6	31.1	32.0	33.0	33.9	34.9	35.9	36.8	37.8	38.8	42.9
50	312	380	399	418	438	458	479	500	522	545	641
	27.2	30.7	31.7	32.6	33.5	34.5	35.4	36.4	37.4	38.4	42.4
35	302	368	386	404	423	443	463	484	505	526	619
	26.9	30.4	31.3	32.2	33.2	34.1	35.1	36.0	37.0	38.0	42.0
25	291	354	371	388	406	425	444	464	484	505	592
	26.6	30.1	31.0	31.9	32.8	33.7	34.7	35.6	36.6	37.5	41.5
20	282	343	359	376	393	411	429	448	468	487	572
	26.4	29.8	30.7	31.6	32.5	33.4	34.3	35.3	36.2	37.2	41.1
15	270	326	341	357	373	390	407	425	443	462	541
	26.0	29.4	30.2	31.1	32.0	32.9	33.8	34.8	35.7	36.6	40.5
10	248	299	312	326	341	356	371	387	403	419	490
	25.4	28.6	29.5	30.3	31.2	32.1	32.9	33.8	34.7	35.6	39.4
5	207	244	255	265	276	287	299	310	322	335	388
	24.1	26.9	27.6	28.4	29.2	29.9	30.7	31.5	32.4	33.2	36.6

h_r, In.^{c}	max A_{ctr}, In.$^{2\ d}$										
0	12.6	8.4	7.8	7.2	6.7	6.3	5.9	5.6	5.3	5.0	4.2
2		35.2	28.2	23.5	20.1	17.6	15.6	14.1	12.8	11.7	8.8
3			160.8	80.4	53.6	40.2	32.2	26.8	23.0	20.1	13.4

[a] Y2 = distance from top of steel beam to concrete centroid
[b] A_{ctr} = transformed effective concrete area = $(b/n)\ t$ for full composite
[c] h_r = height of steel deck ribs
[d] max A_{ctr} = maximum transformed effective concrete area based on full concrete thickness $t = t_o$ for elastic neutral axis to be below the top of the steel deck
[e] Shaded area represents A_{ctr} range wherein elastic neutral axis falls within the concrete slab (h_r = 0)

COMPOSITE BEAM SELECTION TABLE
Transformed Section Properties

$$\overline{I}_{tr}, \text{ In.}^4$$
$$\overline{S}_{tr}, \text{ In.}^3$$

Shape												
W 10×15	$A_s = 4.41$ In.^{2} $I_s = 68.9$ In.^{4} max V_h (steel) = 79 kips for $F_y = 36$ ksi											
	$d = 9.99$ In. $S_s = 13.8$ In.^{3} max V_h (steel) = 110 kips for $F_y = 50$ ksi											

W 10×15

A_{ctr}, In.$^{2\ b,e}$	Y2, In.^{a}										
	2.00	3.00	3.25	3.50	3.75	4.00	4.25	4.50	4.75	5.00	6.00
80	280 *24.1*	343 *27.3*	360 *28.1*	377 *28.9*	395 *29.8*	414 *30.6*	433 *31.5*	452 *32.3*	472 *33.2*	493 *34.1*	581 *37.7*
50	271 *23.8*	332 *26.9*	349 *27.7*	366 *28.6*	383 *29.4*	401 *30.2*	419 *31.1*	438 *32.0*	458 *32.8*	478 *33.7*	563 *37.3*
35	263 *23.5*	322 *26.6*	338 *27.4*	354 *28.3*	371 *29.1*	389 *29.9*	407 *30.8*	425 *31.6*	444 *32.5*	463 *33.4*	545 *36.9*
25	254 *23.3*	311 *26.3*	326 *27.1*	342 *28.0*	358 *28.8*	374 *29.6*	391 *30.4*	409 *31.3*	427 *32.2*	445 *33.0*	524 *36.5*
20	247 *23.1*	302 *26.1*	316 *26.9*	331 *27.7*	347 *28.5*	363 *29.4*	379 *30.2*	396 *31.0*	414 *31.9*	432 *32.7*	507 *36.2*
15	237 *22.8*	288 *25.8*	302 *26.6*	316 *27.3*	331 *28.1*	346 *29.0*	361 *29.8*	377 *30.6*	394 *31.4*	411 *32.3*	482 *35.7*
10	219 *22.3*	265 *25.2*	278 *25.9*	291 *26.7*	304 *27.5*	317 *28.2*	331 *29.0*	346 *29.8*	360 *30.6*	375 *31.5*	440 *34.8*
5	184 *21.1*	219 *23.7*	229 *24.4*	238 *25.1*	249 *25.8*	259 *26.5*	270 *27.2*	281 *27.9*	292 *28.7*	303 *29.4*	353 *32.5*

h_r, In.^{c}	max A_{ctr}, In.$^{2\ d}$										
0	11.0	7.3	6.8	6.3	5.9	5.5	5.2	4.9	4.6	4.4	3.7
2		30.8	24.7	20.6	17.6	15.4	13.7	12.3	11.2	10.3	7.7
3			141.0	70.5	47.0	35.3	28.2	23.5	20.1	17.6	11.8

W 10×12

	$A_s = 3.54$ In.^{2} $I_s = 53.8$ In.^{4} max V_h (steel) = 64 kips for $F_y = 36$ ksi
	$d = 9.87$ In. $S_s = 10.9$ In.^{3} max V_h (steel) = 89 kips for $F_y = 50$ ksi

A_{ctr}, In.$^{2\ b,e}$	Y2, In.^{a}										
	2.00	3.00	3.25	3.50	3.75	4.00	4.25	4.50	4.75	5.00	6.00
80	224 *19.3*	274 *21.9*	288 *22.5*	302 *23.2*	316 *23.9*	331 *24.5*	346 *25.2*	362 *25.9*	378 *26.6*	395 *27.3*	466 *30.2*
50	217 *19.0*	266 *21.6*	279 *22.2*	293 *22.9*	307 *23.6*	322 *24.2*	337 *24.9*	352 *25.6*	368 *26.3*	384 *27.0*	453 *29.9*
35	211 *18.8*	259 *21.3*	272 *22.0*	285 *22.7*	299 *23.3*	313 *24.0*	328 *24.7*	343 *25.4*	358 *26.1*	374 *26.8*	441 *29.7*
25	205 *18.6*	251 *21.1*	264 *21.8*	277 *22.4*	290 *23.1*	303 *23.8*	317 *24.5*	332 *25.1*	347 *25.8*	362 *26.5*	427 *29.4*
20	200 *18.5*	245 *21.0*	257 *21.6*	269 *22.3*	282 *22.9*	296 *23.6*	309 *24.3*	323 *25.0*	338 *25.6*	352 *26.3*	415 *29.2*
15	193 *18.3*	235 *20.7*	247 *21.4*	259 *22.0*	271 *22.7*	284 *23.3*	297 *24.0*	310 *24.7*	324 *25.3*	338 *26.0*	398 *28.8*
10	180 *17.9*	219 *20.3*	230 *20.9*	241 *21.6*	252 *22.2*	263 *22.8*	275 *23.5*	287 *24.1*	300 *24.8*	313 *25.5*	367 *28.2*
5	154 *17.1*	185 *19.3*	193 *19.8*	202 *20.4*	211 *21.0*	220 *21.6*	229 *22.2*	239 *22.8*	249 *23.4*	259 *24.1*	302 *26.6*

h_r, In.^{c}	max A_{ctr}, In.$^{2\ d}$										
0	8.7	5.8	5.4	5.0	4.7	4.4	4.1	3.9	3.7	3.5	2.9
2		24.5	19.6	16.4	14.0	12.3	10.9	9.8	8.9	8.2	6.1
3			112.4	56.2	37.5	28.1	22.5	18.7	16.1	14.0	9.4

[a] $Y2$ = distance from top of steel beam to concrete centroid
[b] A_{ctr} = transformed effective concrete area = $(b/n)\, t$ for full composite
[c] h_r = height of steel deck ribs
[d] max A_{ctr} = maximum transformed effective concrete area based on full concrete thickness $t = t_o$ for elastic neutral axis to be below the top of the steel deck
[e] Shaded area represents A_{ctr} range wherein elastic neutral axis falls within the concrete slab ($h_r = 0$)

COMPOSITE BEAM SELECTION TABLE
Transformed Section Properties

$$\bar{I}_{tr}, \text{ In.}^4$$
$$\bar{S}_{tr}, \text{ In.}^3$$

Shape: W 8×28

$A_s = 8.25$ In.^{2} $\quad I_s = 98.0$ In.^{4} $\quad$ max V_h (steel) = 149 kips for $F_y = 36$ ksi
$d = 8.06$ In. $\quad S_s = 24.3$ In.^{3} $\quad$ max V_h (steel) = 206 kips for $F_y = 50$ ksi

A_{ctr}, In.$^{2\ b,e}$	Y2, In.^{a}										
	2.00	3.00	3.25	3.50	3.75	4.00	4.25	4.50	4.75	5.00	6.00
0	377 / 39.7	474 / 45.6	501 / 47.1	529 / 48.7	557 / 50.3	587 / 51.9	617 / 53.5	649 / 55.2	681 / 56.8	714 / 58.5	857 / 65.3
50	360 / 39.1	452 / 44.9	477 / 46.5	504 / 48.0	531 / 49.6	559 / 51.2	588 / 52.8	617 / 54.4	648 / 56.0	680 / 57.7	815 / 64.4
35	344 / 38.6	431 / 44.3	455 / 45.8	479 / 47.4	505 / 48.9	531 / 50.5	559 / 52.1	587 / 53.7	616 / 55.3	645 / 56.9	773 / 63.6
25	326 / 38.0	407 / 43.7	429 / 45.1	452 / 46.6	476 / 48.1	500 / 49.7	525 / 51.2	551 / 52.8	578 / 54.4	606 / 56.0	724 / 62.6
20	312 / 37.6	388 / 43.1	409 / 44.6	431 / 46.0	453 / 47.5	476 / 49.0	500 / 50.6	525 / 52.1	550 / 53.7	576 / 55.3	687 / 61.7
15	293 / 37.0	362 / 42.3	381 / 43.7	401 / 45.1	421 / 46.6	442 / 48.0	464 / 49.5	487 / 51.0	510 / 52.6	533 / 54.1	635 / 60.4
10	263 / 35.9	322 / 40.9	338 / 42.2	355 / 43.5	372 / 44.9	390 / 46.3	409 / 47.7	428 / 49.1	447 / 50.6	467 / 52.1	554 / 58.1
5	212 / 33.6	252 / 37.7	263 / 38.9	275 / 40.0	287 / 41.2	299 / 42.4	312 / 43.6	325 / 44.8	338 / 46.1	352 / 47.4	412 / 52.7

h_r, In.^{c}	max A_{ctr}, In.$^{2\ d}$										
0	16.6	11.1	10.2	9.5	8.9	8.3	7.8	7.4	7.0	6.6	5.5
2		49.7	39.8	33.2	28.4	24.9	22.1	19.9	18.1	16.6	12.4
3			232.0	116.0	77.3	58.0	46.4	38.7	33.1	29.0	19.3

Shape: W 8×24

$A_s = 7.08$ In.^{2} $\quad I_s = 82.8$ In.^{4} $\quad$ max V_h (steel) = 127 kips for $F_y = 36$ ksi
$d = 7.93$ In. $\quad S_s = 20.9$ In.^{3} $\quad$ max V_h (steel) = 177 kips for $F_y = 50$ ksi

A_{ctr}, In.$^{2\ b,e}$	Y2, In.^{a}										
	2.00	3.00	3.25	3.50	3.75	4.00	4.25	4.50	4.75	5.00	6.00
80	321 / 34.0	405 / 39.1	428 / 40.4	452 / 41.8	477 / 43.1	502 / 44.5	528 / 45.9	556 / 47.3	583 / 48.7	612 / 50.2	735 / 56.0
50	308 / 33.5	388 / 38.5	410 / 39.8	433 / 41.2	456 / 42.5	480 / 43.9	506 / 45.3	531 / 46.7	558 / 48.1	585 / 49.5	703 / 55.4
35	295 / 33.1	371 / 38.1	392 / 39.4	414 / 40.7	436 / 42.0	459 / 43.4	483 / 44.7	508 / 46.1	533 / 47.5	559 / 48.9	670 / 54.7
25	281 / 32.6	353 / 37.5	372 / 38.8	392 / 40.1	413 / 41.4	435 / 42.8	457 / 44.1	480 / 45.5	504 / 46.8	528 / 48.2	633 / 53.9
20	271 / 32.3	338 / 37.1	357 / 38.4	376 / 39.7	396 / 41.0	416 / 42.3	437 / 43.6	459 / 44.9	482 / 46.3	505 / 47.7	604 / 53.3
15	255 / 31.8	317 / 36.5	334 / 37.7	352 / 39.0	370 / 40.2	389 / 41.5	409 / 42.8	429 / 44.1	449 / 45.5	471 / 46.8	562 / 52.3
10	231 / 31.0	285 / 35.4	299 / 36.6	315 / 37.7	330 / 38.9	347 / 40.2	363 / 41.4	381 / 42.7	398 / 43.9	417 / 45.2	495 / 50.5
5	187 / 29.1	225 / 32.9	236 / 33.9	247 / 34.9	258 / 36.0	269 / 37.1	281 / 38.1	293 / 39.3	306 / 40.4	319 / 41.5	374 / 46.3

h_r, In.^{c}	max A_{ctr}, In.$^{2\ d}$										
0	14.0	9.4	8.6	8.0	7.5	7.0	6.6	6.2	5.9	5.6	4.7
2		42.2	33.8	28.2	24.1	21.1	18.8	16.9	15.4	14.1	10.6
3			197.2	98.6	65.7	49.3	39.4	32.9	28.2	24.7	16.4

^{a}Y2 = distance from top of steel beam to concrete centroid
^{b}A_{ctr} = transformed effective concrete area
  = $(b/n)\,t$ for full composite
^{c}h_r = height of steel deck ribs
^{d}max A_{ctr} = maximum transformed effective concrete area based on full concrete thickness $t = t_o$ for elastic neutral axis to be below the top of the steel deck
^{e}Shaded area represents A_{ctr} range wherein elastic neutral axis falls within the concrete slab ($h_r = 0$)

COMPOSITE BEAM SELECTION TABLE
Transformed Section Properties

$$\overline{I}_{tr}, \text{ In.}^4$$
$$\overline{S}_{tr}, \text{ In.}^3$$

Shape												
W 8×21	A_s = 6.16 In.^{2} d = 8.28 In.		I_s = 75.3 In.^{4} S_s = 18.2 In.^{3}			max V_h (steel) = 111 kips for F_y = 36 ksi max V_h (steel) = 154 kips for F_y = 50 ksi						

A_{ctr}, In.$^{2 \text{ b,e}}$	$Y2$, In.^{a}										
	2.00	3.00	3.25	3.50	3.75	4.00	4.25	4.50	4.75	5.00	6.00
80	298 *30.2*	374 *34.7*	394 *35.8*	416 *37.0*	438 *38.2*	461 *39.4*	485 *40.6*	509 *41.8*	534 *43.1*	560 *44.3*	670 *49.4*
50	286 *29.8*	359 *34.2*	379 *35.4*	400 *36.5*	421 *37.7*	443 *38.9*	466 *40.1*	489 *41.3*	513 *42.5*	538 *43.8*	643 *48.9*
35	276 *29.5*	345 *33.8*	364 *34.9*	384 *36.1*	404 *37.3*	425 *38.4*	447 *39.6*	469 *40.8*	492 *42.1*	516 *43.3*	617 *48.3*
25	264 *29.1*	329 *33.4*	347 *34.5*	366 *35.6*	385 *36.8*	405 *37.9*	425 *39.1*	446 *40.3*	468 *41.5*	490 *42.7*	586 *47.7*
20	255 *28.8*	317 *33.0*	334 *34.1*	352 *35.3*	370 *36.4*	389 *37.5*	408 *38.7*	429 *39.9*	449 *41.1*	470 *42.3*	561 *47.2*
15	241 *28.4*	299 *32.5*	315 *33.6*	331 *34.7*	348 *35.8*	366 *36.9*	384 *38.1*	403 *39.2*	422 *40.4*	441 *41.6*	526 *46.4*
10	220 *27.7*	270 *31.6*	284 *32.6*	299 *33.7*	313 *34.7*	329 *35.8*	344 *36.9*	361 *38.0*	377 *39.1*	395 *40.3*	468 *44.9*
5	180 *26.1*	216 *29.5*	226 *30.4*	237 *31.3*	248 *32.3*	259 *33.2*	270 *34.2*	282 *35.2*	294 *36.2*	306 *37.2*	359 *41.4*

h_r, In.^{c}	max A_{ctr}, In.$^{2 \text{ d}}$										
0	12.8	8.5	7.8	7.3	6.8	6.4	6.0	5.7	5.4	5.1	4.3
2		37.8	30.3	25.2	21.6	18.9	16.8	15.1	13.8	12.6	9.5
3			175.9	88.0	58.6	44.0	35.2	29.3	25.1	22.0	14.7

Shape												
W 8×18	A_s = 5.26 In.^{2} d = 8.14 In.		I_s = 61.9 In.^{4} S_s = 15.2 In.^{3}			max V_h (steel) = 95 kips for F_y = 36 ksi max V_h (steel) = 132 kips for F_y = 50 ksi						

A_{ctr}, In.$^{2 \text{ b,e}}$	$Y2$, In.^{a}										
	2.00	3.00	3.25	3.50	3.75	4.00	4.25	4.50	4.75	5.00	6.00
80	250 *25.6*	315 *29.5*	333 *30.4*	351 *31.5*	370 *32.5*	390 *33.5*	410 *34.5*	431 *35.6*	453 *36.7*	475 *37.7*	569 *42.1*
50	241 *25.2*	304 *29.0*	321 *30.0*	339 *31.0*	357 *32.0*	376 *33.1*	396 *34.1*	416 *35.1*	436 *36.2*	458 *37.3*	549 *41.6*
35	233 *25.0*	293 *28.7*	310 *29.7*	327 *30.7*	344 *31.7*	363 *32.7*	381 *33.7*	401 *34.8*	421 *35.8*	441 *36.9*	529 *41.2*
25	224 *24.7*	281 *28.4*	297 *29.3*	313 *30.3*	330 *31.3*	347 *32.3*	365 *33.3*	383 *34.4*	402 *35.4*	421 *36.4*	505 *40.7*
20	217 *24.4*	272 *28.1*	287 *29.1*	302 *30.0*	318 *31.0*	335 *32.0*	352 *33.0*	369 *34.0*	388 *35.1*	406 *36.1*	486 *40.3*
15	207 *24.1*	258 *27.7*	272 *28.6*	286 *29.6*	301 *30.6*	317 *31.5*	333 *32.5*	349 *33.5*	366 *34.5*	384 *35.6*	458 *39.7*
10	190 *23.6*	235 *27.0*	247 *27.9*	260 *28.8*	274 *29.7*	287 *30.7*	301 *31.6*	316 *32.6*	331 *33.6*	346 *34.6*	412 *38.6*
5	157 *22.3*	190 *25.3*	200 *26.1*	209 *27.0*	219 *27.8*	229 *28.6*	240 *29.5*	251 *30.4*	262 *31.3*	273 *32.2*	322 *35.9*

h_r, In.^{c}	max A_{ctr}, In.$^{2 \text{ d}}$										
0	10.7	7.1	6.6	6.1	5.7	5.4	5.0	4.8	4.5	4.3	3.6
2		31.9	25.5	21.3	18.2	16.0	14.2	12.8	11.6	10.6	8.0
3			148.8	74.4	49.6	37.2	29.8	24.8	21.3	18.6	12.4

[a] $Y2$ = distance from top of steel beam to concrete centroid
[b] A_{ctr} = transformed effective concrete area = $(b/n)\,t$ for full composite
[c] h_r = height of steel deck ribs
[d] max A_{ctr} = maximum transformed effective concrete area based on full concrete thickness $t = t_o$ for elastic neutral axis to be below the top of the steel deck
[e] Shaded area represents A_{ctr} range wherein elastic neutral axis falls within the concrete slab ($h_r = 0$)

COMPOSITE BEAM SELECTION TABLE
Transformed Section Properties

$$\overline{I}_{tr}, \text{ In.}^4$$
$$\overline{S}_{tr}, \text{ In.}^3$$

Shape												
W 8×15	A_s = 4.44 In.² $\quad$ I_s = 48.0 In.⁴ $\quad$ max V_h (steel) = 80 kips for F_y = 36 ksi d = 8.11 In. $\quad$ S_s = 11.8 In.³ $\quad$ max V_h (steel) = 111 kips for F_y = 50 ksi											

W 8×15 — A_s = 4.44 In.², I_s = 48.0 In.⁴, max V_h (steel) = 80 kips for F_y = 36 ksi; d = 8.11 In., S_s = 11.8 In.³, max V_h (steel) = 111 kips for F_y = 50 ksi

A_{ctr}, In.² b,e	Y2, In.ᵃ										
	2.00	3.00	3.25	3.50	3.75	4.00	4.25	4.50	4.75	5.00	6.00
80	209 / 21.3	264 / 24.6	279 / 25.4	295 / 26.3	311 / 27.2	328 / 28.0	345 / 28.9	363 / 29.8	381 / 30.7	400 / 31.6	480 / 35.3
50	202 / 21.0	255 / 24.2	270 / 25.1	285 / 25.9	301 / 26.8	317 / 27.7	333 / 28.5	351 / 29.4	368 / 30.3	387 / 31.2	464 / 34.9
35	195 / 20.7	247 / 23.9	261 / 24.8	276 / 25.6	291 / 26.5	307 / 27.4	323 / 28.2	339 / 29.1	356 / 30.0	374 / 30.9	449 / 34.6
25	188 / 20.5	238 / 23.7	251 / 24.5	265 / 25.3	280 / 26.2	295 / 27.1	310 / 27.9	326 / 28.8	342 / 29.7	359 / 30.6	431 / 34.2
20	183 / 20.3	231 / 23.5	244 / 24.3	257 / 25.1	271 / 26.0	285 / 26.8	300 / 27.7	316 / 28.5	331 / 29.4	348 / 30.3	417 / 33.9
15	175 / 20.0	220 / 23.1	232 / 23.9	245 / 24.8	258 / 25.6	272 / 26.4	286 / 27.3	300 / 28.2	315 / 29.0	330 / 29.9	396 / 33.5
10	162 / 19.6	202 / 22.6	213 / 23.4	224 / 24.2	236 / 25.0	248 / 25.8	261 / 26.6	274 / 27.4	287 / 28.3	301 / 29.1	360 / 32.6
5	135 / 18.5	165 / 21.2	174 / 21.9	183 / 22.7	192 / 23.4	201 / 24.2	211 / 24.9	221 / 25.7	231 / 26.5	241 / 27.3	286 / 30.5

h_r, In.ᶜ	max A_{ctr}, In.² d										
0	9.0	6.0	5.5	5.1	4.8	4.5	4.2	4.0	3.8	3.6	3.0
2		26.9	21.5	17.9	15.4	13.4	11.9	10.8	9.8	9.0	6.7
3			125.3	62.6	41.8	31.3	25.1	20.9	17.9	15.7	10.4

W 8×13 — A_s = 3.84 In.², I_s = 39.6 In.⁴, max V_h (steel) = 69 kips for F_y = 36 ksi; d = 7.99 In., S_s = 9.9 In.³, max V_h (steel) = 96 kips for F_y = 50 ksi

A_{ctr}, In.² b,e	Y2, In.ᵃ										
	2.00	3.00	3.25	3.50	3.75	4.00	4.25	4.50	4.75	5.00	6.00
80	178 / 18.3	226 / 21.1	239 / 21.9	252 / 22.6	266 / 23.4	280 / 24.1	295 / 24.9	311 / 25.7	326 / 26.5	343 / 27.2	412 / 30.5
50	172 / 18.0	218 / 20.8	231 / 21.5	244 / 22.3	258 / 23.0	272 / 23.8	286 / 24.6	301 / 25.3	316 / 26.1	332 / 26.9	400 / 30.1
35	167 / 17.8	212 / 20.6	224 / 21.3	237 / 22.0	250 / 22.8	264 / 23.5	278 / 24.3	292 / 25.1	307 / 25.9	322 / 26.7	388 / 29.9
25	161 / 17.6	205 / 20.3	216 / 21.1	229 / 21.8	241 / 22.5	254 / 23.3	268 / 24.0	282 / 24.8	296 / 25.6	311 / 26.4	374 / 29.6
20	157 / 17.4	199 / 20.2	210 / 20.9	222 / 21.6	235 / 22.4	247 / 23.1	260 / 23.9	274 / 24.6	288 / 25.4	302 / 26.2	363 / 29.3
15	151 / 17.2	190 / 19.9	201 / 20.6	213 / 21.3	224 / 22.1	236 / 22.8	249 / 23.6	261 / 24.3	275 / 25.1	288 / 25.8	346 / 29.0
10	140 / 16.8	176 / 19.5	186 / 20.2	196 / 20.9	207 / 21.6	218 / 22.3	229 / 23.0	241 / 23.8	253 / 24.5	265 / 25.2	318 / 28.3
5	118 / 16.0	146 / 18.4	154 / 19.0	162 / 19.7	170 / 20.3	179 / 21.0	188 / 21.7	197 / 22.4	206 / 23.1	216 / 23.8	257 / 26.6

h_r, In.ᶜ	max A_{ctr}, In.² d										
0	7.7	5.1	4.7	4.4	4.1	3.8	3.6	3.4	3.2	3.1	2.6
2		23.0	18.4	15.3	13.2	11.5	10.2	9.2	8.4	7.7	5.8
3			107.4	53.7	35.8	26.9	21.5	17.9	15.3	13.4	9.0

ᵃY2 = distance from top of steel beam to concrete centroid
ᵇA_{ctr} = transformed effective concrete area
= $(b/n)\,t$ for full composite
ᶜh_r = height of steel deck ribs
ᵈmax A_{ctr} = maximum transformed effective concrete area based on full concrete thickness $t = t_o$ for elastic neutral axis to be below the top of the steel deck
ᵉShaded area represents A_{ctr} range wherein elastic neutral axis falls within the concrete slab (h_r = 0)

COMPOSITE BEAM SELECTION TABLE
Transformed Section Properties

$$\overline{I}_{tr}, \text{ In.}^4$$
$$\overline{S}_{tr}, \text{ In.}^3$$

Shape												
W 8×10	$A_s = 2.96$ In.^{2} $\quad I_s = 30.8$ In.^{4} $\quad$ max V_h (steel) = 53 kips for F_y = 36 ksi											
	$d = 7.89$ In. $\quad S_s = 7.8$ In.^{3} $\quad$ max V_h (steel) = 74 kips for F_y = 50 ksi											

A_{ctr}, In.$^{2\ b,e}$	Y2, In.^{a}										
	2.00	3.00	3.25	3.50	3.75	4.00	4.25	4.50	4.75	5.00	6.00
80	138 14.3	175 16.5	185 17.0	196 17.6	206 18.2	218 18.8	229 19.3	241 19.9	253 20.5	266 21.2	320 23.6
50	134 14.0	170 16.2	180 16.7	190 17.3	200 17.9	211 18.5	223 19.1	234 19.7	246 20.3	259 20.9	311 23.4
35	130 13.8	165 16.0	175 16.5	185 17.1	195 17.7	206 18.3	217 18.9	228 19.5	240 20.1	252 20.7	304 23.2
25	126 13.7	161 15.8	170 16.4	180 16.9	190 17.5	200 18.1	211 18.7	222 19.3	233 19.9	245 20.5	295 23.0
20	124 13.5	157 15.7	166 16.2	175 16.8	185 17.4	195 18.0	206 18.6	216 19.1	227 19.7	239 20.3	287 22.8
15	119 13.4	151 15.5	160 16.1	169 16.6	178 17.2	188 17.8	198 18.4	208 18.9	219 19.5	230 20.1	277 22.6
10	112 13.2	142 15.2	150 15.8	158 16.3	167 16.9	176 17.4	185 18.0	195 18.6	204 19.2	214 19.8	258 22.2
5	97 12.6	121 14.6	127 15.1	134 15.6	141 16.1	149 16.6	156 17.2	164 17.7	172 18.3	180 18.8	215 21.1

h_r, In.^{c}	max A_{ctr}, In.$^{2\ d}$										
0	5.8	3.9	3.6	3.3	3.1	2.9	2.7	2.6	2.5	2.3	1.9
2		17.6	14.1	11.7	10.1	8.8	7.8	7.0	6.4	5.9	4.4
3			82.2	41.1	27.4	20.6	16.4	13.7	11.7	10.3	6.9

[a]Y2 = distance from top of steel beam to concrete centroid
[b]A_{ctr} = transformed effective concrete area
$\quad$ = (b/n) t for full composite
[c]h_r = height of steel deck ribs
[d]max A_{ctr} = maximum transformed effective concrete area based on full concrete thickness $t = t_o$ for elastic neutral axis to be below the top of the steel deck
[e]Shaded area represents A_{ctr} range wherein elastic neutral axis falls within the concrete slab ($h_r = 0$)

Notes

BEAM DIAGRAMS AND FORMULAS

Nomenclature

E = Modulus of Elasticity of steel at 29,000 ksi.

I = Moment of Inertia of beam, in.^{4}.

L = Total length of beam between reaction points ft.

M_{max} = Maximum moment, kip in.

M_1 = Maximum moment in left section of beam, kip-in.

M_2 = Maximum moment in right section of beam, kip-in.

M_3 = Maximum positive moment in beam with combined end moment conditions, kip-in.

M_x = Moment at distance x from end of beam, kip-in.

P = Concentrated load, kips

P_1 = Concentrated load nearest left reaction, kips.

P_2 = Concentrated load nearest right reaction, and of different magnitude than P_1, kips.

R = End beam reaction for any condition of symmetrical loading, kips.

R_1 = Left end beam reaction, kips.

R_2 = Right end or intermediate beam reaction, kips.

R_3 = Right end beam reaction, kips.

V = Maximum vertical shear for any condition of symmetrical loading, kips.

V_1 = Maximum vertical shear in left section of beam, kips.

V_2 = Vertical shear at right reaction point, or to left of intermediate reaction point of beam, kips.

V_3 = Vertical shear at right reaction point, or to right of intermediate reaction point of beam, kips.

V_x = Vertical shear at distance x from end of beam, kips.

W = Total load on beam, kips.

a = Measured distance along beam, in.

b = Measured distance along beam which may be greater or less than a, in.

l = Total length of beam between reaction points, in.

w = Uniformly distributed load per unit of length, kips/in.

w_1 = Uniformly distributed load per unit of length nearest left reaction, kips/in.

w_2 = Uniformly distributed load per unit of length nearest right reaction and of different magnitude than w_1, kips/in.

x = Any distance measured along beam from left reaction, in.

x_1 = Any distance measured along overhang section of beam from nearest reaction point, in.

Δ_{max} = Maximum deflection, in.

Δ_α = Deflection at point of laod, in.

Δ_x = Deflection at any point x distance from left reaction, in.

Δ_{x1} = Deflection of overhang section of beam at any distance from nearest reaction point, in.

BEAM DIAGRAMS AND FORMULAS
Frequently used formulas

The formulas given below are frequently required in structural designing. They are included herein for the convenience of those engineers who have infrequent use for such formulas and hence may find reference necessary. Variation from the standard nomenclature on page **2 - 293** is noted.

BEAMS

Flexural stress at extreme fiber:

$$f = Mc/I = M/S$$

Flexural stress at any fiber:

$$f = My/I \qquad y = \text{distance from neutral axis to fiber.}$$

Average vertical shear (for maximum see below):

$$v = V/A = V/dt \text{ (for beams and girders)}$$

Horizontal shearing stress at any section A-A:

$$v = VQ/I\,b \qquad Q = \text{statical moment about the neutral axis of the entire section of that portion of the cross-section lying outside of section A-A,}$$

$$b = \text{width at section A-A}$$

(Intensity of vertical shear is equal to that of horizontal shear acting normal to it at the same point and both are usually a maximum at mid-height of beam.)

Slope and deflection at any point:

$$EI\frac{d^2y}{dx^2} = M \qquad x \text{ and } y \text{ are abscissa and ordinate respectively of a point on the neutral axis, referred to axes of rectangular co-ordinates through a selected point of support.}$$

(First integration gives slopes; second integration gives deflections. Constants of integration must be determined.)

CONTINUOUS BEAMS (THE THEOREM OF THREE MOMENTS)

Uniform load:

$$M_a\frac{l_1}{I_1} + 2M_b\left(\frac{l_1}{I_1} + \frac{l_2}{I_2}\right) + M_c\frac{l_2}{I_2} = -\frac{1}{4}\left(\frac{w_1 l_1^3}{I_1} + \frac{w_2 l_2^3}{I_2}\right)$$

Concentrated loads:

$$M_a\frac{l_1}{I_1} + 2M_b\left(\frac{l_1}{I_1} + \frac{l_2}{I_2}\right) + M_c\frac{l_2}{I_2} = -\frac{P_1 a_1 b_1}{I_1}\left(1 + \frac{a_1}{l_1}\right) - \frac{P_2 a_2 b_2}{I_2}\left(1 + \frac{b_2}{l_2}\right)$$

Considering any two consecutive spans in any continuous structure:

M_a, M_b, M_c = moments at left, center, and right supports respectively, of any pair of adjacent spans.

l_1 and l_2 = length of left and right spans respectively, of the pair.

I_1 and I_2 = moment of inertia of left and right spans respectively.

w_1 and w_2 = load per unit of length on left and right spans respectively.

P_1 and P_2 = concentrated loads on left and right spans respectively.

a_1 and a_2 = distance of concentrated loads from left support in left and right spans respectively.

b_1 and b_2 = distance of concentrated loads from right support in left and right spans respectively.

The above equations are for beams with moment of inertia constant in each span but differing in different spans, continuous over three or more supports. By writing such an equation for each successive pair of spans and introducing the known values (usually zero) of end moments, all other moments can be found.

BEAM DIAGRAMS AND FORMULAS
Table of Concentrated Load Equivalents

n	Loading	Coeff.	Simple Beam	Beam Fixed One End Supported at Other	Beam Fixed Both Ends
∞		a	0.1250	0.0703	0.0417
		b	—	0.1250	0.0833
		c	0.5000	0.3750	—
		d	—	0.6250	0.5000
		e	0.0130	0.0054	0.0026
		f	1.0000	1.0000	0.6667
		g	1.0000	0.4151	0.3000
2		a	0.2500	0.1563	0.1250
		b	—	0.1875	0.1250
		c	0.5000	0.3125	—
		d	—	0.6875	0.5000
		e	0.0208	0.0093	0.0052
		f	2.0000	1.5000	1.0000
		g	0.8000	0.4770	0.4000
3		a	0.3333	0.2222	0.1111
		b	—	0.3333	0.2222
		c	1.0000	0.6667	—
		d	—	1.3333	1.0000
		e	0.0355	0.0152	0.0077
		f	2.6667	2.6667	1.7778
		g	1.0222	0.4381	0.3333
4		a	0.5000	0.2656	0.1875
		b	—	0.4688	0.3125
		c	1.5000	1.0313	—
		d	—	1.9688	1.5000
		e	0.0495	0.0209	0.0104
		f	4.0000	3.7500	2.5000
		g	0.9500	0.4281	0.3200
5		a	0.6000	0.3600	0.2000
		b	—	0.6000	0.4000
		c	2.0000	1.4000	—
		d	—	2.6000	2.0000
		e	0.0630	0.0265	0.0130
		f	4.8000	4.8000	3.2000
		g	1.0080	0.4238	0.3120

Maximum positive moment (kip-ft.):
  a × P × L

Maximum negative moment (kip-ft.):
  b × P × L

Pinned end reaction (kips): c × P

Fixed end reaction (kips): d × P

Maximum deflection (in.): e × Pl^3/EI

Equivalent simple span uniform load (kips):
  f × P

Deflection coeff. for equivalent simple span uniform load: g

Number of equal load spaces: n

Span of beam (ft.): L

Span of beam (in.): l

BEAM DIAGRAMS AND FORMULAS
For various static loading conditions

For meaning of symbols, see page **2 - 293**

---

1. SIMPLE BEAM—UNIFORMLY DISTRIBUTED LOAD

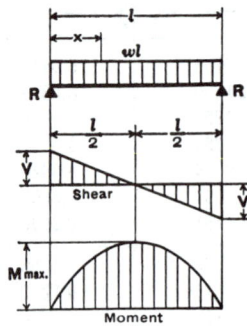

Total Equiv. Uniform Load . . . . $= wl$

$R = V$ . . . . $= \dfrac{wl}{2}$

V_x . . . . $= w\left(\dfrac{l}{2} - x\right)$

M max. $\left(\text{at center}\right)$ . . . . $= \dfrac{wl^2}{8}$

M_x . . . . $= \dfrac{wx}{2}(l - x)$

Δmax. $\left(\text{at center}\right)$ . . . . $= \dfrac{5\,wl^4}{384\,EI}$

Δ_x . . . . $= \dfrac{wx}{24EI}(l^3 - 2lx^2 + x^3)$

---

2. SIMPLE BEAM—LOAD INCREASING UNIFORMLY TO ONE END

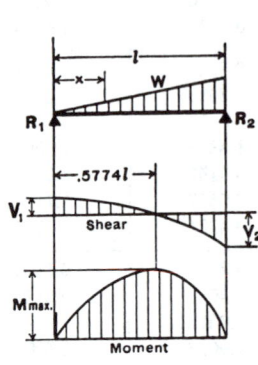

Total Equiv. Uniform Load . . . $= \dfrac{16W}{9\sqrt{3}} = 1.0264W$

$R_1 = V_1$ . . . . $= \dfrac{W}{3}$

$R_2 = V_2$ max. . . . . $= \dfrac{2W}{3}$

V_x . . . . $= \dfrac{W}{3} - \dfrac{Wx^2}{l^2}$

M max. $\left(\text{at } x = \dfrac{l}{\sqrt{3}} = .5774l\right)$ . . $= \dfrac{2Wl}{9\sqrt{3}} = .1283\,Wl$

M_x . . . . $= \dfrac{Wx}{3l^2}(l^2 - x^2)$

Δmax. $\left(\text{at } x = l\sqrt{1 - \sqrt{\dfrac{8}{15}}} = .5193l\right) = .01304\,\dfrac{Wl^3}{EI}$

Δ_x . . . . $= \dfrac{Wx}{180EI\,l^2}(3x^4 - 10l^2x^2 + 7l^4)$

---

3. SIMPLE BEAM—LOAD INCREASING UNIFORMLY TO CENTER

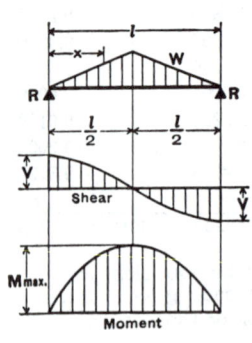

Total Equiv. Uniform Load . . . $= \dfrac{4W}{3}$

$R = V$ . . . . $= \dfrac{W}{2}$

V_x $\left(\text{when } x < \dfrac{l}{2}\right)$ . . . . $= \dfrac{W}{2l^2}(l^2 - 4x^2)$

M max. $\left(\text{at center}\right)$ . . . . $= \dfrac{Wl}{6}$

M_x $\left(\text{when } x < \dfrac{l}{2}\right)$ . . . . $= Wx\left(\dfrac{1}{2} - \dfrac{2x^2}{3l^2}\right)$

Δmax. $\left(\text{at center}\right)$ . . . . $= \dfrac{Wl^3}{60EI}$

Δ_x $\left(\text{when } x < \dfrac{l}{2}\right)$ . . . . $= \dfrac{Wx}{480\,EI\,l^2}(5l^2 - 4x^2)^2$

BEAM DIAGRAMS AND FORMULAS
For various static loading conditions

For meaning of symbols, see page **2 - 293**

---

4. SIMPLE BEAM—UNIFORM LOAD PARTIALLY DISTRIBUTED

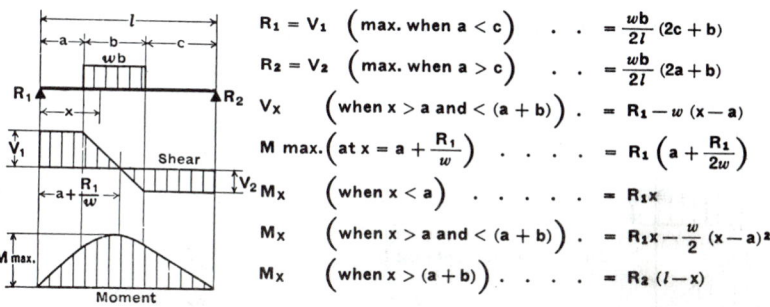

$$R_1 = V_1 \left(\text{max. when } a < c \right) \quad . . \quad = \frac{wb}{2l}(2c + b)$$

$$R_2 = V_2 \left(\text{max. when } a > c \right) \quad . . \quad = \frac{wb}{2l}(2a + b)$$

$$V_x \left(\text{when } x > a \text{ and } < (a + b) \right). \quad = R_1 - w\,(x - a)$$

$$M \text{ max.} \left(\text{at } x = a + \frac{R_1}{w} \right) \; . . . . \quad = R_1 \left(a + \frac{R_1}{2w} \right)$$

$$M_x \left(\text{when } x < a \right) \; . . . . . \quad = R_1 x$$

$$M_x \left(\text{when } x > a \text{ and } < (a + b) \right). \quad = R_1 x - \frac{w}{2}(x - a)^2$$

$$M_x \left(\text{when } x > (a + b) \right). . . . \quad = R_2 (l - x)$$

---

5. SIMPLE BEAM—UNIFORM LOAD PARTIALLY DISTRIBUTED AT ONE END

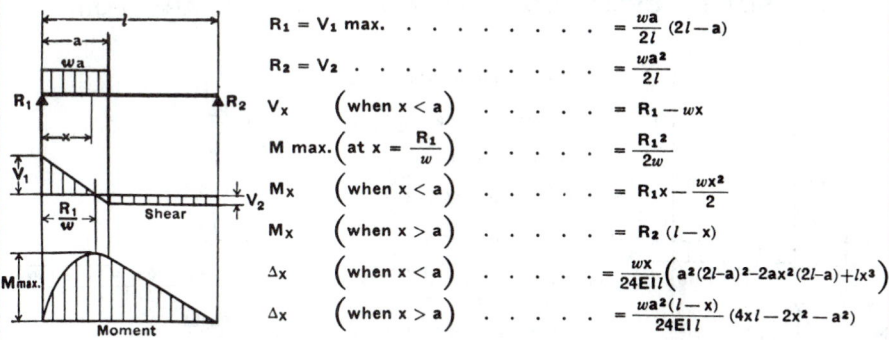

$$R_1 = V_1 \text{ max.} \; . . . . \quad = \frac{wa}{2l}(2l - a)$$

$$R_2 = V_2 \; . . . . \quad = \frac{wa^2}{2l}$$

$$V_x \left(\text{when } x < a \right) \; . . . . . \quad = R_1 - wx$$

$$M \text{ max.} \left(\text{at } x = \frac{R_1}{w} \right) \; . . . . \quad = \frac{R_1{}^2}{2w}$$

$$M_x \left(\text{when } x < a \right) \; . . . . . \quad = R_1 x - \frac{wx^2}{2}$$

$$M_x \left(\text{when } x > a \right) \; . . . . . \quad = R_2 (l - x)$$

$$\Delta_x \left(\text{when } x < a \right) \; . . . . . \quad = \frac{wx}{24EIl}\left(a^2(2l - a)^2 - 2ax^2(2l - a) + lx^3 \right)$$

$$\Delta_x \left(\text{when } x > a \right) \; . . . . . \quad = \frac{wa^2(l - x)}{24EIl}(4xl - 2x^2 - a^2)$$

---

6. SIMPLE BEAM—UNIFORM LOAD PARTIALLY DISTRIBUTED AT EACH END

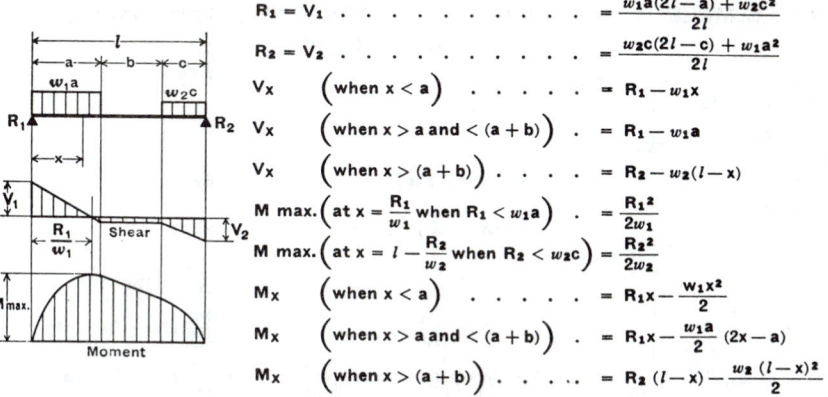

$$R_1 = V_1 \; . . . . \quad = \frac{w_1 a(2l - a) + w_2 c^2}{2l}$$

$$R_2 = V_2 \; . . . . \quad = \frac{w_2 c(2l - c) + w_1 a^2}{2l}$$

$$V_x \left(\text{when } x < a \right) \; . . . . . \quad = R_1 - w_1 x$$

$$V_x \left(\text{when } x > a \text{ and } < (a + b) \right). \quad = R_1 - w_1 a$$

$$V_x \left(\text{when } x > (a + b) \right). . . . \quad = R_2 - w_2(l - x)$$

$$M \text{ max.} \left(\text{at } x = \frac{R_1}{w_1} \text{ when } R_1 < w_1 a \right). \quad = \frac{R_1{}^2}{2w_1}$$

$$M \text{ max.} \left(\text{at } x = l - \frac{R_2}{w_2} \text{ when } R_2 < w_2 c \right) = \frac{R_2{}^2}{2w_2}$$

$$M_x \left(\text{when } x < a \right) \; . . . . . \quad = R_1 x - \frac{w_1 x^2}{2}$$

$$M_x \left(\text{when } x > a \text{ and } < (a + b) \right). \quad = R_1 x - \frac{w_1 a}{2}(2x - a)$$

$$M_x \left(\text{when } x > (a + b) \right). . . . . \quad = R_2 (l - x) - \frac{w_2 (l - x)^2}{2}$$

BEAM DIAGRAMS AND FORMULAS
For various static loading conditions
For meaning of symbols, see page **2 - 293**

---

7. SIMPLE BEAM—CONCENTRATED LOAD AT CENTER

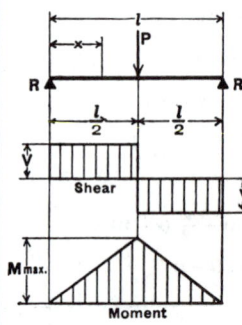

Total Equiv. Uniform Load . . . . . = 2P

$R = V$ . . . . $= \dfrac{P}{2}$

M max. $\left(\text{at point of load}\right)$ . . . . $= \dfrac{Pl}{4}$

M_x $\left(\text{when } x < \dfrac{l}{2}\right)$ . . . . $= \dfrac{Px}{2}$

Δmax. $\left(\text{at point of load}\right)$ . . . . $= \dfrac{Pl^3}{48EI}$

Δ_x $\left(\text{when } x < \dfrac{l}{2}\right)$ . . . . . $= \dfrac{Px}{48EI}(3l^2 - 4x^2)$

---

8. SIMPLE BEAM—CONCENTRATED LOAD AT ANY POINT

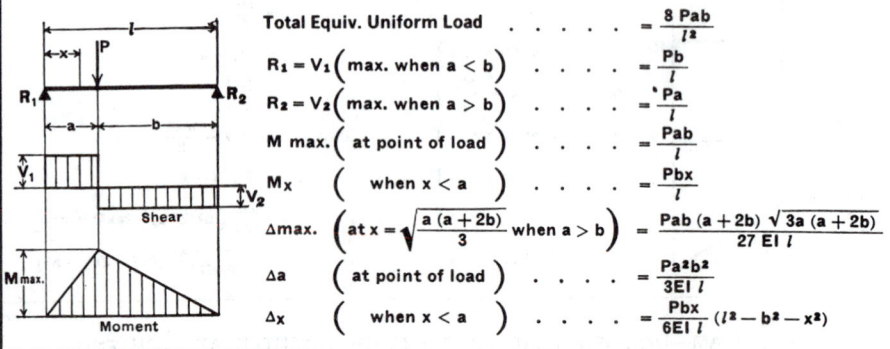

Total Equiv. Uniform Load . . . . . $= \dfrac{8\,Pab}{l^2}$

$R_1 = V_1\left(\text{max. when } a < b\right)$ . . . . $= \dfrac{Pb}{l}$

$R_2 = V_2\left(\text{max. when } a > b\right)$ . . . $= \dfrac{Pa}{l}$

M max. $\left(\text{at point of load}\right)$ . . . . $= \dfrac{Pab}{l}$

M_x $\left(\text{when } x < a\right)$ . . . . $= \dfrac{Pbx}{l}$

Δmax. $\left(\text{at } x = \sqrt{\dfrac{a(a+2b)}{3}} \text{ when } a > b\right) = \dfrac{Pab\,(a+2b)\,\sqrt{3a\,(a+2b)}}{27\,EI\,l}$

Δa $\left(\text{at point of load}\right)$ . . . . $= \dfrac{Pa^2b^2}{3EI\,l}$

Δ_x $\left(\text{when } x < a\right)$ . . . . $= \dfrac{Pbx}{6EI\,l}(l^2 - b^2 - x^2)$

---

9. SIMPLE BEAM—TWO EQUAL CONCENTRATED LOADS SYMMETRICALLY PLACED

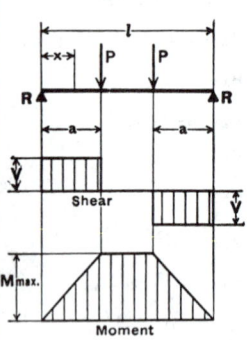

Total Equiv. Uniform Load . . . . . $= \dfrac{8\,Pa}{l}$

$R = V$ . . . . $= P$

M max. $\left(\text{between loads}\right)$ . . . . . $= Pa$

M_x $\left(\text{when } x < a\right)$ . . . . . $= Px$

Δmax. $\left(\text{at center}\right)$ . . . . . . $= \dfrac{Pa}{24EI}(3l^2 - 4a^2)$

Δ_x $\left(\text{when } x < a\right)$ . . . . . $= \dfrac{Px}{6EI}(3la - 3a^2 - x^2)$

Δ_x $\left(\text{when } x > a \text{ and } < (l-a)\right)$ . . $= \dfrac{Pa}{6EI}(3lx - 3x^2 - a^2)$

BEAM DIAGRAMS AND FORMULAS
For various static loading conditions

For meaning of symbols, see page **2 - 293**

---

10. SIMPLE BEAM—TWO EQUAL CONCENTRATED LOADS UNSYMMETRICALLY PLACED

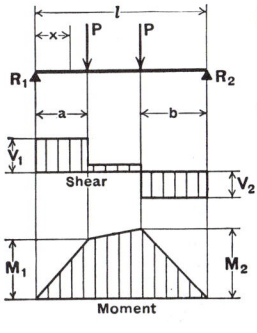

$R_1 = V_1 \left(\text{max. when } a < b \right)$ . . . . $= \dfrac{P}{l}(l - a + b)$

$R_2 = V_2 \left(\text{max. when } a > b \right)$ . . . . $= \dfrac{P}{l}(l - b + a)$

$V_x \left(\text{when } x > a \text{ and } < (l - b) \right)$ . . $= \dfrac{P}{l}(b - a)$

$M_1 \left(\text{max. when } a > b \right)$ . . . . . $= R_1 a$

$M_2 \left(\text{max. when } a < b \right)$ . . . . . $= R_2 b$

$M_x \left(\text{when } x < a \right)$ . . . . $= R_1 x$

$M_x \left(\text{when } x > a \text{ and } < (l - b) \right)$ . . $= R_1 x - P(x - a)$

---

11. SIMPLE BEAM—TWO UNEQUAL CONCENTRATED LOADS UNSYMMETRICALLY PLACED

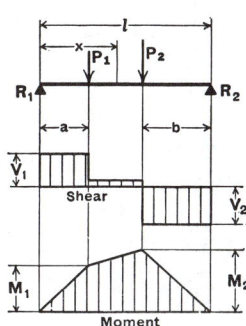

$R_1 = V_1$ . . . . $= \dfrac{P_1(l - a) + P_2 b}{l}$

$R_2 = V_2$ . . . . $= \dfrac{P_1 a + P_2 (l - b)}{l}$

$V_x \left(\text{when } x > a \text{ and } < (l - b) \right)$ . . $= R_1 - P_1$

$M_1 \left(\text{max. when } R_1 < P_1 \right)$ . . . . $= R_1 a$

$M_2 \left(\text{max. when } R_2 < P_2 \right)$ . . . . $= R_2 b$

$M_x \left(\text{when } x < a \right)$ . . . . $= R_1 x$

$M_x \left(\text{when } x > a \text{ and } < (l - b) \right)$ . . $= R_1 x - P_1 (x - a)$

---

12. BEAM FIXED AT ONE END, SUPPORTED AT OTHER— UNIFORMLY DISTRIBUTED LOAD

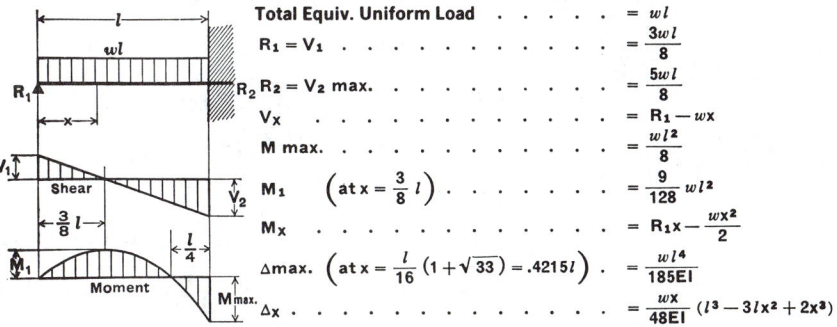

Total Equiv. Uniform Load . . . . . $= wl$

$R_1 = V_1$ . . . . $= \dfrac{3wl}{8}$

$R_2 = V_2$ max. . . . . $= \dfrac{5wl}{8}$

V_x . . . . $= R_1 - wx$

M max. . . . . $= \dfrac{wl^2}{8}$

$M_1 \left(\text{at } x = \dfrac{3}{8} l \right)$ . . . . . . $= \dfrac{9}{128} wl^2$

M_x . . . . $= R_1 x - \dfrac{wx^2}{2}$

$\Delta \text{max.} \left(\text{at } x = \dfrac{l}{16}(1 + \sqrt{33}) = .4215 l \right)$. $= \dfrac{wl^4}{185 EI}$

Δ_x . . . . $= \dfrac{wx}{48 EI}(l^3 - 3lx^2 + 2x^3)$

BEAM DIAGRAMS AND FORMULAS
For various static loading conditions

For meaning of symbols, see page **2 - 293**

---

13. BEAM FIXED AT ONE END, SUPPORTED AT OTHER— CONCENTRATED LOAD AT CENTER

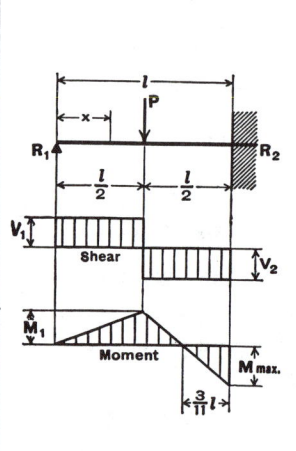

Total Equiv. Uniform Load $\;\ldots\;\ldots\;= \dfrac{3P}{2}$

$R_1 = V_1 \;\ldots\;\ldots\;\ldots\;= \dfrac{5P}{16}$

$R_2 = V_2$ max. $\;\ldots\;\ldots\;= \dfrac{11P}{16}$

M max. $\left(\text{at fixed end}\right) \;\ldots\;= \dfrac{3Pl}{16}$

$M_1 \;\left(\text{at point of load}\right) \;\ldots\;= \dfrac{5Pl}{32}$

$M_x \;\left(\text{when } x < \dfrac{l}{2}\right) \;\ldots\;= \dfrac{5Px}{16}$

$M_x \;\left(\text{when } x > \dfrac{l}{2}\right) \;\ldots\;= P\left(\dfrac{l}{2} - \dfrac{11x}{16}\right)$

Δmax. $\left(\text{at } x = l\sqrt{\dfrac{1}{5}} = .4472l\right) \;\ldots\;= \dfrac{Pl^3}{48EI\sqrt{5}} = .009317\,\dfrac{Pl^3}{EI}$

$\Delta_x \;\left(\text{at point of load}\right) \;\ldots\;= \dfrac{7Pl^3}{768EI}$

$\Delta_x \;\left(\text{when } x < \dfrac{l}{2}\right) \;\ldots\;= \dfrac{Px}{96EI}(3l^2 - 5x^2)$

$\Delta_x \;\left(\text{when } x > \dfrac{l}{2}\right) \;\ldots\;= \dfrac{P}{96EI}(x-l)^2(11x-2l)$

---

14. BEAM FIXED AT ONE END, SUPPORTED AT OTHER— CONCENTRATED LOAD AT ANY POINT

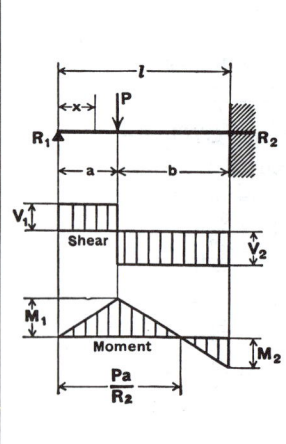

$R_1 = V_1 \;\ldots\;\ldots\;\ldots\;= \dfrac{Pb^2}{2l^3}(a+2l)$

$R_2 = V_2 \;\ldots\;\ldots\;\ldots\;= \dfrac{Pa}{2l^3}(3l^2 - a^2)$

$M_1 \;\left(\text{at point of load}\right) \;\ldots\;= R_1 a$

$M_2 \;\left(\text{at fixed end}\right) \;\ldots\;= \dfrac{Pab}{2l^2}(a+l)$

$M_x \;\left(\text{when } x < a\right) \;\ldots\;= R_1 x$

$M_x \;\left(\text{when } x > a\right) \;\ldots\;= R_1 x - P(x-a)$

Δmax. $\left(\text{when } a < .414l \text{ at } x = l\dfrac{l^2+a^2}{3l^2-a^2}\right) = \dfrac{Pa}{3EI}\dfrac{(l^2-a^2)^3}{(3l^2-a^2)^2}$

Δmax. $\left(\text{when } a > .414l \text{ at } x = l\sqrt{\dfrac{a}{2l+a}}\right) = \dfrac{Pab^2}{6EI}\sqrt{\dfrac{a}{2l+a}}$

$\Delta_a \;\left(\text{at point of load}\right) \;\ldots\;= \dfrac{Pa^2b^3}{12EIl^3}(3l+a)$

$\Delta_x \;\left(\text{when } x < a\right) \;\ldots\;= \dfrac{Pb^2x}{12EIl^3}(3al^2 - 2lx^2 - ax^2)$

$\Delta_x \;\left(\text{when } x > a\right) \;\ldots\;= \dfrac{Pa}{12EIl^3}(l-x)^2(3l^2x - a^2x - 2a^2l)$

BEAM DIAGRAMS AND FORMULAS
For various static loading conditions

For meaning of symbols, see page **2 - 293**

---

15. BEAM FIXED AT BOTH ENDS—UNIFORMLY DISTRIBUTED LOADS

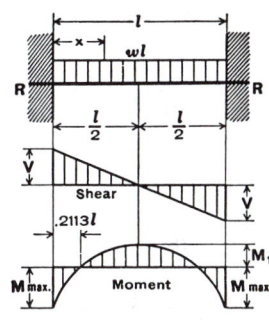

Total Equiv. Uniform Load $\dots\dots = \dfrac{2wl}{3}$

$R = V \dots\dots\dots\dots = \dfrac{wl}{2}$

$V_x \dots\dots\dots\dots = w\left(\dfrac{l}{2}-x\right)$

M max. $\left(\text{at ends}\right) \dots\dots = \dfrac{wl^2}{12}$

$M_1 \quad \left(\text{at center}\right) \dots\dots = \dfrac{wl^2}{24}$

$M_x \dots\dots\dots\dots = \dfrac{w}{12}(6lx - l^2 - 6x^2)$

Δmax. $\left(\text{at center}\right) \dots\dots = \dfrac{wl^4}{384EI}$

$\Delta_x \dots\dots\dots\dots\dots = \dfrac{wx^2}{24EI}(l-x)^2$

---

16. BEAM FIXED AT BOTH ENDS—CONCENTRATED LOAD AT CENTER

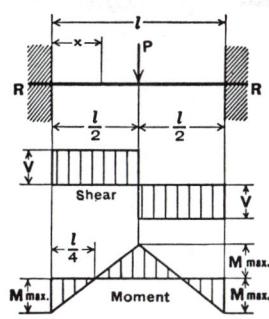

Total Equiv. Uniform Load $\dots\dots = P$

$R = V \dots\dots\dots\dots = \dfrac{P}{2}$

M max. $\left(\text{at center and ends}\right) \dots = \dfrac{Pl}{8}$

$M_x \quad \left(\text{when } x < \dfrac{l}{2}\right) \dots\dots = \dfrac{P}{8}(4x - l)$

Δmax. $\left(\text{at center}\right) \dots\dots = \dfrac{Pl^3}{192EI}$

$\Delta_x \quad \left(\text{when } x < \dfrac{l}{2}\right) \dots\dots = \dfrac{Px^2}{48EI}(3l - 4x)$

---

17. BEAM FIXED AT BOTH ENDS—CONCENTRATED LOAD AT ANY POINT

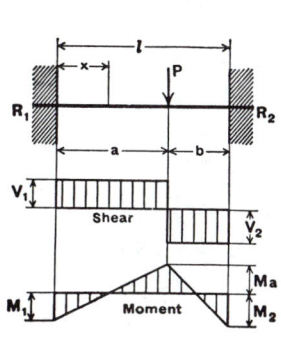

$R_1 = V_1 \left(\text{max. when } a < b\right) \dots = \dfrac{Pb^2}{l^3}(3a + b)$

$R_2 = V_2 \left(\text{max. when } a > b\right) \dots = \dfrac{Pa^2}{l^3}(a + 3b)$

$M_1 \quad \left(\text{max. when } a < b\right) \dots = \dfrac{Pab^2}{l^2}$

$M_2 \quad \left(\text{max. when } a > b\right) \dots = \dfrac{Pa^2b}{l^2}$

$M_a \quad \left(\text{at point of load}\right) \dots = \dfrac{2Pa^2b^2}{l^3}$

$M_x \quad \left(\text{when } x < a\right) \dots\dots = R_1x - \dfrac{Pab^2}{l^2}$

Δmax. $\left(\text{when } a > b \text{ at } x = \dfrac{2al}{3a+b}\right). = \dfrac{2Pa^3b^2}{3EI(3a+b)^2}$

$\Delta_a \quad \left(\text{at point of load}\right) \dots = \dfrac{Pa^3b^3}{3EIl^3}$

$\Delta_x \quad \left(\text{when } x < a\right) \dots\dots = \dfrac{Pb^2x^2}{6EIl^3}(3al - 3ax - bx)$

AMERICAN INSTITUTE OF STEEL CONSTRUCTION

BEAM DIAGRAMS AND FORMULAS
For various static loading conditions

For meaning of symbols, see page **2 - 293**

---

18. CANTILEVER BEAM—LOAD INCREASING UNIFORMLY TO FIXED END

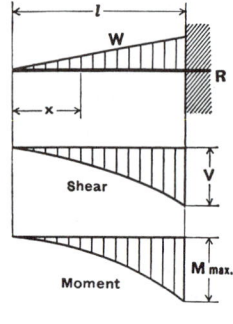

Total Equiv. Uniform Load . . . . $= \dfrac{8}{3} W$

$R = V$ . . . . $= W$

V_x . . . . $= W \dfrac{x^2}{l^2}$

M max. $\left(\text{at fixed end}\right)$ . . . . . $= \dfrac{Wl}{3}$

M_x . . . . $= \dfrac{Wx^3}{3l^2}$

Δmax. $\left(\text{at free end}\right)$ . . . . . $= \dfrac{Wl^3}{15EI}$

Δ_x . . . . $= \dfrac{W}{60EI\,l^2}\,(x^5 - 5l^4x + 4l^5)$

Shear

Moment

M max.

---

19. CANTILEVER BEAM—UNIFORMLY DISTRIBUTED LOAD

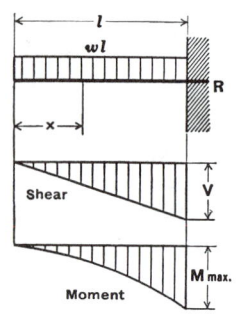

Total Equiv. Uniform Load . . . . $= 4wl$

$R = V$ . . . . $= wl$

V_x . . . . $= wx$

M max. $\left(\text{at fixed end}\right)$ . . . . . $= \dfrac{wl^2}{2}$

M_x . . . . $= \dfrac{wx^2}{2}$

Δmax. $\left(\text{at free end}\right)$ . . . . . $= \dfrac{wl^4}{8EI}$

Δ_x . . . . $= \dfrac{w}{24EI}\,(x^4 - 4l^3x + 3l^4)$

Shear

Moment

M max.

---

20. BEAM FIXED AT ONE END, FREE TO DEFLECT VERTICALLY BUT NOT ROTATE AT OTHER—UNIFORMLY DISTRIBUTED LOAD

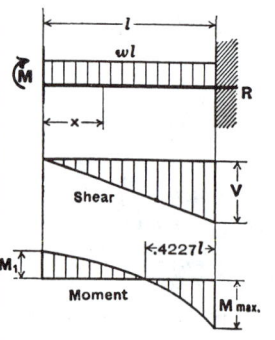

Total Equiv. Uniform Load . . . . $= \dfrac{8}{3} wl$

$R = V$ . . . . $= wl$

V_x . . . . $= wx$

M max. $\left(\text{at fixed end}\right)$ . . . . . $= \dfrac{wl^2}{3}$

M_1 $\left(\text{at deflected end}\right)$ . . . . $= \dfrac{wl^2}{6}$

M_x . . . . $= \dfrac{w}{6}\,(l^2 - 3x^2)$

Δmax. $\left(\text{at deflected end}\right)$ . . . . $= \dfrac{wl^4}{24EI}$

Δ_x . . . . $= \dfrac{w\,(l^2 - x^2)^2}{24EI}$

Shear

Moment

.4227l

M_1

M max.

BEAM DIAGRAMS AND FORMULAS
For various static loading conditions

For meaning of symbols, see page **2 - 293**

---

21. CANTILEVER BEAM—CONCENTRATED LOAD AT ANY POINT

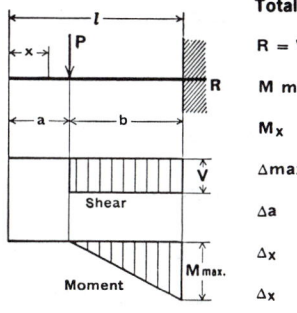

Total Equiv. Uniform Load . . . . $= \dfrac{8Pb}{l}$

$R = V$ . . . . $= P$

M max. $\left(\text{at fixed end}\right)$ . . . . . $= Pb$

$M_x \quad \left(\text{when } x > a\right)$ . . . . . $= P(x-a)$

Δmax. $\left(\text{at free end}\right)$ . . . . . $= \dfrac{Pb^2}{6EI}(3l-b)$

$\Delta a \quad \left(\text{at point of load}\right)$ . . . . $= \dfrac{Pb^3}{3EI}$

$\Delta_x \quad \left(\text{when } x < a\right)$ . . . . . $= \dfrac{Pb^2}{6EI}(3l-3x-b)$

$\Delta_x \quad \left(\text{when } x > a\right)$ . . . . . $= \dfrac{P(l-x)^2}{6EI}(3b-l+x)$

---

22. CANTILEVER BEAM—CONCENTRATED LOAD AT FREE END

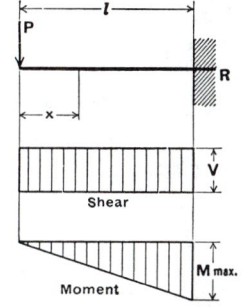

Total Equiv. Uniform Load . . . . $= 8P$

$R = V$ . . . . $= P$

M max. $\left(\text{at fixed end}\right)$ . . . . . $= Pl$

M_x . . . . $= Px$

Δmax. $\left(\text{at free end}\right)$ . . . . $= \dfrac{Pl^3}{3EI}$

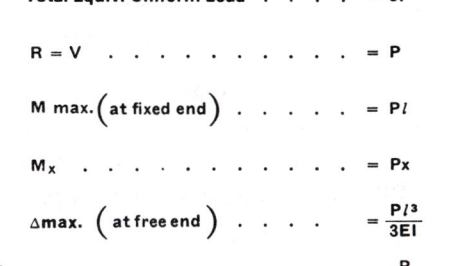

Δ_x . . . . $= \dfrac{P}{6EI}(2l^3-3l^2x+x^3)$

---

23. BEAM FIXED AT ONE END, FREE TO DEFLECT VERTICALLY BUT NOT ROTATE AT OTHER—CONCENTRATED LOAD AT DEFLECTED END

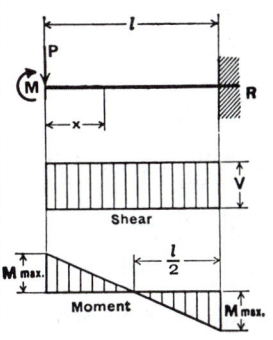

Total Equiv. Uniform Load . . . . $= 4P$

$R = V$ . . . . $= P$

M max. $\left(\text{at both ends}\right)$ . . . . $= \dfrac{Pl}{2}$

M_x . . . . $= P\left(\dfrac{l}{2}-x\right)$

Δmax. $\left(\text{at deflected end}\right)$ . . . . $= \dfrac{Pl^3}{12EI}$

Δ_x . . . . $= \dfrac{P(l-x)^2}{12EI}(l+2x)$

BEAM DIAGRAMS AND FORMULAS
For various static loading conditions

For meaning of symbols, see page 2 - 293

24. BEAM OVERHANGING ONE SUPPORT—UNIFORMLY DISTRIBUTED LOAD

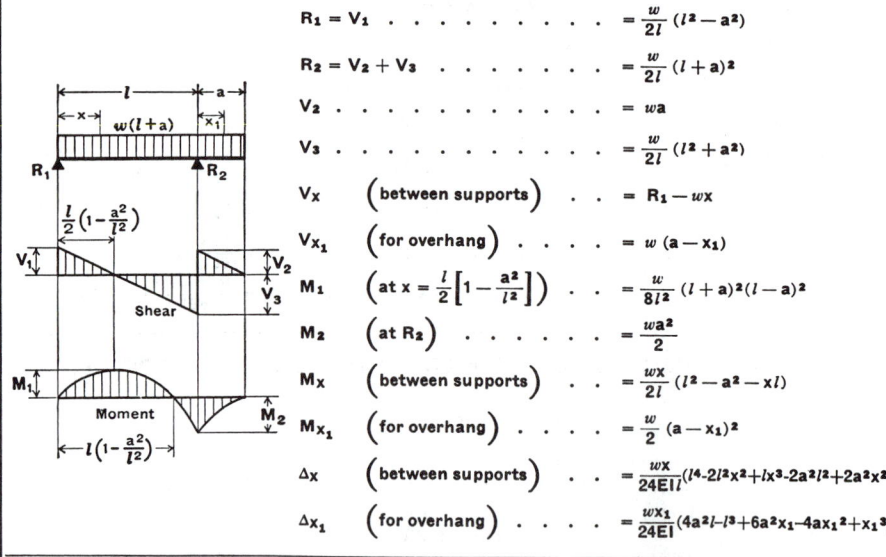

$$R_1 = V_1 \quad \ldots \ldots \ldots \quad = \frac{w}{2l}(l^2 - a^2)$$

$$R_2 = V_2 + V_3 \quad \ldots \ldots \quad = \frac{w}{2l}(l + a)^2$$

$$V_2 \quad \ldots \ldots \ldots \ldots \quad = wa$$

$$V_3 \quad \ldots \ldots \ldots \ldots \quad = \frac{w}{2l}(l^2 + a^2)$$

$$V_x \left(\text{between supports}\right) \quad \ldots \quad = R_1 - wx$$

$$V_{x_1} \left(\text{for overhang}\right) \quad \ldots \ldots \quad = w(a - x_1)$$

$$M_1 \left(\text{at } x = \frac{l}{2}\left[1 - \frac{a^2}{l^2}\right]\right) \ldots \quad = \frac{w}{8l^2}(l + a)^2(l - a)^2$$

$$M_2 \left(\text{at } R_2\right) \quad \ldots \ldots \quad = \frac{wa^2}{2}$$

$$M_x \left(\text{between supports}\right) \quad \ldots \quad = \frac{wx}{2l}(l^2 - a^2 - xl)$$

$$M_{x_1} \left(\text{for overhang}\right) \quad \ldots \ldots \quad = \frac{w}{2}(a - x_1)^2$$

$$\Delta_x \left(\text{between supports}\right) \quad \ldots \quad = \frac{wx}{24EIl}(l^4 - 2l^2x^2 + lx^3 - 2a^2l^2 + 2a^2x^2)$$

$$\Delta_{x_1} \left(\text{for overhang}\right) \quad \ldots \ldots \quad = \frac{wx_1}{24EI}(4a^2l - l^3 + 6a^2x_1 - 4ax_1^2 + x_1^3)$$

25. BEAM OVERHANGING ONE SUPPORT—UNIFORMLY DISTRIBUTED LOAD ON OVERHANG

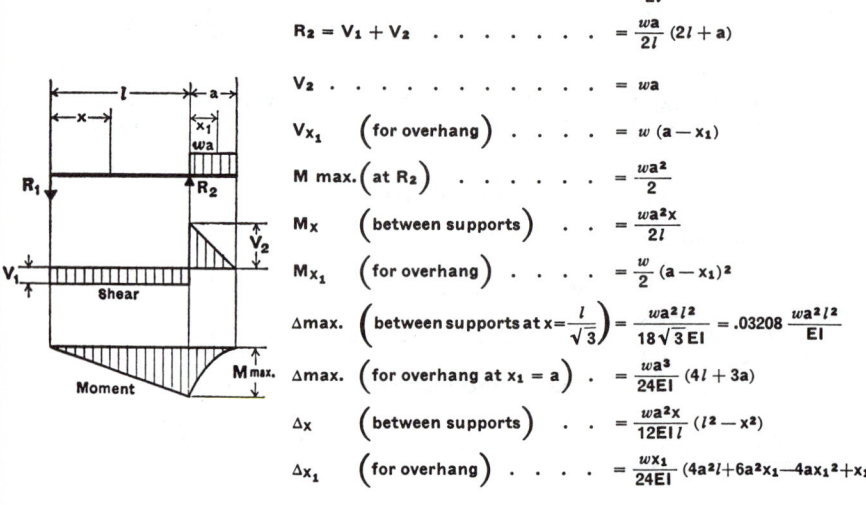

$$R_1 = V_1 \quad \ldots \ldots \ldots \ldots \quad = \frac{wa^2}{2l}$$

$$R_2 = V_1 + V_2 \quad \ldots \ldots \ldots \quad = \frac{wa}{2l}(2l + a)$$

$$V_2 \quad \ldots \ldots \ldots \ldots \ldots \quad = wa$$

$$V_{x_1} \left(\text{for overhang}\right) \quad \ldots \ldots \quad = w(a - x_1)$$

$$M \text{ max.} \left(\text{at } R_2\right) \quad \ldots \ldots \quad = \frac{wa^2}{2}$$

$$M_x \left(\text{between supports}\right) \quad \ldots \quad = \frac{wa^2x}{2l}$$

$$M_{x_1} \left(\text{for overhang}\right) \quad \ldots \quad = \frac{w}{2}(a - x_1)^2$$

$$\Delta \text{max.} \left(\text{between supports at } x = \frac{l}{\sqrt{3}}\right) = \frac{wa^2l^2}{18\sqrt{3}EI} = .03208\frac{wa^2l^2}{EI}$$

$$\Delta \text{max.} \left(\text{for overhang at } x_1 = a\right) \ldots \quad = \frac{wa^3}{24EI}(4l + 3a)$$

$$\Delta_x \left(\text{between supports}\right) \quad \ldots \quad = \frac{wa^2x}{12EIl}(l^2 - x^2)$$

$$\Delta_{x_1} \left(\text{for overhang}\right) \quad \ldots \ldots \quad = \frac{wx_1}{24EI}(4a^2l + 6a^2x_1 - 4ax_1^2 + x_1^3)$$

BEAM DIAGRAMS AND FORMULAS
For various static loading conditions
For meaning of symbols, see page **2 - 293**

---

26. BEAM OVERHANGING ONE SUPPORT—CONCENTRATED LOAD AT END OF OVERHANG

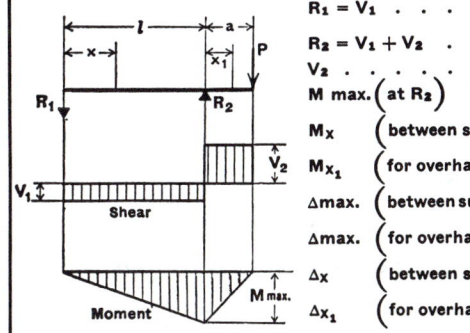

$$R_1 = V_1 \ldots \ldots \ldots = \frac{Pa}{l}$$

$$R_2 = V_1 + V_2 \ldots \ldots = \frac{P}{l}(l+a)$$

$$V_2 \ldots \ldots \ldots \ldots = P$$

$$M \text{ max.} \left(\text{at } R_2\right) \ldots \ldots = Pa$$

$$M_x \left(\text{between supports}\right) \ldots = \frac{Pax}{l}$$

$$M_{x_1} \left(\text{for overhang}\right) \ldots \ldots = P(a - x_1)$$

$$\Delta\text{max.} \left(\text{between supports at } x = \frac{l}{\sqrt{3}}\right) = \frac{Pal^2}{9\sqrt{3}EI} = .06415 \frac{Pal^2}{EI}$$

$$\Delta\text{max.} \left(\text{for overhang at } x_1 = a\right) . = \frac{Pa^2}{3EI}(l+a)$$

$$\Delta_x \left(\text{between supports}\right) \ldots = \frac{Pax}{6EIl}(l^2 - x^2)$$

$$\Delta_{x_1} \left(\text{for overhang}\right) \ldots = \frac{Px_1}{6EI}(2al + 3ax_1 - x_1^2)$$

---

27. BEAM OVERHANGING ONE SUPPORT—UNIFORMLY DISTRIBUTED LOAD BETWEEN SUPPORTS

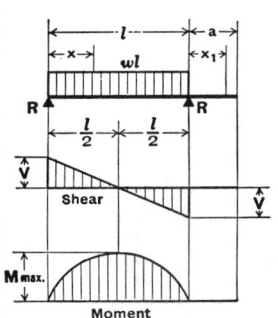

$$\text{Total Equiv. Uniform Load} \ldots = wl$$

$$R = V \ldots \ldots \ldots \ldots = \frac{wl}{2}$$

$$V_x \ldots \ldots \ldots \ldots \ldots = w\left(\frac{l}{2} - x\right)$$

$$M \text{ max.} \left(\text{at center}\right) \ldots \ldots = \frac{wl^2}{8}$$

$$M_x \ldots \ldots \ldots \ldots = \frac{wx}{2}(l - x)$$

$$\Delta\text{max.} \left(\text{at center}\right) \ldots \ldots = \frac{5wl^4}{384EI}$$

$$\Delta_x \ldots \ldots \ldots \ldots = \frac{wx}{24EI}(l^3 - 2lx^2 + x^3)$$

$$\Delta_{x_1} \ldots \ldots \ldots \ldots = \frac{wl^3 x_1}{24EI}$$

---

28. BEAM OVERHANGING ONE SUPPORT—CONCENTRATED LOAD AT ANY POINT BETWEEN SUPPORTS

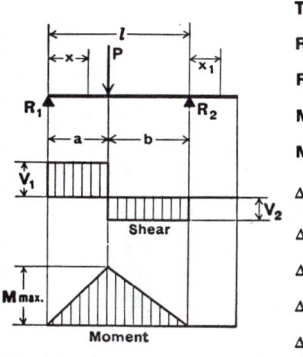

$$\text{Total Equiv. Uniform Load} \ldots = \frac{8Pab}{l^2}$$

$$R_1 = V_1 \left(\text{max. when } a < b\right) \ldots = \frac{Pb}{l}$$

$$R_2 = V_2 \left(\text{max. when } a > b\right) \ldots = \frac{Pa}{l}$$

$$M \text{ max.} \left(\text{at point of load}\right) \ldots = \frac{Pab}{l}$$

$$M_x \left(\text{when } x < a\right) \ldots \ldots = \frac{Pbx}{l}$$

$$\Delta\text{max.} \left(\text{at } x = \sqrt{\frac{a(a+2b)}{3}} \text{ when } a > b\right) = \frac{Pab(a+2b)\sqrt{3a(a+2b)}}{27EIl}$$

$$\Delta a \left(\text{at point of load}\right) \ldots = \frac{Pa^2 b^2}{3EIl}$$

$$\Delta_x \left(\text{when } x < a\right) \ldots \ldots = \frac{Pbx}{6EIl}(l^2 - b^2 - x^2)$$

$$\Delta_x \left(\text{when } x > a\right) \ldots \ldots = \frac{Pa(l-x)}{6EIl}(2lx - x^2 - a^2)$$

$$\Delta_{x_1} \ldots \ldots \ldots \ldots = \frac{Pabx_1}{6EIl}(l + a)$$

BEAM DIAGRAMS AND FORMULAS
For various static loading conditions

For meaning of symbols, see page **2 - 293**

---

29. CONTINUOUS BEAM—TWO EQUAL SPANS—UNIFORM LOAD ON ONE SPAN

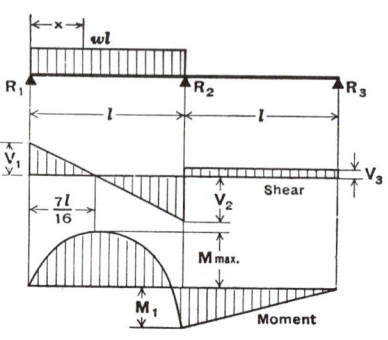

Total Equiv. Uniform Load $= \dfrac{49}{64} wl$

$R_1 = V_1 \ \ldots \ldots \ldots \ldots = \dfrac{7}{16} wl$

$R_2 = V_2 + V_3 \ \ldots \ldots \ = \dfrac{5}{8} wl$

$R_3 = V_3 \ \ldots \ldots \ldots = -\dfrac{1}{16} wl$

$V_2 \ \ldots \ldots \ldots \ldots = \dfrac{9}{16} wl$

M max. $\left(\text{at } x = \dfrac{7}{16} l \right) \ \ldots = \dfrac{49}{512} wl^2$

$M_1 \ \left(\text{at support } R_2 \right) \ . \ = \dfrac{1}{16} wl^2$

$M_x \ \left(\text{when } x < l \right) \ \ldots = \dfrac{wx}{16}(7l - 8x)$

Δ Max. (0.472 l from R_1) $= 0.0092 \ wl^4/EI$

---

30. CONTINUOUS BEAM—TWO EQUAL SPANS—CONCENTRATED LOAD AT CENTER OF ONE SPAN

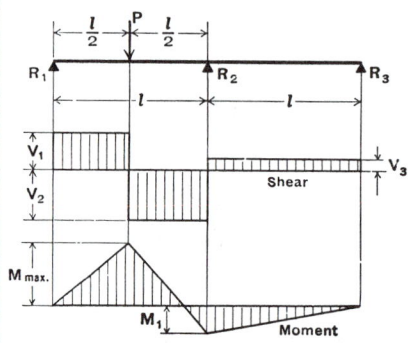

Total Equiv. Uniform Load . $= \dfrac{13}{8} P$

$R_1 = V_1 \ \ldots \ldots \ldots \ldots = \dfrac{13}{32} P$

$R_2 = V_2 + V_3 \ \ldots \ldots \ = \dfrac{11}{16} P$

$R_3 = V_3 \ \ldots \ldots \ldots = -\dfrac{3}{32} P$

$V_2 \ \ldots \ldots \ldots \ldots = \dfrac{19}{32} P$

M max. $\left(\text{at point of load} \right) . \ = \dfrac{13}{64} Pl$

$M_1 \ \left(\text{at support } R_2 \right) \ . \ = \dfrac{3}{32} Pl$

Δ Max. (0.480 l from R_1) $= 0.015 \ Pl^3/EI$

---

31. CONTINUOUS BEAM—TWO EQUAL SPANS—CONCENTRATED LOAD AT ANY POINT

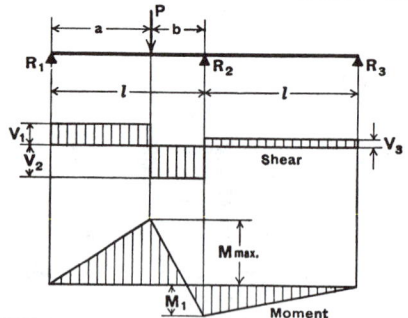

$R_1 = V_1 \ \ldots \ldots \ldots = \dfrac{Pb}{4l^3}\left(4l^2 - a(l+a) \right)$

$R_2 = V_2 + V_3 \ \ldots \ldots = \dfrac{Pa}{2l^3}\left(2l^2 + b(l+a) \right)$

$R_3 = V_3 \ \ldots \ldots \ldots = -\dfrac{Pab}{4l^3}(l+a)$

$V_2 \ \ldots \ldots \ldots \ldots = \dfrac{Pa}{4l^3}\left(4l^2 + b(l+a) \right)$

M max. $\left(\text{at point of load} \right) . \ = \dfrac{Pab}{4l^3}\left(4l^2 - a(l+a) \right)$

$M_1 \ \left(\text{at support } R_2 \right) \ . \ = \dfrac{Pab}{4l^2}(l+a)$

BEAM DIAGRAMS AND FORMULAS
For various static loading conditions

For meaning of symbols, see page **2** - 293

---

32. BEAM—UNIFORMLY DISTRIBUTED LOAD AND VARIABLE END MOMENTS

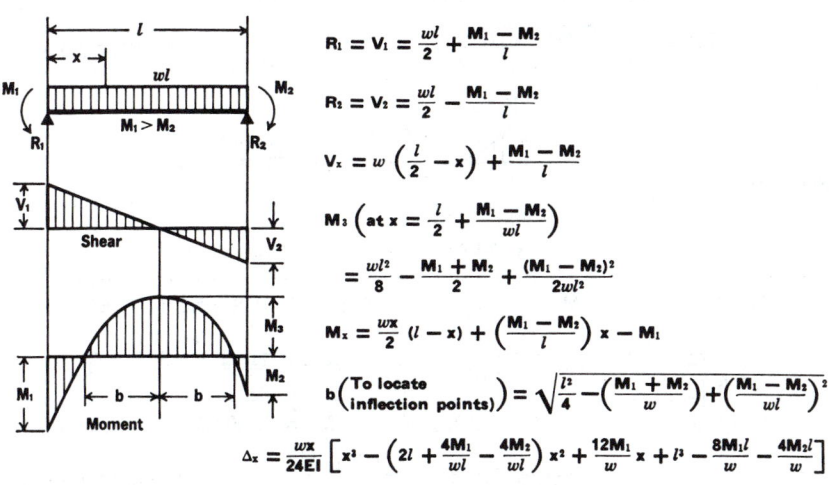

$$R_1 = V_1 = \frac{wl}{2} + \frac{M_1 - M_2}{l}$$

$$R_2 = V_2 = \frac{wl}{2} - \frac{M_1 - M_2}{l}$$

$$V_x = w\left(\frac{l}{2} - x\right) + \frac{M_1 - M_2}{l}$$

$$M_3 \left(\text{at } x = \frac{l}{2} + \frac{M_1 - M_2}{wl}\right)$$

$$= \frac{wl^2}{8} - \frac{M_1 + M_2}{2} + \frac{(M_1 - M_2)^2}{2wl^2}$$

$$M_x = \frac{wx}{2}(l - x) + \left(\frac{M_1 - M_2}{l}\right)x - M_1$$

$$b\left(\begin{smallmatrix}\text{To locate}\\ \text{inflection points}\end{smallmatrix}\right) = \sqrt{\frac{l^2}{4} - \left(\frac{M_1 + M_2}{w}\right) + \left(\frac{M_1 - M_2}{wl}\right)^2}$$

$$\Delta_x = \frac{wx}{24EI}\left[x^3 - \left(2l + \frac{4M_1}{wl} - \frac{4M_2}{wl}\right)x^2 + \frac{12M_1}{w}x + l^3 - \frac{8M_1 l}{w} - \frac{4M_2 l}{w}\right]$$

---

33. BEAM—CONCENTRATED LOAD AT CENTER AND VARIABLE END MOMENTS

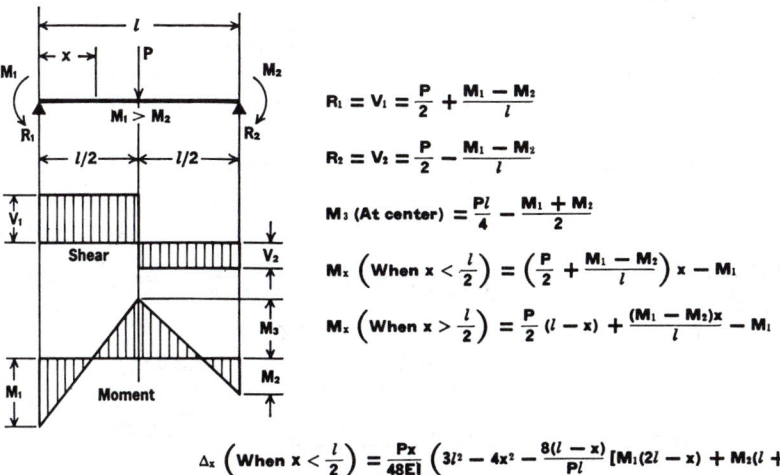

$$R_1 = V_1 = \frac{P}{2} + \frac{M_1 - M_2}{l}$$

$$R_2 = V_2 = \frac{P}{2} - \frac{M_1 - M_2}{l}$$

$$M_3 \text{ (At center)} = \frac{Pl}{4} - \frac{M_1 + M_2}{2}$$

$$M_x \left(\text{When } x < \frac{l}{2}\right) = \left(\frac{P}{2} + \frac{M_1 - M_2}{l}\right)x - M_1$$

$$M_x \left(\text{When } x > \frac{l}{2}\right) = \frac{P}{2}(l - x) + \frac{(M_1 - M_2)x}{l} - M_1$$

$$\Delta_x \left(\text{When } x < \frac{l}{2}\right) = \frac{Px}{48EI}\left(3l^2 - 4x^2 - \frac{8(l - x)}{Pl}[M_1(2l - x) + M_2(l + x)]\right)$$

AMERICAN INSTITUTE OF STEEL CONSTRUCTION

BEAM DIAGRAMS AND DEFLECTIONS
For various static loading conditions

For meaning of symbols, see page **2** - 293

---

34. CONTINUOUS BEAM—THREE EQUAL SPANS—ONE END SPAN UNLOADED

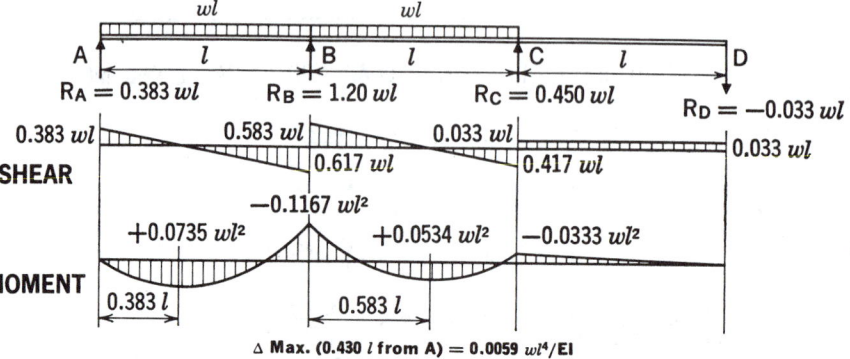

$R_A = 0.383\,wl$ $R_B = 1.20\,wl$ $R_C = 0.450\,wl$ $R_D = -0.033\,wl$

SHEAR

0.383 wl   0.583 wl   0.033 wl   0.033 wl
0.617 wl   0.417 wl

MOMENT

$-0.1167\,wl^2$
$+0.0735\,wl^2$ $+0.0534\,wl^2$ $-0.0333\,wl^2$
0.383 l   0.583 l

Δ **Max. (0.430 _l_ from A) = 0.0059** _wl_⁴/EI

---

35. CONTINUOUS BEAM—THREE EQUAL SPANS—END SPANS LOADED

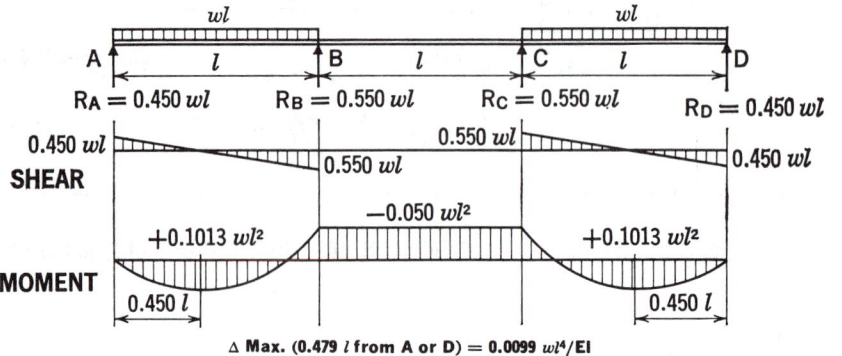

$R_A = 0.450\,wl$ $R_B = 0.550\,wl$ $R_C = 0.550\,wl$ $R_D = 0.450\,wl$

SHEAR

0.450 wl   0.550 wl   0.450 wl
0.550 wl

MOMENT

$-0.050\,wl^2$
$+0.1013\,wl^2$ $+0.1013\,wl^2$
0.450 l   0.450 l

Δ **Max. (0.479 _l_ from A or D) = 0.0099** _wl_⁴/EI

---

36. CONTINUOUS BEAM—THREE EQUAL SPANS—ALL SPANS LOADED

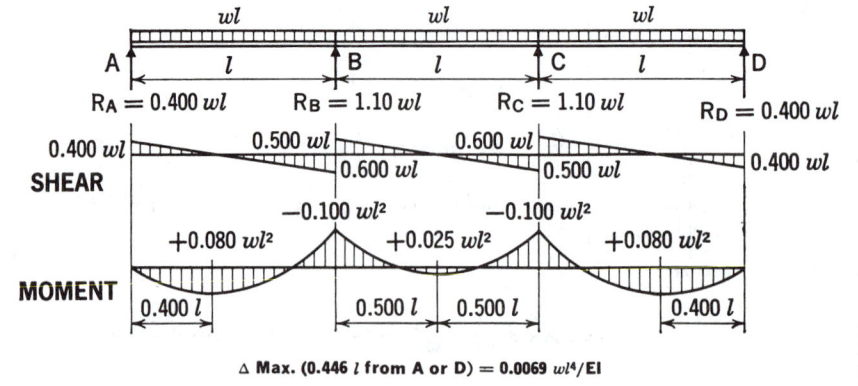

$R_A = 0.400\,wl$ $R_B = 1.10\,wl$ $R_C = 1.10\,wl$ $R_D = 0.400\,wl$

SHEAR

0.400 wl   0.500 wl   0.600 wl   0.400 wl
0.600 wl   0.500 wl

MOMENT

$-0.100\,wl^2$ $-0.100\,wl^2$
$+0.080\,wl^2$ $+0.025\,wl^2$ $+0.080\,wl^2$
0.400 l   0.500 l   0.500 l   0.400 l

Δ **Max. (0.446 _l_ from A or D) = 0.0069** _wl_⁴/EI

BEAM DIAGRAMS AND DEFLECTIONS
For various static loading conditions

For meaning of symbols, see page **2 - 293**

37. CONTINUOUS BEAM—FOUR EQUAL SPANS—THIRD SPAN UNLOADED

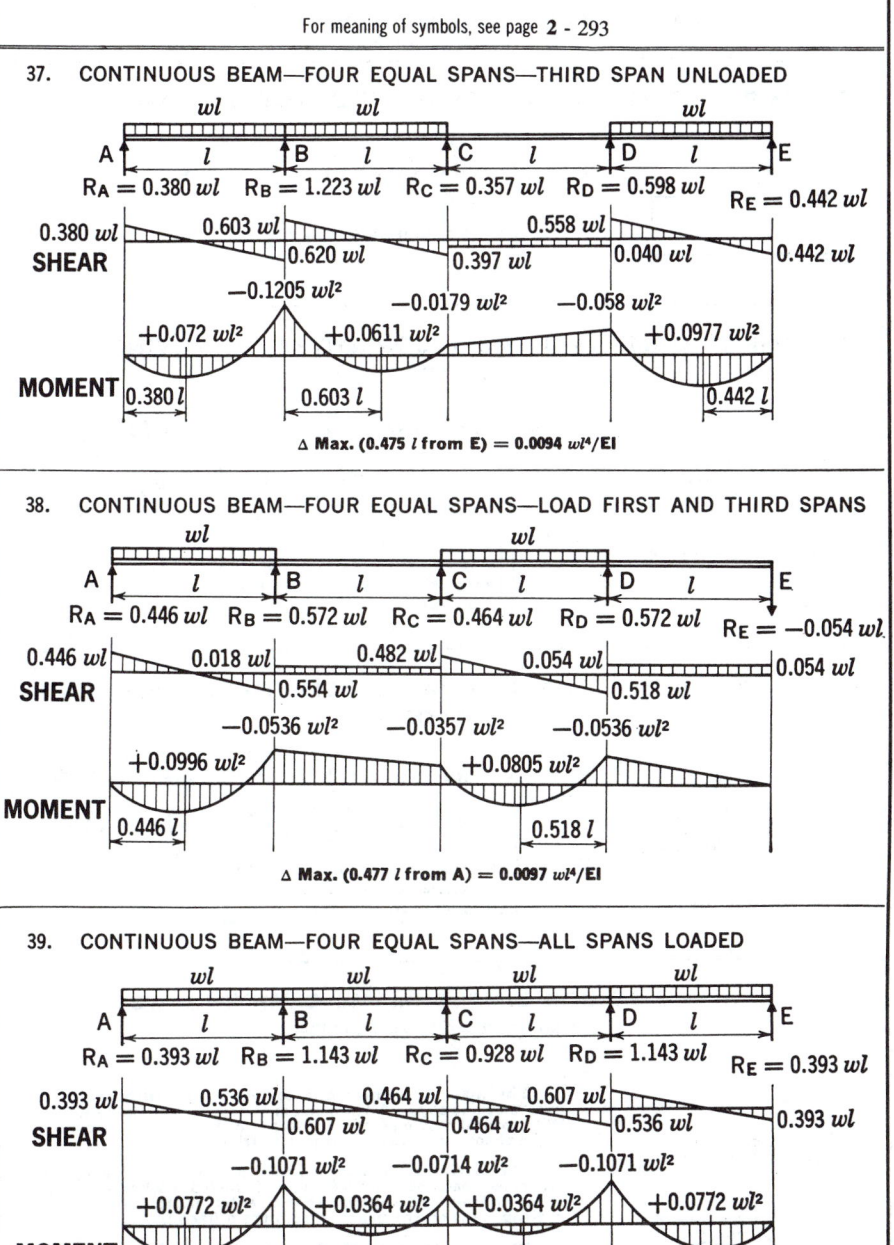

$R_A = 0.380\,wl$ $R_B = 1.223\,wl$ $R_C = 0.357\,wl$ $R_D = 0.598\,wl$ $R_E = 0.442\,wl$

SHEAR

0.380 wl   0.603 wl   0.620 wl   0.397 wl   0.558 wl   0.040 wl   0.442 wl

$-0.1205\,wl^2$ $-0.0179\,wl^2$ $-0.058\,wl^2$

$+0.072\,wl^2$ $+0.0611\,wl^2$ $+0.0977\,wl^2$

MOMENT 0.380 *l*   0.603 *l*   0.442 *l*

Δ **Max. (0.475 *l* from E) = 0.0094 *wl*⁴/EI**

38. CONTINUOUS BEAM—FOUR EQUAL SPANS—LOAD FIRST AND THIRD SPANS

$R_A = 0.446\,wl$ $R_B = 0.572\,wl$ $R_C = 0.464\,wl$ $R_D = 0.572\,wl$ $R_E = -0.054\,wl$

0.446 wl   0.018 wl   0.482 wl   0.054 wl   0.054 wl

SHEAR   0.554 wl   0.518 wl

$-0.0536\,wl^2$ $-0.0357\,wl^2$ $-0.0536\,wl^2$

$+0.0996\,wl^2$ $+0.0805\,wl^2$

MOMENT   0.446 *l*   0.518 *l*

Δ **Max. (0.477 *l* from A) = 0.0097 *wl*⁴/EI**

39. CONTINUOUS BEAM—FOUR EQUAL SPANS—ALL SPANS LOADED

$R_A = 0.393\,wl$ $R_B = 1.143\,wl$ $R_C = 0.928\,wl$ $R_D = 1.143\,wl$ $R_E = 0.393\,wl$

0.393 wl   0.536 wl   0.464 wl   0.607 wl   0.393 wl

SHEAR   0.607 wl   0.464 wl   0.536 wl

$-0.1071\,wl^2$ $-0.0714\,wl^2$ $-0.1071\,wl^2$

$+0.0772\,wl^2$ $+0.0364\,wl^2$ $+0.0364\,wl^2$ $+0.0772\,wl^2$

MOMENT 0.393 *l*   0.536 *l*   0.536 *l*   0.393 *l*

Δ **Max. (0.440 *l* from A and E) = 0.0065 *wl*⁴/EI**

BEAM DIAGRAMS AND FORMULAS
For various concentrated moving loads

The values given in these formulas do not include impact which varies according to the requirements of each case. For meaning of symbols, see page **2 - 293**

---

40. SIMPLE BEAM—ONE CONCENTRATED MOVING LOAD

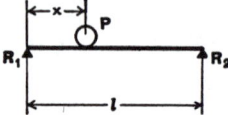

R_1 max. $= V_1$ max. $\left(\text{at } x = 0\right)$ . . . . . $= P$

M max. $\left(\text{at point of load, when } x = \dfrac{l}{2}\right)$. $= \dfrac{Pl}{4}$

---

41. SIMPLE BEAM—TWO EQUAL CONCENTRATED MOVING LOADS

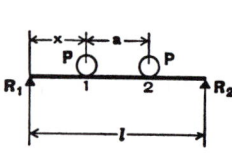

R_1 max. $= V_1$ max. $\left(\text{at } x = 0\right)$ . . . . . $= P\left(2 - \dfrac{a}{l}\right)$

M max. $\begin{cases} \left[\begin{array}{l}\text{when } a < (2 - \sqrt{2})\, l = .586l \\ \text{under load 1 at } x = \frac{1}{2}\left(l - \frac{a}{2}\right)\end{array}\right] = \dfrac{P}{2l}\left(l - \dfrac{a}{2}\right)^2 \\[2em] \left[\begin{array}{l}\text{when } a > (2 - \sqrt{2})\, l = .586l \\ \text{with one load at center of span} \\ \text{(case 40)}\end{array}\right] = \dfrac{Pl}{4} \end{cases}$

---

42. SIMPLE BEAM—TWO UNEQUAL CONCENTRATED MOVING LOADS

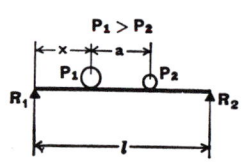

$P_1 > P_2$

R_1 max. $= V_1$ max. $\left(\text{at } x = 0\right)$ . . . . . $= P_1 + P_2\dfrac{l - a}{l}$

M max. $\begin{cases} \left[\text{under } P_1, \text{at } x = \frac{1}{2}\left(l - \frac{P_2 a}{P_1 + P_2}\right)\right] = \left(P_1 + P_2\right)\dfrac{x^2}{l} \\[1.5em] \left[\begin{array}{l}\text{M max. may occur with larger} \\ \text{load at center of span and other} \\ \text{load off span (case 40)}\end{array}\right] = \dfrac{P_1 l}{4} \end{cases}$

GENERAL RULES FOR SIMPLE BEAMS CARRYING MOVING CONCENTRATED LOADS

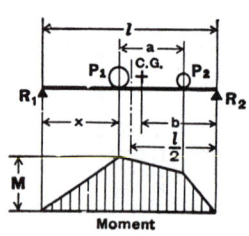

Moment

The maximum shear due to moving concentrated loads occurs at one support when one of the loads is at that support. With several moving loads, the location that will produce maximum shear must be determined by trial.

The maximum bending moment produced by moving concentrated loads occurs under one of the loads when that load is as far from one support as the center of gravity of all the moving loads on the beam is from the other support.

In the accompanying diagram, the maximum bending moment occurs under load P_1 when $x = b$. It should also be noted that this condition occurs when the center line of the span is midway between the center of gravity of loads and the nearest concentrated load.

AMERICAN INSTITUTE OF STEEL CONSTRUCTION

BEAM DIAGRAMS AND FORMULAS
Design properties of cantilevered beams
Equal loads, equally spaced

No. Spans	System
2	
3	
4	
5	
≥6 (even)	
≥7 (odd)	

	n	∞	2	3	4	5
	Typical Span Loading	P	P/2 P P/2	P/2 P P P/2	P/2 P P P P/2	P/2 P P P P P/2
Moments	M_1	0.086 x PL	0.167 x PL	0.250 x PL	0.333 x PL	0.429 x PL
	M_2	0.096 x PL	0.188 x PL	0.278 x PL	0.375 x PL	0.480 x PL
	M_3	0.063 x PL	0.125 x PL	0.167 x PL	0.250 x PL	0.300 x PL
	M_4	0.039 x PL	0.083 x PL	0.083 x PL	0.167 x PL	0.171 x PL
	M_5	0.051 x PL	0.104 x PL	0.139 x PL	0.208 x PL	0.249 x PL
Reactions	A	0.414 x P	0.833 x P	1.250 x P	1.667 x P	2.071 x P
	B	1.172 x P	2.333 x P	3.500 x P	4.667 x P	5.857 x P
	C	0.438 x P	0.875 x P	1.333 x P	1.750 x P	2.200 x P
	D	1.063 x P	2.125 x P	3.167 x P	4.250 x P	5.300 x P
	E	1.086 x P	2.167 x P	3.250 x P	4.333 x P	5.429 x P
	F	1.109 x P	2.208 x P	3.333 x P	4.417 x P	5.557 x P
	G	0.977 x P	1.958 x P	2.917 x P	3.917 x P	4.871 x P
	H	1.000 x P	2.000 x P	3.000 x P	4.000 x P	5.000 x P
Cantilever Dimensions	a	0.172 x L	0.250 x L	0.200 x L	0.182 x L	0.176 x L
	b	0.125 x L	0.200 x L	0.143 x L	0.143 x L	0.130 x L
	c	0.220 x L	0.333 x L	0.250 x L	0.222 x L	0.229 x L
	d	0.204 x L	0.308 x L	0.231 x L	0.211 x L	0.203 x L
	e	0.157 x L	0.273 x L	0.182 x L	0.176 x L	0.160 x L
	f	0.147 x L	0.250 x L	0.167 x L	0.167 x L	0.150 x L

CONTINUOUS BEAMS

MOMENT AND SHEAR CO-EFFICIENTS

EQUAL SPANS, EQUALLY LOADED

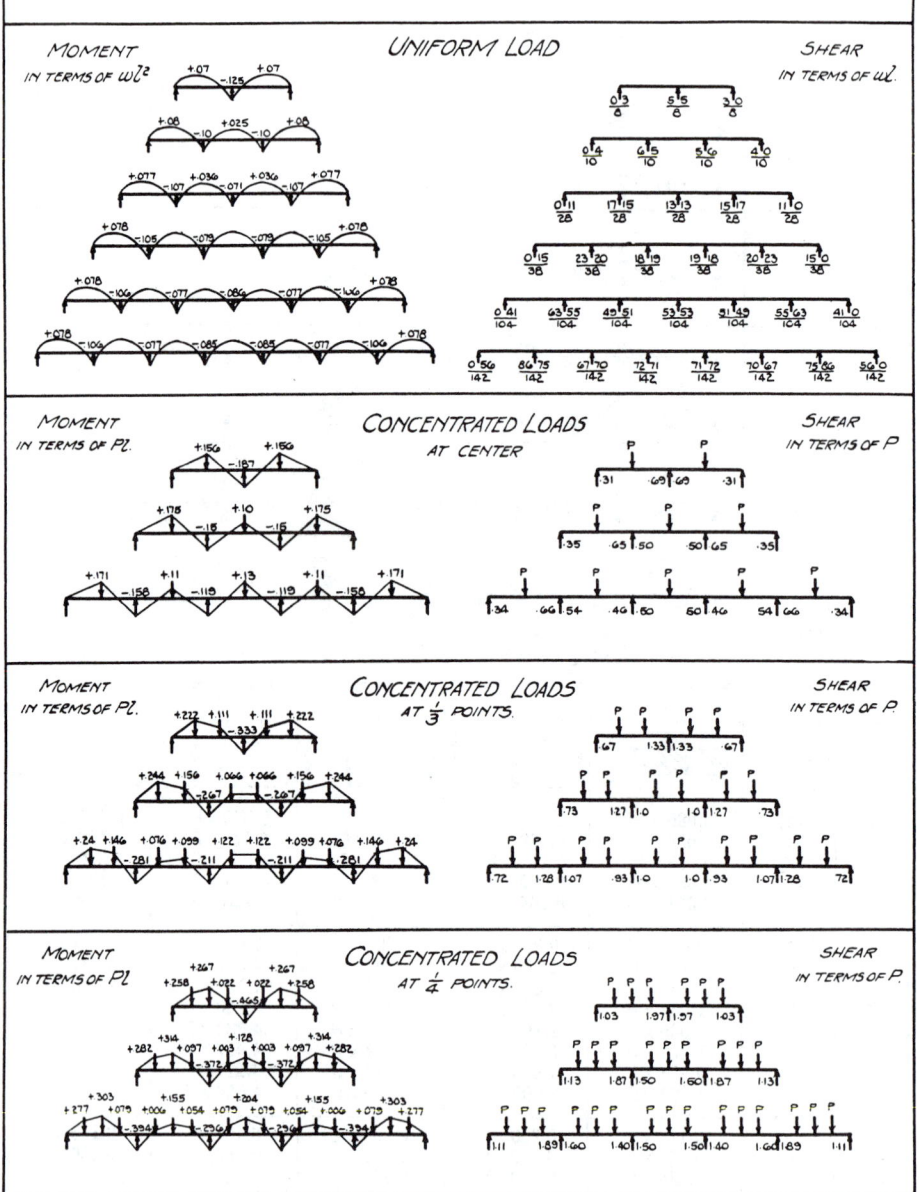

CAMBER AND DEFLECTION
Coefficients
For beams and girders with constant cross section

Given the simple span length, the depth of a beam or girder and the design unit bending stress, the center deflection in inches may be found by multiplying the span length in feet by the tabulated coefficients given in the following table.

For the unit stress values not tabulated, the deflection can be found by the equation $0.00103448 \, (L^2 f_b/d)$ where L is the span in ft, f_b is the fiber stress in kips per sq. in. and d is the depth in inches.

The maximum fiber stresses listed in this table correspond to the allowable unit stresses as provided in Sects. F1.1 and F1.3 of the AISC ASD Specification for steels having yield points ranging between 36 ksi and 65 ksi when $F_b = 0.66F_y$; and between 36 ksi and 100 ksi when $F_b = 0.60F_y$.

The table values, as given, assume a uniformly distributed load. For a single load at center span, multiply these factors by 0.80; for two equal concentrated loads at third points, multiply by 1.02. Likewise, for three equal concentrated loads at quarter points multiply by 0.95.

Ratio of Depth to Span	Maximum Fiber Stress in Kips Per Sq. In.												
	10.0	22.0	24.0	25.2	27.0	28.0	30.0	33.0	36.0	39.0	42.9	54.0	60.0
1/8	.0069	.0152	.0166	.0174	.0186	.0193	.0207	.0228	.0248	.0269	.0296	.0372	.0414
1/9	.0078	.0171	.0186	.0196	.0209	.0217	.0233	.0256	.0279	.0303	.0333	.0419	.0466
1/10	.0086	.0190	.0207	.0217	.0233	.0241	.0259	.0284	.0310	.0336	.0370	.0466	.0517
1/11	.0095	.0209	.0228	.0239	.0256	.0266	.0284	.0313	.0341	.0370	.0407	.0512	.0569
1/12	.0103	.0228	.0248	.0261	.0279	.0290	.0310	.0341	.0372	.0403	.0444	.0559	.0621
1/13	.0112	.0247	.0269	.0282	.0303	.0314	.0336	.0370	.0403	.0437	.0481	.0605	.0672
1/14	.0121	.0266	.0290	.0304	.0326	.0338	.0362	.0398	.0434	.0471	.0518	.0652	.0724
1/15	.0129	.0284	.0310	.0326	.0349	.0362	.0388	.0427	.0466	.0504	.0555	.0698	.0776
1/16	.0138	.0303	.0331	.0348	.0372	.0386	.0414	.0455	.0497	.0538	.0592	.0745	.0828
1/17	.0147	.0322	.0352	.0369	.0396	.0410	.0440	.0484	.0528	.0572	.0629	.0791	.0879
1/18	.0155	.0341	.0372	.0391	.0419	.0434	.0466	.0512	.0559	.0605	.0666	.0838	.0931
1/19	.0164	.0360	.0393	.0413	.0442	.0459	.0491	.0541	.0590	.0639	.0703	.0885	.0983
1/20	.0172	.0379	.0414	.0434	.0466	.0483	.0517	.0569	.0621	.0672	.0740	.0931	.1035
1/21	.0181	.0398	.0434	.0456	.0489	.0507	.0543	.0597	.0652	.0706	.0777	.0978	.1086
1/22	.0190	.0417	.0455	.0478	.0512	.0531	.0569	.0626	.0683	.0740	.0814	.1024	.1138
1/23	.0198	.0436	.0476	.0500	.0535	.0555	.0595	.0654	.0714	.0773	.0851	.1071	.1190
1/24	.0207	.0455	.0497	.0521	.0559	.0579	.0621	.0683	.0745	.0807	.0888	.1117	.1241
1/25	.0216	.0474	.0517	.0543	.0582	.0603	.0647	.0711	.0776	.0841	.0925	.1164	.1293
1/26	.0224	.0493	.0538	.0565	.0605	.0628	.0672	.0740	.0807	.0874	.0962	.1210	.1345
1/27	.0233	.0512	.0559	.0587	.0628	.0652	.0698	.0768	.0838	.0908	.0999	.1257	.1397
1/28	.0241	.0531	.0579	.0608	.0652	.0676	.0724	.0797	.0869	.0941	.1036	.1303	.1448
1/29	.0250	.0550	.0600	.0630	.0675	.0700	.0750	.0825	.0900	.0975	.1073	.1350	.1500
1/30	.0259	.0569	.0621	.0652	.0698	.0724	.0776	.0853	.0931	.1009	.1110	.1397	.1552

PART 3
Column Design

Columns
GENERAL NOTES

COLUMN LOAD TABLES

Column load tables are presented for W, WT and S Shapes, Pipe, Structural Tubing and Double Angles. Tabular loads are computed in accordance with the AISC Specification for Structural Steel Buildings — Allowable Stress Design (ASD), Equations (E2-1) and (E2-2), for axially loaded members having effective unsupported lengths indicated at the left of each table. The effective length KL is the actual unbraced length, in feet, multiplied by the factor K, which depends on the rotational restraint at the ends of the unbraced length and the means available to resist lateral movements.

Table C-C2.1 in the Commentary of the AISC ASD Specification is a guide in selecting the K-factor. Interpolation between the idealized cases is a matter of engineering judgment.

Once sections have been selected for the several framing members, the alignment charts in Fig. 1 (reproduced from the Structural Stability Research Council Guide* and including Fig. C-C2.2 of the AISC ASD Commentary) affords a means of obtaining more precise values for K, if desired. For column behavior in the inelastic range, the values of G as defined in Fig. 1 may be reduced by the values given in Table A, as illustrated in Ex. 3.

Load tables are provided for columns of 36-ksi yield stress steel for all shape categories. In addition, tables for W, WT and S Shapes and for Double Angles are provided for 50-ksi yield stress steel, and tables for Structural Tubing are provided for 46-ksi yield stress steel. All loads are tabulated in kips. Load values are omitted when Kl/r exceeds 200.

The Double Angle and WT tables show loads for effective lengths about both axes. In all other tables allowable loads are given for effective lengths with respect to the minor axis. When the minor axis is braced at closer intervals than the major axis, the capacity of a column must be investigated with reference to both major $(X-X)$ and minor $(Y-Y)$ axes. The ratio r_x/r_y included in these tables provides a convenient method for investigating the strength of a column with respect to its major axis.

To obtain an effective length with respect to the minor axis equivalent in load carrying capacity to the actual effective length about the major axis, divide the major axis effective length by the r_x/r_y ratio. Compare this length with the actual effective length about the minor axis. The longer of the two lengths will control the design and the allowable load may be taken from the table opposite the longer of the two effective lengths with respect to the minor axis.

Properties useful to the designer are listed at the bottom of the column load tables. These properties, and footnotes concerning compact sections, are particularly helpful in the design of members under combined axial and bending stress as discussed below and illustrated in the design examples.

Additional notes relating specifically to the W, WT and S Shape tables, the Steel Pipe and Structural Tubing tables and the Double Angle tables precede each of these groups of tables.

---

*Johnston, Bruce G. (ed.) *Guide to Stability Design Criteria for Metal Structures* Third Edition, John Wiley and Sons, 1976, p. 420.

EXAMPLE 1

Given:

Design the lightest W shape of F_y = 36 ksi steel, to support a concentric load of 670 kips. The effective length with respect to its minor axis is 16 ft. The effective length with respect to its major axis is 31 ft.

Solution:

Enter the appropriate column load table for W shapes at effective length of KL = 16 ft. Since deeper columns are generally more efficient, begin with the W 14 table and work downward, weightwise.

Select W14×132, good for 708 kips > 670 kips.

r_x/r_y = 1.67

Equivalent effective length for *X-X* axis:

31/1.67 = 18.6 ft

Since 18.6 ft > 16 ft, *X-X* axis controls.

Re-enter table for effective length of 18.6 ft to satisfy axial load of 670 kips, select W14×132 with r_x/r_y = 1.67.

By interpolation, the column is good for 679 kips.

Use: W14×132 column

EXAMPLE 2

Given:

Design an 11 ft long W12 interior bay column to support a concentrated concentric axial and roof load of 540 kips. The column is rigidly framed at the top by 30 ft long W30×116 girders connected to each flange. The column is braced normal to its web at top and base so that sidesway is inhibited in this plane. Use F_y = 36 ksi steel.

Solution:

a. Check *Y-Y* axis:

Assume column is pin-connected at top and bottom with sidesway inhibited.

From Table C-C2.1 in the Commentary for condition (d), *K* = 1.0:
Effective length = 11 ft

Enter column load table:
W12×106 good for 593 kips > 540 kips   **o.k.**

b. Check *X-X* axis:

1. Preliminary Selection:

Assume sidesway uninhibited and pin-connected at base.

From Table C-C2.1 for condition (f)*:
K = 2.0.

Approximate effective length relative to *X-X* axis.
2.0 × 11 = 22.0 ft

---

*Table C-C2.1 gives *K* values, in most cases on the conservative side; therefore, final selection may be made by use of Fig. 1 when determining effective length.

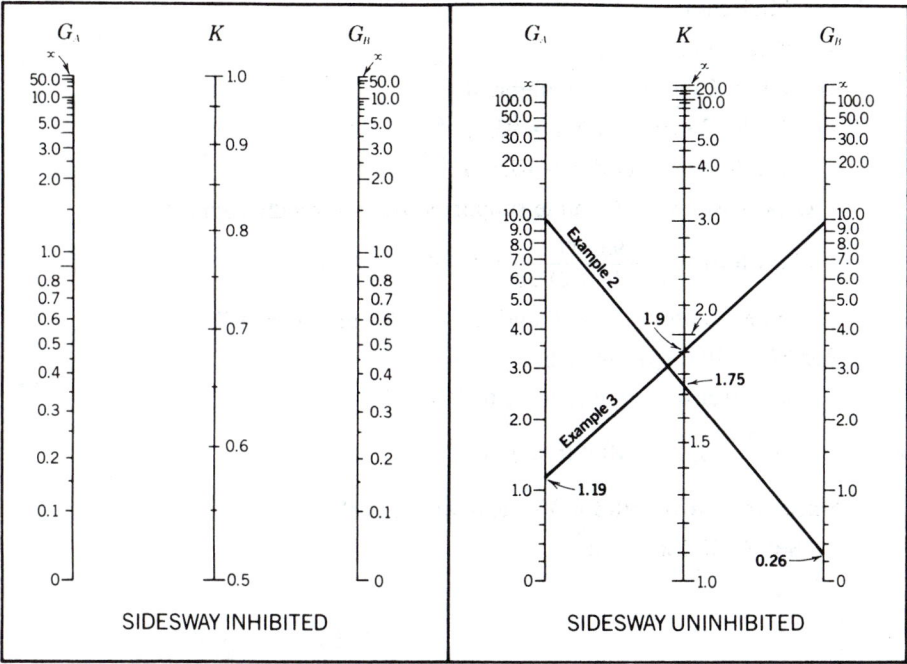

Figure 1. The subscripts A and B refer to the joints at the two ends of the column section being considered. G is defined as

$$G = \frac{\Sigma (I_c/L_c)}{\Sigma (I_g/L_g)}$$

in which Σ indicates a summation of all members rigidly connected to that joint and lying in the plane in which buckling of the column is being considered. I_c is the moment of inertia and L_c the unsupported length of a column section, and I_g is the moment of inertia and L_g the unsupported length of a girder or other restraining member. I_c and I_g are taken about axes perpendicular to the plane of buckling being considered.

For column ends supported by but not rigidly connected to a footing or foundation, G is theoretically infinity, but, unless actually designed as a true friction free pin, may be taken as "10" for practical designs. If the column end is rigidly attached to a properly designed footing, G may be taken as 1.0. Smaller values may be used if justified by analysis.

From properties section in Tables, for **W12** column:

$r_x/r_y \approx 1.76$.

Corresponding effective length relative to the *Y-Y* axis:

$\frac{22.0}{1.76} = 12.5 \text{ ft} > 11.0 \text{ ft}$

∴ Effective length for *X-X* axis is critical.

Enter column load table with an effective length of 12.5 ft:

W12×106 column, by interpolation, good for 577 kips > 540 kips **o.k.**

2. Final Selection:

Try W12×106.

Using Fig. 1 (sidesway uninhibited):

I_x for W12×106 column = 933 in.⁴

I_x for W30×116 girder = 4930 in.⁴

G (at base) = 10 (assume supported but not rigidly connected).

$$G \text{ (at top)} = \frac{933/11}{(4930 \times 2)/30} = 0.258, \text{ say } 0.26.$$

Connect points G_A = 10 and G_B = 0.26, read K = 1.75.

For W12×106, r_x/r_y = 1.76.

Actual effective length relative to Y-Y axis:

$$\frac{1.75}{1.76} \times 11.0 = 10.9 \text{ ft} < 11.0 \text{ ft}$$

Since effective length for Y-Y axis was critical:

Use: W12×106 column

EXAMPLE 3

Given:

Using the alignment chart, Fig. 1 (sidesway uninhib-ited) and Table A, design columns for the bent shown (r.), by the inelastic K-factor procedure. Let F_y = 36 ksi.

Assume continuous support in the transverse direction.

Solution:

The alignment charts in Fig. 1 are applicable to elas-tic columns. By multiplying G-values times the stiff-ness reduction factor E_t/E, the charts may be used for inelastic columns. Since $E_t/E \approx F_a/F'_e$, the rela-tionship may be written as $G_{inelastic} = (F_a/F'_e) G_{elastic}$. By utilizing the actual stress in the reduction factor, instead of the allowable stress (f_a/F'_e instead of F_a/F'_e), a direct solution is possible, using the follow-ing steps:

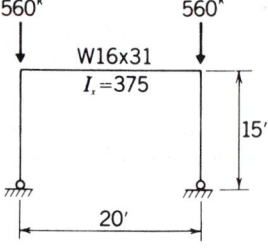

1. For known value of axial load P, select a trial column size:

Assume W12×106:

A = 31.2 in.²; I_x = 933 in.⁴; r_x = 5.47 in.

2. Calculate $f_a = P/A$:

f_a = 560/31.2 = 17.95 ksi

3. From Table A, determine the reduction factor f_a/F_e'. For values of f_a smaller than shown in Table A, the column is elastic, and the reduction factor is 1.0:

$f_a/F_e' = 0.357$, by interpolation

4. Determine $G_{elastic}$ as in Ex. 2:

$G_{elastic}$ (bottom) $= 10.0$

$G_{elastic}$ (top) $= \dfrac{933/15}{375/20} = 3.32$

5. Calculate $G_{inelastic} = (f_a/F_e')\, G_{elastic}$:

$G_{inelastic}$ (top) $= 0.357\,(3.32) = 1.19$

6. Determine K from Fig. 1 using $G_{inelastic}$:

For $G_{top} = 1.19$ and $G_{bot} = 10$:

Read from Fig. 1, $K = 1.9$

7. Calculate Kl/r:

$Kl/r = \dfrac{1.9(15)(12)}{5.47} = 62.5$

8. If $F_a \geq f_a$, column is satisfactory. Check:

From Table C-36, $F_a = 17.19$ ksi < 17.95 ksi   **n.g.**

Try a stronger column.

1. Try W12×120:

$A = 35.3$ in.^{2}, $I_x = 1070$ in.^{4}, $r_x = 5.51$ in.

2. $f_a = 560/35.3 = 15.86$ ksi

3. From Table A: $f_a/F_e' = 0.603$

4. $G_{elastic}$ (top) $= \dfrac{1070/15}{375/20} = 3.80$

5. $G_{inelastic}$ (top) $= 0.603(3.80) = 2.29$

6. $K = 2.13$

7. $Kl/r = \dfrac{2.13(15)(12)}{5.51} = 69.6$

8. $F_a = 16.47$ ksi > 15.86 ksi   **o.k.**

Use: W12×120

Table A
Stiffness Reduction Factors f_a/F_e'

| F_y = 36 ksi |
| F_y = 50 ksi |

f_a	F_y 36 ksi	F_y 50 ksi
28.0	—	0.097
27.9	—	0.104
27.8	—	0.112
27.7	—	0.120
27.6	—	0.127
27.5	—	0.136
27.4	—	0.144
27.3	—	0.152
27.2	—	0.160
27.1	—	0.168
27.0	—	0.177
26.9	—	0.184
26.8	—	0.193
26.7	—	0.202
26.6	—	0.210
26.5	—	0.218
26.4	—	0.227
26.3	—	0.236
26.2	—	0.245
26.1	—	0.253
26.0	—	0.262
25.9	—	0.271
25.8	—	0.280
25.7	—	0.288
25.6	—	0.297
25.5	—	0.306
25.4	—	0.315
25.3	—	0.324
25.2	—	0.333
25.1	—	0.342
25.0	—	0.350
24.9	—	0.359
24.8	—	0.368
24.7	—	0.377
24.6	—	0.386
24.5	—	0.394
24.4	—	0.403
24.3	—	0.412
24.2	—	0.421
24.1	—	0.430
24.0	—	0.439
23.9	—	0.447
23.8	—	0.456
23.7	—	0.465
23.6	—	0.473
23.5	—	0.482
23.4	—	0.490
23.3	—	0.499
23.2	—	0.507
23.1	—	0.516
23.0	—	0.524
22.9	—	0.533
22.8	—	0.541
22.7	—	0.549
22.6	—	0.557
22.5	—	0.565
22.4	—	0.574
22.3	—	0.582
22.2	—	0.590
22.1	—	0.598
22.0	—	0.606

f_a	F_y 36 ksi	F_y 50 ksi
21.9	—	0.614
21.8	—	0.622
21.7	—	0.630
21.6	—	0.637
21.5	—	0.645
21.4	—	0.653
21.3	—	0.660
21.2	—	0.668
21.1	—	0.675
21.0	—	0.683
20.9	—	0.689
20.8	—	0.697
20.7	—	0.704
20.6	—	0.712
20.5	0.064	0.718
20.4	0.074	0.725
20.3	0.083	0.732
20.2	0.093	0.739
20.1	0.102	0.746
20.0	0.114	0.753
19.9	0.125	0.760
19.8	0.136	0.766
19.7	0.147	0.772
19.6	0.158	0.778
19.5	0.169	0.785
19.4	0.181	0.792
19.3	0.193	0.798
19.2	0.204	0.804
19.1	0.216	0.810
19.0	0.228	0.816
18.9	0.241	0.822
18.8	0.252	0.827
18.7	0.264	0.833
18.6	0.277	0.839
18.5	0.288	0.844
18.4	0.301	0.849
18.3	0.314	0.855
18.2	0.326	0.860
18.1	0.338	0.865
18.0	0.350	0.871
17.9	0.363	0.875
17.8	0.375	0.880
17.7	0.387	0.885
17.6	0.400	0.890
17.5	0.411	0.894
17.4	0.424	0.899
17.3	0.436	0.903
17.2	0.448	0.908
17.1	0.460	0.912
17.0	0.472	0.917
16.9	0.484	0.920
16.8	0.496	0.924
16.7	0.508	0.928
16.6	0.519	0.932
16.5	0.531	0.935
16.4	0.543	0.939
16.3	0.554	0.942
16.2	0.565	0.946
16.1	0.577	0.950
16.0	0.588	0.952

f_a	F_y 36 ksi	F_y 50 ksi
15.9	0.599	0.956
15.8	0.610	0.959
15.7	0.621	0.962
15.6	0.632	0.964
15.5	0.643	0.967
15.4	0.653	0.970
15.3	0.664	0.972
15.2	0.675	0.974
15.1	0.684	0.977
15.0	0.695	0.979
14.9	0.704	0.981
14.8	0.715	0.983
14.7	0.724	0.985
14.6	0.734	0.987
14.5	0.743	0.988
14.4	0.753	0.990
14.3	0.762	0.991
14.2	0.770	0.993
14.1	0.780	0.994
14.0	0.789	0.995
13.9	0.797	0.996
13.8	0.805	0.997
13.7	0.814	0.998
13.6	0.822	0.998
13.5	0.830	0.999
13.4	0.838	0.999
13.3	0.845	1.000
13.2	0.853	—
13.1	0.860	—
13.0	0.868	—
12.9	0.874	—
12.8	0.881	—
12.7	0.888	—
12.6	0.895	—
12.5	0.901	—
12.4	0.907	—
12.3	0.913	—
12.2	0.918	—
12.1	0.924	—
12.0	0.929	—
11.9	0.934	—
11.8	0.939	—
11.7	0.944	—
11.6	0.949	—
11.5	0.953	—
11.4	0.958	—
11.3	0.962	—
11.2	0.966	—
11.1	0.970	—
11.0	0.973	—
10.9	0.976	—
10.8	0.979	—
10.7	0.982	—
10.6	0.984	—
10.5	0.987	—
10.4	0.989	—
10.3	0.991	—
10.2	0.993	—
10.1	0.995	—
10.0	0.996	—
9.9	0.997	—
9.8	0.998	—
9.7	0.999	—
9.6	1.000	—

COMBINED AXIAL AND BENDING LOADING (INTERACTION)

Loads given in the column tables are for concentrically loaded columns. For columns subjected to both axial and bending stress, Sect. H1 of the AISC ASD Specification requires that the following equations be satisfied:

$$\frac{f_a}{F_a} + \frac{C_{mx} f_{bx}}{\left(1 - \frac{f_a}{F'_{ex}}\right) F_{bx}} + \frac{C_{my} f_{by}}{\left(1 - \frac{f_a}{F'_{ey}}\right) F_{by}} \le 1.0 \quad \text{Equation (H1-1)}$$

$$\frac{f_a}{0.60 \, F_y} + \frac{f_{bx}}{F_{bx}} + \frac{f_{by}}{F_{by}} \le 1.0 \qquad \text{Equation (H1-2)}$$

Also, when $f_a/F_a \le 0.15$, Equation (H1-3) may be used in lieu of Equations (H1-1) and (H1-2).

$$\frac{f_a}{F_a} + \frac{f_{bx}}{F_{bx}} + \frac{f_{by}}{F_{by}} \le 1.0 \qquad \text{Equation (H1-3)}$$

For convenience, these equations may be written in the following modified forms:

$P + P'_x + P'_y$ = required tabular load

$$= P + \left[B_x M_x C_{mx} \left(\frac{F_a}{F_{bx}} \right) \left(\frac{a_x}{a_x - P(KL)^2} \right) \right] +$$

$$\left[B_y M_y C_{my} \left(\frac{F_a}{F_{by}} \right) \left(\frac{a_y}{a_y - P(KL)^2} \right) \right]$$

Modified Equation (H1-1)

$P + P'_x + P'_y$ = required tabular load

$$= P \left(\frac{F_a}{0.6 \, F_y} \right) + \left[B_x M_x \left(\frac{F_a}{F_{bx}} \right) \right] + \left[B_y M_y \left(\frac{F_a}{F_{by}} \right) \right]$$

Modified Equation (H1-2)

When $f_a/F_a \le 0.15$

$P + P'_x + P'_y$ = required tabular load

$$= P + \left[B_x M_x \left(\frac{F_a}{F_{bx}} \right) \right] + \left[B_y M_y \left(\frac{F_a}{F_{by}} \right) \right]$$

Modified Equation (H1-3)

In Modified Equation (H1-1), for the term $(KL)^2$, K is the effective length factor and L is the actual unbraced length in the plane of bending.

B_x and B_y are, respectively, equal to the area of the column divided by its appropriate section modulus.

Values for the components a_x and a_y, equal to $(0.149 \times 10^6) \, Ar_x^2$ and $(0.149 \times 10^6) \, Ar_y^2$, respectively, are listed at the bottom of the load tables in the form of $a_x/10^6$ and $a_y/10^6$.

Equations for the allowable bending stress F_b are given in Sects. F1, F2 and F3 of the AISC ASD Specification. Sections which are not compact are noted in the load tables by the symbol †, and the lengths L_c and L_u are listed.

AMERICAN INSTITUTE OF STEEL CONSTRUCTION

Values of $F'_{ex} (K_xL_x)^2/10^2$ and $F'_{ey} (K_yL_y)^2/10^2$ are also listed in the column load tables for use in Equation (H1-1), where K_xL_x and K_yL_y are in feet.

The design of a beam-column is a trial and error process in which a trial section is checked for compliance with Equations (H1-1), (H1-2) and (H1-3). A fast method for selecting an economical trial W- or S-shape section, using an equivalent axial load, is illustrated in the example problem, using Table B and the U values listed in the column properties at the bottom of the column load tables.

$F_y = 36$ ksi	TABLE B											
$F_y = 50$ ksi	Values of m*											

F_y	36 ksi						50 ksi					
KL (ft)	10	12	14	16	18	20	22 & over	10	12	14	16	18	20	22 & over

1st Approximation												
All shapes	2.4	2.3	2.2	2.2	2.1	2.0	1.9	2.4	2.3	2.2	2.0	1.9	1.8	1.7

Subsequent Approximations												
W, S 4	3.6	2.6	1.9	1.6	—	—	—	2.7	1.9	1.6	1.6	—	—	—
W, S 5	3.9	3.2	2.4	1.9	1.5	1.4	—	3.3	2.4	1.8	1.6	1.4	1.4	—
W, S 6	3.2	2.7	2.3	2.0	1.9	1.6	1.5	3.0	2.5	2.2	1.9	1.8	1.6	1.5
W 8	3.0	2.9	2.8	2.6	2.3	2.0	2.0	3.0	2.8	2.5	2.2	1.9	1.6	1.6
W 10	2.6	2.5	2.5	2.4	2.3	2.1	2.0	2.5	2.5	2.4	2.3	2.1	1.9	1.7
W 12	2.1	2.1	2.0	2.0	2.0	2.0	2.0	2.0	2.0	2.0	1.9	1.9	1.8	1.7
W 14	1.8	1.7	1.7	1.7	1.7	1.7	1.7	1.8	1.7	1.7	1.7	1.7	1.7	1.7

*Values of m are for $C_m = 0.85$. When C_m is other than 0.85, multiply the tabular value of m by $C_m/0.85$.

The procedure is as follows:

1. With the known value of KL (effective length), select a first approximate value of m from Table B. Let U equal 3.

2. Solve for $P_{eff} = P_0 + M_x m + M_y mU$

   where

   P_0 = actual axial load, kips
   M_x = bending moment about the strong axis, kip-ft
   M_y = bending moment about the weak axis, kip-ft
   m = factor taken from Table B
   U = factor taken from column load table

3. From appropriate column load table, select tentative section to support P_{eff}.

4. Based on the section selected in Step 3, select a "subsequent approximate" value of m from Table B and a U value from the column load table.

5. With the values selected in Step 4, solve for P_{eff}.

6. Repeat Steps 3 and 4 until the values of m and U stabilize.

EXAMPLE 4

Given: $P_0 = 200$ kips
$M_x = 120$ kip-ft
$M_y = 40$ kip-ft
$KL = 14$ ft
$F_y = 36$ ksi
$C_m = 0.85$

Solution:

1. For $KL = 14$ ft, from Table B select a "first trial" value of $m = 2.2$. Let $U = 3.0$.

2. $P_{eff} = P_0 + M_x m + M_y m\, U = 200 + 120(2.2) + 40(2.2)(3.0) = 728$ kips.

3. From column load tables, select W14×132 (730 kips).

4. Select a "second trial" value of $m = 1.7$ from Table B and $U = 2.47$ from column load table.

5. $P_{eff} = 200 + 120(1.7) + 40(1.7)(2.47) = 572$ kips.

6. Repeat steps 3, 4 and 5.

3. Select W14×109 (601 kips).

4. $m = 1.7$, $U = 2.49$.

5. $P_{eff} = 200 + 120(1.7) + 40(1.7)(2.49) = 573$ kips.

   Use: W14×109

EXAMPLE 5

Given:

Check answer in Example 4 by modified equations.

Solution:

For W14×109, from column load table:

$L_c = 15.4$ ft > 14.0 ft

∴ Section is compact.

Use $F_{bx} = 24$ ksi, $F_{by} = 0.75(36) = 27$ ksi

$A = 32.0$ in.^{2}, $P_{all} = 601$ kips, $r_y = 3.73$ in.

$B_x = 0.185$, $B_y = 0.523$

$a_x = 184.5 \times 10^6$, $a_y = 66.3 \times 10^6$

$KL = 14 \times 12 = 168$ in.

$Kl/r_y = 168/3.73 = 45.04$

$F_a = 18.78$ ksi (Table C-36 on page 3-16)

$P(KL)^2 = 200(168)^2 = 2.00(1.68)^2 \times 10^6$

Modified Equation (H1-3) is not used, since $(200/601) > 0.15$.

Req'd tabular load by Modified Equation (H1-1):

$$P + P_x' + P_y' = 200 + \left[0.185(120 \times 12)(0.85) \left(\frac{18.78}{24} \right) \left(\frac{184.5}{184.5 - 2.00(1.68)^2} \right) \right]$$

$$+ \left[0.523(40 \times 12)(0.85) \left(\frac{18.78}{27} \right) \left(\frac{66.3}{66.3 - 2.00(1.68)^2} \right) \right]$$

$$= 200 + 182.78 + 162.23 = 545 \text{ kips}$$

Required tabular load by Modified Equation (H1-2):

$$P + P_x' + P_y' = 200 \left(\frac{18.78}{22} \right) + \left[0.185 (120 \times 12) \left(\frac{18.78}{24} \right) \right]$$

$$+ \left[0.523 (40 \times 12) \left(\frac{18.78}{27} \right) \right]$$

$$= 170.73 + 208.46 + 174.61 = 554 \text{ kips}$$

Modified Equation (H1-2) requires an axial load of 554 kips. W14×109 has an allowable axial load of 601 kips and is satisfactory.

COLUMN WEB STIFFENERS

Values of P_{wo}, P_{wi}, P_{wb}, and P_{fb}, listed in the Properties Section of the column load tables for W- and S-shapes, are useful in determining if a column web requires stiffeners because of forces transmitted into it from the flanges or connecting flange plates of a rigid beam connection to the column flange. When the applied factored beam flange force P_{bf} is equal to or less than the following resisting forces developed within the column section, column web stiffeners are **not** required.

$$P_{bf} \le P_{wb} = \frac{4100 \, t_{wc}^3 \sqrt{F_{yc}}}{d_c} \tag{K1-8}$$

$$P_{bf} \le P_{fb} = \frac{F_{yc} \, t_f^2}{0.16} \tag{K1-1}$$

$$P_{bf} \le P_{wi} t_b + P_{wo} = F_{yc} t_{wc} (t_b + 5k) \tag{K1-9}$$

where

P_{wb} = maximum column web resisting force at beam compression flange, kips

P_{fb} = maximum column web resisting force at beam tension flange, kips

P_{wi} = $F_{yc} t_{wc}$, kips/in.

P_{wo} = $5 F_{yc} t_{wc} k$, kips

F_{yc} = yield strength of column web, ksi

d_c = depth of column web clear of fillets, in.

k = distance from outer face of column flange to web toe of column fillet, in.

t_{wc} = thickness of column web, in.

t_b = thickness of beam flange or connection plate delivering concentrated force, in.

t_f = thickness of column flange, in.

If the factored force P_{bf} transmitted into the column web exceeds any one of the above three resisting forces, stiffeners are required on the column web. Stiffeners must comply with the provisions of AISC ASD Specification Sect. K1.8.

EXAMPLE 6

Given:

The flanges of a W24×84 beam are welded to the flange of a W14×211 column transmitting a moment of 540 kip-ft due to live and dead loads. Determine if stiffeners are required for the columns and, if so, make an appropriate design. Beam and column are F_y = 36 ksi.

Solution:

For W24×84:

d = 24.10 in., t_b = 0.770 in.

$$P_{bf} = 5/3 \left(\frac{540 \times 12}{24.10} \right) = 448 \text{ kips} \qquad \text{Sect. K1.2}$$

For W14×211:

From column load table:

P_{wb} = 2058 kips > 448 kips **o.k.**   AISC ASD Spec. Equation (K1-8)

P_{fb} = 548 kips > 448 kips **o.k.**   AISC ASD Spec. Equation (K1-1)

$P_{wo} + t_b P_{wi}$ = 397 + 0.770(35)   AISC ASD Spec. Equation (K1-9)

= 424 kips < 448 kips **n.g.**

∴ Web stiffeners required

Design web stiffeners according to AISC ASD Specification Sect. K1.8:

Beam data: b_f = 9.020 in. t_b = 0.770 in.

Column data: t = 0.980 in., t_f = 1.560 in., d = 15.72 in.

Req'd stiffener area = $\dfrac{448 - 424}{36}$ = 0.67 in.^{2}

Min. width = $\frac{1}{3} b_f - \dfrac{t}{2}$ = $\frac{1}{3}$ (9.020) − (0.980/2) = 2.517 in.   Sect. K1.8

Min. thickness = $t_b/2$ = 0.770/2 = 0.385 in.   Sect. K1.8

Req'd thickness = 0.67/2.517 = 0.266 in.   Use ½ in.

Req'd width = 0.67/0.5 = 1.34 in.   Use 4 in. for practical detailing considerations.

Min. length = $(d/2) - t_f$

= (15.72/2) − 1.56 = 6.30 in.   Use 7 in.   Sect. K1.8

Use: Two PL4×½×0 ft-7in.

Min. weld size = 5⁄16 in., based on column web thickness of 0.980 in.

(Table J2.4, AISC ASD Specification Sect. J2.2b). Use E70 electrodes.

Req'd length (both PL) $= \dfrac{P}{0.3(70)(.707)(5/16)(1.67)}$

$$= \dfrac{448 - 424}{7.75} = 3.10 \text{ in.}$$

PL 4 x ½ x 0'-7

ALLOWABLE STRESSES FOR COMPRESSION MEMBERS

The allowable stresses F_a in the tables that follow are tabulated for Kl/r from 0 to 200 for $F_y = 36$ ksi and $F_y = 50$ ksi. They are calculated from AISC ASD Specification Equations (E2-1) and (E2-2).

Table C-36
Allowable Stress
For Compression Members of 36-ksi Specified Yield Stress Steel[a]

$F_y = 36$ ksi

$\frac{Kl}{r}$	F_a (ksi)	$\frac{Kl}{r}$	F_a (ksi)	$\frac{Kl}{r}$	F_a (ksi)	$\frac{Kl}{r}$	F_a (ksi)	$\frac{Kl}{r}$	F_a (ksi)
1	21.56	41	19.11	81	15.24	121	10.14	161	5.76
2	21.52	42	19.03	82	15.13	122	9.99	162	5.69
3	21.48	43	18.95	83	15.02	123	9.85	163	5.62
4	21.44	44	18.86	84	14.90	124	9.70	164	5.55
5	21.39	45	18.78	85	14.79	125	9.55	165	5.49
6	21.35	46	18.70	86	14.67	126	9.41	166	5.42
7	21.30	47	18.61	87	14.56	127	9.26	167	5.35
8	21.25	48	18.53	88	14.44	128	9.11	168	5.29
9	21.21	49	18.44	89	14.32	129	8.97	169	5.23
10	21.16	50	18.35	90	14.20	130	8.84	170	5.17
11	21.10	51	18.26	91	14.09	131	8.70	171	5.11
12	21.05	52	18.17	92	13.97	132	8.57	172	5.05
13	21.00	53	18.08	93	13.84	133	8.44	173	4.99
14	20.95	54	17.99	94	13.72	134	8.32	174	4.93
15	20.89	55	17.90	95	13.60	135	8.19	175	4.88
16	20.83	56	17.81	96	13.48	136	8.07	176	4.82
17	20.78	57	17.71	97	13.35	137	7.96	177	4.77
18	20.72	58	17.62	98	13.23	138	7.84	178	4.71
19	20.66	59	17.53	99	13.10	139	7.73	179	4.66
20	20.60	60	17.43	100	12.98	140	7.62	180	4.61
21	20.54	61	17.33	101	12.85	141	7.51	181	4.56
22	20.48	62	17.24	102	12.72	142	7.41	182	4.51
23	20.41	63	17.14	103	12.59	143	7.30	183	4.46
24	20.35	64	17.04	104	12.47	144	7.20	184	4.41
25	20.28	65	16.94	105	12.33	145	7.10	185	4.36
26	20.22	66	16.84	106	12.20	146	7.01	186	4.32
27	20.15	67	16.74	107	12.07	147	6.91	187	4.27
28	20.08	68	16.64	108	11.94	148	6.82	188	4.23
29	20.01	69	16.53	109	11.81	149	6.73	189	4.18
30	19.94	70	16.43	110	11.67	150	6.64	190	4.14
31	19.87	71	16.33	111	11.54	151	6.55	191	4.09
32	19.80	72	16.22	112	11.40	152	6.46	192	4.05
33	19.73	73	16.12	113	11.26	153	6.38	193	4.01
34	19.65	74	16.01	114	11.13	154	6.30	194	3.97
35	19.58	75	15.90	115	10.99	155	6.22	195	3.93
36	19.50	76	15.79	116	10.85	156	6.14	196	3.89
37	19.42	77	15.69	117	10.71	157	6.06	197	3.85
38	19.35	78	15.58	118	10.57	158	5.98	198	3.81
39	19.27	79	15.47	119	10.43	159	5.91	199	3.77
40	19.19	80	15.36	120	10.28	160	5.83	200	3.73

[a]When element width-to-thickness ratio exceeds noncompact section limits of Sect. B5.1, see Appendix B5.

Note: $C_c = 126.1$

AMERICAN INSTITUTE OF STEEL CONSTRUCTION

Table C-50
Allowable Stress
For Compression Members of 50-ksi Specified Yield Stress Steel[a]

$F_y = 50$ ksi

$\frac{Kl}{r}$	F_a (ksi)	$\frac{Kl}{r}$	F_a (ksi)	$\frac{Kl}{r}$	F_a (ksi)	$\frac{Kl}{r}$	F_a (ksi)	$\frac{Kl}{r}$	F_a (ksi)
1	29.94	41	25.69	81	18.81	121	10.20	161	5.76
2	29.87	42	25.55	82	18.61	122	10.03	162	5.69
3	29.80	43	25.40	83	18.41	123	9.87	163	5.62
4	29.73	44	25.26	84	18.20	124	9.71	164	5.55
5	29.66	45	25.11	85	17.99	125	9.56	165	5.49
6	29.58	46	24.96	86	17.79	126	9.41	166	5.42
7	29.50	47	24.81	87	17.58	127	9.26	167	5.35
8	29.42	48	24.66	88	17.37	128	9.11	168	5.29
9	29.34	49	24.51	89	17.15	129	8.97	169	5.23
10	29.26	50	24.35	90	16.94	130	8.84	170	5.17
11	29.17	51	24.19	91	16.72	131	8.70	171	5.11
12	29.08	52	24.04	92	16.50	132	8.57	172	5.05
13	28.99	53	23.88	93	16.29	133	8.44	173	4.99
14	28.90	54	23.72	94	16.06	134	8.32	174	4.93
15	28.80	55	23.55	95	15.84	135	8.19	175	4.88
16	28.71	56	23.39	96	15.62	136	8.07	176	4.82
17	28.61	57	23.22	97	15.39	137	7.96	177	4.77
18	28.51	58	23.06	98	15.17	138	7.84	178	4.71
19	28.40	59	22.89	99	14.94	139	7.73	179	4.66
20	28.30	60	22.72	100	14.71	140	7.62	180	4.61
21	28.19	61	22.55	101	14.47	141	7.51	181	4.56
22	28.08	62	22.37	102	14.24	142	7.41	182	4.51
23	27.97	63	22.20	103	14.00	143	7.30	183	4.46
24	27.86	64	22.02	104	13.77	144	7.20	184	4.41
25	27.75	65	21.85	105	13.53	145	7.10	185	4.36
26	27.63	66	21.67	106	13.29	146	7.01	186	4.32
27	27.52	67	21.49	107	13.04	147	6.91	187	4.27
28	27.40	68	21.31	108	12.80	148	6.82	188	4.23
29	27.28	69	21.12	109	12.57	149	6.73	189	4.18
30	27.15	70	20.94	110	12.34	150	6.64	190	4.14
31	27.03	71	20.75	111	12.12	151	6.55	191	4.09
32	26.90	72	20.56	112	11.90	152	6.46	192	4.05
33	26.77	73	20.38	113	11.69	153	6.38	193	4.01
34	26.64	74	20.10	114	11.49	154	6.30	194	3.97
35	26.51	75	19.99	115	11.29	155	6.22	195	3.93
36	26.38	76	19.80	116	11.10	156	6.14	196	3.89
37	26.25	77	19.61	117	10.91	157	6.06	197	3.85
38	26.11	78	19.41	118	10.72	158	5.98	198	3.81
39	25.97	79	19.21	119	10.55	159	5.91	199	3.77
40	25.83	80	19.01	120	10.37	160	5.83	200	3.73

[a]When element width-to-thickness ratio exceeds noncompact section limits of Sect. B5.1, see Appendix B5.
Note: $C_c = 107.0$

AMERICAN INSTITUTE OF STEEL CONSTRUCTION

ALLOWABLE CONCENTRIC LOADS ON COLUMNS
W and S Shapes

Allowable concentric loads in the tables that follow are tabulated for the effective lengths in feet KL, indicated at the left of each table. They are applicable to axially loaded members with respect to their minor axis in accordance with Section E2 of the AISC ASD Specification. Two strengths are covered, $F_y = 36$ ksi and $F_y = 50$ ksi.

The heavy horizontal lines appearing within the tables indicate $Kl/r = 200$. No values are listed beyond $Kl/r = 200$.

All sections listed satisfy the noncompact section limits of Sect. B5.1 of the AISC ASD Specification with the exception of W14×43 at $F_y = 50$ ksi. For this column, Appendix B5 of the AISC ASD Specification controls the design for effective column lengths KL from zero to approximately 2 ft. Beyond this length, Equations (E2-1) and (E2-2) apply.

For discussion of effective length, range of l/r, strength about the major axis, combined axial and bending stress, and sample problems, see "Columns, General Notes."

Properties and factors are listed at the bottom of the tables for checking strength about the strong axis, combined loading conditions and column stiffener requirements.

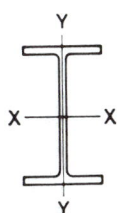

COLUMNS		**F_y = 36 ksi**
W shapes		**F_y = 50 ksi**

Allowable axial loads in kips

Designation		W14									
Wt./ft		730		665		605		550		500	
F_y		36	50	36	50	36	50	36	50	36	50
Effective length in ft KL with respect to least radius of gyration r_y	0	4644	6450	4234	5880	3845	5340	3499	4860	3175	4410
	11	4315	5887	3928	5356	3562	4855	3237	4411	2933	3995
	12	4277	5819	3892	5293	3529	4797	3206	4357	2905	3945
	13	4237	5749	3855	5228	3494	4736	3175	4301	2875	3893
	14	4196	5677	3817	5161	3459	4674	3142	4243	2845	3840
	15	4153	5602	3777	5092	3422	4609	3108	4183	2813	3784
	16	4110	5525	3737	5020	3384	4543	3073	4121	2781	3727
	17	4065	5446	3695	4946	3345	4474	3037	4057	2748	3668
	18	4019	5365	3652	4870	3306	4404	3000	3992	2714	3608
	19	3971	5282	3608	4793	3265	4331	2962	3925	2678	3546
	20	3923	5196	3563	4713	3223	4257	2923	3856	2642	3482
	22	3823	5018	3469	4547	3136	4103	2842	3713	2568	3350
	24	3718	4832	3372	4374	3045	3942	2758	3564	2490	3211
	26	3609	4638	3270	4193	2951	3774	2670	3407	2409	3066
	28	3496	4436	3164	4004	2853	3598	2579	3244	2324	2915
	30	3378	4225	3055	3807	2751	3415	2484	3074	2236	2758
	32	3256	4006	2941	3603	2645	3225	2386	2897	2145	2594
	34	3130	3779	2823	3391	2535	3028	2284	2714	2051	2423
	36	3000	3543	2702	3170	2422	2822	2179	2522	1954	2246
	38	2865	3298	2576	2941	2305	2609	2070	2324	1853	2061
	40	2726	3044	2446	2703	2184	2387	1958	2117	1748	1870
	42	2582	2780	2312	2459	2059	2166	1841	1920	1640	1696
	44	2434	2533	2173	2241	1930	1974	1721	1749	1529	1545
	46	2281	2318	2030	2050	1797	1806	1597	1601	1413	1414
	48	2123	2129	1882	1883	1659	1659	1470	1470	1298	1298
	50	1962	1962	1735	1735	1529	1529	1355	1355	1197	1197

Properties and Reaction Values											
U		2.14	2.14	2.14	2.14	2.16	2.16	2.17	2.17	2.18	2.18
P_{wo} (kips)		3074	4269	2643	3670	2248	3122	1928	2678	1651	2293
P_{wi} (kips/in.)		111	154	102	142	93	130	86	119	79	110
P_{wb} (kips)		63,270	74,560	49,560	58,410	38,210	45,030	29,480	34,740	22,970	27,070
P_{fb} (kips)		5424	7534	4597	6385	3894	5408	3283	4560	2756	3828
L_c (ft)		18.9	16.0	18.6	15.8	18.4	15.6	18.2	15.4	18.0	15.2
L_u (ft)		185.2	133.3	171.5	123.5	159.6	114.9	149.3	107.5	140.3	101.0
A (in.^{2})		215		196		178		162		147	
I_x (in.^{4})		14,300		12,400		10,800		9430		8210	
I_y (in.^{4})		4720		4170		3680		3250		2880	
r_y (in.)		4.69		4.62		4.55		4.49		4.43	
Ratio r_x/r_y		1.74		1.73		1.71		1.70		1.69	
B_x } Bending		0.168		0.170		0.171		0.174		0.175	
B_y } factors		0.408		0.415		0.421		0.429		0.434	
$a_x/10^6$		2138		1860		1614		1405		1225	
$a_y/10^6$		705		623		549		487		430	
$F'_{ex} (K_x L_x)^2/10^2$ (kips)		692		660		631		604		580	
$F'_{ey} (K_y L_y)^2/10^2$ (kips)		228		221		215		209		204	

F_y = 36 ksi										

F_y = 50 ksi

COLUMNS
W shapes
Allowable axial loads in kips

Designation		W14								
Wt./ft		455		426		398		370		342	
F_y		36	50	36	50	36	50	36	50	36	50
	0	2894	4020	2700	3750	2527	3510	2354	3270	2182	3030
	11	2671	3636	2489	3388	2328	3168	2167	2947	2006	2728
	12	2644	3590	2464	3344	2304	3126	2144	2908	1985	2692
	13	2617	3542	2438	3298	2280	3083	2121	2868	1963	2654
	14	2589	3492	2411	3251	2255	3039	2097	2826	1941	2614
	15	2560	3441	2384	3203	2229	2993	2073	2782	1918	2574
	16	2530	3388	2356	3153	2202	2946	2047	2738	1894	2532
	17	2499	3333	2326	3101	2174	2897	2021	2692	1870	2489
	18	2467	3277	2296	3048	2146	2847	1995	2644	1845	2444
	19	2435	3220	2266	2994	2117	2795	1967	2596	1819	2399
	20	2401	3161	2234	2938	2087	2743	1939	2546	1793	2352
	22	2332	3038	2169	2822	2025	2633	1881	2442	1738	2255
	24	2260	2909	2100	2700	1961	2518	1820	2333	1681	2153
	26	2184	2775	2029	2573	1893	2397	1756	2220	1621	2047
	28	2106	2635	1955	2440	1823	2272	1690	2101	1559	1935
	30	2025	2488	1878	2302	1750	2141	1621	1977	1495	1819
	32	1940	2336	1798	2158	1674	2004	1549	1848	1428	1698
	34	1853	2178	1715	2008	1596	1862	1475	1713	1358	1572
	36	1762	2013	1630	1852	1515	1714	1398	1573	1286	1441
	38	1668	1841	1541	1689	1431	1560	1319	1427	1212	1304
	40	1571	1666	1449	1526	1344	1409	1237	1288	1135	1177
	42	1471	1511	1354	1384	1254	1278	1151	1168	1053	1067
	44	1367	1377	1255	1261	1161	1164	1063	1065	972	973
	46	1260	1260	1154	1154	1065	1065	974	974	890	890
	48	1157	1157	1060	1060	978	978	895	895	817	817
	50	1066	1066	977	977	902	902	824	824	753	753

Effective length in ft KL with respect to least radius of gyration r_y

Properties										
U	2.19	2.19	2.20	2.20	2.20	2.20	2.22	2.22	2.23	2.23
P_{wo} (kips)	1405	1952	1245	1729	1115	1549	987	1371	866	1203
P_{wi} (kips/in.)	73	101	68	94	64	89	60	83	55	77
P_{wb} (kips)	17,890	21,080	14,410	16,990	12,130	14,290	9912	11,680	7986	9412
P_{fb} (kips)	2318	3220	2073	2879	1821	2529	1592	2211	1373	1907
L_c (ft)	17.8	15.1	17.6	15.0	17.5	14.9	17.4	14.8	17.3	14.7
L_u (ft)	132.3	95.2	125.1	90.1	118.7	85.5	112.9	81.3	107.7	77.5
A (in.^{2})	134		125		117		109		101	
I_x (in.^{4})	7190		6600		6000		5440		4900	
I_y (in.^{4})	2560		2360		2170		1990		1810	
r_y (in.)	4.38		4.34		4.31		4.27		4.24	
Ratio r_x/r_y	1.67		1.67		1.66		1.66		1.65	
B_x } Bending	0.177		0.177		0.178		0.180		0.181	
B_y } factors	0.441		0.442		0.447		0.452		0.457	
$a_x/10^6$	1073		982		894		812		733	
$a_y/10^6$	383		351		324		296		271	
$F'_{ex}(K_xL_x)^2/10^2$ (kips)	557		547		532		518		505	
$F'_{ey}(K_yL_y)^2/10^2$ (kips)	199		195		193		189		186	

AMERICAN INSTITUTE OF STEEL CONSTRUCTION

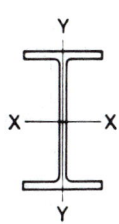

	Fy = 36 ksi
	Fy = 50 ksi

COLUMNS
W shapes
Allowable axial loads in kips

Designation		W14								
Wt./ft		311		283		257		233		211
F_y		36	50	36	50	36	50	36	50	36	50

Effective length in ft KL with respect to least radius of gyration r_y	0	1974	2742	1799	2499	1633	2268	1480	2055	1339	1860
	6	1898	2613	1729	2381	1569	2159	1421	1956	1286	1769
	7	1883	2587	1715	2356	1556	2137	1409	1935	1275	1750
	8	1867	2558	1700	2330	1542	2113	1396	1913	1263	1730
	9	1850	2529	1685	2303	1528	2088	1383	1890	1251	1709
	10	1832	2498	1668	2274	1513	2062	1370	1866	1239	1687
	11	1813	2465	1651	2245	1497	2034	1355	1841	1226	1665
	12	1794	2432	1634	2214	1481	2006	1340	1815	1212	1641
	13	1774	2397	1615	2182	1464	1976	1325	1788	1198	1616
	14	1754	2361	1597	2148	1447	1946	1309	1760	1183	1590
	15	1733	2324	1577	2114	1429	1914	1293	1731	1168	1564
	16	1711	2285	1557	2079	1410	1882	1276	1701	1153	1537
	17	1689	2246	1536	2042	1391	1848	1258	1671	1137	1509
	18	1666	2205	1515	2005	1372	1814	1241	1639	1121	1480
	19	1642	2163	1494	1966	1352	1778	1222	1607	1104	1450
	20	1618	2120	1471	1927	1331	1742	1204	1573	1087	1419
	22	1568	2031	1425	1845	1289	1666	1165	1504	1051	1356
	24	1515	1938	1377	1758	1244	1587	1124	1431	1014	1289
	26	1460	1840	1326	1668	1198	1504	1081	1355	975	1220
	28	1403	1738	1274	1574	1149	1417	1037	1275	934	1147
	30	1344	1631	1219	1476	1099	1326	991	1192	892	1071
	32	1283	1520	1162	1373	1047	1232	943	1106	848	991
	34	1219	1404	1104	1266	993	1133	893	1015	803	908
	36	1153	1283	1043	1155	936	1030	842	921	755	822
	38	1084	1158	980	1040	878	926	788	827	707	738
	40	1013	1045	914	939	818	836	733	746	656	666

Properties										
U	2.24	2.24	2.26	2.26	2.27	2.27	2.28	2.28	2.29	2.29
P_{wo} (kips)	746	1035	639	887	542	753	457	635	397	551
P_{wi} (kips/in.)	51	71	46	65	42	59	39	54	35	49
P_{wb} (kips)	6130	7224	4694	5532	3547	4181	2679	3157	2058	2425
P_{fb} (kips)	1149	1596	964	1339	804	1116	666	925	548	761
L_c (ft)	17.1	14.5	17.0	14.4	16.9	14.3	16.8	14.2	16.7	14.2
L_u (ft)	98.5	70.9	92.6	66.7	85.7	61.7	78.5	56.5	72.3	52.1

A (in.^{2})		91.4		83.3		75.6		68.5		62.0
I_x (in.^{4})		4330		3840		3400		3010		2660
I_y (in.^{4})		1610		1440		1290		1150		1030
r_y (in.)		4.20		4.17		4.13		4.10		4.07
Ratio r_x/r_y		1.64		1.63		1.62		1.62		1.61
B_x ⎱ Bending		0.181		0.181		0.182		0.183		0.183
B_y ⎰ factors		0.459		0.465		0.470		0.472		0.477
$a_x/10^6$		645		572		507		449		396
$a_y/10^6$		240		216		192		172		153
$F'_{ex} (K_x L_x)^2/10^2$ (kips)		491		478		467		456		445
$F'_{ey} (K_y L_y)^2/10^2$ (kips)		183		180		177		174		172

| F_y = 36 ksi |
| F_y = 50 ksi |

COLUMNS
W shapes
Allowable axial loads in kips

Designation		W14							
Wt./ft		193		176		159		145	
F_y		36	50	36	50	36	50	36	50
Effective length in ft KL with respect to least radius of gyration r_y	0	1227	1704	1119	1554	1009	1401	922	1281
	6	1178	1620	1074	1477	968	1331	884	1217
	7	1167	1603	1064	1461	959	1317	877	1203
	8	1157	1584	1054	1444	950	1301	869	1189
	9	1146	1565	1044	1426	941	1285	860	1174
	10	1134	1545	1034	1407	931	1268	851	1159
	11	1122	1524	1022	1388	921	1250	842	1142
	12	1110	1502	1011	1368	911	1232	832	1125
	13	1097	1479	999	1347	900	1213	822	1108
	14	1083	1455	987	1325	889	1193	812	1090
	15	1069	1431	974	1302	877	1173	801	1071
	16	1055	1406	961	1279	865	1152	790	1051
	17	1040	1380	947	1255	853	1130	779	1031
	18	1025	1353	933	1231	840	1107	767	1011
	19	1010	1326	919	1205	827	1085	755	990
	20	994	1298	904	1179	814	1061	743	968
	22	961	1239	874	1125	786	1012	718	923
	24	927	1178	842	1069	758	960	691	875
	26	891	1113	809	1009	727	906	663	825
	28	853	1046	775	947	696	850	634	773
	30	814	976	739	882	663	791	604	719
	32	774	902	701	815	629	729	573	662
	34	732	826	662	744	594	665	540	603
	36	688	745	622	670	558	598	507	541
	38	643	669	580	601	520	537	472	486
	40	596	604	537	543	480	484	435	438

Properties									
U		2.29	2.29	2.31	2.31	2.32	2.32	2.34	2.34
P_{wo} (kips)		340	473	299	415	251	349	214	298
P_{wi} (kips/in.)		32	45	30	42	27	37	24	34
P_{wb} (kips)		1542	1817	1250	1474	904	1066	688	810
P_{fb} (kips)		467	648	386	536	319	443	267	371
L_c (ft)		16.6	14.1	16.5	14.0	16.4	13.9	16.4	13.9
L_u (ft)		68.1	49.0	62.6	45.0	57.2	41.2	52.6	37.9
A (in.^{2})		56.8		51.8		46.7		42.7	
I_x (in.^{4})		2400		2140		1900		1710	
I_y (in.^{4})		931		838		748		677	
r_y (in.)		4.05		4.02		4.00		3.98	
Ratio r_x/r_y		1.60		1.60		1.60		1.59	
B_x } Bending		0.183		0.184		0.184		0.184	
B_y } factors		0.477		0.484		0.485		0.489	
$a_x/10^6$		358		319		283		255	
$a_y/10^6$		139		125		111		101	
$F'_{ex} (K_x L_x)^2/10^2$ (kips)		438		429		422		416	
$F'_{ey} (K_y L_y)^2/10^2$ (kips)		170		168		166		164	

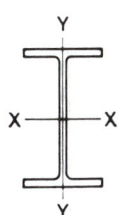

COLUMNS
W shapes
Allowable axial loads in kips

F_y = 36 ksi
F_y = 50 ksi

Designation		\multicolumn{10}{c}{W14}
Wt./ft		\multicolumn{2}{c}{132}	\multicolumn{2}{c}{120}	\multicolumn{2}{c}{109}	\multicolumn{2}{c}{99}	\multicolumn{2}{c}{90}
F_y		36	50	36	50	36	50	36	50†	36	50†
	0	838	1164	762	1059	691	960	629	873	572	795
	6	801	1101	729	1002	661	908	600	825	547	751
	7	794	1088	722	990	654	897	595	815	541	742
	8	786	1074	714	977	647	885	589	805	536	732
	9	777	1060	707	963	640	873	582	793	530	722
	10	768	1044	699	949	633	860	575	782	524	711
	11	759	1028	690	935	626	847	568	769	517	700
	12	750	1011	682	919	618	833	561	757	511	689
	13	740	994	673	903	609	818	554	743	504	676
	14	730	976	663	887	601	803	546	730	497	664
	15	719	958	654	870	592	788	538	715	489	651
	16	708	938	644	852	583	772	529	701	482	637
	17	697	919	633	834	574	755	521	685	474	624
	18	686	898	623	815	564	738	512	670	466	609
	19	674	877	612	796	554	721	503	654	458	595
	20	662	856	601	776	544	703	494	637	449	580
	22	637	811	578	735	523	665	475	603	432	548
	24	610	764	554	692	501	626	454	567	413	515
	26	583	714	528	647	478	585	433	529	394	481
	28	554	663	502	599	454	541	411	489	374	444
	30	524	608	475	549	429	496	388	448	353	406
	32	493	551	446	497	403	449	365	404	331	366
	34	461	492	416	443	376	399	340	359	308	325
	36	427	439	385	395	348	356	314	320	285	290
	38	392	394	353	355	319	320	287	288	260	261

Left label: Effective length in ft KL with respect to least radius of gyration r_y

Properties											
U		2.47	2.47	2.48	2.48	2.49	2.49	2.50	2.28	2.52	2.29
P_{wo} (kips)		196	272	173	240	148	205	125	174	109	151
P_{wi} (kips/in.)		23	32	21	30	19	26	17	24	16	22
P_{wb} (kips)		587	692	449	529	316	373	249	294	186	220
P_{fb} (kips)		239	332	199	276	166	231	137	190	113	158
L_c (ft)		15.5	13.2	15.5	13.1	15.4	13.1	15.4	13.0	15.3	13.0
L_u (ft)		47.7	34.4	44.1	31.7	40.6	29.2	37.0	26.7	34.0	24.5

A (in.^{2})		\multicolumn{2}{c}{38.8}	\multicolumn{2}{c}{35.3}	\multicolumn{2}{c}{32.0}	\multicolumn{2}{c}{29.1}	\multicolumn{2}{c}{26.5}
I_x (in.^{4})		\multicolumn{2}{c}{1530}	\multicolumn{2}{c}{1380}	\multicolumn{2}{c}{1240}	\multicolumn{2}{c}{1110}	\multicolumn{2}{c}{999}
I_y (in.^{4})		\multicolumn{2}{c}{548}	\multicolumn{2}{c}{495}	\multicolumn{2}{c}{447}	\multicolumn{2}{c}{402}	\multicolumn{2}{c}{362}
r_y (in.)		\multicolumn{2}{c}{3.76}	\multicolumn{2}{c}{3.74}	\multicolumn{2}{c}{3.73}	\multicolumn{2}{c}{3.71}	\multicolumn{2}{c}{3.70}
Ratio r_x/r_y		\multicolumn{2}{c}{1.67}	\multicolumn{2}{c}{1.67}	\multicolumn{2}{c}{1.67}	\multicolumn{2}{c}{1.66}	\multicolumn{2}{c}{1.66}
B_x } Bending		\multicolumn{2}{c}{0.186}	\multicolumn{2}{c}{0.186}	\multicolumn{2}{c}{0.185}	\multicolumn{2}{c}{0.185}	\multicolumn{2}{c}{0.185}
B_y } factors		\multicolumn{2}{c}{0.521}	\multicolumn{2}{c}{0.523}	\multicolumn{2}{c}{0.523}	\multicolumn{2}{c}{0.527}	\multicolumn{2}{c}{0.531}
$a_x/10^6$		\multicolumn{2}{c}{228.0}	\multicolumn{2}{c}{204.8}	\multicolumn{2}{c}{184.5}	\multicolumn{2}{c}{165.1}	\multicolumn{2}{c}{148.9}
$a_y/10^6$		\multicolumn{2}{c}{81.7}	\multicolumn{2}{c}{73.6}	\multicolumn{2}{c}{66.3}	\multicolumn{2}{c}{59.7}	\multicolumn{2}{c}{54.1}
$F'_{ex} (K_x L_x)^2/10^2$ (kips)		\multicolumn{2}{c}{409}	\multicolumn{2}{c}{404}	\multicolumn{2}{c}{401}	\multicolumn{2}{c}{395}	\multicolumn{2}{c}{391}
$F'_{ey} (K_y L_y)^2/10^2$ (kips)		\multicolumn{2}{c}{147}	\multicolumn{2}{c}{145}	\multicolumn{2}{c}{144}	\multicolumn{2}{c}{143}	\multicolumn{2}{c}{142}

†Flange is noncompact; see discussion preceding column load tables.

F_y = 36 ksi

F_y = 50 ksi

COLUMNS
W shapes
Allowable axial loads in kips

Designation		W14												
Wt./ft		82		74		68		61		53		48		43
F_y		36	50	36	50	36	50	36	50‡	36	50‡	36	50‡	36‡	50¶
Effective length in ft KL with respect to least radius of gyration r_y	0	521	723	471	654	432	600	387	537	337	468	305	423	272	377
	6	482	657	436	595	400	545	358	487	302	408	273	369	244	329
	7	474	643	429	581	393	533	351	476	295	395	266	356	237	317
	8	465	627	421	567	385	519	345	464	286	380	258	343	230	305
	9	456	610	412	552	377	505	338	452	277	364	250	329	223	292
	10	446	593	403	536	369	491	330	439	268	348	242	313	215	279
	11	435	575	394	520	360	475	322	425	258	330	233	298	207	264
	12	425	555	384	502	351	459	314	410	248	312	224	281	199	249
	14	402	515	363	465	332	425	297	379	226	273	204	245	181	216
	16	377	471	341	426	311	388	278	346	202	229	182	206	161	181
	18	351	423	317	383	289	348	258	310	177	184	159	165	140	144
	20	323	372	292	337	266	305	237	272	149	149	133	133	117	117
	22	293	318	265	287	241	259	214	230	123	123	110	110	96	96
	24	261	267	236	241	214	218	190	193	104	104	93	93	81	81
	26	227	227	206	206	186	186	165	165	88	88	79	79	69	69
	28	196	196	177	177	160	160	142	142	76	76	68	68	60	60
	30	171	171	154	154	139	139	124	124	66	66	59	59	52	52
	31	160	160	145	145	131	131	116	116	62	62	56	56	49	49
	32	150	150	136	136	123	123	109	109	58	58				
	34	133	133	120	120	109	109	96	96						
	36	119	119	107	107	97	97	86	86						
	38	106	106	96	96	87	87	77	77						
Properties														
U		3.69	3.69	3.71	3.71	3.75	3.75	3.77	3.43	4.79	4.35	4.39	4.39	4.44	4.44
P_{wo} (kips)		149	207	127	176	112	156	97	135	96	133	84	117	72	100
P_{wi} (kips/in.)		18	26	16	23	15	21	14	19	13	19	12	17	11	15
P_{wb} (kips)		297	350	204	240	160	188	118	139	113	134	88	104	63	75
P_{fb} (kips)		164	228	139	193	117	162	94	130	98	136	80	111	63	88
L_c (ft)		10.7	9.1	10.6	9.0	10.6	9.0	10.6	9.0	8.5	7.2	8.5	7.2	8.4	7.2
L_u (ft)		28.1	20.2	25.9	18.6	23.9	17.2	21.5	15.5	17.7	12.7	16.0	11.5	14.4	10.4
A (in.^{2})		24.1		21.8		20.0		17.9		15.6		14.1		12.6	
I_x (in.^{4})		882		796		723		640		541		485		428	
I_y (in.^{4})		148		134		121		107		57.7		51.4		45.2	
r_y (in.)		2.48		2.48		2.46		2.45		1.92		1.91		1.89	
Ratio r_x/r_y		2.44		2.44		2.44		2.44		3.07		3.06		3.08	
B_x } Bending		0.196		0.195		0.194		0.194		0.201		0.201		0.201	
B_y } factors		0.823		0.820		0.826		0.833		1.091		1.102		1.115	
$a_x/10^6$		131.4		118.5		107.6		95.4		80.6		71.9		63.6	
$a_y/10^6$		22.1		20.0		18.0		16.0		8.6		7.7		6.7	
F'_{ex} $(K_x L_x)^2/10^2$ (kips)		380		378		375		371		360		355		351	
F'_{ey} $(K_y L_y)^2/10^2$ (kips)		63.8		63.8		62.8		62.2		38.2		37.8		37.0	

‡Web may be noncompact for combined axial and bending stress; see AISC ASD Specification Sect. B5.1.
¶Web exceeds AISC ASD Specification Sect. B5.1 noncompact section limit. See discussion preceding tables.
Note: Heavy line indicates Kl/r of 200.

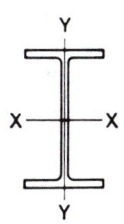

	F_y = 36 ksi
	F_y = 50 ksi

COLUMNS
W shapes
Allowable axial loads in kips

Designation										
Wt./ft	336		305		279		252		230	
F_y	36	50	36	50	36	50	36	50	36	50
0	2134	2964	1935	2688	1769	2457	1601	2223	1462	2031
6	2031	2788	1840	2526	1681	2306	1519	2085	1387	1903
7	2009	2751	1820	2491	1662	2274	1502	2055	1371	1876
8	1986	2711	1799	2454	1642	2240	1484	2023	1355	1847
9	1962	2669	1777	2415	1622	2204	1465	1990	1337	1816
10	1937	2625	1753	2375	1600	2166	1445	1955	1319	1784
11	1911	2579	1729	2332	1578	2126	1425	1919	1300	1750
12	1884	2531	1704	2288	1554	2085	1403	1881	1280	1715
13	1856	2482	1678	2242	1530	2042	1381	1842	1259	1678
14	1827	2430	1651	2194	1505	1998	1358	1801	1238	1641
15	1797	2377	1623	2145	1479	1952	1334	1759	1216	1601
16	1766	2322	1594	2094	1452	1905	1309	1715	1193	1561
17	1733	2265	1565	2041	1425	1856	1284	1670	1169	1519
18	1701	2206	1534	1987	1396	1805	1258	1623	1145	1476
19	1667	2146	1503	1931	1367	1753	1231	1575	1120	1431
20	1632	2084	1471	1873	1337	1699	1203	1526	1095	1386
22	1560	1955	1404	1753	1275	1588	1146	1423	1041	1290
24	1484	1819	1333	1627	1209	1470	1085	1314	985	1190
26	1404	1675	1260	1494	1141	1346	1022	1200	927	1084
28	1321	1525	1183	1354	1069	1216	956	1079	866	972
30	1235	1366	1102	1206	994	1078	887	952	801	855
32	1144	1205	1018	1061	916	948	815	837	734	751
34	1050	1067	930	940	834	839	739	742	664	665
36	951	952	839	839	749	749	661	661	594	594
38	854	854	753	753	672	672	594	594	533	533
40	771	771	679	679	606	606	536	536	481	481

(Leftmost vertical label: Effective length in ft KL with respect to least radius of gyration r_y)

Properties										
U	2.40	2.40	2.41	2.41	2.42	2.42	2.45	2.45	2.46	2.46
P_{wo} (kips)	1178	1636	1005	1396	878	1219	738	1024	636	883
P_{wi} (kips/in.)	64	89	59	81	55	77	50	70	46	64
P_{wb} (kips)	14,480	17,070	11,110	13,100	9274	10,930	7030	8285	5494	6475
P_{fb} (kips)	1965	2729	1646	2287	1373	1907	1139	1582	964	1339
L_c (ft)	14.1	12.0	14.0	11.9	13.9	11.8	13.7	11.6	13.6	11.5
L_u (ft)	107.7	77.5	100.6	72.5	94.5	68.0	87.4	62.9	82.7	59.5
A (in.^{2})	98.8		89.6		81.9		74.1		67.7	
I_x (in.^{4})	4060		3550		3110		2720		2420	
I_y (in.^{4})	1190		1050		937		828		742	
r_y (in.)	3.47		3.42		3.38		3.34		3.31	
Ratio r_x/r_y	1.85		1.84		1.82		1.81		1.80	
B_x ⎱ Bending	0.205		0.206		0.208		0.210		0.211	
B_y ⎰ factors	0.558		0.564		0.573		0.583		0.589	
$a_x/10^6$	605		528		463		405		360	
$a_y/10^6$	177		156		139		123		111	
$F'_{ex} (K_x L_x)^2/10^2$ (kips)	426		410		393		381		370	
$F'_{ey} (K_y L_y)^2/10^2$ (kips)	125		121		118		116		114	

| F_y = 36 ksi |
| F_y = 50 ksi |

COLUMNS
W shapes
Allowable axial loads in kips

Designation												
Wt./ft	210		190		170		152		136		120	
F_y	36	50	36	50	36	50	36	50	36	50	36	50
0	1335	1854	1205	1674	1080	1500	966	1341	862	1197	762	1059
6	1266	1736	1142	1566	1023	1402	914	1253	815	1117	721	987
7	1251	1711	1129	1543	1011	1381	903	1233	805	1100	712	972
8	1236	1684	1115	1518	998	1359	891	1213	795	1082	702	956
9	1219	1655	1100	1492	984	1335	879	1192	784	1062	692	938
10	1202	1625	1084	1465	970	1310	866	1169	772	1042	682	920
11	1185	1594	1068	1437	956	1285	853	1146	760	1021	671	901
12	1166	1562	1051	1407	940	1258	839	1122	747	999	660	881
13	1147	1528	1034	1376	924	1230	825	1096	734	976	648	860
14	1127	1493	1016	1344	908	1200	810	1070	721	952	636	839
15	1107	1457	997	1311	891	1170	794	1042	707	927	624	817
16	1086	1419	978	1276	873	1139	778	1014	693	901	611	794
17	1064	1381	958	1241	855	1107	762	985	678	875	597	770
18	1042	1341	937	1204	837	1074	745	955	662	848	584	746
19	1019	1300	916	1167	817	1039	728	924	647	819	569	720
20	995	1257	894	1128	798	1004	710	892	630	790	555	694
22	946	1169	849	1047	757	931	673	825	597	730	525	640
24	894	1076	802	962	714	853	633	754	561	666	493	583
26	840	977	752	872	668	771	592	680	524	598	460	522
28	783	874	700	776	621	684	549	601	485	527	425	457
30	723	766	645	679	571	597	504	524	444	459	388	398
32	661	673	588	597	519	525	457	461	402	403	349	350
34	596	596	529	529	465	465	408	408	357	357	310	310
36	532	532	472	472	415	415	364	364	319	319	277	277
38	477	477	423	423	372	372	327	327	286	286	248	248
40	431	431	382	382	336	336	295	295	258	258	224	224

Effective length in ft KL with respect to least radius of gyration r_y

Properties												
U	2.47	2.47	2.49	2.49	2.51	2.51	2.53	2.53	2.55	2.55	2.56	2.56
P_{wo} (kips)	558	774	465	646	389	540	333	462	276	383	232	322
P_{wi} (kips/in.)	42	59	38	53	35	48	31	44	28	40	26	36
P_{wb} (kips)	4255	5014	3084	3635	2291	2700	1705	2010	1277	1505	927	1092
P_{fb} (kips)	812	1128	677	941	548	761	441	613	352	488	275	382
L_c (ft)	13.5	11.5	13.4	11.3	13.3	11.3	13.2	11.2	13.1	11.1	13.0	11.0
L_u (ft)	75.9	54.6	71.2	51.3	64.3	46.3	58.6	42.2	53.2	38.3	48.2	34.7

A (in.^{2})	61.8	55.8	50.0	44.7	39.9	35.3
I_x (in.^{4})	2140	1890	1650	1430	1240	1070
I_y (in.^{4})	664	589	517	454	398	345
r_y (in.)	3.28	3.25	3.22	3.19	3.16	3.13
Ratio r_x/r_y	1.80	1.79	1.78	1.77	1.77	1.76
B_x } Bending	0.212	0.212	0.213	0.214	0.215	0.217
B_y } factors	0.594	0.600	0.608	0.614	0.621	0.630
$a_x/10^6$	319.5	281.6	245.5	213.4	185.1	159.7
$a_y/10^6$	99.1	87.8	77.2	67.8	59.4	51.5
F'_{ex} $(K_x L_x)^2/10^2$ (kips)	360	351	342	332	323	315
F'_{ey} $(K_y L_y)^2/10^2$ (kips)	112	110	108	106	104	102

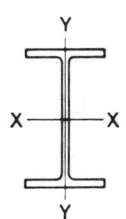

COLUMNS
W shapes
Allowable axial loads in kips

F_y = 36 ksi

F_y = 50 ksi

Designation		W12										
Wt./ft		106		96		87		79		72		65
F_y		36	50	36	50	36	50	36	50	36	50	36	50†
Effective length in ft KL with respect to least radius of gyration r_y	0	674	936	609	846	553	768	501	696	456	633	413	573
	6	637	872	575	788	522	715	473	647	430	589	389	533
	7	629	858	568	775	515	703	467	637	424	579	384	524
	8	620	844	560	762	508	691	460	626	418	569	378	514
	9	611	828	552	748	501	678	453	614	412	558	373	504
	10	602	812	544	733	493	665	446	601	406	547	367	494
	11	593	795	535	718	485	650	439	588	399	535	361	483
	12	583	777	526	701	477	636	431	575	392	522	354	472
	13	572	759	516	685	468	620	423	561	385	509	348	460
	14	561	740	506	667	459	604	415	546	377	496	341	448
	15	550	720	496	649	450	588	407	531	369	482	334	435
	16	539	699	486	630	440	570	398	515	361	468	326	422
	17	527	678	475	611	430	553	389	499	353	453	319	408
	18	514	656	464	591	420	534	379	482	344	438	311	394
	19	502	634	452	570	409	515	370	465	336	422	303	380
	20	489	611	440	549	398	496	360	447	326	406	294	365
	22	462	562	416	505	376	455	339	410	308	372	277	334
	24	433	511	390	458	352	412	317	371	288	336	259	301
	26	404	457	362	408	327	367	294	329	267	297	240	266
	28	372	399	334	356	301	319	270	285	245	258	220	230
	30	340	348	304	310	273	278	245	249	222	225	199	201
	32	305	306	272	273	244	244	219	219	197	197	176	176
	34	271	271	242	242	216	216	194	194	175	175	156	156
	36	241	241	215	215	193	193	173	173	156	156	139	139
	38	217	217	193	193	173	173	155	155	140	140	125	125
	40	196	196	175	175	156	156	140	140	126	126	113	113
Properties												
U		2.59	2.59	2.60	2.60	2.62	2.62	2.63	2.63	2.65	2.65	2.66	2.42
P_{wo} (kips)		185	257	161	223	139	193	122	169	106	148	92	128
P_{wi} (kips/in.)		22	31	20	28	19	26	17	24	15	22	14	20
P_{wb} (kips)		588	693	431	508	354	417	269	317	206	243	154	181
P_{fb} (kips)		221	306	182	253	148	205	122	169	101	140	82	114
L_c (ft)		12.9	10.9	12.8	10.9	12.8	10.9	12.8	10.8	12.7	10.8	12.7	10.7
L_u (ft)		43.3	31.2	39.9	28.7	36.2	26.0	33.3	24.0	30.5	21.9	27.7	20.0
A (in.^{2})		31.2		28.2		25.6		23.2		21.1		19.1	
I_x (in.^{4})		933		833		740		662		597		533	
I_y (in.^{4})		301		270		241		216		195		174	
r_y (in.)		3.11		3.09		3.07		3.05		3.04		3.02	
Ratio r_x/r_y		1.76		1.76		1.75		1.75		1.75		1.75	
B_x } Bending		0.215		0.215		0.217		0.217		0.217		0.217	
B_y } factors		0.633		0.635		0.645		0.648		0.651		0.656	
$a_x/10^6$		139.1		124.3		110.4		98.6		88.6		79.3	
$a_y/10^6$		45.0		40.1		36.0		32.2		29.1		26.0	
F'_{ex} $(K_x L_x)^2/10^2$ (kips)		310		307		300		296		292		289	
F'_{ey} $(K_y L_y)^2/10^2$ (kips)		100		99.0		97.7		96.5		95.8		94.6	

†Flange is noncompact; see discussion preceding column load tables.

F_y = 36 ksi

F_y = 50 ksi

COLUMNS
W shapes
Allowable axial loads in kips

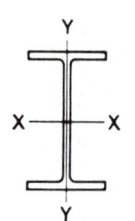

Designation	W12									
Wt./ft	58		53		50		45		40	
F_y	36	50	36	50	36	50	36	50	36	50‡
0	367	510	337	468	318	441	285	396	255	354
6	341	464	312	425	286	386	256	346	229	309
7	335	454	307	416	279	374	250	335	223	299
8	329	443	301	406	271	360	243	322	217	288
9	322	432	295	395	263	346	235	309	210	276
10	315	420	288	384	254	331	228	296	203	264
11	308	407	282	372	246	315	220	281	196	251
12	301	394	275	360	236	298	211	266	188	237
13	293	380	268	347	226	281	202	250	180	222
14	285	365	260	333	216	262	193	233	172	207
15	276	351	252	319	206	243	183	216	163	191
16	268	335	244	305	195	223	173	197	154	175
18	249	302	227	274	171	181	152	159	135	141
20	230	267	209	241	146	146	129	129	114	114
22	209	229	189	206	121	121	106	106	94	94
24	187	193	169	173	102	102	89	89	79	79
26	164	164	147	147	87	87	76	76	67	67
28	142	142	127	127	75	75	66	66	58	58
30	123	123	111	111	65	65	57	57	51	51
32	108	108	97	97	57	57	50	50	45	45
34	96	96	86	86						
38	77	77	69	69						
41	66	66	59	59						

Effective length in ft KL with respect to least radius of gyration r_y

Properties										
U	3.21	3.21	3.24	2.94	4.10	4.10	4.12	3.75	3.77	3.77
P_{wo} (kips)	89	124	78	108	92	127	75	105	66	92
P_{wi} (kips/in.)	13	18	12	17	13	19	12	17	11	15
P_{wb} (kips)	121	142	106	125	131	155	97	115	66	78
P_{fb} (kips)	92	128	74	103	92	128	74	103	60	83
L_c (ft)	10.6	9.0	10.6	9.0	8.5	7.2	8.5	7.2	8.4	7.2
L_u (ft)	24.4	17.5	22.0	15.9	19.6	14.1	17.7	12.8	16.0	11.5
A (in.^{2})	17.0		15.6		14.7		13.2		11.8	
I_x (in.^{4})	475		425		394		350		310	
I_y (in.^{4})	107		95.8		56.3		50.0		44.1	
r_y (in.)	2.51		2.48		1.96		1.94		1.93	
Ratio r_x/r_y	2.10		2.11		2.64		2.65		2.66	
B_x } Bending	0.218		0.221		0.227		0.227		0.227	
B_y } factors	0.794		0.813		1.058		1.065		1.073	
$a_x/10^6$	70.6		63.6		58.8		52.2		46.3	
$a_y/10^6$	16.0		14.3		8.4		7.4		6.5	
F'_{ex} $(K_xL_x)^2/10^2$ (kips)	289		284		278		275		273	
F'_{ey} $(K_yL_y)^2/10^2$ (kips)	65.3		63.8		39.8		39.0		38.6	

‡Web may be noncompact for combined axial and bending stress;
see AISC ASD Specification Sect. B5.1.
Note: Heavy line indicates Kl/r of 200.

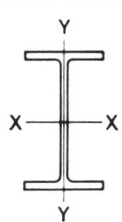

		COLUMNS **W shapes** Allowable axial loads in kips			$F_y = 36$ ksi $F_y = 50$ ksi

Designation		W10									
Wt./ft		112		100		88		77		68	
F_y		36	50	36	50	36	50	36	50	36	50
Effective length in ft KL with respect to least radius of gyration r_y	0	711	987	635	882	559	777	488	678	432	600
	6	663	906	592	808	521	712	454	620	402	548
	7	653	888	583	792	513	697	447	607	395	537
	8	642	869	573	775	504	682	439	593	388	525
	9	631	848	562	756	495	665	431	579	381	512
	10	619	827	551	737	485	648	422	564	373	498
	11	606	805	540	717	475	630	413	548	365	484
	12	593	782	528	696	464	611	404	531	357	469
	13	579	757	516	674	453	591	394	513	348	454
	14	565	732	503	651	442	571	384	495	339	437
	15	550	706	489	627	430	550	373	476	330	421
	16	535	679	476	602	417	528	362	457	320	403
	17	519	651	461	577	405	505	351	437	310	385
	18	503	622	446	550	392	481	339	416	299	366
	19	486	591	431	523	378	457	327	394	289	347
	20	469	560	416	494	364	432	315	371	278	327
	22	433	495	383	435	335	379	289	324	255	285
	24	395	425	348	372	304	323	261	275	230	242
	26	355	362	312	317	271	275	232	234	204	206
	28	313	313	273	273	237	237	202	202	177	177
	30	272	272	238	238	206	206	176	176	155	155
	32	239	239	209	209	181	181	155	155	136	136
	34	212	212	185	185	161	161	137	137	120	120
	36	189	189	165	165	143	143	122	122	107	107
	38	170	170	148	148	129	129	110	110	96	96
	40	153	153	134	134	116	116	99	99	87	87

Properties											
U		2.45	2.45	2.46	2.46	2.49	2.49	2.51	2.51	2.52	2.52
P_{wo} (kips)		255	354	214	298	177	246	143	199	116	162
P_{wi} (kips/in.)		27	38	24	34	22	30	19	27	17	24
P_{wb} (kips)		1388	1636	1014	1196	714	842	480	566	335	395
P_{fb} (kips)		352	488	282	392	221	306	170	237	133	185
L_c (ft)		11.0	9.3	10.9	9.3	10.8	9.2	10.8	9.1	10.7	9.1
L_u (ft)		53.2	38.3	48.2	34.7	43.3	31.2	38.6	27.8	34.8	25.1
A (in.^{2})		32.9		29.4		25.9		22.6		20.0	
I_x (in.^{4})		716		623		534		455		394	
I_y (in.^{4})		236		207		179		154		134	
r_y (in.)		2.68		2.65		2.63		2.60		2.59	
Ratio r_x/r_y		1.74		1.74		1.73		1.73		1.71	
B_x } Bending		0.261		0.263		0.263		0.263		0.264	
B_y } factors		0.726		0.735		0.744		0.751		0.758	
$a_x/10^6$		106.5		92.7		79.5		67.9		58.7	
$a_y/10^6$		35.2		30.8		26.7		22.8		20.0	
$F'_{ex} (K_x L_x)^2/10^2$ (kips)		225		219		214		209		204	
$F'_{ey} (K_y L_y)^2/10^2$ (kips)		74.5		72.8		71.7		70.1		69.6	

F_y = 36 ksi

F_y = 50 ksi

COLUMNS
W shapes
Allowable axial loads in kips

Designation		W10											
Wt./ft		60		54		49		45		39		33	
F_y		36	50	36	50	36	50	36	50	36	50	36	50
	0	380	528	341	474	311	432	287	399	248	345	210	291
	6	353	482	317	433	289	394	260	351	224	303	189	255
	7	348	472	312	423	284	385	253	340	218	293	184	246
	8	341	461	306	414	279	376	247	328	213	283	179	237
	9	335	450	300	403	273	367	240	316	206	272	173	228
	10	328	437	294	392	268	357	232	303	200	260	167	217
	11	321	425	288	381	262	346	224	289	193	248	161	207
	12	313	412	281	369	256	335	216	274	186	235	155	196
	13	306	398	274	356	249	324	208	259	178	221	149	184
	14	297	383	267	343	242	312	199	243	170	207	142	171
	15	289	368	259	330	235	299	190	227	162	193	135	159
	16	280	353	251	316	228	286	180	209	154	177	127	145
	17	271	337	243	301	221	273	170	191	145	161	120	131
	18	262	320	235	286	213	259	160	172	136	144	112	117
	19	253	303	226	271	205	245	149	154	126	130	103	105
	20	243	285	217	255	197	230	138	139	116	117	95	95
	22	222	248	199	221	180	198	115	115	97	97	78	78
	24	201	209	179	186	161	167	97	97	81	81	66	66
	26	177	178	158	159	142	143	82	82	69	69	56	56
	28	154	154	137	137	123	123	71	71	60	60	48	48
	30	134	134	119	119	107	107	62	62	52	52	42	42
	32	118	118	105	105	94	94	54	54	46	46	37	37
	33	111	111	99	99	88	88	51	51	43	43		
	34	104	104	93	93	83	83						
	36	93	93	83	83	74	74						

Effective length in ft KL with respect to least radius of gyration r_y

Properties													
U		2.55	2.55	2.56	2.56	2.57	2.57	3.25	3.25	3.28	3.28	3.35	3.35
P_{wo} (kips)		99	138	83	116	73	101	79	109	64	89	55	77
P_{wi} (kips/in.)		15	21	13	19	12	17	13	18	11	16	10	15
P_{wb} (kips)		239	282	163	193	127	149	138	163	101	119	79	93
P_{fb} (kips)		104	145	85	118	71	98	86	120	63	88	43	59
L_c (ft)		10.6	9.0	10.6	9.0	10.6	9.0	8.5	7.2	8.4	7.2	8.4	7.1
L_u (ft)		31.1	22.4	28.2	20.3	26.0	18.7	22.8	16.4	19.8	14.2	16.5	11.9
A (in.²)		17.6		15.8		14.4		13.3		11.5		9.71	
I_x (in.⁴)		341		303		272		248		209		170	
I_y (in.⁴)		116		103		93.4		53.4		45.0		36.6	
r_y (in.)		2.57		2.56		2.54		2.01		1.98		1.94	
Ratio r_x/r_y		1.71		1.71		1.71		2.15		2.16		2.16	
B_x ⎫ Bending		0.264		0.263		0.264		0.271		0.273		0.277	
B_y ⎭ factors		0.765		0.767		0.770		1.000		1.018		1.055	
$a_x/10^6$		50.5		45.0		40.6		37.2		31.2		25.4	
$a_y/10^6$		17.3		15.4		13.8		8.0		6.7		5.4	
$F'_{ex} (K_x L_x)^2/10^2$ (kips)		200		198		196		194		189		182	
$F'_{ey} (K_y L_y)^2/10^2$ (kips)		68.5		68.0		66.9		41.9		40.7		39.0	

Note: Heavy line indicates Kl/r of 200.

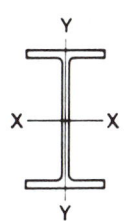

F_y = 36 ksi
F_y = 50 ksi

COLUMNS
W shapes
Allowable axial loads in kips

Designation													
		\multicolumn{12}{W8}											
Wt./ft		67		58		48		40		35		31	
F_y		36	50	36	50	36	50	36	50	36	50	36	50
	0	426	591	369	513	305	423	253	351	222	309	197	274
	6	387	525	336	455	276	375	229	310	201	272	178	241
	7	379	510	328	442	270	363	223	300	197	264	174	234
	8	370	494	320	428	263	352	218	290	191	255	170	226
	9	360	477	312	413	256	339	212	279	186	246	165	217
	10	350	459	303	397	249	326	205	268	180	236	160	208
	11	339	440	293	380	241	312	199	256	174	225	154	199
	12	328	420	283	363	233	297	192	244	168	214	149	189
	13	316	399	273	344	224	282	184	231	162	202	143	179
	14	304	378	263	325	215	266	177	217	155	190	137	168
	15	292	355	251	305	206	249	169	203	148	177	131	156
	16	279	331	240	284	196	232	160	188	141	164	124	145
	17	265	307	228	263	186	214	152	172	133	150	117	132
	18	251	281	216	240	176	195	143	156	125	136	110	119
	19	236	254	203	217	165	175	134	140	117	122	103	107
	20	221	230	190	196	154	158	124	126	109	110	95	97
	22	190	190	162	162	131	131	104	104	91	91	80	80
	24	159	159	136	136	110	110	88	88	76	76	67	67
	26	136	136	116	116	94	94	75	75	65	65	57	57
	28	117	117	100	100	81	81	64	64	56	56	49	49
	30	102	102	87	87	70	70	56	56	49	49	43	43
	32	90	90	76	76	62	62	49	49	43	43	38	38
	33	84	84	72	72	58	58	46	46	40	40	35	35
	34	79	79	68	68	55	55	44	44				
	35	75	75	64	64								

Effective length in ft KL with respect to least radius of gyration r_y

Properties													
U	2.48	2.48	2.50	2.50	2.54	2.54	2.56	2.56	2.59	2.59	2.61	2.61
P_{wo} (kips)	147	205	120	167	86	119	69	96	56	78	48	67
P_{wi} (kips/in.)	21	29	18	26	14	20	13	18	11	16	10	14
P_{wb} (kips)	744	877	533	628	257	303	187	221	120	141	93	110
P_{fb} (kips)	197	273	148	205	106	147	71	98	55	77	43	59
L_c (ft)	8.7	7.4	8.7	7.4	8.6	7.3	8.5	7.2	8.5	7.2	8.4	7.2
L_u (ft)	39.9	28.7	35.3	25.4	30.3	21.8	25.3	18.2	22.6	16.3	20.1	14.5
A (in.²)	\multicolumn{2}{19.7}		\multicolumn{2}{17.1}		\multicolumn{2}{14.1}		\multicolumn{2}{11.7}		\multicolumn{2}{10.3}		\multicolumn{2}{9.13}	
I_x (in.⁴)	\multicolumn{2}{272}		\multicolumn{2}{228}		\multicolumn{2}{184}		\multicolumn{2}{146}		\multicolumn{2}{127}		\multicolumn{2}{110}	
I_y (in.⁴)	\multicolumn{2}{88.6}		\multicolumn{2}{75.1}		\multicolumn{2}{60.9}		\multicolumn{2}{49.1}		\multicolumn{2}{42.6}		\multicolumn{2}{37.1}	
r_y (in.)	\multicolumn{2}{2.12}		\multicolumn{2}{2.10}		\multicolumn{2}{2.08}		\multicolumn{2}{2.04}		\multicolumn{2}{2.03}		\multicolumn{2}{2.02}	
Ratio r_x/r_y	\multicolumn{2}{1.75}		\multicolumn{2}{1.74}		\multicolumn{2}{1.74}		\multicolumn{2}{1.73}		\multicolumn{2}{1.73}		\multicolumn{2}{1.72}	
B_x ⎫ Bending	\multicolumn{2}{0.326}		\multicolumn{2}{0.329}		\multicolumn{2}{0.326}		\multicolumn{2}{0.330}		\multicolumn{2}{0.330}		\multicolumn{2}{0.332}	
B_y ⎭ factors	\multicolumn{2}{0.921}		\multicolumn{2}{0.934}		\multicolumn{2}{0.940}		\multicolumn{2}{0.959}		\multicolumn{2}{0.972}		\multicolumn{2}{0.985}	
$a_x/10^6$	\multicolumn{2}{40.6}		\multicolumn{2}{33.9}		\multicolumn{2}{27.4}		\multicolumn{2}{21.7}		\multicolumn{2}{18.9}		\multicolumn{2}{16.4}	
$a_y/10^6$	\multicolumn{2}{13.2}		\multicolumn{2}{11.2}		\multicolumn{2}{9.1}		\multicolumn{2}{7.3}		\multicolumn{2}{6.3}		\multicolumn{2}{5.6}	
F'_{ex} $(K_x L_x)^2/10^2$ (kips)	\multicolumn{2}{144}		\multicolumn{2}{138}		\multicolumn{2}{135}		\multicolumn{2}{129}		\multicolumn{2}{128}		\multicolumn{2}{125}	
F'_{ey} $(K_y L_y)^2/10^2$ (kips)	\multicolumn{2}{46.6}		\multicolumn{2}{45.7}		\multicolumn{2}{44.9}		\multicolumn{2}{43.2}		\multicolumn{2}{42.7}		\multicolumn{2}{42.3}	

Note: Heavy line indicates Kl/r of 200.

AMERICAN INSTITUTE OF STEEL CONSTRUCTION

F_y = 36 ksi

F_y = 50 ksi

COLUMNS
W shapes
Allowable axial loads in kips

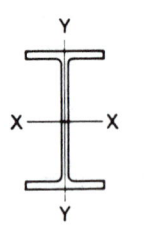

Designation		W8				W6				
Wt./ft		28		24		25		20		15
F_y		36	50	36	50	36	50	36	50	36†	50†
Effective length in ft KL with respect to least radius of gyration r_y	0	178	248	153	212	159	220	127	176	96	133
	6	155	208	133	178	136	182	109	145	81	108
	7	150	198	129	170	131	173	105	137	78	102
	8	144	188	124	161	126	163	100	129	75	96
	9	138	178	118	152	120	152	95	121	71	89
	10	132	166	113	142	114	141	90	112	67	82
	11	125	154	107	132	107	129	85	102	62	74
	12	118	142	101	121	100	117	79	92	58	66
	13	111	128	95	109	93	103	73	81	53	57
	14	103	114	88	97	85	90	67	70	48	49
	15	95	100	81	85	77	78	60	61	43	43
	16	87	88	74	74	69	69	54	54	38	38
	17	78	78	66	66	61	61	47	47	33	33
	18	69	69	59	59	54	54	42	42	30	30
	19	62	62	53	53	49	49	38	38	27	27
	20	56	56	48	48	44	44	34	34	24	24
	22	46	46	39	39	36	36	28	28	20	20
	24	39	39	33	33	31	31	24	24	17	17
	25	36	36	30	30	28	28	22	22		
	26	33	33	28	28						
	27	31	31								

Properties										
U	3.23	3.23	3.27	3.27	2.38	2.07	2.43	1.86	1.93	1.45
P_{wo} (kips)	48	67	39	54	47	65	35	49	26	36
P_{wi} (kips/in.)	10	14	9	12	12	16	9	13	8	12
P_{wb} (kips)	93	110	59	70	170	200	91	107	63	74
P_{fb} (kips)	49	68	36	50	47	65	30	42	15	21
L_c (ft)	6.9	5.9	6.9	5.8	6.4	5.4	6.4	5.4	6.3	5.4
L_u (ft)	17.5	12.6	15.2	10.9	20.0	14.4	16.4	11.8	12.0	8.7

A (in.^{2})	8.25	7.08	7.34	5.87	4.43
I_x (in.^{4})	98.0	82.8	53.4	41.4	29.1
I_y (in.^{4})	21.7	18.3	17.1	13.3	9.32
r_y (in.)	1.62	1.61	1.52	1.50	1.45
Ratio r_x/r_y	2.13	2.12	1.78	1.77	1.77
B_x } Bending	0.340	0.339	0.440	0.438	0.456
B_y } factors	1.244	1.258	1.308	1.331	1.424
$a_x/10^6$	14.63	12.34	7.97	6.19	4.33
$a_y/10^6$	3.23	2.73	2.53	1.97	1.39
$F'_{ex} (K_x L_x)^2/10^2$ (kips)	123	121	75.9	73.1	68.3
$F'_{ey} (K_y L_y)^2/10^2$ (kips)	27.2	26.9	24.0	23.3	21.8

†Flange is noncompact; see discussion preceding column load tables.
Note: Heavy line indicates Kl/r of 200.

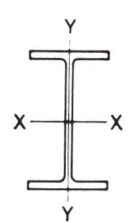

COLUMNS
W shapes
Allowable axial loads in kips

| F_y = 36 ksi |
| F_y = 50 ksi |

Designation		W6						W5				W4	
Wt./ft		16		12		9		19		16		13	
F_y		36	50	36	50	36	50	36	50	36	50	36	50
Effective length in ft KL with respect to least radius of gyration r_y	0	102	142	77	107	58	80	120	166	101	140	83	115
	2	96	132	72	98	54	74	115	158	97	133	78	107
	3	92	124	68	92	51	69	111	152	94	128	75	101
	4	87	116	64	85	48	64	107	145	91	122	71	94
	5	82	106	60	77	45	58	103	138	87	116	67	87
	6	76	95	55	69	41	51	99	129	83	109	62	79
	7	69	83	50	59	37	44	93	120	79	101	57	70
	8	62	70	44	48	33	36	88	111	74	93	52	60
	9	54	57	38	38	28	28	82	100	69	84	46	49
	10	46	46	31	31	23	23	76	89	64	75	39	40
	11	38	38	26	26	19	19	70	77	58	64	33	33
	12	32	32	22	22	16	16	63	65	52	54	28	28
	13	27	27	18	18	13	13	55	56	46	46	24	24
	14	23	23	16	16	12	12	48	48	40	40	20	20
	15	20	20	14	14	10	10	42	42	35	35	18	18
	16	18	18					37	37	31	31	16	16
	17							33	33	27	27		
	18							29	29	24	24		
	19							26	26	22	22		
	20							24	24	20	20		
	21							21	21	18	18		

Properties

	36	50	36	50	36	50	36	50	36	50	36	50
U	3.35	2.55	2.89	2.05	2.26	1.63	2.25	2.01	2.17	1.85	2.30	1.98
P_{wo} (kips)	35	49	26	36	17	24	39	55	32	45	35	48
P_{wi} (kips/in.)	9	13	8	12	6	9	10	14	9	12	10	14
P_{wb} (kips)	91	107	63	74	25	30	138	163	97	115	196	231
P_{fb} (kips)	37	51	18	25	10	14	42	58	29	41	27	37
L_c (ft)	4.3	3.6	4.2	3.6	4.2	3.5	5.3	4.5	5.3	4.5	4.3	3.6
L_u (ft)	12.0	8.7	8.6	6.2	6.7	4.8	19.5	14.0	16.7	12.0	15.6	11.2

	W6-16	W6-12	W6-9	W5-19	W5-16	W4-13
A (in.^{2})	4.74	3.55	2.68	5.54	4.68	3.83
I_x (in.^{4})	32.1	22.1	16.4	26.2	21.3	11.3
I_y (in.^{4})	4.43	2.99	2.20	9.13	7.51	3.86
r_y (in.)	0.966	0.918	0.905	1.28	1.27	1.00
Ratio r_x/r_y	2.69	2.71	2.73	1.70	1.68	1.72
B_x } Bending	0.465	0.486	0.482	0.543	0.550	0.701
B_y } factors	2.155	2.367	2.414	1.526	1.560	2.016
$a_x/10^6$	4.77	3.28	2.44	3.89	3.16	1.69
$a_y/10^6$	0.66	0.45	0.33	1.35	1.12	0.57
$F'_{ex} (K_x L_x)^2/10^2$ (kips)	70.0	64.2	63.3	49.1	47.2	30.7
$F'_{ey} (K_y L_y)^2/10^2$ (kips)	9.68	8.74	8.49	17.0	16.7	10.4

Note: Heavy line indicates Kl/r of 200.

COLUMNS
S shapes
Allowable axial loads in kips

$F_y = 36$ ksi
$F_y = 50$ ksi

Designation	S6				S5				S4				S3			
Wt./ft	17.25		12.5		14.75		10		9.5		7.7		7.5		5.7	
F_y	36	50	36	50	36	50	36	50	36	50	36	50	36	50	36	50
0	110	152	79	110	94	130	64	88	60	84	49	68	48	66	36	50
2	99	134	72	98	84	113	57	77	53	71	43	58	41	55	31	42
3	92	121	67	89	76	100	52	69	48	62	39	51	36	46	28	35
4	83	105	61	78	68	85	47	59	41	51	34	42	31	36	23	28
5	73	87	54	66	58	67	41	48	34	37	28	32	24	24	18	19
6	61	67	47	52	47	48	34	35	26	26	22	22	17	17	13	13
7	49	49	38	39	35	35	26	26	19	19	16	16	12	12	10	10
8	37	37	30	30	27	27	20	20	15	15	12	12	10	10	7	7
9	30	30	23	23	21	21	16	16	12	12	10	10				
10	24	24	19	19	17	17	13	13								
11	20	20	16	16												

Effective length in ft KL with respect to least radius of gyration r_y

Properties

	S6				S5				S4				S3			
U	5.33	3.84	4.99	3.59	4.79	3.45	4.42	3.18	—		—		—		—	
P_{wo} (kips)	73	102	37	51	72	100	31	43	44	61	26	36	43	60	21	29
P_{wi} (kips/in.)	17	23	8	12	18	25	8	11	12	16	7	10	13	17	6	9
P_{wb} (kips)	582	686	72	85	879	1036	71	84	341	402	71	83	644	758	74	88
P_{fb} (kips)	29	40	29	40	24	33	24	33	19	27	19	27	15	21	15	21
L_c (ft)	3.8	3.2	3.5	3.0	3.5	2.9	3.2	2.7	3.0	2.5	2.8	2.4	2.6	2.2	2.5	2.1
L_u (ft)	9.9	7.1	9.2	6.6	9.9	7.2	9.1	6.5	9.5	6.8	9.0	6.5	10.1	7.2	9.4	6.7

	S6		S5		S4		S3	
A (in.^{2})	5.07		3.67		4.34		2.94	
I_x (in.^{4})	26.3		22.1		15.2		12.3	
I_y (in.^{4})	2.31		1.82		1.67		1.22	
r_y (in.)	0.675		0.705		0.620		0.643	
Ratio r_x/r_y	3.38		3.48		3.02		3.19	
B_x } Bending	0.578		0.498		0.713		0.598	
B_y } factors	3.900		3.367		4.297		3.634	
$a_x/10^6$	3.93		3.28		2.26		1.84	
$a_y/10^6$	0.34		0.27		0.25		0.18	
F'_{ex} $(K_x L_x)^2/10^2$ (kips)	54.0		62.4		36.4		43.6	
F'_{ey} $(K_y L_y)^2/10^2$ (kips)	4.72		5.15		3.99		4.29	

	S4		S4		S3		S3	
A (in.^{2})	2.79		2.26		2.21		1.67	
I_x (in.^{4})	6.79		6.08		2.93		2.52	
I_y (in.^{4})	0.903		0.764		0.586		0.455	
r_y (in.)	0.569		0.581		0.516		0.522	
Ratio r_x/r_y	2.74		2.82		2.23		2.36	
B_x } Bending	0.823		0.743		1.133		0.994	
B_y } factors	4.319		3.937		4.722		4.282	
$a_x/10^6$	1.01		0.91		0.44		0.38	
$a_y/10^6$	0.13		0.11		0.09		0.07	
F'_{ex} $(K_x L_x)^2/10^2$ (kips)	25.2		27.8		13.7		15.7	
F'_{ey} $(K_y L_y)^2/10^2$ (kips)	3.36		3.50		2.76		2.83	

Note: Heavy line indicates Kl/r of 200.

AMERICAN INSTITUTE OF STEEL CONSTRUCTION

ALLOWABLE CONCENTRIC LOADS ON COLUMNS
Steel Pipe and Structural Tubing

Allowable concentric loads in the tables that follow are tabulated for the effective lengths in feet KL, indicated at the left of each table. They are applicable to axially loaded members with respect to their minor axis in accordance with Sect. E2 of the AISC ASD Specification.

For discussion of effective length, range of l/r, strength about the major axis, combined axial and bending stress, and sample problems, see "Columns, General Notes."

Properties and factors are listed at the bottom of the tables for checking strength about the strong axis and for checking combined loading conditions.

STEEL PIPE COLUMNS

Allowable loads for unfilled pipe columns are tabulated for $F_y = 36$ ksi. Steel pipe manufactured to ASTM A501 furnishes $F_y = 36$ ksi and ASTM A53, Types E or S, Grade B furnishes $F_y = 35$ ksi and may be designed at stresses allowed for $F_y = 36$ ksi steel.

The heavy horizontal lines within the table indicate $Kl/r = 200$. No values are listed beyond $Kl/r = 200$.

STRUCTURAL TUBING COLUMNS

Allowable loads for square and rectangular structural tubing columns are tabulated for $F_y = 46$ ksi. Structural tubing is manufactured to $F_y = 46$ ksi under ASTM A500, Gr. B.

All tubes listed in the column load tables satisfy the noncompact section limits in Sect. B5.1 of the AISC ASD Specification.

The heavy horizontal lines appearing within the tables indicate $Kl/r = 200$. No values are listed beyond $Kl/r = 200$.

$F_y = 36$ ksi

COLUMNS
Standard steel pipe
Allowable concentric loads in kips

Nominal Dia.		12	10	8	6	5	4	3½	3
Wall Thickness		0.375	0.365	0.322	0.280	0.258	0.237	0.226	0.216
Wt./ft		49.56	40.48	28.55	18.97	14.62	10.79	9.11	7.58
F_y						36 ksi			
	0	315	257	181	121	93	68	58	48
	6	303	246	171	110	83	59	48	38
	7	301	243	168	108	81	57	46	36
	8	299	241	166	106	78	54	44	34
	9	296	238	163	103	76	52	41	31
	10	293	235	161	101	73	49	38	28
	11	291	232	158	98	71	46	35	25
	12	288	229	155	95	68	43	32	22
	13	285	226	152	92	65	40	29	19
	14	282	223	149	89	61	36	25	16
	15	278	220	145	86	58	33	22	14
	16	275	216	142	82	55	29	19	12
	17	272	213	138	79	51	26	17	11
	18	268	209	135	75	47	23	15	10
	19	265	205	131	71	43	21	14	9
	20	261	201	127	67	39	19	12	
	22	254	193	119	59	32	15	10	
	24	246	185	111	51	27	13		
	25	242	180	106	47	25	12		
	26	238	176	102	43	23			
	28	229	167	93	37	20			
	30	220	158	83	32	17			
	31	216	152	78	30	16			
	32	211	148	73	29				
	34	201	137	65	25				
	36	192	127	58	23				
	37	186	120	55	21				
	38	181	115	52					
	40	171	104	47					

Effective length in ft KL with respect to radius of gyration

Properties									
Area A (in.^{2})		14.6	11.9	8.40	5.58	4.30	3.17	2.68	2.23
I (in.^{4})		279	161	72.6	28.1	15.2	7.23	4.79	3.02
r (in.)		4.38	3.67	2.94	2.25	1.88	1.51	1.34	1.16
B } Bending factor		0.333	0.398	0.500	0.657	0.789	0.987	1.12	1.29
$a/10^6$		41.7	23.9	10.8	4.21	2.26	1.08	0.717	0.447

Note: Heavy line indicates Kl/r of 200.

COLUMNS
Extra strong steel pipe
Allowable concentric loads in kips

F_y = 36 ksi

Nominal Dia.		12	10	8	6	5	4	3½	3
Wall Thickness		0.500	0.500	0.500	0.432	0.375	0.337	0.318	0.300
Wt./ft		65.42	54.74	43.39	28.57	20.78	14.98	12.50	10.25
F_y		36 ksi							
	0	415	348	276	181	132	95	79	65
	6	400	332	259	166	118	81	66	52
	7	397	328	255	162	114	78	63	48
	8	394	325	251	159	111	75	59	45
	9	390	321	247	155	107	71	55	41
	10	387	318	243	151	103	67	51	37
	11	383	314	239	146	99	63	47	33
	12	379	309	234	142	95	59	43	28
	13	375	305	229	137	91	54	38	24
	14	371	301	224	132	86	49	33	21
	15	367	296	219	127	81	44	29	18
	16	363	291	214	122	76	39	25	16
	18	353	281	203	111	65	31	20	12
	19	349	276	197	105	59	28	18	11
	20	344	271	191	99	54	25	16	
	21	337	265	185	92	48	22	14	
	22	334	260	179	86	44	21		
	24	323	248	166	73	37	17		
	26	312	236	152	62	32			
	28	301	224	137	54	27			
	30	289	211	122	47	24			
	32	277	197	107	41				
	34	264	183	95	36				
	36	251	168	85	32				
	38	237	152	76					
	40	223	137	69					
Properties									
Area A (in.^{2})		19.2	16.1	12.8	8.40	6.11	4.41	3.68	3.02
I (in.^{4})		362	212	106	40.5	20.7	9.61	6.28	3.89
r (in.)		4.33	3.63	2.88	2.19	1.84	1.48	1.31	1.14
B } Bending factor		0.339	0.408	0.521	0.688	0.822	1.03	1.17	1.36
$a/10^6$		53.6	31.6	15.8	6.00	3.08	1.44	0.941	0.585

Effective length in ft KL with respect to radius of gyration

Note: Heavy line indicates Kl/r of 200.

F_y = 36 ksi

COLUMNS
Double-extra strong
steel pipe
Allowable concentric loads in kips

Nominal Dia.		8	6	5	4	3
Wall Thickness		0.875	0.864	0.750	0.674	0.600
Wt./ft		72.42	53.16	38.55	27.54	18.58
F_y				36 ksi		
	0	461	337	244	175	118
	6	431	306	216	147	91
	7	424	299	209	140	84
	8	417	292	202	133	77
	9	410	284	195	126	69
	10	403	275	187	118	60
	11	395	266	178	109	51
	12	387	257	170	100	43
	13	378	247	160	91	37
	14	369	237	151	81	32
	15	360	227	141	70	28
	16	351	216	130	62	24
	17	341	205	119	55	22
	18	331	193	108	49	
	19	321	181	97	44	
	20	310	168	87	40	
	22	288	142	72	33	
	24	264	119	61		
	26	240	102	52		
	28	213	88	44		
	30	187	76			
	32	164	67			
	34	145	60			
	36	130				
	38	116				
	40	105				

Effective length in ft KL with respect to radius of gyration

Properties						
Area A (in.^{2})		21.3	15.6	11.3	8.10	5.47
I (in.^{4})		162	66.3	33.6	15.3	5.99
r (in.)		2.76	2.06	1.72	1.37	1.05
B } Bending factor		0.567	0.781	0.938	1.19	1.60
$a/10^6$		24.2	9.86	4.98	2.27	0.899

Note: Heavy line indicates Kl/r of 200.

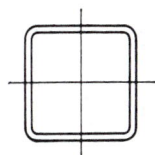

F_y = 46 ksi

COLUMNS
Square structural tubing
Allowable concentric loads in kips

Nominal Size		16 × 16		14 × 14			12 × 12		
Thickness		5/8	1/2	5/8	1/2	3/8	5/8	1/2	3/8
Wt./ft		127.37	103.30	110.36	89.68	68.31	93.34	76.07	58.10
F_y		46 ksi							
Effective length in ft KL with respect to radius of gyration	0	1032	839	894	729	555	756	618	472
	6	1003	815	865	704	536	726	593	453
	7	998	810	859	699	533	720	588	449
	8	992	805	853	694	529	713	583	445
	9	986	800	846	689	525	707	577	441
	10	979	795	840	684	521	699	571	437
	11	973	790	833	678	517	692	566	433
	12	966	785	825	672	513	684	559	428
	13	959	779	818	666	508	676	553	423
	14	951	773	810	660	504	668	546	418
	15	944	767	802	654	499	660	540	413
	16	936	761	794	647	494	651	533	408
	17	928	755	786	641	489	642	526	403
	18	920	748	777	634	484	633	518	397
	19	912	741	769	627	478	623	511	391
	20	903	735	760	619	473	614	503	386
	21	895	728	750	612	467	604	495	380
	22	886	721	741	604	462	594	487	373
	23	877	713	731	597	456	583	478	367
	24	867	706	721	589	450	572	470	361
	25	858	698	711	581	444	562	461	354
	26	848	691	701	573	438	550	452	348
	27	838	683	691	564	432	539	443	341
	28	828	675	680	556	425	528	434	334
	29	818	667	669	547	419	516	424	327
	30	808	658	658	538	412	504	415	320
	32	787	641	635	520	398	479	395	305
	34	765	624	612	501	384	453	374	289
	36	742	606	588	482	370	427	353	274
	38	719	587	563	462	355	399	331	257
	40	695	568	537	441	339	370	308	240
Properties									
A (in.²)		37.40	30.40	32.40	26.40	20.10	27.40	22.40	17.10
I (in.⁴)		1450	1200	952	791	615	580	485	380
r (in.)		6.23	6.29	5.42	5.48	5.54	4.60	4.66	4.72
B } Bending factor		0.205	0.202	0.238	0.233	0.228	0.283	0.276	0.269
$a/10^6$		216	179	142	118	91.7	86.4	72.3	56.7

F_y = 46 ksi

COLUMNS
Square structural tubing
Allowable concentric loads in kips

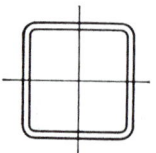

Nominal Size	10 × 10					9 × 9				
Thickness	⅝	9/16	½	⅜	5/16	⅝	9/16	½	⅜	5/16
Wt./ft	76.33	69.48	62.46	47.90	40.35	67.82	61.83	55.66	42.79	36.10
F_y	46 ksi									
0	618	563	508	389	328	549	502	453	348	293
6	588	535	481	370	311	517	473	427	328	276
7	581	529	476	366	308	510	467	421	324	273
8	574	523	471	361	305	503	461	415	320	269
9	567	516	465	357	301	495	454	409	315	266
10	559	509	459	352	297	487	446	403	310	262
11	551	502	452	348	293	479	439	396	305	257
12	543	494	446	343	289	470	431	389	300	253
13	534	487	439	338	285	461	423	382	295	249
14	525	479	432	332	280	452	414	374	289	244
15	516	470	424	327	276	442	406	367	283	239
16	507	462	417	321	271	432	397	359	277	234
17	497	453	409	315	266	422	387	350	271	229
18	487	444	401	309	261	411	378	342	265	224
19	476	435	392	303	256	400	368	333	258	218
20	466	425	384	296	251	389	358	324	251	213
21	455	415	375	290	245	377	347	315	244	207
22	444	405	366	283	240	366	336	305	237	201
23	432	395	357	276	234	353	325	295	230	195
24	420	384	348	269	228	341	314	285	222	189
25	408	373	338	262	222	328	303	275	215	182
26	396	362	328	255	216	315	291	264	207	176
27	383	351	318	247	210	301	279	254	199	169
28	370	339	308	239	203	287	266	243	190	162
29	357	327	297	232	197	273	253	231	182	155
30	344	315	287	224	190	259	240	220	173	148
32	316	290	264	207	176	229	213	195	155	133
34	287	264	241	190	162	203	189	173	138	118
36	257	237	217	172	147	181	168	154	123	106
38	230	213	195	154	132	162	151	139	110	95
40	208	192	176	139	119	146	136	125	99	86

Effective length in ft KL with respect to radius of gyration

Properties										
A (in.^{2})	22.40	20.40	18.40	14.10	11.90	19.90	18.20	16.40	12.60	10.60
I (in.^{4})	321	297	271	214	183	227	211	193	154	132
r (in.)	3.78	3.81	3.84	3.90	3.93	3.37	3.40	3.43	3.49	3.53
B } Bending factor	0.350	0.343	0.339	0.328	0.323	0.395	0.389	0.382	0.370	0.362
$a/10^6$	47.8	44.1	40.4	32.0	27.3	33.7	31.3	28.7	22.9	19.7

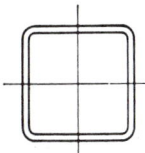

F_y = 46 ksi

COLUMNS
Square structural tubing
Allowable concentric loads in kips

Nominal Size		8 × 8					7 × 7				
Thickness	⅝	%₁₆	½	⅜	%₁₆	¼	%₁₆	½	⅜	%₁₆	¼
Wt./ft	59.32	54.17	48.85	37.60	31.84	25.82	46.51	42.05	32.59	27.59	22.42
F_y						46 ksi					
0	480	439	397	306	258	209	378	342	264	224	182
6	448	409	370	286	242	196	347	314	244	207	168
7	441	403	364	281	238	193	340	308	239	203	165
8	433	396	358	277	234	190	333	301	234	199	162
9	425	389	352	272	230	187	326	294	229	195	158
10	417	381	345	267	226	184	318	287	224	190	155
11	408	373	338	262	222	180	309	280	218	185	151
12	399	365	330	256	217	176	300	272	212	180	147
13	389	356	323	251	212	173	291	264	206	175	143
14	379	348	315	245	207	169	282	255	200	170	139
15	369	338	307	238	202	165	272	246	193	165	135
16	358	329	298	232	197	160	261	237	186	159	130
17	347	319	289	225	191	156	251	228	179	153	125
18	336	309	280	219	186	151	240	218	172	147	120
19	324	298	271	212	180	147	228	208	165	141	115
20	312	288	261	205	174	142	217	197	157	134	110
21	300	276	251	197	168	137	204	187	149	128	105
22	287	265	241	190	162	132	192	176	140	121	99
23	274	253	231	182	155	127	179	164	132	114	94
24	261	241	220	174	148	122	165	152	123	106	88
25	247	229	209	165	141	116	152	140	114	99	82
26	232	216	198	157	134	110	141	130	105	91	76
27	218	203	186	148	127	105	131	120	98	85	70
28	203	189	174	139	120	99	122	112	91	79	65
29	189	176	162	130	112	93	113	104	85	73	61
30	177	165	151	122	105	87	106	98	79	69	57
32	155	145	133	107	92	76	93	86	70	60	50
34	137	128	118	95	82	67	82	76	62	53	44
36	123	115	105	84	73	60	74	68	55	48	40
38	110	103	94	76	65	54	66	61	49	43	35
40	99	93	85	68	59	49	60	55	45	39	32

Effective length in ft KL with respect to radius of gyration

Properties											
A (in.²)	17.40	15.90	14.40	11.10	9.36	7.59	13.70	12.40	9.58	8.11	6.59
I (in.⁴)	153	143	131	106	90.9	75.1	91.4	84.6	68.7	59.5	49.4
r (in.)	2.96	3.00	3.03	3.09	3.12	3.15	2.59	2.62	2.68	2.71	2.74
B } Bending factor	0.455	0.445	0.437	0.420	0.412	0.404	0.525	0.511	0.488	0.477	0.467
$a/10^6$	22.8	21.3	19.6	15.7	13.5	11.2	13.7	12.6	10.2	8.86	7.36

AMERICAN INSTITUTE OF STEEL CONSTRUCTION

F_y = 46 ksi

COLUMNS
Square structural tubing
Allowable concentric loads in kips

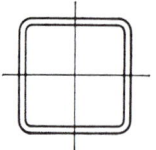

Nominal Size	6 × 6						5 × 5				
Thickness	$\frac{9}{16}$	$\frac{1}{2}$	$\frac{3}{8}$	$\frac{5}{16}$	$\frac{1}{4}$	$\frac{3}{16}$	$\frac{1}{2}$	$\frac{3}{8}$	$\frac{5}{16}$	$\frac{1}{4}$	$\frac{3}{16}$
Wt./ft	38.86	35.24	27.48	23.34	19.02	14.53	28.43	22.37	19.08	15.62	11.97
F_y						46 ksi					
0	315	287	223	189	154	118	231	182	155	127	97
6	283	257	201	171	140	107	200	159	136	111	86
7	275	251	196	167	137	105	193	153	131	108	83
8	268	244	191	163	133	102	186	148	127	104	80
9	259	237	186	158	130	99	178	142	122	100	77
10	251	229	180	154	126	96	169	135	116	96	74
11	242	221	174	149	122	93	160	129	111	92	71
12	232	212	168	143	117	90	151	122	105	87	67
13	222	203	161	138	113	87	141	115	99	82	64
14	212	194	154	132	108	83	131	107	93	77	60
15	201	185	147	126	104	80	120	99	86	72	56
16	190	175	140	120	99	76	109	90	79	66	52
17	178	164	132	113	94	72	97	82	72	60	47
18	166	153	124	107	88	68	87	73	64	54	43
19	153	142	115	100	83	64	78	65	58	49	39
20	140	131	107	93	77	60	70	59	52	44	35
21	127	119	98	85	71	56	64	54	47	40	32
22	116	108	89	78	65	51	58	49	43	36	29
24	98	91	75	65	55	43	49	41	36	31	24
26	83	77	64	56	47	36	41	35	31	26	21
28	72	67	55	48	40	31	36	30	27	22	18
30	62	58	48	42	35	27	31	26	23	20	15
31	58	54	45	39	33	26		25	22	18	14
32	55	51	42	37	31	24				17	14
34	49	45	37	33	27	21					
36	43	40	33	29	24	19					
37			32	27	23	18					
38				26	22	17					
39						16					
Properties											
A (in.^{2})	11.40	10.40	8.08	6.86	5.59	4.27	8.36	6.58	5.61	4.59	3.52
I (in.^{4})	54.1	50.5	41.6	36.3	30.3	23.8	27.0	22.8	20.1	16.9	13.4
r (in.)	2.18	2.21	2.27	2.30	2.33	2.36	1.80	1.86	1.89	1.92	1.95
B } Bending factor	0.633	0.615	0.583	0.567	0.553	0.539	0.773	0.722	0.699	0.677	0.656
$a/10^6$	8.07	7.52	6.20	5.40	4.52	3.54	4.03	3.39	2.99	2.52	2.00

Note: Heavy line indicates Kl/r of 200.

Effective length in ft KL with respect to radius of gyration

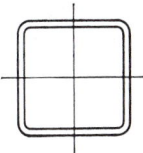

| | | F_y = 46 ksi |

COLUMNS
Square structural tubing
Allowable concentric loads in kips

Nominal Size		4 × 4					3 × 3		
Thickness		½	⅜	⁵⁄₁₆	¼	³⁄₁₆	⁵⁄₁₆	¼	³⁄₁₆
Wt./ft		21.63	17.27	14.83	12.21	9.42	10.58	8.81	6.87
F_y		46 ksi							
Effective length in ft KL with respect to radius of gyration	0	176	140	120	99	76	86	71	56
	2	168	134	115	95	73	80	67	53
	3	162	130	112	92	71	77	64	50
	4	156	126	108	89	69	73	61	48
	5	150	121	104	86	67	68	57	45
	6	143	115	100	83	64	63	53	42
	7	135	110	95	79	61	57	49	39
	8	126	103	90	75	58	51	44	35
	9	117	97	84	70	55	44	38	31
	10	108	89	78	65	51	37	33	27
	11	98	82	72	60	47	31	27	22
	12	87	74	65	55	43	26	23	19
	13	75	65	58	49	39	22	19	16
	14	65	57	51	43	35	19	17	14
	15	57	49	44	38	30	16	15	12
	16	50	43	39	33	27	14	13	11
	17	44	38	34	29	24	13	11	9
	18	39	34	31	26	21		10	8
	19	35	31	28	24	19			
	20	32	28	25	21	17			
	21	29	25	23	19	16			
	22	26	23	21	18	14			
	23	24	21	19	16	13			
	24		19	17	15	12			
	25				14	11			
Properties									
A (in.²)		6.36	5.08	4.36	3.59	2.77	3.11	2.59	2.02
I (in.⁴)		12.3	10.7	9.58	8.22	6.59	3.58	3.16	2.60
r (in.)		1.39	1.45	1.48	1.51	1.54	1.07	1.10	1.13
B } Bending factor		1.04	0.949	0.910	0.874	0.840	1.30	1.23	1.17
$a/10^6$		1.83	1.59	1.43	1.22	0.983	0.533	0.470	0.387

Note: Heavy line indicates Kl/r of 200.

F_y = 46 ksi

COLUMNS
Rectangular structural tubing
Allowable concentric loads in kips

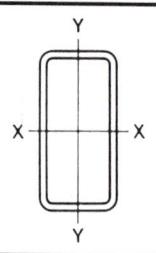

Nominal Size	16 × 12		16 × 8	14 × 10			12 × 8			
Thickness	⅝	½	½	⅝	½	⅜	⅝	9/16	½	⅜
Wt./ft	110.36	89.68	76.07	93.34	76.07	58.10	76.33	69.48	62.46	47.90
F_y	46 ksi									
0	894	729	618	756	618	472	618	563	508	389
6	860	700	580	720	588	450	580	528	475	365
7	853	695	572	712	582	445	571	520	468	360
8	846	689	564	704	576	440	562	512	461	354
9	838	683	555	696	569	435	553	503	453	349
10	830	676	546	687	562	430	543	494	446	343
11	822	670	536	678	555	425	532	485	437	337
12	813	663	526	669	547	419	521	475	429	330
13	805	656	516	659	539	413	510	465	420	323
14	795	648	505	649	531	407	498	455	410	316
15	786	641	494	638	523	401	486	444	401	309
16	776	633	482	627	514	394	474	432	391	302
17	766	625	470	616	505	387	461	421	380	294
18	756	617	458	605	496	381	447	409	370	286
19	745	608	445	593	486	373	434	396	359	278
20	734	600	432	581	477	366	419	384	347	269
22	712	582	405	555	457	351	390	357	324	252
24	688	563	376	529	436	335	359	329	299	233
26	664	543	346	501	413	319	326	300	273	214
28	638	523	314	472	390	302	291	269	245	193
30	612	502	281	442	366	284	255	236	216	172
32	584	480	247	410	341	265	224	208	190	151
34	555	457	219	378	315	245	199	184	168	134
36	526	433	195	343	288	225	177	164	150	119
38	495	409	175	309	260	204	159	147	135	107
39	479	396	166	293	246	194	151	140	128	102
40	463	384	158	278	234	184	143	133	122	97

(Effective length in ft KL with respect to least radius of gyration)

Properties

A (in.^{2})	32.40	26.40	22.40	27.40	22.40	17.10	22.40	20.40	18.40	14.10
I_x (in.^{4})	1160	962	722	728	608	476	418	387	353	279
I_y (in.^{4})	742	618	244	431	361	284	221	205	188	149
r_x/r_y	1.25	1.25	1.72	1.30	1.30	1.29	1.37	1.37	1.37	1.37
r_y (in.)	4.78	4.84	3.30	3.96	4.02	4.08	3.14	3.17	3.20	3.26
B_x } Bending	0.223	0.219	0.248	0.263	0.257	0.251	0.322	0.316	0.312	0.303
B_y } factors	0.261	0.256	0.366	0.318	0.309	0.301	0.405	0.398	0.391	0.377
$a_x/10^6$	173	143	108	108	90.6	70.9	62.3	57.5	52.6	41.6
$a_y/10^6$	110	92.1	36.4	64.0	53.9	42.3	33.0	30.5	28.0	22.2

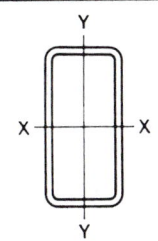

| | F_y = 46 ksi |

COLUMNS
Rectangular structural tubing
Allowable concentric loads in kips

Nominal Size	12 × 6				10 × 8					
Thickness	5/8	9/16	1/2	3/8	5/8	9/16	1/2	3/8	5/16	1/4
Wt./ft	67.82	61.83	55.66	42.79	67.82	61.83	55.66	42.79	36.10	29.23
F_y					46 ksi					
0	549	502	453	348	549	502	453	348	293	237
6	500	457	411	317	513	470	424	326	274	223
7	488	447	402	310	505	463	417	321	270	219
8	476	436	393	303	497	455	411	316	266	216
9	464	425	383	296	488	447	404	311	262	213
10	450	412	372	288	479	439	396	305	257	209
11	436	400	361	280	470	430	389	300	253	205
12	422	387	349	271	460	421	380	294	248	201
13	406	373	337	262	449	412	372	288	243	197
14	390	358	325	253	439	402	363	281	237	193
15	374	344	312	243	427	392	355	274	232	188
16	357	328	298	233	416	382	345	267	226	184
17	339	312	284	222	404	371	336	260	220	179
18	320	295	269	212	392	360	326	253	214	174
19	301	278	254	200	379	348	316	245	208	169
20	281	260	238	189	366	336	305	238	201	164
22	239	223	205	164	339	312	283	221	188	153
24	201	187	173	139	310	286	260	204	173	142
26	171	159	147	118	280	258	236	186	158	130
28	148	138	127	102	248	229	210	167	143	117
30	129	120	111	89	216	200	184	147	126	104
32	113	105	97	78	190	176	162	129	111	91
34	100	93	86	69	168	156	143	114	98	81
36	89	83	77	62	150	139	128	102	87	72
38	80	75	69	55	135	125	115	92	78	65
39	76	71	65	53	128	118	109	87	74	61
40			62	50	122	113	103	83	71	58
Properties										
A (in.^{2})	19.90	18.20	16.40	12.60	19.90	18.20	16.40	12.60	10.60	8.59
I_x (in.^{4})	337	313	287	228	266	247	226	180	154	127
I_y (in.^{4})	112	104	96.0	77.2	187	174	160	127	109	90.2
r_x/r_y	1.74	1.74	1.73	1.72	1.19	1.19	1.19	1.19	1.19	1.19
r_y (in.)	2.37	2.39	2.42	2.48	3.07	3.09	3.12	3.18	3.21	3.24
B_x } Bending	0.354	0.349	0.342	0.330	0.374	0.369	0.363	0.351	0.344	0.338
B_y } factors	0.535	0.524	0.511	0.488	0.425	0.418	0.411	0.396	0.388	0.382
$a_x/10^6$	50.3	46.7	42.8	34.0	39.5	36.7	33.8	26.8	22.9	18.9
$a_y/10^6$	16.6	15.5	14.3	11.5	27.9	25.9	23.8	19.0	16.3	13.4

Effective length in ft KL with respect to least radius of gyration

Note: Heavy line indicates Kl/r of 200.

COLUMNS
Rectangular structural tubing
Allowable concentric loads in kips

F_y = 46 ksi

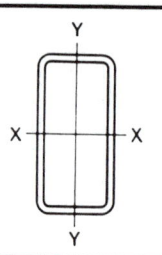

Nominal Size		10 × 6					10 × 5				
Thickness		⅝	⁹⁄₁₆	½	⅜	⁵⁄₁₆	⅝	⁹⁄₁₆	½	⅜	⁵⁄₁₆
Wt./ft		59.32	54.17	48.85	37.69	31.84	55.06	50.34	45.45	35.13	29.72
F_y						46 ksi					
Effective length in ft KL with respect to least radius of gyration	0	480	439	397	306	258	447	408	370	284	241
	6	436	398	360	279	236	394	360	327	253	215
	7	426	389	352	273	231	381	349	317	245	209
	8	415	379	343	266	225	368	337	307	238	202
	9	403	369	334	259	220	354	325	296	230	196
	10	391	358	325	252	214	339	312	284	221	188
	11	379	347	315	245	208	324	298	272	212	181
	12	365	335	304	237	201	308	283	259	202	173
	13	351	322	293	229	194	291	268	245	192	165
	14	337	309	282	220	187	273	252	231	182	156
	15	322	296	270	211	180	254	235	216	171	147
	16	306	282	257	202	172	235	217	201	160	138
	18	274	252	231	183	156	193	180	168	136	118
	20	238	221	203	162	139	156	146	136	111	97
	22	200	187	173	140	120	129	121	113	92	80
	24	168	157	146	118	102	109	101	95	77	67
	25	155	144	134	108	94	100	93	87	71	62
	26	143	134	124	100	87	93	86	81	66	57
	27	133	124	115	93	80	86	80	75	61	53
	28	124	115	107	86	75	80	74	69	57	49
	30	108	100	93	75	65	70	65	61	49	43
	32	95	88	82	66	57	61	57	53	43	38
	34	84	78	73	59	51				38	34
	36	75	70	65	52	45					
	38	67	63	58	47	41					
	39		59	55	45	39					
	40				42	37					

Properties											
A (in.^{2})		17.40	15.90	14.40	11.10	9.36	16.20	14.80	13.40	10.30	8.73
I_x (in.^{4})		211	197	181	145	125	183	171	158	128	110
I_y (in.^{4})		93.5	87.5	80.8	65.4	56.5	60.0	56.5	52.5	42.9	37.2
r_x/r_y		1.50	1.50	1.50	1.49	1.49	1.75	1.74	1.74	1.72	1.71
r_y (in.)		2.32	2.34	2.37	2.43	2.46	1.93	1.95	1.98	2.04	2.07
B_x ⎱ Bending		0.413	0.405	0.397	0.382	0.375	0.441	0.431	0.424	0.404	0.397
B_y ⎰ factors		0.559	0.545	0.533	0.508	0.497	0.675	0.655	0.638	0.602	0.586
$a_x/10^6$		31.4	29.2	26.9	21.6	18.6	27.4	25.5	23.6	18.9	16.4
$a_y/10^6$		13.9	13.0	12.0	9.74	8.42	8.99	8.39	7.83	6.39	5.57

Note: Heavy line indicates Kl/r of 200.

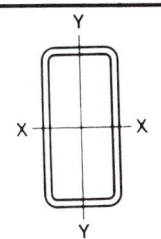

COLUMNS
Rectangular structural tubing
Allowable concentric loads in kips

F_y = 46 ksi

Nominal Size	9 × 7					9 × 6				
Thickness	$\frac{9}{16}$	$\frac{1}{2}$	$\frac{3}{8}$	$\frac{5}{16}$	$\frac{1}{4}$	$\frac{5}{8}$	$\frac{9}{16}$	$\frac{1}{2}$	$\frac{3}{8}$	$\frac{5}{16}$
Wt./ft	54.17	48.85	37.69	31.84	25.82	55.06	50.34	45.45	35.13	29.72
F_y	46 ksi									
0	439	397	306	258	209	447	408	370	284	241
6	405	367	283	239	194	404	370	335	259	220
7	397	360	278	235	191	394	361	328	253	215
8	389	353	273	231	187	384	352	320	247	210
9	381	345	267	226	184	373	342	311	241	204
10	372	338	262	221	180	362	332	302	234	199
11	363	329	255	216	176	350	321	292	227	193
12	353	321	249	211	171	337	310	282	219	187
13	343	312	242	205	167	324	298	272	211	180
14	333	302	235	199	162	310	286	261	203	174
15	322	292	228	193	157	296	273	249	195	167
16	310	282	220	187	152	281	260	238	186	159
17	299	272	213	180	147	266	246	225	177	152
18	287	261	205	174	142	250	232	213	168	144
19	274	250	196	167	137	233	217	200	158	136
20	261	239	188	160	131	216	201	186	148	128
21	248	227	179	153	125	198	185	172	138	119
22	235	214	170	145	119	180	169	157	127	110
23	220	202	160	137	113	165	155	144	116	101
24	206	189	151	129	107	152	142	132	107	93
25	191	175	141	121	100	140	131	122	98	86
26	176	162	131	113	93	129	121	113	91	79
27	164	150	121	104	86	120	112	104	84	73
28	152	140	113	97	80	111	104	97	78	68
29	142	130	105	90	75	104	97	90	73	64
30	133	122	98	85	70	97	91	85	68	59
32	117	107	86	74	62	85	80	74	60	52
34	103	95	76	66	55	76	71	66	53	46
36	92	85	68	59	49	67	63	59	47	41
38	83	76	61	53	44	60	57	53	43	37
40	75	69	55	48	39				38	33
Properties										
A (in.²)	15.90	14.40	11.10	9.36	7.59	16.20	14.80	13.40	10.30	8.73
I_x (in.⁴)	170	157	126	108	89.4	161	150	139	112	96.4
I_y (in.⁴)	115	106	85.1	73.5	60.8	84.5	79.2	73.2	59.4	51.4
r_x/r_y	1.22	1.22	1.22	1.21	1.21	1.38	1.38	1.38	1.37	1.37
r_y (in.)	2.69	2.71	2.77	2.80	2.83	2.28	2.31	2.34	2.40	2.43
B_x Bending	0.420	0.414	0.398	0.390	0.381	0.453	0.443	0.435	0.415	0.408
B_y factors	0.485	0.477	0.457	0.446	0.436	0.890	0.558	0.549	0.520	0.511
$a_x/10^6$	25.3	23.4	18.8	16.1	13.3	24.0	22.4	20.7	16.6	14.3
$a_y/10^6$	17.1	15.8	12.7	10.9	9.06	12.5	11.8	10.9	8.84	7.68

Note: Heavy line indicates Kl/r of 200.

Effective length in ft KL with respect to least radius of gyration

AMERICAN INSTITUTE OF STEEL CONSTRUCTION

COLUMNS
Rectangular structural tubing
Allowable concentric loads in kips

F_y = 46 ksi

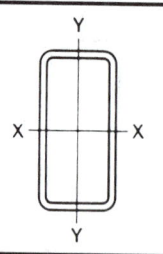

Nominal Size		9 × 5				8 × 6			
Thickness	$^9/_{16}$	$^1/_2$	$^3/_8$	$^5/_{16}$	$^9/_{16}$	$^1/_2$	$^3/_8$	$^5/_{16}$	$^1/_4$
Wt./ft	46.51	42.05	32.58	27.59	46.51	42.05	32.58	27.59	22.42
F_y					46 ksi				

Effective length in ft KL with respect to least radius of gyration

KL	$^9/_{16}$	$^1/_2$	$^3/_8$	$^5/_{16}$	$^9/_{16}$	$^1/_2$	$^3/_8$	$^5/_{16}$	$^1/_4$
0	378	342	264	224	378	342	264	224	182
6	333	302	234	199	342	309	240	204	166
7	322	293	228	193	334	301	235	199	162
8	311	283	220	187	325	294	229	194	158
9	300	273	212	181	316	286	223	189	154
10	287	262	204	174	306	277	216	184	150
11	274	250	196	167	296	268	210	178	145
12	260	238	187	159	285	258	202	172	141
13	246	225	177	152	274	249	195	166	136
14	231	212	167	143	262	238	187	160	131
15	215	198	157	135	250	228	179	153	125
16	198	183	146	126	238	216	171	146	120
18	163	152	123	107	211	193	154	132	108
20	132	123	100	88	183	168	135	116	96
22	109	102	83	72	153	141	115	99	83
24	92	86	70	61	128	118	96	84	70
25	85	79	64	56	118	109	89	77	64
26	78	73	59	52	109	101	82	71	59
27	73	68	55	48	101	93	76	66	55
28	68	63	51	45	94	87	71	61	51
30	59	55	45	39	82	76	62	53	44
32	52	48	39	34	72	66	54	47	39
34				30	64	59	48	42	35
36					57	53	43	37	31
38					51	47	38	33	28
39							36	32	26
40									25

Properties

	$^9/_{16}$	$^1/_2$	$^3/_8$	$^5/_{16}$	$^9/_{16}$	$^1/_2$	$^3/_8$	$^5/_{16}$	$^1/_4$
A (in.^{2})	13.70	12.40	9.58	8.11	13.70	12.40	9.58	8.11	6.59
I_x (in.^{4})	130	121	97.8	84.6	112	103	83.7	72.4	60.1
I_y (in.^{4})	50.9	47.4	38.8	33.8	70.8	65.7	53.5	46.4	38.6
r_x/r_y	1.60	1.59	1.59	1.58	1.25	1.25	1.25	1.25	1.25
r_y (in.)	1.93	1.96	2.01	2.04	2.28	2.31	2.36	2.39	2.42
B_x } Bending	0.472	0.463	0.441	0.431	0.491	0.479	0.458	0.448	0.438
B_y } factors	0.672	0.656	0.618	0.601	0.581	0.565	0.537	0.524	0.512
$a_x/10^6$	19.5	18.0	14.6	12.6	16.7	15.4	12.5	10.8	8.96
$a_y/10^6$	7.60	7.10	5.77	5.03	10.6	9.78	7.97	6.91	5.75

Note: Heavy line indicates Kl/r of 200.

F_y = 46 ksi

COLUMNS
Rectangular structural tubing
Allowable concentric loads in kips

Nominal Size		8 × 4					7 × 5			
Thickness		$\frac{9}{16}$	$\frac{1}{2}$	$\frac{3}{8}$	$\frac{5}{16}$	$\frac{1}{4}$	$\frac{1}{2}$	$\frac{3}{8}$	$\frac{5}{16}$	$\frac{1}{4}$
Wt./ft		38.86	35.24	27.48	23.34	19.02	35.24	27.48	23.34	19.02
F_y						46 ksi				
	0	315	287	223	189	154	287	223	189	154
	6	262	240	188	161	132	252	197	167	137
	7	250	229	180	154	126	244	191	162	133
	8	236	217	172	147	120	235	184	157	129
	9	222	204	162	139	114	226	177	151	124
	10	207	191	153	131	108	216	170	145	119
	11	191	177	142	123	101	206	162	139	114
	12	174	162	131	113	94	196	154	132	109
	13	156	146	120	104	86	184	146	125	103
	14	138	130	108	94	79	173	137	118	98
	15	120	113	95	83	70	160	128	111	92
	16	105	100	83	73	62	147	119	103	85
	18	83	79	66	58	49	120	98	86	72
	20	67	64	53	47	40	97	80	70	59
	22	56	53	44	39	33	80	66	58	48
	24	47	44	37	33	27	68	55	48	41
	25	43	41	34	30	25	62	51	45	37
	26			32	28	23	58	47	41	35
	27				26	22	53	44	38	32
	28						50	41	36	30
	30						43	35	31	26
	32							31	27	23
	33								26	22

Effective length in ft KL with respect to least radius of gyration

Properties										
A (in.^{2})		11.40	10.40	8.08	6.86	5.59	10.40	8.08	6.86	5.59
I_x (in.^{4})		80.5	75.1	61.9	53.9	45.1	63.5	52.2	45.5	38.0
I_y (in.^{4})		26.2	24.6	20.6	18.1	15.3	37.2	30.8	26.9	22.6
r_x/r_y		1.75	1.75	1.73	1.73	1.72	1.31	1.30	1.30	1.30
r_y (in.)		1.51	1.54	1.60	1.62	1.65	1.90	1.95	1.98	2.01
B_x } Bending		0.567	0.552	0.522	0.509	0.496	0.575	0.542	0.528	0.513
B_y } factors		0.870	0.842	0.785	0.758	0.733	0.698	0.657	0.635	0.618
$a_x/10^6$		11.9	11.2	9.22	8.04	6.72	9.53	7.77	6.80	5.67
$a_y/10^6$		3.87	3.66	3.07	2.70	2.27	5.59	4.58	4.01	3.37

Note: Heavy line indicates Kl/r of 200.

COLUMNS
Rectangular structural tubing
Allowable concentric loads in kips

F_y = 46 ksi

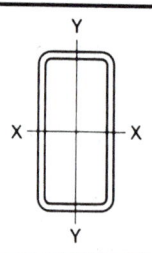

Nominal Size	6 × 5					6 × 4				
Thickness	½	⅜	⁵⁄₁₆	¼	³⁄₁₆	½	⅜	⁵⁄₁₆	¼	³⁄₁₆
Wt./ft	31.84	24.93	21.21	17.32	13.25	28.43	22.37	19.08	15.62	11.97
F_y	46 ksi									
0	258	202	172	140	107	231	182	155	127	97
6	226	178	152	124	95	191	152	130	107	83
7	218	172	147	120	92	182	145	125	103	79
8	210	166	142	116	89	172	138	118	98	75
9	201	160	136	112	86	161	130	112	92	71
10	192	153	131	108	83	150	121	105	87	67
11	183	146	125	103	79	138	112	97	81	63
12	173	138	119	98	76	125	103	90	75	58
13	162	130	112	93	72	112	93	81	68	53
14	151	122	105	87	68	97	83	73	61	48
15	140	114	98	82	63	85	72	64	54	43
16	128	105	91	76	59	75	63	56	48	38
17	115	95	83	70	54	66	56	50	42	33
18	103	86	75	63	50	59	50	44	38	30
19	92	77	67	57	45	53	45	40	34	27
20	83	69	61	51	40	48	41	36	30	24
21	75	63	55	46	37	43	37	33	28	22
22	69	57	50	42	33	39	34	30	25	20
24	58	48	42	36	28	33	28	25	21	17
25	53	44	39	33	26		26	23	19	15
26	49	41	36	30	24			21	18	14
27	46	38	33	28	22					13
29	40	33	29	24	19					
31		29	25	21	17					
32			24	20	16					

Properties

	6 × 5					6 × 4				
A (in.^{2})	9.36	7.33	6.23	5.09	3.89	8.36	6.58	5.61	4.59	3.52
I_x (in.^{4})	42.9	35.6	31.2	26.2	20.6	35.3	29.7	26.2	22.1	17.4
I_y (in.^{4})	32.1	26.8	23.5	19.8	15.6	18.4	15.6	13.8	11.7	9.32
r_x/r_y	1.16	1.16	1.15	1.15	1.15	1.38	1.38	1.37	1.37	1.37
r_y (in.)	1.85	1.91	1.94	1.97	1.97	1.48	1.54	1.57	1.60	1.63
B_x } Bending	0.655	0.616	0.599	0.582	0.566	0.710	0.664	0.643	0.624	0.605
B_y } factors	0.731	0.685	0.663	0.643	0.624	0.907	0.841	0.811	0.782	0.755
$a_x/10^6$	6.39	5.33	4.66	3.91	3.07	5.26	4.43	3.90	3.29	2.60
$a_y/10^6$	4.77	3.98	3.49	2.94	2.32	2.75	2.33	2.06	1.75	1.39

Effective length in ft KL with respect to least radius of gyration

Note: Heavy line indicates Kl/r of 200.

AMERICAN INSTITUTE OF STEEL CONSTRUCTION

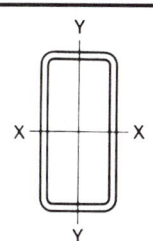

	F_y = 46 ksi

COLUMNS
Rectangular structural tubing
Allowable concentric loads in kips

Nominal Size	6 × 3				5 × 3				
Thickness	³⁄₈	⁵⁄₁₆	¼	³⁄₁₆	½	³⁄₈	⁵⁄₁₆	¼	³⁄₁₆
Wt./ft	19.82	16.96	13.91	10.70	21.23	17.27	14.83	12.21	9.42
F_y					46 ksi				
0	161	137	113	87	176	140	120	99	76
2	152	130	107	82	165	132	114	94	72
3	146	125	103	79	157	126	109	90	70
4	139	119	98	76	149	120	104	86	67
5	131	113	93	72	139	113	98	81	63
6	122	105	87	68	128	105	91	76	59
7	112	98	81	63	117	97	84	70	55
8	102	89	74	58	104	87	77	64	51
9	91	80	67	53	91	77	68	58	46
10	79	70	60	47	76	67	60	51	40
11	67	60	51	41	63	56	50	43	35
12	56	50	43	35	53	47	42	36	29
13	48	43	37	30	45	40	36	31	25
14	41	37	32	26	39	34	31	27	22
15	36	32	28	22	34	30	27	23	19
16	32	28	24	20	30	26	24	20	17
17	28	25	22	17	26	23	21	18	15
18	25	22	19	15		21	19	16	13
19	22	20	17	14			17	15	12
20			16	13					11

Properties									
A (in.²)	5.83	4.98	4.09	3.14	6.36	5.08	4.36	3.59	2.77
I_x (in.⁴)	23.8	21.1	17.9	14.3	16.9	14.7	13.2	11.3	9.06
I_y (in.⁴)	7.78	6.98	6.00	4.83	7.33	6.48	5.85	5.05	4.08
r_x/r_y	1.75	1.74	1.73	1.72	1.52	1.51	1.50	1.50	1.49
r_y (in.)	1.16	1.18	1.21	1.24	1.07	1.13	1.16	1.19	1.21
B_x } Bending	0.735	0.709	0.684	0.661	0.942	0.862	0.827	0.794	0.764
B_y } factors	1.12	1.07	1.02	0.977	1.30	1.17	1.12	1.07	1.02
$a_x/10^6$	3.54	3.14	2.67	2.13	2.51	2.19	1.96	1.68	1.35
$a_y/10^6$	1.16	1.04	0.894	0.719	1.09	0.966	0.871	0.753	0.609

Note: Heavy line indicates Kl/r of 200.

AMERICAN INSTITUTE OF STEEL CONSTRUCTION

Effective length in ft KL with respect to least radius of gyration

F_y = 46 ksi

COLUMNS
Rectangular structural tubing
Allowable concentric loads in kips

Nominal Size		4 × 3			4 × 2			3½ × 2½		3 × 2	
Thickness		$\frac{5}{16}$	$\frac{1}{4}$	$\frac{3}{16}$	$\frac{5}{16}$	$\frac{1}{4}$	$\frac{3}{16}$	$\frac{1}{4}$	$\frac{3}{16}$	$\frac{1}{4}$	$\frac{3}{16}$
Wt./ft		12.70	10.51	8.15	10.58	8.81	6.87	8.81	6.87	7.11	5.59
F_y							46 ksi				
	0	103	85	66	86	71	56	71	56	58	45
	2	97	80	62	77	65	51	66	52	52	41
	3	93	77	60	71	60	47	63	49	48	38
	4	88	73	57	64	54	43	59	46	43	34
	5	83	69	54	56	48	38	54	43	37	30
	6	77	65	51	47	41	33	49	39	31	26
	7	71	60	47	36	32	27	43	34	24	21
	8	64	54	43	28	25	21	37	30	19	16
	9	57	48	38	22	20	16	30	25	15	13
	10	49	42	34	18	16	13	24	20	12	10
	11	40	35	29	15	13	11	20	17	10	8
	12	34	30	24	12	11	9	17	14	8	7
	13	29	25	21			8	14	12		
	14	25	22	18				12	10		
	15	22	19	15				11	9		
	16	19	17	14					8		
	17	17	15	12							
	18	15	13	11							
	19		12	10							

Effective length in ft KL with respect to least radius of gyration

Properties											
A (in.^{2})		3.73	3.09	2.39	3.11	2.59	2.02	2.59	2.02	2.09	1.64
I_x (in.^{4})		7.45	6.45	5.23	5.32	4.69	3.87	3.97	3.26	2.21	1.86
I_y (in.^{4})		4.71	4.10	3.34	1.71	1.54	1.29	2.33	1.93	1.15	0.977
r_x/r_y		1.26	1.25	1.25	1.76	1.75	1.73	1.31	1.30	1.38	1.38
r_y (in.)		1.12	1.15	1.18	0.743	0.770	0.798	0.948	0.977	0.742	0.771
B_x } Bending		1.00	0.957	0.915	1.17	1.10	1.04	1.14	1.09	1.42	1.33
B_y } factors		1.19	1.13	1.07	1.81	1.68	1.57	1.39	1.31	1.81	1.68
$a_x/10^6$		1.11	0.962	0.779	0.792	0.699	0.576	0.593	0.485	0.329	0.277
$a_y/10^6$		0.702	0.611	0.498	0.256	0.229	0.192	0.347	0.287	0.172	0.146

Note: Heavy line indicates Kl/r of 200.

AMERICAN INSTITUTE OF STEEL CONSTRUCTION

ALLOWABLE CONCENTRIC LOADS ON COLUMNS
Double Angles and WT Shapes

For both double angles and WT shapes, the allowable concentric loads are tabulated for the effective length in ft KL with respect to both the X-X and Y-Y axis. Discussion under Sect. C2 of the AISC ASD Specification Commentary points out that for trusses, it is usual practice to take K equal to 1.0. No values are listed beyond Kl/r = 200. Allowable loads about the X-X axis are in accordance with Sect. E2 of the AISC ASD Specification.

For buckling about the Y-Y axis, the allowable loads are based on flexural-torsional buckling. The critical flexural-torsional elastic buckling stress F_e can be obtained from equations in the Load and Resistance Factor Design Specification Appendix E3. It is conservative for C_w to be taken as zero. However, actual values of C_w were used in the tables. Based on this stress, an effective slenderness can be calculated,

$$(Kl/r)_{eff} = \pi \sqrt{E/F_e}$$

This slenderness is inserted into the general column equations of AISC ASD Specification Sect. E2 to determine the Y-Y axis allowable concentric load.

For double angles buckling about the X-X or Y-Y axis, the connectors must be spaced so that the local slenderness a/r_z of the individual member does not exceed ¾ times the governing slenderness ratio of the overall member. Also, at least two intermediate connectors must be used to provide for adequate shear transfer. All connectors must be welded or utilize fully-tightened high-strength bolts.

In designing members fabricated of two angles connected to opposite faces of a gusset plate, Sect. J1.9 of the AISC ASD Specification states that eccentricity between the gage lines and gravity axis may be neglected. In the following tables, eccentricity is neglected.

The tabulated loads for double angles referred to in the Y-Y axis assume a ⅜-in. spacing between angles. These values are conservative when a wider spacing is provided. The following example illustrates a method for determining the allowable load when a ¾-in. gusset plate is used.

Examples 8 and 9 demonstrate how to determine the number of connectors, as well as the allowable loads.

EXAMPLE 7

Given:

Using F_y = 36 ksi steel, determine the maximum allowable concentric load with respect to the Y-Y axis on a double angle member of 8 × 8 × 1 angles with an effective length equal to 12 ft and connected to a ¾-in. thick gusset plate.

Solution:

r_y = 3.53 in. (from the double-angle column load table for 2 L8×8×1 with ⅜-in. plate).

r_y' = 3.67 in. (from Part 1, Properties, Two Equal Angles, 2 L8×8×1 with ¾-in. plate).

$$\frac{r_y}{r_y'} = \frac{3.53}{3.67} = 0.962$$

Equivalent length = 0.962 × 12 ft = 11.5 ft

Enter the column load table for 2 L8×8×1 with reference to the Y-Y axis for effective lengths between 10 and 15 ft, read 564 and 536 kips, respectively.

$$\text{Equivalent allowable load} = 564 - \left[(564 - 536) \times \left(\frac{11.5 - 10}{15 - 10}\right)\right]$$

$$= 556 \text{ kips.}$$

EXAMPLE 8

Given:

Using a double-angle member of 8×6×¾ angles (long legs back-to-back), and 36 ksi steel, with effective lengths $KL_x = KL_y = 16$ ft, determine the allowable load and number of intermediate connectors. Assume $K = 1$.

Solution:

From the double-angle column load tables,

$KL_x = 16$ ft, $P_{cr} = 315$ kips

$KL_y = 16$ ft, $P_{cr} = 300$ kips

Therefore, KL_y governs. The allowable load is 300 kips and the corresponding slenderness ratio is $KL_y/r_y = (12 \times 16) / 2.48 = 77$. The maximum slenderness of the individual angle is

$a/r_z \leq 0.75 \times 77 = 58$

$a = 58 \times 1.29 = 75$ in.

Use 2 intermediate connectors spaced at $(16 \times 12)/3 = 64$ in. This satisfies the requirement that at least 2 intermediate connectors must be used.

EXAMPLE 9

Given:

Using a double-angle member of 5×3×½ angles (short legs back-to-back) and 36 ksi steel, 36 ft long with braced points in the Y direction every 12 ft against buckling about the X-X axis, determine the allowable load and number of intermediate connectors. Assume $K = 1$.

Solution:

From the Double-Angle Column Load Tables,

$KL_x = 12$ ft, $P_{cr} = 37$ kips

$KL_y = 36$ ft, $P_{cr} = 37$ kips

The allowable load is 37 kips and the corresponding slenderness ratio is KL_x/r_x $= (12 \times 12)/0.829 = 174$. The maximum slenderness of the individual angle is

$a/r_z = 0.75 \times 174 = 130$

$a = 130 \times 0.648 = 84$ in.

Place one connector every 6 ft = 72 in. This provides 5 connectors along L_y including the braced points, which satisfies the Specification requirement that at least 2 intermediate connectors be used.

ALLOWABLE CONCENTRIC LOADS ON COLUMNS
SINGLE ANGLE STRUTS

Allowable concentric loads for single angle struts are not tabulated in this Manual because it is virtually impossible to load such struts concentrically. In theory, concentric loading could be accomplished by milling the ends of an angle and loading it through bearing plates. However, in practice, the actual eccentricity of loading is relatively large; and its neglect in design may lead to an underdesigned member.

The design of an equal leg, single angle strut loaded eccentrically must meet the provisions of Chap. H, Combined Stresses. Additional background and design recommendations on single angles are available.*

The following examples illustrate a rational ASD design procedure for an equal leg angle strut based on the mentioned references.

EXAMPLE 10

Given:

An angle $2 \times 2 \times \frac{1}{4}$ is loaded by a gusset plate attached to one leg with geometric axes eccentricities of 0.8 in. and 0.408 in. from the centroid, as shown. Determine the allowable compressive load P_a which may be applied. The maximum effective length KL is 40 in. for all bending axes.

$A = 0.938$ in.^{2}

$r_z = 0.391$ in.

$I_x = I_y = 0.348$ in.^{4}

$\alpha = 45°$

$F_y = 50$ ksi

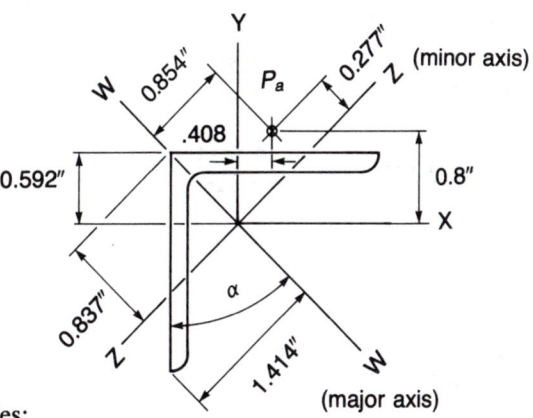

Solution:

Determine principal axis properties:

$$I_z = Ar_z^2 = 0.938 (0.391)^2 = 0.1434 \text{ in.}^4$$

$$I_z + I_w = I_x + I_y$$

$$
\begin{aligned}
I_w &= I_x + I_y - I_z \\
&= 2 (0.348) - 0.1434 = 0.553 \text{ in.}^4 \\
r_w &= \sqrt{I_w / A} = 0.768 \text{ in.} \\
(b/t) &= 2/0.25 = 8 < 76/\sqrt{50} = 10.7 \\
\therefore Q &= 1 \text{ (no local buckling)}
\end{aligned}
$$

Lutz, Leroy M. Behavior and Design of Angle Compression Members. 1988 National Steel Construction Conference Proceedings, AISC, Chicago, IL, June 1988.

American Institute of Steel Construction. Specification for Allowable Stress Design of Single-angle Members, Chicago, IL, 1989.

Compute allowable compression stress F_a without eccentricity for flexural buckling about minor principal axis:

$$\left(\frac{Kl}{r}\right)_z = \frac{40}{0.391} = 102.3$$

By interpolation from Table C-50 of Allowable Stresses for Compression Members ($F_y = 50$ ksi) in Part 3 of the Manual:

$$F_a = 14.17 \text{ ksi}$$

Check flexural-torsional buckling:

$$\left(\frac{Kl}{r}\right)_{max} = 102.3 > \frac{5.4 \, (b/t)}{Q} = \frac{5.4 \, (8)}{1} = 43.2$$

∴. Flexural-torsional buckling does not control, based on an empirically based limit from the first reference above.

Use interaction equation and principal axes considering moments due to load eccentricity:

assume $\left(\dfrac{f_a}{F_a}\right) > 0.15$ and AISC ASD Spec. Equation (H1-1) governs

$$\frac{f_a}{F_a} + \frac{M_w}{S_w F_{bw} \left[1 - (f_a/F'_{ew})\right]} + \frac{M_z}{S_z F_{bz} \left[1 - (f_a/F'_{ez})\right]} \leq 1.0$$

$$M_w = 0.843 P_a; \quad S_w = \frac{0.553}{1.414} = 0.391 \text{ in.}^3; \quad F'_{ew} = 55 \text{ ksi}$$

$$M_z = 0.277 P_a; \quad S_z = \frac{0.1434}{0.837} = 0.171 \text{ in.}^3; \quad F'_{ez} = 14.27 \text{ ksi}$$

Uniform moments: $C_m = C_b = 1.0$

$$\frac{L}{t} = \frac{40}{0.25} = 160 < \frac{9400}{F_y} = \frac{9400}{50} = 188$$

$$\therefore F_{bw} = 0.66 F_y = 33 \text{ ksi}.$$

The $9400/F_y$ limit is from the first reference above and indicates when lateral-torsional buckling will not reduce the allowable stress below $0.66 F_y$. With no local buckling, $F_{bz} = 0.66 F_y = 33$ ksi.

$$\frac{\dfrac{P_a}{0.938}}{14.17} + \frac{0.854 P_a}{(0.391) \, (33) \left[1 - \dfrac{P_a}{0.938 \, (55)}\right]} + \frac{0.277 P_a}{(0.171) \, (33) \left[1 - \dfrac{P_a}{0.938 \, (14.27)}\right]} = 1.0$$

Solve for $P_a = 4.5$ kip

$$\frac{4.5}{(0.938) \, (14.17)} = 0.339 > 0.15 \quad \textbf{o.k.}$$

EXAMPLE 11

Given:

Determine whether the angle strut shown can carry a 7-kip axial compression load. This example includes consideration of local, flexural and flexural-torsional buckling. A36 steel.

$I_x = I_y = 3.04$ in.^{4}

$A = 1.94$ in.^{2}

$x = y = 1.09$ in.

$r_z = 0.795$ in.

$c_t = x\sqrt{2} = 1.54$ in.

$c_c = \dfrac{4 + (0.25/2)}{\sqrt{2}} - c_t = 1.377$ in.

Assume eccentricity of load at 1.45 in.

$I_z = 1.94\,(0.795)^2 = 1.226$ in.^{4}

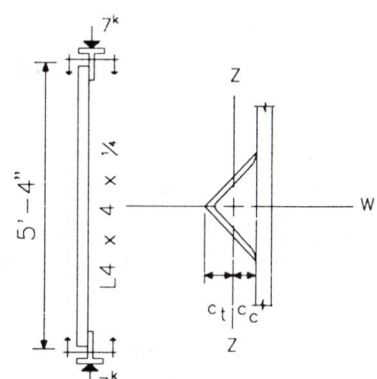

Solution:

Consider unbraced length of 5 ft - 4 in. = 64 in., conservatively use $K_z = 1.0$.

$(Kl/r) = 64 / 0.795 = 80.5$

Check local buckling:

$b/t = 4 / 0.25 = 16 > 76 / \sqrt{36}$

$Q = 1.34 - 0.00447\,(16)\,\sqrt{36} = 0.911$

$C_c' = \sqrt{2\,\pi^2\,(29{,}000) / [0.911(36)]} = 132.1$

For equal leg angles, flexural-torsional buckling will control if:

$(Kl/r)_{max} < 5.4\,(b/t) / Q$

$80.5 < 5.4\,(16) / 0.911 = 95$

Determine $(Kl/r)_{equiv.} = \pi\sqrt{E/F_e}$ using equations for equal leg angles,

$w_o = \sqrt{2}\,(1.09 - 0.25 / 2) = 1.365$ in.

$\bar{r}_o^2 = (1.365)^2 + 2(3.04) / 1.94 = 4.997$ in.

$H = 1 - (1.365)^2/4.996 = 0.627$

$J = At^2/3 = 1.94\,(0.25)^2/3 = 0.04042$ in.^{4}

$I_z + I_w = I_x + I_y$

$I_w = 2\,(3.04) - 1.226 = 4.854$ in.^{4}

$r_w = \sqrt{4.854 / 1.94} = 1.58$ in.

Use $K_w = 0.8$,

$F_{ew} = \pi^2\,(29{,}000)/[0.8(64) / 1.58]^2 = 273$ ksi

$F_{ej} = (11{,}200)\,(0.04042) / [1.94\,(4.996)] = 46.7$ ksi

$$F_e = \frac{273 + 46.7}{2\,(0.627)}\left[1 - \sqrt{1 - \frac{4\,(273)\,(46.7)\,(0.627)}{(273 + 46.7)^2}}\,\right]$$

$$= 43.6 \text{ ksi}$$

$$\left(\frac{Kl}{r}\right)_{equiv.} = \pi\sqrt{\frac{29{,}000}{43.6}} = 81 > 80.5 = \left(\frac{Kl}{r}\right)_z, \text{ or very}$$

minor slenderness increase due to flexural-torsional effects.

Using Table 3,

$$81 \,/\, 132.1 = 0.613,\ C_a = 0.436$$

$$F_a = 0.436\,(0.911)\,(36) = 14.3 \text{ ksi}$$

$$f_a = 7 \,/\, 1.94 = 3.61 \text{ ksi}$$

$$M = P\,(1.45) = 7\,(1.45) = 10.15 \text{ kip-in.}$$

$$f_{bz} = 10.15\,(1.375) \,/\, 1.226 = 11.38 \text{ ksi}$$

With bending only about minor axis Z, there is no flexural-torsional buckling. Check local buckling:

$$F_{bz} = 0.911\,(0.60)\,(36) = 19.68 \text{ ksi}$$

$$F'_{ez} = 22.76 \text{ ksi}$$

$$C_m = 1$$

Check interaction equation:

$$\frac{3.61}{14.3} + \frac{(1)\,11.38}{19.68\,(1 - 3.61/22.76)} = 0.252 + 0.687 = 0.939 < 1 \quad \textbf{o.k.}$$

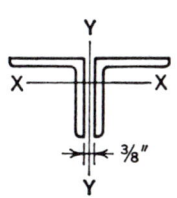

COLUMNS
Double angles
Allowable concentric loads in kips
Equal legs
⅜ in. back-to-back of angles

| F_y = 36 ksi |
| F_y = 50 ksi |

Size				8 x 8									
Thickness		1⅛		1		⅞		¾		⅝		½	
Wt./ft		113.8		102.0		90.0		77.8		65.4		52.8	
F_y		36	50	36	50	36	50	36	50	36	50	36	50
X-X AXIS	0	724	1005	648	900	572	795	495	687	413	539	305	388
	10	616	818	553	734	489	649	423	563	354	447	264	327
	14	552	705	496	634	439	562	381	488	320	392	240	291
	18	479	573	431	518	382	459	332	400	279	327	213	249
	22	395	420	357	383	317	341	277	299	234	253	182	201
	26	301	301	274	274	244	244	214	214	183	183	147	149
	30	226	226	206	206	183	183	161	161	137	137	112	112
	34	176	176	160	160	143	143	125	125	107	107	87	87
	38	141	141	128	128	114	114	100	100	85	85	70	70
	39	134	134	122	122	108	108	95	95	81	81	66	66
	40	127	127	116	116	103	103	91	91	77	77	63	63
	41							86	86	73	73	60	60
Y-Y AXIS	0	724	1005	648	900	572	795	495	687	413	539	305	388
	10	639	859	564	755	488	648	407	535	322	396	221	261
	15	603	795	536	704	467	611	394	510	314	383	217	254
	20	552	704	491	625	429	544	365	459	295	353	208	241
	25	492	596	437	528	382	460	326	389	266	305	192	217
	30	424	473	376	417	328	361	279	304	228	244	169	181
	35	348	352	308	310	268	269	227	227	184	184	140	141
	40	270	270	239	239	207	207	176	176	144	144	111	111
	45	214	214	189	189	165	165	140	140	115	115	89	89
	50	174	174	154	154	134	134	114	114	93	93	73	73
	55	144	144	127	127	111	111	94	94	78	78	61	61
	56	139	139	123	123	107	107	91	91	75	75	59	59
	57	134	134	118	118	103	103	88	88	72	72	57	57
	58	129	129	114	114	100	100	85	85				
	59	125	125										

Effective length in ft KL with respect to indicated axis

Properties of 2 angles — ⅜ in. back-to-back

	1⅛	1	⅞	¾	⅝	½
A (in.²)	33.5	30.0	26.5	22.9	19.2	15.5
r_x (in.)	2.42	2.44	2.45	2.47	2.49	2.50
r_y (in.)	3.55	3.53	3.51	3.49	3.47	3.45

Heavy line indicates Kl/r of 200.

| F_y = 36 ksi |
| F_y = 50 ksi |

COLUMNS
Double angles
Allowable concentric loads in kips
Equal legs
⅜ in. back-to-back of angles

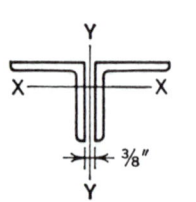

Size		6 x 6										
Thickness		1		⅞		¾		⅝		½		⅜	
Wt./ft		74.8		66.2		57.4		48.4		39.2		29.8	
F_y		36	50	36	50	36	50	36	50	36	50	36	50

Effective length in ft KL with respect to indicated axis

X-X AXIS

KL	36	50	36	50	36	50	36	50	36	50	36	50
0	475	660	421	585	365	507	307	426	248	331	172	218
8	397	524	353	465	306	405	258	341	209	268	147	181
10	369	474	328	421	285	367	240	310	195	245	138	167
12	338	418	300	372	262	326	221	275	180	219	128	152
14	304	357	270	318	236	280	199	237	163	191	117	136
16	267	289	238	259	209	229	176	194	145	159	106	118
18	226	228	202	204	179	181	151	154	125	127	93	99
22	153	153	137	137	121	121	103	103	85	85	66	66
26	109	109	98	98	87	87	74	74	61	61	47	47
30	82	82	74	74	65	65	55	55	46	46	36	36
31									43	43	33	33

Y-Y AXIS

KL	36	50	36	50	36	50	36	50	36	50	36	50
0	475	660	421	585	365	507	307	426	248	331	172	218
10	410	548	361	480	309	409	253	333	195	244	123	145
12	395	520	347	456	298	390	246	320	191	237	121	142
14	377	489	332	429	285	367	236	302	184	227	119	139
16	358	454	315	398	270	341	224	281	176	213	115	133
18	337	418	296	365	254	312	211	257	166	197	111	126
20	316	378	277	330	237	281	197	231	155	178	105	117
22	292	336	256	292	219	248	181	203	143	157	98	106
24	268	291	234	251	200	213	165	174	130	135	90	94
26	242	249	210	215	179	182	148	149	116	117	82	82
28	214	215	186	186	158	158	129	129	101	101	72	72
30	187	187	162	162	138	138	113	113	89	89	64	64
32	165	165	143	143	121	121	100	100	79	79	57	57
34	146	146	126	126	108	108	89	89	70	70	51	51
36	130	130	113	113	96	96	79	79	63	63	46	46
38	117	117	101	101	86	86	71	71	56	56	41	41
40	106	106	92	92	78	78	64	64	51	51	37	37
42	96	96	83	83	71	71	58	58	46	46	34	34
43	92	92	79	79	68	68	56	56	44	44	33	33
44	87	87	76	76	65	65	53	53	42	42		
45	84	84	72	72								

Properties of 2 angles — ⅜ in. back-to-back						
A (in.²)	22.0	19.5	16.9	14.2	11.5	8.72	
r_x (in.)	1.80	1.81	1.83	1.84	1.86	1.88	
r_y (in.)	2.73	2.70	2.68	2.66	2.64	2.62	

Heavy line indicates Kl/r of 200.

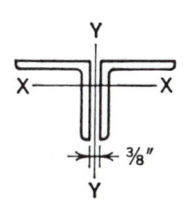

		F_y = 36 ksi
	COLUMNS Double angles Allowable concentric loads in kips **Equal legs** ⅜ in. back-to-back of angles	F_y = 50 ksi

Size		5 x 5								
Thickness		⅞		¾		½		⅜		5⁄16
Wt./ft		54.5		47.2		32.4		24.6		20.6
F_y		36	50	36	50	36	50	36	50	36	50

Effective length in ft KL with respect to indicated axis	X-X AXIS	0	346	480	300	417	205	285	153	199	119	151
		6	296	394	258	343	177	236	133	167	104	129
		8	272	351	237	307	163	212	123	152	97	118
		10	245	302	214	266	148	185	112	134	89	106
		12	214	248	188	219	131	154	99	115	80	92
		14	181	188	160	168	112	119	86	93	70	77
		16	144	144	128	128	91	91	71	71	59	60
		18	114	114	101	101	72	72	56	56	48	48
		20	92	92	82	82	58	58	46	46	39	39
		22	76	76	68	68	48	48	38	38	32	32
		24	64	64	57	57	41	41	32	32	27	27
		25			53	53	37	37	29	29	25	25
		26							27	27	23	23
	Y-Y AXIS	0	346	480	300	417	205	285	153	199	119	151
		6	312	423	268	361	172	227	119	145	87	102
		8	302	404	260	347	169	222	117	143	86	101
		10	288	380	249	327	164	213	115	139	85	99
		12	273	353	236	304	156	199	111	133	83	96
		14	256	323	221	278	147	183	106	124	80	92
		16	238	290	205	249	136	163	98	113	75	85
		18	218	255	188	218	124	142	90	100	70	77
		20	197	216	169	184	112	118	81	85	64	67
		22	175	180	150	153	98	99	71	71	57	57
		24	151	151	129	129	83	83	61	61	49	49
		26	129	129	110	110	71	71	52	52	42	42
		28	111	111	95	95	62	62	45	45	37	37
		30	97	97	83	83	54	54	40	40	32	32
		32	85	85	73	73	48	48	35	35	29	29
		34	76	76	64	64	42	42	31	31	26	26
		36	67	67	58	58	38	38	28	28	23	23
		37	64	64	54	54	36	36	26	26		
		38	61	61	52	52						

Properties of 2 angles — ⅜ in. back-to-back											
A (in.^{2})		16.0		13.9		9.50		7.22		6.05	
r_x (in.)		1.49		1.51		1.54		1.56		1.57	
r_y (in.)		2.30		2.28		2.24		2.22		2.21	

Heavy line indicates Kl/r of 200.

F_y = 36 ksi
F_y = 50 ksi

COLUMNS
Double angles
Allowable concentric loads in kips
Equal legs
3/8 in. back-to-back of angles

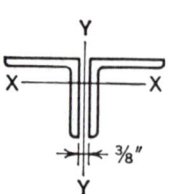

Size		4 x 4										
Thickness		3/4		5/8		1/2		3/8		5/16		1/4
Wt./ft		37.0		31.4		25.6		19.6		16.4		13.2
F_y		36	50	36	50	36	50	36	50	36	50	36	50
X-X AXIS 0		235	327	199	277	162	225	124	172	103	135	76	97
4		209	281	177	238	144	194	110	149	92	118	69	86
6		189	247	161	209	131	172	100	131	84	105	63	77
8		167	206	142	175	116	145	89	111	75	90	57	68
10		140	158	120	136	99	113	76	87	64	73	50	56
12		110	111	95	96	79	80	61	62	52	53	41	44
14		82	82	70	70	59	59	46	46	39	39	32	32
16		63	63	54	54	45	45	35	35	30	30	25	25
18		49	49	42	42	36	36	28	28	24	24	19	19
19		44	44	38	38	32	32	25	25	21	21	17	17
20				34	34	29	29	22	22	19	19	16	16
Y-Y AXIS 0		235	327	199	277	162	225	124	172	103	135	76	97
6		209	281	175	234	139	186	101	132	80	98	55	65
8		198	261	166	219	133	175	98	126	78	95	54	63
10		185	238	155	200	125	159	92	117	75	89	52	61
12		171	213	143	178	115	141	85	104	69	81	50	57
14		155	185	129	154	103	121	77	89	63	70	46	51
16		138	153	115	127	91	99	68	72	55	58	41	44
18		119	122	99	101	78	79	57	58	47	47	36	36
20		99	99	82	82	64	64	47	47	38	38	30	30
22		82	82	68	68	53	53	39	39	32	32	25	25
24		69	69	57	57	45	45	33	33	27	27	21	21
26		59	59	49	49	38	38	28	28	23	23	18	18
28		51	51	42	42	33	33	24	24	20	20	16	16
29		47	47	39	39	31	31	23	23	19	19	15	15
30		44	44	37	37	29	29	21	21	18	18		
31		41	41	34	34								

Effective length in ft KL with respect to indicated axis

Properties of 2 angles — 3/8 in. back-to-back						
A (in.^{2})	10.9	9.22	7.50	5.72	4.80	3.88
r_x (in.)	1.19	1.20	1.22	1.23	1.24	1.25
r_y (in.)	1.88	1.86	1.83	1.81	1.80	1.79

Heavy line indicates Kl/r of 200.

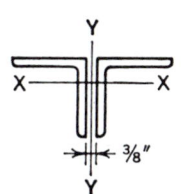

| | | | **COLUMNS** Double angles Allowable concentric loads in kips Equal legs ⅜ in. back-to-back of angles | | | | F_y = 36 ksi F_y = 50 ksi | |

Size		3½ x 3½						
Thickness		⅜		⁵⁄₁₆		¼		
Wt./ft		17.0		14.4		11.6		
F_y		36	50	36	50	36	50	
X-X AXIS	0	107	149	90	124	70	91	
	2	102	139	86	116	67	86	
	4	93	125	79	104	62	78	
	6	83	107	70	89	55	68	
	8	71	84	60	71	48	56	
	10	57	59	48	51	39	42	
	12	41	41	35	35	29	29	
	14	30	30	26	26	21	21	
	16	23	23	20	20	16	16	
	17	20	20	17	17	14	14	
	18			16	16	13	13	
Y-Y AXIS	0	107	149	90	124	70	91	
	6	89	117	72	93	53	64	
	8	84	109	69	87	51	61	
	10	77	97	64	78	48	57	
	12	69	82	57	67	44	50	
	14	60	66	50	53	38	41	
	16	50	51	41	42	32	32	
	18	40	40	33	33	26	26	
	20	33	33	27	27	21	21	
	22	27	27	22	22	18	18	
	24	23	23	19	19	15	15	
	26	20	20	16	16	13	13	

Effective length in ft KL with respect to indicated axis

Properties of 2 angles — ⅜ in. back-to-back			
A (in.²)	4.97	4.21	3.38
r_x (in.)	1.07	1.08	1.09
r_y (in.)	1.61	1.60	1.59

Heavy line indicates Kl/r of 200.

AMERICAN INSTITUTE OF STEEL CONSTRUCTION

| F_y = 36 ksi |
| F_y = 50 ksi |

COLUMNS
Double angles
Allowable concentric loads in kips
Equal legs
⅜ in. back-to-back of angles

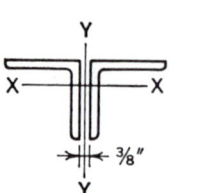

Size		3 x 3								
Thickness		½		⅜		5⁄16		¼		3⁄16	
Wt./ft		18.8		14.4		12.2		9.8		7.42	
F_y		36	50	36	50	36	50	36	50	36	50
X-X AXIS	0	119	165	91	127	77	107	62	83	43	55
	2	111	152	85	116	72	98	58	77	40	51
	4	99	131	76	101	64	85	52	67	37	45
	5	92	118	71	92	60	78	49	61	34	42
	6	84	104	65	81	55	69	45	55	32	38
	7	76	89	59	70	50	59	41	48	29	34
	8	66	72	52	57	44	49	36	40	26	29
	9	56	57	44	45	38	39	31	32	23	25
	10	46	46	36	36	31	31	26	26	20	20
	11	38	38	30	30	26	26	21	21	16	16
	12	32	32	25	25	22	22	18	18	14	14
	13	27	27	22	22	19	19	15	15	12	12
	14	23	23	19	19	16	16	13	13	10	10
	15			16	16	14	14	11	11	9	9
Y-Y AXIS	0	119	165	91	127	77	107	62	83	43	55
	2	109	148	80	108	65	87	50	63	31	37
	4	106	143	79	106	64	85	49	62	31	36
	6	100	132	75	99	62	81	48	60	30	36
	8	91	117	69	88	57	72	45	55	29	34
	10	81	99	61	74	51	61	40	47	27	31
	12	70	79	53	59	44	48	35	37	24	25
	14	58	59	43	44	36	36	28	28	20	20
	16	45	45	33	33	28	28	22	22	16	16
	18	36	36	27	27	22	22	17	17	13	13
	20	29	29	22	22	18	18	14	14	10	10
	22	24	24	18	18	15	15	12	12	9	9
	23	22	22	16	16	14	14	11	11	8	8

Effective length in ft KL with respect to indicated axis

Properties of 2 angles — ⅜ in. back-to-back					
A (in.²)	5.50	4.22	3.55	2.88	2.18
r_x (in.)	0.898	0.913	0.922	0.930	0.939
r_y (in.)	1.43	1.41	1.40	1.39	1.38

Heavy line indicates Kl/r of 200.

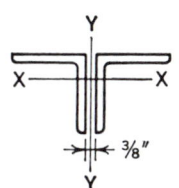

COLUMNS
Double angles
Allowable concentric loads in kips
Equal legs
3/8 in. back-to-back of angles

| F_y = 36 ksi |
| F_y = 50 ksi |

Size		2½ x 2½						
Thickness		3/8		5/16		1/4		3/16
Wt./ft		11.8		10.0		8.2		6.14
F_y		36	50	36	50	36	50	36	50

Effective length in ft KL with respect to indicated axis

X-X AXIS

KL	36	50	36	50	36	50	36	50
0	75	104	63	88	51	71	38	50
2	69	93	58	79	47	64	35	45
3	64	86	54	73	44	59	33	42
4	59	77	50	65	41	53	31	38
5	53	66	45	56	37	46	28	33
6	47	54	40	47	33	38	25	28
7	40	42	34	36	28	30	21	23
8	32	32	27	27	23	23	18	18
9	25	25	22	22	18	18	14	14
10	20	20	18	18	15	15	11	11
11	17	17	15	15	12	12	9	9
12	14	14	12	12	10	10	8	8

Y-Y AXIS

KL	36	50	36	50	36	50	36	50
0	75	104	63	88	51	71	38	50
2	68	91	55	74	43	57	29	36
3	67	90	55	74	43	56	29	35
4	65	87	54	72	42	55	29	35
5	63	83	52	69	41	54	28	34
6	60	78	50	65	40	51	28	33
7	57	72	47	60	38	47	27	32
8	53	65	44	54	35	43	25	29
9	49	58	41	48	33	38	23	26
10	45	51	37	42	30	33	22	23
11	41	43	34	35	27	28	19	20
12	36	36	30	30	23	23	17	17
13	31	31	25	25	20	20	15	15
14	27	27	22	22	17	17	13	13
15	23	23	19	19	15	15	11	11
16	20	20	17	17	13	13	10	10
17	18	18	15	15	12	12	9	9
18	16	16	13	13	11	11	8	8
19	15	15	12	12	10	10	7	7
20	13	13	11	11				

Properties of 2 angles — 3/8 in. back-to-back

	3/8	5/16	1/4	3/16
A (in.^{2})	3.47	2.93	2.38	1.80
r_x (in.)	0.753	0.761	0.769	0.778
r_y (in.)	1.21	1.20	1.19	1.18

Heavy line indicates Kl/r of 200.

$F_y = 36$ ksi
$F_y = 50$ ksi

COLUMNS
Double angles
Allowable concentric loads in kips
Equal legs
⅜ in. back-to-back of angles

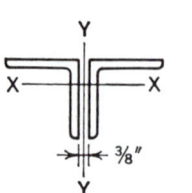

Size		2 x 2									
Thickness		⅜		⁵⁄₁₆		¼		³⁄₁₆		⅛	
Wt./ft		9.4		7.84		6.38		4.88		3.30	
F_y		36	50	36	50	36	50	36	50	36	50
X-X AXIS	0	59	82	50	69	41	56	31	43	19	24
	2	52	70	44	59	36	49	28	37	17	21
	3	47	62	40	52	33	43	25	33	16	19
	4	42	51	35	44	29	36	22	28	14	17
	5	35	39	30	34	25	28	19	22	12	14
	6	27	28	24	24	20	20	15	16	10	11
	7	20	20	18	18	15	15	12	12	8	8
	8	16	16	13	13	11	11	9	9	6	6
	9	12	12	11	11	9	9	7	7	5	5
	10			9	9	7	7	6	6	4	4
Y-Y AXIS	0	59	82	50	69	41	56	31	43	19	24
	2	54	73	45	61	35	47	25	33	13	16
	3	52	71	44	59	35	47	25	33	13	16
	4	50	67	42	56	34	45	25	32	13	15
	5	47	62	40	52	32	41	24	30	13	15
	6	44	56	37	47	30	37	22	27	12	14
	7	41	50	34	41	27	33	20	24	12	13
	8	37	43	31	35	25	28	18	21	11	12
	9	33	35	27	29	22	23	16	17	10	10
	10	28	29	23	24	19	19	14	14	8	8
	11	24	24	20	20	16	16	11	11	7	7
	12	20	20	16	16	13	13	10	10	6	6
	13	17	17	14	14	11	11	8	8	5	5
	14	15	15	12	12	10	10	7	7	5	5
	15	13	13	11	11	8	8	6	6	4	4
	16	11	11	9	9	7	7	5	5	4	4

Effective length in ft KL with respect to indicated axis

Properties of 2 angles — ⅜ in. back-to-back					
A (in.²)	2.72	2.30	1.88	1.43	0.960
r_x (in.)	0.594	0.601	0.609	0.617	0.626
r_y (in.)	1.01	1.00	0.989	0.977	0.965

Heavy line indicates Kl/r of 200.

AMERICAN INSTITUTE OF STEEL CONSTRUCTION

	F_y = 36 ksi
	F_y = 50 ksi

COLUMNS
Double angles
Allowable concentric loads in kips
Unequal legs
Long legs ⅜ in. back-to-back of angles

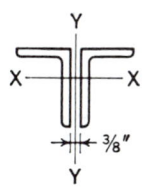

Size	8 x 6							8 x 4					
Thickness	1		¾		½			1		¾		½	
Wt./ft	88.4		67.6		46.0			74.8		57.4		39.2	
F_y	36	50	36	50	36	50		36	50	36	50	36	50

X-X AXIS

KL	36	50	36	50	36	50	KL	36	50	36	50	36	50
0	562	780	430	597	266	338	0	475	660	365	507	226	288
10	481	640	370	492	231	286	10	408	544	314	419	197	245
12	459	600	353	462	222	272	12	389	510	300	394	189	232
14	434	556	334	430	211	256	14	369	474	285	367	181	219
16	408	509	315	394	200	239	16	347	435	268	337	171	205
18	379	458	293	357	188	221	18	324	393	251	305	161	190
20	349	404	271	316	175	201	20	299	347	232	271	151	173
22	317	345	247	272	162	180	22	272	299	212	235	140	156
24	283	290	222	229	148	158	24	244	252	190	198	128	138
26	247	247	195	195	133	136	26	214	214	168	169	115	118
28	213	213	168	168	117	117	28	185	185	145	145	102	102
30	186	186	147	147	102	102	30	161	161	127	127	89	89
32	163	163	129	129	90	90	32	141	141	111	111	78	78
34	145	145	114	114	79	79	34	125	125	99	99	69	69
36	129	129	102	102	71	71	36	112	112	88	88	62	62
38	116	116	91	91	64	64	38	100	100	79	79	55	55
41	99	99	79	79	55	55	40	91	91	71	71	50	50
42			75	75	52	52	42	82	82	65	65	45	45
							43					43	43

Y-Y AXIS

KL	36	50	36	50	36	50	KL	36	50	36	50	36	50
0	562	780	430	597	266	338	0	475	660	365	507	226	288
6	498	670	364	482	202	242	6	401	530	296	387	169	202
8	487	650	358	471	200	239	8	374	483	277	352	160	188
10	470	621	348	454	196	233	10	342	426	252	308	148	168
12	450	586	334	430	191	226	12	306	361	223	256	132	144
14	428	545	318	401	185	215	14	266	288	191	198	114	116
16	402	500	300	368	176	202	16	222	223	155	155	93	93
18	375	451	279	330	166	186	18	177	177	124	124	76	76
20	346	399	257	290	154	168	20	145	145	101	101	63	63
22	315	342	233	246	141	147	22	120	120	84	84	52	52
24	282	289	207	209	126	127	24	101	101	71	71	45	45
25	265	267	193	194	119	119	25	93	93	66	66	41	41
26	247	247	180	180	111	111	26	86	86				
28	214	214	156	156	97	97							
30	187	187	137	137	86	86							
32	165	165	121	121	76	76							
34	146	146	107	107	68	68							
36	130	130	96	96	61	61							
40	106	106	78	78	50	50							
41	101	101	74	74									
42	96	96											

Effective length in ft KL with respect to indicated axis

Properties of 2 angles — ⅜ in. back-to-back						
A (in.²)	26.0	19.9	13.5	22.0	16.9	11.5
r_x (in.)	2.49	2.53	2.56	2.52	2.55	2.59
r_y (in.)	2.52	2.48	2.44	1.61	1.55	1.51

Heavy line indicates Kl/r of 200.

F_y = 36 ksi

F_y = 50 ksi

COLUMNS
Double angles
Allowable concentric loads in kips
Unequal legs
Long legs ⅜ in. back-to-back of angles

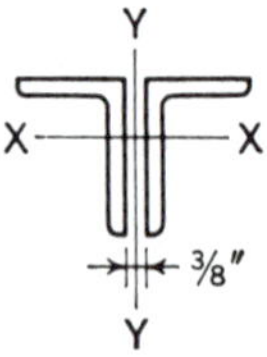

Effective length in ft KL with respect to indicated axis

Size		7 x 4					
Thickness		¾		½		⅜	
Wt./ft		52.4		35.8		27.2	
F_y		36	50	36	50	36	50
X-X AXIS	0	333	462	219	283	144	179
	8	291	391	193	243	129	157
	10	277	365	184	229	124	149
	12	261	337	174	213	118	141
	14	244	306	163	196	111	132
	16	225	272	151	178	105	122
	18	205	236	139	158	97	111
	20	184	197	125	136	89	99
	22	161	163	111	114	81	87
	24	137	137	96	96	72	74
	26	116	116	82	82	63	63
	28	100	100	70	70	54	54
	30	87	87	61	61	47	47
	32	77	77	54	54	42	42
	34	68	68	48	48	37	37
	36	61	61	43	43	33	33
	37	57	57	40	40	31	31
Y-Y AXIS	0	333	462	219	283	144	179
	6	277	366	170	207	104	121
	8	260	335	161	192	100	115
	10	238	296	148	173	94	106
	12	214	252	133	148	86	94
	14	186	201	115	120	76	79
	16	156	156	95	95	65	65
	18	125	125	77	77	53	53
	20	102	102	63	63	44	44
	22	85	85	53	53	37	37
	24	71	71	45	45	32	32
	25	66	66	41	41	30	30
	26	61	61	38	38		
	27	57	57				
Properties of 2 angles — ⅜ in. back-to-back							
A (in.²)		15.4		10.5		7.97	
r_x (in.)		2.22		2.25		2.27	
r_y (in.)		1.62		1.57		1.55	

Size		6 x 4							
Thickness		¾		⅝		½		⅜	
Wt./ft		47.2		40.0		32.4		24.6	
F_y		36	50	36	50	36	50	36	50
X-X AXIS	0	300	417	253	351	205	274	142	181
	8	254	336	214	284	174	223	122	151
	10	237	307	200	259	163	205	115	140
	12	219	274	185	232	151	184	107	128
	14	198	237	168	202	137	162	99	115
	16	177	198	150	169	123	137	89	101
	18	153	157	131	135	107	111	80	85
	20	127	127	110	110	90	90	69	70
	22	105	105	90	90	74	74	58	58
	24	88	88	76	76	62	62	48	48
	26	75	75	65	65	53	53	41	41
	28	65	65	56	56	46	46	36	36
	30	57	57	49	49	40	40	31	31
	31	53	53	46	46	37	37	29	29
	32							27	27
Y-Y AXIS	0	300	417	253	351	205	274	142	181
	6	256	340	212	279	165	208	107	128
	8	241	313	199	258	157	194	103	122
	10	222	280	184	230	145	175	97	112
	12	201	242	166	198	130	151	89	99
	14	178	200	146	162	114	123	79	84
	16	152	156	124	126	96	96	67	67
	18	124	124	101	101	77	77	55	55
	20	101	101	82	82	63	63	45	45
	22	84	84	68	68	53	53	38	38
	24	71	71	58	58	45	45	32	32
	26	60	60	49	49	38	38	28	28
	27	56	56	46	46	36	36	26	26
	28	52	52						
Properties of 2 angles — ⅜ in. back-to-back									
A (in.²)		13.9		11.7		9.50		7.22	
r_x (in.)		1.88		1.90		1.91		1.93	
r_y (in.)		1.69		1.67		1.64		1.62	

Heavy line indicates Kl/r of 200.

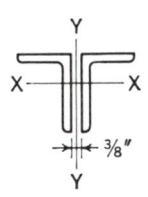

COLUMNS
Double angles
Allowable concentric loads in kips
Unequal legs
Long legs ⅜ in. back-to-back of angles

F_y = 36 ksi
F_y = 50 ksi

Size		6 x 3½				5 x 3½							
Thickness		⅜		5⁄16			¾		½		⅜		5⁄16
Wt./ft		23.4		19.6			39.6		27.2		20.8		17.4
F_y		36	50	36	50		36	50	36	50	36	50	36	50
	0	135	171	102	126	0	251	348	173	240	129	168	101	128
X-X AXIS	8	116	143	89	107	4	231	314	159	217	119	153	94	117
	10	109	133	84	101	6	216	289	150	200	113	142	88	110
	12	102	122	79	93	8	200	260	139	181	105	129	83	101
	14	94	110	74	85	10	181	226	126	158	96	115	76	91
	16	85	96	68	76	12	161	189	113	134	86	99	69	80
	18	76	82	61	67	14	138	147	97	106	75	81	61	67
	20	66	67	54	57	16	113	113	81	81	63	63	52	52
	22	55	55	47	47	18	89	89	64	64	50	50	42	42
	24	46	46	39	39	20	72	72	52	52	40	40	34	34
	26	39	39	33	33	22	60	60	43	43	33	33	28	28
	28	34	34	29	29	24	50	50	36	36	28	28	24	24
	30	30	30	25	25	25	46	46	33	33	26	26	22	22
	32	26	26	22	22	26			31	31	24	24	20	20
	0	135	171	102	126	0	251	348	173	240	129	168	101	128
Y-Y AXIS	4	104	125	75	87	4	225	303	148	196	104	128	77	93
	6	100	119	73	84	6	212	282	141	185	100	123	75	89
	8	94	109	69	78	8	196	253	131	167	94	113	71	83
	10	85	96	63	70	10	178	220	118	144	85	99	65	75
	12	74	79	56	60	12	157	182	104	117	74	81	58	63
	14	61	61	48	47	14	134	141	87	89	62	63	49	49
	16	48	48	38	38	16	108	108	69	69	49	49	40	40
	18	39	39	31	31	18	86	86	55	55	39	39	32	32
	20	32	32	26	26	20	70	70	45	45	32	32	26	26
	22	27	27	22	22	22	58	58	37	37	27	27	22	22
	23	25	25	20	20	24	49	49	31	31	23	23	19	19
						25	45	45						

Effective length in ft KL with respect to indicated axis

Properties of 2 angles — ⅜ in. back-to-back												
A (in.²)		6.84		5.74			11.6		8.00		6.09		5.12
r_x (in.)		1.94		1.95			1.55		1.58		1.60		1.61
r_y (in.)		1.39		1.38			1.53		1.49		1.46		1.45

Heavy line indicates Kl/r of 200.

F_y = 36 ksi
F_y = 50 ksi

COLUMNS
Double angles
Allowable concentric loads in kips
Unequal legs
Long legs ⅜ in. back-to-back of angles

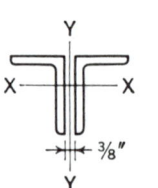

Size		5 x 3							
Thickness		½		⅜		⁵⁄₁₆		¼	
Wt./ft		25.6		19.6		16.4		13.2	
F_y		36	50	36	50	36	50	36	50
X-X AXIS	0	162	225	121	158	94	120	67	82
	2	157	216	117	152	92	116	65	80
	4	149	203	112	144	88	110	63	76
	6	141	188	106	134	83	103	60	72
	8	130	170	98	122	77	95	56	67
	10	119	149	90	109	71	85	52	61
	12	106	126	81	94	64	75	48	55
	14	92	100	70	77	57	63	43	48
	16	76	77	59	60	49	50	38	41
	18	61	61	47	47	40	40	32	33
	20	49	49	38	38	32	32	26	26
	22	41	41	32	32	27	27	22	22
	24	34	34	27	27	22	22	18	18
	26	29	29	23	23	19	19	16	16
	27							14	14
Y-Y AXIS	0	162	225	121	158	94	120	67	82
	2	142	190	100	124	74	90	50	57
	4	136	181	97	119	72	86	48	55
	6	127	164	91	110	68	81	46	52
	8	114	140	82	95	62	72	43	48
	10	97	110	70	77	54	59	38	41
	12	79	80	56	57	44	45	32	32
	14	60	60	43	43	34	34	25	26
	16	46	46	33	33	27	27	20	20
	18	37	37	27	27	22	22	17	17
	20	30	30	22	22	18	18	14	14

Effective length in ft KL with respect to indicated axis

Properties of 2 angles — ⅜ in. back-to-back			
A (in.²)	7.50	5.72	4.80	3.88
r_x (in.)	1.59	1.61	1.61	1.62
r_y (in.)	1.25	1.23	1.22	1.21

Heavy line indicates Kl/r of 200.

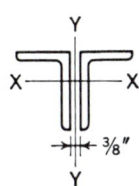

COLUMNS
Double angles
Allowable concentric loads in kips
Unequal legs
Long legs ⅜ in. back-to-back of angles

| F_y = 36 ksi |
| F_y = 50 ksi |

			Size	4 x 3½						
		Size		\multicolumn 4 x 3½						
		Thickness		½		⅜		⁵⁄₁₆		¼
		Wt./ft		23.8		18.2		15.4		12.4
		F_y		36	50	36	50	36	50	36	50
Effective length in ft KL with respect to indicated axis	X-X AXIS	0		151	210	115	160	97	126	71	91
		2		144	198	110	152	92	120	69	87
		4		135	182	103	139	87	110	64	80
		6		123	161	94	123	79	99	59	73
		8		109	136	84	105	71	85	54	64
		10		93	107	72	83	61	69	47	54
		12		75	76	59	60	50	51	39	42
		14		56	56	44	44	38	38	31	31
		16		43	43	34	34	29	29	24	24
		18		34	34	27	27	23	23	19	19
		20		27	27	22	22	18	18	15	15
		21						17	17	14	14
	Y-Y AXIS	0		151	210	115	160	97	126	71	91
		2		135	182	98	130	78	97	54	65
		4		133	179	96	127	77	95	53	64
		6		128	169	94	122	75	92	52	62
		8		119	154	88	113	72	87	51	59
		10		109	135	81	100	66	78	48	55
		12		97	114	72	83	59	66	43	48
		14		84	90	62	65	51	53	38	40
		16		69	69	51	51	41	41	32	32
		18		55	55	40	40	33	33	26	26
		20		45	45	33	33	27	27	21	21
		22		37	37	27	27	22	22	18	18
		24		31	31	23	23	19	19	15	15
		25		29	29	21	21	18	18	14	14
		26		27	27	20	20				

Properties of 2 angles — ⅜ in. back-to-back			
A (in.²)	7.00	5.34	4.49	3.63
r_x (in.)	1.23	1.25	1.26	1.27
r_y (in.)	1.58	1.56	1.55	1.54

Heavy line indicates Kl/r of 200.

AMERICAN INSTITUTE OF STEEL CONSTRUCTION

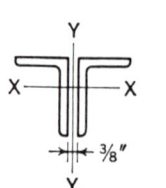

| $F_y = 36$ ksi $F_y = 50$ ksi | COLUMNS Double angles Allowable concentric loads in kips Unequal legs Long legs ⅜ in. back-to-back of angles | | | | | | | |

Size		4 x 3						
Thickness		½		⅜		5⁄16		¼
Wt./ft		22.2		17.0		14.4		11.6
F_y		36	50	36	50	36	50	36	50
X-X AXIS	0	140	195	107	149	90	117	67	85
	2	134	184	103	141	86	111	64	81
	4	126	169	96	130	81	103	60	75
	6	115	150	88	115	74	92	55	68
	8	102	128	78	98	66	80	50	60
	10	88	102	67	78	57	65	44	50
	12	71	73	55	57	47	49	37	40
	14	54	54	42	42	36	36	29	29
	16	41	41	32	32	27	27	22	22
	18	33	33	25	25	22	22	18	18
	20	26	26	20	20	17	17	14	14
	21			19	19	16	16	13	13
Y-Y AXIS	0	140	195	107	149	90	117	67	85
	2	126	170	92	122	74	92	51	62
	4	122	164	90	118	72	89	50	60
	6	114	149	85	110	69	84	49	57
	8	103	130	77	96	63	74	45	52
	10	90	107	67	78	55	62	40	45
	12	76	80	56	58	46	47	34	35
	14	59	59	43	43	35	35	27	27
	16	46	46	34	34	27	27	21	21
	18	36	36	27	27	22	22	17	17
	20	29	29	22	22	18	18	14	14
	21	27	27	20	20	16	16	13	13
	22	24	24						

Effective length in ft KL with respect to indicated axis

Properties of 2 angles — ⅜ in. back-to-back

A (in.²)	6.50	4.97	4.18	3.38
r_x (in.)	1.25	1.26	1.27	1.28
r_y (in.)	1.33	1.31	1.30	1.29

Heavy line indicates Kl/r of 200.

AMERICAN INSTITUTE OF STEEL CONSTRUCTION

F_y = 36 ksi

F_y = 50 ksi

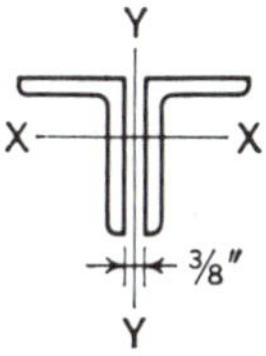

COLUMNS
Double angles
Allowable concentric loads in kips
Unequal legs
Long legs 3/8 in. back-to-back of angles

Effective length in ft KL with respect to indicated axis

Size		3½ x 3					
Thickness		3/8		5/16		1/4	
Wt./ft		15.8		13.2		10.8	
F_y		36	50	36	50	36	50
X-X AXIS	0	99	138	84	114	65	84
	2	94	129	79	107	62	79
	4	87	116	73	97	57	72
	6	77	99	65	83	52	63
	8	66	80	56	67	45	52
	10	54	57	46	49	37	40
	12	39	39	34	34	28	28
	14	29	29	25	25	20	20
	16	22	22	19	19	16	16
	17	20	20	17	17	14	14
	18	17	17	15	15	12	12
Y-Y AXIS	0	99	138	84	114	65	84
	2	86	116	70	92	52	63
	4	85	113	69	89	51	62
	6	80	105	66	84	49	59
	8	73	92	61	75	46	54
	10	65	77	53	62	41	46
	12	55	59	45	48	34	36
	14	44	44	36	36	27	27
	16	34	34	28	28	21	21
	18	27	27	22	22	17	17
	20	22	22	18	18	14	14
	22	18	18	15	15	12	12

Size		3½ x 2½					
Thickness		3/8		5/16		1/4	
Wt./ft		14.4		12.2		9.8	
F_y		36	50	36	50	36	50
X-X AXIS	0	91	127	77	105	60	78
	2	86	119	73	99	57	73
	4	80	107	67	89	53	67
	6	71	92	60	77	48	58
	8	61	74	52	62	41	49
	10	50	53	42	45	34	37
	12	37	37	31	31	26	26
	14	27	27	23	23	19	19
	16	21	21	18	18	15	15
	18	16	16	14	14	12	12
Y-Y AXIS	0	91	127	77	105	60	78
	2	80	107	65	85	48	59
	4	76	101	62	81	46	57
	6	69	88	57	71	43	51
	8	60	72	49	58	38	43
	10	49	51	40	42	31	32
	12	36	36	30	30	23	23
	14	27	27	22	22	17	17
	16	21	21	17	17	13	13
	18	16	16	13	13	11	11

Properties of 2 angles — 3/8 in. back-to-back

	3½ x 3, 3/8	3½ x 3, 5/16	3½ x 3, 1/4	3½ x 2½, 3/8	3½ x 2½, 5/16	3½ x 2½, 1/4
A (in.^{2})	4.59	3.87	3.13	4.22	3.55	2.88
r_x (in.)	1.09	1.10	1.11	1.10	1.11	1.12
r_y (in.)	1.36	1.35	1.33	1.11	1.10	1.09

Heavy line indicates Kl/r of 200.

F_y = 36 ksi
F_y = 50 ksi

COLUMNS
Double angles
Allowable concentric loads in kips
Unequal legs
Long legs ⅜ in. back-to-back of angles

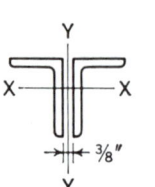

Size	3 x 2½							3 x 2							
Thickness	⅜		¼		³⁄₁₆			⅜		⁵⁄₁₆		¼		³⁄₁₆	
Wt./ft	13.2		9.0		6.77			11.8		10.0		8.2		6.1	
F_y	36	50	36	50	36	50		36	50	36	50	36	50	36	50

X-X AXIS — Effective length in ft KL with respect to indicated axis

KL (3x2½)	⅜ 36	⅜ 50	¼ 36	¼ 50	³⁄₁₆ 36	³⁄₁₆ 50	KL (3x2)	⅜ 36	⅜ 50	⁵⁄₁₆ 36	⁵⁄₁₆ 50	¼ 36	¼ 50	³⁄₁₆ 36	³⁄₁₆ 50
0	83	115	57	76	39	50	0	75	104	63	88	51	69	35	45
2	78	106	53	70	37	46	2	70	96	59	81	48	64	33	42
3	74	100	51	66	35	44	3	67	90	57	77	46	60	32	40
4	70	92	48	62	34	41	4	63	84	54	71	44	56	30	38
5	65	84	45	56	32	38	5	59	77	50	65	41	51	29	35
6	60	75	41	51	29	35	6	55	68	46	58	38	46	27	32
7	54	65	38	44	27	31	7	50	59	42	51	34	41	25	29
8	48	53	34	37	24	27	8	44	49	38	42	31	35	22	25
9	41	42	29	30	22	23	9	38	39	33	34	27	28	20	21
10	34	34	24	24	19	19	10	32	32	27	27	23	23	17	17
11	28	28	20	20	16	16	11	26	26	23	23	19	19	14	14
12	24	24	17	17	13	13	12	22	22	19	19	16	16	12	12
13	20	20	14	14	11	11	13	19	19	16	16	13	13	10	10
14	17	17	12	12	10	10	14	16	16	14	14	12	12	9	9
15	15	15	11	11	8	8	15	14	14	12	12	10	10	8	8
							16							7	7

Y-Y AXIS — Effective length in ft KL with respect to indicated axis

KL (3x2½)	⅜ 36	⅜ 50	¼ 36	¼ 50	³⁄₁₆ 36	³⁄₁₆ 50	KL (3x2)	⅜ 36	⅜ 50	⁵⁄₁₆ 36	⁵⁄₁₆ 50	¼ 36	¼ 50	³⁄₁₆ 36	³⁄₁₆ 50
0	83	115	57	76	39	50	0	75	104	63	88	51	69	35	45
2	74	100	47	59	30	36	2	67	90	55	73	43	54	27	33
3	73	98	46	59	29	35	3	65	86	53	71	42	53	27	32
4	71	95	45	57	29	35	4	61	81	51	66	40	50	26	31
5	68	90	44	55	29	34	5	57	73	48	60	38	46	25	29
6	65	84	42	52	28	33	6	53	65	44	53	35	41	23	27
7	61	77	40	48	27	31	7	48	56	39	46	31	35	21	23
8	57	69	37	43	25	28	8	42	46	35	37	27	28	19	19
9	52	61	34	38	23	26	9	36	36	29	29	23	23	16	16
10	47	52	31	32	21	22	10	30	30	24	24	19	19	13	13
11	42	43	27	27	19	19	11	25	25	20	20	16	16	11	11
12	36	36	23	23	16	16	12	21	21	17	17	13	13	9	9
13	31	31	20	20	14	14	13	18	18	14	14	11	11	8	8
14	27	27	17	17	12	12	14	15	15	12	12	10	10	7	7
15	24	24	15	15	11	11	15	13	13	11	11				
16	21	21	13	13	10	10									
17	18	18	12	12	9	9									
18	16	16	11	11	8	8									
19	15	15													

Properties of 2 angles — ⅜ in. back-to-back

	3x2½ ⅜	3x2½ ¼	3x2½ ³⁄₁₆	3x2 ⅜	3x2 ⁵⁄₁₆	3x2 ¼	3x2 ³⁄₁₆
A (in.^{2})	3.84	2.63	1.99	3.47	2.93	2.38	1.80
r_x (in.)	0.928	0.945	0.954	0.940	0.948	0.957	0.966
r_y (in.)	1.16	1.13	1.12	0.917	0.903	0.891	0.879

Heavy line indicates Kl/r of 200.

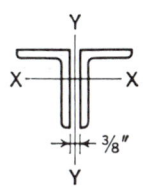

Size		2½ x 2							
Thickness		⅜		⁵⁄₁₆		¼		³⁄₁₆	
Wt./ft		10.6		9.0		7.2		5.5	
F_y		36	50	36	50	36	50	36	50
X-X AXIS	0	67	93	57	79	46	64	34	45
	2	61	83	52	71	42	58	32	41
	3	58	77	49	65	40	53	30	38
	4	53	69	45	59	37	48	28	34
	5	48	60	41	51	34	42	25	30
	6	42	50	36	43	30	35	23	26
	7	36	39	31	33	26	28	20	21
	8	30	30	26	26	21	21	16	17
	9	23	23	20	20	17	17	13	13
	10	19	19	16	16	14	14	11	11
	11	16	16	14	14	11	11	9	9
	12	13	13	11	11	9	9	7	7
	13					8	8	6	6
Y-Y AXIS	0	67	93	57	79	46	64	34	45
	2	60	82	50	67	39	52	27	33
	3	59	79	49	65	38	51	27	33
	4	56	74	47	61	37	48	26	32
	5	52	68	44	56	35	44	25	30
	6	49	61	41	51	32	40	23	27
	7	44	53	37	44	29	34	21	24
	8	40	45	33	37	26	28	19	20
	9	35	36	29	29	22	23	16	16
	10	29	29	24	24	19	19	13	13
	11	24	24	20	20	16	16	11	11
	12	20	20	17	17	13	13	10	10
	13	17	17	14	14	11	11	8	8
	14	15	15	12	12	10	10	7	7
	15	13	13	11	11	8	8	6	6
	16	11	11						

Properties of 2 angles — ⅜ in. back-to-back

	⅜	⁵⁄₁₆	¼	³⁄₁₆
A (in.²)	3.09	2.62	2.13	1.62
r_x (in.)	0.768	0.776	0.784	0.793
r_y (in.)	0.961	0.948	0.935	0.923

Heavy line indicates Kl/r of 200.

Effective length in ft KL with respect to indicated axis

COLUMNS
Double angles
Allowable concentric loads in kips
Unequal legs
Long legs ⅜ in. back-to-back of angles

F_y = 36 ksi F_y = 50 ksi

F_y = 36 ksi
F_y = 50 ksi

COLUMNS
Double angles
Allowable concentric loads in kips
Unequal legs
Short legs ⅜ in. back-to-back of angles

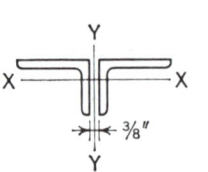

Size	8 x 6							8 x 4					
Thickness	1		¾		½			1		¾		½	
Wt./ft	88.4		67.6		46.9			74.8		57.4		39.2	
F_y	36	50	36	50	36	50		36	50	36	50	36	50

Effective length in ft KL with respect to indicated axis

X-X AXIS

KL	8×6 (36)	(50)	¾(36)	(50)	½(36)	(50)		KL	8×4(36)	(50)	¾(36)	(50)	½(36)	(50)
0	562	780	430	597	266	338		0	475	660	365	507	226	288
8	464	610	357	470	224	276		4	410	547	316	423	199	247
12	390	477	302	371	193	228		6	362	461	280	358	179	216
16	300	315	235	250	156	171		8	304	357	237	281	155	180
20	202	202	160	160	112	112		10	237	242	187	193	128	138
24	140	140	111	111	78	78		12	168	168	134	134	97	97
28	103	103	82	82	57	57		14	123	123	99	99	71	71
29			76	76	53	53		16	95	95	75	75	54	54
								17	84	84	67	67	48	48
								18					43	43

Y-Y AXIS

KL	8×6 (36)	(50)	¾(36)	(50)	½(36)	(50)		KL	8×4(36)	(50)	¾(36)	(50)	½(36)	(50)
0	562	780	430	597	266	338		0	475	660	365	507	226	288
12	493	662	362	479	201	240		12	425	573	311	413	174	209
16	471	622	352	462	199	237		16	409	544	308	408	173	208
20	441	570	333	427	195	231		20	386	504	294	383	172	206
24	408	510	308	383	186	217		24	360	459	275	349	168	200
28	371	443	280	332	172	196		28	333	409	253	310	158	185
32	330	369	249	275	155	170		32	302	354	230	267	145	165
36	287	295	215	219	136	140		36	270	295	204	221	131	142
40	239	239	178	178	115	115		40	235	239	177	179	115	118
44	198	198	148	148	96	96		44	198	198	148	148	98	98
48	166	166	124	124	81	81		48	166	166	124	124	82	82
52	142	142	106	106	69	69		52	142	142	106	106	70	70
56	122	122	91	91	60	60		56	122	122	92	92	61	61
60	107	107	80	80	52	52		60	106	106	80	80	53	53
61	103	103	77	77	51	51		64	94	94	70	70	46	46
62	100	100	75	75				66	88	88	66	66	44	44
63	97	97						67	85	85	64	64		
								68	83	83				

Properties of 2 angles — ⅜ in. back-to-back

	8 x 6				8 x 4		
A (in.²)	26.0	19.9	13.5		22.0	16.9	11.5
r_x (in.)	1.73	1.76	1.79		1.03	1.05	1.08
r_y (in.)	3.78	3.74	3.69		4.10	4.05	4.00

Heavy line indicates Kl/r of 200.

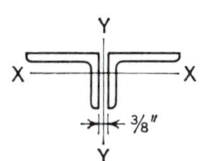

COLUMNS
Double angles
Allowable concentric loads in kips
Unequal legs
Short legs ⅜ in. back-to-back of angles

F_y = 36 ksi
F_y = 50 ksi

Effective length in ft KL with respect to indicated axis

Size 7 x 4

Axis	KL	¾ (36)	¾ (50)	½ (36)	½ (50)	⅜ (36)	⅜ (50)
	Wt./ft	52.4		35.8		27.2	
X-X	0	333	462	219	283	144	179
X-X	4	290	389	192	242	129	157
X-X	6	259	334	173	212	118	141
X-X	8	222	267	150	175	104	121
X-X	10	180	190	123	133	89	99
X-X	12	132	132	93	93	72	73
X-X	14	97	97	68	68	54	54
X-X	16	74	74	52	52	41	41
X-X	18	59	59	41	41	33	33
Y-Y	0	333	462	219	283	144	179
Y-Y	8	292	391	175	215	106	123
Y-Y	12	288	385	174	214	106	123
Y-Y	16	274	360	172	211	105	122
Y-Y	20	254	324	165	199	104	120
Y-Y	24	231	284	151	178	100	115
Y-Y	28	206	238	135	152	93	104
Y-Y	32	179	189	118	124	83	89
Y-Y	36	149	150	98	98	71	73
Y-Y	40	121	121	80	80	59	59
Y-Y	44	100	100	66	66	49	49
Y-Y	48	84	84	56	56	42	42
Y-Y	52	72	72	47	47	35	35
Y-Y	56	62	62	41	41	31	31
Y-Y	57	60	60	40	40	30	30
Y-Y	58	58	58				

Size 6 x 4

Axis	KL	¾ (36)	¾ (50)	⅝ (36)	⅝ (50)	½ (36)	½ (50)	⅜ (36)	⅜ (50)
	Wt./ft	47.2		40.0		32.4		24.6	
X-X	0	300	417	253	351	205	274	142	181
X-X	4	264	353	222	298	181	235	127	158
X-X	6	236	305	200	258	163	205	115	141
X-X	8	204	248	173	211	142	170	102	120
X-X	10	168	181	142	155	118	129	87	97
X-X	12	126	126	108	108	90	90	70	71
X-X	14	92	92	79	79	66	66	52	52
X-X	16	71	71	61	61	51	51	40	40
X-X	18	56	56	48	48	40	40	32	32
X-X	19					36	36	28	28
Y-Y	0	300	417	253	351	205	274	142	181
Y-Y	8	268	361	220	293	170	216	109	131
Y-Y	12	255	338	212	281	167	212	108	129
Y-Y	16	234	301	196	251	157	195	105	125
Y-Y	20	210	258	176	215	141	169	98	114
Y-Y	24	183	210	153	174	123	138	87	97
Y-Y	28	153	158	128	131	102	104	74	76
Y-Y	32	121	121	100	100	80	80	59	59
Y-Y	36	96	96	79	79	63	63	47	47
Y-Y	40	78	78	64	64	51	51	38	38
Y-Y	44	64	64	53	53	43	43	32	32
Y-Y	45	61	61	51	51	41	41	30	30
Y-Y	46	59	59	49	49	39	39	29	29
Y-Y	47	56	56	47	47	37	37	28	28
Y-Y	48	54	54	45	45	36	36		
Y-Y	49	52	52						

Properties of 2 angles — ⅜ in. back-to-back

7 x 4

	¾	½	⅜
A (in.²)	15.4	10.5	7.97
r_x (in.)	1.09	1.11	1.13
r_y (in.)	3.49	3.44	3.42

6 x 4

	¾	⅝	½	⅜
A (in.²)	13.9	11.7	9.50	7.22
r_x (in.)	1.12	1.13	1.15	1.17
r_y (in.)	2.94	2.92	2.90	2.87

Heavy line indicates Kl/r of 200.

F_y = 36 ksi
F_y = 50 ksi

COLUMNS
Double angles
Allowable concentric loads in kips
Unequal legs
Short legs ⅜ in. back-to-back of angles

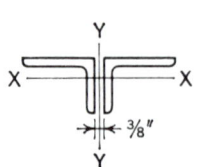

Effective length in ft KL with respect to indicated axis

6 x 3½

Size	6 x 3½			
Thickness	⅜		⁵⁄₁₆	
Wt./ft	23.4		19.6	
F_y	36	50	36	50
X-X AXIS				
0	135	171	102	126
4	116	144	89	108
6	103	123	80	94
8	86	98	69	78
10	68	69	56	59
12	48	48	41	41
14	35	35	30	30
16	27	27	23	23
Y-Y AXIS				
0	135	171	102	126
8	104	125	74	86
12	103	124	74	86
16	101	121	74	85
20	95	111	71	82
24	85	96	66	74
28	73	77	58	62
32	59	59	49	49
36	47	47	39	39
40	38	38	32	32
44	32	32	26	26
48	27	27	22	22
49	26	26	21	21

5 x 3½

Size	5 x 3½							
Thickness	¾		½		⅜		⁵⁄₁₆	
Wt./ft	39.6		27.2		20.8		17.4	
F_y	36	50	36	50	36	50	36	50
X-X AXIS								
0	251	348	173	240	129	168	101	128
2	236	322	163	223	122	157	95	120
4	214	284	149	198	112	140	88	109
6	186	235	130	166	98	120	78	94
8	153	175	109	127	83	95	67	77
10	115	115	84	85	64	66	54	56
12	80	80	59	59	46	46	39	39
14	59	59	43	43	34	34	29	29
16	45	45	33	33	26	26	22	22
17					23	23	19	19
Y-Y AXIS								
0	251	348	173	240	129	168	101	128
8	222	299	148	196	103	128	77	92
10	214	284	144	190	102	126	76	91
12	204	266	138	180	100	123	75	90
14	193	247	131	167	96	116	73	87
16	181	226	123	152	91	107	70	82
18	168	203	114	136	84	97	66	76
20	155	178	104	118	77	86	61	68
22	140	152	94	100	70	74	56	60
24	125	128	83	84	62	62	50	51
26	109	109	72	72	53	53	44	44
28	94	94	62	62	46	46	38	38
30	82	82	54	54	40	40	33	33
32	72	72	48	48	35	35	29	29
34	64	64	42	42	31	31	26	26
36	57	57	38	38	28	28	23	23
38	51	51	34	34	25	25	21	21
39	49	49	32	32	24	24	20	20
40	46	46	31	31	23	23		
41	44	44						

Properties of 2 angles — ⅜ in. back-to-back

	6 x 3½ ⅜	6 x 3½ ⁵⁄₁₆	5 x 3½ ¾	5 x 3½ ½	5 x 3½ ⅜	5 x 3½ ⁵⁄₁₆
A (in.²)	6.84	5.74	11.6	8.00	6.09	5.12
r_x (in.)	0.988	0.996	0.977	1.01	1.02	1.03
r_y (in.)	2.95	2.94	2.48	2.43	2.41	2.39

Heavy line indicates Kl/r of 200.

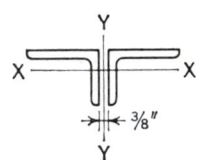

	F_y = 36 ksi
COLUMNS **Double angles** Allowable concentric loads in kips **Unequal legs** Short legs ⅜ in. back-to-back of angles	F_y = 50 ksi

Size		5 x 3						
Thickness		½		⅜		⁵⁄₁₆		¼
Wt./ft		25.6		19.6		16.4		13.2
F_y		36	50	36	50	36	50	36	50

Effective length in ft KL with respect to indicated axis

	X-X AXIS	0	162	225	121	158	94	120	67	82
		2	150	205	113	144	88	111	63	77
		4	132	173	100	124	79	97	57	68
		6	109	132	83	98	67	78	50	57
		8	82	84	64	66	53	56	40	44
		10	53	53	42	42	36	36	30	30
		12	37	37	29	29	25	25	21	21
		13	32	32	25	25	21	21	18	18
		14			22	22	18	18	15	15
	Y-Y AXIS	0	162	225	121	158	94	120	67	82
		8	139	186	97	120	72	86	48	55
		10	137	181	97	120	72	86	48	54
		12	131	172	96	117	71	85	47	54
		14	125	160	92	112	70	83	47	54
		16	117	146	87	104	68	80	47	53
		18	109	132	81	95	64	74	45	51
		20	101	116	75	85	60	67	43	48
		22	92	100	68	74	55	60	41	44
		24	82	84	61	62	49	51	37	40
		26	72	72	53	53	44	44	34	35
		28	62	62	46	46	38	38	30	30
		30	54	54	40	40	33	33	26	26
		32	47	47	35	35	29	29	23	23
		34	42	42	31	31	26	26	21	21
		36	37	37	28	28	23	23	19	19
		38	34	34	25	25	21	21	17	17
		40	30	30	23	23	19	19	15	15
		41	29	29	22	22	18	18	14	14

Properties of 2 angles — ⅜ in. back-to-back								
A (in.²)		7.50		5.72		4.80		3.88
r_x (in.)		0.829		0.845		0.853		0.861
r_y (in.)		2.50		2.48		2.47		2.46

Heavy line indicates Kl/r of 200.

F_y = 36 ksi
F_y = 50 ksi

COLUMNS
Double angles
Allowable concentric loads in kips
Unequal legs
Short legs ⅜ in. back-to-back of angles

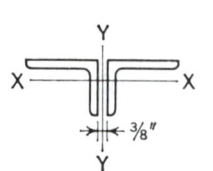

Size		4 x 3½							
Thickness		½		⅜		⁵⁄₁₆		¼	
Wt./ft		23.8		18.2		15.4		12.4	
F_y		36	50	36	50	36	50	36	50
X-X AXIS	0	151	210	115	160	97	126	71	91
	2	143	196	109	150	92	118	68	86
	4	131	175	100	134	84	106	63	78
	6	116	148	89	114	75	92	56	68
	8	97	115	75	90	64	74	49	56
	10	77	79	60	62	51	53	40	43
	12	55	55	43	43	37	37	30	30
	14	40	40	32	32	27	27	22	22
	16	31	31	24	24	21	21	17	17
	17	27	27	22	22	18	18	15	15
Y-Y AXIS	0	151	210	115	160	97	126	71	91
	4	134	180	97	128	77	95	53	63
	6	132	176	96	126	76	94	53	63
	8	126	166	93	122	75	92	52	62
	10	118	152	88	113	72	87	51	60
	12	109	136	82	101	67	80	49	57
	14	99	119	74	88	61	70	45	51
	16	89	99	66	73	55	59	41	45
	18	77	79	57	58	47	48	36	37
	20	64	64	47	47	39	39	30	30
	22	53	53	39	39	32	32	25	25
	24	45	45	33	33	27	27	22	22
	26	38	38	28	28	23	23	19	19
	28	33	33	24	24	20	20	16	16
	30	29	29	21	21	18	18	14	14
	31	27	27	20	20	17	17		

Effective length in ft KL with respect to indicated axis

Properties of 2 angles — ⅜ in. back-to-back				
A (in.²)	7.00	5.34	4.49	3.63
r_x (in.)	1.04	1.06	1.07	1.07
r_y (in.)	1.89	1.87	1.86	1.85

Heavy line indicates Kl/r of 200.

AMERICAN INSTITUTE OF STEEL CONSTRUCTION

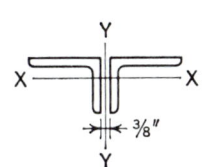

COLUMNS
Double angles
Allowable concentric loads in kips
Unequal legs
Short legs ⅜ in. back-to-back of angles

| F_y = 36 ksi |
| F_y = 50 ksi |

Size		4 x 3							
Thickness		½		⅜		⁵⁄₁₆		¼	
Wt./ft		22.2		17.0		14.4		11.6	
F_y		36	50	36	50	36	50	36	50
X-X AXIS	0	140	195	107	149	90	117	67	85
	2	131	178	100	137	84	108	62	78
	4	116	153	89	117	75	94	56	69
	6	97	119	75	93	63	75	48	57
	8	75	79	59	62	50	53	39	43
	10	50	50	40	40	34	34	28	28
	12	35	35	28	28	24	24	20	20
	14	26	26	20	20	17	17	14	14
Y-Y AXIS	0	140	195	107	149	90	117	67	85
	4	125	169	91	121	73	90	51	60
	6	124	166	90	119	72	90	50	60
	8	119	157	88	116	71	88	50	60
	10	112	145	84	109	69	84	49	58
	12	104	131	78	98	65	78	47	55
	14	95	115	72	86	59	69	44	51
	16	86	98	64	73	53	59	40	45
	18	75	79	56	59	47	48	36	38
	20	64	64	48	48	39	40	31	31
	22	53	53	40	40	33	33	26	26
	24	45	45	33	33	28	28	22	22
	26	38	38	28	28	24	24	19	19
	28	33	33	25	25	20	20	16	16
	30	29	29	21	21	18	18	14	14
	32	25	25	19	19	16	16	12	12

Effective length in ft KL with respect to indicated axis

Properties of 2 angles — ⅜ in. back-to-back				
A (in.²)	6.50	4.97	4.18	3.38
r_x (in.)	0.864	0.879	0.887	0.896
r_y (in.)	1.96	1.94	1.93	1.92

Heavy line indicates Kl/r of 200.

F_y = 36 ksi
F_y = 50 ksi

COLUMNS
Double angles
Allowable concentric loads in kips
Unequal legs
Short legs ⅜ in. back-to-back of angles

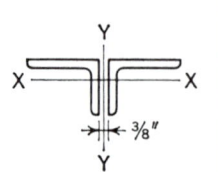

Effective length in ft KL with respect to indicated axis

Size 3½ x 3

Thickness	⅜		5/16		¼	
Wt./ft	15.8		13.2		10.8	
F_y	36	50	36	50	36	50
X-X AXIS						
0	99	138	84	114	65	84
2	93	126	78	105	61	78
4	83	109	70	91	55	69
6	70	87	60	73	47	56
8	55	60	47	51	38	42
10	38	38	33	33	27	27
12	27	27	23	23	19	19
14	20	20	17	17	14	14
15			15	15	12	12
Y-Y AXIS						
0	99	138	84	114	65	84
4	86	114	69	90	51	62
6	84	111	68	89	50	61
8	80	103	66	84	49	59
10	73	93	61	76	47	55
12	66	80	55	66	43	49
14	58	65	49	54	38	41
16	50	51	41	42	32	33
18	40	40	33	33	26	26
20	33	33	27	27	21	21
22	27	27	23	23	18	18
24	23	23	19	19	15	15
26	20	20	16	16	13	13
27	18	18	15	15	12	12

Size 3½ x 2½

Thickness	⅜		5/16		¼	
Wt./ft	14.4		12.2		9.8	
F_y	36	50	36	50	36	50
X-X AXIS						
0	91	127	77	105	60	78
2	83	113	70	94	55	70
4	71	91	60	76	47	58
6	55	62	47	53	37	42
8	35	35	30	30	25	25
10	23	23	19	19	16	16
11	19	19	16	16	13	13
12			14	14	11	11
Y-Y AXIS						
0	91	127	77	105	60	78
4	79	106	64	84	47	58
6	78	104	64	83	47	57
8	75	98	62	79	46	56
10	69	88	58	72	44	53
12	63	77	53	64	41	48
14	56	65	47	53	37	41
16	49	51	40	42	32	33
18	41	41	34	34	27	27
20	33	33	27	27	22	22
22	27	27	23	23	18	18
24	23	23	19	19	15	15
26	20	20	16	16	13	13
28	17	17	14	14	11	11
29	16	16				

Properties of 2 angles — ⅜ in. back-to-back

	3½ x 3			3½ x 2½		
A (in.²)	4.59	3.87	3.13	4.22	3.55	2.88
r_x (in.)	0.897	0.905	0.914	0.719	0.727	0.735
r_y (in.)	1.67	1.66	1.65	1.74	1.73	1.72

Heavy line indicates Kl/r of 200.

	F_y = 36 ksi
COLUMNS	F_y = 50 ksi

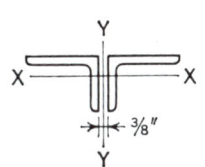

COLUMNS
Double angles
Allowable concentric loads in kips
Unequal legs
Short legs ⅜ in. back-to-back of angles

Size	3 x 2½							3 x 2							
Thickness	⅜		¼		³⁄₁₆			⅜		⁵⁄₁₆		¼		³⁄₁₆	
Wt./ft	13.2		9.0		6.77			11.8		10.0		8.2		6.1	
F_y	36	50	36	50	36	50		36	50	36	50	36	50	36	50

X-X AXIS (Effective length in ft KL with respect to indicated axis)

KL	36	50	36	50	36	50	KL	36	50	36	50	36	50	36	50
0	83	115	57	76	39	50	0	75	104	63	88	51	69	35	45
2	76	103	52	68	36	45	2	66	88	56	75	45	59	32	39
3	71	94	49	63	34	42	3	59	76	50	65	41	51	29	35
4	65	84	45	56	32	38	4	51	62	43	53	36	43	25	30
5	58	72	40	49	29	34	5	42	45	36	39	30	32	22	24
6	51	58	36	41	26	29	6	31	31	27	27	23	23	17	18
7	43	44	30	32	22	24	7	23	23	20	20	17	17	13	13
8	34	34	24	24	19	19	8	18	18	15	15	13	13	10	10
9	27	27	19	19	15	15	9	14	14	12	12	10	10	8	8
10	22	22	15	15	12	12									
11	18	18	13	13	10	10									
12	15	15	11	11	8	8									

Y-Y AXIS

KL	36	50	36	50	36	50	KL	36	50	36	50	36	50	36	50
0	83	115	57	76	39	50	0	75	104	63	88	51	69	35	45
2	74	100	46	59	29	35	2	67	91	55	73	42	54	27	33
4	73	98	46	58	29	35	4	67	90	55	73	42	54	27	32
6	70	92	45	57	29	34	6	64	85	53	71	42	53	27	32
8	64	83	43	53	28	33	8	60	77	50	64	40	50	26	31
10	58	71	39	46	26	30	10	54	67	45	56	36	44	25	29
12	50	58	34	37	23	26	12	48	56	40	47	32	37	23	26
14	42	43	28	28	20	20	14	41	44	34	36	27	29	20	21
16	33	33	22	22	16	16	16	34	34	28	28	22	22	16	16
18	26	26	17	17	13	13	18	27	27	22	22	17	17	13	13
20	21	21	14	14	10	10	20	22	22	18	18	14	14	11	11
22	18	18	12	12	9	9	22	18	18	15	15	12	12	9	9
24	15	15	10	10	7	7	24	15	15	12	12	10	10	7	7
							25	14	14	11	11	9	9	7	7

Properties of 2 angles — ⅜ in. back-to-back

	3 x 2½				3 x 2			
A (in.²)	3.84		2.63	1.99	3.47	2.93	2.38	1.80
r_x (in.)	0.736		0.753	0.761	0.559	0.567	0.574	0.583
r_y (in.)	1.47		1.45	1.44	1.55	1.53	1.52	1.51

Heavy line indicates Kl/r of 200.

F_y = 36 ksi
F_y = 50 ksi

COLUMNS
Double angles
Allowable concentric loads in kips
Unequal legs
Short legs ⅜ in. back-to-back of angles

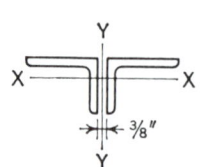

Size			2½ x 2						
Thickness			⅜		⁵⁄₁₆		¼		³⁄₁₆	
Wt./ft			10.6		9.0		7.2		5.5	
F_y			36	50	36	50	36	50	36	50
X-X AXIS	0		67	93	57	79	46	64	34	45
	2		59	79	50	67	41	55	31	39
	3		53	69	45	59	37	48	28	34
	4		46	57	40	49	32	40	25	29
	5		39	43	33	37	27	31	21	23
	6		30	30	26	26	21	22	17	17
	7		22	22	19	19	16	16	12	12
	8		17	17	14	14	12	12	9	9
	9		13	13	11	11	10	10	7	7
	10								6	6
Y-Y AXIS	0		67	93	57	79	46	64	34	45
	2		61	82	50	68	39	52	27	33
	4		59	80	49	66	39	51	27	33
	6		55	72	46	60	37	48	26	32
	8		49	61	41	51	33	41	24	28
	10		42	49	35	41	28	33	21	23
	12		35	36	29	30	23	24	17	17
	14		27	27	22	22	17	17	13	13
	16		20	20	17	17	13	13	10	10
	18		16	16	13	13	11	11	8	8
	20		13	13	11	11	9	9	6	6
	21		12	12	10	10				

Effective length in ft KL with respect to indicated axis

Properties of 2 angles — ⅜ in. back-to-back				
A (in.²)	3.09	2.62	2.13	1.62
r_x (in.)	0.577	0.584	0.592	0.600
r_y (in.)	1.28	1.26	1.25	1.24

Heavy line indicates Kl/r of 200.

AMERICAN INSTITUTE OF STEEL CONSTRUCTION

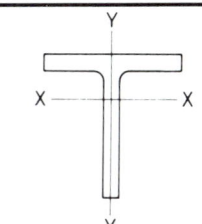

	COLUMNS	F_y = 36 ksi
	Structural Tees Cut from W shapes Allowable axial loads in kips	F_y = 50 ksi

Designation		WT 18								
Wt./ft		150		140		130		122.5		115	
F_y		36	50	36	50	36	50	36	50	36	50
	0	953	1227	890	1071	810	935	733	832	654	725
X-X AXIS	10	901	1149	841	1005	766	880	695	785	620	685
	12	888	1128	829	988	755	866	685	773	612	675
	14	873	1107	816	970	743	851	674	760	603	665
	16	859	1084	802	951	731	835	663	746	593	653
	18	843	1059	787	931	717	818	652	732	583	641
	20	826	1034	771	909	703	800	639	716	573	629
	22	809	1007	755	887	689	781	627	700	562	615
	24	791	979	738	863	674	762	613	684	550	602
	26	772	950	720	839	658	741	599	666	538	587
	28	752	919	702	814	641	720	585	648	526	573
	30	732	888	683	787	624	698	570	630	513	557
	32	711	855	663	760	607	676	555	610	500	541
	34	690	821	643	732	589	652	539	590	486	525
	36	667	786	622	703	570	628	522	570	472	508
	38	644	750	600	673	551	603	505	549	457	490
	40	621	713	578	642	531	577	488	527	442	472
	0	953	1227	890	1071	810	935	733	832	654	725
Y-Y AXIS	10	823	1026	760	892	682	772	610	679	540	589
	12	813	1012	752	881	675	763	605	673	536	584
	14	800	991	740	865	666	752	597	663	530	577
	16	782	964	725	844	653	735	587	650	522	567
	18	762	932	706	818	637	715	574	634	511	555
	20	738	895	685	789	619	691	558	615	498	539
	22	712	855	661	756	598	664	540	592	483	521
	24	685	812	635	720	575	635	520	568	467	501
	26	655	766	608	683	551	603	499	541	449	480
	28	625	718	579	642	525	570	477	513	429	457
	30	592	667	549	600	498	534	453	483	409	432
	32	559	614	517	556	469	497	427	451	387	406
	34	523	558	484	509	439	458	401	418	365	379
	36	487	501	450	461	408	417	374	383	341	350
	38	448	451	413	415	375	376	345	347	316	320
	40	408	408	376	376	341	341	315	315	291	291
Properties										
A (in.²)		44.1		41.2		38.2		36.0		33.8	
r_x (in.)		5.27		5.25		5.26		5.26		5.25	
r_y (in.)		3.83		3.81		3.78		3.75		3.73	

Effective length in ft KL with respect to indicated axis

AMERICAN INSTITUTE OF STEEL CONSTRUCTION

F_y = 36 ksi
F_y = 50 ksi

COLUMNS
Structural Tees
Cut from W shapes
Allowable axial loads in kips

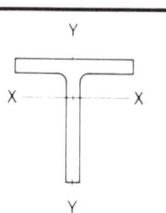

Designation		WT 18												
Wt./ft		105		97		91		85		80		75		67.5	
F_y		36	50	36	50	36	50	36	50	36	50	36	50	36	50
X-X AXIS	0	641	733	546	603	481	510	414	424	365	368	323	322	273	273
	10	610	694	521	573	460	487	396	406	350	352	310	309	262	262
	12	602	684	514	565	454	481	392	401	347	349	307	306	260	260
	14	593	674	507	557	448	474	387	396	343	345	304	303	257	257
	16	584	663	500	549	442	468	382	391	338	340	300	299	254	254
	18	575	651	493	540	436	461	377	386	334	336	297	296	251	251
	20	565	639	485	530	429	453	372	380	329	331	293	292	248	248
	22	555	626	476	520	422	446	366	374	325	326	289	288	245	245
	24	544	612	468	510	415	438	360	368	320	321	284	284	242	242
	26	533	598	459	500	408	429	354	361	314	316	280	279	238	238
	28	522	584	449	488	400	421	348	355	309	311	276	275	235	235
	30	510	569	440	477	392	412	341	348	304	305	271	270	231	231
	32	498	553	430	465	383	402	334	341	298	299	266	265	227	227
	36	472	521	409	440	366	383	320	326	286	287	256	255	219	219
	40	445	486	387	414	347	362	305	310	273	275	245	245	211	211
Y-Y AXIS	0	641	733	546	603	481	510	414	424	365	368	323	322	273	273
	10	507	565	430	465	378	397	324	330	282	284	246	246	197	197
	12	492	546	419	451	369	387	317	323	277	278	242	241	193	193
	14	472	521	404	434	357	374	307	313	269	271	236	235	189	189
	16	449	492	386	412	343	357	296	301	260	261	228	228	183	183
	18	423	458	365	388	325	338	283	287	249	250	219	219	176	176
	20	394	421	342	360	306	317	267	271	236	237	209	208	168	168
	22	362	381	317	330	285	293	250	254	222	223	197	197	158	158
	24	328	337	289	298	262	268	232	234	207	207	184	184	148	148
	26	292	292	260	263	238	241	213	214	190	191	170	170	137	137
	28	255	255	230	230	212	212	192	192	173	173	155	155	124	124
	30	224	224	202	202	187	187	170	170	154	154	139	139	111	111
	32	198	198	179	179	166	166	151	151	137	137	124	124	100	100
	34	177	177	160	160	148	148	135	135	123	123	111	111	90	90
	36	158	158	143	143	133	133	121	121	110	110	100	100	81	81
	39	136	136	123	123	114	114	104	104	95	95	86	86	70	70
	41	123	123	112	112	104	104	95	95	86	86	79	79		
	42	118	118	106	106	99	99	90	90						
	43	112	112												
Properties															
A (in.²)		30.9		28.5		26.8		25.0		23.5		22.1		19.9	
r_x (in.)		5.65		5.62		5.62		5.61		5.61		5.62		5.66	
r_y (in.)		2.58		2.56		2.55		2.53		2.50		2.47		2.38	

Effective length in ft KL with respect to indicated axis

Heavy line indicates Kl/r of 200.

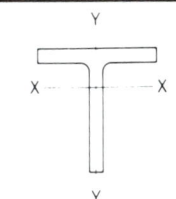

COLUMNS
Structural Tees
Cut from W shapes
Allowable axial loads in kips

F_y = 36 ksi
F_y = 50 ksi

Designation		WT 16.5												
Wt./ft		120.5		110.5		100.5		76		70.5		65		59	
F_y		36	50	36	50	36	50	36	50	36	50	36	50	36	50
X-X AXIS	0	765	920	680	781	571	633	375	386	327	330	284	284	232	232
	10	720	860	641	732	539	596	357	367	312	315	272	271	223	223
	12	708	844	631	719	531	586	353	362	308	311	269	268	220	220
	14	696	827	620	706	523	576	348	357	304	307	265	265	218	218
	16	683	809	609	691	514	565	343	352	300	303	262	261	215	215
	18	669	790	597	676	504	554	337	346	296	298	258	258	212	212
	20	655	770	584	660	494	542	332	340	291	293	254	254	209	209
	22	639	749	571	643	484	529	326	334	286	288	250	250	206	206
	24	624	727	557	626	473	516	320	328	281	283	246	246	203	203
	26	607	704	543	607	461	503	313	321	276	278	242	241	200	200
	28	590	681	528	588	450	488	307	314	270	272	237	237	196	196
	30	572	656	512	569	437	474	300	307	265	266	232	232	193	193
	32	554	630	496	548	425	458	293	299	259	261	228	227	189	189
	34	535	604	480	527	412	443	286	292	253	254	223	222	185	185
	36	515	577	463	505	398	427	278	284	246	248	218	217	182	182
	40	474	519	427	460	370	392	262	267	233	235	207	207	174	174
Y-Y AXIS	0	765	920	680	781	571	633	375	386	327	330	284	284	232	232
	10	643	755	562	632	466	508	296	303	254	255	214	214	170	170
	12	636	744	557	625	462	503	289	296	248	250	210	210	167	167
	14	625	730	549	615	456	497	280	286	241	242	204	204	163	163
	16	611	711	538	601	449	487	269	274	232	233	197	197	158	158
	18	594	687	524	583	439	476	255	260	221	222	189	189	151	151
	20	575	660	507	562	427	461	240	244	209	210	179	179	144	144
	22	553	630	489	539	413	444	224	227	195	196	168	168	136	136
	24	529	597	469	513	397	425	206	208	180	181	156	156	126	126
	26	504	562	447	485	380	405	187	188	164	165	142	142	116	116
	28	477	524	423	455	362	383	167	167	147	148	128	128	105	105
	30	449	485	399	423	342	360	147	147	130	130	114	114	94	94
	34	389	400	346	355	300	309	116	116	104	104	91	91	75	75
	36	357	359	318	319	278	281	104	104	93	93	82	82	68	68
	38	323	323	288	288	254	255	94	94	84	84	74	74	62	62
	39	308	308	274	274	242	242	90	90	80	80	71	71		
	40	293	293	261	261	231	231	85	85	76	76				
	41	279	279	249	249	221	221	81	81						
Properties															
A (in.²)		35.4		32.5		29.5		22.4		20.8		19.2		17.3	
r_x (in.)		4.96		4.96		4.95		5.14		5.15		5.18		5.20	
r_y (in.)		3.63		3.59		3.56		2.47		2.43		2.39		2.32	

Effective length in ft KL with respect to indicated axis

Heavy line indicates Kl/r of 200.

| F_y = 36 ksi |
| F_y = 50 ksi |

COLUMNS
Structural Tees
Cut from W shapes
Allowable axial loads in kips

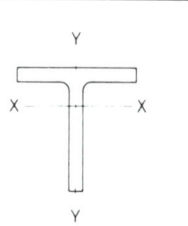

Designation		WT 15														
Wt./ft		105.5		95.5		86.5		66		62		58		54		49.5	
F_y		36	50	36	50	36	50	36	50	36	50	36	50	36	50	36	50
X-X AXIS	0	670	834	596	688	501	560	357	387	315	328	283	290	252	254	214	214
	10	624	769	556	637	469	522	337	363	298	310	268	274	239	241	204	204
	12	613	752	545	624	460	512	332	357	293	305	264	270	235	238	201	201
	14	600	734	534	610	452	501	326	351	288	300	260	266	232	234	198	198
	16	586	714	522	595	442	489	320	344	283	294	256	261	228	230	195	195
	18	572	693	510	579	432	477	314	337	278	289	251	257	224	226	192	192
	20	557	671	497	562	422	464	307	329	272	283	246	252	220	222	189	189
	22	542	648	483	544	411	451	300	321	267	276	241	246	216	218	185	185
	24	525	624	468	525	399	437	293	313	261	270	236	241	211	213	182	181
	26	508	599	453	506	387	422	285	304	254	263	230	235	206	208	178	178
	28	490	573	438	486	375	407	278	295	248	256	225	229	202	203	174	174
	30	472	545	421	465	362	392	269	286	241	249	219	223	196	198	170	170
	32	452	517	404	443	348	375	261	277	234	241	213	217	191	193	166	166
	34	433	487	387	420	334	358	252	267	227	233	206	210	186	187	162	161
	36	412	457	369	397	320	341	243	256	219	225	200	203	180	182	157	157
	40	369	392	331	347	290	304	225	235	203	208	186	189	169	170	148	147
Y-Y AXIS	0	670	834	596	688	501	560	357	387	315	328	283	290	252	254	214	214
	10	567	687	495	559	409	449	276	294	242	250	214	218	185	186	152	152
	12	559	676	489	552	405	444	267	283	235	242	208	211	179	181	148	148
	14	549	660	481	541	399	437	254	269	225	231	199	202	172	173	143	143
	16	535	639	470	527	392	428	240	252	212	218	189	191	164	165	136	136
	18	518	614	456	509	382	416	223	232	198	203	176	179	153	154	128	128
	20	499	586	440	488	370	401	204	211	183	186	163	164	142	142	118	118
	22	478	555	422	465	356	384	184	188	166	168	148	149	129	129	108	108
	24	455	522	402	440	341	366	163	163	147	148	131	132	115	115	96	96
	26	432	486	381	413	324	346	141	141	128	128	115	115	100	100	85	85
	28	406	449	359	384	307	325	123	123	112	112	100	100	88	88	75	75
	30	380	409	336	354	289	302	108	108	98	98	88	88	78	78	66	66
	32	352	367	312	322	269	278	95	95	87	87	78	78	69	69	59	59
	34	323	327	286	289	249	253	85	85	78	78	70	70	62	62	53	53
	35	308	309	273	273	238	240	80	80	74	74	66	66	59	59	50	50
	36	293	293	259	259	227	228	76	76	70	70	63	63				
	37	278	278	246	246	216	216	72	72	66	66						
	38	264	264	234	234	206	206										
	40	239	239	212	212	186	186										
Properties																	
A (in.^{2})		31.0		28.1		25.4		19.4		18.2		17.1		15.9		14.5	
r_x (in.)		4.43		4.42		4.42		4.66		4.66		4.67		4.69		4.71	
r_y (in.)		3.49		3.46		3.43		2.25		2.23		2.19		2.15		2.10	

Effective length in ft KL with respect to indicated axis

Heavy line indicates Kl/r of 200.

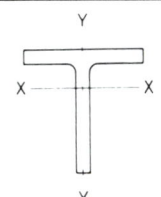

COLUMNS
Structural Tees
Cut from W shapes
Allowable axial loads in kips

| F_y = 36 ksi |
| F_y = 50 ksi |

Designation		WT 13.5												
Wt./ft		89		80.5		73		57		51		47		42	
F_y		36	50	36	50	36	50	36	50	36	50	36	50	36	50
	0	564	734	512	605	436	494	320	353	253	260	217	219	177	177
X-X AXIS	10	520	667	472	553	403	454	299	327	237	244	204	206	168	167
	12	509	649	462	539	395	443	293	321	233	239	201	202	165	164
	14	496	629	450	524	385	431	287	313	228	235	197	199	163	162
	16	483	608	438	507	375	419	280	306	224	230	193	195	160	159
	18	469	586	425	490	365	406	274	298	219	224	189	191	157	156
	20	454	562	412	472	354	392	266	289	213	219	185	187	153	153
	22	439	538	397	453	342	378	259	280	208	213	181	182	150	149
	24	422	512	383	433	330	363	251	271	202	207	176	177	146	146
	26	405	485	367	412	318	347	243	261	196	201	171	172	143	142
	28	388	457	351	390	304	331	234	251	190	194	166	167	139	138
	30	369	427	334	368	291	314	225	240	184	188	161	162	135	135
	32	350	396	317	344	276	296	216	230	177	181	156	157	131	131
	34	330	364	298	319	262	277	206	218	170	174	150	151	127	126
	36	310	330	280	294	246	258	197	206	163	166	144	145	122	122
	38	288	297	260	267	230	238	186	194	156	158	138	139	118	118
	40	266	268	240	241	214	217	176	182	148	150	132	133	113	113
	0	564	734	512	605	436	494	320	353	253	260	217	219	177	177
Y-Y AXIS	10	476	597	424	488	355	394	248	267	195	200	165	166	130	130
	12	468	585	418	480	351	389	238	255	189	193	160	161	127	126
	14	457	567	409	468	345	381	225	240	180	184	153	154	122	122
	16	443	545	397	452	336	370	210	223	170	173	145	146	116	116
	18	426	519	383	433	325	357	194	203	158	160	135	136	109	109
	20	408	489	366	411	312	341	176	181	145	146	125	125	101	101
	22	388	457	349	387	298	323	156	158	130	131	113	113	92	92
	24	367	423	329	361	283	304	135	135	115	115	100	101	83	83
	26	344	387	309	334	266	283	116	116	99	99	87	87	73	73
	28	320	348	287	304	249	261	101	101	87	87	76	76	64	64
	30	295	308	264	273	230	238	88	88	76	76	67	67	56	56
	32	269	272	240	242	211	213	78	78	67	67	60	60	50	50
	34	242	242	215	215	190	190	69	69	60	60	53	53	45	45
	35	228	228	204	204	180	180	66	66	57	57	50	50		
	36	216	216	193	193	171	171	62	62						
	40	176	176	157	157	139	139								
Properties															
A (in.²)		26.1		23.7		21.5		16.8		15.0		13.8		12.4	
r_x (in.)		3.98		3.96		3.95		4.15		4.14		4.16		4.18	
r_y (in.)		3.26		3.24		3.21		2.18		2.15		2.12		2.07	

Effective length in ft KL with respect to indicated axis

Heavy line indicates Kl/r of 200.

| F_y = 36 ksi F_y = 50 ksi | COLUMNS Structural Tees Cut from W shapes Allowable axial loads in kips | 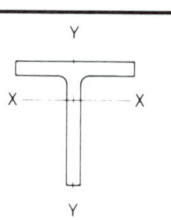 |

Designation		WT 12								
Wt./ft		81		73		65.5		58.5		52	
F_y		36	50	36	50	36	50	36	50	36	50

Effective length in ft KL with respect to indicated axis

X-X AXIS

KL	81 (36)	81 (50)	73 (36)	73 (50)	65.5 (36)	65.5 (50)	58.5 (36)	58.5 (50)	52 (36)	52 (50)
0	516	717	464	611	417	513	357	408	289	317
10	469	636	422	544	379	460	325	369	265	289
12	456	614	411	526	369	446	317	358	258	281
14	443	589	398	506	358	430	307	346	251	273
16	428	563	385	485	346	413	298	334	244	264
18	413	536	371	462	334	395	287	321	236	255
20	396	507	356	439	321	376	276	307	228	245
22	379	476	341	413	307	356	265	293	219	235
24	361	443	325	387	292	335	253	277	210	225
26	342	409	308	359	277	313	240	261	201	213
28	322	373	290	330	261	290	227	244	191	202
30	301	336	271	299	245	266	213	227	180	189
32	280	296	252	267	228	241	199	208	170	177
34	257	263	232	236	210	215	184	189	158	164
36	234	234	210	211	191	191	168	170	147	150
38	210	210	189	189	172	172	152	152	135	135
40	190	190	171	171	155	155	137	137	122	122

Y-Y AXIS

KL	81 (36)	81 (50)	73 (36)	73 (50)	65.5 (36)	65.5 (50)	58.5 (36)	58.5 (50)	52 (36)	52 (50)
0	516	717	464	611	417	513	357	408	289	317
10	443	590	392	495	343	408	287	321	228	246
12	432	571	383	481	336	398	283	315	225	242
14	418	546	371	462	327	385	276	307	221	237
16	402	517	357	439	315	368	267	295	215	230
18	384	485	341	413	301	347	256	281	207	221
20	364	450	323	385	285	325	243	265	199	211
22	344	412	305	354	268	300	229	247	188	199
24	321	372	284	321	250	274	214	228	177	186
26	298	330	263	286	231	245	198	207	165	171
28	274	286	241	249	210	215	181	185	152	156
30	248	250	217	218	189	189	163	163	139	140
32	221	221	192	192	167	167	144	144	124	124
34	196	196	171	171	148	148	128	128	111	111
36	175	175	153	153	133	133	115	115	99	99
38	157	157	138	138	120	120	104	104	90	90
40	142	142	124	124	108	108	94	94	81	81

Properties					
A (in.²)	23.9	21.5	19.3	17.2	15.3
r_x (in.)	3.50	3.50	3.52	3.51	3.51
r_y (in.)	3.05	3.01	2.97	2.94	2.91

AMERICAN INSTITUTE OF STEEL CONSTRUCTION

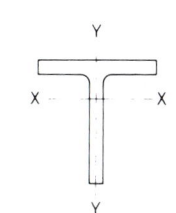

	COLUMNS	F_y = 36 ksi
	Structural Tees	F_y = 50 ksi
	Cut from W shapes	
	Allowable axial loads in kips	

Designation							WT 12						
Wt./ft			47		42		38		34		31		27.5
F_y			36	50	36	50	36	50	36	50	36	50	36	50

Effective length in ft KL with respect to indicated axis

X-X AXIS

KL	47 (36)	47 (50)	42 (36)	42 (50)	38 (36)	38 (50)	34 (36)	34 (50)	31 (36)	31 (50)	27.5 (36)	27.5 (50)
0	267	296	217	227	179	182	147	147	142	144	109	109
10	246	271	201	210	167	169	137	137	133	134	103	103
12	240	264	196	205	163	165	135	134	130	131	101	101
14	234	257	192	200	160	162	132	132	128	129	99	99
16	227	249	187	195	156	158	129	129	125	126	97	97
18	220	241	182	189	152	154	126	126	122	123	95	95
20	213	232	176	183	148	149	123	123	119	119	93	93
22	206	223	170	177	143	145	119	119	115	116	91	91
24	198	213	164	170	139	140	116	116	112	113	88	88
26	189	203	158	163	134	135	112	112	108	109	86	86
28	180	193	151	156	129	130	108	108	105	105	83	83
30	171	182	145	149	124	125	104	104	101	101	81	81
32	162	170	138	141	118	119	100	100	97	97	78	78
34	152	158	130	133	112	113	96	96	92	92	75	75
36	142	146	123	125	107	107	91	91	88	88	72	72
38	131	133	115	116	101	101	87	87	84	84	69	69
40	120	120	107	107	94	95	82	82	79	79	66	66

Y-Y AXIS

KL	47 (36)	47 (50)	42 (36)	42 (50)	38 (36)	38 (50)	34 (36)	34 (50)	31 (36)	31 (50)	27.5 (36)	27.5 (50)
0	267	296	217	227	179	182	147	147	142	144	109	109
10	205	222	165	171	135	136	107	107	91	92	69	69
12	194	209	157	162	129	130	103	103	83	83	63	63
14	180	192	147	152	122	123	98	98	72	72	56	56
16	165	174	136	139	113	114	91	91	60	60	48	48
18	148	153	123	126	103	104	84	84	49	49	40	40
20	129	131	109	110	92	93	75	75	41	41	33	33
22	110	110	94	94	81	81	66	66	34	34	28	28
23	101	101	87	87	74	74	61	61	32	32		
24	93	93	80	80	69	69	57	57				
26	80	80	69	69	59	59	49	49				
28	69	69	60	60	52	52	43	43				
30	61	61	52	52	45	45	38	38				
31	57	57	49	49	43	43	36	36				
32	53	53	46	46	40	40						
33	50	50										

Properties						
A (in.^{2})	13.8	12.4	11.2	10.0	9.11	8.10
r_x (in.)	3.67	3.67	3.68	3.70	3.79	3.80
r_y (in.)	1.98	1.95	1.92	1.87	1.38	1.34

Heavy line indicates Kl/r of 200.

F_y = 36 ksi F_y = 50 ksi	COLUMNS Structural Tees Cut from W shapes Allowable axial loads in kips								

Designation		WT 10.5								
Wt./ft		73.5		66		61		55.5		50.5	
F_y		36	50	36	50	36	50	36	50	36	50
	0	467	648	419	582	387	533	352	448	319	369
X-X AXIS	10	416	561	373	503	344	461	313	390	283	324
	12	402	537	361	481	333	440	303	374	274	312
	14	388	510	347	457	320	418	291	357	263	298
	16	372	482	333	432	306	395	279	338	252	284
	18	355	452	318	404	292	369	266	318	240	268
	20	337	420	301	375	277	343	252	296	228	252
	22	318	385	284	344	261	314	237	274	214	235
	24	298	349	266	311	244	284	222	250	200	216
	26	277	311	247	276	226	252	205	225	186	197
	28	255	271	227	240	208	219	188	198	170	177
	30	232	236	206	209	188	191	170	172	154	156
	32	207	208	184	184	168	168	152	152	137	137
	34	184	184	163	163	148	148	134	134	121	121
	36	164	164	145	145	132	132	120	120	108	108
	38	147	147	130	130	119	119	107	107	97	97
	40	133	133	118	118	107	107	97	97	87	87
Y-Y AXIS	0	467	648	419	582	387	533	352	448	319	369
	10	402	536	357	473	326	429	292	358	260	294
	12	390	515	347	457	318	415	286	349	255	288
	14	376	490	335	436	307	397	277	336	248	279
	16	360	462	321	411	295	375	266	319	239	267
	18	343	431	306	383	281	349	253	299	228	252
	20	324	397	289	352	265	322	239	277	215	236
	22	304	361	271	320	248	292	224	254	202	219
	24	283	323	252	285	231	260	208	229	188	200
	26	261	282	231	248	212	226	191	202	172	179
	28	237	244	210	215	192	196	173	175	156	158
	30	212	213	188	188	172	172	153	153	138	138
	32	188	188	166	166	151	151	135	135	122	122
	34	166	166	147	147	134	134	120	120	109	109
	36	149	149	131	131	120	120	108	108	97	97
	38	134	134	118	118	108	108	97	97	88	88
	40	121	121	107	107	98	98	87	87	79	79
Properties										
A (in.^{2})		21.6		19.4		17.9		16.3		14.9	
r_x (in.)		3.08		3.06		3.04		3.03		3.01	
r_y (in.)		2.95		2.93		2.92		2.90		2.89	

Effective length in ft KL with respect to indicated axis

AMERICAN INSTITUTE OF STEEL CONSTRUCTION

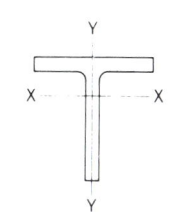

	COLUMNS	F_y = 36 ksi
	Structural Tees Cut from W shapes Allowable axial loads in kips	F_y = 50 ksi

Designation		WT 10.5															
Wt./ft		46.5		41.5		36.5		34		31		28.5		25		22	
F_y		36	50	36	50	36	50	36	50	36	50	36	50	36	50	36	50
X-X AXIS	0	296	398	264	313	210	234	184	199	155	160	143	149	116	118	89	89
	6	280	373	250	295	199	222	175	189	148	152	137	142	111	113	86	86
	8	274	361	243	287	195	216	171	185	144	149	134	139	109	110	85	85
	10	266	349	237	278	190	210	167	180	141	145	131	136	107	108	83	83
	12	258	336	229	268	184	204	163	175	138	142	128	132	105	106	81	81
	14	249	321	222	257	178	196	158	169	134	137	124	128	102	103	79	79
	16	240	305	213	246	172	189	152	163	129	133	121	124	99	100	77	77
	18	230	289	204	234	165	181	147	156	125	128	117	120	96	97	75	75
	20	220	271	195	221	158	172	141	150	120	123	112	116	93	94	73	73
	22	208	252	185	207	151	163	135	142	115	118	108	111	89	90	71	71
	24	197	233	174	193	143	154	128	135	110	113	103	106	86	87	68	68
	26	185	212	163	178	135	144	121	127	105	107	99	101	82	83	66	66
	28	172	190	152	162	126	133	114	119	99	101	94	96	78	79	63	63
	30	158	167	139	145	117	122	106	110	94	95	88	90	74	75	60	60
	32	144	147	127	128	108	111	99	101	87	89	83	84	70	71	58	58
	34	130	130	113	113	98	99	90	92	81	82	77	78	66	66	55	55
	36	116	116	101	101	88	88	82	82	74	75	71	72	61	62	51	51
	38	104	104	91	91	79	79	74	74	68	68	65	65	57	57	48	48
	40	94	94	82	82	71	71	66	66	61	61	59	59	52	52	45	45
Y-Y AXIS	0	296	398	264	313	210	234	184	199	155	160	143	149	116	118	89	89
	6	247	317	216	249	169	185	148	158	107	110	110	113	85	86	63	63
	8	237	300	208	238	164	179	143	153	105	107	104	106	81	81	60	60
	10	223	276	196	222	156	169	137	145	102	104	95	97	74	74	56	56
	12	206	247	181	202	145	156	128	136	97	99	84	85	65	66	50	50
	14	186	214	163	178	133	141	118	123	91	92	71	71	55	56	43	43
	16	164	176	144	152	118	124	106	110	83	84	57	57	45	45	35	35
	18	140	142	123	123	103	104	93	94	73	74	46	46	36	36	29	29
	20	116	116	101	101	86	86	78	78	63	63	38	38	30	30	24	24
	21	105	105	92	92	78	78	72	72	58	58	34	34	27	27	22	22
	22	96	96	84	84	72	72	66	66	54	54	31	31				
	24	81	81	71	71	61	61	56	56	46	46						
	26	70	70	61	61	52	52	48	48	40	40						
	28	60	60	53	53	45	45	41	41	35	35						
	29	56	56	49	49	42	42	39	39	33	33						
	30	53	53	46	46	39	39	36	36								

Effective length in ft KL with respect to indicated axis

Properties								
A (in.^{2})	13.7	12.2	10.7	10.0	9.13	8.37	7.36	6.49
r_x (in.)	3.25	3.22	3.21	3.20	3.21	3.29	3.30	3.31
r_y (in.)	1.84	1.83	1.81	1.80	1.77	1.35	1.30	1.26

Heavy line indicates Kl/r of 200.

AMERICAN INSTITUTE OF STEEL CONSTRUCTION

F_y = 36 ksi	COLUMNS
F_y = 50 ksi	Structural Tees Cut from W shapes Allowable axial loads in kips

Designation			WT 9								
Wt./ft			59.5		53		48.5		43		38	
F_y			36	50	36	50	36	50	36	50	36	50
		0	378	525	337	468	309	429	274	357	239	278
	X-X AXIS	10	327	436	291	389	266	355	236	298	206	235
		12	313	411	278	366	254	334	226	281	197	223
		14	297	383	264	341	241	311	214	263	187	210
		16	280	354	250	315	227	286	202	243	176	196
		18	263	322	234	286	213	259	188	222	164	181
		20	244	288	217	255	197	231	174	199	152	165
		22	224	251	199	222	180	200	159	174	139	148
		24	202	213	180	188	162	169	143	149	125	129
		26	180	181	159	161	143	144	126	127	110	111
		28	156	156	138	138	124	124	109	109	96	96
		30	136	136	121	121	108	108	95	95	83	83
		34	106	106	94	94	84	84	74	74	65	65
		38	85	85	75	75	67	67	59	59	52	52
		42	70	70	62	62	55	55	49	49	42	42
		43	66	66	59	59						
	Y-Y AXIS	0	378	525	337	468	309	429	274	357	239	278
		10	323	430	285	378	259	343	226	282	193	217
		12	311	409	275	360	251	327	219	272	188	211
		14	297	384	263	338	240	308	210	257	181	203
		16	282	356	249	314	227	286	200	240	172	191
		18	265	327	234	287	214	261	188	221	162	178
		20	248	295	218	258	199	235	175	200	151	163
		22	229	260	201	227	183	206	161	177	139	148
		24	209	224	183	194	167	176	146	153	126	131
		26	188	191	164	166	149	151	130	131	112	113
		28	165	165	144	144	130	130	113	113	98	98
		30	144	144	125	125	114	114	99	99	86	86
		34	113	113	98	98	89	89	78	78	67	67
		38	90	90	79	79	71	71	62	62	54	54
		42	74	74	64	64	59	59	51	51	44	44
		43	71	71	61	61	56	56	49	49	42	42
		44	67	67	59	59	53	53				
Properties											
A (in.^{2})			17.5		15.6		14.3		12.7		11.2	
r_x (in.)			2.60		2.59		2.56		2.55		2.54	
r_y (in.)			2.69		2.66		2.65		2.63		2.61	

Effective length in ft KL with respect to indicated axis

Heavy line indicates Kl/r of 200.

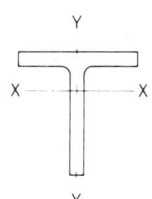

COLUMNS
Structural Tees
Cut from W shapes
Allowable axial loads in kips

F_y = 36 ksi
F_y = 50 ksi

Designation		WT 9															
Wt./ft		35.5		32.5		30		27.5		25		23		20		17.5	
F_y		36	50	36	50	36	50	36	50	36	50	36	50	36	50	36	50
X-X AXIS	0	225	300	206	251	184	211	160	179	130	137	122	129	88	87	71	71
	10	196	254	180	215	161	182	140	155	116	121	108	114	79	79	65	65
	12	188	241	173	204	154	174	135	149	112	117	105	110	77	77	63	63
	14	180	227	165	193	148	165	129	142	107	112	101	106	74	74	61	61
	16	171	212	156	181	140	156	123	135	102	107	96	101	71	71	59	59
	18	161	196	147	168	132	146	117	126	97	101	92	96	69	69	57	57
	20	151	178	138	155	124	135	110	118	92	96	87	91	66	66	54	54
	22	140	160	127	140	115	124	102	109	87	90	82	85	62	62	52	52
	24	128	140	117	125	106	112	95	99	81	83	77	79	59	59	50	50
	26	116	120	105	108	96	99	86	89	75	76	71	73	56	56	47	47
	28	103	103	93	93	85	86	78	79	68	69	65	67	52	52	44	44
	30	90	90	81	81	75	75	69	69	61	62	59	60	48	48	41	41
	32	79	79	72	72	66	66	60	60	54	54	53	53	44	44	38	38
	34	70	70	63	63	58	58	53	53	48	48	47	47	40	40	35	35
	36	62	62	57	57	52	52	48	48	43	43	42	42	36	36	32	32
	38	56	56	51	51	47	47	43	43	38	38	37	37	32	32	29	29
	40	51	51	46	46	42	42	39	39	35	35	34	34	29	29	26	26
	42	46	46	42	42	38	38	35	35	31	31	31	31	26	26	24	24
	43	44	44	40	40	36	36	33	33	30	30	29	29	25	25	22	22
	45	40	40	36	36	33	33	30	30	27	27	27	27	23	23	21	21
	46											25	25	22	22	20	20
Y-Y AXIS	0	225	300	206	251	184	211	160	179	130	137	122	129	88	87	71	71
	10	165	202	150	172	134	148	116	125	95	99	78	81	59	59	46	46
	12	150	176	136	153	122	133	106	114	88	91	68	69	52	52	40	40
	14	132	147	120	130	109	115	95	100	79	82	55	55	44	44	34	34
	16	113	116	103	105	93	96	82	84	70	71	43	43	36	36	28	28
	18	93	93	84	84	77	77	68	68	59	60	34	34	29	29	22	22
	20	76	76	68	68	63	63	56	56	49	49	28	28	23	23	19	19
	21	69	69	62	62	57	57	51	51	45	45	26	26	21	21		
	22	63	63	57	57	52	52	47	47	41	41						
	24	53	53	48	48	44	44	39	39	35	35						
	26	45	45	41	41	38	38	34	34	30	30						
	27	42	42	38	38	35	35	31	31	28	28						
	28	39	39	36	36	33	33										

Properties																	
A (in.^{2})		10.4		9.55		8.82		8.10		7.33		6.77		5.88		5.15	
r_x (in.)		2.74		2.72		2.71		2.71		2.70		2.77		2.76		2.79	
r_y (in.)		1.70		1.69		1.69		1.67		1.65		1.29		1.27		1.22	

Effective length in ft KL with respect to indicated axis

Heavy line indicates Kl/r of 200.

F_y = 36 ksi	COLUMNS
F_y = 50 ksi	Structural Tees

COLUMNS
Structural Tees
Cut from W shapes
Allowable axial loads in kips

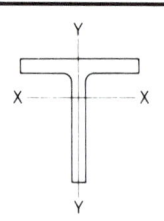

Designation			WT 8						
Wt./ft			50		44.5		38.5		33.5	
F_y			36	50	36	50	36	50	36	50
		0	318	441	283	393	244	335	213	254
		10	266	352	237	313	204	266	177	206
		12	252	326	224	290	192	246	167	192
		14	236	298	210	264	180	224	156	177
		16	219	267	194	237	166	200	144	161
		18	200	234	178	207	152	175	131	143
		20	181	198	160	175	136	147	117	124
		22	160	164	142	145	120	121	103	104
	X-X AXIS	24	138	138	122	122	102	102	87	87
		26	117	117	104	104	87	87	74	74
		28	101	101	89	89	75	75	64	64
		30	88	88	78	78	65	65	56	56
		32	77	77	68	68	57	57	49	49
		34	69	69	61	61	51	51	44	44
		36	61	61	54	54	45	45	39	39
		37	58	58	51	51	43	43	37	37
		38	55	55						
		0	318	441	283	393	244	335	213	254
		10	269	357	238	315	203	265	173	201
		12	257	336	228	296	195	251	167	193
		14	244	312	216	276	185	233	159	182
		16	229	286	203	252	174	214	150	169
		18	214	258	189	227	161	192	139	155
		20	197	228	174	200	148	169	128	139
		22	179	195	158	171	135	144	116	122
	Y-Y AXIS	24	161	165	141	144	120	122	103	104
		26	140	141	123	123	104	104	89	89
		28	121	121	106	106	90	90	77	77
		30	106	106	93	93	78	78	67	67
		32	93	93	82	82	69	69	59	59
		34	83	83	72	72	61	61	53	53
		36	74	74	65	65	55	55	47	47
		38	66	66	58	58	49	49	42	42
		41	57	57	50	50	42	42	36	36
Properties										
A (in.^{2})			14.7		13.1		11.3		9.84	
r_x (in.)			2.28		2.27		2.24		2.22	
r_y (in.)			2.51		2.49		2.47		2.46	

Effective length in ft KL with respect to indicated axis

Heavy line indicates Kl/r of 200.

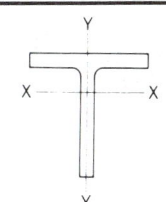

COLUMNS
Structural Tees
Cut from W shapes
Allowable axial loads in kips

F_y = 36 ksi
F_y = 50 ksi

Designation		WT 8												
Wt./ft		28.5		25		22.5		20		18		15.5		13	
F_y		36	50	36	50	36	50	36	50	36	50	36	50	36	50
X-X AXIS	0	181	237	158	183	129	144	100	103	86	88	66	66	47	47
	6	167	215	146	167	120	133	93	96	80	82	62	62	44	44
	8	161	205	140	160	116	128	90	93	78	79	60	60	43	43
	10	154	194	134	152	111	122	87	89	75	77	58	58	42	42
	12	146	182	127	144	106	116	83	85	72	74	56	56	41	41
	14	138	169	120	134	100	109	79	81	69	70	54	54	40	40
	16	129	154	112	124	94	102	75	77	66	67	52	52	38	38
	18	119	139	104	114	88	94	71	72	62	63	49	49	37	37
	20	109	122	95	102	81	86	66	67	58	59	47	47	35	35
	22	98	104	86	90	74	77	61	62	54	55	44	44	33	33
	24	87	88	76	76	66	68	55	56	50	50	41	41	31	31
	26	75	75	65	65	58	58	50	50	45	45	38	38	29	29
	28	64	64	56	56	50	50	44	44	40	40	35	35	27	27
	30	56	56	49	49	44	44	38	38	35	35	31	31	25	25
	32	49	49	43	43	38	38	34	34	31	31	28	28	23	23
	34	44	44	38	38	34	34	30	30	28	28	25	25	21	21
	36	39	39	34	34	30	30	26	26	25	25	22	22	19	19
	38	35	35	30	30	27	27	24	24	22	22	20	20	17	17
	39	33	33	29	29	26	26	23	23	21	21	19	19	16	16
	40	32	32	28	28					20	20	18	18	15	15
	41													14	14
Y-Y AXIS	0	181	237	158	183	129	144	100	103	86	88	66	66	47	47
	6	150	188	128	145	106	116	80	82	66	67	50	50	34	34
	8	141	174	121	136	101	110	77	79	64	65	47	47	32	32
	10	129	155	111	123	93	101	73	74	61	61	42	42	30	30
	12	116	133	100	108	84	90	67	68	56	56	37	37	26	26
	14	100	108	87	91	74	77	60	61	50	50	30	30	22	22
	16	83	84	72	72	62	63	52	53	43	44	24	24	18	18
	18	67	67	57	57	50	50	44	44	36	36	19	19	14	14
	19	60	60	52	52	45	45	40	40	33	33	17	17		
	20	54	54	47	47	41	41	36	36	30	30				
	22	45	45	39	39	34	34	30	30	25	25				
	24	38	38	33	33	29	29	25	25	21	21				
	25	35	35	30	30	27	27	23	23	19	19				
	26	32	32	28	28	25	25	22	22						
Properties															
A (in.²)		8.38		7.37		6.63		5.89		5.28		4.56		3.84	
r_x (in.)		2.41		2.40		2.39		2.37		2.41		2.45		2.47	
r_y (in.)		1.60		1.59		1.57		1.57		1.52		1.17		1.12	

Effective length in ft KL with respect to indicated axis

Heavy line indicates Kl/r of 200.

$F_y = 36$ ksi	COLUMNS

$F_y = 50$ ksi

COLUMNS
Structural Tees
Cut from W shapes
Allowable axial loads in kips

Designation		WT 7									
Wt./ft		66		60		54.5		49.5		45	
F_y		36	50	36	50	36	50	36	50	36	50
X-X AXIS	0	419	582	382	531	346	480	315	438	285	396
	2	406	561	371	511	335	462	306	421	276	381
	4	390	532	355	485	321	437	292	399	264	360
	6	370	497	337	452	303	407	276	371	250	335
	8	346	455	315	414	283	371	258	338	233	305
	10	320	408	290	370	261	331	237	301	214	271
	12	291	356	263	321	235	286	214	259	193	233
	14	259	298	234	267	208	235	188	213	169	190
	16	224	235	201	210	177	183	160	165	144	147
	18	186	186	166	166	145	145	130	130	116	116
	20	151	151	134	134	117	117	106	106	94	94
	22	124	124	111	111	97	97	87	87	78	78
	24	105	105	93	93	81	81	73	73	65	65
	26	89	89	79	79	69	69	62	62	56	56
	27	83	83	74	74	64	64	58	58	52	52
	28	77	77	68	68	60	60				
Y-Y AXIS	0	419	582	382	531	346	480	315	438	285	396
	6	376	507	338	455	301	403	270	359	239	316
	8	375	506	338	454	301	402	269	358	239	315
	10	374	504	337	452	300	401	269	357	238	314
	12	370	497	335	448	299	399	268	356	238	313
	14	362	484	329	439	296	393	266	352	236	311
	16	353	466	321	424	289	381	262	344	234	307
	18	342	447	311	406	280	366	254	332	229	297
	20	330	426	300	387	271	349	246	317	221	284
	22	317	404	289	367	260	331	237	300	213	270
	24	304	380	277	345	250	311	227	282	204	254
	26	291	356	264	323	238	291	216	263	195	237
	28	276	330	251	299	226	269	205	244	185	219
	30	261	303	237	274	214	247	194	223	175	200
	32	246	274	223	248	201	223	182	201	164	181
	34	230	245	208	221	187	198	170	179	153	161
	36	213	219	193	197	173	177	157	160	141	143
	38	195	196	176	177	159	159	143	143	129	129
	40	177	177	160	160	144	144	130	130	116	116
Properties											
A (in.^{2})		19.4		17.7		16.0		14.6		13.2	
r_x (in.)		1.73		1.71		1.68		1.67		1.66	
r_y (in.)		3.76		3.74		3.73		3.71		3.70	

Effective length in ft KL with respect to indicated axis

Heavy line indicates Kl/r of 200.

AMERICAN INSTITUTE OF STEEL CONSTRUCTION

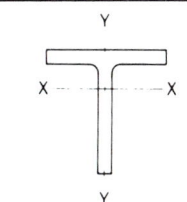

COLUMNS		F_y = 36 ksi
Structural Tees Cut from W shapes Allowable axial loads in kips		F_y = 50 ksi

Designation			WT 7											
Wt./ft		41		37		34		30.5		26.5		24		21.5	
F_y		36	50	36	50	36	50	36	50	36	50	36	50	36	50
X-X AXIS	0	259	360	235	327	216	300	194	261	169	224	153	187	129	147
	4	243	332	220	301	202	275	181	241	158	208	143	174	121	137
	6	231	312	210	282	192	258	172	226	151	196	137	164	116	130
	8	218	289	197	261	181	238	162	209	143	182	129	154	110	123
	10	203	262	184	236	168	216	150	189	133	167	120	141	103	114
	12	187	233	168	209	154	191	138	168	123	150	111	128	95	104
	14	169	201	152	179	139	163	124	144	112	131	101	113	86	93
	16	150	166	134	146	122	133	109	117	99	110	89	97	77	82
	18	129	131	114	116	104	105	92	93	86	88	77	79	67	69
	20	106	106	94	94	85	85	75	75	72	72	64	64	57	57
	22	88	88	77	77	70	70	62	62	59	59	53	53	47	47
	24	74	74	65	65	59	59	52	52	50	50	45	45	39	39
	26	63	63	55	55	50	50	45	45	42	42	38	38	34	34
	28	54	54	48	48	43	43	38	38	37	37	33	33	29	29
	30	47	47	42	42	38	38	33	33	32	32	28	28	25	25
	31									30	30	27	27	24	24
Y-Y AXIS	0	259	360	235	327	216	300	194	261	169	224	153	187	129	147
	8	228	306	206	276	187	250	165	214	140	178	126	149	106	118
	10	220	291	199	264	181	240	161	208	132	165	119	139	100	111
	12	210	274	190	248	173	226	155	196	123	149	110	127	94	103
	14	199	254	180	230	164	209	147	183	112	131	100	113	86	93
	16	187	232	169	211	154	191	138	168	100	111	90	98	77	82
	18	173	209	157	189	143	172	128	151	87	90	78	81	67	69
	20	160	184	145	167	132	150	117	133	73	73	66	66	57	57
	22	145	156	131	142	119	128	106	113	61	61	54	54	47	47
	24	129	132	117	119	106	108	94	95	51	51	46	46	40	40
	26	112	112	102	102	92	92	81	81	44	44	39	39	34	34
	28	97	97	88	88	79	79	70	70	38	38	34	34	29	29
	30	85	85	77	77	69	69	61	61	33	33	29	29	26	26
	31	79	79	72	72	65	65	58	58	31	31	28	28	24	24
	32	74	74	68	68	61	61	54	54	29	29				
	34	66	66	60	60	54	54	48	48						
	36	59	59	53	53	48	48	43	43						
	40	48	48	43	43	39	39	35	35						
	41	45	45	41	41	37	37								
Properties															
A (in.²)		12.0		10.9		9.99		8.96		7.81		7.07		6.31	
r_x (in.)		1.85		1.82		1.81		1.80		1.88		1.87		1.86	
r_y (in.)		2.48		2.48		2.46		2.45		1.92		1.91		1.89	

Effective length in ft KL with respect to indicated axis

Heavy line indicates Kl/r of 200.

			WT 7							

Fy = 36 ksi
Fy = 50 ksi

COLUMNS
Structural Tees
Cut from W shapes
Allowable axial loads in kips

Designation			WT 7							
Wt./ft		19		17		15		13		11	
Fy		36	50	36	50	36	50	36	50	36	50
X-X AXIS	0	113	127	93	100	77	81	61	62	44	44
	2	110	124	90	98	76	79	60	61	43	43
	4	106	120	88	95	74	77	59	59	42	42
	6	102	115	85	91	71	74	57	57	41	41
	8	98	109	81	87	68	71	55	55	39	39
	10	92	102	77	82	65	68	52	53	38	38
	12	87	95	72	77	61	64	50	50	37	37
	14	80	87	68	72	58	60	47	48	35	35
	16	73	79	62	66	54	55	44	45	33	33
	18	66	70	57	59	49	50	41	41	31	31
	20	58	60	51	52	44	45	38	38	29	29
	22	50	50	44	45	39	40	34	34	27	27
	24	42	42	37	37	34	34	31	31	25	25
	26	36	36	32	32	29	29	27	27	22	22
	28	31	31	28	28	25	25	23	23	20	20
	30	27	27	24	24	22	22	20	20	17	17
	32	24	24	21	21	19	19	18	18	15	15
	34	21	21	19	19	17	17	16	16	13	13
	35							15	15	13	13
Y-Y AXIS	0	113	127	93	100	77	81	61	62	44	44
	6	91	101	74	79	59	62	47	47	32	32
	8	87	95	71	75	57	59	42	42	30	30
	10	80	87	66	70	54	55	36	37	26	26
	12	72	77	60	63	49	50	30	30	22	22
	14	62	65	52	54	43	44	22	22	17	17
	16	52	52	44	45	36	37	17	17	13	13
	17	46	46	40	40	33	33	16	16	12	12
	18	41	41	36	36	29	29	14	14		
	20	34	34	29	29	24	24				
	22	28	28	24	24	20	20				
	24	24	24	21	21	17	17				
	25	22	22	19	19						

Effective length in ft KL with respect to indicated axis

Properties										
A (in.²)		5.58		5.00		4.42		3.85		3.25	
rx (in.)		2.04		2.04		2.07		2.12		2.14	
ry (in.)		1.55		1.53		1.49		1.08		1.04	

Heavy line indicates Kl/r of 200.

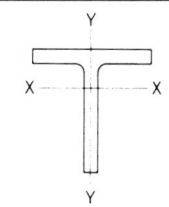

COLUMNS
Structural Tees
Cut from W shapes
Allowable axial loads in kips

$F_y = 36$ ksi
$F_y = 50$ ksi

Designation		WT 6									
Wt./ft		29		26.5		25		22.5		20	
F_y		36	50	36	50	36	50	36	50	36	50
X-X AXIS	0	184	256	168	233	159	220	143	198	127	157
	2	178	245	162	223	153	211	138	190	123	151
	4	169	229	154	210	146	199	132	179	117	143
	6	158	210	144	192	138	184	124	165	110	133
	8	145	188	133	172	128	167	115	149	102	121
	10	131	162	120	149	117	147	104	131	93	107
	12	115	133	105	123	104	124	93	110	82	92
	14	97	101	89	94	91	99	80	87	71	76
	16	78	78	72	72	75	76	67	67	59	59
	18	61	61	57	57	60	60	53	53	46	46
	20	50	50	46	46	49	49	43	43	38	38
	22	41	41	38	38	40	40	35	35	31	31
	24	35	35	32	32	34	34	30	30	26	26
	25	32	32	29	29	31	31	27	27	24	24
	26					29	29	25	25	22	22
Y-Y AXIS	0	184	256	168	233	159	220	143	198	127	156
	6	161	215	144	192	139	186	124	165	108	130
	8	159	213	143	190	133	176	119	157	105	125
	10	156	206	140	185	126	163	112	145	100	117
	12	149	195	135	176	117	147	104	131	93	107
	14	142	181	128	164	107	129	96	115	85	95
	16	133	166	121	150	96	110	86	97	76	83
	18	124	150	112	135	85	89	75	78	66	69
	20	115	133	103	119	72	72	64	64	56	56
	22	104	114	94	101	60	60	53	53	46	46
	24	93	96	84	85	50	50	44	44	39	39
	26	82	82	73	73	43	43	38	38	33	33
	28	71	71	63	63	37	37	33	33	29	29
	30	62	62	55	55	32	32	29	29	25	25
	32	54	54	48	48	28	28	25	25	22	22
	34	48	48	43	43						
	36	43	43	38	38						
	38	38	38	34	34						
	41	33	33	29	29						

Effective length in ft KL with respect to indicated axis

Properties					
A (in.^{2})	8.52	7.78	7.34	6.61	5.89
r_x (in.)	1.50	1.51	1.60	1.58	1.57
r_y (in.)	2.51	2.48	1.96	1.94	1.93

Heavy line indicates Kl/r of 200.

| F_y = 36 ksi |
| F_y = 50 ksi |

COLUMNS
Structural Tees
Cut from W shapes
Allowable axial loads in kips

Designation		WT 6													
Wt./ft		17.5		15		13		11		9.5		8		7	
F_y		36	50	36	50	36	50	36	50	36	50	36	50	36	50
X-X AXIS	0	112	133	85	94	63	65	62	69	48	50	38	38	28	28
	2	108	129	82	91	62	63	61	67	47	49	37	37	28	28
	4	104	123	79	87	60	61	59	65	45	47	36	36	27	27
	6	99	116	76	83	57	58	56	62	44	45	35	35	26	26
	8	93	107	71	78	54	55	54	59	42	43	33	34	25	25
	10	86	98	67	72	51	52	50	55	40	41	31	32	24	24
	12	78	88	61	66	48	48	47	51	37	38	30	30	23	23
	14	70	77	55	59	44	44	43	46	34	35	28	28	21	21
	16	61	64	49	51	39	40	39	41	32	32	26	26	20	20
	18	51	51	42	43	35	35	35	36	28	29	23	24	19	19
	20	42	42	35	35	30	30	30	30	25	25	21	21	17	17
	22	34	34	29	29	25	25	25	25	22	22	18	18	15	15
	24	29	29	24	24	21	21	21	21	18	18	16	16	14	14
	26	25	25	21	21	18	18	18	18	15	15	13	13	12	12
	28	21	21	18	18	15	15	15	15	13	13	12	12	10	10
	29	20	20	17	17	14	14	14	14	12	12	11	11	9	9
	30							13	13	12	12	10	10	9	9
	31							13	13	11	11	9	9	8	8
	32											9	9	8	8
Y-Y AXIS	0	112	133	85	94	63	65	62	69	48	50	38	38	28	28
	2	96	112	71	77	52	53	51	55	38	39	27	28	20	20
	4	95	110	70	76	52	53	48	52	36	37	26	26	19	19
	6	92	106	69	75	51	52	42	44	32	32	23	23	17	17
	8	86	99	65	71	49	50	33	34	25	26	18	18	14	14
	10	79	88	60	65	46	47	23	23	18	18	13	13	10	10
	12	70	76	54	57	42	43	16	16	13	13	9	9	7	7
	13	65	69	51	53	40	41	14	14	11	11				
	14	60	62	47	49	38	38	12	12						
	16	48	48	39	40	32	33								
	18	38	38	32	32	27	27								
	20	31	31	26	26	22	22								
	22	26	26	21	21	18	18								
	24	22	22	18	18	15	15								
	25	20	20	17	17	14	14								

Effective length in ft KL with respect to indicated axis

Properties							
A (in.^{2})	5.17	4.40	3.82	3.24	2.79	2.36	2.08
r_x (in.)	1.76	1.75	1.75	1.90	1.90	1.92	1.92
r_y (in.)	1.54	1.52	1.51	0.847	0.822	0.773	0.753

Heavy line indicates Kl/r of 200.

AMERICAN INSTITUTE OF STEEL CONSTRUCTION

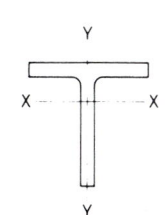

COLUMNS
Structural Tees
Cut from W shapes
Allowable axial loads in kips

| F_y = 36 ksi |
| F_y = 50 ksi |

Designation		WT 5											
Wt./ft		22.5		19.5		16.5		15		13		11	
F_y		36	50	36	50	36	50	36	50	36	50	36	50
X-X AXIS	0	143	199	124	172	105	146	95	133	82	103	70	81
	2	137	188	118	163	100	138	92	127	79	99	67	78
	4	128	172	111	149	94	127	87	118	75	93	64	74
	6	117	153	101	132	86	113	81	108	70	85	60	68
	8	104	129	90	112	77	96	74	96	64	76	55	61
	10	89	102	77	88	66	77	67	82	57	66	49	54
	12	72	73	62	63	54	55	58	66	49	54	43	46
	14	54	54	47	47	41	41	48	49	41	42	36	37
	16	41	41	36	36	31	31	38	38	32	32	28	28
	18	33	33	28	28	25	25	30	30	25	25	22	22
	20	26	26	23	23	20	20	24	24	20	20	18	18
	21					18	18	22	22	19	19	16	16
	22							20	20	17	17	15	15
	24							17	17	14	14	12	12
Y-Y AXIS	0	143	199	124	172	105	146	95	133	82	103	70	81
	2	129	175	110	147	90	119	84	113	71	87	57	65
	4	129	174	109	146	89	118	83	111	70	85	57	64
	6	127	171	108	144	88	117	78	103	67	80	54	61
	8	122	162	104	138	86	114	71	90	61	72	50	56
	10	115	150	99	128	82	106	63	75	54	61	44	48
	12	107	136	92	116	77	96	54	58	46	49	37	39
	14	99	120	84	102	70	84	43	43	36	36	29	29
	16	89	104	76	87	63	71	33	33	28	28	23	23
	18	79	85	67	71	55	57	26	26	22	22	18	18
	20	68	69	57	58	47	47	21	21	18	18	15	15
	22	57	57	48	48	39	39	18	18	15	15	12	12
	24	48	48	40	40	33	33						
	26	41	41	34	34	28	28						
	28	35	35	30	30	24	24						
	30	31	31	26	26	21	21						
	32	27	27	23	23	18	18						
	33	25	25	21	21								
Properties													
A (in.^{2})		6.63		5.73		4.85		4.42		3.81		3.24	
r_x (in.)		1.24		1.24		1.26		1.45		1.44		1.46	
r_y (in.)		2.01		1.98		1.94		1.37		1.36		1.33	

Effective length in ft KL with respect to indicated axis

Heavy line indicates Kl/r of 200.

F_y = 36 ksi	COLUMNS

F_y = 50 ksi

COLUMNS
Structural Tees
Cut from W shapes
Allowable axial loads in kips

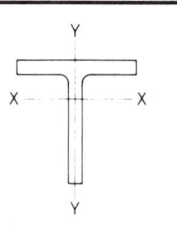

Designation			WT 5						
Wt./ft			9.5		8.5		7.5		6	
F_y			36	50	36	50	36	50	36	50
		0	61	73	54	63	47	54	30	31
		2	59	71	52	61	45	52	29	30
		4	56	67	50	58	43	49	28	29
		6	52	62	47	54	40	46	27	28
		8	48	56	43	49	38	42	25	26
		10	44	50	39	44	34	38	23	24
	X-X AXIS	12	39	43	35	38	31	33	21	22
		14	33	35	30	32	26	28	19	19
		16	27	27	25	25	22	22	17	17
		18	21	21	19	19	17	17	14	14
		20	17	17	16	16	14	14	11	11
		22	14	14	13	13	12	12	9	9
		24	12	12	11	11	10	10	8	8
		25	11	11	10	10	9	9	7	7
		26			9	9	8	8	7	7
		0	61	73	54	63	47	54	30	31
		2	51	60	43	49	35	39	22	22
		4	47	55	40	45	33	36	21	21
		6	41	45	34	37	28	30	18	18
		8	31	32	26	26	20	20	14	14
		10	21	21	17	17	14	14	10	10
	Y-Y AXIS	12	15	15	12	12	10	10	7	7
		13	13	13	11	11	8	8	6	6
		14	11	11	9	9				

Effective length in ft KL with respect to indicated axis

Properties				
A (in.^{2})	2.81	2.50	2.21	1.77
r_x (in.)	1.54	1.56	1.57	1.57
r_y (in.)	0.874	0.844	0.810	0.785

Heavy line indicates Kl/r of 200.

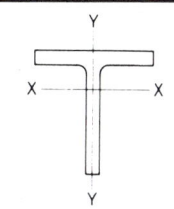

	COLUMNS
	Structural Tees
	Cut from W shapes
	Allowable axial loads in kips

$F_y = 36$ ksi
$F_y = 50$ ksi

Designation		WT 4												
Wt./ft		14		12		10.5		9		7.5		6.5		5	
F_y		36	50	36	50	36	50	36	50	36	50	36	50	36	50
X-X AXIS	0	89	124	76	106	67	92	57	79	48	67	41	58	29	33
	2	84	115	72	99	63	87	54	74	46	63	40	54	28	31
	3	80	109	69	93	61	83	52	71	44	60	38	52	27	30
	4	76	102	66	87	58	78	50	67	43	58	37	50	26	29
	5	72	94	62	80	56	73	48	63	41	54	35	47	25	28
	6	67	85	57	73	52	68	45	58	39	51	34	44	24	26
	8	56	65	48	55	45	55	39	48	34	43	30	37	21	23
	10	43	44	36	37	37	40	32	35	29	33	26	29	18	19
	12	30	30	25	25	28	28	25	25	23	24	21	21	15	15
	14	22	22	19	19	20	20	18	18	17	17	15	15	11	11
	16	17	17	14	14	16	16	14	14	13	13	12	12	9	9
	18					12	12	11	11	11	11	9	9	7	7
	19							10	10	9	9	8	8	6	6
	20									9	9	8	8	6	6
Y-Y AXIS	0	89	124	76	106	67	92	57	79	48	67	41	58	29	33
	2	80	107	67	89	58	78	48	64	40	54	33	44	22	24
	4	79	106	66	89	57	76	47	62	38	49	31	40	21	23
	6	77	102	65	86	54	70	45	58	32	39	27	31	19	20
	8	72	93	61	79	48	60	40	49	25	26	20	20	15	15
	10	66	82	56	70	41	48	34	39	17	17	13	13	10	10
	12	59	70	50	60	34	35	27	28	12	12	9	9	7	7
	14	51	56	44	48	26	26	21	21	9	9	7	7	5	7
	16	43	43	36	37	20	20	16	16						
	18	34	34	29	29	16	16	13	13						
	20	28	28	24	24	13	13	10	10						
	21	25	25	21	21	11	11								
	22	23	23	20	20										
	24	19	19	16	16										
	26	17	17	14	14										
	27	15	15												

Effective length in ft KL with respect to indicated axis

Properties							
A (in.^{2})	4.12	3.54	3.08	2.63	2.22	1.92	1.48
r_x (in.)	1.01	0.999	1.12	1.14	1.22	1.23	1.20
r_y (in.)	1.62	1.61	1.26	1.23	0.876	0.843	0.841

Heavy line indicates Kl/r of 200.

COLUMN BASE PLATES
Design Procedure

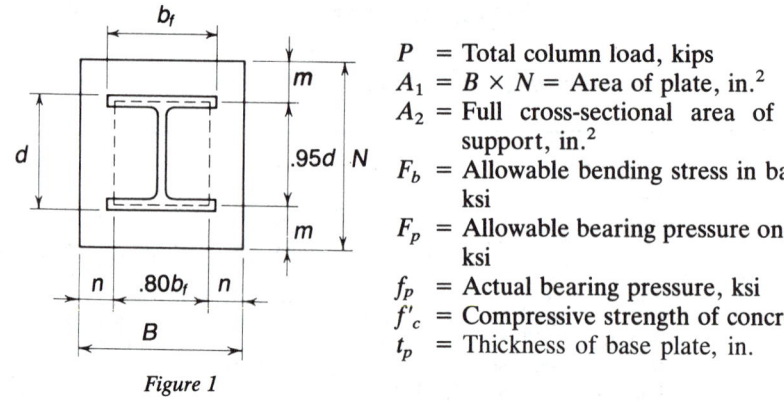

P = Total column load, kips
A_1 = $B \times N$ = Area of plate, in.^{2}
A_2 = Full cross-sectional area of concrete support, in.^{2}
F_b = Allowable bending stress in base plate, ksi
F_p = Allowable bearing pressure on support, ksi
f_p = Actual bearing pressure, ksi
f'_c = Compressive strength of concrete, ksi
t_p = Thickness of base plate, in.

Figure 1

Steel base plates are generally used under columns for distribution of the column load over a sufficient area of the concrete pier or foundation.

Unless the m and n dimensions are small, the base plate is designed as a cantilever beam, fixed at the edges of a rectangle whose sides are $0.80b_f$ and $0.95d$. The column load P is assumed to be distributed uniformily over the base plate within the rectangle. Letting F_b equal $0.75F_y$, the required thickness is found from the formulas

$$t_p = 2m \sqrt{\frac{f_p}{F_y}} \quad \text{and} \quad t_p = 2n \sqrt{\frac{f_p}{F_y}}$$

Dimensions of the base plate are optimized if $m = n$. This condition is approached when $N \approx \sqrt{A_1} + \Delta$, where $\Delta = 0.5 (0.95d - 0.80b_f)$ and $B = A_1/N$.

When the values of m and n are small (the base plate is just large enough in area to accommodate the column profile), a different model is used. For light loads with this type of base plate, the column load is assumed to be distributed to the concrete area, as shown by cross-hatching in Fig. 2 where L is the cantilever distance subjected to the maximum bearing pressure, F_p.

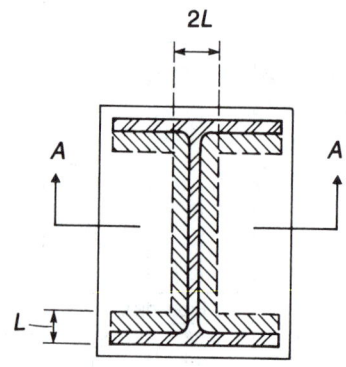

Figure 2

For more heavily loaded small plates, a yield-line solution,* assuming a fixed edge along the column web and simple supports along the flanges, is closely approximated by

$$t_p = 2n'\sqrt{\frac{f_p}{F_y}}$$

where

$$n' = \frac{\sqrt{db_f}}{4}$$

d = depth of column section, in.
b_f = flange width of column section, in.

A smooth transition between the lightly and heavily loaded condition for small plates can be derived as the coefficient λ,** where:

$$\lambda = \frac{2[1-\sqrt{1-q}}{\sqrt{q}} \le 1.0$$

$$q = \frac{4f_p db_f}{(d+b_f)^2 F_p} < 1.0$$

and

$$t_p = 2(\lambda n')\sqrt{\frac{f_p}{F_y}}$$

When λ is less than 1.0, or, equivalently when q is less than 0.64, the design for lightly loaded plates governs, as shown in Fig. 2. The L distance is already factored into the expression for λ. The upper bound of $\lambda = 1.0$ represents the yield-line solution which is conservative to always use for simplicity. Because the above small plate equation is in the same form as the required thickness for large plates, the largest distance m, n or $\lambda n'$ controls.

The allowable bearing strength F_p of the concrete depends on f_c' and the percent of support area occupied by the base plate. From AISC ASD Specification Sect. J9,

$$F_p = 0.35 f_c'$$

when the entire area of a concrete support is covered, and

$$F_p = 0.35 f_c' \sqrt{A_2/A_1} \le 0.7 f_c'$$

when less than the full area is covered. By substituting $P/A_1 \le F_p$, this formula may be rewritten as:

$$\left(\frac{P}{0.35 f_c'}\right)^2 \le A_1 A_2 \le 4A_1^2$$

The first two terms give the general equation:

$$A_1 \ge \frac{1}{A_2}\left(\frac{P}{0.35 f_c'}\right)^2$$

The first and third terms give the equation for the minimum base plate area for the upper concrete bearing limit:

$$A_1 \ge \frac{P}{0.7 f_c'}$$

---

* Thornton, W. A., *Design of Small Base Plates for Wide Flange Columns*, Engineering Journal, AISC, Vol. 27, No. 3, 3rd Quarter 1990, pp. 108–110.
** Thornton, W. A., *Design of Base Plates for Wide Flange Columns—A Concatenation of Methods*, Engineering Journal, AISC, Vol. 27, No. 4, 4th Quarter 1990.

From the second and third terms, it may be noted the condition exists for the lightest base plate when $A_2 = 4A_1$. Substituting into the general equation, the required pedestal area for this condition is:

$$A_2 \geq \frac{P}{0.175 \, f_c'}$$

If conditions permit, the pedestal should be made at least this size for optimum concrete bearing stress.

Steps in the design of a base plate are:

1. Find $A_1 = \frac{1}{A_2}\left(\frac{P}{0.35f_c'}\right)^2$, $A_1 = \frac{P}{0.7f_c'}$, or $A_1 = b_f d$
   Use larger value.

2. Determine $N \approx \sqrt{A_1} + \Delta \geq d$ and $B = A_1/N \geq b_f$

3. Determine uniform and allowable bearing pressure on concrete and check $f_p \leq F_p$:

$$f_p = P/(B \times N)$$

$$F_p = 0.35f_c'\sqrt{\frac{A_2}{A_1}} \leq 0.7f_c'$$

4. Determine $m = (N - 0.95d)/2$ and $n = (B - 0.80b)/2$.

5. Compute q and λ, or, conservatively set $\lambda = 1.0$ and $n' = \dfrac{\sqrt{db_f}}{4}$

6. Determine t_p by formula:

$$c = \max(m, n, \lambda n')$$

$$t_p = 2c\sqrt{\frac{f_p}{F_y}}$$

EXAMPLE 12

Given:

A W10×100 column ($d = 11.10$ in., $b_f = 10.34$ in.) has a reaction of 525 kips, and bears on a 28-in.× 28-in. pier. $f_c' = 3$ ksi, $F_y = 36$ ksi.

Solution:

$A_2 = 28 \times 28 = 784$ in.^{2}

1. $A_1 = \dfrac{1}{A_2}\left(\dfrac{P}{0.35 \, f_c'}\right)^2 = \dfrac{1}{784}\left(\dfrac{525}{0.35 \, (3)}\right)^2$

   $= 319$ in.^{2} **governs**

   $A_1 = \dfrac{P}{0.7 \, f_c'} = \dfrac{525}{0.7 \, (3)} = 250$ in.^{2}

   $A_1 = b_f d = (10.34)(11.1) = 114.8$ in.^{2}

2. $\Delta = 0.5 [(0.95 \times 11.10) - (0.8 \times 10.34)] = 1.14$ in.

$N = \sqrt{A_1} + \Delta = \sqrt{319} + 1.14 = 19.0$

$B = A_1/N = 319/19.0 = 16.8$ in. (use 17 in.)

$A_1 = 19 \times 17 = 323$ in.^{2}

3. $f_p = P / (B \times N) = 525/323 = 1.63$ ksi

$F_p = 0.35 f_c' \sqrt{A_2/A_1} \leq 0.70 f_c'$

$F_p = 0.35(3) \sqrt{784/323} \leq 0.7(3)$

$F_p = 1.64 \leq 2.1$. Use 1.64 ksi

$f_p < F_p$ o.k.

4. $m = (N - 0.95d) / 2 = [19 - (0.95 \times 11.1)]/2$
$= 4.23$ in.

$n = (B - 0.80b) / 2 = [17 - (0.8 \times 10.34)]/2$
$= 4.36$ in.

5. $q = \dfrac{4(1.63)(114.8)}{(11.1+10.34)^2 (1.64)}$

$= 0.993 > 0.64, \therefore \lambda = 1.0$

$n' = \dfrac{\sqrt{11.1 (10.34)}}{4} = 2.68$ in.

6. $c = \max(4.23, 4.36, 2.68)$

$c = 4.36$

$t_p = 2c\sqrt{\dfrac{f_p}{F_y}} = (2)4.36\sqrt{\dfrac{1.63}{36}}$

$= 1.86$ in. (use 2 in.)

Use: Base plate $17 \times 2 \times 1$ ft-7 in.

EXAMPLE 13

Given:

A W12×106 column ($d = 12.89$ in. and $b_f = 12.22$ in.) has a reaction of 600 kips. Select the dimensions of the pier ($f_c' = 3$ ksi) and design the base plate for the smallest nominal area possible. $F_y = 36$ ksi.

Solution:

For maximum F_p, use $A_2 = \dfrac{P}{0.175 f_c'} = \dfrac{600}{0.175 (3)} = 1143$ in.^{2}

Use $A_2 = 34 \times 34 = 1156$ in.^{2}

1. $A_1 = \dfrac{1}{1156} \left(\dfrac{600}{0.35 (3)}\right)^2 = 282$ in.^{2}

$A_1 = \dfrac{600}{0.7 (3)} = 286$ in.^{2} **governs**

$A_1 = b_f d = (12.22)(12.89) = 157.5$ in.^{2}

2. $\Delta = 0.5 [(0.95 \times 12.89) - (0.8 \times 12.22)] = 1.235$ in.

   $N \approx \sqrt{286} + 1.235 = 18.1$ in. (use 19 in.)

   $B = 286 / 19 = 15.6$ in. (use 16 in.)

   $A_1 = 19 \times 16 = 304$ in.^{2}

3. $f_p = 600 / 304 = 1.97$ ksi

   $F_p = 0.35 f_c' \sqrt{A_2/A_1} \leq 0.7 f_c'$

   $F_p = 0.35(3) \sqrt{1156/304} \leq 0.7(3) = 2.1$ ksi

   $\quad = 2.1 \therefore F_p = 2.1$ ksi

   $f_p < F_p$ **o.k.**

4. $m = [19 - (0.95 \times 12.89)]/2 = 3.38$ in.

   $n = [16 - (0.8 \times 12.22)]/2 = 3.11$ in.

5. $q = \dfrac{4(1.97)(157.5)}{(12.22+12.89)^2 \ (2.1)}$

   $\quad = 0.937 > 0.64, \therefore \lambda = 1.0$

   $n' = \dfrac{\sqrt{157.5}}{4} = 3.14$ in.

6. $c = \max(3.38, 3.11, 3.14)$

   $\quad = 3.38$

   $t_p = 2c\sqrt{\dfrac{f_p}{F_y}} = (2)3.38\sqrt{\dfrac{1.97}{36}} = 1.58$ in. (use 1¾ in.)

Use: Base plate 16 × 1¾ × 1 ft-7in.

COLUMN BASE PLATES
Finishing

Rolled steel plates are extensively used for column bases. So that the base plates function properly in transmitting loads to masonry supports, finishing is regulated by specification.

In AISC ASD Specification, Sect. M2.8, it is stated:

"Column bases and base plates shall be finished in accordance with the following requirements:

a. Rolled steel bearing plates 2 in. or less in thickness may be used without milling,* provided a satisfactory contact bearing is obtained; rolled steel bearing plates over 2 in. but not over 4 in. in thickness may be straightened by pressing or, if presses are not available, by milling for all bearing surfaces (except as noted in Subparagraphs c and d of this section), to obtain a satisfactory contact bearing; rolled steel bearing plates over 4 in. thick shall be milled for all bearing surfaces (except as noted in Subparagraphs c and d of this section).

b. Column bases other than rolled steel bearing plates shall be milled for all bearing surfaces (except as noted in Subparagraphs c and d of this section).

c. The bottom surfaces of bearing plates and column bases which are grouted to insure full bearing contact on foundations need not be milled.

d. The top surfaces of base plates with columns full-penetration welded need not be pressed or milled."

---

*See Commentary Sect. J8.

PART 4
Connections

Part 4
CONNECTIONS

Preface

Part 4 provides engineering data and tabular information for the design of connections. Examples illustrate the use of the information for specific applications. More detailed design calculations and pre-engineered connections are given in two companion publications available from AISC:

1. *Engineering for Steel Construction* (1st Ed. 1984)
   This source book of connections contains suggested design and detailing procedures for use by advanced detailers and engineers.
   AISC Publication No. M014*

2. *Detailing for Steel Construction* (1st Ed. 1983)
   This text book directed to detailers contains instruction, explanations, problem solutions and many typical shop details and drawings.
   AISC Publication No. M013*

---

*Keyed to the 8th Edition *Manual of Steel Construction*. Specific designs should be checked against procedures and specifications of the 9th Edition.

BOLTS, THREADED PARTS AND RIVETS
Tension
Allowable loads in kips

TABLE I-A. BOLTS AND RIVETS
Tension on gross (nominal) area

ASTM Designation	F_t Ksi	5/8	3/4	7/8	1	1 1/8	1 1/4	1 3/8	1 1/2
		\multicolumn{8}{Nominal Diameter d, In.}							
		0.3068	0.4418	0.6013	0.7854	0.9940	1.227	1.485	1.767
A307 bolts	20.0	6.1	8.8	12.0	15.7	19.9	24.5	29.7	35.3
A325 bolts	44.0	13.5	19.4	26.5	34.6	43.7	54.0	65.3	77.7
A490 bolts	54.0	16.6	23.9	32.5	42.4	53.7	66.3	80.2	95.4
A502-1 rivets	23.0	7.1	10.2	13.8	18.1	22.9	28.2	34.2	40.6
A502-2,3 rivets	29.0	8.9	12.8	17.4	22.8	28.8	35.6	43.1	51.2

The above table lists ASTM specified materials that generally are intended for use as structural fasteners.

For dynamic and fatigue loading, only A325 or A490 high-strength bolts should be specified. See AISC Specification. Appendix K4.

For allowable combined shear and tension loads, see AISC ASD Specification Sects. J3.5 and J3.6.

TABLE I-B. THREADED FASTENERS
Tension on gross (nominal) area

ASTM Designation	F_y Ksi	F_u ksi	F_t ksi	5/8	3/4	7/8	1	1 1/8	1 1/4	1 3/8	1 1/2
				Nominal Diameter d, In.							
				0.3068	0.4418	0.6013	0.7854	0.9940	1.227	1.485	1.767
A36	36	58	19.1	5.9	8.4	11.5	15.0	19.0	23.4	28.4	33.7
A572, Gr. 50	50	65	21.5	6.6	9.5	12.9	16.9	21.4	26.4	31.9	38.0
A588	50	70	23.1	7.1	10.2	13.9	18.1	23.0	28.3	34.3	40.8
A449 $d \leq 1$	92	120	39.6	12.1	17.5	23.8	31.1	—	—	—	—
$1 < d \leq 1 1/2$	81	105	34.7	—	—	—	—	34.5	42.6	51.5	61.3

The above table lists ASTM specified materials available in round bar stock that are generally intended for use in threaded applications such as tie rods, cross bracing and similar uses. The tensile capacity of the threaded portion of an upset rod shall be larger than the body area times $0.6F_y$.

F_u = specified minimum tensile strength of the fastener material.

$F_t = 0.33F_u$ = allowable tensile stress in threaded fastener.

BOLTS AND THREADED PARTS
ASTM Specifications

TABLE I-C. MATERIAL FOR ANCHOR BOLTS AND TIE RODS

| | ASTM Specification | Strength, Ksi | | | Maximum Diameter In. | Type of Material[b] | Headed or Unheaded |
		Proof Load	Yield (Min.)	Tensile (Min.)			
Bolts and Studs	A307	—	—	60	4	C	H
	A325[a]	85	92	120	½ to 1, incl.	C, QT	H
		74	81	105	1⅛ to 1½ incl.		
	A354 Gr. BD	120	130	150	¼ to 2½ incl.	A, QT	H, U
		105	115	140	over 2½ to 4 incl.		
	A354 Gr. BC	105	109	125	¼ to 2½ incl.	A, QT	H, U
		95	99	115	over 2½ to 4 incl.		
	A449	85	92	120	¼ to 1 incl.	C, QT	H, U
		74	81	105	1⅛ to 1½ incl.		
		55	58	90	1¾ to 3 incl.		
	A490	120	—	150	½ to 1½ incl.	A, QT	H
	A687	—	105	150[c]	⅝ to 3 incl.	A, QT, NT	U
Threaded Round Stock	A36	—	36	58	8	C	U
	A572 Gr. 50	—	50	65	2	HSLA	U
	A572 Gr. 42	—	42	60	6	HSLA	U
	A588	—	50	70	To 4 incl.	HSLA, ACR	U
		—	46	67	over 4 to 5 incl.		
		—	42	63	over 5 to 8 incl.		

[a]Available with weathering (atmospheric corrosion resistance) characteristics comparable to ASTM A242 and A588 steel.

[b]C = carbon
QT = quenched and tempered
A = alloy
NT = notch tough (Charpy V-notch 15 ft-lb. @ −20°F)
HSLA = high-strength low alloy
ACR = atmospheric corrosion-resistant

[c]Maximum (ultimate tensile strength)

Notes:
ASTM specified material for anchor bolts, tie rods and similar applications can be obtained from either specifications for threaded bolts and studs normally used as connectors or for structural material available in round stock that may then be threaded. The material supplier should be consulted for availability of size and length.

Suitable nuts by grade may be obtained from ASTM Specification A563.

Anchor bolt material that is quenched and tempered should not be welded or heated.

Threaded rod with properties meeting A325, A490 or A449 Specifications may be obtained by the use of an appropriate steel (such as AISI C1040 or C4140), quenched and tempered after fabrication.

4 - 5

BOLTS, THREADED PARTS AND RIVETS
Shear
Allowable load in kips

TABLE I-D. SHEAR

ASTM Designation	Connection Type[a]	Hole Type[b]	F_v ksi	Loading[c]	5/8 .3068	3/4 .4418	7/8 .6013	1 .7854	1 1/8 .9940	1 1/4 1.227	1 3/8 1.485	1 1/2 1.767
Bolts												
A307	—	STD NSL	10.0	S	3.1	4.4	6.0	7.9	9.9	12.3	14.8	17.7
				D	6.1	8.8	12.0	15.7	19.9	24.5	29.7	35.3
A325	SC[a] Class A	STD	17.0	S	5.22	7.51	10.2	13.4	16.9	20.9	25.2	30.0
				D	10.4	15.0	20.4	26.7	33.8	41.7	50.5	60.1
		OVS, SSL	15.0	S	4.60	6.63	9.02	11.8	14.9	18.4	22.3	26.5
				D	9.20	13.3	18.0	23.6	29.8	36.8	44.6	53.0
		LSL	12.0	S	3.68	5.30	7.22	9.42	11.9	14.7	17.8	21.2
				D	7.36	10.6	14.4	18.8	23.9	29.4	35.6	42.4
	N	STD, NSL	21.0	S	6.4	9.3	12.6	16.5	20.9	25.8	31.2	37.1
				D	12.9	18.6	25.3	33.0	41.7	51.5	62.4	74.2
	X	STD, NSL	30.0	S	9.2	13.3	18.0	23.6	29.8	36.8	44.5	53.0
				D	18.4	26.5	36.1	47.1	59.6	73.6	89.1	106.0
A490	SC[a] Class A	STD	21.0	S	6.44	9.28	12.6	16.5	20.9	25.8	31.2	37.1
				D	12.9	18.6	25.3	33.0	41.7	51.5	62.4	74.2
		OVS, SSL	18.0	S	5.52	7.95	10.8	14.1	17.9	22.1	26.7	31.8
				D	11.0	15.9	21.6	28.3	35.8	44.2	53.5	63.6
		LSL	15.0	S	4.60	6.63	9.02	11.8	14.9	18.4	22.3	26.5
				D	9.20	13.3	18.0	23.6	29.8	36.8	44.6	53.0
	N	STD, NSL	28.0	S	8.6	12.4	16.8	22.0	27.8	34.4	41.6	49.5
				D	17.2	24.7	33.7	44.0	55.7	68.7	83.2	99.0
	X	STD, NSL	40.0	S	12.3	17.7	24.1	31.4	39.8	49.1	59.4	70.7
				D	24.5	35.3	48.1	62.8	79.5	98.2	119.0	141.0
Rivets												
A502-1	—	STD	17.5	S	5.4	7.7	10.5	13.7	17.4	21.5	26.0	30.9
				D	10.7	15.5	21.0	27.5	34.8	42.9	52.0	61.8
A502-2 A502-3	—	STD	22.0	S	6.7	9.7	13.2	17.3	21.9	27.0	32.7	38.9
				D	13.5	19.4	26.5	34.6	43.7	54.0	65.3	77.7
Threaded Parts												
A36 (F_u=58 ksi)	N	STD	9.9	S	3.0	4.4	6.0	7.8	9.8	12.1	14.7	17.5
				D	6.1	8.7	11.9	15.6	19.7	24.3	29.4	35.0
	X	STD	12.8	S	3.9	5.7	7.7	10.1	12.7	15.7	19.0	22.6
				D	7.9	11.3	15.4	20.1	25.4	31.4	38.0	45.2
A572, Gr. 50 (F_u=65 ksi)	N	STD	11.1	S	3.4	4.9	6.7	8.7	11.0	13.6	16.5	19.6
				D	6.8	9.8	13.3	17.4	22.1	27.2	33.0	39.2
	X	STD	14.3	S	4.4	6.3	8.6	11.2	14.2	17.5	21.2	25.3
				D	8.8	12.6	17.2	22.5	28.4	35.1	42.5	50.5
A588 (F_u=70 ksi)	N	STD	11.9	S	3.7	5.3	7.2	9.3	11.8	14.6	17.7	21.0
				D	7.3	10.5	14.3	18.7	23.7	29.2	35.3	42.1
	X	STD	15.4	S	4.7	6.8	9.3	12.1	15.3	18.9	22.9	27.2
				D	9.4	13.6	18.5	24.2	30.6	37.8	45.7	54.4

[a]SC = Slip critical connection.
N: Bearing-type connection with threads *included* in shear plane.
X: Bearing-type connection with threads *excluded* from shear plane.
[b]STD: Standard round holes (d + 1/16 in.)   OVS: Oversize round holes
LSL: Long-slotted holes normal to load direction   SSL: Short-slotted holes
NSL: Long-or short-slotted hole normal to load direction (required in bearing-type connection).
[c]S: Single shear   D: Double shear.
For threaded parts of materials not listed, use $F_v = 0.17F_u$ when threads are included in a shear plane, and $F_v = 0.22F_u$ when threads are excluded from a shear plane.
To fully pretension bolts 1⅛-in. dia. and greater, special impact wrenches may be required.
When bearing-type connections used to splice tension members have a fastener pattern whose length, measured parallel to the line of force, exceeds 50 in., tabulated values shall be reduced by 20%. See AISC ASD Commentary Sect. J3.4.

AMERICAN INSTITUTE OF STEEL CONSTRUCTION

BOLTS AND THREADED PARTS
Bearing
Allowable loads in kips

TABLE I-E. BEARING
Slip-critical and Bearing-type Connections

Material Thickness	$F_u = 58$ ksi Bolt dia.			$F_u = 65$ ksi Bolt dia.			$F_u = 70$ ksi Bolt dia.			$F_u = 100$ ksi Bolt dia.		
	3/4	7/8	1	3/4	7/8	1	3/4	7/8	1	3/4	7/8	1
1/8	6.5	7.6	8.7	7.3	8.5	9.8	7.9	9.2	10.5	11.3	13.1	15.0
3/16	9.8	11.4	13.1	11.0	12.8	14.6	11.8	13.8	15.8	16.9	19.7	22.5
1/4	13.1	15.2	17.4	14.6	17.1	19.5	15.8	18.4	21.0	22.5	26.3	30.0
5/16	16.3	19.0	21.8	18.3	21.3	24.4	19.7	23.0	26.3	28.1	32.8	37.5
3/8	19.6	22.8	26.1	21.9	25.6	29.3	23.6	27.6	31.5	33.8	39.4	45.0
7/16	22.8	26.6	30.5	25.6	29.9	34.1	27.6	32.2	36.8		45.9	52.5
1/2	26.1	30.5	34.8	29.3	34.1	39.0	31.5	36.8	42.0			60.0
9/16	29.4	34.3	39.2	32.9	38.4	43.9		41.3	47.3			
5/8	32.6	38.1	43.5		42.7	48.8		45.9	52.5			
11/16		41.9	47.9		46.9	53.6			57.8			
3/4		45.7	52.2			58.5						
13/16			56.6									
7/8			60.9									
15/16												
1	52.2	60.9	69.6	58.5	68.3	78.0	63.0	73.5	84.0	90.0	105.0	120.0

Notes:
This table is applicable to all mechanical fasteners in both slip-critical and bearing-type connections utilizing standard holes. Standard holes shall have a diameter nominally 1/16-in. larger than the nominal bolt diameter ($d + 1/16$ in.).
Tabulated bearing values are based on $F_p = 1.2 F_u$.
F_u = specified minimum tensile strength of the connected part.
In connections transmitting axial force whose length between extreme fasteners measured parallel to the line of force exceeds 50 in., tabulated values shall be reduced 20%.
Connections using high-strength bolts in slotted holes with the load applied in a direction other than approximately normal (between 80 and 100 degrees) to the axis of the hole and connections with bolts in oversize holes shall be designed for resistance against slip at working load in accordance with AISC ASD Specification Sect. J3.8.
Tabulated values apply when the distance l parallel to the line of force from the center of the bolt to the edge of the connected part is not less than 1 1/2 d and the distance from the center of a bolt to the center of an adjacent bolt is not less than 3d. See AISC ASD Commentary J3.8.
Under certain conditions, values greater than the tabulated values may be justified under Specification Sect. J3.7.
Values are limited to the double-shear bearing capacity of A490-X bolts.
Values for decimal thicknesses may be obtained by multiplying the decimal value of the unlisted thickness by the value given for a 1-in. thickness.

BOLTS AND RIVETS
Bearing
Allowable loads in kips

TABLE I-F. EDGE DISTANCE

spacing = 3
n = no. of bolts

COPED

Edge Distance[b] l_v In.	Allowable Loads, Kips[a] (for one fastener, 1-in. thick material)			
	$F_u = 58$	$F_u = 65$	$F_u = 70$	$F_u = 100$
1	29.0	32.5	35.0	50.0
1⅛	32.6	36.6	39.4	56.3
1¼	36.3	40.6	43.8	62.5
<1½	43.5	48.8	52.5	75.0

Bolt Dia.	1½ d In.	Values when edge distance is 1½ d or greater [c]			
1	1½	69.6	78.0	84.0	120
⅞	1⁵⁄₁₆	60.9	68.3	73.5	105
¾	1⅛	52.2	58.5	63.0	90.0

[a] Total allowable load $= \Sigma\left[(\text{tabular value}) \times n\right] t$
where

   t = thickness of critical connected part, in.
   n = number of fasteners.

[b] $l_v \geq 2P/F_u t$ (AISC ASD Spec. J3.9) distance center of hole to free edge of connected part in direction of force, in.

where

   F_u = specified minimum tensile strength of material, ksi
   P = force transmitted by one fastener to the critical connected part, kips
[c] $P = 1.2 F_u d$ (AISC ASD Spec. Sect. J3.7).

BOLTS AND RIVETS
Bearing

TABLE I-G. COEFFICIENTS FOR WEB TEAR-OUT (BLOCK SHEAR)
Based on standard holes and 3-in. fastener spacing

Coefficient C_1

l_v In.	l_h, In.												
	1	1⅛	1¼	1⅜	1½	1⅝	1¾	1⅞	2	2¼	2½	2¾	3
1¼	.88	.94	1.00	1.06	1.13	1.19	1.25	1.31	1.38	1.50	1.63	1.75	1.88
1⅜	.91	.98	1.04	1.10	1.16	1.23	1.29	1.35	1.41	1.54	1.66	1.79	1.91
1½	.95	1.01	1.08	1.14	1.20	1.26	1.33	1.39	1.45	1.58	1.70	1.83	1.95
1⅝	.99	1.05	1.11	1.18	1.24	1.30	1.36	1.43	1.49	1.61	1.74	1.86	1.99
1¾	1.03	1.09	1.15	1.21	1.28	1.34	1.40	1.46	1.53	1.65	1.78	1.90	2.03
1⅞	1.06	1.13	1.19	1.25	1.31	1.38	1.44	1.50	1.56	1.69	1.81	1.94	2.06
2	1.10	1.16	1.23	1.29	1.35	1.41	1.48	1.54	1.60	1.73	1.85	1.98	2.10
2¼	1.18	1.24	1.30	1.36	1.43	1.49	1.55	1.61	1.68	1.80	1.93	2.05	2.18
2½	1.25	1.31	1.38	1.44	1.50	1.56	1.63	1.69	1.75	1.88	2.00	2.13	2.25
2¾	1.33	1.39	1.45	1.51	1.58	1.64	1.70	1.76	1.83	1.95	2.08	2.20	2.33
3	1.40	1.46	1.53	1.59	1.65	1.71	1.78	1.84	1.90	2.03	2.15	2.28	2.40

Coefficient C_2

n	Bolt Dia., In.		
	¾	⅞	1
2	.33	.24	.16
3	.99	.86	.74
4	1.64	1.48	1.32
5	2.30	2.10	1.90
6	2.96	2.72	2.48
7	3.61	3.34	3.06
8	4.27	3.96	3.64
9	4.93	4.58	4.23
10	5.58	5.19	4.81

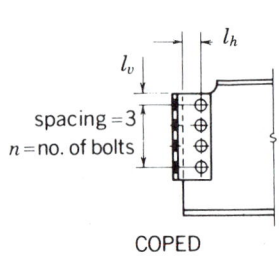

spacing = 3
n = no. of bolts

COPED

Notes:

R_{BS} = Resistance to block shear, kips

$\quad = 0.30\,A_v F_u + 0.50\,A_t F_u$ (from AISC ASD Sect. J4)

$\quad = \{(0.30\,l_v + 0.5\,l_h) + 0.30\,[(n-1)(s-d_h) - d_h/2] - d_h/4\} \times F_u t$

$\quad = (C_1 + C_2) F_u t$

where

A_v = net shear area, in.²
A_t = net tension area, in.²
F_u = specified min. tensile strength, ksi
d_h = dia. of hole (dia. of fastener + ¹⁄₁₆), in.
l_h = distance from center of hole to beam end, in.
l_v = distance from center of hole to edge of web, in.
n = number of fasteners
s = fasteners spacing, in.

Tabular values are based on the following:
AISC ASD Specification Sects. D1, J4, J3.9, J5.2.
AISC ASD Commentary Sects. D1, J4, J3.9.

FRAMED BEAM CONNECTIONS
Bolted
TABLE II

BEAM REACTIONS

For economical connections, beam reactions should be shown on the contract draw-ings. If these reactions are not shown, connections *must* be selected to support one-half the total uniform load capacity shown in the Allowable Uniform Load Tables, Part 2 of this Manual, for the given beam, span and grade of steel specified. The ef-fects of any concentrated loads must be taken into account.

Beam reactions *must* be shown on contract drawings for composite construction and continuous framing.

TYPE OF CONNECTION

Tables are developed for allowable reactions for simply supported beams (Type 2 framing). No eccentricity or moment resistance is considered in determining the tab-ulated values. The inherent rigidity of the connections is a factor the designer should be aware of and consider where critical. The thickness limitation of ⅝ in. for framing angles in the tables was selected to assure flexibility.

In applying the provisions of AISC *Specification for Structural Steel Buildings—Allowable Stress Design (ASD)* Sect. J4, the calculation of area effective in resisting failure due to shear (and tension, when present) is accomplished by deducting the area of the standard holes along the minimum net failure surface. A standard hole is defined as the nominal fastener diameter plus ¹⁄₁₆ in. Attention is called to the dif-ference between this net area provision and that described in AISC ASD Specifica-tion Sects. B2 and B3, which apply to the minimum net *tensile* failure planes.

BOLTS

Bolts approved for use in steel building structures are listed in AISC ASD Specifica-tion Sect. A3.4, Bolts, Washers and Nuts. Application should comply with Sect. J3, Bolts, Threaded Parts and Rivets; Sect. J1.12, Limitations on Bolted and Welded Connections; Sect. J3.7, Size and Use of Holes; and Sect. M2.5, High-strength-bolted Construction—Assembling.

The type of application of a high-strength bolt is indicated as follows:

A325-SC and A490-SC: Slip-critical connection.

A325-N and A490-N: Bearing-type connection with threads included in a shear plane.

A325-X and A490-X: Bearing-type connection with threads excluded from the shear planes.

Note that SC-N-X is not part of the ASTM Specification designation, but is descrip-tive of the design assumptions.

TABLE II-A

This table is for bolts in bearing-type connections having standard or slotted holes, and for bolts in slip-critical connections having standard holes and class A, coated or clean mill scale surface condition (see the *Specification for Structural Joints Using ASTM A325 or A490 bolts* as approved by the Research Council on Structural Connections of the Engineering Foundation, i.e. RCSC Specification).

Allowable loads in the table are based on:

1. Shear capacity of the type and number of bolts in the connection.

2. Bearing capacity of the angles. Footnote b notes that for angle length L the shear on the net section of the angles governs (Table II-C).

3. Shear capacity based on net section of the angles. Footnote c references Table II-C in which these values are given.

TABLE II-B

This table is for bolts in slip-critical connections having oversize, short-slotted, or long-slotted holes and class A, clean mill scale surface condition.

Allowable loads in the table are based on:

1. Shear capacity of the type and number of bolts in the connection and type of hole.

2. Shear capacity based on net section of angles, as noted in footnote b.

For allowable shear values of other classes of surface conditions in slip-critical connections, see the RCSC Specification.

TABLE II-C

This table is for checking the shear capacity of the net section of the connection angles.

Allowable loads are based on $0.3F_u$ (17.4 ksi for A36 material) on the net section vertically through the bolt holes, per AISC ASD Specification Sect. J4, using:

$$R_v = 2t[(L \text{ or } L') - n(d + \tfrac{1}{16})] \, 0.3F_u, \text{ kips}$$

WEB TEAR-OUT (BLOCK SHEAR)

When high fastener values are used on relatively thin material, failure can occur by a combination of shear along the vertical plane through the fasteners plus tension along a horizontal plane on the area effective in resisting tearing failure (see Figs. C-J4.1 through C-J4.4 in the AISC ASD Commentary). The effective area is the minimum net failure surface bounded by the bolt holes. This condition is critical and should be investigated when beam flanges are coped, or in any other similar situation. The net shearing area of the connection angles may also be critical and should be investigated. Refer to Commentary on the AISC ASD Specification Sects. J3.9 and J4 for additional discussion. This failure mechanism is referred to as web tear-out or block shear.

Table I-G lists coefficients to be used for determining block shear capacity for several common beam framing conditions of l_v and l_h and three different bolt diameters in standard size holes at 3-in. spacing.

The block shear or web tear-out allowable reaction is determined by adding coefficient C_1, for the given l_v and l_h, to coefficient C_2 for number and size of bolts, then multiplying this sum by the given value of F_u and the web thickness. For conditions that differ from those tabulated, the general equation shown in Table I-G can be utilized.

For oversize and slotted holes, refer to AISC ASD Specification Sects. J3.9 and J4.

SELECTION OF CONNECTION

The 1989 AISC ASD Specification requires that several conditions be checked to determine the allowable capacity for a framed beam connection. These conditions are: bolt shear, bolt bearing on connecting material, beam web tear-out (block shear), shear on the net area of the connection angles or connection plate and local bending stresses.

In Tables II-A and II-B, connection angle thicknesses have been established to conform with an end distance of 1¼ in. for material with $F_u = 58$ ksi. Note that certain values are governed by the bearing capacity of the angles. Other values are limited by net shear on the angles. The footnotes identify the values so limited and refer to Table II-C.

When the beam flange is not coped and standard size holes at 3-in. spacing are used, the allowable load in kips is obtained from Table II-A, or if the footnotes in Table II-A indicate, from Table II-C.

When the beam flange is coped and standard size holes at 3-in. spacing are used, the allowable load in kips is obtained from Table II-A or II-C, as described above. In addition, block shear, determined by using the coefficients in Table I-G, must be checked.

For bolts in slip-critical connections having oversize or slotted holes and Class A, clean mill scale surface condition, the allowable load is given in Table II-B. If the beam is coped, the block shear must be checked using the coefficients in Table I-G and appropriate adjustments must be made to comprehend the length of slot as it affects l_v or l_h.

For bearing-type connections using slotted holes, it is required that the long direction of the slot be perpendicular to the direction of the load. For an uncoped beam with slots in the web, the bolt shear allowable values are determined from Table II-A and the bolt bearing allowable value may be determined from Table I-F by entering the table with the actual l_h reduced by the increment C_2, as defined in AISC ASD Specification Table J3.6. For coped beams, beam web tear-out must also be checked using Table I-G by reducing l_h by the increment C_2. Intermediate values in Tables I-F and I-G can be obtained by proportioning. Use Table II-C to check the net shear capacity.

For slip-critical connections using slotted holes or oversized holes in the connection angles, allowable bolt shear values are determined from Table II-B. When slots are used in the beam web and the long direction of the slot is perpendicular to the direction of the load, bolt bearing values and beam web tear-out may be determined as given above for the bearing-type connection. For slots with the long direction of the slot parallel to the direction of the load and for oversize holes, see the AISC ASD Specification.

Connection angle lengths L (L' for staggered holes) and number of bolts at 3-in. spacing with 1¼-in. end distance are tabulated in Tables II-A and II-B.

DETAILS

1. In Tables II-A and II-B, connection angle lengths vary from 5½ in. to 29½ in. (7 in. to 31 in. for staggered hole arrangement) in multiples of 3 in. The length of the angles for a connection must be compatible with the beam T-dimension for uncoped beams. It is recommended that the minimum length of connection angle be at least one-half the T-dimension, to provide stability during erection.

2. Vertical fastener spacing is arbitrarily chosen as 3 in. for these tables. This may be varied within the parameters established by the AISC ASD Specification.

3. End distance on angles is set at 1¼ in., as permitted by AISC ASD Specification Table J3.5. When using oversize or slotted holes, the end distance must be adjusted over that for standard holes as required by AISC ASD Specification Sect. J3.9.

4. Gages for supporting members should be selected to meet the requirements of AISC ASD Specification Sects. J3.8, J3.9 and J3.10.

5. Clearance for assembly is essential in all cases. Framing angles with staggered holes are permitted as alternates to provide clearance and to permit smaller gages on legs of connection angles.

COMBINATION OF WELDED AND BOLTED FRAMED BEAM CONNECTIONS

See Framed Beam Connections—Welded, Table III, for appropriate Case I or Case II combination welded and bolted connections with tabulated reaction values.

OTHER FRAMED CONNECTIONS

These tables are not intended to preclude the use of other adequately designed connections.

EXAMPLE 1

Given: Beam: W 18 × 50, t_w = 0.355 in.
ASTM A36 (F_y = 36 ksi and F_u = 58 ksi)
Top flange coped 2 in. deep
Beam reaction = 38 kips
Bolts: ¾ in. dia. A325-N in ¹³⁄₁₆-in. dia. holes

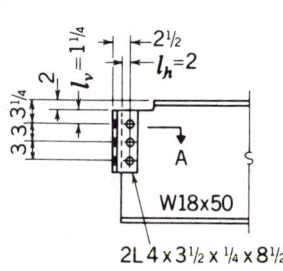

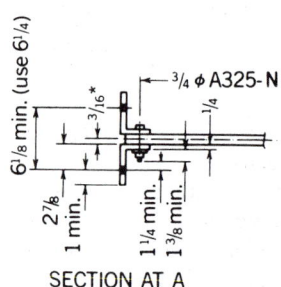

SECTION AT A

---

* This dimension (see sketch, section at A)is determined to be one-half the decimal web thickness rounded to the next higher ¹⁄₁₆ in. Example: 0.355 ÷ 2 = 0.1775; use ³⁄₁₆ in. This will produce a spacing of holes slightly smaller than detailed to permit spreading, rather than closing, at time of erection to supporting member.

Solution:

From Table II-A, select a connection with 3 rows ($n = 3$) of ¾ in. dia. A325-N bolts and shear value of 55.7 kips using ⁵⁄₁₆-in. angles. This exceeds the 38-kip reaction.

The net shear in the connection angles may be checked using Table II-C. ¼-in. connection angles with ¾-in. dia. bolts have a net shear value of 52.7 kips, so may be used.

Since the connection could slip into bearing, (although an extremely remote possibility), the beam web and connecting material should be checked for bearing stress. See AISC ASD Commentary Sect. J3.7.

Using the normal 1¼-in. edge distance to the cope ($l_v = 1¼$) and a 3-in. fastener spacing of ¾-in. bolts, enter Table I-F at 58 ksi and read coefficient 52.2 kips per in. of thickness for all three bolts. The allowable load is then (52.2×3 bolts) $\times 0.355$ $= 55.6$ kips > 38 kips and 1¼-in. edge distance is adequate for beam web and angles.

The maximum load for a ¾-in. bolt bearing on a 0.355-in. web is calculated by 1.2 $F_u dt$, when the edge distance (and equivalent spacing) $\geq 1.5\, d = 1.125$.

AISC ASD Specification Table J3.5 edge requirements are also met.

Since this beam is coped, beam web tear-out (block shear) must be checked:

Enter Table I-G with $l_v = 1¼$ in. and $l_h = 2$ in.

Read coefficient $C_1 = 1.38$.

Coefficient C_2 for ¾-in. bolts and $n = 3$ is 0.99.

Allowable reaction $R_{BS} = (1.38 + 0.99) \times 58 \times 0.355 = 48.8$ kips > 38 kips **o.k.**

The connection as sketched is adequate. Attention must also be given to the following:

a. Insertion and tightening clearances; see Table of Assembling Clearances (Manual, Part 4).

b. Need for reinforcement at the coped section in bending.

c. Adequacy of this connection with respect to the supporting member.

d. If a smaller gage on the outstanding leg of the connection angles is required, the bolts must be staggered with those in the beam web. An even stagger will permit the use of a 5-in. gage using angles 3½ × 3½ × ¼ × 0'-10 . The stagger portion of the Table of Stagger for Tightening (Manual, Part 4) is helpful in verifying clearances.

EXAMPLE 2

Given: Beam: W 36 × 230, $t_w = 0.760$ in.
        A36 ($F_y = 36$ ksi and $F_u = 58$ ksi)
        Beam reaction = 340 kips
        Bolts: ⅞-in. dia. A490-X in ¹⁵⁄₁₆-in. dia. holes

Solution:

From Table II-A, select a connection with 8 rows ($n = 8$) of ⅞-in. dia. A490-X bolts and load value of 363 kips, which exceeds the 340-kip reaction. Note that this value is footnoted (c) and is therefore limited by shear on the net section of the ⅝-in. thick

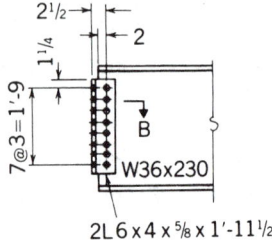

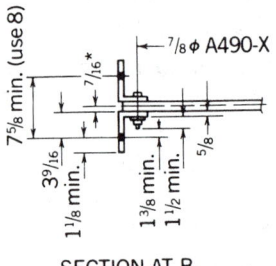

SECTION AT B

angles at length $L = 23\frac{1}{2}$ in. From Table II-C, the net shear capacity with $\frac{7}{8}$-in. dia. bolts and $n = 8$ on angles $\frac{5}{8}$-in. thick and $23\frac{1}{2}$-in. long is 348 kips > 340 kips **o.k.**

Bearing capacity of the beam may be checked from Table I-E, where the value 60.9 kips/in. thickness is read for $F_u = 58$ ksi and $\frac{7}{8}$-in. dia. bolts at 3-in. spacing. Allowable bearing $R = 60.9 \times 0.760 \times 8 = 370$ kips > 340 kips **o.k.**

Bearing on the angle material is checked with Table I-F. The $1\frac{1}{4}$-in. vertical edge distance on the top bolt is less than $1\frac{1}{2}d$ ($1\frac{5}{16}$ in.), so the bearing value for the top bolt is 36.3 kips/in. The value for the remaining seven bolts is 60.9 kips/in. The capacity of the connection angles is $\frac{5}{8}$ [36.3 + (7 × 60.9)] = 289 kips > 340/2 = 170 kips **o.k.**

Since this beam is not coped, beam web tear-out does not require checking.

The connection as sketched is adequate. Attention must also be given to:

  a. Insertion and driving clearance; see Table of Assembling Clearances (Manual, Part 4).

  b. Adequacy of this connection with respect to the supporting member.

  c. If a smaller gage is required on the outstanding legs of the connection angles, the bolts must be staggered with those in the beam web. An even stagger will permit the gage to be reduced to 7 in. and still maintain driving clearances while maintaining the 3-in. pitch. See Table of Stagger for Tightening (Manual, Part 4).

EXAMPLE 3

Given:

  Beam: W24 × 76, $t_w = 0.440$ in.
      A572 Gr. 50 ($F_y = 50$ ksi and $F_u = 65$ ksi)
  Connection angles: A36 ($F_y = 36$ ksi and $F_u = 58$ ksi)
  Beam reaction = 156 kips
  Bolts: 1-in. dia. A325-N
  Holes: $1\frac{1}{16} \times 1\frac{5}{16}$ (short slots), long axis perpendicular to transmitted force

---

* This dimension (see sketch, Section at B) is one-half the decimal web thickness rounded to the next higher $\frac{1}{16}$ in., as in Ex. 1.

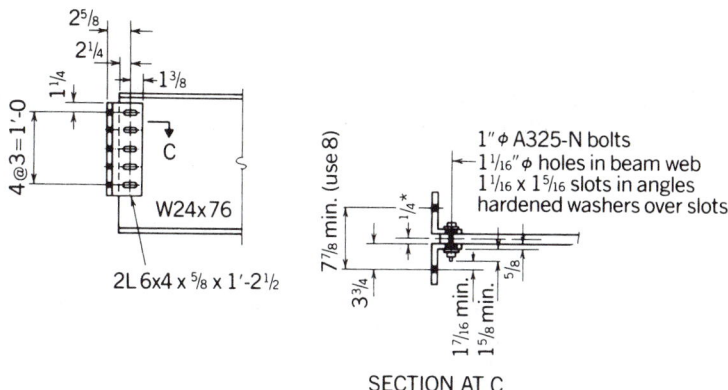

SECTION AT C

Solution:

From Table II-A, select a connection with 5 rows ($n = 5$) of 1-in. dia. A325-N bolts and shear value of 165 kips using ⅝-in. thick angles. This exceeds the 156-kip reaction.

Bearing capacity of the beam may be checked from Table I-E, where the value 78.0 kips/in. thickness is read for $F_u = 65$ ksi and 1-in. dia. bolts at 3-in. spacing. Allowable bearing $R = 78.0 \times 0.440 \times 5 = 172$ kips > 156 kips **o.k.**

Check capacity of the A36 connection angles using Table I-F with $F_u = 58$ ksi and $l_v = 1¼$-in. edge distance. The top bolt has a material bearing value of 36.3 kips/in. Allowable bearing on the connection angles then is ⅝ $\times [36.3 + (4 \times 69.6)] = 197$ kips $> 156/2 = 78$ kips. **o.k.**

Since this beam is not coped, beam web tear-out (block shear) does not require checking.

The net shear capacity of the angles need not be checked, using Table II-C, for the reason given in Ex. 1.

Note that short slots are permitted in any and all plies of slip-critical or bearing-type connections using high-strength bolts. To simplify the fabrication, slots are put in the angles, rather than the beam web. Hardened washers are required over these slots because they are in an outer ply.

The connection as sketched is adequate. Attention must also be given to the following:

a. Insertion and driving clearances; see Table of Assembling Clearances (Manual, Part 4).

b. Adequacy of this connection with respect to the supporting member.

c. An even stagger of web bolts with those in the outstanding legs will permit use of a 7-in. gage and still maintain driving clearances while maintaining the 3-in. pitch. See the Table of Stagger for Tightening (Manual, Part 4).

---

* This dimension (see sketch, Section at C) is one-half the decimal web thickness rounded to the next higher ¹⁄₁₆ in., as in Ex. 1.

EXAMPLE 4

Given:

The same conditions as Ex. 3, except the beam is coped 1¾ in. deep at the top flange. Find the allowable reaction on the modified beam.

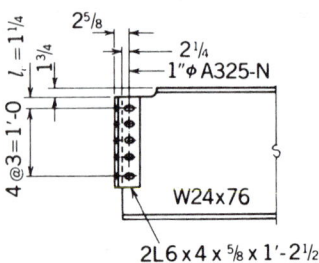

2L6 x 4 x ⅝ x 1'-2½

Solution:

The allowable reaction of the coped beam will be the least value determined by the bolt shear, bolt bearing, beam web tear-out (block shear) or shear capacity of the connection angles.

1. Bolt shear: From Table II-A, R = 165 kips.

2. Bolt bearing of beam web:

   From Table I-F, for l_v = 1¼ in. and F_u = 65 ksi:

   Allowable R = 0.440 in. × [(1 bolt × 40.6) + (4 bolts × 78.0)] = 155 kips

3. Bolt bearing on connection angles:

   From Table I-F, for l_v = 1¼ in. and F_u = 58 ksi:

   Allowable R = ⅝ in. × 2 angles × [(1 bolt × 36.3) + (4 bolts × 69.6)] = 393 kips

4. Web tear-out:

   From Table I-G, for l_v = 1¼ in., l_h = 2¼ in., t = 0.440 in. and F_u = 65 ksi:

   Coefficient C_1 = 1.50
   Coefficient C_2 = 1.90
   Total   3.40

   $R_{BS} = (C_1 + C_2)F_u t$
        = 3.40 × 65 × 0.440 = 97.2 kips

5. Shear capacity of angles:

   From Table II-C, for 1-in. bolts, angle thickness = ⅝ in., L = 14½ in. (n = 5):

   Shear capacity = 200 kips

The allowable reaction is limited to 97.2 kips and the conditions assumed would require a redesign of this connection.

EXAMPLE 5

Given: Beam: W 36 × 300, $t_w = 0.945$ in.
        A36 ($F_y = 36$ ksi and $F_u = 58$ ksi)
        Top flange coped 3 in. deep
        Beam reaction = 300 kips
        Bolts: 1⅛-in. dia. A490-N in 1³⁄₁₆-in. dia. holes

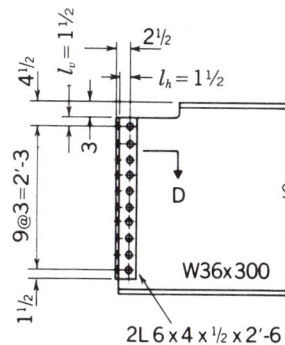

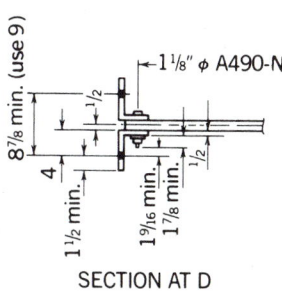

SECTION AT D

Solution:

Since values are not tabulated in this Manual for 1⅛-in. dia. fasteners, they must be derived from the AISC ASD Specification. The following is offered as a guide and outline. References are to sections of the AISC ASD Specification and Commentary.

	Specification and Commentary Reference
1. Bolt shear:	
Area of 1⅛-in. dia. bolt = 0.994 in.^{2}	
$F_v = 28.0$ ksi	
Single shear capacity = 0.994 × 28 = 27.8 kips/bolt	Table J3.2
No. bolts req'd = 300/(27.8 × 2) = 5.4.	
Since this beam is coped, judgement and experience suggest trying 10 bolts because web tear-out is likely to be critical.	
2. Bolt spacing:	J3.8
Minimum spacing = 2⅔d = 3 in.	
Preferred spacing = 3d = 3⅜ in.	
To satisfy bearing on web:	
Spacing = $2P/F_u t + d/2$	
= (2 × 300/10)/(58 × 0.945) + (1⅛)/2 = 1.66 in.	
Use 3-in. spacing, which is standard but less than the preferred spacing.	
3. Edge distance:	J3.9
Minimum distance in the direction of reactive force:	
$l_v = 2P/F_u t$ = (2 × 300/10)/(58 × 0.945) = 1.09 in.	
Use 1½ in. (minimum for gas cut edge)	Table J3.5

4. Since this beam is coped, failure by shear and tension through the fastener holes must be checked.

ASD Specification Reference

Allowable block shear $R_{BS} = 0.30A_v F_u + 0.50A_t F_u$

J4

This equation expands to the general expression shown at Table I-G

$$R_{BS} = \{(0.3\, l_v + 0.5\, l_h) + 0.3\,[(n-1)(s-d_h) - d_h/2] - d_h/4\}\, F_u t$$
$$= \{[(0.3 \times 1\frac{1}{2}) + (0.5 \times 1\frac{1}{2})]$$
$$+ 0.3[(10-1)(3 - 1\frac{3}{16}) - \frac{19}{32}] - \frac{19}{64}\} \, 58 \times 0.945$$
$$= 308 \text{ kips}$$

5. Connection angles:

   Try 2 L 6 × 4 × ½ × 2′- 6, A36.

   Allowable bearing:

   From AISC ASD Specification Table J3.5 for 1⅛-in. dia. bolts, $l_v = 1\frac{1}{2}$ in. < $1\frac{1}{2}\, d = 1.69$,

   Top Bolt: $F_u \times t/2 \times$ edge distance
   $$58 \times 0.5/2 \times 1\frac{1}{2} = 21.8 \text{ kips}$$

   Remaining Bolts: $1.2 \times F_u \times d \times t$
   $$1.2 \times 58 \times 1\frac{1}{8} \times \frac{1}{2} = 39.2 \text{ kips}$$

   R = (1 bolt × 21.8) + (9 bolts × 39.2) = 374 kips **o.k.**

   Shear on plane through fasteners:

   Allowable shear is $0.4F_y = 14.5$ ksi on the gross area or $0.3F_u = 17.4$ ksi on the net area. Shear on the *net* area governs when $(d + \frac{1}{16} \text{ in.}) > L/6n$. Then $30/(6 \times 10) = 0.5$ in. and *net* area governs.

   Net area $= 2t\,[L - n(d + \frac{1}{16})]$
   $$= (2 \times \frac{1}{2})\,[30 - (10 \times 1\frac{3}{16})] = 18.1 \text{ in.}^2$$

   $f_v = 300/18.1 = 16.6$ ksi < 17.4 ksi

6. The connection as sketched is adequate. Attention must also be given to the following:

   a. Insertion and driving clearances; see Table of Assembling Clearances (Manual, Part 4).

   b. Need for reinforcement at the coped section in bending.

   c. Adequacy of this connection with respect to the supporting member.

   d. If a smaller gage is required on the outstanding legs, a new design is required, since the maximum number of fasteners has been used in the beam web. Consideration can be given to larger diameter bolts and to the use of a second gage line in the web. The effect of web tear-out (block shear) on a connection with two gage lines has not been determined at this time.

FRAMED BEAM CONNECTIONS
Bolted
TABLE II Allowable loads in kips

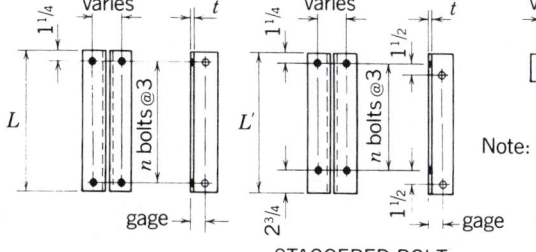

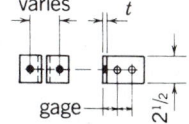

Note: For $L=2\frac{1}{2}$ use one half
the tabular load value
shown for $L=5\frac{1}{2}$, for the
same bolt type, diameter,
and thickness.

STAGGERED BOLT
ALTERNATE

TABLE II-A Bolt Shear[a]

For A307 bolts in standard or slotted holes and for A325 and A490 bolts in **slip-critical** connections with standard holes and Class A, clean mill scale surface condition.

Bolt Type	A307			A325-SC			A490-SC			Note:
F_v, Ksi	10.0			17.0			21.0			For slip-critical
Bolt Dia., d In.	$\frac{3}{4}$	$\frac{7}{8}$	1	$\frac{3}{4}$	$\frac{7}{8}$	1	$\frac{3}{4}$	$\frac{7}{8}$	1	connections with oversize
Angle Thickness t, In.	$\frac{1}{4}$	$\frac{1}{4}$	$\frac{1}{4}$	$\frac{1}{4}$	$\frac{5}{16}$	$\frac{1}{2}$	$\frac{5}{16}$	$\frac{1}{2}$	$\frac{5}{8}$	or slotted holes, see
L In. / L' In. / n										Table II-B.
29½ 31 10	88.4	120	157	150	204	267	186	253	330	
26½ 28 9	79.5	108	141	135	184	240	167	227	297	
23½ 25 8	70.7	96.2	126	120	164	214	148	202	264	
20½ 22 7	61.9	84.2	110	105	143	187	130	177	231	
17½ 19 6	53.0	72.2	94.2	90.1	123	160	111	152	198	
14½ 16 5	44.2	60.1	78.5	75.1	102	134	92.8	126	165	
11½ 13 4	35.3	48.1	62.8	60.1	81.8	107	74.2	101	132	
8½ 10 3	26.5	36.1	47.1[b]	45.1	61.3	80.1	55.7	75.8	99.0	
5½ 7 2	17.7	24.1	31.4[b]	30.0	40.9	53.4	37.1	50.5	66.0	

Notes:

[a]Tabulated load values are based on double shear of bolts unless noted. See RCSC Specification for other surface conditions.

[b]Capacity shown is based on double shear of the bolts; however, for length L, net shear on the angle thickness specified is critical. See Table II-C.

FRAMED BEAM CONNECTIONS
Bolted
TABLE II Allowable loads in kips

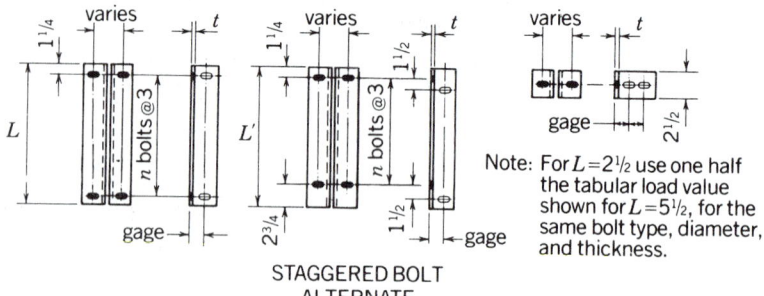

Note: For $L=2^{1}/_{2}$ use one half the tabular load value shown for $L=5^{1}/_{2}$, for the same bolt type, diameter, and thickness.

STAGGERED BOLT
ALTERNATE

TABLE II-A Bolt Shear[a]
For bolts in **bearing-type** connections with standard or slotted holes.

Bolt Type			A325-N			A490-N			A325-X			A490-X		
F_v, Ksi			21.0			28.0			30.0			40.0		
Bolt Dia., d In.			$^{3}/_{4}$	$^{7}/_{8}$	1	$^{3}/_{4}$	$^{7}/_{8}$	1	$^{3}/_{4}$	$^{7}/_{8}$	1	$^{3}/_{4}$	$^{7}/_{8}$	1
Angle Thickness t, In.			$^{5}/_{16}$	$^{3}/_{8}$	$^{5}/_{8}$	$^{3}/_{8}$	$^{1}/_{2}$	$^{5}/_{8}$	$^{3}/_{8}$	$^{5}/_{8}$	$^{5}/_{8}$	$^{1}/_{2}$	$^{5}/_{8}$	$^{5}/_{8}$
L In.	L' In.	n												
$29^{1}/_{2}$	31	10	186	253	330	247	337	440[b]	265	361	c	353	481	c
$26^{1}/_{2}$	28	9	167	227	297	223	303	396[b]	239	325	c	318	433	c
$23^{1}/_{2}$	25	8	148	202	264	198	269	352[b]	212	289	c	283	385	c
$20^{1}/_{2}$	22	7	130	177	231	173	236	308[b]	186	253	c	247	337	c
$17^{1}/_{2}$	19	6	111	152	198	148	202	264[b]	159	216	283	212	289	377
$14^{1}/_{2}$	16	5	92.8	126	165	124	168	220[b]	133	180	236	177	242	314
$11^{1}/_{2}$	13	4	74.2	101	132	99.0	135	176[b]	106	144	188	141	192	251
$8^{1}/_{2}$	10	3	55.7	75.8[b]	99.0	74.2	101[b]	132[b]	79.5[b]	108	141	106[b]	144	188
$5^{1}/_{2}$	7	2	37.1	50.5[b]	66.0	49.5	67.3[b]	88.0[b]	53.0[b]	72.2	94	70.7[b]	96	126

[a]Tabulated load values are based on double shear of bolts unless noted. See RCSC Specification for other surface conditions.

[b]Capacity shown is based on double shear of the bolts; however, for length L, net shear on the angle thickness specified is critical. See Table II-C.

[c]Capacity is governed by net shear on angles for lengths L and L'. See Table II-C.

FRAMED BEAM CONNECTIONS
Bolted
TABLE II Allowable loads in kips

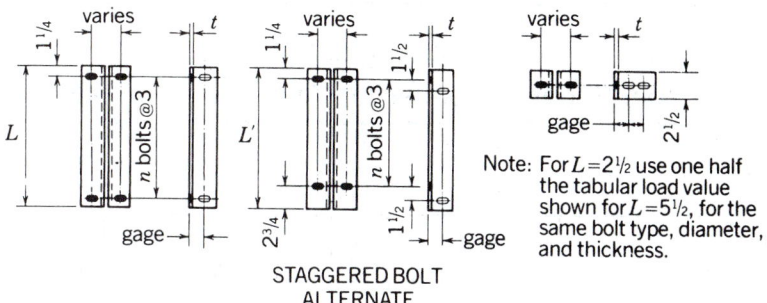

Note: For $L=2\frac{1}{2}$ use one half
the tabular load value
shown for $L=5\frac{1}{2}$, for the
same bolt type, diameter,
and thickness.

STAGGERED BOLT
ALTERNATE

TABLE II-B Bolt Shear[a]

For bolts in **slip-critical** connections with oversize, short-slotted, or long-slotted holes
and Class A, clean mill scale surface condition.

Hole Type			Long-slotted Holes						Oversize and Short-slotted Holes					
Bolt Type			A325-SC			A490-SC			A325-SC			A490-SC		
F_v, Ksi			12.0			15.0			15.0			18.0		
Bolt Dia., d, In.			¾	⅞	1	¾	⅞	1	¾	⅞	1	¾	⅞	1
Angle Thickness, t, In.			¼	¼	5/16	¼	5/16	½	¼	5/16	½	5/16	⅜	⅝
L In.	L' In.	n												
29½	31	10	106	144	188	133	180	236	133	180	236	159	216	283
26½	28	9	95.4	130	170	119	162	212	119	162	212	143	195	254
23½	25	8	84.8	115	151	106	144	188	106	144	188	127	173	226
20½	22	7	74.2	101	132	92.8	126	165	92.8	126	165	111	152	198
17½	19	6	63.6	86.6	113	79.5	108	141	79.5	108	141	95.4	130	170
14½	16	5	53.0	72.2	94.2	66.3	90.2	118	66.3	90.2	118	79.5	108	141
11½	13	4	42.4	57.7	75.4	53.0	72.2	94.2	53.0	72.2	94.2	63.6	86.6	113
8½	10	3	31.8	43.3	56.5	39.8	54.1	70.7	39.8	54.1	70.7	47.7	64.9	84.8
5½	7	2	21.2	28.9	37.7[b]	26.5	36.1	47.1	26.5	36.1	47.1	31.8	43.3[c]	56.5

[a]Tabulated load values are based on double shear of bolts unless noted. See RCSC Specification, for other surface conditions. Slotted holes are parallel to beam flange. When slotted holes are not parallel to beam flanges, the above connection angle values and details will differ. See AISC ASD Specification, Sects. J3.8 and J3.9.

[b]Capacity shown is based on double shear of the bolts; however, for length L, net shear on the angle thickness specified is critical. See Table II-C.

[c]Capacity shown is based on double shear of bolts; however, for $L = 5\frac{1}{2}$ in., with oversize holes, net shear in the angle thickness specified is critical and tabulated load reduces to 44.0 kips.

FRAMED BEAM CONNECTIONS
BOLTED
TABLE II Allowable loads in kips

TABLE II-C Allowable Shear in Connection Angles for A36 Material														
Bolt Dia., d In.		3/4			7/8					1				
Angle Thick-ness, t In.	1/4	5/16	3/8	1/2	1/4	5/16	3/8	1/2	5/8	1/4	5/16	3/8	1/2	5/8

L In.	n														
29½	10	186	232	279	372	175	219	263	350	438	164	205	246	328	411
26½	9	167	209	250	334	157	196	236	314	393	147	184	221	295	368
23½	8	148	185	222	296	139	174	209	278	348	131	163	196	261	326
20½	7	129	161	193	258	121	152	182	243	303	114	142	170	227	284
17½	6	110	137	165	220	103	129	155	207	258	96.8	121	145	194	242
14½	5	90.8	114	136	182	85.4	107	128	171	213	79.9	99.9	120	160	200
11½	4	71.8	89.7	108	144	67.4	84.3	101	135	169	63.1	78.8	94.6	126	158
8½	3	52.7	65.9	79.1	105	49.5	61.9	74.2	99.0	124	46.2	57.8	69.3	92.4	116
5½	2	33.7	42.1	50.6	67.4	31.5	39.4	47.3	63.1	78.8	29.4	36.7	44.0	58.7	73.4

L' In.	n														
31	10	199	249	299	398	188	235	282	376	470	177	222	266	355	443
28	9	180	225	270	360	170	213	255	340	425	160	201	241	321	401
25	8	161	201	241	322	152	190	228	305	381	144	179	215	287	359
22	7	142	177	213	284	134	168	201	269	336	127	158	190	253	317
19	6	123	154	184	246	116	145	175	233	291	110	137	165	220	275
16	5	104	130	156	208	98.4	123	148	197	246	93.0	116	139	186	232
13	4	84.8	106	127	170	80.5	101	121	161	201	76.1	95.2	114	152	190
10	3	65.8	82.2	98.7	132	62.5	78.2	93.8	125	156	59.3	74.1	88.9	119	148
7	2	46.8	58.5	70.1	93.5	44.6	55.7	66.9	89.2	111	42.4	53.0	63.6	84.8	106

NOTES: Table based on an allowable shear of $0.3F_u$ (17.4 ksi for A36 angles) of the net section of two angles.

Net section based on diameter of fastener + 1/16 in.

FRAMED BEAM CONNECTIONS
Welded–E70XX electrodes for
combination with Table II connections
TABLE III

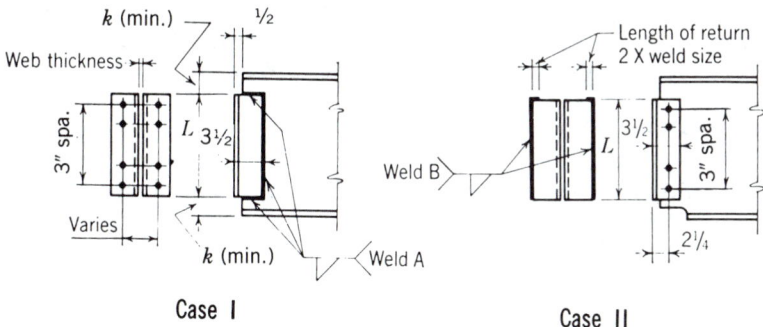

Case I

Case II

Table III is arranged to permit substitution of welds for bolts in the connections shown in Table II which fall within the weld capacities. Welds A replace fasteners in the beam web legs (Case I). Welds B replace fasteners in the outstanding legs (Case II).

To accommodate usual gages, angle leg widths will generally be 4 × 3½, with the 4-in. leg outstanding. Width of web legs in Case I may be reduced optionally from 3½ to 3 in. Width of outstanding legs in Case II may be reduced optionally from 4 to 3 in. for values of $L = 5\frac{1}{2}$ in. through 1'-5½.

Angle thickness is equal to weld size plus ¹⁄₁₆ in., or thickness of angle from applicable Table II, whichever is greater.

Angle length L must be as tabulated in Table III.

Holes for erection bolts may be placed as required in legs to be field-welded (optional).

When bolts are used, investigate bearing capacity of supporting member.

Although it is permissible to use Welds A and B in combination to obtain all-welded connections, it is recommended such connections be chosen from Table IV. This table will usually provide greater economy and allow increased flexibility in selection of angle lengths and connection capacities.

Allowable capacity for Weld A utilizes instantaneous center solutions based on the same criteria developed for Tables XIX to XXVI, Eccentric Loads on Weld Groups. However, capacity for Weld B is computed using traditional vector analysis techniques.

CASE I

EXAMPLE 6

Given:

Beam:	W 36 × 150 (not coped); $t_w = 0.625$ in.
	$F_y = 36$ ksi; $F_v = 14.5$ ksi
Reaction:	200 kips
Bolts:	⅞ in. dia., ASTM A325-N
Welds:	E70XX

Solution:

Enter Table III under Weld A and note that a value that satisfies the reaction is 222 kips. This requires ⁵⁄₁₆-in. welds and 23½-in. long angles. Use ⅜-in. thick angles to meet the weld requirement stipulated in AISC ASD Specification Sect. J2.2b. The 0.625-in. web thickness is less than the minimum required 0.64 in., so the reduction in capacity is $(0.625/0.64) \times 222$ kips = 217 kips.

Note, in Table II-A, the angle length provides for 8 rows of ⅞-in. dia. ASTM A325-N bolts with a capacity of 202 kips. The ⅜-in. required angle thickness is the same as the ⅜-in. angle thickness required due to Weld A.

Detail data: Two L 4 × 3½ × ⅜ × 1′−11½

F_y = 36 ksi

Sixteen ⅞-in. dia. ASTM A325-N bolts (threads included in shear plane)

⁵⁄₁₆-in fillet weld, E70XX

EXAMPLE 7

Given:

Beam:	W 16 × 26 (not coped); t_w = 0.25 in.
	F_y = 36 ksi; F_v = 14.5 ksi
Reaction:	46 kips
Bolts:	⅞-in. dia. ASTM A307
Welds:	E70XX

Solution:

See Table II-A and note 4 rows of bolts with 11½-in. long angles are compatible with a 16-in. deep section. Capacity of the ⅞-in. dia. ASTM A307 bolts with ¼-in. thick angles is 48.1 kips.

Note in Table III that 72.7 kips capacity is designated for ³⁄₁₆-in. Weld A and 11½-in. long angles. The 0.25-in. web thickness is less than the minimum 0.38 in. listed. The reduced capacity is $(0.25/0.38) \times 72.7$ kips = 47.8 kips. The ¼-in. angle thickness required for bolts is satisfactory for the ³⁄₁₆-in. weld.

Detail data: Two L 4 × 3½ × ¼ × 0′−11½

F_y = 36 ksi

Eight ⅞-in. dia. ASTM A307 bolts

³⁄₁₆-in fillet weld, E70XX

CASE II

EXAMPLE 8

Given:

Beam:	W 36 × 150 (coped); t_w = 0.625 in.
	F_y = 36 ksi; F_v = 14.5 ksi
Reaction:	155 kips
Bolts:	⅞-in. dia., ASTM A490-N
Welds:	E70XX

Solution:

Enter Table III under Weld B and note the value most nearly satisfying the reaction is 156 kips. This requires ⁵/₁₆-in. Weld B and 20½-in. long, ⅜-in thick angles. Table II-A shows a bolt capacity for a 7 fastener connection of 236 kips, which is more than the 155 kips required. Therefore, a 20½-in. long angle is selected from Table III.

Check bearing on beam web = 155/7 = 22.1 kips per bolt in double shear bearing. From Table I-E, the allowable bearing 60.9 kips for 1-in. thickness and 3-in. spacing; 60.9 × 0.625 in. = 38.1 kips per bolt allowable. Assuming the beam is coped, and using 1¼-in. edge distance for the top bolt, the single shear bearing capacity of the ⅜-in. connection angle (Table I-F) is,

⅜ [(1 bolt × 36.3) + (6 bolts × 60.9 kips)] = 151 kips   **o.k.**

Check block shear:

$$(C_1 + C_2) F_u\, t = (1.25 + 3.34)\ 58 \times 0.625 = 166.4 \text{ kips} \quad \textbf{o.k.}$$

Detail data: Two L 4 × 3½ × ⅜ × 1′−8½
F_y = 36 ksi
Seven ⅞-in. dia. ASTM A490-N bolts
⁵/₁₆-in. fillet weld, E70XX

EXAMPLE 9

Given:

Beam: W 16 × 31 (not coped); t_w = 0.275 in.
           F_y = 50 ksi; F_v = 20 ksi
Reaction: 39 kips
Bolts: ¾-in. dia., ASTM A325-N
Welds: E70XX

Solution:

Enter Table III under Weld B and note the value most nearly satisfying the reaction is 40.3 kips. This requires ⁵/₁₆-in. Weld B and 8½-in. long, ⅜-in. thick angles.

Enter Table II-A for 3 rows of fasteners and note that the angle length is compatible with beam size. Capacity of three ¾-in. dia. ASTM A325-N bolts is 55.7 kips.

Check bearing on beam web = 39/3 = 13 kips per bolt in double shear. From Table I-E, the allowable bearing is 58.5 kips for 1-in. thickness at 3-in. spacing; 58.5 × 0.275 = 16.1 kips per bolt allowable. Assuming 1¼-in. edge distance for the top bolt, the single shear bearing capacity of ⅜-in. connecting angle (Table I-F) is,

$$3/8(3 \times 52.2) = 58.7 \text{ kips} \quad \textbf{o.k.}$$

Detail data: Two L 4 × 3½ × ⅜ × 0′−8½
F_y = 36 ksi
Three ¾-in. dia. ASTM A325-N bolts
⁵/₁₆-in. fillet weld, E70XX

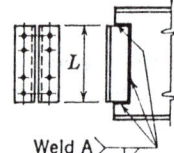

FRAMED BEAM CONNECTIONS
Welded—E70XX electrodes for combination with Table II connections
TABLE III Allowable loads in kips

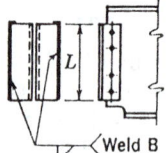

Weld A

Weld B

Weld A		Weld B		Angle Length	[a]Minimum Web Thickness for Welds A		Maximum Number of Fasteners in One Vertical Row (Table II)
Capacity, Kips	[b]Size, In.	[c]Capacity, Kips	Size, In.	L In.	F_y = 36 ksi F_v = 14.5 ksi	F_y = 50 ksi F_v = 20 ksi	
266	5/16	296	3/8	29½	.64	.46	
213	1/4	247	5/16	29½	.51	.37	10
160	3/16	197	1/4	29½	.38	.28	
245	5/16	261	3/8	26½	.64	.46	
196	1/4	217	5/16	26½	.51	.37	9
147	3/16	173	1/4	26½	.38	.28	
222	5/16	223	3/8	23½	.64	.46	
178	1/4	186	5/16	23½	.51	.37	8
133	3/16	149	1/4	23½	.38	.28	
198	5/16	187	3/8	20½	.64	.46	
158	1/4	156	5/16	20½	.51	.37	7
119	3/16	125	1/4	20½	.38	.28	
174	5/16	152	3/8	17½	.64	.46	
139	1/4	126	5/16	17½	.51	.37	6
104	3/16	101	1/4	17½	.38	.28	
148	5/16	115	3/8	14½	.64	.46	
118	1/4	95.7	5/16	14½	.51	.37	5
88.7	3/16	76.6	1/4	14½	.38	.28	
121	5/16	80.1	3/8	11½	.64	.46	
97.0	1/4	66.9	5/16	11½	.51	.37	4
72.7	3/16	53.4	1/4	11½	.38	.28	
92.1	5/16	48.2	3/8	8½	.64	.46	
73.7	1/4	40.3	5/16	8½	.51	.37	3
55.3	3/16	32.2	1/4	8½	.38	.28	
61.8	5/16	21.9	3/8	5½	.64	.46	
49.5	1/4	18.3	5/16	5½	.51	.37	2
37.1	3/16	14.6	1/4	5½	.38	.28	

[a]When the beam web thickness is less than the minimum, multiply the connection capacity furnished by Weld A by the ratio of the actual web thickness to the tabulated minimum thickness. Thus, if 5/16 in. Weld A, with a connection capacity of 148 kips and a 14½-in. long angle, is considered for a beam of web thickness of 0.375 in. with F_y = 36 ksi, the connection capacity must be multiplied by 0.375/0.64, giving 86.7 kips.

[b]Should the thickness of material to which connection angles are welded exceed the limits set by AISC ASD Specification Sects. J2.1b or J2.2b for weld sizes specified, increase the weld size as required, but not to exceed the angle thickness.

[c]When welds are used on outstanding legs, connection capacity may be limited by the shear capacity of the supporting member as stipulated by AISC ASD Specification Sect. F4. See Ex. 13 and 14 for Table IV.

Note 1: Connection angles: Two L 4 × 3½ × thickness × L; F_y = 36 ksi. See discussion preceding examples for Table III for limiting values of thickness and optional width of legs.

Note 2: Capacities shown in this table apply only when the material welded is F_y = 36 ksi or F_y = 50 ksi steel.

FRAMED BEAM CONNECTIONS
Welded–E70XX electrodes
TABLE IV

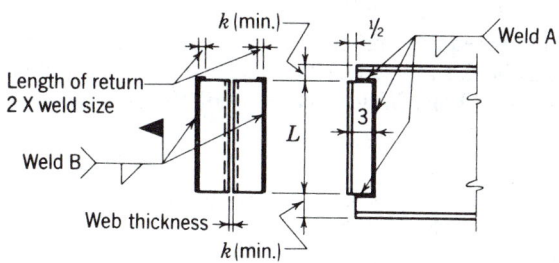

Table IV lists capacities and details for angle connections welded to both the beam web and the supporting member.

Holes for erection bolts may be placed as required in legs that are to be field welded (optional).

Allowable capacity for Weld A utilizes instantaneous center solutions based on the same criteria developed for Tables XIX to XXVI, Eccentric Loads on Weld Groups. However, Weld B capacity is computed using traditional vector analysis technique.

EXAMPLE 10

Given:

Beam:	W 36 × 150; t_w = 0.625 in.; T = 32⅛ in.
	F_y = 36 ksi; F_v = 14.5 ksi
Weld:	E70XX
Reaction:	170 kips

Solution:

Enter Table IV and select a Weld A capacity of 174 kips (weld size = ¼ in.). Weld B has a capacity of 191 kips and is satisfactory. The angle length (24 in.) is less than T for the W 36 × 150 and is satisfactory. The beam web thickness (0.625 in.) exceeds the minimum web thickness (0.51 in.), so no reduction in Weld A capacity is required.

Detail data: Two L 4 × 3 × ⅜ × 2'-0 ; F_y = 36 ksi
Weld A = ¼ in., E70XX
Weld B = ⁵⁄₁₆ in., E70XX

AMERICAN INSTITUTE OF STEEL CONSTRUCTION

EXAMPLE 11

Given:

Same data as Ex. 10, except the reaction is 144 kips.

Solution:

Enter Table IV and select a Weld A capacity of 151 kips (weld size = ¼ in.). Weld B has a capacity of 152 kips and is satisfactory. The angle length (20 in.) is less than *T* and is satisfactory. The beam web thickness (0.625 in.) exceeds the minimum web thickness (0.51 in.), so no reduction in Weld A capacity is required.

Unless framing details require this short angle length, longer angles with less deposited weld metal may be desirable. The 28-in. long angles with Weld A capacity of 149 kips (weld size = 3/16 in.) and Weld B capacity of 185 kips are also satisfactory and may be selected.

Detail data: Two L 4 × 3 × 5/16 × 2'- 4 ; F_y = 36 ksi
Weld A = 3/16 in., E70XX
Weld B = ¼ in., E70XX

EXAMPLE 12

Given:

Beam: W 16 × 26; t_w = 0.25 in.; *T* = 13⅝ in.
F_y = 50 ksi; F_v = 20 ksi
Weld: E70XX
Reaction: 35 kips

Solution:

Enter Table IV and select a Weld B capacity of 35.5 kips (weld size = ¼ in.). Angle length (8 in.) is less than *T* and is satisfactory. Weld A has a capacity of 51.5 kips using 3/16-in. weld size. However, the beam web thickness (0.25 in.) is less than the minimum web thickness (0.28 in.), so Weld A capacity is reduced to .25/.28 times 51.5 or 46.0 kips.

Detail data: Two L 3 × 3 × 5/16 × 0'- 8 ; F_y = 36 ksi
Weld A = 3/16 in., E70XX
Weld B = ¼ in., E70XX

WELDS TO SUPPORTING MEMBERS

Selection of connections tabulated herein is based on and limited by the requirement that Weld B will be applied in accordance with AISC ASD Specification Sects. J2.1b and J2.2b, which stipulates minimum welds for various material thicknesses.

With respect to Weld B, it should be noted that supporting members with limited shear capacity, or which support opposed connections, may be subject to a reduction in connection capacity. See AISC ASD Specification Sect. J2.4.

EXAMPLE 13

Given:

Weld B = ⁵⁄₁₆-in. fillet weld, E70XX, fully loaded on one side of ¼-in. thick supporting member web of F_y = 36 ksi steel.

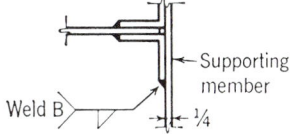

Solution:

Shear value of one ⁵⁄₁₆-in. fillet weld = 0.3125 in. × 0.707 × 21.0 ksi = 4.64 kips/lin. in. Shear value of ¼-in. thick web = 0.25 in. × 14.5 ksi = 3.63 kips/lin. in.

Because of this deficiency in web shear capacity, the total capacity selected from the Weld B column for ⁵⁄₁₆-in. weld must be reduced by the ratio 3.63/4.64.

EXAMPLE 14

Given:

Two floor beams with end reactions of 15.0 kips each are to be supported by a beam of F_y = 36 ksi steel with a ⁵⁄₁₆-in. thick web.

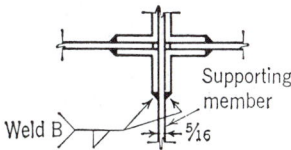

Solution:

¼-in. Weld B with 5-in. long angles has a capacity of 15.7 kips when fully stressed. Maximum shear developed in the two ¼-in. fillet Welds B on opposite sides of the supporting beam web = 2 × 0.25 × 0.707 × 21.0 × 15/15.7 = 7.09 kips/lin. in. Shear capacity of ⁵⁄₁₆-in. web = 0.3125 × 14.5 = 4.53 kips/lin. in. A longer connection is required to reduce the web shear. Required Weld B capacity is 15.7 kips × 7.09/4.53 = 24.6. Two 7-in. long angles with ¼-in. Weld B have a tabulated capacity of 28.3 kips and are adequate.

FRAMED BEAM CONNECTIONS
Welded—E70XX electrodes
Table IV

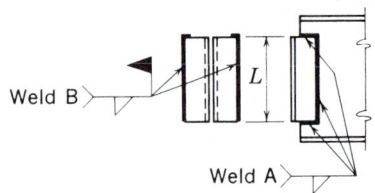

Weld A		Weld B		Angle Length L In.	Angle Size $(F_y = 36$ ksi$)$	^aMinimum Web Thickness for Weld A	
Capacity Kips	^bSize In.	^cCapacity Kips	^bSize In.			$F_y = 36$ ksi $F_v = 14.5$ ksi	$F_y = 50$ ksi $F_v = 20$ ksi
277	5⁄16	326	3⁄8	32	4 × 3 × 1⁄2	.64	.46
221	1⁄4	271	5⁄16	32	4 × 3 × 3⁄8	.51	.37
166	3⁄16	217	1⁄4	32	4 × 3 × 5⁄16	.38	.28
262	5⁄16	302	3⁄8	30	4 × 3 × 1⁄2	.64	.46
210	1⁄4	251	5⁄16	30	4 × 3 × 3⁄8	.51	.37
157	3⁄16	201	1⁄4	30	4 × 3 × 5⁄16	.38	.28
248	5⁄16	278	3⁄8	28	4 × 3 × 1⁄2	.64	.46
198	1⁄4	231	5⁄16	28	4 × 3 × 3⁄8	.51	.37
149	3⁄16	185	1⁄4	28	4 × 3 × 5⁄16	.38	.28
234	5⁄16	254	3⁄8	26	4 × 3 × 1⁄2	.64	.46
187	1⁄4	211	5⁄16	26	4 × 3 × 3⁄8	.51	.37
140	3⁄16	169	1⁄4	26	4 × 3 × 5⁄16	.38	.28
218	5⁄16	230	3⁄8	24	4 × 3 × 1⁄2	.64	.46
174	1⁄4	191	5⁄16	24	4 × 3 × 3⁄8	.51	.37
131	3⁄16	153	1⁄4	24	4 × 3 × 5⁄16	.38	.28
204	5⁄16	206	3⁄8	22	4 × 3 × 1⁄2	.64	.46
163	1⁄4	171	5⁄16	22	4 × 3 × 3⁄8	.51	.37
122	3⁄16	137	1⁄4	22	4 × 3 × 5⁄16	.38	.28
188	5⁄16	181	3⁄8	20	4 × 3 × 1⁄2	.64	.46
151	1⁄4	152	5⁄16	20	4 × 3 × 3⁄8	.51	.37
113	3⁄16	121	1⁄4	20	4 × 3 × 5⁄16	.38	.28
172	5⁄16	157	3⁄8	18	4 × 3 × 1⁄2	.64	.46
138	1⁄4	131	5⁄16	18	4 × 3 × 3⁄8	.51	.37
103	3⁄16	105	1⁄4	18	4 × 3 × 5⁄16	.38	.28
156	5⁄16	148	3⁄8	16	3 × 3 × 1⁄2	.64	.46
125	1⁄4	123	5⁄16	16	3 × 3 × 3⁄8	.51	.37
94.0	3⁄16	98.8	1⁄4	16	3 × 3 × 5⁄16	.38	.28
139	5⁄16	124	3⁄8	14	3 × 3 × 1⁄2	.64	.46
112	1⁄4	103	5⁄16	14	3 × 3 × 3⁄8	.51	.37
83.7	3⁄16	82.5	1⁄4	14	3 × 3 × 5⁄16	.38	.28
122	5⁄16	99.6	3⁄8	12	3 × 3 × 1⁄2	.64	.46
97.2	1⁄4	83.1	5⁄16	12	3 × 3 × 3⁄8	.51	.37
72.9	3⁄16	66.5	1⁄4	12	3 × 3 × 5⁄16	.38	.28

For footnotes, see next page.

FRAMED BEAM CONNECTIONS
Welded—E70XX electrodes
Table IV

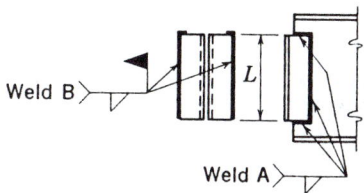

Weld B

L

Weld A

Weld A		Weld B		Angle Length	Angle Size	aMinimum Web Thickness for Weld A	
Capacity Kips	Sizeb In.	Capacityc Kips	Sizeb In.	L In.	F_y = 36 ksi	F_y = 36 ksi F_v = 14.5 ksi	F_y = 50 ksi F_v = 20 ksi
104	5/16	75.9	3/8	10	3 × 3 × 1/2	.64	.46
83.2	1/4	63.3	5/16	10	3 × 3 × 3/8	.51	.37
62.4	3/16	50.5	1/4	10	3 × 3 × 5/16	.38	.28
94.6	5/16	64.3	3/8	9	3 × 3 × 1/2	.64	.46
75.7	1/4	53.7	5/16	9	3 × 3 × 3/8	.51	.37
56.7	3/16	42.9	1/4	9	3 × 3 × 5/16	.38	.28
85.8	5/16	53.2	3/8	8	3 × 3 × 1/2	.64	.46
68.6	1/4	44.4	5/16	8	3 × 3 × 3/8	.51	.37
51.5	3/16	35.5	1/4	8	3 × 3 × 5/16	.38	.28
74.8	5/16	42.5	3/8	7	3 × 3 × 1/2	.64	.46
59.8	1/4	35.5	5/16	7	3 × 3 × 3/8	.51	.37
44.9	3/16	28.3	1/4	7	3 × 3 × 5/16	.38	.28
64.9	5/16	32.6	3/8	6	3 × 3 × 1/2	.64	.46
51.9	1/4	27.1	5/16	6	3 × 3 × 3/8	.51	.37
38.9	3/16	21.7	1/4	6	3 × 3 × 5/16	.38	.28
54.0	5/16	23.4	3/8	5	3 × 3 × 1/2	.64	.46
43.2	1/4	19.5	5/16	5	3 × 3 × 3/8	.51	.37
32.4	3/16	15.7	1/4	5	3 × 3 × 5/16	.38	.28
44.0	5/16	15.5	3/8	4	3 × 3 × 1/2	.64	.46
35.2	1/4	12.9	5/16	4	3 × 3 × 3/8	.51	.37
26.4	3/16	10.4	1/4	4	3 × 3 × 5/16	.38	.28

aWhen the beam web thickness is less than the minimum, multiply the connection capacity furnished by Welds A by the ratio of the actual thickness to the tabulated minimum thickness. Thus, if 5/16-in. Weld A, with a connection capacity of 85.8 kips and an 8-in. long angle, is considered for a beam of web thickness 0.305 in. and F_y = 36 ksi, the connection capacity must be multiplied by 0.305/0.64, giving 40.9 kips.

bShould the thickness of material to which connection angles are welded exceed the limits set by AISC ASD Specification Sects. J2.1b and J2.2b for weld sizes specified, increase the weld size as required, but not to exceed the angle thickness.

cFor welds on outstanding legs, connection capacity may be limited by the shear capacity of the supporting members, as stipulated by AISC ASD Specification Sect. F4. See Ex. 13 and 14 for Table IV.

Note 1: Capacities shown in this table apply only when connection angles are F_y = 36 ksi steel and the material to which they are welded is either F_y = 36 ksi or F_y = 50 ksi steel.

SEATED BEAM CONNECTIONS
Bolted
TABLE V

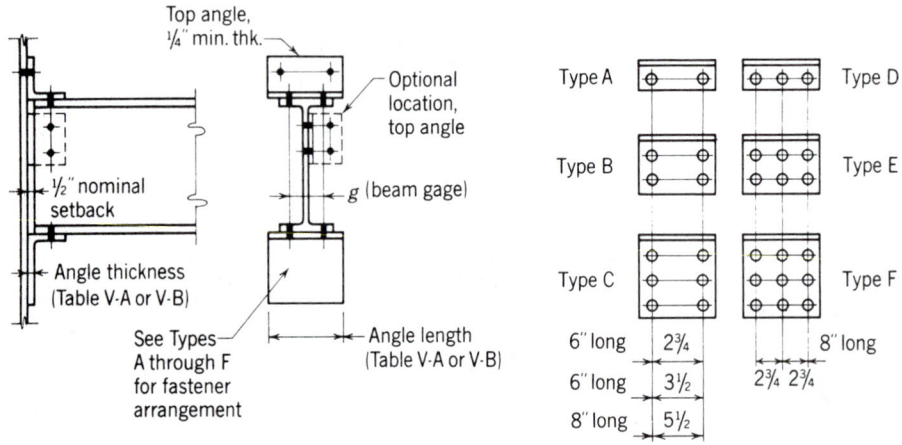

Seat Angle Types

Seated connections should be used only when the beam is supported by a top angle placed as shown above, or in the optional location as indicated.

Nominal beam setback is ½ in. Allowable loads in Tables V-A and V-B are based on ¾-in. setback, which provides for possible mill underrun in beam length.

ASTM A307 bolts may be used in seated connections, provided the stipulations of AISC ASD Specification Sect. J1.12 are observed.

Allowable loads in Table V-A are based on F_y = 36 ksi steel in both beam and seat angle. These values will be conservative when used with beams of F_y greater than 36 ksi. For beams with F_y equal to or greater than 50 ksi, use Table V-B.

Allowable loads in Table V-B are based on F_y = 36 ksi steel in the seat angle and F_y = 50 ksi steel in the beams. For beams with F_y greater than 50 ksi, these values will be conservative.*

Vertical spacing of fasteners and gages in seat angles may be arranged to suit conditions, provided they conform to AISC ASD Specification Sects. J3.8 and J3.9 with regard to minimum spacing and minimum edge distances. Where thick angles are used, driving clearances may require an increase in the outstanding leg gage.

In the event the thin web of a supporting member limits its bearing capacity, it will be necessary to reduce values listed in Table V-C.

For the most economical connection, the reaction values of the beams should be shown on the contract drawings. If the reactions are not shown, the connections must be selected to support the beam end reaction calculated from the Allowable Uniform Load Tables for the given shape, span and steel specified for the beam in question. The effect of concentrated loads near an end connection *must* also be considered.

---

* For the static model of the connections, see *Brockenbrough, R. L. and J. H. Garrett, Jr.* Design Loads For Seated-beam Connections in LRFD. AISC *Engineering Journal, 2nd Qtr., 1986, Chicago, IL.*

It should be noted that certain beams with slender webs might have allowable reactions less than the seat capacities tabulated in Table V. Since the values in Table V result from a model that includes Equation (K1-3), beams need only be checked by Equation (K1-5) or the allowable end reaction calculated by $R_3 + (N \times R_4)$.

EXAMPLE 15

Given:

Beam:	W 16 × 50, (⅜-in. web)
	F_y = 36 ksi material
Reaction:	30 kips
Bolts:	⅞-in. dia. A325-N
Column gage:	5½ in. in column web

Solution:

Enter Table V-A under 8-in. angle length, for a ⅜-in. beam web; select a ¾-in. angle thickness (capacity = 34.4 kips). Enter Table V-C opposite ⅞-in. dia. A325-N; note that a Type D connection (capacity = 37.9 kips) is required. From Table V-D, with a Type D connection, a 4 × 4 angle is available in ¾-in. thickness. Check beam web crippling. From the Allowable Uniform Load Tables, R_3 = 37.9 and R_4 = 3.28, R_3 + NR_4 = 37.9 + (4 × 3.28) = 51.0 kips **o.k.**

Detail data:

Seat: One L 4 × 4 × ¾ × 0'- 8 with three ⅞-in. dia. A325-N bolts.
Top angle or side support: To be chosen to suit conditions.

EXAMPLE 16

Given:

Same as Ex. 15 except connect to a column flange with column gage = 5½-in.

Solution:

As in Ex. 15, a ¾-in. angle thickness is adequate. Enter Table V-C opposite ⅞-in. dia. A325-N; note that a Type B connection (capacity = 50.5 kips) is required. From Table V-D, with a Type B connection, a 6 × 4 angle is available in ¾-in. thickness.

Detail data: Seat: One L 6 × 4 × ¾ × 0'- 8 (4-in. OSL) with four ⅞-in. dia. A325-N bolts.
Top angle or side support: To be chosen to suit conditions.

SEATED BEAM CONNECTIONS
Bolted
TABLE V Allowable loads in kips

TABLE V-A. Outstanding Leg Capacity, kips (based on OSL = 4 in.)

			Angle Length		6 In.					8 In.			
			Angle Thickness, In.	³⁄₈	½	⁵⁄₈	¾	1	³⁄₈	½	⁵⁄₈	¾	1
F_y (ksi)	36	Beam Web Thickness, In.	³⁄₁₆	8.23	11.5	14.1	16.7	19.7	9.50	12.7	15.5	18.4	19.7
			¼	9.50	14.2	18.9	21.9	28.1	11.0	16.2	20.4	23.8	28.6
			⁵⁄₁₆	10.6	16.1	21.8	27.5	36.3	12.3	18.3	24.5	30.7	39.0
			³⁄₈	11.6	17.9	24.3	30.9	41.1	13.4	20.3	27.3	34.4	47.6
			⁷⁄₁₆	12.6	19.5	26.8	34.1	48.9	14.5	22.2	29.9	37.8	53.8
			½	13.4	21.1	29.1	37.2	53.6	15.5	23.9	32.5	41.2	58.8
			⁹⁄₁₆	14.2	22.6	31.4	40.2	58.2	16.5	25.5	34.9	44.4	63.6

TABLE V-B. Outstanding Leg Capacity, kips (based on OSL = 4 in.)

			Angle Length		6 In.					8 In.			
			Angle Thickness, In.	³⁄₈	½	⁵⁄₈	¾	1	³⁄₈	½	⁵⁄₈	¾	1
F_y (ksi)	50	Beam Web Thickness, In.	³⁄₁₆	9.69	14.5	17.7	20.9	27.2	11.2	15.8	19.3	22.9	27.4
			¼	11.2	17.1	23.2	28.0	35.7	12.9	19.4	26.0	30.2	38.6
			⁵⁄₁₆	12.5	19.5	26.6	33.9	46.9	14.5	22.0	29.8	37.6	49.9
			³⁄₈	13.7	21.6	29.9	38.2	55.2	15.8	24.4	33.3	42.3	60.4
			⁷⁄₁₆	14.8	23.7	32.9	42.4	61.5	17.1	26.7	36.6	46.7	67.1
			½	15.8	25.6	35.9	46.4	67.6	18.3	28.8	39.8	51.0	73.6
			⁹⁄₁₆	16.8	27.5	38.8	50.3	73.6	19.4	30.9	42.9	55.1	80.5

TABLE V-C Fastener Capacity, kips

Fastener Designation[a]	Fastener Dia., In.	Connection Type					
		A	B	C	D	E	F
A307	¾	8.84	17.7	26.5	13.3	26.5	39.8
	⁷⁄₈	12.0	24.0	36.1	18.0	36.1	54.1
	1	15.7	31.4	47.1	23.6	47.1	70.7
A325-SC	¾	15.0	30.0	45.1	22.5	45.1	67.6
	⁷⁄₈	20.4	40.8	61.2	30.6	61.2	91.8
	1	26.8	53.6	80.4	40.2	80.4	—
A325-N	¾	18.6	37.1	55.7	27.8	55.7	83.5
	⁷⁄₈	25.3	50.5	75.8	37.9	75.8	—
	1	33.0	66.0	99.0	49.5	99.0	—
A325-X	¾	26.5	53.0	79.5	39.8	79.5	119
	⁷⁄₈	36.1	72.2	—	54.1	—	—
	1	47.1	94.2	—	70.7	—	—
A490-SC	¾	18.6	37.1	55.7	27.8	55.6	83.5
	⁷⁄₈	25.2	50.4	75.6	37.8	75.6	—
	1	33.0	66.0	—	49.5	—	—
A490-N	¾	24.7	49.5	74.2	37.1	74.2	111
	⁷⁄₈	33.7	67.3	101	50.5	101	—
	1	44.0	88.0	—	66.0	—	—
A490-X	¾	35.3	70.7	106	53.0	106	159
	⁷⁄₈	48.1	96.2	—	72.2	—	—
	1	62.8	—	—	94.3	—	—

TABLE V-D Available Seat Angle and Thickness Range

Type	Angle Size In.	t In.
A, D	4 × 3	³⁄₈– ½
	4 × 3½	³⁄₈– ½
	4 × 4	³⁄₈– ¾
B, E	6 × 4	³⁄₈– ¾
	7 × 4	³⁄₈– ¾
	8 × 4	½–1
C, F	[b]8 × 4	½–1

[b]Suitable for use with ¾-in. and ⁷⁄₈-in. fasteners only.

[a]A325-SC and A490-SC: Slip-critical connections with standard holes.
A325-N and A490-N: Bearing-type connections with threads included in the shear plane.
A325-X and A490-X: Bearing type connections with threads excluded from shear plane.

AMERICAN INSTITUTE OF STEEL CONSTRUCTION

SEATED BEAM CONNECTIONS
Welded–E70XX electrodes
TABLE VI

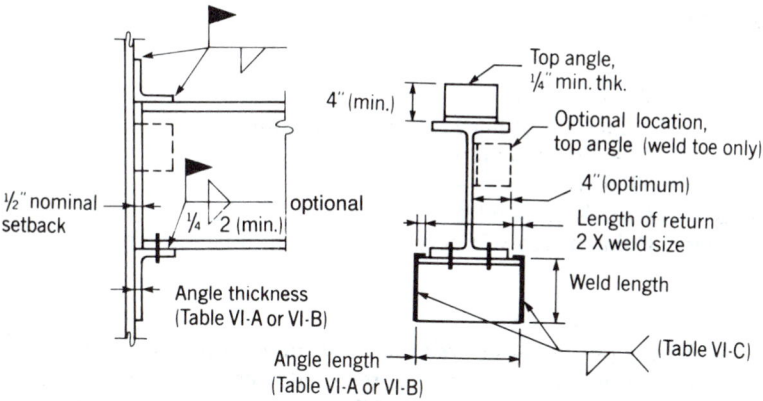

Seated connections should be used only when the beam is supported by a top angle placed as shown above, or in the optional location as indicated.

Allowable loads in Table VI are based on the use of E70XX electrodes. The table may be used for other electrodes, provided the tabular values are adjusted for the electrodes used (e.g., for E60XX electrodes, multiply tabular values by 18/21, or 0.86, etc.) and the welds and base metal meet the required strength level provisions of AISC ASD Specification Sect. J2.4.

Welds attaching beams to seat or top angles may be replaced by bolts. ASTM A307 bolts may be used in seated connections, provided the stipulations of AISC ASD Specification Sect. J1.12, are observed.

In addition to the welds shown, temporary erection bolts may be used to attach beams to seats (optional).

Nominal beam setback is ½ in. Allowable loads in Tables VI-A and VI-B are based on ¾-in. setback, which provides for possible mill underrun in beam length.

Allowable loads in Table VI-A are based on F_y = 36 ksi material in both beam and seat angle. These values will be conservative when used for beams with F_y greater than 36 ksi.

Allowable loads in Table VI-B are based on F_y = 36 ksi material in the seat angle with beam material of F_y = 50 ksi. These values will be conservative when used with beams of F_y greater than 50 ksi.

Allowable weld capacities in Table VI-C are computed using traditional vector analysis.

Should combinations of material thickness and weld size selected from Tables VI-A or VI-B and VI-C exceed the limits set by AISC ASD Specification Sects. J2.1b, J2.2b and J2.4 increase the weld size or material thickness as required.

No reduction of the tabulated weld capacity is required when unstiffened seats line up on opposite sides of the supporting web.

For the most economical connection, the reaction values of the beams should be shown on the contract drawings. If the reactions are not shown, the connections must be selected to support the beam end reaction calculated from the Allowable Uniform Load Tables for the given shape, span and steel specified for the beam in question. The effect of concentrated loads near an end connection *must* also be considered.

It should be noted that certain beams, with slender webs, might have allowable reactions less than the seat capacities tabulated in Table V. Since the values in Table V result from a model that includes Equation (K1-3), beams need only be checked by Equation (K1-5) or the allowable end reaction calculated by $R_3 + (N \times R_4)$.

EXAMPLE 17

Given:

Beam:	W 21 $\times$ 62, (3⁄8-in. web).
	Attach beam flange to seat with bolts.
	$F_y = 36$ ksi material
Reaction:	30 kips
Welds:	E70XX electrodes
Column:	Column web will permit use of 8-in. long seat angle.

Solution:

Enter Table VI-A opposite 3⁄8-in. web thickness; under 8-in. angle length, read 34.4 kips. Note that a 3⁄4-in. angle thickness is required. Enter Table VI-C and note that satisfactory weld capacities appear under 5 through 8 inch leg angles, all of which are shown to be available in 3⁄4-in. thickness. In this case, the 5 $\times$ 3½ and 6 $\times$ 4 angles are ruled out because of the rather heavy welds required. (½ in. and 7⁄16 in. respectively).

Angles 8 $\times$ 4 (capacity = 35.6 kips, ¼-in. weld) and 7 $\times$ 4 (capacity = 35.6 kips, 5⁄16-in. weld) are equally suitable. Angle 7 $\times$ 4 is chosen because the material savings will usually offset the cost differential between welds of 1⁄16-in. thickness differential, provided each weld can be made with the same number of passes (5⁄16-in. welds and smaller are single pass welds). Check beam web crippling. From the Allowable Uniform Load Tables, $R_3 = 40.5$ and $R_4 = 3.03$, $R_3 + NR_4 = 40.5 + (3.5 \times 3.03) = 51.1$ kips **o.k.**

Detail data: One L 7 $\times$ 4 $\times$ 3⁄4 $\times$ 0'- 8 with 5⁄16-in. welds (E70XX).

Top or side angle, as required.

Had it been required to weld the beam to the seat, the 3⁄4-in. seat angle thickness would dictate a ¼-in. weld, which is compatible with the 5⁄8-in. beam flange thickness (see AISC ASD Specification Sect. J2.2). Block beam flange to permit welding to the 8-in. seat angle or use a longer seat angle if space permits.

SEATED BEAM CONNECTIONS
Welded—E70XX electrodes
TABLE VI Allowable loads in kips

TABLE VI-A. Outstanding Leg Capacity, kips (based on OSL = 3½ or 4 in.)

		Angle Length		6 In.					8 In.			
		Angle Thickness, In.	⅜	½	⅝	¾	1	⅜	½	⅝	¾	1
F_y, ksi	36	Beam Web Thickness, In.										
		³⁄₁₆	8.23	11.5	14.1	16.7	19.7	9.50	12.7	15.5	18.4	19.7
		¼	9.50	14.2	18.9	21.9	28.1	11.0	16.2	20.4	23.8	28.6
		⁵⁄₁₆	10.6	16.1	21.8	27.5	36.3	12.3	18.3	24.5	30.7	39.0
		⅜	11.6	17.9	24.3	30.9	41.1	13.4	20.3	27.3	34.4	47.6
		⁷⁄₁₆	12.6	19.5	26.8	34.1	48.9	14.5	22.2	29.9	37.8	53.8
		½	13.4	21.1	29.1	37.2	53.6	15.5	23.9	32.5	41.2	58.8
		⁹⁄₁₆	14.2	22.6	31.4	40.2	58.2	16.5	25.5	34.9	44.4	63.6

Note: Values above heavy lines apply only for 4-in. outstanding legs.

TABLE VI-B. Outstanding Leg Capacity, kips (based on OSL = 3½ or 4 in.)

		Angle Length		6 In.					8 In.			
		Angle Thickness, In.	⅜	½	⅝	¾	1	⅜	½	⅝	¾	1
F_y, ksi	50	Beam Web Thickness, In.										
		³⁄₁₆	9.69	14.5	17.7	20.9	27.2	11.2	15.8	19.3	22.9	27.4
		¼	11.2	17.1	23.2	28.0	35.7	12.9	19.4	26.0	30.2	38.6
		⁵⁄₁₆	12.5	19.5	26.6	33.9	46.9	14.5	22.0	29.8	37.6	49.9
		⅜	13.7	21.6	29.9	38.2	55.2	15.8	24.4	33.3	42.3	60.4
		⁷⁄₁₆	14.8	23.7	32.9	42.4	61.5	17.1	26.7	36.6	46.7	67.1
		½	15.8	25.6	35.9	46.4	67.6	18.3	28.8	39.8	51.0	73.6
		⁹⁄₁₆	16.8	27.5	38.8	50.3	73.6	19.4	30.9	42.9	55.1	80.5

Note: Values above heavy lines apply only for 4-in. outstanding legs.

TABLE VI-C. Weld Capacity, kips

Weld Size In.	E70XX Electrodes				
	Seat Angle Size (long-leg vertical)				
	4×3½	5×3½	6×4	7×4	8×4
¼	11.5	17.2	21.8	28.5	35.6
⁵⁄₁₆	14.3	21.5	27.3	35.6	44.5
⅜	17.2	25.8	32.7	42.7	53.4
⁷⁄₁₆	20.1	30.1	38.2	49.8	62.3
½	—	34.4	43.6	56.9	71.2
⅝	—	43.0	54.5	71.2	89.0
¹¹⁄₁₆	—	47.3	60.0	78.3	—
¾	—	—	—	—	—
Range of Available Seat Angle Thicknesses					
Minimum	⅜	⅜	⅜	⅜	½
Maximum	½	¾	¾	¾	1

AMERICAN INSTITUTE OF STEEL CONSTRUCTION

Notes

STIFFENED SEATED BEAM CONNECTIONS
Bolted
TABLE VII

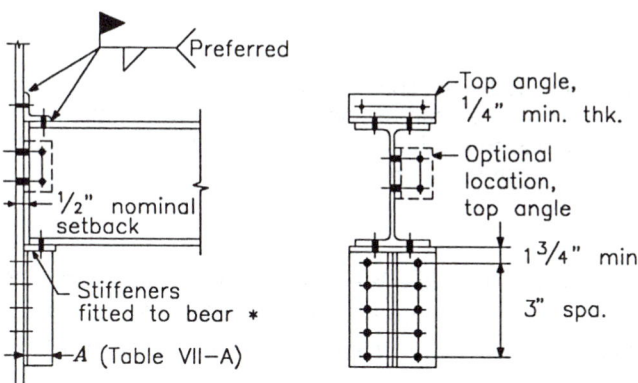

Seated connections should be used only when the beam is supported by a top angle placed as shown above, or in the optional location as indicated.

Allowable capacities in Table VII-A are based on allowable bearing using steel of F_y = 36 ksi or F_y = 50 ksi in the stiffener angles. Capacities of fastener groups in Table VII-B are based on single shear. Connection capacity is based on the lesser of these two values in conjunction with the local web-yielding and web-crippling value of the supported beam.

Effective length of stiffener bearing is assumed ½-in. less than the length of the outstanding leg.

Maximum gage in legs of stiffeners connected to columns is 2½ in.

ASTM A307 bolts may be used in seated connections, provided the stipulations of ASD Specification Sect. J1.12 are observed.

Vertical spacing of fasteners in stiffener angles may be arranged to suit conditions, provided they conform to Sects. J3.8 and J3.9 with respect to minimum spacing and minimum edge distances.

Beam seats, traditionally provided in column webs for simple beams, permit advantages:

1. Beams require only routine punching
2. Ample erection clearance is provided
3. Erection is fast, safe and simple
4. Accuracy of bay size is easy to maintain.

To permit selection of the most economical beam connection, the beam reactions should be shown on the contract drawings. If they are not shown, the connections must be selected to support the beam end reaction from the Allowable Uniform Load Tables for the given shape, span, and steel of the beam in question. The

---

* A structural tee may be used instead of a pair of angles.

effect of concentrated loads near an end connection *must* also be considered. This is done most easily by checking the beam web for yielding: $R = R_1 + (N \times R_2)$ and for crippling: $R = R_3 + (N \times R_4)$.

For loads in excess of tabulated capacities, or for thin webs of supporting members, it is necessary to design special seated connections.

EXAMPLE 18

Given:

Design a stiffened seated beam connection of $F_y = 36$ ksi steel to support a W 30 × 99, also $F_y = 36$ ksi, with an end reaction of 84 kips. Use ⅞-in. dia. ASTM A325-N bolts to attach the seat to a column web with a 5½-in. gage and thickness of ½ in. Assume that a top angle is required.

Solution:

1. From the $F_y = 36$ ksi Allowable Uniform Load Table: under W30×99, note $R_1 = 44.4$ kips, $R_2 = 12.4$ kips/in., $R_3 = 62.6$ kips and $R_4 = 4.33$ kips/in.

   For yielding N, req'd $= (R - R_1)/R_2 = (84 - 44.4)/12.4 = 3.19$ in.

   For crippling N, req'd $= (R - R_3)/R_4 = (84 - 62.6)/4.33 = 4.94$ in. governs.

   From Table VII-A, under $F_y = 36$ ksi, it will be seen that a 4.94-in. length of bearing requires 5-in. bearing and the 84 kip reaction requires stiffener angles of ⁵⁄₁₆-in. thickness. Use a seat plate of ⅜-in. thickness extending beyond the stiffener angle; this requires a 6-in. leg outstanding.

2. In Table VII-B for a ⅞-in. dia. A325-N fastener, 4 rows of bolts with a capacity of 101 kips will be required for an 84 kip reaction.

3. Bearing on web $= 84/8 = 10.5$ kips/bolt < 30.5 kips (Table I-E, $F_u = 58$ ksi, fastener spacing $= 3$ in.)

Detail data:

Steps 1 and 2 indicate the use of a connection with 4 rows of ⅞-in. dia. A325-N bolts. Assuming it is possible to employ the suggested spacing of fasteners, detail material will be as follows:

Steel:	$F_y = 36$ ksi
2 Stiffeners:	L 5 × 5 × ⁵⁄₁₆ × 0'-11⅝
1 Seat plate:	PL ⅜ × 6 × 0'-11
1 Top angle:	L 5 × 5 × ⅜ × 0'-8

STIFFENED SEATED BEAM CONNECTIONS
Bolted
TABLE VII

TABLE VII-A Stiffener Angle Capacity, kips						
Stiffener Material		$F_y = 36$ ksi ($F_p = 32.4$ ksi)			$F_y = 50$ ksi ($F_p = 45$ ksi)	
Stiffener Outstanding Leg, A, In.		3½	4	5	3½	4	5
Max. Length Beam Bearing, In.		3½	4	5	3½	4	5
Thickness of Stiffener Outstanding Legs, In.	5/16	60.8	70.9	91.1	84.4	98.4	127
	3/8	72.9	85.1	109	101	118	152
	½	97.2	113	146	135	158	203
	5/8	122	142	182	169	197	253
	¾	146	170	219	203	236	304

Use ⅜-in. thick seat plate wide enough to extend beyond outstanding legs of stiffener.

TABLE VII-B Fastener Capacity, kips						
Fastener Specification[a]	Fastener Diameter In.	Number of Fasteners in One Vertical Row				
		3	4	5	6	7
A307	¾	26.5	35.3	44.2	53.0	61.9
	7/8	36.1	48.1	60.1	72.2	84.2
	1	47.1	62.8	78.5	94.2	110
A325-N	¾	55.7	74.2	92.8	111	130
	7/8	75.8	101	126	152	177
	1	99.0	132	165	198	231
A325-X	¾	79.5	106	133	159	186
	7/8	108	144	180	216	253
	1	141	188	236	283	330
A490-N	¾	74.2	99.0	124	148	173
	7/8	101	135	168	202	236
	1	132	176	220	264	308
A490-X	¾	106	141	177	212	247
	7/8	144	192	241	289	337
	1	188	251	314	377	440

[a]A325-N and A490-N: Bearing-type connections with threads included in shear plane.
A325-X and A490-X: Bearing-type connections with threads excluded from shear plane.

STIFFENED SEATED BEAM CONNECTIONS
Welded–E70XX Electrodes
TABLE VIII

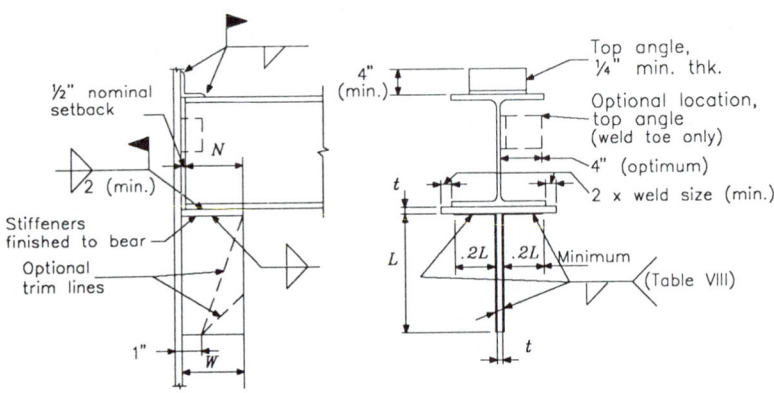

Seated connections should be used only when the beam is supported by a top angle placed as shown above, or in the optional location as indicated.

Allowable loads in Table VIII are based on the use of E70XX electrodes. The table may be used for other electrodes, provided the tabular values are adjusted for the electrodes used (e.g., for E60XX electrodes, multiply tabular values by $\frac{18}{21}$, or 0.86, etc.) and the welds and base metal meet the provisions of AISC ASD Specification Sect. J2.4.

Allowable weld capacities in Table VIII are computed using traditional vector analysis.

Based on $F_y = 36$ ksi bracket material, minimum stiffener plate thickness, t, for supported beams with unstiffened webs should not be less than the supported beam web thickness for $F_y = 36$ ksi beams, and not less than 1.4 times the beam web thickness for beams with $F_y = 50$ ksi. Based on bracket material of $F_y = 50$ ksi or greater, the minimum stiffener plate thickness, t, for supported beams with unstiffened webs should be the beam web thickness multiplied by the ratio of F_y of the beam to F_y of the bracket (e.g., F_y (beam) = 65 ksi; F_y (bracket) = 50 ksi; $t = t_w$ (beam) $\times$ 65/50, minimum). The minimum stiffener plate thickness, t, should be at least two times the required E70XX weld size when F_y of the bracket is 36 ksi, and should be at least 1.5 times the required E70XX weld size when F_y of the bracket is 50 ksi.

Thickness t of the horizontal seat plate, or tee flange, should not be less than $\frac{3}{8}$ in.

If seat and stiffener are separate plates, finish stiffener to bear against seat. Welds connecting the two plates should have a strength equal to, or greater than, the horizontal welds to the support under the seat plate.

Welds attaching beam to seat may be replaced by bolts.

ASTM A307 bolts may be used in seated connections, if the stipulations of AISC ASD Specification Sect. J1.12 are observed.

For stiffener seats in line on opposite sides of a column web of $F_y = 36$ ksi material, select E70XX weld size no greater than 0.50 of column web thickness. For column web of F_y = 50 ksi, select E70XX weld size no greater than 0.67 of column web thickness.

Should combinations of material thickness and weld size selected from Table VIII exceed the limits set by AISC ASD Specification Sects. J2.2 and J2.4, increase the weld size or material thickness as required.

In addition to the welds shown, temporary erection bolts may be used to attach beams to seats (optional).

To permit selection of the most economical connection, the reaction values should be given on the contract drawings. If the reaction values are not given, the connections must be selected to support the beam end reaction calculated from the Allowable Uniform Load Tables for the given shape, span, and steel specification of the beam in question. The effect of concentrated loads near an end connection *must* also be considered.

EXAMPLE 19

Given:

Beam: W 30 × 116 (flange = 10.495 in. × 0.85 in.; web = 0.565 in.)
ASTM A36 steel (F_y = 36 ksi)

Welds: E70XX

Reaction: 100 kips

Design a two-plate welded stiffener seat using ASTM A36 steel.

Solution:

From the F_y = 36 ksi, Allowable Uniform Load Table for W30 × 116, note that R_1, = 54.5 kips, R_2 = 13.4 kips/in., R_3 = 79.9 kips, R_4 = 4.33 kips/in.

For yielding N, req'd = $(R - R_1)/R_2$

$$= (100 - 54.5)/13.4 = 3.40 \text{ in.}$$

For buckling N, req'd = $(R - R_3)/R_4$

$$= (100 - 79.9)/4.33 = 4.64 \text{ in.}$$

Stiffener width = 4.64 + 0.5 (setback) = 5.14 in.

Use W = 6 in.

Enter Table VIII with W = 6 in. and a reaction of 100 kips; select a $5/16$-in. weld with L = 15 in., which has a capacity of 103 kips. From this, the minimum length of weld between seat plate and support is 2 × 0.2L = 6 in. This also establishes the minimum weld between the seat plate and the stiffener as 6 in. total, or 3 in. on each side of stiffener.

Stiffener plate thickness t to develop welds is 2 × $5/16$ = $5/8$ in., or 0.625 in. This is greater than the beam web thickness of 0.565 in.; thus, the stiffener plate thickness need not be increased.

Use: $5/8$-in. plate for the stiffener and $3/8$-in. plate for seat.

Welds attaching the beam flange to the seat must be $\frac{5}{16}$ in. to conform to AISC ASD Specification Sect. J2.2, due to the 0.85-in. flange thickness of the W 30 × 116 beam.

Seat plate width to permit field welding of beam to seat = flange width + (4 × weld size) = 10.5 + (4 × $\frac{5}{16}$) = 11.75 in.

Use: $\frac{3}{8}$ × 6 × 1'-0 plate

This is adequate for the required minimum weld length.

Detail data:

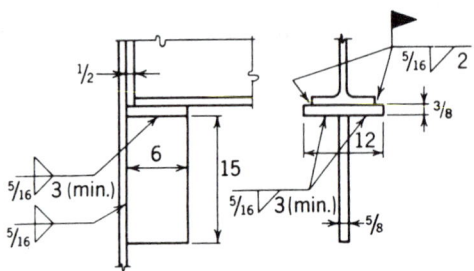

Use: L 4 × 4 × $\frac{3}{8}$ × 0'-4 top angle (F_y = 36 ksi) with $\frac{5}{16}$-in. welds along toes of angle only.

EXAMPLE 20

Given:

Beam:	W 21 × 68 (flange = 8.27 in. × 0.685 in.; web = 0.43 in.)
	ASTM A572, Grade 50 steel (F_y = 50 ksi)
Welds:	E70XX electrodes
Reaction:	83 kips

Design a two-plate welded stiffener seat using ASTM A36 steel.

Solution:

From the F_y = 50 ksi, Allowable Uniform Load Table for W21 × 68, note R_1 = 51.0 kips, R_2 = 14.2 kips/in., R_3 = 56.1 kips, R_4 = 3.96 kips/in.

For yielding N, req'd = $(R - R_1)/R_2$

$$= (83 - 51.0)/14.2 = 2.25 \text{ in.}$$

For buckling N, req'd = $(R - R_3)/R_4$

$$= (83 - 56.1)/3.96 = 6.79 \text{ in.}$$

Stiffener width = 6.79 + 0.5 (setback) = 7.29 in.

Use W = 8 in.

Enter Table VIII with W = 8 in. and a reaction of 83 kips; satisfying these requirements are a $\frac{5}{16}$-in. weld, L = 15 in. (84.4 kips), or a $\frac{3}{8}$-in. weld, L = 14 in. (89.8 kips), or an even larger weld size. Generally, the $\frac{5}{16}$-in. weld is the better selection, as this can be made in one pass using manual welding. Select ⁵⁄₁₆-in. weld. From this, the minimum length of ⁵⁄₁₆-in. weld between seat plate and support is 2 × 0.2L = 6 in.

Use: 6 in. weld length. This also establishes the minimum weld between the seat plate and the stiffener as 6 in. total, or 3 in. on each side.

Stiffener plate thickness t to develop welds is 2 × ⁵⁄₁₆ = ⅝ in., or 0.625 in. The minimum thickness t for a bracket of $F_y = 36$ ksi with a beam of $F_y = 50$ ksi is 1.4 times the beam web thickness = 1.4 × 0.43 = 0.602 in.

Use: ⅝-in. plate for the stiffener and ⅜-in. plate for the seat.

Welds attaching the beam flange to the seat can be ¼ in. for a flange of 0.685 in., per AISC ASD Specification Sect. J2.2.

Seat plate width, to permit field welding of beam to seat = flange width + (4 × weld size) = 8.27 + (4 × ¼) = 9.27 in.

Use: ⅜ × 8 × 0'-10 plate

This is adequate for the required minimum weld length.

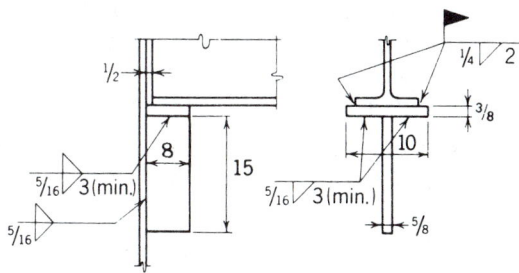

Detail data:

Use: L 4 × 4 × ⅜ × 0'-4 top angle ($F_y = 36$ ksi) with ⁵⁄₁₆-in. welds along toes of angle only.

STIFFENED SEATED BEAM CONNECTIONS
Welded—E70XX electrodes
TABLE VIII Allowable loads in kips

L In.	Width of Seat W, In.														
	4					5						6			
	Weld Size, In.					Weld Size, In.						Weld Size, In.			
	3/16	1/4	5/16	3/8	7/16	3/16	1/4	5/16	3/8	7/16	1/2	5/16	3/8	7/16	1/2
6	17.0	22.7	28.4	34.0	39.7	14.1	18.8	23.5	28.1	32.8	37.5	19.9	23.9	27.9	31.9
7	22.4	29.9	37.4	44.9	52.4	18.7	25.0	31.2	37.5	43.7	50.0	26.7	32.0	37.3	42.7
8	28.3	37.8	47.2	56.7	66.1	23.9	31.9	39.8	47.8	55.8	63.7	34.3	41.1	48.0	54.8
9	34.6	46.1	57.6	69.2	80.7	29.5	39.3	49.1	59.0	68.8	78.6	42.5	51.1	59.6	68.1
10	41.1	54.9	68.6	82.3	96.0	35.4	47.2	59.0	70.8	82.6	94.4	51.4	61.7	72.0	82.3
11	47.9	63.9	79.8	95.8	112	41.6	55.5	69.4	83.3	97.1	111	60.9	73.1	85.2	97.4
12	54.8	73.1	91.4	110	128	48.1	64.1	80.2	96.2	112	128	70.8	85.0	99.2	113
13	61.9	82.5	103	124	144	54.8	73.0	91.3	110	128	146	81.1	97.4	114	130
14	69.0	92.0	115	138	161	61.6	82.1	103	123	144	164	91.9	110	129	147
15	76.2	101	127	152	178	68.5	91.4	114	137	160	183	103	123	144	165
16	83.5	111	139	167	195	75.6	100	126	151	176	202	115	138	160	183
17	90.7	121	151	181	212	82.7	110	138	165	193	221	126	151	176	201
18	98.0	131	163	196	229	89.9	119	150	180	210	240	137	164	192	219
19	105	140	175	211	246	97.1	129	162	194	227	259	149	179	208	238
20	112	150	188	225	263	104	139	174	209	243	278	161	193	225	257
21	119	160	200	240	280	111	148	189	223	260	298	173	207	242	276
22	127	169	212	254	296	118	158	198	238	277	317	185	221	258	295
23	134	179	224	269	313	126	168	210	252	294	336	197	236	275	315
24	141	189	236	283	330	133	177	222	267	311	356	209	250	292	334
25	148	198	248	297	347	140	187	234	281	328	375	221	265	309	353
26	155	208	260	312	364	148	197	247	296	345	394	233	279	326	373
27	163	217	272	326	380	155	206	259	310	362	414	245	294	343	392

Note: Loads shown are for E70XX electrodes. For E60XX electrodes, multiply tabular loads by 0.86, or enter table with 1.17 times the given reaction. For E80XX electrodes, multiply tabular loads by 1.14 or enter table with 0.875 times the given reaction.

STIFFENED SEATED BEAM CONNECTIONS
Welded—E70XX electrodes

TABLE VIII Allowable loads in kips

L In.	Width of Seat W, In.											
	7				8				9			
	Weld Size, In.				Weld Size, In.				Weld Size, In.			
	5/16	3/8	7/16	1/2	5/16	3/8	1/2	5/8	5/16	3/8	1/2	5/8
11	54.0	64.8	75.6	86.3	48.4	58.0	77.3	96.6	43.7	52.4	69.9	87.4
12	63.1	75.7	88.4	101	56.7	68.1	90.7	113	51.4	61.7	82.2	103
13	72.7	87.2	102	117	65.5	78.7	105	131	59.6	71.5	95.3	119
14	82.6	99.1	116	132	74.8	89.8	120	149	68.2	81.8	109	136
15	92.9	112	130	149	84.4	101	135	169	77.2	92.6	123	154
16	104	124	145	166	94.4	113	151	189	86.5	104	138	173
17	114	137	160	183	105	126	167	209	96.2	115	154	192
18	126	151	176	201	115	138	184	230	106	127	170	212
19	137	164	192	219	126	151	202	252	117	140	186	233
20	148	178	208	237	137	165	219	274	127	152	203	254
21	160	192	224	256	148	178	237	296	138	165	220	276
22	172	206	240	274	159	192	255	319	149	178	238	297
23	183	220	257	293	171	205	274	342	160	192	256	320
24	195	234	274	312	183	219	292	365	171	205	274	342
25	207	249	290	331	195	233	311	389	182	219	292	365
26	219	263	307	351	206	248	330	412	194	233	310	388
27	231	278	324	370	218	262	349	436	206	247	329	411
28	243	292	341	389	230	276	368	460	217	261	348	435
29	256	307	358	409	242	291	387	484	229	275	367	458
30	268	321	375	428	254	305	406	508	241	289	386	482
31	280	336	392	447	266	319	426	532	253	303	405	506
32	292	350	409	467	278	334	445	556	265	318	424	530

Note: Loads shown are for E70XX electrodes. For E60XX electrodes, multiply tabular loads by 0.86, or enter table with 1.17 times the given reaction. For E80XX electrodes, multiply tabular loads by 1.14 or enter table with 0.875 times the given reaction.

AMERICAN INSTITUTE OF STEEL CONSTRUCTION

Notes

END-PLATE SHEAR CONNECTIONS
TABLE IX

This type of connection consists of a plate, less than the beam depth in length, perpendicular to the longitudinal axis of the beam, welded to the beam web with fillet welds each side of the beam web. The end-plate connection compares favorably to the double-angle connection and, for like thicknesses, gage lines and length of connection will furnish end rotation capacity and strength of connection closely approximating that of the double-angle framing connection, within the range listed in the table.

Fabrication of this type of connection requires close control in cutting the beam to length; adequate consideration must be given to squaring the beam ends such that both end plates are parallel and the effect of beam camber does not result in out-of-square end plates, which make erection and field fit-up difficult. Shims may be required on runs of beams to compensate for mill and shop tolerances.

For adequate end-rotation capacity, it is suggested end plates be designed for a plate thickness range of ¼ in. to ⅜ in., inclusive. To develop full capacity of the fasteners and welds, the end plate and web thicknesses must equal or exceed the values listed in the table. If the material thickness supplied by either the plate or the web is less than required, the fastener or weld capacity must be reduced by the ratio of thickness supplied to thickness required.

The gage g should be 3½ in. to 5½ in. for average plate thicknesses, with an edge distance of 1¼ in. Lesser values of edge distance should be avoided. Plates ¼-in. thick, of $F_y = 36$ ksi steel and a gage of 3 in. should provide adequate end rotation capacity in the connection. All end-plate material thicknesses listed in the table are for $F_y = 36$ ksi. Use of higher values of F_y should be based on an engineering investigation that confirms that adequate end rotation capacity is available.

Weld values listed are for two fillet welds and are based on the use of E70XX electrodes. These weld values have been reduced by considering the effective weld length equal to the plate length minus twice the weld size. Welds should not be returned across the web at the top or bottom of the end plates.

EXAMPLE 21

Given:

Select an end-plate connection for a W14 × 30 beam.

$F_y = 36$ ksi and end reaction = 24 kips.

Solution:

Beam web thickness: 0.270 in.

Gage: 3½ in.

From Table IX, for beam depth limits 12 in. through 18 in., select a plate length of 8½ in. with three ¾-in. dia. A307 bolts per vertical row, with a listed capacity of 26.5 kips and a required minimum plate thickness of $t = 0.13$ in.

Select a ¼-in. plate: 0.25 in. > 0.13 in. **o.k.**

From Table IX:

Weld capacity: 8½ in. of ³⁄₁₆-in. fillet = 45.2 kips

AMERICAN INSTITUTE OF STEEL CONSTRUCTION

Minimum web thickness = 0.38 in.

$$\frac{0.270}{0.38} \times 45.2 = 32.1 \text{ kips} > 24 \text{ kips} \quad \textbf{o.k.}$$

Use:

End plate 6 in. wide × 8½-in. long × ¼-in. thick, with six ¾-in. dia. A307 bolts on 3½-in. gage. Weld the plate to the beam web with ³⁄₁₆-in. fillet welds on each side of the web.

EXAMPLE 22

Given:

Select an end-plate connection for two W 12 × 58 beams framing into both sides of a W 30 × 191 girder. Beam reaction = 34 kips for each of the W 12 × 58 beams and F_y = 50 ksi for both beams and girder.

Solution:

Beam web thickness: 0.360 in.

Gage: 5½ in.

Girder web thickness: 0.710 in.

From Table IX for beam depth limits 8 in. through 12 in., select a plate length of 5½ in. with two ¾-in. dia. A325-N bolts per vertical row, with a listed capacity of 37.1 kips and minimum t = 0.28 in.

Use a ⁵⁄₁₆-in. end-plate thickness t = 0.3125 in.

From Table IX:

   Weld capacity: 5½ in. of ¼-in. fillet weld = 37.1 kips

   Minimum web t = 0.37 in.

$$\frac{0.360}{0.37} \times 37.1 = 36.1 \text{ kips} > 34 \text{ kips} \quad \textbf{o.k.}$$

Girder web thickness must be checked separately, using fastener spacing bearing criteria:

   Total load on girder web in double shear = 2 × 34 = 68 kips

   Using Table I-E, based on ¾-in. dia. bolts, 3 in. pitch and F_u = 65 ksi:

$$t = \frac{68}{4 \times 58.5} = 0.291 \text{ in.} < 0.710 \text{ in.}$$

   ∴ Girder web is adequate

Use:

End plate 8 in. wide × 5½-in. long × ⁵⁄₁₆-in. thick, with four ¾-in. dia., A325-N bolts on 5½-in. gage. Weld the plate to the beam web with ¼-in. fillet welds on each side of the web.

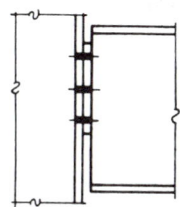

END PLATE SHEAR
CONNECTIONS

Welded—E70XX electrodes

TABLE IX

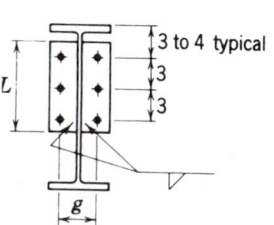

Bolts per Vertical Line	Bolt Designation	¾-in. Diameter		⅞-in. Diameter		Plate Length L In.	Beam Depth Limits In.
		Total Capacity Kips	Min[a] Plate Thickness t In.	Total Capacity Kips	Min[a] Plate Thickness t In.		
1	A307	8.8	.12	12.0	.17		
	A325-N	18.6	.26*	25.3	.35	3	5–8
	A325-X	26.5	.37*	—	—		
2	A307	17.7	.13	24.1	.19		
	A325-N	37.1	.28	—	—	5½	8–12
3	A307	26.5	.13	36.1	.18		
	A325-N	55.7	.26	—	—	8½	12–18
	A325-X	79.5	.38	—	—		
4	A307	35.3	.12	48.1	.18		
	A325-N	74.2	.26	101	.37	11½	15–24
	A325-X	106	.37	—	—		
5	A307	44.2	.12	60.1	.18		
	A325-N	92.8	.26	126	.37	14½	18–30
	A325-X	133	.37	—	—		
6	A307	53.0	.12	72.2	.18		
	A325-N	111	.26*	152	.37	17½	21–36
	A325-X	159	.37*	—	—		

[a]Net shear controls thickness unless noted by asterisk (*).
*Indicates edge distance bearing controls.

WELD CAPACITY, Kips

Weld Size	Thickness, In.		Plate Length, L, In.					
	$F_y = 36$	$F_y = 50$	3	5½	8½	11½	14½	17½
3/16	.38	.28	14.6	28.5	45.2	61.9	78.6	95.4
¼	.51	.37	18.6	37.1	59.4	81.7	104	126
5/16	.64	.46	22.0	45.2	73.1	101	129	157
3/8	.77	.56	25.1	52.9	86.3	120	153	187

SINGLE-PLATE SHEAR CONNECTIONS
TABLE X

The Allowable Loads tabulated in Table X are based on recent research at the University of California-Berkeley* incorporating some simplified and conservative assumptions. Research on these connections has also been conducted at the University of Arizona.**

NOMENCLATURE

d_b = Bolt diameter, in.
n = Number of bolts, in.
t = Plate thickness, in.
L = Plate length, in.

DESIGN ASSUMPTIONS

1. Single bolt row. $2 \leq n \leq 7$
2. Bolt pitch = 3 in.
3. Vertical edge distance = 1½ in.
4. Distance from the weld line to the bolt line = 3 in.
5. Shear plates are F_y = 36 ksi
6. Fillet welds are E70XX electrodes
7. $t \leq d_b/2 + \frac{1}{16}$ in.
8. $L \geq \frac{1}{2} T$ (see Part 1, Dimensions and Properties for T of beam)
9. Tabulated values are valid for composite and non-composite beams, standard or short-slotted holes, fully-tightened or snug-tight, and for all grades of beam steels and all loadings. These values may be overly conservative for snug-tight bolts in slotted holes. Currently, research is being conducted on the subject.

DESIGN PROCEDURE

1. Calculate plate capacity in yielding, R_o, kips.

$$R_o = 0.4 (36) \times L \times t$$

2. Calculate fillet weld size, D in sixteenths, to develop R_o.

$$D = R_o/(L \times C)$$

where,

C is taken from Table XIX, assuming al = 3 in. or $n \times 1$ in., whichever is larger, and k = 0.

---

* *Astaneh-Asl, A., K. M. McMullin and S. M. Call* Design of Single-plate Framing Connections, *Report No. UCB/SEMM University of California–Berkeley, July 1988.*

** *Richard, R. M., P. E. Gillett, J. D. Kriegh and B. A. Lewis* The Analysis and Design of Single-plate Framing Connections, *AISC Engineering Journal, Vol. 17, No. 2, 2nd Qtr., 1980.*

Richard, R. M., J. D. Kriegh and D. E. Hormby Design of Single-Plate Framing Connections with A307 Bolts *AISC Engineering Journal, Vol. 19, No. 4, 4th Qtr., 1982.*

Hormby, D. E., R. M. Richard and J. D. Kriegh Single-plate Framing Connections with Grade 50 Steel and Composite Construction, *AISC Engineering Journal, Vol. 21, No. 3, 3rd Qtr., 1984.*

Young, N. W. and R. O. Disque Design Aids for Single-plate Framing Connections, *AISC Engineering Journal, Vol. 18, No. 4, 4th Qtr., 1981.*

In addition, research indicates that fillet weld size w need not exceed $0.75\,t$

3. Calculate bolt group capacity, R_b, kips.

$$R_b = C \times r_v$$

where,

C is taken from Table XI assuming $l = 3$ in.

r_v = Single shear bolt capacity, kips

4. Calculate net shear fracture capacity of plate R_{ns}, kips

$$R_{ns} = 0.3\,(58)\,[L - n\,(d_b + \tfrac{1}{16})]\,t$$

Tabulated loads are the lesser of the bolt capacity and the net shear capacity of the plate. Yielding of the plate does not govern any of the values tabulated.

5. Check plate bearing capacity, P_b

and, if beam is coped, check block shear.

EXAMPLE 23

Given:

Beam:	W27 $\times$ 114, $t_w = 0.570$ in., A36 steel
Reaction:	100 kips (service load)
Bolts:	$\frac{7}{8}$ in., A490-N, snug-tight, with 3-in. spacing
Welds:	E70XX fillet welds

Solution:

From Table X:

7 A490-N bolts ($L = 21$) with $\frac{7}{16}$-in. plate and $\frac{3}{8}$-in. fillet welds has a capacity of 102 kips.

SINGLE-PLATE SHEAR CONNECTIONS

TABLE X-A Allowable loads in kips

n = 2   L = 6

	Plate Thickness, t In.	Bolt Size, In.					
		¾		⅞		1	
		Load	Weld	Load	Weld	Load	Weld
A325-N	¼	8.2	³⁄₁₆	11.1	³⁄₁₆	14.5	³⁄₁₆
	⁵⁄₁₆	8.2	¼	11.1	¼	14.5	¼
	⅜	8.2	⁵⁄₁₆	11.1	⁵⁄₁₆	14.5	⁵⁄₁₆
	⁷⁄₁₆	8.2	⅜	11.1	⅜	14.5	⅜
	½	—	—	11.1	⅜	14.5	⅜
	⁹⁄₁₆	—	—	—	—	14.5	⁷⁄₁₆
A490-N	¼	10.9	³⁄₁₆	14.8	³⁄₁₆	19.4	³⁄₁₆
	⁵⁄₁₆	10.9	¼	14.8	¼	19.4	¼
	⅜	10.9	⁵⁄₁₆	14.8	⁵⁄₁₆	19.4	⁵⁄₁₆
	⁷⁄₁₆	10.9	⅜	14.8	⅜	19.4	⅜
	½	—	—	14.8	⅜	19.4	⅜
	⁹⁄₁₆	—	—	—	—	19.4	⁷⁄₁₆

TABLE X-B Allowable loads in kips

n = 3   L = 9

	Plate Thickness, t In.	Bolt Size, In.					
		¾		⅞		1	
		Load	Weld	Load	Weld	Load	Weld
A325-N	¼	16.3	³⁄₁₆	22.1	³⁄₁₆	25.3	³⁄₁₆
	⁵⁄₁₆	16.3	¼	22.1	¼	28.9	¼
	⅜	16.3	⁵⁄₁₆	22.1	⁵⁄₁₆	28.9	⁵⁄₁₆
	⁷⁄₁₆	16.3	⅜	22.1	⅜	28.9	⅜
	½	—	—	22.1	⅜	28.9	⅜
	⁹⁄₁₆	—	—	—	—	28.9	⁷⁄₁₆
A490-N	¼	21.7	³⁄₁₆	26.9	³⁄₁₆	25.3	³⁄₁₆
	⁵⁄₁₆	21.7	¼	29.4	¼	31.6	¼
	⅜	21.7	⁵⁄₁₆	29.4	⁵⁄₁₆	37.9	⁵⁄₁₆
	⁷⁄₁₆	21.7	⅜	29.4	⅜	38.5	⅜
	½	—	—	29.4	⅜	38.5	⅜
	⁹⁄₁₆	—	—	—	—	38.5	⁷⁄₁₆

SINGLE-PLATE SHEAR CONNECTIONS

TABLE X-C Allowable loads in kips

n = 4   L = 12

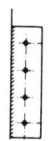

	Plate Thickness, t In.	Bolt Size, In.					
		³⁄₄		⁷⁄₈		1	
		Load	Weld	Load	Weld	Load	Weld
A325-N	¼	26.1	³⁄₁₆	35.4	³⁄₁₆	33.7	³⁄₁₆
	⁵⁄₁₆	26.1	¼	35.4	¼	42.1	¼
	³⁄₈	26.1	⁵⁄₁₆	35.4	⁵⁄₁₆	46.4	⁵⁄₁₆
	⁷⁄₁₆	26.1	³⁄₈	35.4	³⁄₈	46.4	³⁄₈
	½	—	—	35.4	³⁄₈	46.4	³⁄₈
	⁹⁄₁₆	—	—	—	—	46.4	⁷⁄₁₆
A490-N	¼	34.8	³⁄₁₆	35.9	³⁄₁₆	33.7	³⁄₁₆
	⁵⁄₁₆	34.8	¼	44.9	¼	42.1	¼
	³⁄₈	34.8	⁵⁄₁₆	47.2	⁵⁄₁₆	50.6	⁵⁄₁₆
	⁷⁄₁₆	34.8	³⁄₈	47.2	³⁄₈	59.0	³⁄₈
	½	—	—	47.2	³⁄₈	61.8	³⁄₈
	⁹⁄₁₆	—	—	—	—	61.8	⁷⁄₁₆

TABLE X-D Allowable loads in kips

n = 5   L = 15

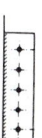

	Plate Thickness, t In.	Bolt Size, In.					
		³⁄₄		⁷⁄₈		1	
		Load	Weld	Load	Weld	Load	Weld
A325-N	¼	36.3	³⁄₁₆	44.9	³⁄₁₆	42.1	³⁄₁₆
	⁵⁄₁₆	36.3	¼	49.1	¼	52.7	¼
	³⁄₈	36.3	⁵⁄₁₆	49.1	⁵⁄₁₆	63.2	⁵⁄₁₆
	⁷⁄₁₆	36.3	³⁄₈	49.1	³⁄₈	64.4	³⁄₈
	½	—	—	49.1	³⁄₈	64.4	³⁄₈
	⁹⁄₁₆	—	—	—	—	64.4	⁷⁄₁₆
A490-N	¼	47.6	³⁄₁₆	44.9	³⁄₁₆	42.1	³⁄₁₆
	⁵⁄₁₆	48.4	¼	56.1	¼	52.7	¼
	³⁄₈	48.4	⁵⁄₁₆	65.5	⁵⁄₁₆	63.2	⁵⁄₁₆
	⁷⁄₁₆	48.4	³⁄₈	65.5	³⁄₈	73.7	³⁄₈
	½	—	—	65.5	³⁄₈	84.3	³⁄₈
	⁹⁄₁₆	—	—	—	—	85.8	⁷⁄₁₆

SINGLE-PLATE SHEAR CONNECTIONS

TABLE X-E Allowable loads in kips

n = 6   L = 18

	Plate Thickness, *t* In.	Bolt Size, In.					
		3/4		7/8		1	
		Load	Weld	Load	Weld	Load	Weld
A325-N	1/4	46.3	3/16	53.8	3/16	50.6	3/16
	5/16	46.3	1/4	62.7	1/4	63.2	1/4
	3/8	46.3	5/16	62.7	5/16	75.9	5/16
	7/16	46.3	3/8	62.7	3/8	82.2	3/8
	1/2	—	—	62.7	3/8	82.2	3/8
	9/16	—	—	—	—	82.2	7/16
A490-N	1/4	57.1	3/16	53.8	3/16	50.6	3/16
	5/16	61.8	1/4	67.3	1/4	63.2	1/4
	3/8	61.8	5/16	80.7	5/16	75.9	5/16
	7/16	61.8	3/8	83.7	3/8	88.5	3/8
	1/2	—	—	83.7	3/8	101	3/8
	9/16	—	—	—	—	110	7/16

TABLE X-F Allowable loads in kips

n = 7   L = 21

	Plate Thickness, *t* In.	Bolt Size, In.					
		3/4		7/8		1	
		Load	Weld	Load	Weld	Load	Weld
A325-N	1/4	56.4	3/16	62.8	3/16	59.0	3/16
	5/16	56.4	1/4	76.4	1/4	73.7	1/4
	3/8	56.4	5/16	76.4	5/16	88.5	5/16
	7/16	56.4	3/8	76.4	3/8	100	3/8
	1/2	—	—	76.4	3/8	100	3/8
	9/16	—	—	—	—	100	7/16
A490-N	1/4	66.6	3/16	62.8	3/16	59.0	3/16
	5/16	75.1	1/4	78.5	1/4	73.7	1/4
	3/8	75.1	5/16	94.2	5/16	88.5	5/16
	7/16	75.1	3/8	102	3/8	103	3/8
	1/2	—	—	102	3/8	118	3/8
	9/16	—	—	—	—	133	7/16

ECCENTRIC LOADS ON FASTENER GROUPS
TABLES XI–XVIII

ULTIMATE STRENGTH METHOD*

When fastener groups are loaded in shear by an external load that does not act through the center of gravity of the group, the load is eccentric and will tend to cause a relative rotation and translation of the connected material. This condition is equivalent to that of pure rotation about a single point. This point is called the *instantaneous center of rotation.*

The individual resistance force of each fastener can then be assumed to act on a line perpendicular to a ray passing through the instantaneous center and that fastener's location (see Fig. 1).

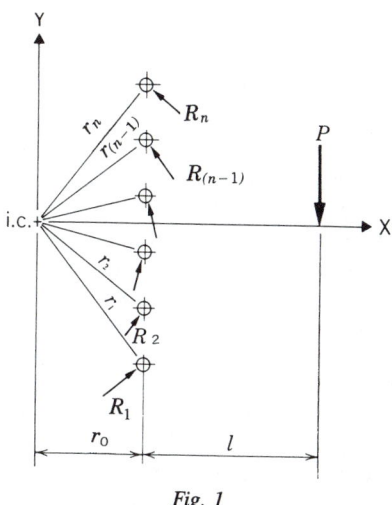

Fig. 1

The ultimate shear strength of fastener groups can be obtained from the load deformation relationship of a single fastener, which is expressed as:

$$R = R_{ult}(1 - e^{-10\Delta})^{0.55}$$

where

R = Shear force in a single fastener at any given deformation (see Fig. 2)
R_{ult} = Ultimate shear load of a single fastener
Δ = Total deformation of a fastener, including shearing, bending, and bearing deformation, plus local bearing deformation of the plate
e = Base of natural logarithm ≈ 2.718

By applying a maximum deformation Δ_{max} to the fastener (or fasteners) most remote from the instantaneous center, the maximum shear force for that fastener can be computed. For fasteners in less remote locations, deformations are computed to vary linearly from the instantaneous center and shear forces can be obtained from the above relationship.

---

* *Crawford, S. F. and G. L. Kulak Eccentrically Loaded Bolted Connections ASCE Journal of the Structural Division, Vol. 97, No. ST3, March 1971, New York (pp. 765-783).*

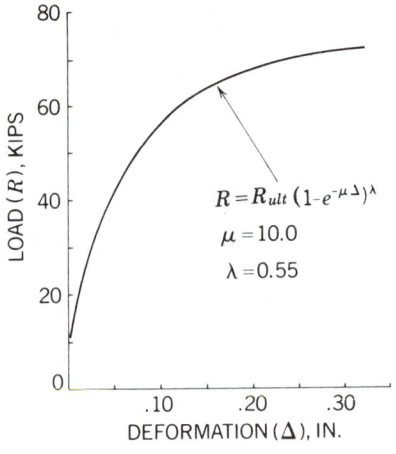

$$R = R_{ult}\left(1 - e^{-\mu \Delta}\right)^{\lambda}$$
$$\mu = 10.0$$
$$\lambda = 0.55$$

Fig. 2

The total resistance force of all fasteners then combine to resist the eccentric external load and if the correct location of the instantaneous center has been selected, the three equations of statics will be satisfied.

Although the development of fastener load-deformation relationships as expressed above is based on connections which may experience slip under load, both load tests and analytical studies* indicate that extending these procedures to slip-resistant connections is conservative.

Tables XI to XVIII for vertical loads have been based on the solution of the instantaneous center problem for each fastener pattern and each eccentric condition. These Manual tables may be easily extended to inclined eccentric loads through Alternate Method 2 described later. The load-deformation relationship data is based upon the following values obtained experimentally for ¾-in. dia. ASTM A325 bolts:

$$R_{ult} = 74 \text{ kips}$$
$$\Delta_{max} = 0.34 \text{ in.}$$

The non-dimensional coefficient C is obtained by dividing the ultimate load P by R_{ult}. The values derived can be safely used with any fastener diameter and are conservative when used with ASTM A490 bolts. In using these tables, margins of safety are provided equivalent to those bolts used in joints less than 50-in. long, subject to shear produced by concentric load only in both slip-critical or bearing-type connections.

For any fastener group listed in these tables, the coefficient C times the allowable value of one fastener equals the total load P located at an eccentric distance from the centroid of the fastener group ($P = C \times r_v$). Thus, by dividing P by the allowable fastener value r_v the minimum coefficient is obtained and a fastener group can be selected for which the coefficient C is of that magnitude or greater.

---

* *Kulak, G. L.* Eccentrically Loaded Slip-resistant Connections *AISC Engineering Journal, Vol. 12, No. 2, 1975, New York.*

ALTERNATE METHOD 1—ELASTIC

Recognition of the elastic method is continued in this edition for unusual fastener groups not conforming to the tables. Each fastener is assumed to support:

1. An equal share of the vertical component of the load
2. An equal share of the horizontal component (if any) of the load
3. A proportional share (depending on the fastener's distance from the centroid of the group) of the eccentric moment portion of the load.

The maximum load is calculated from the vectorial resolution of these stresses at the fastener most remote from the group's centroid. This method, although providing a simplified and conservative approach, does not render a consistent factor of safety and, in some cases, results in excessively conservative designs of connections.

ALTERNATE METHOD 2

Since the Manual tables are intended for eccentric loads that are vertical, they are not applicable directly to eccentric loads that are inclined at an angle Θ from the vertical (see Fig. 3). If the preferred ultimate strength re-analysis is not feasible for this condition, Alternate Method 1 (elastic) could be used.

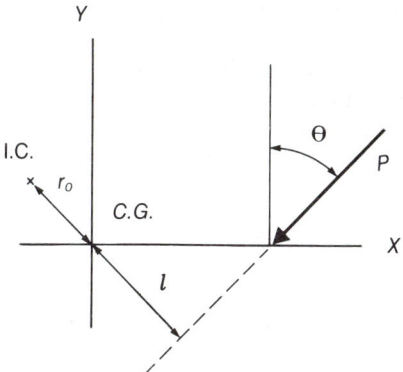

Fig. 3. Inclined Eccentric Load

In addition, a new, easy and conservative method to approximate eccentricity coefficients for inclined loads on connector patterns covered by the Manual tables for vertical loads has been developed* to overcome this limitation. It is based on arithmetic, rather than vectorial, addition of connector strength as an exaggerated load effect or, equivalently, on a linear interaction between eccentric and direct shear. Alternate Method 2 is computationally easy and has some of the redistribution benefits of the ultimate strength method.

---

* *Iwankiw, Nestor R.* Design for Eccentric and Inclined Loads on Bolt and Weld Groups. *AISC Engineering Journal, 4th Qtr., 1987, Chicago, Ill.*

Iwankiw, Nestor R. Addendum/Closure on Design for Eccentric and Inclined Loads on Bolt and Weld Groups. *AISC Engineering Journal, 3rd Qtr., 1988, Chicago, Ill.*

First, define C_{max} as the total number of bolts. Next, let

$$A = \frac{C_{max}}{C_o} \geq 1.0$$

where C_o is the tabulated C for a given vertical load case. For a particular connector pattern and load eccentricity distance, A is a constant relative to the load angle Θ; it serves as the single characteristic input property of the connector geometry.

The approximate eccentricity coefficient C_a for the inclined load is

$$\frac{C_a}{C_o} = \frac{A}{(\sin\Theta + A\cos\Theta)} \geq 1.0$$

Only this one equation is necessary to represent capacity as a function of load angle. The allowable inclined eccentric load is then defined as

$$P = C_a r_v$$

The minimum $\dfrac{C_a}{C_o}$ limit of 1.0 is based on ultimate strength solutions which demonstrate that the worst case occurs when the applied load is vertical.

Figure 4 shows a graph of $\dfrac{C_a}{C_o}$ for specific A values and for load angles from 0 to 90°. It gives a clear overall picture of the C_a sensitivity to both A and Θ. The end values at $\Theta = 0°$ and 90° are 1.0 and A, respectively. Thus, for a vertical load ($\Theta = 0°$), C_a equals C_o while for a horizontal load ($\Theta = 90°$), $C_a = C_{max}$, as expected. The C_a/C_o ordinates and curve slopes increase with higher A values since the curvature is concave upward, such that linear interpolation between Θ can be greatly misleading, especially for larger Θ's and larger A's.

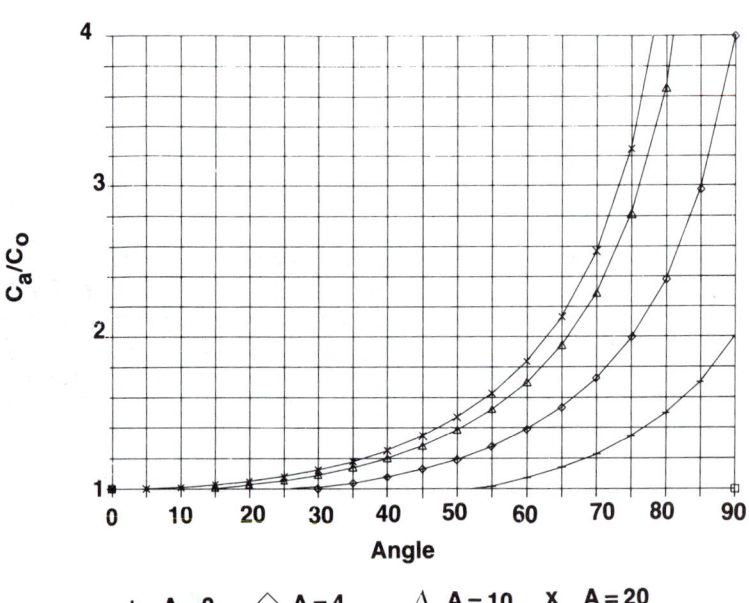

Fig. 4.   Coefficient C_a for inclined eccentric load.

EXAMPLE 24

Given:

Find the maximum load ($\Theta = 0$) that can be supported by the bracket shown in Fig. 5. Column and brackets are $F_y = 36$ ksi. Use ⅞-in. dia. A325-SC bolts in standard holes and assume that the column flange is at least adequately thick so that bolt shear will govern.

$n = 6, \quad b = 3 \text{ in.}, \quad D = 5\frac{1}{2} \text{ in.}, \quad l = 16 \text{ in.}$

Solution:

From Table XIII, with $n = 6$, $b = 3$ in. and $l = 16$ in.:

$C = 3.55$

Using $r_v = 10.2$ kips:

$P = 3.55 \times 10.2 = 36.2$ kips

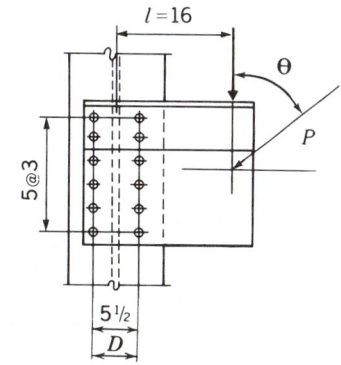

Fig. 5

EXAMPLE 25

Given:

Same conditions as Ex. 24, except that an additional horizontal load is applied as shown so that the resultant, P, with a vertical force forms an angle $\Theta = 60°$. Use Alternate Method 2 and Alternate Method 1.

Solution:

From Table XIII, with $n = 6$, $b = 3$ in. and $l = 16$ in.:

$C_o = C = 3.55$ (same as Ex. 24)

Use Alternate Method 2:

$C_{max} = 2 \times 6 = 12$

$A = \dfrac{12}{3.55} = 3.38$

$\dfrac{C_a}{C_o} = \dfrac{3.38}{(\sin 60° + 3.38\cos 60°)} = 1.322 \geq 1.0. \quad \textbf{o.k.}$

$C_a = 1.322(3.55) = 4.69$

using $r_v = 10.2$ kips

$P = C_a r_v = 4.69 \times 10.2 = 47.8$ kips

Using Alternate Method 1 (elastic), $C_e = 4.1 < 4.69$

ECCENTRIC LOADS ON FASTENER GROUPS
TABLE XI Coefficients C

Required minimum $C = \dfrac{P}{r_v}$

$P = C \times r_v$

n = Total number of fasteners in one vertical row
P = Allowable load acting with lever arm l, in.
r_v = Allowable load on one fastener by Specification
C = Coefficients tabulated below.

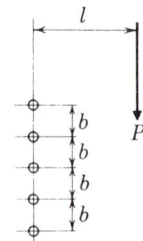

	l					n						
	In.	2	3	4	5	6	7	8	9	10	11	12
	2	1.18	2.23	3.32	4.40	5.45	6.48	7.51	8.52	9.53	10.54	11.54
	3	.88	1.77	2.83	3.91	4.99	6.07	7.13	8.17	9.21	10.24	11.26
	4	.69	1.40	2.38	3.42	4.49	5.58	6.66	7.73	8.79	9.85	10.89
	5	.56	1.15	2.01	2.97	4.01	5.08	6.16	7.24	8.32	9.39	10.46
	6	.48	1.00	1.73	2.59	3.57	4.60	5.66	6.74	7.82	8.90	9.98
	7	.41	.83	1.51	2.28	3.18	4.16	5.19	6.24	7.32	8.40	9.48
	8	.36	.73	1.34	2.04	2.86	3.76	4.75	5.77	6.83	7.89	8.97
	9	.32	.65	1.21	1.83	2.59	3.43	4.35	5.32	6.36	7.41	8.48
b = 3 In.	10	.29	.59	1.09	1.66	2.36	3.14	4.00	4.93	5.91	6.95	8.00
	12	.24	.49	.92	1.40	2.00	2.68	3.44	4.27	5.16	6.10	7.08
	14	.21	.42	.79	1.21	1.74	2.33	3.01	3.75	4.56	5.42	6.32
	16	.18	.37	.70	1.06	1.53	2.06	2.67	3.33	4.06	4.85	5.69
	18	.16	.33	.62	.95	1.37	1.84	2.39	3.00	3.66	4.38	5.15
	20	.15	.30	.56	.85	1.24	1.67	2.17	2.72	3.33	3.99	4.70
	24	.12	.25	.47	.71	1.03	1.40	1.82	2.29	2.81	3.37	3.99
	28	.11	.21	.40	.61	.89	1.20	1.57	1.97	2.42	2.92	3.45
	32	.09	.19	.35	.54	.78	1.05	1.38	1.73	2.13	2.57	3.04
	36	.08	.17	.31	.48	.69	.94	1.22	1.54	1.90	2.29	2.72
	2	1.63	2.72	3.75	4.77	5.77	6.77	7.76	8.75	9.74	10.73	11.72
	3	1.39	2.48	3.56	4.60	5.63	6.65	7.65	8.66	9.66	10.65	11.64
	4	1.18	2.23	3.32	4.40	5.45	6.48	7.51	8.52	9.53	10.54	11.54
	5	1.01	1.99	3.07	4.16	5.23	6.29	7.33	8.36	9.39	10.40	11.41
	6	.88	1.77	2.83	3.91	4.99	6.07	7.13	8.17	9.21	10.24	11.26
	7	.77	1.57	2.59	3.66	4.75	5.83	6.90	7.96	9.01	10.05	11.09
	8	.69	1.40	2.38	3.42	4.49	5.58	6.66	7.73	8.79	9.85	10.89
	9	.62	1.26	2.18	3.19	4.25	5.33	6.41	7.49	8.56	9.63	10.69
b = 6 In.	10	.56	1.15	2.01	2.97	4.01	5.08	6.16	7.24	8.32	9.39	10.46
	12	.48	1.00	1.73	2.59	3.57	4.60	5.66	6.74	7.82	8.90	9.98
	14	.41	.83	1.51	2.28	3.18	4.16	5.19	6.24	7.32	8.40	9.48
	16	.36	.73	1.34	2.04	2.86	3.76	4.75	5.77	6.83	7.89	8.97
	18	.32	.65	1.21	1.83	2.59	3.43	4.35	5.32	6.36	7.41	8.48
	20	.29	.59	1.09	1.66	2.36	3.14	4.00	4.93	5.91	6.95	8.00
	24	.24	.49	.92	1.40	2.00	2.68	3.44	4.27	5.16	6.10	7.08
	28	.21	.42	.79	1.21	1.74	2.33	3.01	3.75	4.56	5.42	6.32
	32	.18	.37	.70	1.06	1.53	2.06	2.67	3.33	4.06	4.85	5.69
	36	.16	.33	.62	.95	1.37	1.84	2.39	3.00	3.66	4.38	5.15

ECCENTRIC LOADS ON FASTENER GROUPS
TABLE XII Coefficients *C*

Required minimum $C = \dfrac{P}{r_v}$

$P = C \times r_v$
n = Total number of fasteners in one vertical row
P = Allowable load acting with lever arm l, in.
r_v = Allowable load on one fastener by Specification
C = Coefficients tabulated below.

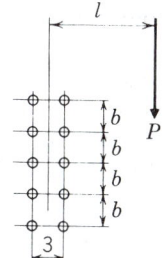

l In.						n						
	1	2	3	4	5	6	7	8	9	10	11	12
b = 3 In.												
2	.84	2.54	4.50	6.62	8.74	10.84	12.92	14.97	17.00	19.03	21.04	23.04
3	.65	2.03	3.68	5.69	7.80	9.94	12.07	14.19	16.29	18.36	20.42	22.47
4	.54	1.67	3.06	4.87	6.87	8.97	11.11	13.26	15.40	17.52	19.64	21.73
5	.45	1.42	2.59	4.21	6.03	8.04	10.14	12.28	14.43	16.58	18.72	20.85
6	.39	1.22	2.25	3.69	5.33	7.20	9.21	11.30	13.44	15.59	17.74	19.90
7	.35	1.08	1.98	3.27	4.75	6.48	8.37	10.38	12.47	14.59	16.74	18.90
8	.31	.96	1.78	2.93	4.27	5.87	7.63	9.55	11.56	13.63	15.75	17.90
9	.28	.86	1.61	2.65	3.87	5.34	6.97	8.76	10.72	12.73	14.80	16.92
10	.26	.79	1.46	2.42	3.53	4.90	6.42	8.10	9.92	11.89	13.90	15.97
12	.22	.67	1.24	2.06	3.01	4.19	5.51	7.01	8.64	10.39	12.24	14.24
14	.19	.58	1.08	1.78	2.62	3.66	4.82	6.15	7.61	9.20	10.90	12.69
16	.17	.51	.95	1.57	2.32	3.24	4.27	5.47	6.79	8.23	9.78	11.43
18	.15	.45	.85	1.41	2.07	2.90	3.83	4.92	6.11	7.43	8.85	10.37
20	.14	.41	.77	1.27	1.88	2.63	3.48	4.46	5.55	6.76	8.07	9.48
24	.12	.34	.65	1.07	1.58	2.21	2.93	3.76	4.69	5.73	6.84	8.06
28	.10	.29	.56	.92	1.36	1.90	2.53	3.25	4.05	4.96	5.93	7.00
32	.09	.26	.49	.80	1.19	1.67	2.22	2.86	3.56	4.36	5.23	6.18
36	.08	.23	.43	.72	1.06	1.49	1.98	2.55	3.18	3.89	4.67	5.52
b = 6 In.												
2	.84	3.25	5.40	7.48	9.51	11.52	13.52	15.51	17.49	19.47	21.45	23.43
3	.65	2.79	4.94	7.08	9.18	11.24	13.27	15.29	17.29	19.29	21.28	23.27
4	.54	2.41	4.45	6.62	8.76	10.87	12.94	14.99	17.02	19.05	21.06	23.06
5	.45	2.10	3.98	6.13	8.29	10.43	12.54	14.63	16.69	18.74	20.78	22.80
6	.39	1.85	3.56	5.64	7.80	9.96	12.10	14.22	16.31	18.39	20.45	22.49
7	.35	1.64	3.19	5.19	7.30	9.46	11.62	13.76	15.88	17.99	20.07	22.14
8	.31	1.47	2.87	4.77	6.82	8.96	11.12	13.28	15.43	17.55	19.66	21.76
9	.28	1.34	2.61	4.40	6.37	8.47	10.62	12.79	14.94	17.09	19.22	21.34
10	.26	1.22	2.39	4.07	5.95	8.00	10.13	12.28	14.45	16.60	18.75	20.88
12	.22	1.04	2.04	3.52	5.21	7.14	9.18	11.29	13.44	15.60	17.77	19.92
14	.19	.90	1.78	3.10	4.61	6.37	8.32	10.36	12.46	14.60	16.76	18.92
16	.17	.80	1.57	2.75	4.12	5.74	7.53	9.50	11.53	13.63	15.76	17.91
18	.15	.71	1.41	2.48	3.72	5.21	6.88	8.70	10.68	12.71	14.80	16.92
20	.14	.64	1.28	2.25	3.38	4.77	6.31	8.02	9.86	11.86	13.89	15.97
24	.12	.54	1.07	1.90	2.86	4.06	5.40	6.91	8.56	10.33	12.20	14.15
28	.10	.46	.93	1.64	2.47	3.52	4.70	6.05	7.52	9.13	10.84	12.65
32	.09	.41	.81	1.44	2.18	3.11	4.16	5.36	6.69	8.15	9.71	11.38
36	.08	.36	.73	1.29	1.94	2.78	3.72	4.81	6.02	7.34	8.78	10.31

ECCENTRIC LOADS ON FASTENER GROUPS
TABLE XIII Coefficients C

Required minimum $C = \dfrac{P}{r_v}$

$P = C \times r_v$

n = Total number of fasteners in one vertical row
P = Allowable load acting with lever arm l, in.
r_v = Allowable load on one fastener by Specification
C = Coefficients tabulated below.

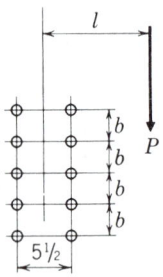

	l In.	n											
		1	2	3	4	5	6	7	8	9	10	11	12
b = 3 In.	2	1.14	2.75	4.63	6.70	8.77	10.84	12.90	14.94	16.98	19.00	21.01	23.02
	3	.94	2.32	3.94	5.84	7.88	9.97	12.07	14.17	16.26	18.33	20.39	22.44
	4	.80	1.99	3.39	5.10	7.01	9.05	11.15	13.27	15.38	17.50	19.60	21.69
	5	.70	1.74	2.97	4.51	6.25	8.18	10.22	12.31	14.44	16.57	18.70	20.82
	6	.62	1.54	2.64	4.03	5.61	7.40	9.34	11.38	13.48	15.60	17.74	19.88
	7	.55	1.38	2.36	3.63	5.07	6.73	8.55	10.51	12.54	14.64	16.76	18.90
	8	.50	1.25	2.14	3.30	4.61	6.15	7.86	9.71	11.67	13.71	15.80	17.92
	9	.46	1.14	1.96	3.01	4.22	5.66	7.24	8.99	10.86	12.83	14.87	16.96
	10	.42	1.04	1.80	2.78	3.89	5.23	6.71	8.35	10.13	12.02	14.00	16.03
	12	.37	.90	1.55	2.39	3.36	4.53	5.82	7.29	8.88	10.61	12.43	14.35
	14	.32	.79	1.37	2.10	2.96	3.99	5.13	6.44	7.87	9.44	11.09	12.85
	16	.29	.70	1.22	1.87	2.64	3.55	4.58	5.76	7.05	8.48	9.99	11.62
	18	.26	.63	1.10	1.68	2.38	3.20	4.14	5.21	6.38	7.68	9.08	10.58
	20	.24	.58	1.00	1.53	2.16	2.91	3.77	4.75	5.82	7.02	8.31	9.69
	24	.20	.49	.84	1.29	1.83	2.46	3.19	4.03	4.94	5.97	7.07	8.28
	28	.18	.42	.73	1.12	1.58	2.13	2.77	3.49	4.29	5.19	6.15	7.21
	32	.16	.38	.64	.98	1.39	1.88	2.44	3.08	3.79	4.58	5.44	6.38
	36	.14	.33	.58	.88	1.24	1.68	2.18	2.75	3.39	4.10	4.87	5.72
b = 6 In.	2	1.14	3.28	5.40	7.46	9.50	11.51	13.51	15.50	17.48	19.46	21.44	23.42
	3	.94	2.86	4.95	7.07	9.16	11.22	13.25	15.27	17.28	19.28	21.27	23.26
	4	.80	2.52	4.49	6.62	8.74	10.84	12.92	14.97	17.01	19.03	21.04	23.05
	5	.70	2.24	4.05	6.14	8.28	10.41	12.52	14.61	16.67	18.72	20.76	22.78
	6	.62	2.00	3.66	5.68	7.80	9.94	12.08	14.19	16.29	18.37	20.43	22.47
	7	.55	1.81	3.32	5.24	7.32	9.46	11.60	13.74	15.86	17.97	20.05	22.12
	8	.50	1.64	3.02	4.85	6.86	8.97	11.11	13.26	15.40	17.53	19.64	21.73
	9	.46	1.50	2.77	4.50	6.42	8.49	10.62	12.77	14.92	17.06	19.19	21.30
	10	.42	1.38	2.56	4.18	6.02	8.03	10.13	12.28	14.43	16.58	18.73	20.86
	12	.37	1.19	2.21	3.66	5.31	7.19	9.20	11.30	13.43	15.59	17.75	19.90
	14	.32	1.05	1.95	3.24	4.72	6.46	8.36	10.38	12.46	14.59	16.75	18.91
	16	.29	.93	1.74	2.90	4.24	5.83	7.59	9.54	11.55	13.63	15.75	17.90
	18	.26	.84	1.57	2.62	3.84	5.31	6.95	8.75	10.71	12.72	14.80	16.92
	20	.24	.76	1.43	2.39	3.50	4.87	6.39	8.09	9.91	11.88	13.90	15.97
	24	.20	.64	1.21	2.02	2.98	4.16	5.49	6.99	8.62	10.37	12.23	14.23
	28	.18	.55	1.05	1.76	2.59	3.63	4.80	6.13	7.59	9.18	10.88	12.68
	32	.16	.49	.93	1.55	2.29	3.21	4.25	5.45	6.77	8.21	9.76	11.42
	36	.14	.44	.83	1.38	2.05	2.88	3.81	4.90	6.09	7.41	8.83	10.36

ECCENTRIC LOADS ON FASTENER GROUPS
TABLE XIV Coefficients *C*

Required minimum $C = \dfrac{P}{r_v}$

$P = C \times r_v$

n = Total number of fasteners in one vertical row
P = Allowable load acting with lever arm l, in.
r_v = Allowable load on one fastener by Specification
C = Coefficients tabulated below.

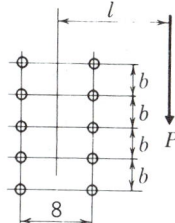

	l In.	n											
		1	2	3	4	5	6	7	8	9	10	11	12
$b = 3$ In.	2	1.31	2.91	4.74	6.85	8.85	10.88	12.91	14.94	16.97	18.99	21.00	23.00
	3	1.16	2.54	4.15	5.99	8.02	10.06	12.12	14.19	16.26	18.32	20.37	22.42
	4	.98	2.24	3.66	5.33	7.20	9.18	11.23	13.32	15.41	17.50	19.59	21.67
	5	.92	2.20	3.27	4.80	6.50	8.37	10.35	12.40	14.49	16.60	18.71	20.81
	6	.79	1.80	2.96	4.35	5.91	7.65	9.53	11.51	13.57	15.66	17.77	19.89
	7	.71	1.63	2.70	3.97	5.40	7.02	8.79	10.69	12.68	14.73	16.82	18.93
	8	.65	1.50	2.46	3.65	4.97	6.48	8.13	9.93	11.84	13.83	15.89	17.98
	9	.60	1.38	2.27	3.37	4.59	6.01	7.56	9.26	11.08	13.00	14.99	17.05
	10	.56	1.28	2.11	3.13	4.27	5.59	7.05	8.65	10.38	12.22	14.15	16.15
	12	.49	1.11	1.84	2.73	3.73	4.90	6.19	7.63	9.18	10.86	12.65	14.52
	14	.44	.99	1.64	2.42	3.32	4.36	5.51	6.80	8.20	9.73	11.37	13.11
	16	.39	.89	1.47	2.17	2.98	3.91	4.95	6.13	7.40	8.80	10.30	11.90
	18	.36	.81	1.33	1.97	2.70	3.55	4.50	5.57	6.73	8.01	9.39	10.87
	20	.33	.74	1.22	1.80	2.47	3.25	4.12	5.10	6.17	7.35	8.62	10.00
	24	.28	.63	1.04	1.54	2.11	2.77	3.51	4.35	5.28	6.30	7.39	8.59
	28	.25	.55	.91	1.34	1.83	2.41	3.06	3.79	4.60	5.50	6.46	7.51
	32	.22	.49	.81	1.18	1.62	2.13	2.71	3.36	4.08	4.87	5.73	6.67
	36	.20	.44	.73	1.06	1.46	1.91	2.43	3.01	3.66	4.37	5.15	5.99
$b = 6$ In.	2	1.31	3.37	5.42	7.46	9.49	11.50	13.50	15.49	17.47	19.46	21.43	23.41
	3	1.16	2.94	4.99	7.08	9.15	11.21	13.24	15.26	17.27	19.27	21.26	23.25
	4	.98	2.63	4.55	6.64	8.74	10.83	12.90	14.95	16.99	19.01	21.03	23.03
	5	.92	2.37	4.15	6.18	8.29	10.41	12.51	14.59	16.66	18.70	20.74	22.77
	6	.79	2.15	3.78	5.74	7.82	9.95	12.07	14.18	16.27	18.35	20.41	22.45
	7	.71	1.97	3.47	5.33	7.36	9.47	11.60	13.73	15.84	17.94	20.03	22.10
	8	.65	1.81	3.19	4.96	6.92	8.99	11.12	13.26	15.39	17.51	19.61	21.71
	9	.60	1.67	2.95	4.63	6.50	8.53	10.64	12.77	14.91	17.05	19.17	21.28
	10	.56	1.55	2.75	4.33	6.11	8.09	10.16	12.28	14.43	16.57	18.71	20.84
	12	.49	1.35	2.41	3.82	5.43	7.27	9.25	11.33	13.44	15.59	17.74	19.89
	14	.44	1.20	2.14	3.41	4.87	6.57	8.44	10.42	12.49	14.60	16.74	18.90
	16	.39	1.08	1.92	3.07	4.40	5.97	7.71	9.60	11.59	13.65	15.76	17.90
	18	.36	1.01	1.75	2.79	4.00	5.46	7.08	8.86	10.77	12.76	14.82	16.93
	20	.33	.89	1.60	2.56	3.67	5.02	6.53	8.21	10.02	11.93	13.93	15.99
	24	.28	.76	1.37	2.19	3.14	4.32	5.63	7.11	8.72	10.45	12.29	14.27
	28	.25	.66	1.19	1.90	2.75	3.78	4.93	6.26	7.70	9.27	10.96	12.74
	32	.22	.59	1.06	1.69	2.44	3.35	4.38	5.58	6.88	8.31	9.85	11.49
	36	.20	.52	.95	1.51	2.19	3.01	3.94	5.02	6.21	7.52	8.93	10.44

ECCENTRIC LOADS ON FASTENER GROUPS
TABLE XV Coefficients C

Required minimum $C = \dfrac{P}{r_v}$

$P = C \times r_v$

n = Total number of fasteners in one vertical row
P = Allowable load acting with lever arm l, in.
r_v = Allowable load on one fastener by Specification
C = Coefficients tabulated below.

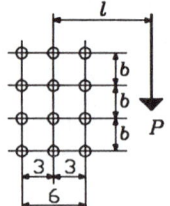

	l In.	n											
		1	2	3	4	5	6	7	8	9	10	11	12
b = 3 In.	2	1.72	4.08	6.89	9.96	13.09	16.21	19.31	22.39	25.44	28.48	31.50	34.51
	3	1.42	3.40	5.79	8.66	11.73	14.89	18.06	21.21	24.35	27.47	30.56	33.63
	4	1.21	2.90	4.97	7.54	10.41	13.49	16.65	19.84	23.03	26.20	29.36	32.51
	5	1.06	2.51	4.35	6.64	9.26	12.16	15.23	18.40	21.59	24.80	28.00	31.19
	6	.92	2.21	3.85	5.91	8.28	10.98	13.91	16.98	20.14	23.34	26.55	29.77
	7	.81	1.96	3.45	5.31	7.46	9.96	12.71	15.66	18.73	21.88	25.07	28.29
	8	.72	1.77	3.11	4.80	6.78	9.09	11.65	14.45	17.40	20.47	23.62	26.81
	9	.64	1.60	2.83	4.38	6.20	8.34	10.73	13.37	16.19	19.15	22.22	25.36
	10	.58	1.46	2.59	4.02	5.71	7.70	9.92	12.40	15.08	17.93	20.90	23.97
	12	.49	1.24	2.21	3.44	4.91	6.65	8.59	10.80	13.20	15.79	18.55	21.43
	14	.42	1.08	1.92	3.01	4.30	5.83	7.57	9.53	11.68	14.01	16.51	19.17
	16	.37	.95	1.70	2.66	3.82	5.19	6.75	8.51	10.45	12.58	14.87	17.32
	18	.33	.85	1.52	2.39	3.43	4.67	6.08	7.68	9.44	11.40	13.50	15.75
	20	.29	.77	1.37	2.16	3.11	4.24	5.53	6.99	8.61	10.40	12.34	14.43
	24	.25	.65	1.15	1.82	2.62	3.57	4.67	5.92	7.30	8.84	10.50	12.31
	28	.21	.56	.99	1.57	2.26	3.08	4.03	5.12	6.33	7.67	9.13	10.71
	32	.18	.49	.87	1.38	1.98	2.71	3.55	4.51	5.58	6.77	8.06	9.47
	36	.16	.43	.77	1.23	1.77	2.42	3.17	4.03	4.98	6.05	7.21	8.48
b = 6 In.	2	1.72	4.88	8.07	11.18	14.23	17.26	20.26	23.24	26.22	29.19	32.16	35.12
	3	1.42	4.24	7.39	10.59	13.73	16.81	19.87	22.90	25.91	28.91	31.90	34.89
	4	1.21	3.72	6.69	9.90	13.10	16.25	19.37	22.44	25.50	28.54	31.56	34.57
	5	1.06	3.29	6.02	9.18	12.40	15.60	18.76	21.90	25.00	28.07	31.13	34.17
	6	.92	2.93	5.43	8.47	11.67	14.89	18.10	21.27	24.42	27.54	30.63	33.70
	7	.81	2.63	4.91	7.81	10.94	14.16	17.38	20.59	23.78	26.94	30.06	33.17
	8	.72	2.38	4.47	7.22	10.24	13.42	16.65	19.87	23.09	26.28	29.45	32.58
	9	.64	2.17	4.09	6.68	9.58	12.70	15.90	19.13	22.37	25.58	28.77	31.95
	10	.58	2.00	3.78	6.21	8.97	12.01	15.17	18.39	21.62	24.86	28.07	31.27
	12	.49	1.71	3.27	5.42	7.91	10.74	13.77	16.92	20.13	23.36	26.60	29.83
	14	.42	1.49	2.87	4.78	7.03	9.64	12.50	15.53	18.67	21.87	25.10	28.34
	16	.37	1.32	2.55	4.28	6.29	8.69	11.34	14.26	17.29	20.42	23.61	26.83
	18	.33	1.19	2.30	3.86	5.70	7.91	10.38	13.08	16.03	19.06	22.17	25.35
	20	.29	1.08	2.09	3.51	5.20	7.25	9.54	12.09	14.82	17.79	20.82	23.93
	24	.25	.91	1.76	2.97	4.42	6.19	8.18	10.44	12.89	15.52	18.31	21.32
	28	.21	.78	1.52	2.57	3.84	5.39	7.14	9.15	11.35	13.73	16.29	18.99
	32	.18	.69	1.33	2.27	3.39	4.76	6.33	8.13	10.11	12.28	14.61	17.10
	36	.16	.61	1.19	2.03	3.03	4.27	5.68	7.31	9.10	11.08	13.22	15.51

ECCENTRIC LOADS ON FASTENER GROUPS
TABLE XVI Coefficients *C*

Required minimum $C = \dfrac{P}{r_v}$

$P = C \times r_v$

n = Total number of fasteners in one vertical row
P = Allowable load acting with lever arm l, in.
r_v = Allowable load on one fastener by Specification
C = Coefficients tabulated below.

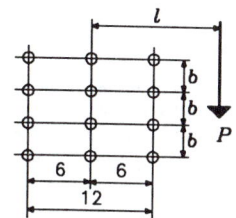

	l In.						n						
		1	2	3	4	5	6	7	8	9	10	11	12
b = 3 in.	2	2.29	4.62	7.24	10.41	13.36	16.35	19.37	22.39	25.41	28.44	31.45	34.45
	3	1.92	4.06	6.43	9.11	12.16	15.15	18.20	21.27	24.35	27.42	30.49	33.55
	4	1.72	3.65	5.80	8.24	10.92	13.89	16.90	19.98	23.09	26.20	29.32	32.43
	5	1.55	3.31	5.27	7.51	9.99	12.72	15.63	18.65	21.74	24.86	28.00	31.15
	6	1.42	3.02	4.82	6.88	9.16	11.71	14.46	17.37	20.39	23.48	26.62	29.77
	7	1.31	2.78	4.44	6.34	8.46	10.83	13.41	16.18	19.10	22.13	25.23	28.37
	8	1.21	2.56	4.10	5.88	7.85	10.07	12.48	15.11	17.91	20.84	23.87	26.96
	9	1.13	2.38	3.81	5.47	7.32	9.39	11.66	14.15	16.81	19.63	22.57	25.60
	10	1.06	2.22	3.56	5.10	6.84	8.79	10.93	13.29	15.82	18.52	21.35	24.30
	12	.92	1.95	3.12	4.48	6.04	7.78	9.70	11.81	14.09	16.56	19.18	21.94
	14	.81	1.72	2.78	4.00	5.39	6.95	8.69	10.61	12.68	14.93	17.33	19.89
	16	.72	1.54	2.49	3.59	4.85	6.27	7.85	9.60	11.50	13.56	15.77	18.13
	18	.64	1.39	2.25	3.25	4.40	5.70	7.15	8.75	10.50	12.41	14.44	16.63
	20	.58	1.26	2.05	2.97	4.02	5.22	6.56	8.03	9.65	11.42	13.31	15.34
	24	.49	1.06	1.74	2.52	3.43	4.45	5.60	6.88	8.29	9.82	11.47	13.24
	28	.42	.92	1.50	2.19	2.98	3.87	4.88	6.00	7.24	8.59	10.05	11.62
	32	.37	.81	1.32	1.93	2.63	3.42	4.32	5.32	6.42	7.62	8.93	10.33
	36	.33	.72	1.18	1.72	2.35	3.06	3.87	4.77	5.76	6.84	8.02	9.29
b = 6 in.	2	2.29	5.15	8.15	11.18	14.21	17.22	20.22	23.21	26.19	29.16	32.13	35.09
	3	1.92	4.48	7.53	10.61	13.70	16.77	19.82	22.85	25.86	28.87	31.86	34.84
	4	1.72	4.08	6.89	9.96	13.09	16.21	19.31	22.39	25.44	28.48	31.50	34.51
	5	1.55	3.71	6.31	9.29	12.42	15.57	18.71	21.83	24.93	28.01	31.06	34.11
	6	1.42	3.40	5.79	8.66	11.73	14.89	18.06	21.21	24.35	27.47	30.56	33.63
	7	1.31	3.14	5.35	8.07	11.05	14.18	17.36	20.54	23.71	26.86	29.99	33.09
	8	1.21	2.90	4.97	7.54	10.41	13.49	16.65	19.84	23.03	26.20	29.36	32.51
	9	1.13	2.69	4.64	7.07	9.81	12.81	15.93	19.12	22.32	25.51	28.70	31.86
	10	1.06	2.51	4.35	6.64	9.26	12.16	15.23	18.40	21.59	24.80	28.00	31.19
	12	.92	2.21	3.85	5.91	8.28	10.98	13.91	16.98	20.14	23.34	26.55	29.77
	14	.81	1.96	3.45	5.31	7.46	9.96	12.71	15.66	18.73	21.88	25.07	28.29
	16	.72	1.77	3.11	4.80	6.78	9.09	11.65	14.45	17.40	20.47	23.62	26.81
	18	.64	1.60	2.83	4.38	6.20	8.34	10.73	13.37	16.19	19.15	22.22	25.36
	20	.58	1.46	2.59	4.02	5.71	7.70	9.92	12.40	15.08	17.93	20.90	23.97
	24	.49	1.24	2.21	3.44	4.91	6.65	8.59	10.80	13.20	15.79	18.55	21.43
	28	.42	1.08	1.92	3.01	4.30	5.83	7.57	9.53	11.68	14.01	16.51	19.17
	32	.37	.95	1.70	2.66	3.82	5.19	6.75	8.51	10.45	12.58	14.87	17.32
	36	.33	.85	1.52	2.39	3.43	4.67	6.08	7.68	9.44	11.40	13.50	15.75

ECCENTRIC LOADS ON FASTENER GROUPS
TABLE XVII Coefficients C

Required minimum $C = \dfrac{P}{r_v}$

$P = C \times r_v$

n = Total number of fasteners in one vertical row
P = Allowable load acting with lever arm l, in.
r_v = Allowable load on one fastener by Specification
C = Coefficients tabulated below.

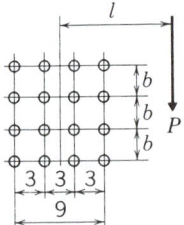

	l In.	n											
		1	2	3	4	5	6	7	8	9	10	11	12
$b = 3$ In.	2	2.60	5.71	9.37	13.42	17.50	21.61	25.71	29.79	33.86	37.91	41.94	45.95
	3	2.23	4.92	8.05	11.77	15.75	19.88	24.05	28.23	32.39	36.53	40.66	44.76
	4	1.94	4.30	7.09	10.39	14.09	18.09	22.23	26.43	30.65	34.86	39.06	43.24
	5	1.69	3.79	6.31	9.29	12.66	16.42	20.42	24.56	28.77	33.01	37.26	41.49
	6	1.49	3.37	5.65	8.37	11.45	14.94	18.73	22.74	26.88	31.10	35.35	39.61
	7	1.32	3.03	5.10	7.59	10.43	13.67	17.22	21.05	25.06	29.20	33.42	37.67
	8	1.19	2.74	4.64	6.93	9.56	12.57	15.88	19.51	23.36	27.38	31.52	35.73
	9	1.07	2.50	4.24	6.36	8.81	11.61	14.70	18.13	21.80	25.68	29.71	33.84
	10	1.00	2.29	3.90	5.86	8.15	10.77	13.67	16.90	20.39	24.10	28.00	32.03
	12	.83	1.96	3.34	5.06	7.06	9.37	11.96	14.83	17.96	21.35	24.95	28.72
	14	.73	1.72	2.92	4.44	6.22	8.27	10.59	13.17	15.99	19.07	22.37	25.87
	16	.65	1.52	2.59	3.95	5.54	7.39	9.48	11.82	14.39	17.19	20.20	23.43
	18	.58	1.37	2.33	3.56	4.99	6.67	8.57	10.70	13.05	15.62	18.38	21.37
	20	.53	1.24	2.12	3.23	4.53	6.07	7.81	9.77	11.93	14.30	16.85	19.61
	24	.45	1.05	1.79	2.73	3.83	5.14	6.62	8.30	10.15	12.19	14.41	16.80
	28	.39	.90	1.55	2.36	3.31	4.45	5.73	7.20	8.82	10.61	12.55	14.66
	32	.34	.79	1.36	2.07	2.92	3.92	5.05	6.35	7.79	9.38	11.11	12.98
	36	.31	.71	1.21	1.85	2.60	3.50	4.51	5.68	6.96	8.39	9.95	11.64
$b = 6$ In.	2	2.60	6.59	10.75	14.88	18.95	22.97	26.98	30.96	34.94	38.90	42.86	46.81
	3	2.23	5.77	9.87	14.09	18.26	22.38	26.45	30.49	34.51	38.51	42.50	46.48
	4	1.94	5.12	8.96	13.18	17.42	21.62	25.77	29.88	33.95	38.00	42.03	46.05
	5	1.69	4.58	8.13	12.24	16.49	20.75	24.96	29.14	33.28	37.38	41.46	45.51
	6	1.49	4.13	7.38	11.34	15.54	19.81	24.07	28.30	32.50	36.66	40.78	44.88
	7	1.32	3.74	6.74	10.50	14.59	18.84	23.13	27.40	31.64	35.85	40.03	44.17
	8	1.19	3.42	6.20	9.74	13.69	17.88	22.15	26.45	30.72	34.97	39.19	43.38
	9	1.07	3.14	5.73	9.06	12.84	16.94	21.18	25.47	29.76	34.04	38.30	42.53
	10	1.00	2.89	5.32	8.45	12.06	16.04	20.21	24.48	28.78	33.08	37.36	41.62
	12	.83	2.50	4.63	7.43	10.68	14.39	18.38	22.54	26.80	31.10	35.41	39.71
	14	.73	2.20	4.09	6.60	9.53	12.97	16.72	20.72	24.88	29.12	33.41	37.73
	16	.65	1.95	3.65	5.93	8.59	11.76	15.26	19.06	23.07	27.21	31.44	35.73
	18	.58	1.76	3.29	5.37	7.81	10.73	13.99	17.58	21.41	25.41	29.55	33.77
	20	.53	1.60	2.99	4.90	7.15	9.85	12.86	16.21	19.83	23.75	27.76	31.88
	24	.45	1.35	2.53	4.16	6.10	8.44	11.07	14.05	17.28	20.77	24.46	28.44
	28	.39	1.17	2.19	3.61	5.31	7.37	9.69	12.34	15.25	18.41	21.79	25.37
	32	.34	1.03	1.93	3.19	4.69	6.53	8.61	10.99	13.60	16.48	19.57	22.87
	36	.31	.92	1.72	2.85	4.20	5.85	7.73	9.89	12.26	14.89	17.72	20.76

ECCENTRIC LOADS ON FASTENER GROUPS
TABLE XVIII Coefficients C

Required minimum $C = \dfrac{P}{r_v}$

$P = C \times r_v$
n = Total number of fasteners in one vertical row
P = Allowable load acting with lever arm l, in.
r_v = Allowable load on one fastener by Specification
C = Coefficients tabulated below.

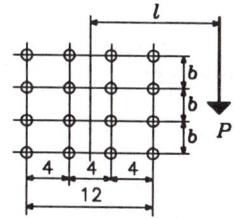

l In.	\multicolumn{12}{c}{n}
	1	2	3	4	5	6	7	8	9	10	11	12
b = 3 In.												
2	2.82	5.99	9.58	13.73	17.69	21.71	25.75	29.80	33.84	37.88	41.90	45.91
3	2.50	5.31	8.46	12.05	16.04	20.06	24.15	28.27	32.40	36.51	40.61	44.71
4	2.23	4.74	7.57	10.81	14.46	18.35	22.39	26.52	30.69	34.86	39.03	43.19
5	2.01	4.27	6.86	9.82	13.12	16.77	20.66	24.72	28.86	33.05	37.26	41.46
6	1.81	3.87	6.24	8.96	11.99	15.39	19.07	22.98	27.04	31.19	35.39	39.61
7	1.64	3.52	5.71	8.22	11.05	14.21	17.66	21.38	25.30	29.36	33.51	37.71
8	1.49	3.22	5.24	7.58	10.22	13.17	16.40	19.92	23.68	27.61	31.67	35.83
9	1.36	2.96	4.83	7.01	9.48	12.25	15.29	18.62	22.19	25.98	29.92	33.99
10	1.25	2.74	4.48	6.51	8.83	11.44	14.31	17.45	20.84	24.47	28.28	32.24
12	1.07	2.37	3.89	5.68	7.74	10.06	12.64	15.47	18.52	21.83	25.34	29.04
14	.95	2.08	3.42	5.03	6.86	8.95	11.28	13.84	16.62	19.63	22.85	26.28
16	.83	1.86	3.05	4.50	6.15	8.04	10.15	12.49	15.04	17.80	20.75	23.92
18	.75	1.67	2.75	4.06	5.56	7.29	9.22	11.36	13.70	16.25	18.98	21.90
20	.68	1.52	2.50	3.70	5.07	6.66	8.43	10.40	12.57	14.92	17.46	20.18
24	.58	1.29	2.12	3.14	4.30	5.66	7.18	8.88	10.75	12.79	15.00	17.38
28	.50	1.12	1.85	2.72	3.74	4.92	6.24	7.73	9.37	11.17	13.12	15.22
32	.45	1.00	1.63	2.40	3.30	4.34	5.51	6.84	8.29	9.90	11.64	13.51
36	.40	.88	1.46	2.15	2.95	3.89	4.94	6.13	7.43	8.88	10.40	12.14
b = 6 In.												
2	2.82	6.77	10.81	14.88	18.93	22.95	26.95	30.94	34.91	38.88	42.84	46.79
3	2.50	5.92	9.96	14.10	18.24	22.35	26.42	30.46	34.48	38.48	42.48	46.46
4	2.23	5.34	9.10	13.22	17.41	21.59	25.73	29.84	33.91	37.97	42.00	46.01
5	2.01	4.84	8.31	12.32	16.51	20.73	24.93	29.10	33.23	37.34	41.41	45.47
6	1.81	4.42	7.61	11.45	15.58	19.81	24.05	28.26	32.45	36.61	40.73	44.83
7	1.64	4.05	7.02	10.66	14.67	18.86	23.11	27.36	31.59	35.79	39.98	44.12
8	1.49	3.73	6.51	9.94	13.80	17.92	22.15	26.42	30.68	34.92	39.13	43.33
9	1.36	3.45	6.06	9.30	12.98	17.01	21.19	25.45	29.73	34.00	38.25	42.47
10	1.25	3.21	5.66	8.73	12.23	16.13	20.25	24.48	28.76	33.04	37.32	41.57
12	1.07	2.80	4.98	7.74	10.91	14.54	18.47	22.58	26.80	31.08	35.37	39.67
14	.95	2.48	4.43	6.92	9.81	13.17	16.86	20.80	24.91	29.12	33.39	37.69
16	.83	2.22	3.98	6.25	8.90	12.00	15.43	19.18	23.14	27.24	31.45	35.71
18	.75	2.01	3.60	5.68	8.13	10.99	14.19	17.72	21.50	25.47	29.57	33.77
20	.68	1.83	3.29	5.21	7.47	10.13	13.11	16.43	20.02	23.83	27.81	31.91
24	.58	1.55	2.79	4.45	6.40	8.72	11.33	14.28	17.46	20.91	24.65	28.50
28	.50	1.35	2.42	3.87	5.59	7.64	9.96	12.59	15.45	18.58	21.93	25.48
32	.45	1.19	2.14	3.43	4.95	6.79	8.87	11.23	13.83	16.67	19.73	23.00
36	.40	1.06	1.92	3.07	4.44	6.10	7.98	10.12	12.48	15.09	17.90	20.92

ECCENTRIC LOADS ON WELD GROUPS
TABLES XIX–XXVI

ULTIMATE STRENGTH METHOD*

When weld groups are loaded in shear by an external load that does not act through the center of gravity of the group, the load is eccentric and will tend to cause a relative rotation and translation between the parts connected by the weld. The point about which rotation tends to take a place is called the *instantaneous center of rotation*. Its location is dependent upon the eccentricity, geometry of the weld group, and deformation of the weld at different angles of the resultant elemental force relative to the weld axis.

The individual resistance force of each unit weld element can then be assumed to act on a line perpendicular to a ray passing through the instantaneous center and that element's location (see Fig. 1).

The ultimate shear strength of weld groups can be obtained from the load deformation relationship of a single unit weld element which is expressed as:

$$R = R_{ult}(1 - e^{-\mu\Delta})^\lambda$$

where

R = Shear force in a single element at any given deformation
R_{ult} = Ultimate shear load of a single element
μ, λ = Regression coefficients
Δ = Deformation of a weld element
e = Base of natural logarithm ≈ 2.718

Unlike the load-deformation relationship for bolts, strength and deformation performance in welds are dependent on the angle Θ that the resultant elemental force makes with the axis of the weld element (see Fig. 1).

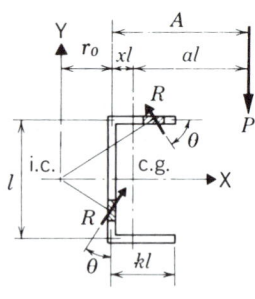

Fig.1

The critical weld element is usually (but not always) the weld element furthest from the instantaneous center. The critical deformation can be calculated as:

$$\Delta_{max} = 0.225 (\Theta + 5)^{-0.47}, \text{ where } \Theta \text{ is expressed in degrees}$$

* *Butler, Pal and Kulak* Eccentrically Loaded Weld Connections. *ASCE Journal of the Structural Division, Vol. 98, No. ST5, May 1972 (pp. 989–1005).*

The deformation of other weld elements can then be calculated as:

$$\Delta = \frac{r}{r_{max}} \Delta_{max}$$

The values of R_{ult}, μ and Δ depend on the value of the angle Θ and can be obtained from the following relations:

$$R_{ult} = \frac{10 + \Theta}{0.92 + 0.0603\Theta}$$

$$\mu = 75\,e^{0.0114\,\Theta}$$

$$\lambda = 0.4\,e^{0.0146\,\Theta}$$

The total resistance of all the weld elements combine to resist the eccentric ultimate load, and if the correct location of the instantaneous center has been selected, the three equations of statics will be satisfied. General performance curves for values of $\Theta = 0°$, $10°$, $30°$ and $90°$ are shown in Fig. 2.

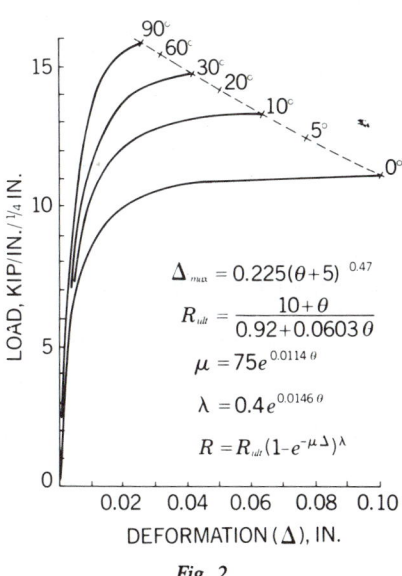

Fig. 2

TABLES XIX-XXVI

To obtain reliable coefficients based on ultimate strength analysis that would replace the traditional elastic C-value in eccentric load design tables, several intermediate steps are required. These include proper correlation factors applied to research data, the application of an acceptable factor of safety, and the use of upper bound limits at points of critical stress in the group to prevent overstress in the weld metal.

Tests were performed on eccentrically loaded ¼-in. fillet weld groups made with E60XX electrodes. To obtain C-tables for E70XX electrode series and a base weld size of ¹⁄₁₆ in., the ultimate capacities were adjusted by the factor of ¼ × 70/60. The resulting value was then converted to an allowable stress by multiplying it by 0.30.

Additionally, this value was reduced by a factor that would prevent the stress in any element of weld to exceed the allowable stress for fillet weld metal as required in AISC ASD Specification Sect. J2.4. Tests have demonstrated that the fusion face of weld metal and base material is not critical in determining weld strength. The Manual tables are, therefore, valid for weld metal with a strength level that matches the base material.

To obtain the capacity of a weld group carrying an eccentric load:

$$P = CC_1Dl$$

where

P = Allowable load, kips
C = Tabular value
C_1 = Coefficient for electrode used (see table below)
   = 1.0 for E70XX electrodes
l = Length of vertical weld, in.
D = Number of sixteenths of an inch, weld size

Electrode	E60	E70	E80	E90	E100	E110
F_v (ksi)	18.0	21.0	24.0	27.0	30.0	33.0
C_1	0.857	1.0	1.14	1.29	1.43	1.57

Tables XIX through XXVI are based on welds made with E70XX electrodes and matching base metal (AWS Table 4.1.1). They also recognize that for equal leg fillet welds the area of the fusion surface is always larger than the leg dimension times the weld length; therefore, the values are based upon the strength through the throat of the weld ($0.3 \times F_u \times 0.707 \times \frac{1}{16}$). When electrodes other than E70XX are used with matching or stronger base metals, multiply by C_1 values tabulated in the table above.

These Manual tables may be easily extended to inclined eccentric loads through Alternate Method 2 described later.

ALTERNATE METHOD 1—ELASTIC

In addition to the ultimate strength method previously described, the elastic method may be used to design/analyze eccentrically loaded weld groups not conforming to the AISC Manual tables. By assuming each weld element as a line coincident with the edge of a fillet weld, each unit element is assumed to support:

1. An equal share of the vertical component of the load.
2. An equal share of the horizonal component of the load.
3. A proportional share (dependent on the element's distance from the centroid of the group) of the eccentric moment portion of the load.

The maximum load is determined from the vectorial resolution of these stresses at the element most remote from the group's centroid. This elastic method, although providing a simplified and conservative approach, does not render a consistent factor of safety and in some cases results in excessively conservative designs of connections.

ALTERNATE METHOD 2

As discussed in the eccentrically loaded bolt section, this new method permits extension of the published Manual weld tables to eccentric loads that are inclined at an angle Θ from the vertical. It is based on arithmetic, rather than vectorial, addition of connector strength as an exaggerated load effect or, equivalently, on a linear interaction between eccentric and direct shear.*

First, define C_{max} as the maximum concentric weld coefficient (e.g. 0.928 (1 + 2k) for C-welds).

Next, let

$$A = \frac{C_{max}}{C_o} \geq 1.0$$

where C_o is the AISC Manual tabulated C for a given vertical load case. For a particular connector pattern and load eccentricity distance, A is a constant relative to the load angle Θ; it serves as the single characteristic input property of the connector geometry.

The approximate eccentricity coefficient C_a for the inclined load is

$$\frac{C_a}{C_o} = \frac{A}{(\sin\Theta + A\cos\Theta)} \geq 1.0$$

Only this one equation (the same as for bolts) is necessary to represent capacity as a function of load angle. The allowable load P is determined as previously described except that the computed C_a replaces the tabulated C value. For more guidance and a graphical design aid on application of this approximate method, see bolt discussion on Alternate Method 2.

EXAMPLE 26

Given:

C-Weld group as shown with $l = 10$ in., $kl = 5$ in. and $xl + al = 10$ in. Find the maximum allowable load P ($\Theta = 0$) for a ⅜-in. weld using E70XX electrodes by using Table XXIII.

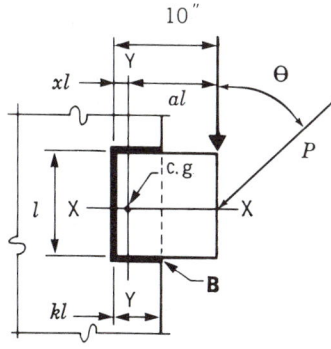

---

* Iwankiw, Nestor R.   Design for Eccentric and Inclined Loads on Bolt and Weld Groups   *AISC Engineering Journal, 4th Qtr., 1987, Chicago, Ill.*

** Iwankiw, Nestor R.   Addendum/Closure on Design for Eccentric and Inclined Loads on Bolt and Weld Groups   *AISC Engineering Journal, 3rd Qtr., 1988, Chicago, Ill.*

Solution:

$$k = \frac{kl}{l} = \frac{5}{10} = 0.5;$$

Enter Table XXIII: For $k = 0.5$, $x = 0.125$

$xl = 0.125 \times 10 = 1.25$ in.
$al = 10 - xl = 8.75$ in.; $a = 0.875$

Interpolating between $a = 0.8$ and $a = 0.9$ for $k = 0.5$:

$C = 0.704$
$D = 6$ (⅜-in. weld)
$C_1 = 1.0$ for E70XX electrodes (see table above)
$P = C_1CDl = 1.0 \times 0.704 \times 6 \times 10 = 42.2$ kips

EXAMPLE 27

Given:

C-Weld as shown in Ex. 26, with $l = 10$ in., $kl = 5$ in., and $al = 0.875$, except eccentric service load $P = 90$ kips at 75.° Determine minimum required weld size.

Use Alternate Method 2:

Solution:

From Table XXIII (as in Ex. 26)

$C_o = C = 0.704$

$C_{max} = 0.928\,(1 + 2(.5)) = 1.856$

$A = \dfrac{1.856}{0.704} = 2.64$

$\dfrac{C_a}{C_o} = \dfrac{2.64}{[.966 + 2.64\,(.259)]} = 1.6 \geq 1.0,$ **o.k.**

$C_a = 1.6\,(0.704) = 1.13$

$D = \dfrac{90}{(1.13)\,10} = 7.96$, say 8, use ½-in weld

Using Alternate Method 1 (elastic), $C_e = 1.09 < 1.13$

ECCENTRIC LOADS ON WELD GROUPS
TABLE XIX Coefficients C

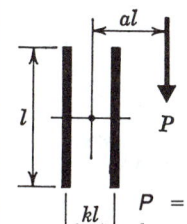

Required Minimum $C = \dfrac{P}{C_1 Dl}$

"   " $D = \dfrac{P}{CC_1 l}$

"   " $l = \dfrac{P}{CC_1 D}$

P = Allowable eccentric load in kips
l = Length of each weld in in.
D = Number of sixteenths of an in. in fillet weld size
C = Coefficients tabulated below
C_1 = Coefficient for electrode used (see Table on p. 4-72).
  = 1.0 for E70XX electrodes.

$P = CC_1 Dl$

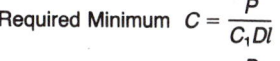

SPECIAL CASE*
(Load not in plane of weld group)
Use C-values given in column headed $k = 0$

a	k															
	0	0.1	0.2	0.3	0.4	0.5	0.6	0.7	0.8	0.9	1.0	1.2	1.4	1.6	1.8	2.0
.06	1.67	1.67	1.68	1.68	1.68	1.69	1.69	1.69	1.69	1.70	1.70	1.70	1.71	1.71	1.71	1.71
.08	1.64	1.65	1.65	1.65	1.66	1.66	1.66	1.66	1.67	1.67	1.67	1.67	1.68	1.68	1.69	1.69
.10	1.61	1.61	1.62	1.62	1.62	1.63	1.63	1.63	1.63	1.64	1.64	1.65	1.65	1.66	1.66	1.67
.15	1.51	1.51	1.52	1.52	1.53	1.53	1.54	1.54	1.55	1.56	1.56	1.57	1.58	1.59	1.60	1.61
.20	1.39	1.39	1.40	1.41	1.42	1.43	1.44	1.45	1.46	1.47	1.48	1.50	1.52	1.53	1.54	1.56
.25	1.26	1.27	1.28	1.30	1.31	1.33	1.35	1.36	1.38	1.39	1.41	1.43	1.45	1.47	1.49	1.50
.30	1.14	1.15	1.17	1.19	1.21	1.24	1.26	1.28	1.30	1.32	1.33	1.36	1.39	1.41	1.43	1.45
.40	.939	.951	.976	1.01	1.04	1.07	1.10	1.13	1.16	1.18	1.20	1.24	1.28	1.31	1.33	1.36
.50	.787	.792	.813	.865	.903	.941	.976	1.01	1.04	1.07	1.09	1.14	1.18	1.21	1.25	1.27
.60	.673	.679	.701	.734	.795	.834	.872	.907	.940	.970	.998	1.05	1.09	1.13	1.17	1.20
.70	.585	.592	.615	.647	.708	.748	.787	.823	.857	.888	.918	.971	1.02	1.06	1.10	1.13
.80	.517	.524	.546	.579	.636	.676	.714	.751	.786	.818	.848	.903	.952	.995	1.03	1.07
.90	.463	.469	.491	.524	.576	.615	.654	.690	.725	.757	.788	.844	.893	.938	.978	1.02
1.00	.419	.425	.446	.478	.527	.565	.602	.638	.672	.704	.735	.791	.842	.887	.928	.965
1.20	.351	.357	.377	.406	.448	.484	.519	.553	.586	.617	.647	.702	.752	.798	.840	.878
1.40	.302	.307	.326	.352	.390	.423	.455	.488	.519	.548	.577	.631	.680	.725	.766	.805
1.60	.265	.270	.287	.311	.344	.375	.405	.435	.465	.493	.520	.572	.619	.664	.704	.743
1.80	.236	.241	.256	.278	.308	.336	.365	.393	.421	.448	.474	.523	.569	.612	.652	.689
2.00	.213	.217	.231	.251	.279	.305	.331	.358	.384	.410	.434	.481	.526	.567	.606	.642
2.20	.193	.198	.211	.229	.254	.279	.303	.328	.353	.377	.401	.446	.488	.528	.566	.602
2.40	.177	.181	.194	.211	.234	.256	.280	.303	.327	.350	.372	.415	.456	.495	.531	.566
2.60	.164	.168	.179	.195	.216	.237	.259	.282	.304	.326	.347	.388	.428	.465	.500	.534
2.80	.152	.156	.166	.181	.201	.221	.242	.263	.284	.305	.325	.365	.402	.438	.472	.505
3.00	.142	.145	.155	.169	.188	.207	.226	.246	.266	.286	.306	.344	.380	.415	.448	.479

*Valid only when the connection material between the welds is solid and does not bend in the plane of the welds.

ECCENTRIC LOADS ON WELD GROUPS
TABLE XX Coefficients C

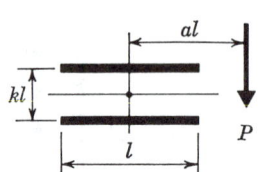

P = Allowable eccentric load in kips
l = Length of each weld in in.
D = Number of sixteenths of an in. in fillet weld size
C = Coefficients tabulated below
C_1 = Coefficient for electrode used (see Table on p. 4-72)
= 1.0 for E70XX electrodes

$$P = CC_1Dl$$

Required Minimum $C = \dfrac{P}{C_1Dl}$

"   " $D = \dfrac{P}{CC_1l}$

"   " $l = \dfrac{P}{CC_1D}$

a	k															
	0	0.1	0.2	0.3	0.4	0.5	0.6	0.7	0.8	0.9	1.0	1.2	1.4	1.6	1.8	2.0
.06	1.56	1.57	1.61	1.66	1.72	1.76	1.79	1.80	1.82	1.83	1.83	1.84	1.84	1.85	1.85	1.85
.08	1.48	1.49	1.54	1.60	1.66	1.71	1.75	1.77	1.79	1.81	1.82	1.83	1.84	1.84	1.85	1.85
.10	1.41	1.43	1.47	1.53	1.60	1.66	1.70	1.74	1.77	1.79	1.80	1.82	1.83	1.84	1.84	1.84
.15	1.26	1.28	1.33	1.39	1.46	1.53	1.59	1.65	1.69	1.72	1.74	1.78	1.80	1.81	1.82	1.83
.20	1.13	1.15	1.20	1.27	1.34	1.42	1.49	1.55	1.60	1.64	1.68	1.73	1.76	1.78	1.80	1.81
.25	1.02	1.04	1.09	1.16	1.23	1.31	1.38	1.45	1.51	1.57	1.61	1.67	1.72	1.75	1.77	1.79
.30	.934	.953	1.00	1.06	1.14	1.21	1.29	1.36	1.43	1.49	1.54	1.62	1.67	1.71	1.74	1.76
.40	.789	.806	.850	.909	.977	1.05	1.12	1.20	1.27	1.34	1.40	1.50	1.57	1.63	1.67	1.70
.50	.680	.695	.734	.789	.853	.920	.989	1.06	1.13	1.20	1.27	1.38	1.47	1.54	1.59	1.63
.60	.595	.608	.644	.695	.753	.816	.881	.947	1.01	1.08	1.15	1.27	1.36	1.44	1.51	1.56
.70	.528	.540	.573	.619	.673	.731	.792	.854	.915	.977	1.04	1.17	1.27	1.36	1.43	1.49
.80	.473	.484	.515	.557	.607	.661	.718	.776	.834	.892	.950	1.07	1.18	1.27	1.35	1.41
.90	.428	.439	.467	.506	.552	.603	.656	.710	.765	.819	.874	.988	1.10	1.19	1.27	1.34
1.00	.391	.401	.426	.463	.506	.553	.603	.654	.705	.757	.808	.913	1.02	1.12	1.20	1.28
1.20	.333	.341	.363	.395	.433	.474	.518	.563	.609	.655	.702	.794	.891	.990	1.08	1.15
1.40	.289	.296	.316	.344	.377	.414	.453	.494	.535	.577	.618	.702	.786	.878	.965	1.04
1.60	.255	.262	.279	.304	.334	.367	.402	.439	.476	.514	.552	.628	.704	.784	.869	.946
1.80	.228	.234	.250	.273	.300	.330	.362	.395	.429	.464	.498	.568	.638	.709	.785	.862
2.00	.207	.212	.226	.247	.272	.299	.328	.359	.390	.422	.454	.518	.582	.647	.715	.788
2.20	.189	.194	.207	.226	.248	.273	.300	.328	.357	.387	.416	.476	.535	.595	.657	.723
2.40	.173	.178	.190	.208	.229	.252	.277	.303	.330	.357	.384	.440	.495	.551	.608	.667
2.60	.161	.165	.176	.192	.212	.233	.257	.281	.306	.331	.357	.409	.461	.513	.566	.620
2.80	.149	.153	.164	.179	.197	.217	.239	.262	.285	.309	.333	.382	.431	.480	.529	.580
3.00	.140	.143	.153	.168	.184	.203	.224	.245	.267	.289	.312	.358	.404	.451	.497	.544

ECCENTRIC LOADS ON WELD GROUPS
TABLE XXI Coefficients C

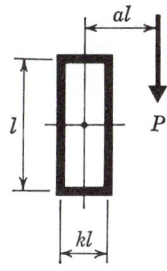

P = Allowable eccentric load in kips
l = Length of longer welds in in.
D = Number of sixteenths of an in. in fillet weld size
C = Coefficients tabulated below
C_1 = Coefficient for electrode used (see Table on p. 4-72)
  = 1.0 for E70XX electrodes.

$$P = CC_1Dl$$

Required Minimum $C = \dfrac{P}{C_1Dl}$

"   " $D = \dfrac{P}{CC_1l}$

"   " $l = \dfrac{P}{CC_1D}$

Note: When load P is perpendicular to longer side l use Table XXII.

a	k										
	0	0.1	0.2	0.3	0.4	0.5	0.6	0.7	0.8	0.9	1.0
.06	1.67	1.61	1.77	1.93	2.10	2.27	2.45	2.62	2.80	2.97	3.15
.08	1.64	1.63	1.79	1.95	2.12	2.28	2.45	2.63	2.80	2.97	3.14
.10	1.61	1.64	1.80	1.96	2.12	2.28	2.45	2.62	2.79	2.95	3.12
.15	1.51	1.63	1.78	1.93	2.09	2.24	2.40	2.56	2.72	2.88	3.04
.20	1.39	1.58	1.72	1.87	2.02	2.17	2.32	2.47	2.62	2.78	2.93
.25	1.26	1.46	1.64	1.78	1.92	2.07	2.21	2.36	2.51	2.66	2.81
.30	1.14	1.33	1.52	1.68	1.82	1.96	2.10	2.25	2.39	2.54	2.69
.40	.939	1.11	1.29	1.47	1.62	1.75	1.89	2.03	2.16	2.30	2.44
.50	.787	.925	1.10	1.26	1.43	1.56	1.69	1.82	1.96	2.09	2.23
.60	.673	.793	.929	1.10	1.26	1.40	1.52	1.65	1.77	1.90	2.03
.70	.585	.691	.813	.971	1.12	1.26	1.38	1.50	1.62	1.74	1.87
.80	.517	.611	.721	.843	1.00	1.14	1.25	1.37	1.48	1.60	1.72
.90	.463	.546	.647	.758	.907	1.04	1.15	1.26	1.37	1.48	1.60
1.00	.419	.494	.586	.690	.827	.951	1.06	1.16	1.27	1.38	1.49
1.20	.351	.414	.493	.584	.702	.811	.915	1.01	1.11	1.20	1.31
1.40	.302	.356	.426	.506	.609	.706	.805	.891	.979	1.07	1.16
1.60	.265	.312	.374	.447	.537	.624	.716	.796	.877	.960	1.05
1.80	.236	.278	.334	.399	.480	.559	.643	.720	.794	.870	.949
2.00	.213	.250	.301	.361	.434	.506	.582	.656	.725	.796	.869
2.20	.193	.228	.274	.329	.395	.462	.532	.603	.667	.733	.801
2.40	.177	.209	.252	.302	.363	.425	.490	.557	.617	.678	.742
2.60	.164	.193	.233	.279	.336	.393	.454	.518	.574	.632	.691
2.80	.152	.180	.216	.260	.312	.366	.422	.482	.536	.591	.647
3.00	.142	.168	.202	.243	.292	.342	.395	.451	.503	.555	.608

ECCENTRIC LOADS ON WELD GROUPS
TABLE XXII Coefficients C

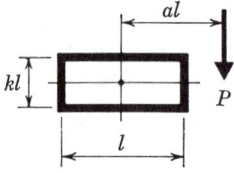

P = Allowable eccentric load in kips
l = Length of longer welds in in.
D = Number of sixteenths of an in. in fillet weld size
C = Coefficients tabulated below
C_1 = Coefficient for electrode used (see Table on p. 4-72)
= 1.0 for E70XX electrodes.

$$P = CC_1Dl$$

Required Minimum $C = \dfrac{P}{C_1Dl}$

" " $D = \dfrac{P}{CC_1l}$

" " $l = \dfrac{P}{CC_1D}$

Note: When load P is parallel to longer side l use Table XXI.

a	k										
	0	0.1	0.2	0.3	0.4	0.5	0.6	0.7	0.8	0.9	1.0
.06	1.56	1.69	1.85	2.03	2.21	2.39	2.55	2.71	2.86	3.01	3.15
.08	1.48	1.61	1.78	1.96	2.14	2.33	2.51	2.68	2.84	2.99	3.14
.10	1.41	1.54	1.71	1.89	2.08	2.27	2.45	2.63	2.80	2.97	3.12
.15	1.26	1.40	1.55	1.73	1.92	2.11	2.31	2.50	2.68	2.87	3.04
.20	1.13	1.27	1.43	1.59	1.78	1.97	2.16	2.36	2.55	2.74	2.93
.25	1.02	1.16	1.31	1.48	1.65	1.84	2.03	2.22	2.42	2.62	2.81
.30	.934	1.06	1.21	1.37	1.54	1.72	1.90	2.10	2.29	2.49	2.69
.40	.789	.903	1.04	1.20	1.35	1.52	1.69	1.87	2.06	2.25	2.44
.50	.680	.782	.911	1.05	1.20	1.35	1.51	1.68	1.86	2.04	2.23
.60	.595	.687	.806	.938	1.08	1.22	1.37	1.52	1.69	1.86	2.03
.70	.528	.612	.720	.843	.972	1.11	1.25	1.39	1.54	1.70	1.87
.80	.473	.550	.650	.764	.885	1.01	1.14	1.28	1.42	1.57	1.72
.90	.428	.499	.592	.697	.811	.929	1.05	1.18	1.31	1.45	1.60
1.00	.391	.456	.542	.641	.747	.859	.975	1.09	1.22	1.35	1.49
1.20	.333	.389	.464	.551	.645	.744	.848	.955	1.07	1.18	1.31
1.40	.289	.339	.405	.482	.566	.655	.749	.846	.947	1.05	1.16
1.60	.255	.300	.359	.428	.504	.584	.669	.758	.850	.946	1.05
1.80	.228	.268	.322	.385	.453	.527	.605	.686	.770	.858	.949
2.00	.207	.243	.292	.349	.412	.480	.551	.626	.703	.784	.869
2.20	.189	.222	.267	.319	.377	.440	.506	.575	.647	.722	.801
2.40	.173	.204	.246	.294	.348	.406	.467	.532	.599	.669	.742
2.60	.161	.189	.228	.273	.323	.377	.434	.494	.557	.623	.691
2.80	.149	.176	.212	.254	.301	.352	.406	.462	.521	.583	.647
3.00	.140	.165	.198	.238	.282	.330	.380	.433	.489	.547	.608

ECCENTRIC LOADS ON WELD GROUPS
TABLE XXIII Coefficients C

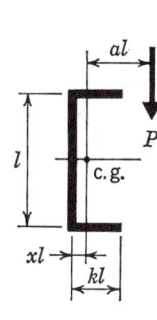

P = Allowable eccentric load in kips
l = Length of weld parallel to load P, in.
D = Number of sixteenths of an in. in fillet weld size
C = Coefficients tabulated below
C_1 = Coefficient for electrode used (see Table on p. 4-72)
   = 1.0 for E70XX electrodes
xl = Distance from vertical weld to center of gravity of weld group

$$P = CC_1Dl$$

Required Minimum $\quad C = \dfrac{P}{C_1Dl}$

" " $\quad D = \dfrac{P}{CC_1l}$

" " $\quad l = \dfrac{P}{CC_1D}$

a	k															
	0	0.1	0.2	0.3	0.4	0.5	0.6	0.7	0.8	0.9	1.0	1.2	1.4	1.6	1.8	2.0
.06	.835	.883	1.05	1.22	1.40	1.58	1.76	1.94	2.12	2.30	2.48	2.84	3.21	3.58	3.95	4.32
.08	.820	.895	1.06	1.23	1.41	1.58	1.76	1.94	2.12	2.30	2.48	2.85	3.21	3.58	3.94	4.31
.10	.804	.902	1.07	1.24	1.41	1.59	1.76	1.94	2.12	2.30	2.48	2.83	3.20	3.56	3.92	4.28
.15	.753	.895	1.06	1.22	1.39	1.56	1.73	1.90	2.07	2.24	2.42	2.76	3.11	3.46	3.81	4.16
.20	.693	.865	1.02	1.18	1.34	1.50	1.67	1.83	1.99	2.16	2.32	2.65	2.99	3.32	3.66	4.00
.25	.630	.823	.972	1.12	1.28	1.43	1.59	1.74	1.90	2.06	2.21	2.53	2.85	3.18	3.51	3.84
.30	.570	.750	.917	1.06	1.21	1.35	1.50	1.65	1.80	1.95	2.10	2.41	2.72	3.04	3.36	3.68
.40	.469	.627	.803	.934	1.07	1.20	1.33	1.47	1.61	1.74	1.89	2.17	2.47	2.77	3.08	3.39
.50	.393	.529	.666	.819	.937	1.06	1.18	1.30	1.43	1.56	1.69	1.96	2.24	2.53	2.83	3.13
.60	.336	.453	.574	.721	.829	.939	1.05	1.17	1.28	1.40	1.53	1.78	2.05	2.32	2.61	2.90
.70	.293	.395	.502	.611	.739	.839	.942	1.05	1.16	1.27	1.39	1.63	1.88	2.14	2.41	2.69
.80	.259	.349	.444	.543	.664	.756	.852	.950	1.05	1.16	1.27	1.49	1.73	1.98	2.24	2.51
.90	.232	.312	.398	.488	.602	.687	.775	.867	.962	1.06	1.16	1.38	1.60	1.84	2.09	2.36
1.00	.209	.282	.360	.442	.550	.629	.711	.796	.885	.978	1.07	1.28	1.49	1.72	1.96	2.21
1.20	.176	.236	.302	.372	.445	.536	.608	.683	.762	.844	.929	1.11	1.31	1.52	1.74	1.97
1.40	.151	.203	.260	.320	.384	.466	.530	.597	.667	.741	.818	.985	1.17	1.36	1.56	1.78
1.60	.132	.178	.228	.281	.338	.412	.469	.529	.593	.660	.731	.883	1.05	1.22	1.41	1.61
1.80	.118	.158	.203	.250	.301	.369	.420	.475	.533	.595	.660	.799	.951	1.11	1.29	1.47
2.00	.106	.142	.182	.225	.272	.334	.381	.431	.484	.541	.601	.730	.870	1.02	1.18	1.35
2.20	.097	.129	.166	.205	.247	.305	.348	.394	.444	.496	.552	.671	.802	.942	1.09	1.25
2.40	.089	.119	.152	.188	.227	.280	.320	.363	.409	.458	.510	.621	.743	.874	1.01	1.16
2.60	.082	.110	.140	.174	.210	.259	.297	.337	.380	.425	.474	.578	.692	.815	.946	1.09
2.80	.076	.102	.130	.161	.195	.242	.277	.314	.354	.397	.442	.540	.647	.763	.886	1.02
3.00	.071	.095	.122	.151	.182	.226	.259	.294	.332	.372	.415	.507	.608	.717	.834	.958
x	0	.008	.028	.056	.088	.125	.163	.204	.246	.289	.333	423	.515	.609	.704	.800

ECCENTRIC LOADS ON WELD GROUPS
TABLE XXIV Coefficients C

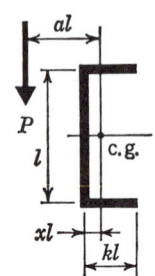

P = Allowable eccentric load in kips
l = Length of weld parallel to load P, in.
D = Number of sixteenths of an in. in fillet weld size
C = Coefficients tabulated below
C_1 = Coefficient for electrode used (see Table on p. 4-72)
 = 1.0 for E70XX electrodes
xl = Distance from vertical weld to center of gravity of weld group

$$P = CC_1Dl$$

Required Minimum $C = \dfrac{P}{C_1Dl}$

" " $D = \dfrac{P}{CC_1l}$

" " $l = \dfrac{P}{CC_1D}$

a	k																
	0	0.1	0.2	0.3	0.4	0.5	0.6	0.7	0.8	0.9	1.0	1.2	1.4	1.6	1.8	2.0
.06	.834	.899	1.07	1.25	1.42	1.60	1.78	1.95	2.13	2.30	2.48	2.82	3.17	3.51	3.85	4.20
.08	.820	.907	1.08	1.25	1.42	1.60	1.77	1.94	2.12	2.29	2.46	2.80	3.14	3.48	3.81	4.15
.10	.804	.911	1.08	1.25	1.42	1.59	1.76	1.93	2.10	2.27	2.44	2.77	3.11	3.44	3.77	4.10
.15	.753	.904	1.07	1.23	1.39	1.56	1.72	1.88	2.05	2.21	2.37	2.69	3.01	3.33	3.66	3.98
.20	.692	.877	1.03	1.19	1.35	1.51	1.67	1.82	1.98	2.13	2.29	2.60	2.91	3.22	3.54	3.85
.25	.630	.820	.993	1.15	1.30	1.45	1.60	1.75	1.90	2.05	2.20	2.50	2.81	3.11	3.42	3.72
.30	.570	.752	.932	1.09	1.24	1.39	1.53	1.68	1.82	1.97	2.11	2.41	2.70	3.00	3.30	3.60
.40	.469	.628	.790	.956	1.12	1.25	1.39	1.53	1.66	1.80	1.94	2.21	2.49	2.78	3.07	3.36
.50	.393	.528	.668	.815	.967	1.12	1.25	1.38	1.50	1.63	1.76	2.03	2.30	2.57	2.85	3.13
.60	.336	.450	.568	.696	.834	.979	1.12	1.24	1.37	1.49	1.61	1.87	2.12	2.39	2.66	2.93
.70	.293	.390	.492	.608	.734	.868	1.01	1.13	1.25	1.37	1.48	1.72	1.97	2.22	2.48	2.75
.80	.259	.343	.434	.539	.655	.777	.905	1.04	1.15	1.26	1.37	1.60	1.84	2.08	2.33	2.58
.90	.232	.305	.389	.484	.590	.703	.821	.944	1.06	1.16	1.27	1.49	1.71	1.95	2.19	2.43
1.00	.209	.276	.352	.439	.536	.640	.750	.865	.983	1.08	1.18	1.39	1.61	1.83	2.06	2.30
1.20	.176	.231	.295	.370	.452	.541	.638	.739	.844	.945	1.04	1.23	1.42	1.63	1.84	2.06
1.40	.151	.199	.255	.318	.390	.468	.553	.642	.736	.834	.919	1.09	1.27	1.46	1.66	1.86
1.60	.132	.174	.223	.280	.343	.412	.487	.566	.651	.739	.824	.982	1.15	1.32	1.50	1.69
1.80	.118	.155	.199	.249	.305	.367	.434	.506	.582	.663	.746	.891	1.04	1.21	1.38	1.55
2.00	.106	.140	.180	.225	.275	.331	.392	.457	.536	.604	.674	.815	.957	1.11	1.27	1.43
2.20	.097	.127	.163	.205	.250	.301	.357	.417	.488	.549	.614	.750	.883	1.02	1.17	1.33
2.40	.089	.117	.150	.188	.230	.277	.328	.393	.447	.504	.563	.689	.819	.950	1.09	1.23
2.60	.082	.108	.139	.173	.212	.255	.302	.362	.412	.465	.520	.637	.763	.886	1.02	1.15
2.80	.076	.100	.129	.161	.197	.237	.281	.336	.382	.431	.483	.592	.710	.830	.953	1.08
3.00	.071	.094	.120	.150	.184	.221	.262	.314	.357	.402	.450	.553	.664	.780	.897	1.02
x	0	.008	.028	.056	.088	.125	.163	.204	.246	.289	.333	.423	.515	.609	.704	.800

ECCENTRIC LOADS ON WELD GROUPS
TABLE XXV Coefficients *C*

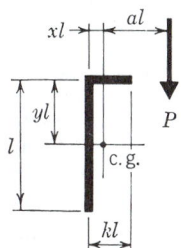

P = Allowable eccentric load in kips
l = Length of weld parallel to load P, in.
D = Number of sixteenths of an in. in fillet weld size
C = Coefficients tabulated below
C_1 = Coefficient for electrode used (see Table on p. 4-72)
  = 1.0 for E70XX electrodes
xl = Distance from vertical weld to center of gravity of weld group
yl = Distance from vertical weld to center of gravity of weld group

$$P = CC_1Dl$$

Required Minimum $\quad C = \dfrac{P}{C_1Dl}$

$\qquad$ " $\qquad$ " $\quad D = \dfrac{P}{CC_1l}$

$\qquad$ " $\qquad$ " $\quad l = \dfrac{P}{CC_1D}$

a	\multicolumn{17}{c}{k}
	0.0	0.1	0.2	0.3	0.4	0.5	0.6	0.7	0.8	0.9	1.0	1.2	1.4	1.6	1.8	2.0
.06	.835	.801	.882	.965	1.05	1.14	1.22	1.31	1.40	1.48	1.57	1.75	1.93	2.11	2.29	2.47
.08	.820	.814	.892	.974	1.06	1.14	1.23	1.32	1.40	1.49	1.58	1.74	1.92	2.10	2.28	2.47
.10	.804	.818	.895	.976	1.06	1.14	1.23	1.31	1.40	1.48	1.57	1.73	1.91	2.10	2.28	2.47
.15	.753	.810	.882	.957	1.03	1.11	1.19	1.27	1.34	1.42	1.51	1.67	1.84	2.09	2.28	2.46
.20	.693	.780	.844	.915	.985	1.06	1.13	1.20	1.27	1.35	1.42	1.58	1.74	1.91	2.28	2.26
.25	.630	.714	.795	.862	.926	.990	1.06	1.12	1.19	1.26	1.34	1.49	1.64	1.80	1.97	2.46
.30	.570	.649	.724	.798	.864	.923	.984	1.05	1.11	1.18	1.25	1.39	1.55	1.70	1.87	2.04
.40	.469	.538	.602	.665	.729	.797	.851	.908	.967	1.03	1.09	1.23	1.38	1.53	1.69	1.85
.50	.393	.452	.507	.562	.618	.676	.740	.791	.846	.904	.960	1.09	1.23	1.38	1.53	1.68
.60	.336	.387	.435	.483	.532	.584	.640	.697	.747	.801	.858	.981	1.11	1.25	1.40	1.55
.70	.293	.337	.379	.421	.465	.512	.563	.620	.667	.718	.771	.887	1.01	1.15	1.29	1.43
.80	.259	.297	.335	.373	.413	.455	.501	.558	.602	.649	.699	.808	.926	1.05	1.19	1.33
.90	.232	.266	.300	.334	.370	.409	.452	.498	.547	.591	.638	.741	.853	.975	1.10	1.24
1.00	.209	.240	.271	.303	.335	.371	.410	.453	.501	.542	.586	.683	.790	.906	1.03	1.16
1.20	.176	.201	.227	.254	.282	.313	.347	.384	.428	.464	.504	.591	.688	.793	.905	1.02
1.40	.151	.173	.196	.219	.243	.270	.300	.333	.373	.406	.441	.520	.607	.703	.806	.915
1.60	.132	.152	.172	.192	.213	.237	.264	.294	.330	.360	.392	.464	.543	.630	.725	.825
1.80	.118	.135	.153	.171	.190	.212	.236	.263	.296	.324	.353	.418	.491	.571	.658	.751
2.00	.106	.122	.138	.154	.172	.191	.213	.238	.269	.294	.321	.380	.447	.521	.602	.688
2.20	.097	.111	.125	.140	.156	.174	.194	.217	.242	.269	.294	.349	.411	.480	.554	.635
2.40	.089	.101	.115	.128	.143	.160	.179	.200	.223	.248	.271	.322	.380	.444	.513	.589
2.60	.082	.094	.106	.119	.132	.148	.165	.185	.207	.230	.251	.299	.353	.413	.478	.548
2.80	.076	.087	.098	.110	.123	.137	.154	.172	.192	.214	.234	.279	.330	.386	.447	.513
3.00	.071	.081	.092	.103	.115	.128	.144	.161	.180	.200	.219	.262	.309	.362	.420	.482
x	0	.004	.016	.034	.057	.083	.112	.144	.177	.213	.250	.327	.408	.492	.578	.666
y	.500	.454	.416	.384	.357	.333	.312	.294	.277	.263	.250	.227	.208	.192	.178	.166

ECCENTRIC LOADS ON WELD GROUPS
TABLE XXVI Coefficients C

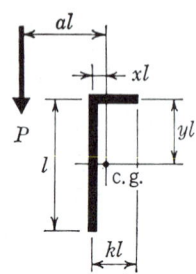

P = Allowable eccentric load in kips
l = Length of weld parallel to load P, in.
D = Number of sixteenths of an in. in fillet weld size
C = Coefficients tabulated below
C_1 = Coefficient for electrode used (see Table on p. 4-72)
$\quad$ = 1.0 for E70XX electrodes
xl = Distance from vertical weld to center of gravity of weld group
yl = Distance from horizontal weld to center of gravity of weld group

$$P = CC_1 Dl$$

Required Minimum $\quad C = \dfrac{P}{C_1 Dl}$

" $\quad$ " $\quad D = \dfrac{P}{CC_1 l}$

" $\quad$ " $\quad l = \dfrac{P}{CC_1 D}$

a	k															
	0	0.1	0.2	0.3	0.4	0.5	0.6	0.7	0.8	0.9	1.0	1.2	1.4	1.6	1.8	2.0
.06	.835	.807	.894	.983	1.07	1.16	1.25	1.34	1.42	1.50	1.58	1.73	1.88	2.03	2.19	2.34
.08	.820	.820	.904	.990	1.08	1.16	1.25	1.33	1.41	1.48	1.56	1.70	1.85	2.00	2.15	2.31
.10	.804	.826	.908	.992	1.08	1.16	1.24	1.31	1.39	1.46	1.53	1.68	1.82	1.97	2.12	2.27
.15	.753	.821	.898	.976	1.05	1.13	1.20	1.27	1.33	1.40	1.47	1.60	1.74	1.89	2.03	2.19
.20	.693	.781	.866	.946	1.02	1.10	1.18	1.21	1.27	1.33	1.40	1.53	1.66	1.81	1.95	2.10
.25	.630	.715	.795	.871	.945	1.02	1.09	1.16	1.24	1.26	1.33	1.46	1.59	1.73	1.88	2.03
.30	.570	.650	.725	.797	.867	.936	1.01	1.08	1.15	1.23	1.26	1.39	1.52	1.66	1.80	1.95
.40	.469	.538	.602	.663	.727	.794	.864	.936	1.01	1.09	1.16	1.26	1.39	1.53	1.67	1.81
.50	.393	.450	.503	.559	.620	.683	.749	.818	.889	.962	1.04	1.19	1.28	1.41	1.55	1.69
.60	.336	.384	.429	.480	.536	.595	.657	.721	.789	.858	.931	1.08	1.18	1.31	1.44	1.58
.70	.293	.333	.374	.420	.470	.524	.581	.642	.705	.771	.840	.985	1.09	1.21	1.34	1.48
.80	.259	.294	.331	.372	.418	.467	.520	.576	.636	.698	.763	.901	1.05	1.13	1.26	1.39
.90	.232	.263	.296	.334	.375	.420	.469	.522	.577	.636	.697	.828	.968	1.06	1.18	1.30
1.00	.209	.237	.268	.302	.340	.381	.427	.475	.528	.583	.641	.764	.897	.991	1.11	1.23
1.20	.176	.199	.225	.254	.286	.321	.360	.403	.449	.497	.549	.660	.781	.910	.986	1.10
1.40	.151	.171	.194	.218	.246	.277	.311	.348	.389	.432	.479	.579	.688	.807	.886	.992
1.60	.132	.150	.170	.192	.216	.243	.273	.307	.343	.382	.423	.514	.614	.723	.802	.901
1.80	.118	.134	.151	.171	.192	.216	.243	.273	.306	.341	.379	.462	.553	.653	.761	.825
2.00	.106	.120	.136	.154	.173	.195	.219	.247	.276	.308	.343	.418	.503	.577	.667	.759
2.20	.097	.110	.124	.140	.158	.177	.200	.225	.252	.281	.313	.382	.450	.527	.610	.700
2.40	.089	.101	.114	.128	.145	.163	.183	.206	.231	.258	.287	.352	.414	.485	.562	.646
2.60	.082	.093	.105	.119	.134	.150	.169	.190	.213	.239	.266	.326	.382	.448	.521	.599
2.80	.076	.086	.098	.110	.124	.140	.157	.177	.198	.222	.247	.300	.355	.417	.485	.558
3.00	.071	.081	.091	.103	.116	.130	.147	.165	.185	.207	.231	.280	.332	.390	.453	.522
x	0	.004	.016	.034	.057	.083	.112	.144	.177	.213	.250	.327	.408	.492	.578	.666
y	.500	.454	.416	.384	.357	.333	.312	.294	.277	.263	.250	.227	.208	.192	.178	.166

SINGLE-ANGLE CONNECTIONS

Single-angle connections are usually shop-fastened to the supporting members, either by bolting or welding, and the field connection is made by bolting into the supported member.

In designing a single-angle connection, it is customary to consider vertical shear or bearing in all fasteners; the shear in the connection angle through the least net section; the effect of eccentricity in the bolts or weld for the angle leg fastened to the supporting member; and the bending of the connection angle at the critical section for this leg of the angle.

This type of connection is categorized as an AISC Type 2 simple-framed connection. The minimum angle thickness for ¾ and ⅞-in. dia. bolts should be ⅜-in. and for 1-in. dia. bolts a ½-in. minimum is suggested. Table II-C is useful for shear investigations and the equation on p. 4-88 is useful for determining the section modulus of the angle at the net section. The following table will be useful in investigating the eccentricity of the bolt group. If welding is used Table XXV on page 4-81 is helpful.

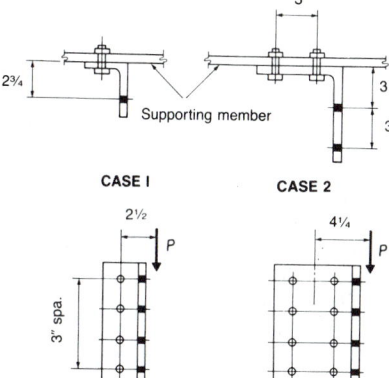

	Coefficient, C	
n	Case I	Case II
1	—	0.50
2	1.02	1.55
3	2.00	2.83
4	3.08	4.54
5	4.16	6.45
6	5.22	8.51
7	6.28	10.6
8	7.32	10.7
9	8.35	14.8
10	9.37	25.8

$$P = Cr_v \text{ or } C = \frac{P}{r_v}$$

where

n = Total no. of fasteners in one vertical row
C = Coefficient
P = Allowable load, kips
r_v = Allowable shear or bearing value for one fastener, kips
l = Distance between centerline of connected beam web and center of gravity of fasteners, in.

When using the coefficient table shown above, do not exceed the eccentricities shown for the angle leg which attaches to the *supporting* member. Eccentricities less than those given will produce conservative results. Only standard holes should be used in this leg. The connection may be designed as a bearing-type or a slip-critical connection. If larger gages are necessary, coefficients should be interpolated from Tables XI and XII.

Do not exceed the gages shown for the angle leg which attaches to the supported member. Normal diameter, round holes or short, horizontal slots may be used in this leg to facilitate erection. A325, A490 or A307 bolts may be used.

Where possible, the distance between the centers of the top and bottom connecting bolts should equal or exceed, one-half the T-distance of the supported member to guard against overturning of the beam. It is permissible to design a connection with certain combinations of leg widths, bolt types, diameters and holes or slots within the bounds noted above.

Case I is by far the most common type of single-angle connection and is usually the first choice. Case II is used as an alternate in case the allowable loads for Case I are exceeded by the design loads.

If it is desired to weld the single-angle connection to the supporting member, the weld should take the form of an L, as shown in Ex. 30.

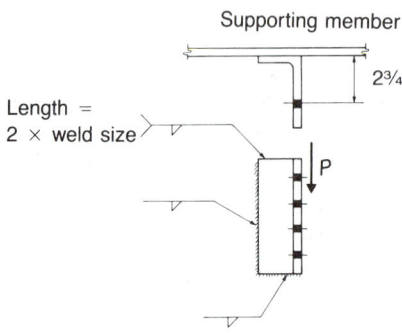

To provide the necessary flexibility, there should be no weld at the top or heel of the angle. A 4 × 3 angle is normally selected for this welded connection with the 3-in. leg being the welded leg.

Other types of one-sided connections using tees and shear plates are shown elsewhere in this manual.

When two filler beams frame to a girder beam directly opposite each other and share the same bolts, or have welds directly opposite each other, the beam web of the girder must be checked for bolt bearing capacity or maximum permitted weld size.

EXAMPLE 28

Given:

W18 × 35 filler beam, F_y = 36 ksi
End reaction = 26 kips
Connection angle is Case I type, F_y = 36 kips
Fasteners are ¾-in. dia. A325-N
W21 × 62 supporting beam (F_y = 36 ksi)

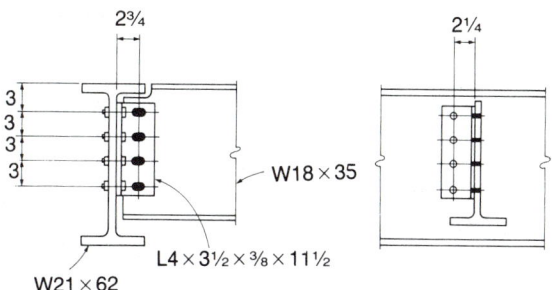

Solution:

Assume an angle 4 × 3½ × ⅜.

1. For leg against the W21 × 62:

   r_v = 9.3 kips/bolt from Table I -D

   Req'd C = 26/9.3 = 2.8   Try 4 bolts (C = 3.08)

   Distance center-to-center extreme bolts = 9 in. ≥ half T-distance of beam.   **o.k.**

   Angle length = 11½ in.

   Shear = 54 kips > 26 kips from Table II-C.   **o.k.**

   Section modulus at net section (from p. 4-88)

   $$S = \frac{0.375 \, (11.5)^2}{6} - \frac{3^2 \, (4) \, (4^2-1) \, [0.375 \times (.75 + 0.125)]}{6 \, (11.5)}$$

   $$= 5.7 \text{ in.}^3$$

   e = 2.25 + 0.15 = 2.40 in. < 2.5 in.   **o.k.**

   M = 26 × 2.40 = 62.4 kip-in.

   $$f_b = \frac{62.4}{5.7} = 10.95 \text{ ksi} < 0.6 \, F_y. \quad \textbf{o.k.}$$

2. For leg against the W18 × 35:

   Short slots are used in this leg

   Allowable P = 4 (9.3) = 37.2 kips > 26.   **o.k.**

   Use L4 × 3½ × ⅜ × 0'-11½

3. Because W18 × 35 is coped check block shear per p. 4-8.

EXAMPLE 29

Given:

Same information as Ex. 28, except end reaction = 45 kips

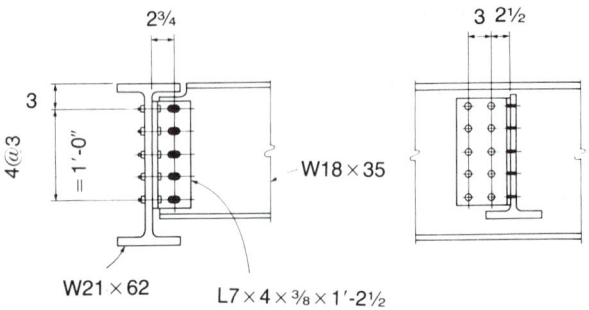

W18 × 35

W21 × 62   L7 × 4 × ⅜ × 1'-2½

Solution:

Assume L4 × 3½ × ⅜.

1. For leg against the W21 × 62:

   Req'd C = 45/9.3 = 4.8. This would require 6 bolts which will not fit in the web of a W18 beam. Therefore, we investigate a Case II connection.

   Assume an angle 8 × 8 × ½

   l = 2.5 + 1.5 + 0.15 = 4.15 < 4.25

   From Table XII req'd n = 5 rows

   Angle length = 1 ft-2½ in.

   By inspection shear at net section is **o.k.**

   Section modulus at net section (from p. 4-88)

   $$S = \frac{0.5\ (14.5)^2}{6} - \frac{3^2\ (5)\ (5^2-1)\ [0.5 \times (0.75 + 0.125)]}{6\ (14.5)}$$

   $$= 12.1\ \text{in.}^3$$

   M = 45 (2.5 + 0.15) = 119 kip-in.

   f_b = 119/12.1 = 9.83 ksi < 0.6 F_y **o.k.**

2. For leg against the W18 × 35:

   Req'd no. of bolts = 45/9.3 = 4.8. Since it is not required to consider eccentricity in this leg, one vertical row of 5 bolts could be used in which case an angle size of L7 × 4 × ⅜ would be substituted with the 4-in. leg against the W18 × 35. From Table II-C, allowable shear in 2 connection angles = 136 kips. For 1 angle, allowable shear = 68 kips > 45   **o.k.**

   Recheck f_b = (9.83) (0.50)/(0.375) = 13.11 ksi < 0.6 F_y **o.k.**

   Use L7 × 4 × ⅜ × 1'-2½"

3. Check W18 × 35 for block shear per p. 4-8.

EXAMPLE 30

Given:

Same information as Ex. 28 except use an angle shop welded to the W21 × 62 with E70XX electrodes.

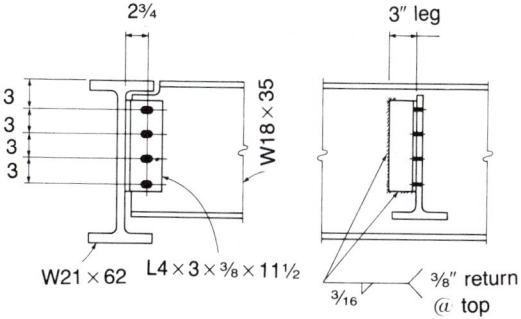

Solution:

Assume an angle 4 × 3 × ⅜ with the 3-in. leg against the W21 × 62.

1. No. of bolts at field connection = 26/9.3 = 2.8 required, but this would not provide a deep enough connection. For this requirement, 4 bolts are necessary to give an angle length of 11½ in. Eccentric capacity of this weld is found from Table XXV on p. 4-81.

$$e = 3 + 0.15 = 3.15 \text{ in.}$$

$$k = \frac{kl}{l} = \frac{3}{11.5} = 0.26$$

$$x = 0.027 \text{ for } k = 0.26$$

$$xl = 0.026 \, (11.5) = 0.3$$

$$al = 3.15 - 0.3 = 2.85$$

$$a = \frac{2.85}{11.5} = 0.25$$

$$C = 0.84$$

$$C_1 = 1 \text{ for E70XX electrodes}$$

$$D = \frac{P}{CC_1 l} = \frac{26}{0.84 \, (1) \, (11.5)} = 2.7 \text{ sixteenths min.}$$

Use ³⁄₁₆-in. weld which meets the requirements of Table J2.4. Use L4 × 3 × ⅜ × 0'-11½" with 4 bolts.

2. Check block shear of web of W18 × 35 — p. 4-8.

BRACKET PLATES
Net section moduli

Diameter of holes assumed $\frac{1}{8}$ in. larger than nominal diameter of fastener

Section moduli taken along this line

Fasteners spaced 3 in. vertically

No. of Fasteners in One Vertical Line	Depth of Plate In.	3/4-In. Fasteners					7/8-In. Fasteners					1-In. Fasteners				
		Thickness of Plate, In.					Thickness of Plate, In.					Thickness of Plate, In.				
		1/4	3/8	1/2	5/8	3/4	3/8	1/2	5/8	3/4	7/8	1/2	5/8	3/4	7/8	1
2	6	1.2	1.8	2.3	2.9	3.5	1.7	2.3	2.9	3.4	4.0	2.2	2.7	3.2	3.8	4.3
3	9	2.5	3.8	5.0	6.3	7.5	3.6	4.8	5.9	7.1	8.3	4.5	5.6	6.8	7.9	9.0
4	12	4.4	6.3	8.7	11	13	6.2	8.2	10	12	14	7.8	9.7	12	14	16
5	15	6.8	10	14	17	20	10	13	16	19	22	12	15	18	21	24
6	18	9.6	15	19	24	29	14	18	23	27	32	17	21	26	30	34
7	21	13	20	26	33	39	19	25	31	37	43	23	29	35	41	47
8	24	17	26	34	43	51	24	32	40	48	56	30	38	45	53	61
9	27	22	32	43	54	65	31	41	51	61	71	38	48	57	67	77
10	30	27	40	53	67	80	38	50	63	75	88	47	59	71	83	94
12	36	38	58	77	96	115	54	72	90	108	126	68	85	102	119	136
14	42	52	78	104	130	157	74	98	123	147	172	92	115	138	161	184
16	48	68	102	136	170	204	96	128	160	192	224	120	150	180	211	241
18	54	86	129	172	215	259	122	162	203	243	284	152	190	228	266	304
20	60	106	160	213	266	319	150	200	250	300	350	188	235	282	329	376
22	66	129	193	257	322	386	182	242	303	363	424	227	284	341	398	454
24	72	153	230	306	383	459	216	288	360	432	504	270	338	406	473	541
26	78	180	270	359	449	539	254	338	423	507	592	317	397	476	555	634
28	84	208	313	417	521	625	294	392	490	588	686	368	460	552	644	736
30	90	240	359	478	598	718	338	450	563	675	788	422	528	633	739	845
32	96	272	408	544	680	816	384	512	640	768	896	480	600	721	841	961
34	102	308	461	614	768	922	434	578	723	867	1012	542	678	813	949	1085
36	108	344	517	689	861	1033	486	648	810	972	1134	608	760	912	1064	1216

Interpolate for intermediate thickness of plates.

General equation for net section modulus of bracket plates:

$$S_{net} = \frac{t_p d^2}{6} - \frac{b^2 n (n^2 - 1) [t_p \times (\text{bolt dia.} + 0.125)]}{6d}$$

where

t_p = Plate thickness, in.
d = Plate depth, in.
n = Number of fasteners in one vertical row
b = Fastener spacing vertically, in.

HANGER-TYPE CONNECTIONS
Fasteners loaded in tension

In the design of hanger-type connections, prying action must be considered. The table below is useful for making a preliminary selection of a trial fitting, using $F_y = 36$ ksi. The fitting must then be checked for bending stresses and tension in the bolts, using a procedure which includes prying action.

PRELIMINARY SELECTION TABLE

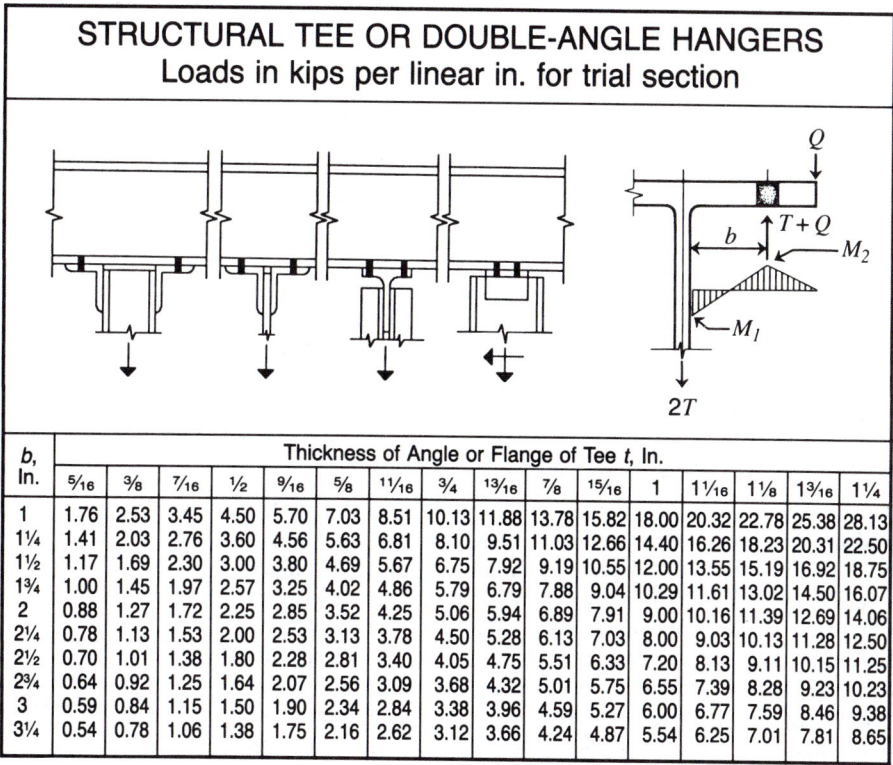

STRUCTURAL TEE OR DOUBLE-ANGLE HANGERS
Loads in kips per linear in. for trial section

b, In.	Thickness of Angle or Flange of Tee t, In.														
	5/16	3/8	7/16	1/2	9/16	5/8	11/16	3/4	13/16	7/8	15/16	1	1 1/16	1 1/8	1 3/16	1 1/4
1	1.76	2.53	3.45	4.50	5.70	7.03	8.51	10.13	11.88	13.78	15.82	18.00	20.32	22.78	25.38	28.13
1 1/4	1.41	2.03	2.76	3.60	4.56	5.63	6.81	8.10	9.51	11.03	12.66	14.40	16.26	18.23	20.31	22.50
1 1/2	1.17	1.69	2.30	3.00	3.80	4.69	5.67	6.75	7.92	9.19	10.55	12.00	13.55	15.19	16.92	18.75
1 3/4	1.00	1.45	1.97	2.57	3.25	4.02	4.86	5.79	6.79	7.88	9.04	10.29	11.61	13.02	14.50	16.07
2	0.88	1.27	1.72	2.25	2.85	3.52	4.25	5.06	5.94	6.89	7.91	9.00	10.16	11.39	12.69	14.06
2 1/4	0.78	1.13	1.53	2.00	2.53	3.13	3.78	4.50	5.28	6.13	7.03	8.00	9.03	10.13	11.28	12.50
2 1/2	0.70	1.01	1.38	1.80	2.28	2.81	3.40	4.05	4.75	5.51	6.33	7.20	8.13	9.11	10.15	11.25
2 3/4	0.64	0.92	1.25	1.64	2.07	2.56	3.09	3.68	4.32	5.01	5.75	6.55	7.39	8.28	9.23	10.23
3	0.59	0.84	1.15	1.50	1.90	2.34	2.84	3.38	3.96	4.59	5.27	6.00	6.77	7.59	8.46	9.38
3 1/4	0.54	0.78	1.06	1.38	1.75	2.16	2.62	3.12	3.66	4.24	4.87	5.54	6.25	7.01	7.81	8.65

To select a preliminary size of tee or double angle hanger, the above table assumes equal critical moments at the fastener line and at the face of the tee stem or angle leg. Then

$$M_1 = M_2 = Pb/2 = 27t^2/6$$

Hanger capacity $= 2P = 18t^2/b$

where

$2P$ = Allowable load on two angles or a structural tee, in kips per linear inch, using maximum allowable bending stress of 27 ksi ($.75F_y$)

t = Thickness of angle or tee flange, in.

b = Distance from fastener line to the face of outstanding leg of angle or tee stem, in.

AMERICAN INSTITUTE OF STEEL CONSTRUCTION

DESIGN AND ANALYSIS METHODS

Nomenclature

T = Applied tension per bolt (exclusive of initial tightening and prying force), kips

Q = Prying force per bolt at design load, kips

B = Allowable tension per bolt, kips

t_c = Flange or angle thickness required to develop B in bolts with no prying action, in.

$$= \sqrt{\frac{8Bb'}{pF_y}}$$

F_y = Yield strength of the flange material, ksi

p = Length of flange, parallel to stem or leg, tributary to each bolt, in.

a = Distance from bolt centerline to edge of tee flange or angle leg but not more than $1.25b$, in.

d = Bolt dia., in.

d' = Width of bolt hole parallel to tee stem or angle leg, in.

b' = $b - d/2$, in.

a' = $a + d/2$, in.

ρ = b'/a'

α = $(0 \le \alpha \le 1.0)$. Ratio of moment at bolt line to moment at stem line

$\quad = M_2/\delta M_1$

α' = Value of α for which required thickness $(t_{req'd})$ is a minimum or allowable applied tension per bolt (T_{all}) is a maximum.

δ = Ratio of net area (at bolt line) and the gross area (at the face of the stem or angle leg)

$\quad = 1 - d'/p$

DESIGN CONSIDERATIONS

The actual distribution of stress in the tee flange or angle leg and the extent of prying action are extremely complex. Significant deformation of the fitting under design load is seldom tolerable. Flange stiffness, rather than bending strength, is the key to satisfactory performance. Therefore, dimension b should be made as small as the wrench clearance will permit. Since dimension b is only slightly larger than the thickness of the fitting, the classical moment diagram (right) does not truly represent all the restraining forces at the bolt line, and overestimates the actual prying force. In addition, local de-

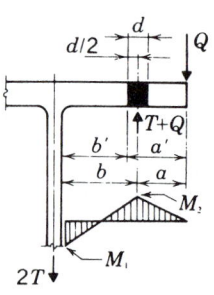

formation of the fitting ("quilting") under the pretension force of high-strength bolts also accounts for a less critical prying force than indicated by earlier investigations.

Good correlation between estimated connection strength and observed test results has been obtained using theory of Ref. 1. Design and analysis procedures based on this theory have been developed in Refs. 2 and 3.

1. *Kulak, G. L., Fisher, J. W. and Struik, J. H. A.* Guide to Design Criteria for Bolted and Riveted Joints, *2nd Ed., John Wiley and Sons, New York, 1987, pp. 277-282.*

2. *Astaneh, A.* Procedure for Design and Analysis of Hanger-Type Connections. *AISC Engineering Journal, Vol. 22, No. 2, 2nd Quarter, 1985, pp. 63-66.*

3. *Thornton, W. A.* Prying Action—A General Treatment. *AISC Engineering Journal, Vol. 22, No. 2, 2nd Quarter, 1985, pp. 67-75.*

SOLUTION PROCEDURES

Method 1 (Design)

Given: p, F_y, B, T
Find: $t_{req'd}$

Step 1. Determine the number and size of bolts required so that $T \leq B$

Step 2. Using preliminary selection table to estimate required flange thickness, choose a trial section with flange thickness t, and calculate b, a ($\leq 1.25b$), b', a', ρ, δ

Step 3. Calculate $\beta = \dfrac{1}{\rho}\left(\dfrac{B}{T} - 1\right)$

if $\beta \geq 1$ set $\alpha' = 1$

if $\beta < 1$ set $\alpha' =$ lesser of $\dfrac{1}{\delta}\left(\dfrac{\beta}{1-\beta}\right)$ and 1.0

Step 4. Calculate

$$t_{req'd,} = \sqrt{\frac{8Tb'}{pF_y\,(1+\delta\alpha')}}$$

if $t_{req'd} \leq t$, design is satisfactory

if $t_{req'd} > t$, choose a heavier section with $t \geq t_{req'd}$, or change geometry (b and p), and repeat steps 3 and 4

Step 5. If the prying force Q is required, it can be calculated as follows:

$$\alpha = \frac{1}{\delta}\left[\frac{T/B}{(t/t_c)^2} - 1\right]; \text{ if } \alpha < 0 \text{ set } \alpha = 0$$

$$Q = B\delta\alpha\rho\left(\frac{t}{t_c}\right)^2$$

Step 6. In applications where the prying force Q must be reduced to an insignificant amount, skip Steps 3, 4 and 5, set $\alpha = 0$, and calculate

$$t_{req'd} = \sqrt{\frac{8Tb'}{pF_y}}$$

Method 2 (Analysis)

Given: t, a, b, p, B, T

Find: T_{all}

Step 1. Check $T \leq B$. If true, proceed; if not, use more or stronger bolts

Step 2. Then, calculate

$$\alpha' = \frac{1}{\delta(1 + \rho)}\left[\left(\frac{t_c}{t}\right)^2 - 1\right]$$

if $\alpha' > 1$, $T_{all} = B\left(\dfrac{t}{t_c}\right)^2 (1 + \delta)$

if $0 \leq \alpha' \leq 1$, $T_{all} = B\left(\dfrac{t}{t_c}\right)^2 (1 + \delta\,\alpha')$

if $\alpha' < 0$, $T_{all} = B$

Step 3. If $T_{all} \geq T$ design is satisfactory

If $T_{all} < T$, choose section with thicker flange, or change geometry (b,p) and repeat Step 2.

Step 4. If the prying force is required, use formulas of Step 5 of Method 1.

EXAMPLE 31

Given:

Select a WT section hanger using A36 steel and ¾-in. dia. A325-N bolts to support 44 kips suspended from the bottom flange of a W36×160. Bolts to be located on a 4-in. beam gage and the fitting length is 9 in. max.

Solution: (Method 1)

1. $B = 19.4$ kips (from Table I-A)

   No. of bolts required $= \dfrac{44}{19.4} = 2.27$; try 4 bolts

   $T = \dfrac{44}{4} = 11$ kips < 19.4 kips **o.k.**

2. From preliminary selection table with

   $\dfrac{2 \times 11}{4.5} = 4.89$ kips/in.

   and $b = 2 - 0.25 = 1.75$ in., choose preliminary thickness $t_{pre} = {}^{11}\!/_{16}$ in. to ¾ in. Tentatively select a WT9×30 with $t = 0.695$ in.

   $b = \dfrac{4 - 0.415}{2} = 1.792$ in. $> 1¼$ in. wrench clearance

   $a = \dfrac{7.555 - 4}{2} = 1.778$

   check $1.25 \, b = 1.25 \times 1.792 = 2.240$

   since $2.240 > 1.778$, use $a = 1.778$ in.

   $b' = 1.792 - 0.375 = 1.417$

   $a' = 1.778 + 0.375 = 2.153$

   $p = 4.5$

   $d' = {}^{13}\!/_{16} = 0.8125$

   $\delta = 1 - 0.8125/4.5 = 0.819$

   $\rho = 1.417/2.153 = 0.658$

3. Calculate $\beta = \dfrac{1}{0.658} \left(\dfrac{19.4}{11} - 1 \right) = 1.16$

   Since $\beta > 1$, $\alpha' = 1$

4. Calculate $t_{req'd} = \sqrt{\dfrac{8 \times 11 \times 1.417}{4.5 \times 36 \times 1.819}} = 0.651$ in.

Since 0.651 in. $<$ 0.695 in., WT9×30 **o.k.**

5. If the prying force is required:

$$t_c = \sqrt{\dfrac{8 \times 19.4 \times 1.417}{4.5 \times 36}} = 1.165$$

$$\alpha = \dfrac{1}{0.819}\left[\dfrac{11/19.4}{(0.695/1.165)^2} - 1\right] = 0.724$$

$$Q = 19.4 \times 0.819 \times 0.724 \times 0.658 \times \left(\dfrac{0.695}{1.165}\right)^2 = 2.69 \text{ kips}$$

EXAMPLE 32

Given:

If the connection of Ex. 31 is subjected to fatigue loading of more than 500,000 cycles, the following design check using AISC ASD Specification Appendix K4 must be made.

Solution:

Using Method 2:

1. Maximum allowable tension per bolt B = 31 × 0.4418 = 13.7 kips
   T = 11 kips $<$ B = 13.7 kips **o.k.**

2. $t_c = \sqrt{\dfrac{8 \times 13.7 \times 1.417}{4.5 \times 36}} = 0.979$

$$\alpha' = \dfrac{1}{0.819 \times 1.658}\left[\left(\dfrac{0.979}{0.695}\right)^2 - 1\right] = 0.725$$

$$T_{all} = 13.7\left(\dfrac{0.695}{0.979}\right)^2 [1 + (.819 \times 0.725)] = 11.004 \text{ kips} > 11 \quad \textbf{o.k.}$$

3. $\alpha = \dfrac{1}{0.819}\left[\dfrac{11.0/13.7}{(.695/.979)^2} - 1\right] = 0.724$

$$Q = 13.7 \times 0.819 \times .724 \times .658 \times \left(\dfrac{0.695}{0.979}\right)^2$$

$$= 2.69 \text{ kips} < .6 \times 11 = 6.60 \text{ kips} \quad \textbf{o.k.}$$

EXAMPLE 33

Given:

Redesign the hanger fitting of Ex. 31 for a loading condition where virtual elimination of the prying force is required.

Solution:

The solution is given by Step 6 of Method 1. Note that because $\alpha = 0$, the solution does not depend on a. Assuming a WT hanger stem thickness of ½ in.

$b = 2 - 0.25 = 1.75, \; b' = 1.75 - 0.375 = 1.375$

$$t_{req'd} = \sqrt{\frac{8 \times 11 \times 1.375}{4.5 \times 36}} = 0.864 \text{ in.}$$

choose a WT8×44.5

$t = 0.875 \text{ in.} > 0.864 \text{ in.}$ **o.k.**

$t_w = 0.525 \text{ in.} > 0.5 \text{ in.}$ (assumed) **o.k.**

Use: WT8×44.5

EXAMPLE 34

Given:

Select a double-angle connection using A36 steel and ¾-in. dia. A325-N bolts to carry a load of 60 kips at a bevel of 6 to 12. Fasteners to be located on a 5½-in. column gage.

Solution: (Method 2)

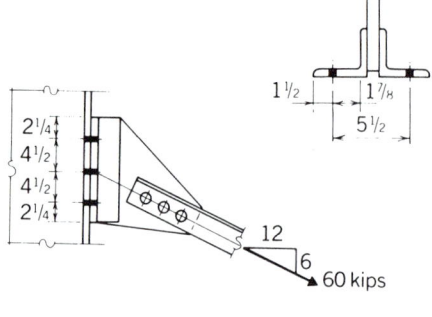

$\text{Shear} = \dfrac{1}{\sqrt{5}} \times 60 = 26.8 \text{ kips}$

$\text{Tension} = \dfrac{2}{\sqrt{5}} \times 60 = 53.7 \text{ kips}$

Assume ½-in. gusset plate

1. Try 6 bolts.

    shear per bolt $V = \dfrac{26.8}{6} = 4.47 \text{ kips}$

    tension per bolt $T = \dfrac{53.7}{6} = 8.95 \text{ kips}$

From Table I-D:

Bolt allowable shear $= 0.4418 \times 21 = 9.3 \text{ kips}$

Since 4.47 kips < 9.3 kips, bolts **o.k.** for shear

From interaction (Table J3.3)

$F_t = \sqrt{(44)^2 - 4.39 \, (4.47/0.4418)^2} = 38.6 \text{ ksi}$

$B = 38.6 \times 0.4418 = 17 \text{ kips} < 19.4 \text{ kips}$ **o.k.**

Since 8.95 kips < 17 kips, bolts **o.k.** for tension

2. Assume $p = 4.5$ in.

    Enter preliminary selection table with

$\dfrac{2T}{p} = \dfrac{2 \times 8.95}{4.5} = 3.98 \text{ kips/in.,}$ and

b equal to approximately 2 in.

Select $t_{pre} = \text{⅝ to ¹¹⁄₁₆}$

Try two angles 4 × 4 × ⅝ × 1'-1½

$b = 1.875 \text{ in.}$

$b' = 1.875 - 0.375 = 1.50 \text{ in.}$

$a = 1.5$ in. $< 1.25b$

$a' = 1.5 + 0.375 = 1.875$ in.

$\rho = 1.50/1.875 = 0.800$

$\delta = 0.819$

$$t_c = \sqrt{\frac{8 \times 17.0 \times 1.50}{4.5 \times 36}} = 1.122 \text{ in.}$$

Calculate $\alpha' = \dfrac{1}{0.819 \times 1.800} \left[\left(\dfrac{1.122}{.625}\right)^2 - 1 \right] = 1.508$; use $\alpha' = 1$

3. $T_{all} = 17.0 \times \left(\dfrac{0.625}{1.122}\right)^2 (1 + 0.819) = 9.60$ kips

Since 9.60 kips $>$ 8.95 kips, angles $4 \times 4 \times \frac{5}{8}$ are **o.k.**

TENSION MEMBERS
Net areas

	Two angles—net area, In.²											
	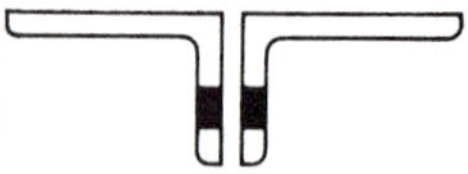 2 Holes out				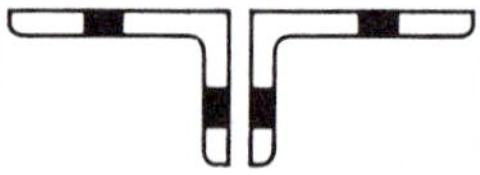 4 Holes out				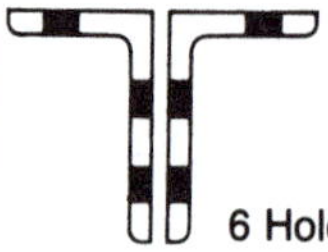 6 Holes out			
Angle Designation	Fastener Dia., In.				Fastener Dia., In.				Fastener Dia., In.			
	¾	⅞	1	1⅛	¾	⅞	1	1⅛	¾	⅞	1	1⅛
L8×8×1⅛	31.5	31.2	30.9	30.7	29.5	29.0	28.4	27.8	27.6	26.7	25.9	25.0
1	28.3	28.0	27.8	27.5	26.5	26.0	25.5	25.0	24.8	24.0	23.3	22.5
⅞	24.9	24.7	24.5	24.3	23.4	23.0	22.5	22.1	21.9	21.2	20.6	19.9
¾	21.6	21.4	21.2	21.0	20.3	19.9	19.5	19.1	18.9	18.4	17.8	17.3
⅝	18.1	18.0	17.8	17.7	17.0	16.7	16.4	16.1	15.9	15.5	15.0	14.5
½	14.6	14.5	14.4	14.3	13.8	13.5	13.3	13.0	12.9	12.5	12.1	11.8
L8×6×1	24.3	24.0	23.8	23.5	22.5	22.0	21.5	21.0	20.8	20.0	19.3	18.5
¾	18.6	18.4	18.2	18.0	17.3	16.9	16.5	16.1	15.9	15.4	14.8	14.3
½	12.6	12.5	12.4	12.3	11.8	11.5	11.3	11.0	10.9	10.5	10.1	9.75
L8×4×1	20.3	20.0	19.8	19.5	18.5	18.0	17.5	17.0	16.8	16.0	15.3	14.5
¾	15.6	15.4	15.2	15.0	14.3	13.9	13.5	13.1	12.9	12.4	11.8	11.3
½	10.6	10.5	10.4	10.3	9.75	9.50	9.25	9.00	8.87	8.50	8.13	7.75
L7×4× ¾	14.1	13.9	13.7	13.5	12.8	12.4	12.0	11.6	11.4	10.9	—	—
½	9.62	9.50	9.37	9.25	8.75	8.50	8.25	8.00	7.87	7.50	—	—
⅜	7.32	7.23	7.14	7.04	6.67	6.48	6.27	6.08	6.01	5.73	—	—
L6×6×1	20.3	20.0	19.8	19.5	18.5	18.0	17.5	17.0	16.8	16.0	—	—
⅞	17.9	17.7	17.5	17.3	16.4	16.0	15.5	15.1	14.9	14.2	—	—
¾	15.6	15.4	15.2	15.0	14.3	13.9	13.5	13.1	12.9	12.4	—	—
⅝	13.1	13.0	12.8	12.7	12.0	11.7	11.4	11.1	10.9	10.5	—	—
½	10.6	10.5	10.4	10.3	9.75	9.50	9.25	9.00	8.87	8.50	—	—
⅜	8.06	7.97	7.88	7.78	7.41	7.22	7.03	6.84	6.75	6.47	—	—
L6×4× ¾	12.6	12.4	12.2	12.0	11.3	10.9	10.5	10.1	9.94	—	—	—
⅝	10.6	10.5	10.3	10.2	9.53	9.22	8.91	8.60	8.44	—	—	—
½	8.62	8.50	8.37	8.25	7.75	7.50	7.25	7.00	6.88	—	—	—
⅜	6.56	6.47	6.38	6.28	5.91	5.72	5.53	5.34	5.25	—	—	—

Net areas are computed in accordance with AISC Specification Sect. B2.

AMERICAN INSTITUTE OF STEEL CONSTRUCTION

TENSION MEMBERS
Net areas

Angle Designation	Two angles—net area, In.²											
	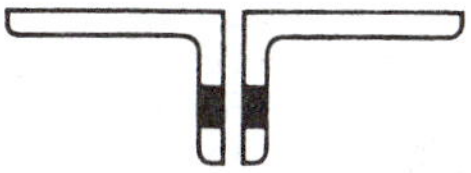 2 Holes out				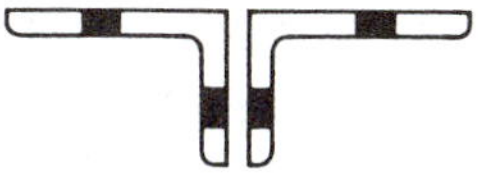 4 Holes out				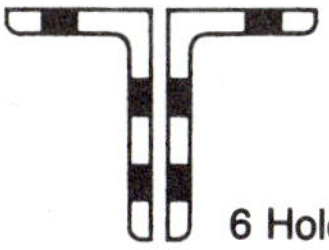 6 Holes out			
	Fastener Dia., In.				Fastener Dia., In.				Fastener Dia., In.			
	³⁄₄	⁷⁄₈	1	1⅛	³⁄₄	⁷⁄₈	1	1⅛	³⁄₄	⁷⁄₈	1	1⅛
L6×3½×⅜	6.18	6.09	6.00	—	5.53	5.34	—	—	4.87	—	—	—
⁵⁄₁₆	5.19	5.11	5.04	—	4.65	4.49	—	—	4.10	—	—	—
L5×5 ×⅞	14.4	14.2	14.0	13.8	12.9	12.5	12.0	11.6	—	—	—	—
³⁄₄	12.6	12.4	12.2	12.0	11.3	10.9	10.5	10.1	—	—	—	—
½	8.62	8.50	8.37	8.25	7.75	7.50	7.25	7.00	—	—	—	—
⅜	6.56	6.47	6.38	6.28	5.91	5.72	5.53	5.34	—	—	—	—
L5×3½×¾	10.3	10.1	9.93	—	8.99	8.62	—	—	—	—	—	—
½	7.12	7.00	6.87	—	6.25	6.00	—	—	—	—	—	—
⅜	5.44	5.35	5.25	—	4.79	4.60	—	—	—	—	—	—
⁵⁄₁₆	4.57	4.49	4.42	—	4.03	3.87	—	—	—	—	—	—
L5×3 ×½	6.62	6.50	—	—	5.75	5.50	—	—	—	—	—	—
⅜	5.06	4.97	—	—	4.41	4.22	—	—	—	—	—	—
⁵⁄₁₆	4.25	4.17	—	—	3.71	3.55	—	—	—	—	—	—
L4×4 ×¾	9.57	9.38	9.19	9.00	8.26	7.88	—	—	—	—	—	—
⅝	8.13	7.97	7.81	7.66	7.03	6.72	—	—	—	—	—	—
½	6.62	6.50	6.37	6.25	5.75	5.50	—	—	—	—	—	—
⅜	5.06	4.97	4.88	4.78	4.41	4.22	—	—	—	—	—	—
⁵⁄₁₆	4.25	4.17	4.10	4.02	3.71	3.55	—	—	—	—	—	—
L4×3½×½	6.12	6.00	5.87	—	5.25	5.00	—	—	—	—	—	—
⅜	4.68	4.59	4.50	—	4.03	3.84	—	—	—	—	—	—
⁵⁄₁₆	3.95	3.87	3.80	—	3.41	3.25	—	—	—	—	—	—
L4×3 ×½	5.62	5.50	—	—	4.75	4.50	—	—	—	—	—	—
⅜	4.30	4.21	—	—	3.65	3.46	—	—	—	—	—	—
⁵⁄₁₆	3.63	3.55	—	—	3.09	2.93	—	—	—	—	—	—
¼	2.94	2.88	—	—	2.50	2.38	—	—	—	—	—	—

Net areas are computed in accordance with AISC Specification Sect. B2.

REDUCTION OF AREA FOR HOLES
Area in square inches = diameter of hole × thickness of material

Thickness In.	Dia. of Hole, In.							
	3/4	13/16	7/8	15/16	1	1 1/16	1 1/8	1 3/16
3/16	.141	.152	.164	.176	.188	.199	.211	.223
1/4	.188	.203	.219	.234	.250	.266	.281	.297
5/16	.234	.254	.273	.293	.313	.332	.352	.371
3/8	.281	.305	.328	.352	.375	.398	.422	.445
7/16	.328	.355	.383	.410	.438	.465	.492	.520
1/2	.375	.406	.438	.469	.500	.531	.563	.594
9/16	.422	.457	.492	.527	.563	.598	.633	.668
5/8	.469	.508	.547	.586	.625	.664	.703	.742
11/16	.516	.559	.602	.645	.688	.730	.773	.816
3/4	.563	.609	.656	.703	.750	.797	.844	.891
13/16	.609	.660	.711	.762	.813	.863	.914	.965
7/8	.656	.711	.766	.820	.875	.930	.984	1.04
15/16	.703	.762	.820	.879	.938	.996	1.05	1.11
1	.750	.813	.875	.938	1.00	1.06	1.13	1.19
1/16	.797	.863	.930	.996	1.06	1.13	1.20	1.26
1/8	.844	.914	.984	1.05	1.13	1.20	1.27	1.34
3/16	.891	.965	1.04	1.11	1.19	1.26	1.34	1.41
1/4	.938	1.02	1.09	1.17	1.25	1.33	1.41	1.48
5/16	.984	1.07	1.15	1.23	1.31	1.39	1.48	1.56
3/8	1.03	1.12	1.20	1.29	1.38	1.46	1.55	1.63
7/16	1.08	1.17	1.26	1.35	1.44	1.53	1.62	1.71
1/2	1.13	1.22	1.31	1.41	1.50	1.59	1.69	1.78
9/16	1.17	1.27	1.37	1.46	1.56	1.66	1.76	1.86
5/8	1.22	1.32	1.42	1.52	1.63	1.73	1.83	1.93
11/16	1.27	1.37	1.48	1.58	1.69	1.79	1.90	2.00
3/4	1.31	1.42	1.53	1.64	1.75	1.86	1.97	2.08
13/16			1.59	1.70	1.81	1.93	2.04	2.15
7/8			1.64	1.76	1.88	1.99	2.11	2.23
15/16			1.70	1.82	1.94	2.06	2.18	2.30
2			1.75	1.88	2.00	2.13	2.25	2.38
1/16			1.80	1.93	2.06	2.19	2.32	2.45
1/8			1.86	1.99	2.13	2.26	2.39	2.52
3/16			1.91	2.05	2.19	2.32	2.46	2.60
1/4			1.97	2.11	2.25	2.39	2.53	2.67
5/16			2.02	2.17	2.31	2.46	2.60	2.75
3/8			2.08	2.23	2.38	2.52	2.67	2.82
7/16			2.13	2.29	2.44	2.59	2.74	2.89
1/2			2.19	2.34	2.50	2.66	2.81	2.97
5/8			2.30	2.46	2.63	2.79	2.95	3.12
3/4			2.41	2.58	2.75	2.92	3.09	3.27
7/8			2.52	2.70	2.88	3.05	3.23	3.41
3			2.63	2.81	3.00	3.19	3.38	3.56

NET SECTION OF TENSION MEMBERS

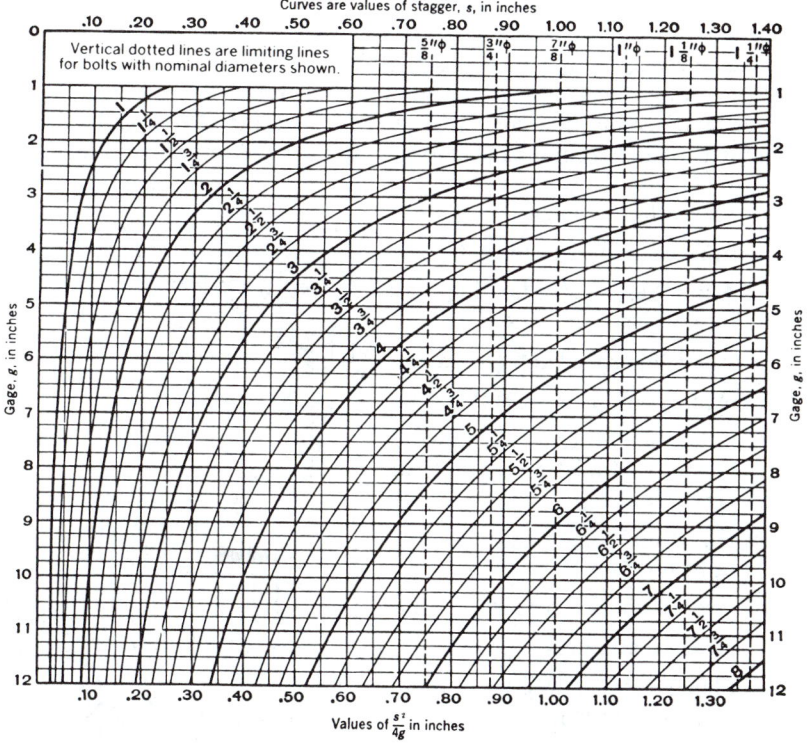

Curves are values of stagger, s, in inches

Vertical dotted lines are limiting lines for bolts with nominal diameters shown.

Gage, g, in inches

Values of $\frac{s^2}{4g}$ in inches

The above chart will simplify the application of the rule for net width, Sects. B1, B2 and B3 of the AISC ASD Specification. Entering the chart at left or right with the gage g and proceeding horizontally to intersection with the curve for the pitch s, then vertically to top or bottom, the value of $s^2/4g$ may be read directly.

Step 1 of the example below illustrates the application of the rule and the use of the chart. Step 2 illustrates the application of the 85% of gross area limitation applicable to connection fittings.

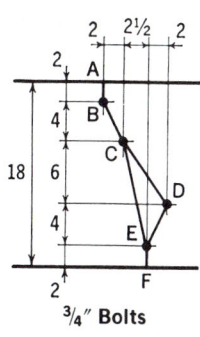

$\frac{3}{4}''$ Bolts

Step 1: Chain A B C E F

Deduct for 3 holes @ ($\frac{3}{4}$ + $\frac{1}{8}$)	= −2.625
BC, g = 4, s = 2; add $s^2/4g$	= +0.25
CE, g = 10, s = 2½; add $s^2/4g$	= +0.16
Total Deduction	= $\overline{-2.215}$ in.

Chain A B C D E F

Deduct for 4 holes @ ($\frac{3}{4}$ + $\frac{1}{8}$)	= −3.50
BC, as above, add	= +0.25
CD, g = 6, s = 4½; add $s^2/4g$	= +0.85
DE, g = 4, s = 2; add $s^2/4g$	= +0.25
Total Deduction	= $\overline{-2.15}$ in.

Net Width = 18.0 − 2.215 = 15.785 in.

Step 2: Net Width = 18.0 × 0.85 = 15.3 in.
(Governs in this example)

In comparing the path CDE with the path CE, it is seen that if the sum of the two values of $s^2/4g$ for CD and DE exceed the single value of $s^2/4g$ for CE, by more than the deduction for one hole, then the path CDE is not critical compared to CE.

Evidently if the value of $s^2/4g$ for one leg CD of the path CDE is greater than the deduction for one hole, the path CDE cannot be critical as compared with CE. The vertical dotted lines in the chart serve to indicate, for the respective bolt diameters noted at the top thereof, that any value of $s^2/4g$ to the right of such line is derived from a non-critical chain which need not be further considered.

MOMENT CONNECTIONS
Welded

Many moment connections can be designed to resist wind moments only and, at the same time, to rotate sufficiently to accommodate the simple beam gravity rotation as a Type 2 (simple) connection. See AISC ASD Specification Sect. A2.2. The following example is one of several connections recommended for such a design.

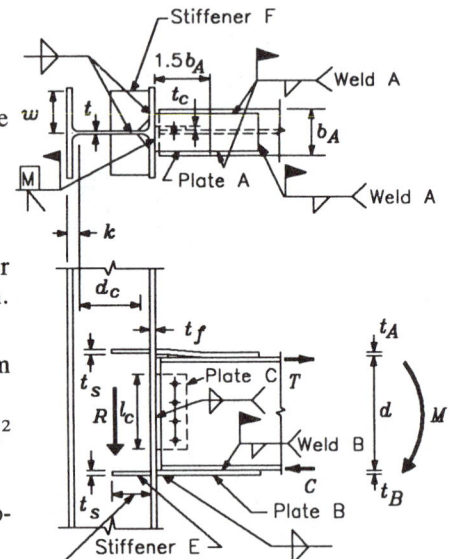

M	= Connection moment, kip-ft
R	= End reaction of beam, kips
r_v	= Allowable shear or bearing value for one fastener, kips
F_{vp}	= Allowable shear in plate, ksi
D	= Number of $\frac{1}{16}$-in. in weld size
$0.928D$	= Value of E70XX weld, kips per linear in. of fillet weld per $\frac{1}{16}$-in. leg
T	= Horizontal force top and bottom of beam, kips
A_p	= Area of plate, top or bottom, in.^{2}
A_{st}	= Area of stiffeners, in.^{2}
A_{bc}	= Planar area of web at beam-to-column connection, in.^{2}

If stiffeners are required they need not exceed one-half the column depth when beam is on one flange only.

The moment is assumed to be resisted by plates A and B welded to the top and bottom of beam and to the column. The shear is assumed to be transferred to the column by the vertical web plate C, using fasteners in the beam web and shop welds to the column.

An unwelded length of 1.5 times the top flange width b_A is assumed in this example to permit the elongation of the plate that is necessary to obtain the desired semi-rigid action.

DESIGN PROCEDURE

A. Determine horizontal force $T = (M \times 12)/d$

B. Design top plate A; determine length and size of weld A.

$$A_p = T/F_t$$

Length of weld = $T/0.928D$ (for E70XX Electrodes)

Select bottom plate B and determine length and size of weld B.
Area of plate B should be ≥ area of plate A.

C. Design the web connection:

1. No. of fasteners required for shear = R/r_v (Table I-D).

2. Check bearing on beam web (Table I-E or I-F).

3. Design shear plate C:

   a. Check net section in shear

              J4

   b. Check bearing on plate C (Table I-E).

   c. Determine weld size.

$$\text{Min. } D = \frac{R}{2 \times 0.928l} \text{ (for full length welds both sides)}$$

If intermittent or less than full length welds are used, min. plate thickness and weld size must be adjusted to satisfy AISC ASD Specification Sects. J2.2b and J2.4.

   d. Check minimum plate thickness for weld used.

D. Investigate column web shear:

Column must be reinforced in the panel zone with a doubler plate if the column web is not thick enough to resist the shear through the web in the plane below the top beam flange. Doubler plates are expensive and, in most cases, it would be more economical to choose a column section with a thicker web and avoid them.

E. Check column for web stiffeners

Column web stiffeners are required (for nomenclature, see Part 3, Columns, General Notes on Column Web Stiffeners):

At both flanges if $P_{bf} > t_b P_{wi} + P_{wo}$

At compression flange if $P_{bf} > P_{wb}$

At tension flange if $P_{bf} > P_{fb}$

Determine area of stiffeners.

$$A_{st} = \frac{P_{bf} - F_{yc}t_{wc}(t_b + 5k)}{F_{yst}} = \frac{P_{bf} - t_b P_{wi} - P_{wo}}{F_{yst}}$$

Required stiffeners must comply with the provisions of Sect. K1.8.

Selected stiffeners must comply with width-thickness ratio limitation.

F. Determine weld requirements of stiffeners to column web and flange.

EXAMPLE 35

Given:

Design a semi-rigid connection of a W18×50 beam framed to the flange of an exterior W14×109 column. The end moment of 150 kip-ft and end reaction of 50 kips are from live and dead load only. Material is A36, bolts are ¾-in. dia. A325-X, weld is E70XX.

From Properties Tables, Part 1:

Beam (W18×50): d = 17.99 in., b_f = 7½ in.
t_w = 0.355 in., t_f = 0.570 in.

Column (W14×109): d = 14.32 in., b_f = 14⅝ in.
t_{wc} = 0.525 in., t_f = 0.860 in.

Solution:

A. Horizontal force at beam flange:

$T = (M \times 12)/d = (150 \times 12)/18 = 100$ kips

B. $A_p = T/F_t = 100/22 = 4.55$ in.^{2}

Top flange plate A:

Select 6-in. wide plate:

$t_A = 4.55/6 = 0.76$ in. (use ¾ × 0'-6 plate)

Bottom flange plate B:

Select 9-in. wide plate:

$t_B = 4.55/9 = 0.51$ in. (use ½ × 0'-9 plate)

Design Welds A and B:

Select ⁵⁄₁₆-in. fillet weld, E70XX.

Length of weld = $100/(5 \times 0.928) = 22$-in. minimum.

Weld A: Use 6 in. across and 8 in. along each side.

Weld B: Use 11 in. along each side.

C. Design web connection:

1. Connect beam web to column shear plate with ¾-in. dia. A325-X bolts.

No. bolts required = $R/r_v = 50/13.3 = 3.8$; use 4 bolts

2. Check bearing on beam web:

Assume $l_v \geq 1\frac{1}{2}\, d$.

For F_u = 58 ksi and t_w = 0.355 in. (use Table I-F):

Allowable load = $52.2 \times 0.355 \times 4 = 74.1$ kips > 50 kips

3. Design shear plate C:

a. Try a shear plate with 3-in. pitch, 1¼-in. end distance, l_c = 12 in.:

$l_{net} = 12 - 4\ (¾ + \frac{1}{16}) = 8.75$ in.

$F_{vp} = 0.30 \times 58 = 17.4$ ksi

$t_c = 50/(17.4 \times 8.75) = 0.33$ in.

Try a ⅜-in. plate.

b. Check bearing:

From Table I-E, for F_u = 58 ksi, l_v = 1¼ in. > 1.5 × ¾ = 1⅛ in.

Allowable load: 19.6 kips/bolt at 3-in. spacing

19.6 kips/bolt > $50/4 = 12.5$ kips/bolt **o.k.**

c. Determine required weld to column flange:

$$D_{min} = \frac{50}{2 \times 0.928 \times 12} = 2.25$$

Since the column flange thickness is over ¾ in., use ⁵⁄₁₆-in. fillet weld (Specification Table J2.4).

Use: Shear connection plate ⅜ × 5 × 1'-0 welded to the column flange with ⁵⁄₁₆-in. fillet welds full length each side.

D. Investigate column web shear: Story shear, $V_s = 0.0$ kips

From Commentary Sect. E6:

$$\text{Web panel shear} = \Sigma F = \frac{M_1}{0.95 \, d_1} + \frac{M_2}{0.95 \, d_2} - V_s$$

$$\Sigma F = \frac{150 \times 12}{0.95 \times 17.99} + 0 - 0 = 105 \text{ kips}$$

Web resisting force $= 0.4 \times F_y \times t_w \times d$
$$= 0.4 \times 36 \times 0.525 \times 14.32 = 108 \text{ kips}$$

108 kips > 105 kips **o.k.**

∴ column web need not be reinforced

E. Determine need for column web stiffeners:

$$\text{Horizontal force at stiffeners} = \frac{M \times 12}{d + \frac{1}{2}(t_A + t_B)}$$

$$= \frac{150 \times 12}{17.99 + \frac{1}{2}(\frac{3}{4} + \frac{1}{2})}$$

$$= 96.7 \text{ kips}$$

From Column Load Tables for W14×109:

$P_{wo} = 148$ kips

$P_{wi} = 19$ kips/in.

$P_{wb} = 316$ kips

$P_{fb} = 166$ kips

$P_{bf} = \frac{5}{3} \times 96.7 = 161.2$ kips

Column web stiffeners with cross-sectional area A_{st} are required opposite both the tension and compression flanges of the beam whenever Eq. (K1-9) gives a positive answer:

$$A_{st} = \frac{P_{bf} - F_{yc}t_{wc} \, (t_b + 5k)}{F_{yst}}$$

$$= \frac{161.2 - 36(0.525)[\frac{1}{2} + (5 \times 1\frac{9}{16})]}{36}$$

$$= 0.114 \text{ in.}^2$$

Alternatively, this calculation could have been made as follows:

$$A_{st} = \frac{P_{bf} - t_b P_{wi} - P_{wo}}{F_{yst}}$$

$$= \frac{161.2 - (\frac{1}{2} \times 19) - 148}{36} = 0.103 \text{ in.}^{2*}$$

Note that stiffeners are required. If this formula had a negative answer, it would require a further check by Eqs. (K1-1) and (K1-8), as follows:

At compression flange, stiffeners are required if

$$d_c > \frac{4100 t_{wc}{}^3 \sqrt{F_{yc}}}{P_{bf}} \text{ or } P_{bf} > P_{wb}$$

At tension flange, stiffeners are required if

$$t_f < 0.4 \sqrt{\frac{P_{bf}}{F_{yc}}} \text{ or } P_{bf} > P_{fb}$$

The engineer may judge that the required A_{st} is so small that the stiffeners can be omitted. If stiffeners are provided, they must be proportioned to comply with AISC ASD Specification criteria in Sect. K1.8:

1. $w + t/2 \geq b/3$

   where

   w = stiffener width
   t = thickness of column web
   b = width of connecting plate or beam flange

   $w = (b/3) - (t/2) = (9/3) - (0.525/2)$

   $= 2\frac{3}{4} \text{ in. (min.)}$

2. Stiffener thickness $t_s \geq t_b/2$

   where t_b = thickness of connecting plate or beam flange

   $t_s = 0.75/2 = 0.375$ in. at top flange

   $= 0.50/2 = 0.25$ in. at bottom flange

   For practical detailing considerations use a stiffener $\frac{1}{2} \times 4$ in. on each side at top and bottom flange. Clip corner $\frac{3}{4}$ in. $\times \frac{3}{4}$ in.

3. Length of stiffener need not exceed one-half the column web depth [unless they are required by Sects. K1.4 or K1.6. See Sect. K1.8]:

   Check stiffeners for slenderness:

   Area furnished = $4 \times \frac{1}{2} \times 2 = 4.0 \text{ in.}^2 > 0.114 \text{ in.}^2$ **o.k.**

   Width-thickness ratio (Sect. B5.1):

   $w/t = 4/0.5 = 8.0 < 95/\sqrt{F_y} = 15.8$ **o.k.**

   Length = $(14\frac{3}{8} / 2) - \frac{7}{8} = 6\frac{5}{16}$ in. (use $6\frac{1}{2}$ in.)

---

* The minor discrepancy from the Eq. (K1-9) result is due to rounding of the terms.

F. Stiffener weld requirements:

Use fillet welds, E70XX.

From AISC ASD Specification Table J2.4:

Min. weld size to web: ¼ in.

Min. weld size to flange: ⁵⁄₁₆ in.

Weld lengths:

Length of weld must be sufficient to develop the following (see Step E).

$$P_{bf} - F_{yc}\, t\,(t_b + 5k) = 161.2 - 157.1 = 4.1 \text{ kips}$$

or

$$P_{bf} - t_b P_{wi} - P_{wo} = 161.2 - (½ \times 19) - 148 = 3.7 \text{ kips}$$

Minimum length of ⁵⁄₁₆-in. fillet weld to tension flange:

4.1/(2 × 5 × 0.928 × 1.67) = 0.3 in.

Minimum length of ¼-in. fillet weld to web:

4.1/(2 × 4 × 0.928 × 1.67) = 0.3 in.

Since the problem, as stated, does not have reversible end moments, it would be permissible to finish the two compression flange stiffeners to bear instead of welding to the flange, using $F_p = 0.9 \times 36 = 32.4$ ksi (Sect. J8).

$$f_p = \frac{4.1}{2\,(4 - ¾) \times ½} = 1.3 \text{ ksi} < 32.4 \text{ ksi} \quad \textbf{o.k.}$$

If welding is used, it should be the same at tension and compression stiffeners.

Note that even though compression stiffeners are not indicated by Eq. (K1-1) or (K1-8), they must be furnished as a requirement of Eq. (K1-9).

MOMENT CONNECTIONS
Field-welded, field-bolted

Many framing systems are designed as Type 1 (rigid-frame) and the connections must be designed to develop the frame moments. The following example illustrates the design of a moment connection that may be used in rigid-frame construction. For nomenclature, see Moment Connections, Welded.

The full plastic moment of the beam can be developed by welding the flanges. The shear is assumed to be transferred to the column by a vertical plate shop welded to the column and field connected to the beam web.

DESIGN PROCEDURE

A. Check beam strength.

B. Check beam web connection.

C. Investigate column web high shear.

D. Determine need for stiffeners.

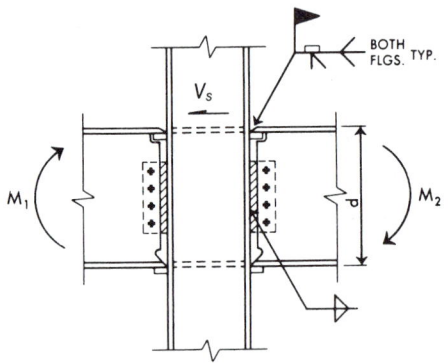

EXAMPLE 36

Given:

Design a moment connection for a W24×55 beam framed to both flanges of a W14×99 column. The design wind moments are $M_1 = M_2 = 105$ kip-ft and the end reaction is 33 kips. Column web shear is 18.0 kips. All material is ASTM A36 steel. Use ⅞-dia. A325-N bolts and E70 electrodes. Gravity load moment = 125 kip-ft.

Solution:

From Properties and Dimensions Tables, Part 1:

Beam (W24×55): d = 23.57 in., b_f = 7.005 in.
t_w = 0.395 in., t_f = 0.505 in.

Column (W14×99): d = 14.16 in., b_f = 14.57 in.
t_{wc} = 0.485 in., t_f = 0.78 in.
k = 1⁷⁄₁₆ in., T = 11¼ in.

Research performed at the University of California and Lehigh University has demonstrated the full plastic moment capacity of the girder can be developed by welding only the flanges. Therefore, a full-penetration weld will be used to connect the girder flanges to the column. A plate will be used to transfer the shear.

A. Check beam strength:

$$S_{req'd} = (230)(12)/0.66(36)(1.33) = 87.3 \text{ in.}^3 < 114 \text{ in.}^3 \quad \textbf{o.k.}$$

B. Design web connection:

$$33/12.6 = 2.6, \text{ try 4 bolts}$$

Check bearing on beam web

From Table I-E

$4 \times 60.9 \times 0.395 = 96$ kips > 33 kips   **o.k.**

Design shear plate

Try $\frac{5}{16}$-plate, 3-in. pitch, $l = 12$ in.

$A_{net} = .3125 \times [12 - 4(\frac{7}{8} + \frac{1}{16})] = 2.58$ in.^{2}

$0.30 \times 58 \times 2.58 = 44.9$ kips > 33 kips   **o.k.**

Check bearing on shear plate

From Table I-E

$4 \times 19.0 = 76.0$ kips > 33 kips   **o.k.**

Weld to column flange

$D_{min} = 33/(2 \times 0.928 \times 12) = 1.5$ in.

Use $\frac{5}{16}$-in. fillet. See Table J2.4.

C. Investigate column web high shear, $V_s = 18.0$ kips

$M_{1L} = M_{2L} = 105$ kip-ft, wind: net $= 105 + 105$

$M_{1G} = M_{2G} = 125$ kip-ft, gravity (dead load): net $= 0$

Web panel shear $= \Sigma F = \dfrac{M_1}{0.95 \, d_1} + \dfrac{M_2}{0.95 \, d_2} - V_s$

$\qquad = \dfrac{(105 + 125) \times 12}{0.95 \times 23.57} + \dfrac{(105 - 125) \times 12}{0.95 \times 23.57} - 18.0$

$\qquad = 94.6$ kips

Web resisting force $= 0.4 \times F_y \times t_w \times d$

$\qquad\qquad = 0.4 \times 36 \times 0.485 \times 14.16 \times 1.33 = 132$ kips > 94.6

Doubler plate not required.

D. Determine column-web stiffener requirements

Horizontal force at stiffeners:

$\dfrac{M \times 12}{d - t_f} = \dfrac{230 \times 12}{23.57 - 0.505} = 120$ kips

From Column Load Tables for W14×99

$P_{wo} = 125$ kips (yielding)
$P_{wi} = 17$ kips/in. (yielding)
$P_{wb} = 249$ kips (buckling)
$P_{fb} = 137$ kips (tension)
$P_{bf} = 120 \times 4/3 = 160$ kips

$P_{wo} + t_b P_{wi} = 125 + (0.505 \times 17) = 134$ kips < 160 kips; stiffeners required
$P_{wb} = 249$ kips > 160 kips; stiffeners not required
$P_{fb} = 137$ kips < 160 kips; stiffeners required

Stiffeners are required at both the compression flange and the tension flange. For stiffener design guidelines, see AISC ASD Specification Sect. K1.8.

Stiffener width = 7.005/3 − 0.485/2 = 2.09 in. use 2½ in.
Stiffener thickness = 0.505/2 = 0.25 in. use ¼ in.
Stiffener length = (14.16/2) − 0.78 = 6.3 in. use 6½ in.
Stiffener area req'd = (160 − 134)/36 = 0.72 in.^{2}
Stiffener area supplied = 2 × (2½ × ¼) = 1.25 in.^{2} **o.k.**

E. Stiffener weld requirements
   Use fillet welds, E70 electrodes.
   From Table J2.4:
      Min. weld size to column web: ³⁄₁₆ in.
      Min. weld size to column flange: ⁵⁄₁₆ in.

Weld length calculation:
   Force on compression flange: 160 − 134 = 26.0 kips
   Force on tension flange: 160 − 137 = 23.0 kips
      Compression flange governs.
   Load in stiffener = 26/2 = 13 kips
   Remove load factor of 4/3 to determine the weld length (web),
   13/(2 × 3 × 0.928 × (4/3)) = 1.8 in.
      Make weld full length of stiffener:
      6½ − ¾ (clip) = 5¾ in., both sides.
   Remove load factor of 4/3 to determine the weld length (flange),
   13/(2 × 5 × 0.928 × (4/3)) = 1.05
      2½ − ¾ (clip) = 1¾ in., both sides.
Check shear stress in stiffener base metal:

$$f_v = \frac{13 \times \frac{3}{4}}{\frac{1}{4} \times 5\frac{3}{4}}$$

$$= 6.78 \text{ ksi} < 0.4 \times 36 \times 1.33 = 19.2 \text{ ksi} \quad \textbf{o.k.}$$

It may be permissible to furnish the two compression flange stiffeners to bear instead of welding to the column flange. See AISC ASD Specification Sect. J8.

13 × ¾ kips < 0.9 × 36 × ¼ × 2½ × 1.33 = 27 kips **o.k.**

MOMENT CONNECTIONS
Shop-welded, field-bolted

Many framing systems are designed as Type 1 (rigid-frame) and the connections must be designed to develop the frame moments. The following example illustrates the design of a moment connection that may be used in rigid-frame construction. For nomenclature, see Moment Connections, Welded.

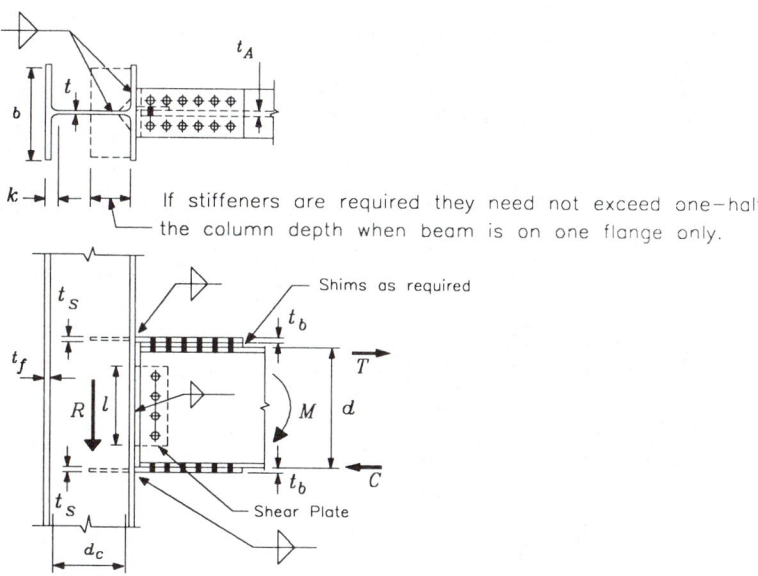

If stiffeners are required they need not exceed one-half the column depth when beam is on one flange only.

The moment is assumed to be resisted by flange plates shop welded to the column and field-connected to the beam flanges. The shear is assumed to be transferred to the column by a vertical plate shop welded to the column and field-connected to the beam web.

	ASD Specification Reference
DESIGN PROCEDURE	
A. Determine beam flange area reduction for fastener holes, if req'd.	**B2** **B10**
B. Determine horizontal force at beam flange: $T = (M \times 12)/d$	
C. Design flange plates: Gross section: $f_t = T/A_{gross} \le 0.6\,F_y$ Net section: $f_t = T/A_{net} \le 0.5\,F_u$	**B3** **D1** **J3**
D. Determine the number of fasteners required to develop the horizontal force in the flanges: No. of fasteners = T/r_v	

E. Design the web connection:

No. of fasteners required for shear = R/r_v (Table I-D).

Check bearing on beam web (Table I-E).

Check shear on plate.

Check bearing on plate.

Determine weld requirements (AISC ASD Specification Table J2.4).

F. Check column web shear:

Column must be reinforced in the panel zone with a doubler plate if the column web is not thick enough to resist the shear through the web in the plane below the top beam flange. Doubler plates are expensive and, in most cases, it would be more economical to choose a column section with a thicker web and avoid them.

G. Check column for web stiffeners:

Column web stiffeners are required (for nomenclature, see Part 3, Columns, General Notes on Column Web Stiffeners):

At both flanges if $P_{bf} > t_b P_{wi} + P_{wo}$

At compression flange if $P_{bf} > P_{wb}$

At tension flange if $P_{bf} > P_{fb}$

Determine area of stiffeners, if required:

$$A_{st} = \frac{P_{bf} - F_{yc} t_{wc}(t_b + 5k)}{F_{yst}}$$

$$= \frac{P_{bf} - t_b P_{wi} - P_{wo}}{F_{yst}}$$

where

P_{bf} = factored beam flange or connection plate force in a restrained connection, kips

Required stiffeners must comply with the provisions of Sect. K1.8.

Selected stiffeners must comply with width-thickness ratio limitation.

H. Determine weld requirements of stiffeners to column web and flange.

Note: Oversize holes in the moment flange plates assist in the field assembly of this type connection by compensating for the rolling, fabrication, and erection tolerances.

ASD Specification Reference

J2

Commentary E6

K1.2
K1.6
K1.8

Equation (K1-9)

B5.1

J2

EXAMPLE 37

Given:

Design a moment connection for a W18×55 beam framing into each side of a W14×99 column. The design moment of 153 kip-ft and the end reaction of 40 kips are results of dead and live load only. All material is ASTM A36 steel with $F_t = 22$ ksi (for beam, $F_b = 24$ ksi). Use A325 bolts and E70XX electrodes. Oversize holes are permitted in the flange connection plates.

From Properties and Dimensions Tables, Part 1:

Beam (W18×55): d = 18.11 in., b_f = 7.530 in.
t_w = 0.390 in., t_f = 0.630 in.

Column (W14×99): d = 14.16 in., b_f = 14.565 in.
t_{wc} = 0.485 in., t_f = 0.780 in.
k = 1⅟₁₆ in., T = 11¼ in.

Solution:

$M = 153$ kip-ft

$S_{req} = (153 \times 12)/24.0 = 76.5$ in.$^3 < 98.3$ in.^{2} **o.k.**

Assume 2 rows of ⅞-in. dia. A325 bolts in a slip-critical connection.

A. Beam-flange area reduction:

A_{fg} (gross) $= 7.530 \times 0.630 = 4.74$ in.

A_{fn} (net) $= 4.74 - 2(0.875 + 0.125)(0.63) = 3.48$ in.

$0.5 F_u A_{fn} = 0.5 (58)(3.48) = 100.9$ kips

$0.6 F_y A_{fg} = 0.6 (36)(4.74) = 102.4$ kips

Since 100.9 kips < 102.4 kips, the effective tension flange area is,

$$A_{fe} = \frac{5}{6}\left(\frac{58}{36}\right)(3.48) = 4.67 \text{ in.}^2$$

$$I \text{ (net)} = 890 - \left[(4.74 - 4.67)\left(\frac{18.11 - 0.63}{2}\right)^2\right] = 884 \text{ in.}^4$$

S (net) $= 884/9.06 = 97.6$ in.$^3 > 76.5$ in.^{3} **beam is o.k.**

B. Horizontal force at beam flange:

$T = (M \times 12)/d = (153 \times 12)/18.11 = 101.4$ kips

C. Design flange plates:

Try plate ⅞ × 7¾

Gross section

$$f_t = \frac{101.4}{⅞ \times 7¾} = 15.0 < 22.0 \text{ ksi} \quad \textbf{o.k.}$$

Net section

Net area $= ⅞ [(7¾) - 2 (1⅟₁₆ \text{ oversize} + ⅟₁₆ \text{ net})]$ (Table J3.1)
$= 4.81 < 0.85 \times 6.78 = 5.76$ in.^{2}

$$f_t = \frac{101.4}{4.81} = 21.1 < 29.0 \text{ ksi} \quad \textbf{o.k.}$$

D. Flange connection:

Assume ⅞-in. dia. A325 bolts in a slip-critical connection in oversize holes, clean mill scale.

No. bolts required:

$T/r_v = 101.4/9.02 = 11.2$ (Table I-D)

Use: 12 – $\frac{7}{8}$-in. dia. A325-SC bolts.

E. Web connection:

Assume $\frac{7}{8}$-in. dia. A325 bolts in a bearing-type connection with threads included in the shear plane.

No. of bolts required for shear:

$R/r_v = 40/12.6 = 3.17$ (Table I-D)

Try 4 bolts.

Check bearing on beam web:

Assume $1\frac{1}{2}$ in. end distance.

$1\frac{1}{2}$ in. $> 1\frac{1}{2}$ ($\frac{7}{8}$) = 1.31 in. Use Table I-E.

From Table I-E for $F_u = 58$ ksi, $t = 0.39$ in.,

Allowable load = $60.9 \times 0.39 \times 4 = 95$ kips

Use: Four $\frac{7}{8}$-in. dia. A325-N bolts.

Design shear plate:

Try a shear plate with 3-in. pitch, $1\frac{1}{2}$-in. end distance, $l = 12$ in.

$l_{net} = 12 - 4 (\frac{7}{8} + \frac{1}{8}) = 8$ in.

$F_{vp} = 0.30 \times 58 = 17.4$ ksi

$t_A = 40/(17.4 \times 8) = 0.29$ in.

Try a $\frac{5}{16}$-in. plate

Check bearing:

From Table I-E for $F_u = 58$ ksi, $t = \frac{5}{16}$ in.

Allowable load = 19 kips/bolt

19 kips/bolt $> 40/4 = 10$ kips/bolt

Determine required weld to column flange:

$$D_{min} = \frac{40}{2 \times 0.928 \times 12} = 1.80$$

Since the column flange thickness is over $\frac{3}{4}$ in., use $\frac{5}{16}$-in. fillet weld (AISC ASD Specification Table J2.4).

Use: Shear connector plate $\frac{5}{16} \times 5 \times 12$ in. welded to the column flange with $\frac{5}{16}$-in. fillet welds full length each side. Designate on the drawings that weld is to be built out to obtain $\frac{5}{16}$-in. throat thickness.

F. Investigate column web shear:

Connection is balanced so web shear is zero.

G. Determine need for column web stiffeners:

Horizontal force at stiffeners $= \dfrac{M \times 12}{d + t_b}$

$$= \dfrac{153 \times 12}{18.11 + \tfrac{7}{8}}$$

$$= 96.7 \text{ kips}$$

$P_{bf} = \tfrac{5}{3} \times 96.7 = 161 \text{ kips}$

Column web stiffeners with cross-sectional area A_{st} are required opposite both the tension and compression flange connection of the beam whenever Eq. (K1-9) gives a positive answer:

$$A_{st} = \dfrac{P_{bf} - F_{yc}t_{wc}\,(t_b + 5k)}{F_{yst}}$$

$$= \dfrac{161 - 36(0.485)[\tfrac{7}{8} + (5 \times 1\tfrac{7}{16})]}{36}$$

$$= \dfrac{161 - 140.8}{36} = 0.56 \text{ in.}^2$$

Alternatively, this calculation could have been made as follows:

$$A_{st} = \dfrac{P_{bf} - t_b P_{wi} - P_{wo}}{F_{yst}}$$

$$= \dfrac{161 - (\tfrac{7}{8} \times 17) - 125}{36}$$

$$= \dfrac{161 - 139.9}{36} = 0.59 \text{ in.}^{2*}$$

Note, since the answer is positive, stiffeners are required at both the tension and compression flanges. If this answer had been negative, it would require a further check by Eqs. (K1-1) and (K1-8), as follows:

Stiffeners are required opposite the compression flange if:

$$d_c > \dfrac{4100 t_{wc}{}^3 \sqrt{F_{yc}}}{P_{bf}} \text{ or } P_{bf} > P_{wb}$$

Stiffeners are required opposite the tension flange if:

$$t_f < 0.4 \sqrt{\dfrac{P_{bf}}{F_{yc}}} \text{ or } P_{bf} > P_{fb}$$

In either case the stiffeners must be proportioned to comply with Specification criteria in Sect. K1.8:

Stiffener width:

$$w + t_{wc}/2 \geq b/3$$

where

$w = $ stiffener width

$t_{wc}= $ thickness of column web

$b = $ width of connecting plate or beam flange

---

* The minor discrepancy from the Eq. (K1-9) result is due to rounding of the terms.

$w = (b/3) - (t_{wc}/2) = (7.75/3) - (0.485/2) = 2.34$ in. (min.)

Stiffener thickness:

$t_s \geq t_b/2$

where t_b = thickness of connecting plate or beam flange

$t_s = 0.875/2 = 0.4375$ in. (min.) at top and bottom flanges

For practical detailing considerations, use a stiffener ½ × 4 in. on each side at top and bottom flanges. Clip corner ¾ in. × ¾ in.

Check stiffeners:

Area furnished $= 4 \times \frac{1}{2} \times 2 = 4$ in.$^2 > 0.56$ in.^{2} **o.k.**

Width-thickness ratio [Sect. B5.1]:

$w/t = 4/0.5 = 8.0 \leq 95/\sqrt{F_y} = 95/\sqrt{36} = 15.8$ **o.k.**

Length of stiffener should be not less than one-half the column depth [Sect. K1.8]:

Length $= (14.16/2) - \frac{7}{8} = 6.2$ in. (use 6½ in.)

H. Stiffener weld requirements:

Use fillet welds, E70XX.

From AISC ASD Specification Table J2.4:

Min. weld size to column web: ³⁄₁₆ in.

Min. weld size to column flange: ⁵⁄₁₆ in.

Weld lengths:

Determine forces at column web to be resisted by stiffener welds (see Step G):

$P_{bf} - F_{yc}t_{wc}(t_b + 5k) = 161 - 140.8 = 20.2$ kips

or

$P_{bf} - t_bP_{wi} - P_{wo} = 161 - 139.9 = 21.1$ kips

Determine forces at tension flange to be resisted by stiffener welds:

Using Eq. (K1-1):

$$P_{bf} - \frac{t_f^2 F_{yc}}{0.16} = 161 - \frac{(0.780)^2 \times 36}{0.16} = 24 \text{ kips}$$

or, using Properties from Column Load Tables,

$P_{bf} - P_{fb} = 161 - 137 = 24$ kips

Tension flange force governs.

Weld requirements:

Load in stiffener $= 24/2 = 12$ kips each side

$$l_w = \frac{12}{2 \times 3 \times 0.928 \times 1.67} = 1.3 \text{ in. (min. length of } \frac{3}{16}\text{-in. fillet weld to web)}$$

Make weld full length of stiffener: $6\frac{1}{2} - \frac{3}{4} = 5\frac{3}{4}$ in. both sides.

$$l_f = \frac{12}{2 \times 5 \times 0.928 \times 1.67}$$

$= 0.8$ in. (min. length of ⁵⁄₁₆-in. fillet weld to tension flange)

Make weld full width of stiffener: $4 - \frac{3}{4} = 3\frac{1}{4}$ in both sides.

Minimum thickness of plate to develop $\frac{5}{16}$ fillet welds at maximum shear stress $= 0.64$ in.

Check shear stress in weld metal:

$$f_v = \frac{12}{2 \times \frac{5}{16} \times 0.707 \times 3\frac{1}{4}} = 8.4 \text{ ksi} < 0.3 \times 70 \times 1.67 = 35.1 \text{ ksi} \quad \textbf{o.k.}$$

Check shear stress in stiffener base metal:

$$f_v = \frac{12}{\frac{1}{2} \times 3\frac{1}{4}} = 7.4 \text{ ksi} < 0.4 \times 36 \times 1.67 = 24.0 \text{ ksi} \quad \textbf{o.k.}$$

Since the problem, as stated, does not have reversible end moments, it is permissible to finish the two compression flange stiffeners to bear instead of welding to the flange.

$$F_p = 0.90F_y = 0.90 \times 36 \times 1.67 = 54.1 \text{ ksi} \quad [\text{Sect. J8}]$$

$$f_p = \frac{24}{2(4 - \frac{3}{4}) \times \frac{1}{2}} = 7.4 \text{ ksi} < 54.1 \text{ ksi} \quad \textbf{o.k.}$$

If welding is used, it would be the same at tension and compression stiffeners.

Note that even though compression stiffeners are not indicated by Eq. (K1-8), they must be furnished as a requirement of Eq. (K1-9).

MOMENT CONNECTIONS
End Plate (Static Loading Only)

GENERAL NOTES

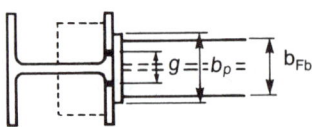

Section A-A

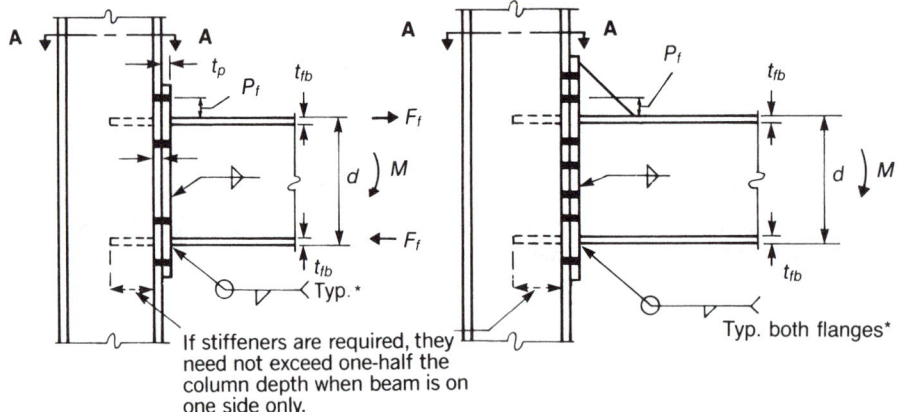

4-tension Bolts **8-tension Bolts**

If stiffeners are required, they need not exceed one-half the column depth when beam is on one side only.

*For fillet welds size greater than ½ in, full penetration welds with reinforcement should be considered.

1. Both four-tension bolt and eight-tension bolt end plates are discussed on the following pages. The eight-bolt type will carry greater moments than the four bolt and therefore a wider range of beam sizes may be used. Additionally, even though the applied moment is not beyond the limit of four-tension bolts, an eight-bolt type might be chosen because it may eliminate the need for stiffeners in the column opposite the tension flange (see item 6C under Notes for Four Tension Bolt Type). Stiffeners in columns are an item of expense. More importantly, they interfere with connections into the column web and should be avoided if at all possible.

2. Bolts must be fully-tightened for end-plate connections.

3. Tension bolts in slip-critical, end-plate connections are fully effective for shear (see Sect. C5 Commentary on Specification for Structural Joints Using ASTM A325 or A490 Bolts, Manual, Part 5). In bearing type end-plate connections, bolts subjected to both tension and shear can be designed using a shear-tension interaction equation. For end plate connections, 2 to 8 or more bolts are always provided in addition to the tension bolts. These additional bolts are fully effective for shear. Also, fully stressed tension bolts can be additionally stressed in shear to at least ¼ of their normal shear value.

End reaction shear is seldom a problem in end-plate connections and calculations for bolt shear are not shown in the following examples. However, a designer may want to check bolt shear, particularly in cases of deep short span beams subjected primarily to concentrated loads.

4. End-plate material for custom fabricators should preferably be A36.

5. Attention should be given to the possibility of reverse, or partially reversed moments and the design adjusted accordingly.

6. To allow for bolt driving clearances, the minimum value of P_f must be equal to the bolt diameter plus ½ in. Thus, for 1½-in. dia. bolts, P_f must be at least 2 in. It may be practical and economical to standardize on a value of $P_f = 2$ in. to minimize drafting and fabrication costs.*

NOTES FOR FOUR-TENSION BOLT TYPE

1. When the high-strength bolts in an end-plate moment connection are located at a distance P_f above and approximately the same distance below the beam tension flange, the force applied to each bolt by the beam end moment can be considered equal.

2. When the applied moment is less than the moment capacity of the beam, the bolts and end plate may be designed for the applied moment only. However, the beam web weld to the end plate should develop the capacity of the web in tension at $0.60F_y$ for a distance of approximately $2P_f$ below the beam tension flange. The balance of the web to end-plate weld should be continuous, but may usually be a minimum size as required by the end-plate thickness.

3. Where possible, the end-plate width should be approximately 1 in. greater than the width of the beam flange for proper welding. This width of $b_f + 1$ is also the maximum effective width which may be used in design calculations.

4. Maximum bolt size is 1½-in. dia. No research on end plates has been done with larger bolts.

5. A325 or A490 bolts may be used.

6. Column stiffeners are required for the following cases:

   A. For column web yielding opposite the compression flange, when

   $$P_{bf} \geq F_{yc}t_{wc}(t_{fb} + 6k + 2t_p + 2w)$$

   B. For column-web buckling, when

   $$P_{bf} \geq \frac{4100t_{wc}^3 \sqrt{F_{yc}}}{d_c}$$

   C. Opposite the beam tension flange, when $t_{fc} < t_p$ (req'd) using the four tension bolt end-plate design procedure** with the following assumptions:

   1. Effective column flange length (analogous to end-plate width):

   $$b_p = 2.5 (P_f + t_{fb} + P_f)$$

---

* Griffiths, John D. End-plate Moment Connections—Their Use and Misuse. *AISC Engineering Journal*, Vol. 21, 1st Qtr., 1984, Chicago, Ill.

** Krishnamurthy, N. A Fresh Look at Bolted End-plate Behavior and Design. *AISC Engineering Journal*, Vol. 15, 2nd Qtr., 1978, Chicago, Ill.

Note: For $P_f = 2$ in. and with the inner tension bolts located 3 in. below the top of the beam flange the effective column flange length is $2.5 (2 + 3) = 12.5$ in.

2. $P_{ec} = (g/2) - k_1 - (d_b/4)$

Where P_{ec} = effective horizontal bolt distance analogous to P_e for end plates.*

3. $C_b = 1.0$
$A_f/A_w = 1.0$

7. In the preliminary design procedure which follows:

The calculated bolt diameters obtained from the preliminary design should not vary from the four-tension bolt end-plate design procedure by more than 3 to 4%. There will be a greater variation for end plate thicknesses—generally not more than 10% undersize or oversize. Ten percent should be added to the preliminary design end plate thicknesses for beams with depths of 14 in. and under.

A point of inflection is assumed halfway between the beam flange or weld and the centerline of the upper bolt. Thus the lever arm is reduced from P_f to $P_f/2$. Note that nominal beam depths and flange widths are used to simplify the preliminary design.

The P_f value selected may be reduced by the leg size of the fillet weld or the leg size of the reinforcing fillet (if used) for groove welds.

The thickness of the end plate is determined by limiting the calculated extreme fiber bending stress to $0.75F_y$, which for A36 steel is 27 ksi.

$$t_p = \sqrt{\frac{6 \times M_e}{27 \times b_p}}$$

EXAMPLE 38

Preliminary Design

Given:

Design a four-tension bolt end-plate connection for a W18×50 (7½ × 9/16-in. flange) with a 160 kip-ft moment and a 40-kip end reaction framing into the flange of a W14×132 column, no reverse moment. A36 steel. Assume $P_f = 2.0$ in.

Solution:

End Plate Width $b_p = b_{fb} + 1 = 7.5 + 1 = 8.5$ in.

Flange Force $F_f = \dfrac{M}{\text{Nominal depth}} = \dfrac{160 \times 12}{18} = 107$ kips

Bolt force $T = \dfrac{107}{4} = 26.8$ kips

Use 1-in. dia. A325 bolts (allowable tension = 34.6 kips).

Flange Weld $D = \dfrac{107}{0.928 \, [2 \, (7.495 + 0.570) - 0.355]} = 7.31$

---

* Murray, T. M. and L. E. Curtis   Column-flange Strength at End-plate Connections   AISC Engineering Journal, 2nd Qtr., 1989, Chicago, Ill.

Use ½-in. fillets.

End Plate Moment $M_e = \dfrac{26.8 \times 2 \times (2 - 0.5)}{2} = 40.2$ kip-in.

End plate thickness $t_p = \sqrt{\dfrac{6 \times 40.2}{27 \times 8.5}} = 1.03$ in.

Use 1⅛-in. plate

Column Stiffeners

Opposite tension flange, effective column length = 12.5 in.

Assume $g = 5.5$ in.

For preliminary design, k_1 can be taken conservatively as 0.75; and $d_b/4$ may be omitted.

$P_{ec} = (g/2) - 0.75$

$\qquad = (5.5/2) - 0.75 = 2$ in.

Column flange moment $= (26.8 \times 2 \times 2)/2$

$\qquad\qquad\qquad\qquad = 53.6$ kip-in.

Req'd column-flange thickness:

$t_{fc} = \sqrt{\dfrac{6 \times 53.6}{27 \times 12.5}} = 0.98$ in.

$0.98 < 1.03$ in. = column flange thickness **o.k.**

See Ex. 39 for further information on column stiffeners.

Summary: Six 1-in. dia. A325-X Bolts (4-in tension)
$\qquad\qquad$ 1⅛ × 8½ × 2'-1 plate
$\qquad\qquad$ Welds as normally selected according to Note 2 above.

FOUR TENSION BOLT END-PLATE DESIGN PROCEDURE

The effective bolt distance P_e used to compute the bending moment in the end plate may be taken as:

$P_e = P_f - (d_b/4) - 0.707w$

where

P_f = distance from centerline of bolt to nearer surface of the tension flange, in.
w = fillet weld size or reinforcement of groove weld
d_b = nominal bolt diameter, in.

The end plate is designed to resist the moment M_e using $F_b = 0.75\,F_y = 27$ ksi (A36 steel):

$M_e = \alpha_m F_f P_e / 4$

where

$\alpha_m = C_a C_b (A_f/A_w)^{1/3} (P_e/d_b)^{1/4}$
$C_a = 1.13$ for beam $F_y = 36$ ksi; 1.11 for beam $F_y = 50$ ksi
$C_b = \sqrt{b_{fb}/b_p}$
A_f = area of tension flange, in.^{2}
A_w = web area, clear of flanges, in.^{2}

Recommended plate width:

$b_p = b_{fb} + 1$

Required plate thickness, t_p:

$$t_p = \sqrt{\frac{6 \times M_e}{27 \times b_p}}$$

Required flange fillet weld:

$$D_{Req'd} = \frac{F_f}{[2 \, (b_f + t_f) - t_w] \, 0.928}$$

Required weld to develop maximum web bending stress near flanges:

$$D = \frac{0.60F_y \times t_{wb}}{2 \times 0.928}$$

Check adequacy of column size and need for web stiffeners or reinforcement. See Note 6.

EXAMPLE 39

Four-Tension Bolt End-Plate Design

Given:

Design an end plate for a W16 × 40 beam having an end reaction of 30 kips and a fixed-end moment of 120 kip-ft (due to gravity load only) framed to the flange of a W14 × 193 column. Use minimum P_f (bolt dia. + ½ in.). Use A325 bolts, A36 steel for all members.

Solution:

A. W16×40 $S_x = 64.7$ in.^{3}
   $d = 16.01$ in., $b_{fb} = 6.995$ in.
   $t_{fb} = 0.505$ in., $t_{wb} = 0.305$ in.

B. Bolt design:

$$F_f = \frac{M}{(d - t_{fb})} = \frac{120 \times 12}{(16.01 - 0.505)} = 92.9 \text{ kips}$$

Req'd $T = 92.9/4 = 23.2$ kips

Use ⅞-in. dia. A325 bolts

From Table I-A, $T_{allow} = 26.5$ kips   **o.k.**

C. Top flange to end plate weld

$$D_{Req'd} = \frac{92.9}{0.928 \, [2 \, (6.995 + 0.505) - 0.305]} = 6.8$$

Use ⁷⁄₁₆-in. fillet welds

D. End-plate design:

Maximum design plate width:

$b_p = 6.995 + 1 = 7.995$

Use 8-in. plate with 5½-in gage

Effective bolt distance:

$P_e = (\tfrac{7}{8} + \tfrac{1}{2}) - (\tfrac{1}{4} \times \tfrac{7}{8}) - (0.707 \times \tfrac{7}{16}) = 0.847$ in.

$C_a = 1.13$ for $F_y = 36$ ksi

$C_b = \sqrt{\dfrac{6.995}{7.995}} = 0.935$

$\dfrac{A_f}{A_w} = \dfrac{6.995 \times 0.505}{[16.01 - (2 \times 0.505)] \times 0.305} = 0.772$

$\dfrac{P_e}{d_b} = \dfrac{0.847}{0.875} = 0.968$

$\alpha_m = 1.13 \times 0.935 \times (0.772)^{1/3} \times (0.968)^{1/4} = 0.961$

$M_e = 0.961 \times 92.9 \times (0.847/4) = 18.9$ kip-in.

Req'd plate thickness:

$$t_p = \sqrt{\dfrac{6 \times 18.9}{27 \times 8}} = 0.725 \text{ in.}$$

Use: ¾ in. × 0'-8 A36 plate

E. Beam web to end-plate weld

Minimum size fillet weld is ¼ in.

Required weld to develop maximum web tension stress ($0.60F_y = 21.6$ ksi) in web near flanges.

$$D = \dfrac{21.6 \times 0.305}{2 \times 0.928} = 3.5 < 4$$

Use ¼-in. fillet welds continuous on both sides of beam web

F. Column stiffeners

Check column yielding $k = 2\tfrac{1}{8}$

$P_{bf} = 5/3 \times$ flange force $= 5/3 \times 92.9 = 154.8$ kips

P_{bf} (allow) $= F_{yc} \times t_{wc} \times (t_{fb} + 6k + 2t_p + 2w)$

$\qquad = 36 \times .89 \times [.505 + (6 \times 2.125) + (2 \times 0.75) + (2 \times 0.4375)]$

$\qquad = 500.8$ kips > 154.8 kips **o.k.**

Check column web buckling.

$$P_{bf} \text{ (allow)} = \dfrac{4100 t_{wc}^3 \sqrt{F_{yc}}}{d_c}$$

$$= \dfrac{4100 (0.890)^3 \sqrt{36}}{11.25} = 1542 \text{ kips} > 154.8 \text{ kips} \quad \textbf{o.k.}$$

Check column flange bending opposite tension flange.

$k_1 \qquad = 1\tfrac{1}{16}$ in.

$b_p \qquad = 2.5 (1.375 + 0.505 + 1.375) = 8.14$ in.

$P_{ec} \qquad = (5.5/2) - 1.0625 - (0.875/4) = 1.47$ in.

C_b = 1.0

A_f/A_w = 1.0

α_m = 1.13 × 1.0 × (1.0)$^{1/3}$ × (1.47/0.875)$^{1/4}$

= 1.29

M_e = 1.29 × 92.9 × (1.47/4) = 44.04 kip-in.

Required flange thickness:

$$t_{fc} = \sqrt{\frac{6 \times 44.04}{27 \times 8.14}} = 1.10 \text{ in.} < 1.44 \text{ in.} \quad \textbf{o.k.}$$

Column web stiffeners are not required.

NOTES FOR EIGHT TENSION BOLT TYPE[*]

1. The connected beam section must be hot-rolled and included in the Allowable Stress Design Selection Table, pg. 2-4.

2. The vertical pitch P_f from the face of the beam tension flange to the first row of bolts must not exceed 2½ in.

   The vertical spacing between bolt rows P_b must not exceed $3d_b$ except it is common to use P_b = 3 in. when the bolt dia. is less than 1 in.

3. The horizontal gage g must be between 5½ and 7½ in.

4. The thickness of stiffeners between the beam flanges and the end plate should be equal to or greater than the thickness of the beam web.

5. For the shortened 8-bolt design (which follows), six of the eight bolts are considered to be fully effective.

6. The web to end-plate welds should develop the beam web in tension for a distance of P_f below the innermost bolt.

7. Column stiffener requirements may be determined as for the four-tension bolt type except that the effective column-flange length, which is the equivalent end-plate width, is

   $b_p = t_{fb} + 2P_f + 3.5P_b$[**]

8. Research on this connection has been limited to A325 bolts; therefore, only A325 bolts may be used in 8-tension bolt end-plate connections.

SHORTENED EIGHT TENSION BOLT END PLATE DESIGN PROCEDURE

1. Compute beam flange force F_f:

$$F_f = \frac{M}{(d - t_{fb})}$$

---

[*] *Murray, T. M. and A. R. Kukreti.* Design of 8-Bolt Stiffened Moment End Plates. *AISC Engineering Journal, 2nd Qtr., 1988, Chicago, Ill.*

[**] *Murray, T. M. and L. E. Curtis.* Column-flange Strength at End-plate Connections. *AISC Engineering Journal, 2nd Qtr., 1989, Chicago, Ill.*

Where:

M = beam end moment

d = beam depth

t_{fb} = beam flange thickness

2. Determine req'd bolt force T:

$$T = \frac{F_f}{6}$$

3. Select bolt size, A325:

Select bolt size from Table I-A, $T \leq T_{allow}$

4. Compute tee-stub analogy moment M_a:

M_a = moment in plate caused by two bolts with inflection point at $P_f/2$.

$$= 2 \times T \times (P_f/2) = T \times P_f$$

5. Compute tee stub analogy plate thickness t_{pa}:

$$t_{pa} = \sqrt{\frac{6 \times M_a}{27 \times b_p}}$$

6. Compute correction factor CF:

$$(CF)^2 = \frac{\sqrt{g^2 + P_f^2}}{5}$$

7. Compute final thickness t_p:

$$t_p = CF \times t_{pa}$$

EXAMPLE 40

Shortened Eight-tension Bolt End-plate Design

Given:

Design an eight-tension bolt moment end plate to develop the moment capacity of a W24×94, A36 steel, beam under gravity loading. The beam frames into a W14×311. Gage = 5.5 in., P_{bf} = bolt dia. + ½ in.

Solution:

W24×94 b_{fb} = 9.065 in. d = 24.31 in.

               t_{fb} = 0.875 in. t_{wb} = 0.515 in.

$M = M_R$ = 440 kip-ft (Manual, Part 2, p. 2-9)

Flange Force: $F_f = \dfrac{440 \times 12}{24.31 - 0.875}$

$$= 225.3 \text{ kips}$$

Single Bolt Force: $T = \dfrac{225.3}{6} = 37.6$ kips

Bolt Size: Select 1⅛-in. dia. A325 bolts

T_{allow} = 43.7 kips (Table I-A) > 37.6 kips   **o.k.**

End-plate Geometry:

$d_b = 1\frac{1}{8}$ in., $\qquad P_f = 1\frac{1}{8} + \frac{1}{2} = 1\frac{5}{8}$ in.

$g = 5\frac{1}{2}$ in., $\qquad b_p = 10$ in. $< b_{fb} + 1 = 10.065$ in.

$P_b = 3d_b = 3\frac{3}{8}$ in.

Tee-stub Analogy End-plate Moment:

$M_a = T \times P_f = 37.6 \times 1.625 = 61.1$ kip-in.

Tee-stub Analogy End-plate Thickness:

$$t_{pa} = \sqrt{\frac{6 \times 61.1}{27 \times 10}} = 1.17 \text{ in.}$$

Correction Factor CF

$$(CF)^2 = \frac{\sqrt{(5.5)^2 + (1.625)^2}}{5} = 1.15$$

$CF = 1.07$

Final End-plate Thickness:

$t_p = t_{pa} \times CF = 1.17 \times 1.07 = 1.25$ in.

Top Flange to End-plate Weld:

$$D_{Req'd} = \frac{(225.3)}{0.928 \, [2 \, (9.065 + 0.875) - 0.515]} = 12.5 \text{ sixteenths}$$

Use full-penetration groove weld.

Beam-web to End-plate Weld:

Minimum size fillet weld is $\frac{5}{16}$-in.

Required weld to develop maximum web tension stress $(0.60F_y = 21.6 \text{ ksi})$ in web near flange:

$$D_{Req'd} = \frac{0.515 \times 21.6}{2 \times 0.928} = 5.99 \text{ sixteenths}$$

Use $\frac{3}{8}$-in. fillet welds on both sides of beam web for a distance P_f below the inner most bolt. A $\frac{5}{16}$-in. weld may be used on the remaining plate to web weld.

Stiffener to End-plate and Beam-flange Weld:

Use same weld as for beam web near tension flange, e.g. $\frac{3}{8}$ in. both sides.

Column Stiffeners:

Check column yielding: $k = 2\frac{15}{16}$ in.

$P_{bf} = 5/3 \times$ flange force $= 5/3 \times 225.3$

$\qquad = 375.5$ kips

P_{bf} (all) $= F_{yc} \times t_{wc} \times (t_{fb} + 6k + 2t_p)$

$= 36 \times 1.410 \times [0.875 + (6 \times 2.9375) + (2 \times 1.25)]$

$= 1066$ kips > 375.5 kips   **o.k.**

Check column web buckling $T = 11.25$ in.

$$T_{Req'd} = \frac{4100 t_{wc}^3 \sqrt{F_{yc}}}{P_{bf}}$$

$$= \frac{4100 (1.410)^3 \sqrt{36}}{375.5}$$

$$= 183.6 \text{ in.} > 11.25 \text{ in.} \quad \textbf{o.k.}$$

Check column-flange bending

$$
\begin{aligned}
b_p &= t_{fb} + 2 P_f + 3.5 P_b \\
&= 0.875 + (2 \times 1.625) + (3.5 \times 3.375) \\
&= 15.94 \text{ in.} \\
P_e &= (g/2) - k_1 - (d_b/4) \\
&= (5.5/2) - 1.3125 - (1.125/4) \\
&= 1.16 \text{ in.} \\
C_b &= 1.0 \\
A_f/A_w &= 1.0 \\
\alpha_m &= 1.13 \times 1.0 \times (1.0)^{1/3} \times (1.16/1.125)^{1/4} \\
&= 1.14 \\
M_e &= 1.14 \times 225.3 \times (1.16/4) = 74.48 \text{ kip-in.} \\
t_{fc} &= \sqrt{\frac{6 \times 74.48}{15.94 \times 27}} = 1.02 \text{ in.} < 2.26 \text{ in.} \quad \textbf{o.k.}
\end{aligned}
$$

Summary:

Eight 1⅛-in. dia. tension A325 bolts

End plate: 1¼ × 9 in.

Stiffener: ½-in.

SUGGESTED DETAILS
Beam framing

SKEWED AND SLOPED CONNECTIONS

Details on this and succeeding pages are suggested treatments only and are not intended to limit the use of other connections not illustrated.

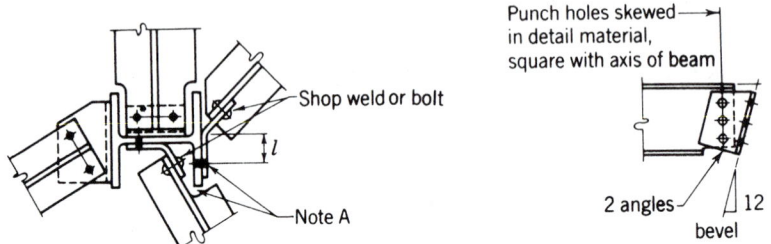

Note A: If a combination of several connections occurs at one level, provide field and driving clearance.

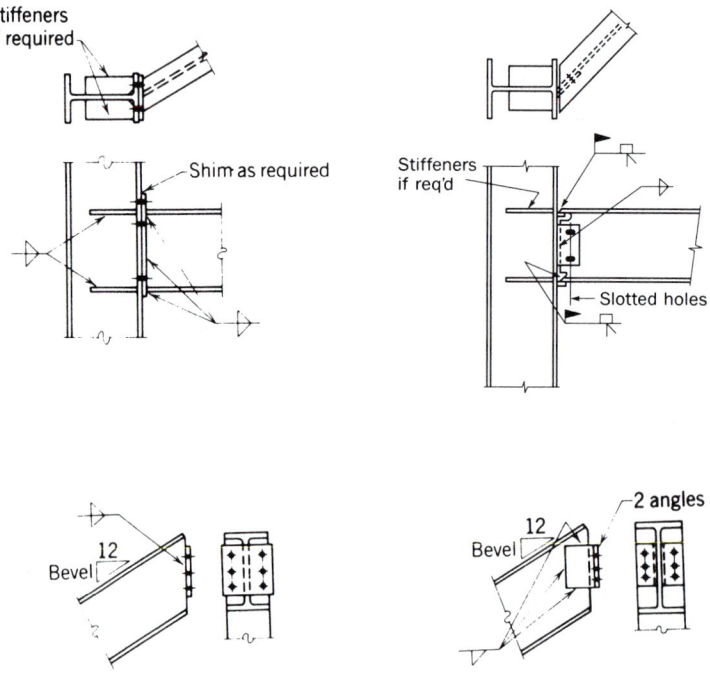

SUGGESTED DETAILS
Beam framing

MOMENT CONNECTIONS

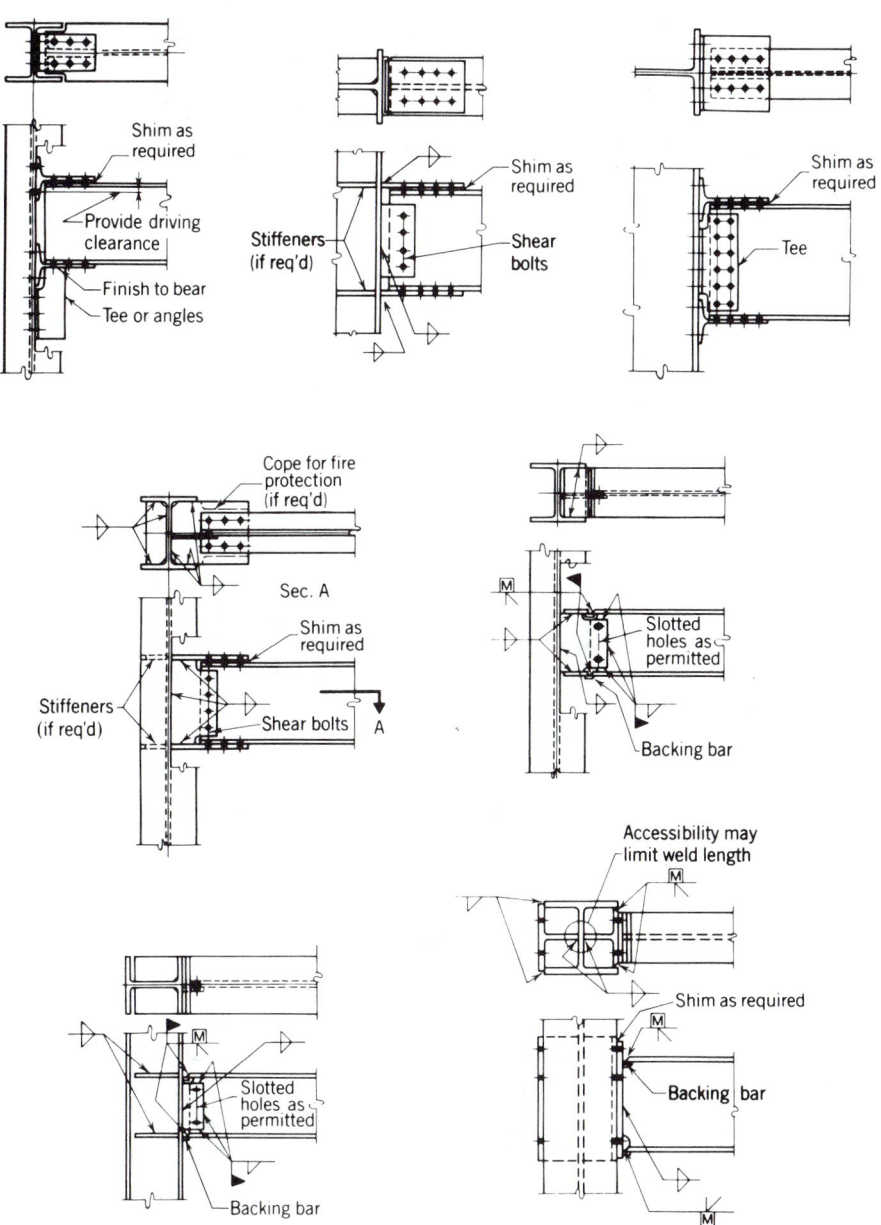

For additional information on moment connections of beams to column webs, refer to *AISC Engineering for Steel Construction*.

SUGGESTED DETAILS
Beam framing

SHEAR CONNECTIONS

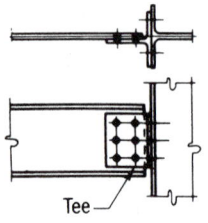

Tee

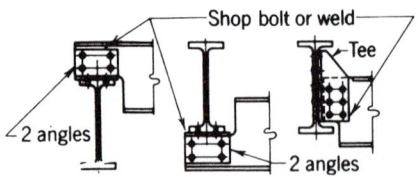

Shop bolt or weld

Tee

2 angles

2 angles

Note: Check web shear and moment in coped beam.

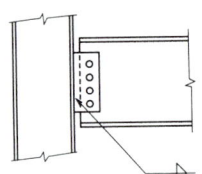

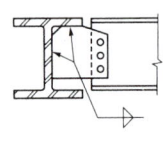

SHEAR SPLICES

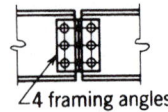

4 framing angles

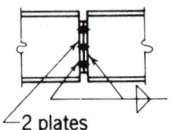

2 plates

Note: Of the above types, 4 framing angles is most flexible.

BOLTED MOMENT SPLICES

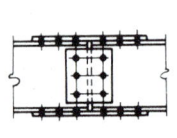

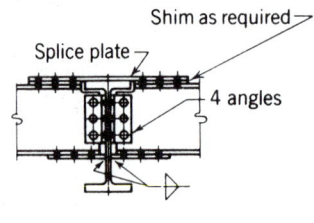

Shim as required

Splice plate

4 angles

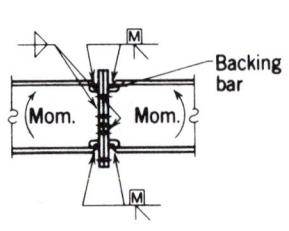

M

Backing bar

Mom.   Mom.

M

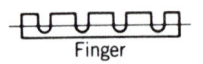

Finger

Strip

TYPICAL SHIMS

SUGGESTED DETAILS
Beam framing

WELDED MOMENT SPLICES

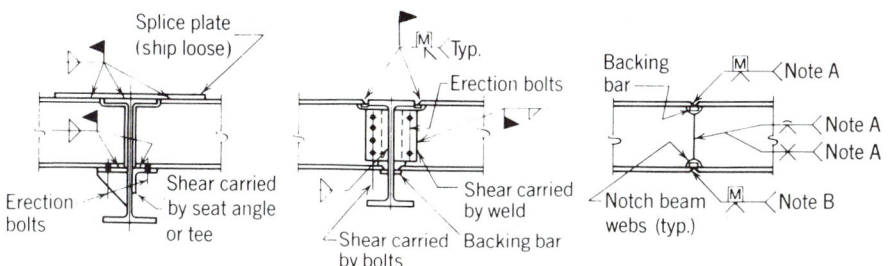

Note A: Joint preparation depends on thickness of material and welding process.
Note B: Invert this joint preparation if beam cannot be turned over.

MOMENT SPLICE AT RIDGE
(Field-bolted)

*BEAM OVER COLUMN
(With continuity)

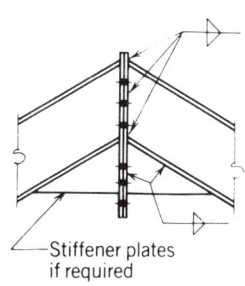

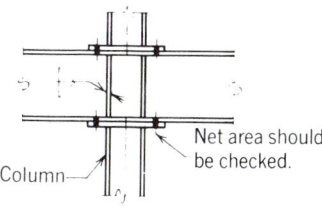

*For Plastic Design see Spec. Sect. N6.

SUGGESTED DETAILS
Column base plates

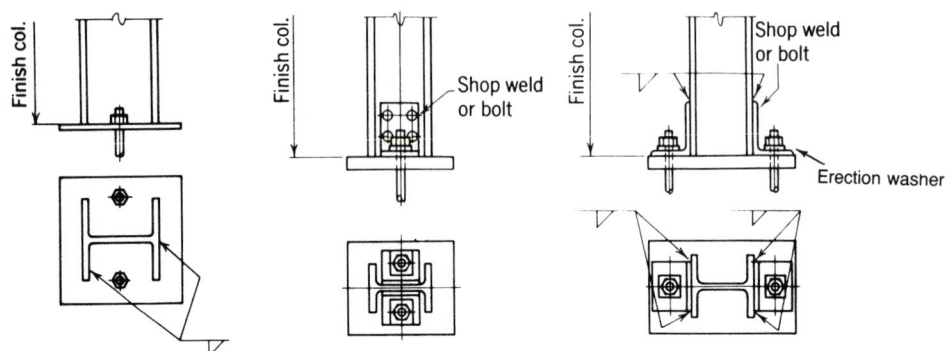

Base plate detailed and shipped loose when required.

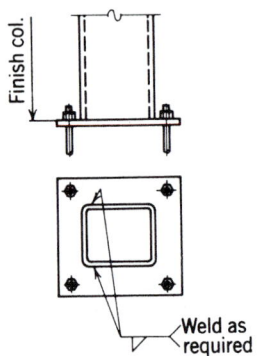

Note: Hole sizes for anchor bolts are normally
made oversize to facilitate erection as follows:
Bolts ¾ to 1 in.—⁵⁄₁₆ in. oversize
Bolts 1 to 2 in.—½ in. oversize
Bolts over 2 in.—1 in. oversize
Larger holes are permitted
if plate washers are used.

SUGGESTED DETAILS
Column base plates

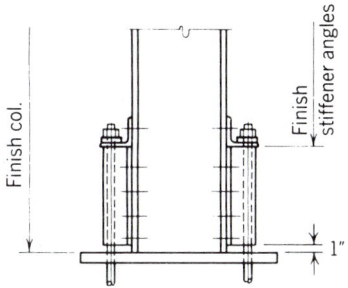

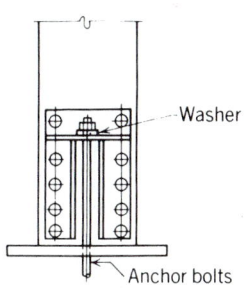

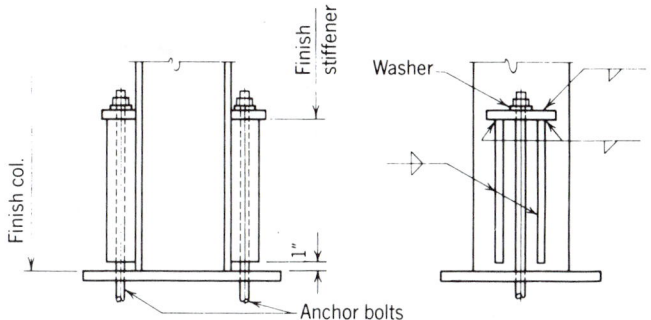

Base plates are normally detailed and shipped loose.

Note: Hole sizes for anchor bolts are normally
made oversize to facilitate erection as follows:
Bolts ¾ to 1 in.—⁵⁄₁₆ in. oversize
Bolts 1 to 2 in.—½ in. oversize
Bolts over 2 in.—1 in. oversize
Larger holes are permitted
if plate washers are used.

SUGGESTED DETAILS
Column splices

RIVETED AND BOLTED

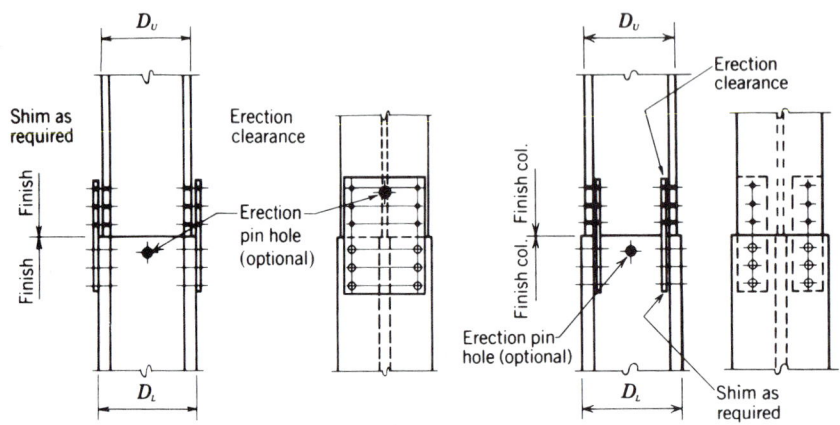

DEPTH OF D_U AND D_L
NOMINALLY THE SAME

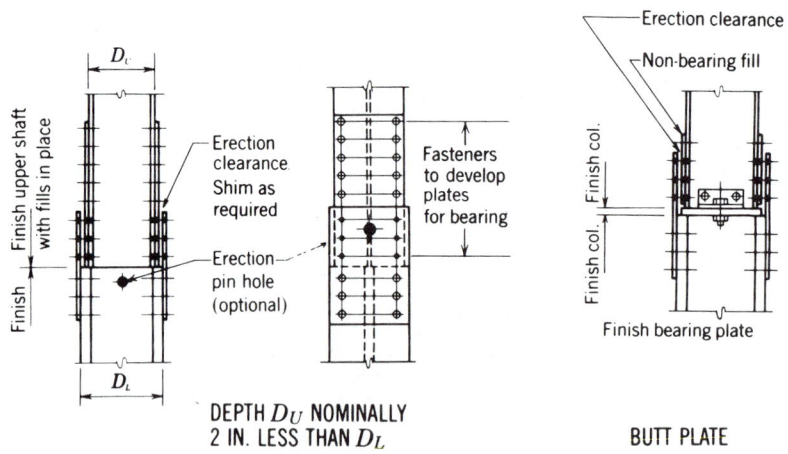

DEPTH D_U NOMINALLY
2 IN. LESS THAN D_L

BUTT PLATE

Note: Erection clearance = ⅛ in.

SUGGESTED DETAILS
Column splices

WELDED

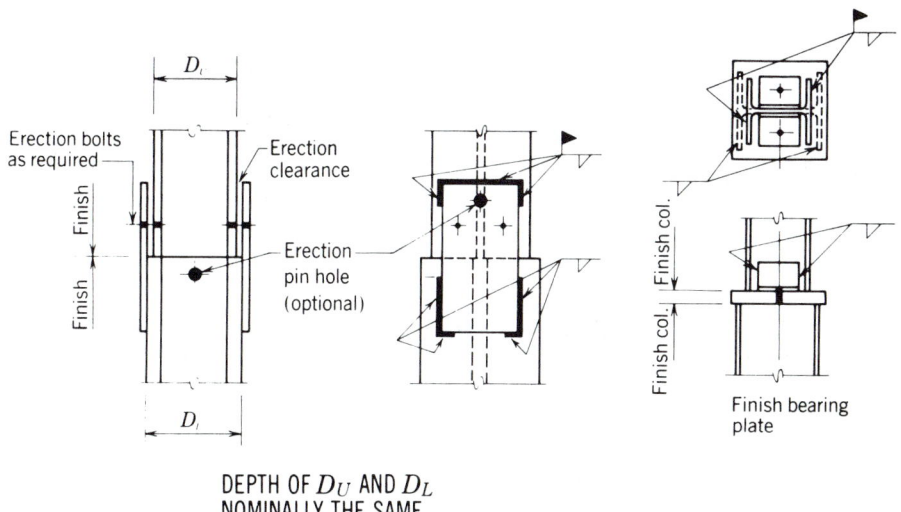

DEPTH OF D_U AND D_L
NOMINALLY THE SAME

BUTT PLATE

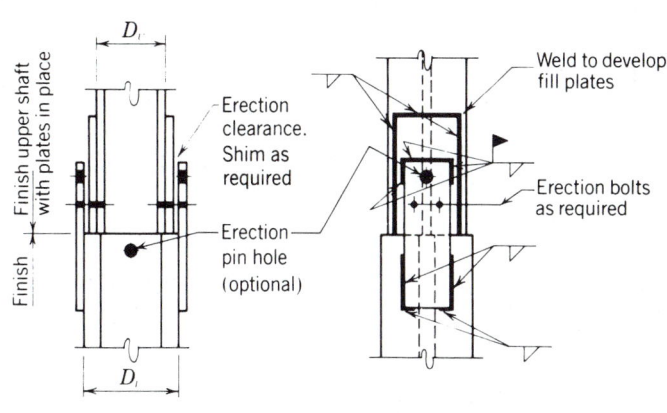

DEPTH D_U NOMINALLY
2 IN. LESS THAN D_L

Notes: Erection clearance = ¹⁄₁₆ in.

When D_U and D_L are nominally the same and thin fills are required, shop may attach splice plate to upper section and provide field clearance over lower section.

Stability of upper shaft, with its loading, should be considered until the final welding is completed.

AMERICAN INSTITUTE OF STEEL CONSTRUCTION

SUGGESTED DETAILS
Column splices

WELDED

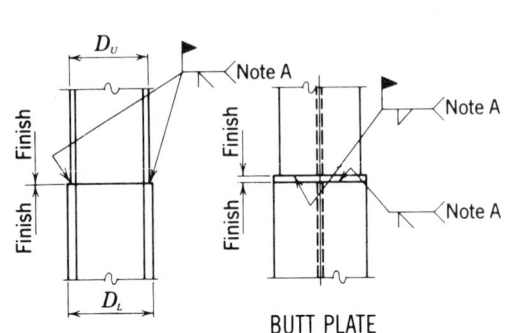

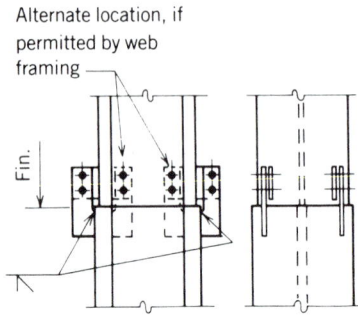

BUTT PLATE

DEPTH OF D_U AND D_L
NOMINALLY THE SAME

DEPTH D_U NOMINALLY
2 IN. LESS THAN D_L

ERECTION AID AND
STABILITY DEVICE

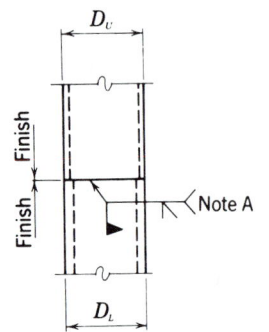

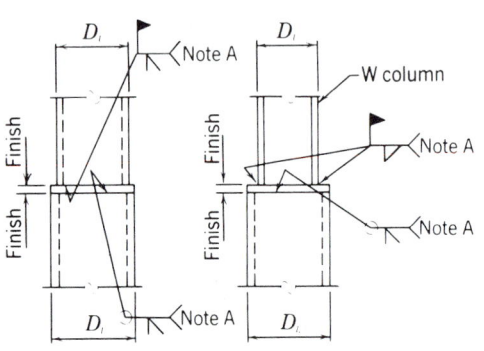

BUTT PLATE

DEPTH OF D_U AND D_L
NOMINALLY THE SAME

DEPTH D_U NOMINALLY
2 IN. LESS THAN D_L

Note A: Use fillet welds or partial penetration
weld whenever possible.

Finish bearing plates in accordance with AISC Spec. Sect. M2.8.

SUGGESTED DETAILS
Miscellaneous

STRUCTURAL TUBING AND PIPE
BEAM-TO-COLUMN CONNECTIONS

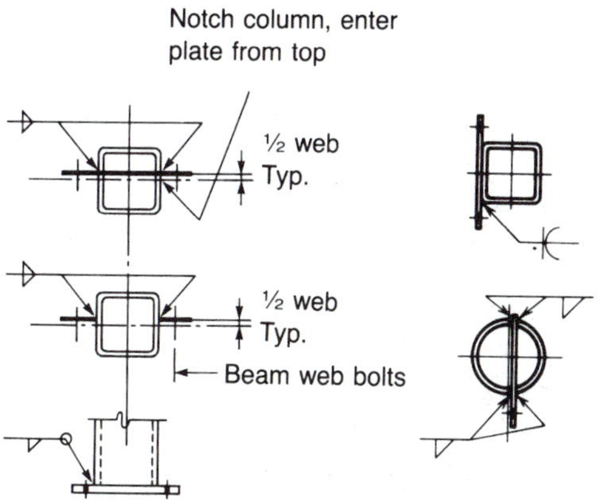

Notch column, enter
plate from top

½ web
Typ.

½ web
Typ.

Beam web bolts

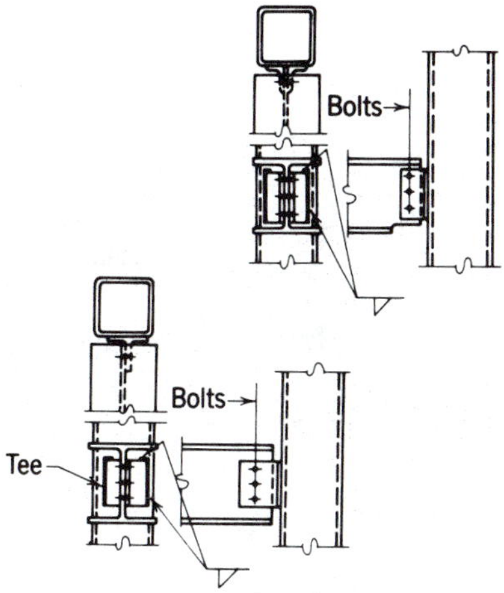

Bolts→

Bolts→

Tee

Note: Details similar for pipe and tubing.

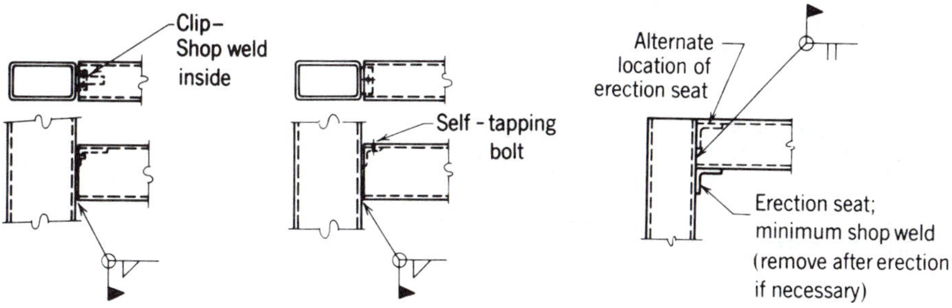

Clip–
Shop weld
inside

Self - tapping
bolt

Alternate
location of
erection seat

Erection seat;
minimum shop weld
(remove after erection
if necessary)

Note: Connections within tubes and pipe may be difficult or impossible to erect.

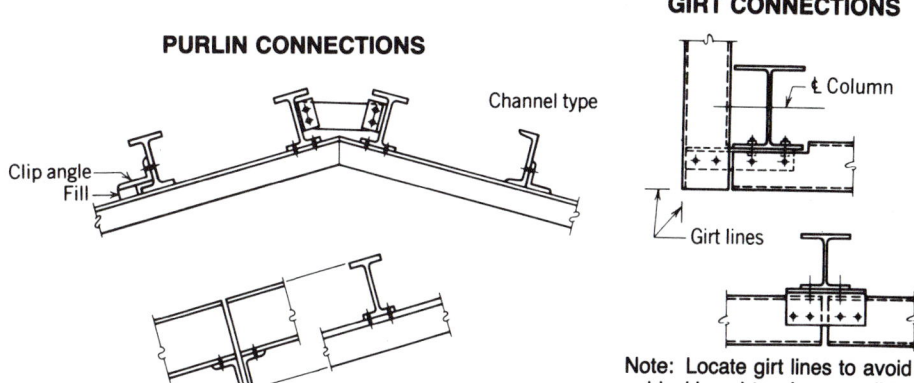

GIRT CONNECTIONS

PURLIN CONNECTIONS

Channel type

₡ Column

Clip angle
Fill

Girt lines

Note: Locate girt lines to avoid
blocking girts when possible.

SUGGESTED DETAILS
Miscellaneous

TIE RODS AND ANCHORS

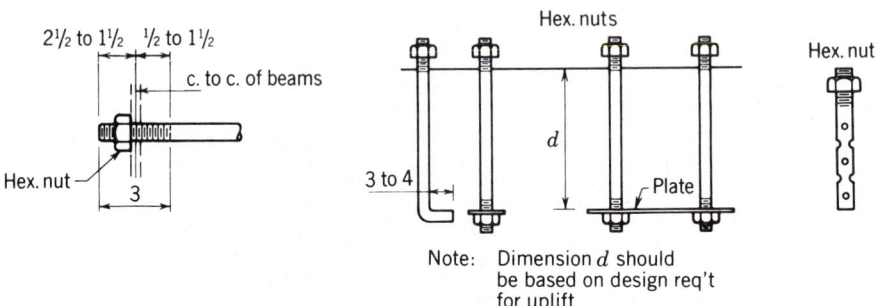

Tie Rods

Note: Dimension d should be based on design req't for uplift

Anchor Bolts

Swedge Bolts

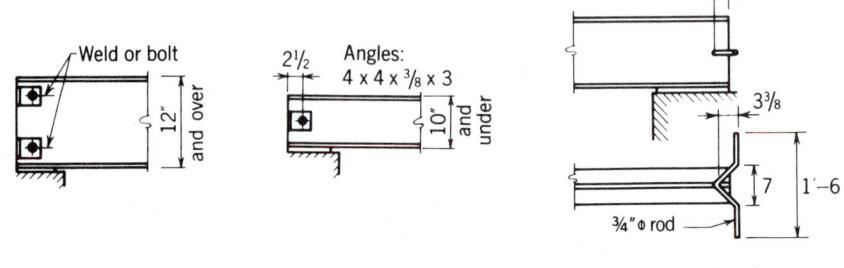

Angle Wall Anchors

Government Anchor

THREADED FASTENERS

Assembling clearances

A325 AND A490 HIGH-STRENGTH BOLTS ENTERING AND TIGHTENING CLEARANCES

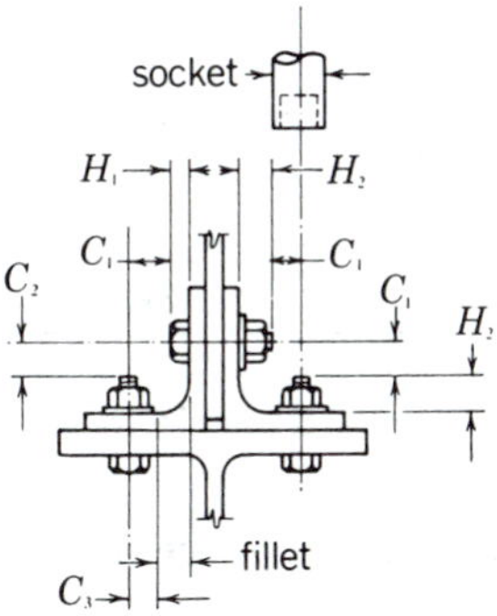

Bolt Dia.	Std. Socket	H_1	H_2	C_1	C_2	C_3 Round	C_3 Clipped
5/8	1 3/4	25/64	1 1/4	1	11/16	11/16	9/16
3/4	2 1/4	15/32	1 3/8	1 1/4	3/4	3/4	11/16
7/8	2 1/2	36/64	1 1/2	1 3/8	7/8	7/8	13/16
1	2 5/8	39/64	1 5/8	1 7/16	15/16	1	7/8
1 1/8	2 7/8	11/16	1 7/8	1 9/16	1 1/16	1 1/8	1
1 1/4	3 1/8	25/32	2	1 11/16	1 1/8	1 1/4	1 1/8
1 3/8	3 1/4	27/32	2 1/8	1 3/4	1 1/4	1 3/8	1 1/4
1 1/2	3 1/2	15/16	2 1/4	1 7/8	1 5/16	1 1/2	1 5/16

H_1 = height of head
H_2 = shank extension, max., based on one flat washer
C_1 = clearance for tightening
C_2 = clearance for entering
C_3 = clearance for fillet, based on std. hardened washer

A325 AND A490 HIGH-STRENGTH BOLTS STAGGER FOR IMPACT WRENCH TIGHTENING

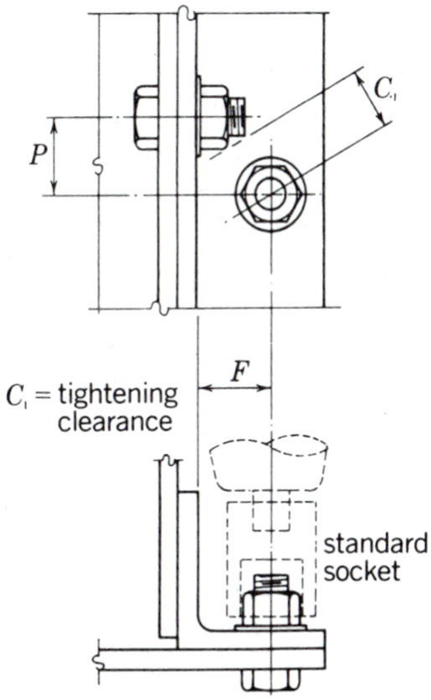

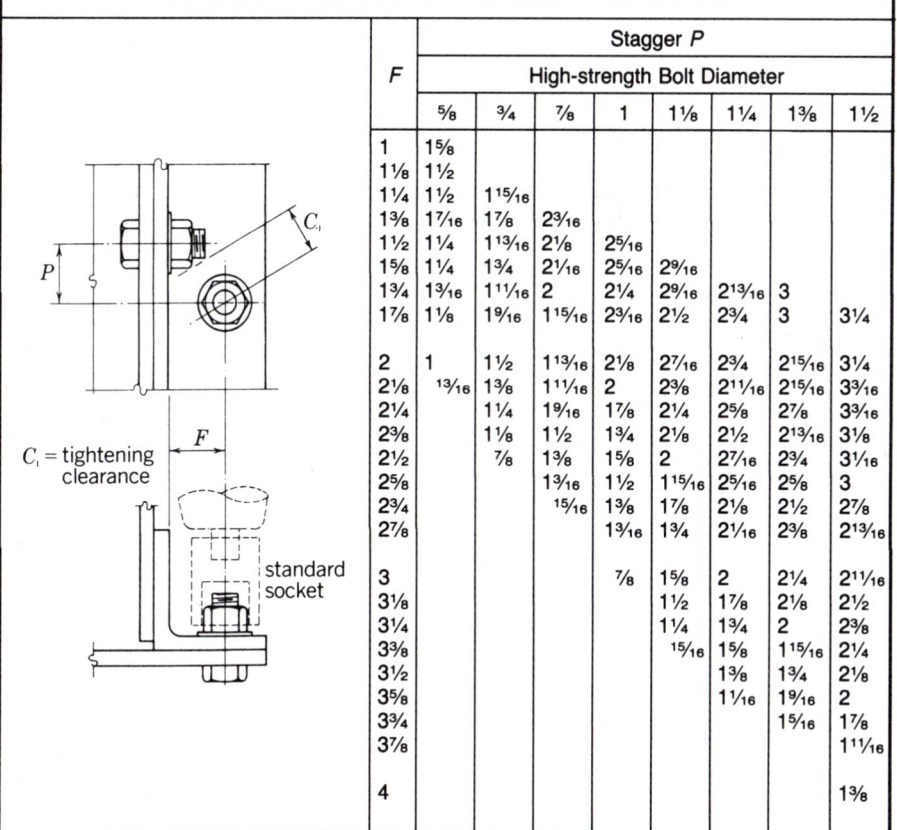

F	Stagger P — High-strength Bolt Diameter 5/8	3/4	7/8	1	1 1/8	1 1/4	1 3/8	1 1/2
1	1 5/8							
1 1/8	1 1/2							
1 1/4	1 1/2	1 15/16						
1 3/8	1 7/16	1 7/8	2 3/16					
1 1/2	1 1/4	1 13/16	2 1/8	2 5/16				
1 5/8	1 1/4	1 3/4	2 1/16	2 5/16	2 9/16			
1 3/4	1 3/16	1 11/16	2	2 1/4	2 9/16	2 13/16	3	
1 7/8	1 1/8	1 9/16	1 15/16	2 3/16	2 1/2	2 3/4	3	3 1/4
2	1	1 1/2	1 13/16	2 1/8	2 7/16	2 3/4	2 15/16	3 1/4
2 1/8	13/16	1 3/8	1 11/16	2	2 3/8	2 11/16	2 15/16	3 3/16
2 1/4		1 1/4	1 9/16	1 7/8	2 1/4	2 5/8	2 7/8	3 3/16
2 3/8		1 1/8	1 1/2	1 3/4	2 1/8	2 1/2	2 13/16	3 1/8
2 1/2		7/8	1 3/8	1 5/8	2	2 7/16	2 3/4	3 1/16
2 5/8			1 3/16	1 1/2	1 15/16	2 5/16	2 5/8	3
2 3/4			15/16	1 3/8	1 7/8	2 1/8	2 1/2	2 7/8
2 7/8				1 3/16	1 3/4	2 1/16	2 3/8	2 13/16
3				7/8	1 5/8	2	2 1/4	2 11/16
3 1/8					1 1/2	1 7/8	2 1/8	2 1/2
3 1/4					1 1/4	1 3/4	2	2 3/8
3 3/8					15/16	1 5/8	1 15/16	2 1/4
3 1/2						1 3/8	1 3/4	2 1/8
3 5/8						1 1/16	1 9/16	2
3 3/4							1 5/16	1 7/8
3 7/8								1 11/16
4								1 3/8

THREADED FASTENERS
Assembling clearances

A325 AND A490 HIGH-STRENGTH TENSION CONTROL BOLTS
ENTERING AND TIGHTENING CLEARANCES
SMALL INSTALLATION TOOL

Bolt Dia.	Tool Dia.	H_1	H_2	C_1	C_2	C_3 Round
5/8	3	7/16	1 1/4	1 5/8	13/16	11/16
3/4	3	1/2	1 3/8	1 5/8	7/8	3/4
7/8	3	9/16	1 1/2	1 5/8	1	7/8
5/8	2 3/16	7/16	1 1/4	1 1/8	13/16	11/16
3/4	2 3/16	1/2	1 3/8	1 1/8	7/8	3/4
7/8	2 3/16	9/16	1 1/2	1 1/8	1	7/8

H_1 = height of head
H_2 = shank extension, max., based on one flat washer
C_1 = clearance for tightening
C_2 = clearance for entering
C_3 = clearance for fillet, based on std. hardened washer

A325 AND A490 HIGH-STRENGTH TENSION CONTROL BOLTS
STAGGER FOR INSTALLATION TOOL

F	Stagger P Bolt Diameter			
	5/8	3/4	7/8	1
1 1/4	1 13/16			
1 3/8	1 3/4	2 1/16	2 1/4	2 7/16
1 1/2	1 11/16	2	2 3/16	2 3/8
1 5/8	1 9/16	1 7/8	2 1/16	2 1/4
1 3/4	1 1/2	1 13/16	2	2 3/16
1 7/8	1 7/16	1 3/4	1 7/8	2 1/8
2	1 5/16	1 5/8	1 3/4	2
2 1/8	1 1/4	1 9/16	1 11/16	1 15/16
2 1/4	1 3/16	1 1/2	1 9/16	1 7/8
2 3/8	1 1/8	1 3/8	1 1/2	1 3/4
2 1/2	1	1 5/16	1 3/8	1 11/16
2 5/8		1 3/16	1 5/16	1 9/16
2 3/4		1 1/8	1 3/16	1 1/2
2 7/8			1 1/8	1 3/8
3				1 5/16
3 3/8				1 5/16

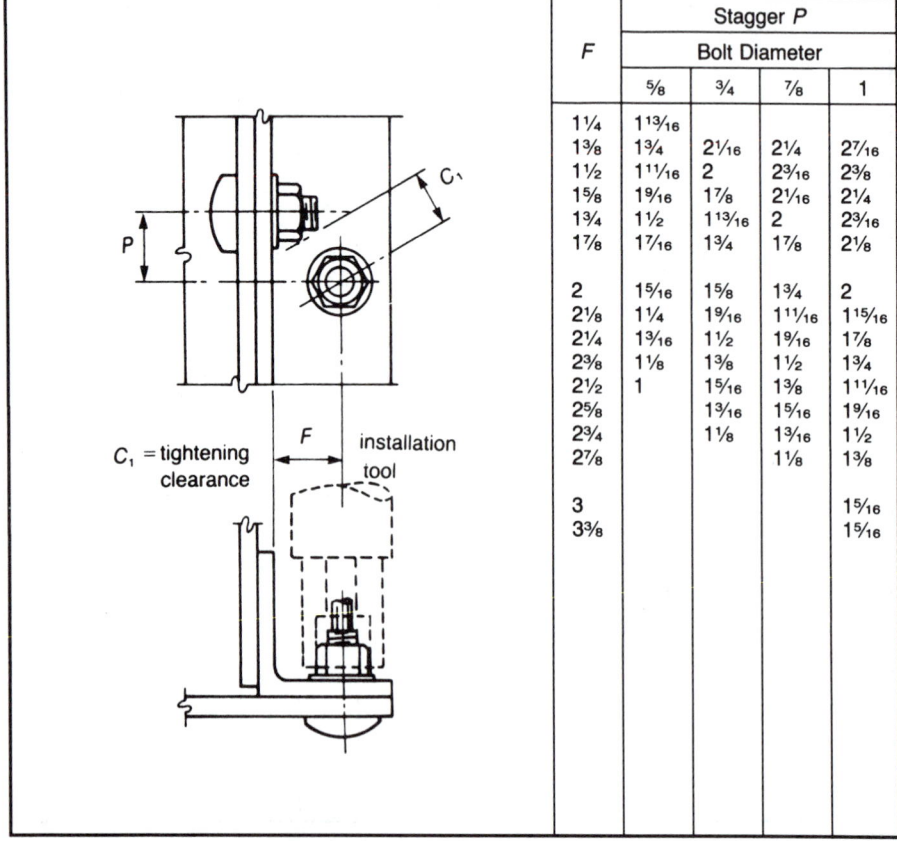

C_1 = tightening clearance

F = installation tool

AMERICAN INSTITUTE OF STEEL CONSTRUCTION

THREADED FASTENERS
Assembling clearances

A325 AND A490 HIGH-STRENGTH TENSION CONTROL BOLTS ENTERING AND TIGHTENING CLEARANCES LARGE INSTALLATION TOOL

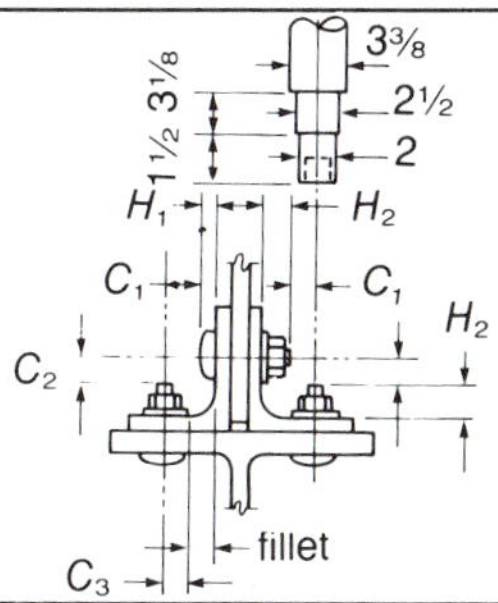

Bolt Dia.	Tool Dia.	H_1	H_2	C_1	C_2	C_3 Round	
¾	3⅜	½	1⅜	1⅞	⅞	¾	
⅞	3⅜	9/16	1½	1⅞	1	⅞	
1	3⅜	⅝	1¾	1⅞	1⅛	1	
¾	2½	½	1⅜	1⅜	⅞	¾	
⅞	2½	9/16	1½	1⅜	1	⅞	
1	2½	⅝	1¾	1⅜	1⅛	1	

H_1 = height of head
H_2 = shank extension, max., based on one flat washer
C_1 = clearance for tightening
C_2 = clearance for entering
C_3 = clearance for fillet, based on std. hardened washer

A325 AND A490 HIGH-STRENGTH TENSION CONTROL BOLTS ENTERING AND TIGHTENING CLEARANCES LARGE INSTALLATION TOOL

Bolt Dia.	H_1	H_2	C_1	C_2	C_3 Round	
⅝	—	—	—	—	—	
¾	½	1⅜	1⅜	⅞	¾	
⅞	9/16	1½	1⅜	1	⅞	
1	⅝	—	1⅜	1⅛	1	

			Tool			
Socket		C	D	E		
2	1½	2½	3½	3⅜		

H_1 = height of head
H_2 = shank extension, max., based on one flat washer
C_1 = clearance for tightening
C_2 = clearance for entering
C_3 = clearance for fillet, based on std. hardened washer

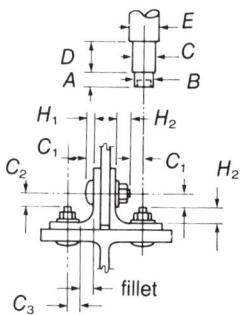

Notes

RIVETS AND THREADED FASTENERS
Field erection clearances

RIVET CLEARANCE—W COLUMNS

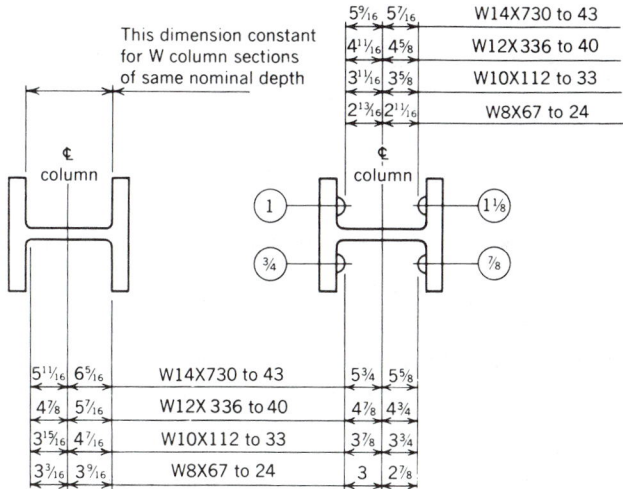

This dimension constant for W column sections of same nominal depth

5 9/16	5 7/16	W14X730 to 43
4 11/16	4 5/8	W12X336 to 40
3 11/16	3 5/8	W10X112 to 33
2 13/16	2 11/16	W8X67 to 24

5 11/16	6 5/16	W14X730 to 43	5 3/4	5 5/8
4 7/8	5 7/16	W12X336 to 40	4 7/8	4 3/4
3 15/16	4 7/16	W10X112 to 33	3 7/8	3 3/4
3 3/16	3 9/16	W8X67 to 24	3	2 7/8

FLANGE CUTS FOR COLUMN WEB CONNECTIONS

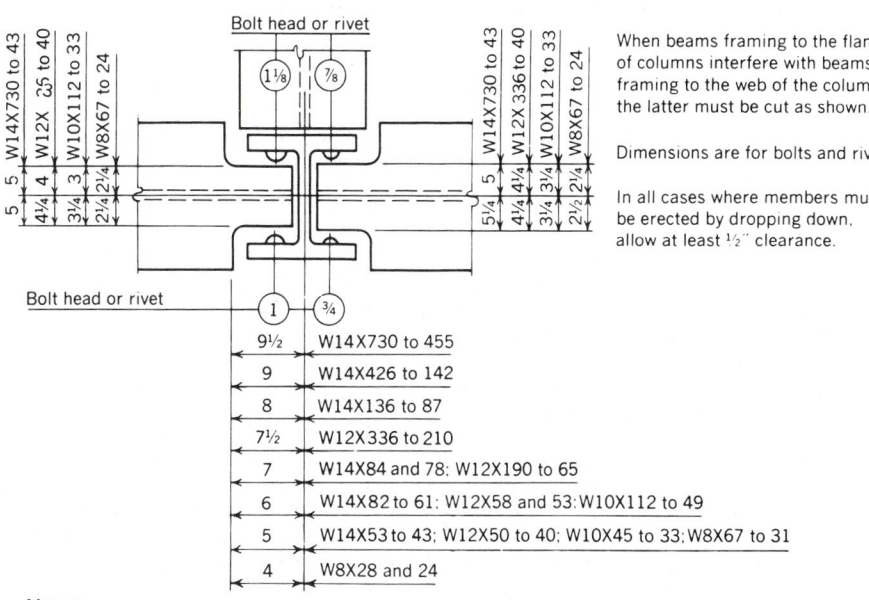

When beams framing to the flanges of columns interfere with beams framing to the web of the column, the latter must be cut as shown.

Dimensions are for bolts and rivets.

In all cases where members must be erected by dropping down, allow at least ½" clearance.

9 1/2	W14X730 to 455
9	W14X426 to 142
8	W14X136 to 87
7 1/2	W12X336 to 210
7	W14X84 and 78; W12X190 to 65
6	W14X82 to 61; W12X58 and 53;W10X112 to 49
5	W14X53 to 43; W12X50 to 40; W10X45 to 33;W8X67 to 31
4	W8X28 and 24

Notes:
1. Information shown on these clearance diagrams applies to both the old and new WF and W series. Maximum clearances are shown to accommodate the slight differences in dimensions.
2. Values shown for clearances over rivet heads are applicable when applied to bolt heads, but not to the nut and stick-through. See Table of Assembling Clearances for high-strength bolt clearance dimensions.
3. Based on Table of Dimensions of Structural Rivets.

AMERICAN INSTITUTE OF STEEL CONSTRUCTION

THREADED FASTENERS

Bolt heads

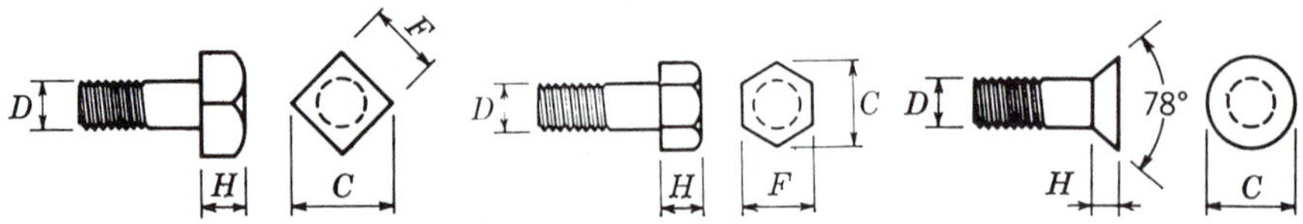

Square Hex Countersunk

Bolt head dimensions, rounded to nearest 1/16 in., are in accordance with ANSI B18.2.1—1981 (Square and Hex) and ANSI 18.5—1978 (Countersunk).

Standard Dimensions for Bolt Heads

Dia. of Bolt D	Square			Hex			Heavy Hex			Countersunk	
	Width F	Width C	Height H	Width F	Width C	Height H	Width F	Width C	Height H	Dia. C	Height H
In.	In.	In.	In.	In.	In.	In.	In.	In.	In.	In.	In.
1/4	3/8	1/2	3/16	7/16	1/2	3/16	—	—	—	1/2	1/8
3/8	9/16	13/16	1/4	9/16	5/8	1/4	—	—	—	11/16	3/16
1/2	3/4	1 1/16	5/16	3/4	7/8	3/8	7/8	1	3/8	7/8	1/4
5/8	15/16	1 5/16	7/16	15/16	1 1/16	7/16	1 1/16	1 1/4	7/16	1 1/8	5/16
3/4	1 1/8	1 9/16	1/2	1 1/8	1 5/16	1/2	1 1/4	1 7/16	1/2	1 3/8	3/8
7/8	1 5/16	1 7/8	5/8	1 5/16	1 1/2	9/16	1 7/16	1 11/16	9/16	1 9/16	7/16
1	1 1/2	2 1/8	11/16	1 1/2	1 3/4	11/16	1 5/8	1 7/8	11/16	1 13/16	1/2
1 1/8	1 11/16	2 3/8	3/4	1 11/16	1 15/16	3/4	1 13/16	2 1/16	3/4	2 1/16	9/16
1 1/4	1 7/8	2 5/8	7/8	1 7/8	2 3/16	7/8	2	2 5/16	7/8	2 1/4	5/8
1 3/8	2 1/16	2 15/16	15/16	2 1/16	2 3/8	15/16	2 3/16	2 1/2	15/16	2 1/2	11/16
1 1/2	2 1/4	3 3/16	1	2 1/4	2 5/8	1	2 3/8	2 3/4	1	2 11/16	3/4
1 3/4	—	—	—	2 5/8	3	1 3/16	2 3/4	3 3/16	1 3/16	—	—
2	—	—	—	3	3 7/16	1 3/8	3 1/8	3 5/8	1 3/8	—	—
2 1/4	—	—	—	3 3/8	3 7/8	1 1/2	3 1/2	4 1/16	1 1/2	—	—
2 1/2	—	—	—	3 3/4	4 5/16	1 11/16	3 7/8	4 1/2	1 11/16	—	—
2 3/4	—	—	—	4 1/8	4 3/4	1 13/16	4 1/4	4 15/16	1 13/16	—	—
3	—	—	—	4 1/2	5 3/16	2	4 5/8	5 5/16	2	—	—
3 1/4	—	—	—	4 7/8	5 5/8	3 3/16	—	—	—	—	—
3 1/2	—	—	—	5 1/4	6 1/16	2 5/16	—	—	—	—	—
3 3/4	—	—	—	5 5/8	6 1/2	2 1/2	—	—	—	—	—
4	—	—	—	6	6 15/16	2 11/16	—	—	—	—	—

For dimensions for high strength bolts, refer to *Allowable Stress Design Specification for Structural Joints Using ASTM A325 or A490 Bolts* in Part 5 of this manual.
Countersunk head bolts may be ordered with slotted or socket head.

THREADED FASTENERS
Nuts

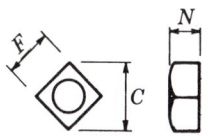

Square

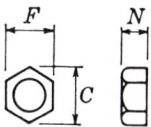

Hex

Nut dimensions, rounded to nearest ¹⁄₁₆ in., are in accordance with ANSI B18.2.2—1972 (R1983).

Dimensions for Nuts

Nut Size	Square			Hex			Heavy Square			Heavy Hex		
	Width F	Width C	Height N	Width F	Width C	Height N	Width F	Width C	Height N	Width F	Width C	Height N
In.	In.	In.	In.	In.	In.	In.	In.	In.	In.	In.	In.	In.
¼	⁷⁄₁₆	⅝	¼	⁷⁄₁₆	½	¼	½	¹¹⁄₁₆	¼	½	⁹⁄₁₆	¼
⅜	⅝	⅞	⁵⁄₁₆	⁹⁄₁₆	⅝	⁵⁄₁₆	¹¹⁄₁₆	1	⅜	¹¹⁄₁₆	¹³⁄₁₆	⅜
½	¹³⁄₁₆	1⅛	⁷⁄₁₆	¾	⅞	⁷⁄₁₆	⅞	1¼	½	⅞	1	½
⅝	1	1⁷⁄₁₆	⁹⁄₁₆	¹⁵⁄₁₆	1¹⁄₁₆	⁹⁄₁₆	1¹⁄₁₆	1½	⅝	1¹⁄₁₆	1¼	⅝
¾	1⅛	1⁹⁄₁₆	¹¹⁄₁₆	1⅛	1⁵⁄₁₆	⅝	1¼	1¾	¾	1¼	1⁷⁄₁₆	¾
⅞	1⁵⁄₁₆	1⅞	¾	1⁵⁄₁₆	1½	¾	1⁷⁄₁₆	2¹⁄₁₆	⅞	1⁷⁄₁₆	1¹¹⁄₁₆	⅞
1	1½	2⅛	⅞	1½	1¾	⅞	1⅝	2⁵⁄₁₆	1	1⅝	1⅞	1
1⅛	1¹¹⁄₁₆	2⅜	1	1¹¹⁄₁₆	1¹⁵⁄₁₆	1	1¹³⁄₁₆	2⁹⁄₁₆	1⅛	1¹³⁄₁₆	2¹⁄₁₆	1⅛
1¼	1⅞	2⅝	1⅛	1⅞	2³⁄₁₆	1¹⁄₁₆	2	2¹³⁄₁₆	1¼	2	2⁵⁄₁₆	1¼
1⅜	2¹⁄₁₆	2¹⁵⁄₁₆	1¼	2¹⁄₁₆	2⅜	1³⁄₁₆	2³⁄₁₆	3⅛	1⅜	2³⁄₁₆	2½	1⅜
1½	2¼	3³⁄₁₆	1⁵⁄₁₆	2¼	2⅝	1⁵⁄₁₆	2⅜	3⅜	1½	2⅜	2¾	1½
1¾	—	—	—	—	—	—	—	—	—	2¾	3³⁄₁₆	1¾
2	—	—	—	—	—	—	—	—	—	3⅛	3⅝	2
2¼	—	—	—	—	—	—	—	—	—	3½	4¹⁄₁₆	2³⁄₁₆
2½	—	—	—	—	—	—	—	—	—	3⅞	4½	2⁷⁄₁₆
2¾	—	—	—	—	—	—	—	—	—	4¼	4¹⁵⁄₁₆	2¹¹⁄₁₆
3	—	—	—	—	—	—	—	—	—	4⅝	5⁵⁄₁₆	2¹⁵⁄₁₆
3¼	—	—	—	—	—	—	—	—	—	5	5¾	3³⁄₁₆
3½	—	—	—	—	—	—	—	—	—	5⅜	6³⁄₁₆	3⁷⁄₁₆
3¾	—	—	—	—	—	—	—	—	—	5¾	6⅝	3¹¹⁄₁₆
4	—	—	—	—	—	—	—	—	—	6⅛	7¹⁄₁₆	3¹⁵⁄₁₆

For dimensions for high strength bolts, refer to *Allowable Stress Design Specification for Structural Joints Using ASTM A325 or A490 Bolts* in Part 5 of this manual.

THREADED FASTENERS
Weight of bolts
With square heads and hexagon nuts in pounds per 100

Length Under Head In.	Diameter of Bolts, In.								
	1/4	3/8	1/2	5/8	3/4	7/8	1	1 1/8	1 1/4
1	2.38	6.11	13.0	24.1	38.9	—	—	—	—
1 1/4	2.71	6.71	14.0	25.8	41.5	—	—	—	—
1 1/2	3.05	7.47	15.1	27.6	44.0	67.3	95.1	—	—
1 3/4	3.39	8.23	16.5	29.3	46.5	70.8	99.7	—	—
2	3.73	8.99	17.8	31.4	49.1	74.4	104	143	—
2 1/4	4.06	9.75	19.1	33.5	52.1	77.9	109	149	—
2 1/2	4.40	10.5	20.5	35.6	55.1	82.0	114	155	206
2 3/4	4.74	11.3	21.8	37.7	58.2	86.1	119	161	213
3	5.07	12.0	23.2	39.8	61.2	90.2	124	168	221
3 1/4	5.41	12.8	24.5	41.9	64.2	94.4	129	174	229
3 1/2	5.75	13.5	25.9	44.0	67.2	98.5	135	181	237
3 3/4	6.09	14.3	27.2	46.1	70.2	103	140	188	246
4	6.42	15.1	28.6	48.2	73.3	107	145	195	254
4 1/4	6.76	15.8	29.9	50.3	76.3	111	151	202	262
4 1/2	7.10	16.6	31.3	52.3	79.3	115	156	208	271
4 3/4	7.43	17.3	32.6	54.4	82.3	119	162	215	279
5	7.77	18.1	33.9	56.5	85.3	123	167	222	288
5 1/4	8.11	18.9	35.3	58.6	88.4	127	172	229	296
5 1/2	8.44	19.6	36.6	60.7	91.4	131	178	236	304
5 3/4	8.78	20.4	38.0	62.8	94.4	136	183	242	313
6	9.12	21.1	39.3	64.9	97.4	140	188	249	321
6 1/4	9.37	21.7	40.4	66.7	100	143	193	255	329
6 1/2	9.71	22.5	41.8	68.7	103	147	198	262	337
6 3/4	10.1	23.3	43.1	70.8	106	151	204	269	345
7	10.4	24.0	44.4	72.9	109	156	209	275	354
7 1/4	10.7	24.8	45.8	75.0	112	160	214	282	362
7 1/2	11.0	25.5	47.1	77.1	115	164	220	289	371
7 3/4	11.4	26.3	48.5	79.2	118	168	225	296	379
8	11.7	27.0	49.8	81.3	121	172	231	303	387
8 1/2	—	28.6	52.5	85.5	127	180	241	316	404
9	—	30.1	55.2	89.7	133	189	252	330	421
9 1/2	—	31.6	57.9	93.9	139	197	263	343	438
10	—	33.1	60.6	98.1	145	205	274	357	454
10 1/2	—	34.6	63.3	102	151	213	284	371	471
11	—	36.2	66.0	106	157	221	295	384	488
11 1/2	—	37.7	68.7	110	163	230	306	398	505
12	—	39.2	71.3	115	170	238	316	411	522
12 1/2	—	—	74.0	119	176	246	327	425	538
13	—	—	76.7	123	182	254	338	439	556
13 1/2	—	—	79.4	127	188	263	349	452	572
14	—	—	82.1	131	194	271	359	466	589
14 1/2	—	—	84.8	135	200	279	370	479	605
15	—	—	87.5	140	206	287	381	493	622
15 1/2	—	—	90.2	144	212	296	392	507	639
16	—	—	92.9	148	218	304	402	520	656
Per Inch Additional	1.3	3.0	5.4	8.4	12.1	16.5	21.4	27.2	33.6

Bolt is square bolt, ANSI B18.2.1—81 and nut is hex nut, ANSI B18.2.2—72 (R1983). This table conforms to weight standards adopted by the Industrial Fasteners Institute.

THREADED FASTENERS
Weight of bolts
Special cases in pounds per 100

VARIATIONS IN BOLT AND NUT TYPES

Weights for combinations of bolt heads and nuts, other than square heads and hex nuts, may be determined by making the appropriate additions and deductions tabulated below from the weight per 100 shown on the previous page.

Combination	Add or Subtract	Diameter of Bolt in Inches								
		¼	⅜	½	⅝	¾	⅞	1	1⅛	1¼
Square bolt with square nut	+	0.1	1.0	2.0	3.4	3.5	5.5	8.0	12.2	16.3
Square bolt with heavy square nut	+	0.6	2.1	4.1	7.0	11.6	17.2	23.2	32.1	41.2
Square bolt with heavy hex nut	+	0.4	1.5	2.8	4.6	7.6	10.7	14.2	18.9	24.3
Hex bolt with square nut	+	0.1	0.6	1.1	1.4	0.2	0.5	−0.2	−0.1	−1.7
Hex bolt with hex nut	−	0.0	0.4	0.9	2.0	3.3	5.0	8.2	12.3	18.0
Hex bolt with heavy square nut	+	0.6	1.7	3.2	5.0	8.3	12.2	15.0	19.8	23.2
Hex bolt with heavy hex nut	+	0.4	1.1	1.9	2.6	4.3	5.7	6.0	6.6	6.3
Heavy hex bolt with heavy square nut	+	—	—	4.7	7.3	11.3	16.5	20.7	27.0	33.6
Heavy hex bolt with heavy hex nut	+	—	—	3.4	4.9	7.3	10.0	11.7	13.8	16.7

LARGE DIAMETER BOLTS

Weights of bolts over 1¼ inches in diameter may be calculated from the following data. Standard practice is hex head bolts with heavy hex nut. Square head bolts and square nuts are not standard in sizes over 1½ inches.

Weight of 100 Each	Diameter of Bolt in Inches											
	1⅜	1½	1¾	2	2¼	2½	2¾	3	3¼	3½	3¾	4
Square heads	105	130	—	—	—	—	—	—	—	—	—	—
Hex heads	84	112	178	259	369	508	680	900	1120	1390	1730	2130
Heavy hex heads	95	124	195	280	397	541	720	950	—	—	—	—
Square nuts	94.5	122	—	—	—	—	—	—	—	—	—	—
Heavy square nuts	125	161	—	—	—	—	—	—	—	—	—	—
Heavy hex nuts	102	131	204	299	419	564	738	950	1190	1530	1810	2180
Linear inch of threaded shank	35.0	42.5	57.4	75.5	97.4	120	147	178	210	246	284	325
Linear inch of unthreaded shank	42.0	50.0	68.2	89.0	113	139	168	200	235	272	313	356

THREADED FASTENERS

Weight of ASTM A325 or A490 high strength bolts

Heavy hex structural bolts with heavy hex nuts in pounds per 100

Length Under Head In.	Diameter of Bolts, In.								
	½	⅝	¾	⅞	1	1⅛	1¼	1⅜	1½
1	16.5	29.4	47.0	—	—	—	—	—	—
1¼	17.8	31.1	49.6	74.4	104	—	—	—	—
1½	19.2	33.1	52.2	78.0	109	148	197	—	—
1¾	20.5	35.3	55.3	81.9	114	154	205	261	333
2	21.9	37.4	58.4	86.1	119	160	212	270	344
2¼	23.3	39.8	61.6	90.3	124	167	220	279	355
2½	24.7	41.7	64.7	94.6	130	174	229	290	366
2¾	26.1	43.9	67.8	98.8	135	181	237	300	379
3	27.4	46.1	70.9	103	141	188	246	310	391
3¼	28.8	48.2	74.0	107	146	195	255	321	403
3½	30.2	50.4	77.1	111	151	202	263	332	416
3¾	31.6	52.5	80.2	116	157	209	272	342	428
4	33.0	54.7	83.3	120	162	216	280	353	441
4¼	34.3	56.9	86.4	124	168	223	289	363	453
4½	35.7	59.0	89.5	128	173	230	298	374	465
4¾	37.1	61.2	92.7	133	179	237	306	384	478
5	38.5	63.3	95.8	137	184	244	315	395	490
5¼	39.9	65.5	98.9	141	190	251	324	405	503
5½	41.2	67.7	102	146	196	258	332	416	515
5¾	42.6	69.8	105	150	201	265	341	426	527
6	44.0	71.9	108	154	207	272	349	437	540
6¼	—	74.1	111	158	212	279	358	447	552
6½	—	76.3	114	163	218	286	367	458	565
6¾	—	78.5	118	167	223	293	375	468	577
7	—	80.6	121	171	229	300	384	479	589
7¼	—	82.8	124	175	234	307	392	489	602
7½	—	84.9	127	179	240	314	401	500	614
7¾	—	87.1	130	183	246	321	410	510	626
8	—	89.2	133	187	251	328	418	521	639
8¼	—	—	—	192	257	335	427	531	651
8½	—	—	—	196	262	342	435	542	664
8¾	—	—	—	—	—	—	444	552	676
9	—	—	—	—	—	—	453	563	689
Per inch additional add	5.5	8.6	12.4	16.9	22.1	28.0	34.4	42.5	49.7
For each 100 plain round washers add	2.1	3.6	4.8	7.0	9.4	11.3	13.8	16.8	20.0
For each 100 beveled square washers add	23.1	22.4	21.0	20.2	19.2	34.0	31.6	—	—

This table conforms to weight standards adopted by the Industrial Fasteners Institute, 1965, updated for washer weights.

THREADED FASTENERS

SCREW THREADS

Unified Standard Series—UNC/UNRC and 4UN/4UNR

ANSI B1.1–1982

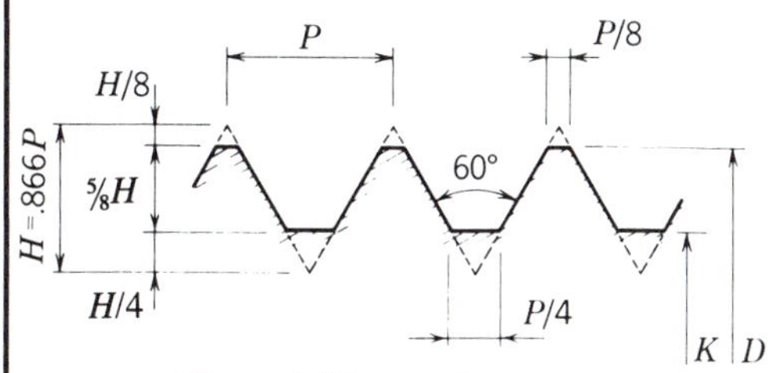
P
P/8
H/8
⅝H
H=.866P
60°
H/4
P/4
K
D

Thread Dimensions

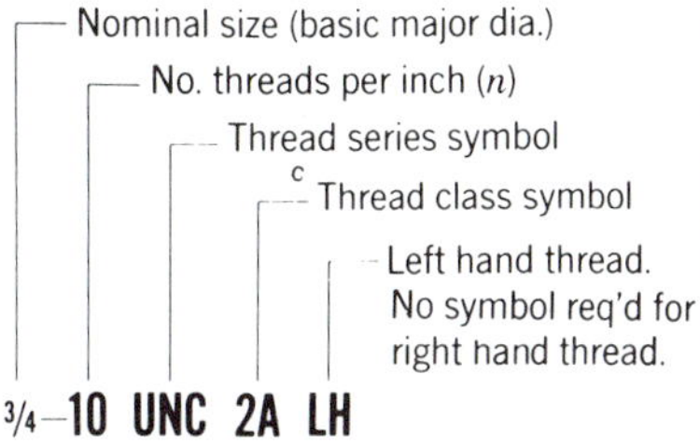
Nominal size (basic major dia.)
No. threads per inch (n)
Thread series symbol
c Thread class symbol
Left hand thread. No symbol req'd for right hand thread.
¾–10 UNC 2A LH

Standard Designations

Dia. Basic Major D In.	Dia. Min. Root K In.	Area Gross A_D In.^{2}	Area Min. Root A_K[a] In.^{2}	Area Tensile[a] Stress In.^{2}	Th'ds[b] per In. n	Dia. Basic Major D In.	Dia. Min. Root K In.	Area Gross A_D In.^{2}	Area Min. Root A_K[a] In.^{2}	Area Tensile[a] Stress In.^{2}	Th'ds[b] per In. n
¼	.196	.049	.027	.032	20	2¾	2.479	5.940	4.62	4.93	4
⅜	.307	.110	.068	.078	16	3	2.729	7.069	5.62	5.97	4
½	.417	.196	.126	.142	13	3¼	2.979	8.296	6.72	7.10	4
⅝	.527	.307	.202	.226	11	3½	3.229	9.621	7.92	8.33	4
¾	.642	.442	.302	.334	10	3¾	3.479	11.045	9.21	9.66	4
⅞	.755	.601	.419	.462	9	4	3.729	12.566	10.61	11.1	4
1	.865	.785	.551	.606	8	4¼	3.979	14.186	12.1	12.6	4
1⅛	.970	.994	.693	.763	7	4½	4.229	15.904	13.7	14.2	4
1¼	1.095	1.227	.890	.969	7	4¾	4.479	17.721	15.4	15.9	4
1⅜	1.195	1.485	1.05	1.16	6	5	4.729	19.635	17.2	17.8	4
1½	1.320	1.767	1.29	1.41	6	5¼	4.979	21.648	19.1	19.7	4
1¾	1.534	2.405	1.74	1.90	5	5½	5.229	23.758	21.0	21.7	4
2	1.759	3.142	2.30	2.50	4½	5¾	5.479	25.967	23.1	23.8	4
2¼	2.009	3.976	3.02	3.25	4½	6	5.729	28.274	25.3	26.0	4
2½	2.229	4.909	3.72	4.00	4						

[a]Tensile stress area = $0.7854\left(D - \frac{.9743}{n}\right)^2$. $A_K = 0.7854\left(D - \frac{1.3}{n}\right)^2$

[b]For basic major diameters of ¼ to 4 in. incl., thread series is UNC (coarse); for 4¼ in. dia. and larger, thread series is 4UN.

[c]2A denotes Class 2A fit applicable to external threads, 2B denotes corresponding Class 2B fit for internal threads.

MINIMUM LENGTH OF THREAD ON BOLTS

ANSI B18.2.1–1972

Length of Bolt	Diameter of Bolt D, In.																
	¼	⅜	½	⅝	¾	⅞	1	1⅛	1¼	1⅜	1½	1¾	2	2¼	2½	2¾	3
To 6 in. Incl.	¾	1	1¼	1½	1¾	2	2¼	2½	2¾	3	3¼	3¾	4¼	4¾	5¼	5¾	6¼
Over 6 in.	1	1¼	1½	1¾	2	2¼	2½	2¾	3	3¼	3½	4	4½	5	5½	6	6½

Thread length for bolts up to 6 in. long is $2D$ + ¼. For bolts over 6-in. long, thread length is $2D$ + ½. These proportions may be used to compute thread length for diameters not shown in the table. Bolts which are too short for listed or computed thread lengths are threaded as close to the head as possible.

For thread lengths for high-strength bolts, refer to *Allowable Stress Design Specification for Structural Joints Using ASTM A325 or A490 Bolts.*

CLEVISES

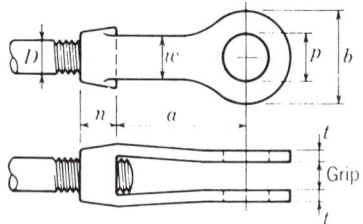

Thread: UNC Class 2B

Grip = thickness
plate + $\frac{1}{4}$"

Clevis Number	Dimensions, In.							Weight Pounds	Safe Working Load, Kips[a]
	Max. D	Max. p	b	n	a	w	t		
2	$\frac{5}{8}$	$\frac{3}{4}$	$1\frac{7}{16}$	$\frac{5}{8}$	$3\frac{7}{8}$	$1\frac{1}{16}$	$\frac{5}{16}(+\frac{1}{32}-0)$	1.0	3.5
$2\frac{1}{2}$	$\frac{7}{8}$	$1\frac{1}{2}$	$2\frac{1}{2}$	$1\frac{1}{8}$	4	$1\frac{1}{4}$	$\frac{5}{16}(+\frac{1}{32}-0)$	2.0	7.5
3	$1\frac{3}{8}$	$1\frac{3}{4}$	3	$1\frac{5}{16}$	5	$1\frac{1}{2}$	$\frac{1}{2}(+\frac{1}{32}-0)$	4.0	15
$3\frac{1}{2}$	$1\frac{1}{2}$	2	$3\frac{1}{2}$	$1\frac{5}{8}$	6	$1\frac{3}{4}$	$\frac{1}{2}(+\frac{1}{32}-0)$	6.0	18
4	$1\frac{3}{4}$	$2\frac{1}{4}$	4	$1\frac{3}{4}$	6	2	$\frac{1}{2}(+\frac{1}{32}-0)$	8.0	21
5	2	$2\frac{1}{2}$	5	$2\frac{1}{4}$	7	$2\frac{1}{2}$	$\frac{5}{8}(+\frac{1}{16}-0)$	16.0	37.5
6	$2\frac{1}{2}$	3	6	$2\frac{3}{4}$	8	3	$\frac{3}{4}(+\frac{3}{32}-0)$	26.0	54
7	3	$3\frac{3}{4}$	7	3	9	$3\frac{1}{2}$	$\frac{7}{8}(+\frac{1}{8}-0)$	36.0	68.5
8	4	4	8	4	10	4	$1\frac{1}{2}(+\frac{1}{8}-0)$	80.0	135

[a]Safe working load based on 5:1 safety factor using maximum pin diameter.

CLEVIS NUMBERS FOR VARIOUS RODS AND PINS

Diameter of Tap, In.	Diameter of Pin, In.															
	$\frac{5}{8}$	$\frac{3}{4}$	$\frac{7}{8}$	1	$1\frac{1}{4}$	$1\frac{1}{2}$	$1\frac{3}{4}$	2	$2\frac{1}{4}$	$2\frac{1}{2}$	$2\frac{3}{4}$	3	$3\frac{1}{4}$	$3\frac{1}{2}$	$3\frac{3}{4}$	4
$\frac{5}{8}$	2	2	$2\frac{1}{2}$	$2\frac{1}{2}$	$2\frac{1}{2}$	$2\frac{1}{2}$										
$\frac{3}{4}$	—	$2\frac{1}{2}$	$2\frac{1}{2}$	$2\frac{1}{2}$	$2\frac{1}{2}$	$2\frac{1}{2}$										
$\frac{7}{8}$	—	—	$2\frac{1}{2}$	$2\frac{1}{2}$	$2\frac{1}{2}$	$2\frac{1}{2}$										
1	—	—	—	3	3	3	3									
$1\frac{1}{4}$	—	—	—	3	3	3	3	$3\frac{1}{2}$								
$1\frac{3}{8}$	—	—	—	3	3	3	$3\frac{1}{2}$	$3\frac{1}{2}$	4							
$1\frac{1}{2}$	—	—	—	$3\frac{1}{2}$	$3\frac{1}{2}$	$3\frac{1}{2}$	4	4	5							
$1\frac{3}{4}$	—	—	—	—	4	4	5	5	5	5						
2	—	—	—	—	—	5	5	5	5	5	6	6				
$2\frac{1}{4}$	—	—	—	—	—	—	—	6	6	6	6	6	7	7		
$2\frac{1}{2}$	—	—	—	—	—	—	—	6	6	6	7	7	7	7	7	
$2\frac{3}{4}$	—	—	—	—	—	—	—	—	—	7	7	7	7	8	8	
3	—	—	—	—	—	—	—	—	—	7	8	8	8	8	8	8
$3\frac{1}{4}$	—	—	—	—	—	—	—	—	—	—	8	8	8	8	8	8
$3\frac{1}{2}$	—	—	—	—	—	—	—	—	—	—	8	8	8	8	8	8
$3\frac{3}{4}$	—	—	—	—	—	—	—	—	—	—	8	8	8	8	8	8
4	—	—	—	—	—	—	—	—	—	—	8	8	8	8	8	8

Above Table of Clevis Sizes is based on the Net Area of Clevis through Pin Hole being equal to or greater than 125% of Net Area of Rod. Table applies to round rods without upset ends. Pins are sufficient for shear but must be investigated for bending. For other combinations of pin and rod or net area ratios, required clevis size can be calculated by reference to the tabulated dimensions.

Weights and dimensions of clevises are typical. Products of all suppliers are similar and essentially the same.

TURNBUCKLES

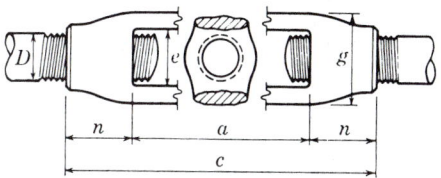

Threads: UNC and 4 UN Class 2B

Dia. D In.	Standard Turnbuckles					Weight of Turnbuckles, Pounds						Turnbuckle Safe Working
	Dimensions, In.					Length a, In.						Load, Kips*
	a	n	c	e	g	6	9	12	18	24	26	
3/8	6	9/16	7 1/8	9/16	1 1/32	.41						1.2
1/2	6	3/4	7 1/2	11/16	1 5/16	.75	.80	1.00				2.2
5/8	6	29/32	7 13/16	13/16	1 1/2	1.00	1.38	1.50	2.43			3.5
3/4	6	1 1/16	8 1/8	15/16	1 23/32	1.45	1.63	2.13	3.06	4.25		5.2
7/8	6	1 7/32	8 7/16	1 3/32	1 7/8	1.85		2.83	4.20	5.43		7.2
1	6	1 3/8	8 3/4	1 9/32	2 1/32	2.60		3.20	4.40	6.85	10.0	9.3
1 1/8	6	1 9/16	9 1/8	1 13/32	2 9/32	2.72		4.70	6.10			11.6
1 1/4	6	1 3/4	9 1/2	1 9/16	2 17/32	3.58		4.70	7.13	11.30	13.1	15.2
1 3/8	6	1 15/16	9 7/8	1 11/16	2 3/4	4.50						17.4
1 1/2	6	2 1/8	10 1/4	1 27/32	3 1/32	5.50		8.00	9.13	16.80	19.4	21.0
1 5/8	6	2 1/4	10 1/2	1 31/32	3 9/32	7.50						24.5
1 3/4	6	2 1/2	11	2 1/8	3 9/16	9.50		15.25	16.00	19.50		28.3
1 7/8	6	2 3/4	11 1/2	2 3/8	4	11.50						37.2
2	6	2 3/4	11 1/2	2 3/8	4	11.50		15.25		27.50		37.2
2 1/4	6	3 3/8	12 3/4	2 11/16	4 5/8	18.00		35.25		43.50		48.0
2 1/2	6	3 3/4	13 1/2	3	5	23.25		33.60		42.38		60.0
2 3/4	6	4 1/8	14 1/4	3 1/4	5 5/8	31.50				54.00		75.0
3	6	4 1/2	15	3 5/8	6 1/8	39.50						96.7
3 1/4	6	5 1/4	16 1/2	3 7/8	6 3/4	60.50						122.2
3 1/2	6	5 1/4	16 1/2	3 7/8	6 3/4	60.50						122.2
3 3/4	6	6	18	4 5/8	8 1/2	95.00						167.8
4	6	6	18	4 5/8	8 1/2	95.00						167.8
4 1/4	9	6 3/4	22 1/2	5 1/4	9 3/4		152.0					233.8
4 1/2	9	6 3/4	22 1/2	5 1/4	9 3/4		152.0					233.8
4 3/4	9	6 3/4	22 1/2	5 1/4	9 3/4		152.0					233.8
5	9	7 1/2	24	6	10		200.0					294.7

* Tabulated loads for Clevises and Turnbuckles have been supplied to AISC and are claimed to be based on a 5:1 safety factor. The higher-than-usual safety factor is because these devices are most often used for rigging and may be subject to dynamic and impact loading. Users should check with individual suppliers to verify the specified breaking load.

SLEEVE NUTS

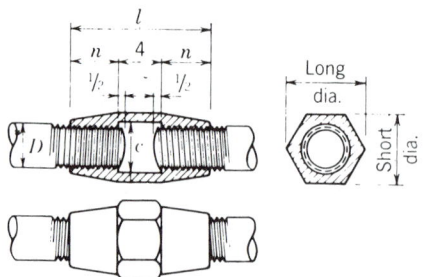

Thread: UNC and 4 UN Class 2B

Diameter of Screw D In.	Dimensions, In.					Weight Lbs.
	Short Diameter	Long Diameter	Length l	Nut n	Clear c	
3/8	11/16	25/32	4	—	—	.27
7/16	25/32	7/8	4	—	—	.34
1/2	7/8	1	4	—	—	.43
9/16	15/16	1 1/16	5	—	—	.64
5/8	1 1/16	1 7/32	5	—	—	.93
3/4	1 1/4	1 7/16	5	—	—	1.12
7/8	1 7/16	1 5/8	7	1 7/16	1	1.75
1	1 5/8	1 13/16	7	1 7/16	1 1/8	2.46
1 1/8	1 13/16	2 1/16	7 1/2	1 5/8	1 1/4	3.10
1 1/4	2	2 1/4	7 1/2	1 5/8	1 3/8	4.04
1 3/8	2 3/16	2 1/2	8	1 7/8	1 1/2	4.97
1 1/2	2 3/8	2 11/16	8	1 7/8	1 5/8	6.16
1 5/8	2 9/16	2 15/16	8 1/2	2 1/16	1 3/4	7.36
1 3/4	2 3/4	3 1/8	8 1/2	2 1/16	1 7/8	8.87
1 7/8	2 15/16	3 5/16	9	2 5/16	2	10.42
2	3 1/8	3 1/2	9	2 5/16	2 1/8	12.24
2 1/4	3 1/2	3 15/16	9 1/2	2 1/2	2 3/8	16.23
2 1/2	3 7/8	4 3/8	10	2 3/4	2 5/8	21.12
2 3/4	4 1/4	4 13/16	10 1/2	2 15/16	2 7/8	26.71
3	4 5/8	5 1/4	11	3 3/16	3 1/8	33.22
3 1/4	5	5 5/8	11 1/2	3 3/8	3 3/8	40.62
3 1/2	5 3/8	6	12	3 5/8	3 5/8	49.07
3 3/4	5 3/4	6 3/8	12 1/2	3 13/16	3 7/8	58.57
4	6 1/8	6 7/8	13	4 1/16	4 1/8	69.22
4 1/4	6 1/2	7 1/2	13 1/2	4 3/4	4 3/8	75.00
4 1/2	6 7/8	7 15/16	14	5	4 3/4	90.00
4 3/4	7 1/4	8 3/8	14 1/2	5 1/4	5	98.00
5	7 5/8	8 7/8	15	5 1/2	5 1/4	110.0
5 1/4	8	9 1/4	15 1/2	5 3/4	5 1/2	122.0
5 1/2	8 3/8	9 3/4	16	6	5 3/4	142.0
5 3/4	8 3/4	10 1/8	16 1/2	6 1/4	6	157.0
6	9 1/8	10 5/8	17	6 1/2	6 1/4	176.0

Strengths are greater than the corresponding connecting rod when same material is used. Weights and dimensions are typical. Products of all suppliers are similar and essentially the same.

RECESSED PIN NUTS AND COTTER PINS

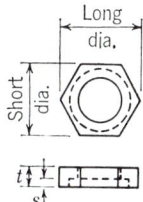

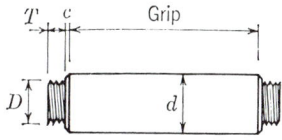

Diameter of Pin d		Pin			Nut (suggested dimensions)					Weight, Lbs.
		Thread		c	Thickness t	Diameter		Recess		
		D	T			Short Dia.	Long Dia.	Rough Dia.	s	
	2	2¼	1½	1	⅛	⅞	3	3⅜	2⅝	¼	1
	2½	2¾	2	1⅛	⅛	1	3⅝	4⅛	3⅛	¼	2
3	3¼	3½	2½	1¼	⅛	1⅛	4⅜	5	3⅞	⅜	3
	3¾	4	3	1⅜	¼	1¼	4⅞	5⅝	4⅜	⅜	4
4¼	4½	4¾	3½	1½	¼	1⅜	5¾	6⅝	5¼	½	5
	5	5¼	4	1⅝	¼	1½	6¼	7¼	5¾	½	6
5½	5¾	6	4½	1¾	¼	1⅝	7	8⅛	6½	⅝	8
	6¼	6½	5	1⅞	⅜	1¾	7⅝	8⅞	7	⅝	10
	6¾	7	5½	2	⅜	1⅞	8⅛	9⅜	7½	¾	12
	7¼	7½	5½	2	⅜	1⅞	8⅝	10	8	¾	14
7¾	8	8¼	6	2¼	⅜	2⅛	9⅜	10⅞	8¾	¾	19
8½	8¾	9	6	2¼	⅜	2⅛	10¼	11⅞	9⅝	¾	24
	9¼	9½	6	2⅜	⅜	2¼	11¼	13	10⅝	¾	32
	9¾	10	6	2⅜	⅜	2¼	11¼	13	10⅝	¾	32

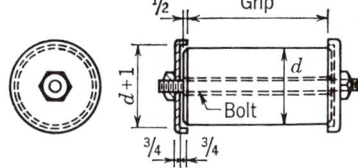

Typical Pin Cap Detail for Pins over 10 Inches in Diameter

Although nuts may be used on all sizes of pins as shown above, for pins over 10-in. dia. the preferred practice is a detail similar to that shown at the left, in which the pin is held in place by a recessed cap at each end and secured by a bolt passing completely through the caps and pin. Suitable provision must be made for attaching pilots and driving nuts.

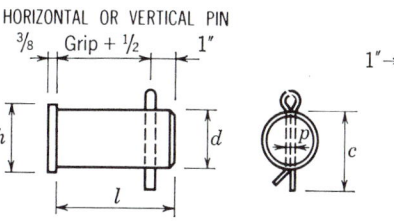

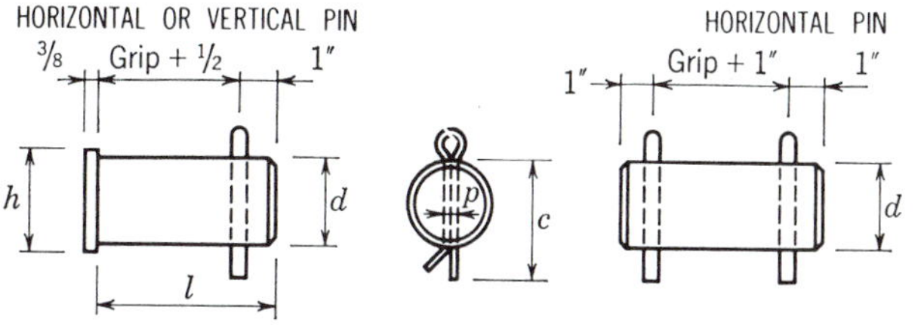

l = Length of pin, in inches.

Pin Dia. d	Pins with Heads		Cotter			Pin Dia. d	Pins with Heads		Cotter		
	Head Dia. h	Weight of One Lb.	Length c	Dia. p	Wt. Per 100 Lb.		Head Dia. h	Weight of One Lb.	Length c	Dia. p	Wt. Per 100 Lb.
1¼	1½	.19 + .35l	2	¼	2.64	2¾	3⅛	.82 + 1.68l	4	⅜	11.4
1½	1¾	.26 + .50l	2½	¼	3.10	3	3½	1.02 + 2.00l	5	½	28.5
1¾	2	.33 + .68l	2¾	¼	3.50	3¼	3¾	1.17 + 2.35l	5	½	28.5
2	2⅜	.47 + .89l	3	⅜	9.00	3½	4	1.34 + 2.73l	6	½	33.8
2¼	2⅝	.58 + 1.13l	3¼	⅜	9.40	3¾	4¼	1.51 + 3.13l	6	½	33.8
2½	2⅞	.70 + 1.39l	3¾	⅜	10.9						

WELDED JOINTS
Requirements

The AISC Specification and the Structural Welding Code of the American Welding Society exempt from tests and qualification most of the common welded joints used in steel structures. Such exempt joints are designated *prequalified*. AWS prequalification of a weld joint is based upon experience that sound weld metal with appropriate mechanical properties can be deposited, provided work is performed in accordance with all applicable provisions of the Structural Welding Code. Among the applicable provisions are requirements for joint form and geometry, which are reproduced for convenience on the following pages.

Prequalification is intended only to mean that sound weld metal can be deposited and fused to the base metal. Suitability of particular joints for specific applications is not assured merely by the selection of a prequalified joint form. The design and detailing for successful welded construction require consideration of factors which include, but are not limited to, magnitude, type and distribution of forces to be transmitted, accessibility, restraint to weld metal contraction, thickness of connected material, effect of residual welding stresses on connected material and distortion.

In general, all fillet welds are deemed prequalified, whether illustrated or not, provided they conform to requirements of the AWS Code and the AISC ASD Specification.

These prequalified joints are limited to those made by the shielded metal arc, submerged arc, gas metal arc (except short circuiting transfer) and flux-cored arc welding procedures. Small deviations from dimensions, angles of grooves, and variation in the depth of groove joints are permissible within the tolerances given. Other joint forms and welding procedures may be employed, provided they are tested and qualified in accordance with AWS D1.1-88.

Most prequalified joints illustrated are also applicable for bridge construction. (See notes to Prequalified Welded Joints, immediately preceding the tables, and to prohibited types in Sect. 9 of AWS D1.1-88.)

For information on the subject of highly restrained welded joints, refer to the article "Commentary on Highly Restrained Welded Connections," AISC Engineering Journal, Vol. 10, No. 3, 3rd Quarter 1973, pp. 61–73.

The designations such as B-L1a, B-U2, B-P3, which are given on the following pages, are used in the AWS standards. Groove welds are classified using the following convention:

1. *Symbols for Joint Types*

    B   —butt joint
    C   —corner joint
    T   —T-joint
    BC —butt or corner joint
    TC —T or corner joint
    BTC—butt, T or corner joint

2. *Symbols for Base Metal Thickness and Penetration*

    L—limited thickness, complete joint penetration
    U—unlimited thickness, complete joint penetration
    P—partial joint penetration

3. *Symbols for Weld Types*

    1—square groove
    2—single-V groove
    3—double-V groove
    4—single-bevel groove
    5—double-bevel groove
    6—single-U groove
    7—double-U groove
    8—single-J groove
    9—double-J groove
   10—flare-bevel groove

4. *Symbols for Welding Processes*

If not shielded metal arc (SMAW):
S—submerged arc welding (SAW)
G—gas metal arc welding (GMAW)
F—flux-cored arc welding (FCAW)

5. *Symbols for Welding Positions*

   F—flat
   H—horizontal
   V—vertical
OH—overhead

6. The lower case letters, e.g., a, b, c, etc., are used to differentiate between joints that would otherwise have the same joint designation.

NOTES TO PREQUALIFIED WELD JOINTS

  A: Not prequalified for gas metal arc welding using short circuiting transfer.
  B: Joints welded from one side.
 Br: Bridge application limits the use of these joints to the horizontal position.
  C: Gouge root of joint to sound metal before welding second side.
  E: Minimum effective throat E as shown in AISC Specification, Table J2.3. *S* as specified on drawings.
  J: If fillet welds are used in buildings to reinforce groove welds in corner and T-joints, they shall be equal to ¼ T_1 but need not exceed ⅜ in. Groove welds in corner and T-joints in bridges shall be made with fillet welds equal to ¼ T_1, but not more than ⅜ in.
 J2: If fillet welds are used in buildings to reinforce groove welds in corner and T-joints, they shall be equal to ¼ T_1, but need not exceed ⅜ in.
  L: Butt and T-joints are not prequalified for bridges.
  M: Double-groove welds may have grooves of unequal depth, but the depth of the shallower groove shall be no less than one-fourth of the thickness of the thinner part joined.
 Mp: Double-groove welds may have grooves of unequal depth, provided they conform to the limitations of Note E. Also, the effective throat E, less any reduction, applies individually to each groove.
  N: The orientation of the two members in the joints may vary from 135° to 180°, provided the basic joint configuration (groove angle, root face, root opening) remain the same and that the design throat thickness is maintained.

Q: For corner and T-joints, the member orientation may be changed, provided the groove angle is maintained as specified.

Q2: The member orientation may be changed, provided the groove dimensions are maintained as specified.

R: The orientation of two members in the joint may vary from 45° to 135° for corner joints and from 45° to 90° for T-joints, provided the basic joint configuration (groove angle, root face, root opening) remain the same and the design throat thickness is maintained.

V: For corner joints, the outside groove preparation may be in either or both members, provided the basic groove configuration is not changed and adequate edge distance is maintained to support the welding operations without excessive edge melting.

Z: Effective throat E is based on joints welded flush.

Note: Data on welded prequalified joints are reproduced on the following pages by courtesy of the American Welding Society.

WELDED JOINTS
Standard symbols

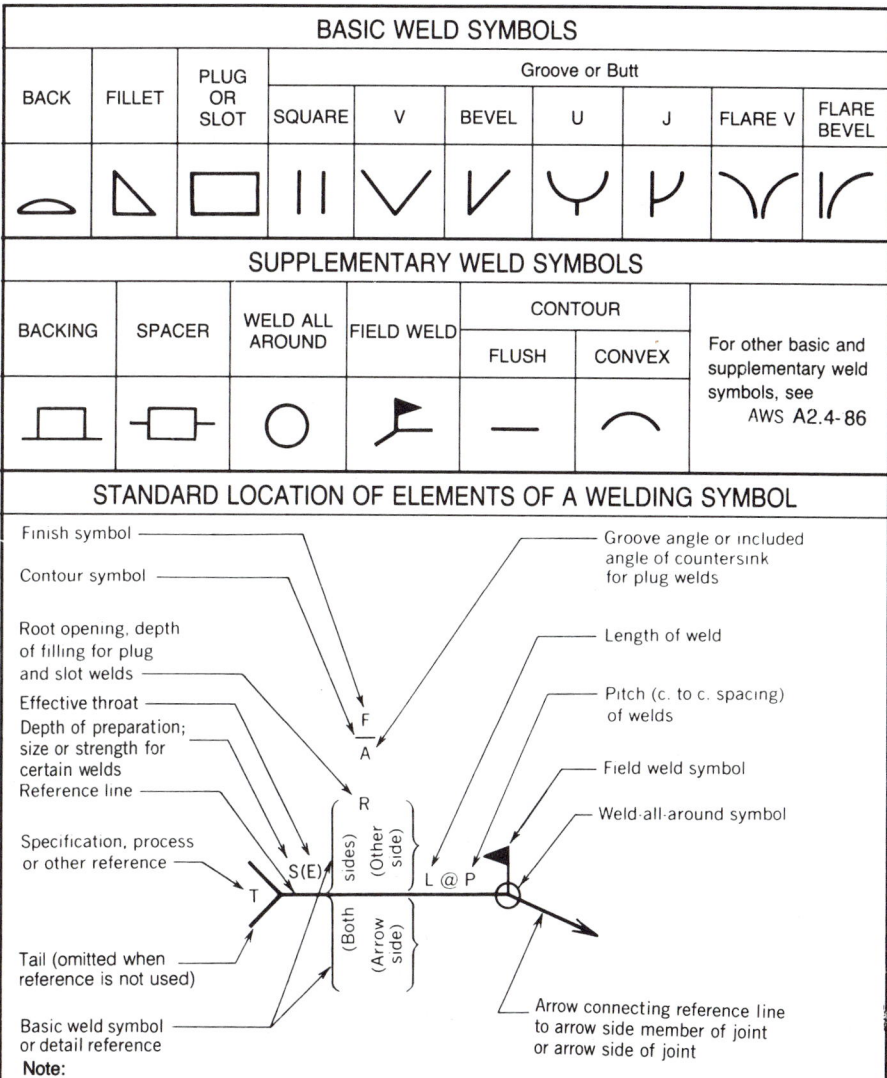

BASIC WELD SYMBOLS

BACK	FILLET	PLUG OR SLOT	Groove or Butt						
			SQUARE	V	BEVEL	U	J	FLARE V	FLARE BEVEL

SUPPLEMENTARY WELD SYMBOLS

BACKING	SPACER	WELD ALL AROUND	FIELD WELD	CONTOUR		For other basic and supplementary weld symbols, see AWS A2.4-86
				FLUSH	CONVEX	

STANDARD LOCATION OF ELEMENTS OF A WELDING SYMBOL

Finish symbol

Contour symbol

Root opening, depth of filling for plug and slot welds

Effective throat

Depth of preparation; size or strength for certain welds

Reference line

Specification, process or other reference

Tail (omitted when reference is not used)

Basic weld symbol or detail reference

Groove angle or included angle of countersink for plug welds

Length of weld

Pitch (c. to c. spacing) of welds

Field weld symbol

Weld-all-around symbol

Arrow connecting reference line to arrow side member of joint or arrow side of joint

Note:

Size, weld symbol, length of weld and spacing must read in that order from left to right along the reference line. Neither orientation of reference line nor location of the arrow alters this rule.

The perpendicular leg of $\triangle$, V, $\vdash$, $\vdash$ weld symbols must be at left.

Arrow and Other Side welds are of the same size unless otherwise shown. Dimensions of fillet welds must be shown on both the Arrow Side and the Other Side Symbol.

Flag of field-weld symbol shall be placed above and at right angle to reference line of junction with the arrow.

Symbols apply between abrupt changes in direction of welding unless governed by the "all around" symbol or otherwise dimensioned.

These symbols do not explicitly provide for the case that frequently occurs in structural work, where duplicate material (such as stiffeners) occurs on the far side of a web or gusset plate. The fabricating industry has adopted this convention: that when the billing of the detail material discloses the existence of a member on the far side as well as on the near side, the welding shown for the near side shall be duplicated on the far side.

PREQUALIFIED WELDED JOINTS
Fillet welds

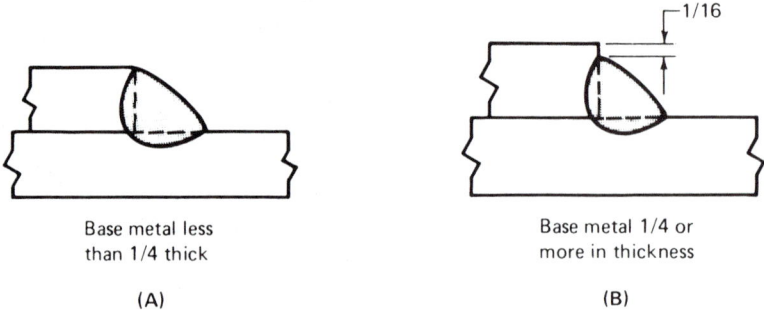

Base metal less than 1/4 thick	Base metal 1/4 or more in thickness
(A)	(B)

Maximum detailed size of fillet weld along edges

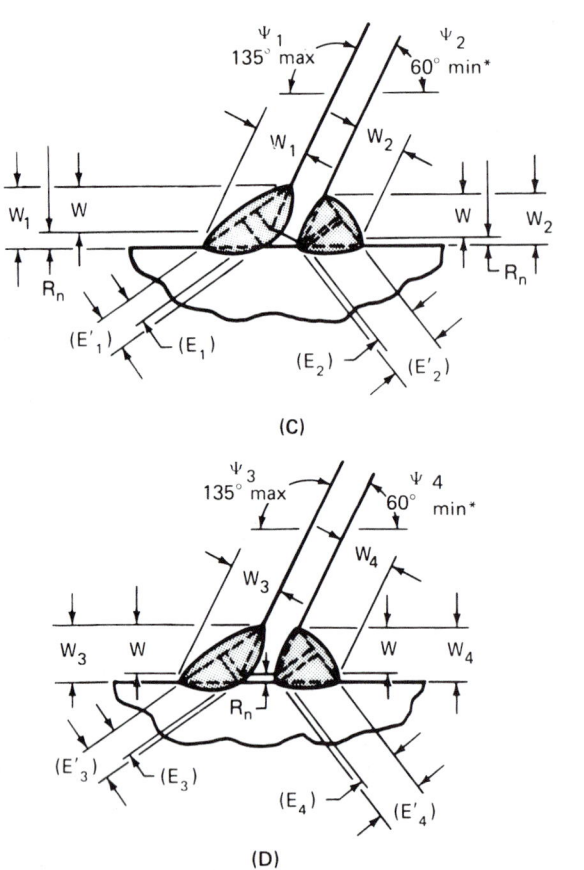

(C)

(D)

Skewed T-joints

Note: $(E)_{(n)}$, $(E')_{(n)}$ = effective throats dependent on magnitude of root opening (R_n). See AWS 3.3.1. Subscript (n) represents 1, 2, 3 or 4.

* Angles smaller than 60 degrees are permitted; however, in such cases, the weld is considered to be a partial joint penetration groove weld.

For additional requirements for skewed T-joints, see 2.3.2.4 of AWS D1.1-88.

AMERICAN INSTITUTE OF STEEL CONSTRUCTION

PREQUALIFIED WELDED JOINTS
Complete-penetration groove welds

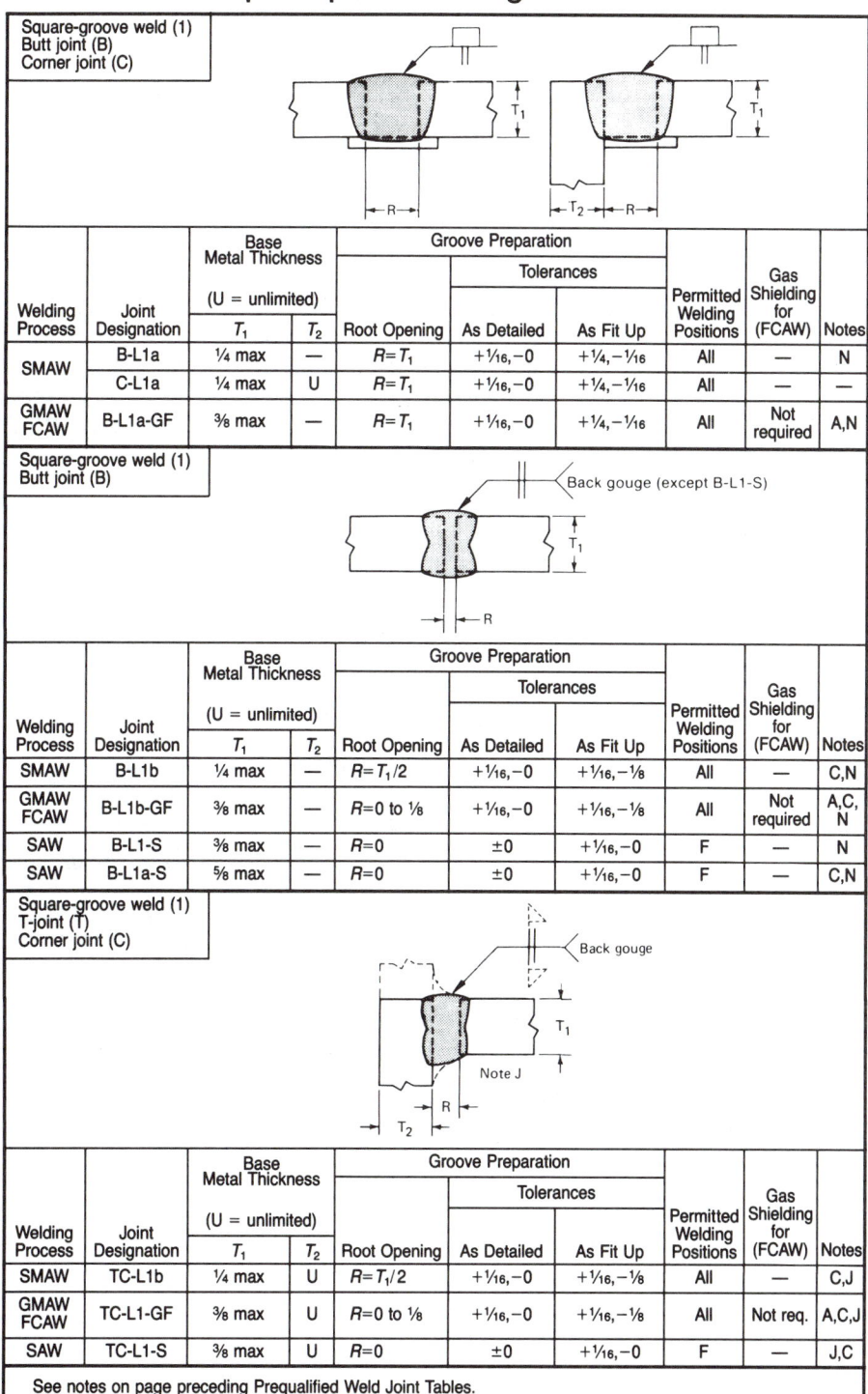

Square-groove weld (1)
Butt joint (B)
Corner joint (C)

Welding Process	Joint Designation	Base Metal Thickness (U = unlimited)		Groove Preparation			Permitted Welding Positions	Gas Shielding for (FCAW)	Notes
		T_1	T_2	Root Opening	As Detailed	As Fit Up			
SMAW	B-L1a	¼ max	—	$R=T_1$	+¹⁄₁₆, −0	+¼, −¹⁄₁₆	All	—	N
SMAW	C-L1a	¼ max	U	$R=T_1$	+¹⁄₁₆, −0	+¼, −¹⁄₁₆	All	—	—
GMAW FCAW	B-L1a-GF	⅜ max	—	$R=T_1$	+¹⁄₁₆, −0	+¼, −¹⁄₁₆	All	Not required	A,N

Square-groove weld (1)
Butt joint (B)

Back gouge (except B-L1-S)

Welding Process	Joint Designation	Base Metal Thickness (U = unlimited)		Groove Preparation			Permitted Welding Positions	Gas Shielding for (FCAW)	Notes
		T_1	T_2	Root Opening	As Detailed	As Fit Up			
SMAW	B-L1b	¼ max	—	$R=T_1/2$	+¹⁄₁₆, −0	+¹⁄₁₆, −⅛	All	—	C,N
GMAW FCAW	B-L1b-GF	⅜ max	—	$R=0$ to ⅛	+¹⁄₁₆, −0	+¹⁄₁₆, −⅛	All	Not required	A,C,N
SAW	B-L1-S	⅜ max	—	$R=0$	±0	+¹⁄₁₆, −0	F	—	N
SAW	B-L1a-S	⅝ max	—	$R=0$	±0	+¹⁄₁₆, −0	F	—	C,N

Square-groove weld (1)
T-joint (T)
Corner joint (C)

Back gouge — Note J

Welding Process	Joint Designation	Base Metal Thickness (U = unlimited)		Groove Preparation			Permitted Welding Positions	Gas Shielding for (FCAW)	Notes
		T_1	T_2	Root Opening	As Detailed	As Fit Up			
SMAW	TC-L1b	¼ max	U	$R=T_1/2$	+¹⁄₁₆, −0	+¹⁄₁₆, −⅛	All	—	C,J
GMAW FCAW	TC-L1-GF	⅜ max	U	$R=0$ to ⅛	+¹⁄₁₆, −0	+¹⁄₁₆, −⅛	All	Not req.	A,C,J
SAW	TC-L1-S	⅜ max	U	$R=0$	±0	+¹⁄₁₆, −0	F	—	J,C

See notes on page preceding Prequalified Weld Joint Tables.

PREQUALIFIED WELDED JOINTS
Complete-penetration groove welds

Single V-groove weld (2)
Butt joint (B)

Tolerances	
As detailed	As fit up
$R=+\frac{1}{16},-0$	$+\frac{1}{4},-\frac{1}{16}$
$\alpha=+10°,-0°$	$+10°,-5°$

Welding Process	Joint Designation	Base Metal Thickness (U = unlimited) T_1	T_2	Groove Preparation Root Opening	Groove Angle	Permitted Welding Positions	Gas Shielding for (FCAW)	Notes
SMAW	B-U2a	U	—	$R=\frac{1}{4}$	$\alpha=45°$	All	—	N
				$R=\frac{3}{8}$	$\alpha=30°$	F,V,OH	—	N
				$R=\frac{1}{2}$	$\alpha=20°$	F,V,OH	—	N
GMAW FCAW	B-U2a-GF	U	—	$R=\frac{3}{16}$	$\alpha=30°$	F,V,OH	Required	A,N
				$R=\frac{3}{8}$	$\alpha=30°$	F,V,OH	Not req.	A,N
				$R=\frac{1}{4}$	$\alpha=45°$	F,V,OH	Not req.	A,N
SAW	B-L2a-S	2 max	—	$R=\frac{1}{4}$	$\alpha=30°$	F	—	N
SAW	B-U2-S	U	—	$R=\frac{5}{8}$	$\alpha=20°$	F	—	N

Single V-groove weld (2)
Corner joint (C)

Tolerances	
As detailed	As fit up
$R=+\frac{1}{16},-0$	$+\frac{1}{4},-\frac{1}{16}$
$\alpha=+10°,-0°$	$+10°,-5°$

Welding Process	Joint Designation	Base Metal Thickness (U = unlimited) T_1	T_2	Groove Preparation Root Opening	Groove Angle	Permitted Welding Positions	Gas Shielding for (FCAW)	Notes
SMAW	C-U2a	U	U	$R=\frac{1}{4}$	$\alpha=45°$	All	—	Q
				$R=\frac{3}{8}$	$\alpha=30°$	F,V,OH	—	Q
				$R=\frac{1}{2}$	$\alpha=20°$	F,V,OH	—	Q
GMAW FCAW	C-U2a-GF	U	U	$R=\frac{3}{16}$	$\alpha=30°$	F,V,OH	Required	A
				$R=\frac{3}{8}$	$\alpha=30°$	F,V,OH	Not req.	A,Q
				$R=\frac{1}{4}$	$\alpha=45°$	F,V,OH	Not req.	A,Q
SAW	C-L2a-S	2 max	U	$R=\frac{1}{4}$	$\alpha=30°$	F	—	Q
SAW	C-U2-S	U	U	$R=\frac{5}{8}$	$\alpha=20°$	F	—	Q

See notes on page preceding Prequalified Weld Joint Tables.

PREQUALIFIED WELDED JOINTS
Complete-penetration groove welds

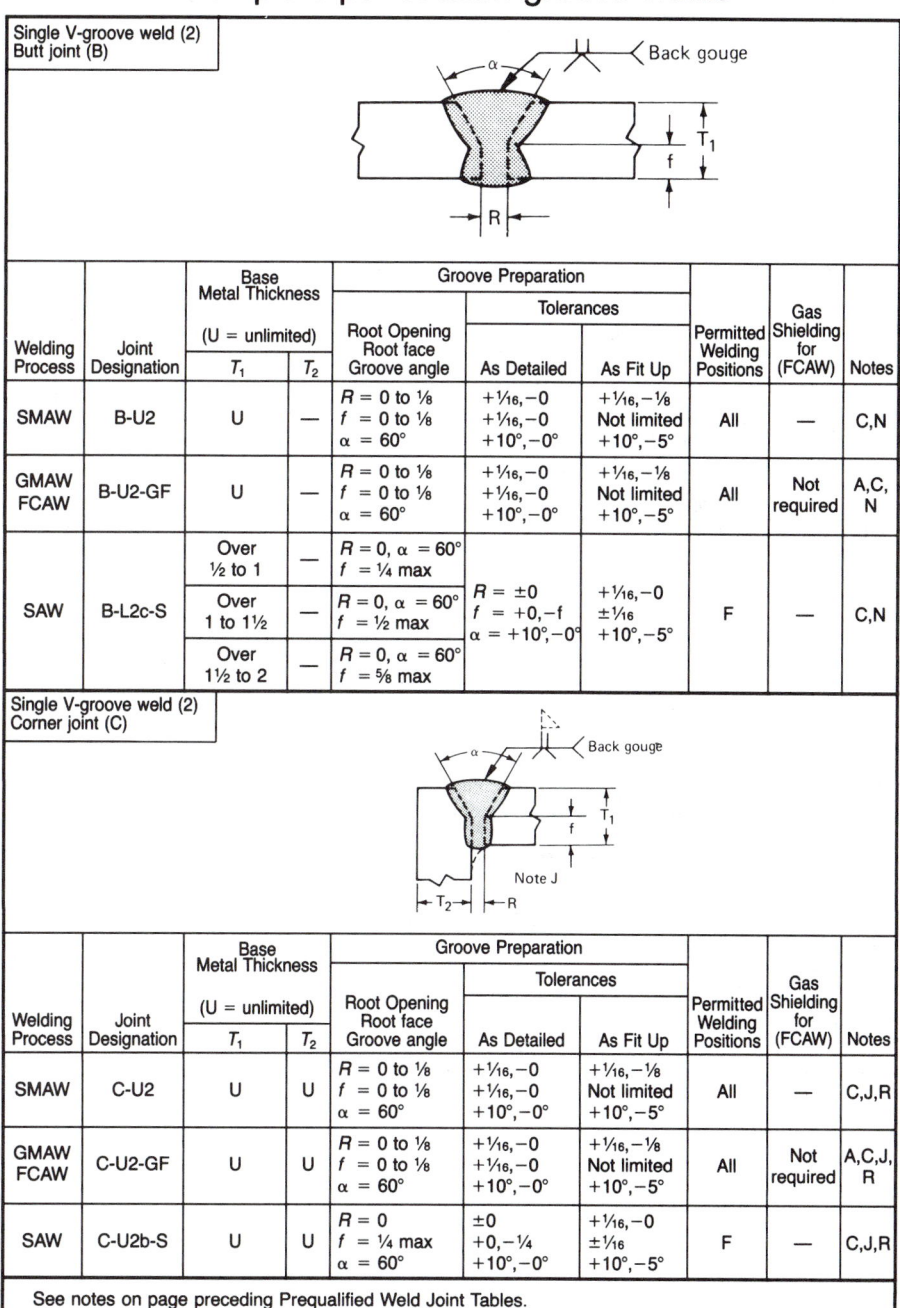

Single V-groove weld (2)
Butt joint (B)

Welding Process	Joint Designation	Base Metal Thickness (U = unlimited) T₁	T₂	Groove Preparation Root Opening Root face Groove angle	Tolerances As Detailed	As Fit Up	Permitted Welding Positions	Gas Shielding for (FCAW)	Notes
SMAW	B-U2	U	—	R = 0 to ⅛ f = 0 to ⅛ α = 60°	+1/16, −0 +1/16, −0 +10°, −0°	+1/16, −⅛ Not limited +10°, −5°	All	—	C,N
GMAW FCAW	B-U2-GF	U	—	R = 0 to ⅛ f = 0 to ⅛ α = 60°	+1/16, −0 +1/16, −0 +10°, −0°	+1/16, −⅛ Not limited +10°, −5°	All	Not required	A,C, N
SAW	B-L2c-S	Over ½ to 1	—	R = 0, α = 60° f = ¼ max					
		Over 1 to 1½	—	R = 0, α = 60° f = ½ max	R = ±0 f = +0,−f α = +10°,−0°	+1/16,−0 ±1/16 +10°, −5°	F	—	C,N
		Over 1½ to 2	—	R = 0, α = 60° f = ⅝ max					

Single V-groove weld (2)
Corner joint (C)

Note J

Welding Process	Joint Designation	Base Metal Thickness (U = unlimited) T₁	T₂	Groove Preparation Root Opening Root face Groove angle	Tolerances As Detailed	As Fit Up	Permitted Welding Positions	Gas Shielding for (FCAW)	Notes
SMAW	C-U2	U	U	R = 0 to ⅛ f = 0 to ⅛ α = 60°	+1/16, −0 +1/16, −0 +10°, −0°	+1/16, −⅛ Not limited +10°, −5°	All	—	C,J,R
GMAW FCAW	C-U2-GF	U	U	R = 0 to ⅛ f = 0 to ⅛ α = 60°	+1/16, −0 +1/16, −0 +10°, −0°	+1/16, −⅛ Not limited +10°, −5°	All	Not required	A,C,J, R
SAW	C-U2b-S	U	U	R = 0 f = ¼ max α = 60°	±0 +0, −¼ +10°, −0°	+1/16, −0 ±1/16 +10°, −5°	F	—	C,J,R

See notes on page preceding Prequalified Weld Joint Tables.

PREQUALIFIED WELDED JOINTS
Complete-penetration groove welds

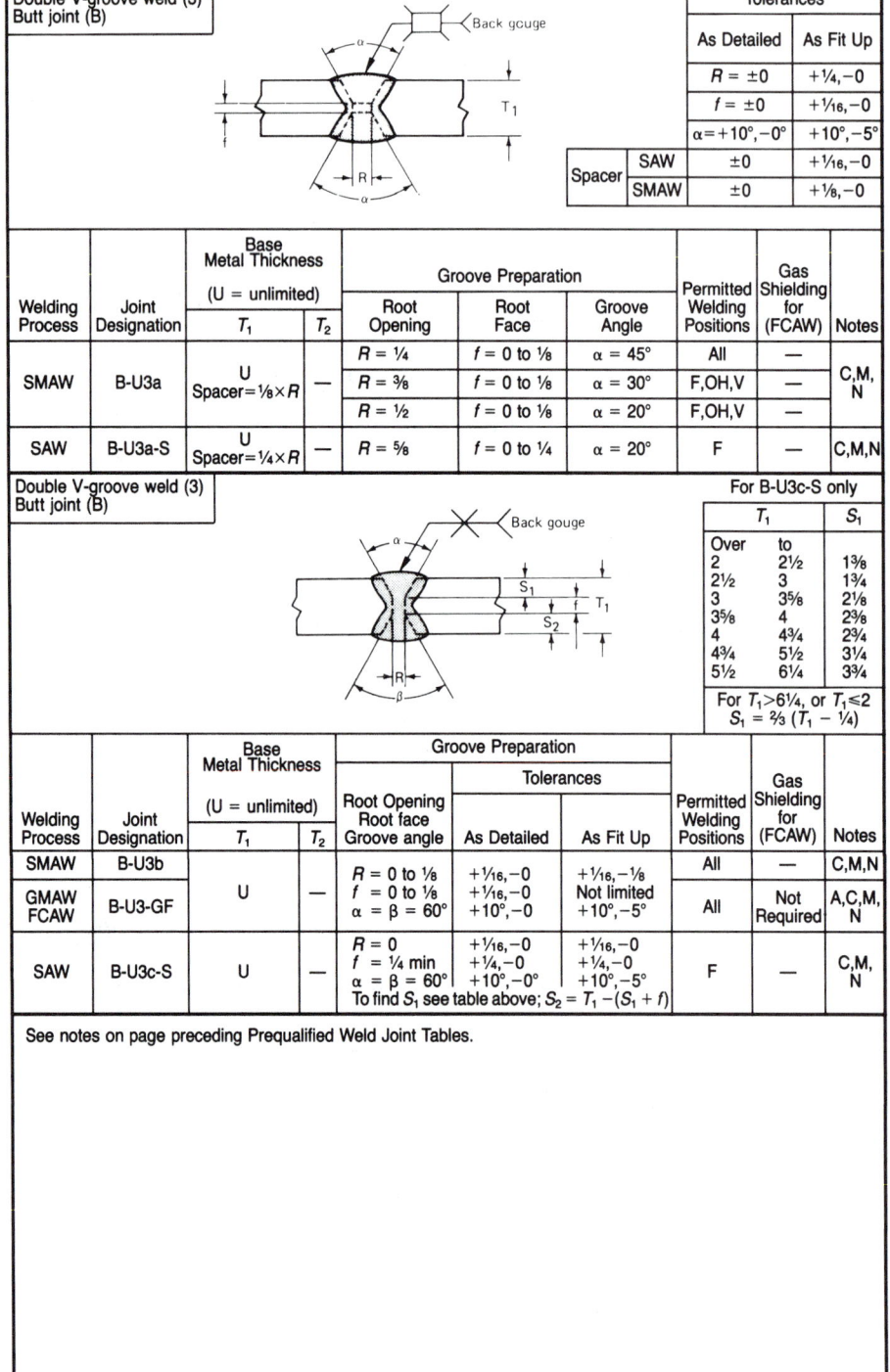

Double V-groove weld (3)
Butt joint (B)

Tolerances		
	As Detailed	As Fit Up
$R = \pm0$		$+\frac{1}{4},-0$
$f = \pm0$		$+\frac{1}{16},-0$
$\alpha = +10°,-0°$		$+10°,-5°$
Spacer — SAW	±0	$+\frac{1}{16},-0$
Spacer — SMAW	±0	$+\frac{1}{8},-0$

Welding Process	Joint Designation	Base Metal Thickness (U = unlimited) T_1	T_2	Groove Preparation Root Opening	Root Face	Groove Angle	Permitted Welding Positions	Gas Shielding for (FCAW)	Notes
SMAW	B-U3a Spacer=$\frac{1}{8}\times R$	U	—	$R = \frac{1}{4}$	$f = 0$ to $\frac{1}{8}$	$\alpha = 45°$	All	—	C,M, N
				$R = \frac{3}{8}$	$f = 0$ to $\frac{1}{8}$	$\alpha = 30°$	F,OH,V	—	
				$R = \frac{1}{2}$	$f = 0$ to $\frac{1}{8}$	$\alpha = 20°$	F,OH,V	—	
SAW	B-U3a-S Spacer=$\frac{1}{4}\times R$	U	—	$R = \frac{5}{8}$	$f = 0$ to $\frac{1}{4}$	$\alpha = 20°$	F	—	C,M,N

Double V-groove weld (3)
Butt joint (B)

For B-U3c-S only

T_1		S_1
Over	to	
2	2½	1⅜
2½	3	1¾
3	3⅝	2⅛
3⅝	4	2⅜
4	4¾	2¾
4¾	5½	3¼
5½	6¼	3¾

For $T_1>6\frac{1}{4}$, or $T_1\leqslant2$
$S_1 = \frac{2}{3}(T_1 - \frac{1}{4})$

Welding Process	Joint Designation	Base Metal Thickness (U = unlimited) T_1	T_2	Groove Preparation Root Opening Root face Groove angle	Tolerances As Detailed	As Fit Up	Permitted Welding Positions	Gas Shielding for (FCAW)	Notes
SMAW	B-U3b	U	—	$R = 0$ to $\frac{1}{8}$	$+\frac{1}{16},-0$	$+\frac{1}{16},-\frac{1}{8}$	All	—	C,M,N
				$f = 0$ to $\frac{1}{8}$	$+\frac{1}{16},-0$	Not limited			
GMAW FCAW	B-U3-GF			$\alpha = \beta = 60°$	$+10°,-0$	$+10°,-5°$	All	Not Required	A,C,M, N
SAW	B-U3c-S	U	—	$R = 0$	$+\frac{1}{16},-0$	$+\frac{1}{16},-0$	F	—	C,M, N
				$f = \frac{1}{4}$ min	$+\frac{1}{4},-0$	$+\frac{1}{4},-0$			
				$\alpha = \beta = 60°$	$+10°,-0°$	$+10°,-5°$			
				To find S_1 see table above; $S_2 = T_1-(S_1 + f)$					

See notes on page preceding Prequalified Weld Joint Tables.

PREQUALIFIED WELDED JOINTS
Complete-penetration groove welds

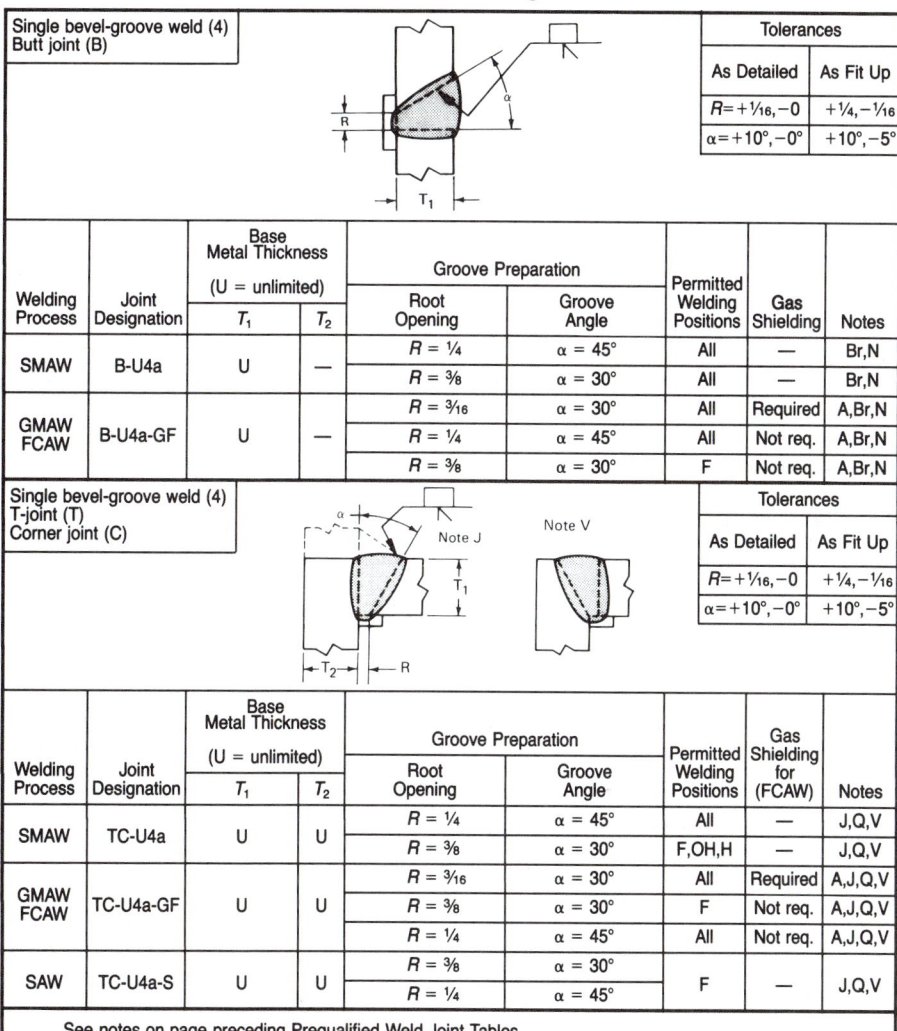

Single bevel-groove weld (4)
Butt joint (B)

Tolerances		
	As Detailed	As Fit Up
$R = +\frac{1}{16}, -0$		$+\frac{1}{4}, -\frac{1}{16}$
$\alpha = +10°, -0°$		$+10°, -5°$

Welding Process	Joint Designation	Base Metal Thickness (U = unlimited) T_1	T_2	Groove Preparation Root Opening	Groove Angle	Permitted Welding Positions	Gas Shielding	Notes
SMAW	B-U4a	U	—	$R = \frac{1}{4}$	$\alpha = 45°$	All	—	Br,N
				$R = \frac{3}{8}$	$\alpha = 30°$	All	—	Br,N
GMAW FCAW	B-U4a-GF	U	—	$R = \frac{3}{16}$	$\alpha = 30°$	All	Required	A,Br,N
				$R = \frac{1}{4}$	$\alpha = 45°$	All	Not req.	A,Br,N
				$R = \frac{3}{8}$	$\alpha = 30°$	F	Not req.	A,Br,N

Single bevel-groove weld (4)
T-joint (T)
Corner joint (C)

Note J   Note V

Tolerances		
	As Detailed	As Fit Up
$R = +\frac{1}{16}, -0$		$+\frac{1}{4}, -\frac{1}{16}$
$\alpha = +10°, -0°$		$+10°, -5°$

Welding Process	Joint Designation	Base Metal Thickness (U = unlimited) T_1	T_2	Groove Preparation Root Opening	Groove Angle	Permitted Welding Positions	Gas Shielding for (FCAW)	Notes
SMAW	TC-U4a	U	U	$R = \frac{1}{4}$	$\alpha = 45°$	All	—	J,Q,V
				$R = \frac{3}{8}$	$\alpha = 30°$	F,OH,H	—	J,Q,V
GMAW FCAW	TC-U4a-GF	U	U	$R = \frac{3}{16}$	$\alpha = 30°$	All	Required	A,J,Q,V
				$R = \frac{3}{8}$	$\alpha = 30°$	F	Not req.	A,J,Q,V
				$R = \frac{1}{4}$	$\alpha = 45°$	All	Not req.	A,J,Q,V
SAW	TC-U4a-S	U	U	$R = \frac{3}{8}$	$\alpha = 30°$	F	—	J,Q,V
				$R = \frac{1}{4}$	$\alpha = 45°$			J,Q,V

See notes on page preceding Prequalified Weld Joint Tables.

PREQUALIFIED WELDED JOINTS
Complete-penetration groove welds

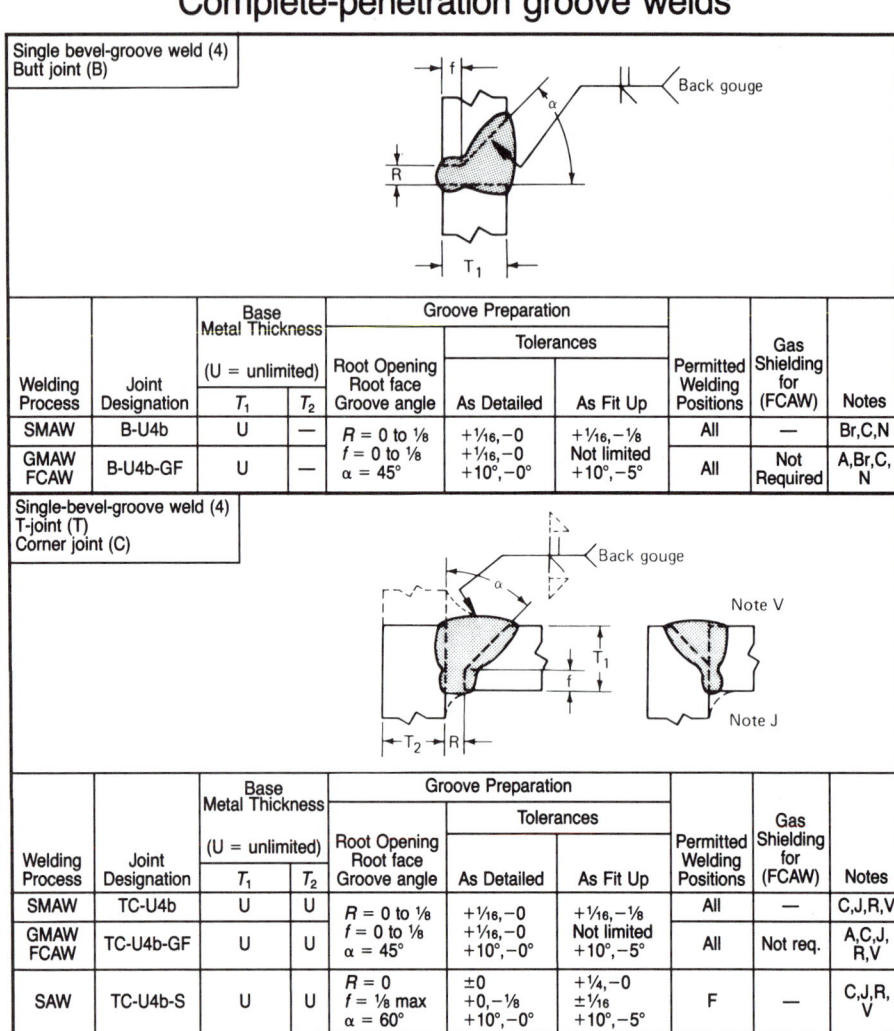

Single bevel-groove weld (4)
Butt joint (B)

Welding Process	Joint Designation	Base Metal Thickness (U = unlimited) T_1	T_2	Groove Preparation — Root Opening Root face Groove angle	Tolerances — As Detailed	Tolerances — As Fit Up	Permitted Welding Positions	Gas Shielding for (FCAW)	Notes
SMAW	B-U4b	U	—	R = 0 to ⅛ / f = 0 to ⅛ / α = 45°	+1/16,−0 / +1/16,−0 / +10°,−0°	+1/16,−⅛ / Not limited / +10°,−5°	All	—	Br,C,N
GMAW FCAW	B-U4b-GF	U	—				All	Not Required	A,Br,C, N

Single-bevel-groove weld (4)
T-joint (T)
Corner joint (C)

Welding Process	Joint Designation	Base Metal Thickness (U = unlimited) T_1	T_2	Groove Preparation — Root Opening Root face Groove angle	Tolerances — As Detailed	Tolerances — As Fit Up	Permitted Welding Positions	Gas Shielding for (FCAW)	Notes
SMAW	TC-U4b	U	U	R = 0 to ⅛ / f = 0 to ⅛ / α = 45°	+1/16,−0 / +1/16,−0 / +10°,−0°	+1/16,−⅛ / Not limited / +10°,−5°	All	—	C,J,R,V
GMAW FCAW	TC-U4b-GF	U	U				All	Not req.	A,C,J, R,V
SAW	TC-U4b-S	U	U	R = 0 / f = ⅛ max / α = 60°	±0 / +0,−⅛ / +10°,−0°	+¼,−0 / ±1/16 / +10°,−5°	F	—	C,J,R, V

See notes on page preceding Prequalified Weld Joint Tables.

AMERICAN INSTITUTE OF STEEL CONSTRUCTION

PREQUALIFIED WELDED JOINTS
Complete-penetration groove welds

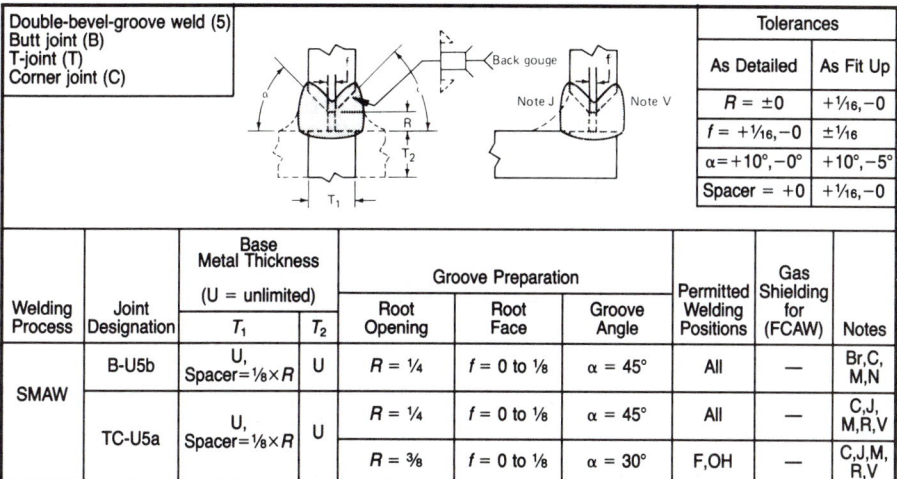

Double-bevel-groove weld (5) Butt joint (B) T-joint (T) Corner joint (C)		Tolerances	
		As Detailed	As Fit Up
		$R = \pm0$	$+\frac{1}{16}, -0$
		$f = +\frac{1}{16}, -0$	$\pm\frac{1}{16}$
		$\alpha = +10°, -0°$	$+10°, -5°$
		Spacer $= +0$	$+\frac{1}{16}, -0$

Welding Process	Joint Designation	Base Metal Thickness (U = unlimited) T_1	T_2	Groove Preparation Root Opening	Root Face	Groove Angle	Permitted Welding Positions	Gas Shielding for (FCAW)	Notes
SMAW	B-U5b	U, Spacer$=\frac{1}{8}\times R$	U	$R = \frac{1}{4}$	$f = 0$ to $\frac{1}{8}$	$\alpha = 45°$	All	—	Br,C, M,N
	TC-U5a	U, Spacer$=\frac{1}{8}\times R$	U	$R = \frac{1}{4}$	$f = 0$ to $\frac{1}{8}$	$\alpha = 45°$	All	—	C,J, M,R,V
				$R = \frac{3}{8}$	$f = 0$ to $\frac{1}{8}$	$\alpha = 30°$	F,OH	—	C,J,M, R,V

See notes on page preceding Prequalified Weld Joint Tables.

PREQUALIFIED WELDED JOINTS
Complete-penetration groove welds

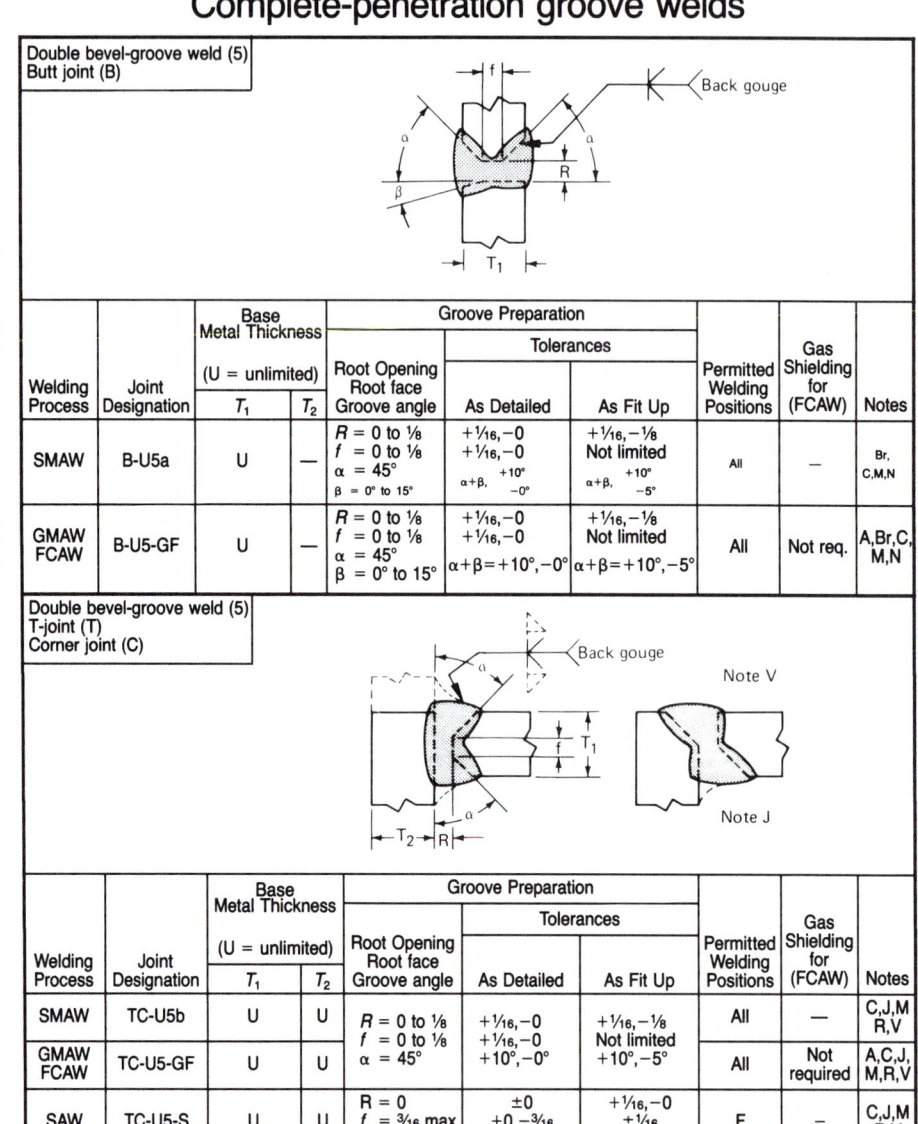

Double bevel-groove weld (5)
Butt joint (B)

Welding Process	Joint Designation	Base Metal Thickness (U = unlimited) T_1	T_2	Groove Preparation Root Opening Root face Groove angle	Tolerances As Detailed	As Fit Up	Permitted Welding Positions	Gas Shielding for (FCAW)	Notes
SMAW	B-U5a	U	—	$R = 0$ to $\frac{1}{8}$ $f = 0$ to $\frac{1}{8}$ $\alpha = 45°$ $\beta = 0°$ to $15°$	$+\frac{1}{16}, -0$ $+\frac{1}{16}, -0$ $\alpha+\beta, \begin{smallmatrix}+10°\\-0°\end{smallmatrix}$	$+\frac{1}{16}, -\frac{1}{8}$ Not limited $\alpha+\beta, \begin{smallmatrix}+10°\\-5°\end{smallmatrix}$	All	—	Br, C,M,N
GMAW FCAW	B-U5-GF	U	—	$R = 0$ to $\frac{1}{8}$ $f = 0$ to $\frac{1}{8}$ $\alpha = 45°$ $\beta = 0°$ to $15°$	$+\frac{1}{16}, -0$ $+\frac{1}{16}, -0$ $\alpha+\beta = +10°, -0°$	$+\frac{1}{16}, -\frac{1}{8}$ Not limited $\alpha+\beta = +10°, -5°$	All	Not req.	A,Br,C, M,N

Double bevel-groove weld (5)
T-joint (T)
Corner joint (C)

Note V

Note J

Welding Process	Joint Designation	Base Metal Thickness (U = unlimited) T_1	T_2	Groove Preparation Root Opening Root face Groove angle	Tolerances As Detailed	As Fit Up	Permitted Welding Positions	Gas Shielding for (FCAW)	Notes
SMAW	TC-U5b	U	U	$R = 0$ to $\frac{1}{8}$ $f = 0$ to $\frac{1}{8}$ $\alpha = 45°$	$+\frac{1}{16}, -0$ $+\frac{1}{16}, -0$ $+10°, -0°$	$+\frac{1}{16}, -\frac{1}{8}$ Not limited $+10°, -5°$	All	—	C,J,M R,V
GMAW FCAW	TC-U5-GF	U	U				All	Not required	A,C,J, M,R,V
SAW	TC-U5-S	U	U	$R = 0$ $f = \frac{3}{16}$ max $\alpha = 60°$	± 0 $+0, -\frac{3}{16}$ $+10°, -0°$	$+\frac{1}{16}, -0$ $\pm\frac{1}{16}$ $+10°, -5°$	F	–	C,J,M R,V

See notes on page preceding Prequalified Weld Joint Tables.

PREQUALIFIED WELDED JOINTS
Complete-penetration groove welds

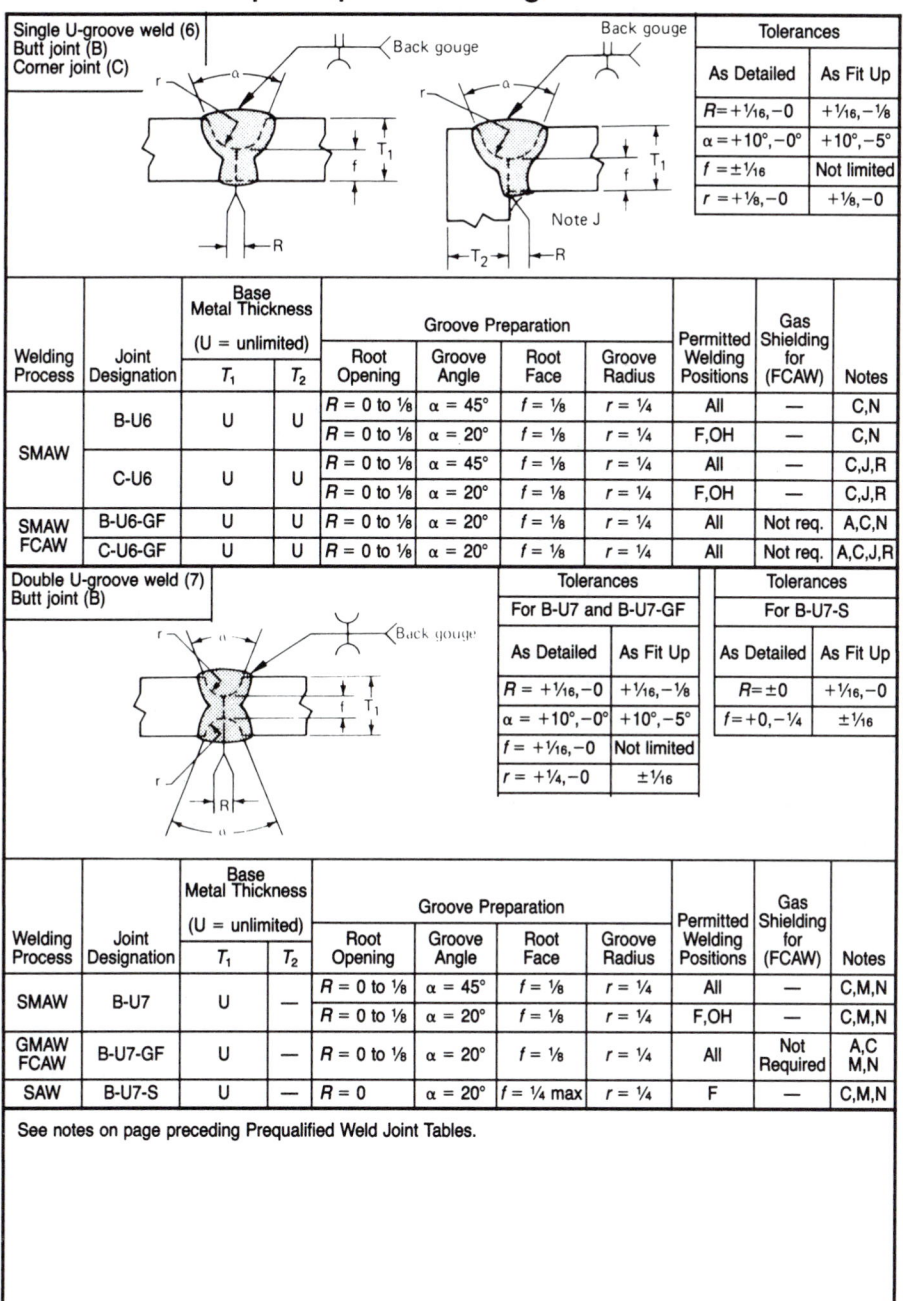

Single U-groove weld (6)
Butt joint (B)
Corner joint (C)

Tolerances	As Detailed	As Fit Up
$R = +\frac{1}{16}, -0$		$+\frac{1}{16}, -\frac{1}{8}$
$\alpha = +10°, -0°$		$+10°, -5°$
$f = \pm\frac{1}{16}$		Not limited
$r = +\frac{1}{8}, -0$		$+\frac{1}{8}, -0$

Welding Process	Joint Designation	Base Metal Thickness (U = unlimited) T_1	T_2	Groove Preparation Root Opening	Groove Angle	Root Face	Groove Radius	Permitted Welding Positions	Gas Shielding for (FCAW)	Notes
SMAW	B-U6	U	U	$R = 0$ to $\frac{1}{8}$	$\alpha = 45°$	$f = \frac{1}{8}$	$r = \frac{1}{4}$	All	—	C,N
				$R = 0$ to $\frac{1}{8}$	$\alpha = 20°$	$f = \frac{1}{8}$	$r = \frac{1}{4}$	F,OH	—	C,N
	C-U6	U	U	$R = 0$ to $\frac{1}{8}$	$\alpha = 45°$	$f = \frac{1}{8}$	$r = \frac{1}{4}$	All	—	C,J,R
				$R = 0$ to $\frac{1}{8}$	$\alpha = 20°$	$f = \frac{1}{8}$	$r = \frac{1}{4}$	F,OH	—	C,J,R
SMAW FCAW	B-U6-GF	U	U	$R = 0$ to $\frac{1}{8}$	$\alpha = 20°$	$f = \frac{1}{8}$	$r = \frac{1}{4}$	All	Not req.	A,C,N
	C-U6-GF	U	U	$R = 0$ to $\frac{1}{8}$	$\alpha = 20°$	$f = \frac{1}{8}$	$r = \frac{1}{4}$	All	Not req.	A,C,J,R

Double U-groove weld (7)
Butt joint (B)

Tolerances For B-U7 and B-U7-GF	As Detailed	As Fit Up
$R = +\frac{1}{16}, -0$		$+\frac{1}{16}, -\frac{1}{8}$
$\alpha = +10°, -0°$		$+10°, -5°$
$f = +\frac{1}{16}, -0$		Not limited
$r = +\frac{1}{4}, -0$		$\pm\frac{1}{16}$

Tolerances For B-U7-S	As Detailed	As Fit Up
$R = \pm 0$		$+\frac{1}{16}, -0$
$f = +0, -\frac{1}{4}$		$\pm\frac{1}{16}$

Welding Process	Joint Designation	Base Metal Thickness (U = unlimited) T_1	T_2	Groove Preparation Root Opening	Groove Angle	Root Face	Groove Radius	Permitted Welding Positions	Gas Shielding for (FCAW)	Notes
SMAW	B-U7	U	—	$R = 0$ to $\frac{1}{8}$	$\alpha = 45°$	$f = \frac{1}{8}$	$r = \frac{1}{4}$	All	—	C,M,N
				$R = 0$ to $\frac{1}{8}$	$\alpha = 20°$	$f = \frac{1}{8}$	$r = \frac{1}{4}$	F,OH	—	C,M,N
GMAW FCAW	B-U7-GF	U	—	$R = 0$ to $\frac{1}{8}$	$\alpha = 20°$	$f = \frac{1}{8}$	$r = \frac{1}{4}$	All	Not Required	A,C M,N
SAW	B-U7-S	U	—	$R = 0$	$\alpha = 20°$	$f = \frac{1}{4}$ max	$r = \frac{1}{4}$	F	—	C,M,N

See notes on page preceding Prequalified Weld Joint Tables.

PREQUALIFIED WELDED JOINTS
Complete-penetration groove welds

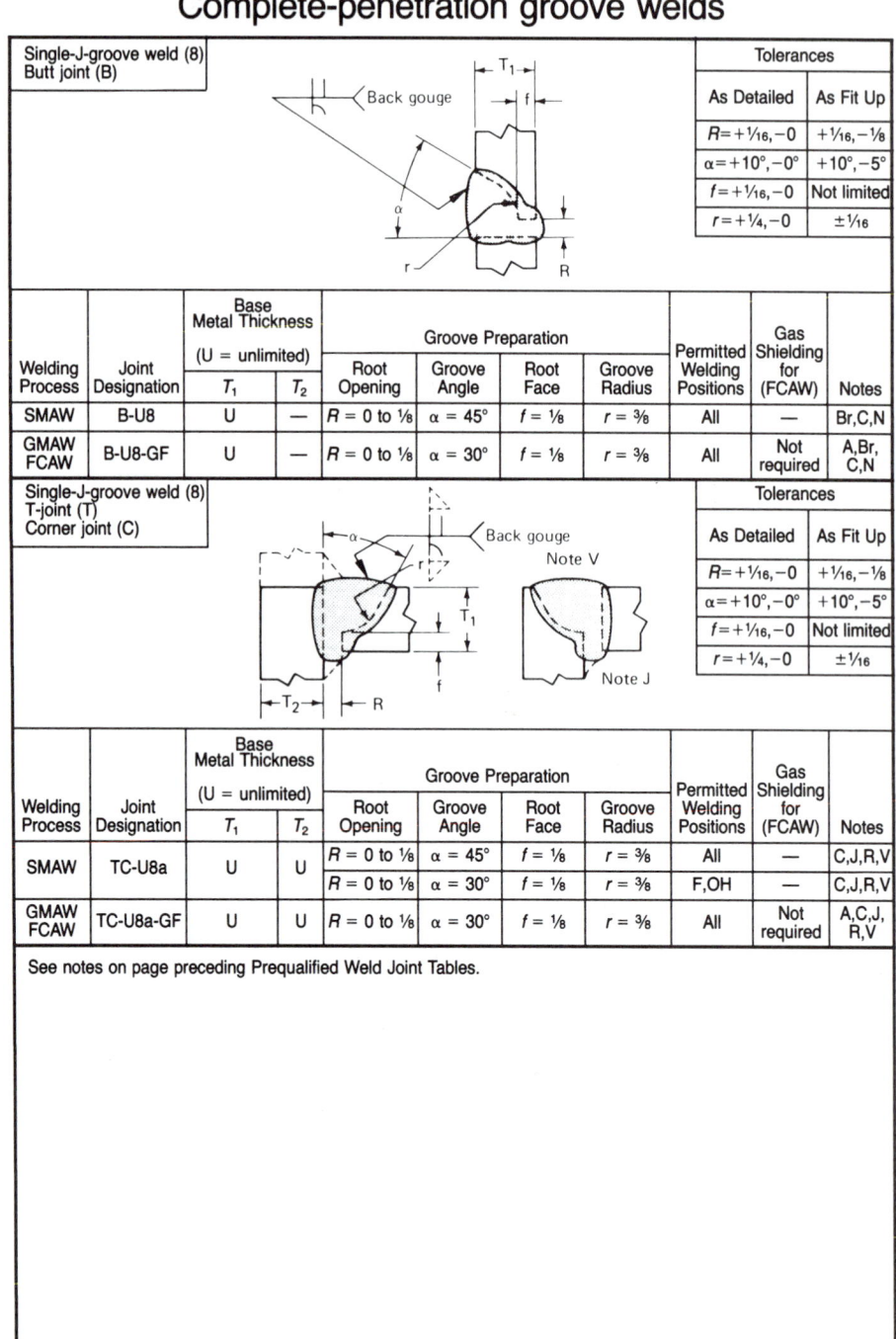

Single-J-groove weld (8)
Butt joint (B)

Tolerances		
	As Detailed	As Fit Up
$R = +\frac{1}{16}, -0$	$+\frac{1}{16}, -\frac{1}{8}$
$\alpha = +10°, -0°$	$+10°, -5°$
$f = +\frac{1}{16}, -0$	Not limited
$r = +\frac{1}{4}, -0$	$\pm\frac{1}{16}$

Welding Process	Joint Designation	Base Metal Thickness (U = unlimited)		Groove Preparation				Permitted Welding Positions	Gas Shielding for (FCAW)	Notes
		T_1	T_2	Root Opening	Groove Angle	Root Face	Groove Radius			
SMAW	B-U8	U	—	R = 0 to $\frac{1}{8}$	α = 45°	f = $\frac{1}{8}$	r = $\frac{3}{8}$	All	—	Br,C,N
GMAW FCAW	B-U8-GF	U	—	R = 0 to $\frac{1}{8}$	α = 30°	f = $\frac{1}{8}$	r = $\frac{3}{8}$	All	Not required	A,Br, C,N

Single-J-groove weld (8)
T-joint (T)
Corner joint (C)

Tolerances		
	As Detailed	As Fit Up
$R = +\frac{1}{16}, -0$	$+\frac{1}{16}, -\frac{1}{8}$
$\alpha = +10°, -0°$	$+10°, -5°$
$f = +\frac{1}{16}, -0$	Not limited
$r = +\frac{1}{4}, -0$	$\pm\frac{1}{16}$

Welding Process	Joint Designation	Base Metal Thickness (U = unlimited)		Groove Preparation				Permitted Welding Positions	Gas Shielding for (FCAW)	Notes
		T_1	T_2	Root Opening	Groove Angle	Root Face	Groove Radius			
SMAW	TC-U8a	U	U	R = 0 to $\frac{1}{8}$	α = 45°	f = $\frac{1}{8}$	r = $\frac{3}{8}$	All	—	C,J,R,V
				R = 0 to $\frac{1}{8}$	α = 30°	f = $\frac{1}{8}$	r = $\frac{3}{8}$	F,OH	—	C,J,R,V
GMAW FCAW	TC-U8a-GF	U	U	R = 0 to $\frac{1}{8}$	α = 30°	f = $\frac{1}{8}$	r = $\frac{3}{8}$	All	Not required	A,C,J, R,V

See notes on page preceding Prequalified Weld Joint Tables.

PREQUALIFIED WELDED JOINTS
Complete-penetration groove welds

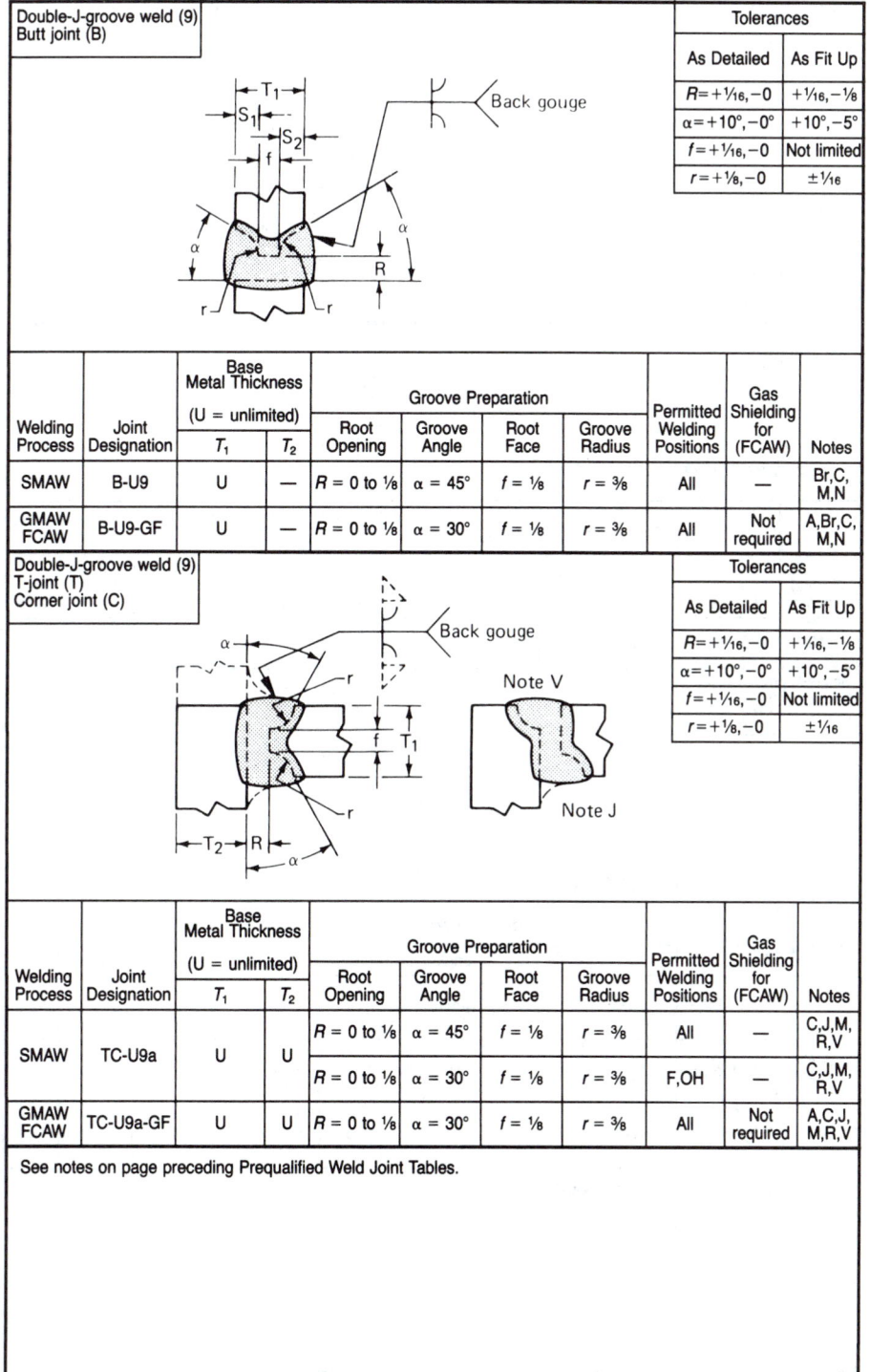

Double-J-groove weld (9)
Butt joint (B)

Back gouge

Tolerances		
	As Detailed	As Fit Up
$R = +\frac{1}{16}, -0$		$+\frac{1}{16}, -\frac{1}{8}$
$\alpha = +10°, -0°$		$+10°, -5°$
$f = +\frac{1}{16}, -0$		Not limited
$r = +\frac{1}{8}, -0$		$\pm\frac{1}{16}$

Welding Process	Joint Designation	Base Metal Thickness (U = unlimited)		Groove Preparation				Permitted Welding Positions	Gas Shielding for (FCAW)	Notes
		T_1	T_2	Root Opening	Groove Angle	Root Face	Groove Radius			
SMAW	B-U9	U	—	$R = 0$ to $\frac{1}{8}$	$\alpha = 45°$	$f = \frac{1}{8}$	$r = \frac{3}{8}$	All	—	Br,C, M,N
GMAW FCAW	B-U9-GF	U	—	$R = 0$ to $\frac{1}{8}$	$\alpha = 30°$	$f = \frac{1}{8}$	$r = \frac{3}{8}$	All	Not required	A,Br,C, M,N

Double-J-groove weld (9)
T-joint (T)
Corner joint (C)

Back gouge

Note V

Note J

Tolerances		
	As Detailed	As Fit Up
$R = +\frac{1}{16}, -0$		$+\frac{1}{16}, -\frac{1}{8}$
$\alpha = +10°, -0°$		$+10°, -5°$
$f = +\frac{1}{16}, -0$		Not limited
$r = +\frac{1}{8}, -0$		$\pm\frac{1}{16}$

Welding Process	Joint Designation	Base Metal Thickness (U = unlimited)		Groove Preparation				Permitted Welding Positions	Gas Shielding for (FCAW)	Notes
		T_1	T_2	Root Opening	Groove Angle	Root Face	Groove Radius			
SMAW	TC-U9a	U	U	$R = 0$ to $\frac{1}{8}$	$\alpha = 45°$	$f = \frac{1}{8}$	$r = \frac{3}{8}$	All	—	C,J,M, R,V
				$R = 0$ to $\frac{1}{8}$	$\alpha = 30°$	$f = \frac{1}{8}$	$r = \frac{3}{8}$	F,OH	—	C,J,M, R,V
GMAW FCAW	TC-U9a-GF	U	U	$R = 0$ to $\frac{1}{8}$	$\alpha = 30°$	$f = \frac{1}{8}$	$r = \frac{3}{8}$	All	Not required	A,C,J, M,R,V

See notes on page preceding Prequalified Weld Joint Tables.

PREQUALIFIED WELDED JOINTS
Partial-penetration groove welds

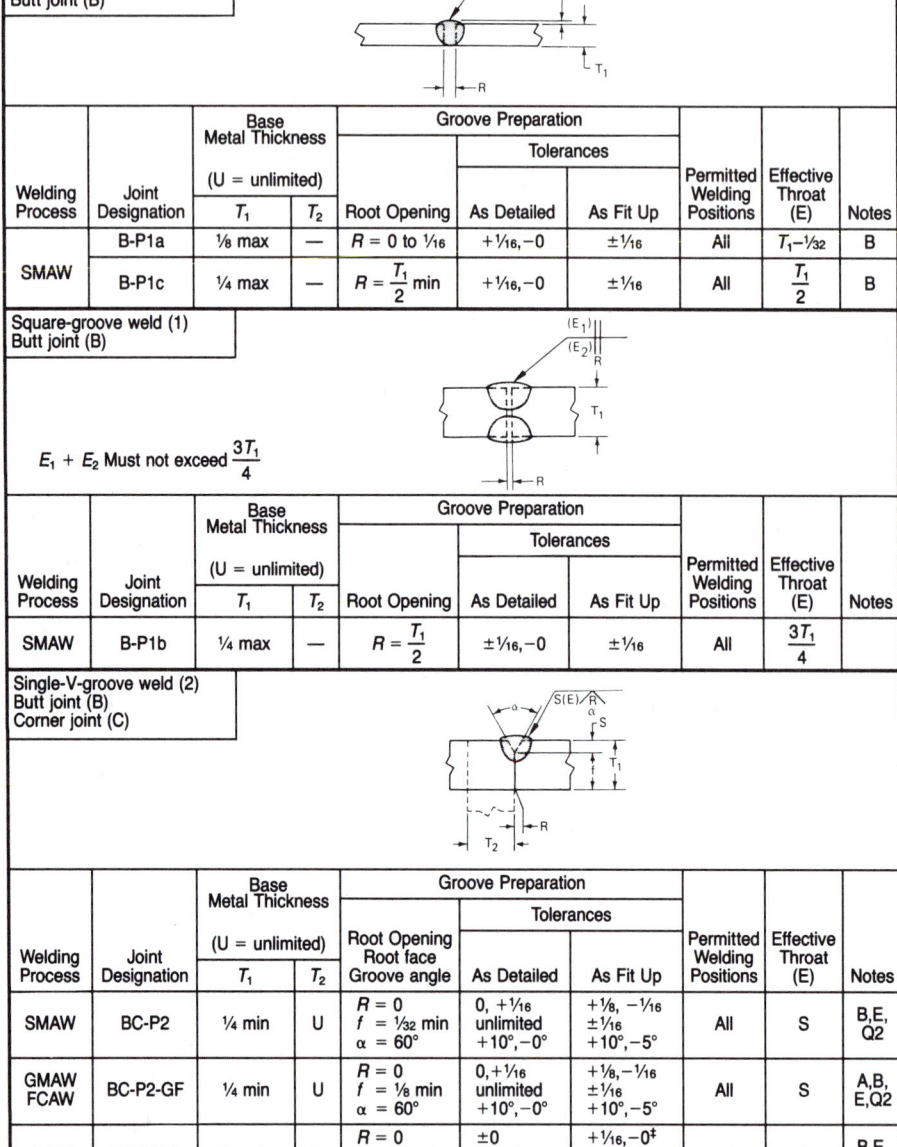

Square-groove weld (1)
Butt joint (B)

Reinforcement 1/32 to 1/8, no tolerance

Welding Process	Joint Designation	Base Metal Thickness (U = unlimited)		Groove Preparation			Permitted Welding Positions	Effective Throat (E)	Notes
		T_1	T_2	Root Opening	Tolerances				
					As Detailed	As Fit Up			
SMAW	B-P1a	⅛ max	—	R = 0 to ¹⁄₁₆	$+¹⁄₁₆, -0$	$±¹⁄₁₆$	All	$T_1 - ¹⁄₃₂$	B
SMAW	B-P1c	¼ max	—	$R = \dfrac{T_1}{2}$ min	$+¹⁄₁₆, -0$	$±¹⁄₁₆$	All	$\dfrac{T_1}{2}$	B

Square-groove weld (1)
Butt joint (B)

$E_1 + E_2$ Must not exceed $\dfrac{3T_1}{4}$

Welding Process	Joint Designation	Base Metal Thickness (U = unlimited)		Groove Preparation			Permitted Welding Positions	Effective Throat (E)	Notes
		T_1	T_2	Root Opening	Tolerances				
					As Detailed	As Fit Up			
SMAW	B-P1b	¼ max	—	$R = \dfrac{T_1}{2}$	$±¹⁄₁₆, -0$	$±¹⁄₁₆$	All	$\dfrac{3T_1}{4}$	

Single-V-groove weld (2)
Butt joint (B)
Corner joint (C)

Welding Process	Joint Designation	Base Metal Thickness (U = unlimited)		Groove Preparation			Permitted Welding Positions	Effective Throat (E)	Notes
		T_1	T_2	Root Opening Root face Groove angle	Tolerances				
					As Detailed	As Fit Up			
SMAW	BC-P2	¼ min	U	$R = 0$ $f = ¹⁄₃₂$ min $α = 60°$	$0, +¹⁄₁₆$ unlimited $+10°, -0°$	$+⅛, -¹⁄₁₆$ $±¹⁄₁₆$ $+10°, -5°$	All	S	B,E, Q2
GMAW FCAW	BC-P2-GF	¼ min	U	$R = 0$ $f = ⅛$ min $α = 60°$	$0, +¹⁄₁₆$ unlimited $+10°, -0°$	$+⅛, -¹⁄₁₆$ $±¹⁄₁₆$ $+10°, -5°$	All	S	A,B, E,Q2
SAW	BC-P2-S	⁷⁄₁₆ min	U	$R = 0$ $f = ¼$ min $α = 60°$	$±0$ unlimited $+10°, -0°$	$+¹⁄₁₆, -0‡$ $±¹⁄₁₆$ $+10°, -5°$	F	S	B,E, Q2

See notes on page preceding Prequalified Weld Joint Tables.

‡Fit-up tolerance, SAW: See AWS 3.3.2; for rolled shape R may be ⁵⁄₁₆ in. in thick plates if backing is provided.

PREQUALIFIED WELDED JOINTS
Partial-penetration groove welds

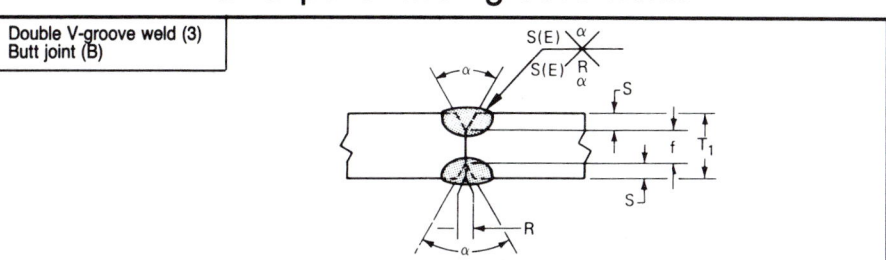

| Double V-groove weld (3) Butt joint (B) | | | | | | | | | |

Welding Process	Joint Designation	Base Metal Thickness (U = unlimited) T_1	T_2	Groove Preparation Root Opening Root face Groove angle	Tolerances As Detailed	As Fit Up	Permitted Welding Positions	Effective Throat (E)	Notes
SMAW	B-P3	½ min	—	$R = 0$ f = ⅛ min $\alpha = 60°$	+1/16, −0 unlimited +10°, −0°	+⅛, −1/16 ±1/16 +10°, −5°	All	S	E,Mp, Q2
GMAW FCAW	B-P3-GF	½ min	—	$R = 0$ f = ⅛ min $\alpha = 60°$	+1/16, −0 unlimited +10°, −0°	+⅛, −1/16 ±1/16 +10°, −5°	All	S	A,E, Mp, Q2
SAW	B-P3-S	¾ min	—	$R = 0$ f = ¼ min $\alpha = 60°$	±0 unlimited +10°, −0°	+1/16, −0‡ ±1/16 +10°, −5°	F	S	E,Mp, Q2

‡Fit-up tolerance, SAW: See AWS 3.3.2; for rolled shapes R may be 5/16 inches in thick plates if backing is provided.

See notes on page preceding Prequalifed Weld Joint Tables.

PREQUALIFIED WELDED JOINTS
Partial-penetration groove welds

| Single-bevel-groove weld (4) |
| Butt joint (B) |
| T-joint (T) |
| Corner joint (C) |

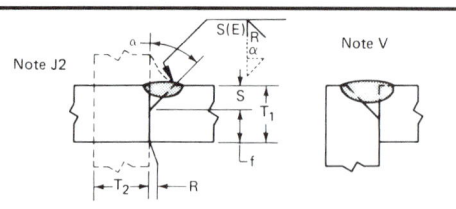

Welding Process	Joint Designation	Base Metal Thickness (U = unlimited) T_1	T_2	Groove Preparation Root Opening Root face Groove angle	Tolerances As Detailed	As Fit Up	Permitted Welding Positions	Effective Throat (E)	Notes
SMAW	BTC-P4	U	U	$R = 0$ $f = \frac{1}{8}$ min $\alpha = 45°$	$+\frac{1}{16}, -0$ unlimited $+10°, -0°$	$+\frac{1}{8}, -\frac{1}{16}$ $\pm\frac{1}{16}$ $+10°, -5°$	All	$S - \frac{1}{8}$	B,E,J2, Q2,V
GMAW FCAW	BTC-P4-GF	$\frac{1}{4}$ min	U	$R = 0$ $f = \frac{1}{8}$ min $\alpha = 45°$	$+\frac{1}{16}, -0$ unlimited $+10°, -0°$	$+\frac{1}{8}, -\frac{1}{16}$ $\pm\frac{1}{16}$ $+10°, -5°$	F,H V,OH	S $S - \frac{1}{8}$	A,B,E, J2,Q2, V
SAW	TC-P4-S	$\frac{7}{16}$ min	U	$R = 0$ $f = \frac{1}{4}$ min $\alpha = 60°$	± 0 unlimited $+10°, -0°$	$+\frac{1}{16}, -0‡$ $\pm\frac{1}{16}$ $+10°, -5°$	F	S	B,E,J2, Q2,V

‡Fit-up tolerance, SAW: See AWS 3.3.2; for rolled shapes, R may be $\frac{5}{16}$ inches in thick plates if backing is provided.

| Double-bevel-groove weld (5) |
| Butt joint (B) |
| T-joint (T) |
| Corner joint (C) |

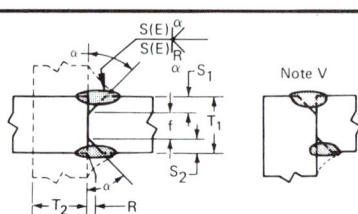

Welding Process	Joint Designation	Base Metal Thickness (U = unlimited) T_1	T_2	Groove Preparation Root Opening Root face Groove angle	Tolerances As Detailed	As Fit Up	Permitted Welding Positions	Effective Throat (E)	Notes
SMAW	BTC-P5	$\frac{5}{16}$ min	U	$R = 0$ $f = \frac{1}{8}$ min $\alpha = 45°$	$+\frac{1}{16}, -0$ unlimited $+10°, -0°$	$+\frac{1}{8}, -\frac{1}{16}$ $\pm\frac{1}{16}$ $+10°, -5°$	All	$(S_1 + S_2)$ $-\frac{1}{4}$	E,J2, L,Mp, Q2,V
GMAW FCAW	BTC-P5-GF	$\frac{1}{2}$ min	U	$R = 0$ $f = \frac{1}{8}$ min $\alpha = 45°$	$+\frac{1}{16}, -0$ unlimited $+10°, -0°$	$+\frac{1}{8}, -\frac{1}{16}$ $\pm\frac{1}{16}$ $+10°, -5°$	F,H V,OH	$(S_1 + S_2)$ $(S_1 + S_2)$ $-\frac{1}{4}$	A,E,J2, L,Mp, Q2,V
SAW	TC-P5-S	$\frac{3}{4}$ min	U	$R = 0$ $f = \frac{1}{4}$ min $\alpha = 60°$	± 0 unlimited $+10°, -0°$	$+\frac{1}{16}, -0‡$ $\pm\frac{1}{16}$ $+10°, -5°$	F	$S_1 + S_2$	E,J2, L,MP, Q2,V

‡Fit-up tolerance, SAW: See AWS 3.3.2; for rolled shapes, R may be $\frac{5}{16}$ inches in thick plates if backing is provided.

See notes on page preceding Prequalified Weld Joint Tables.

PREQUALIFIED WELDED JOINTS
Partial-penetration groove welds

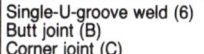

Single-U-groove weld (6)
Butt joint (B)
Corner joint (C)

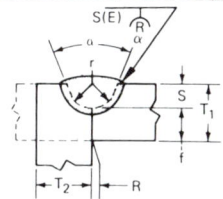

Welding Process	Joint Designation	Base Metal Thickness (U = unlimited)		Groove Preparation			Permitted Welding Positions	Effective Throat (E)	Notes
		T_1	T_2	Root Opening Groove face Groove radius Groove angle	Tolerances As Detailed	As Fit Up			
SMAW	BC-P6	¼ min	U	$R = 0$ $f = \frac{1}{32}$ min $r = \frac{1}{4}$ $\alpha = 45°$	$+\frac{1}{16}, -0$ unlimited $+\frac{1}{4}, -0$ $+10°, -0°$	$+\frac{1}{8}, -\frac{1}{16}$ $\pm\frac{1}{16}$ $\pm\frac{1}{16}$ $+10°, -5°$	All	S	B,E,Q2
GMAW FCAW	BC-P6-GF	¼ min	U	$R = 0$ $f = \frac{1}{8}$ min $r = \frac{1}{4}$ $\alpha = 20°$	$+\frac{1}{16}, -0$ unlimited $+\frac{1}{4}, -0$ $+10°, -0°$	$+\frac{1}{8}, -\frac{1}{16}$ $\pm\frac{1}{16}$ $\pm\frac{1}{16}$ $+10°, -5°$	All	S	A,B, E,Q2
SAW	BC-P6-S	⁷⁄₁₆ min	U	$R = 0$ $f = \frac{1}{4}$ min $r = \frac{1}{4}$ $\alpha = 20°$	± 0 unlimited $+\frac{1}{4}, -0$ $+10°, -0°$	$+\frac{1}{16}, -0‡$ $\pm\frac{1}{16}$ $\pm\frac{1}{16}$ $+10°, -5°$	F	S	B,E,Q2

‡Fit-up tolerance, SAW: See AWS 3.3.2; for rolled shapes, R may be ⁵⁄₁₆ inches in thick plates if backing is provided.

Double-V-groove weld (7)
Butt joint (B)

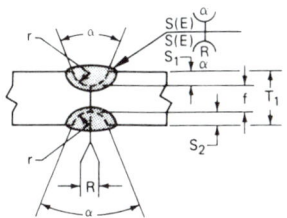

Welding Process	Joint Designation	Base Metal Thickness (U = unlimited)		Groove Preparation			Permitted Welding Positions	Effective Throat (E)	Notes
		T_1	T_2	Root Opening Groove face Groove radius Groove angle	Tolerances As Detailed	As Fit Up			
SMAW	B-P7	½ min	—	$R = 0$ $f = \frac{1}{8}$ min $r = \frac{1}{4}$ $\alpha = 45°$	$+\frac{1}{16}, -0$ unlimited $+\frac{1}{4}, -0$ $+10°, -0°$	$+\frac{1}{8}, -\frac{1}{16}$ $\pm\frac{1}{16}$ $\pm\frac{1}{16}$ $+10°, -5°$	All	$S_1 + S_2$	E, Mp, Q2
GMAW FCAW	B-P7-GF	½ min	—	$R = 0$ $f = \frac{1}{8}$ min $r = \frac{1}{4}$ $\alpha = 20°$	$+\frac{1}{16}, -0$ unlimited $+\frac{1}{4}, -0$ $+10°, -0°$	$+\frac{1}{8}, -\frac{1}{16}$ $\pm\frac{1}{16}$ $\pm\frac{1}{16}$ $+10°, -5°$	All	$S_1 + S_2$	A,E, Mp, Q2
SAW	B-P7-S	¾ min	—	$R = 0$ $f = \frac{1}{4}$ min $r = \frac{1}{4}$ $\alpha = 20°$	± 0 unlimited $+\frac{1}{4}, -0$ $+10°, -0°$	$+\frac{1}{16}, -0‡$ $\pm\frac{1}{16}$ $\pm\frac{1}{16}$ $+10°, -5°$	F	$S_1 + S_2$	E, MP, Q2

‡Fit-up tolerance, SAW: See AWS 3.2.2; for rolled shapes, R may be ⁵⁄₁₆ inches in thick plates if backing is provided.

See notes on page preceding Prequalified Weld Joint Tables.

PREQUALIFIED WELDED JOINTS
Partial penetration groove welds

Single-J-groove weld (8)
Butt joint (B)
T-joint (T)
Corner joint (C)

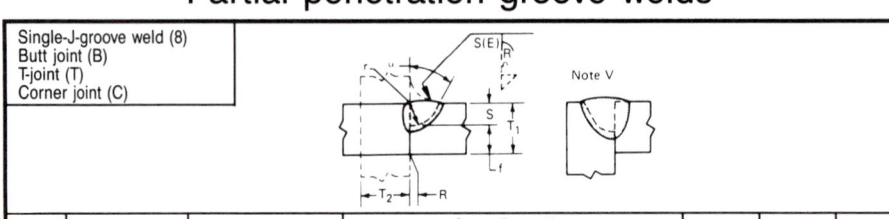

Note V

Welding Process	Joint Designation	Base Metal Thickness (U = unlimited)		Groove Preparation			Permitted Welding Positions	Effective Throat (E)	Notes
		T_1	T_2	Root Opening Root face Groove radius Groove angle	Tolerances				
					As Detailed	As Fit Up			
SMAW	TC-P8*	¼ min	U	$R=0$ $f=\frac{1}{8}$ min $r=\frac{3}{8}$ $\alpha=45°$	$+\frac{1}{16}, -0$ Not limited $+\frac{1}{4}, -0$ $+10°, -0°$	$+\frac{1}{8}, -\frac{1}{16}$ $\pm\frac{1}{16}$ $\pm\frac{1}{16}$ $+10°, -5°$	All	S	E,J2, Q2,V
SMAW	BC-P8**	¼ min	U	$R=0$ $f=\frac{1}{8}$ min $r=\frac{3}{8}$ $\alpha=30°$	$+\frac{1}{16}, -0$ Not limited $+\frac{1}{4}, -0$ $+10°, -0°$	$+\frac{1}{8}, -\frac{1}{16}$ $\pm\frac{1}{16}$ $\pm\frac{1}{16}$ $+10°, -5°$	All	S	E,J2, Q2,V
GMAW FCAW	TC-P8-GF*	¼ min	U	$R=0$ $f=\frac{1}{8}$ min $r=\frac{3}{8}$ $\alpha=45°$	$+\frac{1}{16}, -0$ Not limited $+\frac{1}{4}, -0$ $+10°, -0°$	$+\frac{1}{8}, -\frac{1}{16}$ $\pm\frac{1}{16}$ $\pm\frac{1}{16}$ $+10°, -5°$	All	S	A,E, J2,Q2, V
GMAW FCAW	BC-P8-GF**	¼ min	U	$R=0$ $f=\frac{1}{8}$ min $r=\frac{3}{8}$ $\alpha=30°$	$+\frac{1}{16}, -0$ Not limited $+\frac{1}{4}, -0$ $+10°, -0°$	$+\frac{1}{8}, -\frac{1}{16}$ $\pm\frac{1}{16}$ $\pm\frac{1}{16}$ $+10°, -5°$	All	S	A,E, J2,Q2, V
SAW	TC-P8-S*	$\frac{7}{16}$ min	U	$R=0$ $f=\frac{1}{4}$ min $r=\frac{1}{2}$ $\alpha=45°$	±0 Not limited $+\frac{1}{4}, -0$ $+10°, -0°$	$+\frac{1}{16}, -0‡$ $\pm\frac{1}{16}$ $\pm\frac{1}{16}$ $+10°, -5°$	F	S	E,J2, Q2,V
SAW	C-P8-S**	$\frac{7}{16}$ min	U	$R=0$ $f=\frac{1}{4}$ min $r=\frac{1}{2}$ $\alpha=20°$	±0 Not limited $+\frac{1}{4}, -0°$ $+10°, -0°$	$+\frac{1}{16}, -0‡$ $\pm\frac{1}{16}$ $\pm\frac{1}{16}$ $+10°, -5°$	F	S	E,J2, Q2,V

Double-J-groove weld (9)
Butt joint (B)
T-joint (T)
Corner joint (C)

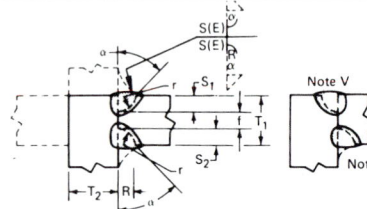

Note V

Note J

Welding Process	Joint Designation	Base Metal Thickness (U = unlimited)		Groove Preparation			Permitted Welding Positions	Effective Throat (E)	Notes
		T_1	T_2	Root Opening Root face Groove radius Groove angle	Tolerances				
					As Detailed	As Fit Up			
SMAW	BTC-P9*	½ min	U	$R=0$ $f=\frac{1}{8}$ min $r=\frac{3}{8}$ $\alpha=45°$	$+\frac{1}{16}, -0$ -0 $+\frac{1}{4}, -0$ $+10°, -0°$	$+\frac{1}{8}, -\frac{1}{16}$ $\pm\frac{1}{16}$ $\pm\frac{1}{16}$ $+10°, -5°$	All	$S_1 + S_2$	E,J2, Mp,Q2, V
GMAW FCAW	BTC-P9-GF*	½ min	U	$R=0$ $f=\frac{1}{8}$ min $r=\frac{3}{8}$ $\alpha=30°$	$+\frac{1}{16}, -0$ Not limited $+\frac{1}{4}, -0$ $+10°, -0°$	$+\frac{1}{8}, -\frac{1}{16}$ $\pm\frac{1}{16}$ $\pm\frac{1}{16}$ $+10°, -5°$	All	$S_1 + S_2$	A,J2, Mp,Q2, V
SAW	C-P9-S*	¾ min	U	$R=0$ $f=\frac{1}{4}$ min $r=\frac{1}{2}$ $\alpha=45°$	±0 Not limited $+\frac{1}{4}, -0$ $+10°, -0°$	$+\frac{1}{16}, -0‡$ $\pm\frac{1}{16}$ $\pm\frac{1}{16}$ $+10°, -5°$	F	$S_1 + S_2$	E,J2, Mp,Q2 V
SAW	C-P9-S**	¾ min	U	$R=0$ $f=\frac{1}{4}$ min $r=\frac{1}{2}$ $\alpha=20°$	±0 Not limited $+\frac{1}{4}, -0$ $+10°, -0°$	$+\frac{1}{16}, -0‡$ $\pm\frac{1}{16}$ $\pm\frac{1}{16}$ $+10°, -5°$	F	$S_1 + S_2$	E,J2, Mp,Q2, V
SAW	T-P9-S	¾ min	U	$R=0$ $f=\frac{1}{4}$ min $r=\frac{1}{2}$ $\alpha=45°$	±0 Not limited $+\frac{1}{4}, -0$ $+10°, -0°$	$+\frac{1}{16}, -0‡$ $\pm\frac{1}{16}$ $\pm\frac{1}{16}$ $+10°, -5°$	F	$S_1 + S_2$	E,J2, Mp,Q2

*Applies to inside corner joints.
**Applies to outside corner joints.
‡ Fit-up tolerance, SAW. See AWS 3.3.2; for rolled shapes R may be $\frac{9}{16}$ in. in thick plates if backing is provided.
See notes on page preceding Prequalified Welded Joint Tables.

PREQUALIFIED WELDED JOINTS
Partial-penetration groove welds

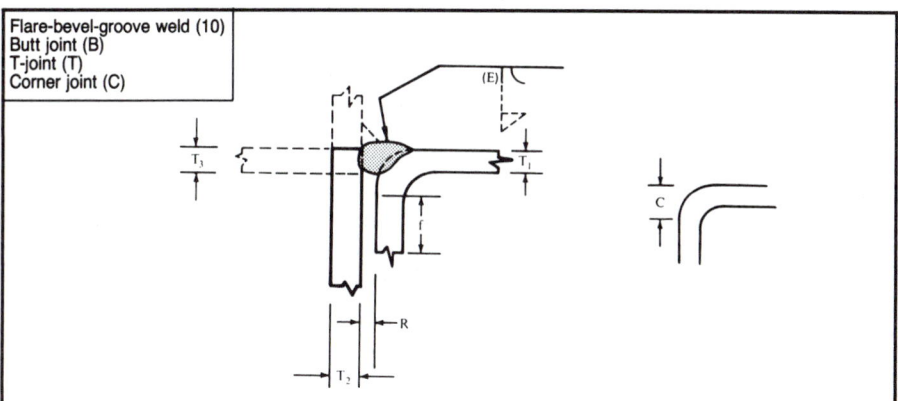

Flare-bevel-groove weld (10)
Butt joint (B)
T-joint (T)
Corner joint (C)

Welding Process	Joint Designation	Base Metal Thickness (U = unlimited)			Groove Preparation			Permitted Welding Positions	Effective Throat (E)	Notes
		T_1	T_2	T_3	Root Opening Root face Bend Radius*	Tolerances				
						As Detailed	As Fit Up			
SMAW	BTC-P10	3/16 min	U	T_1 min	$R = 0$ f = 3/16 min $C = 3\ T_1$ min 2	+1/16, −0 Not Limited −0, +Not Limited	+1/8, −1/16 +U, −1/16 −0, +Not Limited	All	5/8 T_1	J2 Q2 Z
GMAW FCAW	BTC-P10-GF	3/16 min	U	T_1 min	$R = 0$ f = 3/16 min $C = 3\ T_1$ min 2	+1/16, −0 Not Limited −0, +Not Limited	+1/8, −1/16 +U, −1/16 −0, +Not Limited	All	5/8 T_1	A J2 Q2 Z
SAW	T-P10-S	1/2 min	1/2 min	N/A	$R = 0$ f = 1/2 min $C = 3\ T_1$ min 2	±0 Not Limited −0, +Not Limited	+1/16, −0 +U, −1/16 Limited	F	5/8 T_1	J2 Q2 Z

* For cold formed (A500) rectangular tubes, C dimension is not limited (see AWS commentary).

BENT PLATES

Minimum radius for cold bending

The following table gives the generally accepted minimum inside radii of bends in terms of thickness *t* for various steels listed. Values are for bend lines transverse to the direction of final rolling. When bend lines are parallel to the direction of final rolling, the values may have to be approximately doubled. When bend lines are longer than 36 in., all radii may have to be increased if problems in bending are encountered.

Before bending, special attention should be paid to the condition of plate edges transverse to the bend lines. Flame-cut edges of hardenable steels should be machined or softened by heat treatment. Nicks should be ground out. Sharp corners should be rounded.

ASTM Designation		Thickness, in.				
		Up to ¼	Over ¼ to ½	Over ½ to 1	Over 1 to 1½	Over 1½ to 2
A36		1½*t*	1½*t*	2*t*	3*t*	4*t*
A242		2*t*	3*t*	5*t*	—[a]	—[a]
A441		2*t*	3*t*	5*t*	—[a]	—[a]
A529		2*t*	2*t*	—	—	—
A572[c]	Gr. 42	2*t*	2*t*	3*t*	4*t*	5*t*
	Gr. 50	2½*t*	2½*t*	4*t*	—[a]	—[a]
	Gr. 60	3½*t*	3½*t*	6*t*	—[a]	—
	Gr. 65	4*t*	4*t*	—[a]	—[a]	—
A588		2*t*	3*t*	5*t*	—[a]	—[a]
A852[b]		2*t*	2*t*	3*t*	3*t*	3*t*
A514[b]		2*t*	2*t*	2*t*	3*t*	3*t*

[a]It is recommended that steel in this thickness range be bent hot. Hot bending, however, may result in a decrease in the as-rolled mechanical properties.

[b]The mechanical properties of A852 and ASTM A514 steel results from a quench-and-temper operation. Hot bending may adversely affect these mechanical properties. If necessary to hot-bend, fabricator should discuss procedure with the steel supplier.

[c]Thickness may be restricted because of columbium content. Consult supplier.

FABRICATING PRACTICES

Maximum efficiency in the fabrication of structural steel by modern shops is entirely dependent upon close cooperation between designing office, drafting room and shop. Designs should be favorable to, the drafting room should recognize and call for, and the shop should adapt its equipment to, the use of recurrent details which have been standardized.

Consideration should be given to duplication of details and multiple punching or drilling. Utilization of standard jigs and machine set-ups eliminates unnecessary handling of material and facilitates the drilling or punching of holes. Gage lines should conform to standard machine set ups. Once determined, they should be duplicated as far as possible throughout any one job. Gages and hole sizes on an individual member should not be varied throughout the length of that member.

Keep gages and longitudinal spacing alike to permit maximum economy in either drilling or punching operations. Longitudinal spacing should preferably be 3 in., or multiples of 3 in., since most shops consider this to be standard.

Copes, blocks and cuts

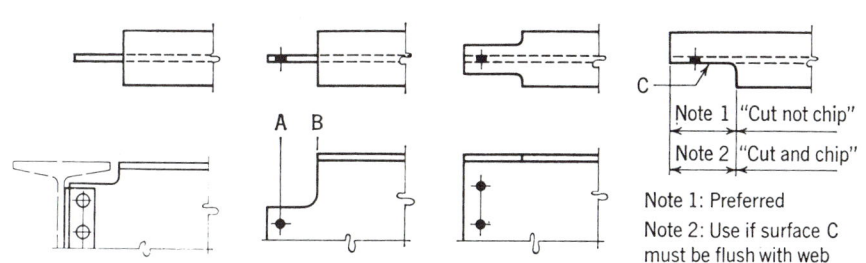

All re-entrant corners shall be shaped, notch-free, to a radius.

The above sketches indicate standard methods of providing clearance for beams connecting to beams or columns. Where possible, a minimum clearance of ½ in. is to be provided. Fabricators may vary in designation and dimensions of copes and blocks. Some fabricators designate all of the operations pictured above by the term "cuts." Note recommended cutting practice in sketch below.

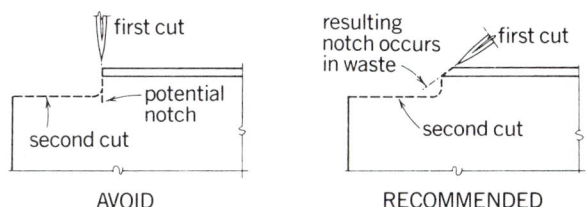

For economy, coping or blocking of beams should be avoided if possible. When construction will permit, the elevation of the top of filler beams should be established a sufficient distance below the top of girders to clear the girder fillet. Unusually long or deep copes and blocks, or blocks in beams with thin webs, may materially affect the capacity of the beam. Such beams must be investigated for both shear and moment at lines A and B and, when necessary, adequate reinforcement provided.

PART 5
Specifications and Codes

SPEC.

Cross Reference to the 1978 AISC Specification for the Design, Fabrication and Erection of Structural Steel for Buildings

This table provides a cross-reference of the 1989 *Specification for Structural Steel Buildings—Allowable Stress Design and Plastic Design* alphanumeric section designations and their headings, to the 1978 Specification section numbers. A "—" indicates there was no specific section in the 1978 Specification corresponding to that 1989 Specification section.

1989 Spec.		1978 Spec.
Chapter A GENERAL PROVISIONS		
A1	Scope	
A2	Limits of Applicability	
A2.1	Structural Steel Defined	—
A2.2	Types of Construction	1.2
A3	Material	1.4
A3.1	Structural Steel	
A3.1a	ASTM designations	1.4.1.1
A3.1b	Unidentified steel	1.4.1.2
A3.1c	Heavy shapes	new
A3.2	Steel Castings and Forgings	1.4.2, 1.5.4
A3.3	Rivets	1.4.3
A3.4	Bolts, Washers and Nuts	1.4.4
A3.5	Anchor Bolts and Threaded Rods	—
A3.6	Filler Metal and Flux for Welding	1.4.5
A3.7	Stud Shear Connectors	1.4.6
A4	Loads and Forces	1.3.7
A4.1	Dead Load and Live Load	1.3.1/2
A4.2	Impact	1.3.3
A4.3	Crane Runway Horizontal Forces	1.3.4
A4.4	Wind	1.3.5
A4.5	Other Forces	1.3.6
A5	Design Basis	
A5.1	Allowable Stresses	1.5*
A5.2	Wind and Seismic Stresses	1.5.6
A5.3	Structural Analysis	—
A5.4	Design for Serviceability and other Considerations	—
A6	Referenced Codes and Standards	—
A7	Design Documents	
A7.1	Plans	1.1.1
A7.2	Standard Symbols and Nomenclature	1.1.4*
A7.3	Notation for Welding	1.1.3

* first paragraph only

1989 Spec.		1978 Spec.
Chapter B DESIGN REQUIREMENTS		
B1	Gross Area	1.14.1*/3
B2	Net Area	1.14.1**/4
		1.14.2.1
B3	Effective Net Area	1.14.2.2/3
B4	Stability	1.8.1+
B5	Local Buckling	1.9
B5.1	Classification of Steel Sections	1.9.1++/2
B5.2	Slender Compression Elements	1.9.1+++
B6	Rotational Restraint at Points of Support	1.10.11
B7	Limiting Slenderness Ratios	1.8.4
B8	Simple Spans	1.12.1
B9	End Restraint	1.12.2
B10	Proportions of Beams and Girders	1.10.1/3/4
B11	Proportioning of Crane Girders	1.10.9
Chapter C FRAMES AND OTHER STRUCTURES		
C1	General	—
C2	Frame Stability	
C2.1	Braced Frames	1.8.2
C2.2	Unbraced Frames	1.8.3
Chapter D TENSION MEMBERS		
D1	Allowable Stress	1.5.1.1
D2	Built-up Members	1.18.3
D3	Pin-Connected Members	—
D3.1	Allowable Stress	1.5.1.1
D3.2	Pin-connected Plates	1.14.5***
D3.3	Eyebars	1.14.5++++

    * except net
   ** except gross
  *** except first two paragraphs
   + first paragraph only
  ++ except last paragraph
 +++ last paragraph only
++++ first two paragraphs only

	1989 Spec.	1978 Spec.
Chapter E	**COLUMNS AND OTHER COMPRESSION MEMBERS**	
E1	Effective Length and Slenderness Ratio	1.8.1**
E2	Allowable Stress	1.5.1.3.1/2
E3	Flexural-torsional Buckling	—
E4	Built-up Members	1.18.2
E5	Pin-connected Compression Members	—
E6	Column Web Shear	1.15.5.5
Chapter F	**BEAMS AND OTHER FLEXURAL MEMBERS**	
F1	Allowable Stress: Strong Axis Bending of I-Shaped Members and Channels	—
F1.1	Members with Compact Sections	1.5.1.4.1+++
F1.2	Members with Noncompact Sections	1.5.1.4.2/5(2b)
F1.3	Members with Compact or Noncompact Sections with Unbraced Length Greater than L_c	1.5.1.4.5*
F2	Allowable Stress: Weak Axis Bending of I-Shaped Members, Solid Bars and Rectangular Plates	1.5.1.4.3 1.10.5.2*
F2.1	Members with Compact Sections	1.5.1.4.3+
F2.2	Members with Noncompact Sections	1.5.1.4.3++/5(2b)
F3	Allowable Stress: Bending of Box Members, Rectangular Tubes and Circular Tubes	—
F3.1	Members with Compact Sections	1.5.1.4.1***/3**
F3.2	Members with Noncompact Sections	1.5.1.4.4
F4	Allowable Shear Stress	1.5.1.2.1, 1.10.5.2+
F5	Transverse Stiffeners	1.10.5.3*
F6	Built-up Members	1.18.1
F7	Web-tapered Members	Appendix D
Chapter G	**PLATE GIRDERS**	1.10
G1	Web Slenderness Limitations	1.10.2

   * except last paragraph
  ** last paragraph only
 *** first paragraph and Item 6. only
   + first paragraph only
  ++ second paragraph only
 +++ first and last paragraph only

	1989 Spec.	1978 Spec.
G2	Allowable Bending Stress	1.10.6
G3	Allowable Shear Stress with Tension Field Action	1.10.5.2**
G4	Transverse Stiffeners	1.10.5.3*/4
G5	Combined Shear and Tension Stress	1.10.7
Chapter H	**COMBINED STRESSES**	1.6
H1	Axial Compression and Bending	1.6.1
H2	Axial Tension and Bending	1.6.2
Chapter I	**COMPOSITE CONSTRUCTION**	1.11
I1	Definition	1.11.1
I2	Design Assumptions	1.11.2
I3	End Shear	1.11.3
I4	Shear Connectors	1.11.4
I5	Composite Beams or Girders with Formed Steel Deck	1.11.5
I5.1	General	1.11.5.1
I5.2	Deck Ribs Oriented Perpendicular to Steel Beam or Girder	1.11.5.2
I5.3	Deck Ribs Oriented Parallel to Steel Beam or Girder	1.11.5.3
I6	Special Cases	1.11.6
Chapter J	**CONNECTIONS, JOINTS AND FASTENERS**	1.15
J1	General Provisions	—
J1.1	Design Basis	—
J1.2	Simple Connections	1.15.4
J1.3	Moment Connections	1.15.5.1
J1.4	Compression Members with Bearing Joints	1.15.8
J1.5	Connections of Tension and Compression Members in Trusses	1.15.7
J1.6	Minimum Connections	1.15.1
J1.7	Splices in Heavy Sections	new
J1.8	Beam Copes and Weld Access Holes	new
J1.9	Placement of Welds, Bolts and Rivets	1.15.3
J1.10	Bolts in Combination with Welds	1.15.10

* last paragraph only
** first paragraph only

	1989 Spec.	1978 Spec.
J1.11	High-Strength Bolts in Slip-Critical Connections in Combination with Rivets	1.15.11
J1.12	Limitations on Bolted and Welded Connections	1.15.12
J2	Welds	1.17.1
J2.1	Groove Welds	—
J2.1a	Effective Area	1.14.6.1
J2.1b	Limitations	1.17.2*
J2.2	Fillet Welds	—
J2.2a	Effective Area	1.14.6.2
J2.2b	Limitations	1.17.2**
		1.17.3
		1.17.4
		1.17.5
		1.17.6
		1.17.7
		1.17.8
J2.3	Plug and Slot Welds	—
J2.3a	Effective Area	1.14.6.3
J2.3b	Limitations	1.17.9
J2.4	Allowable Stresses	1.5.3
J2.5	Combination of Welds	1.15.9
J2.6	Mixed Weld Metal	new
J2.7	Preheat for Heavy Shapes	new
J3	Bolts, Threaded Parts and Rivets	1.16
J3.1	High-strength Bolts	1.16.1
J3.2	Size and Use of Holes	1.23.4.1/2+/ 3/4/5
J3.3	Effective Bearing Area	1.16.2
J3.4	Allowable Tension and Shear	1.5.2.1/2
J3.5	Combined Tension and Shear in Bearing-type Connections	1.6.3+++
J3.6	Combined Tension and Shear in Slip-critical Joints	1.6.3++
J3.7	Allowable Bearing at Bolt Holes	1.5.1.5.3
J3.8	Minimum Spacing	1.16.4
J3.9	Minimum Edge Distance	1.16.5
J3.10	Maximum Edge Distance & Spacing	1.16.6
J3.11	Long Grips	1.16.3
J4	Allowable Shear Rupture	1.5.1.2.2

 * excluding fillet weld references
 ** excluding groove weld references
 + first paragraph only
 ++ last paragraph only
 +++ except last paragraph

	1989 Spec.	1978 Spec.
J5	Connecting Elements	
J5.1	Eccentric Connections	1.15.2
J5.2	Allowable Shear Rupture	1.5.1.2.2
J6	Fillers	1.15.6
J7	Splices	1.10.8
J8	Allowable Bearing Stress	1.5.1.5.1/2
J9	Column Bases and Bearing on Masonry and Concrete	1.5.5, 1.21.1
J10	Anchor Bolts	1.22
Chapter K	**SPECIAL DESIGN CONSIDERATIONS**	
K1	Webs and Flanges Under Concentrated Forces	—
K1.1	Design Basis	—
K1.2	Local Flange Bending	1.15.5.3*
K1.3	Local Web Yielding	1.10.10.1
K1.4	Web Crippling	1.10.10.2
K1.5	Sidesway Web Buckling	—
K1.6	Compression Buckling of the Web	1.15.5.3**
K1.7	Compression Members with Web Panels Subject to High Shear	1.5.1.2.1
K1.8	Stiffener Requirements for Concentrated Loads	1.10.5.1, 1.15.5.4, 1.15.5.2
K2	Ponding	1.13.3
K3	Torsion	—
K4	Fatigue	1.7.1*** 1.7.2
Chapter L	**SERVICEABILITY DESIGN CONSIDERATIONS**	
L1	Camber	1.19
L2	Expansion and Contraction	1.20
L3	Deflection, Vibration and Drift	
L3.1	Deflection	1.13.1
L3.2	Vibration	1.13.2
L4	Connection Slip	—
L5	Corrosion	—
Chapter M	**FABRICATION, ERECTION AND QUALITY CONTROL**	

  * only information pertinent to Equation (K1-1)
 ** only information pertinent to Equation (K1-8)
*** last paragraph only

	1989 Spec.	1978 Spec.
M1	Shop Drawings	1.1.2
M2	Fabrication	1.23
M2.1	Cambering, Curving and Straightening	1.23.1
M2.2	Thermal Cutting	1.23.2
M2.3	Planing of Edges	1.23.3
M2.4	Welded Construction	1.23.6
M2.5	High-strength Bolted Construction-Assembly	1.23.4.2+ 1.23.5++
M2.6	Compression Joints	1.23.7
M2.7	Dimensional Tolerances	1.23.8
M2.8	Finishing of Column Bases	1.21.3
M3	Shop Painting	1.24
M3.1	General Requirements	1.24.1
M3.2	Inaccessible Surfaces	1.24.2
M3.3	Contact Surfaces	1.24.3
M3.4	Finished Surfaces	1.24.4
M3.5	Surfaces Adjacent to Field Welds	1.24.5
M4	Erection	1.25
M4.1	Alignment of Column Bases	1.21.2
M4.2	Bracing	1.25.1
M4.3	Alignment	1.25.3
M4.4	Fit of Column Compression Joints	1.25.4
M4.5	Field Welding	1.25.5
M4.6	Field Painting	1.25.6
M4.7	Field Connections	1.25.2
M5	Quality Control	1.26.1
M5.1	Cooperation	1.26.2
M5.2	Rejections	1.26.3
M5.3	Inspection of Welding	1.26.4
M5.4	Inspection of Slip-critical, High-strength Bolted Connections	—
M5.5	Identification of Steel	1.26.5
Chapter N	**PLASTIC DESIGN**	Part 2
N1	Scope	2.1
N2	Structural Steel	2.2
N3	Basis for Maximum Strength Determination	2.3
N3.1	Stability of Braced Frames	2.3.1
N3.2	Stability of Unbraced Frames	2.3.2
N4	Columns	2.4

+ last paragraph
++ except last paragraph

	1989 Spec.	1978 Spec.
N5	Shear	2.5
N6	Web Crippling	2.6
N7	Minimum Thickness (Width-Thickness Ratios)	2.7
N8	Connections	2.8
N9	Lateral Bracing	2.9
N10	Fabrication	2.10
APPENDIX B5	Local Buckling	App. C
APPENDIX F7	Web-tapered Members	App. D
APPENDIX K4	Fatigue	1.7.1* App. B
* first paragraph only		

Specification for Structural Steel Buildings

Allowable Stress Design and
Plastic Design
June 1, 1989

with Commentary

AMERICAN INSTITUTE OF STEEL CONSTRUCTION, INC.
One East Wacker Drive, Suite 3100
Chicago, IL 60601-2001

AMERICAN INSTITUTE OF STEEL CONSTRUCTION

PREFACE

The AISC *Specification for Structural Steel Buildings*—Allowable Stress Design (ASD) and Plastic Design has evolved through numerous versions from the 1st Edition, published June 1, 1923. Each succeeding edition has been based upon past successful usage, advances in the state of knowledge and changes in design practice. The data included has been developed to provide a uniform practice in the design of steel-framed buildings. The intention of the Specification is to provide design criteria for routine use and not to cover infrequently encountered problems which occur in the full range of structural design.

The AISC Specification is the result of the deliberations of a committee of structural engineers with wide experience and high professional standing, representing a wide geographical distribution throughout the U. S. The committee includes approximately equal numbers of engineers in private practice, engineers involved in research and teaching and engineers employed by steel fabricating companies.

To avoid reference to proprietary steels, which may have limited availability, only those steels which can be identified by ASTM specifications are listed as approved under this Specification. However, some steels covered by ASTM specifications, but subject to more costly manufacturing and inspection techniques than deemed essential for structures covered by this Specification, are not listed, even though they may provide all of the necessary characteristics of less expensive steels which are listed. Approval of such steels is left to the owner's representative.

The Appendices to this Specification are an integral part of the Specification.

A Commentary has been included to provide background for these and other provisions.

This edition of the Specification has been developed primarily upon the basis of the criteria in the Specification dated November 1, 1978. That Specification, as well as earlier editions, was arranged essentially on the basis of type of stress with special or supplementary requirements for different kinds of members and details contained in succeeding sections. The provisions of the 1978 Specification have been reorganized using decision table logic techniques to provide an allowable stress design specification that is more logically arranged on the basis of type of member.

This arrangement is more convenient to the user because general design requirements are presented first, followed by chapters containing the information required to design members of each type. This organization is consistent with that used in the *Load and Resistance Factor Design Specification for Structural Steel Buildings*.

The principal changes incorporated in this edition of the Specification include:

- Reorganization of provisions to be consistent with LRFD format.
- New provisions for built-up compression members.
- New provisions for the design of webs under concentrated forces.
- Updated provisions for slender web girders.
- Updated provisions for design for fatigue.
- Recommendations for the use of heavy rolled shapes and welded members made up of thick plates.

The reader is cautioned that independent professional judgment must be exercised when data or recommendations set forth in this Specification are applied. The publication of the material contained herein is not intended as a representation or warranty on the part of the American Institute of Steel Construction, Inc.—or any other person named herein—that this information is suitable for general or particular use, or freedom from infringement of any patent or patents. Anyone making use of this information assumes all liability arising from such use. The design of structures is within the scope of expertise of a competent licensed structural engineer, architect, or other licensed professional for the application of principles to a particular structure.

By the Committee,

A. P. Arndt, Chairman	A. L. Johnson
E. W. Miller,	Donald L. Johnson
Vice Chairman	L. A. Kloiber
Horatio Allison	William J. LeMessurier
Lynn S. Beedle	Stanley D. Lindsey
Reidar Bjorhovde	Richard W. Marshall
Omer W. Blodgett	William McGuire
Roger L. Brockenbrough	William A. Milek
John H. Busch	Walter P. Moore
Wai-Fah Chen	William E. Moore, II
Duane S. Ellifritt	Thomas M. Murray
Bruce Ellingwood	Clarkson W. Pinkham
Shu-Jin Fang	Egor P. Popov
Steven J. Fenves	Donald R. Sherman
Richard F. Ferguson	Frank Sowokinos
James M. Fisher	Sophus A. Thompson
John W. Fisher	William A. Thornton
Theodore V. Galambos	Raymond H. R. Tide
Geerhard Haaijer	Ivan M. Viest
Mark V. Holland	Lyle L. Wilson
Ira Hooper	Joseph A. Yura
Jerome S. B. Iffland	Charles Peshek, Secretary

June 1989

TABLE OF CONTENTS

COMMENTARY

CHAPTER A

GENERAL PROVISIONS

A1. SCOPE

The *Specification for Structural Steel Buildings—Allowable Stress Design and Plastic Design* is intended as an alternate to the currently approved *Load and Resistance Factor Design Specification for Structural Steel Buildings* of the American Institute of Steel Construction., Inc.

A2. LIMITS OF APPLICABILITY

1. Structural Steel Defined

As used in this Specification, the term *structural steel* refers to the steel elements of the structural steel frame essential to the support of the design loads. Such elements are generally enumerated in Sect. 2.1 of the AISC *Code of Standard Practice for Steel Buildings and Bridges*. For the design of cold-formed steel structural members, whose profiles contain rounded corners and slender flat elements, the provisions of the American Iron and Steel Institute *Specification for the Design of Cold-formed Steel Structural Members* are recommended.

2. Types of Construction

Three basic types of construction and associated design assumptions are permissible under the respective conditions stated herein, and each will govern in a specific manner the size of members and the types and strength of their connections:

Type 1, commonly designated as "rigid-frame" (continuous frame), assumes that beam-to-column connections have sufficient rigidity to hold virtually unchanged the original angles between intersecting members.

Type 2, commonly designated as "simple framing" (unrestrained, free-ended), assumes that, insofar as gravity loading is concerned, ends of beams and girders are connected for shear only and are free to rotate under gravity load.

Type 3, commonly designated as "semi-rigid framing" (partially restrained), assumes that the connections of beams and girders possess a dependable and known moment capacity intermediate in degree between the rigidity of Type 1 and the flexibility of Type 2.

The design of all connections shall be consistent with the assumptions as to type of construction called for on the design drawings.

Type 1 construction is unconditionally permitted under this Specification. Two different methods of design are recognized. Within the limitations laid down in Sect. N1, members of continuous frames or continuous portions of frames may be proportioned, on the basis of their maximum predictable strength, to resist the specified design loads multiplied by the prescribed load factors. Otherwise, Type 1 construction shall be designed, within the limitations of Chapters A through M, to resist the stresses produced by the specified design loads, assuming moment distribution in accordance with the elastic theory.

Type 2 construction is permitted under this Specification, subject to the stipulations of the following paragraph, wherever applicable.

In buildings designed as Type 2 construction (i.e., with beam-to-column connections other than wind connections assumed flexible under gravity loading) the wind moments may be distributed among selected joints of the frame, provided:

1. Connections and connected members have adequate capacity to resist wind moments.
2. Girders are adequate to carry full gravity load as "simple beams."
3. Connections have adequate inelastic rotation capacity to avoid overstress of the fasteners or welds under combined gravity and wind loading.

Type 3 (semi-rigid) construction is permitted upon evidence the connections to be used are capable of furnishing, as a minimum, a predictable proportion of full end restraint. The proportioning of main members joined by such connections shall be predicated upon no greater degree of end restraint than this minimum.

Types 2 and 3 construction may necessitate some nonelastic, but self- limiting, deformation of a structural steel part.

A3. MATERIAL

1. Structural Steel

a. ASTM designations

Material conforming to one of the following standard specifications is approved for use under this Specification:

Structural Steel, ASTM A36
Pipe, Steel, Black and Hot-dipped, Zinc-coated Welded and Seamless Steel Pipe, ASTM A53, Gr. B
High-strength Low-alloy Structural Steel, ASTM A242
High-strength Low-alloy Structural Manganese Vanadium Steel, ASTM A441
Cold-formed Welded and Seamless Carbon Steel Structural Tubing in Rounds and Shapes, ASTM A500
Hot-formed Welded and Seamless Carbon Steel Structural Tubing, ASTM A501
High-yield Strength, Quenched and Tempered Alloy-Steel Plate, Suitable for Welding, ASTM A514
Structural Steel with 42 ksi Minimum Yield Point, ASTM A529
Steel, Sheet and Strip, Carbon, Hot-rolled, Structural Quality, ASTM A570 Gr. 40, 45 and 50

High-strength, Low-alloy Columbium-Vanadium Steels of Structural Quality, ASTM A572

High-strength Low-alloy Structural Steel with 50 ksi Minimum Yield Point to 4-in. Thick, ASTM A588

Steel, Sheet and Strip, High-strength, Low-alloy, Hot-rolled and Cold-rolled, with Improved Atmospheric Corrosion Resistance, ASTM A606

Steel, Sheet and Strip, High-strength, Low-alloy, Columbium or Vanadium, or both, Hot-rolled and Cold-rolled, ASTM A607

Hot-formed Welded and Seamless High-strength Low-alloy Structural Tubing, ASTM A618

Structural Steel for Bridges, ASTM A709

Quenched and Tempered Low-alloy Structural Steel Plate with 70 ksi Minimum Yield Strength to 4 in. thick, ASTM A852

Certified mill test reports or certified reports of tests made by the fabricator or a testing laboratory in accordance with ASTM A6 or A568, as applicable, and the governing specification shall constitute sufficient evidence of conformity with one of the above ASTM standards. Additionally, the fabricator shall, if requested, provide an affidavit stating the structural steel furnished meets the requirements of the grade specified.

b. Unidentified steel

Unidentified steel, if free from surface imperfections, is permitted for parts of minor importance, or for unimportant details, where the precise physical properties of the steel and its weldability would not affect the strength of the structure.

c. Heavy shapes

For ASTM A6 Groups 4 and 5 rolled shapes to be used as members subject to primary tensile stresses due to tension or flexure, toughness need not be specified if splices are made by bolting. If such members are spliced using full penetration welds, the steel shall be specified in the contract documents to be supplied with Charpy V-Notch testing in accordance with ASTM A6, Supplementary Requirement S5. The impact test shall meet a minimum average value of 20 ft-lbs. absorbed energy at +70°F and shall be conducted in accordance with ASTM A673 with the following exceptions:

a. The center longitudinal axis of the specimens shall be located as near as practical to midway between the inner flange surface and the center of the flange thickness at the intersection with the web mid-thickness.

b. Tests shall be conducted by the producer on material selected from a location representing the top of each ingot or part of an ingot used to produce the product represented by these tests.

For plates exceeding 2-in. thick used for built-up members with bolted splices and subject to primary tensile stresses due to tension or flexure, material toughness need not be specified. If such members are spliced using full penetration welds, the steel shall be specified in the contract documents to be supplied with Charpy V-Notch testing in accordance with ASTM A6, Supplementary Requirement S5. The impact test shall be conducted by the producer in accordance with ASTM A673, Frequency P, and shall meet a minimum average value of 20 ft-lbs. absorbed energy at +70°F.

The above supplementary toughness requirements shall also be considered for welded full-penetration joints other than splices in heavy rolled and built-up members subject to primary tensile stresses.

Additional requirements for joints in heavy rolled and built-up members are given in Sects. J1.7, J1.8, J2.6, J2.7 and M2.2.

2. Steel Castings and Forgings

Cast steel shall conform to one of the following standard specifications:

> Mild-to-medium-strength Carbon-steel Castings for General Applications, ASTM A27, Gr. 65-35
> High-strength Steel Castings for Structural Purposes, ASTM A148, Gr. 80-50

Steel forgings shall conform to the following standard specification:

> Steel Forgings Carbon and Alloy for General Industrial Use, ASTM A668

Certified test reports shall constitute sufficient evidence of conformity with the standards.

Allowable stresses shall be the same as those provided for other steels, where applicable.

3. Rivets

Steel rivets shall conform to the following standard specification:

> Steel Structural Rivets, ASTM A502

Manufacturer's certification shall constitute sufficient evidence of conformity with the standard.

4. Bolts, Washers and Nuts

Steel bolts shall conform to one of the following standard specifications:

> Carbon Steel Bolts and Studs, 60,000 psi Tensile Strength, ASTM A307
> High-strength Bolts for Structural Steel Joints, ASTM A325
> Quenched and Tempered Steel Bolts and Studs, ASTM A449
> Heat-treated Steel Structural Bolts, 150 ksi Min. Tensile Strength, ASTM A490
> Carbon and Alloy Steel Nuts, ASTM A563
> Hardened Steel Washers, ASTM F436

A449 bolts are permitted only in connections requiring bolt diameters greater than 1½ in. and shall not be used in slip-critical connections.

Manufacturer's certification shall constitute sufficient evidence of conformity with the standards.

5. Anchor Bolts and Threaded Rods

Anchor bolt and threaded rod steel shall conform to one of the following standard specifications:

> Structural Steel, ASTM A36
>
> Carbon and Alloy Steel Nuts for Bolts for High-pressure and High-temperature Service, ASTM A194, Gr.7
>
> Quenched and Tempered Alloy Steel Bolts, Studs and other Externally Threaded Fasteners, ASTM A354
>
> Quenched and Tempered Steel Bolts and Studs, ASTM A449
>
> High-Strength Low-Alloy Columbium-Vanadium Steels of Structural Quality, ASTM A572
>
> High-strength Low-alloy Structural Steel with 50,000 psi Minimum Yield Point to 4 in. Thick, ASTM A588
>
> High-strength Non-headed Steel Bolts and Studs, ASTM A687

Threads on bolts and rods shall conform to Unified Standard Series of latest edition of ANSI B18.1 and shall have Class 2A tolerances.

Steel bolts conforming to other provisions of Sect. A3 are permitted as anchor bolts. A449 material is acceptable for high-strength anchor bolts and threaded rods of any diameter.

Manufacturer's certification shall constitute sufficient evidence of conformity with the standards.

6. Filler Metal and Flux for Welding

Welding electrodes and fluxes shall conform to one of the following specifications of the American Welding Society:*

> Specification for Covered Carbon Steel Arc Welding Electrodes, AWS A5.1
>
> Specification for Low-alloy Steel Covered Arc Welding Electrodes, AWS A5.5
>
> Specification for Carbon Steel Electrodes and Fluxes for Submerged-Arc Welding, AWS A5.17
>
> Specification for Carbon Steel Filler Metals for Gas-Shielded Arc Welding, AWS A5.18
>
> Specification for Carbon Steel Electrodes for Flux-Cored Arc Welding, AWS A5.20
>
> Specification for Low-alloy Steel Electrodes and Fluxes for Submerged-arc Welding, AWS A5.23
>
> Specification for Low-alloy Steel Filler Metals for Gas-shielded Arc Welding, AWS A5.28
>
> Specification for Low-alloy Steel Electrodes for Flux-cored Arc Welding, AWS A5.29

Manufacturer's certification shall constitute sufficient evidence of conformity with the standards.

---

*Approval of these welding electrode specifications is given without regard to weld metal notch toughness requirements, which are generally not critical for building construction. See Commentary, Sect. A3.

7. Stud Shear Connectors

Steel stud shear connectors shall conform to the requirements of *Structural Welding Code—Steel*, AWS D1.1.

Manufacturer's certification shall constitute sufficient evidence of conformity with the code.

A4. LOADS AND FORCES

The nominal loads shall be the minimum design loads stipulated by the applicable code under which the structure is designed or dictated by the conditions involved. In the absence of a code, the loads and load combinations shall be those stipulated in the American National Standard *Minimum Design Loads for Buildings and Other Structures*, ANSI A58.1.

1. Dead Load and Live Load

The dead load to be assumed in design shall consist of the weight of steelwork and all material permanently fastened thereto or supported thereby.

The live load, including snow load if any, shall be that stipulated by the applicable code under which the structure is being designed or that dictated by the conditions involved. Snow load shall be considered as applied either to the entire roof area or to a part of the roof area, and any probable arrangement of loads resulting in the highest stresses in the supporting members shall be used in the design.

2. Impact

For structures carrying live loads* which induce impact, the assumed live load shall be increased sufficiently to provide for same.
If not otherwise specified, the increase shall be not less than:

For supports of elevators	100%
For cab-operated traveling crane support girders and their connections	25%
For pendant-operated traveling crane support girders and their connections	10%
For supports of light machinery, shaft or motor driven	20%
For supports of reciprocating machinery or power driven units ..	50%
For hangers supporting floors and balconies	33%

3. Crane Runway Horizontal Forces

The lateral force on crane runways to provide for the effect of moving crane trolleys shall be not less than 20% of the sum of weights of the lifted load and of the crane trolley, but exclusive of other parts of the crane. The force shall

---

*Live loads on crane support girders shall be taken as the maximum crane wheel loads.

be assumed to be applied at the top of the rails, acting in either direction normal to the runway rails, and shall be distributed with due regard for lateral stiffness of the structure supporting the rails.

The longitudinal tractive force shall be not less than 10% of the maximum wheel loads of the crane applied at the top of the rail, unless otherwise specified.

The crane runway shall also be designed for crane stop forces.

4. Wind

Proper provision shall be made for stresses caused by wind, both during erection and after completion of the building.

5. Other Forces

Structures in localities subject to earthquakes, hurricanes and other extraordinary conditions shall be designed with due regard for such conditions.

A5. DESIGN BASIS

1. Allowable Stresses

Except as provided in Chapter N, all structural members, connections and connectors shall be proportioned so the stresses due to the working loads do not exceed the allowable stresses specified in Chapters D through K. The allowable stresses specified in these chapters do not apply to peak stresses in regions of connections (see also Sect. B9), provided requirements of Chapter K are satisfied.

For provisions pertaining to plastic design, refer to Chapter N.

2. Wind and Seismic Stresses

Allowable stresses may be increased ⅓ above the values otherwise provided when produced by wind or seismic loading, acting alone or in combination with the design dead and live loads, provided the required section computed on this basis is not less than that required for the design dead and live load and impact (if any) computed without the ⅓ stress increase, and further provided that stresses are not otherwise* required to be calculated on the basis of reduction factors applied to design loads in combinations. The above stress increase does not apply to allowable stress ranges provided in Appendix K4.

3. Structural Analysis

The stresses in members, connections and connectors shall be determined by structural analysis for the loads defined in Sect. A4. Selection of the method of analysis is the prerogative of the responsible engineer.

---

*For example, see ANSI A58.1, Sect. 2.3.3.

4. Design for Serviceability and Other Considerations

The overall structure and the individual members, connections and connectors shall be checked for serviceability in accordance with Chapter L.

A6. REFERENCED CODES AND STANDARDS

Where codes and standards are referenced in this Specification, the editions of the following listed adoption dates are intended:

American National Standards Institute
   ANSI B18.1-72
   ANSI A58.1-82

American Society of Testing and Materials

ASTM A6-87d	ASTM A27-87	ASTM A36-87
ASTM A53-88	ASTM A148-84	ASTM A242-87
ASTM A307-86a	ASTM A325-86	ASTM A354-86
ASTM A441-85	ASTM A449-87	ASTM A490-85
ASTM A500-84	ASTM A501-84	ASTM A514-87a
ASTM A529-85	ASTM A563-84	ASTM A570-85
ASTM A572-85	ASTM A588-87	ASTM A606-85
ASTM A607-85	ASTM A618-84	ASTM A668-85a
ASTM A687-84	ASTM C33-86	ASTM C330-87
ASTM F436-86	ASTM A502-83A	ASTM A709-87b
ASTM A852-85		

American Welding Society

AWS D1.1-88	AWS A5.1-81	AWS A5.5-81
AWS A5.17-80	AWS A5.18-79	AWS A5.20-79
AWS A5.23-80	AWS A5.28-79	AWS A5.29-80

Research Council on Structural Connections
   Specification for Structural Joints Using ASTM A325 or A490 Bolts, 1985

A7. DESIGN DOCUMENTS

1. Plans

The design plans shall show a complete design with sizes, sections and relative locations of the various members. Floor levels, column centers and offsets shall be dimensioned. Drawings shall be drawn to a scale large enough to show the information clearly.

Design documents shall indicate the type or types of construction as defined in Sect. A2.2 and shall include the loads and design requirements necessary for preparation of shop drawings including shears, moments and axial forces to be resisted by all members and their connections.

Where joints are to be assembled with high-strength bolts, design documents shall indicate the connection type (slip-critical, tension or bearing).

Camber of trusses, beams and girders, if required, shall be called for in the design documents. The requirements for stiffeners and bracing shall be shown on the design documents.

2. Standard Symbols and Nomenclature

Welding and inspection symbols used on plans and shop drawings shall preferably be the American Welding Society symbols. Other adequate welding symbols are permitted, provided a complete explanation thereof is shown in the design documents.

3. Notation for Welding

Notes shall be made in the design documents and on the shop drawings of those joints or groups of joints in which the welding sequence and technique of welding shall be carefully controlled to minimize distortion.

Weld lengths called for in the design documents and on the shop drawings shall be the net effective lengths.

CHAPTER B

DESIGN REQUIREMENTS

This chapter contains provisions which are common to the Specification as a whole.

B1. GROSS AREA

The gross area of a member at any point shall be determined by summing the products of the thickness and the gross width of each element as measured normal to the axis of the member.

For angles, the gross width shall be the sum of the widths of the legs less the thickness.

B2. NET AREA

The net area A_n of a member is the sum of the products of the thickness and the net width of each element computed as follows:

The width of a bolt or rivet hole shall be taken as $\frac{1}{16}$ in. greater than the nominal dimension of the hole.

For a chain of holes extending across a part in any diagonal or zigzag line, the net width of the part shall be obtained by deducting from the gross width the sum of the diameters or slot dimensions as provided in Sect. J3.2, of all holes in the chain, and adding, for each gage space in the chain, the quantity

$$s^2/4g$$

where

    s = longitudinal center-to-center spacing (pitch) of any two consecutive holes, in.

    g = transverse center-to-center spacing (gage) between fastener gage lines, in.

For angles, the gage for holes in opposite adjacent legs shall be the sum of the gages from the back of the angles less the thickness.

The critical net area A_n of the part is obtained from that chain which gives the least net width.

In determining the net area across plug or slot welds, the weld metal shall not be considered as adding to the net area.

B3. EFFECTIVE NET AREA

When the load is transmitted directly to each of the cross-sectional elements by connectors, the effective net area A_e is equal to the net area A_n.

When the load is transmitted by bolts or rivets through some but not all of the cross-sectional elements of the member, the effective net area A_e shall be computed as:

$$A_e = U A_n \qquad \text{(B3-1)}$$

where

A_n = net area of the member, in.^{2}
U = reduction coefficient

When the load is transmitted by welds through some but not all of the cross-sectional elements of the member, the effective net area A_e shall be computed as:

$$A_e = U A_g \qquad \text{(B3-2)}$$

where

A_g = gross area of member, in.^{2}

Unless a larger coefficient is justified by tests or other criteria, the following values of U shall be used:

    a. W, M or S shapes with flange widths not less than ⅔ the depth, and structural tees cut from these shapes, provided the connection is to the flanges. Bolted or riveted connections shall have no fewer than three fasteners per line in the direction of stress .... $U = 0.90$
    b. W, M or S shapes not meeting the conditions of subparagraph a, structural tees cut from these shapes and all other shapes, including built-up cross sections. Bolted or riveted connections shall have no fewer than three fasteners per line in the direction of stress .... $U = 0.85$
    c. All members with bolted or riveted connections having only two fasteners per line in the direction of stress .... $U = 0.75$

When load is transmitted by transverse welds to some but not all of the cross-sectional elements of W, M or S shapes and structural tees cut from these shapes, A_e shall be taken as the area of the directly connected elements.

When the load is transmitted to a plate by longitudinal welds along both edges at the end of the plate, the length of the welds shall not be less than the width of the plate. The effective net area A_e shall be computed by Equation (B3-2).

Unless a larger coefficient can be justified by tests or other criteria, the following values of U shall be used:

    a. When $l > 2w$ .... $U = 1.0$
    b. When $2w > l > 1.5w$ .... $U = 0.87$
    c. When $1.5w > l > w$ .... $U = 0.75$

    where

        l = weld length, in.
        w = plate width (distance between welds), in.

Bolted and riveted splice and gusset plates and other connection fittings subject to tensile force shall be designed in accordance with the provisions of Sect. D1, where the effective net area shall be taken as the actual net area, except that, for the purpose of design calculations, it shall not be taken as greater than 85% of the gross area.

B4. STABILITY

General stability shall be provided for the structure as a whole and for each compression element.

Consideration shall be given to significant load effects resulting from the deflected shape of the structure or of individual elements of the lateral load resisting system, including effects on beams, columns, bracing, connections and shear walls.

B5. LOCAL BUCKLING

1. Classification of Steel Sections

Steel sections are classified as compact, noncompact and slender element sections. For a section to qualify as compact, its flanges must be continuously connected to the web or webs and the width-thickness ratios of its compression elements must not exceed the applicable limiting width-thickness ratios from Table B5.1. Steel sections that do not qualify as compact are classified as noncompact if the width-thickness ratios of the compression elements do not exceed the values shown for noncompact in Table B5.1. If the width-thickness ratios of any compression element exceed the latter applicable value, the section is classified as a slender element section.

For unstiffened elements which are supported along only one edge, parallel to the direction of the compression force, the width shall be taken as follows:
   a. For flanges of I-shaped members and tees, the width b is half the full nominal width.
   b. For legs of angles and flanges of channels and zees, the width b is the full nominal dimension.
   c. For plates, the width b is the distance from the free edge to the first row of fasteners or line of welds.
   d. For stems of tees, d is taken as the full nominal depth.

For stiffened elements, i.e., supported along two edges parallel to the direction of the compression force, the width shall be taken as follows:
   a. For webs of rolled, built-up or formed sections, h is the clear distance between flanges.
   b. For webs of rolled, built-up or formed sections, d is the full nominal depth.
   c. For flange or diaphragm plates in built-up sections, the width b is the distance between adjacent lines of fasteners or lines of welds.
   d. For flanges of rectangular hollow structural sections, the width b is the clear distance between webs less the inside corner radius on each side. If the corner radius is not known, the flat width may be taken as the total section width minus three times the thickness.

For tapered flanges of rolled sections, the thickness is the nominal value halfway between the free edge and the corresponding face of the web.

2. Slender Compression Elements

For the design of flexural and compressive sections with slender compressive elements see Appendix B5.

TABLE B5.1
Limiting Width-Thickness Ratios
for Compression Elements

Description of Element	Width-Thickness Ratio	Limiting Width-Thickness Ratios	
		Compact	Noncompact[c]
Flanges of I-shaped rolled beams and channels in flexure[a]	b/t	$65/\sqrt{F_y}$	$95/\sqrt{F_y}$
Flanges of I-shaped welded beams in flexure	b/t	$65/\sqrt{F_y}$	$95/\sqrt{F_{yf}/k_c}$ [e]
Outstanding legs of pairs of angles in continuous contact; angles or plates projecting from rolled beams or columns; stiffeners on plate girders	b/t	NA	$95/\sqrt{F_y}$
Angles or plates projecting from girders, built-up columns or other compression members; compression flanges of plate girders	b/t	NA	$95/\sqrt{F_y/k_c}$
Stems of tees	d/t	NA	$127/\sqrt{F_y}$
Unstiffened elements simply supported along one edge, such as legs of single-angle struts, legs of double-angle struts with separators and cross or star-shaped cross sections	b/t	NA	$76/\sqrt{F_y}$
Flanges of square and rectangular box and hollow structural sections of uniform thickness subject to bending or compression[d]; flange cover plates and diaphragm plates between lines of fasteners or welds	b/t	$190/\sqrt{F_y}$	$238/\sqrt{F_y}$
Unsupported width of cover plates perforated with a succession of access holes[b]	b/t	NA	$317/\sqrt{F_y}$
All other uniformly compressed stiffened elements, i.e., supported along two edges	b/t h/t_w	NA	$253/\sqrt{F_y}$
Webs in flexural compression[a]	d/t	$640/\sqrt{F_y}$	—
	h/t_w	—	$760/\sqrt{F_b}$
Webs in combined flexural and axial compression	d/t_w	for $f_a/F_y \leq 0.16$ $\dfrac{640}{\sqrt{F_y}}\left(1 - 3.74\dfrac{f_a}{F_y}\right)$ for $f_a/F_y > 0.16$ $257/\sqrt{F_y}$	—
	h/t_w	—	$760/\sqrt{F_b}$
Circular hollow sections In axial compression In flexure	D/t	$3,300/F_y$ $3,300/F_y$	— —

[a]For hybrid beams, use the yield strength of the flange F_{yf} instead of F_y.
[b]Assumes net area of plate at widest hole.
[c]For design of slender sections that exceed the noncompact limits see Appendix B5.
[d]See also Sect. F3.1.
[e]$k_c = \dfrac{4.05}{(h/t)^{0.46}}$ if $h/t > 70$, otherwise $k_c = 1.0$.

B6. ROTATIONAL RESTRAINT AT POINTS OF SUPPORT

At points of support, beams, girders and trusses shall be restrained against rotation about their longitudinal axis.

B7. LIMITING SLENDERNESS RATIOS

For members whose design is based on compressive force, the slenderness ratio Kl/r preferably should not exceed 200. If this limit is exceeded, the allowable stress shall not exceed the value obtained from Equation (E2-2).

For members whose design is based on tensile force, the slenderness ratio L/r preferably should not exceed 300. The above limitation does not apply to rods in tension. Members which have been designed to perform as tension members in a structural system, but experience some compression loading, need not satisfy the compression slenderness limit.

B8. SIMPLE SPANS

Beams, girders and trusses designed on the basis of simple spans shall have an effective length equal to the distance between centers of gravity of the members to which they deliver their end reactions.

B9. END RESTRAINT

When designed on the assumption of full or partial end restraint due to continuous, semi-continuous or cantilever action, the beams, girders and trusses, as well as the sections of the members to which they connect, shall be designed to carry the shears and moments so introduced, as well as all other forces, without exceeding at any point the unit stresses prescribed in Chapters D through F, except that some non-elastic but self-limiting deformation of a part of the connection is permitted when this is essential to avoid overstressing of fasteners.

B10. PROPORTIONS OF BEAMS AND GIRDERS

Rolled or welded shapes, plate girders and cover-plated beams shall, in general, be proportioned by the moment of inertia of the gross section. No deduction shall be made for shop or field bolt or rivet holes in either flange provided that

$$0.5F_u\, A_{fn} \geq 0.6F_y\, A_{fg} \qquad (B10\text{-}1)$$

where A_{fg} is the gross flange area and A_{fn} is the net flange area, calculated in accordance with the provisions of Sects. B1 and B2.

If

$$0.5F_u\, A_{fn} < 0.6F_y\, A_{fg} \qquad (B10\text{-}2)$$

the member flexural properties shall be based on an effective tension flange area A_{fe}

$$A_{fe} = \frac{5}{6}\frac{F_u}{F_y}\, A_{fn} \qquad (B10\text{-}3)$$

Hybrid girders may be proportioned by the moment of inertia of their gross section,* subject to the applicable provisions in Sect. G1, provided they are not required to resist an axial force greater than $0.15F_y$ times the area of the gross section, where F_y is the yield stress of the flange material. To qualify as hybrid girders, the flanges at any given section shall have the same cross-sectional area and be made of the same grade of steel.

Flanges of welded beams or girders may be varied in thickness or width by splicing a series of plates or by the use of cover plates.

The total cross-sectional area of cover plates of bolted or riveted girders shall not exceed 70% of the total flange area.

High-strength bolts, rivets or welds connecting flange to web, or cover plate to flange, shall be proportioned to resist the total horizontal shear resulting from the bending forces on the girder. The longitudinal distribution of these bolts, rivets or intermittent welds shall be in proportion to the intensity of the shear. However, the longitudinal spacing shall not exceed the maximum permitted for compression or tension members in Sect. D2 or E4, respectively. Bolts, rivets or welds connecting flange to web shall also be proportioned to transmit to the web any loads applied directly to the flange, unless provision is made to transmit such loads by direct bearing.

Partial length cover plates shall be extended beyond the theoretical cutoff point and the extended portion shall be attached to the beam or girder by high-strength bolts in a slip-critical connection, rivets or fillet welds adequate, at the applicable stresses allowed in Sects. J2.4, J3.4, or K4, to develop the cover plate's portion of the flexural stresses in the beam or girder at the theoretical cutoff point.

In addition, for welded cover plates, the welds connecting the cover plate termination to the beam or girder in the length a', defined below, shall be adequate, at the allowed stresses, to develop the cover plate's portion of the flexural stresses in the beam or girder at the distance a' from the end of the cover plate. The length a', measured from the end of the cover plate, shall be:

1. A distance equal to the width of the cover plate when there is a continuous weld equal to or larger than ¾ of the plate thickness across the end of the plate and continuous welds along both edges of the cover plate in the length a'.
2. A distance equal to 1½ times the width of the cover plate when there is a continuous weld smaller than ¾ of the plate thickness across the end of the plate and continuous welds along both edges of the cover plate in the length a'.
3. A distance equal to 2 times the width of the cover plate when there is no weld across the end of the plate, but continuous welds along both edges of the cover plate in the length a'.

B11. PROPORTIONING OF CRANE GIRDERS

The flanges of plate girders supporting cranes or other moving loads shall be proportioned to resist the horizontal forces produced by such loads.

---

*No limit is placed on the web stresses produced by the applied bending moment for which a hybrid girder is designed, except as provided in Sect. K4 and Appendix K4.

CHAPTER   C

FRAMES AND OTHER STRUCTURES

This chapter specifies general requirements to assure stability of the structure as a whole.

C1. GENERAL

In addition to meeting the requirements of member strength and stiffness, frames and other continous structures shall be designed to provide the needed deformation capacity and to assure over-all frame stability.

C2. FRAME STABILITY

1. Braced Frames

In trusses and in those frames where lateral stability is provided by adequate attachment to diagonal bracing, to shear walls, to an adjacent structure having adequate lateral stability or to floor slabs or roof decks secured horizontally by walls or bracing systems parallel to the plane of the frame, the effective length factor K for the compression members shall be taken as unity, unless analysis shows that a smaller value is permitted.

2. Unbraced Frames

In frames where lateral stability is dependent upon the bending stiffness of rigidly connected beams and columns, the effective length Kl of compression members shall be determined by analysis and shall not be less than the actual unbraced length.

CHAPTER D

TENSION MEMBERS

This section applies to prismatic members subject to axial tension caused by forces acting through the centroidal axis. For members subject to combined axial tension and flexure, see Sect. H2. For members subject to fatigue, see Sect. K4. For tapered members, see Appendix F7. For threaded rods see Sect. J3.

D1. ALLOWABLE STRESS

The allowable stress F_t shall not exceed $0.60F_y$ on the gross area nor $0.50F_u$ on the effective net area. In addition, pin-connected members shall meet the requirements of Sect. D3.1 at the pin hole.

Block shear strength shall be checked at end connections of tension members in accordance with Sect. J4.

Eyebars shall meet the requirements of Sect. D3.1.

D2. BUILT-UP MEMBERS

The longitudinal spacing of connectors between elements in continuous contact consisting of a plate and a shape or two plates shall not exceed:

> 24 times the thickness of the thinner plate, nor 12 in. for painted members or unpainted members not subject to corrosion.
> 14 times the thickness of the thinner plate, nor 7 in. for unpainted members of weathering steel subject to atmospheric corrosion.

In a tension member the longitudinal spacing of fasteners and intermittent welds connecting two or more shapes in contact shall not exceed 24 inches. Tension members composed of two or more shapes or plates separated by intermittent fillers shall be connected to one another at these fillers at intervals such that the slenderness ratio of either component between the fasteners does not exceed 300.

Either perforated cover plates or tie plates without lacing are permitted on the open sides of built-up tension members. Tie plates shall have a length not less than ⅔ the distance between the lines of welds or fasteners connecting them to the components of the member. The thickness of such tie plates shall not be less than ¹⁄₅₀ of the distance between these lines. The longitudinal spacing of intermittent welds or fasteners at tie plates shall not exceed 6 in.

The spacing of tie plates shall be such that the slenderness ratio of any component in the length between tie plates should preferably not exceed 300.

D3. PIN-CONNECTED MEMBERS

1. Allowable Stress

The allowable stress on the net area of the pin hole for pin-connected members is 0.45 F_y. The bearing stress on the projected area of the pin shall not exceed the stress allowed in Sect. J8.

The allowable stress on eyebars meeting the requirements of Sect. D3.3 is 0.60 F_y on the body area.

2. Pin-connected Plates

The minimum net area beyond the pin hole, parallel to the axis of the member, shall not be less than ⅔ of the net area across the pin hole.

The distance used in calculations, transverse to the axis of pin-connected plates or any individual element of a built-up member, from the edge of the pin hole to the edge of the member or element shall not exceed 4 times the thickness at the pin hole. For calculation purposes, the distance from the edge of the pin hole to the edge of the plate or to the edge of a separated element of a built-up member at the pin hole, shall not be assumed to be more than 0.8 times the diameter of the pin hole.

For pin-connected members in which the pin is expected to provide for relative movement between connected parts while under full load, the diameter of the pin hole shall not be more than ¹⁄₃₂ in. greater than the diameter of the pin.

The corners beyond the pin hole may be cut at 45° to the axis of the member, provided the net area beyond the pin hole, on a plane perpendicular to the cut, is not less than that perpendicular to the direction of the applied load.

3. Eyebars

Eyebars shall be of uniform thickness, without reinforcement at the pin holes, and have circular heads whose periphery is concentric with the pin hole. The radius of the transition between the circular head and the eyebar body shall not be less than the diameter of the head.

For calculation purposes, the width of the body of an eyebar shall not exceed 8 times its thickness.

The thickness may be less than ½-in. only if external nuts are provided to tighten pin plates and filler plates into snug contact. For calculation purposes, the distance from the hole edge to plate edge perpendicular to the direction of the applied load shall not be less than ⅔ nor greater than ¾ times the width of the eyebar body.

The pin diameter shall be not less than ⅞ times the eyebar width.

The pin-hole diameter shall be no more than ¹⁄₃₂-in. greater than the pin diameter.

For steel having a yield stress greater than 70 ksi, the hole diameter shall not exceed 5 times the plate thickness and the width of the eyebar shall be reduced accordingly.

CHAPTER E

COLUMNS AND OTHER COMPRESSION MEMBERS

This section applies to prismatic members with compact and noncompact sections subject to axial compression through the centroidal axis. For members with slender elements, see Appendix B5.2. For members subject to combined axial compression and flexure, see Chap. H. For tapered members, see Appendix F7.

E1. EFFECTIVE LENGTH AND SLENDERNESS RATIO

The effective-length factor K shall be determined in accordance with Sect. C2.

In determining the slenderness ratio of an axially loaded compression member, the length shall be taken as its effective length Kl and r as the corresponding radius of gyration. For limiting slenderness ratios, see Sect. B7.

E2. ALLOWABLE STRESS

On the gross section of axially loaded compression members whose cross sections meet the provisions of Table B5.1, when Kl/r, the largest effective slenderness ratio of any unbraced segment is less than C_c, the allowable stress is:

$$F_a = \frac{\left[1 - \dfrac{(Kl/r)^2}{2C_c^2}\right] F_y}{\dfrac{5}{3} + \dfrac{3(Kl/r)}{8C_c} - \dfrac{(Kl/r)^3}{8C_c^3}} \tag{E2-1}$$

where

$$C_c = \sqrt{\frac{2\pi^2 E}{F_y}}$$

On the gross section of axially loaded compression members, when Kl/r exceeds C_c, the allowable stress is:

$$F_a = \frac{12\pi^2 E}{23(Kl/r)^2} \tag{E2-2}$$

E3. FLEXURAL-TORSIONAL BUCKLING

Singly symmetric and unsymmetric columns, such as angles or tee-shaped columns, and doubly symmetric columns such as cruciform or built-up columns with very thin walls, may require consideration of flexural-torsional and torsional buckling.

E4. BUILT-UP MEMBERS

All parts of built-up compression members and the transverse spacing of their lines of fasteners shall meet the requirements of Sect. B7.

For spacing and edge distance requirements for weathering steel members, see Sect. J3.10.

At the ends of built-up compression members bearing on base plates or milled surfaces, all components in contact with one another shall be connected by rivets or bolts spaced longitudinally not more than 4 diameters apart for a distance equal to $1\frac{1}{2}$ times the maximum width of the member, or by continuous welds having a length not less than the maximum width of the member.

The longitudinal spacing for intermediate bolts, rivets or intermittent welds in built-up members shall be adequate to provide for the transfer of calculated stress. The maximum longitudinal spacing of bolts, rivets or intermittent welds connecting two rolled shapes in contact shall not exceed 24 in. In addition, for painted members and unpainted members not subject to corrosion where the outside component consists of a plate, the maximum longitudinal spacing shall not exceed:

$127/\sqrt{F_y}$ times the thickness of the outside plate nor 12 in. when fasteners are not staggered along adjacent gage lines.

$190/\sqrt{F_y}$ times the thickness of the outside plate nor 18 in. when fasteners are staggered along adjacent gage lines.

Compression members composed of two or more rolled shapes separated by intermittent fillers shall be connected at these fillers at intervals such that the slenderness ratio Kl/r of either shape, between the fasteners, does not exceed $\frac{3}{4}$ times the governing slenderness ratio of the built-up member. The least radius of gyration r shall be used in computing the slenderness ratio of each component part. At least two intermediate connectors shall be used along the length of the built-up member.

All connections, including those at the ends, shall be welded or shall utilize high-strength bolts tightened to the requirements of Table J3.7.

Open sides of compression members built up from plates or shapes shall be provided with lacing having tie plates at each end and at intermediate points if the lacing is interrupted. Tie plates shall be as near the ends as practicable. In main members carrying calculated stress, the end tie plates shall have a length of not less than the distance between the lines of fasteners or welds connecting them to the components of the member. Intermediate tie plates shall have a length not less than $\frac{1}{2}$ of this distance. The thickness of tie plates shall not be less than $\frac{1}{50}$ of the distance between the lines of fasteners or welds connecting them to the components of the member. In bolted and riveted construction, the spacing in the direction of stress in tie plates shall not be more than 6 diameters and the tie plates shall be connected to each component by at least 3 fasteners. In welded construction, the welding on each line connecting a tie plate shall aggregate not less than $\frac{1}{3}$ the length of the plate.

Lacing, including flat bars, angles, channels or other shapes employed as lacing, shall be so spaced that the ratio l/r of the flange included between their connections shall not exceed $\frac{3}{4}$ times the governing ratio for the member as a whole. Lacing shall be proportioned to resist a shearing stress normal to the axis of the member equal to 2% of the total compressive stress in the member. The ratio l/r for lacing bars arranged in single systems shall not exceed 140. For double lacing this ratio shall not exceed 200. Double lacing bars shall be joined at their intersections. For lacing bars in compression the unsupported length of the lacing bar shall be taken as the distance between fasteners or welds connecting it to the components of the built-up member for single lacing, and 70% of that distance for double lacing. The inclination of lacing bars to the axis of the member shall preferably be not less than 60° for single lacing and 45° for double lacing. When the distance between the lines of fasteners or welds in the flanges is more than 15 in., the lacing preferably shall be double or be made of angles.

The function of tie plates and lacing may be performed by continuous cover plates perforated with access holes. The unsupported width of such plates at access holes, as defined in Sect. B5, is assumed available to resist axial stress, provided that: the width-to-thickness ratio conforms to the limitations of Sect. B5; the ratio of length (in direction of stress) to width of holes shall not exceed 2; the clear distance between holes in the direction of stress shall be not less than the transverse distance between nearest lines of connecting fasteners or welds; and the periphery of the holes at all points shall have a minimum radius of $1\frac{1}{2}$ in.

E5. PIN-CONNECTED COMPRESSION MEMBERS

Pin-connections of pin-connected compression members shall conform to the requirements of Sect. D3.

E6. COLUMN WEB SHEAR

Column connections must be investigated for concentrated force introduction in accordance with Sect. K1.

CHAPTER F

BEAMS AND OTHER FLEXURAL MEMBERS

Beams shall be distinguished from plate girders on the basis of the web slenderness ratio h/t_w. When this value is greater than $970/\sqrt{F_y}$ the allowable bending stress is given in Chapter G. The allowable shear stresses and stiffener requirements are given in Chapter F unless tension field action is used, then the allowable shear stresses are given in Chapter G.

This chapter applies to singly or doubly symmetric beams including hybrid beams and girders loaded in the plane of symmetry. It also applies to channels loaded in a plane passing through the shear center parallel to the web or restrained against twisting at load points and points of support. For members subject to combined flexural and axial force, see Sect. H1.

F1. ALLOWABLE STRESS: STRONG AXIS BENDING OF I-SHAPED MEMBERS AND CHANNELS

1. Members with Compact Sections

For members with compact sections as defined in Sect. B5.1 (excluding hybrid beams and members with yield points greater than 65 ksi) symmetrical about, and loaded in, the plane of their minor axis the allowable stress is

$$F_b = 0.66 \, F_y \tag{F1-1}$$

provided the flanges are connected continuously to the web or webs and the laterally unsupported length of the compression flange L_b does not exceed the value of L_c, as given by the smaller of:

$$\frac{76b_f}{\sqrt{F_y}} \text{ or } \frac{20{,}000}{(d/A_f) \, F_y} \tag{F1-2}$$

Members (including composite members and excluding hybrid members and members with yield points greater than 65 ksi) which meet the requirements for compact sections and are continuous over supports or rigidly framed to columns, may be proportioned for $9/10$ of the negative moments produced by gravity loading when such moments are maximum at points of support, provided that, for such members, the maximum positive moment is increased by $1/10$ of the average negative moments. This reduction shall not apply to moments produced by loading on cantilevers. If the negative moment is resisted by a column rigidly framed to the beam or girder, the $1/10$ reduction is permitted in proportioning the column for the combined axial and bending loading, provided that the stress f_a due to any concurrent axial load on the member, does not exceed $0.15F_a$.

2. Members with Noncompact Sections

For members meeting the requirements of Sect. F1.1 except that their flanges are noncompact (excluding built-up members and members with yield points greater than 65 ksi), the allowable stress is

$$F_b = F_y \left[0.79 - 0.002 \frac{b_f}{2t_f} \sqrt{F_y} \right] \qquad (F1\text{-}3)$$

For built-up members meeting the requirements of Sect. F1.1 except that their flanges are noncompact and their webs are compact or noncompact, (excluding hybrid girders and members with yield points greater than 65 ksi) the allowable stress is

$$F_b = F_y \left[0.79 - 0.002 \frac{b_f}{2t_f} \sqrt{\frac{F_y}{k_c}} \right] \qquad (F1\text{-}4)$$

where

$$k_c = \frac{4.05}{(h/t_w)^{0.46}} \text{ if } h/t_w > 70, \text{ otherwise } k_c = 1.0.$$

For members with a noncompact section (Sect. B5), but not included above, and loaded through the shear center and braced laterally in the region of compression stress at intervals not exceeding $76b_f/\sqrt{F_y}$, the allowable stress is

$$F_b = 0.60 \, F_y \qquad (F1\text{-}5)$$

3. Members with Compact or Noncompact Sections with Unbraced Length Greater than L_c

For flexural members with compact or noncompact sections as defined in Sect. B5.1, and with unbraced lengths greater than L_c as defined in Sect. F1.1, the allowable bending stress in tension is determined from Equation (F1-5).

For such members with an axis of symmetry in, and loaded in the plane of their web, the allowable bending stress in compression is determined as the larger value from Equations (F1-6) or (F1-7) and (F1-8), except that Equation (F1-8) is applicable only to sections with a compression flange that is solid and approximately rectangular in cross section and that has an area not less than the tension flange. Higher values of the allowable compressive stress are permitted if justified by a more precise analysis. Stresses shall not exceed those permitted by Chapter G, if applicable.

For channels bent about their major axis, the allowable compressive stress is determined from Equation (F1-8).

AMERICAN INSTITUTE OF STEEL CONSTRUCTION

When

$$\sqrt{\frac{102 \times 10^3 C_b}{F_y}} \le \frac{l}{r_T} \le \sqrt{\frac{510 \times 10^3 C_b}{F_y}}:$$

$$F_b = \left[\frac{2}{3} - \frac{F_y \, (l/r_T)^2}{1530 \times 10^3 C_b}\right] F_y \le 0.60 \, F_y \qquad \text{(F1-6)}$$

When

$$\frac{l}{r_T} \ge \sqrt{\frac{510 \times 10^3 C_b}{F_y}}:$$

$$F_b = \frac{170 \times 10^3 C_b}{(l/r_T)^2} \le 0.60 \, F_y \qquad \text{(F1-7)}$$

For any value of l/r_T:

$$F_b = \frac{12 \times 10^3 C_b}{ld/A_f} \le 0.60 \, F_y \qquad \text{(F1-8)}$$

where

l = distance between cross sections braced against twist or lateral displacement of the compression flange, in. For cantilevers braced against twist only at the support, l may conservatively be taken as the actual length.

r_T = radius of gyration of a section comprising the compression flange plus $\frac{1}{3}$ of the compression web area, taken about an axis in the plane of the web, in.

A_f = area of the compression flange, in.^{2}

C_b = $1.75 + 1.05 \, (M_1/M_2) + 0.3 \, (M_1/M_2)^2$, but not more than 2.3,* where M_1 is the smaller and M_2 the larger bending moment at the ends of the unbraced length, taken about the strong axis of the member, and where M_1/M_2, the ratio of end moments, is positive when M_1 and M_2 have the same sign (reverse curvature bending) and negative when they are of opposite signs (single curvature bending). When the bending moment at any point within an unbraced length is larger than that at both ends of this length, the value of C_b shall be taken as unity. When computing F_{bx} to be used in Equation (H1-1), C_b may be computed by the equation given above for frames subject to joint translation, and it shall be taken as unity for frames braced against joint translation. C_b may conservatively be taken as unity for cantilever beams.**

---

*It is conservative to take C_b as unity. For values smaller than 2.3, see Table 6 in the Numerical Values Section.

**For the use of larger C_b values, see Galambos (1988).

For hybrid plate girders, F_y for Equations (F1-6) and (F1-7) is the yield stress of the compression flange. Equation (F1-8) shall not apply to hybrid girders.

Sect. F1.3 does not apply to tee sections if the stem is in compression anywhere along the unbraced length.

F2. ALLOWABLE STRESS: WEAK AXIS BENDING OF I-SHAPED MEMBERS, SOLID BARS AND RECTANGULAR PLATES

Lateral bracing is not required for members loaded through the shear center about their weak axis nor for members of equal strength about both axes.

1. Members With Compact Sections

For doubly symmetrical I- and H-shape members with compact flanges (Sect. B5) continuously connected to the web and bent about their weak axes (except members with yield points greater than 65 ksi); solid round and square bars; and solid rectangular sections bent about their weaker axes, the allowable stress is

$$F_b = 0.75 \, F_y \tag{F2-1}$$

2. Members With Noncompact Sections

For members not meeting the requirements for compact sections of Sect. B5 and not covered in Sect. F3, bent about their minor axis, the allowable stress is

$$F_b = 0.60 \, F_y \tag{F2-2}$$

Doubly symmetrical I- and H-shape members bent about their weak axes (except members with yield points greater than 65 ksi) with noncompact flanges (Sect. B5) continuously connected to the web may be designed on the basis of an allowable stress of

$$F_b = F_y \left[1.075 - 0.005 \left(\frac{b_f}{2t_f} \right) \sqrt{F_y} \right] \tag{F2-3}$$

F3. ALLOWABLE STRESS: BENDING OF BOX MEMBERS, RECTANGULAR TUBES AND CIRCULAR TUBES

1. Members With Compact Sections

For members bent about their strong or weak axes, members with compact sections as defined in Sect. B5 and flanges continuously connected to the webs, the allowable stress is

$$F_b = 0.66 \, F_y \tag{F3-1}$$

To be classified as a compact section, a box-shaped member shall have, in addition to the requirements in Sect. B5, a depth not greater than 6 times the width, a flange thickness not greater than 2 times the web thickness and a laterally unsupported length L_b less than or equal to

$$L_c = \left(1{,}950 + 1{,}200 \, \frac{M_1}{M_2} \right) \frac{b}{F_y} \tag{F3-2}$$

except that it need not be less than 1,200 (b/F_y), where M_1 is the smaller and M_2 the larger bending moment at the ends of the unbraced length, taken about the strong axis of the member, and where M_1/M_2, the ratio of end moments, is positive when M_1 and M_2 have the same sign (reverse curvature bending) and negative when they are of opposite signs (single curvature bending).

2. **Members With Noncompact Sections**

For box-type and tubular flexural members that meet the noncompact section requirements of Sect. B5, the allowable stress is

$$F_b = 0.60 \, F_y \tag{F3-3}$$

Lateral bracing is not required for a box section whose depth is less than 6 times its width. Lateral-support requirements for box sections of larger depth-to-width ratios must be determined by special analysis.

F4. ALLOWABLE SHEAR STRESS

For $h/t_w \leq 380/\sqrt{F_y}$, on the overall depth times the web thickness, the allowable shear stress is

$$F_v = 0.40 \, F_y \tag{F4-1}$$

For $h/t_w > 380/\sqrt{F_y}$, the allowable shear stress is on the clear distance between flanges times the web thickness is

$$F_v = \frac{F_y}{2.89} \, (C_v) \leq 0.40 F_y \tag{F4-2}$$

where

$$C_v = \frac{45,000 k_v}{F_y (h/t_w)^2} \text{ when } C_v \text{ is less than } 0.8$$

$$= \frac{190}{h/t_w} \sqrt{\frac{k_v}{F_y}} \text{ when } C_v \text{ is more than } 0.8$$

$$k_v = 4.00 + \frac{5.34}{(a/h)^2} \text{ when } a/h \text{ is less than } 1.0$$

$$= 5.34 + \frac{4.00}{(a/h)^2} \text{ when } a/h \text{ is more than } 1.0$$

t_w = thickness of web, in.
a = clear distance between transverse stiffeners, in.
h = clear distance between flanges at the section under investigation, in.

For shear rupture on coped beam end connections see Sect. J4.

Maximum h/t_w limits are given in Chapter G.

An alternative design method for plate girders utilizing tension field action is given in Chapter G.

F5. TRANSVERSE STIFFENERS

Intermediate stiffeners are required when the ratio h/t_w is greater than 260 and the maximum web shear stress f_v is greater than that permitted by Equation (F4-2).

The spacing of intermediate stiffeners, when required, shall be such that the web shear stress will not exceed the value for F_v given by Equation (F4-2) or (G3-1), as applicable, and

$$\frac{a}{h} \leq \left[\frac{260}{(h/t_w)} \right]^2 \text{ and } 3.0 \qquad \text{(F5-1)}$$

F6. BUILT-UP MEMBERS

Where two or more rolled beams or channels are used side-by-side to form a flexural member, they shall be connected together at intervals of not more than 5 ft. Through-bolts and separators are permitted, provided that, in beams having a depth of 12 in. or more, no fewer than 2 bolts shall be used at each separator location. When concentrated loads are carried from one beam to the other, or distributed between the beams, diaphragms having sufficient stiffness to distribute the load shall be riveted, bolted or welded between the beams.

F7. WEB-TAPERED MEMBERS

See Appendix F7.

CHAPTER G

PLATE GIRDERS

Plate girders shall be distinguished from beams on the basis of the web slenderness ratio h/t_w. When this value is greater than $970/\sqrt{F_y}$, the provisions of this chapter shall apply for allowable bending stress, otherwise Chapter F is applicable.

For allowable shear stress and transverse stiffener design see appropriate sections in Chapter F or this chapter if tension field action is utilized.

G1. WEB SLENDERNESS LIMITATIONS

When no transverse stiffeners are provided or when transverse stiffeners are spaced more than $1\frac{1}{2}$ times the distance between flanges

$$\frac{h}{t_w} \leq \frac{14,000}{\sqrt{F_{yf}\,(F_{yf} + 16.5)}} \qquad \text{(G1-1)}$$

When transverse stiffeners are provided, spaced not more than $1\frac{1}{2}$ times the distance between flanges

$$\frac{h}{t_w} \leq \frac{2,000}{\sqrt{F_{yf}}} \qquad \text{(G1-2)}$$

G2. ALLOWABLE BENDING STRESS

When the web depth-to-thickness ratio exceeds $970/\sqrt{F_y}$, the maximum bending stress in the compression flange shall not exceed

$$F_b' \leq F_b\, R_{PG}\, R_e \qquad \text{(G2-1)}$$

where

F_b = applicable bending stress given in Chapter F, ksi

$$R_{PG} = 1 - 0.0005\, \frac{A_w}{A_f} \left(\frac{h}{t} - \frac{760}{\sqrt{F_b}}\right) \leq 1.0$$

$$R_e = \frac{12 + \left(\dfrac{A_w}{A_f}\right)(3\alpha - \alpha^3)}{12 + 2\left(\dfrac{A_w}{A_f}\right)} \leq 1.0$$

(non-hybrid girders, $R_e = 1.0$)

A_w = area of web at the section under investigation, in.^{2}

A_f = area of compression flange, in.^{2}

α = $0.6\, F_{yw}/F_b \leq 1.0$

AMERICAN INSTITUTE OF STEEL CONSTRUCTION

G3. ALLOWABLE SHEAR STRESS WITH TENSION FIELD ACTION

Except as herein provided, the largest average web shear, f_v, in kips per sq. in., computed for any condition of complete or partial loading, shall not exceed the value given by Equation (F4-2).

Alternatively, for girders other than hybrid girders, if intermediate stiffeners are provided and spaced to satisfy the provisions of Sect. G4 and if $C_v \leq 1$, the allowable shear including tension field action given by Equation (G3-1) is permitted in lieu of the value given by Equation (F4-2).

$$F_v = \frac{F_y}{2.89}\left[C_v + \frac{1 - C_v}{1.15\sqrt{1 + (a/h)^2}}\right] \leq 0.40F_y \qquad (G3-1)*$$

G4. TRANSVERSE STIFFENERS

Transverse stiffeners shall meet the requirements of Sect. F5.

In girders designed on the basis of tension field action, the spacing between stiffeners at end panels, at panels containing large holes, and at panels adjacent to panels containing large holes shall be such that f_v does not exceed the value given by Equation (F4-2).

Bolts and rivets connecting stiffeners to the girder web shall be spaced not more than 12 in. o.c. If intermittent fillet welds are used, the clear distance between welds shall not be more than 16 times the web thickness nor more than 10 in.

The moment of inertia, I_{st}, of a pair of intermediate stiffeners, or a single intermediate stiffener, with reference to an axis in the plane of the web, shall be limited as follows

$$I_{st} \geq \left(\frac{h}{50}\right)^4 \qquad (G4-1)$$

The gross area (*total* area, when stiffeners are furnished in pairs), in sq. in., of intermediate stiffeners spaced as required for Equation (G3-1) shall be not less than

$$A_{st} = \frac{1 - C_v}{2}\left[\frac{a}{h} - \frac{(a/h)^2}{\sqrt{1 + (a/h)^2}}\right] YDht \qquad (G4-2)$$

where

C_v, a, h, and t are as defined in Sect. F4
Y = ratio of yield stress of web steel to yield stress of stiffener steel
D = 1.0 for stiffeners furnished in pairs
= 1.8 for single angle stiffeners
= 2.4 for single plate stiffeners

---

*Equation (G3-1) recognizes the contribution of tension field action.

When the greatest shear stress f_v in a panel is less than that permitted by Equation (G3-1), the reduction of this gross area requirement is permitted in like proportion.

Intermediate stiffeners required by Equation (G3-1) shall be connected for a total shear transfer, in kips per linear inch of single stiffener or pair of stiffeners, not less than

$$f_{vs} = h \sqrt{\left(\frac{F_y}{340}\right)^3} \qquad\qquad \text{(G4-3)}$$

where F_y = yield stress of web steel.

This shear transfer may be reduced in the same proportion that the largest computed shear stress f_v in the adjacent panels is less than that permitted by Equation (G3-1). However, rivets and welds in intermediate stiffeners which are required to transmit to the web an applied concentrated load or reaction shall be proportioned for not less than the applied load or reaction.

Intermediate stiffeners may be stopped short of the tension flange, provided bearing is not needed to transmit a concentrated load or reaction. The weld by which intermediate stiffeners are attached to the web shall be terminated not closer than 4 times nor more than 6 times the web thickness from the near toe of the web-to-flange weld. When single stiffeners are used, they shall be attached to the compression flange, if it consists of a rectangular plate, to resist any uplift tendency due to torsion in the plate. When lateral bracing is attached to a stiffener, or a pair of stiffeners, in turn, these shall be connected to the compression flange to transmit 1% of the total flange stress, unless the flange is composed only of angles.

G5. COMBINED SHEAR AND TENSION STRESS

Plate girder webs which depend upon tension field action, as provided in Equation (G3-1), shall be so proportioned that bending tensile stress, due to moment in the plane of the girder web, shall not exceed $0.60F_y$ nor

$$\left(0.825 - 0.375 \frac{f_v}{F_v}\right) F_y \qquad\qquad \text{(G5-1)}$$

where
    f_v = computed average web shear stress (total shear divided by web area), ksi
    F_v = allowable web shear stress according to Equation (G3-1), ksi

The allowable shear stress in the webs of girders having flanges and webs with yield point greater than 65 ksi shall not exceed the values given by Equation (F4-2) if the flexural stress in the flange f_b exceeds $0.75F_b$.

CHAPTER H

COMBINED STRESSES

The strength of members subjected to combined stresses shall be determined according to the provisions of this chapter.

This chapter pertains to doubly and singly symmetrical members only. See Chapter E for determination of F_a and Chapter F for determination of F_{bx} and F_{by}.

H1. AXIAL COMPRESSION AND BENDING

Members subjected to both axial compression and bending stresses shall be proportioned to satisfy the following requirements:

$$\frac{f_a}{F_a} + \frac{C_{mx}f_{bx}}{\left(1 - \dfrac{f_a}{F'_{ex}}\right)F_{bx}} + \frac{C_{my}f_{by}}{\left(1 - \dfrac{f_a}{F'_{ey}}\right)F_{by}} \le 1.0 \qquad (\text{H1-1})$$

$$\frac{f_a}{0.60F_y} + \frac{f_{bx}}{F_{bx}} + \frac{f_{by}}{F_{by}} \le 1.0 \qquad (\text{H1-2})$$

When $f_a/F_a \le 0.15$, Equation (H1-3) is permitted in lieu of Equations (H1-1) and (H1-2):

$$\frac{f_a}{F_a} + \frac{f_{bx}}{F_{bx}} + \frac{f_{by}}{F_{by}} \le 1.0 \qquad (\text{H1-3})$$

In Equations (H1-1), (H1-2) and (H1-3), the subscripts x and y, combined with subscripts b, m and e, indicate the axis of bending about which a particular stress or design property applies, and

F_a = axial compressive stress that would be permitted if axial force alone existed, ksi

F_b = compressive bending stress that would be permitted if bending moment alone existed, ksi

$$F'_e = \frac{12\,\pi^2 E}{23(Kl_b/r_b)^2}$$

= Euler stress divided by a factor of safety, ksi (In the expression for F'_e, l_b is the actual unbraced length *in the plane of bending* and r_b is the corresponding radius of gyration. K is the effective length factor *in the plane of bending*.) As in the case of F_a, F_b and $0.60F_y$, F'_e may be increased $\frac{1}{3}$ in accordance with Sect. A5.2.

f_a = computed axial stress, ksi

f_b = computed compressive bending stress at the point under consideration, ksi

C_m = Coefficient whose value shall be taken as follows:

   a. For compression members in frames subject to joint translation (sidesway), $C_m = 0.85$.

   b. For rotationally restrained compression members in frames braced against joint translation and not subject to transverse loading between their supports in the plane of bending,

$$C_m = 0.6 - 0.4\,(M_1/M_2)$$

   where M_1/M_2 is the ratio of the smaller to larger moments at the ends of that portion of the member unbraced in the plane of bending under consideration. M_1/M_2 is positive when the member is bent in reverse curvature, negative when bent in single curvature.

   c. For compression members in frames braced against joint translation in the plane of loading and subjected to transverse loading between their supports, the value of C_m may be determined by an analysis. However, in lieu of such analysis, the following values are permitted:

      i. For members whose ends are restrained against rotation in the plane of bending .... $C_m = 0.85$

      ii. For members whose ends are unrestrained against rotation in the plane of bending .... $C_m = 1.0$

H2.   AXIAL TENSION AND BENDING

Members subject to both axial tension and bending stresses shall be proportioned at all points along their length to satisfy the following equation:

$$\frac{f_a}{F_t} + \frac{f_{bx}}{F_{bx}} + \frac{f_{by}}{F_{by}} \leq 1.0 \qquad \text{(H2-1)}$$

where f_b is the computed bending tensile stress, f_a is the computed axial tensile stress, F_b is the allowable bending stress and F_t is the governing allowable tensile stress defined in Sect. D1.

However the computed bending compressive stress arising from an independent load source relative to the axial tension, taken above, shall not exceed the applicable value required in Chapter F.

CHAPTER I

COMPOSITE CONSTRUCTION

This chapter applies to steel beams supporting a reinforced concrete slab* so interconnected that the beams and the slab act together to resist bending. Simple and continuous composite beams with shear connectors and concrete-encased beams, constructed with or without temporary shores, are included.

I1. DEFINITION

Two cases of composite members are recognized: Totally encased members which depend upon natural bond for interaction with the concrete and those with shear connectors (mechanical anchorage to the slab) with the steel member not necessarily encased.

A beam totally encased in concrete cast integrally with the slab may be assumed to be connected to the concrete by natural bond, without additional anchorage, provided that:

1. Concrete cover over beam sides and soffit is at least 2 in.
2. The top of the beam is at least 1½ in. below the top and 2 in. above bottom of the slab.
3. Concrete encasement contains adequate mesh or other reinforcing steel throughout the whole depth and across the soffit of the beam to prevent spalling of the concrete.

Shear connectors must be provided for composite action if the steel member is not totally encased in concrete. The portion of the effective width of the concrete slab on each side of the beam centerline shall not exceed:

a. One-eighth of the beam span, center-to-center of supports;
b. One-half the distance to the centerline of the adjacent beam; or
c. The distance from the beam centerline to the edge of the slab.

I2. DESIGN ASSUMPTIONS

1. Encased beams shall be proportioned to support, unassisted, all dead loads applied prior to the hardening of the concrete (unless these loads are supported temporarily on shoring) and, acting in conjunction with the slab, to support all dead and live loads applied after hardening of the concrete, without exceeding a computed bending stress of $0.66F_y$, where F_y is the yield stress of the steel beam. The bending stress produced by loads after the concrete has hardened shall be computed on the basis of the section properties of the composite section. Concrete tension stresses shall be ne-

---

*See Commentary Sect. I2.

glected. Alternatively, the steel beam alone may be proportioned to resist, unassisted, the positive moment produced by all loads, live and dead, using a bending stress equal to $0.76F_y$, in which case temporary shoring is not required.

2. When shear connectors are used in accordance with Sect. I4, the composite section shall be proportioned to support all of the loads without exceeding the allowable stress prescribed in Sect. F1.1, even when the steel section is not shored during construction. In positive moment areas, the steel section is exempt from compact flange criteria (Sect. B5) and there is no limit on the unsupported length of the compression flange.

Reinforcement parallel to the beam within the effective width of the slab, when anchored in accordance with the provisions of the applicable building code, may be included in computing the properties of composite sections, provided shear connectors are furnished in accordance with the requirements of Sect. I4. The section properties of the composite section shall be computed in accordance with the elastic theory. Concrete tension stresses shall be neglected. For stress computations, the compression area of lightweight or normal weight concrete shall be treated as an equivalent area of steel by dividing it by the modular ratio n for normal weight concrete of the strength specified when determining the section properties. For deflection calculations, the transformed section properties shall be based on the appropriate modular ratio n for the strength and weight concrete specified, where $n = E/E_c$.

In cases where it is not feasible or necessary to provide adequate connectors to satisfy the horizontal shear requirements for full composite action, the effective section modulus shall be determined as

$$S_{eff} = S_s + \sqrt{\frac{V_h'}{V_h}} (S_{tr} - S_s) \qquad (I2\text{-}1)$$

where

V_h and V_h' are as defined in Sect. I4

S_s = section modulus of the steel beam referred to its bottom flange, in.^{3}

S_{tr} = section modulus of the transformed composite section referred to its bottom flange, based upon maximum permitted effective width of concrete flange (Sect. I1), in.^{3}

For composite beams constructed without temporary shoring, stresses in the steel section shall not exceed $0.90F_y$. Stresses shall be computed assuming the steel section alone resists all loads applied before the concrete has reached 75% of its required strength and the effective composite section resists all loads applied after that time.

The actual section modulus of the transformed composite section shall be used in calculating the concrete flexural compression stress and, for construction without temporary shores, this stress shall be based upon loading applied after the concrete has reached 75% of its required strength. The stress in the concrete shall not exceed $0.45f_c'$.

I3. END SHEAR

The web and the end connections of the steel beam shall be designed to carry the total reaction.

I4. SHEAR CONNECTORS

Except in the case of encased beams, as defined in Sect. I2.1, the entire horizontal shear at the junction of the steel beam and the concrete slab shall be assumed to be transferred by shear connectors welded to the top flange of the beam and embedded in the concrete. For full composite action with concrete subject to flexural compression, the total horizontal shear to be resisted between the point of maximum positive moment and points of zero moment shall be taken as the smaller value using Equations (I4-1) and (I4-2):

$$V_h = 0.85f'_c A_c/2 \qquad (\text{I4-1})^*$$

and

$$V_h = F_y A_s/2 \qquad (\text{I4-2})$$

where
f'_c = specified compression strength of concrete, ksi
A_c = actual area of effective concrete flange defined in Sect. I1, in.^{2}
A_s = area of steel beam, in.^{2}

In continuous composite beams where longitudinal reinforcing steel is considered to act compositely with the steel beam in the negative moment regions, the total horizontal shear to be resisted by shear connectors between an interior support and each adjacent point of contraflexure shall be taken as

$$V_h = F_{yr} A_{sr}/2 \qquad (\text{I4-3})$$

where
A_{sr} = total area of longitudinal reinforcing steel at the interior support located within the effective flange width specified in Sect. I1, in.^{2}
F_{yr} = specified minimum yield stress of the longitudinal reinforcing steel, ksi

For full composite action, the number of connectors resisting the horizontal shear, V_h, each side of the point of maximum moment, shall not be less than that determined by the relationship V_h/q, where q, the allowable shear load for one connector, is given in Table I4.1 for flat soffit concrete slabs made with ASTM C33 aggregates. For flat soffit concrete slabs made with rotary kiln produced aggregates, conforming to ASTM C330 with concrete unit weight not less than 90 pcf, the allowable shear load for one connector is obtained by multiplying the values from Table I4.1 by the coefficient from Table I4.2.

For partial composite action with concrete subject to flexural compression, the horizontal shear V'_h to be used in computing S_{eff} shall be taken as the product

---

*The term $\frac{1}{2} F_{yr} A'_s$ shall be added to the right-hand side of Equation (I4-1) if longitudinal reinforcing steel with area A'_s located within the effective width of the concrete flange is included in the properties of the composite section.

of q times the number of connectors furnished between the point of maximum moment and the nearest point of zero moment.

The value of V_h' shall not be less than ¼ the smaller value of Equation (I4-1), using the maximum permitted effective width of the concrete flange, or Equation (I4-2). The effective moment of inertia for deflection computations shall be determined by:

$$I_{eff} = I_s + \sqrt{\frac{V_h'}{V_h}}\,(I_{tr} - I_s) \qquad (I4\text{-}4)$$

where

I_s = moment of inertia of the steel beam, in.[4]
I_{tr} = moment of inertia of the transformed composite section, in.[4]

The connectors required each side of the point of maximum moment in an area of positive bending may be uniformly distributed between that point and adjacent points of zero moment, except that N_2, the number of shear connectors

Table I4.1
Allowable Horizontal
Shear Load for One Connector (q), kips[a]

Connector[b]	Specified Compressive Strength of Concrete (f_c'), ksi		
	3.0	3.5	≥4.0
½″ dia. × 2″ hooked or headed stud	5.1	5.5	5.9
⅝″ dia. × 2½″ hooked or headed stud	8.0	8.6	9.2
¾″ dia. × 3″ hooked or headed stud	11.5	12.5	13.3
⅞″ dia. × 3½″ hooked or headed stud	15.6	16.8	18.0
Channel C3 × 4.1	$4.3w^c$	$4.7w^c$	$5.0w^c$
Channel C4 × 5.4	$4.6w^c$	$5.0w^c$	$5.3w^c$
Channel C5 × 6.7	$4.9w^c$	$5.3w^c$	$5.6w^c$

[a]Applicable only to concrete made with ASTM C33 aggregates.
[b]The allowable horizontal loads tabulated are also permitted for studs longer than shown.
[c]w = length of channel, in.

Table I4.2
Coefficients for Use with Concrete Made with
C330 Aggregates

Specified Compressive Strength of Concrete (f_c')	Air Dry Unit Weight of Concrete, pcf						
	90	95	100	105	110	115	120
≤4.0 ksi	0.73	0.76	0.78	0.81	0.83	0.86	0.88
≥5.0 ksi	0.82	0.85	0.87	0.91	0.93	0.96	0.99

required between any concentrated load in that area and the nearest point of zero moment, shall be not less than that determined by Equation (I4-5).

$$N_2 = \frac{N_1 \left[\dfrac{M\beta}{M_{max}} - 1 \right]}{\beta - 1} \qquad (I4\text{-}5)$$

where

M = moment (less than the maximum moment) at a concentrated load point

N_1 = number of connectors required between point of maximum moment and point of zero moment, determined by the relationship V_h/q or V_h'/q, as applicable

β = $\dfrac{S_{tr}}{S_s}$ or $\dfrac{S_{eff}}{S_s}$, as applicable

For a continuous beam, connectors required in the region of negative bending may be uniformly distributed between the point of maximum moment and each point of zero moment.

Shear connectors shall have at least 1 in. of lateral concrete cover, except for connectors installed in the ribs of formed steel decks. Unless located directly over the web, the diameter of studs shall not be greater than 2½ times the thickness of the flange to which they are welded. The minimum center-to-center spacing of stud connectors shall be 6 diameters along the longitudinal axis of the supporting composite beam and 4 diameters transverse to the longitudinal axis of the supporting composite beam. The maximum center-to-center spacing of stud connectors shall not exceed 8 times the total slab thickness.

I5. COMPOSITE BEAMS OR GIRDERS WITH FORMED STEEL DECK

Composite construction of concrete slabs on formed steel deck connected to steel beams or girders shall be designed by the applicable portions of Sects. I1 through I4, with the following modifications.

1. General

1. Section I5 is applicable to decks with nominal rib height not greater than 3 inches.
2. The average width of concrete rib or haunch w_r shall be not less than 2 in., but shall not be taken in calculations as more than the minimum clear width near the top of the steel deck. See Sect. I5.3, subparagraphs 2 and 3, for additional provisions.
3. The concrete slab shall be connected to the steel beam or girder with welded stud shear connectors ¾ in. or less in diameter (AWS D1.1, Sect. 7, Part F). Studs may be welded through the deck or directly to the steel member.
4. Stud shear connectors shall extend not less than 1½ in. above the top of the steel deck after installation.
5. The slab thickness above the steel deck shall be not less than 2 in.

2. Deck Ribs Oriented Perpendicular to Steel Beam or Girder

1. Concrete below the top of the steel deck shall be neglected when determining section properties and in calculating A_c for Equation (I4-1).

2. The spacing of stud shear connectors along the length of a supporting beam or girder shall not exceed 36 in.

3. The allowable horizontal shear load per stud connector q shall be the value stipulated in Sect. I4 (Tables I4.1 and I4.2) multiplied by the following reduction factor:

$$\left(\frac{0.85}{\sqrt{N_r}}\right) \left(\frac{w_r}{h_r}\right) \left(\frac{H_s}{h_r} - 1.0\right) \leq 1.0 \tag{I5-1}$$

where

h_r = nominal rib height, in.

H_s = length of stud connector after welding, in., not to exceed the value $(h_r + 3)$ in computations, although the actual length may be greater

N_r = number of stud connectors on a beam in one rib, not to exceed 3 in computations, although more than 3 studs may be installed.

w_r = average width of concrete rib, in. (see Sect. I5.1, subparagraph 2)

4. To resist uplift, the steel deck shall be anchored to all compositely designed steel beams or girders at a spacing not to exceed 16 in. Such anchorage may be provided by stud connectors, a combination of stud connectors and arc spot (puddle) welds, or other devices specified by the designer.

3.   Deck Ribs Oriented Parallel to Steel Beam or Girder

1. Concrete below the top of the steel deck may be included when determining section properties and shall be included in calculating A_c for Equation (I4-1).

2. Steel deck ribs over supporting beams or girders may be split longitudinally and separated to form a concrete haunch.

3. When the nominal depth of steel deck is 1½ in. or greater, the average width w_r of the supported haunch or rib shall be not less than 2 in. for the first stud in the transverse row plus 4 stud diameters for each additional stud.

4. The allowable horizontal shear load per stud connector q shall be the value stipulated in Sect. I4 (Tables I4.1 and I4.2), except when the ratio w_r/h_r is less than 1.5, the allowable load shall be multiplied by the following reduction factor:

$$0.6 \left(\frac{w_r}{h_r}\right) \left(\frac{H_s}{h_r} - 1.0\right) \leq 1.0 \tag{I5-2}$$

where h_r and H_s are as defined in Sect. I5.2 and w_r is the average width of concrete rib or haunch (see Sect. I5.1, subparagraph 2, and Sect. I5.3, subparagraph 3).

I6.   SPECIAL CASES

When composite construction does not conform to the requirements of Sects. I1 through I5, the allowable load per shear connector must be established by a suitable test program.

CHAPTER J

CONNECTIONS, JOINTS AND FASTENERS

This chapter applies to connections consisting of connecting elements (plates, stiffeners, gussets, angles, brackets) and connectors (welds, bolts, rivets).

J1. GENERAL PROVISIONS

1. Design Basis

Connections shall be proportioned so that the calculated stress is less than the allowable stress determined (1) by structural analysis for loads acting on the structure or (2) as a specified proportion of the strength of the connected members, whichever is appropriate.

2. Simple Connections

Except as otherwise indicated in the design documents, connections of beams, girders or trusses shall be designed as flexible and ordinarily may be proportioned for the reaction shears only. Flexible beam connections shall accommodate end rotations of unrestrained (simple) beams. To accomplish this, inelastic deformation in the connection is permitted.

3. Moment Connections

End connections of restrained beams, girders and trusses shall be designed for the combined effect of forces resulting from moment and shear induced by the rigidity of the connections.

4. Compression Members with Bearing Joints

When columns bear on bearing plates or are finished to bear at splices, there shall be sufficient connectors to hold all parts securely in place.

When other compression members are finished to bear, the splice material and its connectors shall be arranged to hold all parts in line and shall be proportioned for 50% of the strength of the member.

All compression joints shall be proportioned to resist any tension developed by the specified lateral loads acting in conjunction with 75% of the calculated dead-load stress and no live load.

5. Connections of Tension and Compression Members in Trusses

The connections at ends of tension or compression members in trusses shall develop the force due to the design load, but not less than 50% of the effective strength of the member, unless a smaller percentage is justified by engineering analysis that considers other factors including handling, shipping and erection.

6. Minimum Connections

Connections carrying calculated stresses, except for lacing, sag bars and girts, shall be designed to support not less than 6 kips.

7. Splices in Heavy Sections

This section applies to ASTM A6 Group 4 and 5 rolled shapes, or shapes built-up by welding plates more than 2 in. thick together to form the cross section*, and where the cross section is to be spliced and subject to primary tensile stresses due to tension or flexure.

When tensile forces in these sections are to be transmitted through splices by full-penetration groove welds, material notch-toughness requirements as given in Sect. A3.1c, weld access holes details as given in Sect. J1.8, compatible welding procedures as given in Sect. J2.6, welding preheat requirements as given in Sect. J2.7 and thermal cut surface preparation and inspection requirements as given in Sect. M2.2 apply.

At tension splices in these sections, weld tabs and backing shall be removed and the surfaces ground smooth.

When splicing these sections, and where the section is to be used as a primary compression member, all weld access holes required to facilitate groove welding operations shall satisfy the provisions of Sect. J1.8.

Alternatively, splicing of such members subject to compression, including members which are subject to tension due to wind or seismic loads, may be accomplished using splice details which do not induce large weld shrinkage strains such as partial-penetration flange groove welds with fillet-welded surface lap plate splices on the web, or with bolted or combination bolted/fillet-welded lap plate splices.

8. Beam Copes and Weld Access Holes

All weld access holes required to facilitate welding operations shall have a length from the toe of the weld preparation not less than 1½ times the thickness of the material in which the hole is made. The height of the access hole shall be adequate for deposition of sound weld metal in the adjacent plates and provide clearance for weld tabs. In hot rolled shapes and built-up shapes, all beam copes and weld access holes shall be shaped free of notches or sharp re-entrant corners except that, when fillet web-to-flange welds are used in built-up shapes, access holes are permitted to terminate perpendicular to the flange.

For Group 4 and 5 shapes and built-up shapes of material more than 2 in. thick, the thermally cut surfaces of beam copes and weld access holes shall be ground

---

*When the individual elements of the cross section are spliced prior to being joined to form the cross section in accordance with AWS D1.1, Article 3.4.6, the applicable provisions of AWS D1.1 apply in lieu of the requirements of this Section.

to bright metal and inspected by either magnetic particle or dye penetrant methods. If the curved transition portion of weld access holes and beam copes are formed by predrilled or sawed holes, that portion of the access hole or cope need not be ground. Weld access holes and beam copes in other shapes need not be ground nor inspected by dye penetrant or magnetic particle.

9. Placement of Welds, Bolts and Rivets

Groups of welds, bolts or rivets at the ends of any member which transmit axial stress into that member shall be sized so the center of gravity of the group coincides with the center of gravity of the member, unless provision is made for the eccentricity. The foregoing provision is not applicable to end connections of statically loaded single-angle, double-angle and similar members. Eccentricity between the gravity axes of such members and the gage lines for their riveted or bolted end connections may be neglected in statically loaded members, but shall be considered in members subject to fatigue loading.

See Sect. J3.10 for placement of fasteners in built-up members made of weathering steel.

10. Bolts in Combination with Welds

In new work, A307 bolts or high-strength bolts used in bearing-type connections shall not be considered as sharing the stress in combination with welds. Welds, if used, shall be provided to carry the entire stress in the connection. High-strength bolts proportioned for slip-critical connections may be considered as sharing the stress with the welds.

In making welded alterations to structures, existing rivets and high-strength bolts tightened to the requirements for slip-critical connections are permitted for carrying stresses resulting from loads present at the time of alteration, and the welding need be adequate to carry only the additional stress.

11. High-strength Bolts in Slip-Critical Connections in Combination with Rivets

In both new work and alterations, high-strength bolts in slip-critical connections may be considered as sharing the load with rivets.

12. Limitations on Bolted and Welded Connections

Fully-tensioned high-strength bolts (see Table J3.7) or welds shall be used for the following connections:

Column splices in all tier structures 200 ft or more in height

Column splices in tier structures 100 to 200 ft in height, if the least horizontal dimension is less than 40% of the height

Column splices in tier structures less than 100 ft in height, if the least horizontal dimension is less than 25% of the height

Connections of all beams and girders to columns and of any other beams and girders on which the bracing of columns is dependent, in structures over 125 ft in height

In all structures carrying cranes of over 5-ton capacity: roof truss splices and connections of trusses to columns, column splices, column bracing, knee braces and crane supports

Connections for supports of running machinery or of other live loads which produce impact or reversal of stress

Any other connections stipulated on the design plans.

In all other cases, connections may be made with high-strength bolts tightened to the snug-tight condition or with A307 bolts.

For the purpose of this section, the height of a tier structure shall be taken as the vertical distance from the curb level to the highest point of the roof beams in the case of flat roofs, or to the mean height of the gable in the case of roofs having a rise of more than 2⅔ in 12. Where the curb level has not been established, or where the structure does not adjoin a street, the mean level of the adjoining land shall be used instead of curb level. Penthouses may be excluded in computing the height of the structure.

J2. WELDS

All provisions of the American Welding Society *Structural Welding Code— Steel,* AWS D1.1, except Sects. 2.3.2.4, 2.5, 8.13.1, 9, and 10, apply to work performed under this Specification.

1. Groove Welds

a. Effective Area

The effective area of groove welds shall be considered as the effective length of the weld times the effective throat thickness.

The effective length of a groove weld shall be the width of the part joined.

The effective throat thickness of a complete-penetration groove weld shall be the thickness of the thinner part joined.

The effective throat thickness of a partial-penetration groove weld shall be as shown in Table J2.1.

The effective throat thickness of a flare groove weld when flush to the surface of a bar or 90° bend in a formed section shall be as shown in Table J2.2. Random sections of production welds for each welding procedure, or such test sections as may be required by design documents, shall be used to verify that the effective throat is consistently obtained.

Larger effective throat thicknesses than those in Table J2.2 are permitted, provided the fabricator can establish by qualification that he can consistently provide such larger effective throat thicknesses. Qualification shall consist of sectioning the weld normal to its axis, at mid-length and terminal ends. Such sectioning shall be made on a number of combinations of material sizes representative of the range to be used in the fabrication or as required by the designer.

b. Limitations

The minimum effective throat thickness of a partial-penetration groove weld shall be as shown in Table J2.3. Minimum effective throat thickness is determined by the thicker of the two parts joined, except that the weld size need not exceed the thickness of the thinnest part joined. For this exception, particular care shall be taken to provide sufficient preheat for soundness of the weld.

TABLE J2.1
Effective Throat Thickness of Partial-penetration Groove Welds

Welding Process	Welding Position	Included Angle at Root of Groove	Effective Throat Thickness
Shielded metal arc Submerged arc Gas metal arc Flux-cored arc	All	J or U joint	Depth of chamfer
		Bevel or V joint ≥ 60°	Depth of chamfer
		Bevel or V joint < 60° but ≥ 45°	Depth of chamfer minus ⅛-in.

TABLE J2.2
Effective Throat Thickness of Flare Groove Welds

Type of Weld	Radius (R) of Bar or Bend	Effective Throat Thickness
Flare bevel groove	All	$\frac{5}{16}R$
Flare V-groove	All	$\frac{1}{2}R$[a]

[a]Use ⅜R for Gas Metal Arc Welding (except short circuiting transfer process) when $R \geq$ ½-in.

TABLE J2.3
Minimum Effective Throat Thickness of Partial-penetration Groove Welds

Material Thickness of Thicker Part Joined (in.)	Minimum Effective Throat Thickness[a] (in.)
To ¼ inclusive	⅛
Over ¼ to ½	3/16
Over ½ to ¾	¼
Over ¾ to 1½	5/16
Over 1½ to 2¼	⅜
Over 2¼ to 6	½
Over 6	⅝

[a]See Sect. J2.

2. Fillet Welds

a. Effective Area

The effective area of fillet welds shall be taken as the effective length times the effective throat thickness.

The effective length of fillet welds, except fillet welds in holes and slots, shall be the overall length of full-size fillets, including returns.

The effective throat thickness of a fillet weld shall be the shortest distance from the root of the joint to the face of the diagrammatic weld, except that for fillet welds made by the submerged arc process, the effective throat thickness shall be taken equal to the leg size for $\frac{3}{8}$-in. and smaller fillet welds, and equal to the theoretical throat plus 0.11-in. for fillet welds larger than $\frac{3}{8}$-in.

For fillet welds in holes and slots, the effective length shall be the length of the centerline of the weld along the center of the plane through the throat. In the case of overlapping fillets, the effective area shall not exceed the nominal cross-sectional area of the hole or slot in the plane of the faying surface.

b. Limitations

The *minimum size of fillet welds* shall be as shown in Table J2.4. Minimum weld size is dependent upon the thicker of the two parts joined, except that the weld size need not exceed the thickness of the thinner part. For this exception, particular care shall be taken to provide sufficient preheat for soundness of the weld. Weld sizes larger than the thinner part joined are permitted if required by calculated strength. In the as-welded condition, the distance between the edge of the base metal and the toe of the weld may be less than $\frac{1}{16}$-in. provided the weld size is clearly verifiable.

The *maximum size of fillet welds* that is permitted along edges of connected parts shall be:

- Material less than $\frac{1}{4}$-in. thick, not greater than the thickness of the material.
- Material $\frac{1}{4}$-in. or more in thickness, not greater than the thickness of the material minus $\frac{1}{16}$-in., unless the weld is especially designated on the drawings to be built out to obtain full-throat thickness.

TABLE J2.4
Minimum Size of Fillet Welds

Material Thickness of Thicker Part Joined (in.)	Minimum Size of Fillet Weld[a] (in.)
To $\frac{1}{4}$ inclusive	$\frac{1}{8}$
Over $\frac{1}{4}$ to $\frac{1}{2}$	$\frac{3}{16}$
Over $\frac{1}{2}$ to $\frac{3}{4}$	$\frac{1}{4}$
Over $\frac{3}{4}$	$\frac{5}{16}$
[a]Leg dimension of fillet welds. Single-pass welds must be used.	

The *minimum effective length of fillet welds* designed on the basis of strength shall be not less than 4 times the nominal size, or else the size of the weld shall be considered not to exceed $\frac{1}{4}$ of its effective length. If longitudinal fillet welds are used alone in end connections of flat bar tension members, the length of each fillet weld shall be not less than the perpendicular distance between them. The transverse spacing of longitudinal fillet welds used in end connections of tension members shall not exceed 8 in., unless the member is designed on the basis of effective net area in accordance with Sect. B3.

Intermittent fillet welds are permitted to transfer calculated stress across a joint or faying surfaces when the strength required is less than that developed by a continuous fillet weld of the smallest permitted size, and to join components of built-up members. The effective length of any segment of intermittent fillet welding shall be not less than 4 times the weld size, with a minimum of $1\frac{1}{2}$ in.

In *lap joints,* the minimum lap shall be 5 times the thickness of the thinner part joined, but not less than 1 in. Lap joints joining plates or bars subjected to axial stress shall be fillet welded along the end of both lapped parts, except where the deflection of the lapped parts is sufficiently restrained to prevent opening of the joint under maximum loading.

Fillet welds in holes or slots are permitted to transmit shear in lap joints or to prevent the buckling or separation of lapped parts and to join components of built-up members. Such fillet welds may overlap, subject to the provisions of Sect. J2. Fillet welds in holes or slots are not to be considered plug or slot welds.

Side or end fillet welds terminating at ends or sides, respectively, of parts or members shall, wherever practicable, be returned continuously around the corners for a distance not less than 2 times the nominal size of the weld. This provision shall apply to side and top fillet welds connecting brackets, beam seats and similar connections, on the plane about which bending moments are computed. For framing angles and simple end-plate connections which depend upon flexibility of the outstanding legs for connection flexibility, end returns shall not exceed four times the nominal size of the weld. Fillet welds which occur on opposite sides of a common plane shall be interrupted at the corner common to both welds. End returns shall be indicated on the design and detail drawings.

3. Plug and Slot Welds

a. Effective Area

The effective shearing area of plug and slot welds shall be considered as the nominal cross-sectional area of the hole or slot in the plane of the faying surface.

b. Limitations

Plug or slot welds are permitted to transmit shear in lap joints or to prevent buckling of lapped parts and to join component parts of built-up members.

The diameter of the hole for a plug weld shall not be less than the thickness of the part containing it plus ⁵⁄₁₆-in., rounded to the next larger odd ¹⁄₁₆-in., nor greater than the minimum diameter plus ⅛-in. or 2¼ times the thickness of the weld.

The minimum c.-to-c. spacing of plug welds shall be four times the diameter of the hole.

The minimum spacing of lines of slot welds in a direction transverse to their length shall be 4 times the width of the slot. The minimum c.-to-c. spacing in a longitudinal direction on any line shall be 2 times the length of the slot.

The length of slot for a slot weld shall not exceed 10 times the thickness of the weld. The width of the slot shall be not less than the thickness of the part containing it plus ⁵⁄₁₆-in., rounded to the next larger odd ¹⁄₁₆-in., nor shall it be larger than 2¼ times the thickness of the weld. The ends of the slot shall be semicircular or shall have the corners rounded to a radius not less than the thickness of the part containing it, except those ends which extend to the edge of the part.

The thickness of plug or slot welds in material ⅝-in. or less in thickness shall be equal to the thickness of the material. In material over ⅝-in. thick, the thickness of the weld shall be at least ½ the thickness of the material but not less than ⅝-in.

4. Allowable Stresses

Except as modified by the provisions of Sect. K4, welds shall be proportioned to meet the stress requirements given in Table J2.5.

5. Combination of Welds

If two or more of the general types of weld (groove, fillet, plug, slot) are combined in a single joint, the effective capacity of each shall be separately computed with reference to the axis of the group in order to determine the allowable capacity of the combination.

6. Mixed Weld Metal

When notch-toughness is specified, the process consumables for all weld metal, tack welds, root pass and subsequent passes, deposited in a joint shall be compatible to assure notch-tough composite weld metal.

7. Preheat for Heavy Shapes

For ASTM A6 Group 4 and 5 shapes and welded built-up members made of plates more than 2 in. thick, a preheat equal to or greater than 350°F shall be used when making groove weld splices.

TABLE J2.5
Allowable Stress on Welds[f]

Type of Weld and Stress[a]	Allowable Stress	Required Weld Strength Level[b,c]
Complete-penetration Groove Welds		
Tension normal to effective area	Same as base metal	"Matching" weld metal shall be used.
Compression normal to effective area	Same as base metal	Weld metal with a strength level equal to or less than "matching" weld metal is permitted.
Tension or compression parallel to axis of weld	Same as base metal	
Shear on effective area	0.30 × nominal tensile strength of weld metal (ksi)	
Partial-penetration Groove Welds[d]		
Compression normal to effective area	Same as base metal	Weld metal with a strength level equal to or less than "matching" weld metal is permitted.
Tension or compression parallel to axis of weld[e]	Same as base metal	
Shear parallel to axis of weld	0.30 × nominal tensile strength of weld metal (ksi)	
Tension normal to effective area	0.30 × nominal tensile strength of weld metal (ksi), except tensile stress on base metal shall not exceed 0.60 × yield stress of base metal	
Fillet Welds		
Shear on effective area	0.30 × nominal tensile strength of weld metal (ksi)	Weld metal with a strength level equal to or less than "matching" weld metal is permitted.
Tension or compression Parallel to axis of weld[e]	Same as base metal	
Plug and Slot Welds		
Shear parallel to faying surfaces (on effective area)	0.30 × nominal tensile strength of weld metal (ksi)	Weld metal with a strength level equal to or less than "matching" weld metal is permitted.

[a]For definition of effective area, see Sect. J2.

[b]For "matching" weld metal, see Table 4.1.1, AWS D1.1.

[c]Weld metal one strength level stronger than "matching" weld metal will be permitted.

[d]See Sect. J2.1b for a limitation on use of partial-penetration groove welded joints.

[e]Fillet welds and partial-penetration groove welds joining the component elements of built-up members, such as flange-to-web connections, may be designed without regard to the tensile or compressive stress in these elements parallel to the axis of the welds.

[f]The design of connected material is governed by Chapters D through G. Also see Commentary Sect. J2.4.

J3. BOLTS, THREADED PARTS AND RIVETS

1. High-strength Bolts

Except as otherwise provided in this Specification, use of high-strength bolts shall conform to the provisions of the *Specification for Structural Joints Using ASTM A325 or A490 Bolts* approved by the Research Council on Structural Connections of the Engineering Foundation (RCSC).

If required to be tightened to more than 50% of their minimum specified tensile strength, ASTM A449 bolts in tension and bearing-type shear connections shall have an ASTM F436 hardened washer installed under the bolt head, and the nuts shall meet the requirements of ASTM A563.

2. Size and Use of Holes

a. The *maximum sizes* of holes for bolts are given in Table J3.1, except that larger holes, required for tolerance on location of anchor bolts in concrete foundations, are permitted in column base details.

b. *Standard holes* shall be provided in member-to-member connections, unless oversized, short-slotted or long-slotted holes in bolted connections are approved by the designer. Finger shims up to $\frac{1}{4}$ in. may be introduced into slip-critical connections designed on the basis of standard holes without reducing the allowable shear stress of the fastener.

c. *Oversized holes* are permitted in any or all plies of slip-critical connections, but they shall not be used in bearing-type connections. Hardened washers shall be installed over oversized holes in an outer ply.

d. *Short-slotted holes* are permitted in any or all plies of slip-critical or bearing-type connections. The slots are permitted without regard to direction of loading in slip-critical connections, but the length shall be normal to the direction of the load in bearing-type connections. Washers shall be installed over short-slotted holes in an outer ply; when high-strength bolts are used, such washers shall be hardened.

e. *Long-slotted holes* are permitted in only one of the connected parts of either a slip-critical or bearing-type connection at an individual faying surface. Long-slotted holes are permitted without regard to direction of loading in slip-critical connections, but shall be normal to the direction of load in bearing-type connections. Where long-slotted holes are used in an outer ply, plate washers or a continuous bar with standard holes, having a size suf-

TABLE J3.1
Nominal Hole Dimensions

Bolt Dia.	Hole Dimensions			
	Standard (Dia.)	Oversize (Dia.)	Short-slot (Width × length)	Long-slot (Width × length)
$\frac{1}{2}$	$\frac{9}{16}$	$\frac{5}{8}$	$\frac{9}{16} \times \frac{11}{16}$	$\frac{9}{16} \times 1\frac{1}{4}$
$\frac{5}{8}$	$\frac{11}{16}$	$\frac{13}{16}$	$\frac{11}{16} \times \frac{7}{8}$	$\frac{11}{16} \times 1\frac{9}{16}$
$\frac{3}{4}$	$\frac{13}{16}$	$\frac{15}{16}$	$\frac{13}{16} \times 1$	$\frac{13}{16} \times 1\frac{7}{8}$
$\frac{7}{8}$	$\frac{15}{16}$	$1\frac{1}{16}$	$\frac{15}{16} \times 1\frac{1}{8}$	$\frac{15}{16} \times 2\frac{3}{16}$
1	$1\frac{1}{16}$	$1\frac{1}{4}$	$1\frac{1}{16} \times 1\frac{5}{16}$	$1\frac{1}{16} \times 2\frac{1}{2}$
$\geq 1\frac{1}{8}$	$d + \frac{1}{16}$	$d + \frac{5}{16}$	$(d + \frac{1}{16}) \times (d + \frac{3}{8})$	$(d + \frac{1}{16}) \times (2.5 \times d)$

ficient to completely cover the slot after installation, shall be provided. In high-strength bolted connections, such plate washers or continuous bars shall be not less than $\frac{5}{16}$-in. thick and shall be of structural grade material, but need not be hardened. If hardened washers are required for use of high-strength bolts, the hardened washers shall be placed over the outer surface of the plate washer or bar.

  f. When A490 bolts over 1-in. dia. are used in slotted or oversize holes in external plies, a single hardened washer conforming to ASTM F436, except with $\frac{5}{16}$-in. minimum thickness, shall be used in lieu of the standard washer.

3. Effective Bearing Area

The effective bearing area of bolts, threaded parts and rivets shall be the diameter multiplied by the length in bearing, except that for countersunk bolts and rivets $\frac{1}{2}$ the depth of the countersink shall be deducted.

4. Allowable Tension and Shear

Allowable tension and shear stresses on bolts, threaded parts and rivets shall be as given in Table J3.2, in ksi of the nominal body area of rivets (before driving) or the unthreaded nominal body area of bolts and threaded parts other than upset rods (see footnote c, Table J3.2). High-strength bolts supporting applied load by direct tension shall be so proportioned that their average tensile stress, computed on the basis of nominal bolt area and independent of any initial tightening force, will not exceed the appropriate stress given in Table J3.2. The applied load shall be the sum of the external load and any tension resulting from prying action produced by deformation of the connected parts.

When specified by the designer, the nominal slip resistance for connections having special faying surface conditions may be increased to the applicable values in the *RCSC Specification for Structural Joints Using ASTM A325 or A490 Bolts.*

Finger shims up to $\frac{1}{4}$-in. may be introduced into slip-critical connections designed on the basis of standard holes without reducing the allowable shear stress of the fastener to that specified for slotted holes.

Design for bolts, threaded parts and rivets subject to fatigue loading shall be in accordance with Appendix K4.3.

5. Combined Tension and Shear in Bearing-type Connections

Bolts and rivets subject to combined shear and tension shall be so proportioned that the tension stress F_t in ksi on the nominal body area A_b produced by forces applied to the connected parts, shall not exceed the values computed from the equations in Table J3.3, where f_v, the shear stress produced by the same forces, shall not exceed the value for shear given in Table J3.2. When allowable stresses are increased for wind or seismic loads in accordance with Sect. A5.2, the constants in the equations listed in Table J3.3 shall be increased by $\frac{1}{3}$, but the coefficient applied to f_v shall not be increased.

TABLE J3.2
Allowable Stress on Fasteners, ksi

Description of Fasteners	Allowable Tension[g] (F_t)	Allowable Shear[g] (F_v)					
		Slip-critical Connections[e,i]					Bearing-type Connections[i]
		Standard size Holes	Oversized and Short-slotted Holes	Long-slotted holes			
				Transverse[j] Load	Parallel[j] Load		
A502, Gr. 1, hot-driven rivets	23.0[a]						17.5[f]
A502, Gr. 2 and 3, hot-driven rivets	29.0[a]						22.0[f]
A307 bolts	20.0[a]						10.0[b,f]
Threaded parts meeting the requirements of Sects. A3.1 and A3.4 and A449 bolts meeting the requirements of Sect. A3.4, when threads are not excluded from shear planes	$0.33F_u$[a,c,h]						$0.17F_u$[h]
Threaded parts meeting the requirements of Sects. A3.1 and A3.4, and A449 bolts meeting the requirements of Sect. A3.4, when threads are excluded from shear planes	$0.33F_u$[a,h]						$0.22F_u$[h]
A325 bolts, when threads are not excluded from shear planes	44.0[d]	17.0	15.0	12.0	10.0		21.0[f]
A325 bolts, when threads are excluded from shear planes	44.0[d]	17.0	15.0	12.0	10.0		30.0[f]
A490 bolts, when threads are not excluded from shear planes	54.0[d]	21.0	18.0	15.0	13.0		28.0[f]
A490 bolts, when threads are excluded from shear planes	54.0[d]	21.0	18.0	15.0	13.0		40.0[f]

[a]Static loading only.

[b]Threads permitted in shear planes.

[c]The tensile capacity of the threaded portion of an upset rod, based upon the cross-sectional area at its major thread diameter A_b shall be larger than the nominal body area of the rod before upsetting times $0.60F_y$.

[d]For A325 and A490 bolts subject to tensile fatigue loading, see Appendix K4.3.

[e]Class A (slip coefficient 0.33). Clean mill scale and blast-cleaned surfaces with Class A coatings. When specified by the designer, the allowable shear stress, F_v, for slip-critical connections having special faying surface conditions may be increased to the applicable value given in the RCSC Specification.

[f]When bearing-type connections used to splice tension members have a fastener pattern whose length, measured parallel to the line of force, exceeds 50 in., tabulated values shall be reduced by 20%.

[g]See Sect. A5.2

[h]See Table 2, Numerical Values Section for values for specific ASTM steel specifications.

[i]For limitations on use of oversized and slotted holes, see Sect. J3.2.

[j]Direction of load application relative to long axis of slot.

6. Combined Tension and Shear in Slip-critical Joints

For A325 and A490 bolts used in slip-critical connections, the maximum shear stress allowed by Table J3.2 shall be multiplied by the reduction factor $(1 - f_t A_b / T_b)$, where f_t is the average tensile stress due to a direct load applied to all of the bolts in a connection and T_b is the pretension load of the bolt specified in Table J3.7. When allowable stresses are increased for wind or seismic loads in accordance with the provisions of Sect. A5.2, the reduced allowable shear stress shall be increased by $\frac{1}{3}$.

7. Allowable Bearing at Bolt Holes

On the projected area of bolts and rivets in shear connections with the end distance in the line of force not less than $1\frac{1}{2}\,d$ and the distance c. to c. of bolts not less than $3d$:

1. In standard-or short-slotted holes with two or more bolts in the line of force,

$$F_p = 1.2\, F_u \qquad\qquad (J3\text{-}1)$$

where

F_p = allowable bearing stress, ksi

2. In long-slotted holes with the axis of the slot perpendicular to the direction of load and with two or more bolts in the line of force,

$$F_p = 1.0\, F_u \qquad\qquad (J3\text{-}2)$$

On the projected area of the bolt or rivet closest to the edge in standard or short-slotted holes with the edge distance less than $1\frac{1}{2}d$ and in all connections with a single bolt in the line of force:

$$F_p = L_e F_u / 2d \le 1.2\, F_u \qquad\qquad (J3\text{-}3)$$

TABLE J3.3
Allowable Tension Stress F_t for Fasteners in Bearing-type Connections

Description of Fasteners	Threads Included in Shear Plane	Threads Excluded from Shear Plane
A307 bolts	$26 - 1.8f_v \le 20$	
A325 bolts	$\sqrt{(44)^2 - 4.39f_v^2}$	$\sqrt{(44)^2 - 2.15f_v^2}$
A490 bolts	$\sqrt{(54)^2 - 3.75f_v^2}$	$\sqrt{(54)^2 - 1.82f_v^2}$
Threaded parts, A449 bolts over 1½-in. dia.	$0.43F_u - 1.8f_v \le 0.33F_u$	$0.43F_u - 1.4f_v \le 0.33F_u$
A502 Gr. 1 rivets	$30 - 1.3f_v \le 23$	
A502 Gr. 2 rivets	$38 - 1.3f_v \le 29$	

where,

L_e = distance from the free edge to center of the bolt, in.
d = bolt dia., in.

If deformation around the hole is not a design consideration and adequate spacing and edge distance is as required by Sects. J3.8 and J3.9, the following equation is permitted in lieu of Equation (J3-1):

$$F_p = 1.5\ F_u \qquad (J3\text{-}4)$$

and the limit in Equation (J3-3) shall be increased to $1.5F_u$.

8.   Minimum Spacing

The distance between centers of standard, oversized or slotted fastener holes shall not be less than $2\frac{2}{3}$ times the nominal diameter of the fastener* nor less than that required by the following paragraph, if applicable.

Along a line of transmitted forces, the distance between centers of holes s shall be not less than $3d$ when F_p is determined by Equations (J3-1) and (J3-2). Otherwise, the distance between centers of holes shall not be less than the following:

a. For standard holes:

$$s \geq 2P/F_u t + d/2 \qquad (J3\text{-}5)$$

where

P = force transmitted by one fastener to the critical connected part, kips
F_u = specified minimum tensile strength of the critical connected part, ksi
t = thickness of the critical connected part, in.

b. For oversized and slotted holes, the distance required for standard holes in subparagraph a, (above), plus the applicable increment C_1 from Table J3.4, but the clear distance between holes shall not be less than one bolt diameter.

9.   Minimum Edge Distance

The distance from the center of a standard hole to an edge of a connected part shall be not less than the applicable value from Table J3.5 nor the value from Equation (J3-6), as applicable.

Along a line of transmitted force, in the direction of the force, the distance from the center of a standard hole to the edge of the connected part L_e shall be not less than $1\frac{1}{2}d$ when F_p is determined by Equations (J3-1) or (J3-2). Otherwise the edge distance shall be not less than

$$L_e \geq 2P/F_u t \qquad (J3\text{-}6)$$

where P, F_u, t are defined in Sect. J3.8.

---

*A distance of $3d$ is preferred.

TABLE J3.4
Values of Spacing Increment C_1, in.

Nominal Dia. of Fastener	Oversize Holes	Slotted Holes		
		Perpendicular to Line of Force	Parallel to Line of Force	
			Short-slots	Long-slots[a]
$\leq 7/8$	$1/8$	0	$3/16$	$1\frac{1}{2}d - 1/16$
1	$3/16$	0	$1/4$	$1^{7}/16$
$\geq 1\frac{1}{8}$	$1/4$	0	$5/16$	$1\frac{1}{2}d - 1/16$

[a]When length of slot is less than maximum allowed in Table J3.1, C_1 may be reduced by the difference between the maximum and actual slot lengths.

TABLE J3.5
Minimum Edge Distance, in.
(Center of Standard Hole[a] to Edge of Connected Part)

Nominal Bolt or Rivet Dia. (in.)	At Sheared Edges	At Rolled Edges of Plates, Shapes or Bars, Gas Cut or Saw-cut Edges[b]
$1/2$	$7/8$	$3/4$
$5/8$	$1\frac{1}{8}$	$7/8$
$3/4$	$1\frac{1}{4}$	1
$7/8$	$1\frac{1}{2}$[c]	$1\frac{1}{8}$
1	$1\frac{3}{4}$[c]	$1\frac{1}{4}$
$1\frac{1}{8}$	2	$1\frac{1}{2}$
$1\frac{1}{4}$	$2\frac{1}{4}$	$1\frac{5}{8}$
Over $1\frac{1}{4}$	$1\frac{3}{4} \times$ Dia.	$1\frac{1}{4} \times$ Dia.

[a]For oversized or slotted holes, see Table J3.6.
[b]All edge distances in this column may be reduced $1/8$-in. when the hole is at a point where stress does not exceed 25% of the maximum design strength in the element.
[c]These may be $1\frac{1}{4}$ in. at the ends of beam connection angles.

TABLE J3.6
Values of Edge Distance Increment C_2, in.

Nominal Dia. of Fastener (in.)	Oversized Holes	Slotted Holes		
		Perpendicular to Edge		Parallel to Edge
		Short Slots	Long Slots[a]	
$\leq 7/8$	$1/16$	$1/8$		
1	$1/8$	$1/8$	$3/4d$	0
$\leq 1\frac{1}{8}$	$1/8$	$3/16$		

[a]When length of slot is less than maximum allowable (see Table J3.1), C_2 may be reduced by one-half the difference between the maximum and actual slot lengths.

The distance from the center of an oversized or slotted hole to an edge of a connected part shall be not less than required for a standard hole plus the applicable increment C_2 from Table J3.6.

10. Maximum Edge Distance and Spacing

The maximum distance from the center of any rivet or bolt to the nearest edge of parts in contact shall be 12 times the thickness of the connected part under consideration, but shall not exceed 6 in. Bolted joints in unpainted steel exposed to atmospheric corrosion require special limitations on pitch and edge distance.

For unpainted, built-up members made of weathering steel which will be exposed to atmospheric corrosion, the spacing of fasteners connecting a plate and a shape or two-plate components in contact shall not exceed 14 times the thickness of the thinnest part nor 7 in., and the maximum edge distance shall not exceed eight times the thickness of the thinnest part, or 5 in.

11. Long Grips

A307 bolts which carry calculated stress, with a grip exceeding five diameters, shall have their number increased 1% for each additional $\frac{1}{16}$ in. in the grip.

J4. ALLOWABLE SHEAR RUPTURE

At beam end connections where the top flange is coped, and in similar situations where failure might occur by shear along a plane through the fasteners, or by a combination of shear along a plane through the fasteners plus tension along a perpendicular plane:

$$F_v = 0.30F_u \tag{J4-1}$$

TABLE J3.7
Minimum Pretension for
Fully-tightened Bolts, kips[a]

Bolt Size, in.	A325 Bolts	A490 Bolts
$\frac{1}{2}$	12	15
$\frac{5}{8}$	19	24
$\frac{3}{4}$	28	35
$\frac{7}{8}$	39	49
1	51	64
$1\frac{1}{8}$	56	80
$1\frac{1}{4}$	71	102
$1\frac{3}{8}$	85	121
$1\frac{1}{2}$	103	148

[a]Equal to 0.70 of minimum tensile strength of bolts, rounded off to nearest kip, as specified in ASTM specifications for A325 and A490 bolts with UNC threads.

acting on the net shear area A_v and,

$$F_t = 0.50F_u \qquad (J4-2)$$

acting on the net tension area A_t.

The minimum net failure path on the periphery of welded connections shall be checked.[*]

J5.   CONNECTING ELEMENTS

This section applies to the design of connecting elements, such as stiffeners, gussets, angles and brackets and the panel zones of beam-to-column connections.

1.   Eccentric Connections

Intersecting axially stressed members shall have their gravity axes intersect at one point, if practicable; if not, provision shall be made for bending and shearing stresses due to the eccentricity.

2.   Allowable Shear Rupture

For situations where failure might occur by shear along a plane through the fasteners, or by a combination of shear along a plane through the fasteners plus tension along a perpendicular plane, see Sect. J4.

J6.   FILLERS

In welded construction, any filler $\frac{1}{4}$-in. or more in thickness shall extend beyond the edges of the splice plate and shall be welded to the part on which it is fitted with sufficient weld to transmit the splice plate stress, applied at the surface of the filler as an eccentric load. The welds joining the splice plate to the filler shall be sufficient to transmit the splice plate stress and shall be long enough to avoid overstressing the filler along the toe of the weld. Any filler less than $\frac{1}{4}$-in. thick shall have its edges flush with the edges of the splice plate and the weld size shall be the sum of the size necessary to carry the splice plate stress plus the thickness of the filler plate.

When bolts or rivets carrying computed stress pass through fillers thicker than $\frac{1}{4}$-in., except in slip-critical connections assembled with high-strength bolts, the fillers shall be extended beyond the splice material and the filler extension shall be secured by enough bolts or rivets to distribute the total stress in the member uniformly over the combined section of the member and the filler, or an equivalent number of fasteners shall be included in the connection. Fillers between $\frac{1}{4}$-in. and $\frac{3}{4}$-in. thick, inclusive, need not be extended and developed, provided the allowable shear stress in the bolts is reduced by the factor, $0.4 \, (t - 0.25)$, where t is the total thickness of the fillers, up to $\frac{3}{4}$ in.

---

[*]See Sects. B2 and Commentary Figs. C-J4.1, C-J4.2, C-J4.3 and C-J4.4.

J7. SPLICES

Groove welded splices in plate girders and beams shall develop the full strength of the smaller spliced section. Other types of splices in cross sections of plate girders and beams shall develop the strength required by the stresses at the point of splice.

J8. ALLOWABLE BEARING STRESS

On contact area of milled surfaces and ends of fitted bearing stiffeners; on projected area of pins in reamed, drilled or bored holes:

$$F_p = 0.90F_y* \tag{J8-1}$$

Expansion rollers and rockers, kips per lin. in.:

$$F_p = \left(\frac{F_y - 13}{20}\right) 0.66d \tag{J8-2}$$

where d is the diameter of roller or rocker, in.

J9. COLUMN BASES AND BEARING ON MASONRY AND CONCRETE

Proper provision shall be made to transfer the column loads and moments to the footings and foundations.

In the absence of code regulations the following stresses apply:

On sandstone and limestone .... $F_p = 0.40$ ksi
On brick in cement mortar .... $F_p = 0.25$ ksi
On the full area of a concrete support .... $F_p = \ \ 0.35f'_c$
On less than the full area of a
 concrete support................ $F_p = 0.35f'_c \sqrt{A_2/A_1} \leq 0.70f'_c$

where

f'_c = specified compressive strength of concrete, ksi
A_1 = area of steel concentrically bearing on a concrete support, in.^{2}
A_2 = maximum area of the portion of the supporting surface that is geometrically similar to and concentric with the loaded area, in.^{2}

J10. ANCHOR BOLTS

Anchor bolts shall be designed to provide resistance to all conditions on completed structures of tension and shear at the bases of columns, including the net tensile components of any bending moment which may result from fixation or partial fixation of columns.

---

*When parts in contact have different yield stresses, F_y shall be the smaller value.

CHAPTER K

SPECIAL DESIGN CONSIDERATIONS

This chapter covers member strength design considerations related to concentrated forces, ponding, torsion, and fatigue.

K1. WEBS AND FLANGES UNDER CONCENTRATED FORCES

1. Design Basis

Members with concentrated loads applied normal to *one flange* and symmetric to the web shall have a flange and web proportioned to satisfy the local flange bending, web yielding strength, web crippling and sidesway web buckling criteria of Sects. K1.2, K1.3, K1.4 and K1.5. Members with concentrated loads applied to *both flanges* shall have a web proportioned to satisfy the web yielding, web crippling and column web buckling criteria of Sects. K1.3, K1.4 and K1.6.

Where pairs of stiffeners are provided on opposite sides of the web, at concentrated loads, and extend at least half the depth of the member, Sects. K1.2 and K1.3 need not be checked.

For column webs subject to high shears, see Sect. K1.7; for bearing stiffeners, see Sect. K1.8.

2. Local Flange Bending

A pair of stiffeners shall be provided opposite the tension flange or flange plate of the beam or girder framing into the member when the thickness of the member flange t_f is less than

$$0.4 \sqrt{\frac{P_{bf}}{F_{yc}}} \qquad (K1\text{-}1)$$

where

F_{yc} = column yield stress, ksi

P_{bf} = the computed force delivered by the flange or moment connection plate multiplied by $\frac{5}{3}$, when the computed force is due to live and dead load only, or by $\frac{4}{3}$,* when the computed force is due to live and dead load in conjunction with wind or earthquake forces, kips

If the length of loading measured across the member flange is less than $0.15b$, where b is the member flange width, Equation (K1-1) need not be checked.

---

*Except where other codes may govern. For example, see Section 4(D) "Recommended Lateral Force Requirements and Commentary," Structural Engineers Assoc. of California, 1975.

3. Local Web Yielding

Bearing stiffeners shall be provided if the compressive stress at the web toe of the fillets resulting from concentrated loads exceeds $0.66F_y$.

a. When the force to be resisted is a concentrated load producing tension or compression, applied at a distance from the member end that is greater than the depth of the member,

$$\frac{R}{t_w(N + 5k)} \leq 0.66F_y \tag{K1-2}$$

b. When the force to be resisted is a concentrated load applied at or near the end of the member,

$$\frac{R}{t_w(N + 2.5k)} \leq 0.66F_y \tag{K1-3}$$

where

R = concentrated load or reaction, kips
t_w = thickness of web, in.
N = length of bearing (not less than k for end reactions), in.
k = distance from outer face of flange to web toe of fillet, in.

4. Web Crippling

Bearing stiffeners shall be provided in the webs of members under concentrated loads, when the compressive force exceeds the following limits:

a. When the concentrated load is applied at a distance not less than $d/2$ from the end of the member:

$$R = 67.5t_w^2 \left[1 + 3\left(\frac{N}{d}\right)\left(\frac{t_w}{t_f}\right)^{1.5}\right]\sqrt{F_{yw}\, t_f/t_w} \tag{K1-4}$$

b. When the concentrated load is applied less than a distance $d/2$ from the end of the member:

$$R = 34t_w^2 \left[1 + 3\left(\frac{N}{d}\right)\left(\frac{t_w}{t_f}\right)^{1.5}\right]\sqrt{F_{yw}\, t_f/t_w} \tag{K1-5}$$

where

F_{yw} = specified minimum yield stress of beam web, ksi
d = overall depth of the member, in.
t_f = flange thickness, in.

If stiffeners are provided and extend at least one-half the web depth, Equations (K1-4) and (K1-5) need not be checked.

5. Sidesway Web Buckling

Bearing stiffeners shall be provided in the webs of members with flanges not restrained against relative movement by stiffeners or lateral bracing and subject to concentrated compressive loads, when the compressive force exceeds the following limits:

a. If the loaded flange is restrained against rotation and $(d_c/t_w)/(l/b_f)$ is less than 2.3:

$$R = \frac{6{,}800 t_w^3}{h}\left[1 + 0.4\left(\frac{d_c/t_w}{l/b_f}\right)^3\right] \tag{K1-6}$$

b. If the loaded flange is not restrained against rotation and $(d_c/t_w)/(l/b_f)$ is less than 1.7:

$$R = \frac{6{,}800 t_w^3}{h}\left[0.4\left(\frac{d_c/t_w}{l/b_f}\right)^3\right] \tag{K1-7}$$

where
l = largest laterally unbraced length along either flange at the point of load, in.
b_f = flange width, in.
$d_c = d - 2k$ = web depth clear of fillets, in.

Equations (K1-6) and (K1-7) need not be checked providing $(d_c/t_w)/(l/b_f)$ exceeds 2.3 or 1.7, respectively, or for webs subject to uniformly distributed load.

6. Compression Buckling of the Web

For unstiffened portions of webs of members under concentrated loads to both flanges, a stiffener or a pair of stiffeners shall be provided opposite the compression flange when the web depth clear of fillets d_c is greater than

$$\frac{4100\, t_{wc}^3\sqrt{F_{yc}}}{P_{bf}} \tag{K1-8}$$

where
t_{wc} = thickness of column web, in.

7. Compression Members with Web Panels Subject to High Shear

Members subject to high shear stress in the web should be checked for conformance with Sect. F4.*

8. Stiffener Requirements for Concentrated Loads

Stiffeners shall be placed in pairs at unframed ends or at points of concentrated loads on the interior of beams, girders or columns if required by Sect. K1.2 through K1.6, as applicable.

If required by Sects. K1.2, K1.3 or Equation (K1-9), stiffeners need not extend more than one-half the depth of the web, except as follows:

If stiffeners are required by Sects. K1.4 or K1.6, the stiffeners shall be designed as axially compressed members (columns) in accordance with requirements of Sect. E2 with an effective length equal to 0.75h, a cross section composed of two stiffeners and a strip of the web having a width of $25t_w$ at interior stiffeners and $12t_w$ at the ends of members.

When the load normal to the flange is tensile, the stiffeners shall be welded to

---

*See Commentary Sect. E6.

the loaded flange. When the load normal to the flange is compressive, the stiffeners shall either bear on or be welded to the loaded flange.

When flanges or moment connection plates for end connections of beams and girders are welded to the flange of an I- or H-shape column, a pair of column-web stiffeners having a combined cross-sectional area A_{st} not less than that computed from Equation (K1-9) shall be provided whenever the calculated value of A_{st} is positive.

$$A_{st} = \frac{P_{bf} - F_{yc} t_{wc} (t_b + 5k)}{F_{yst}} \qquad \text{(K1-9)}$$

where

F_{yst} = stiffener yield stress, ksi

k = distance between outer face of column flange and web toe of its fillet, if column is a rolled shape, or equivalent distance if column is a welded shape, in.

t_b = thickness of flange or moment connection plate delivering concentrated force, in.

Stiffeners required by the provisions of Equation (K1-9) and Sects. K1.2 and K1.6 shall comply with the following criteria:

1. The width of each stiffener plus ½ the thickness of the column web shall be not less than ⅓ the width of the flange or moment connection plate delivering the concentrated force.
2. The thickness of stiffeners shall be not less than one-half the thickness of the flange or plate delivering the concentrated load.*
3. The weld joining stiffeners to the column web shall be sized to carry the force in the stiffener caused by unbalanced moments on opposite sides of the column.

K2. PONDING

The roof system shall be investigated by structural analysis to assure adequate strength and stability under ponding conditions, unless the roof surface is provided with sufficient slope toward points of free drainage or adequate individual drains to prevent the accumulation of rainwater.

The roof system shall be considered stable and not requiring further investigation if:

$$C_p + 0.9C_s \leq 0.25 \qquad \text{(K2-1)}$$
$$\text{and } I_d \geq 25 \, (S^4) 10^{-6} \qquad \text{(K2-2)}$$

where

$$C_p = \frac{32 L_s L_p^4}{10^7 I_p}$$

$$C_s = \frac{32 S L_s^4}{10^7 I_s}$$

L_p = column spacing in direction of girder (length of primary members), ft

---

*See Commentary Sect. K1 for comment on width-thickness ratio for stiffeners.

L_s = column spacing perpendicular to direction of girder (length of secondary members), ft

S = spacing of secondary members, ft

I_p = moment of inertia of primary members, in.^{4}

I_s = moment of inertia of secondary members, in.^{4}

I_d = moment of inertia of the steel deck supported on secondary members, in.^{4} per ft

For trusses and steel joists, the moment of inertia I_s shall be decreased 15% when used in the above equation. A steel deck shall be considered a secondary member when it is directly supported by the primary members.

Total bending stress due to dead loads, gravity live loads (if any) and ponding shall not exceed $0.80F_y$ for primary and secondary members. Stresses due to wind or seismic forces need not be included in a ponding analysis.

K3. TORSION

The effects of torsion shall be considered in the design of members and the normal and shearing stresses due to torsion shall be added to those from all other loads, with the resultants not exceeding the allowable values.

K4. FATIGUE

Members and their connections subject to fatigue loading shall be proportioned in accordance with the provisions of Appendix K4.

Few members or connections in conventional buildings need to be designed for fatigue, since most load changes in such structures occur only a small number of times or produce only minor stress fluctuations. The occurrence of full design wind or earthquake loads is too infrequent to warrant consideration in fatigue design. However, crane runways and supporting structures for machinery and equipment are often subject to fatigue loading conditions.

CHAPTER   L

SERVICEABILITY DESIGN CONSIDERATIONS

This chapter provides design guidance for serviceability considerations not covered elsewhere. Serviceability is a state in which the function of a building, its appearance, maintainability, durability and comfort of its occupants are preserved under normal usage.

Limiting values of structural behavior to ensure serviceability (e.g., maximum deflections, accelerations, etc.) shall be chosen with due regard to the intended function of the structure.

L1.   CAMBER

If any special camber requirements are necessary to bring a loaded member into proper relation with the work of other trades, the requirements shall be set forth in the design documents.

Trusses of 80 ft or greater span generally shall be cambered for approximately the dead-load deflection. Crane girders of 75 ft or greater span generally shall be cambered for approximately the dead-load deflection plus ½ the live-load deflection.

Beams and trusses detailed without specified camber shall be fabricated so that after erection any camber due to rolling or shop assembly shall be upward. If camber involves the erection of any member under a preload, this shall be noted in the design documents.

L2.   EXPANSION AND CONTRACTION

Provision shall be made for expansion and contraction appropriate to the service conditions of the structure.

L3.   DEFLECTION, VIBRATION AND DRIFT

1.   Deflection

Beams and girders supporting floors and roofs shall be proportioned with due regard to the deflection produced by the design loads. Beams and girders supporting plastered ceilings shall be so proportioned that the maximum live-load deflection does not exceed ¹⁄₃₆₀ of the span.

2.   Vibration

Beams and girders supporting large open floor areas free of partitions or other sources of damping shall be designed with due regard for vibration.

L4. CONNECTION SLIP

For the design of slip-resistant connections see Sect. J3.

L5. CORROSION

When appropriate, structural components shall be designed to tolerate corrosion or shall be protected against corrosion that impairs the strength or serviceability of the structure.

Where beams are exposed they shall be sealed against corrosion of interior surfaces or spaced sufficiently far apart to permit cleaning and painting.

CHAPTER M

FABRICATION, ERECTION AND QUALITY CONTROL

M1. SHOP DRAWINGS

Shop drawings giving complete information necessary for the fabrication of the component parts of the structure, including the location, type and size of all welds, bolts and rivets, shall be prepared in advance of the actual fabrication. These drawings shall clearly distinguish between shop and field welds and bolts and shall clearly identify type of high-strength bolted connection (snug-tight or fully-tightened bearing, or slip-critical).

Shop drawings shall be made in conformity with the best practice and with due regard to speed and economy in fabrication and erection.

M2. FABRICATION

1. Cambering, Curving and Straightening

Local application of heat or mechanical means are permitted to introduce or correct camber, curvature and straightness. The temperature of heated areas, as measured by approved methods, shall not exceed 1050°F for A852 steel, 1100°F for A514 steel nor 1200°F for other steels. The same limits apply for equivalent grades of A709 steels.

2. Thermal Cutting

Thermally cut free edges which will be subject to substantial tensile stress shall be free of gouges greater than $\frac{3}{16}$-in. Gouges greater than $\frac{3}{16}$-in. deep and sharp notches shall be removed by grinding or repaired by welding. Thermally cut edges which are to have weld deposited upon them, shall be reasonably free of notches or gouges.

All reentrant corners shall be shaped to a smooth transition. If specific contour is required, it must be shown on the contract documents.

Beam copes and weld access holes shall meet the geometrical requirements of Sect. J1.8. Beam copes and weld access holes in ASTM A6 Group 4 and 5 shapes and welded built-up shapes with material thickness greater than 2 in. shall be preheated to a temperature of not less than 150°F prior to thermal cutting.

3. Planing of Edges

Planing or finishing of sheared or thermally cut edges of plates or shapes will not be required unless specifically called for in the design documents or included in a stipulated edge preparation for welding.

4. Welded Construction

The technique of welding, the workmanship, appearance and quality of welds and the methods used in correcting nonconforming work shall be in accordance with "Sect. 3—Workmanship" and "Sect. 4—Technique" of the AWS *Structural Welding Code—Steel,* D1.1.

5. High-strength Bolted Construction—Assembly

All parts of bolted members shall be pinned or bolted and held together rigidly while assembling. Use of a drift pin in bolt holes during assembling shall not distort the metal or enlarge the holes. Poor matching of holes shall be cause for rejection.

If the thickness of the material is not greater than the nominal diameter of the bolt plus $\frac{1}{8}$-in., the holes may be punched. If the thickness of the material is greater than the nominal diameter of the bolt plus $\frac{1}{8}$-in., the holes shall be either drilled or sub-punched and reamed. The die for all sub-punched holes and the drill for all sub-drilled holes shall be at least $\frac{1}{16}$-in. smaller than the nominal diameter of the bolt. Holes in A514 steel plates over $\frac{1}{2}$-in. thick shall be drilled.

Surfaces of high-strength-bolted parts in contact with the bolt head and nut shall not have a slope of more than 1:20 with respect to a plane normal to the bolt axis. Where the surface of a high-strength-bolted part has a slope of more than 1:20, a beveled washer shall be used to compensate for the lack of parallelism. High-strength-bolted parts shall fit solidly together when assembled and shall not be separated by gaskets or any other interposed compressible materials.

The orientation of fully inserted finger shims, with a total thickness of not more than $\frac{1}{4}$-in. within a joint, is independent of the direction of application of the load.

When assembled, all joint surfaces, including surfaces adjacent to the bolt head and nut, shall be free of scale, except tight mill scale and shall be free of dirt or other foreign material. Burrs that would prevent solid seating of the connected parts in the snug-tight condition shall be removed. Contact surfaces within slip-critical connections shall be free of oil, paint, lacquer or other coatings, except as listed in Table 3 of the RCSC *Specification for Structural Joints Using ASTM A325 or A490 Bolts.*

The use of high-strength bolts shall conform to the requirements of the RCSC *Specification for Structural Joints Using ASTM A325 or A490 Bolts.*

6. Compression Joints

Compression joints which depend on contact bearing as part of the splice capacity shall have the bearing surfaces of individual fabricated pieces prepared by milling, sawing or other suitable means.

7. Dimensional Tolerances

Dimensional tolerances shall be as permitted in the *Code of Standard Practice* of the American Institute of Steel Construction, Inc.

8. Finishing of Column Bases

Column bases and base plates shall be finished in accordance with the following requirements:

    a. Rolled steel bearing plates 2 in. or less in thickness are permitted without milling,* provided a satisfactory contact bearing is obtained; rolled steel bearing plates over 2 in. but not over 4 in. in thickness may be straightened by pressing, or if presses are not available, by milling for all bearing surfaces (except as noted in subparagraphs c. and d. of this section), to obtain a satisfactory contact bearing; rolled steel bearing plates over 4 in. thick shall be milled for all bearing surfaces (except as noted in subparagraphs c. and d. of this section).

    b. Column bases other than rolled steel bearing plates shall be milled for all bearing surfaces (except as noted in subparagraphs c. and d. of this section).

    c. The bottom surfaces of bearing plates and column bases which are grouted to insure full bearing contact on foundations need not be milled.

    d. The top surfaces of base plates with columns full-penetration welded need not be pressed or milled.

M3. SHOP PAINTING

1. General Requirements

Shop painting and surface preparation shall be in accordance with the provisions of the *Code of Standard Practice* of the American Institute of Steel Construction, Inc.

Unless otherwise specified, steelwork which will be concealed by interior building finish or will be in contact with concrete need not be painted. Unless specifically excluded, all other steelwork shall be given one coat of shop paint.

2. Inaccessible Surfaces

Except for contact surfaces, surfaces inaccessible after shop assembly shall be cleaned and painted prior to assembly, if required by the design documents.

3. Contact Surfaces

Paint is permitted unconditionally in bearing-type connections. For slip-critical connections, the faying surface requirements shall be in accordance with the RCSC *Specification for Structural Joints Using ASTM A325 or A490 Bolts,* paragraph 3.(b).

---

*See Commentary Sect. J8.

4. Finished Surfaces

Machine-finished surfaces shall be protected against corrosion by a rust-inhibiting coating that can be removed prior to erection, or which has characteristics that make removal prior to erection unnecessary.

5. Surfaces Adjacent to Field Welds

Unless otherwise specified in the design documents, surfaces within 2 in. of any field weld location shall be free of materials that would prevent proper welding or produce toxic fumes during welding.

M4. ERECTION

1. Alignment of Column Bases

Column bases shall be set level and to correct elevation with full bearing on concrete or masonry.

2. Bracing

The frame of steel skeleton buildings shall be carried up true and plumb within the limits defined in the *Code of Standard Practice* of the American Institute of Steel Construction. Temporary bracing shall be provided, in accordance with the requirements of the *Code of Standard Practice*, wherever necessary to take care of all loads to which the structure may be subjected, including equipment and the operation of same. Such bracing shall be left in place as long as may be required for safety.

Wherever piles of material erection equipment or other loads are supported during erection, proper provision shall be made to take care of stresses resulting from such loads.

3. Alignment

No permanent bolting or welding shall be performed until as much of the structure as will be stiffened thereby has been properly aligned.

4. Fit of Column Compression Joints

Lack of contact bearing not exceeding a gap of $\frac{1}{16}$-in., regardless of the type of connection used (partial-penetration, groove-welded or bolted), shall be acceptable. If the gap exceeds $\frac{1}{16}$-in., but is less than $\frac{1}{4}$-in., and if an engineering investigation shows sufficient contact area does not exist, the gap shall be packed with non-tapered steel shims. Shims may be of mild steel, regardless of the grade of the main material.

5. Field Welding

Shop paint on surfaces adjacent to welds shall be wire-brushed to reduce paint film to a minimum.

6. Field Painting

Responsibility for touch-up painting, cleaning and field-painting shall be allocated in accordance with accepted local practices, and this allocation shall be set forth explicitly in the design documents.

7. Field Connections

As erection progresses, the work shall be securely bolted or welded to take care of all dead load, wind and erection stresses.

M5. QUALITY CONTROL

The fabricator shall provide quality control procedures to the extent that he deems necessary to assure that all work is performed in accordance with this Specification. In addition to the fabricator's quality control procedures, material and workmanship at all times may be subject to inspection by qualified inspectors representing the purchaser. If such inspection by representatives of the purchaser will be required, it shall be so stated in design documents.

1. Cooperation

As far as possible, all inspection by representatives of the purchaser shall be made at the fabricator's plant. The fabricator shall cooperate with the inspector, permitting access for inspection to all places where work is being done. The purchaser's inspector shall schedule his work for minimum interruption to the work of the fabricator.

2. Rejections

Material or workmanship not in reasonable conformance with the provisions of this Specification may be rejected at any time during the progress of the work. The fabricator shall receive copies of all reports furnished to the purchaser by the inspection agency.

3. Inspection of Welding

The inspection of welding shall be performed in accordance with the provisions of Sect. 6 of the AWS *Structural Welding Code—Steel,* D1.1.

When nondestructive testing is required, the process, extent and standards of acceptance shall be defined clearly in the design documents.

4. Inspection of Slip-critical, High-strength Bolted Connections

The inspection of slip-critical, high-strength bolted connections shall be in accordance with the provisions of the RCSC *Allowable Stress Design Specification for Structural Joints Using ASTM A325 or A490 Bolts.*

5. Identification of Steel

The fabricator shall be able to demonstrate by a written procedure and by actual practice a method of material application and identification, visible at least through the "fit-up" operation, of the main structural elements of a shipping piece.

The identification method shall be capable of verifying proper material application as it relates to:

1. Material specification designation
2. Heat number, if required
3. Material test reports for special requirements

CHAPTER N

PLASTIC DESIGN

N1. SCOPE

Subject to the limitations contained herein, simple and continuous beams, braced and unbraced planar rigid frames, and similar parts of structures rigidly constructed so as to be continuous over at least one interior support,* are permitted to be proportioned on the basis of plastic design, i.e., on the basis of their maximum strength. This strength, as determined by rational analysis, shall be not less than that required to support a factored load equal to 1.7 times the given live load and dead load, or 1.3 times these loads acting in conjunction with 1.3 times any specified wind or earthquake forces.

Rigid frames shall satisfy the requirements for Type 1 construction in the plane of the frame, as provided in Sect. A2.2. This does not preclude the use of some simple connections, provided provisions of Sect. N3 are satisfied. Type 2 construction is permitted for members between rigid frames. Connections joining a portion of a structure designed on the basis of plastic behavior with a portion not so designed need be no more rigid than ordinary seat-and-top-angle or ordinary web connections.

Where plastic design is used as the basis for proportioning continuous beams and structural frames, the provisions relating to allowable stress are waived. Except as modified by these rules, however, all other pertinent provisions of Chapters A through M shall govern.

It is not recommended that crane runways be designed continuous over interior vertical supports on the basis of maximum strength. However, rigid frame bents supporting crane runways may be considered as coming within the scope of the rules.

N2. STRUCTURAL STEEL

Structural steel shall conform to one of the following specifications:

Structural Steel, ASTM A36
High-strength Low-alloy Structural Steel, ASTM A242
High-strength Low-alloy Structural Manganese Vanadium Steel, ASTM A441
Structural Steel with 42 ksi Minimum Yield Point, ASTM A529
High-strength Low-alloy Columbium-Vanadium Steels of Structural Quality, ASTM A572
High-strength Low-alloy Structural Steel with 50 ksi Minimum Yield Point to 4-in. Thick, ASTM A588

---

*As used here, "interior support" includes a rigid-frame knee formed by the junction of a column and a sloping or horizontal beam or girder.

N3. BASIS FOR MAXIMUM STRENGTH DETERMINATION

For one- or two-story frames, the maximum strength is permitted to be determined by a routine plastic analysis procedure and ignore the frame instability effect ($P\Delta$). For braced multi-story frames, provisions shall be made to include the frame instability effect in the design of bracing system and frame members. For unbraced multi-story frames, the frame instability effect shall be included directly in the calculations for maximum strength.

1. Stability of Braced Frames

The vertical bracing system for a plastically designed braced multi-story frame shall be adequate, as determined by an analysis, to:

1. Prevent buckling of the structure under factored gravity loads
2. Maintain the lateral stability of the structure, including the overturning effects of drift, under factored gravity plus factored horizontal loads

It is permitted to consider that the vertical bracing system functions together with in-plane shear-resisting exterior and interior walls, floor slabs and roof decks, if these walls, slabs and decks are secured to the structural frames. The columns, girders, beams and diagonal members, when used as the vertical bracing system, could be considered to comprise a vertical-cantilever, simply connected truss in the analyses for frame buckling and lateral stability. Axial deformation of all members in the vertical bracing system shall be included in the lateral stability analysis. The axial force in these members caused by factored gravity plus factored horizontal loads shall not exceed $0.85P_y$, where P_y is the product of yield stress times the profile area of the member.

Girders and beams included in the vertical bracing system of a braced multi-story frame shall be proportioned for axial force and moment caused by the concurrent factored horizontal and gravity loads, in accordance with Equation (N4-2), with P_{cr} taken as the maximum axial strength of the beam, based on the actual slenderness ratio between braced points in the plane of bending.

2. Stability of Unbraced Frames

The strength of an unbraced multi-story frame shall be determined by an analysis which includes the effect of frame instability and column axial deformation. Such a frame shall be designed to be stable under (1) factored gravity loads and (2) factored gravity loads plus factored horizontal loads. The axial force in the columns at factored load levels shall not exceed $0.75P_y$.

N4. COLUMNS

In the plane of bending of columns which would develop a plastic hinge at ultimate loading, the slenderness ratio l/r shall not exceed C_c, defined in Sect. E2.

The maximum strength of an axially loaded compression member shall be taken as

$$P_{cr} = 1.7F_a A \qquad (N4-1)$$

where A is the gross area of the member and F_a, as defined by Equation (E2-1), is based upon the applicable slenderness ratio.

Members subject to combined axial load and bending moment shall be proportioned to satisfy the following interaction formulas:

$$\frac{P}{P_{cr}} + \frac{C_m M}{\left(1 - \dfrac{P}{P_e}\right) M_m} \le 1.0 \tag{N4-2}$$

$$\frac{P}{P_y} + \frac{M}{1.18 M_p} \le 1.0; \quad M \le M_p \tag{N4-3}$$

in which

M = maximum factored moment, kip-ft
P = factored axial load, kips
P_e = Euler buckling load, kips
    = $(23/12)F'_e A$, where F'_e is as defined in Sect. H1.
C_m = coefficient defined in Sect. H1.
M_m = maximum moment that can be resisted by the member in the absence of axial load, kip-ft
M_p = plastic moment, kip-ft
    = $F_y Z$
Z = plastic section modulus, in.

For columns braced in the weak direction:

$$M_m = M_{px} \tag{N4-4}$$

For columns unbraced in the weak direction:

$$M_m = \left[1.07 - \frac{(l/r_y)\,\sqrt{F_y}}{3160}\right] M_{px} \le M_{px} \tag{N4-5}$$

N5. SHEAR

Unless reinforced by diagonal stiffeners or a doubler plate, the webs of columns, beams and girders, including areas within the boundaries of the connections, shall be so proportioned that

$$V \le 0.55 F_y t_w d \tag{N5-1}$$

where

V = shear that would be produced by the required factored loading, kips
d = depth of the member, in.
t_w = web thickness, in.

N6. WEB CRIPPLING

Web stiffeners are required on a member at a point of load application where a plastic hinge would form.

At points on a member where the concentrated load delivered by the flanges of a member framing into it would produce web crippling opposite the compression flange or high-tensile stress in the connection of the tension flange, web stiffeners are required in accordance with the provisions of Sect. K1.

N7. MINIMUM THICKNESS (WIDTH-THICKNESS RATIOS)

The width-thickness ratio for flanges of rolled W, M or S shapes and similar built-up, single-web shapes subjected to compression involving hinge rotation under ultimate loading shall not exceed the following values:

F_y	$b_f/2t_f$
36	8.5
42	8.0
45	7.4
50	7.0
55	6.6
60	6.3
65	6.0

It is permitted to take the thickness of sloping flanges as their average thickness.

The width-thickness ratio of similarly compressed flange plates in box sections and cover plates shall not exceed $190/\sqrt{F_y}$. For this purpose, the width of a cover plate shall be taken as the distance between longitudinal lines of connecting rivets, high-strength bolts or welds.

The depth-thickness ratio of webs of members subject to plastic bending shall not exceed the value given by Equation (N7-1) or (N7-2), as applicable.

$$\frac{d}{t} = \frac{412}{\sqrt{F_y}}\left(1 - 1.4\,\frac{P}{P_y}\right) \quad \text{when} \quad \frac{P}{P_y} \le 0.27 \qquad \text{(N7-1)}$$

$$\frac{d}{t} = \frac{257}{\sqrt{F_y}} \quad \text{when} \quad \frac{P}{P_y} > 0.27 \qquad \text{(N7-2)}$$

N8. CONNECTIONS

All connections, the rigidity of which is essential to the continuity assumed as the basis of the analysis, shall be capable of resisting the moments, shears and axial loads to which they would be subjected by the full factored loading, or any probable partial distribution thereof.

Corner connections (haunches) that are tapered or curved for architectural reasons shall be so proportioned that the full plastic bending strength of the section adjacent to the connection can be developed, if required.

Stiffeners shall be used, as required, to preserve the flange continuity of interrupted members at their junction with other members in a continuous frame. Such stiffeners shall be placed in pairs on opposite sides of the web of the member which extends continuously through the joint.

High-strength bolts, A307 bolts, rivets and welds shall be proportioned to resist the forces produced at factored load, using stresses equal to 1.7 times those given in Chapters A through M. In general, groove welds are preferable to fillet welds, but their use is not mandatory.

High-strength bolts are permitted in joints having painted contact surfaces when these joints are of such size that the slip required to produce bearing would not interfere with the formation, at factored loading, of the plastic hinges assumed in the design.

N9. LATERAL BRACING

Members shall be braced adequately to resist lateral and torsional displacements at the plastic hinge locations associated with the failure mechanism. The laterally unsupported distance l_{cr} from such braced hinge locations to similarly braced adjacent points on the member or frame shall not exceed the value determined from Equation (N9-1) or (N9-2), as applicable.

$$\frac{l_{cr}}{r_y} = \frac{1375}{F_y} + 25 \quad \text{when} \quad +1.0 > \frac{M}{M_p} > -0.5 \qquad \text{(N9-1)}$$

$$\frac{l_{cr}}{r_y} = \frac{1375}{F_y} \quad \text{when} \quad -0.5 \geq \frac{M}{M_p} > -1.0 \qquad \text{(N9-2)}$$

where

r_y = radius of gyration of the member about its weak axis, in.
M = lesser of the moments at the ends of the unbraced segment, kip-ft
M/M_p = end moment ratio, positive when the segment is bent in reverse curvature and negative when bent in single curvature.

The foregoing provisions need not apply in the region of the last hinge to form in the failure mechanism assumed as the basis for proportioning a given member, nor in members oriented with their weak axis normal to the plane of bending. However, in the region of the last hinge to form, and in regions not adjacent to a plastic hinge, the maximum distance between points of lateral support shall be such as to satisfy the requirements of Equations (F1-5), (F1-6), or (F1-7), as well as Equations (H1-1) and (H1-2). For this case, the values of f_a and f_b shall be computed from the moment and axial force at factored loading, divided by the applicable load factor.

Members built into a masonry wall and having their web perpendicular to this wall can be assumed to be laterally supported with respect to their weak axis of bending.

N10. FABRICATION

The provisions of Chapters A through M with respect to workmanship shall govern the fabrication of structures, or portions of structures, designed on the basis of maximum strength, subject to the following limitations:

1. The use of sheared edges shall be avoided in locations subject to plastic hinge rotation at factored loading. If used, they shall be finished smooth by grinding, chipping or planing.
2. In locations subject to plastic hinge rotation at factored loading, holes for rivets or bolts in the tension area shall be sub-punched and reamed or drilled full size.

APPENDIX B

DESIGN REQUIREMENTS

B5. LOCAL BUCKLING

2. Slender Compression Elements

Axially loaded members and flexural members containing elements subject to compression which have a width-thickness ratio in excess of the applicable noncompact value, as stipulated in Sect. B5.1 shall be proportioned according to this Appendix.

a. Unstiffened Compression Elements

The allowable stress of unstiffened compression elements whose width-thickness ratio exceeds the applicable noncompact value as stipulated in Sect. B5.1 shall be subject to a reduction factor Q_s. The value of Q_s shall be determined by Equations (A-B5-1) through (A-B5-6), as applicable, where b is the width of the unstiffened element as defined in Sect. B5.1. When such elements comprise the compression flange of a flexural member, the maximum allowable bending stress shall not exceed $0.60\,F_yQ_s$ nor the applicable value as provided in Sect. F1.3. The allowable stress of axially loaded compression members shall be modified by the appropriate reduction factor Q, as provided in paragraph c.

For single angles:

When $76.0/\sqrt{F_y} < b/t < 155/\sqrt{F_y}$:

$$Q_s = 1.340 - 0.00447(b/t)\sqrt{F_y} \tag{A-B5-1}$$

When $b/t \geq 155/\sqrt{F_y}$:

$$Q_s = 15,500/[F_y(b/t)^2] \tag{A-B5-2}$$

For angles or plates projecting from columns or other compression members, and for projecting elements of compression flanges of beams and girders:

When $95.0/\sqrt{F_y/k_c} < b/t < 195/\sqrt{F_y/k_c}$

$$Q_s = 1.293 - 0.00309(b/t)\sqrt{F_y/k_c} \tag{A-B5-3}$$

When $b/t > 195/\sqrt{F_y/k_c}$

$$Q_s = 26,200\,k_c/[F_y(b/t)^2] \tag{A-B5-4}$$

For stems of tees:

When $127/\sqrt{F_y} < b/t < 176/\sqrt{F_y}$:

$$Q_s = 1.908 - 0.00715(b/t)\sqrt{F_y} \tag{A-B5-5}$$

When $b/t \geq 176/\sqrt{F_y}$:

$$Q_s = 20,000/[F_y(b/t)^2] \tag{A-B5-6}$$

AMERICAN INSTITUTE OF STEEL CONSTRUCTION

where
   b = width of unstiffened compression element as defined in Sect. B5.1
   t = thickness of unstiffened element, in.
   F_y = specified minimum yield stress, ksi

$$k_c = \frac{4.05}{(h/t)^{.46}} \text{ if } h/t > 70, \text{ otherwise } k_c = 1.0.$$

Unstiffened elements of tees whose proportions exceed the limits of Sect. B5.1 shall conform to the limits given in Table A-B5.1.

b. **Stiffened Compression Elements**

When the width-thickness ratio of uniformly compressed stiffened elements (except perforated cover plates) exceeds the noncompact limit stipulated in Sect. B5.1, a reduced effective width b_e shall be used in computing the design properties of the section containing the element, except that the ratio b_e/t need not be taken as less than the applicable value permitted in Sect. B5.1.

For the flanges of square and rectangular sections of uniform thickness:

$$b_e = \frac{253t}{\sqrt{f}} \left[1 - \frac{50.3}{(b/t)\sqrt{f}} \right] \leq b \qquad \text{(A-B5-7)}$$

For other uniformly compressed elements:

$$b_e = \frac{253t}{\sqrt{f}} \left[1 - \frac{44.3}{(b/t)\sqrt{f}} \right] \leq b \qquad \text{(A-B5-8)}$$

where

   b = actual width of a stiffened compression element, as defined in Sect. B5.1, in.

   b_e = reduced width, in.

   t = element thickness, in.

   f = computed compressive stress (axial plus bending stresses) in the stiffened elements, based on the design properties as specified in Appendix B5.2, ksi. If unstiffened elements are included in the

Table A-B5.1
Limiting Proportions for Channels and Tees

Shape	Ratio of full flange width to profile depth	Ratio of flange thickness to web or stem thickness
Built-up or rolled channels	≤0.25	≤3.0
	≤0.50	≤2.0
Built-up tees	≥0.50	≥1.25
Rolled tees	≥0.50	≥1.10

total cross section, f for the stiffened element must be such that the maximum compressive stress in the unstiffened element does not exceed F_aQ_s or F_bQ_s, as applicable.

When the allowable stresses are increased due to wind or seismic loading in accordance with the provisions of Sect. A5.2, the effective width b_e shall be determined on the basis of 0.75 times the stress caused by wind or seismic loading acting alone or in combination with the design dead and live loading.

For axially loaded circular sections:

Members with diameter-to-thickness ratios D/t greater than $3,300/F_y$, but having a diameter-to-thickness ratio of less than $13,000/F_y$, shall not exceed the smaller value determined by Sect. E2 nor

$$F_a = \frac{662}{D/t} + 0.40F_y \qquad \text{(A-B5-9)}$$

where

D = outside diameter, in.

t = wall thickness, in.

c. **Design Properties**

Properties of sections shall be determined using the full cross section, except as follows:

In computing the moment of inertia and section modulus of flexural members, the effective width of uniformly compressed stiffened elements, as determined in Appendix B5.2b, shall be used in determining effective cross-sectional properties.

For stiffened elements of the cross section

$$Q_a = \frac{\text{effective area}}{\text{actual area}} \qquad \text{(A-B5-10)}$$

For unstiffened elements of the cross section, Q_s is determined from Appendix B5.2a.

For axially loaded compression members the gross cross-sectional area and the radius of gyration r shall be computed on the basis of the actual cross section.

The allowable stress for axially loaded compression members containing unstiffened or stiffened elements shall not exceed

$$F_a = \frac{Q\left[1 - \frac{(Kl/r)^2}{2C_c'^2}\right]F_y}{\frac{5}{3} + \frac{3(Kl/r)}{8C_c'} - \frac{(Kl/r)^3}{8C_c'^3}} \qquad \text{(A-B5-11)}$$

when Kl/r is less than C_c', where

$$C_c' = \sqrt{\frac{2\pi^2 E}{Q F_y}}$$

and

$$Q = Q_s Q_a$$

a. Cross sections composed entirely of unstiffened elements,
$$Q = Q_s \text{ i.e. } (Q_a = 1.0)$$

b. Cross sections composed entirely of stiffened elements,
$$Q = Q_a \text{ i.e. } (Q_s = 1.0)$$

c. Cross sections composed of both stiffened and unstiffened elements,
$$Q = Q_s Q_a$$

When Kl/r exceeds C_c':

$$F_a = \frac{12\pi^2 E}{23(Kl/r)^2} \tag{A-B5-12}$$

d. Combined Axial and Flexural Stress

In applying the provisions of Chapter H to members subject to combined axial and flexural stress and containing stiffened elements whose width-thickness ratio exceeds the applicable noncompact limit given in Sect. B5.1, the stresses F_a, f_{bx} and f_{by} shall be calculated on the basis of the section properties as provided in Appendix B5.2c, as applicable. The allowable bending stress F_b for members containing unstiffened elements whose width-thickness ratio exceeds the noncompact limit given in Sect. B5.1 shall be the smaller value, $0.60F_y Q_s$ or that provided in Sect. F1.3. The term $f_a/0.60F_y$ in Equations (H1-2) and (A-F7-13) shall be replaced by $f_a/0.60F_y Q$.

APPENDIX F

BEAMS AND OTHER FLEXURAL MEMBERS

F7. WEB-TAPERED MEMBERS

The design of tapered members meeting the requirements of this section shall be governed by the provisions of Chapter F, except as modified by this Appendix.

1. General Requirements

In order to qualify under this Specification, a tapered member must meet the following requirements:

    a. It shall possess at least one axis of symmetry which shall be perpendicular to the plane of bending if moments are present.

    b. The flanges shall be of equal and constant area.

    c. The depth shall vary linearly as

$$d = d_o \left(1 + \gamma \frac{z}{L}\right) \tag{A-F7-1}$$

where

d_o = depth at smaller end of member, in.
d_L = depth at larger end of member, in.
γ = $(d_L - d_o)/d_o \leq$ the smaller of $0.268(L/d_o)$ or 6.0
z = distance from the smaller end of member, in.
L = unbraced length of member measured between the center of gravity of the bracing members, in.

2. Allowable Tensile Stress

The allowable tensile stress of tapered tension members shall be determined in accordance with Sect. D1.

3. Allowable Compressive Stress

On the gross section of axially loaded tapered compression members, the allowable compressive stress, in kips per sq. in., shall not exceed the following:

When the effective slenderness ratio S is less than C_c:

$$F_{a\gamma} = \frac{\left(1.0 - \dfrac{S^2}{2C_c{}^2}\right) F_y}{\dfrac{5}{3} + \dfrac{3S}{8C_c} - \dfrac{S^3}{8C_c^3}} \tag{A-F7-2}$$

When the effective slenderness ratio S exceeds C_c:

$$F_{a\gamma} = \frac{12\pi^2 E}{23S^2}$$ (A-F7-3)

where

S = Kl/r_{oy} for weak axis bending and $K_\gamma l/r_{ox}$ for strong axis bending
K = effective length factor for a prismatic member
K_γ = effective length factor for a tapered member as determined by an analysis*
l = actual unbraced length of member, in.
r_{ox} = strong axis radius of gyration at the smaller end of a tapered member, in.
r_{oy} = weak axis radius of gyration at the smaller end of a tapered member, in.

4. Allowable Flexural Stress**

Tension and compression stresses on extreme fibers of tapered flexural members, in kips per sq. in., shall not exceed the following values:

$$F_{b\gamma} = \frac{2}{3}\left[1.0 - \frac{F_y}{6B\sqrt{F_{s\gamma}^2 + F_{w\gamma}^2}}\right] F_y \le 0.60F_y$$ (A-F7-4)

unless $F_{b\gamma} \le F_y/3$, in which case

$$F_{b\gamma} = B\sqrt{F_{s\gamma}^2 + F_{w\gamma}^2}$$ (A-F7-5)

In the above equations,

$$F_{s\gamma} = \frac{12 \times 10^3}{h_s L d_o/A_f}$$ (A-F7-6)

$$F_{w\gamma} = \frac{170 \times 10^3}{(h_w L/r_{To})^2}$$ (A-F7-7)

where
h_s = factor equal to $1.0 + 0.0230\gamma\sqrt{Ld_o/A_f}$
h_w = factor equal to $1.0 + 0.00385\gamma\sqrt{L/r_{To}}$
r_{To} = radius of gyration of a section at the smaller end, considering only the compression flange plus ⅓ of the compression web area, taken about an axis in the plane of the web, in.
A_f = area of the compression flange, in.^{2}

and where B is determined as follows:

a. When the maximum moment M_2 in three adjacent segments of approximately equal unbraced length is located within the central segment and

---

*See Commentary Appendix F7.3.
**See Commentary Appendix F7.4.

M_1 is the larger moment at one end of the three-segment portion of a member:*

$$B = 1.0 + 0.37\left(1.0 + \frac{M_1}{M_2}\right) + 0.50\gamma\left(1.0 + \frac{M_1}{M_2}\right) \geq 1.0 \qquad \text{(A-F7-8)}$$

b. When the largest computed bending stress f_{b2} occurs at the larger end of two adjacent segments of approximately equal unbraced lengths and f_{b1} is the computed bending stress at the smaller end of the two-segment portion of a member:*

$$B = 1.0 + 0.58\left(1.0 + \frac{f_{b1}}{f_{b2}}\right) - 0.70\gamma\left(1.0 + \frac{f_{b1}}{f_{b2}}\right) \geq 1.0 \qquad \text{(A-F7-9)}$$

c. When the largest computed bending stress f_{b2} occurs at the smaller end of two adjacent segments of approximately equal unbraced length and f_{b1} is the computed bending stress at the larger end of the two-segment portion of a member:**

$$B = 1.0 + 0.55\left(1.0 + \frac{f_{b1}}{f_{b2}}\right) + 2.20\gamma\left(1.0 + \frac{f_{b1}}{f_{b2}}\right) \geq 1.0 \qquad \text{(A-F7-10)}$$

In the foregoing, $\gamma = (d_L - d_o)/d_o$ is calculated for the unbraced length containing the maximum computed bending stress.

d. When the computed bending stress at the smaller end of a tapered member or segment thereof is equal to zero:

$$B = \frac{1.75}{1.0 + 0.25\sqrt{\gamma}} \qquad \text{(A-F7-11)}$$

where $\gamma = (d_L - d_o)/d_o$, calculated for the unbraced length adjacent to the point of zero bending stress.

5. Allowable Shear

The allowable shear stress of tapered flexural members shall be in accordance with Sect. F4.

6. Combined Flexure and Axial Force

Tapered members and unbraced segments thereof subjected to both axial compression and bending stresses shall be proportioned to satisfy the following requirement:

$$\left(\frac{f_{ao}}{F_{a\gamma}}\right) + \frac{C_m'}{\left(1 - \frac{f_{ao}}{F_{e\gamma}'}\right)}\left(\frac{f_{b1}}{F_{b\gamma}}\right) \leq 1.0 \qquad \text{(A-F7-12)}$$

---

* M_1/M_2 is considered as negative when producing single curvature. In the rare case where M_1/M_2 is positive, it is recommended it be taken as zero.
** f_{b1}/f_{b2} is considered as negative when producing single curvature. If a point of contraflexure occurs in one of two adjacent unbraced segments, f_{b1}/f_{b2} is considered as positive. The ratio $f_{b1}/f_{b2} \neq 0$.

and

$$\frac{f_a}{0.60F_y} + \frac{f_b}{F_{b\gamma}} \leq 1.0 \qquad \text{(A-F7-13)}$$

When $f_{ao}/F_{a\gamma} \leq 0.15$, Equation (A-F7-14) is permitted in lieu of Equations (A-F7-12) and (A-F7-13).

$$\left(\frac{f_{ao}}{F_{a\gamma}}\right) + \left(\frac{f_{bl}}{F_{b\gamma}}\right) \leq 1.0 \qquad \text{(A-F7-14)}$$

where

$F_{a\gamma}$ = axial compressive stress permitted in the absence of bending moment, ksi

$F_{b\gamma}$ = bending stress permitted in the absence of axial force, ksi

$F_{e\gamma}'$ = Euler stress divided by factor of safety, ksi, equal to

$$\frac{12\pi^2 E}{23(K_\gamma l_b/r_{bo})^2}$$

where l_b is the actual unbraced length in the plane of bending and r_{bo} is the corresponding radius of gyration at its smaller end

f_{ao} = computed axial stress at the smaller end of the member or unbraced segment thereof, as applicable, ksi

f_{bl} = computed bending stress at the larger end of the member or unbraced segment thereof, as applicable, ksi

C_m' = coefficient applied to bending term in interaction equation

$$= 1.0 + 0.1 \left(\frac{f_{ao}}{F_{e\gamma}'}\right) + 0.3 \left(\frac{f_{ao}}{F_{e\gamma}'}\right)^2$$

when the member is subjected to end moments which cause single curvature bending and approximately equal computed bending stresses at the ends

$$= 1.0 - 0.9 \left(\frac{f_{ao}}{F_{e\gamma}'}\right) + 0.6 \left(\frac{f_{ao}}{F_{e\gamma}'}\right)^2$$

when the computed bending stress at the smaller end of the unbraced length is equal to zero.

When $Kl/r \geq C_c$ and combined stresses are checked incrementally along the length, f_{ao} may be replaced by f_a, and f_{bl} may be replaced by f_b, in Equations (A-F7-12) and (A-F7-14).

APPENDIX K

STRENGTH DESIGN CONSIDERATIONS

K4. FATIGUE

Members and connections subject to fatigue loading shall be proportioned in accordance with the provisions of this Appendix.

Fatigue, as used in this Specification, is defined as the damage that may result in fracture after a sufficient number of fluctuations of stress. Stress range is defined as the magnitude of these fluctuations. In the case of a stress reversal, the stress range shall be computed as the numerical sum of maximum repeated tensile and compressive stresses or the sum of maximum shearing stresses of opposite direction at a given point, resulting from differing arrangement of live load.

1. Loading Conditions; Type and Location of Material

In the design of members and connections subject to repeated variation of live load, consideration shall be given to the number of stress cycles, the expected range of stress and the type and location of member or detail.

Loading conditions shall be classified according to Table A-K4.1.

The type and location of material shall be categorized according to Table A-K4.2.

2. Allowable Stress Range

The maximum stress shall not exceed the basic allowable stress provided in Chapters A through M of this Specification and the maximum range of stress shall not exceed that given in Table A-K4. 3.

TABLE A-K4.1
Number of Loading Cycles

Loading Condition	From	To
1	20,000[a]	100,000[b]
2	100,000	500,000[c]
3	500,000	2,000,000[d]
4	Over 2,000,000	

[a]Approximately equivalent to two applications every day for 25 years.
[b]Approximately equivalent to 10 applications every day for 25 years.
[c]Approximately equivalent to 50 applications every day for 25 years.
[d]Approximately equivalent to 200 applications every day for 25 years

AMERICAN INSTITUTE OF STEEL CONSTRUCTION

3. Tensile Fatigue

When subject to tensile fatigue loading, the tensile stress in A325 or A490 bolts due to the combined applied load and prying forces shall not exceed the following values, and the prying force shall not exceed 60% of the externally applied load.

Number of Cycles	A325	A490
Not more than 20,000	44	54
From 20,000 to 500,000	40	49
More than 500,000	31	38

Bolts must be tensioned to the requirements of Table J3.7.

The use of other bolts and threaded parts subjected to tensile fatigue loading is not recommended.

TABLE A-K4.2
Stress Category Classifications

General Condition	Situation	Kind of Stress[a]	Stress Category (see Table A-K4.3)	Illus-trative Example Nos. (see Fig. A-K4.1)[b]
Plain Material	Base metal with rolled or cleaned surface. Flame-cut edges with ANSI smoothness of 1,000 or less	T or Rev.	A	1,2
Built-up Members	Base metal in members without attachments, built-up plates or shapes connected by continuous full-penetration groove welds or by continuous fillet welds parallel to the direction of applied stress	T or Rev.	B	3,4,5,6
	Base metal in members without attachments, built-up plates, or shapes connected by full-penetration groove welds with backing bars not removed, or by partial-penetration groove welds parallel to the direction of applied stress	T or Rev.	B′	3,4,5,6
	Base metal at toe welds on girder webs or flanges adjacent to welded transverse stiffeners	T or Rev.	C	7
	Base metal at ends of partial length welded cover plates narrower than the flange having square or tapered ends, with or without welds across the ends or wider than flange with welds across the ends			
	Flange thickness ≤ 0.8 in.	T or Rev.	E	5
	Flange thickness > 0.8 in.	T or Rev.	E′	5
	Base metal at end of partial length welded cover plates wider than the flange without welds across the ends		E′	5

[a] "T" signifies range in tensile stress only; "Rev." signifies a range involving reversal of tensile or compressive stress; "S" signifies range in shear, including shear stress reversal.

[b] These examples are provided as guidelines and are not intended to exclude other reasonably similar situations.

[c] Allowable fatigue stress range for transverse partial-penetration and transverse fillet welds is a function of the effective throat, depth of penetration and plate thickness. See Frank and Fisher (1979).

TABLE A-K4.2 (cont'd)
Type and Location of Material

General Condition	Situation	Kind of Stress[a]	Stress Category (see Table A-K4.3)	Illustrative Example Nos. (see Fig. A-K4.1)[b]
Groove Welds	Base metal and weld metal at full-penetration groove welded splices of parts of similar cross section ground flush, with grinding in the direction of applied stress and with weld soundness established by radiographic or ultrasonic inspection in accordance with the requirements of 9.25.2 or 9.25.3 of AWS D1.1	T or Rev.	B	10,11
	Base metal and weld metal at full-penetration groove welded splices at transitions in width or thickness, with welds ground to provide slopes no steeper than 1 to 2½ with grinding in the direction of applied stress, and with weld soundness established by radiographic or ultrasonic inspection in accordance with the requirements of 9.25.2 or 9.25.3 of AWS D1.1			
	A514 base metal	T or Rev.	B'	12,13
	Other base metals	T or Rev.	B	12,13
	Base metal and weld metal at full-penetration groove welded splices, with or without transitions having slopes no greater than 1 to 2½ when reinforcement is not removed but weld soundness is established by radiographic or ultrasonic inspection in accordance with requirements of 9.25.2 or 9.25.3 of AWS D1.1	T or Rev.	C	10,11,12,13
Partial-Penetration Groove Welds	Weld metal of partial-penetration transverse groove welds, based on effective throat area of the weld or welds	T or Rev.	F[c]	16

TABLE A-K4.2 (cont'd)
Type and Location of Material

General Condition	Situation	Kind of Stress[a]	Stress Category (see Table A-K4.3)	Illustrative Example Nos. (see Fig. A-K4.1)[b]
Fillet-welded Connections	Base metal at intermittent fillet welds	T or Rev.	E	
	Base metal at junction of axially loaded members with fillet-welded end connections. Welds shall be disposed about the axis of the member so as to balance weld stresses $b \le 1$ in. $b > 1$ in.	T or Rev. T or Rev.	E E′	17,18 17,18
	Base metal at members connected with transverse fillet welds $b \le \frac{1}{2}$ in. $b > \frac{1}{2}$ in.	T or Rev.	C See Note c	20,21
Fillet Welds	Weld metal of continuous or intermittent longitudinal or transverse fillet welds	S	F[c]	15,17,18 20,21
Plug or Slot Welds	Base metal at plug or slot welds	T or Rev.	E	27
	Shear on plug or slot welds	S	F	27
Mechanically Fastened Connections	Base metal at gross section of high-strength bolted slip-critical connections, except axially loaded joints which induce out-of-plane bending in connected material	T or Rev.	B	8
	Base metal at net section of other mechanically fastened joints	T or Rev.	D	8,9
	Base metal at net section of fully tensioned high-strength, bolted-bearing connections	T or Rev.	B	8,9

TABLE A-K4.2 (cont'd)
Type and Location of Material

General Condition	Situation	Kind of Stress[a]	Stress Category (see Table A-K4.3)	Illustrative Example Nos. (see Fig. A-K4.1)[b]
Attachments	Base metal at details attached by full-penetration groove welds subject to longitudinal and/or transverse loading when the detail embodies a transition radius R with the weld termination ground smooth and for transverse loading, the weld soundness established by radiographic or ultrasonic inspection in accordance with 9.25.2 or 9.25.3 of AWS D1.1			
	Longitudinal loading			
	$R > 24$ in.	T or Rev.	B	14
	24 in. $> R > 6$ in.	T or Rev.	C	14
	6 in. $> R > 2$ in.	T or Rev.	D	14
	2 in. $> R$	T or Rev.	E	14
	Detail base metal for transverse loading: equal thickness and reinforcement removed			
	$R > 24$ in.	T or Rev.	B	14
	24 in. $> R > 6$ in.	T or Rev.	C	14
	6 in. $> R > 2$ in.	T or Rev.	D	14
	2 in. $> R$	T or Rev.	E	14,15
	Detail base metal for transverse loading: equal thickness and reinforcement not removed			
	$R > 24$ in.	T or Rev.	C	14
	24 in. $> R > 6$ in.	T or Rev.	C	14
	6 in. $> R > 2$ in.	T or Rev.	D	14
	2 in. $> R$	T or Rev.	E	14,15

TABLE A-K4.2 (cont'd)
Type and Location of Material

General Condition	Situation	Kind of Stress[a]	Stress Category (see Table A-K4.3)	Illustrative Example Nos. (see Fig. A-K4.1)[b]
Attachments (cont'd)	Detail base metal for transverse loading: unequal thickness and reinforcement removed			
	$R > 2$ in.	T or Rev.	D	14
	2 in. $> R$	T or Rev.	E	14,15
	Detail base metal for transverse loading: unequal thickness and reinforcement not removed			
	all R	T or Rev.	E	14,15
	Detail base metal for transverse loading			
	$R > 6$ in.	T or Rev.	C	19
	6 in. $> R > 2$ in.	T or Rev.	D	19
	2 in. $> R$	T or Rev.	E	19
	Base metal at detail attached by full-penetration groove welds subject to longitudinal loading			
	$2 < a < 12b$ or 4 in.	T or Rev.	D	15
	$a > 12b$ or 4 in. when $b \leq 1$ in.	T or Rev.	E	15
	$a > 12b$ or 4 in. when $b > 1$ in.	T or Rev.	E'	15
	Base metal at detail attached by fillet welds or partial-penetration groove welds subject to longitudinal loading			
	$a < 2$ in.	T or Rev.	C	15,23,24, 25,26
	2 in. $< a < 12b$ or 4 in.	T or Rev.	D	15,23, 24,26
	$a > 12b$ or 4 in. when $b \leq 1$ in.	T or Rev.	E	15,23, 24,26
	$a > 12b$ or 4 in. when $b > 1$ in.	T or Rev.	E'	15,23, 24,26

TABLE A-K4.2 (cont'd)
Type and Location of Material

General Condition	Situation	Kind of Stress[a]	Stress Category (see Table A-K4.3)	Illustrative Example Nos. (see Fig. A-K4.1)[b]

Attachments (cont'd)	Base metal attached by fillet welds or partial-penetration groove welds subjected to longitudinal loading when the weld termination embodies a transition radius with the weld termination ground smooth:
	    $R > 2$ in.
	    $R \leq 2$ in.
		T or Rev.
		T or Rev.
			D
			E
				19
				19

	Fillet-welded attachments where the weld termination embodies a transition radius, weld termination ground smooth, and main material subject to longitudinal loading:
	    Detail base metal for transverse loading:
	    $R > 2$ in.
	    $R < 2$ in.
		T or Rev.
		T or Rev.
			D
			E
				19
				19

	Base metal at stud-type shear connector attached by fillet weld or automatic end weld	T or Rev.	C	22
	Shear stress on nominal area of stud-type shear connectors	S	F	

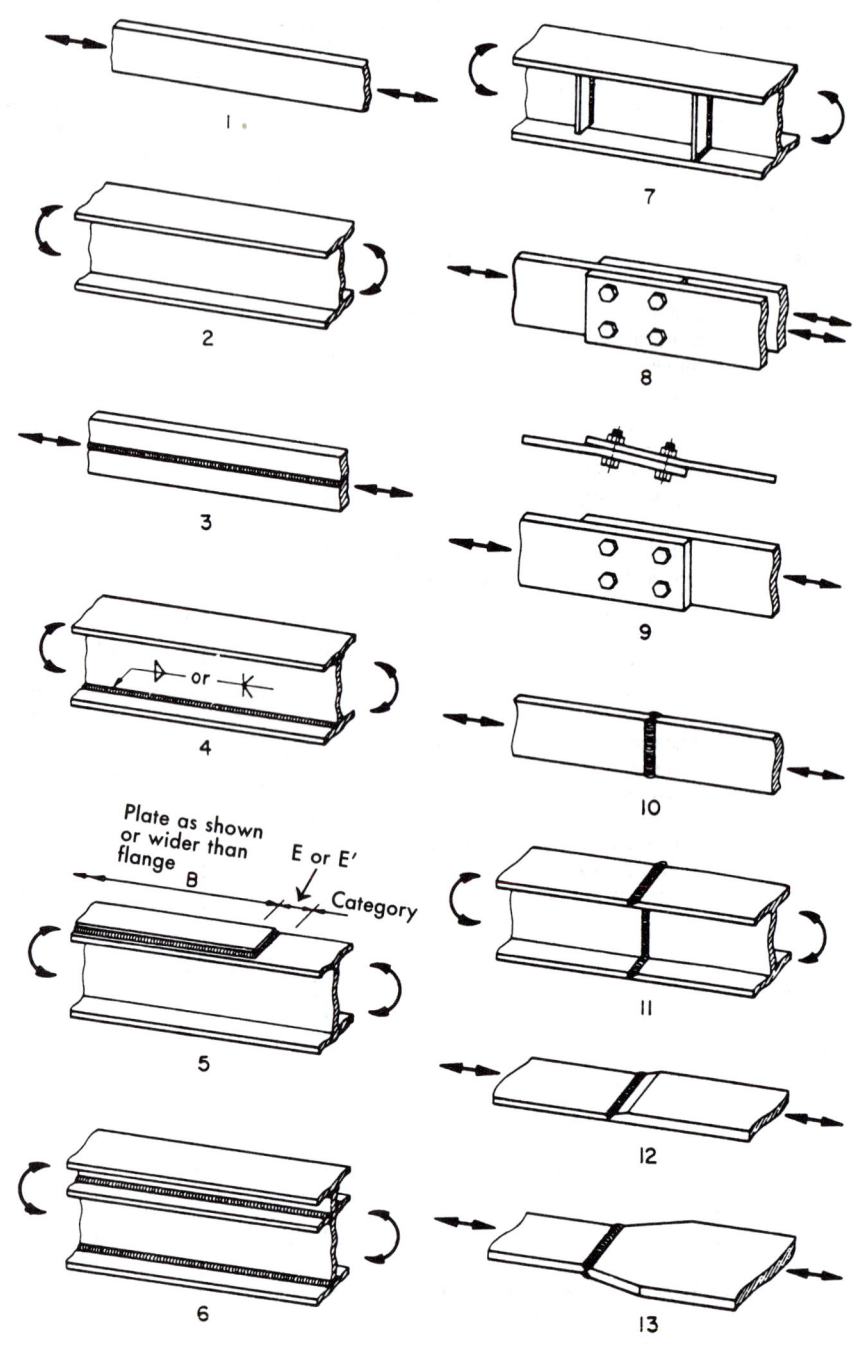

Illustrative examples

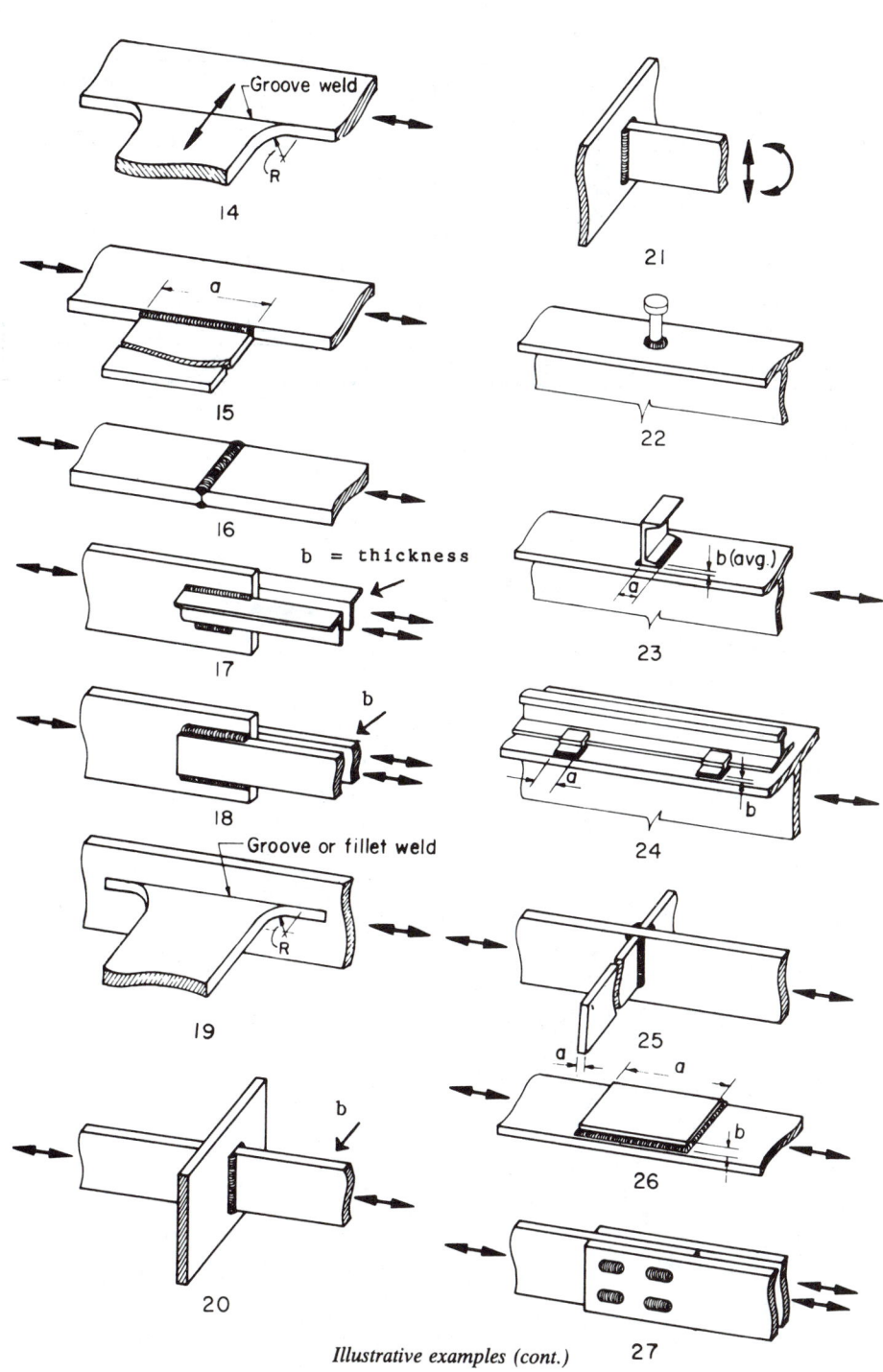

Illustrative examples (cont.)

TABLE A-K4.3
Allowable Stress Range, Ksi

Category (from Table A-K4.2)	Loading Condition 1	Loading Condition 2	Loading Condition 3	Loading Condition 4
A	63	37	24	24
B	49	29	18	16
B′	39	23	15	12
C	35	21	13	10[a]
D	28	16	10	7
E	22	13	8	5
E′	16	9	6	3
F	15	12	9	8

[a] Flexural stress range of 12 ksi permitted at toe of stiffener welds on flanges

NUMERICAL VALUES

TABLE 1
Allowable Stress as a Function of F_y

F_y (ksi)	Allowable Stress (ksi)					
	$0.40F_y$ [b,e]	$0.45F_y$ [a]	$0.60F_y$ [a,c]	$0.66F_y$ [c]	$0.75F_y$ [c]	$0.90F_y$ [d]
33	13.2	14.9	19.8	21.8	24.8	29.7
35	14.0	15.8	21.0	23.1	26.3	31.5
36	14.5	16.2	21.6	23.8	27.0	32.4
40	16.0	18.0	24.0	26.4	30.0	36.0
42	16.8	18.9	25.2	27.7	31.5	37.8
45	18.0	20.3	27.0	29.7	33.8	40.5
46	18.4	20.7	27.6	30.4	34.5	41.4
50	20.0	22.5	30.0	33.0	37.5	45.0
55	22.0	24.8	33.0	36.3	41.3	49.5
60	24.0	27.0	36.0	39.6	45.0	54.0
65	26.0	29.3	39.0	42.9	48.8	58.5
70	28.0	31.5	42.0			63.0
90	36.0	40.5	54.0			81.0
100	40.0	45.0	60.0			90.0

[a]See Sect. D1, D3 Tension
[b]See Sect. D3, F4, K1 Shear
[c]See Sect. F1, F2 Bending
[d]See Sect. J8 Bearing
[e]See Sect. G3 Shear in Plate Girders

TABLE 2
Allowable Stresses as a Function of F_u

Item	ASTM Designation	F_y (ksi)	F_u (ksi)	Connected Part of Designated Steel		Bolt or Threaded Part of Designated Steel		
				Tension $0.5F_u$[a]	Bearing $1.2F_u$[b]	Tension $0.33F_u$[c]	Shear $0.17F_u$[d]	Shear $0.22F_u$[e]
Shapes, Plates, Bars, Sheet and Tubing, or Threaded Parts	A36	36	58—80	29.0	69.6	19.1	9.9	12.8
	A53	35	60	30.0	72.0	—	—	—
	⌈A242⌉	50	70	35.0	84.0	23.1	11.9	15.4
	A441	46	67	33.5	80.4	22.1	11.4	14.7
	⌊A588⌋	42	63	31.5	75.6	20.8	10.7	13.9
		40[f]	60	30.0	72.0	19.8	10.2	13.2
	A500	33/39[g]	45	22.5	54.0	—	—	—
		42/46[g]	58	29.0	69.6	—	—	—
		46/50[g]	62	31.0	74.4	—	—	—
	A501	36	58	29.0	69.6	—	—	—
	A529	42	60—85	30.0	72.0	19.8	10.2	13.2
	A570	40	55	27.5	66.0	—	—	—
		42	58	29.0	69.6	—	–	–
	A572	42	60	30.0	72.0	19.8	10.2	13.2
		50	65	32.5	78.0	21.5	11.1	14.3
		60	75	37.5	90.0	24.8	12.8	16.5
		65	80	40.0	96.0	26.4	13.6	17.6
	A514	100	110—130	55.0	132	36.3	18.7	24.2
		90	100—130	50.0	120	33.0	17.0	22.0
	A606	45	65	32.5	78.0	—	—	—
		50	70	35.0	84.0	—	—	—
	A607	45	60	30.0	72.0	—	—	—
		50	65	32.5	78.0	—	—	—
		55	70	35.0	84.0	—	—	—
		60	75	37.5	90.0	—	—	—
		65	80	40.0	96.0	—	—	—
		70	85	42.5	102	—	—	—
	A618	50	70	35.0	84.0	—	—	—
		50	65	32.5	78.0	—	—	—
	A852	70	90—110	45.0	108	—	—	—
Bolts	A449	92	120	—	—	39.6	20.4	26.4
		81	105	—	—	34.7	17.9	23.1
		58	90	—	—	29.7	15.3	19.8

[a] On effective net area, see Sects. D1, J4.
[b] Produced by fastener in shear, see Sect. J3.7. Note that smaller maximum design bearing stresses, as a function of hole spacing, may be required by Sects. J3.8 and J3.9.
[c] On nominal body area, see Table J3.2.
[d] Threads not excluded from shear plane, see Table J3.2.
[e] Threads excluded from shear plane, see Table J3.2.
[f] For A441 material only.
[g] Smaller value for circular shapes, larger for square or rectangular shapes.
Note: For dimensional and size limitations, see the appropriate ASTM Specification.

TABLE 3
VALUES OF C_a

For Determining Allowable Stress When $Kl/r \leq C_c$
for Steel of Any Yield Stress (by Eq. $F_a = C_aF_y$)[a]

$\dfrac{Kl/r}{C_c}$	C_a	$\dfrac{Kl/r}{C_c}$	C_a	$\dfrac{Kl/r}{C_c}$	C_a	$\dfrac{Kl/r}{C_c}$	C_a
.01	.599	.26	.548	.51	.472	.76	.375
.02	.597	.27	.546	.52	.469	.77	.371
.03	.596	.28	.543	.53	.465	.78	.366
.04	.594	.29	.540	.54	.462	.79	.362
.05	.593	.30	.538	.55	.458	.80	.357
.06	.591	.31	.535	.56	.455	.81	.353
.07	.589	.32	.532	.57	.451	.82	.348
.08	.588	.33	.529	.58	.447	.83	.344
.09	.586	.34	.527	.59	.444	.84	.339
.10	.584	.35	.524	.60	.440	.85	.335
.11	.582	.36	.521	.61	.436	.86	.330
.12	.580	.37	.518	.62	.432	.87	.325
.13	.578	.38	.515	.63	.428	.88	.321
.14	.576	.39	.512	.64	.424	.89	.316
.15	.574	.40	.509	.65	.420	.90	.311
.16	.572	.41	.506	.66	.416	.91	.306
.17	.570	.42	.502	.67	.412	.92	.301
.18	.568	.43	.499	.68	.408	.93	.296
.19	.565	.44	.496	.69	.404	.94	.291
.20	.563	.45	.493	.70	.400	.95	.286
.21	.561	.46	.489	.71	.396	.96	.281
.22	.558	.47	.486	.72	.392	.97	.276
.23	.556	.48	.483	.73	.388	.98	.271
.24	.553	.49	.479	.74	.384	.99	.266
.25	.551	.50	.476	.75	.379	1.00	.261

All grades of steel

[a]When ratios exceed the noncompact section limits of Sect. B5.1, use $\dfrac{Kl/r}{C_c'}$ in lieu of $\dfrac{Kl/r}{C_c}$ values and equation $F_a = C_aQ_aQ_sF_y$ (Appendix Sect. B5).

NUMERICAL VALUES

TABLE 4
VALUES OF C_c

For Use with Equations (E2-1) and (E2-2) and in Table 3

F_y (ksi)	C_c	F_y (ksi)	C_c
33	131.7	46	111.6
35	127.9	50	107.0
36	126.1	55	102.0
39	121.2	60	97.7
40	119.6	65	93.8
42	116.7	90	79.8
45	112.8	100	75.7

TABLE 5
Slenderness Ratios of Elements as a Function of F_y

Specification Section and Ratios	F_y (ksi)					
	36	42	46	50	60	65
Table B5.1						
$65/\sqrt{F_y}$	10.8	10.0	9.6	9.2	8.4	8.1
$190/\sqrt{F_y}$	31.7	29.3	28.0	26.9	24.5	23.6
$640/\sqrt{F_y}$	106.7	98.8	94.4	90.5	82.6	79.4
$257/\sqrt{F_y}$	42.8	39.7	37.9	36.3	33.2	31.9
Sect. F1.2						
$\sqrt{\dfrac{102 \times 10^3 C_b}{F_y}}$	$53\sqrt{C_b}$	$49\sqrt{C_b}$	$47\sqrt{C_b}$	$45\sqrt{C_b}$	$41\sqrt{C_b}$	$40\sqrt{C_b}$
$\sqrt{\dfrac{510 \times 10^3 C_b}{F_y}}$	$119\sqrt{C_b}$	$110\sqrt{C_b}$	$105\sqrt{C_b}$	$101\sqrt{C_b}$	$92\sqrt{C_b}$	$89\sqrt{C_b}$
Table B5.1						
$76/\sqrt{F_y}$	12.7	11.7	11.2	10.7	9.8	9.4
$95/\sqrt{F_y}$	15.8	14.7	14.0	13.4	12.3	11.8
$127/\sqrt{F_y}$	21.2	19.6	18.7	18.0	16.4	15.8
Table B5.1						
$238/\sqrt{F_y}$	39.7	36.7	35.1	33.7	30.7	29.5
$317/\sqrt{F_y}$	52.8	48.9	46.7	44.8	40.9	39.3
$253/\sqrt{F_y}$	42.2	39.0	37.3	35.8	32.7	31.4
Table B5.1—Appendix B5.2b						
$3300/F_y$	91.7	78.6	71.7	66.0	55.0	50.8
$13000/F_y$	361	310	283	260	217	200
Sect. G1						
$\dfrac{14000}{\sqrt{F_y(F_y + 16.5)}}$	322	282	261	243	207	192
$2000/\sqrt{F_y}$	333	309	295	283	258	248

AMERICAN INSTITUTE OF STEEL CONSTRUCTION

TABLE 6
Values of C_b

For Use in Equations (F1-6), (F1-7) and (F1-8)

$\dfrac{M_1}{M_2}$	C_b	$\dfrac{M_1}{M_2}$	C_b	$\dfrac{M_1}{M_2}$	C_b
−1.00	1.00	−0.45	1.34	0.10	1.86
−0.95	1.02	−0.40	1.38	0.15	1.91
−0.90	1.05	−0.35	1.42	0.20	1.97
−0.85	1.07	−0.30	1.46	0.25	2.03
−0.80	1.10	−0.25	1.51	0.30	2.09
−0.75	1.13	−0.20	1.55	0.35	2.15
−0.70	1.16	−0.15	1.60	0.40	2.22
−0.65	1.19	−0.10	1.65	0.45	2.28
−0.60	1.23	−0.05	1.70	≥0.47	2.30
−0.55	1.26	0	1.75		
−0.50	1.30	0.05	1.80		

Note 1: $C_b = 1.75 + 1.05(M_1/M_2) + 0.3\,(M_1/M_2)^2 \leq 2.3$
Note 2: M_1/M_2 positive for reverse curvature and negative for single curvature.

TABLE 7
Values of C_m

For Use in Equation (H1-1)

$\dfrac{M_1}{M_2}$	C_m	$\dfrac{M_1}{M_2}$	C_m	$\dfrac{M_1}{M_2}$	C_m
−1.00	1.00	−0.45	0.78	0.10	0.56
−0.95	0.98	−0.40	0.76	0.15	0.54
−0.90	0.96	−0.35	0.74	0.20	0.52
−0.85	0.94	−0.30	0.72	0.25	0.50
−0.80	0.92	−0.25	0.70	0.30	0.48
−0.75	0.90	−0.20	0.68	0.35	0.46
−0.70	0.88	−0.15	0.66	0.40	0.44
−0.65	0.86	−0.10	0.64	0.45	0.42
−0.60	0.84	−0.05	0.62	0.50	0.40
				0.60	0.36
−0.55	0.82	0	0.60	0.80	0.28
−0.50	0.80	0.05	0.58	1.00	0.20

Note 1: $C_m = 0.6 - 0.4(M_1/M_2)$
Note 2: M_1/M_2 is positive for reverse curvature and negative for single curvature.

NUMERICAL VALUES

TABLE 8
Values of F_e'

For Use in Equation (H1-1), for Steel of Any Yield Stress

All grades of steel

$\dfrac{Kl_b}{r_b}$	F_e' (ksi)	$\dfrac{Kl_b}{r_b}$	F_e' (ksi)	$\dfrac{Kl_b}{r_b}$	F_e' (ksi)	$\dfrac{Kl_b}{r_b}$	F_e' (ksi)	$\dfrac{Kl_b}{r_b}$	F_e' (ksi)	$\dfrac{Kl_b}{r_b}$	F_e' (ksi)
21	338.62	51	57.41	81	22.76	111	12.12	141	7.51	171	5.11
22	308.54	52	55.23	82	22.21	112	11.90	142	7.41	172	5.05
23	282.29	53	53.16	83	21.68	113	11.69	143	7.30	173	4.99
24	259.26	54	51.21	84	21.16	114	11.49	144	7.20	174	4.93
25	238.93	55	49.37	85	20.67	115	11.29	145	7.10	175	4.88
26	220.90	56	47.62	86	20.19	116	11.10	146	7.01	176	4.82
27	204.84	57	45.96	87	19.73	117	10.91	147	6.91	177	4.77
28	190.47	58	44.39	88	19.28	118	10.72	148	6.82	178	4.71
29	177.56	59	42.90	89	18.85	119	10.55	149	6.73	179	4.66
30	165.92	60	41.48	90	18.44	120	10.37	150	6.64	180	4.61
31	155.39	61	40.13	91	18.03	121	10.20	151	6.55	181	4.56
32	145.83	62	38.85	92	17.64	122	10.03	152	6.46	182	4.51
33	137.13	63	37.62	93	17.27	123	9.87	153	6.38	183	4.46
34	129.18	64	36.46	94	16.90	124	9.71	154	6.30	184	4.41
35	121.90	65	35.34	95	16.55	125	9.56	155	6.22	185	4.36
36	115.22	66	34.28	96	16.20	126	9.41	156	6.14	186	4.32
37	109.08	67	33.27	97	15.87	127	9.26	157	6.06	187	4.27
38	103.42	68	32.29	98	15.55	128	9.11	158	5.98	188	4.23
39	98.18	69	31.37	99	15.24	129	8.97	159	5.91	189	4.18
40	93.33	70	30.48	100	14.93	130	8.84	160	5.83	190	4.14
41	88.83	71	29.62	101	14.64	131	8.70	161	5.76	191	4.09
42	84.65	72	28.81	102	14.35	132	8.57	162	5.69	192	4.05
43	80.76	73	28.02	103	14.08	133	8.44	163	5.62	193	4.01
44	77.13	74	27.27	104	13.81	134	8.32	164	5.55	194	3.97
45	73.74	75	26.55	105	13.54	135	8.19	165	5.49	195	3.93
46	70.57	76	25.85	106	13.29	136	8.07	166	5.42	196	3.89
47	67.60	77	25.19	107	13.04	137	7.96	167	5.35	197	3.85
48	64.81	78	24.54	108	12.80	138	7.84	168	5.29	198	3.81
49	62.20	79	23.93	109	12.57	139	7.73	169	5.23	199	3.77
50	59.73	80	23.33	110	12.34	140	7.62	170	5.17	200	3.73

Note: $F_e' = \dfrac{12\pi^2 E}{23(Kl_b/r_b)^2}$

Commentary

ON THE SPECIFICATION FOR STRUCTURAL STEEL BUILDINGS—
ALLOWABLE STRESS DESIGN AND PLASTIC DESIGN
(June 1, 1989)

INTRODUCTION

This Commentary provides information on the basis and limitations of various provisions of the Specification, so that designers, fabricators and erectors (users) can make more efficient use of the Specification. The Commentary and Specification, termed as documents, do not attempt to anticipate and/or set forth all the questions or possible problems that may be encountered, or situations in which special consideration and engineering judgment should be exercised in using and applying the documents. Such a recitation could not be made complete and would make the documents unduly lengthy and cumbersome.

Warning is given that AISC assumes the users of its documents are competent in their fields of endeavor and are informed on current developments and findings related to their fields.

CHAPTER A

GENERAL PROVISIONS

A2. LIMITS OF APPLICABILITY

2. Types of Construction

In order that adequate instructions can be issued to shop and erection personnel, the basic assumptions underlying the design must be thoroughly understood by all concerned. These assumptions are classified under three separate but generally recognized types of construction.

For better clarity, provisions covering tier buildings of Type 2 construction designed for wind loading were reworded in the 1969 Specification, but without change in intent. Justification for these provisions has been discussed by Disque (1964 and 1975) and Ackroyd (1987).

A3. MATERIAL

1. Structural Steel

a. ASTM Designations

The grades of structural steel approved for use under the Specification, covered by ASTM standard specifications, extend to a yield stress of 100 ksi. Some of these ASTM standards specify a minimum yield point, while others specify a minimum yield strength. The term "yield stress" is used in the Specification as a generic term to denote either the yield point or the yield strength. When requested, the fabricator must provide an affidavit that all steel specified has been provided in accordance with the plans and Specification.

In keeping with the inclusion of steels of several strength grades, a number of corresponding ASTM standards for cast steel forgings and other materials such as rivets, bolts and welding electrodes are also included.

Provisions of the Specification are based on providing a factor of safety against reaching yield stress in primary connected material at allowable loads. The direction parallel to the direction of rolling is the direction of principal interest in the design of steel structures. Hence, yield stress as determined by the standard tensile test is the principal mechanical property recognized in the selection of the steels approved for use under the Specification. It must be recognized that other mechanical and physical properties of rolled steel, such as anisotropy, ductility, notch toughness, formability, and corrosion resistance may also be important to the satisfactory performance of a structure. In such situations, the user of the Specification is advised to make use of reference material contained in the literature on the specific properties of concern and to specify supplementary material production or quality requirements as provided for in ASTM material specifications. One such situation, for example, is the design of highly restrained welded connections (AISC, 1973). Rolled steel is anisotropic, especially insofar as ductility is concerned; therefore weld contraction strains in the region of highly restrained welded connections may exceed the capabilities of the material if special attention is not given to material selection, details, workmanship and inspection. Another special situation is that of fracture control design for certain types of service conditions (Rolfe, 1977). The relatively warm temperatures of steel in buildings, the essentially static strain rates, the stress intensity and the number of cycles of full allowable stress make the probability of fracture in building structures extremely remote. Good details which incorporate joint geometry that avoids severe stress concentrations and good workmanship are generally the most effective means to provide fracture-resistant construction. However, for especially demanding service conditions, such as low temperatures with impact loading, the specification of steels with superior notch toughness should be specified.

c. Heavy Shapes

The web-to-flange intersection and the web center of heavy hot-rolled shapes, as well as the interior portions of heavy plates, may contain a coarser grain structure and/or lower toughness than other areas of these products. This is probably caused by ingot segregation, as well as somewhat less deformation during hot rolling, higher finishing temperature and a slower cooling rate after rolling. This characteristic is not detrimental to suitability for service as compression members or non-welded members. However when heavy sections are fabricated using full-penetration welds, tensile strains induced by weld shrinkage may result in cracking. For critical applications such as primary tension members, material should be produced to provide adequate toughness. Because of differences in the strain rate between the Charpy V-Notch (CVN) impact test and the strain rate experienced in actual structures, the CVN test is conducted at a temperature higher than the anticipated service temperature.

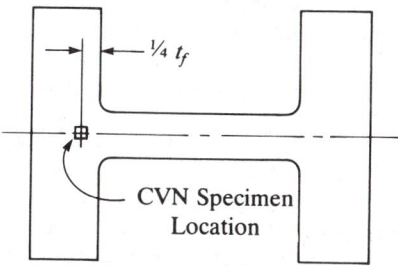

Fig. C-A3.1c Location from which charpy impact specimen shall be taken.

The toughness requirements of Sect. A3.1c are intended only to provide material of necessary toughness for ordinary service application. For unusual applications and/or low temperature service, more restrictive requirements and/or toughness requirements for other section sizes and thickness would be appropriate.

To minimize the potential for fracture, the notch toughness requirements of Sect. A3.1c must be used in conjunction with good design and fabrication procedures. Specific requirements are given in Sects. J1.7, J1.8, J2.6, J2.7 and M2.2.

4. Bolts, Washers and Nuts

The ASTM standard for A307 bolts covers two grades of fasteners. Either grade may be used under the Specification; however, Grade B is intended for pipe flange bolting. Grade A is used for structural applications.

6. Filler Metal and Flux for Welding

When specifying filler metal and/or flux by AWS designation, the applicable standard specifications should be carefully reviewed to assure a complete understanding of the electrode designation. This is necessary because the AWS designation systems are not consistent. For example, in the case of electrodes for shielded metal arc welding (AWS A5.1), the first two or three digits indicate the nominal tensile strength classification, (in ksi) of the weld material and the final two digits indicate the type coating. However, in the case of carbon steel electrodes for submerged arc welding (AWS A5.17), the first one or two digits times 10 indicates the nominal tensile strength classification, and the final digit or digits times − 10 indicates the testing temperature, in degrees F, for weld metal impact tests. In the case of low-alloy, steel-covered arc welding electrodes (AWS A5.5), certain portions of the designation indicate a requirement for stress relief, while others indicate no stress relief requirement.

A4. LOADS AND FORCES

The specification does not presume to establish the loading requirements for which structures should be designed. In most cases these are adequately covered in the applicable local building codes. Where this is not the case, the generally recognized standards of the American National Standards Institute (ANSI) are recommended as the basis for design.

2. Impact

A mass of the total moving load (wheel load) is used as the basis for impact loads on crane runway girders, because maximum impact load results when cranes travel while supporting lifted loads.

The increase in load, in recognition of random impacts, is not required to be applied to supporting columns, because the impact load effects (increase in eccentricities or increase in out-of-straightness) will not develop or will be negligible during the short duration of impact.

Association of Iron and Steel Engineers (AISE, 1979) gives more stringent requirements for crane girder and crane runway design.

3. Crane Runway Horizontal Forces

Minimum crane horizontal and longitudinal forces are provided in the Specification. Some cranes may require that the runway be designed for larger forces.

The magnitude and point of application of the crane stop forces should be provided by the owner. For additional information on runway forces, see AISE (1979).

A5. DESIGN BASIS

1. Allowable Stresses

The allowable stresses contained within the Specification are to be compared with stresses determined by analysis of the effects of design loads upon the structure. The factor of safety inherent in the allowable stresses provide for the uncertainties that are associated with typical simplifying assumptions and the use of nominal or average calculated stresses as the basis for manual methods of analysis. It is not intended that highly localized peak stresses that may be determined by sophisticated computer-aided methods of analysis, and which may be blunted by confined yielding, must be less than the stipulated allowable stresses. The exercise of engineering judgment is required.

In keeping with the inclusion of high strength low-alloy steels, the Specification recognizes high strength steel castings. Allowable stresses are expressed in terms of the specified minimum yield stress for castings.

CHAPTER B

DESIGN REQUIREMENTS

B3. EFFECTIVE NET AREA

Section B3 deals with the effect of shear lag. The inclusion of welded members acknowledges that shear lag is also a factor in determining the effective area of welded connections where the welds are so distributed as to directly connect some, but not all, of the elements of a tension member. However, since welds are applied to the unreduced cross-sectional area, the reduction coefficient U is applied to the gross area A_g. With this modification the values of U are the same as for similar shapes connected by bolts and rivets except that: (1) the provisions for members having only two fasteners per line in the direction of stress have no application to welded connections; and (2) tests (Kulak, Fisher and Struik, 1987) have shown that flat plates , or bars axially loaded in tension and connected only by longitudinal fillet welds, may fail prematurely by shear lag at their corners if the welds are separated by too great a distance. Therefore, the values of U are specified unless the member is designed on the basis of effective net area as discussed below.

As the length of a connection l is increased the intensity of shear lag is diminished. The concept can be expressed empirically as:

$$U = 1 - \bar{x}/l \qquad \text{(C-B1-1)}$$

where:
$\quad \bar{x} =$ the distance from the centroid of the shape profile to the shear plane of the connection, in.
$\quad l \ =$ length

Munse and Chesson have shown, using this expression to compute an effective net area, that with few exceptions, the estimated strength of some 1,000 test specimens correlated with observed test results with a scatterband of $\pm10\%$ (Kulak, Fisher, and Struik, 1987; Munse and Chesson, 1963; Gaylord and Gaylord, 1972). For any given profile and connected elements, length l is dependent upon the number of fasteners or length of weld required to develop the given tensile force, and this in turn is dependent upon the mechanical properties of the member and the capacity of the fasteners or weld used. The values of U, given as the reduction coefficients in Sect. B3, are reasonable lower bounds for the profile types and connections described, based upon the use of the above expression.

The restriction that the net area shall in no case be considered as comprising more than 85% of the gross area is limited to relatively short fittings, such as splice plates, gusset plates or beam-to-column fittings.

B4. STABILITY

The stability of structures must be considered from the standpoint of the structures as a whole, including not only the compression members, but also the beams, bracing system and connections. The stability of individual elements must also be provided. Considerable attention has been given to this subject in the technical literature, and various methods of analysis are available to assure stability. The SSRC *Guide to Design Criteria for Metal Compression Members* (Galambos, 1988) devotes several chapters to the stability of different types of members considered as individual elements, and then considers the effects of individual elements on the stability of the structure as a whole.

B5. LOCAL BUCKLING

For the purposes of the ASD Specification, steel sections are divided into compact sections, noncompact sections and sections with slender compression elements.

When the width-thickness ratio of the compressed elements in a member does not exceed the noncompact section limit specified in Table B5.1, no reduction in allowable stress is necessary in order to prevent local buckling. Appendix B provides a design procedure for those infrequent situations where width-thickness ratios in excess of the limits given in Sect. B5.1 are involved.

Equations (A-B5-1), (A-B5-2), (A-B5-5) and (A-B5-6) are based upon the following expression for critical buckling stress σ_c for a plate supported against lateral deflection along one or both edges (Galambos, 1988), with or without torsional restraint along these edges and subject to in-plane compressive force:

$$\sigma_c = k_c \left[\frac{\pi^2 E \sqrt{\eta}}{12 \, (1-v^2)(b/t)^2} \right] \qquad \text{(C-B5-1)}$$

where:

η = the ratio of the tangent modulus to the elastic modulus, E_t/E
v = Poisson's ratio

The assumption of nothing more than knife-edge lateral support applied along one edge of the unstiffened element under a uniformly distributed stress (the most critical case) would give a value of $k_c = 0.425$. Some increase in this value is warranted because of the torsional restraint provided by the supporting element and because of the difference between b, as defined in Sect. B5.1, and the theoretical width b.

Equations (A-B5-3) and (A-B5-4) have been revised for this AISC ASD Specification. In the 1978 AISC Specification, these formulas assumed partial end restraint from the beam web in rolled shapes for compression flange stability. However, with more slender girder webs that may have already buckled, this beneficial effect is diminished and the previous Q factors have been reduced to account for this local buckling interaction. Research by Butler Manufacturing resulted in new provisions, given in Appendix B5.2, which are also reflected in Sects. B5.1 and G2 (Johnson, 1985).

In the interest of simplification, when $\sqrt{\eta} < 1.0$, a linear formula is substituted for the theoretical expression. Its agreement with the latter may be judged by the comparison shown in Fig. C-B5.1.

Equation (A-B5-5) recognizes that the torsional restraint characteristics of tees cut from rolled shapes might be of quite different proportions than those of tees formed by welding two plates together.

It has been shown that singly symmetrical members whose cross section consists of elements having large width-thickness ratios may fail by twisting under a smaller axial load than associated with general column failure (Chajes and Winter, 1965). Such is not generally the case with hot-rolled shapes. To guard against this type of failure, particularly when relatively thin-walled members are fabricated from plates, Table A-B5.1 in Appendix B places an upper limit on the proportions permissible for channels and tees.

With both edges parallel to the applied load supported against buckling, stiffened compression elements can support a load producing an average stress σ_c greater than that given in the above expression for critical plate buckling stress. This is true even when k_c is taken as 4.0, applicable to the case where both edges are simply supported, or a value between 4.0 and 6.97, applicable when some torsional restraint is also provided along these edges.

A better estimate of the compressive strength of stiffened elements, based upon an "effective width" concept was first proposed by von Karman, Sechler and Donnell (1932). This was later modified by Winter (1947) to provide a transition between very slender elements and stockier elements shown by tests to be fully effective.

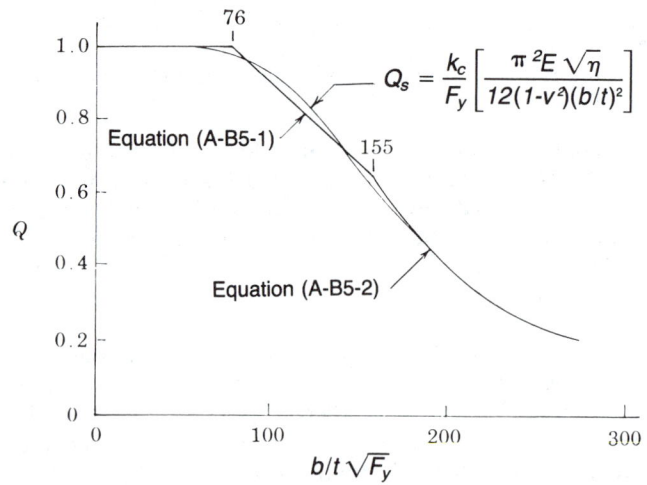

$$Q_s = \frac{k_c}{F_y}\left[\frac{\pi^2 E \sqrt{\eta}}{12(1-v^2)(b/t)^2}\right]$$

Fig. C-B5.1

As modified, the ratio of effective width to actual width increases as the level of compressive stress applied to a stiffened element in a member is decreased and takes the form

$$\frac{b_e}{t} = 1.9 \sqrt{\frac{E}{f}} \left[1 - \frac{C}{(b/t)} \sqrt{\frac{E}{f}} \right] \qquad \text{(C-B5-2)}$$

where f is the level of uniformly distributed stress to which the element would be subjected based upon the design of the member, and C is an arbitrary constant based on test results (Winter, 1947).

Holding the effective width of a stiffened element to no greater value than that given by the limits provided in Sect. B5.1 is unnecessarily conservative when the maximum uniformly distributed design stress is substantially less than $0.60F_y$, or when the ratio b/t is considerably in excess of the limit given in Sect. B5.1.

For the case of square and rectangular box sections, the sides of which in their buckled condition afford negligible torsional restraint for one another along their corner edges, the value of C reflected in Equation (A-B5-7) is higher than for the other case, thereby providing a slightly more conservative evaluation of effective width. For cases where appreciable torsional restraint is provided, as for example the web of an I-shape column, the value of C implicit in Equation (A-B5-8) is decreased slightly. As in earlier editions of the AISC Specification, for such cases no reduction from actual width is required when the width-thickness ratio does not exceed $253/\sqrt{F_y}$, and for greater widths the effective width may be taken as equal to $253t/\sqrt{F_y}$. If the actual width-thickness ratio is substantially greater than $253/\sqrt{F_y}$, however, a larger effective width can be obtained using Equation (A-B5-8) rather than the earlier provisions.

In computing the section modulus of a member subject to bending, the area of stiffened elements parallel to the axis of bending and subject to compressive stress must be based upon their effective, rather than actual, width. In computing the effective area of a member subject to axial loading, the effective, rather than actual, area of all stiffened elements must be used. However, the radius of gyration of the *actual* cross section together with the form factor Q_a may be used to determine the allowable axial stress. If the cross section contains an unstiffened element, the allowable axial stress must be modified by the reduction factor Q_s.

The classical theory of longitudinally compressed cylinders overestimates the actual buckling strength, often by 200% or more. Inevitable imperfections of shape and the axiality of load are responsible for the reduction in actual strength below theoretical strength. The limits of B5.1 are based upon test evidence (Sherman, 1976), rather than theoretical calculations, that local buckling will not occur if the D/t ratio is equal to or less than $3300/F_y$ when the applied stress is equal to F_y. When D/t exceeds $3300/F_y$, but is less than $13,000/F_y$, Equation (A-B5-9) provides for a reduction in allowable stress with a factor of safety against local buckling of at least 1.67. The Specification contains no recommendations for allowable stresses when D/t exceeds $13,000/F_y$.

B6. ROTATIONAL RESTRAINT AT POINTS OF SUPPORT

Slender beams and girders resting on top of columns and stayed laterally only in the plane of their top flanges may become unstable due to the flexibility of the column. Unless lateral support is provided for the bottom flange, either by bracing or continuity at the beam-to-column connection, lateral displacement at the top of the column, accompanied by rotation of the beam about its longitudinal axis, may lead to collapse of the framing.

B7. LIMITING SLENDERNESS RATIOS

The slenderness limitations recommended for tension members are not essential to the structural integrity of such members; they merely afford a degree of stiffness such that undesirable lateral movement ("slapping" or vibration) will be avoided. These limitations are not mandatory.

See Commentary E4.

B10. PROPORTIONS OF BEAMS AND GIRDERS

As in earlier editions of the Specification, it is provided that flexural members be proportioned to resist bending on the basis of moment of inertia of their gross cross section. However, the 15% flange area allowance for holes in previous specifications (Lilly and Carpenter, 1940), has been replaced by an improved criterion based on a direct comparison of tensile fracture and yield. For the fracture calculation, no hole deduction need be made until $A_{net}/A_{gross} = 6/5$ (F_y/F_u). This is equivalent to a hole allowance of 25.5% for A36 and 7.7% for $F_y = 50$ ksi material. This provision includes the design of hybrid flexural members whose flanges are fabricated from a stronger grade of steel than that in their web. As in the case of flexural members having the same grade of steel throughout their cross section, their bending strength is defined by the product of the section modulus of the gross cross section multiplied by the allowable bending stress. On this basis, the stress in the web at its junction with the flanges may even exceed the yield stress of the web material, but under strains controlled by the elastic state of stress in the stronger flanges. Numerous tests have shown that, with only minor adjustment in the basic allowable bending stress as provided in Equation (G2-1), the bending strength of a hybrid member is predictable within the same degree of accuracy as is that of a homogeneous member (ASCE-AASHO, 1968).

If a partial length cover plate is to function as an integral part of a beam or girder at the theoretical cutoff point beyond which it is not needed, it must be developed in an extension beyond this point by high-strength bolts or welding to develop its portion of the flexural stresses (i.e., the stresses which the plate would have received had it been extended the full length of the member). The cover plate force to be developed by the fasteners in the extension is equal to

$$\frac{MQ}{I} \qquad \text{(C-B10-1)}$$

where

M = moment at theoretical cutoff
Q = statical moment of cover plate area about neutral axis of coverplated section
I = moment of inertia of coverplated section

When the nature of the loading is such as to produce fatigue, the fasteners must be proportioned in accordance with the provisions of Appendix K4.

In the case of welded cover plates, it is further provided that the amount of stress that may be carried by a partial length of cover plate, at a distance a' in from its actual end, may not exceed the capacity of the terminal welds deposited along its edges and optionally across its end within this distance a'. If the moment, computed by equating MQ/I to the capacity of the welds in this distance, is less than the value at the theoretical cutoff point, either the size of the welds must be increased or the end of the cover plate must be extended to a point such that the moment on the member at the distance a' from the end of the cover plate is equal to that which the terminal welds will support.

CHAPTER C

FRAMES AND OTHER STRUCTURES

C2. FRAME STABILITY

The stability of structures as a whole must be considered from the standpoint of the structure, including not only the columns, but also the beams, bracing system and connections. The stability of individual elements must also be provided. Considerable attention has been given in the technical literature to this subject, and various methods of analysis are available to assure stability. The SSRC *Guide to Design Criteria for Metal Compression Members* devotes several chapters to the stability of different types of members considered as individual elements, and then considers the effects of individual elements on the stability of the structure as a whole (Galambos, 1988).

The effective length concept is one method for estimating the interaction effects of the total frame on a column being considered. This concept uses K-factors to equate the strength of a framed compression element of length L to an equivalent pin-ended member of length KL subject to axial load only. Other methods are available for evaluating the stability of frames subject to gravity and lateral loading and individual compression members subject to axial load and moments. The effective length concept is one tool available for handling several cases which occur in practically all structures, and it is an essential part of many analysis procedures. Although the concept is completely valid for ideal structures, its practical implementation involves several assumptions of idealized conditions which will be mentioned later.

Two conditions, opposite in their effect upon column strength under axial loading, must be considered. If enough axial load is applied to the columns in an unbraced frame dependent entirely on its own bending stiffness for resistance to lateral deflection of the tops of the columns with respect to their bases (see Fig. C-C2.1), the effective length of these columns will exceed the actual length. On the other hand, if the same frame were braced to resist such lateral movement, the effective length would be less than the actual length, due to the restraint (resistance to joint rotation) provided by the bracing or other lateral support. The ratio K, effective column length to actual unbraced length, may be greater or less than 1.0.

The theoretical K-values for six idealized conditions in which joint rotation and translation are either fully realized or nonexistent are tabulated in Table C-C2.1. Also shown are suggested design values recommended by the Structural Stability Research Council for use when these conditions are approximated in actual design. In general, these suggested values are slightly higher than their theoretical equivalents, since joint fixity is seldom fully realized.

If the column base in Case f of Table C-C2.1 were truly pinned, K would actually exceed 2.0 for a frame such as that pictured in Fig. C-C2.1, because the flexibility of the horizontal member would prevent realization of full fixity at the top of the column. On the other hand, the restraining influence of foundations, even where these footings are designed only for vertical load, can be very substantial in the case of flat-ended column base details with ordinary anchorage (Stang and Jaffe, 1948). For this condition, a design *K*-value of 1.5 would generally be conservative in Case f.

While in some cases the existence of masonry walls provides enough lateral support for their building frames to control lateral deflection, the increasing use of light curtain wall construction and wide column spacing for high-rise structures not provided with a positive system of diagonal bracing can create a situation where only the bending stiffness of the frame itself provides this support.

Table C-C2.1

	(a)	(b)	(c)	(d)	(e)	(f)
Buckled shape of column is shown by dashed line						
Theoretical *K* value	0.5	0.7	1.0	1.0	2.0	2.0
Recommended design value when ideal conditions are approximated	0.65	0.80	1.2	1.0	2.10	2.0
End condition code		Rotation fixed and translation fixed				
		Rotation free and translation fixed				
		Rotation fixed and translation free				
		Rotation free and translation free				

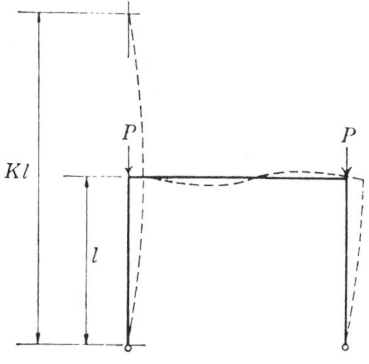

Figure C-C2.1

AMERICAN INSTITUTE OF STEEL CONSTRUCTION

In this case, the effective length factor K for an unbraced length of column L is dependent on the amount of bending stiffness provided by the other in-plane members entering the joint at each end of the unbraced segment. If the combined stiffness provided by the beams is sufficiently small, relative to that of the unbraced column segments, KL could exceed two or more story heights (Bleich, 1952).

Several methods are available for estimating the effective length of columns in an unbraced frame. These range from simple interpolation between the idealized cases shown in Table C-C2.1 to very complex analytical procedures. Once a trial selection of framing members has been made, the use of the alignment chart in Fig. C-C2.2 affords a fairly rapid method for determining adequate K-values.

However, this alignment chart is based upon assumptions of idealized conditions which seldom exist in real structures (Galambos, 1988). These assumptions are as follows:

1. Behavior is purely elastic.
2. All members have constant cross section.
3. All joints are rigid.
4. For braced frames, rotations at opposite ends of beams are equal in magnitude, producing single curvature bending.
5. For unbraced frames, rotation at opposite ends of the restraining beams are equal in magnitude, producing reverse curvature bending.
6. The stiffness parameters $L\sqrt{P/EI}$ of all columns are equal.
7. Joint restraint is distributed to the column above and below the joint in proportion to I/L of the two columns.
8. All columns buckle simultaneously.

Where the actual conditions differ from these assumptions, unrealistic designs may result. There are design procedures available which may be used in the calculation of G for use in Fig. C-C2.2 to give results more truly representative of conditions in real structures (Yura, 1971 and Disque, 1973).

Research at Lehigh University on the load-carrying capacity of regular rectangular rigid frames has shown that it is not always necessary to directly account for the $P\Delta$ effect for a certain class of adequately stiff rigid frames (Ozer et al, 1974 and Cheong-Sait Moy, Ozer and Lu, 1977). In the research, second-order analyses using different load sequences to failure were used to confirm the adequacy of alternate allowable stress design procedures. The loading sequences used in the second order analysis were:

1. Constant gravity load at a load factor of 1.0 while the lateral load was progressively increased.
2. Constant gravity load at a load factor of 1.3 while the lateral load was progressively increased.
3. Both the lateral and gravity loads were progressively increased proportionately.

The seven frames included in the study were 10 to 40 stories high and in-plane column slenderness ratios h/r_x ranged from 18 to 42. The live load, including partitions, varied from 40 to 100 psf and the dead load from 50 to 75 psf. A uniform wind load of 20 psf was specified throughout. All beams and column sections were compact. The axial load ratios f_a/F_a and $f_a/0.60F_y$ were limited to not more than 0.75.

The results of the second order analyses showed that adequate strength and stability were assured under combined gravity and lateral loads or gravity load alone, when the rigid frames were designed by either a stress design procedure according to AISC ASD Specification requirements or by a modified stress design procedure. The modified allowable stress design procedure incorporated a stiffness parameter[*] which assured adequate frame stiffness, while the effective length factor K was assumed to be unity in calculations of f_a and F'_e, and the coefficient C_m was computed as for a braced frame.

Several other references[**] are available concerning alternatives to effective

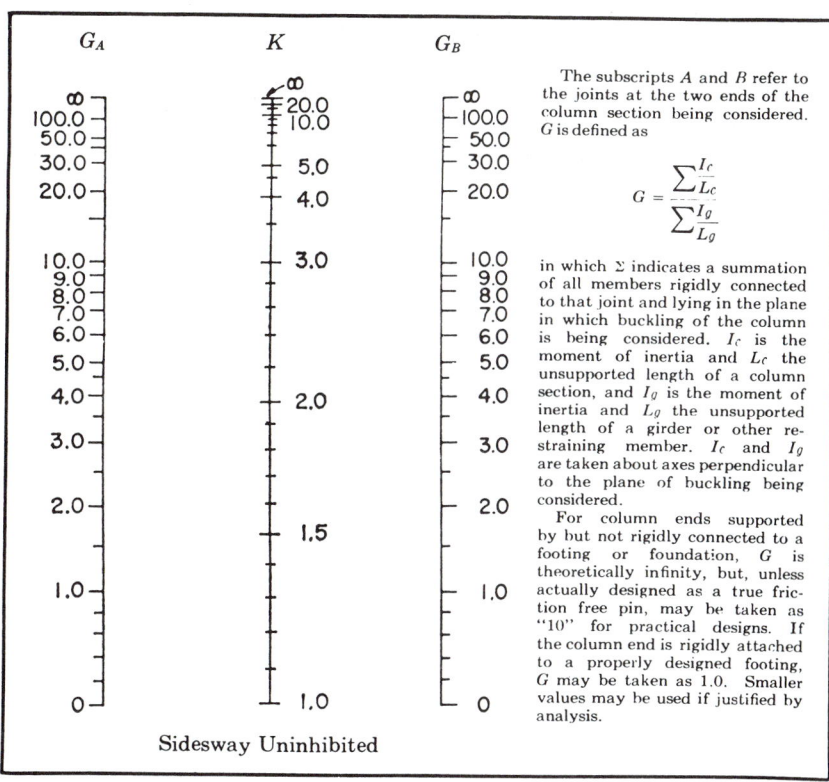

Alignment Chart for Effective Length of Columns in Continuous Frames

Fig. C-C2.2

---

[*]A design procedure based only upon a first order drift index may not assure frame stability.

[**]Yura, 1971; Springfield and Adams, 1972; Liapunov, 1974 (pp 1643-1655); Daniels and Lu, 1972; LeMessurier, 1976; and LeMessurier, 1977.

length factors for multistory frames under combined loads or gravity loads alone.

In frames which depend upon their own bending stiffness for stability, the amplified moments are accounted for in the design of columns by means of the interaction equations of Sect. H1. However, moments are also induced in the beams which restrain the columns; thus, consideration must be given to the amplification of those portions of the beam moments that are increased when the frame drifts. The effect may be particularly important in frames in which the contribution to individual beam moments from story shears becomes small as a result of distribution to many bays, but in which the $P\Delta$ moment in individual columns and beams is not diminished and becomes dominant.

If roof decks and floor slabs, anchored to shear walls or vertical plane bracing systems, are counted upon to provide lateral support for individual columns in a building frame, due consideration must be given to their stiffness when functioning as a horizontal diaphragm (Winter, 1958).

Although translation of the joints in the plane of a truss is inhibited and, due to end restraint, the effective length of compression members might therefore be assumed to be less than the distance between panel points, it is usual practice to take K as equal to 1.0, since, if all members of the truss reached their ultimate load capacity simultaneously, the restraints at the ends of the compression members would disappear or, at least, be greatly reduced. For K less than unity in trusses, see Galambos (1988).

CHAPTER D

TENSION MEMBERS

D1. ALLOWABLE STRESS

Due to strain hardening, a ductile steel bar loaded in axial tension can resist, without fracture, a force greater than the product of its gross area and its coupon yield stress. However, excessive elongation of a tension member due to uncontrolled yielding of its gross area not only marks the limit of its usefulness, but can precipitate failure of the structural system of which it is a part. On the other hand, depending upon the scale of reduction of gross area and the mechanical properties of the steel, the member can fail by fracture of the net area at a load smaller than required to yield the gross area. Hence, general yielding of the gross area and fracture of the net area both constitute failure limit states.

To prevent failure of a member loaded in tension, Sect. D1 has imposed a factor of 1.67 against yielding of the entire member and of 2.0 against fracture of its weakest effective net area.

The part of the member occupied by the net area at fastener holes has a negligible length relative to the total length of the member; thus, yielding of the net area at fastener holes does not constitute a limit state of practical significance. For the very rare case where holes or slots, other than rivet or bolt holes, are located in a tension member, it is conceivable that they could have an appreciable length in the direction of the tensile force. The failure limit states of general yielding on the gross area and fracture on the reduced area are still the principal limit states of concern. However, when the length of the reduced area exceeds the member depth or constitutes an appreciable portion of the member's length, yielding of the net area may become a serviceability limit state meriting special consideration and exercise of engineering judgment.

The mode of failure is dependent upon the ratio of effective net area to gross area and the mechanical properties of the steel. The boundary between these modes, according to the provisions of Sect. D1, is defined by the equation $A_e/A_g = 0.6F_y/0.5F_u$. When $A_e/A_g \geq F_y/0.833F_u$, general yielding of the member will be the failure mode. When $A_e/A_g < F_y/0.833F_u$, fracture at the weakest net area will be the failure mode.

In the case of short fittings used to transfer tensile force, an upper limit of 0.85 is placed on the ratio A_e/A_g. See B3.

D3. PIN-CONNECTED MEMBERS

Forged eyebars have generally been replaced by pin-connected plates or eyebars thermally cut from plates. Provisions for the proportioning of eyebars are based upon standards evolved from long experience with forged eyebars. Through extensive destructive testing, eyebars have been found to provide balanced designs when they are thermally cut instead of forged. The somewhat more conservative rules for pin-connected members of nonuniform cross section and those not having enlarged "circular" heads are likewise based on the results of experimental research (Johnston, 1939).

Somewhat stockier proportions are provided for eyebars and pin-connected members fabricated from steel having yield stress greater than 70 ksi to eliminate any possibility of their "dishing" under the higher working stress.

CHAPTER E

COLUMNS AND OTHER COMPRESSION MEMBERS

E1. EFFECTIVE LENGTH AND SLENDERNESS RATIO

The Commentary on Sect. C2 regarding frame stability and effective length factors applies here. Further analytical methods, formulas, charts and references for the determination of effective length are provided in the SSRC *Guide to Stability Design Criteria for Metal Structures* (Galambos, 1988).

E2. ALLOWABLE STRESS*

Equations (E2-1) and (E2-2) are founded upon the basic column strength estimate suggested by the Structural Stability Research Council. This estimate assumes that the upper limit of elastic buckling failure is defined by an average column stress equal to ½ of yield stress. The slenderness ratio C_c corresponding to this limit, can be expressed in terms of the yield stress of a given grade of structural steel as

$$C_c = \sqrt{\frac{2\pi^2 E}{F_y}} \qquad \text{(C-E2-1)}$$

A variable factor of safety has been applied to the column strength estimate to obtain allowable stresses. For very short columns, this factor has been taken as equal to, or only slightly greater than, that required for members axially loaded in tension, and can be justified by the insensitivity of such members to accidental eccentricities. For longer columns, entering the Euler slenderness range, the factor is increased 15% to approximately the value provided in the AISC Specification since it was first published.

To provide a smooth transition between these limits, the factor of safety has been defined arbitrarily by the algebraic equivalent of a quarter sine curve whose abscissas are the ratio of given Kl/r values to the limiting value C_c, and whose ordinates vary from 5/3 when Kl/r equals 0 to 23/12 when Kl/r equals C_c.

Equation (E2-2) covering slender columns (Kl/r greater than C_c) which fail by elastic buckling, is based upon a constant factor of safety of 23/12 with respect to the elastic (Euler) column strength.

E3. FLEXURAL-TORSIONAL BUCKLING

Torsional buckling of symmetric shapes and flexural-torsional buckling of unsymmetric shapes are failure modes usually not considered in the design of hot-rolled columns. They generally do not govern or the critical load differs very little from the weak axis planar buckling load. Such buckling loads may, however, control the capacity of symmetric columns made from relatively thin plate elements and of unsymmetric columns.

---

*For tapered members, also see Commentary Appendix F7.

Appendix E3 of the LRFD Specification (AISC, 1986) may be used to establish the effect of flexural-torsional buckling. The critical elastic buckling stress F_e can be obtained directly from the equations in LRFD Appendix E3. The effective slenderness is then given by

$$\left(\frac{KL}{r}\right)_e = \pi \sqrt{\frac{E}{F_e}} \qquad \text{(C-E2-2)}$$

The allowable stress is then obtained from Equations (E2-1) or (E2-2).

E4. BUILT-UP MEMBERS

Requirements for detailing of built-up members, which cannot be stated in terms of calculated stress, are based upon judgment, tempered by experience.

The longitudinal spacing of fasteners connecting components of built-up compression members must be such that the effective slenderness ratio K_a/r of the individual shape does not exceed 75% of the slenderness ratio Kl/r of the entire member. In addition, at least two intermediate connectors must be used along the length of the built-up member. To minimize the possibility of slip, the connectors must be welded or use high-strength bolts tightened to the requirements of Table J3.7. However, maximum fastener spacing less than that necessary to prevent local buckling may be needed to ensure a close fit-up over the entire faying surface of components designed to be in contact.

Provisions based on this latter consideration are of little structural significance. Hence, some latitude is warranted in relating them to the given dimensions of a particular member.

The provisions governing the proportioning of perforated cover plates are based on extensive experimental research (Stang and Jaffe, 1948).

E6. COLUMN WEB SHEAR

The column web shear stresses may be high within the boundaries of the rigid connection of two or more members whose webs lie in a common plane. Such webs should be reinforced when the calculated stress along plane $A\text{-}A$ in Fig. C-E6.1 exceeds the allowable shear stress

$$\Sigma F = \frac{M_1}{0.95d_1} + \frac{M_2}{0.95d_2} - V_s \qquad \text{(C-E6-1)}$$

$$\Sigma F/(d_c \times t_w) \le F_v \qquad \text{(C-E6-2)}$$

where:
$M_1 = M_{1L} + M_{1G} =$ sum of the moments due to the lateral load M_{1L} and the moments due to gravity load M_{1G} on the leeward side of the connection, kip-in.

$M_2 = M_{2L} - M_{2G} =$ difference between the moments due to lateral load M_{2L} and the moments due to gravity load M_{2G} on the windward side of the connection, kip-in.

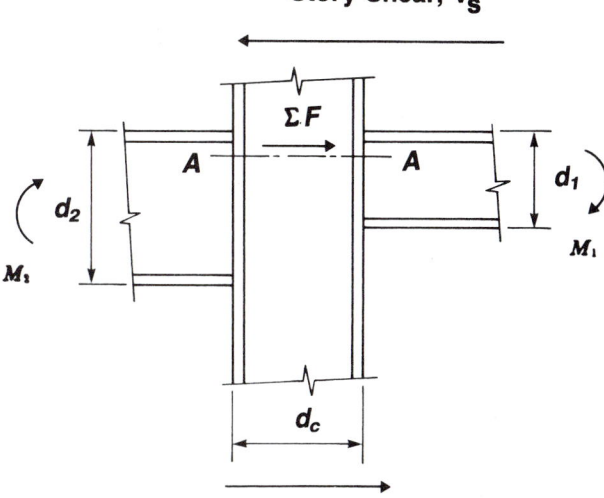

Fig. C-E6.1

CHAPTER F

BEAMS AND OTHER FLEXURAL MEMBERS

When flexural members, loaded to produce bending about their strong axis, are proportioned with width-thickness ratios not exceeding the noncompact section limits of Sect. B5, and are adequately braced to prevent the lateral displacement of the compression flange, they provide bending resistance equal at least to the product of their section modulus and yield stress, even when the width-thickness ratio of compressed elements of their profile is such that local buckling may be imminent. Lateral buckling of members bent about their strong axis may be prevented by bracing which either restrains the compression flange against lateral displacement or restrains the cross section against twisting which would induce bending about the weaker axis. Members bent solely about their minor axis, and members having approximately the same strength about both axes, do not buckle laterally and therefore may be stressed to the full allowable bending stress, consistent with the width-thickness proportions of their compression elements, without bracing.

F1. ALLOWABLE STRESS: STRONG AXIS BENDING OF I-SHAPED MEMBERS AND CHANNELS

1. Members with Compact Sections

Research in plastic design has demonstrated that local buckling will not occur in homogeneous sections meeting the requirements of Sect. F1.1 before the full plastic moment is reached. Practically all W and S shapes of A36 steel and a large proportion of these shapes having a yield stress of 50 ksi meet these provisions and are termed "compact" sections. It is obvious that the possibility of overload failure in bending of such rolled shapes must involve a higher level of stress (computed on the basis of M/S) than members having more slender compression elements. Since the shape factor of W and S beams is generally in excess of 1.12, the allowable bending stress for such members has been raised 10% from $0.60F_y$ to $0.66F_y$.

The further provisions permitting the arbitrary redistribution of 10% of the moment at points of support, due to gravity loading, gives partial recognition to the philosophy of plastic design. Subject to the restrictions provided in Sect. F1.1, continuous framing consisting of compact members may be proportioned on the basis of the allowable stress provisions of Chaps. D through K of the Specification when the moments, before redistribution, are determined on the basis of an elastic analysis. Fig. C-F1.1 illustrates the application of this provision by comparing calculated moment diagrams with the diagrams as altered by this provision.

2. Members with Noncompact Sections

Equation (F1-3) avoids an abrupt transition between an allowable bending stress of $0.66F_y$ when the half-flange width-to-thickness ratio of laterally supported compression flanges exceeds $65/\sqrt{F_y}$ and when this ratio is no more than $95/\sqrt{F_y}$. The assured hinge rotation capacity in this range is too small to permit redistribution of computed moment. Equation (F1-4) performs a similar function for homogeneous plate girders. See Commentary Sect. B5.

The allowable bending stress for all other flexural members is given as $0.60F_y$, provided the member is braced laterally at relatively close intervals ($l/b_f \leq 76/\sqrt{F_y}$).

3. **Members with Compact or Noncompact Sections With Unbraced Length Greater than L$_c$**

Members bent about their major axis and having an axis of symmetry in the plane of loading may be braced laterally at intervals greater than $76b_f/\sqrt{F_y}$ or $20{,}000/(d/A_f)F_y$ if the maximum bending stress is reduced sufficiently to prevent premature buckling of the compression flange.

The combination of Equations (F1-6) or (F1-7) and (F1-8) provides a reasonable design criterion in convenient form. Equations (F1-6) and (F1-7) are based on the assumption that only the bending stiffness of the compression flange will prevent the lateral displacement of that element between bracing points.

Equation (F1-8) is a convenient approximation which assumes the presence of both lateral bending resistance and St. Venant torsional resistance. Its agreement with more exact expressions for the buckling strength of intermittently braced flexural members (Galambos, 1988) is closest for homogeneous sections having substantial resistance to St. Venant torsion, identifiable in the case of

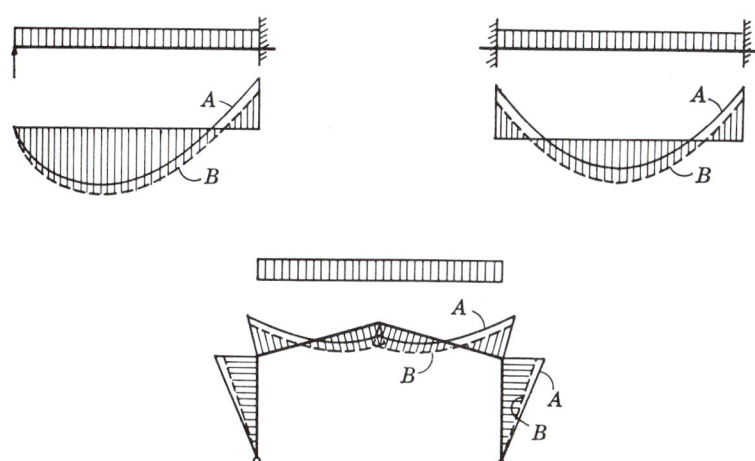

A = Actual moment diagram
B = Modified diagram corresponding to 10 percent moment reduction allowance at interior supports

Fig. C-F1.1

doubly symmetrical sections by a relatively low d/A_f ratio. Due to the difference between flange and web yield strength of a hybrid girder, it is desirable to base the lateral buckling resistance solely on warping torsion of the flange. Hence, use of Equation (F1-8) is not permitted for such members.

For some sections having a compression flange area distinctly smaller than the tension flange area, Equation (F1-8) may be unconservative; for this reason, its use is limited to sections whose compression flange area is at least as great as the tension flange. In plate girders, which usually have a much higher d/A_f ratio than rolled W shapes, Equation (F1-8) may err grossly on the conservative side. For such members, the larger stress permitted by Equation (F1-6), and at times by Equation (F1-7), affords the better estimate of buckling strength. Although these latter equations underestimate the buckling strength somewhat because they ignore the St. Venant torsional rigidity profile, this rigidity for such sections is relatively small and the margin of overconservatism, therefore, is likewise small.

Equation (F1-8) is written for the case of elastic buckling. A transition is not provided for this formula in the inelastic stress range because, when actual conditions of load application and variation in bending moment are considered, any unconservative error without the transition will be small.

Singly symmetrical, built-up, I-shape members, such as some crane girders, often have an increased compression flange area in order to resist bending due to lateral loading acting in conjunction with the vertical loads. Such members usually can be proportioned for the full allowable bending stress when the stress is produced by the combined vertical and horizontal loading. Where the failure mode of a singly symmetrical I-shape member having a larger compression than tension flange would be by lateral buckling, the allowable bending stress can be obtained by using Equations (F1-6), (F1-7) or (F1-8).

Through the introduction of the modifier C_b, some liberalization in stress is permissible when there is moment gradient over the unbraced length, except that C_b must be taken as unity when computing F_{bx} for use in Equation (H1-1) for frames braced against joint translation (Galambos, 1988).

Equations (F1-6) and (F1-7) may be refined to include both St. Venant and warping torsion by substituting a derived value for r_T. The equivalent radius of gyration, r_{Tequiv}, can be obtained by equating the appropriate expression giving the critical elastic bending stress for the compression flange of a beam with that of an axially loaded column (Galambos, 1988).

For the case of a doubly-symmetrical I-shape beam,

$$(r_{Tequiv})^2 = \frac{I_y}{2S_x}\sqrt{d^2 + \frac{0.156l^2J}{I_y}} \qquad \text{(C-F1-1)}$$

where

I_y = minor axis moment of inertia of the member
S_x = major axis section modulus

$$J = \frac{2b_f t_f^3}{3} + \frac{dt^3}{3} \qquad \text{(C-F1-2)}$$

Closer approximations of Equations (F1-7) and (F1-8) are given in Galambos, 1988.

F2. ALLOWABLE STRESS: WEAK AXIS BENDING OF I-SHAPED MEMBERS, SOLID BARS AND RECTANGULAR PLATES

The 25% increase in allowable bending stress for compact sections and solid rectangular bars bent about their weak axis, as well as for square and rectangular bars, is based upon the favorable shape factor present when these sections are bent about their weaker axis, and the fact that, in this position, they are not subject to lateral-torsional buckling. While the plastic bending strength of these shapes, bent in this direction, is considerably more than 25% in excess of their elastic bending strength, full advantage is not taken of this fact in order to provide elastic behavior at service loading.

Equation (F2-3), like Equation (F1-3), is a transition between the allowable bending stress of $0.75F_y$ at $b_f/2t_f = 65/\sqrt{F_y}$ and the lower stress of $0.60F_y$ at $b_f/2t_f = 95/\sqrt{F_y}$.

F3. ALLOWABLE STRESS: BENDING OF BOX MEMBERS, RECTANGULAR TUBES AND CIRCULAR TUBES

The provision for compact circular members is given in Table B5.1 (Sherman, 1976).

Supplement No. 3 (1974) to the 1969 Specification added Equation (F3-2), an unsupported length criteria for compact tubular members with rectangular cross sections. The equation recognizes the effect of moment gradient, and tests have shown it to be conservative (Sherman, 1976).

Box-type members are torsionally very stiff (Galambos, 1988). The critical flexural stress due to lateral-torsional buckling, for the compression flange of a box-type beam loaded in the plane of its minor axis so as to bend about its major axis, can be obtained using Equation (E2-1) with an equivalent slenderness ratio, by the expression

$$\left(\frac{l}{r}\right)_{equiv} = \sqrt{\frac{5.1lS_x}{\sqrt{J}I_y}} \qquad \text{(C-F3-1)}$$

where:
 l = distance between points of lateral support, in.
 S_x = elastic section modulus about major axis, in.[3]
 I_y = moment of inertia about minor axis, in.[4]
 J = torsional constant for a section, in.[4]

It can be shown that, when $d < 10b$ and $l/b > 2500/F_y$, the allowable compression flange stress indicated by the above equation will approximate $0.60F_y$. Beyond this limit, deflection rather than stress is likely to be the design criterion.

F4. ALLOWABLE SHEAR STRESS

Although the shear yield stress of structural steel has been variously estimated as between $\frac{1}{2}$ and $\frac{5}{8}$ of the tension and compression yield stress and is frequently taken as $F_y/\sqrt{3}$, it will be noted that the allowable value is given as $\frac{2}{3}$ the recommended basic allowable tensile stress, substantially as it has been since the first edition of the AISC Specification published in 1923. This apparent reduction in factor of safety is justified by the minor consequences of shear yielding, as compared with those associated with tension and compression yielding, and by the effect of strain hardening.

Although the allowable stress of $0.40F_y$ may be applied over the full area of the beam web, judgment should be used in cases where a connection length is considerably less than the depth of the beam.

When the computed average shear stress in the web is less than that permitted by Equation (F4-2), intermediate stiffeners are not required, provided the depth of the girders is limited to 260 times the web thickness. Such girders do not depend upon tension field action.

F5. TRANSVERSE STIFFENERS

In order to facilitate handling during fabrication and erection, when intermediate stiffeners are required the panel aspect ratio a/h is arbitrarily limited by Equation (F5-1) to $[260/(h/t_w)]^2$, with a maximum spacing of 3 times the girder depth.

CHAPTER   G

PLATE GIRDERS

G1. WEB SLENDERNESS LIMITATIONS

The limiting web depth-thickness ratio to prevent vertical buckling of the compression flange into the web, before attainment of yield stress in the flange due to flexure, may be increased when transverse stiffeners are provided, spaced not more than 1½ times the girder depth on centers.

The provision $h/t_w \leq 2000/\sqrt{F_y}$ is based upon tests on both homogeneous and hybrid girders with flanges having a specified yield stress of 100 ksi and a web of similar or weaker steel (ASCE-AASHO, 1968).

G2. ALLOWABLE BENDING STRESS

In regions of maximum bending moment, a portion of a thin web may deflect enough laterally on the compression side of the neutral axis that it does not provide the full bending resistance assumed in proportioning the girder on the basis of its moment of inertia. The compression stress which the web would have resisted is therefore shifted to the compression flange. But because the relative bending strength of this flange is so much greater than that of the laterally displaced portion of the web, the resulting increase in flange stress is at most only a few percent. The allowable design stress in the compression flange is reduced by the plate girder factor R_{PG} to ensure sufficient bending capacity is provided in the flange to compensate for any loss of bending strength in the web due to its lateral displacement.

To compensate for the slight loss of bending resistance when portions of the web or a hybrid flexural member are strained beyond their yield stress limit, the hybrid girder factor R_e reduces the allowable flange bending stress applicable to both flanges. The extent of the reduction is dependent upon the ratio of web area to flange area and of $.6F_{yw}$ to F_b. This is changed due to the reduction of F_{yf} based on local or lateral buckling. These reduction factors are multiplicable in the determination of the allowable bending stress for hybrid girders (Equation (G2-1)). This is to reflect the fact that the web continues to contribute some strength beyond the point of theoretical web buckling ($h/t_w = 760/\sqrt{F_b}$).

G3. ALLOWABLE SHEAR STRESS WITH TENSION FIELD ACTION

Unlike columns, which actually are on the verge of collapse as their buckling stage is approached, the panels of the plate girder web, bounded on all sides by the girder flanges or transverse stiffeners, are capable of carrying loads far in excess of their "web buckling" load. Upon reaching the theoretical buckling limit, very slight lateral displacements will have developed in the web. Nevertheless, they are of no structural significance, because other means are still present to assist in resisting further loading.

When transverse stiffeners are properly spaced and strong enough to act as compression struts, membrane stresses, due to shear forces greater than those associated with the theoretical buckling load, form diagonal tension fields. The resulting combination in effect provides a Pratt truss which, without producing yield stress in the steel, furnishes the capacity to resist applied shear forces unaccounted for by the linear buckling theory.

Analytical methods based upon this action have been developed (Basler and Thurlimann, 1963 and Basler, 1961) and corroborated in an extensive program of tests (Basler et al, 1960). These methods form the basis for Equation (G3-1). Use of tension field action is not counted upon when $0.60F_y\sqrt{3} \le F_u \le 0.40F_y$, nor when $a/h > 3.0$. Until further research is completed, it is not recommended for hybrid girders.

To provide adequate lateral support for the web, all stiffeners are required to have a moment of inertia at least equal to $(h/50)^4$. In many cases, however, this provision will be overshadowed by the gross area requirement. The amount of stiffener area necessary to develop the tension field, which is dependent upon the ratios a/h and h/t_w, is given by Equation (G4-1). Larger gross areas are required for one-sided stiffeners than for pairs of stiffeners, because of the eccentric nature of their loading.

The amount of shear to be transferred between web and stiffeners is not affected by the eccentricity of loading and generally is so small that it can be taken care of by the minimum sized fillet weld. The specified Equation (G4-3) affords a conservative estimate of required shear transfer under any condition of stress permitted by Equation (G3-1). The shear transfer between web and stiffener due to tension field action and that due to a concentrated load or reaction in line with the stiffeners are not additive. The stiffener need only be connected for the larger of the two shears.

G4. TRANSVERSE STIFFENERS

See Commentary G3.

G5. COMBINED SHEAR AND TENSION STRESS

Unless a flexural member is designed on the basis of tension field action, no stress reduction is required due to the interaction of concurrent bending and shear stress.

It has been shown that plate girder webs subject to tension field action can be proportioned on the basis of (Basler, 1979):
1. The allowable bending stress F_b, when the concurrent shear stress f_v is not greater than 0.60 of the allowable shear stress F_v or
2. The allowable shear stress F_v when the concurrent bending stress f_b is not greater than 0.75 of the allowable bending stress F_b.

Beyond these limits a linear interaction formula is provided in the AISC ASD Specification by Equation (G5-1).

However, because the webs of homogeneous girders of steel with yield points greater than 65 ksi loaded to their full capacity in bending develop more waviness than less-heavily-stressed girder webs of lower strength grades of steel, use of tension field action is limited in the case of webs with yield stress greater than 65 ksi to regions where the concurrent bending stress is no more than $0.75F_b$.

CHAPTER H

COMBINED STRESSES

H1. AXIAL COMPRESSION AND BENDING

The application of moment along the unbraced length of axially loaded members, with its attendant axial displacement in the plane of bending, generates a second-ary moment equal to the product of resulting eccentricity and the applied axial load, which is not reflected in the computed stress f_b. To provide for this added moment in the design of members subject to combined axial and bending stress, Equation (H1-1) requires that f_b be amplified by the factor

$$\frac{1}{\left(1 - \dfrac{f_a}{F'_e}\right)} \qquad \text{(C-H1-1)}$$

Depending upon the shape of the applied moment diagram (and, hence, the critical location and magnitude of the induced eccentricity), this factor may overestimate the extent of the secondary moment. To take care of this condi-tion the amplification factor is modified, as required, by a reduction factor C_m.

When bending occurs about both the x- and y-axes, the bending stress calcu-lated about each axis is adjusted by the value of C_m and F'_e corresponding to the distribution of moment and the slenderness ratio in its plane of bending. It is then taken as a fraction of the stress permitted for bending about that axis, with due regard to the unbraced length of compression flange where this is a factor.

When the computed axial stress is no greater than 15% of the permissible axial stress, the influence of

$$\frac{C_m}{\left(1 - \dfrac{f_a}{F'_e}\right)} \qquad \text{(C-H1-2)}$$

is generally small and may be neglected, as provided in Equation (H1-3). How-ever, its use in Equation (H1-1) is not intended to permit a value of f_b greater than F_b when the value of C_m and f_a are both small.

Depending upon the slenderness ratio of the given unbraced length of a mem-ber in the plane of bending, the combined stress computed at one or both ends of this length may exceed the combined stress at all intermediate points where lateral displacement is created by the applied moments. The limiting value of the combined stress in this case is established by Equation (H1-2).

The classification of members subject to combined axial compression and bending stresses is dependent upon two conditions: the stability against sides-way of the frame of which they are an integral part, and the presence or ab-sence of transverse loading between points of support in the plane of bending.

Note that f_b is defined as the computed bending stress *at the point under consideration*. In the absence of transverse loading between points of support, f_b is computed from the larger of the moments at these points of support. When intermediate transverse loading is present, the larger moment at one of the two supported points is used to compute f_b for use in the Equation (H1-2). However, to investigate the possibility of buckling failure, the maximum moment between points of support is used to compute f_b in Equation (H1-1).

In Equations (H1-1), (H1-2) and (H1-3), F_{bx} includes lateral-torsional buckling effects as provided in Equations (F1-6), (F1-7) and (F1-8).

Three categories are to be considered in computing values of C_m:

Category a covers columns in frames subject to sidesway, i.e., frames which depend upon the bending stiffness of their several members for overall lateral stability. For determining the value of F_a and F'_e, the effective length of such members, as discussed under C2, is never less than the actual unbraced length in the plane of bending, and may be greater than this length. The actual length is used in computing moments. For this case the value of C_m can be taken as

$$C_m = 1 - 0.18f_a/F'_e \qquad \text{(C-H1-3)}$$

However, under the combination of compression stress and bending stress most affected by the amplification factor, a value of 0.15 can be substituted for $0.18f_a/F'_e$. Hence, a constant value of 0.85 is recommended for C_m here.[*]

Category b applies to columns not subject to transverse loading in frames where sidesway is prevented. For determining the value of F_a and F'_e, the effective length of such members is never greater than the actual unbraced length and may be somewhat less. The actual length is used in computing moments.

For this category, the greatest eccentricity, and hence the greatest amplification, occurs when the end moments, M_1 and $-M_2$[**] are numerically equal and cause single curvature. It is least when they are numerically equal and of a direction to cause reverse curvature.

To properly evaluate the relationship between end moment and amplified moment, the concept of an equivalent moment M_e to be used in lieu of the numerically smaller end moment, has been suggested. M_e can be defined as the value of equal end moments of opposite signs which would cause failure at the same concurrent axial load as would the given unequal end

---

[*]See Commentary Sect. C2 for cases where C_m for unbraced frames 10 to 40 stories high may be computed as for braced frames.
[**]The sign convention for moments here and in Chap. H is that generally used in frame analysis. It should not be confused with the beam sign convention used in many textbooks. Moments are considered positive when acting clockwise about a fixed point, negative when acting counterclockwise.

moments. Then, M_e/M_2 can be written in terms of $\pm M_1/M_2$ as (Galambos, 1988):

$$\frac{M_e}{M_2} = C_m = \sqrt{0.3 \left(\frac{M_1}{M_2}\right)^2 - 0.4 \left(\pm \frac{M_1}{M_2}\right) + 0.3} \qquad \text{(C-H1-4)}$$

It has been noted that the simpler formulation (Austin, 1961):

$$C_m = 0.6 - 0.4 \left(\pm\frac{M_1}{M_2}\right) \geq 0.4 \qquad \text{(C-H1-5)}$$

affords a good approximation to this expression. The 0.4 limit on C_m corresponding to a M_1/M_2 ratio of 0.5, was included in the 1978 AISC Specification. The limit was intended to apply to lateral-torsional buckling and not to second-order, in-plane bending strength. As in the 1978 AISC Specification and the 1986 AISC LRFD Specification, this AISC ASD Specification uses a modification factor C_b as given in Sect. F1.3 for lateral-torsional buckling. C_b which is limited to 2.3, is approximately the inverse of C_m as presented in Austin (1961) with a 0.4 limit. In Zandonini (1985) it was pointed out this C_m equation could be used for in-plane second order moments if the 0.4 limit was eliminated. This adjustment has been made here, as it is in the 1986 AISC LRFD Specification.

Category c is exemplified by the compression chord of a truss subject to transverse loading between panel points, or by a simply supported column subjected to transverse loads between supports. For such cases, the value of C_m can be approximated using the equation:

$$C_m = 1 + \psi \frac{f_a}{F'_e} \qquad \text{(C-H1-6)}$$

Values of ψ for several conditions of transverse loading and end restraint (simulating continuity at panel points) are given in Table C-H1.1, together with two cases of simply supported beam-columns. In the case of continuity at panel points, f_b is maximum at the restrained ends or end, and the value of C_m for usual f_a/F'_e ratios is only slightly less than unity (a value of 0.85 is suggested in the Specification in the final paragraph of H1). For determinate (simply supported) beam-columns, f_b is maximum at or near midspan, depending upon the pattern of transverse loading. For this case

$$\psi = \frac{\pi^2 \delta_o E I}{M_o L^2} - 1 \qquad \text{(C-H1-7)}$$

where
  δ_o = maximum deflection due to tranverse loading
  M_o = maximum moment between supports due to transverse loading

If, as in the case of a derrick boom, such a beam-column is subject to transverse (gravity) load and a calculable amount of end moment, δ_o should include the deflection between supports produced by this moment.

It should be noted that, for amplified end moments in indeterminate members, stress alone is critical and is controlled by Equation (H1-2). For determinate members, where the amplified bending stress is maximum between supports, buckling-type failure is also of concern.

Note that F_a is governed by the maximum slenderness ratio, regardless of the plane of bending. F'_e, on the other hand, is always governed by the slenderness ratio in the plane of bending. Thus, when flexure is about the strong axis only, two different values of slenderness ratio may be involved in solving a given problem.

H2. Axial Tension and Bending

Contrary to the behavior in compression members, axial tension tends to *reduce* the bending stress because the secondary moment, which is the product of the deflection and the axial tension, is opposite in sense to the applied moment; thus, the secondary moment diminishes, rather than amplifies, the primary moment.

TABLE C-H1.1
Amplification Factors ψ and C_m

Case	ψ	C_m
	0	1.0
	-0.4	$1 - 0.4\dfrac{f_a}{F'_e}$
	-0.4	$1 - 0.4\dfrac{f_a}{F'_e}$
	-0.2	$1 - 0.2\dfrac{f_a}{F'_e}$
L/2	-0.3	$1 - 0.3\dfrac{f_a}{F'_e}$
	-0.2	$1 - 0.2\dfrac{f_a}{F'_e}$

CHAPTER I

COMPOSITE CONSTRUCTION

I1. DEFINITION

When the dimensions of a concrete slab supported by steel beams are such that the slab can effectively serve as the flange of a composite T-beam, and the concrete and steel are adequately tied together so as to act as a unit, the beam can be proportioned on the assumption of composite action.

Two cases are recognized: fully encased steel beams, which depend upon natural bond for interaction with the concrete, and beams with mechanical anchorage to the slab (shear connectors), which do not have to be encased.

For composite beams with formed steel deck, studies have demonstrated that total slab thickness, including ribs, can be used in determining effective slab width (Grant, Fisher and Slutter, 1977 and Fisher, 1970).

I2. DESIGN ASSUMPTIONS

Unless temporary shores are used, beams encased in concrete and interconnected only by a natural bond must be proportioned to support all of the dead load, unassisted by the concrete, plus the superimposed live load in composite action, without exceeding the allowable bending stress for steel provided in Chap. F.

Because the completely encased steel section is restrained from both local and lateral buckling, an allowable stress of $0.66F_y$, rather than $0.60F_y$, can be applied when the analysis is based on the properties of the transformed section. The alternate provision to be used in designs where a fully encased beam is proportioned, on the basis of the steel beam alone, to resist all loads at a stress not greater than $0.76F_y$, reflects a common engineering practice where it is desired to eliminate the calculation of composite section properties.

When shear connectors are used to obtain composite action, this action may be assumed, within certain limits, in proportioning the beam for the moments created by the sum of live and dead loads, even for unshored construction (Fisher, 1970). This liberalization is based upon an ultimate strength concept, although the provisions for proportioning of the member are based upon the elastic section modulus of the transformed cross section.

The flexural capacity of composite steel-concrete beams designed for complete composite action is the same for either lightweight or normal weight concrete, given the same area of concrete slab and concrete strength, but with the number of shear connectors appropriate to the type of concrete. The same concrete design stress level can be used for both types of concrete.

For unshored construction, so the steel beam under service loading will remain elastic, the superposition of precomposite and composite stresses is limited to $0.9F_y$. This direct stress check replaces the derived equivalent maximum transformed section modulus used in the past. The $0.9F_y$ stress limit only prevents permanent deformation under service loads and has no effect on the ultimate moment capacity of the composite beam.

On the other hand, to avoid excessively conservative slab-to-beam proportions, it is required that the flexural stress in the concrete slab, due to composite action, be computed on the basis of the transformed section modulus, referred to the top of concrete, and limited to the generally accepted working stress limit.

For a given beam and concrete slab, the increase in bending strength intermediate between no composite action and full composite action is dependent upon the shear resistance developed between the steel and concrete, i.e., the number of shear connectors provided between these limits (Slutter and Driscoll, 1965). Usually, it is not necessary, and occasionally it may not be feasible, to provide full composite action. Therefore, the AISC ASD Specification recognizes two conditions: full and partial composite action.

For the case where total shear V'_h developed between steel and concrete on each side of the point of maximum moment is less than V_h, Equation (I2-1) can be used to derive an effective section modulus S_{eff} having a value less than the section modulus for fully effective composite action S_{tr}, but more than that of the steel beam alone.

I4.　SHEAR CONNECTORS

Composite beams in which the longitudinal spacing of shear connectors has been varied according to the intensity of shear, and duplicate beams where the required number of connectors were uniformly spaced, have exhibited the same ultimate strength and the same amount of deflection at normal working loads. Only a slight deformation in the concrete and the more heavily stressed shear connectors is needed to redistribute the horizontal shear to the other less heavily stressed connectors. The important consideration is that the total number of connectors be sufficient to develop the shear V_h either side of the point of maximum moment. The provisions of the AISC ASD Specification are based upon this concept of composite action.

In computing the section modulus at points of maximum negative bending, reinforcement parallel to the steel beam and lying within the effective width of slab may be included, provided such reinforcement is properly anchored beyond the region of negative moment. However, enough shear connectors are required to transfer, from slab to the steel beam, one-half of the ultimate tensile strength of the reinforcement.

Studies have defined stud shear connection strength Q_u in terms of normal weight and lightweight aggregate concretes, as a function of both concrete modulus of elasticity and concrete strength (McGarraugh and Baldwin, 1971 and Ollgaard, Slutter and Fisher, 1971):

$$Q_u = 0.5A_s\sqrt{f_c' E_c}$$ (C-I4-1)

where

A_s = cross-sectional area of stud, in.^{2}
f_c' = concrete compressive strength, ksi
E_c = concrete modulus of elasticity, ksi

Tests have shown that fully composite beams designed using the values in Tables I4.1 and/or I4.2 as appropriate, and concrete meeting the requirements of Part 3, Chap. 4, "Concrete Quality", of ACI Standard 318-83 made with ASTM C33 or C330 aggregates, develop their full flexural capacity (Ollgaard, Slutter and Fisher, 1971). For normal weight concrete, compressive strengths greater than 4.0 ksi do not increase the shear capacity of the connectors, as is reflected in Table I4.1. For lightweight concrete, compressive strengths greater than 5 ksi do not increase the shear capacity of the connectors. The reduction coefficients in Table I4.2 are applicable to both stud and channel shear connectors and provide comparable margins of safety.

When partial composite action is counted upon to provide flexural capacity, the restriction on the minimum value of V_h' is to prevent excessive slip as well as substantial loss in beam stiffness. Studies indicate that Equations (I2-1) and (I4-4) adequately reflect the reduction in strength and beam stiffness, respectively, when fewer connectors than required for full composite action are used.

Where adequate flexural capacity is provided by the steel beam alone, that is, composite action to any degree is not required for flexural strength, but where it is desired to provide interconnection between the steel frame and the concrete slab for other reasons, such as to increase frame stiffness or to take advantage of diaphragm action, the minimum requirement that V_h be not less than $V_h/4$ does not apply.

The required shear connectors can generally be spaced uniformly between the points of maximum and zero moment (Slutter and Driscoll, 1965). However, certain loading patterns can produce a condition where closer connector spacing is required over part of this distance.

For example, consider the case of a uniformly loaded simple beam also required to support two equal concentrated loads, symmetrically disposed about midspan, of such magnitude that the moment at the concentrated loads is only slightly less than the maximum moment at midspan. The number of shear connectors N_2 required between each end of the beam and the adjacent concentrated load would be only slightly less than the number N_1 required between each end and midspan.

Equation (I4-5) is provided to determine the number of connectors, N_2, required between one of the concentrated loads and the nearest point of zero moment. It is based upon the following requirement:

$$\frac{N_2}{N_1} = \frac{S - S_s}{S_{eff} - S_s} = \frac{\left[\dfrac{S}{S_{eff}} \times \dfrac{S_{eff}}{S_s}\right] - 1}{\dfrac{S_{eff}}{S_s} - 1} \qquad \text{(C-I4-2)}$$

where

S = section modulus required at the concentrated load at which location moment equals M, in.^{3}

S_{eff} = section modulus required at M_{max} (equal to S_{tr} for fully composite case), in.^{3}

S_s = section modulus of steel beam, in.^{3}

N_1 = number of studs required from M_{max} to zero moment

N_2 = number of studs required from M to zero moment

M = moment at the concentrated load point

M_{max} = maximum moment in the beam

Noting that $S/S_{eff} = M/M_{max}$, and defining β as S_{eff}/S_s, the above equation is equivalent to Equation (I4-5).

With the issuance of Supplement No. 3 to the 1969 AISC Specification, the requirement for 1-in. cover over the tops of studs was eliminated. Only the concrete surrounding the stud below the head contributes to the strength of the stud in resistance to shear. When stud shear connectors are installed on beams with formed steel deck, concrete cover at the sides of studs adjacent to sides of steel ribs is not critical. Tests have shown that studs installed as close as is permitted to accomplish welding of studs does not reduce the composite beam capacity.

Stud welds not located directly over the web of a beam tend to tear out of a thin flange before attaining their full shear-resisting capacity. To guard against this contingency, the size of a stud not located over the beam web is limited to 2½ times the flange thickness.

I5. COMPOSITE BEAMS OR GIRDERS WITH FORMED STEEL DECK

The 6-diameter minimum center-to-center spacing of studs in the longitudinal direction is based upon observation of concrete shear failure surfaces in sectioned flat soffit concrete slab composite beams which had been tested to full ultimate strength. The reduction in connection capacity of more closely spaced shear studs within the ribs of formed steel decks oriented perpendicular to beam or girder, is accounted for by the parameter $0.85/\sqrt{N_r}$ in Equation (I5-1).

When studs are used on beams with formed steel deck, they may be welded directly through the deck or through prepunched or cut-in-place holes in the deck. The usual procedure is to install studs by welding directly through the deck; however, when the deck thickness is greater than 16 ga. for single thickness, or 18 ga. for each sheet of double thickness, or when the total thickness of galvanized coating is greater than 1.25 ounces per sq. ft., special precautions and procedures recommended by the stud manufacturer should be followed.

Fig. C-I5.1 is a graphic presentation of the terminology used in Sect. I5.1.

The design rules which have been added for composite construction with formed steel deck are based upon a study of all available test results (Grant, Fisher and Slutter, 1977). The limiting parameters listed in Sect. I5.1 were established to keep composite construction with formed steel deck within the available research data.

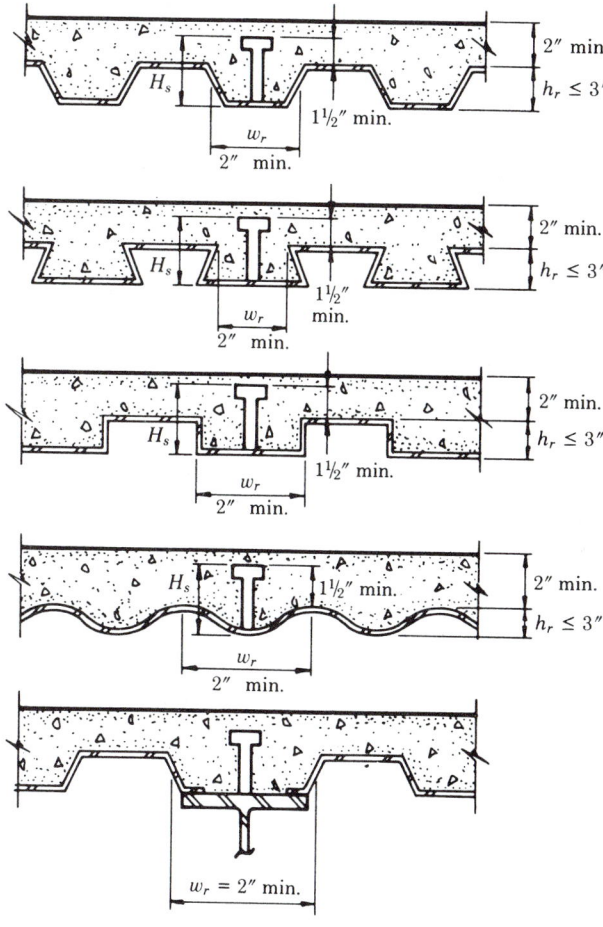

Figure C-I5.1

Seventeen full-sized composite beams with concrete slab on formed steel deck were tested at Lehigh University and the results supplemented by the results of 58 tests performed elsewhere. The range of stud and steel deck dimensions encompassed by the 75 tests were limited to:

1. Stud dimensions: ¾-in dia. × 3.00 to 7.00 in.
2. Rib width: 1.94 in. to 7.25 in.
3. Rib height: 0.88 in. to 3.00 in.
4. Ratio w_r/h_r: 1.30 to 3.33
5. Ratio H_s/h_r: 1.50 to 3.41
6. Number of studs in any one rib: 1, 2, or 3

Based upon all tests, the strength of stud connectors in flat soffit composite slab beams, determined in previous test programs, when multiplied by values computed from Equation (I5-1), reasonably approximates the strength of stud connectors installed in the ribs of concrete slabs on formed steel deck with the ribs oriented perpendicular to the steel beam (Ollgaard, Slutter and Fisher, 1971). Hence, Equation (I5-1) provides a reasonable reduction factor to be applied to the allowable design stresses in Tables I4.1 and I4.2.

Testing has shown that the maximum spacing of shear connectors can be increased to 36 in. instead of the previous value of 32 in. (Klyce, 1988).

For the case where ribs run parallel to the beam, limited testing has shown that shear connection is not significantly affected by the ribs (Grant, Fisher and Slutter, 1977). However, for narrow ribs, where the ratio w_r/h_r is less than 1.5, a shear stud reduction factor, Equation (I5-2), has been suggested in view of lack of test data.

The Lehigh study also indicated that Equation (I2-1) for effective section modulus and Equation (I4-4) for effective moment of inertia were valid for composite construction with formed steel deck (Grant, Fisher and Slutter, 1977).

When metal deck includes units for carrying electrical wiring, crossover headers are commonly installed over the cellular deck, perpendicular to the ribs, in effect creating trenches which completely or partially replace sections of the concrete slab above the deck. These trenches, running parallel to or transverse to a composite beam, may reduce the effectiveness of the concrete flange. Without special provisions to replace the concrete displaced by the trench, the trench should be considered as a complete structural discontinuity in the concrete flange. When trenches are parallel to the composite beam, the effective flange width should be determined from the known position of the trench.

Trenches oriented transverse to the composite beam should, if possible, be located in areas of low bending moment and the full required number of studs should be placed between the trench and the point of maximum positive moment. Where the trench cannot be located in an area of low moment, the beam should be designed as non-composite.

CHAPTER J

CONNECTIONS, JOINTS AND FASTENERS

J1. GENERAL PROVISIONS

7. Splices in Heavy Sections

Solidified but still hot weld metal contracts significantly as it cools to ambient temperature. Shrinkage of large welds between elements which are not free to move to accommodate the shrinkage causes strains in the material adjacent to the weld that can exceed the yield point strain. In thick material, the weld shrinkage is restrained in the thickness direction as well as in the width and length directions causing triaxial stresses to develop that may inhibit the ability of ductile steel to deform in a ductile manner. Under these conditions, the possibility of brittle fracture increases.

When splicing ASTM Group 4 and 5 rolled sections or heavy welded built-up members, the potentially harmful weld shrinkage strains can be avoided by use of bolted splices or fillet welded lap splices or a splice using a combination welded and bolted detail (Fig. C-J1.1). Details and techniques, that perform well for materials of modest thickness usually must be changed or supplemented by more demanding requirements when welding thick material. Also, the provisions of Structural Welding Code AWS D1.1 are minimum requirements that apply to most structural welding situations; however, when designing and fabricating welded splices of Group 4 and 5 shapes and similar built-up

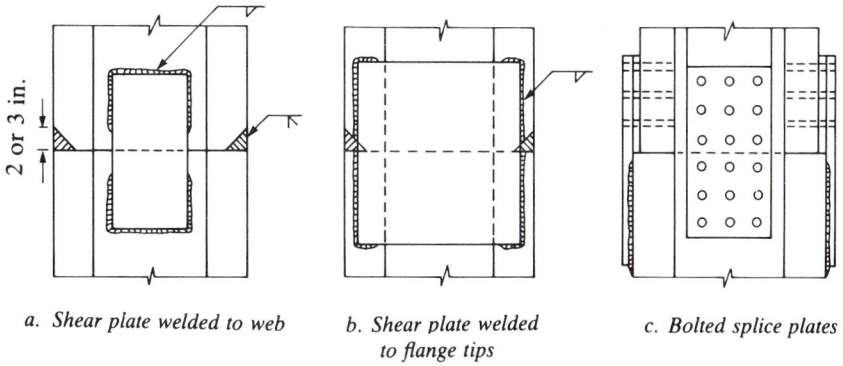

a. Shear plate welded to web

b. Shear plate welded to flange tips

c. Bolted splice plates

Fig. C-J1.1. Alternative splices that minimize weld resistant tensile stresses

cross sections special consideration must be given to all aspects of the welded splice detail. These are as follows:

- Notch-tough requirements should be specified for tension members. See Commentary A3.1c.
- Generously sized weld access holes (Fig. C-J1.2) are required to provide increased relief from concentrated weld shrinkage strains, to avoid close juncture of welds in orthogonal directions, and to provide adequate clearance for the exercise of high quality workmanship in hole preparation, welding and ease of inspection.
- Preheating for thermal cutting is required to minimize the formation of a hard surface layer.
- Grinding to bright metal and inspection using magnetic particle or dye penetrant methods is required to remove the hard surface layer and to assure smooth transitions free of notches or cracks.

In addition to tension splices of truss chord members and tension flanges of flexural members, other joints fabricated of heavy sections subject to tension should be given special consideration during design and fabrication.

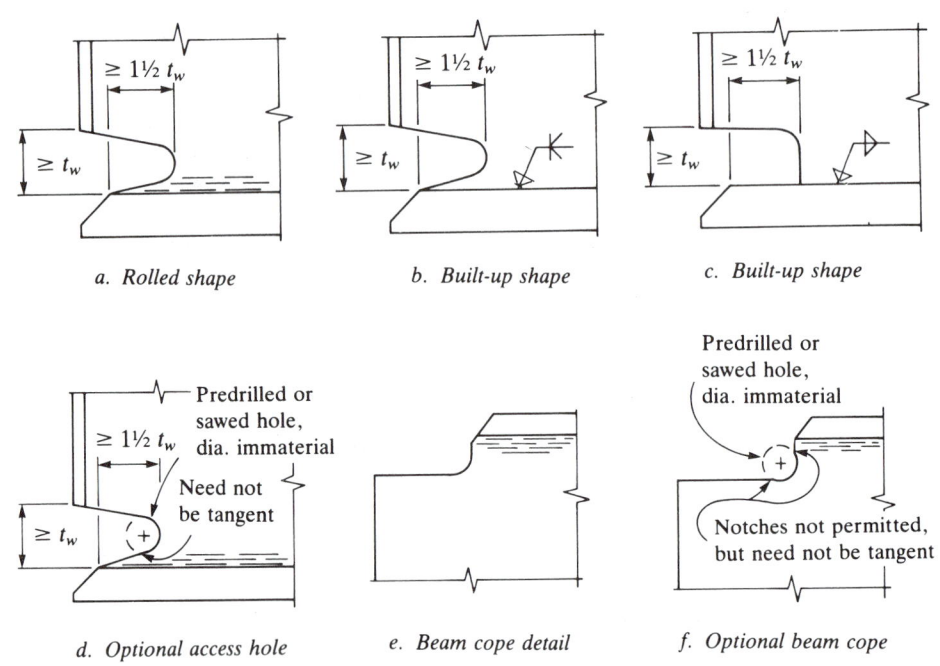

a. Rolled shape   *b. Built-up shape*   *c. Built-up shape*

d. Optional access hole   *e. Beam cope detail*   *f. Optional beam cope*

Note: For Group 4 and 5 shapes and welded built-up members made of material more than 2-in. thick, preheat prior to thermal cutting, grind and inspect thermally cut edges using magnetic particle or dye penetrant methods.

Fig. C-J1.2. Weld access hole and beam cope geometry

9. Placement of Welds, Bolts and Rivets

Slight eccentricities between the gravity axis of single- and double-angle members and the center of gravity of their connecting bolts or rivets have long been ignored as having negligible effect upon the static strength of such members. Tests have shown that similar practice is warranted in the case of welded members in statically loaded structures (Gibson and Wake, 1942). However, the fatigue life of single angles, loaded in tension or compression, has been shown to be very short (Koppel and Seeger, 1964).

10. Bolts In Combination with Welds

High-strength bolts used in bearing-type connections should not be required to share load with welds. High-strength bolts used in slip-critical connections, however, because of the rigidity of the connection, may be proportioned to function in conjunction with welds in resisting the transfer of stress across faying surfaces. Because the welds, if installed prior to final tightening of the bolts, might interfere with the development of the high contact pressure between faying surfaces that is counted upon in slip-critical connections, it is advisable that the welds be made after the bolts are tightened. At the location of the fasteners, the heat of welding the connected parts will not alter the mechanical properties of the fasteners.

In making alterations to existing structures, it is assumed whatever slip is likely to occur in high-strength bolted, bearing-type connections will have already taken place. Hence, in such cases the use of welding to resist all stresses other than those produced by existing dead load present at the time of making the alteration is permitted.

J2. WELDS

The requirements of the AWS Code have been adopted by reference, with four exceptions and most requirements governing welding workmanship have been deleted. For convenience of the designer, provisions for allowable design stresses and proportioning of welds have been retained, even though the AISC and AWS provisions are consistent.

The provisions of the AWS *Structural Welding Code* to which exception is taken in the AISC ASD Specification are as follows:

1. Section 2.3.2.4 of the AWS Code and Sect. J2.2a of the AISC ASD Specification both define the effective throat of fillet welds as the shortest distance from the root to the face of the diagrammatic weld. However, for fillet welds made by the submerged arc process, Sect. J2.2a additionally recognizes the deep penetration that is provided by this automatic process at the root of the weld beyond the limits of the diagrammatic weld.

2. Section 2.5 of the AWS Code prohibits the use of partial-penetration welds subject to cyclic tension normal to the longitudinal axis of the weld, whereas the AISC ASD Specification Appendix K4 recognizes partial-penetration welds subject to fatigue loading, but only at the same severely limited stress ranges of Category F that are appropriate to fillet welds.

3. Section 8.13 of the AWS Code provides criteria for the flatness of girder webs, which are arbitrary and based upon a concern for possible cyclic secondary stresses resulting from breathing action of thin girder webs subject to fatigue loading. The AISC ASD Specification does not include such criteria, because lateral deflection or out-of-flatness of webs of girders subject to static loading is of no structural significance. If architectural appearance of exposed girders is of importance, then tolerances based upon specific consideration of architectural requirements, rather than tolerances based upon unrelated consideration of fatigue effects should be provided in the project specification.

4. Section 9 of the AWS Welding Code is applicable to bridges, which are outside the scope of the AISC ASD Specification. Therefore, no comparable provisions are included in the AISC ASD Specification.

5. Section 10 of the AWS Welding Code is applicable to offshore construction, which is outside the scope of the AISC ASD Specification. Therefore, no comparable provisions are included in the AISC ASD Specification.

As in earlier editions, the Specification accepts, without further procedure qualification, numerous weld and joint details executed in accordance with the provisions of the AWS Code. Other welding procedures may be used, provided they are qualified to the satisfaction of the designer and the building code authority and are executed in accordance with the provisions of the AWS Code.

4. Allowable Stresses

The strength of welds is governed by the strength of either the base material or the deposited weld metal.

It should be noted that in Table J2.5 the allowable stress of fillet welds is determined from the effective throat area, whereas the design of the connected parts is governed by their respective thicknesses. Fig. C-J2.1 illustrates the shear planes for fillet welds and base material:

a. Plane 1-1, in which the design is governed by the shear strength for material A

b. Plane 2-2, in which the design is governed by the shear strength of the weld metal

c. Plane 3-3, in which the design is governed by the shear strength of material B

The design of the welded joint is governed by the weakest plane of shear transfer. Note that planes 1-1 and 3-3 are positioned away from the fusion areas between the weld and the base material. Tests have demonstrated that the stress on this fusion area is normally not critical in determining the shear strength of fillet welds. (Preece, 1968) However, if the weld metal is overstrength as might occur when materials with two different strength levels are connected, then the shear plane of the lower strength material at the fusion area may govern. The allowable shear stress on the leg of the weld at the lower strength base metal will be 0.3 times the tensile strength of the base metal.

As in the past, the allowable stresses for statically loaded full-penetration welds are the same as those permitted for the base metal, provided the me-

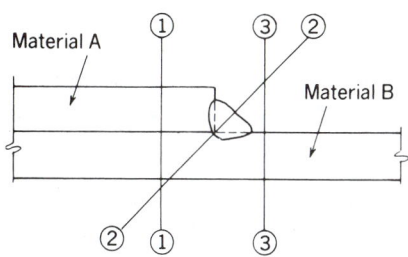 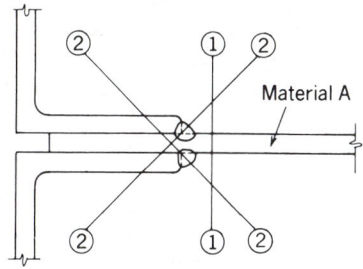

Fig. C-J2.1 Alternative column splices that minimize weld restraint tensile stresses (From: Fisher, J.W. and Pense, A.W. 1987)

chanical properties of the electrodes used are such as to match or exceed those of the weakest grade of base metal being joined.

On the basis of physical tests, the allowable stress on fillet welds deposited on "matching" base metal, or on steel having mechanical properties higher than those specified for such base metal, has been given in terms of the nominal tensile strength* of the weld metal since the 1969 edition of the Specification (Higgins and Preece, 1968).

As in the past, the same allowable value is given to a transverse as to a longitudinal weld, even though the force the former can resist is substantially greater than that of the latter. In the case of tension on the throat of partial-penetration groove welds normal to their axis (more nearly analogous to that of transverse than longitudinal fillets), the allowable stress is conservatively taken the same as for fillet welds.

When partial-penetration groove welds are so disposed that they are stressed in tension parallel to the longitudinal axis of the groove, or primarily in compression or bearing, they may be proportioned to resist such stress at the same unit value permitted in the base metal.

6. Mixed Weld Metal

Instances have been reported in which tack welds deposited using a self-shielded process with aluminum deoxidizers (which by itself provided notch-tough weld metal) were subsequently covered by weld passes using a submerged arc process (which by itself provided notch-tough weld metal) resulted in composite weld metal with low notch-toughness (Terashima and Hart, 1984; Kotecki and Moll, 1970; and Kotecki and Moll, 1972).

J3. BOLTS, THREADED PARTS AND RIVETS

The provisions for mechanical fasteners are based on an extended review and reexamination of the large body of data growing out of voluminous research, which has been completed in the past two decades. In order to consolidate and

---

*See Commentary Sect. A3.

organize this material and, for the convenience of the engineering profession, to present concise, rational and well balanced conclusions within the covers of a single volume, the Research Council on Structural Connections sponsored the preparation of the 2nd Edition of *Guide to Design Criteria for Bolted and Riveted Joints*, (Kulak, Fisher and Struik, 1987) (in subsequent references this publication will be noted as the "Guide").

The first edition of the Guide was published in 1974 and has provided the background for two revisions of the *Specification for Structural Joints Using ASTM A325 or A490 Bolts* of the Research Council on Structural Connections. The most recent version was approved Sept. 1, 1986. Likewise, it has been the basis for the revision of AISC ASD Specification provisions concerning mechanically fastened structural connections.

At the outset, the Guide notes a distinction between a factor of safety adequate to prevent loss of usefulness of a structure, member, or connection, and one needed to insure against complete failure of these entities. In the latter case, it notes that, under the long-standing misconception of "balanced design," when the weakest element of a joint has a factor of safety of 2, other elements may be grossly overdesigned, with attendant loss in economy (Kulak, Fisher and Struik, 1987).

The balanced design concept may have been valid when there was but one grade of structural steel and but one grade of fastener. However, it has lost its meaning with today's multiplicity of both fastener and connected material strengths.

Based on the earlier criteria, the weakest component in some of the largest and most important joints of existing structures have a factor of safety no greater than 2, yet they have proven with time to be entirely satisfactory. The Guide has adopted this value as basic with respect to failure, increasing it somewhat in rounding off to even working stress values or, as in the case of slip-resistance, reducing it somewhat when impairment of usefulness alone is at stake. With considerable accumulation of data now available as to the effectiveness of joint components under various loading conditions, probabilistic methods of statistical analysis have been used in determining the critical stress to which the factor of safety should be applied (Kulak, Fisher and Struik, 1987).

Provision for the limited use of A449 bolts, in lieu of A325 bolts, is predicated on the fact that the provisions of ASTM A449 concerning quality control are less stringent that those contained in ASTM A325. These bolts differ from A325 bolts only as to reduced size of head and increased length of threading.

4. Allowable Tension and Shear

Allowable stresses for rivets are given in terms applicable to the nominal cross-sectional area of the rivet before driving. For convenience in the proportioning of high-strength bolted connections, allowable stresses for bolts and threaded items are given in terms applicable to their nominal body area, i.e., the area of the threaded part based on its major diameter.

Except as provided in Appendix K4.3, any additional fastener tension resulting from prying action due to distortion of the connection details should be added to the stress calculated directly from the applied tension in proportioning fasteners for an applied tensile force, using the specified allowable stresses. Depending upon the relative stiffness of the fasteners and the connection material, this prying action may be negligible or it may be a substantial part of the total tension in the fasteners (Kulak, Fisher and Struik, 1987).

Mechanically fastened connections which transmit load by means of shear in their fasteners are categorized as either slip-critical or bearing type. The former depend upon sufficiently high clamping force to prevent slip of the connected parts under anticipated service conditions. The latter depend upon contact of the fasteners against the sides of their holes to transfer the load from one connected part to another.

The amount of clamping force developed by shrinkage of a rivet after cooling and by A307 bolts is unpredictable and generally insufficient to prevent complete slippage at the allowable stress. Riveted connections and connections made with A307 and A449 bolts for shear are treated as bearing-type. The high clamping force produced by properly tightened A325 and A490 bolts is sufficient to assure that slip will not occur at full allowable stress in slip-critical connections and probably will not occur at service loads in bearing-type connections.

The working values given in Table J3.2 for slip-critical and bearing-type shear connections are, with only minor modifications based on reliability analysis of existing data, equivalent to those in previous editions of the AISC ASD Specification for use with A325 and A490 bolts in standard or slotted holes with tight mill scale surfaces.

The requirement of footnote f in Table J3.2, which calls for a 20% reduction in allowable fastener shear stress, as noted in the Guide, is based upon tests on butt-type splice specimens where all connected parts were loaded in tension. This footnote provision would not apply to connection angles at the ends of plate girders which transmit the girder reaction to the supporting member by means of shear in the connection angles. Nor would the distance between extreme fasteners in tension members connected at opposite edges of a gusset plate govern; instead, the length of the connection for each tension member would control the design.

Bearing-type connections are intended for use where service conditions are such that cyclic loading approaching complete stress reversal will not occur, and deformation of the structural frame or a component thereof, due to slip of the connection into bearing, can be tolerated. The allowable stresses in this case are based upon a factor of safety of 2 or more, which over a long period has been found to be adequate. This is substantially higher than that which is basic to the design of the connected members.

The efficiency of threaded fasteners in resisting shear in bearing-type connections is reduced when the threading extends into the shear plane between the connected parts. Except in the case of A307 bolts, two allowable shear stress values are given: one when threading is excluded from the shear plane and one when it is not. In selecting appropriate allowable shear stresses, it was deemed an unwarranted refinement to make a distinction between threads in a single shear plane and threads in two planes (double shear of an enclosed part). Therefore, the allowable stresses were established on the conservative assumption of threads in two planes. Because it is not customary to control this feature in the case of A307 bolts, and because the length of threading on A307 and A449 bolts is greater than on A325 and A490 bolts, it is assumed threading may extend into the shear plane and the allowable shear value, applicable to the gross area, is reduced accordingly.

5. Combined Tension and Shear in Bearing-type Connections

The strength of fasteners subject to combined tension and shear is provided by elliptical interaction curves in Table J3.3 for A325 and A490 bolts, which account for the connection length effect on bolts loaded in shear, the ratio of shear strength to tension strength of threaded fasteners and the ratios of root area to nominal body area and tensile stress area to nominal body area (Yura, 1987). The elliptical interaction curve provides the best estimate of the strength of bolts subject to combined shear and tension and thus is used in this Specification.

6. Combined Tension and Shear in Slip-critical Joints

In the case of slip-critical connections subject to combined tension and shear at the contact surface common to a beam connection and the supporting member, where the fastener tension f_t is produced by moment in the plane of the beam web, the shear component may be neglected in proportioning the fasteners for tension. This is because the shear component assigned to the fasteners subject to direct tensile stress is picked up by the increase in compressive force on the compression side of the beam axis, resulting in no actual shear force on the fasteners in tension.

However, when a slip-critical connection must resist an axially applied tensile force, the clamping force is reduced and F_v must be reduced in proportion to the loss of pretension.

7. Allowable Bearing at Bolt Holes

Bearing values are provided, not as a protection to the fastener (because it needs no such protection) but for the protection of the connected parts. Therefore, the same bearing value applies to joints assembled by bolts, regardless of fastener shear strength or the presence or absence of threads in the bearing area.

It should be noted that the value for bearing stress $1.5F_u$ is the *maximum* allowable value provided deformation around the bolt hole is not a design consideration. As explained under Sects. J3.8 and J3.9 of this Commentary, this maximum value is permitted only if the end distance and intermediate spacing of fasteners, measured in the direction of applied force, are adequate to prevent failure by splitting of a connected part parallel to the line of force at a load less than required to cause transverse fracture through the net area of the part.

Tests have demonstrated that hole elongation greater than 0.25 in. will begin to develop as the bearing stress is increased beyond the values given in Equations (J3-1) and (J3-2), especially if it is combined with high-tensile stress on the net section, even though rupture does not occur. Equation (J3-4) considers the effect of hole ovalization.

Although the possibility of a slip-critical connection slipping into bearing under anticipated service conditions is extremely remote, such connections should comply with the provisions of Sects. J3.4 and J3.8 to insure the usual minimum factor of safety of 2 against complete connection failure.

8. Minimum Spacing

Critical bearing stress is a function of the material tensile strength, the spacing of fasteners, and the distance from the edge of the part to the centerline of the nearest fastener. Tests have shown that a linear relationship exists between the ratio of critical bearing stress to tensile strength of the connected material and the ratio of fastener spacing (in the line of force) to fastener diameter (Kulak, Fisher and Struik, 1987). The following equation affords a good lower bound to published test data for single-fastener connections with standard holes, and is conservative for adequately spaced multi-fastener connections:

$$\frac{F_{pcr}}{F_u} = \frac{l_e}{d} \qquad \text{(C-J3-1)}$$

where

F_{pcr} = critical bearing stress
F_u = tensile strength of the connected material
l_e = distance, along a line of transmitted force, from the center of a fastener to the nearest edge of an adjacent fastener or to the free edge of a connected part (in the direction of stress), in.
d = diameter of a fastener, in.

This equation, modified by a safety factor of 2, is the basis for Equations (J3-5) and (J3-6).

Along a line of transmitted force, the required spacing center-to-center of standard holes is found from Equation (J3-5). For oversized and slotted holes, this spacing is increased by an increment C_1, given in Table J3.4, providing the same clear distance between holes as for standard holes.

The required edge distance in the direction of stress is found from Equation (J3-6) as the distance from the center of a standard hole to the edge of a connected part. For oversized and slotted holes, this distance is increased by an increment C_2, given in Table J3.6, providing the same clear distance from the edge of the hole as for a standard hole.

The provisions of Sect. J3.8 are concerned with l_e as hole spacings, whereas Sect. J3.9 is concerned with l_e as edge distance L_e in the direction of stress, and Sect. J3.7 establishes a maximum allowable bearing stress. Spacing and/or edge distance may be increased to provide for a required bearing stress, or bearing force may be reduced to satisfy a spacing and/or edge distance limitation.

The critical bearing stress of a single fastener connection is more dependent upon a given edge distance than multi-fastener connections (Jones, 1940). For this reason, longer edge distances (in the direction of force) are required for connections with one fastener in the line of transmitted force than required for those having two or more.

9. Minimum Edge Distance

See Commentary Sect. J3.8.

10. Maximum Edge Distance and Spacing

See Brockenbrough (1983).

11. Long Grips

Provisions requiring a decrease in calculated stress for A307 bolts having long grips (by arbitrarily increasing the required number an amount in proportion to the grip length) are not required for high-strength bolts. Tests have demonstrated the ultimate shearing strength of high-strength bolts having a grip of 8 or 9 diameters is no less than that of similar bolts with much shorter grips (Bendigo, Hansen and Rumpf, 1963).

J4. ALLOWABLE SHEAR RUPTURE

Tests have shown high-strength-bolted beam end connections which subject a coped web to high bearing stresses may cause a tearing failure mode where a portion of the beam web tears out along the perimeter of the holes (Birkemoe and Gilmor, 1978). The tests demonstrated the failure load can be predicted using an analytical model which combines ultimate shear strength of the net section subject to shear stress with the ultimate tensile strength of the net section subject to tensile stress. More recent research has suggested an alternative approach (Ricles and Yura, 1983 and Hardash and Bjorhovde, 1985).

The block shear failure mode is not limited to coped ends of beams (Fig. C-J4.1). Other examples are shown in Figs. C-J4.2, C-J4.3 and C-J4.4.

There may be similar connections, such as thin bolted gusset plates in double shear, where this type of failure could occur. Such situations should be investigated.

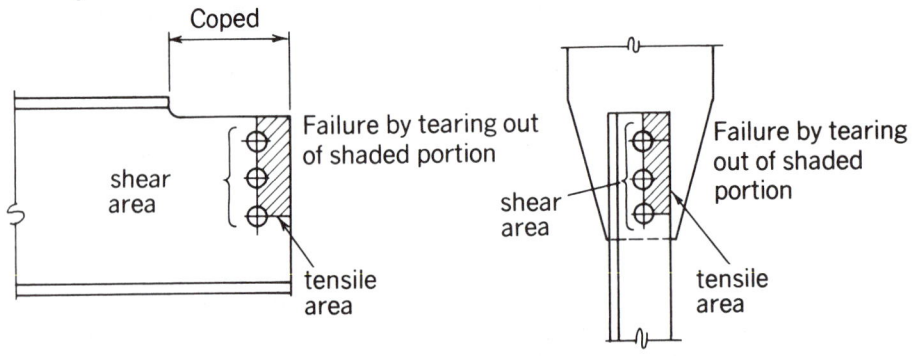

Fig. C-J4.1   Fig. C-J4.2

J6. FILLERS

The practice of securing fillers by means of additional fasteners, so that they are in effect an integral part of a shear-connected component, is not required where a connection is designed as a slip-critical connection using high-strength bolts. In such connections the resistance to slip between filler and either connected part is comparable to that which would exist between the connected parts if no fill were required.

J8. ALLOWABLE BEARING STRESS

As used throughout the AISC ASD Specification, the terms milled surface, milled, or milling are intended to include surfaces which have been accurately sawed or finished to a true plane by any suitable means. The recommended bearing stress on pins is not the same as for bolts and rivets. The lower value, $\frac{9}{10}$ of the yield stress of the part containing the pin hole, provides a safeguard against instability of the plate beyond the hole and high bearing stress concentration which might interfere with operation of the pin, but which is of no concern with bolts and rivets (Johnston, 1939).

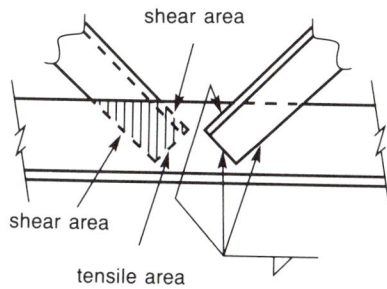

Fig. C-J4.3

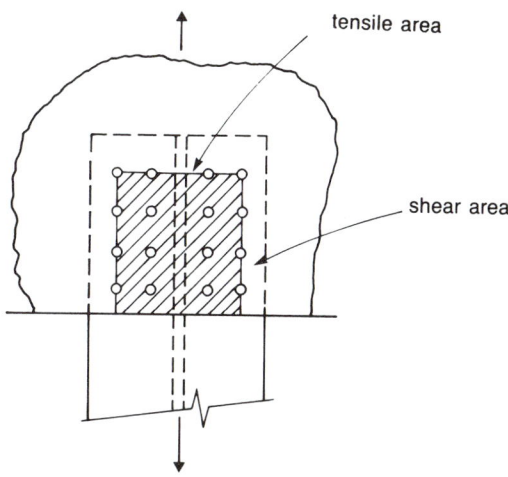

Fig. C-J4.4

J9.   COLUMN BASES AND BEARING ON MASONRY AND CONCRETE

It is not the intent of this Specification to prescribe bearing values for masonry materials. The values specified are included to permit a complete design within the scope of this specification, if desired.

The provisions given were derived from ACI Standard 318-83 ultimate strength criteria, using a load factor of 1.7 applied to both live and dead load. These provisions are more conservative than the ACI ultimate strength provisions, wherein a load factor of 1.4 is permitted for dead load.

J10.   ANCHOR BOLTS

Shear at the base of a column resisted by bearing of the column base details against the anchor bolts is seldom, if ever, critical. Even considering the lowest conceivable slip coefficient, the vertical load on a column is generally more than sufficient to result in the transfer of any likely amount of shear from column base to foundation by frictional resistance, so that the anchor bolts usually experience only tensile stress. Generally, the largest tensile force for which anchor bolts should be designed is that produced by bending moment at the column base, at times augmented by uplift caused by the overturning tendency of a building under lateral load.

Hence, the use of oversized holes required to accommodate the tolerance in setting anchor bolts cast in concrete, permitted in Sect. J3.2, is not detrimental to the integrity of the supported structure.

CHAPTER K

SPECIAL DESIGN CONSIDERATIONS

K1. WEBS AND FLANGES UNDER CONCENTRATED FORCES

1. Design Basis

Whether or not transverse stiffeners are required on the web of a member opposite the flanges of members rigidly connected to its flanges, as in Fig. C-K1.1, depends on the proportions of these members.

Equation (K1-1) limits the bending stress in the flange of the supporting member. Equation (K1-8) limits the slenderness ratio of an unstiffened web of the supporting member, in order to avoid possibility of its buckling.

When Equation (K1-1) and/or Equation (K1-8) indicate the need for stiffeners; the required area of stiffeners is not given. However, minimum stiffener dimensions are given in Sect. K1.8 and their width-to-thickness ratio must satisfy Sect. B5.

Equation (K1-9), giving the required area of stiffeners when stiffeners are needed, is based on tests supporting the concept that, in the absence of transverse stiffeners, the web and flange thickness of member A should be such that these elements will not yield inelastically under concentrated forces delivered by member B (Graham et al, 1959).

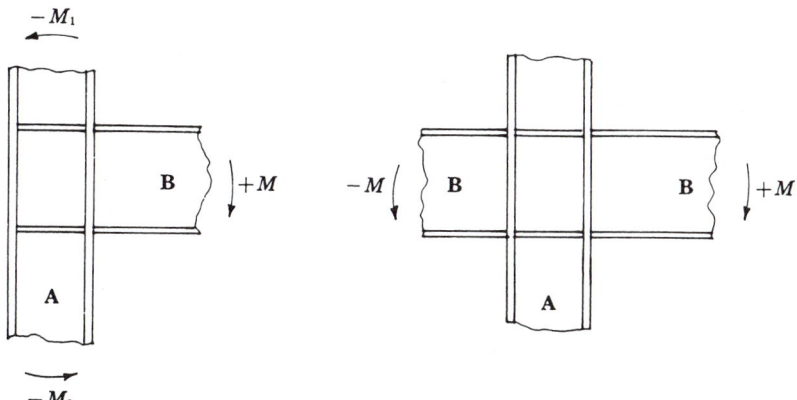

Fig. C-K1.1

AMERICAN INSTITUTE OF STEEL CONSTRUCTION

3. Local Web Yielding

This web strength criteria has been established to limit the stress in the web member into which a force is being transmitted. The stress at the toe of the flange fillet, assumed to be distributed longitudinally a distance no greater than the length of the bearing plus 2.5 or 5 times the k-distance of the flange, depending upon the location of the load, is limited by Equation (K1-2) or (K1-3) to $0.66F_y$. This represents a change from the past web yield criteria that is consistent with AISC (1986).

4. Web Crippling

The expression for resistance to web crippling at a concentrated load is a departure from previous specifications (IABSE, 1968; Bergfelt, 1971; Hoglund, 1971; and Elgaaly, 1983). Equations (K1-4) and (K1-5) are based on research by Roberts (1981).

5. Sidesway Web Buckling

The sidesway web buckling criteria were developed after observing several unexpected failures in beams (Yura, 1982). In these tests, the compression flanges were braced at the concentrated load, the web was squeezed into compression and the tension flange buckled. (see Fig. C-K1.2).

Sidesway web buckling will not occur in the following cases:
For flanges restrained against rotation:

$$\frac{d_c/t_w}{\ell/b_f} > 2.3 \qquad\qquad \text{(C-K1-1)}$$

For flange rotation not restrained:

$$\frac{d_c/t_w}{\ell/b_f} > 1.7 \qquad\qquad \text{(C-K1-2)}$$

Sidesway web buckling can also be prevented by the proper design of lateral bracing or stiffeners at the load point. It is suggested that local bracing at both flanges be designed for 1% of the concentrated load applied to that point. Stiffeners must extend from the load point through at least one-half the girder depth. In addition, the pair of stiffeners should be designed to carry the full load. If flange rotation is permitted at the loaded flange, stiffeners will not be effective.

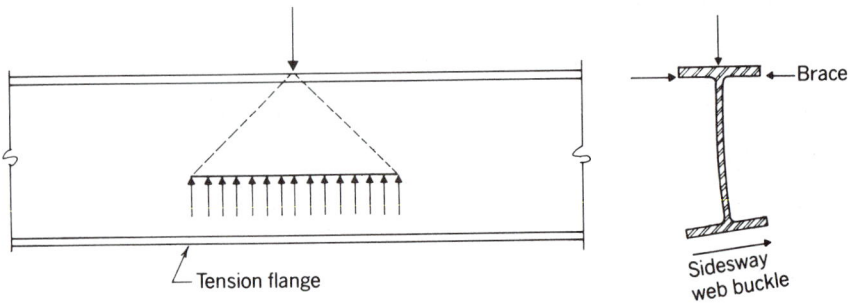

Fig. C-K1.2 Sidesway web buckling

K2. PONDING

As used in the Specification, ponding refers to the retention of water due solely to the deflection of flat roof framing. The amount of this water is dependent upon the flexibility of the framing. If the roof framing members have insufficient stiffness, the water can accumulate and collapse the roof.

Representing the deflected shape of the primary and critical secondary member as a half-sine wave, the weight and distribution of the ponded water can be estimated and, from this, the contribution that the deflection each of these members makes to the total ponding can be expressed (Marino, 1966) as

$$\Delta_w = \frac{\alpha_p \Delta_o \left[1 + \frac{\pi}{4} \alpha_s + \frac{\pi}{4} \rho \, (1 + \alpha_s) \right]}{1 - \frac{\pi}{4} \alpha_p \alpha_s} \qquad \text{(C-K2-1)}$$

for the primary member, and

$$\delta_w = \frac{\alpha_s \delta_o \left[1 + \frac{\pi^3}{32} \alpha_p + \frac{\pi^2}{8\rho} (1 + \alpha_p) + 0.185 \alpha_s \alpha_p \right]}{1 - \frac{\pi}{4} \alpha_p \alpha_s} \qquad \text{(C-K2-2)}$$

for the secondary member. In these expressions Δ_o and δ_o are, respectively, the primary and secondary beam deflections due to loading present at the initiation of ponding, $\alpha_p = C_p/(1-C_p)$, $\alpha_s = C_s/(1-C_s)$, and $\rho = \delta_o/\Delta_o = C_s/C_p$.

Using the above expression for Δ_w and δ_w, the ratios Δ_w/Δ_o and δ_w/δ_o can be computed for any given combination of primary and secondary beam framing using, respectively, the computed value of parameters C_p and C_s defined in the AISC ASD Specification.

Even on the basis of unlimited elastic behavior, it is seen that the ponding deflections would become infinitely large unless

$$\left(\frac{C_p}{1 - C_p} \right) \left(\frac{C_s}{1 - C_s} \right) < \frac{4}{\pi} \qquad \text{(C-K2-3)}$$

Because elastic behavior is limited, the effective bending strength available in each member to resist the stress caused by ponding action is restricted to the difference between the yield stress of the member and the stress, f_o, produced by the total load supported by it before consideration of ponding is included.

Noting that elastic deflection is directly proportional to stress, and providing a factor of safety of 1.25 with respect to stress due to ponding, the admissible amount of ponding deflection in either the primary or critical (midspan) secondary member, in terms of the applicable ratio Δ_w/Δ_o or δ_w/δ_o, can be represented as $(0.8F_y - f_o)/f_o$. Substituting this expression for Δ_w/Δ_o and δ_w/δ_o and combining with the foregoing expressions for Δ_w and δ_w, the relationship between critical values for C_p and C_s and the available elastic bending strength to resist ponding is obtained. The curves presented in Figs. C-K2.1 and C-K2.2

are based upon this relationship. They constitute a design aid for use when a more exact determination of required flat roof framing stiffness is needed than given by the Specification provision that $C_p + 0.9C_s \leq 0.25$.

Given any combination of primary and secondary framing, the stress index is computed as

$$U_p = \left(\frac{0.8F_y - f_o}{f_o}\right)_p \quad \text{for the primary member} \qquad \text{(C-K2-4)}$$

$$U_s = \left(\frac{0.8F_y - f_o}{f_o}\right)_s \quad \text{for the secondary member} \qquad \text{(C-K2-5)}$$

where f_o, in each case, is the computed bending stress in the member due to the supported loading, neglecting ponding effect. Depending upon geographic location, this loading should include such amount of snow as might also be present, although ponding failures have occurred more frequently during torrential summer rains, when the rate of precipitation exceeded the rate of drainage runoff and the resulting hydraulic gradient over large roof areas caused substantial accumulation of water some distance from the eaves.

Given the size, spacing, and span of a tentatively selected combination of primary and secondary beams, for example, one may enter Fig. C-K2.1 at the level of the computed stress index U_p determined for the primary beam; move horizontally to the computed C_s-value of the secondary beams; and, thence, downward to the abscissa scale. The combination stiffness of the primary and secondary framing is sufficient to prevent ponding if the flexibility constant read from this latter scale is more than the value of C_p computed for the given primary member; if not, a stiffer primary or secondary beam, or combination of both, is required.

If the roof framing consists of a series of equally spaced wall-bearing beams, they would be considered as secondary members, supported on an infinitely stiff primary member. For this case, one would enter Fig. C-K2.2. The limiting value of C_s would be determined by the intercept of a horizontal line representing the U_s-value and the curve for $C_p = 0$.

The ponding deflection contributed by a metal deck is usually such a small part of the total ponding deflection of a roof panel, that it is sufficient merely to limit its moment of inertia (per foot of width normal to its span) to 0.000025 times the fourth power of its span length, as provided in the Specification. However, the stability against ponding of a roof consisting of a metal roof deck of relatively slender depth-span ratio, spanning between beams supported directly on columns, may need to be checked. This can be done using Fig. C-K2.1 or C-K2.2 with the following computed values:

U_p, the stress index for the supporting beam
U_s, the stress index for the roof deck
C_p, the flexibility constant for the supporting beams
C_s, the flexibility constant for one foot width of the roof deck ($S = 1.0$)

Since the shear rigidity of their web system is less than that of a solid plate, the moment of inertia of steel joists and trusses should be taken as somewhat *less* than that of their chords.

K3. TORSION

See AISC (1983).

K4. FATIGUE

Because most members in building frames are not subject to a large enough number of cycles of full allowable stress application to require design for fatigue, the provisions covering such designs have been placed in Appendix K4.

When fatigue is a design consideration, its severity is most significantly affected by the number of load applications, the magnitude of the stress range, and the severity of the stress concentrations associated with the particular details. These factors are not encountered in normal building designs; however, when encountered and when fatigue is of concern, all provisions of Appendix K4 must be satisfied.

Members or connections subject to less than 20,000 cycles of loading will not involve a fatigue condition, except in the case of repeated loading involving large ranges of stress. For such conditions, the admissible range of stress can conservatively be taken as 1½ times the applicable value given in Table A-K4.3 for Loading Condition 1.

Fluctuation in stress which does not involve tensile stress does not cause crack propagation and is not considered to be a fatigue situation. On the other hand, in elements of members subject solely to calculated compression stress, fatigue cracks may initiate in regions of high tensile residual stress. In such situations, the cracks generally do not propagate beyond the region of the residual tensile stress, because the residual stress is relieved by the crack. For this reason, stress ranges that are completely in compression are not included in the column headed by "Kind of Stress" in Table A-K4.2 of Appendix K4. This is also true of comparable tables of the current AASHTO and AREA Specifications.

When fabrication details involving more than one category occur at the same location in a member, the stress range at that location must be limited to that of the most restrictive category. By locating notch-producing fabrication details in regions subject to a small range of stress, the need for a member larger than required by static loading will often be eliminated.

Extensive test programs using full size specimens, substantiated by theoretical stress analysis, have confirmed the following general conclusions (Fisher et al, 1970; Fisher et al, 1974):

1. Stress range and notch severity are the dominant stress variables for welded details and beams.
2. Other variables such as minimum stress, mean stress and maximum stress are not significant for design purposes.

3. Structural steels with yield points of 36 to 100 ksi do not exhibit significantly different fatigue strength for given welded details fabricated in the same manner.

Allowable stress ranges can be read directly from Table A-K4.3 for a particular category and loading condition. The values are based on recent research (Keating and Fisher, 1985). Provisions for A325 and A490 bolts subjected to tension are given in Appendix K4.3.

Tests have uncovered dramatic differences in fatigue life, not completely pre-

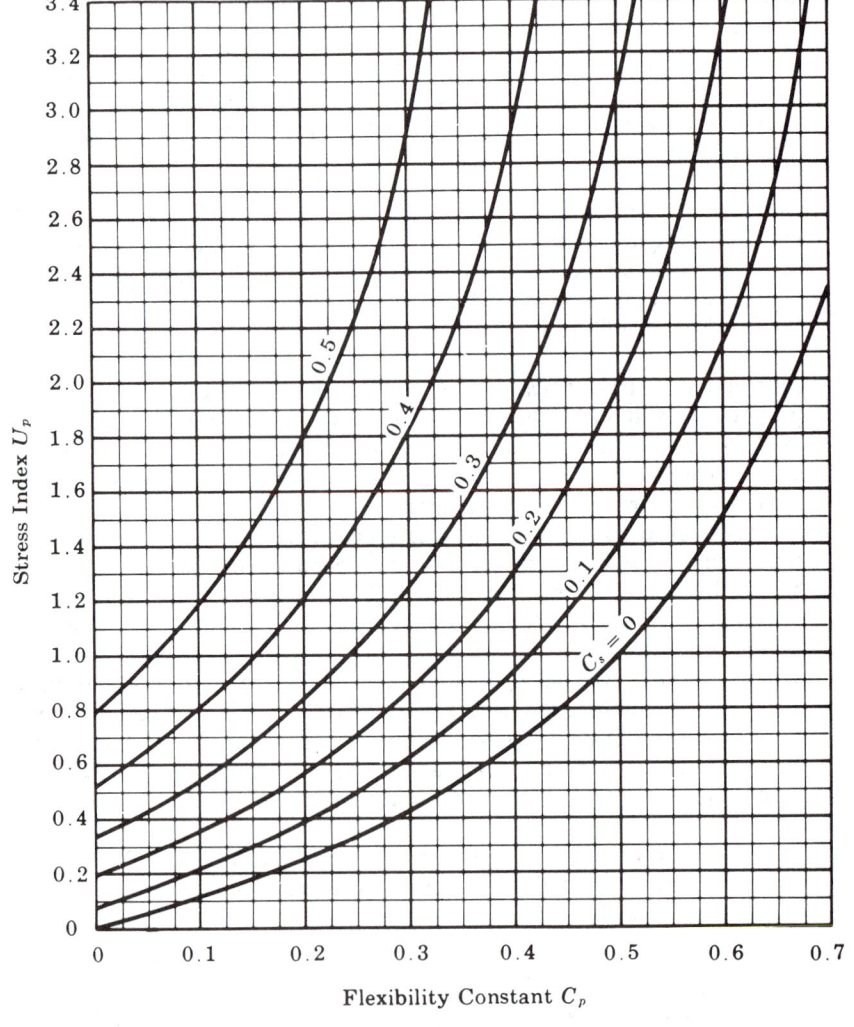

Fig. C-K2.1

dictable from the various published formulas for estimating the actual magnitude of prying force (Kulak, Fisher and Struik, 1987).

The use of other types of mechanical fasteners to resist applied cyclic loading in tension is not permitted. Lacking a high degree of assured pretension, the range of stress is generally too great to resist such loading for long. However, all types of mechanical fasteners survive unharmed when subject to cyclic stresses sufficient to fracture the connected parts, which is provided for elsewhere in Appendix K4.

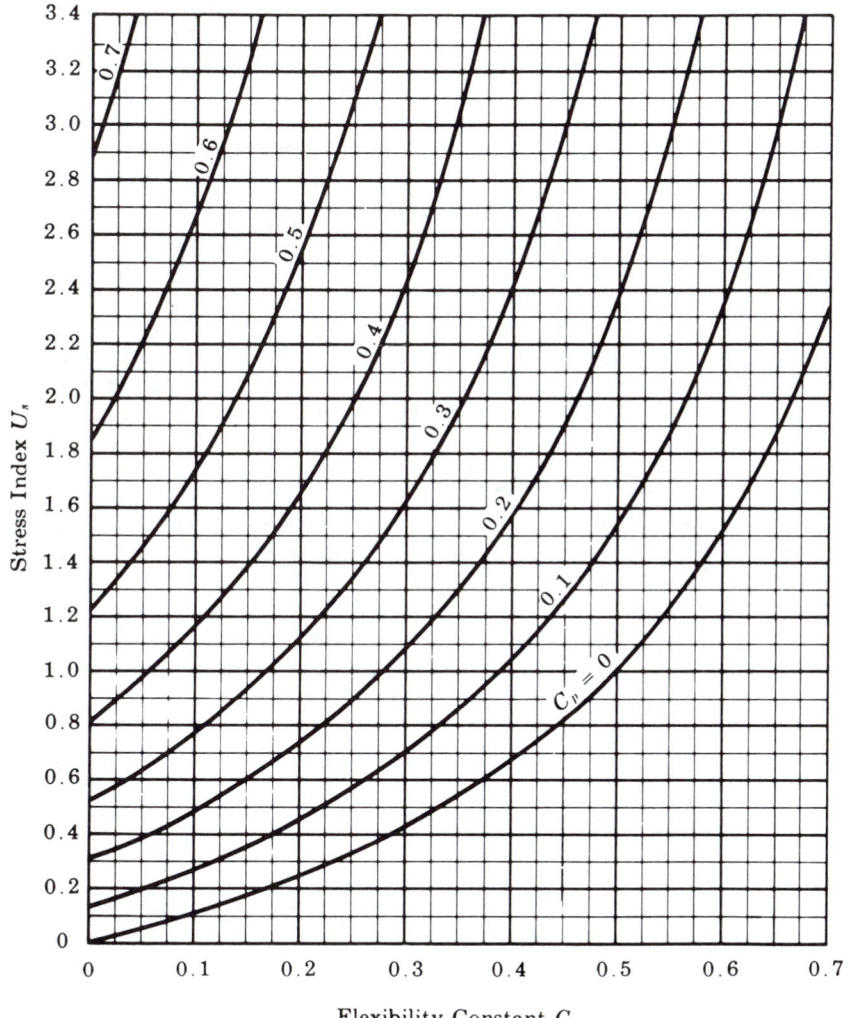

Fig. C-K2.2

CHAPTER L

SERVICEABILITY DESIGN CONSIDERATIONS

L1. CAMBER

The cambering of flexural members, to eliminate the appearance of sagging or to match the elevation of adjacent building components when the member is loaded, is accomplished in various ways. In the case of trusses and girders, the desired curvature can be built in during assembly of the component parts. Within limits, rolled beams can be cold-cambered at the producing mill.

The local application of heat has come into common use as a means of straightening or cambering beams and girders. The method depends upon an ultimate shortening of the heated zones. A number of such zones, on the side of the member that would be subject to compression during cold-cambering or gagging, are heated enough to be upset by the restraint provided by surrounding unheated areas. Shortening takes place upon cooling.

Although the final curvature of camber produced by any of these methods can be controlled to a degree, it must be realized that some tolerance, to cover workmanship error and permanent change due to handling, is inevitable.

L2. EXPANSION AND CONTRACTION

As in the case of deflections, the satisfactory control of expansion cannot be reduced to a few simple rules, but must depend largely upon the good judgment of qualified engineers.

The problem is more serious in buildings having masonry wall enclosures than where the walls consist of prefabricated units. Complete divorcement of the framing, at widely spaced expansion joints, is generally more satisfactory than more frequently located devices dependent upon the sliding of parts in bearing, and usually less expensive than rocker or roller expansion bearings.

L3. DEFLECTION, VIBRATION AND DRIFT

1. Deflection

Although deflection, rather than stress, is sometimes the criterion of satisfactory design, there is no single scale by which the limit of tolerable deflection can be defined. Where limitations on flexibility are desirable, they are often dictated by the nature of collateral building components, such as plastered walls and ceilings, rather than by considerations of human comfort and safety. The admissible amount of movement varies with the type of component.

The most satisfactory solution must rest upon the sound judgment of qualified engineers. As a guide, the following rules are suggested:

1. The depth of fully stressed beams and girders in floors should, if practicable , be not less than $(F_y/800)$ times the span. If members of less depth are used, the unit stress in bending should be decreased in the same ratio as the depth is decreased from that recommended above.
2. The depth of fully stressed roof purlins should, if practicable, be not less than $(F_y/1000)$ times the span, except in the case of flat roofs.

2. Vibration

Where human comfort is the criterion for limiting motion, as in the case of perceptible vibrations, the limit of tolerable amplitude is dependent on both the frequency of the vibration and the damping effect provided by components of the construction. At best, the evaluation of these criteria is highly subjective, although mathematical models do exist which may be useful (Murray, 1975). When such vibrations are caused by running machinery, they should be isolated by damping devices or by the use of independent foundations.

The depth of a steel beam supporting large open floor areas free of partitions or other sources of damping should not be less than 1/20 of the span to minimize perceptible transient vibration due to pedestrian traffic.

L5.   CORROSION

Steel members may deteriorate in particular service environments. This deterioration may appear either in external corrosion, which would be visible upon inspection, or in undetected changes in the material that would reduce its load-carrying capacity. The designer should recognize these problems by either factoring a specific amount of damage tolerance into his design or providing adequate protection systems (e.g., coating, cathodic protection) and/or planned maintenance programs so that such problems do not occur.

CHAPTER M

FABRICATION, ERECTION AND QUALITY CONTROL

M2. FABRICATION

2. Thermal Cutting

Thermal cutting should preferably be done by machine. The requirements for a positive preheat of 150°F minimum when thermal cutting beam copes and weld access holes in ASTM A6 Group 4 and 5 shapes and in built-up shapes made of material more than 2-in. thick tends to minimize the hard surface layer and the possible initiation of cracks.

5. High-strength Bolted Construction - Assembly

In the past, all ASTM A325 and A490 bolts in both slip-critical and bearing-type connections were required to be tightened to a specified tension. The requirement was changed in 1985 to permit some bearing-type connections to be tightened to only a snug-tight condition.

To qualify as a snug-tight bearing connection, the bolts are not subject to tension loads, slip is permitted and loosening or fatigue due to vibration or load fluctuations are not design considerations.

It is suggested that snug-tight bearing-type connections be used in applications when A307 bolts would be permitted. Sect. J1.12 serves as a guide to these applications.

In other cases, A325 and A490 bolts are required to be tightened to 0.7 of their tensile strength. This may be done either by the turn-of-nut method, by a calibrated wrench or by using direct tension indicators (RCSC, 1985). Since fewer fasteners and stiffer connected parts are involved than is generally the case with A307 bolts, the greater clamping force is recommended to ensure solid seating of the connected parts. However, because the performance of bolts in bearing is not dependent on an assured minimum level of pretension, thorough inspection requirements to assure full compliance with pretightening criteria are not warranted. This is especially true regarding the arbitration inspection requirements of Sect 9b of the RCSC Specification (1985). Visual evidence of solid seating of the connected parts, and of wrench impacting to assure that the nut has been tightened sufficiently to prevent it from loosening, is adequate.

M3. SHOP PAINTING

The surface condition of steel framing disclosed by the demolition of long-standing buildings has been found to be unchanged from the time of its erection except at isolated spots where leakage may have occurred. Where such leakage is not eliminated, the presence or absence of a shop coat is of minor influence (Bigos et al, 1954).

The Specification does not define the type of paint to be used when a shop coat is required. Conditions of exposure and individual preferences with regard to finish paint are factors which have a bearing on the selection of the proper primer. Hence, a single formulation would not suffice.*

M4. ERECTION

4. Fit of Column Compression Joints

Tests on spliced, full-sized columns with joints that had been intentionally milled out-of-square, relative to either strong or weak axis, demonstrated their load-carrying capacity was the same as for a similar unspliced column (Popov and Stephen, 1977). In the tests, gaps of $\frac{1}{16}$-in. were not shimmed; gaps of $\frac{1}{4}$-in. were shimmed with non-tapered mild steel shims. Minimum size partial-penetration welds were used in all tests. No tests were performed on specimens with gaps greater than $\frac{1}{4}$-in.

The criteria for fit of column compression joints are equally applicable to joints at column splices and joints between columns and base plates.

---

*For a comprehensive treatment of the subject, see Ref. 54.

CHAPTER N

PLASTIC DESIGN

N1. SCOPE

The Specification recognizes three categories of profiles, classified according to the ability to resist local buckling of elements of the cross section when subject to compressive stress. The categories are : (1) noncompact, (2) compact, and (3) plastic design. The elements of *noncompact* sections (B5) will not buckle locally when subject to elastic limit strains. Elements of *compact* sections (B5) are proportioned so that the cross section may be strained in bending to the degree necessary to achieve full plastification of the cross section; however, the reserve for inelastic strains is adequate only to achieve modest redistribution of moments. The elements of *plastic design* sections (N7) are proportioned so they will not only achieve full plastification of the cross section, but also will remain stable while being bent through an appreciable angle at a constant plastic moment up to the point where strain hardening is initiated. Thus, plastic design cross sections are capable of providing the hinge rotations that are counted upon in the plastic method of analysis.

Superior bending strength of compact sections is recognized in F1.1 of the Specification by increasing the allowable bending stress to $0.66F_y$ and by permitting 10% redistribution of moment. By the same token, the logical load factor for plastically designed beams is given by the equation

$$F = \frac{F_y}{0.66F_y} \times (\text{shape factor}) \qquad \text{(C-N1-1)}$$

For such shapes listed in the AISC *Manual of Steel Construction*, the variation of shape factor is from 1.10 to 1.23, with a mode of 1.12 for the most commonly used shapes. Then, the corresponding load factor must vary from 1.67 to 1.86, with a mode of 1.70. Such a load factor is consistent and in better balance with that inherent in the allowable working stresses for tension members and deep plate girders.

Research on the ultimate strength of heavily loaded columns subjected to concurrent bending moments has provided data which justifies a load factor, for such members, that is the same as that provided for members subject to bending only, namely 1.7. Consistent with the 1/3 increase in allowable stress permitted in Sect. A5.2 of the Specification, the load factor to be used in designing for gravity loading combined with wind or seismic loading is 1.3 (Van Kuren and Galambos, 1964).

Based on research on multi-story framing, application of the Specification provisions includes the complete design of braced and unbraced planar frames in high-rise buildings (Driscoll et al, 1965; Driscoll, 1966). Systematic procedures for application of plastic design in proportioning the members of such frames have been developed (AISI, 1968; Lu, 1967).

N2. STRUCTURAL STEEL

Research testing has demonstrated the suitability of all of the steels listed in this section for use in plastic design(Adams, Lay and Galambos, 1965; ASCE, 1971).

N3. BASIS FOR MAXIMUM STRENGTH DETERMINATION

Although resistance to wind and seismic loading can be provided in moderate height buildings by means of concrete and masonry shear walls, which also provide for overall frame stability at factored gravity loading, taller building frames must provide this resistance acting alone. This can be achieved in one of two ways: either by a system of bracing or by a moment-resisting frame.

For one- and two-story unbraced frames with Type 1 construction throughout, where the column axial loads are generally modest, the frame instability effect is small and $P\Delta$ effects* may be safely ignored. However, where such frames are designed with a mixture of rigid connections and simple or semi-rigid connections (Type 2 and Type 3 construction), it may be necessary to consider the frame instability effect $P\Delta$. In this case, stability is dependent upon a reduced number of rigid connections and the effect of frame drift may be a significant consideration in the design.

1. Stability of Braced Frames

The limitation on axial force $0.85P_y$ was inserted as a simple means to compensate for three possible effects (Douty and McGuire, 1965):

1. Loss of stiffness due to residual stress
2. Effect of secondary $P\Delta$ moments on the vertical bracing system
3. Lateral-torsional buckling effect

N4. COLUMNS

Equations (N4-2) and (N4-3) will be recognized as similar in type to Equations (H1-1) and (H1-2), except they are written in terms of factored loads and moments, instead of allowable stresses at service loading. As in the case of Equations (H1-1) and (H1-2), P_{cr} is computed on the basis of l/r_x or l/r_y, whichever is larger, for any given unbraced length (Driscoll et al, 1965).

A column is considered to be fully braced if the slenderness ratio l/r_y between the braced points is less than or equal to that specified in Sect. N9. When the unbraced length ratio of a member bent about its strong axis exceeds the limit specified in Sect. N9, the rotation capacity of the member may be impaired, due to the combined influence of lateral and torsional deformation, to such an extent that plastic hinge action within the member cannot be counted upon. However, if the computed value of M is small enough so limitations of Equations (N4-2) and (N4-3) are met, the member will be strong enough to function at a joint where the required hinge action is provided in another member entering the joint. An assumed reduction in moment-resisting capacity is provided by using the value M_m, computed from Equation (N4-4), in Equation (N4-2).

---

*See Commentary C2 for discussion of $P\Delta$ effects.

Equation (N4-4) was developed empirically on the basis of test observations and provides an estimate of the critical lateral buckling moment, in the absence of axial load, for the case where $M_1/M_2 = -1.0$ (single curvature bending) (Driscoll et al, 1965). For other values of M_1/M_2, adjustment is provided by using the appropriate C_m value as defined in Sect. H1.

Equation (N4-4) is to be used only in connection with Equation (N4-2).

Space frames containing plastically designed planar rigid frames are assumed to be supported against sidesway normal to these frames. Depending upon other conditions of restraint, the basis for determination of proper values for P_{cr} and P_e and M_m, for a plastically designed column oriented to resist bending about its strong axis, is outlined in Table C-N4.1. In each case l is the distance between points of lateral support corresponding to r_x or r_y, as applicable. When K is indicated, its value is governed by the provisions of Sect. C2.2.

N5. SHEAR

Using the von Mises criterion, the average stress at which an unreinforced web would be fully yielded in pure shear can be expressed as $F_y/\sqrt{3}$. It has been observed that the plastic bending strength of an I-shaped beam is not reduced appreciably until shear yielding occurs over the full effective depth, which may be taken as the distance between the centroids of its flanges (approximately 0.95 times its actual depth) (ASCE, 1971). Thus,

$$V_u = \frac{F_y}{\sqrt{3}} \times 0.95dt = 0.55F_y dt_w \qquad \text{(C-N5-1)}$$

Shear stresses are generally high within the boundaries of a rigid connection of two or more members whose webs lie in a common plane. Assuming the moment $+M$, in Fig. C-N5.1, expressed in kip-ft, to be resisted by a force couple acting at the centroid of the beam flanges, the shear, in kips, produced in beam-to-column connections web abcd can be computed as

$$V = \frac{12M}{0.95d_b} - V_s \qquad \text{(C-N5-2)}$$

when $V = 0.55F_y d_c t_w$

$$\text{req'd } t_w = \frac{1}{0.55F_y d_c}\left[\frac{12M}{0.95d_b} - V_s\right] \qquad \text{(C-N5-3)}$$

TABLE C-N4.1

	Braced Planar Frames	One- and Two-story Unbraced Planar Frames
P_{cr}	Use larger ratio, $\dfrac{l}{r_y}$ or $\dfrac{l}{r_x}$	[1]Use larger ratio, $\dfrac{l}{r_y}$ or $\dfrac{Kl}{r_x}$
P_e	Use l/r_x	[1]Use Kl/r_x
M_m	Use l/r_y	Use l/r_y
[1]Webs of columns assumed to be in plane of frame.		

where A_{bc} is the planar area *abcd* and F_y is expressed in ksi. If the thickness of the web panel is less than that given by this formula, the deficiency may be compensated for by a pair of diagonal stiffeners or by a reinforcing plate in contact with the web panel and welded around its boundary to the column flanges and horizontal stiffeners.

N6. WEB CRIPPLING

Usually stiffeners are needed, as *ab* and *dc* in Fig. C-N5.1, in line with the flanges of a beam rigidly connected to the flange of a second member so located that their webs lie in the same plane to prevent crippling of the web of the latter opposite the compression flange of the former. A stiffener may also be required opposite the tension flange to protect the weld joining the two flanges; otherwise the stress in the weld might be too great in the region of the beam web, because of the lack of bending stiffness in the flange to which the beam is connected. Since their design is based on equating the plastic resisting capacity of the supporting member to the plastic moment delivered by the supported member, Equations (K1-1), (K1-8) and (K1-11) are equally applicable to allowable stress design and plastic design.

When stiffeners are required, as an alternative to the usual pair of horizontal plates, vertical plates parallel to but separated from the web, as shown in Fig. C-N6.1, may prove advantageous. See Graham et al., 1959.

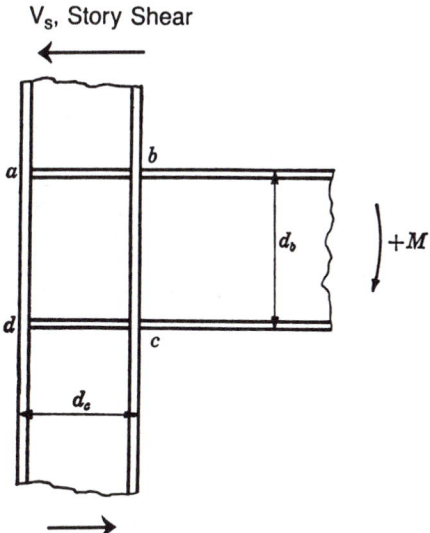

Fig. C-N5.1

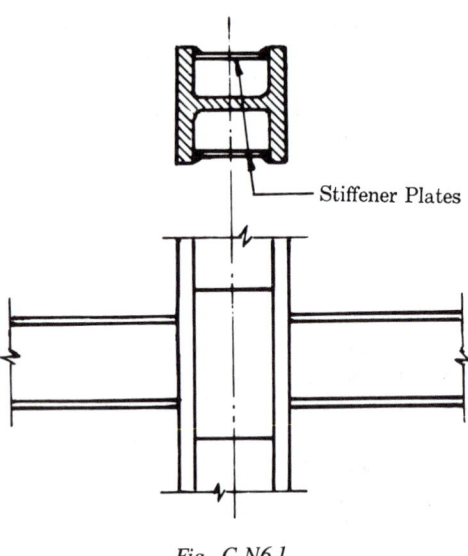

Fig. C-N6.1

N7. MINIMUM THICKNESS (WIDTH-THICKNESS RATIOS)

Research has shown the limiting flange and web width-thickness ratios, below which ample plastic hinge rotations could be relied upon without reduction in the M_p value due to local buckling, are not exactly proportional to $1/\sqrt{F_y}$, although the discrepancy using such a relationship, within the range of yield stress presently permitted by the Specification, is not large (ASCE, 1971). Expressions including other pertinent factors are complex and involve use of mechanical properties that have not been defined clearly. Tabular values for limiting flange width-thickness ratios are given in the Specification for the approved grades of steel.

No change in basic philosophy was involved in extending the earlier expression for limiting web depth-thickness ratio to stronger steels. Equations (N7-1) and (N7-2) were derived, with minor adjustments for better correlation with observed test results, by multiplying Equation (25) of the 1963 Specification by the factor $\sqrt{36/F_y}$ to cover the accepted range in yield point stress. Equation (N7-1) is identical to Equation (1.5-4) in Part 1 of the 1969 Specification, except that it is written in terms of factored loads instead of allowable stresses at service loading. Equation (1.5-4) in the 1969 Specification was liberalized in 1974 and redesignated as Equation (1.5-4a) in the 1978 Specification; in this Specification it is included in Table B5.1. However, this liberalization was not extended to plastic design sections, which require greater rotational capacity than compact sections.

N8. CONNECTIONS

Connections located outside of regions where hinges would have formed at ultimate load can be treated in the same manner as similar connections in frames designed in accordance with the provisions of Chaps. A through M. Since the moments and forces to be resisted will be those corresponding to the factored loading, the allowable stresses to be used in proportioning parts of the connections can be taken as 1.7 times those given in J.

The same procedure is valid in proportioning connections located in the region of a plastic hinge. Connections required to resist moments and forces due to wind and earthquake loads combined with gravity loading factored to 1.3, and proportioned on the basis of limiting stresses equal to 1.7 times those given in J, provide a balance between frame strength and connection strength, provided they are adequate to resist gravity loading alone, factored to 1.7.

The width-thickness ratio and unbraced length of all parts of the connection that would be subject to compression stresses in the region of a hinge must meet the requirements given in N, and sheared edges and punched holes must not be used in portions of the connection subject to tension.

When a haunched connection is proportioned elastically for the moments that would exist within its length, the continuous frame can be analyzed as a mechanism having a hinge at the small end of the haunch, rather than at the intersection point between connected members, with some attendant economy (ASCE, 1971).

Tests have shown that splices assembled with high-strength bolts are capable of developing the M_p value of the gross cross section of the connected part (Douty and McGuire, 1965). It has also been demonstrated that beam-to-column connections involving use of welded or mechanically fastened fittings, instead of full-penetration groove welds matching the full member cross section, not only are capable of developing the M_p value of the member, but that the resulting hinge rotation can be reversed several times without failure (Popov and Pinkney, 1968).

N9. LATERAL BRACING

Portions of members that would be required to rotate inelastically as a plastic hinge in reducing a continuous frame to a mechanism at ultimate load, need more bracing than similar parts of a continuous frame designed in accordance with the elastic theory. Not only must they reach yield point at a load factor of 1.7, they must also strain inelastically to provide the necessary hinge rotation. This is not true at the last hinge to form, since the factored load is assumed to have been reached when this hinge starts to rotate. When bending takes place about the strong axis, any I-shape member tends to buckle out of the plane of bending. For this reason, lateral bracing is needed. The same tendency exists with highly stressed members in elastically designed frames, and in portions of plastically designed frames outside of the hinge areas, but the problem is less severe, since hinge rotation is not involved.

The Specification provisions governing unbraced length are based upon research on members with moment gradients (ASCE, 1971).

In keeping with similar usage of the parameter M/M_p in Chap. H of the Specification, the sign convention adopted in Equations (N9-1) and (N9-2) is generally found more convenient in frame analysis, namely that clockwise moments about a fixed point are positive and counterclockwise moments are negative.

APPENDIX   B

DESIGN REQUIREMENTS

B5.   LOCAL BUCKLING

2.   Slender Compression Elements*

---

*See Commentary Chap. B for the discussion of provisions for Slender Compression Elements.

APPENDIX F

BEAMS AND OTHER FLEXURAL MEMBERS

F7. WEB-TAPERED MEMBERS

The provisions contained in Appendix F7 cover only those aspects of the design of tapered members that are unique to tapered members. For other criteria of design not covered specifically in Appendix F7, see the appropriate portions of Chaps. A through M.

3. Allowable Compressive Stress

The approach in formulating $F_{a\gamma}$ of tapered columns is based on the concept that the critical stress for an axially loaded tapered column is equal to that of a prismatic column of different length, but of the same cross section as the smaller end of the tapered column. This resulted in an equivalent effective length factor K_γ for a tapered member subjected to axial compression (Lee, Morrell and Ketter, 1972). This factor, used to determine the value of S in Equations (A-F7-2) and (A-F7-3), can be determined accurately for a symmetrical rectangular rigid frame composed of prismatic beams and tapered columns.

With modifying assumptions, such a frame can be used as a mathematical model to determine, with sufficient accuracy, the influence of the stiffness $\Sigma(I/b)_g$ of beams and rafters which afford restraint at the ends of a tapered column in other cases, such as those shown in Fig. C-A-F7.1. From Equations (A-F7-2) and (A-F7-3), the critical load P_{cr} can be expressed as $\pi^2 EI_o/(K_\gamma l)^2$. The value of K_γ can be obtained by interpolation, using the appropriate chart (Figs. C-A-F7.2 to C-A-F7.17), and restraint modifiers G_T and G_B. In each of these modifiers the tapered column, treated as a prismatic member having a moment of inertia I_o computed at the smaller end, and its actual length l, is assigned the stiffness I_o/l, which is then divided by the stiffness of the restraining members at the end of the tapered column under consideration. Such an approach is well documented.

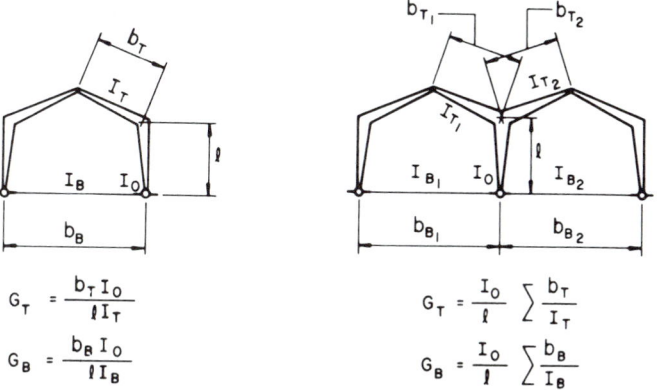

$$G_T = \frac{b_T I_o}{l I_T}$$

$$G_B = \frac{b_B I_o}{l I_B}$$

$$G_T = \frac{I_o}{l} \Sigma \frac{b_T}{I_T}$$

$$G_B = \frac{I_o}{l} \Sigma \frac{b_B}{I_B}$$

Fig. C-A-F7.1

AMERICAN INSTITUTE OF STEEL CONSTRUCTION

4.   Allowable Flexural Stress

The development of the allowable flexural stress for tapered beams follows closely with that for prismatic beams. The basic concept is to replace a tapered beam by an equivalent prismatic beam with a different length, but with a cross section identical with that of the smaller end of the tapered beam (Lee, Morrell and Ketter, 1972). This has led to the modified length factors h_s and h_w in Equations (A-F7-4) and (A-F7-5).

Equations (A-F7-4) and (A-F7-5) are based on total resistance to lateral buckling using both St. Venant and warping resistance. The factor B modifies the basic $F_{b\gamma}$ to account for moment gradient and lateral restraint offered by adjacent segments. For members which are continuous past lateral supports, categories a, b and c of Appendix F7.4 usually apply; however, note they apply only when the axial force is small and adjacent unbraced segments are approximately equal in length. For a single member, or segments which do not fall into category a, b, c or d, the recommended value for B is unity. The value of B should also be taken as unity when computing the value of $F_{b\gamma}$ to be used in Equation (A-F7-12), since the effect of moment gradient is provided for by the factor C_m. The background material is given in WRC Bulletin No. 192 (Morrell and Lee, 1974).

Thus, note that in these charts the values of K_γ represent the combined effects of end restraints and tapering. For the case $\gamma = 0$, K_γ becomes K, which can also be determined from the alignment chart for effective length of columns in continuous frames (Fig. C-C2.2). For cases when the restraining beams are also tapered, the procedure used in WRC Bulletin No. 173 can be followed, or appropriate estimation of K_γ can be made based on these charts (Lee et al. 1972).

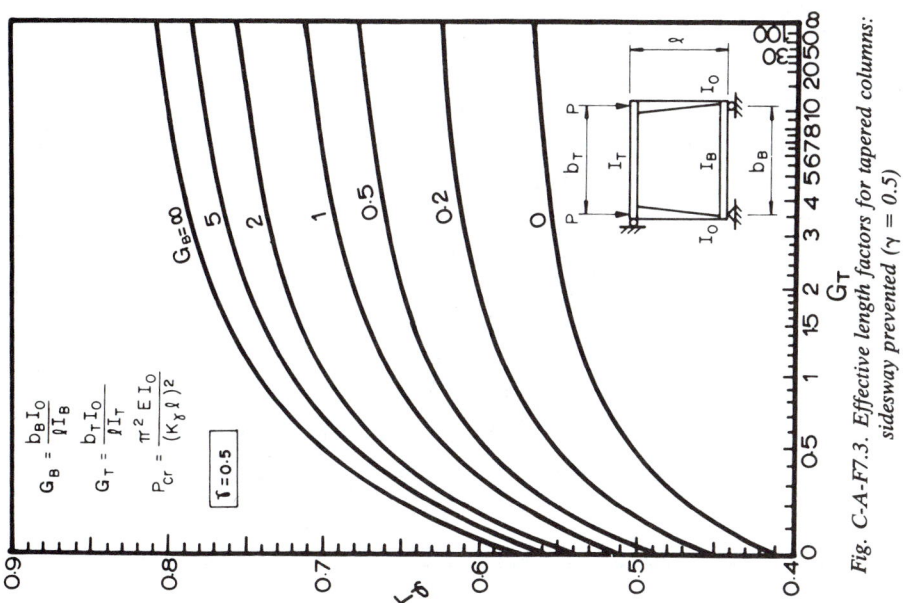

Fig. C-A-F7.3. Effective length factors for tapered columns: sideway prevented ($\gamma = 0.5$)

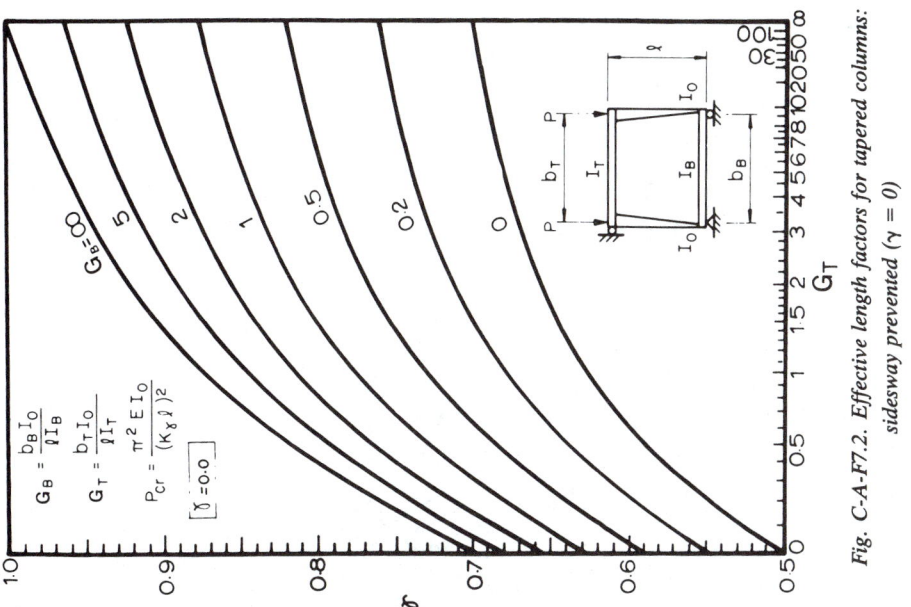

Fig. C-A-F7.2. Effective length factors for tapered columns: sideway prevented ($\gamma = 0$)

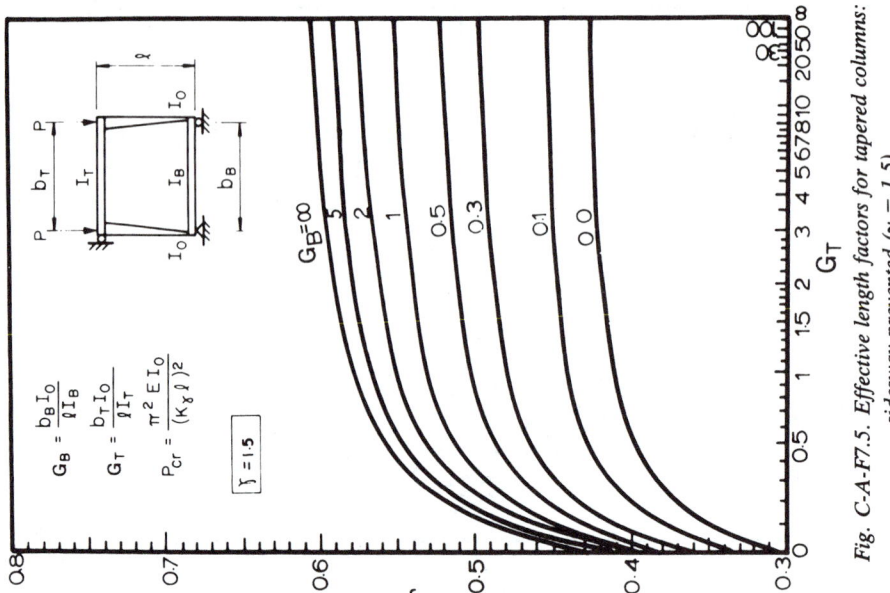

Fig. C-A-F7.5. Effective length factors for tapered columns:
sidesway prevented (γ = 1.5)

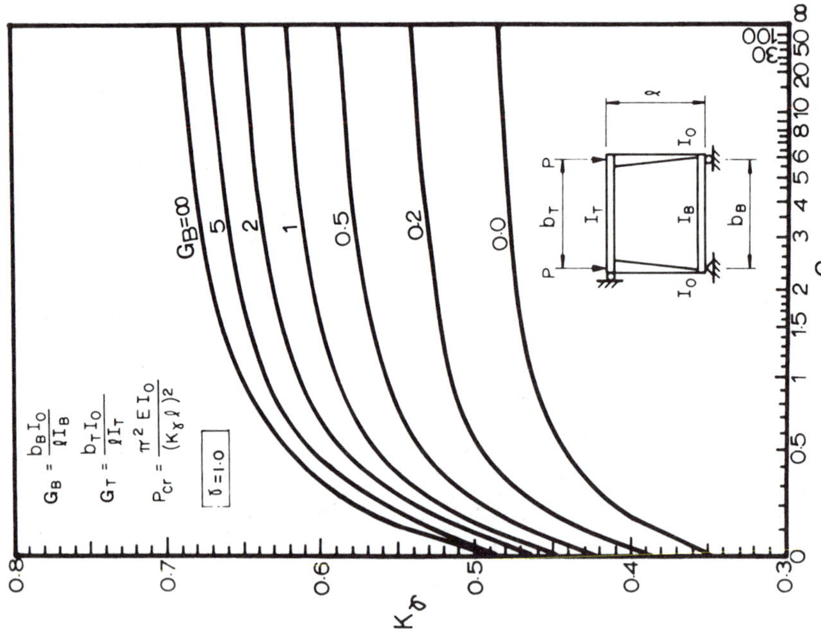

Fig. C-A-F7.4. Effective length factors for tapered columns:
sidesway prevented (γ = 1.0)

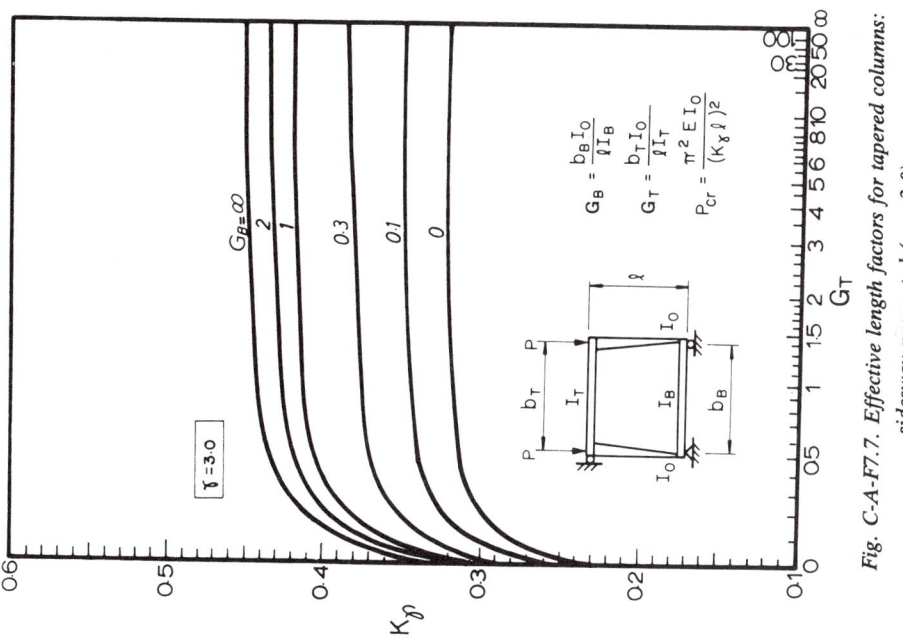

Fig. C-A-F7.7. Effective length factors for tapered columns: sideway prevented ($\gamma = 3.0$)

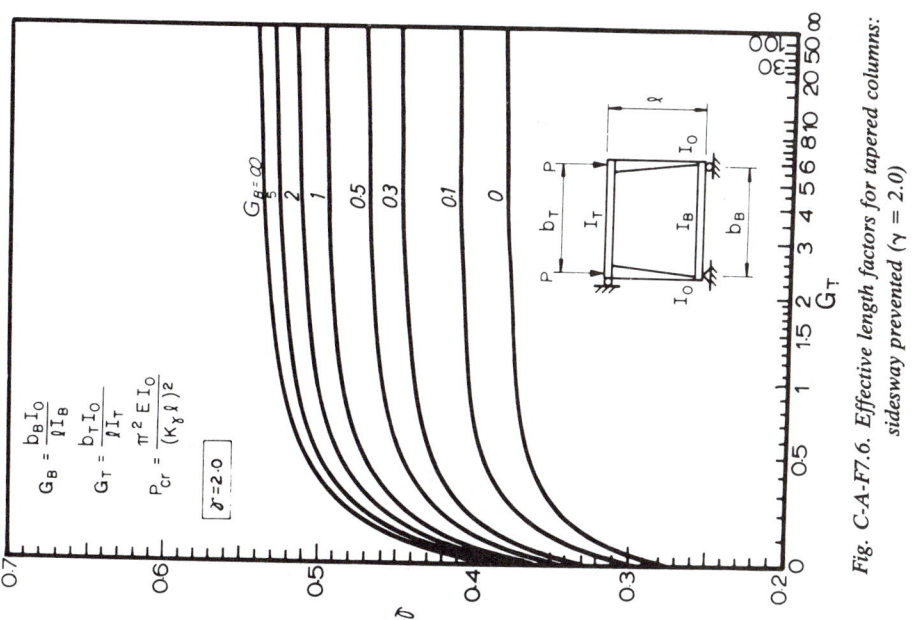

Fig. C-A-F7.6. Effective length factors for tapered columns: sideway prevented ($\gamma = 2.0$)

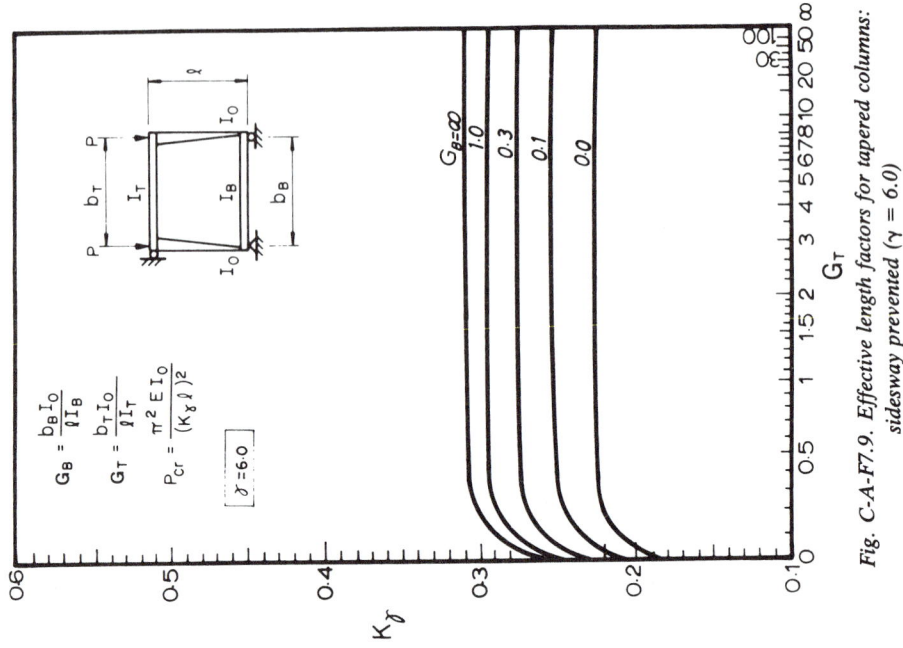

Fig. C-A-F7.9. *Effective length factors for tapered columns: sidesway prevented* ($\gamma = 6.0$)

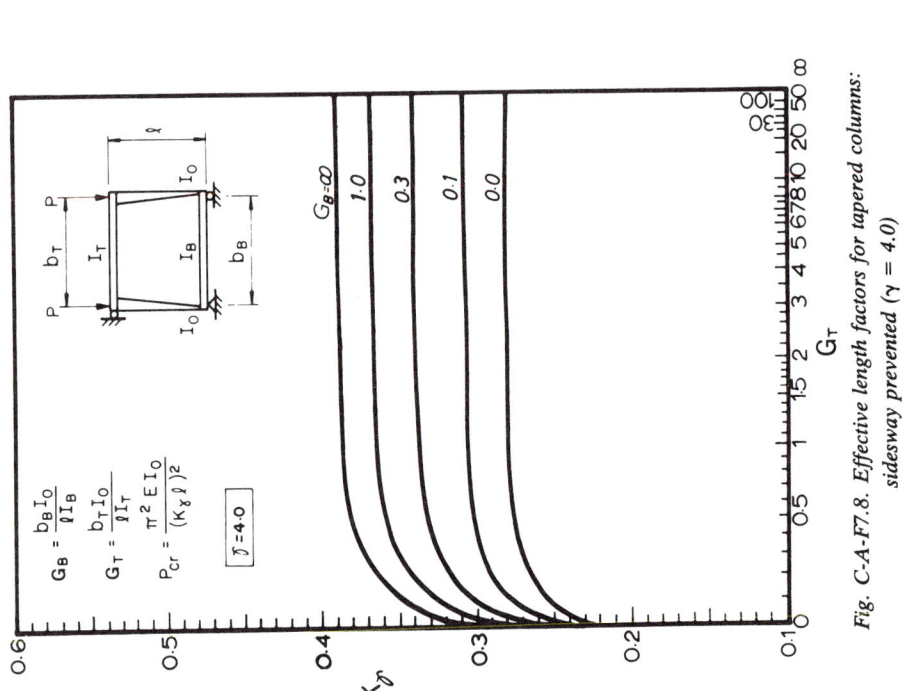

Fig. C-A-F7.8. *Effective length factors for tapered columns: sidesway prevented* ($\gamma = 4.0$)

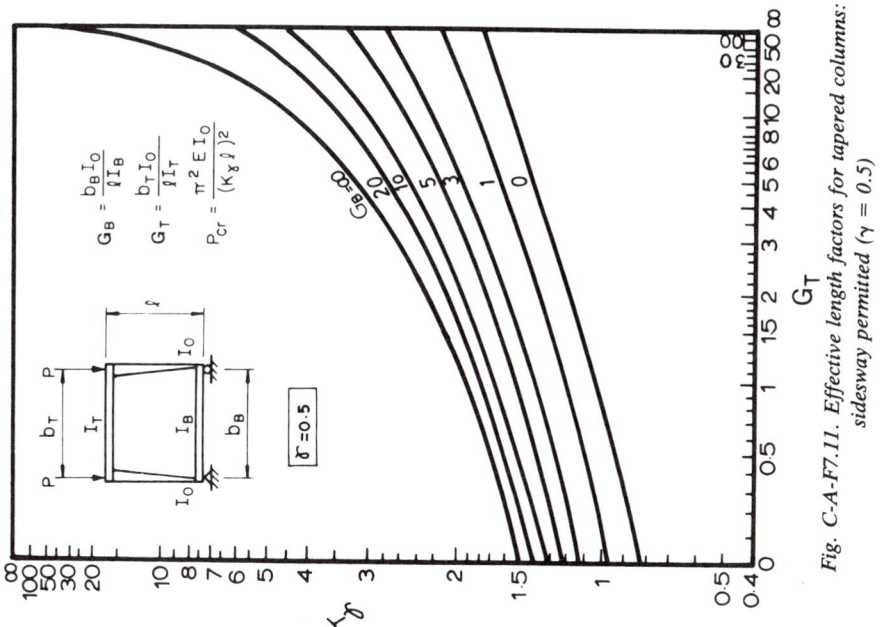

Fig. C-A-F7.11. Effective length factors for tapered columns: sidesway permitted ($\gamma = 0.5$)

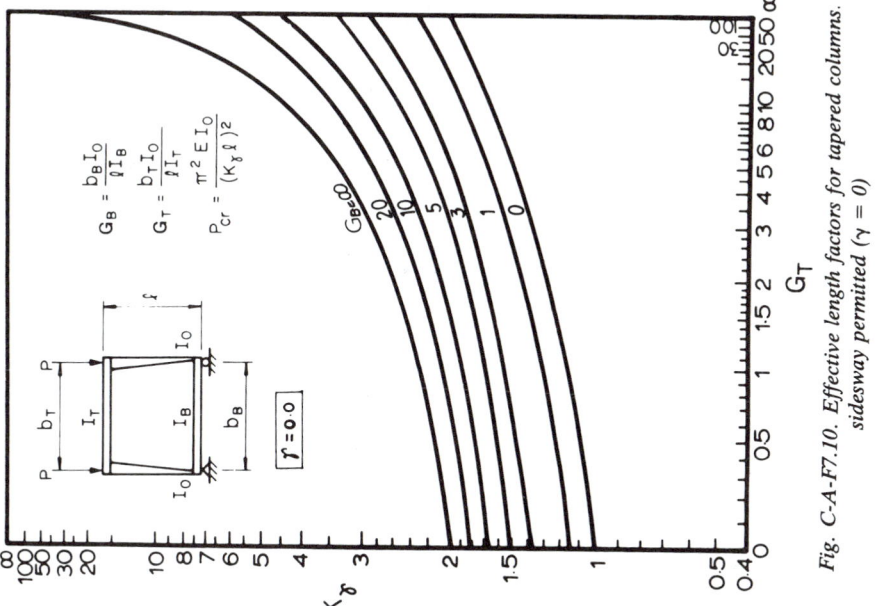

Fig. C-A-F7.10. Effective length factors for tapered columns. sidesway permitted ($\gamma = 0$)

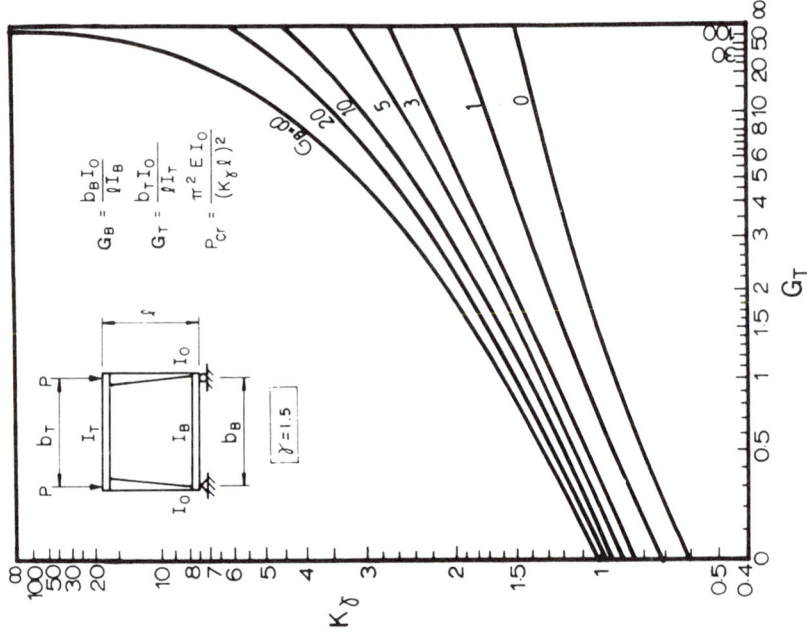

Fig. C-A-F7.13. Effective length factors for tapered columns: sideway permitted ($\gamma = 1.5$)

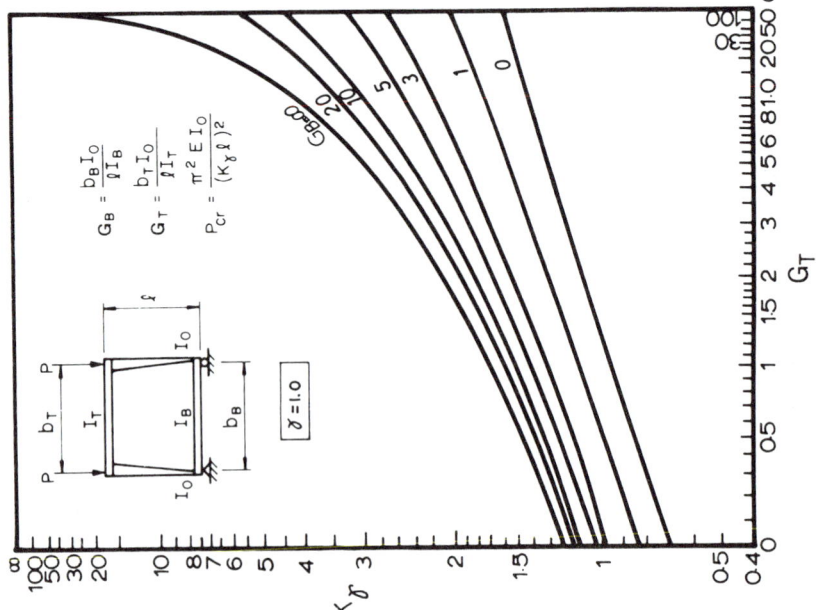

Fig. C-A-F7.12. Effective length factors for tapered columns: sideway permitted ($\gamma = 1.0$)

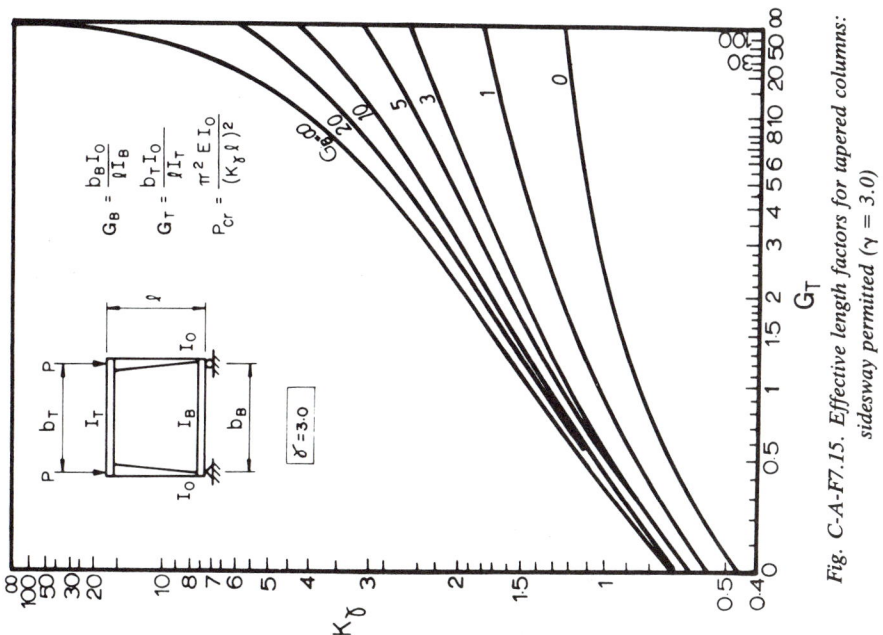

Fig. C-A-F7.15. Effective length factors for tapered columns: sideway permitted ($\gamma = 3.0$)

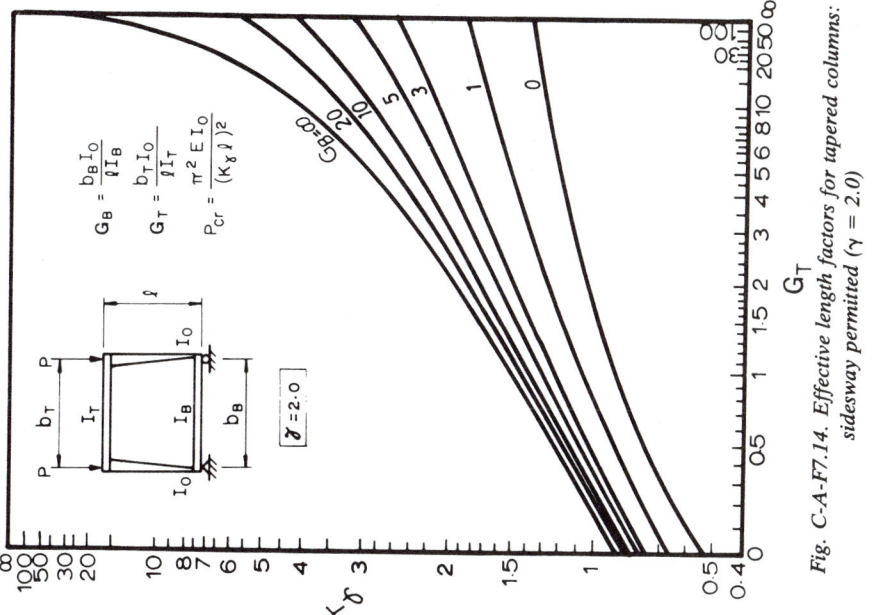

Fig. C-A-F7.14. Effective length factors for tapered columns: sideway permitted ($\gamma = 2.0$)

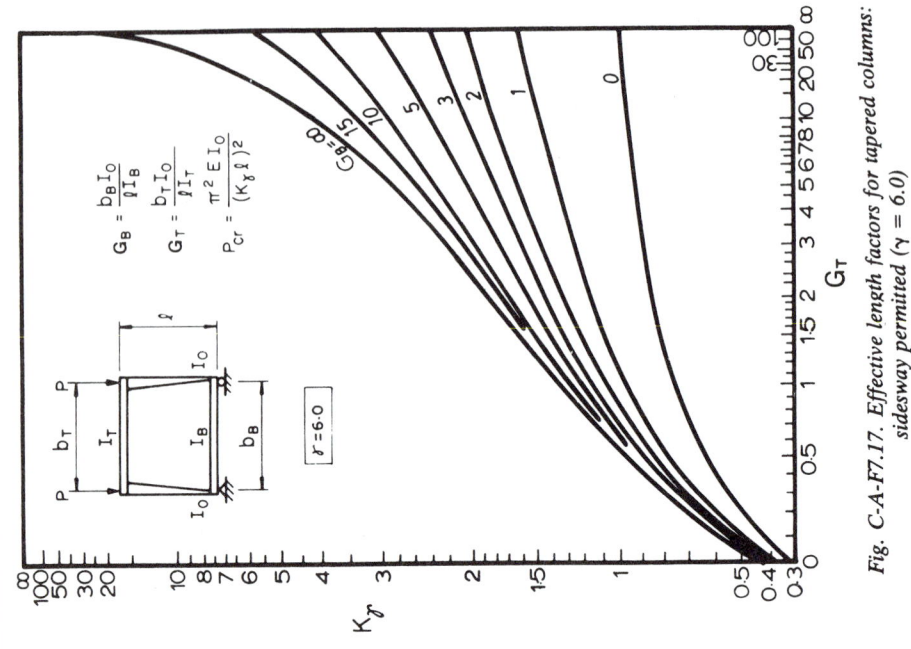

Fig. C-A-F7.17. Effective length factors for tapered columns: sidesway permitted ($\gamma = 6.0$)

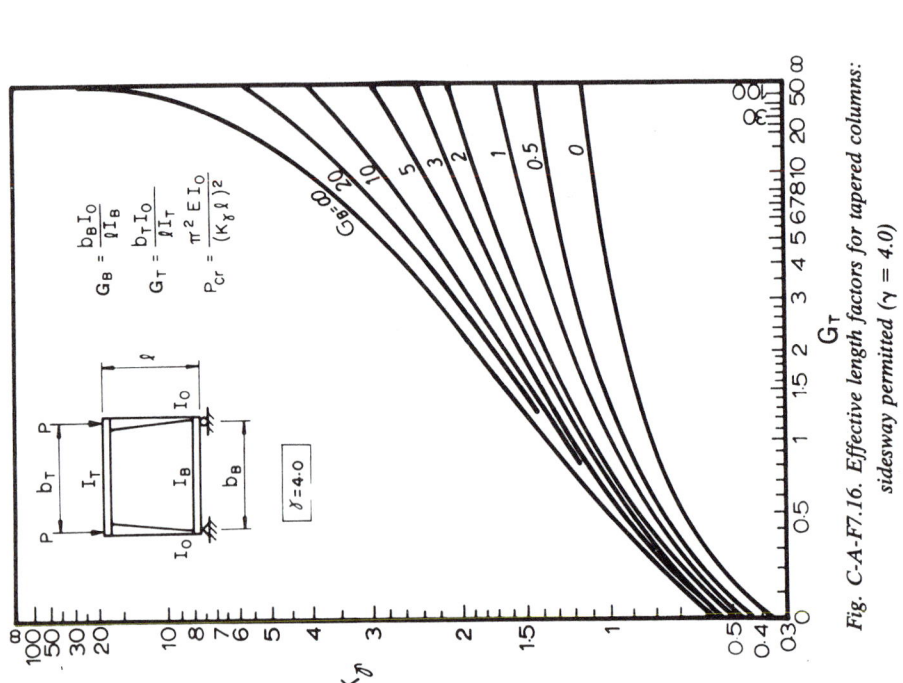

Fig. C-A-F7.16. Effective length factors for tapered columns: sidesway permitted ($\gamma = 4.0$)

SYMBOLS

The section numbers in parentheses after the definition of a symbol refers to the section where the symbol is first used.

A Gross area of an axially loaded compression member, in.^{2} (N4)

A_b Nominal body area of a fastener, in.^{2} (J3.5); area of an upset rod based upon the major diameter of its threads, i.e., the diameter of a coaxial cylinder which would bound the crests of the threads, in.^{2} (J3.4)

A_c Actual area of effective concrete flange in composite design, in.^{2} (I4)

A_e Effective net area of an axially loaded tension member, in.^{2} (B3)

A_f Area of compression flange, in.^{2} (F1.1)

A_{fe} Effective tension flange area, in.^{2} (B10)

A_{fg} Gross beam flange area, in.^{2} (B10)

A_{fn} Net beam flange area, in.^{2} (B10)

A_g Gross area of member, in.^{2} (B3)

A_n Net area of an axially loaded tension member, in.^{2} (B2)

A_s Area of steel beam in composite design, in.^{2} (I4)

A_s' Area of compressive reinforcing steel, in.^{2} (I4)

A_{sr} Area of reinforcing steel providing composite action at point of negative moment, in.^{2} (I4)

A_{st} Cross-sectional area of stiffener or pair of stiffeners, in.^{2} (G4)

A_t Net tension area, in.^{2} (J4)

A_v Net shear area, in.^{2} (J4)

A_w Area of girder web, in.^{2} (G2)

A_1 Area of steel concentrically bearing on a concrete support, in.^{2} (J9)

A_2 Maximum area of the portion of the supporting surface that is geometrically similar to and concentric with the loaded area, in.^{2} (J9)

B Bending coefficient dependent upon computed moment or stress at the ends of unbraced segments of a tapered member (Appendix F7.4)

C_a Coefficient used in Table 4 of Numerical Values

C_b Bending coefficient dependent upon moment gradient (F1.3)

C_c Column slenderness ratio separating elastic and inelastic buckling (E2)

C_c' Slenderness ratio of compression elements (Appendix B5.2)

C_h Coefficient used in Table 12 of Numerical Values

C_m Coefficient applied to bending term in interaction equation for prismatic members and dependent upon column curvature caused by applied moments (H1)

C_m' Coefficient applied to bending term in interaction equation for tapered members and dependent upon axial stress at the small end of the member (Appendix F7.6)

C_p Stiffness factor for primary member in a flat roof (K2)

C_s Stiffness factor for secondary member in a flat roof (K2)

C_v Ratio of "critical" web stress, according to the linear buckling theory, to the shear yield stress of web material (F4)

C_1 Increment used in computing minimum spacing of oversized and slotted holes (J3.8)

C_2 Increment used in computing minimum edge distance for oversized and slotted holes (J3.9)

D Factor depending upon type of transverse stiffeners (G4); outside diameter of tubular member, in. (Appendix B5.2)

E Modulus of elasticity of steel (29,000 ksi) (E2)

E_c Modulus of elasticity of concrete, ksi (I2)

F_a Axial compressive stress permitted in a prismatic member in the absence of bending moment, ksi (E2)

$F_{a\gamma}$ Axial compressive stress permitted in a tapered member in the absence of bending moment, ksi (Appendix F7.3)

F_b Bending stress permitted in a prismatic member in the absence of axial force, ksi (F1.1)

F_b' Allowable bending stress in compression flange of plate girders as reduced for hybrid girders or because of large web depth-to-thickness ratio, ksi (G2)

$F_{b\gamma}$ Bending stress permitted in a tapered member in the absence of axial force, ksi (Appendix F7.6)

F_e' Euler stress for a prismatic member divided by factor of safety, ksi (H1)

$F_{e\gamma}'$ Euler stress for a tapered member divided by factor of safety, ksi (Appendix F7.6)

F_p Allowable bearing stress, ksi (J3.7)

$F_{s\gamma}$ St. Venant torsion resistance bending stress in a tapered member, ksi (Appendix F7.4)

F_t Allowable axial tensile stress, ksi (D1)

F_u Specified minimum tensile strength of the type of steel or fastener being used, ksi (B10)

F_v Allowable shear stress, ksi (F4)

$F_{w\gamma}$ Flange warping torsion resistance bending stress in a tapered member, ksi (Appendix F7.4)

F_y Specified minimum yield stress of the type of steel being used, ksi (B5.1). As used in this Specification, "yield stress" denotes either the specified minimum yield point (for those steels with a yield point) or specified minimum yield strength (for those steels without a yield point)

F_{yc} Specified minimum column yield stress, ksi (K1.2)

F_{yf} Specified minimum yield stress of flange, ksi (Table B5.1).

F_{yr} Specified minimum yield stress of the longitudinal reinforcing steel, ksi (I4)

F_{yst} Specified minimum stiffener yield stress, ksi (K1.8)

F_{yw} Specified minimum yield stress of beam web, ksi (B5.1)

H_s Length of a stud shear connector after welding, in. (I5.2)

I_d Moment of inertia of steel deck supported on secondary members, in.[4] (K2)

I_{eff} Effective moment of inertia of composite sections for deflection computations, in.[4] (I4)

I_p Moment of inertia of primary member in flat roof framing, in.^{4} (K2)

I_s Moment of inertia of secondary member in flat roof framing, in.^{4} (K2); moment of inertia of steel beam in composite construction, in.^{4} (I4)

I_{tr} Moment of inertia of transformed composite section, in.^{4} (I4)

K Effective length factor for a prismatic member (B7)

K_γ Effective length factor for a tapered member (Appendix F7.3)

L Unbraced length of tensile members, in. (B7); actual unbraced length of a column, in. (C2); unbraced length of member measured between the centers of gravity of the bracing members, in. (Appendix F7.1)

L_c Maximum unbraced length of the compression flange at which the allowable bending stress may be taken at $0.66F_y$ or as determined by AISC Specification Equation (F1-3) or Equation (F2-3), when applicable, ft (F1)

L_e Distance from free edge to center of the bolt, in. (J3.6)

L_p Length of primary member in flat roof framing, ft (K2)

L_s Length of secondary member in flat roof framing, ft (K2)

M Moment, kip-ft. (I4); maximum factored bending moment, kip-ft, (N4)

M_1 Smaller moment at end of unbraced length of beam-column (F3.1); larger moment at one end of three-segment portion of a tapered member (Appendix F7.4)

M_2 Larger moment at end of unbraced length of beam-column (F3.1); maximum moment in three adjacent segments of a tapered member (Appendix F7.4)

M_m Critical moment that can be resisted by a plastically designed member in the absence of axial load, kip-ft (N4)

M_p Plastic moment, kip-ft (N4)

N Length of bearing of applied load, in. (K1.3)

N_r Number of stud shear connectors on a beam in one transverse rib of a metal deck, not to exceed 3 in calculations (I5.2)

N_1 Number of shear connectors required between point of maximum moment and point of zero moment (I4)

N_2 Number of shear connectors required between concentrated load and point of zero moment (I4)

P Force transmitted by a fastener, kips (J3.8); factored axial load, kips (N3); normal force, kips (J10.2); axial load, kips (C1)

P_{bf} Factored beam flange or connection plate force in a restrained connection, kips (K1.2)

P_{cr} Maximum strength of an axially loaded compression member or beam, kips (N3.1)

P_e Euler buckling load, kips (N4)

P_y Plastic axial load, equal to profile area times specified minimum yield stress, kips (N3.1)

Q Full reduction factor for slender compression elements (Appendix B5.2)

Q_a Ratio of effective profile area of an axially loaded member to its total profile area (Appendix B5.2)

Q_s Axial stress reduction factor where width-thickness ratio of unstiffened elements exceeds noncompact section limits given in Sect. B5 (Appendix B5.2)

R Reaction or concentrated load applied to beam or girder, kips (K1.3); radius, in. (J2.1)

R_{PG} Plate girder bending strength reduction factor (G2)

R_e Hybrid girder factor (G2)

S Spacing of secondary members in a flat roof, ft (K2); governing slenderness ratio of a tapered member (Appendix F7.3)

S_{eff} Effective section modulus corresponding to partial composite action, in.^{3} (I2)

S_s Section modulus of steel beam used in composite design, referred to the bottom flange, in.^{3} (I2)

S_{tr} Section modulus of transformed composite cross section, referred to the bottom flange; based upon maximum permitted effective width of concrete flange, in.^{3} (I2)

T_b Specified pretension of a high-strength bolt, kips (J3.6)

U Reduction coefficient used in calculating effective net area (B3)

V Shear produced by factored loading, kips (N5); friction force, kips (J10.2)

V_h Total horizontal shear to be resisted by connectors under full composite action, kips (I2)

V_h' Total horizontal shear provided by the connectors providing partial composite action, kips (I2)

Y Ratio of yield stress of web steel to yield stress of stiffener steel (G4)

Z Plastic section modulus, in.^{3} (N4)

a Clear distance between transverse stiffeners, in. (F4); dimension parallel to the direction of stress, in. (Appendix K4)

a' Distance beyond theoretical cut-off point required at ends of welded partial length cover plate to develop stress, in. (B10)

b Actual width of stiffened and unstiffened compression elements as defined in Sect. B5.1, in.; dimension normal to the direction of stress, in. (Appendix K4)

b_e Effective width of stiffened compression element, in (Appendix B5.2)

b_f Flange width of rolled beam or plate girder, in. (F1.1)

d Depth of beam or girder, in. (B5.1); diameter of a roller or rocker bearing, in. (J8); nominal diameter of a fastener, in. (J3.7)

d_L Depth at the larger end of a tapered member, in. (Appendix F7.1)

d_c Web depth clear of fillets, in. (K1.5)

d_o Depth at the smaller end of a tapered member or unbraced segment thereof, in. (Appendix F7.1)

f Axial compression stress on member based on effective area, ksi (Appendix B5.2)

f_a Computed axial stress, ksi (B5.1)

f_{ao} Computed axial stress at the smaller end of a tapered member or unbraced segment thereof, ksi (Appendix F7.6)

f_b Computed bending stress, ksi (H1)

f_{b1} Smallest computed bending stress at one end of a tapered segment, ksi (Appendix F7.4)

f_{b2} Largest computed bending stress at one end of a tapered segment, ksi (Appendix F7.4)

f_{bl} Computed bending stress at the larger end of a tapered member or unbraced segment thereof, ksi (Appendix F7.6)

f'_c Specified compression strength of concrete, ksi (I2)

f_t Computed tensile stress, ksi (J3.6)

f_v Computed shear stress, ksi (F5)

f_{vs} Shear between girder web and transverse stiffeners, kips per linear in. of single stiffener or pair of stiffeners (G4)

g Transverse spacing between fastener gage lines, in. (B2)

h Clear distance between flanges of a beam or girder at the section under investigation, in. (B5)

h_r Nominal rib height for steel deck, in. (I5.2)

h_s Factor applied to the unbraced length of a tapered member (Appendix F7.4)

h_w Factor applied to the unbraced length of a tapered member (Appendix F7.4)

k Distance from outer face of flange to web toe of fillet of rolled shape or equivalent distance on welded section, in. (K1.3)

k_c Compression element restraint coefficient (B5)

k_v Shear buckling coefficient for girder webs (F4)

l For beams, distance between cross sections braced against twist or lateral displacement of the compression flange, in. (F1.3); for columns, actual unbraced length of member, in. (B7); unsupported length of a lacing bar, in. (E4); weld length, in. (B3); largest laterally unbraced length along either flange at the point of load, in. (K1.5)

l_b Actual unbraced length in plane of bending, in. (H1)

l_{cr} Critical unbraced length adjacent to plastic hinge, in. (N9)

n Modular ratio (E/E_c) (I2)

q Allowable horizontal shear to be resisted by a shear connector, kips (I4)

r Governing radius of gyration, in. (B7)

r_T Radius of gyration of a section comprising the compression flange plus $\frac{1}{3}$ of the compression web area, taken about an axis in the plane of the web, in. (F1.3)

r_{To} Radius of gyration at the smaller end of a tapered member or unbraced segment thereof, considering only the compression flange plus $\frac{1}{3}$ of the compression web area, taken about an axis in the plane of the web, in. (Appendix F7.4)

r_b Radius of gyration about axis of concurrent bending, in. (H1)

r_{bo} Radius of gyration about axis of concurrent bending at the smaller end of a tapered member or unbraced segment thereof, in. (Appendix F7.6)

r_o Radius of gyration at the smaller end of a tapered member, in. (Appendix F7.3)

s Longitudinal center-to-center spacing (pitch) of any two consecutive holes, in. (B2)

t Thickness of a connected part, in. (J3.9); wall thickness of a tubular member, in. (Appendix B5); compression element thickness, in. (B5.1); filler thickness, in. (J6)

t_b Thickness of beam flange or moment connection plate at rigid beam-to-column connection, in. (K1.8)

t_f	Flange thickness, in. (F1.1)
t_w	Web thickness, in. (B5.1)
t_{wc}	Column web thickness, in. (K1.6)
w	Length of channel shear connectors, in. (I4); plate width (distance between welds), in. (B3)
w_r	Average width of rib or haunch of concrete slab on formed steel deck, in. (I5.1)
x	Subscript relating symbol to strong axis bending
y	Subscript relating symbol to weak axis bending
z	Distance from the smaller end of a tapered member, in. (Appendix F7.3)
α	$= 0.6\,F_{yw}/F_b < 1.0$ (G2)
β	Ratio S_{tr}/S_s or S_{eff}/S_s (I4)
γ	Tapering ratio of a tapered member or unbraced segment of a tapered member (Appendix F7.1); subscript relating symbol to tapered members
Δ	Displacement of the neutral axis of a loaded member from its position when the member is not loaded, in. (C1)
μ	Coefficient of friction (J10.2)

LIST OF REFERENCES

Ackroyd, M. (1987) Simplified Frame Design of Type PR Construction *AISC Engineering Journal, 4th Quarter, 1987, Chicago, IL.*

Adams, P.F., Lay, M.G., and Galambos, T.V. (1965) Experiments on High-strength Steel Members *Welding Research Council Bulletin No. 110, November 1965.*

American Concrete Institute (1983) Building Code Requirements for Reinforced Concrete *ACI 318-83, Detroit MI, 1983.*

American Institute of Steel Construction (1973) Commentary on Highly Restrained Welded Connections *AISC Engineering Journal, Vol. 10, No. 3, 3rd Quarter, 1973.*

American Institute of Steel Construction, Inc. (1983) Torsional Analysis of Steel Members *Chicago, IL., 1983.*

American Institute of Steel Construction, Inc. (1986) Load and Resistance Factor Design Specification for Structural Steel Buildings *Chicago, IL., 1986.*

American Iron and Steel Institute (1968) Plastic Design of Braced Multistory Steel Frames *New York, NY, 1968.*

American Society of Civil Engineers (1971) Plastic Design in Steel *ASCE Manual of Engineering Practice No. 41, 2nd Edition, 1971.*

Joint ASCE-AASHO Committee on Flexural Members (1968) Design of Hybrid Steel Beams *Report of Subcommittee 1, Journal of the Structural Division, ASCE, Vol. 94, No. ST6, June 1968.*

Joint ASCE-ACI Committee on Composite Construction (1960) Progress Report *ASCE Journal of the Structural Division, December 1960.*

Association of Iron and Steel Engineers (1979) Technical Report No. 13 *Pittsburgh, PA, 1979.*

Austin, W.J. (1961) Strength and Design of Metal Beam-Columns *ASCE Journal of the Structural Division, April 1961.*

Australian Institute of Steel Construction (1975) Australian Standard A51250-1975.

Basler, K. (1961) New Provisions for Plate Girder Design Appendix C *1961 Proceedings AISC National Engineering Conference.*

Basler, K. (1961) Strength of Plate Girders in Shear *Journal of the Structural Division, ASCE, Vol.87, No. ST7, October 1961.*

Basler, K. (1961a) Strength of Plate Girders Under Combined Bending and Shear *Journal of the Structural Division, ASCE, Vol. 87, No. ST7, October 1961.*

Basler, K. and Thurlimann, B. (1963) Strength of Plate Girders in Bending *Journal of the Structural Division, ASCE, Vol. 89, No. ST4, August 1963.*

Basler, K., Yen, B.T., Mueller, J. A. and Thurlimann, B. (1960) Web-buckling Tests on Welded Plate Girders *Welding Research Council Bulletin No. 64, September 1960.*

Bendigo, R.A., Hansen, R.M. and Rumpf, J.L. (1963) Long Bolted Joints *Journal of the Structural Division, ASCE, Vol. 89, No. ST6, December 1963.*

Bergfelt, A. (1971) Studies and Tests on Slender Plate Girders without Stiffeners March 1971.

Bigos, J., Smith, G.W., Ball, E.F. and Foehl, P.J. (1954) Shop Paint and Painting Practice *1954 Proceedings, AISC National Engineering Conference.*

Birkemoe, P.C. and Gilmor, M.I. (1978) Behavior of Bearing-critical Double-angle Beam Connections *AISC Engineering Journal, 4th Quarter, 1978.*

Bleich, F. (1952) Buckling Strength of Metal Structures *McGraw-Hill Book Co., New York, 1952.*

Brockenbrough, R.L. (1983) Considerations in the Design of Bolted Joints for Weathering Steel *AISC Engineering Journal, 1st Quarter, 1983, (p. 40) Chicago, IL.*

Chajes, A. and Winter G. (1965) Torsional Flexural Buckling of Thin-walled Members *Journal of the Structural Division, ASCE, Vol. 91, No. ST4, August 1965.*

Cheong-Siat Moy., F, Ozer, E. and Lu, L.W. (1977) Strength of Steel Frames Under Gravity Loads *Journal of the Structural Division, ASCE, Vol. 103, No. ST6, June 1977.*

Daniels, J.H. and Lu. L.W. (1972) Plastic Subassemblage Analysis for Unbraced Frames *Journal of the Structural Division, ASCE, Vol. 98, No. ST8, August 1972.*

Disque, R.O. (1964) Wind Connections with Simple Framing *AISC Engineering Journal, Vol. 1, No. 3, July 1964.*

Disque, R.O. (1973) Inelastic K-factor in Design *AISC Engineering Journal, 2nd Quarter, 1973.*

Disque, R.O. (1975) Directional Moment Connections–A Proposed Design Method for Unbraced Steel Frames *AISC Engineering Journal, 1st Quarter, 1975, Chicago, IL.*

Douty, R.T. and McGuire, W. (1965) High-strength Bolted Moment Connections *Journal of the Structural Division, ASCE, Vol. 91, No. ST2, April 1965.*

Driscoll, G.C. (1966) Lehigh Conference on Plastic Design of Multi-story Frames—a Summary *AISC Engineering Journal, April 1966.*

Driscoll, G.C. et al. (1965) Plastic Design of Multi-story Frames—Lecture Notes *Fritz Engineering Laboratory Report No. 273.20, Lehigh University, August 1965.*

Elgaaly, M. (1983) Web Design under Compressive Edge Loads *AISC Engineering Journal, 4th Quarter, 1983.*

Fisher, J. W. (1970) Design of Composite Beams with Formed Metal Deck *AISC Engineering Journal, Vol.7, No. 3, July 1970.*

Fisher, J.W., Albrecht, P.A., Yen, B.T., Klingerman, D.J. and McNamee, B.M. (1974) Fatigue Strength of Steel Beams With Welded Stiffeners and Attachments *National Cooperative Highway Research Program, Report 147, 1974.*

Fisher, J.W., Frank, K.H., Hirt, M.A. and McNamee, B.M. (1970) Effect of Weldments on the Fatigue Strength of Beams *National Cooperative Highway Research Program, Report 102, 1970.*

Fisher, J.W. and Pense, A.W. (1987) Experience with Use of Heavy W-shapes in Tension *AISC Engineering Journal, 2nd Quarter, 1987, Chicago, IL.*

Frank, K.H. and Fisher, J.W. (1979) Fatigue Strength of Fillet Welded Cruciform Joints *Journal of the Structural Division, ASCE, Vol. 105, No. ST9, September, 1979.*

Frank, K.H. and J.A. Yura (1981) An Experimental Study of Bolted Shear Connections *FHWA/RD-81/148, December 1981.*

Galambos, T.V. (1960) Influence of Partial Base Fixity on Frame Stability *ASCE Journal of the Structural Division, May 1960.*

Galambos, T.V. (1968) Structural Members and Frames *Prentice-Hall, Englewood Cliffs, NJ, 1968.*

Galambos, T.V. (Ed.) (1988) Structural Stability Research Council Guide to Stability Design Criteria for Metal Structures *4th Edition. John Wiley & Sons, 1988.*

Gaylord, E. H. and Gaylord, C.N. (1972) Design of Steel Structures *2nd Edition, McGraw-Hill Book Co., New York, 1972.*

Gibson, G.T. and Wake, B.T. (1942) An Investigation of Welded Connections for Angle Tension Members *The Welding Journal, American Welding Society, January 1942.*

Gjelsvik, A. (1981) The Theory of Thin-walled Bars *John Wiley and Sons, New York, 1981.*

Graham, J.D., Sherbourne, A.N., Knabbaz, R.N. and Jensen, C.D. (1959) Welded Interior Beam-to-Column Connections *American Institute of Steel Construction, 1959.*

Grant, J.A., Fisher, J. W. and Slutter, R.O. (1977) Composite Beams with Formed Steel Deck *AISC Engineering Journal, Vol.14, No.1, 1st Quarter, 1977.*

Hardash, S.G. and Bjorhovde, R. (1985) New Design Criteria for Gusset Plates in Tension *AISC Engineering Journal, 2nd Quarter, 1985.*

Higgins, T.R. and Preece, F.R. (1968) Proposed Working Stresses for Fillet Welds in Building Construction *Welding Journal Research Supplement, Oct., 1968.*

Hoglund, T. (1971) Simply-supported Long Thin Plate I-Girders without Web Stiffeners, Subjected to Distributed Transverse Load *Dept. of Building Statics and Structural Engineering of the Royal Institute of Technology, Stockholm, Sweden, 1971.*

International Association of Bridge and Structural Engineering (1968) Final Report of the Eighth Congress *September 1968.*

Johnson, D.L. (1985)   An Investigation into the Interaction of Flanges and Webs in Wide-flange Shapes   *1985 Proceedings SSRC Annual Technical Session.*

Johnston, B.G. (1939)   Pin-Connected Plate Links   *1939 ASCE Transactions.*

Jones, J. (1940)   Static Tests on Riveted Joints   *Civil Engineering, May, 1940.*

Keating, P.B. and J.W. Fisher (1985)   Review of Fatigue Tests and Design Criteria on Welded Details   *NCHRP Project 12-15(50), October 1985.*

Ketter, R.L. (1961)   Further Studies of the Strength of Beam Columns   *ASCE Journal of the Structural Division, Vol. 87, No. ST6, August 1961.*

Kloppel, K. and Seeger, T. (1964)   Dauerversuche Mit Einsohnittigen Hv-Verbindurgen Aus ST37   *Der Stahlbau, Vol. 33, No. 8, August 1964, pp. 225–245 and Vol. 33, No. 11, November 1964. pp. 335-346.*

Klyce, David C. (1988)   Shear Connector Spacing in Composite Members with Formed Steel Deck   *Lehigh University, May 1988.*

Kotecki, D.S. and Moll, R.A. (1970)   A Toughness Study of Steel Weld Metal from Self-shielded, Flux-cored Electrodes, Part 1   *Welding Journal Vol. 49, April 1970.*

Kotecki, D.S. and Moll, R.A. (1972)   A Toughness Study of Steel Weld Metal from Self-shielded, Flux-cored Electrodes, Part 2   *Welding Journal, Vol. 51, March 1972.*

Krishnamurthy, N. (1978)   A Fresh Look at Bolted End-Plate Behavior and Design   *AISC Engineering Journal, Vol. 15, No. 2, 2nd Quarter, 1978.*

Kulak, G.L., Fisher, J.W. and Struik, J.H.A. (1987)   Guide to Design Criteria for Bolted and Riveted Joints, 2nd Edition   *John Wiley & Sons, New York 1987.*

Lay, M.G. and Galambos, T.V. (1967)   Inelastic Beams Under Moment Gradient   *Journal of the Structural Division, ASCE, Vol. 93, No. ST1, February 1967.*

Lee, G.C., Morrell, M.L. and Ketter, R.L. (1972)   Design of Tapered Members   *WRC Bulletin No. 173, June 1972.*

Leigh, J.M. and M.G. Lay (1978)   Laterally Unsupported Angles with Equal and Unequal Legs   *Report MRL 22/2 July 1978, Melbourne Research Laboratories, Clayton.*

Leigh, J.M. and M.G. Lay (1984)   The Design of Laterally Unsupported Angles, in   *Steel Design Current Practice, Sect. 2, Bending Members, American Institute of Steel Construction, January 1984.*

LeMessurier, W.J. (1976)   A Practical Method of Second Order Analysis/Part 1—Pin-jointed Frames   *AISC Engineering Journal, Vol. 13, No. 4, 4th Quarter, 1976.*

LeMessurier, W.J. (1977)   A Practical Method of Second Order Analysis/Part 2—Rigid Frames   *AISC Engineering Journal, Vol. 14, No. 2, 2nd Quarter, 1977.*

Liapunov, S. (1974)   Ultimate Load Studies of Plane Multi-story Steel Rigid Frames   *Journal of the Structural Division, ASCE, Vol. 100, No. ST8, Proc. Paper 10750, August 1974.*

Lilly, S.B. and Carpenter, S.T. (1940)   Effective Moment of Inertia of a Riveted Plate Girder   *1940 ASCE Transactions.*

Lu, Le-Wu (1967) Design of Braced Multi-story Frames by the Plastic Method *AISC Engineering Journal, January 1967.*

Madugula, M.K.S. and J.B. Kennedy (1985) Single and Compound Angle Members *Elsevier Applied Science, New York, 1985.*

Marino, F.J. (1966) Ponding of Two-way Roof Systems *AISC Engineering Journal, Vol. 3, No. 3, July 1966.*

McGarraugh, J.B. and Baldwin, J.W. (1971) Lightweight Concrete-on-steel Composite Beams *AISC Engineering Journal, Vol. 8, No. 3, July 1971.*

Morrell, M.L. and Lee, G.C. (1974) Allowable Stress for Web-tapered Members *WRC Bulletin 192, February 1974.*

Munse, W.H. and Chesson, E. Jr. (1963) Riveted and Bolted Joints: Net Section Design *Journal of the Structural Division, ASCE, Vol. 89, No. ST1, February 1963.*

Murray, T.M. (1975) Design to Prevent Floor Vibration *AISC Engineering Journal, Vol. 12, No. 3, 3rd Quarter, 1975.*

Ollgaard, J.G., Slutter, R.G. and Fisher, J.W. (1971) Shear Strength of Stud Shear Connections in Lightweight and Normal Weight Concrete *AISC Engineering Journal, Vol. 8, No. 2, April 1971.*

Ozer, E., Okten, O.S., Morino, S., Daniels, J.H. and Lu, L.W. (1974) Frame Stability and Design of Columns in Unbraced Multi-story Steel Frames *Fritz Engineering Laboratory Report No. 375.2, Lehigh University, November 1974.*

Popov, E.P. and Pinkney, R.B. (1968) Behavior of Steel Building Connections Subjected to Inelastic Strain Reversals *Bulletin Nos. 13 and 14, American Iron and Steel Institute, November 1968.*

Popov, E.P. and Stephen, R.M. (1977) Capacity of Columns with Splice Imperfections *AISC Engineering Journal, Vol. 14, No. 1, 1st Quarter, 1977.*

Preece, F.R. (1968) AWS-AISC Fillet Weld Study Longitudinal and Transverse Shear Tests *Testing Engineers, Inc., Los Angeles, May 31, 1968.*

Ravindra, M.K. and Galambos, T.V. (1978) Load and Resistance Factor Design for Steel *ASCE Journal of the Structural Division, Vol. 104, No. ST9, September 1978.*

Research Council on Structural Connections (1985) Specification for Structural Joints Using ASTM A325 or A490 Bolts *1985.*

Ricles, J.M. and Yura, J.A. (1983) Strength of Double-row Bolted Web Connections *ASCE Journal of the Structural Division, Vol. 109, No. ST1, January 1983.*

Roberts, T.M. (1981) Slender Plate Girders Subjected to Edge Loading *Proceedings of Institute of Civil Engineers, Part 2, 71, September 1981.*

Rolfe, S.T. (1977) Fracture and Fatigue Control in Steel Structures *AISC Engineering Journal, Vol. 14, No. 1, 1st Quarter, 1977.*

Rolfe, S.T. and J.M. Barsom (1987) Fracture and Fracture Control in Structures, 2nd Edition, *Prentice-Hall, Inc., Englewood Cliffs, NJ, 1987.*

Sherman, D.R. (1976) Tentative Criteria for Structural Applications of Steel Tubing and Pipe *American Iron and Steel Institute, Washington, D.C., August, 1976.*

Slutter, R.G. and Driscoll, G. C. (1965) Flexural Strength of Steel-Concrete Composite Beams *Journal of the Structural Division, ASCE, Vol. 91, No. ST2, April 1965.*

Springfield, J. and Adams, P.F. (1972) Aspects of Column Design in Tall Steel Buildings *Journal of the Structural Division, ASCE, Vol. 98, No. ST5, May 1972.*

Stang, A.H. and Jaffe, B.S. (1948) Perforated Cover Plates for Steel Columns *Research Paper RP1861, National Bureau of Standards, 1948.*

Steel Structures Painting Council (1982) Steel Structures Painting Manual, Vol. 2, Systems and Specifications *Pittsburgh, PA., 1982*

Terashima, H. and Hart, P.H.M. (1984) Effect of Aluminum on Carbon, Manganese, Niobium Steel Submerged Arc Weld Metal Properties *Welding Journal, Vol. 63, June 1984.*

Timoshenko, S.P. and J.M. Gere (1952) Theory of Elastic Stability *McGraw-Hill Book Company, 1952.*

Van Kuren, R.C. and Galambos, T.V. (1964) Beam-Column Experiments *Journal of the Structural Division, ASCE, Vol. 90, No. ST2, April 1964.*

v. Karman, T., Sechler, E.E. and Donnell, L.H. (1932) The Strength of Thin Plates in Compression *1932 ASME Transactions, Vol. 54, APM-54-5.*

Winter, G. (1947) Strength of Steel Compression Flanges. *1947 ASCE Transactions.*

Winter, G. (1958) Lateral Bracing of Columns and Beams *Journal of the Structural Division, ASCE, Vol. 84, No. ST2, March 1958.*

Yura, J.A. (1971) The Effective Length of Columns in Unbraced Frames *AISC Engineering Journal, Vol. 8, No. 2, April 1971.*

Yura, J.A. (1982) The Behavior of Beams Subjected to Concentrated Loads *PMFSEL Report No. 82-5, August 1982, University of Texas-Austin*

Yura, J.A. Frank, K.H. and Polyzois, D. (1987) High Strength Bolts for Bridges, *PMFSEL Report No. 87-3, May 1987, University of Texas-Austin.*

Zahn, C.J. (1987) Plate Girder Design Using LRFD *AISC Engineering Journal, 1st Quarter, 1987.*

Zandonini, R. (1985) Stability of Compact Built-up Struts: Experimental Investigation and Numerical Simulation *Construction Metalliche, No. 4, 1985.*

GLOSSARY

Alignment chart for columns. A nomograph for determining the effective length factor K for some types of columns

Amplification factor. A multiplier of the value of moment or deflection in the unbraced length of an axially loaded member to reflect the secondary values generated by the eccentricity of the applied axial load within the member

Aspect ratio. In any rectangular configuration, the ratio of the lengths of the sides

Batten plate. A plate element used to join two parallel components of a built-up column, girder or strut rigidly connected to the parallel components and designed to transmit shear between them

Beam. A structural member whose primary function is to carry loads transverse to its longitudinal axis

Beam-column. A structural member whose primary function is to carry loads both transverse and parallel to its longitudinal axis

Bent. A plane framework of beam or truss members which support loads and the columns which support these members

Biaxial bending. Simultaneous bending of a member about two perpendicular axes

Bifurcation. The phenomenon whereby a perfectly straight member under compression may either assume a deflected position or may remain undeflected, or a beam under flexure may either deflect and twist out of plane or remain in its in-plane deflected position.

Braced frame. A frame in which the resistance to lateral load or frame instability is primarily provided by a diagonal, a K-brace or other auxiliary system of bracing

Brittle fracture. Abrupt cleavage with little or no prior ductile deformation

Buckling load. The load at which a perfectly straight member under compression assumes a deflected position

Built-up member. A member made of structural metal elements that are welded, bolted or riveted together

Cladding. The exterior covering of the structural components of a building

Cold-formed members. Structural members formed from steel without the application of heat

Column. A structural member whose primary function is to carry loads parallel to its longitudinal axis

Column curve. A curve expressing the relationship between axial column strength and slenderness ratio

Combined mechanism. A mechanism determined by plastic analysis procedure which combines elementary beam, panel and joint mechanisms

Compact section. Compact sections are capable of developing a fully plastic stress distribution and possess rotation capacity of approximately 3 before the onset of local buckling

Composite beam. A steel beam structurally connected to a concrete slab so that the beam and slab respond to loads as a unit. See also Concrete-encased beam

Composite column. A steel column fabricated from rolled or built-up steel shapes and encased in structural concrete or fabricated from steel pipe or tubing and filled with structural concrete

Concrete-encased beam. A beam totally encased in concrete cast integrally with the slab

Connection. Combination of joints used to transmit forces between two or more members. Categorized by the type and amount of force transferred (moment, shear, end reaction). See also splices

Critical load. The load at which bifurcation occurs as determined by a theoretical stability analysis

Curvature. The rotation per unit length due to bending

Design documents. See structural design documents

Design strength. Resistance (force, moment, stress, as appropriate) provided by element or connection; the product of the nominal strength and the resistance factor

Diagonal bracing. Inclined structural members carrying primarily axial load employed to enable a structural frame to act as a truss to resist horizontal loads

Diaphragm. Floor slab, metal wall or roof panel possessing a large in-plane shear stiffness and strength adequate to transmit horizontal forces to resisting systems

Diaphragm action. The in-plane action of a floor system (also roofs and walls) such that all columns framing into the floor from above and below are maintained in their same position relative to each other

Double curvature. A bending condition in which end moments on a member cause the member to assume an S-shape

Drift. Lateral deflection of a building

Drift index. The ratio of lateral deflection to the height of the building

Ductility factor. The ratio of the total deformation at maximum load to the elastic-limit deformation

Effective length. The equivalent length KL used in compression formulas and determined by a bifurcation analysis

Effective length factor K. The ratio between the effective length and the unbraced length of the member measured between the centers of gravity of the bracing members

Effective moment of inertia. The moment of inertia of the cross section of a member that remains elastic when partial plastification of the cross section takes place, usually under the combination of residual stress and applied stress. Also, the moment of inertia based on effective widths of elements that buckle locally. Also, the moment of inertia used in the design of partially composite members

Effective stiffness. The stiffness of a member computed using the effective moment of inertia of its cross section

Effective width. The reduced width of a plate or slab which, with an assumed uniform stress distribution, produces the same effect on the behavior of a structural member as the actual plate width with its nonuniform stress distribution

Elastic analysis. Determination of load effects (force, moment, stress as appropriate) on members and connections based on the assumption that material deformation disappears on removal of the force that produced it

Elastic-perfectly plastic. A material which has an idealized stress-strain curve that varies linearly from the point of zero strain and zero stress up to the yield point of the material, and then increases in strain at the value of the yield stress without any further increases in stress

Embedment. A steel component cast in a concrete structure which is used to transmit externally applied loads to the concrete structure by means of bearing, shear, bond, friction or any combination thereof. The embedment may be fabricated of structural-steel plates, shapes, bars, bolts, pipe, studs, concrete reinforcing bars, shear connectors or any combination thereof

Encased steel structure. A steel-framed structure in which all of the individual frame members are completely encased in cast-in-place concrete

Euler formula. The mathematical relationship expressing the value of the Euler load in terms of the modulus of elasticity, the moment of inertia of the cross section and the length of a column

Euler load. The critical load of a perfectly straight, centrally loaded, pin-ended column

Eyebar. A particular type of pin-connected tension member of uniform thickness with forged or flame cut head of greater width than the body proportioned to provide approximately equal strength in the head and body

Factored load. The product of the nominal load and a load factor

Fastener. Generic term for welds, bolts, rivets or other connecting device

Fatigue. A fracture phenomenon resulting from a fluctuating stress cycle

First-order analysis. Analysis based on first-order deformations in which equilibrium conditions are formulated on the undeformed structure

Flame-cut plate. A plate in which the longitudinal edges have been prepared by oxygen cutting from a large plate

Flat width. For a rectangular tube, the nominal width minus twice the outside corner radius. In absence of knowledge of the corner radius, the flat width may be taken as the total section width minus three times the thickness

Flexible connection. A connection permitting a portion, but not all, of the simple beam rotation of a member end

Floor system. The system of structural components separating the stories of a building

Force. Resultant of distribution of stress over a prescribed area. A reaction that develops in a member as a result of load (formerly called total stress or stress). Generic term signifying axial loads, bending moment, torques and shears

Fracture toughness. Measurement of the ability to absorb energy without fracture. Generally determined by impact loading of specimens containing a notch having a prescribed geometry

Frame buckling. A condition under which bifurcation may occur in a frame

Frame instability. A condition under which a frame deforms with increasing lateral deflection under a system of increasing applied monotonic loads until a maximum value of the load called the stability limit is reached, after which the frame will continue to deflect without further increase in load

Fully composite beam. A composite beam with sufficient shear connectors to develop the full flexural strength of the composite section

High-cycle fatigue. Failure resulting from more than 20,000 applications of cyclic stress

Hybrid beam. A fabricated steel beam composed of flanges with a greater yield strength that that of the web. Whenever the maximum flange stress is less than or equal to the web yield stress the girder is considered homogeneous

Hysteresis loop. A plot of force versus displacement of a structure or member subjected to reversed, repeated load into the inelastic range, in which the path followed during release and removal of load is different from the path for the addition of load over the same range of displacement

Inclusions. Nonmetallic material entrapped in otherwise sound metal

Incomplete fusion. Lack of union by melting of filler and base metal over entire prescribed area

Inelastic action. Material deformation that does not disappear on removal of the force that produced it

Instability. A condition reached in the loading of an element or structure in which continued deformation results in a decrease of load-resisting capacity

Joint. Area where two or more ends, surfaces, or edges are attached. Categorized by type of fastener or weld used and method of force transfer

K-bracing. A system of struts used in a braced frame in which the pattern of the struts resembles the letter *K*, either normal or on its side

Lamellar tearing. Separation in highly restrained base metal caused by through-thickness strains induced by shrinkage of adjacent weld metal

Lateral bracing member. A member utilized individually or as a component of a lateral bracing system to prevent buckling of members or elements and/or to resist lateral loads

Lateral (or lateral-torsional) buckling. Buckling of a member involving lateral deflection and twist

Limit state. A condition in which a structure or component becomes unfit for service and is judged either to be no longer useful for its intended function *(serviceability limit state)* or to be unsafe *(strength limit state)*

Limit states. Limits of structural usefulness, such as brittle fracture, plastic collapse, excessive deformation, durability, fatigue, instability and serviceability

Load factor. A factor that accounts for unavoidable deviations of the actual load from the nominal value and for uncertainties in the analysis that transform the load into a load effect

Loads. Forces or other actions that arise on structural systems from the weight of all permanent construction, occupants and their possessions, environmental effects, differential settlement and restrained dimensional changes. *Permanent* loads are those loads in which variations in time are rare or of small magnitude. All other loads are *variable* loads. See *Nominal loads.*

LRFD (Load and Resistance Factor Design). A method of proportioning structural components (members, connectors, connecting elements and assemblages) such that no applicable limit state is exceeded when the structure is subjected to all appropriate load combinations

Local buckling. The buckling of a compression element which may precipitate the failure of the whole member

Low-cycle fatigue. Fracture resulting from a relatively high stress range resulting in a relatively small number of cycles to failure

Lower bound load. A load computed on the basis of an assumed equilibrium moment diagram in which the moments are not greater than M_p, that is, less than or at best equal to the true ultimate load

Mechanism. An articulated system able to deform without an increase in load, used in the special sense that the linkage may include real hinges or plastic hinges, or both

Mechanism method. A method of plastic analysis in which equilibrium between external forces and internal plastic hinges is calculated on the basis of an assumed mechanism. The failure load so determined is an upper bound

Nominal loads. The magnitudes of the loads specified by the applicable code

Nominal strength. The capacity of a structure or component to resist the effects of loads, as determined by computations using specified material strengths and dimensions and formulas derived from accepted principles of structural mechanics or by field tests or laboratory tests of scaled models, allowing for modeling effects and differences between laboratory and field conditions

Noncompact section. Noncompact sections can develop yield stress in compression elements before local buckling occurs, but will not resist inelastic local buckling at strain levels required for a fully plastic stress distribution

P-Delta effect. Secondary effect of column axial loads and lateral deflection on the moments in members

Panel zone. The zone in a beam-to-column connection that transmits moments by a shear panel

Partially composite beam. A composite beam for which the shear strength of shear connectors governs the flexural strength

Plane frame. A structural system assumed for the purpose of analysis and design to be two-dimensional

Plastic analysis. Determination of load effects (force, moment, stress, as appropriate) on members and connections based on the assumption of rigid-plastic behavior, i.e., that equilibrium is satisfied throughout the structure and yield is not exceeded anywhere. Second order effects may need to be considered

Plastic design section. The cross section of a member which can maintain a full plastic moment through large rotations so that a mechanism can develop; the section suitable for plastic design

Plastic hinge. A yielded zone which forms in a structural member when the plastic moment is attained. The beam is assumed to rotate as if hinged, except that it is restrained by the plastic moment M_p

Plastic-limit load. The maximum load that is attained when a sufficient number of yield zones have formed to permit the structure to deform plastically without further increase in load. It is the largest load a structure will support, when perfect plasticity is assumed and when such factors as instability, second-order effects, strain hardening and fracture are neglected

Plastic mechanism. See mechanism

Plastic modulus. The section modulus of resistance to bending of a completely yielded cross-section. It is the combined static moment about the neutral axis of the cross-sectional areas above and below that axis

Plastic moment. The resisting moment of a fully yielded cross section

Plastic strain. The difference between total strain and elastic strain

Plastic zone. The yielded region of a member

Plastification. The process of successive yielding of fibers in the cross section of a member as bending moment is increased

Plate girder. A built-up structural beam

Post-buckling strength. The load that can be carried by an element, member or frame after buckling

Redistribution of moment. A process which results in the successive formation of plastic hinges so that less highly stressed portions of a structure may carry increased moments

Required strength. Load effect (force, moment, stress, as appropriate) acting on an element or connection determined by structural analysis from the factored loads (using most appropriate critical load combinations)

Residual stress. The stress that remains in an unloaded member after it has been formed into a finished product. (Examples of such stresses include, but are not limited to, those induced by cold bending, cooling after rolling, or welding.)

Resistance. The capacity of a structure or component to resist the effects of loads. It is determined by computations using specified material strengths, dimensions and formulas derived from accepted principles of structural mechanics, or by field tests or laboratory tests of scaled models, allowing for modeling effects and differences between laboratory and field conditions. Resistance is a generic term that includes both strength and serviceability limit states

Resistance factor. A factor that accounts for unavoidable deviations of the actual strength from the nominal value and the manner and consequences of failure

Rigid frame. A structure in which connections maintain the angular relationship between beam and column members under load

Root of the flange. Location on the web of the corner radius termination point or the toe of the flange-to-web weld. Measured as the k-distance from the far side of the flange

Rotation capacity. The incremental angular rotation that a given shape can accept prior to local failure defined as $R = (\theta_u/\theta_p) - 1$ where θ_u is the overall rotation attained at the factored load state and θ_p is the idealized rotation corresponding to elastic theory applied to the case of $M = M_p$

St. Venant torsion. That portion of the torsion in a member that induces only shear stresses in the member

Second-order analysis. Analysis based on second-order deformations, in which equilibrium conditions are formulated on the deformed structure

Service load. Load expected to be supported by the structure under normal usage; often taken as the nominal load

Serviceability limit state. Limiting condition affecting the ability of a structure to preserve its appearance, maintainability, durability or the comfort of its occupants or function of machinery under normal usage.

Shape factor. The ratio of the plastic moment to the yield moment, or the ratio of the plastic modulus to the section modulus for a cross section

Shear-friction. Friction between the embedment and the concrete that transmits shear loads. The relative displacement in the plane of the shear load is considered to be resisted by shear-friction anchors located perpendicular to the plane of the shear load

Shear lugs. Plates, welded studs, bolts and other steel shapes that are embedded in the concrete and located transverse to the direction of the shear force and that transmit shear loads introduced into the concrete by local bearing at the shear lug-concrete interface

Shear wall. A wall that in its own plane resists shear forces resulting from applied wind, earthquake or other transverse loads or provides frame stability. Also called a structural wall

Sidesway. The lateral movement of a structure under the action of lateral loads, unsymmetrical vertical loads or unsymmetrical properties of the structure

Sidesway buckling. The buckling mode of a multistory frame precipitated by the relative lateral displacements of joints, leading to failure by sidesway of the frame

Simple plastic theory. See Plastic design

Single curvature. A deformed shape of a member having one smooth continuous arc, as opposed to double curvature which contains a reversal

Slender section. The cross section of a member which will experience local buckling in the elastic range

Slenderness ratio. The ratio of the effective length of a column to the radius of gyration of the column, both with respect to the same axis of bending

Slip-critical joint. A bolt joint in which the slip resistance of the connection is required

Space frame. A three-dimensional structural framework (as contrasted to a plane frame)

Splice. The connection between two structural elements joined at their ends to form a single, longer element

Stability-limit load. Maximum (theoretical) load a structure can support when second-order instability effects are included

Stepped column. A column with changes from one cross section to another occurring at abrupt points within the length of the column

Stiffener. A member, usually an angle or plate, attached to a plate or web of a beam or girder to distribute load, to transfer shear or to prevent buckling of the member to which it is attached

Stiffness. The resistance to deformation of a member or structure measured by the ratio of the applied force to the corresponding displacement

Story drift. The difference in horizontal deflection at the top and bottom of a story

Strain hardening. Phenomenon wherein ductile steel, after undergoing considerable deformation at or just above yield point, exhibits the capacity to resist substantially higher loading than that which caused initial yielding

Strain-hardening strain. For structural steels that have a flat (plastic) region in the stress-strain relationship, the value of the strain at the onset of strain hardening

Strength design. A method of proportioning structural members using load factors and resistance factors such that no applicable limit state is exceeded (also called load and resistance factor design)

Strength limit state. Limiting condition affecting the safety of the structure, in which the ultimate load-carrying capacity is reached

Stress. Force per unit area

Stress concentration. Localized stress considerably higher than average (even in uniformly loaded cross sections of uniform thickness) due to abrupt changes in geometry or localized loading

Strong axis. The major principal axis of a cross section

Structural design documents. Documents prepared by the designer (plans, design details and job specifications)

Structural system. An assemblage of load-carrying components which are joined together to provide regular interaction or interdependence

Stub column. A short compression-test specimen, long enough for use in measuring the stress-strain relationship for the complete cross section, but short enough to avoid buckling as a column in the elastic and plastic ranges

Subassemblage. A truncated portion of a structural frame

Supported frame. A frame which depends upon adjacent braced or unbraced frames for resistance to lateral load or frame instability. (This transfer of load is frequently provided by the floor or roof system through diaphragm action or by horizontal cross bracing in the roof.)

Tangent modulus. At any given stress level, the slope of the stress-strain curve of a material in the inelastic range as determined by the compression test of a small specimen under controlled conditions.

Temporary structure. A general term for anything that is built or constructed (usually to carry construction loads) that will eventually be removed before or after completion of construction and does not become part of the permanent structural system

Tensile strength. The maximum tensile stress that a material is capable of sustaining

Tension field action. The behavior of a plate girder panel under shear force in which diagonal tensile stresses develop in the web and compressive forces develop in the transverse stiffeners in a manner analogous to a Pratt truss

Toe of the fillet. Termination point of fillet weld or of rolled section fillet

Torque-tension relationship Term applied to the wrench torque required to produce specified pre-tension in high-strength bolts

Turn-of-nut method. Procedure whereby the specified pre-tension in high-strength bolts is controlled by rotation of the wrench a predetermined amount after the nut has been tightened to a snug fit

Unbraced frame. A frame in which the resistance to lateral load is provided by the bending resistance of frame members and their connections

Unbraced length. The distance between braced points of a member, measured between the centers of gravity of the bracing members

Undercut. A notch resulting from the melting and removal of base metal at the edge of a weld

Universal-mill plate. A plate in which the longitudinal edges have been formed by a rolling process during manufacture. Often abbreviated as UM plate

Upper bound load. A load computed on the basis of an assumed mechanism which will always be at best equal to or greater than the true ultimate load

Vertical bracing system. A system of shear walls, braced frames or both, extending throughout one or more floors of a building

Von Mises yield criterion. A theory which states that inelastic action at any point in a body under any combination of stresses begins only when the strain energy of distortion per unit volume absorbed at the point is equal to the strain energy of distortion absorbed per unit volume at any point in a simple tensile bar stressed to the elastic limit under a state of uniaxial stress. It is often called the maximum strain-energy-of-distortion theory. Accordingly, shear yield occurs at 0.58 times yield strength

Warping torsion. That portion of the total resistance to torsion that is provided by resistance to warping of the cross section

Weak axis. The minor principal axis of a cross section

Weathering steel. A type of high-strength, low-alloy steel which can be used in normal environments (not marine) and outdoor exposures without protective paint covering. This steel develops a tight adherent rust at a decreasing rate with respect to time

Web buckling. The buckling of a web plate

Web crippling. The local failure of a web plate in the immediate vicinity of a concentrated load or reaction

Working load. Also called service load. The actual load assumed to be acting on the structure.

Yield moment. In a member subjected to bending, the moment at which an outer fiber first attains the yield stress

Yield plateau. The portion of the stress-strain curve for uniaxial tension or compression in which the stress remains essentially constant during a period of substantially increased strain

Yield point. The first stress in a material at which an increase in strain occurs without an increase in stress, the yield point less than the maximum attainable stress

Yield strength. The stress at which a material exhibits a specified limiting deviation from the proportionality of stress to strain. Deviation expressed in terms of strain

Yield stress. Yield point, yield strength or yield-stress level as defined

Yield-stress level. The average stress during yielding in the plastic range, the stress determined in a tension test when the strain reaches 0.005 in. per in.

Code of Standard Practice

FOR STEEL BUILDINGS AND BRIDGES

Adopted Effective September 1, 1986
American Institute of Steel Construction, Inc.

AMERICAN INSTITUTE OF STEEL CONSTRUCTION, INC.
1 East Wacker Drive, Suite 3100, Chicago, IL 60601-2001

PREFACE

When contractual documents do not contain specific provisions to the contrary, existing trade practices are considered to be incorporated into the relationships between the parties to a contract. As in any industry, trade practices have developed among those involved in the purchase, design, fabrication and erection of structural steel. The American Institute of Steel Construction has continuously surveyed the structural steel fabrication industry to determine standard practices and, commencing in 1924, published its *Code of Standard Practice*. Since that date, the Code has been periodically updated to reflect new and changing technology and practices of the industry.

It is the Institute's intention to provide to owners, architects, engineers, contractors and others associated with construction, a useful framework for a common understanding of acceptable standards when contracting for structural steel construction.

This edition is the third complete revision of the Code since it was first published. It includes a number of new sections covering new subjects not included in the previous Code, but which are an integral part of the relationship of the parties to a contract.

The Institute acknowledges the valuable information and suggestions provided by trade associations and other organizations associated with construction and the fabricating industry in developing this current *Code of Standard Practice*.

While every precaution has been taken to insure that all data and information presented is as accurate as possible, the Institute cannot assume responsibility for errors or oversights in the information published herein, or the use of the information published or incorporation of such information in the preparation of detailed engineering plans. The Code should not replace the judgment of an experienced architect or engineer who has the responsibility of design for a specific structure.

Code of Standard Practice
for Steel Buildings and Bridges

Adopted Effective September 1, 1986
American Institute of Steel Construction, Inc.

SECTION 1.   GENERAL PROVISIONS

1.1.   Scope

The practices defined herein have been adopted by the AISC as the commonly accepted standards of the structural steel fabricating industry. In the absence of other instructions in the contract documents, the trade practices defined in this *Code of Standard Practice*, as revised to date, govern the fabrication and erection of structural steel.

1.2.   Definitions

AISC Specification—The *Specification for the Design, Fabrication and Erection of Structural Steel for Buildings* as adopted by the American Institute of Steel Construction.

ANSI—American National Standards Institute.

Architect/Engineer—The owner's designated representative with full responsibility for the design and integrity of the structure.

ASTM—The material standard of the American Society for Testing and Materials.

AWS Code—The *Structural Welding Code* of the American Welding Society.

Code—The *Code of Standard Practice* as adopted by the American Institute of Steel Construction.

Contract Documents—The documents which define the responsibilities of the parties involved in bidding, purchasing, supplying and erecting structural steel. Such documents normally consist of a contract, plans and specifications.

Drawings—Shop and field erection drawings prepared by the fabricator and erector for the performance of the work.

Erector—The party responsible for the erection of the structural steel.

Fabricator—The party responsible for furnishing fabricated structural steel.

General Contractor—The owner's designated representative with full responsibility for the construction of the structure.

MBMA—Metal Building Manufacturers Association.

Mill Material—Steel mill products ordered expressly for the requirements of a specific project.

Owner—The owner of the proposed structure or his designated representatives, who may be the architect, engineer, general contractor, public authority or others.

AMERICAN INSTITUTE OF STEEL CONSTRUCTION

Plans—Design drawings furnished by the party responsible for the design of the structure.

Release for Construction—The release by the owner permitting the fabricator to commence work under the contract, including ordering material and the preparation of shop drawings.

SSPC—The Steel Structures Painting Council, publishers of the *Steel Structures Painting Manual*, Vol. 2, "Systems and Specifications."

Tier—The word Tier used in Sect. 7.11 is defined as a column shipping piece.

1.3. Design Criteria for Buildings and Similar Type Structures

In the absence of other instructions, the provisions of the AISC Specification govern the design of the structural steel.

1.4. Design for Bridges

In the absence of other instructions, the following provisions govern, as applicable:

> *Standard Specifications for Highway Bridges* of American Association of State Highway and Transportation Officials
>
> *Specifications for Steel Railway Bridges* of American Railway Engineering Association
>
> *Structural Welding Code* of American Welding Society

1.5. Responsibility for Design

1.5.1. When the owner provides the design, plans and specifications, the fabricator and erector are not responsible for the suitability, adequacy or legality of the design. The fabricator is not responsible for the practicability or safety of erection if the structure is erected by others.

1.5.2. If the owner desires the fabricator or erector to prepare the design, plans and specifications, or to assume any responsibility for the suitability, adequacy or legality of the design, he clearly states his requirements in the contract documents.

1.6. Patented Devices

Except when the contract documents call for the design to be furnished by the fabricator or erector, the fabricator and erector assume that all necessary patent rights have been obtained by the owner and that the fabricator or erector will be fully protected in the use of patented designs, devices or parts required by the contract documents.

SECTION 2.0. CLASSIFICATION OF MATERIALS

2.1. Definition of Structural Steel

"Structural Steel," as used to define the scope of work in the contract documents, consists of the steel elements of the structural steel frame essential to support the design loads. Unless otherwise specified in the contract documents, these elements consist of material as shown on the structural steel plans and described as:

Anchor bolts for structural steel
Base or bearing plates
Beams, Girders, Purlins and Girts
Bearings of steel for girders, trusses or bridges
Bracing
Columns, posts
Connecting materials for framing structural steel to structural steel
Crane rails, splices, stops, bolts and clamps
Door frames constituting part of the steel frame
Expansion joints connected to steel frame
Fasteners for connecting structural steel items:
    Shop rivets
    Permanent shop bolts
    Shop bolts for shipment
    Field rivets for permanent connections
    Field bolts for permanent connections
    Permanent pins
Floor Plates (checkered or plain) attached to steel frame
Grillage beams and girders
Hangers essential to the structural steel frame
Leveling plates, wedges, shims & leveling screws
Lintels, if attached to the structural steel frame
Marquee or canopy framing
Machinery foundations of rolled steel sections and/or plate attached to the structural frame
Monorail elements of standard structural shapes when attached to the structural frame
Roof frames of standard structural shapes
Shear connectors—if specified shop attached
Struts, tie rods and sag rods forming part of the structural frame
Trusses

2.2. Other Steel or Metal Items

The classification "Structural Steel," does not include steel, iron or other metal items not generally described in Paragraph 2.1, even when such items are shown on the structural steel plans or are attached to the structural frame. These items include but are not limited to:

Cables for permanent bracing or suspension systems
Chutes and hoppers
Cold-formed steel products
Door and corner guards
Embedded steel parts in precast or poured concrete
Flagpole support steel
Floor plates (checkered or plain) not attached to the steel frame
Grating and metal deck
Items required for the assembly or erection of materials supplied by trades other than structural steel fabricators or erectors
Ladders and safety cages
Lintels over wall recesses
Miscellaneous metal

Non-steel bearings
Open-web, long-span joists and joist girders
Ornamental metal framing
Shear connectors field installed
Stacks, tanks and pressure vessels
Stairs, catwalks, handrail and toeplates
Trench or pit covers.

SECTION 3.   PLANS AND SPECIFICATIONS

3.1.   Structural Steel

In order to insure adequate and complete bids, the contract documents provide complete structural steel design plans clearly showing the work to be performed and giving the size, section, material grade and the location of all members, floor levels, column centers and offsets, camber of members, with sufficient dimensions to convey accurately the quantity and nature of the structural steel to be furnished. Structural steel specifications include any special requirements controlling the fabrication and erection of the structural steel.

3.1.1.   Wind bracing, connections, column stiffeners, bearing stiffeners on beams and girders, web reinforcement, openings for other trades, and other special details where required are shown in sufficient detail so that they may be readily understood.

3.1.2.   Plans include sufficient data concerning assumed loads, shears, moments and axial forces to be resisted by members and their connections, as may be required for the development of connection details on the shop drawings and the erection of the structure.

3.1.3.   Where connections are not shown, the connections are to be in accordance with the requirements of the AISC Specification.

3.1.4.   When loose lintels and leveling plates are required to be furnished as part of the contract requirements, the plans and specifications show the size, section and location of all pieces.

3.2.   Architectural, Electrical and Mechanical

Architectural, electrical and mechanical plans may be used as a supplement to the structural steel plans to define detail configurations and construction information, provided all requirements for the structural steel are noted on the structural steel plans.

3.3.   Discrepancies

In case of discrepancies between plans and specifications for buildings, the specifications govern. In case of discrepancies between plans and specifications for bridges, the plans govern. In case of discrepancies between scale dimensions on the plans and figures written on them, the figures govern. In case of discrepancies between the structural steel plans and plans for other trades, the structural steel plans govern.

3.4. Legibility of Plans

Plans are clearly legible and made to a scale not less then ⅛-inch to the foot. More complex information is furnished to an adequate scale to convey the information clearly.

3.5. Special Conditions

When it is required that a project be advertised for bidding before the requirements of Article 3.1 can be met, the owner must provide sufficient information in form of scope, drawings, weights, outline specifications, and other descriptive data to enable the fabricator and erector to prepare a knowledgeable bid.

SECTION 4.   SHOP AND ERECTION DRAWINGS

4.1. Owner Responsibility

To enable the fabricator and erector to properly and expeditiously proceed with the work, the owner furnishes, in a timely manner and in accordance with the contract documents, complete structural steel plans and specifications released for construction. "Released for construction" plans and specifications are required by the fabricator for ordering the mill material and for the preparation and completion of shop and erection drawings.

4.2. Approval

When shop drawings are made by the fabricator, prints thereof are submitted to the owner for his examination and approval. The fabricator includes a maximum allowance of fourteen (14) calendar days in his schedule for the return of shop drawings. Return of shop drawings is noted with the owner's approval, or approval subject to corrections as noted. The fabricator makes the corrections, furnishes corrected prints to the owner, and is released by the owner to start fabrication.

4.2.1.   Approval by the owner of shop drawings prepared by the fabricator indicates that the fabricator has correctly interpreted the contract requirements, and is released by the owner to start fabrication. This approval constitutes the owner's acceptance of all responsibility for the design adequacy of any detail configuration of connections developed by the fabricator as part of his preparation of these shop drawings. Approval does not relieve the fabricator of the responsibility for accuracy of detail dimensions on shop drawings, nor the general fit-up of parts to be assembled in the field.

4.2.2.   Unless specifically stated to the contrary, any additions, deletions or changes indicated on the approval of shop and erection drawings are authorizations by the owner to release the additions, deletions or revisions for construction.

4.3. Drawings Furnished by Owner

When the shop drawings are furnished by the owner, he must deliver them to the fabricator in time to permit material procurement and fabrication to proceed in an orderly manner in accordance with the prescribed time schedule. The owner prepares these shop drawings, insofar as practicable, in accordance with the shop and drafting room standards of the fabricator. The owner is responsible for the completeness and accuracy of shop drawings so furnished.

SECTION 5.   MATERIALS

5.1.   Mill Materials

5.1.1.   Mill tests are performed to demonstrate material conformance to ASTM specifications in accordance with the contract requirements. Unless special requirements are included in the contract documents, mill testing is limited to those tests required by the applicable ASTM material specifications. Mill test reports are furnished by the fabricator only if requested by the owner, either in the contract documents or in separate written instructions prior to the time the fabricator places his material orders with the mill.

5.1.2.   When material received from the mill does not satisfy ASTM A6 tolerances for camber, profile, flatness or sweep, the fabricator is permitted to perform corrective work by the use of controlled heating and mechanical straightening, subject to the limitations of the AISC Specification.

5.1.3.   Corrective procedures described in ASTM A6 for reconditioning the surface of structural steel plates and shapes before shipment from the producing mill may also be performed by the fabricator, at his option, when variations described in ASTM A6 are discovered or occur after receipt of the steel from the producing mill.

5.1.4.   When special requirements demand tolerances more restrictive than allowed by ASTM A6, such requirements are defined in the contract documents and the fabricator has the option of corrective measures as described above.

5.2.   Stock Materials

5.2.1.   Many fabricators maintain stocks of steel products for use in their fabricating operations. Materials taken from stock by the fabricator for use for structural purposes must be of a quality at least equal to that required by the ASTM specifications applicable to the classification covering the intended use.

5.2.2.   Mill test reports are accepted as sufficient record of the quality of materials carried in stock by the fabricator. The fabricator reviews and retains the mill test reports covering the materials he purchases for stock, but he does not maintain records that identify individual pieces of stock material against individual mill test reports. Such records are not required if the fabricator purchases for stock under established specifications as to grade and quality.

5.2.3.   Stock materials purchased under no particular specifications or under specifications less rigid than those mentioned above, or stock materials which have not been subject to mill or other recognized test reports, are not used without the express approval of the owner, except where the quality of the material could not affect the integrity of the structure.

SECTION 6.   FABRICATION AND DELIVERY

6.1.   Identification of Material

6.1.1.   High strength steel and steel ordered to special requirements is marked by the supplier, in accordance with ASTM A6 requirements, prior to delivery to the fabricator's shop or other point of use.

6.1.2.   High strength steel and steel ordered to special requirements that has not been marked by the supplier in accordance with Sect. 6.1.1 is not used until its identification is established by means of tests as specified in Sect. 1.4.1.1 of the AISC Specification, and until a fabricator's identification mark, as described in Sect. 6.1.3, has been applied.

6.1.3.   During fabrication, up to the point of assembling members, each piece of high strength steel and steel ordered to special requirements carries a fabricator's identification mark or an original supplier's identification mark. The fabricator's identification mark is in accordance with the fabricator's established identification system, which is on record and available for the information of the owner or his representative, the building commissioner and the inspector, prior to the start of fabrication.

6.1.4.   Members made of high strength steel and steel ordered to special requirements are not given the same assembling or erecting mark as members made of other steel, even though they are of identical dimensions and detail.

6.2.   Preparation of Material

6.2.1.   Thermal cutting of structural steel may be by hand or mechanically guided means.

6.2.2.   Surfaces noted as "finished" on the drawings are defined as having a maximum ANSI roughness height value of 500. Any fabricating technique, such as friction sawing, cold sawing, milling, etc., that produces such a finish may be used.

6.3.   Fitting and Fastening

6.3.1.   Projecting elements of connection attachments need not be straightened in the connecting plane if it can be demonstrated that installation of the connectors or fitting aids will provide reasonable contact between faying surfaces.

6.3.2.   Runoff tabs are often required to produce sound welds. The fabricator or erector does not remove them unless specified in the contract documents. When their removal is required, they may be hand flame-cut close to the edge of the finished member with no further finishing required, unless other finishing is specifically called for in the contract documents.

6.3.3.   All high-strength bolts for shop attached connection material are to be installed in the shop in accordance with *Specification for Structural Joints Using A325 or A490 Bolts*, unless otherwise noted on the shop drawings.

6.4.   Dimensional Tolerances

6.4.1.   A variation of 1/32-inch is permissible in the overall length of members with both ends finished for contact bearing as defined in Sect. 6.2.2.

6.4.2. Members without ends finished for contact bearing, which are to be framed to other steel parts of the structure, may have a variation from the detailed length not greater than $\frac{1}{16}$-inch for members 30 ft or less in length, and not greater than $\frac{1}{8}$-inch for members over 30 ft in length.

6.4.3. Unless otherwise specified, structural members, whether of a single-rolled shape or built-up, may vary from straightness within the tolerances allowed for wide-flange shapes by ASTM Specification A6, except that the tolerance on deviation from straightness of compression members is $\frac{1}{1000}$ of the axial length between points which are to be laterally supported.

Completed members should be free from twists, bends and open joints. Sharp kinks or bends are cause for rejection of material.

6.4.4. Beams and trusses detailed without specified camber are fabricated so that after erection any camber due to rolling or shop fabrication is upward.

6.4.5. Any permissible deviation in depths of girders may result in abrupt changes in depth at splices. Any such difference in depth at a bolted joint, within the prescribed tolerances, is taken up by fill plates. At welded joints the weld profile may be adjusted to conform to the variation in depth, provided that the minimum cross section of required weld is furnished and that the slope of the weld surface meets AWS Code requirements.

6.5. Shop Painting

6.5.1. The contract documents specify all the painting requirements, including members to be painted, surface preparation, paint specifications, manufacturer's product identification and the required dry film thickness, in mils, of the shop coat.

6.5.2. The shop coat of paint is the prime coat of the protective system. It protects the steel for only a short period of exposure in ordinary atmospheric conditions, and is considered a temporary and provisional coating. The fabricator does not assume responsibility for deterioration of the prime coat that may result from extended exposure to ordinary atmospheric conditions, nor from exposure to corrosive conditions more severe than ordinary atmospheric conditions.

6.5.3. In the absence of other requirements in the contract documents, the fabricator hand cleans the steel of loose rust, loose mill scale, dirt and other foreign matter, prior to painting, by means of wire brushing or by other methods elected by the fabricator, to meet the requirements of SSPC-SP2. The fabricator's workmanship on surface preparation is considered accepted by the owner unless specifically disapproved prior to paint application.

6.5.4. Unless specifically excluded, paint is applied by brush, spray, roller coating, flow coating or dipping, at the election of the fabricator. When the term "shop coat" or "shop paint" is used with no paint system specified, the fabricator's standard paint shall be applied to a minimum dry film thickness of one mil.

6.5.5 Steel not requiring shop paint is cleaned of oil or grease by solvent cleaners and cleaned of dirt and other foreign material by sweeping with a fiber brush or other suitable means.

6.5.6. Abrasions caused by handling after painting are to be expected. Touch-up of these blemished areas is the responsibility of the contractor performing field touch-up or field painting.

6.6. Marking and Shipping of Materials

6.6.1. Erection marks are applied to the structural steel members by painting or other suitable means, unless otherwise specified in the contract documents.

6.6.2. Rivets and bolts are commonly shipped in separate containers according to length and diameter; loose nuts and washers are shipped in separate containers according to sizes. Pins and other small parts, and packages of rivets, bolts, nuts and washers are usually shipped in boxes, crates, kegs or barrels. A list and description of the material usually appears on the outside of each closed container.

6.7. Delivery of Materials

6.7.1 Fabricated structural steel is delivered in such sequence as will permit the most efficient and economical performance of both shop fabrication and erection. If the owner wishes to prescribe or control the sequence of delivery of materials, he reserves such right and defines the requirements in the contract documents. If the owner contracts separately for delivery and erection, he must coordinate planning between contractors.

6.7.2. Anchor bolts, washers and other anchorage or grillage materials to be built into masonry should be shipped so that they will be on hand when needed. The owner must give the fabricator sufficient time to fabricate and ship such materials before they are needed.

6.7.3. The quantities of material shown by the shipping statement are customarily accepted by the owner, fabricator and erector as correct. If any shortage is claimed, the owner or erector should immediately notify the carrier and the fabricator in order that the claim may be investigated.

6.7.4. The size and weight of structural steel assemblies may be limited by shop capabilities, the permissible weight and clearance dimensions of available transportation and the job site conditions. The fabricator limits the number of field splices to those consistent with minimum project cost.

6.7.5. If material arrives at its destination in damaged condition, it is the responsibility of the receiving party to promptly notify the fabricator and carrier prior to unloading the material, or immediately upon discovery.

SECTION 7.   ERECTION

7.1.   Method of Erection

When the owner wishes to control the method and sequence of erection, or when certain members cannot be erected in their normal sequence, the owner so specifies in the contract documents. In the absence of such restrictions, the erector will proceed using the most efficient and economical method and sequence available to him consistent with the contract documents. When the owner contracts separately for fabrication and erection services, the owner is responsible for coordinating planning between contractors.

7.2. Site Conditions

The owner provides and maintains adequate access roads into and through the site for the safe delivery of derricks, cranes, other necessary equipment, and the material to be erected. The owner affords the erector a firm, properly graded, drained, convenient and adequate space at the site for the operation of his equipment, and removes all overhead obstructions such as power lines, telephone lines, etc., in order to provide a safe working area for erection of the steelwork. The erector provides and installs the safety protection required for his own work. Any protection for other trades not essential to the steel erection activity is the responsibility of the owner. When the structure does not occupy the full available site, the owner provides adequate storage space to enable the fabricator and erector to operate at maximum practicable speed.

7.3. Foundations, Piers and Abutments

The accurate location, strength, suitability and access to all foundations, piers and abutments is the sole responsibility of the owner.

7.4. Building Lines and Bench Marks

The owner is responsible for accurate location of building lines and bench marks at the site of the structure, and for furnishing the erector with a plan containing all such information.

7.5. Installation of Anchor Bolts and Embedded Items

7.5.1. Anchor bolts and foundation bolts are set by the owner in accordance with an approved drawing. They must not vary from the dimensions shown on the erection drawings by more than the following:

(a) ⅛-inch center to center of any two bolts within an anchor bolt group, where an anchor bolt group is defined as the set of anchor bolts which receive a single fabricated steel shipping piece.

(b) ¼-inch center to center of adjacent anchor bolt groups.

(c) Elevation of the top of anchor bolts ± ½-inch

(d) Maximum accumulation of ¼-inch per hundred feet along the established column line of multiple anchor bolt groups, but not to exceed a total of 1 in., where the established column line is the actual field line most representative of the centers of the as-built anchor bolt groups along a line of columns.

(e) ¼-inch from the center of any anchor bolt group to the established column line through that group.

(f) The tolerances of paragraphs b, c and d apply to offset dimensions shown on the plans, measured parallel and perpendicular to the nearest established column line for individual columns shown on the plans to be offset from established column lines.

7.5.2. Unless shown otherwise, anchor bolts are set perpendicular to the theoretical bearing surface.

7.5.3. Other embedded items or connection materials between the structural steel and the work of other trades are located and set by the owner in accordance with approved location or erection drawings. Accuracy of these items must satisfy the erection tolerance requirements of Sect. 7.11.3.

7.5.4. All work performed by the owner is completed so as not to delay or interfere with the erection of the structural steel.

7.6. Bearing Devices

The owner sets to line and grade all leveling plates and loose bearing plates which can be handled without a derrick or crane. All other bearing devices supporting structural steel are set and wedged, shimmed or adjusted with leveling screws by the erector to lines and grades established by the owner. The fabricator provides the wedges, shims or leveling screws that are required, and clearly scribes the bearing devices with working lines to facilitate proper alignment. Promptly after the setting of any bearing devices, the owner checks lines and grades, and grouts as required. The final location and proper grouting of bearing devices are the responsibility of the owner. Tolerance on elevation relative to established grades of bearing devices, whether set by the owner or by the erector, is plus or minus ⅛-inch.

7.7 Field Connection Material

7.7.1 The fabricator provides field connection details consistent with the requirements of the contract documents which will, in his opinion, result in the most economical fabrication and erection cost.

7.7.2. When the fabricator erects the structural steel, the fabricator supplies all materials required for temporary and permanent connection of the component parts of the structural steel.

7.7.3 When the erection of the structural steel is performed by someone other than the fabricator, the fabricator furnishes the following field connection material:

(a) Bolts of required size and in sufficient quantity for all field connections of steel to steel which are to be permanently bolted. Unless high-strength bolts or other special types of bolts and washers are specified, common bolts are furnished. An extra 2 percent of each bolt size (diameter and length) are furnished.

(b) Rivets of required size and in sufficient quantity for all field connections of steel to steel which are to be riveted field connections. An extra 10 percent of each rivet size are furnished.

(c) Shims shown as necessary for make-up of permanent connections of steel to steel.

(d) Back-up bars or run-off tabs that may be required for field welding.

7.7.4. When the erection of the structural steel is performed by someone other than the fabricator, the erector furnishes all welding electrodes, fit-up bolts and drift pins used for erection of the structural steel.

7.7.5. Field-installed shear connectors are supplied by the shear connector applicator.

7.7.6. Metal deck support angles are the responsibility of the metal deck supplier.

7.8. Loose Material

Loose items of structural steel not connected to the structural frame are set by the owner without assistance from the erector, unless otherwise specified in the contract documents.

7.9. Temporary Support of Structural Steel Frames.

7.9.1. General

Temporary supports, such as temporary guys, braces, falsework, cribbing or other elements required for the erection operation will be determined and furnished and installed by the erector. These temporary supports will secure the steel framing, or any partly assembled steel framing, against loads comparable in intensity to those for which the structure was designed, resulting from wind, seismic forces and erection operations, but not the loads resulting from the performance of work by or the acts of others, nor such unpredictable loads as those due to tornado, explosion or collision.

7.9.2. Self-supporting Steel Frames

A self-supporting steel frame is one that provides the required stability and resistance to gravity loads and design wind and seismic forces without interaction with other elements of the structure. The erector furnishes and installs only those temporary supports that are necessary to secure any element or elements of the steel framing until they are made stable without external support.

7.9.3. Non-Self-supporting Steel Frames

A non-self-supporting steel frame is one that requires interaction with other elements not classified as Structural Steel to provide the required stability or resistance to wind and seismic forces. Such frames shall be clearly identified in the contract documents. The contract documents specify the sequence and schedule of placement of such elements and the effect of loads imposed by these partially or completely installed interacting elements on the bare steel frame. The erector determines the need and furnishes and installs the temporary supports in accordance with this information. The owner is responsible for the installation and timely completion of all elements that are required for stability of the frame.

7.9.4. Special Erection Conditions

When the design concept of a structure is dependent upon use of shores, jacks or loads which must be adjusted as erection progresses to set or maintain camber or prestress, such requirement is specifically stated in the contract documents.

7.9.5. Removal of Temporary Supports

The temporary guys, braces, falsework, cribbing and other elements required for the erection operation, which are furnished and installed by the erector, are not the property of the owner.

In *self-supporting structures*, temporary supports are not required after the structural steel for a self-supporting element is located and finally fastened within the required tolerances. After such final fastening, the erector is no longer responsible for temporary support of the self-supporting element and may remove the temporary supports.

In *non-self-supporting structures*, the erector may remove temporary supports when the necessary non-structural steel elements are complete. Temporary supports are not to be removed without the consent of the erector. At completion of steel erection, any temporary supports that are required to be left in place are removed by the owner and returned to the erector in good condition.

7.9.6. Temporary Supports for Other Work

Should temporary supports beyond those defined as the responsibility of the erector in Sects. 7.9.1, 7.9.2 and 7.9.3 be required, either during or after the erection of the structural steel, responsibility for the supply and installation of such supports rests with the owner.

7.10. Temporary Floors and Handrails for Buildings

The erector provides floor coverings, handrails and walkways as required by law and applicable safety regulations for protection of his own personnel. As work progresses, the erector removes such facilities from units where the erection operations are completed, unless other arrangements are included in the contract documents. The owner is responsible for all protection necessary for the work of other trades. When permanent steel decking is used for protective flooring and is installed by the owner, all such work is performed so as not to delay or interfere with erection progress and is scheduled by the owner and installed in a sequence adequate to meet all safety regulations.

7.11. Frame Tolerances

7.11.1. Overall Dimensions

Some variation is to be expected in the finished overall dimensions of structural steel frames. Such variations are deemed to be within the limits of good practice when they do not exceed the cumulative effect of rolling tolerances, fabricating tolerances and erection tolerances.

7.11.2. Working Points and Working Lines

Erection tolerances are defined relative to member working points and working lines as follows:

(a) For members other than horizontal members, the member work point is the actual center of the member at each end of the shipping piece.

(b) For horizontal members, the working point is the actual center line of the top flange or top surface at each end.

(c) Other working points may be substituted for ease of reference, providing they are based upon these definitions.

(d) The member working line is a straight line connecting the member working points.

7.11.3. Position and Alignment

The tolerances on position and alignment of member working points and working lines are as follows:

7.11.3.1. Columns

Individual column shipping pieces are considered plumb if the deviation of the working line from a plumb line does not exceed 1:500, subject to the following limitations:

(a) The member working points of column shipping pieces adjacent to elevator shafts may be displaced no more than 1 in. from the established column line in the first 20 stories; above this level, the displacement may be increased ⅟₃₂-inch for each additional story up to a maximum of 2 in.

(b) The member working points of exterior column shipping pieces may be displaced from the established column line no more than 1 in. toward nor 2 in. away from the building line in the first 20 stories; above the 20th story, the displacement may be increased ⅟₁₆-inch for each additional story, but may not exceed a total displacement of 2 in. toward nor 3 in. away from the building line.

(c) The member working points of exterior column shipping pieces at any splice level for multi-tier buildings and at the tops of columns for single tier buildings may not fall outside a horizontal envelope, parallel to the building line, 1½-inch wide for buildings up to 300 ft in length. The width of the envelope may be increased by ½-inch for each additional 100 ft in length, but may not exceed 3 in.

(d) The member working points of exterior column shipping pieces may be displaced from the established column line, in a direction parallel to the building line, no more than 2 in. in the first 20 stories; above the 20th story, the displacement may be increased ⅟₁₆-inch for each additional story, but may not exceed a total displacement of 3 in. parallel to the building line.

7.11.3.2. Members Other Than Columns

(a) Alignment of members which consist of a single straight shipping piece containing no field splices, except cantilever members, is considered acceptable if the variation in alignment is caused solely by the variation of column alignment and/or primary supporting member alignment within the permissible limits for fabrication and erection of such members.

(b) The elevation of members connecting to columns is considered acceptable if the distance from the member working point to the upper milled splice line of the column does not deviate more than plus ³⁄₁₆-inch or minus ⁵⁄₁₆-inch from the distance specified on the drawings.

(c) The elevation of members which consist of a single shipping piece, other than members connected to columns, is considered acceptable if the variation in actual elevation is caused solely by the variation in elevation of the supporting members which are within permissible limits for fabrication and erection of such members.

(d) Individual shipping pieces which are segments of field assembled units containing field splices between points of support are considered plumb, level and aligned if the angular variation of the working line of each shipping piece relative to the plan alignment does not exceed 1:500.

(e) The elevation and alignment of cantilever members shall be considered plumb, level and aligned if the angular variation of the working line from a straight line extended in the plan direction from the working point at its supported end does not exceed 1:500.

(f) The elevation and alignment of members which are of irregular shape shall be considered plumb, level and aligned if the fabricated member is within its tolerance and its supporting member or members are within the tolerances specified in this Code.

7.11.3.3. Adjustable Items

The alignment of lintels, wall supports, curb angles, mullions and similar supporting members for the use of other trades, requiring limits closer than the foregoing tolerances, cannot be assured unless the owner's plans call for adjustable connections of these members to the supporting structural frame. When adjustable connections are specified, the owner's plans must provide for the total adjustment required to accommodate the tolerances on the steel frame for the proper alignment of these supports for other trades. The tolerances on position and alignment of such adjustable items are as follows:

(a) Adjustable items are considered to be properly located in their vertical position when their location is within ⅜-inch of the location established from the upper milled splice line of the nearest column to the support location as specified on the drawings.

(b) Adjustable items are considered to be properly located in their horizontal position when their location is within ⅜-inch of the proper location relative to the established finish line at any particular floor.

7.11.4. Responsibility for Clearances

In the design of steel structures, the owner is responsible for providing clearances and adjustments of material furnished by other trades to accommodate all of the foregoing tolerances of the structural steel frame.

7.11.5. Acceptance of Position and Alignment

Prior to placing or applying any other materials, the owner is responsible for determining that the location of the structural steel is acceptable for plumbness, level and alignment within tolerances. The erector is given timely notice of acceptance by the owner or a listing of specific items to be corrected in order to obtain acceptance. Such notice is rendered immediately upon completion of any part of the work and prior to the start of work by other trades that may be supported, attached or applied to the structural steelwork.

7.12. Correction of Errors

Normal erection operations include the correction of minor misfits by moderate amounts of reaming, chipping, welding or cutting, and the drawing of elements into line through the use of drift pins. Errors which cannot be corrected by the foregoing means or which require major changes in member configuration are reported immediately to the owner and fabricator by the erector, to enable whoever is responsible either to correct the error or to approve the most efficient and economic method of correction to be used by others.

7.13. Cuts, Alterations and Holes for Other Trades

Neither the fabricator nor the erector will cut, drill or otherwise alter his work, or the work of other trades, to accommodate other trades, unless such work is clearly specified in the contract documents. Whenever such work is specified, the owner is responsible for furnishing complete information as to materials, size, location and number of alterations prior to preparation of shop drawings.

7.14. Handling and Storage

The erector takes reasonable care in the proper handling and storage of steel during erection operations to avoid accumulation of unnecessary dirt and foreign matter. The erector is not responsible for removal from the steel of dust, dirt or other foreign matter which accumulates during the erection period as the result of exposure to the elements.

7.15. Field Painting

The erector does not paint field bolt heads and nuts, field rivet heads and field welds, nor touch up abrasions of the shop coat, nor perform any other field painting.

7.16 Final Cleaning Up

Upon completion of erection and before final acceptance, the erector removes all of his falsework, rubbish and temporary buildings.

SECTION 8. QUALITY CONTROL

8.1. General

8.1.1. The fabricator maintains a quality control program to the extent deemed necessary so that the work is performed in accordance with this Code, the AISC Specification and contract documents. The fabricator has the option to use the AISC Quality Certification Program in establishing and administering the quality control program.

8.1.2. The erector maintains a quality control program to the extent the erector deems necessary so that all of the work is performed in accordance with this Code, the AISC Specification and the contract documents. The erector shall be capable of performing the erection of the structural steel, and shall provide the equipment, personnel and management for the scope, magnitude and required quality of each project.

8.1.3. When the owner requires more extensive quality control or independent inspection by qualified personnel, or requires the fabricator to be certified by the AISC Quality Certification Program, this shall be clearly stated in the contract documents, including a definition of the scope of such inspection.

8.2. Mill Material Inspection

The fabricator customarily makes a visual inspection, but does not perform any material tests, depending upon mill reports to signify that the mill product satisfies material order requirements. The owner relies on mill tests required by contract and on such additional tests as he orders the fabricator to have made at the owner's expense. If mill inspection operations are to be monitored, or if tests other than mill tests are desired, the owner so specifies in the contract documents and should arrange for such testing through the fabricator to assure coordination.

8.3. Non-destructive Testing

When non-destructive testing is required, the process, extent, technique and standards of acceptance are clearly defined in the contract documents.

8.4. Surface Preparation and Shop Painting Inspection

Surface preparation and shop painting inspection must be planned for acceptance of each operation as completed by the fabricator. Inspection of the paint system, including material and thickness, is made promptly upon completion of the paint application. When wet film thickness is inspected, it must be measured immediately after application.

8.5. Independent Inspection

When contract documents specify inspection by other than the fabricator's and erector's own personnel, both parties to the contract incur obligations relative to the performance of the inspection.

8.5.1. The fabricator and erector provide the inspector with access to all places where work is being done. A minimum of 24 hours notification is given prior to commencement of work.

8.5.2. Inspection of shop work by the owner or his representative is performed in the fabricator's shop to the fullest extent possible. Such inspections should be in sequence, timely, and performed in such a manner as to minimize disruptions in operations and to permit the repair of all non-conforming work while the material is in process in the fabricating shop.

8.5.3. Inspection of field work is completed promptly, so that corrections can be made without delaying the progress of the work.

8.5.4. Rejection of material or workmanship not in conformance with the contract documents may be made at any time during the progress of the work. However, this provision does not relieve the owner of his obligation for timely, in-sequence inspections.

8.5.5. The fabricator and erector receive copies of all reports prepared by the owner's inspection representative.

SECTION 9.   CONTRACTS

9.1.   Types of Contracts

9.1.1. For contracts stipulating a lump sum price, the work required to be performed by the fabricator and erector is completely defined by the contract documents.

9.1.2. For contracts stipulating a price per pound, the scope of work, type of materials, character of fabrication, and conditions of erection are based upon the contract documents which must be representative of the work to be performed.

9.1.3. For contracts stipulating a price per item, the work required to be performed by the fabricator and erector is based upon the quantity and the character of items described in the contract documents.

9.1.4. For contracts stipulating unit prices for various categories of structural steel, the scope of the work required to be performed by the fabricator and erector is based upon the quantity, character and complexity of the items in each category as described in the contract documents. The contract documents must be representative of the work to be done in each category.

9.2. Calculation of Weights

Unless otherwise set forth in the contract, on contracts stipulating a price per pound for fabricated structural steel delivered and/or erected, the quantities of materials for payment are determined by the calculation of gross weight of materials as shown on the shop drawings.

9.2.1. The unit weight of steel is assumed to be 490 pounds per cubic foot. The unit weight of other materials is in accordance with the manufacturer's published data for the specific product.

9.2.2. The weights of shapes, plates, bars, steel pipe and structural tubing are calculated on the basis of shop drawings showing actual quantities and dimensions of material furnished, as follows:

(a) The weight of all structural shapes, steel pipe and structural tubing is calculated using the nominal weight per foot and the detailed overall length.

(b) The weight of plates and bars is calculated using the detailed overall rectangular dimensions.

(c) When parts can be economically cut in multiples from material of larger dimensions, the weight is calculated on the basis of the theoretical rectangular dimensions of the material from which the parts are cut.

(d) When parts are cut from structural shapes, leaving a non-standard section not useable on the same contract, the weight is calculated on the basis of the nominal unit weight of the section from which the parts are cut.

(e) No deductions are to be made for material removed by cuts, copes, clips, blocks, drilling, punching, boring, slot milling, planing or weld joint preparation.

9.2.3. The calculated weights of castings are determined from the shop drawings of the pieces. An allowance of 10 percent is added for fillets and overrun. Scale weights of rough castings may be used if available.

9.2.4. The items for which weights are shown in tables in the AISC *Manual of Steel Construction* are calculated on the basis of tabulated unit weights.

9.2.5. The weight of items not included in the tables in the AISC *Manual of Steel Construction* shall be taken from the manufacturers' catalog and the manufacturers' shipping weight shall be used.

9.2.6. The weight of shop or field weld metal and protective coatings is not included in the calculated weight for pay purposes.

9.3. Revisions to Contract Documents

9.3.1. Revisions to the contract are made by the issuance of new documents or the reissuance of existing documents. In either case, all revisions are clearly indicated and the documents are dated.

9.3.2. A revision to the requirements of the contract documents are made by change orders, extra work orders, or notations on the shop and erection drawings when returned upon approval.

9.3.3. Unless specifically stated to the contrary, the issuance of a revision is authorization by the owner to release these documents for construction.

9.4. Contract Price Adjustment

9.4.1. When the scope of work and responsibilities of the fabricator and erector are changed from those previously established by the contract documents, an appropriate modification of the contract price is made. In computing the contract price adjustment, the fabricator and erector consider the quantity of work added or deleted, modifications in the character of the work, and the timeliness of the change with respect to the status of material ordering, detailing, fabrication and erection operations.

9.4.2. Requests for contract price adjustments are presented by the fabricator and erector in a timely manner and are accompanied by a description of the change in sufficient detail to permit evaluation and timely approval by the owner.

9.4.3. Price per pound and price per item contracts generally provide for additions or deletions to the quantity of work prior to the time work is released for construction. Changes to the character of the work, at any time, or additions and/or deletions to the quantity of the work after it is released for construction, may require a contract price adjustment.

9.5. Scheduling

9.5.1. The contract documents specify the schedule for the performance of the work. This schedule states when the "released for construction" plans will be issued and when the job site, foundations, piers and abutments will be ready, free from obstructions and accessible to the erector, so that erection can start at the designated time and continue without interference or delay caused by the owner or other trades.

9.5.2. The fabricator and erector have the responsibility to advise the owner, in a timely manner, of the effect any revision has on the contract schedule.

9.5.3. If the fabrication or erection is significantly delayed due to design revisions, or for other reasons which are the owner's responsibility, the fabricator and erector are compensated for additional costs incurred.

9.6 Terms of Payment. The terms of payment for the contract shall be outlined in the contract documents.

SECTION 10. ARCHITECTURALLY EXPOSED STRUCTURAL STEEL

10.1. Scope

This section of the Code defines additional requirements which apply only to members specifically designated by the contract documents as "Architecturally Exposed Structural Steel" (AESS). All provisions of Sects. 1 through 9 of the Code apply unless specifically modified in this section. AESS members or components are fabricated and erected with the care and dimensional tolerances indicated in this section.

10.2. Additional Information Required in Contract Documents

(a) Specific identification of members or components which are to be AESS.
(b) Fabrication and erection tolerances which are more restrictive than provided for in this section.
(c) Requirements, if any, of a test panel or components for inspection and acceptance standards prior to the start of fabrication.

10.3. Fabrication

10.3.1. Rolled Shapes

Permissible tolerances for out-of-square or out-of-parallel, depth, width and symmetry of rolled shapes are as specified in ASTM Specification A6. No attempt to match abutting cross-sectional configurations is made unless specifically required by the contract documents. The as-fabricated straightness tolerances of members are one-half of the standard camber and sweep tolerances in ASTM A6.

10.3.2. Built-up Members

The tolerances on overall profile dimensions of members made up from a series of plates, bars and shapes by welding are limited to the accumulation of permissible tolerances of the component parts as provided by ASTM Specification A6. The as-fabricated straightness tolerances for the member as a whole are one-half the standard camber and sweep tolerances for rolled shapes in ASTM A6.

10.3.3. Weld Show-through

It is recognized that the degree of weld show-through, which is any visual indication of the presence of a weld or welds on the side away from the viewer, is a function of weld size and material thickness. The members or components will be acceptable as produced unless specific acceptance criteria for weld show-through are included in the contract documents.

10.3.4. Joints

All copes, miters and butt cuts in surfaces exposed to view are made with uniform gaps of ⅛-inch if shown to be open joints, or in reasonable contact if shown without gap.

10.3.5. Welding

Reasonably smooth and uniform as-welded surfaces are acceptable on all welds exposed to view. Butt and plug welds do not project more than ¹⁄₁₆-inch above the exposed surface. No finishing or grinding is required except where clearances or fit of other components may necessitate, or when specifically required by the contract documents.

10.3.6. Weathering Steel

Members fabricated of weathering steel which are to be AESS shall not have erection marks or other painted marks on surfaces that are to be exposed in the completed structure. If cleaning other than SSPC-SP6 is required, these requirements shall be defined in the contract documents.

10.4. Delivery of Materials

The fabricator uses special care to avoid bending, twisting or otherwise distorting individual members.

10.5 Erection

10.5.1. General

The erector uses special care in unloading, handling and erecting the steel to avoid marking or distorting the steel members. Care is also taken to minimize damage to any shop paint. If temporary braces or erection clips are used, care is taken to avoid unsightly surfaces upon removal. Tack welds are ground smooth and holes are filled with weld metal or body solder and smoothed by grinding or filing. The erector plans and executes all operations in such a manner that the close fit and neat appearance of the structure will not be impaired.

10.5.2. Erection Tolerances

Unless otherwise specifically designated in the contract documents, members and components are plumbed, leveled and aligned to a tolerance not to exceed one-half the amount permitted for structural steel. These erection tolerances for AESS require that the owner's plans specify adjustable connections between AESS and the structural steel frame or the masonry or concrete supports, in order to provide the erector with means for adjustment.

10.5.3. Components with Concrete Backing

When AESS is backed with concrete, it is the general contractor's responsibility to provide sufficient shores, ties and strongbacks to assure against sagging, bulging, etc., of the AESS resulting from the weight and pressure of the wet concrete.

Commentary
on the
Code of Standard Practice

PREFACE

This Commentary has been prepared to assist those who use the *Code of Standard Practice* in understanding the background, basis and intent of its provisions.

Each section in the Commentary is referenced to the corresponding section or subsection in the Code. Not all sections of the Code are discussed; sections are covered only if it is believed that additional explanation may be helpful.

While every precaution has been taken to insure that all data and information presented is as accurate as possible, the Institute cannot assume responsibility for errors or oversights in the information published herein or the use of the information published or incorporating such information in the preparation of detailed engineering plans. The figures are for illustrative purposes only and are not intended to be applicable to any actual design. The information should not replace the judgment of an experienced architect or engineer who has the responsibility of design for a specific structure.

Commentary

ON THE CODE OF STANDARD PRACTICE
FOR STEEL BUILDINGS AND BRIDGES

(Adopted Effective September 1, 1986)

SECTION 1. GENERAL PROVISIONS

1.1. Scope

This Code is not applicable to metal building systems, which are the subject of standards published by the Metal Building Manufacturers Association in their *Metal Building Systems Manual*. AISC has not participated in the development of the MBMA code and, therefore, takes no position and is not responsible for any of its provisions.

This Code is not applicable to standard steel joists, which are the subject of *Recommended Code of Standard Practice for Steel Joists*, published by the Steel Joist Institute. AISC has not participated in the development of the SJI code and, therefore, takes no position and is not responsible for any of its provisions.

SECTION 2. CLASSIFICATION OF MATERIALS

2.2. Other Steel or Metal Items

These items include materials which may be supplied by the steel fabricator which require coordination between other material suppliers and trades. If they are to be supplied by the fabricator, they must be specifically called for and detailed in contract documents.

SECTION 3. PLANS AND SPECIFICATIONS

3.1. Structural Steel

Project specifications vary greatly in complexity and completeness. There is a benefit to the owner if the specifications leave the contractor reasonable latitude in performing his work. However, critical requirements affecting the integrity of the structure or necessary to protect the owner's interest must be covered in the contract documents. The following checklist is included for reference:

Standard codes and specifications governing structural steelwork
Material specifications
Mill test reports
Welded joint configuration
Weld procedure qualification
Bolting specifications
Special requirements for work of other trades
Runoff tabs
Surface preparation and shop painting
Shop inspection
Field inspection
Non-destructive testing, including acceptance criteria
Special requirements on delivery
Special erection limitations
Temporary bracing for non-self-supporting structures
Special fabrication and erection tolerances for AESS
Special pay weight provisions

SECTION 4.   SHOP AND ERECTION DRAWINGS

4.1.   Owner's Responsibility

The owner's responsibility for the proper planning of the work and the communication of all facts of his particular project is a requirement of the Code, not only at the time of bidding, but also throughout the term of any project. The contract documents, including the plans and specification, are for the purpose of communication. It is the owner's responsibility to properly define the scope of work, and to define information or items required and outlined in the plans and specifications. When the owner releases plans and specifications for construction, the fabricator and erector rely on the fact that these are the owner's requirements for his project.

The Code defines the owner as including a designated representative such as the architect, engineer or project manager, and when these representatives direct specific action to be taken, they are acting as and for the owner.

On phased construction projects, to insure the orderly flow of material procurement, detailing, fabrication and erection activities, it is essential that designs are not continuously revised after progressive releases for construction are made. In essence, once a portion of a design is released for construction, the essential elements of that design should be "frozen" to assure adherence to the construction schedule or all parties should reach an understanding on the effects of future changes as they affect scheduled deliveries and added costs, if any.

4.2.   Approval

4.2.1.   In those instances where a fabricator develops the detail configuration of connections during the preparation of shop drawings, he does not thereby become responsible for the design of that part of the overall structure. The Engineer-of-Record has the final and total responsibility for the adequacy and safety of a structure, and is the only individual who has all the information necessary to evaluate the total impact of the connection details on the structural design. The structural steel fabricator is in no position to accept such design responsibility, for two practical reasons:

(a) The structural steel plans may be released for construction with incomplete or preliminary member reaction data, forcing a review by the Engineer at the time of approval.

(b) Few fabricators have engineers registered in all of the states in which they do business.

In practice, the fabricator develops connection details which satisfy two basic criteria:

(a) The connections must be of suitable strength and rigidity to meet the requirements of the design information provided by the engineer-of-record.

(b) The detail configuration accommodates the fabricator's shop equipment and procedures.

Since each shop has different equipment and skills, the fabricator is best suited to develop connection details which satisfy the second requirement. However, the overriding first requirement necessitates acceptance of responsibility and approval by the engineer.

SECTION 5.   MATERIALS

5.1.   Mill Materials

5.1.2.   Mill dimensional tolerances are completely set forth as part of ASTM Specification A6. Variation in cross section geometry of rolled members must be recognized by the designer, the fabricator and erector (see Fig. 1). Such tolerances are

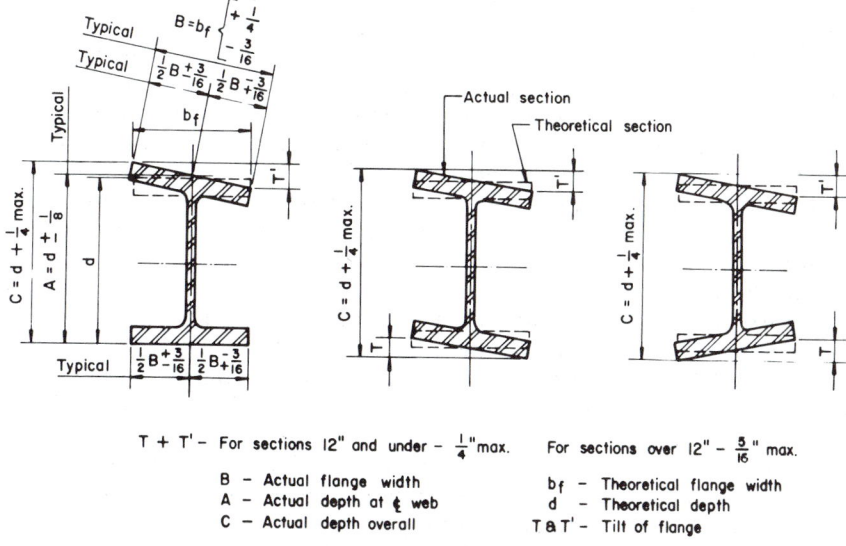

$T + T'$ – For sections 12" and under – $\frac{1}{4}$" max.   For sections over 12" – $\frac{5}{16}$" max.

B – Actual flange width b_f – Theoretical flange width
A – Actual depth at ⊄ web   d – Theoretical depth
C – Actual depth overall   T & T' – Tilt of flange

Fig. 1.   Mill tolerances on cross-section dimensions

mandatory because roll wear, thermal distortions of the hot cross section immediately after leaving the forming rolls, and differential cooling distortions that take place on the cooling beds are economically beyond precise control. Absolute perfection of cross section geometry is not of structural significance and, if the tolerances are recognized and provided for, also not of architectural significance. ASTM A6 also stipulates straightness and camber tolerances which are adequate for most conventional construction; however, these characteristics may be controlled or corrected to closer tolerances during the fabrication process when the unique demands of a particular project justify the added cost.

SECTION 6.   FABRICATION AND DELIVERY

6.4.   Dimensional Tolerances

Fabrication tolerances are stipulated in several specification documents, each applicable to a special area of construction. Basic fabrication tolerances are stipulated in Sects. 6.4 and 10 of the Code and Sect. 1.23.8.1 of the AISC Specification. Other specifications and codes frequently incorporated by reference in the contract documents are the AWS *Structural Welding Code* and AASHTO *Standard Specifications for Highway Bridges*.

6.5.   Shop Painting

6.5.2., 6.5.3.   The selection of a paint system is a design decision involving many factors, including owner's preference, service life of the structure, severity of environmental exposure, the cost of both initial application and future renewals, and the compatibility of the various components comprising the paint system, i.e., surface preparation, prime coat and subsequent coats.

Because inspection of shop painting needs to be concerned with workmanship at each stage of the operation, the fabricator provides notice of the schedule of operations and affords access to the work site to inspectors. Inspection must be coordinated with that schedule in such a way as to avoid delay of the scheduled operations.

Acceptance of the prepared surface must be made prior to application of the prime coat, because the degree of surface preparation cannot be readily verified after painting. Also, time delay between surface preparation and application of the prime coat can, especially with blast-cleaned surfaces, result in unacceptable deterioration of a properly prepared surface, necessitating a repetition of surface preparation. Therefore, to avoid potential deterioration of the surface it is assumed that surface preparation is accepted unless it is inspected and rejected prior to the scheduled application of the prime coat.

The prime coat in any paint system is designed to maximize the wetting and adherence characteristics of the paint, usually at the expense of its weathering capabilities. Consequently, extended exposure of the prime coat to weather or to a corrosive atmosphere will lead to its deterioration and may necessitate repair, possibly including repetition of surface preparation and primer application in limited areas. With the introduction of high performance paint systems in the recent past, delay in the application of the prime coat has become more critical. High performance paint systems generally require a greater degree of surface preparation, as well as early application of weathering protection for the prime coat.

Since the fabricator does not control the selection of the paint system, the compatibility of the various components of the total paint system, nor the length of exposure of the prime coat, he cannot guarantee the performance of the prime coat or any other part of the system. Rather, he performs specific operations to the requirements established in the contract documents.

Section 6.5.3 stipulates cleaning the steel to the requirements of SSPC-SP2. This is not meant as an exclusive cleaning level, but rather that level of surface preparation which will be furnished if the steel is to be painted and if the job specifications are silent or do not require more stringent surface preparation requirements.

Further information regarding shop painting is available in *A Guide to Shop Painting of Structural Steel*, published jointly by the Steel Structures Painting Council and the American Institute of Steel Construction.

6.5.5. Extended exposure of unpainted steel which has been cleaned for subsequent fire protection material application can be detrimental to the fabricated product. Most levels of cleaning require the removal of all loose mill scale, but permit some amount of "tightly adhering mill scale." When a piece of structural steel which has been cleaned to an acceptable level is left exposed to a normal environment, moisture can penetrate behind the scale, and some "lifting" of the scale by the oxidation products is to be expected. Cleanup of "lifted" mill scale is not the responsibility of the fabricator, but is assigned by contract requirement to an appropriate contractor.

Section 6.5.5 of the Code is not applicable to weathering steel, for which special cleaning specifications are always required in the contract documents.

SECTION 7. ERECTION

7.5. Installation of Anchor Bolts and Embedded Items

7.5.1. While the general contractor must make every effort to set anchor bolts accurately to theoretical drawing dimensions, minor errors may occur. The tolerances set forth in this section were compiled from data collected from general contractors and erectors. They can be attained by using reasonable care and will ordinarily allow the steel to be erected and plumbed to required tolerances. If special conditions require closer tolerances, the contractor responsible for setting the anchor bolts should be so informed by the contract documents. When anchor bolts are set in sleeves, the adjustment provided may be used to satisfy the required anchor bolt setting tolerances.

The tolerances established in this section of the Code have been selected to be compatible with oversize holes in base plates, as recommended in the AISC textbook *Detailing for Steel Construction*.

An *anchor bolt group* is the set of anchor bolts which receive a single fabricated steel shipping piece.

The *established column line* is the actual field line most representative of the centers of the as-built anchor bolt groups along a line of columns. It must be straight or curved as shown on the plans.

7.6. Bearing Devices

The ⅛-inch tolerance on elevation of bearing devices relative to established grades is provided to permit some variation in setting bearing devices and to account for attainable accuracy with standard surveying instruments.

The use of leveling plates larger than 12 in. × 12 in. is discouraged and grouting is recommended with larger sizes.

7.9.3. Non Self-supporting Steel Frames

To rationally provide temporary supports and/or bracing, the erector must be informed by the owner of the sequence of installation and the effect of loads imposed by such elements at various stages during the sequence until they become effective. For example, precast tilt-up slabs or channel slab facia elements which depend upon attachment to the steel frame for stability against overturning due to eccentricity of their gravity load, may induce significant unbalanced lateral forces on the bare steel frame when partially installed.

7.11.   Framing Tolerances

The erection tolerances defined in this section of the Code have been developed through long-standing usage as practical criteria for the erection of structural steel. Erection tolerances were first defined by AISC in its *Code of Standard Practice* of October, 1924 in Paragraph 7 (f), "Plumbing Up." With the changes that took place in the types and use of materials in building construction after World War II, and the increasing demand by architects and owners for more specific tolerances, AISC adopted new standards for erection tolerances in Paragraph 7 (h) of the March 15, 1959 edition of the Code. Experience has proven that those tolerances can be economically obtained.

The current requirements were first published in the October 1, 1972 edition of the Code. They provide an expanded set of criteria over earlier Code editions. The basic premise that the final accuracy of location of any specific point in a structural steel frame results from the combined mill, fabrication and erection tolerances, rather than from the erection tolerances alone, remains unchanged in this edition of the Code. However, to improve clarity, pertinent standard fabrication tolerances are now stipulated in Sect. 7.11, rather than by reference to the AISC Specification as in previous editions. Additionally, expanded coverage has been given to definition of working points and working lines governing measurements of the actual steel location. Illustrations for defining and applying the applicable Code tolerances are provided in this Commentary.

The recent trend in building work is away from built-in-place construction wherein compatibility of the frame and the facade or other collateral materials is automatically provided for by the routine procedures of the crafts. Building construction today frequently incorporates prefabricated components wherein large units are developed with machine-like precision to dimensions that are theoretically correct for a perfectly aligned steel frame with ideal member cross sections. This type of construction has made the magnitude of the tolerances allowed for structural steel building frames increasingly of concern to owners, architects and engineers. This has led to the inclusion in job specifications of unrealistically small tolerances, which indicate a general lack of recognition of the effects of the accumulation of dead load, temperature effects and mill, fabrication and erection tolerances. Such tolerances are not economically feasible and do not measurably increase the structure's functional value. This edition of the Code incorporates tolerances previously found to be practical and presents them in a precise and clear manner. Actual application methods have been considered and the application of the tolerance limitations to the actual structure defined.

7.11.3. Position and Alignment

The limitations described in Sect. 7.11.3.1 and illustrated in Figs. 2 and 3 make it possible to maintain built-in-place or prefabricated facades in a true vertical plane up to the 20th story, if connections which provide for 3-in. adjustment are used. Above the 20th story, the facade may be maintained within $\frac{1}{16}$-inch per story with a maximum total deviation of 1 in. from a true vertical plane, if the 3-in. adjustment is provided.

Section 7.11.3.1(c) limits the position of exterior column working points at any given splice elevation to a narrow horizontal envelope parallel to the building line (see Fig. 4). This envelope is limited to a width of 1½ in., normal to the building line, in up to 300 ft of building length. The horizontal location of this envelope is not necessarily directly above or below the corresponding envelope at the adjacent splice elevations, but should be within the limitation of the 1:500 allowable tolerance in plumbness of the controlling columns (see Fig. 3).

Connections permitting adjustments of plus 2 in. to minus 3 in. (5 in. total) will be necessary in cases where the architect or owner insists upon attempting to construct the facade to a true vertical plane above the 20th story.

Usually there is a differential shortening of the internal versus the external columns during construction, due to non-uniform rate of accumulation of dead load stresses (see Fig. 5). The amount of such differential shortening is indeterminate because it varies dependent upon construction sequence from day to day as the construction progresses, and does not reach its maximum shortening until the building is in service. When floor concrete is placed while columns are supporting different percentages of their full design loads, the floor must be finished to slopes established by measurements from the tops of beams at column connections. The effects of differential shortening, plus mill camber and deflections, all become very important when there is little cover over the steel, when there are electrical fittings mounted on

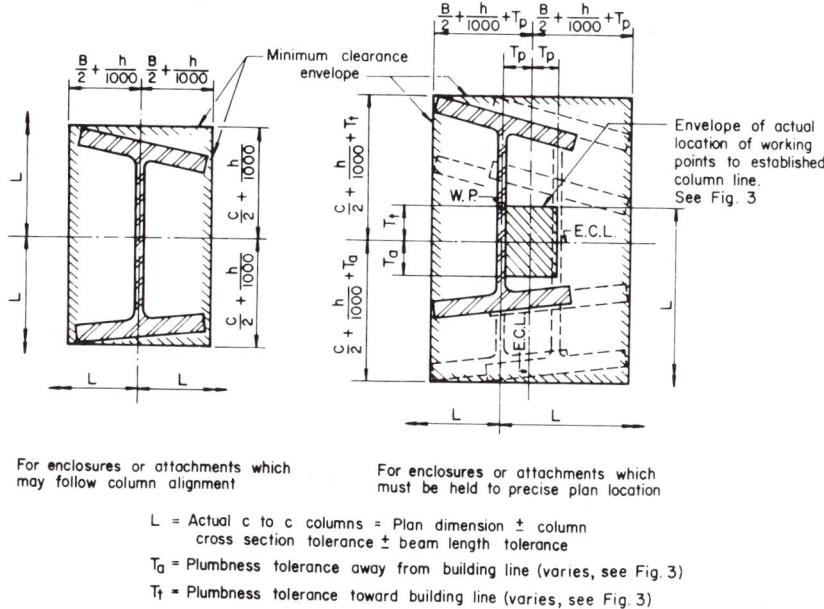

For enclosures or attachments which may follow column alignment

For enclosures or attachments which must be held to precise plan location

L = Actual c to c columns = Plan dimension ± column cross section tolerance ± beam length tolerance

T_a = Plumbness tolerance away from building line (varies, see Fig. 3)

T_t = Plumbness tolerance toward building line (varies, see Fig. 3)

T_p = Plumbness tolerance parallel to building line (= T_a)

Fig. 2. Clearance required to accommodate accumulated column tolerances

AMERICAN INSTITUTE OF STEEL CONSTRUCTION

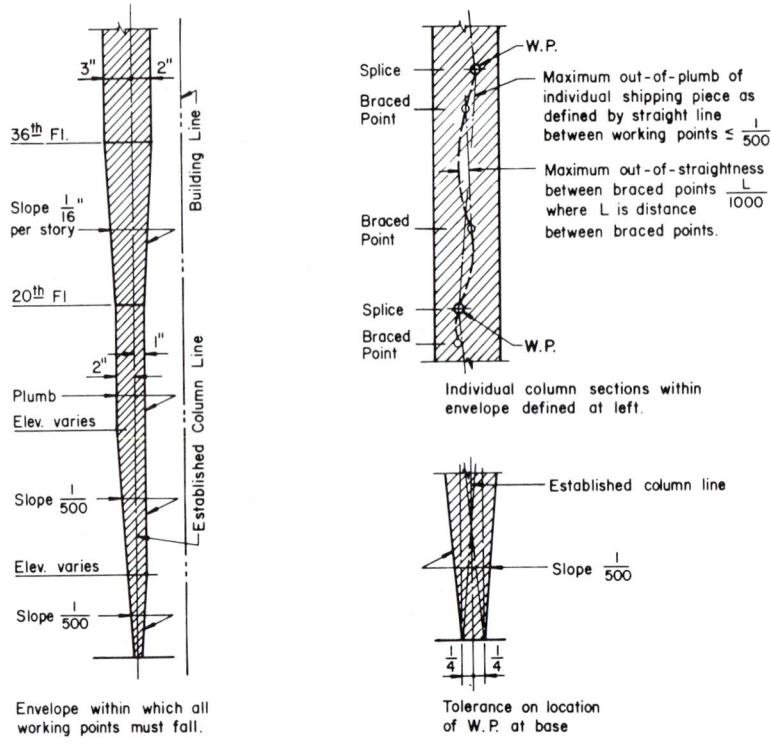

Envelope within which all
working points must fall.

Tolerance on location
of W. P. at base

NOTE: The plumb line thru the base working point for an individual column is not necessarily the precise plan location because Sect. 7.11.3.1 deals only with plumbness tolerance and does not include inaccuracies in location of established column line, foundations and anchor bolts beyond the erector's control.

Fig. 3.   Exterior column plumbness tolerances normal to building line

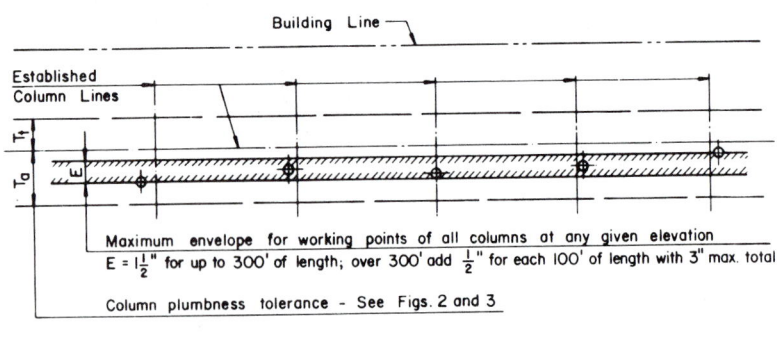

Maximum envelope for working points of all columns at any given elevation
$E = 1\frac{1}{2}"$ for up to 300' of length; over 300' add $\frac{1}{2}"$ for each 100' of length with 3" max. total

Column plumbness tolerance - See Figs. 2 and 3

ϕ — Indicates column working points.

At any splice elevation, envelope "E" is located within the limits T_a and T_t.
At any splice elevation, envelope "E" may be located offset from the corresponding envelope at the adjacent splice elevations, above and below, by an amount not greater than $\frac{1}{500}$ of the column length.

Fig. 4.   Tolerances in plan at any splice elevation of exterior columns

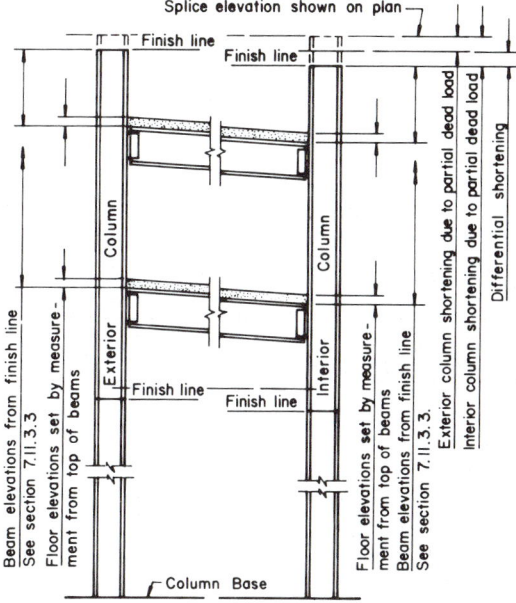

On a particular date during the erection of structural steel and placement of other material, (floor concrete, facade etc.) the interior columns will be carrying a higher percentage of their final loads than the exterior columns Therefore, for equal design unit stresses, the actual stress on that date for interior columns will be greater than the actual stresses on exterior columns. When all dead loads have been applied, stresses and shortening in all columns will be approximately equal.

Fig. 5. *Effect of differential column shortening*

the steel flooring whose tops are supposed to be flush with the finished floor, when there is small clearance between bottom of beams and top of door frames, etc., and when there is little clearance around ductwork. To finish floors to precise level plane, for example by the use of laser leveling techniques, can result in significant differential floor thicknesses, different increases above design dead loads for individual columns and, thus, permanent differential column shortening and out-of-level completed floors.

Similar considerations make it infeasible to attempt to set the elevation of a given floor in a multistory building by reference to a bench mark at the base of the structure. Columns are fabricated to a length tolerance of plus or minus $\frac{1}{32}$-inch while under a zero state of stress. As dead loads accumulate, the column shortening which takes place is negligible within individual stories and in low buildings, but will accumulate to significant magnitude in tall buildings; thus, the upper floors of tall buildings will be excessively thick and the lower floors will be below the initial finish elevation if floor elevations are established relative to a ground level bench mark.

If foundations and base plates are accurately set to grade and the lengths of individual column sections are checked for accuracy prior to erection, and if floor elevations are established by reference to the elevation of the top of beams, the effect of column shortening due to dead load will be minimized.

Since a long unencased steel frame will expand or contract $\frac{1}{8}$-in. per 100 feet for each change of 15°F in temperature, and since the change in length can be assumed to act about the center of rigidity, the end columns anchored to foundations will be plumb only when the steel is at normal temperature (see Fig. 6). It is, therefore, necessary to correct field measurements of offsets to the structure from established baselines for the expansion or contraction of the exposed steel frame. For example, a building 200-ft long that is plumbed up at 100°F should have working points at the tops of end columns positioned $\frac{1}{2}$-inch out from the working point at the base in order for the column to be

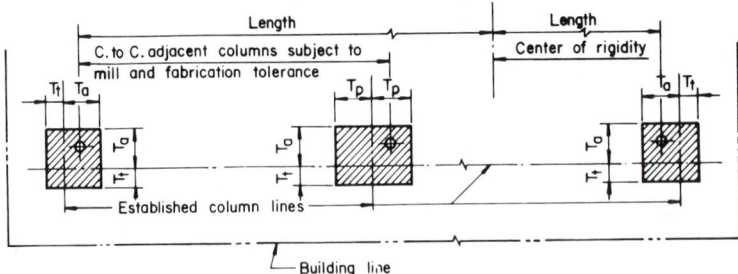

Fig. 6. Tolerances in plan location of columns

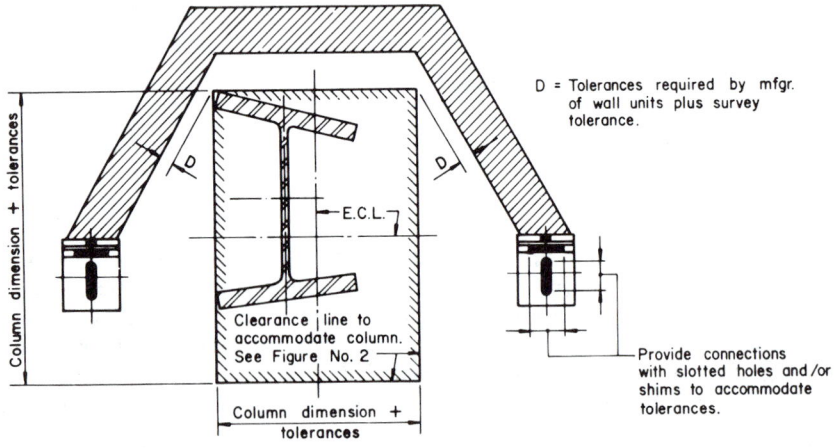

Fig. 7. Clearance required to accommodate fascia

plumb at 60°F. Differential temperature effects on column length should also be taken into account in plumbing surveys when tall steel frames are subject to strong sun exposure on one side.

The alignment of lintels, spandrels, wall supports and similar members used to connect other building construction units to the steel frame should have an adjustment of sufficient magnitude to allow for the accumulative effect of mill, fabrication and erection tolerances on the erected steel frame (see Fig. 7).

7.11.3.2. Alignment Tolerance for Members with Field Splices

The angular misalignment of the working line of all fabricated shipping pieces relative to the line between support points of the member as a whole in erected position must not exceed 1 in 500. Note that the tolerance is not stated in terms of a

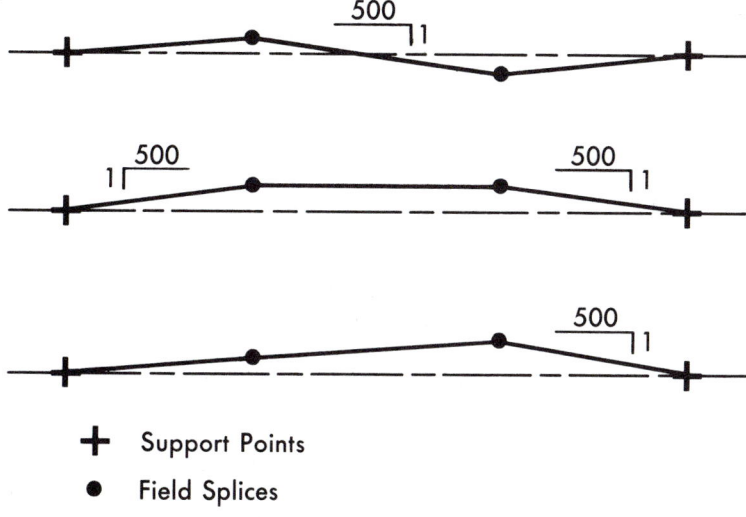

+   Support Points

●   Field Splices

Fig. 8. Alignment tolerances for members with field splices

linear displacement at any point and is not to be taken as the overall length between supports divided by 500. Typical examples are shown in Fig. 8. Numerous conditions within tolerance for these and other cases are possible. This condition applies to both plan and elevation tolerances.

7.11.4. Responsibility for Clearances

In spite of all efforts to minimize inaccuracies, deviations will still exist; therefore, in addition, the designs of prefabricated wall panels, partition panels, fenestrations, floor-to-ceiling door frames and similar elements must provide for clearance and details for adjustment, as described in Sect. 7.11.4. Designs must provide for adjustment in the vertical dimension of prefabricated facade panels supported by the steel frame, because the accumulation of shortening of stressed steel columns will result in the unstressed facade supported at each floor level being higher than the steel frame connections to which it must attach. Observations in the field have shown that where a heavy facade is erected to a greater height on one side of a multistory building than on the other, the steel framing will be pulled out of alignment. Facades should be erected at a relatively uniform rate around the perimeter of the structure.

SECTION 8. QUALITY CONTROL

8.1.1. The AISC Quality Certification Program confirms to the construction industry that a certified structural steel fabricating plant has the capability by reason of commitment, personnel, organization, experience, procedures, knowledge and equipment to produce fabricated structural steel of the required quality for a given category of structural steelwork. The AISC Quality Certification Program is not intended to involve inspection and/or judgement of product quality on individual projects. Neither is it intended to guarantee the quality of specific fabricated steel products.

9.2. Calculation of Weights

The standard procedure for calculation of weights that is described in the Code meets the need for a universally acceptable system for defining "pay weights" in contracts based on the weight of delivered and/or erected materials. This procedure permits owners to easily and accurately evaluate price per pound proposals from potential suppliers and enables both parties to a contract to have a clear understanding of the basis for payment.

The Code procedure affords a simple, readily understood method of calculation which will produce pay weights which are consistent throughout the industry and which may be easily verified by the owner. While this procedure does not produce actual weights, it can be used by purchasers and suppliers to define a widely accepted basis for bidding and contracting for structural steel. However, any other system can be used as the basis for a contractual agreement. When other systems are used, both supplier and purchaser should clearly understand how the alternate procedure is handled.

9.3. Revisions to Contract Documents

9.3.1. Revisions to the contract are proposed by the issuance of new documents or the re-issuance of existing documents. Individual revisions are noted where they occur and documents are dated with latest issue date and reasons for issuing are identified.

9.3.2. Revisions to the contract are also proposed by change order, extra work order or notations on the shop and erection drawings when returned from approval. However, revisions proposed in this manner are incorporated subsequently as revisions to the plans and/or specifications and re-issued in accordance with Article 9.3.1.

9.3.3. Unless specifically stated to the contrary in the contract documents, the issuance of revisions authorizes the fabricator and erector to incorporate the revisions in the work. This authorization obligates the owner to pay the fabricator and erector for costs associated with changed and/or additional work.

When authorization for revisions is not granted to the fabricator and erector by issuance of new or revised documents, revisions affecting contract price and/or schedule are only incorporated by issuance of (1) change order, (2) extra work order, or (3) other documents expressing the agreement of all contract parties to such revisions. The fabricator and erector must promptly notify the owner of the effect and cost of proposed revisions to contract price and schedule, enabling orderly progress of the work.

9.6. Terms of Payment

These terms include such items as progress payments for material, fabrication, erection, retainage, performance and payment bonds and final payment. If a performance or payment bond, paid for by the owner, is required by contract, then no retainage shall be required.

SECTION 10.   ARCHITECTURALLY EXPOSED STRUCTURAL STEEL

The rapidly increasing use of exposed structural steel as a medium of architectural expression has given rise to a demand for closer dimensional tolerances and smoother finished surfaces than required for ordinary structural steel framing.

This section of the Code establishes standards for these requirements which take into account both the desired finished appearance and the abilities of the fabrication shop to produce the desired product. These requirements were previously contained in the AISC *Specification for Architecturally Exposed Structural Steel* which Architects and Engineers have specified in the past. It should be pointed out that the term "Architecturally Exposed Structural Steel" (AESS), as covered in this section, must be specified in the contract documents if the fabricator is required to meet the fabricating standards of Sect. 10, and applies only to that portion of the structural steel so identified. In order to avoid misunderstandings and to hold costs to a minimum, only those steel surfaces and connections which will remain exposed and subject to normal view by pedestrians or occupants of the completed structure should be designated AESS.

ALLOWABLE STRESS DESIGN

Specification for Structural Joints Using ASTM A325 or A490 Bolts

Approved by the Research Council on Structural Connections
of the Engineering Foundation, November 13, 1985
Endorsed by American Institute of Steel Construction
Endorsed by Industrial Fasteners Institute

AMERICAN INSTITUTE OF STEEL CONSTRUCTION, INC.
1 East Wacker Drive, Suite 3100, Chicago, IL 60601-2001

PREFACE

The purpose of the Research Council on Structural Connections is to stimulate and support such investigation as may be deemed necessary and valuable to determine the suitability and capacity of various types of structural connections, to promote the knowledge of economical and efficient practices relating to such structural connections, and to prepare and publish related standards and such other documents as necessary to achieving its purpose.

The Council membership consists of qualified structural engineers from the academic and research institutions, practicing design engineers, suppliers and manufacturers of threaded fasteners, fabricators and erectors and code writing authorities. Each version of the Specification is based upon deliberations and letter ballot of the full Council membership.

The first *Specification for Assembly of Structural Joints Using High Tensile Steel Bolts* approved by the Council was published in January 1951. Since that time the Council has published 11 succeeding editions each based upon past successful usage, advances in the state of knowledge and changes in engineering design practice. This twelfth version of the Council's *Allowable Stress Design Specification* is significantly reorganized and revised from earlier versions.

The intention of the Specifications is to cover the design criteria and normal usage and practices involved in the everyday use of high-strength bolts in steel-to-steel structural connections. It is not intended to cover the full range of structural connections using threaded fasteners nor the use of high-strength bolts other than those included in ASTM A325 or ASTM A490 Specifications nor the use of ASTM A325 or A490 Bolts in connections with material other than steel within the grip.

A Commentary has been prepared to accompany these Specifications to provide background and aid the user to better understand and apply the provisions.

The user is cautioned that independent professional judgement must be exercised when data or recommendations set forth in these Specifications are applied. The design and the proper installation and inspection of bolts in structural connections is within the scope of expertise of a competent licensed architect, structural engineer or other licensed professional for the application of the principles to a particular case.

ALLOWABLE STRESS DESIGN

Specification for Structural Joints Using ASTM A325 or A490 Bolts

Approved by Research Council on Structural Connections of the
Engineering Foundation, November 13, 1985.

Endorsed by American Institute of Steel Construction
Endorsed by Industrial Fasteners Institute

1. Scope

This Specification relates to the allowable stress design for strength and slip resistance of structural joints using ASTM A325 high-strength bolts, ASTM A490 heat treated high-strength bolts or equivalent fasteners, and for the installation of such bolts in connections of structural steel members. The Specification relates only to those aspects of the connected materials that bear upon the performance of the fasteners.

Construction shall conform to an applicable code or specification for structures of carbon, high strength low alloy steel or quenched and tempered structural steel.

The attached Commentary provides background information in order that the user may better understand the provisions of the Specification.

2. Bolts, Nuts, Washers and Paint

(a) **Bolt Specifications.** Bolts shall conform to the requirements of the current edition of the Specifications of the American Society for Testing and Materials for High-strength Bolts for Structural Steel Joints, ASTM A325, or Heat Treated Steel Structural Bolts, 150 ksi Minimum Tensile Strength, ASTM A490, except as provided in paragraph (d) of this section. The designer shall specify the type of bolts to be used.

(b) **Bolt Geometry.** Bolt dimensions shall conform to the current requirements of the American National Standards Institute for Heavy Hex Structural Bolts, ANSI Standard B18.2.1, except as provided in paragraph (d) of this section. The length of bolts shall be such that the end of the bolt will be flush with or outside the face of the nut when properly installed.

(c) **Nut Specifications.** Nuts shall conform to the current chemical and mechanical requirements of the American Society for Testing and Materials Standard Specification for Carbon and Alloy Steel Nuts, ASTM A563 or Standard Specification for Carbon and Alloy Steel Nuts for Bolts for High Pressure and High Temperatures Service, ASTM A194. The grade and surface finish of nuts for each type shall be as follows:

A325 Bolt Type	Nut Specification, Grade and Finish
1 and 2, plain (uncoated)	A563 C, C3, D, DH, and D3 or A194 2 and 2H; plain
1 and 2, galvanized	A563 DH or A194 2H; galvanized
3 plain	A563 C3 and DH3

A490 Bolt Type	Nut Specification, Grade and Finish
1 and 2, plain	A563 DH and DH3 or A194 2H; plain
3 plain	A563 DH3

Nut dimensions shall conform to the current requirements of the American National Standards Institute for Heavy Hex Nuts, ANSI Standard B18.2.2., except as provided in paragraph (d) of this section.

(d) **Alternate Fastener Designs.** Other fasteners or fastener assemblies which meet the materials, manufacturing and chemical composition requirements of ASTM Specification A325 or A490 and which meet the mechanical property requirements of the same specifications in full-size tests, and which have a body diameter and bearing areas under the head and nut not less than those provided by a bolt and nut of the same nominal dimensions prescribed by paragraphs 2(b) and 2(c), may be used subject to the approval of the responsible Engineer. Such alternate fasteners may differ in other dimensions from those of the specified bolts and nuts. Their installation procedure and inspection may differ from procedures specified for regular high-strength bolts in Sections 8 and 9. When a different installation procedure or inspection is used, it shall be detailed in a supplemental specification applying to the alternate fastener and that specification must be approved by the engineer responsible for the design of the structure.

(e) **Washers.** Flat circular washers and square or rectangular beveled washers shall conform to the current requirements of the American Society for Testing and Materials Standard Specification for Hardened Steel Washers, ASTM F436.

(f) **Load Indicating Devices.** Load indicating devices may be used in conjunction with bolts, nuts and washers specified in 2(a) through 2(e) provided they satisfy the requirements of 8(d)(4). Their installation procedure and inspection shall be detailed in supplemental specifications provided by the manufacturer and subject to the approval of the engineer responsible for the design of the structure.

(g) **Faying Surface Coatings.** Paint, if used on faying surfaces of connections which are not specified to be slip critical, may be of any formulation. Paint, used on the faying surfaces of connections specified to be slip critical, shall be qualified by test in accordance with *Test Method to Determine the Slip Coefficient for Coatings Used in Bolted Joints* as adopted by the Research Council on Structural Connections, see Appendix A. Manufacturer's certification shall include a certified copy of the test report.

3. Bolted Parts

(a) **Connected Material.** All material within the grip of the bolt shall be steel. There shall be no compressible material such as gaskets or insulation within the grip. Bolted steel parts shall fit solidly together after the bolts are tightened, and may be coated or noncoated. The slope of the surfaces of parts in contact with the bolt head or nut shall not exceed 1:20 with respect to a plane normal to the bolt axis.

(b) **Surface Conditions.** When assembled, all joint surfaces, including surfaces adjacent to the bolt head and nut, shall be free of scale, except tight mill scale, and shall be free of dirt or other foreign material. Burrs that would prevent solid seating of the connected parts in the snug tight condition shall be removed.

Paint is permitted on the faying surfaces unconditionally in connections except in slip critical connections as defined in Section 5(a).

The faying surfaces of slip critical connections shall meet the requirements of the following paragraphs, as applicable.

(1) In noncoated joints, paint, including any inadvertent overspray, shall be excluded from areas closer than one bolt diameter but not less than one inch from the edge of any hole and all areas within bolt pattern.

(2) Joints specified to have painted faying surfaces shall be blast cleaned and coated with a paint which has been qualified as class A or B in accordance with the requirements of paragraph 2(g), except as provided in 3(b)(3).

(3) Subject to the approval of the Engineer, coatings providing a slip coefficient less than 0.33 may be used provided the mean slip coefficient is established by test in accordance with the requirements of paragraph 2(g), and the allowable slip load per unit area established. The allowable slip load per unit area shall be taken as equal to the allowable slip load per unit area from Table 3 for Class A coatings as appropriate for the hole type and bolt type times the slip coefficient determined by test divided by 0.33.

(4) Coated joints shall not be assembled before the coatings have cured for the minimum time used in the qualifying test.

(5) Galvanized faying surfaces shall be hot dip galvanized in accordance with ASTM Specification A123 and shall be roughened by means of hand wire brushing. Power wire brushing is not permitted.

(c) **Hole Types.** Hole types recognized under this specification are standard holes, oversize holes, short slotted holes and long slotted holes. The nominal dimensions for each type hole shall be not greater than those shown in Table 1. Holes not more than $1/32$ inch larger in diameter than the true decimal equivalent of the nominal diameter that may result from a drill or reamer of the nominal diameter are considered acceptable. The slightly conical hole that naturally results from punching operations is considered acceptable. The width of slotted holes which are produced by flame cutting or a combination of drilling or punching and flame cutting shall generally be not more than $1/32$ inch greater than the nominal width except that gouges not more than $1/16$ inch deep shall be permitted. For statically loaded connections, the flame cut surface need not be ground. For dynamically loaded connections, the flame cut surface shall be ground smooth.

Table 1. Nominal Hole Dimensions

Bolt Dia.	Hole Dimensions			
	Standard (Dia.)	Oversize (Dia.)	Short Slot (Width x Length)	Long Slot (Width x Length)
$1/2$	$9/16$	$5/8$	$9/16 \times 11/16$	$9/16 \times 1 1/4$
$5/8$	$11/16$	$13/16$	$11/15 \times 7/8$	$11/16 \times 1 9/16$
$3/4$	$13/16$	$15/16$	$13/16 \times 1$	$13/16 \times 1 7/8$
$7/8$	$15/16$	$1 1/16$	$15/16 \times 1 1/8$	$15/16 \times 2 3/16$
1	$1 1/16$	$1 1/4$	$1 1/16 \times 1 5/16$	$1 1/16 \times 2 1/2$
$\geq 1 1/8$	$d + 1/16$	$d + 5/16$	$(d + 1/16) \times (d + 3/8)$	$(d + 1/16) \times (2.5 \times d)$

4. Design for Strength of Bolted Connections

(a) **Allowable Strength.** The allowable working stress in shear and bearing, independent of the method of tightening, for A325 and A490 bolts is given in Table 2. Also given in Table 2 is the allowable working stress in axial tension for A325 and A490 bolts which are tightened to the minimum fastener tension specified in Table 4. The allowable working stresses in Table 2 are to be used in conjunction with the cross sectional area of the bolt corresponding to the nominal diameter.

(b) **Bearing Force.** The computed bearing force shall be assumed to be distributed over an area equal to the nominal bolt diameter times the thickness of the connected part.

A value of allowable bearing pressure on the connected material at a bolt greater than permitted by Table 2 can be justified provided deformation around the bolt hole is not a design consideration and adequate pitch and end distance L is provided according to:

$$F_p = LF_u/2d \leq 1.5F_u$$

Table 2.   Allowable Working Stress[a] on Fasteners or Connected Material (ksi)

Load Condition	A325	A490
Applied Static Tension [b,c].	44	54
Shear on bolt with threads in shear plane.	21[d]	28[d]
Shear on bolt without threads in shear plane.	30[d]	40[d]
Bearing on connected material with single bolt in line of force in a standard or short slotted hole.	$1.0F_u$ [e,f,g,h]	
Bearing on connected material with 2 or more bolts in line of force in standard or short slotted holes.	$1.2F_u$ [e,f,g,h]	
Bearing on connected material in long slotted holes.	$1.0F_u$ [e,f,g,h]	

[a]Ultimate failure load divided by factor of safety.

[b]Bolts must be tensioned to requirements of Table 4.

[c]See 4 (d) for bolts subject to tensile fatigue.

[d]In connections transmitting axial force whose length between extreme fasteners measured parallel to the line of force exceeds 50 inches, tabulated values shall be reduced 20 percent.

[e]F_u = specified minimum tensile strength of connected part.

[f]Connections using high strength bolts in slotted holes with the load applied in a direction other than approximately normal (between 80 and 100 degrees) to the axis of the hole and connections with bolts in oversize holes shall be designed for resistance against slip at working load in accordance with Section 5.

[g]Tabulated values apply when the distance L parallel to the line of force from the center of the bolt to the edge of the connected part is not less than 1 1/2d and the distance from the center of a bolt to the center of an adjacent bolt is not less than 3d. When either of these conditions is not satisfied, the distance L requirement of 4(b) determines allowable bearing stress. See Commentary.

[h]Except as may be justified under provision 4(b).

Where

d = bolt diameter

F_p = allowable bearing pressure at a bolt

F_u = specified minimum tensile strength of connected part

(c) **Prying Action.**   The force in bolts required to support loads by means of direct tension shall be calculated considering the effects of the external load and any tension resulting from prying action produced by deformation of the connected parts.

(d) **Tensile Fatigue.**   When subject to tensile fatigue loading, the tensile stress in the bolt due to the combined applied load and prying forces shall not exceed the following values dependent upon the bolt grade and number of cycles, and the prying force shall not exceed 60 percent of the externally applied load.

Number of Cycles	A325	A490
Not more than 20,000	44	54
From 20,000 to 500,000	40	49
More than 500,000	31	38

Bolts must be tensioned to requirements of Table 4.

5. Design Check for Slip Resistance

(a) **Slip-Critical Joints.** Slip-critical joints are defined as joints in which slip would be detrimental to the serviceability of the structure. They include:

(1) Joints subject to fatigue loading.

(2) Joints with bolts installed in oversized holes.

(3) Except where the Engineer intends otherwise and so indicates in the contract documents, joints with bolts installed in slotted holes where the force on the joint is in a direction other than normal (between approximately 80 and 100 degrees) to the axis of the slot.

(4) Joints subject to significant load reversal.

(5) Joints in which welds and bolts share in transmitting load at a common faying surface. See Commentary.

(6) Joints in which, in the judgement of the Engineer, any slip would be critical to the performance of the joint or the structure and so designated on the contract plans and specifications.

(b) **Allowable Slip Load.** In addition to the requirements of Section 4, the force on a slip-critical joint shall not exceed the allowable resistance (P_s) of the connection (See Commentary) according to:

$$P_s = F_s A_b N_b N_s$$

Where

F_s = allowable slip load per unit area of bolt from Table 3
A_b = area corresponding to the nominal body area of the bolt
N_b = number of bolts in the joint
N_s = number of slip planes

Class A, B or C surface conditions of the bolted parts as defined in Table 3 shall be used in joints designated as slip-critical except as permitted in 3(b)(3).

6. Increase in Allowable Stresses

When the applicable code or specification for design of connected members permits an increase in working stress for loads in combination with wind or seismic forces, the permitted increases in working stresses may be applied with wind or seismic forces, the permitted increases in working stresses may be applied to the allowable stresses in Sections 4 and 5. When the effect of loads in combination with wind or seismic forces are accounted for by reduction in the load factors, the allowable stresses in Sections 4 and 5 may not be increased.

7. Design Details of Bolted Connections

(a) **Standard Holes.** In the absence of approval by the engineer for use of other hole types, standard holes shall be used in high strength bolted connections.

Table 3.   Allowable Load for Slip-critical Connections
(Slip Load per Unit of Bolt Area, ksi)

Contact Surface of Bolted Parts	Hole Type and Direction of Load Application							
	Any Direction				Transverse		Parallel	
	Standard		Oversize & Short Slot		Long Slots		Long Slots	
	A325	A490	A325	A490	A325	A490	A325	A490
Class A (Slip Coefficient 0.33) Clean mill scale and blast-cleaned surfaces with Class A coatings[a]	17	21	15	18	12	15	10	13
Class B (Slip Coefficient 0.50) Blast-cleaned surfaces and blast-cleaned surfaces with Class B coatings[a]	28	34	24	29	20	24	17	20
Class C (Slip Coefficient 0.40) Hot dip Galvanized and rough-ened surfaces	22	27	19	23	16	19	14	16

[a]Coatings classified as Class A or Class B includes those coatings which provide a mean slip coefficient not less than 0.33 or 0.50, respectively, as determined by Testing Method to Determine the Slip Coefficient for Coatings Used in Bolted Joints, see Appendix A.

(b) **Oversize and Slotted Holes.**   When approved by the Engineer, oversize, short slotted holes or long slotted holes may be used subject to the following joint detail requirements:

(1)   Oversize holes may be used in all plies of connections in which the allowable slip resistance of the connection is greater than the applied load.

(2)   Short slotted holes may be used in any or all plies of connections designed on the basis of allowable stress on the fasteners in Table 2 provided the load is applied approximately normal (between 80 and 100 degrees) to the axis of the slot. Short slotted holes may be used without regard for the direction of applied load in any or all plies of connections in which the allowable slip resistance is greater than the applied force.

(3)   Long slotted holes may be used in one of the connected parts at any individual faying surface in connections designed on the basis of allowable stress on the fasteners in Table 2 provided the load is applied approximately normal (between 80 and 100 degrees) to the axis of the slot. Long slotted holes may be used in one of the connected parts at any individual faying surface without regard for the direction of applied load on connections in which the allowable slip resistance is greater than the applied force.

(4)   Fully inserted finger shims between the faying surfaces of load transmitting elements of connections are not to be considered a long slot element of a connection.

(c) **Washer Requirements.** Design details shall provide for washers in high strength bolted connections as follows:

(1) Where the outer face of the bolted parts has a slope greater than 1:20 with respect to a plane normal to the bolt axis, a hardened beveled washer shall be used to compensate for the lack of parallelism.

(2) Hardened washers are not required for connections using A325 and A490 bolts except as required in paragraphs 7(c)(3) through 7(c)(7) for slip-critical connections and connections subject to direct tension or as required by paragraph 8(c) for shear/bearing connections.

(3) Hardened washers shall be used under the element turned in tightening when the tightening is to be performed by calibrated wrench method.

(4) Irrespective of the tightening method, hardened washers shall be used under both the head and the nut when A490 bolts are to be installed and tightened to the tension specified in Table 4 in material having a specified yield point less than 40 ksi.

(5) Where A325 bolts of any diameter or A490 bolts equal to or less than 1 inch in diameter are to be installed and tightened in an oversize or short slotted hole in an outer ply, a hardened washer conforming to ASTM F436 shall be used.

(6) When A490 bolts over 1 inch in diameter are to be installed and tightened in an oversize or short slotted hole in an outer ply, hardened washers conforming to ASTM F436 except with $5/16$ inch minimum thickness shall be used under both the head and the nut in lieu of standard thickness hardened washers. Multiple hardened washers with combined thickness equal to or greater than $5/16$ inch do not satisfy this requirement.

(7) Where A325 bolts of any diameter or A490 bolts equal to or less than 1 inch in diameter are to be installed and tightened in a long slotted hole in an outer ply, a plate washer or continuous bar of at least $5/16$ inch thickness with standard holes shall be provided. These washers or bars shall have a size sufficient to completely cover the slot after installation and shall be of structural grade material, but need not be hardened except as follows. When A490 bolts over 1 inch in diameter are to be used in long slotted holes in external plies, a single hardened washer conforming to ASTM F436 but with $5/16$ inch minimum thickness shall be used in lieu of washers or bars of structural grade material. Multiple hardened washers with combined thickness equal to or greater than $5/16$ inch do not satisfy this requirement.

(8) Alternate design fasteners meeting the requirements of 2(d) with a geometry which provides a bearing circle on the head or nut with a diameter equal to or greater than the diameter of hardened washers meeting the requirements ASTM F436 satisfy the requirements for washers specified in paragraphs 7(c)(4) and 7(c)(5).

8. Installation and Tightening

(a) **Handling and Storage of Fasteners.** Fasteners shall be protected from dirt and moisture at the job site. Only as many fasteners as are anticipated to be installed and tightened during a work shift shall be taken from protected storage. Fasteners not used shall be returned to protected storage at the end of the shift. Fasteners shall not be cleaned of lubricant that is present in as-delivered condition. Fasteners for slip critical connections which must be cleaned of accumulated rust or dirt resulting from job site conditions, shall be cleaned and relubricated prior to installation.

(b) Tension Calibrator. A tension measuring device shall be required at all job sites where bolts in slip-critical joints or connections subject to direct tension are being installed and tightened. The tension measuring device shall be used to confirm: (1) the suitability to satisfy the requirements of Table 4 of the complete fastener assembly, including lubrication if required to be used in the work, (2) calibration of wrenches, if applicable, and (3) the understanding and proper use by the bolting crew of the method to be used. The frequency of confirmation testing, the number of tests to be performed and the test procedure shall be as specified in 8(d), as applicable. The accuracy of the tension measuring device shall be confirmed through calibration by an approved testing agency at least annually.

(c) Joint Assembly and Tightening of Shear/Bearing Connections. Bolts in connections not within the slip-critical category as defined in Section 5(a) nor subject to tension loads nor required to be fully tensioned bearing-type connections shall be installed in properly aligned holes, but need only be tightened to the snug tight condition. The snug tight condition is defined as the tightness that exists when all plies in a joint are in firm contact. This may be attained by a few impacts of an impact wrench or the full effort of a man using an ordinary spud wrench. See Commentary. If a slotted hole occurs in an outer ply, a flat hardened washer or common plate washer shall be installed over the slot. Bolts which may be tightened only to a snug tight condition shall be clearly identified on the drawings.

(d) Joint Assembly and Tightening of Connections Requiring Full Pretensioning. In slip-critical connections, connections subject to direct tension, and fully pre-tensioned bearing connections, fasteners, together with washers of size and quality specified, located as required by Section 7(c), shall be installed in properly aligned holes and tightened by one of the methods described in Subsections 8(d)(1) through 8(d)(4) to at least the minimum tension specified in Table 4 when all the fasteners are tight. Tightening may be done by turning the bolt while the nut is prevented from rotating when it is impractical to turn the nut. Impact wrenches, if used, shall be of adequate capacity and sufficiently supplied with air to perform the required tightening of each bolt in approximately 10 seconds.

(1) Turn-of-nut Tightening. When turn-of-nut tightening is used, hardened washers are not required except as may be specified in 7(c).

A representative sample of not less than three bolts and nuts of each diameter, length and grade to be used in the work shall be checked at the start of work in a device capable of indicating bolt tension. The test shall demonstrate that the method of estimating the snug-tight condition and controlling turns from snug tight to be used by the bolting crews develops a tension not less than five percent greater than the tension required by Table 4.

Bolts shall be installed in all holes of the connection and brought to a snug-tight condition. Snug tight is defined as the tightness that exist when the plies of the joint are in firm contact. This may be attained by a few

Table 4.   Fastener Tension Required for Slip-critical Connections and Connections Subject to Direct Tension

Nominal Bolt Size, Inches	Minimum Tension[a] in 1000's of Pounds (kips)	
	A325 Bolts	A490 Bolts
1/2	12	15
5/8	19	24
3/4	28	35
7/8	39	49
1	51	64
1 1/8	56	80
1 1/4	71	102
1 3/8	85	121
1 1/2	103	148

[a]Equal to 70 percent of specified minimum tensile strengths of bolts (as specified in ASTM Specifications for tests of full size A325 and A490 bolts with UNC threads loaded in axial tension) rounded to the nearest kip.

impacts of an impact wrench or the full effort of a man using an ordinary spud wrench. Snug tightening shall progress systematically from the most rigid part of the connection to the free edges, and then the bolts of the connection shall be retightened in a similar systematic manner as neccessary until all bolts are simultaneously snug tight and the connection is fully compacted. Following this initial operation all bolts in the connection shall be tightened further by the applicable amount of rotation specified in Table 5. During the tightening operation there shall be no rotation of the part not turned by the wrench. Tightening shall progress systematically from the most rigid part of the joint to its free edges.

(2)   Calibrated Wrench Tightening.   Calibrated wrench tightening may be used only when installation procedures are calibrated on a daily basis and when a hardened washer is used under the element turned in tightening. See the Commentary to this Section. This specification does not recognize standard torques determined from tables or from formulas which are assumed to relate torque to tension.

When calibrated wrenches are used for installation, they shall be set to provide a tension not less than 5 percent in excess of the minimum tension specified in Table 4. The installation procedures shall be calibrated at least once each working day for each bolt diameter, length and grade using fastener assemblies that are being installed in the work. Calibration shall be accomplished in a device capable of indicating actual bolt tension by tightening three typical bolts of each diameter, length and grade from the bolts being installed and with a hardened washer from the washers being used in the work under the element turned in tightening. Wrenches shall be recalibrated when significant difference is noted in the surface condition of the bolts threads, nuts or washers. It shall be verified during actual installation in the assembled steelwork that the wrench adjustment selected by the calibration does not produce a nut or bolt head rotation from snug tight greater than that permitted in Table 5. If manual torque wrenches are used, nuts shall be turned in the tightening direction when torque is measured.

Table 5. Nut Rotation from Snug Tight Condition[a,b]

Bolt length (Under side of head to end of bolt)	Disposition of Outer Face of Bolted Parts		
	Both faces normal to bolt axis	One face normal to bolt axis and other sloped not more than 1:20 (beveled washer not used)	Both faces sloped not more than 1:20 from normal to the bolt axis (beveled washer not used)
Up to and including 4 diameters	1/3 turn	1/2 turn	2/3 turn
Over 4 diameters but not exceeding 8 dia.	1/2 turn	2/3 turn	5/6 turn
Over 8 diameters but not exceeding 12 dia.[c]	2/3 turn	5/6 turn	1 turn

[a]Nut rotation is relative to bolt regardless of the element (nut or bolt) being turned. For bolts installed by 1/2 turn and less, the tolerance should be plus or minus 30 degrees; for bolts installed by 2/3 turn and more, the tolerance should be plus or minus 45 degrees.

[b]Applicable only to connections in which all material within the grip of the bolt is steel.

[c]No research has been performed by the Council to establish the turn-of-nut procedure for bolt lengths exceeding 12 diameters. Therefore, the required rotation must be determined by actual test in a suitable tension measuring device which simulates conditions of solidly fitted steel.

When calibrated wrenches are used to install and tension bolts in a connection, bolts shall be installed with hardened washers under the element turned in tightening bolts in all holes of the connection and brought to a snug tight condition. Following this initial tightening operation, the connection shall be tightened using the calibrated wrench. Tightening shall progress systematically from the most rigid part of the joint to its free edges. The wrench shall be returned to "touch up" previously tightened bolts which may have been relaxed as a result of the subsequent tightening of adjacent bolts until all bolts are tightened to the prescribed amount.

(3) Installation of Alternate Design Bolts. When fasteners which incorporate a design feature intended to indirectly indicate the bolt tension or to automatically provide the tension required by Table 4 and which have been qualified under Section 2(d) are to be installed, a representative sample of not less than three bolts of each diameter, length and grade shall be checked at the job site in a device capable of indicating bolt tension. The test assembly shall include flat hardened washers, if required in the actual connection, arranged as in the actual connections to be tensioned. The calibration test shall demonstrate that each bolt develops a tension not less than five percent greater than the tension required by Table 4. Manufacturer's installation procedure as required by Section 2(d) shall be followed for installation of bolts in the calibration device and in all connections.

When alternate design features of the fasteners involve an irreversible mechanism such as yield or twist-off of an element, bolts shall be installed in all holes of the connection and initially brought to a snug tight condition. All fasteners shall then be tightened, progressing systematically from the most rigid part of the connection to the free edges in a manner that will minimize relaxation of previously tightened fasteners prior to final twist-off or yielding of the control or indicator element of the individual fasteners. In some cases, proper tensioning of the bolts may require more than a single cycle of systematic tightening.

(4) Direct Tension Indicator Tightening. Tightening of bolts using direct tension indicator devices is permitted provided the suitability of the device can be demonstrated by testing a representative sample of not less than three devices for each diameter and grade of fastener in a calibration device capable of indicating bolt tension. The test assembly shall include flat hardened washers, if required in the actual connection, arranged as those in the actual connections to be tensioned. The calibration test shall demonstrate that the device indicates a tension not less than five percent greater than that required by Table 4. Manufacturer's installation procedure as required by Section 2(d) shall be followed for installation of bolts in the calibration device and in all connections. Special attention shall be given to proper installation of flat hardened washers when load indicating devices are used with bolts installed in oversize or slotted holes and when the load indicating devices are used under the turned element.

When the direct tension indicator involves an irreversible mechanism such as yielding or fracture of an element, bolts shall be installed in all holes of the connection and brought to snug tight condition. All fasteners shall then be tightened, progressing systematically from the most rigid part of the connection to the free edges in a manner that will minimize relaxation of previously tightened fasteners prior to final twist-off or yielding of the control or indicator element of the individual devices. In some cases, proper tensioning of the bolts may require more than a single cycle of systematic tightening.

(e) **Reuse of Bolts.** A490 bolts and galvanized A325 bolts shall not be reused. Other A325 bolts may be reused if approved by the Engineer responsible. Touching up or retightening previously tightened bolts which may have been loosened by the tightening of adjacent bolts shall not be considered as reuse provided the snugging up continues from the initial position and does not require greater rotation, including the tolerance, than that required by Table 5.

9. Inspection

(a) **Inspector Responsibility.** While the work is in progress, the Inspector shall determine that the requirements of Sections 2, 3 and 8 of this Specification are met in the work. The Inspector shall observe the calibration procedures when such procedures are required by contract documents and shall monitor the installation of bolts to determine that all plies of connected material have been drawn together and that the selected procedure is properly used to tighten all bolts.

In addition to the requirement of the foregoing paragraph, for all connections specified to be slip critical or subject to axial tension, the Inspector shall assure that the specified procedure was followed to achieve the pretension specified in Table 4. Bolts installed by procedures in Section 8(d) may reach tensions substantially greater than values given in Table 4, but this shall not be cause for rejection.

Bolts in connections identified as not being slip-critical nor subject to direct tension need not be inspected for bolt tension other than to ensure that the plies of the connected elements have been brought into snug contact.

(b) **Arbitration Inspection.** When high strength bolts in slip-critical connections and connections subject to direct tension have been installed by any of the tightening methods in Section 8(d) and inspected in accordance with Section 9(a) and a disagreement exists as to the minimum tension of the installed bolts, the following arbitration procedure may be used. Other methods for arbitration inspection may be used if approved by the engineer.

(1) The Inspector shall use a manual torque wrench which indicates torque by means of a dial or which may be adjusted to give an indication that the job inspecting torque has been reached.

(2) This Specifcation does not recognize standard torques determined from tables or from formulas which are assumed to relate torque to tension. Testing using such standard torques shall not be considered valid.

(3) A representative sample of five bolts from the diameter, length and grade of the bolts used in the work shall be tightened in the tension measuring device by any convenient means to an initial condition equal to approximately 15 percent of the required fastener tension and then to the minimum tension specified in Table 4. Tightening beyond the initial condition must not produce greater nut rotation than 1 1/2 times that permitted in Table 5. The job inspecting torque shall be taken as the average of three values thus determined after rejecting the high and low values. The inspecting wrench shall then be applied to the tightened bolts in the work and the torque necessary to turn the nut or head 5 degrees (approximately 1 inch at 12 inch radius) in the tightening direction shall be determined.

(4) Bolts represented by the sample in the foregoing paragraph which have been tightened in the structure shall be inspected by applying, in the tightening direction, the inspecting wrench and its job torque to 10 percent of the bolts, but not less than 2 bolts, selected at random in each connection in question. If no nut or bolt head is turned by application of the job inspecting torque, the connection shall be accepted as properly tightened. If any nut or bolt is turned by the application of the job inspecting torque, all bolts in the connection shall be tested, and all bolts whose nut or head is turned by the job inspecting torque shall be tightened and reinspected. Alternatively, the fabricator or erector, at his option, may retighten all of the bolts in the connection and then resubmit the connection for the specified inspection.

(c) **Delayed Verification Inspection.** The procedure specified in Sections 9(a) and (b) are intended for inspection of bolted connections and verification of pretension at the time of tensioning the joint. If verification of bolt tension is required after a passage of a period of time and exposure of the completed joints, the procedures of Section 9(b) will provide indication of bolt tension which is of questionable accuracy. Procedures appropriate to the specific situation should be used for verification of bolt tension. This might involve use of the arbitration inspection procedure contained herein, or might require the development and use of alternate procedures.

See Commentary.

APPENDIX A

Testing Method To Determine the Slip Coefficient for Coatings Used in Bolted Joints

Reprinted from *Engineering Journal*
American Institute of Steel Construction, Third Quarter, 1985.

JOSEPH A. YURA and KARL H. FRANK

In 1975, the Steel Structures Painting Council (SSPC) contacted the Research Council In Riveted and Bolted Structural Joints (RCRBSJ), now the Research Council on Structural Connections (RCSC), regarding the difficulties and costs which steel fabricators encounter with restrictions on coatings of contact surfaces for friction-type structural joints. The SSPC also expressed the need for a "standardized test which can be conducted by any certified testing agency at the initiative and expense of any interested party, including the paint manufacturer."And finally, the RCSC was requested to "prepare and promulgate a specification for the conduct of such a standard test for slip coefficients."

The following Testing Method is the answer of Research Council on Structural Connections to the SSPC request. The test method was developed by Professors Joseph A. Yura and Karl H. Frank of The University of Texas at Austin under a grant from the Federal Highway Administration. The Testing Method was approved by the RCSC on June 14, 1984.

1.0 GENERAL PROVISIONS

1.1 Purpose and Scope

The purpose of the testing procedure is to determine the slip coefficient of a coating for use in high-strength bolted connections. The testing specification ensures that the creep deformation of the coating due to both the clamping force of the bolt and the service load joint shear are such that the coating will provide satisfactory performance under sustained loading.

Joseph A. Yura, M. ASCE, is Warren S. Bellows Centennial Professor in Civil Engineering, University of Texas at Austin, Austin, Texas.

Karl H. Frank, A.M. ASCE, is Associate Professor, Department of Civil Engineering, University of Texas at Austin, Austin, Texas.

1.2 Definition of Essential Variables

Essential variables mean those variables which, if changed, will require retesting of the coating to determine its slip coefficient. The essential variables are given below. The relationship of these variables to the limitation of application of the coating for structural joints is also given.

The *time interval* between application of the coating and the time of testing is an essential variable. The time interval must be recorded in hours and any special curing procedures detailed. Curing according to published manufacturer's recommendations would not be considered a special curing procedure. The coatings are qualified for use in structural connections which are assembled after coating for a time equal to or greater than the interval used in the test specimens. Special curing conditions used in the test specimens will also apply to the use of the coating in the structural connections.

The *coating thickness* is an essential variable. The maximum average coating thickness allowed on the bolted structure will be the average thickness, rounded to the nearest whole mil, of the coating used on the creep test specimens minus 2 mils.

The *composition of the coating,* including the thinners used, and its method of manufacture are essential variables. Any change will require retesting of the coating.

1.3 Retesting

A coating which fails to meet the creep or the post-creep slip test requirements given in Sect. 4 may be retested in accordance with methods in Sect. 4 at a lower slip coefficient, without repeating the static short-term tests specified in Sect. 3. Essential variables must remain unchanged in the retest.

2.0 TEST PLATES AND COATING OF THE SPECIMENS

2.1 Test Plates

The test specimen plates for the short-term static tests are shown in Fig. 1. The plates are 4×4 in. (102×102 mm) plates, ⁵/₈-in. (16 mm) thick, with a 1-in. (25 mm) dia. hole drilled 1¹/₂ in. ± ¹/₁₆ in. (38 mm ± 1.6 mm) from one edge. The specimen plates for the creep specimen are shown in Fig. 2. The plates are 4×7 in. (102×178 mm), ⁵/₈-in. (16 mm) thick, with two 1-in. (25 mm) holes, 1¹/₂ in. ± ¹/₁₆ in. (38 mm ± 1.6 mm) from each end. The edges of the plates may be milled, as rolled or saw cut.

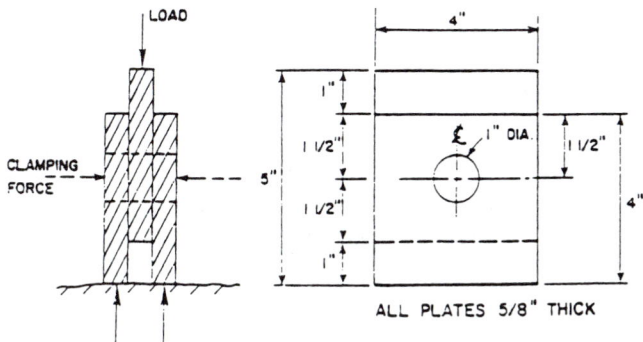

Fig. 1.   Compression test specimen

Flame cut edges are not permitted. The plates should be flat enough to ensure they will be in reasonably full contact over the faying surface. Any burrs, lips or rough edges should be filed or milled flat. The arrangement of the specimen plates for the testing is shown in Figs. 2 and 3. The plates are to be fabricated from a steel with a minimum yield strength between 36 to 50 ksi (250 to 350 MPa).

If specimens with more than one bolt are desired, the contact surface per bolt should be 4×3 in. (102×76.5 mm) as shown for the single bolt specimen in Fig. 1.

2.2 Specimen Coating

The coatings are to be applied to the specimens in a manner consistent with the actual intended structural application. The method of applying the coating and the surface preparation should be given in the test report. The specimens are to be coated to an average thickness 2 mils (0.05 mm) greater than average thickness to be used in the structure. The thickness of the total coating and the primer, if used, shall be measured on the contact surface of the specimens. The thickness should be measured in accordance with the Steel Structures Painting Council specification SSPC-PA2, Measurement of Dry Paint Thickness with Magnetic Gages.[1] Two spot readings (six gage readings) should be made for each contact surface. The overall average thickness from the three plates comprising a specimen is the average thickness for the specimen. This value should be reported for each specimen. The average coating thickness of the three creep specimens will be calculated and reported. The average thickness of the creep specimen minus two mils rounded to the nearest whole mil is the maximum average thickness of the coating to be used in the faying surface of a structure.

The time between painting and specimen assembly is to be the same for all specimens within ± 4 hours. The average time is to be calculated and reported. The two coating applications required in Sect. 3 are to use the same equipment and procedures.

3.0 SLIP TESTS

The methods and procedures described herein are used to determine experimentally the slip coefficient (sometimes called the coefficient of friction) under short-term static loading for high-strength bolted connections. The slip coefficient will be determined by testing two sets of five specimens. The two sets are to be coated at different times at least one week apart.

3.1 Compression Test Setup

The test setup shown in Fig. 3 has two major loading components, one to apply a clamping force to the specimen plates and another to apply a compressive load to the specimen so that the load is transferred across the faying surfaces by friction.

Clamping Force System. The clamping force system consists of a 7/8-in. (22 mm) dia. threaded rod which passes through the specimen and a centerhole compression ram. A 2H nut is used at both ends of the rod, and a hardened washer is used at each side of the test specimen. Between the ram and the specimen is a specially fabricated 7/8-in. (22 mm) 2H nut in which the threads have been drilled out so that it will slide with little resistance along the rod. When oil is pumped into the centerhole ram, the piston rod extends, thus forcing the special nut against one of the outside plates of the specimen. This action puts tension in the threaded rod and applies a clamping force to

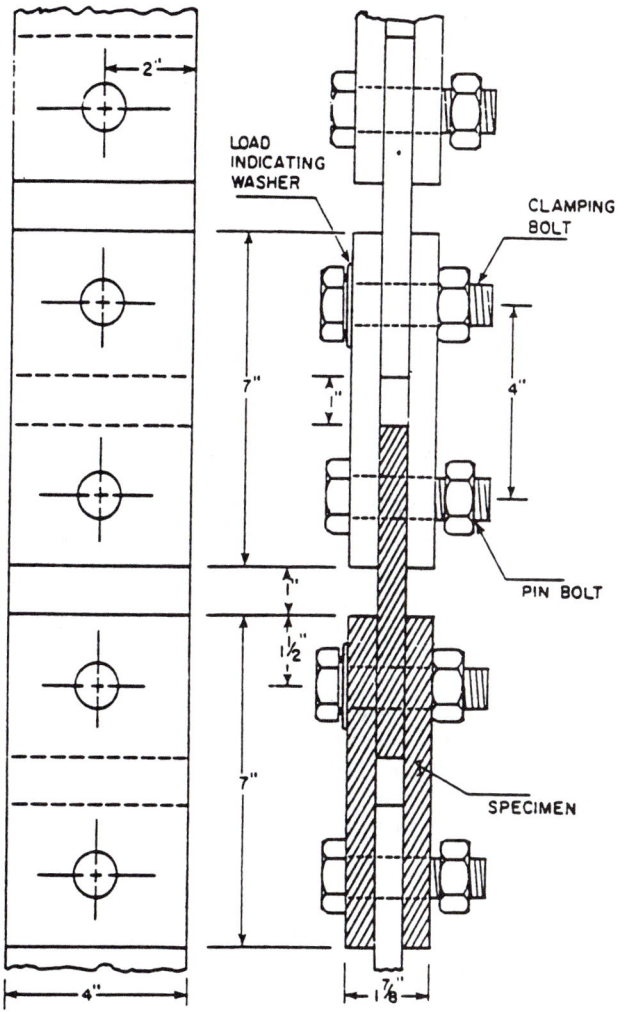

Fig. 2.   Creep test specimens

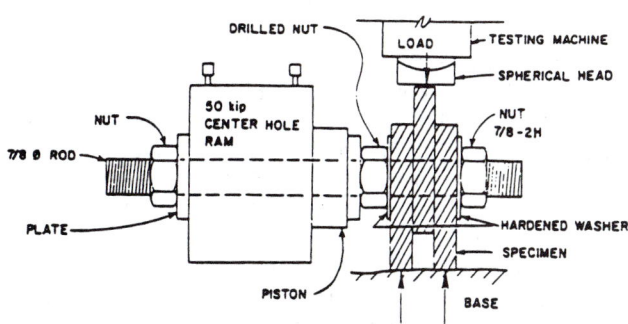

Fig. 3.   Test setup

the specimen which simulates the effect of a tightened bolt. If the diameter of the centerhole ram is greater than 1 in. (25 mm), additional plate washers will be necessary at the ends of the ram. The clamping force system must have a capability to apply a load of at least 49 kips (219 kN) and maintain this load during the test with an accuracy of ±1%.

Compressive Load System. A compressive load is applied to the specimen until slip occurs. This compressive load can be applied by a compression test machine or compression ram. The machine, ram and the necessary supporting elements should be able to support a force of 90 kips (400 kN).

The compression loading system should have an accuracy of 1.0% of the slip load.

3.2 Instrumentation

Clamping Force. The clamping force must be measured within 0.5 kips (2.2 kN). This may be accomplished by measuring the pressure in the calibrated ram or placing a load cell in series with the ram.

Compression Load. The compression load must be measured during the test. This may be accomplished by direct reading from a compression testing machine, a load cell in series with the specimen and the compression loading device, or pressure readings on a calibrated compression ram.

Slip Deformation. The relative displacement of the center plate and the two outside plates must be measured. This displacement, called slip for simplicity, should be the average which occurs at the centerline of the specimen. This can be accomplished by using the average of two gages placed on the two exposed edges of the specimen or by monitoring the movement of the loading head relative to the base. If the latter method is used, due regard must be taken for any slack that may be present in the loading system prior to application of the load. Deflections can be measured by dial gages or any other calibrated device which has an accuracy of 0.001 in. (0.20 mm).

3.3 Test Procedure

The specimen is installed in the test setup as shown in Fig. 3. Before the hydraulic clamping force is applied, the individual plates should be positioned so that they are in, or are close to, bearing contact with the 7/8-in. (22 mm) threaded rod in a direction opposite to the planned compressive loading to ensure obvious slip deformation. Care should be taken in positioning the two outside plates so that the specimen will be straight and both plates are in contact with the base.

After the plates are positioned, the centerhole ram is engaged to produce a clamping force of 49 kips (219 kN). The applied clamping force should be maintained within ±0.5 kips (2.2 kN) during the test until slip occurs.

The spherical head of the compression loading machine should be brought in contact with the center plate of the specimen after the clamping force is applied. The spherical head or other appropriate device ensures uniform contact along the edge of the plate, thus eliminating eccentric loading. When 1 kip (4.45 kN) or less of compressive load is applied, the slip gages should be engaged or attached. The purpose of engaging the deflection gage(s), after a slight load is applied, is to eliminate initial specimen settling deformation from the slip readings.

When the slip gages are in place, the compression load is applied at a rate not exceeding 25 kips (109 kN) per minute, or 0.003 in. (0.07 mm) of slip displacement per minute until the slip load is reached. The test should be terminated when a slip of

0.05 in. (1.3 mm) or greater is recorded. The load-slip relationship should preferably be monitored continuously on an *X-Y* plotter throughout the test, but in lieu of continuous data, sufficient load-slip data must be recorded to evaluate the slip load defined below.

3.4 Slip Load

Typical load-slip response is shown in Fig. 4. Three types of curves are usually observed and the slip load associated with each type is defined as follows:

Curve (a). Slip load is the maximum load, provided this maximum occurs before a slip of 0.02 in. (0.5 mm) is recorded.

Curve (b). Slip load is the load at which the slip rate increases suddenly.

Curve (c). Slip load is the load corresponding to a deformation of 0.02 in. (0.5 mm). This definition applies when the load vs. slip curves show a gradual change in response.

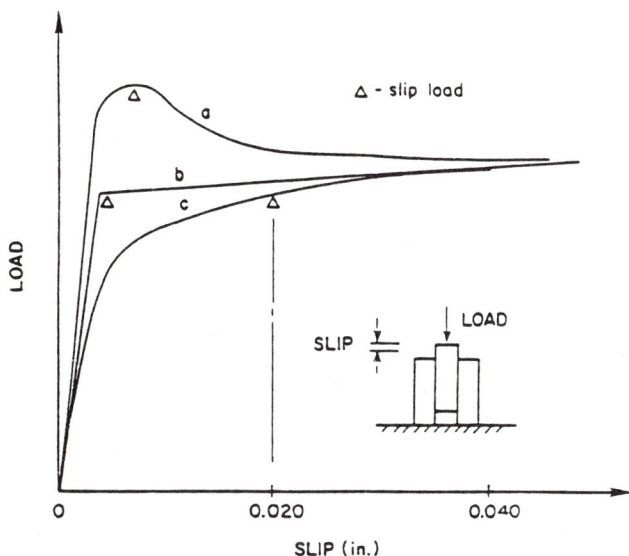

Fig. 4. Definition of slip load

3.5 Coefficient of Slip

The slip coefficient k_s for an individual specimen is calculated as follows:

$$k_s = \frac{\text{slip load}}{2 \times \text{clamping force}}$$

The mean slip coefficient for both sets of five specimens must be compared. If the two means differ by more than 25%, using the smaller mean as the base, a third five-specimen set must be tested. The mean and standard deviation of the data from all specimens tested define the slip coefficient of the coating.

3.6 Alternate Test Methods

Other test meethods to determine slip may be used provided the accuracy of load measurement and clamping satisfies the conditions presented in the previous sections. For example, the slip load may be determined from a tension-type test setup rather than the compression-type as long as the contact surface area per fastener of the test specimen is the same as shown in Fig. 1. The clamping force of at least 49 kips (219 kN) may be applied by any means provided the force can be established within ±1%. Strain-gaged bolts can usually provide the desired accuracy. However, bolts installed by turn-of-nut method, tension indicating fasteners and load indicator washers usually show too much variation to be used in the slip test.

4.0 TENSION CREEP TESTS

The test method outlined is intended to ensure the coating will not undergo significant creep deformation under service loading. The test also determines the loss in clamping force in the fastener due to the compression or creep of the paint. Three replicate specimens are to be tested.

4.1 Test Setup

Tension-type specimens, as shown in Fig. 2, are to be used. The replicate specimens are to be linked togther in a single chain-like arrangement, using loose pin bolts, so the same load is applied to all specimens. The specimens shall be assembled so the specimen plates are bearing against the bolt in a direction opposite to the applied tension loading. Care should be taken in the assembly of the specimens to ensure the centerline of the holes used to accept the pin bolts is in line with the bolts used to assemble the joint. The load level, specified in Sect. 4.2, shall be maintained constant within ±1% by springs, load maintainers, servo controllers, dead weights or other suitable equipment. The bolts used to clamp the specimens together shall be $^7/_8$-in. (22 mm) dia. A490 bolts. All bolts should come from the same lot.

 The clamping force in the bolts should be a minimum of 49 kips (219 kN). The clamping force is to be determined by calibrating the bolt force with bolt elongation, if standard bolts are used. Special fasteners which control the clamping force by other means such as bolt torque or strain gages may be used. A minimum of three bolt calibrations must be performed using the technique selected for bolt force determination. The average of the three-bolt calibration is to be calculated and reported. The method of measuring bolt force must ensure the clamping force is within ±2 kips (9 kN) of the average value.

 The relative slip between the outside plates and the center plates shall be measured to an accuracy of 0.001 in. (0.02 mm). This is to be measured on both sides of each specimen.

4.2 Test Procedure

The load to be placed on the creep specimens is the service load permitted for $^7/_8$-in. A490 bolts in slip-critical connections by the latest edition of the *Specification for Structural Joints Using ASTM A325 or A490 Bolts*[2] for the particular slip coefficient category under consideration. The load is to be placed on the specimen and held for 1,000 hours. The creep deformation of a specimen is calculated using the average reading of the two dispacements on each side of the specimen. The difference between the average after 1,000 hours and the initial average reading taken within one-half hour after loading the specimens is defined as the creep deformation of the specimen. This value is to be reported for each specimen. If the creep deformation of any specimen exceeds 0.005 in. (0.12 mm), the coating has failed the test for the slip coefficient used. The coating may be retested using new specimens in accordance with this section at a load corresponding to a lower value of slip coefficient.

If the value of creep deformation is less than 0.005 in. (0.12 mm) for all specimens, the specimens are to be loaded in tension to a load calculated as

$$P_u = \text{average clamping force} \times \text{design slip coefficient} \times 2$$

since there are two slip planes. The average slip deformation which occurs at this load must be less than 0.015 in. (0.38 mm) for the three specimens. If the deformation is greater than this value, the coating is considered to have failed to meet the requirements for the particular slip coefficient used. The value of deformation for each specimen is to be reported.

COMMENTARY

The slip coefficient under short-term static loading has been found to be independent of clamping force, paint thickness and hole diameter.[3] The slip coefficient can be easily determined using the hydraulic bolt test setup included in this specification. The sip load measured in this setup yields the slip coefficient directly since the clamping force is controlled. The slip coefficient k_s is given by

$$k_s = \frac{\text{slip load}}{2 \times \text{clamping force}}$$

The resulting slip coefficient has been found to correlate with both tension and compression tests of bolted specimens. However, tests of bolted specimens revealed that the clamping force may not be constant but decreases with time due to the compressive creep of the coating on the faying surfaces and under the nut and bolt head. The reduction of the clamping force can be considerable for joints with high clamping force and thick coatings, as much as a 20% loss. This reduction in clamping force causes a corresponding reduction in the slip load. The resulting reduction in slip load must be considered in the procedure used to determine the design allowable slip loads for the coating.

The loss in clamping force is a characteristic of the coating. Consequently, it cannot be accounted for by an increase in the factor of safety or a reduction in the clamping force used for design without unduly penalizing coatings which do not exhibit this behavior.

The creep deformation of the bolted joint under the applied shear loading is also an important characteristic and a function of the coating applied. Thicker coatings tend to creep more than thinner coatings. Rate of creep deformation increases as the applied load approaches the slip load. Extensive testing has shown the rate of creep is not constant with time, rather it decreases with time. After 1,000 hours of loading, the additional creep deformation is negligible.

The proposed test methods are designed to provide the necessary information to evaluate the suitability of a coating for slip critical bolted connections and to determine the slip coefficient to be used in the design of the connections. The initial testing of the compression specimens provides a measure of the scatter of the slip coefficient. In order to get better statistical information, a third set of specimens must be tested whenever the means of the initial two sets differ by more than 25%.

The creep tests are designed to measure the paint's creep behavior under the service loads determined by the paint's slip coefficient based on the compression test results. The slip test conducted at the conclusion of the creep test is to ensure the loss of clamping force in the bolt does not reduce the slip load below that associated with the design slip coefficient. A490 bolts are specified, since the loss of clamping force is larger for these bolts than A325 bolts. Qualifying of the paint for use in a structure at an averge thickness of 2 mils less than the test specimen is to ensure that a casual buildup of paint due to overspray, etc., does not jeopardize the coating's performance.

The use of 1-in. (25 mm) holes in the specimens is to ensure that adequate clearance is available for slip. Fabrication tolerances, coating buildup on the holes and assembly tolerances reduce the apparent clearances.

REFERENCES

1. *Steel Structures Painting Council* Steel Structures Painting Manual *Vols. 1 and 2. Pittsburgh, Pa., 1982.*
2. *Research Council on Structural Connections* Specification for Structural Joints Using ASTM A325 or A490 Bolts *American Institute of Steel Construction., Inc., Chicago, Ill., November 1985.*
3. *Frank, K.H., and J. A. Yura* An Experimental Study of Bolted Shear Connections *FHWA/RD-81-148, Federal Highway Administration, Washington, D.C., December 1981.*

---

*Presently identified as A502, Grade 1.

Commentary on Specifications for Structural Joints Using ASTM A325 or A490 Bolts

November 13, 1985

PREFACE

This Commentary to *Specifications for Structural Joints Using ASTM A325 or A490 Bolts* is equally applicable to the *Allowable Stress Design* version or the *Load and Resistance Factor Design* version. It is provided as an aid to the user of either Specification. By providing background information, references to source information and discussion relative to questions raised over the years by users of earlier versions of the Council Specifications, it is intended to provide a record of the reasoning behind the requirements and understanding of the intent of the Specification provisions.

Historical Notes

When first approved by the Research Council on Structural Connections of the Engineering Foundation, January 1951, the *Specification for Assembly of Structural Joints Using High-Strength Bolts* merely permitted the substitution of a like number of A325 high-strength bolts for hot driven ASTM A141* steel rivets of the same nominal diameter. It was required that all contact surfaces be free of paint. As revised in 1954, the omission of paint was required to apply only to "joints subject to stress reversal, impact or vibration, or to cases where stress redistribution due to joint slippage would be undesirable." This relaxation of the earlier provision recognized the fact that, in a great many cases, movement of the connected parts that brings the bolts into bearing against the sides of their holes is in no way detrimental.

In the first edition of the Specification published in 1951, a table of torque to tension relationships for bolts of various diameters was included. It was soon demonstrated in research that a variation in the torque to tension relationship of as high as plus or minus 40 percent must be anticipated unless the relationship is established individually for each bolt lot, diameter and fastener condition. Hence, by the 1954 edition of the Specification, recognition of standard torque to tension relationships in the form of tabulated values or formulas was withdrawn. Recognition of the calibrated wrench method of tightening was retained however until 1980, but with the requirement that the torque required for installation or inspection be determined specifically for the bolts being installed on a daily basis. Recognition of the method was withdrawn in 1980 because of continuing controversy resulting from failure of users to adhere to the detailed requirements for valid use of the method both during installation and inspection. With this version of the Specification, the calibrated wrench method has been reinstated, but with more detailed requirements which should be carefully followed.

The increasing use of high-strength steels created the need for bolts substantially stronger than A325, in order to resist the much greater forces they support without resort to very large connections. To meet this need, a new ASTM specification, A490, was developed. When provisions for the use of these bolts were included in this Specification in 1964, it was required that they be tightened to their specified proof load, as was required for the installation of A325 bolts. However, the ratio of proof load to specified minimum tensile strength is approximately 0.7 for A325 bolts, whereas it is 0.8 for A490 bolts. Calibration studies have shown that high-strength bolts have ultimate load capacities in torqued tension which vary from about 80 to 90 percent of the pure-tension tensile strength.[1] Hence, if minimum strength A490 bolts were supplied and they experienced the maximum reduction due to torque required to induce the tension, there is a possibility that these bolts could not be tightened to proof load by any method of installation. Also, statistical studies have shown that tightening to the 0.8 times tensile strength under calibrated wrench control may result in some "twist-off" bolt failures during installation or in some cases a slight amount of under tightening.[2] Therefore, the required installed tension for A490 bolts was reduced to 70 percent of the specified minimum tensile strength. For consistency, but with only minor change, the initial tension required for A325 bolts was also set at 70 percent of their specified minimum tensile strength and at the same time the values for minimum required pretension were rounded off to the nearest kip.

C1 Scope

This Specification deals only with two types of high-strength bolts, namely, ASTM A325 and A490, and to their installation in structural steel joints. The provisions may not be relied upon for high-strength fasteners of other chemical composition or mechanical properties or size. The provisions do not apply to ASTM A325 or A490 fasteners when material other than steel is included in the grip. The provisions do not apply to high-strength anchor bolts.

The Specification relates only to the performance of fasteners in structural steel connections and those few aspects of the connected material that affect the performance of the fasteners in connections. Many other aspects of connection design and fabrication are of equal importance and must not be overlooked. For information on questions of design of connected material, not covered herein, the user is directed to standard textbooks on design of structural steel and also to "Fisher, J.W. and J.H.A. Struik," Guide to Design Criteria for Bolted and Riveted Joints, John Wiley & Sons, New York, 1974. (Hereinafter referred to as the Guide.)

C2 Bolts, Nuts, Washers and Paint

Complete familiarity with the referenced ASTM Specification requirements is necessary for the proper application of this Specification. Discussion of referenced specifications in this Commentary is limited to only a few frequently overlooked or little understood items.

In this Specification a single style of fastener (heavy hex structural bolts with heavy hex nuts), available in two strength grades (A325 and A490) is specified as a principal style, but conditions for acceptance of other types of fasteners are provided.

Bolt Specifications ASTM A325 and A490 bolts are manufactured to dimensions specified in ANSI Standard B18.2.1 for Heavy Hex Structural Bolts. The basic dimensions as defined in Figure C1 are shown in Table C1. The principal geometric features of heavy hex structural bolts that distinguish them from bolts for general application are the size of the head and the body length. The head of the heavy hex

---

[1]Christopher, R.J., G. L. Kulak, and J. W. Fisher, "Calibration of Alloy Steel Bolts," *Journal of the Structural Division*, ASCE, Vol. 92, No. ST2, Proc. Paper 4768, April, 1966, pp. 19–40.

[2]Gill, P.J., "Specifications of Minimum Preloads for Structural Bolts," Memorandum 30, G. K. N. Group Research Laboratory, England, 1966 (unpublished report).

structural bolt is specified to be the same size as a heavy hex nut of the same nominal diameter in order that the ironworker may use a single size wrench or socket on both the head and the nut. Heavy hex structural bolts have shorter thread length than bolts for general application. By making the body length of the bolt the control dimension, it has been possible to exclude the thread from all shear planes, except in the case of thin outside parts adjacent to the nut. Depending upon the amount of bolt length added to adjust for incremental stock lengths, the full thread may extend into the grip by as much as $3/8$ inch for $1/2$ in., $5/8$ in., $3/4$ in., $7/8$ in., $1 1/4$ in., and $1 1/2$ in. diameter bolts and as much as $1/2$ inch for 1 in., $1 1/8$ in. and $1 3/8$ in. diameter bolts. Inclusion of some thread run-out in the plane of shear is permissible. Of equal or even greater importance is exercise of care to provide sufficient thread for nut tightening to keep the nut threads from jamming into the thread run-out. When the thickness of an outside part is less than the amount the threads may extend into the grip tabulated above, it may be necessary to call for the next increment of bolt length together with sufficient flat washers to insure full tightening of the nut without jamming nut threads on the thread run-out.

Table C1

Nominal Bolt Size, Inches D	Bolt Dimensions, Inches Heavy Hex Structural Bolts			Nut Dimensions, Inches Heavy Hex nuts	
	Width across flats, F	Height, H	Thread length	Width across flats, W	Height, H
$1/2$	$7/8$	$5/16$	1	$7/8$	$31/64$
$5/8$	$1 1/16$	$25/64$	$1 1/4$	$1 1/16$	$39/64$
$3/4$	$1 1/4$	$15/32$	$1 3/8$	$1 1/4$	$47/64$
$7/8$	$1 7/16$	$35/64$	$1 1/2$	$1 7/16$	$55/64$
1	$1 5/8$	$39/64$	$1 3/4$	$1 5/8$	$63/64$
$1 1/8$	$1 13/16$	$11/16$	2	$1 13/16$	$1 7/64$
$1 1/4$	2	$25/32$	2	2	$1 7/32$
$1 3/8$	$2 3/16$	$27/32$	$2 1/4$	$2 3/16$	$1 11/32$
$1/2$	$2 3/8$	$15/16$	$2 1/4$	$2 3/8$	$1 15/32$

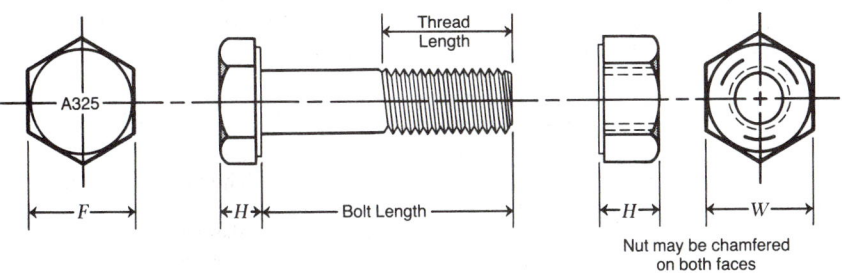

Nut may be chamfered on both faces

Fig. C1. Heavy hex structural bolt and heavy hex nut

There is an exception to the short thread length requirements for ASTM A325 bolts discussed in the foregoing. Beginning with ASTM A325-83, supplementary requirements have been added to the ASTM A325 Specification which permit the purchaser, when the bolt length is equal to or shorter than four times the nominal diameter, to specify that the bolt be threaded for the full length of the shank. This exception to the requirements for thread length of heavy hex structural bolts was provided in the Specification in order to increase economy through simplified ordering and inventory control in the fabrication and erection of structures using relatively thin materials where strength of the connection is not dependent upon shear strength of the bolt, whether threads are in the shear plane or not. The Specification requires that bolts ordered to such supplementary requirements be marked with the symbol A325T.

In order to determine the required bolt length, the value shown in Table C2 should be added to the grip (i.e., the total thickness of all connected material, exclusive of washers). For each hardened flat washer that is used, add $5/32$ inch, and for each beveled washer add $5/16$ inch. The tabulated values provide appropriate allowances for manufacturing tolerances, and also provide for full thread engagement* with an installed heavy hex nut. The length determined by the use of Table C2 should be adjusted to the next longer $1/4$ inch length.

Table C2

Nominal Bolt Size, Inches	To Determine Required Bolt Length, Add to Grip, in Inches
$1/2$	$11/16$
$5/8$	$7/8$
$3/4$	1
$7/8$	$1 1/8$
1	$1 1/4$
$1 1/8$	$1 1/2$
$1 1/4$	$1 5/8$
$1 3/8$	$1 3/4$
$1 1/2$	$1 7/8$

ASTM A325 and ASTM A490 currently provide for three types (according to metallurgical classification) of high-strength structural bolts, supplied in sizes $1/2$ inch to $1 1/2$ inch inclusive except for A490 Type 2 bolts which are available in diameters from $1/2$ inch to 1 inch inclusive:

Type 1. Medium carbon steel for A325 bolts, alloy steel for A490 bolts.

Type 2. Low carbon martensitic steel for both A325 and A490 bolts.

Type 3. Bolts having improved atmospheric corrosion resistance and weathering characteristics for both A325 and A490 bolts.

---

*Defined as: Having the end of the bolt at least flush with the face of the nut.

When the bolt type is not specified, either Type 1, Type 2 or Type 3 may be supplied at the option of the manufacturer. Special attention is called to the requirement in ASTM A325 that where elevated temperature applications are involved, Type 1 bolts shall be specified by the purchaser. This is because the chemistry of Type 2 bolts permits heat treatment at sufficiently low temperatures that subsequent heating to elevated temperatures may affect the mechanical properties.

Heavy Hex Nuts.   Heavy hex nuts for use with A325 bolts may be manufactured to the requirements of ASTM A194 for grades 2 or 2H or the requirements of ASTM A563 for grades C, C3, D, DH or DH3 except that nuts to be galvanized for use with galvanized bolts must be hardened nuts meeting the requirements for 2H, DH or DH3.

The heavy hex nuts for use with A490 bolts may be manufactured to the requirements of ASTM A194 for grade 2H or the requirements of ASTM A563 for grade DH or DH3.

Galvanized High-strength Bolts.   Galvanized high-strength bolts and nuts must be considered as a manufactured matched assembly; hence, comments relative to them have not been included in the foregoing paragraphs where bolts and nuts were considered separately. Insofar as the galvanized bolt and nut assembly, per se, is concerned, four principal factors need be discussed in order that the provisions of the Specification may be understood and properly applied. They are (1) the effect of the galvanizing process on the mechanical properties of high-strength steels, (2) the effect of galvanized coatings on the nut stripping strength, (3) the effect of galvanizing upon the torque involved in the tightening operation and (4) shipping requirements.

Effect of Galvanizing on the Strength of Steels.   Steels in the 200 ksi and higher tensile strength range are subject to embrittlement if hydrogen is permitted to remain in the steel and the steel is subjected to high tensile stress. The minimum tensile strength of A325 bolts is 105 or 120 ksi, depending upon the size, comfortably below the critical range. The required minimum tensile strength for A490 bolts was set at 170 ksi in order to provide a little more than a ten percent margin below 200 ksi; however, because manufacturers must target their production slightly higher than the required minimum, A490 bolts close to the critical range of tensile strength must be anticipated. For black bolts this is not a cause for concern, but, if the bolt is galvanized, a hazard of delayed brittle fracture in service exists because of the real possibility of introduction of hydrogen into the steel during the pickling operation of the galvanizing process and the subsequent "sealing-in" of the hydrogen by the zinc coating. ASTM specifications provide for the galvanizing of A325 bolts but not A490 bolts. Galvanizing of A490 bolts is not permitted.

The heat treatment temperatures for Type 2 ASTM A325 bolts is in the range of the molten zinc temperatures for hot-dip galvanizing; therefore there is a potential for diminishing the heat treated mechanical properties of Type 2 A325 bolts by the galvanizing process. For this reason, the Specifications require that such fasteners be tension tested after galvanizing to check the mechanical properties. Special attention should be given to specifying only Type 1 bolts for hot-dip galvanizing or to assuring that this requirement has not been overlooked if galvanized A325 bolts with Type 2 head markings are supplied. Because it is recommended that A490 bolts not be hot-dip galvanized, a similar requirement is not part of the ASTM Specification for Type 2 A490 bolts.

Nut Stripping Strength. Hot-dip galvanizing affects the stripping strength of the nut-bolt assembly primarily because to accommodate the relatively thick zinc coatings on bolt threads, it is usual practice to galvanize the blank nut and then to tap the nut oversize after galvanizing. This overtapping results in a reduction in the amount of engagement between the steel portions of the male and female threads with a consequent approximately 25 percent reduction in the stripping strength. Only the stronger hardened nuts have adequate strength to meet specification requirements even with the reduction due to overtapping; therefore, ASTM A325 specifies that only Grades DH and 2H be used for galvanized nuts. This requirement should not be overlooked if non-galvanized nuts are purchased and then sent to a local galvanizer for hot-dip galvanizing.

Effect of Galvanizing upon Torque Involved in Tightening. Research[3] has shown that, as galvanized, hot-dip galvanizing both increases the friction between the bolt and nut threads and also makes the torque induced tension much more variable. Lower torque and more consistent results are provided if the nuts are lubricated; thus, ASTM A325 requires that a galvanized bolt and a tapped oversize lubricated galvanized nut intended to be used with the bolt shall be assembled in a steel joint with a galvanized washer and tested in accordance with ASTM A536 by the manufacturer prior to shipment to assure that the galvanized nut with the lubricant provided may be rotated from the snug tight condition well in excess of the rotation required for full tensioning of the bolts without stripping.

Shipping Requirements for Galvanized Bolts and Nuts. The above requirements clearly indicate that galvanized bolts and nuts are to be treated as an assembly and shipped together. Purchase of galvanized bolts and galvanized nuts from separate sources is not recommended because the amount of over tapping appropriate for the bolt and the testing and application of lubricant would cease to be under the control of a single supplier and the responsibility for proper performance of the nut/bolt assembly would become obscure. Because some of the lubricants used to meet the requirements of ASTM Specification are water soluble, it is advisable that galvanized bolts and nuts be shipped and stored in plastic bags in wood or metal containers.

Washers. The primary function of washers is to provide a hardened non-galling surface under the element turned in tightening for those installation procedures which depend upon torque for control. Circular hardened washers meeting the requirements of ASTM F436 provide an increase in bearing area of 45 to 55 percent over the area provided by a heavy hex bolt head or nut; however, tests have shown that standard thickness washers play only a minor role in distributing the pressure induced by the bolt pretension, except where oversize or short slotted holes are used. Hence, consideration is given to this function only in the case of oversize and short slotted holes. The requirement for standard thickness hardened washers, when such washers are specified as an aid in the distribution of pressure, is waived for alternate design fasteners which incorporate a bearing surface under the head of the same diameter as the hardened washer; however, the requirements for hardened washers to satisfy the principal requirement of providing a non-galling surface under the element turned in tightening is not waived. The maximum thickness is the same for all standard washers up to and including 1 1/2 inch bolt diameter in order that washers may be produced from a single stock of material.

---

[3]Birkemoe, P. C., and D. C. Herrschaft, "Bolted Galvanized Bridges—Engineering Acceptance Near," *Civil Engineering*, ASCE, April 1970.

The requirement that heat-treated washers not less than 5/16 inch thick be used to cover oversize and slotted holes in external plies, when A490 bolts of 1 1/8 inch or larger diameter are used, was found necessary to distribute the high clamping pressure so as to prevent collapse of the hole perimeter and enable development of the desired clamping force. Preliminary investigation has shown that a similar but less severe deformation occurs when oversize or slotted holes are in the interior plies. The reduction in clamping force may be offset by "keying" which tends to increase the resistance to slip. These effects are accentuated in joints of thin plies.

Marking. Heavy hex structural bolts and heavy hex nuts are required by ASTM Specifications to be distinctively marked. Certain markings are mandatory. In addition to the mandatory markings the manufacturer may apply other distinguishing markings. The mandatory and optional markings are shown in Figure C2.

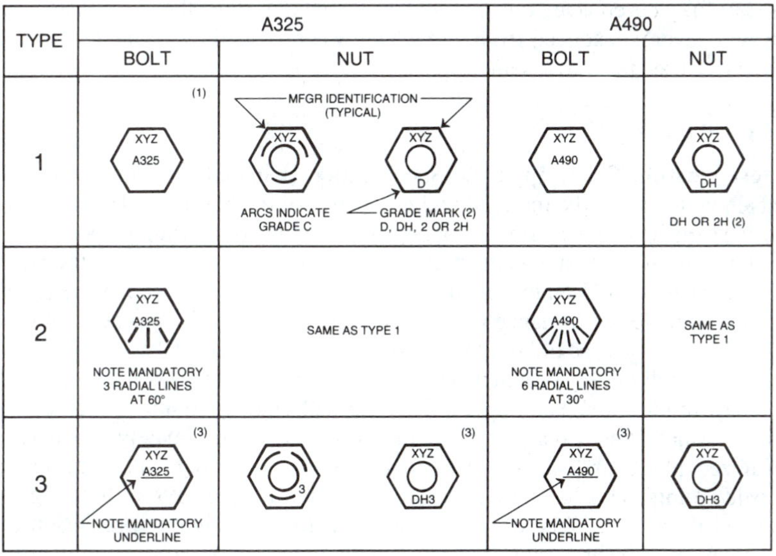

Fig. C2.   *Required marking for acceptable bolt and nut assemblies*

(1) ADDITIONAL OPTIONAL 3 RADIAL LINES AT 120° MAY BE ADDED.
(2) TYPE 3 ALSO ACCEPTABLE.
(3) ADDITIONAL OPTIONAL MARK INDICATING WEATHERING GRADE MAY BE ADDED.

Paint. In the previous edition of the Specification, generic names for paints applied to faying surfaces was the basis for categories of allowable working stresses in "friction" type connections. Research[4] completed since the adoption of the 1980 Specification has demonstrated that the slip coefficients for paints described by a generic type are not single values but depend also upon the type vehicle used. Small differences in formulation from manufacturer to manufacturer or from lot to lot with a single manufacturer, if certain essential variables within a generic type were changed, significantly affected slip coefficients; hence it is unrealistic to assign paints to categories with relatively small incremental differences between categories based solely upon a generic description. As a result of the research, a test method was developed

---

[4]Frank, Karl H. and J. A. Yura, "An Experimental Study of Bolted Shear Connections," FHWA/RD-81/148, Dec. 1981.

and adopted by the Council titled "Test Method to Determine the Slip Coefficient for Coatings Used in Bolted Joints." A copy of this document is appended to this Specification as Appendix A. The method, which requires requalification if an essential variable is changed, is the sole basis for qualification of any paint to be used under this Specification. Further, normally only 2 categories of slip coefficient for paints to be used in slip critical joints are recognized, Class A for coatings which do not reduce the slip coefficient below that provided by clean mill scale, and Class B for paints which do not reduce the slip coefficient below that of blast-cleaned steel surfaces.

The research cited in the preceding paragraph also investigated the effect of varying the time from coating the faying surfaces to assembly of the connection and tightening the bolts. The purpose was to ascertain if partially cured paint continued to cure within the assembled joint over a period of time. It was learned that all curing ceased at the time the joint was assembled and tightened and that paint coatings that were not fully cured acted much as a lubricant would; thus, the slip resistance of the joint was severely reduced from that which was provided by faying surfaces which were fully cured prior to assembly.

C3 Bolted Parts

Material within the Grip. The Specification is intended to apply to structural joints in which all of the material within the grip of the bolt is steel because predictable and satisfactory performance of slip critical joints is dependent upon predictable and stable installed tension in the bolts. The Test Method to Determine the Slip Coefficient for Coatings Used in Bolted Joints includes long term creep test requirements to assure reliable performance for qualified paint coatings. However, it must be recognized that in the case of hot dip galvanized coatings, especially if the joint consists of many plies of thickly coated material, relaxation of bolt tension may be significant and may require retensioning of the bolts subsequent to the initial tightening. Research[5] has shown that a loss of pretension of approximately 6.5 percent for galvanized plates and bolts due to relaxation as compared with 2.5 percent for uncoated joints. This loss of bolt tension occurred in five days with negligible loss recorded thereafter. This loss can be allowed for in design or pretension may be brought back to the prescribed level by retightening the bolts after an initial period of "settling-in."

This Specification has permitted the use of bolt holes $1/16$ inch larger than the bolts installed in them since it was first published. Research[6] has shown that, where greater latitude is needed in meeting dimensional tolerances during erection, somewhat larger holes can be permitted for bolts $5/8$ inch diameter and larger without adversely affecting the performance of shear connections assembled with high-strength bolts. The oversize and slotted hole provisions of this Specification are based upon these findings. Because an increase in hole size generally reduces the net area of a connected part, the use of oversize holes is subject to approval by the Engineer.

---

[5]Munse, W. H., "Structural Behavior of Hot Galvanized Bolted Connections," 8th International Conference on Hot-dip Galvanizing," London, England, June 1967.

[6]Allen, R. N., and J. W. Fisher, "Bolted Joints With Oversize or Slotted Holes," *Journal of the Structural Division*, ASCE, Vol. 94, No. ST9, September 1968.

[7]Polyzois, D. and J. A. Yura, "Effect of Burrs on Bolted Friction Connections," AISC *Engineering Journal*, Vol. 22, No. 3, Third Quarter 1985.

Burrs.   Based upon tests,[7] which demonstrated that the slip resistance of joints was unchanged or slightly improved by the presence of burrs, burrs which do not prevent solid seating of the connected parts in the snug tight condition need not be removed. On the other hand, parallel tests in the same program demonstrated that large burrs can cause a small increase in the required turns from snug tight condition to achieve specified pretension with turn-of-nut method of tightening.

Unqualified Paint on Faying Surfaces.   An extension to the research on the slip resistance of shear connections cited in footnote 4 investigated the effect of ordinary paint coatings on limited portions of the contact area within joints and the effect of overspray over the total contact area. The tests demonstrated that the effective area for transfer of shear by friction between contact surfaces was concentrated in an annular ring around and close to the bolts. Paint on the contact surfaces approximately one inch but not less then the bolt diameter away from the edge of the hole did not reduce the slip resistance. Because in connections of thick material involving a number of bolts on multiple gage lines, bolt pretension might not be adequate to completely flatten and pull thick material into tight contact around every bolt, the Specification requires that all areas between bolts also be free of paint. See Figure C3. The new requirements have a potential for increased economy because the paint free area may easily be protected using masking tape located relative to the hole pattern, and further, the narrow paint strip around the perimeter of the faying surface will minimize uncoated material outside the connection requiring field touch up.

This research also investigated the effect of various degrees of inadvertent overspray on slip resistance. It was found that even the smallest amount of overspray of ordinary paint (that is, not qualified as Class A) within the specified paint free area on clean mill scale reduced the slip resistance significantly. On blast cleaned surfaces, the presence of a small amount of overspray was not as detrimental. For simplicity, the Specification prohibits any overspray from areas required to be free of paint in slip-critical joints regardless of whether the surface is clean mill scale or blast cleaned.

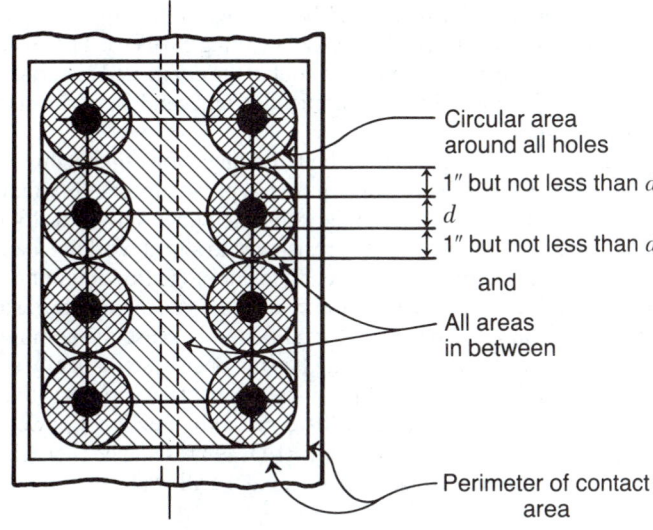

Circular area
around all holes

1″ but not less than d

d

1″ but not less than d

and

All areas
in between

Perimeter of contact
area

Areas outside the defined area need not
be free of paint.

Figure C3

Galvanized Faying Surfaces. The slip factor for initial slip with clean hot-dip galvanized surfaces is of the order of 0.19 as compared with a factor of about 0.35 for clean mill scale. However, research[3] has shown that the slip factor of galvanized surfaces is significantly improved by treatments such as hand wire brushing or light "brush-off" grit blasting. In either case, the treatment must be controlled in order to achieve the necessary roughening or scoring. Power wire brushing is unsatisfactory because it tends to polish rather than roughen the surface.

Field experience and test results have indicated that galvanized members may have a tendency to continue to slip under sustained loading.[8] Tests of hot-dip galvanized joints subject to sustained loading show a creep type behavior. Treatments to the galvanized faying surfaces prior to assembly of the joint which caused an increase in the slip resistance under short duration loads did not significantly improve the slip behavior under sustained loading.

C4 Design for Strength of Bolted Connections

Background for Design Stresses. With this edition of the Specification, the arbitrary designations "friction type" and "bearing type" connections used in former editions, and which were frequently misinterpreted as implying an actual difference in the manner of performance or strength of the two types of connection, are discontinued in order to focus attention more upon the real manner of performance of bolted connections.

In bolted connections subject to shear type loading, the load is transferred between the connected parts by friction up to a certain level of force which is dependent upon the total clamping force on the faying surfaces and the coefficient of friction of the faying surfaces. The connectors are not subject to shear nor is the connected material subject to bearing stress. As loading is increased to a level in excess of the frictional resistance between the faying surfaces, slip occurs, but failure in the sense of rupture does not occur. As even higher levels of load are applied, the load is resisted by shear upon the fastener and bearing upon the connected material plus some uncertain amount of friction between the faying surfaces. The final failure will be by shear failure of the connectors or tear out of the connected material or unacceptable ovalization of the holes. Final failure load is independent of the clamping force provided by the bolts.[9]

Thus, the design of high-strength bolted connections under this Specification begins with consideration of strength required to prevent premature failure by shear of the connectors or bearing failure of the connected material. Next, for connections which are defined as "slip-critical" the resistance to slip at working load is checked. Because high clamping force with high coefficient of friction might mathematically exceed the ultimate shear or bearing of the fasteners, even though the fasteners would not be subject to shear or bearing prior to slip, and because the combined effect of frictional resistance with shear or bearing has not been systematically studied and is uncertain, the allowable force for slip critical connections conservatively must not exceed the lesser of the allowable loads determined by Section 4 or Section 5 of the allowable stress design specification. For LRFD, slip resistance is checked at service load. The resistance at service load is identical to the allowable stress for ASD.

---

[8]Fisher, J. W. and J. H. A. Struik, *Guide to Design Criteria for Bolted and Riveted Joints*, John Wiley & Sons, New York, 1974, Pg. 205 and 206. (Hereinafter referred to as the Guide.)

[9]Ibid., pp. 49–52.

Connection Slip. There are practical cases in the design of structures where slip of the connection is desirable in order to permit rotation in a joint or to minimize the transfer of moment. Additionally there are cases where, because of the number of fasteners in a joint, the probability of slip is extremely small or where, if slip did occur, it would not be detrimental to the serviceability of the structure. In order to provide for such cases while at the same time making use of the higher shear strength of high-strength bolts, as contrasted to ASTM A307 bolts, the Specification now permits joints tightened only to the snug tight condition.

The maximum amount of slip that can occur in connections that are not classified as slip-critical under the Specification rules is limited theoretically to $^1/_{16}$ inch. In most practical cases, however, the real magnitude of slip would probably be much less because the acceptable inaccuracies in the location of holes within a pattern of bolts would usually cause one or more bolts to be in bearing in the initial unloaded condition. Further, in statically loaded structures, even with perfectly positioned holes, the usual method of erection would cause the weight of the connected elements to put the bolts into direct bearing at the time the member is supported on loose bolts and the lifting crane is unhooked. Subsequent additional gravity loading could not cause additional connection slip.

Connections classified as slip-critical include those cases where slip could theoretically exceed $^1/_{16}$ inch, and thus, possibly affect the suitability for service of the structure by excessive distortion or reduction in strength or stability even though the resistance to fracture of the connection, per se, may be adequate. Also included are those cases where slip of any magnitude must be prevented, for example, joints subject to fatigue loading.

Shear and Bearing on Fasteners. Several interrelated parameters influence the shear and bearing strength of connections. These include such geometric parameters as the net-to-gross-area ratio of the connected parts, the ratio of the net area of the connected parts to the total shear-resisting area of the fasteners, and the ratio of transverse fastener spacing to fastener diameter and the ratio of transverse fastener spacing to connected part thickness. In addition, the ratio of yield strength to tensile strength of the steel comprising the connected parts, as well as the total distance between extreme fasteners, measured parallel to the line of direct tensile force, play a part.

In the past, a balanced design concept had been sought in developing criteria for mechanically fastened joints to resist shear between connected parts by means of bearing of the fasteners against the sides of the holes. This philosophy resulted in wide variations in the factor of safety for the fasteners, because the ratio of yield to tensile strength increases significantly with increasingly stronger grades of steel. It had no application at all in the case of very long joints used to transfer direct tension, because the end fasteners "unbutton" before the plate can attain its full strength or before the interior fasteners can be loaded to their rated shear capacity.

By means of a mathematical model it was possible to study the interrelationship of the previously mentioned parameters.[10,11] It has been shown that the factor of safety against shear failure ranged from 3.3 for compact (short) joints to approximately 2.0 for joints with an overall length in excess of 50 inches. It is of interest to

---

[10]Fisher, J.W., and L. S. Beedle, "Analysis of Bolted Butt Joints," *Journal of the Structural Division,* ASCE, Vol. 91, No. ST5, October 1965.

[11]Guide, pp. 84–107 and 123–127.

note that the longest (and often the most important) joints had the lowest factor, indicating that a factor of safety of 2.0 has proven satisfactory in service.

The absence of any working stress or design strength provisions for the case where a bolt in double shear has a nonthreaded shank in one shear plane and a threaded section in the other shear plane recognizes that knowledge as to the bolt placement (which might leave both shear planes in the threaded section) is not ordinarily available to the detailer.

The allowable working stresses and corresponding design strengths for fasteners subject to applied tension or shear given in Table 2 are unchanged from the 1980 edition of the Specification. The values are based upon the research and recommendations reported in the Guide. With the wealth of data available, it was possible, through statistical analyses, to adjust allowable working stresses to provide uniform reliability for all loading and joint types. The design of connections is more conservative than that of the connected members of buildings and bridges by a substantial margin, in the sense that the failure load of the fasteners is substantially in excess of the maximum serviceability limit (yield) of the connected material.

Design for Tension.   The allowable working stresses and design strengths specified for applied tension[12] are intended to apply to the external bolt load plus any tension resulting from prying action produced by deformation of the connected parts. When stressed in tension to the recommended working value (approximately equal to two-thirds of the initial tightening force), high-strength bolts tensioned to the requirements of Table 4 will experience little if any actual change in stress. For this reason, bolts in connections in which the applied loads subject the bolts to axial tension are required to be fully tensioned even though the connection may not be subject to fatigue loading nor classified as slip-critical.

Properly tightened A325 and A490 bolts are not adversely affected by repeated application of the recommended working tensile stress, provided the fitting material is sufficiently stiff, so that the prying force is a relatively small part of the applied tension.[13] The provisions covering bolt tensile fatigue are based upon study of test reports of bolts that were subjected to repeated tensile load to failure.

Design for Shear.   The strength in shear is based upon the assumption of a ratio of shear strength to tensile strength of 0.6.[14] In the allowable stress design specification, the allowable shear is based upon a factor of safety of approximately 2.35. Load and Resistance Factor Design uses nominal shear strength with a resistance factor of 0.65 or 0.75, depending on the position of the shear plane relative to the bolt threads to establish the design shear strength.

Design for Bearing   Bearing stress produced by a high-strength bolt pressing against the side of the hole in a connected part is important only as an index to behavior of the connected part. It is of no significance to the bolt. The critical value can be derived from the case of a single bolt at the end of a tension member.

---

[12]Ibid., pp. 257–276.

[13]Ibid., pg. 266.

[14]Ibid., pp. 46–53.

It has been shown,[15] using finger-tight bolts, that a connected plate will not fail by tearing through the free edge of the material if the distance L, measured parallel to the line of applied force from a single bolt to the free edge of the member toward which the force is directed, is not less than the diameter of the bolt multiplied by the ratio of the bearing stress to the tensile strength of the connected part.

Providing a factor of safety of 2.0, the working stress design criterion given in 4(b) is

$$L/d \geq 2(F_p/F_u).$$

where

F_p = allowable working stress in bearing.
F_u = specified minimum tensile strength of the connected part.

When using factored loads, the load and resistance factor criterion given in 4(b) is

$$L/d \geq R_n/F_u$$

where

R_n = nominal bearing pressure.
F_u = specified minimum tensile strength of the connected part.

As a practical consideration, a lower limit of 1.5 is placed on the ratio L/d and an upper limit of 1.5 on the ratio F_p/F_u and an upper limit of 3.0 on the ratio R_n/F_u.

The foregoing leads to the rules governing bearing strength in both versions of the specification. The permitted bearing pressure in the 1980 Specification and the current provisions are fully justifiable from the standpoint of strength of the connected material. However, recent tests have demonstrated that severe ovalization of the hole will begin to develop, even though rupture does not occur, as bearing stress is increased beyond the previously permitted stess, especially if it is combined with high-tensile stress on the net section. Thus, the rules for bearing strength have been revised to provide increased conservatism, unless special consideration is given to the effect of possible hole ovalization.

For connections with more than a single bolt in the direction of force, the resistance may be taken as the sum of the resistances of the individual bolts.

C5 Design Check for Slip Resistance

The Specification recognizes that, for a significant number of cases, slip of the joint would be undesirable or must be precluded. Such joints are termed "slip-critical" joints. This is somewhat different from the previous term "friction type" connection because it recognizes that all tensioned high-strength bolted joints resist load by friction between faying surfaces up to the slip load and subsequently are able to resist even greater loads by shear and bearing. The Specification requires that, in addition to assuring that the connection has adequate strength, the design of slip-critical joints be checked to assure that slip will not occur at working load.

It must be recognized that the formula for P_s in 5(b) is for connections subject to a linear load. For cases in which the load tends to rotate the connection in the plane of the faying surface, a modified formula accounting for the placement of bolts relative to the center of rotation should be used.

The safety index for serviceability considerations has traditionally been less than that required for strength considerations. In the consideration of the consequences of slip of bolts into bearing, a single criterion cannot apply. In the case of bolts in holes

---

[15]Ibid., pg. 137.

affording only small clearance as in standard holes, oversize holes, slotted holes loaded transverse to the axis of the slot and short slotted holes loaded parallel to the axis of the slot, the consequences of slip are trivial except for fatigue applications. In the case of slip critical connections in which load is applied to bolts parallel to the axis of a long slot in which they are installed, it is conceivable that failure by slip could lead to critical geometric distortions of the frame, endangering the strength of the structure, even though danger of strength failure of the connection does not exist.

Extensive test data developed through research sponsored by the Council and others have made possible a statistical analysis of the slip probability of connections tensioned to the requirements of Table 4. The frequency distribution and mean value of clamping force for bolts tightened by the calibrated wrench method[16] is due to variation in the torque-tension ratio from bolt to bolt, the tolerance on wrench performance, hole type and human error. Both the variation in the slip coefficient (or degree of surface roughness) and the frequency distribution of the magnitude of clamping force provided by A325 and A490 bolts in the connection were considered.[17]

In the 1978 edition of this Specification, nine classes of faying surface conditions were introduced and significant increases were made in the recommended working stresses for proportioning connections which function by transfer of shear between connected parts by friction. These classes and the stresses were adopted on the basis of statistical evaluation of the information then available.

The allowable stresses or nominal resistances for bolts in standard holes in Table 3 of these Specifications were developed for a 10 percent probability of slip considering only faying surface treatment and bolt clamping force. This is analogous to the product of the resistance factor and the nominal strength (ϕR) used in reliability concepts.[18] The bolt clamping force was based upon *calibrated wrench* method of installation using the clamping force variation shown in Figure 5.8 of the Guide. An examination of the slip coefficient for a wide range of surface conditions, including additional data developed during the past ten years,[4] indicate that the variability for each surface class was about the same ($\sigma = 0.007$ to 0.09). Rather than providing a separate class for each of the individual surface conditions, the surface conditions with approximately the same mean values were grouped into three classes, with mean slip coefficients of 0.33, 0.40 and 0.50. The revised edition of the Guide is scheduled for printing in 1986. Table 5.5 of the revised edition provides the equivalent shear stress values for mean slip coefficients between 0.2 and 0.6, with calibrated wrench installation. Mean values for turn-of-nut method of installation are provided in Table 5.4.

Because of the effects of oversize and slotted holes on the induced tension in bolts using any of the specified installation methods, lower loads are provided in Table 3 for bolts in these hole types. In the case of bolts in long slotted holes, even though the slip load is the same for bolts loaded transverse or parallel to the axis of the slot, the load for bolts loaded parallel to the axis has been further reduced in recognition of the greater consequences of slip.

---

[16]Guide, pg. 82, Fig. 5.8.

[17]The probability distribution function of the product of two independent random variables can be determined using standard statistical techniques as outlined, for example, in "Introductory Probability and Statistical Applications," by P. L. Meyer, Addison Publishing Company, 1965.

[18]Fisher, J. W., etal "Load and Resistance Design Criteria for Connectors," *Journal of the Structural Division*, ASCE, Vol. 104, No. ST9, September 1978.

The frequency distribution and mean value of clamping force for bolts tightened by turn-of-nut method are higher, due to the elimination of variables which affect torque-tension ratios and to higher-than-specified minimum strength of production bolts. Because properly applied turn-of-nut installation induces yield point strain in the bolt, the higher-than-specified yield strength of production bolts will be mobilized and result in higher clamping force by the method. On the other hand, with the calibrated wrench method, which is dependent upon the calibration of wrenches in a tension indicating device independent of the actual bolt properties, any additional strength of production bolts will not be mobilized. High clamping force might be achieved by the calibrated wrench method if the wrench was set to a higher torque value. However, this would require more attention to the degrees of rotation (because the limitation on the maximum rotation of the nut specified in the second paragraph of 8(d)(2) to prevent excessive deformation of the bolt would control more often), otherwise a torsional bolt failure might result. Because of the increased clamping force, connections having bolts installed by turn-of-nut method provide a greater resistance to slip (lower probability of slip).

Connections of the type shown in Figure C4(a), in which some of the bolts (A) lose a part of their clamping force due to applied tension, suffer no overall loss of frictional resistance. The bolt tension produced by the moment is coupled with a compensating compressive force (C) on the other side of the axis of bending. In a connection of the type shown in Figure C4(b), however, all fasteners (B) receive applied tension which reduces the initial compression force at the contact surface. If slip under load cannot be tolerated, the design slip load value of the bolts in shear should be reduced in proportion to the ratio of residual axial force to initial tension. If slip of the joint can be tolerated, the bolt shear stress should be reduced according to the tension-shear interaction as outlined in the Guide page 69. Because the bolts are subject to applied axial tension, they are required to be pretensioned in either case.

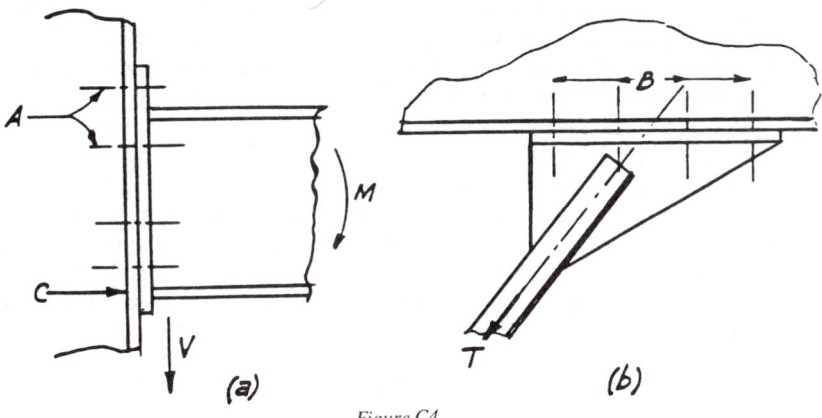

Figure C4

While connections with bolts pretensioned to the levels specified in Table 4 do not ordinarily slip into bearing against the sides of the hole when subjected to the allowable loads of Table 3, it is required that they meet the requirements for allowable stress in Table 2 in order to maintain the factor of safety of 2 against fracture in the event that the bolts do slip into bearing as a result of large unforeseen loads.

To cover those cases where a coefficient of friction less than 0.33 might be adequate for a given situation the Specification provides that, subject to the approval of the Engineer, and provided the mean slip coefficient is determined by the specified test procedure allowable slip loads or nominal resistances less than those provided by Class A faying surface coating may be used.

It should be noted that both Class A and Class B coatings are required to be applied to blast cleaned steel.

High-strength Bolts in Combination with Weld or Rivets. For high-strength bolts in combination with welds in statically loaded conditions, the allowable load or nominal strength may be taken as the sum of two contributions.[19] One results from the slip resistance of the bolted parts and may be determined in accordance with Section 5(b). The second results from the resistance of the welds and may be determined on the basis of the allowable stresses for welds given in the applicable specifications.

For high-strength bolts in combination with welds in fatigue loaded applications, data available are not sufficient to develop general design recommendations at this time. High-strength bolts in combination with rivets are rarely encountered in modern practice. If need arises, guidance may be found in the Guide.

C7 Design Details of Bolted Connections

A new section has been added with this edition of the Specification in order to bring together a number of requirements for proper design and detailing of high-strength bolted connections. The material covered in the Specification, and Section 7 in particular, is not intended to provide comprehensive coverage of the design of high-strength bolted connections. For example, other design considerations of importance to the satisfactory performance of the connected material such as block shear, shear lag, prying action, connection stiffness, effect on the performance of the structure and others are beyond the scope of this Specification and Commentary.

Proper location of hardened washers is as important as other elements of a detail to the peformance of the fasteners. Drawings and details should clearly reflect the number and disposition of washers, especially the thick hardened washers that are required for several slotted hole applications. Location of washers is a design consideration which should not be left to the experience of the iron worker.

Finger shims are a necessary device or tool of the trade to permit adjusting alignment and plumbing of structures. When these devices are fully and properly inserted, they do not have the same effect on bolt tension relaxation or the connection performance as do long slotted holes in an outer ply which is the basis for allowable design stresses. When fully inserted, the shim provides support around approximately 75 percent of the perimeter of the bolt in contrast to the greatly reduced area that exists with a bolt centered in a long slot. Further, finger shims would always be enclosed on either side by the connected material which would be fully effective in bridging the space between the fingers.

C8 Installation and Tightening

Several methods for installation and tensioning of high-strength bolts, when tensioning is required, are provided without preference in the Specification. Each method recognized in Section 8, when properly used as specified, may be relied upon to provide satisfactory results. All methods may be misused or abused.

At the expense of redundancy, the provisions stipulating the manner in which each method is intended to be used is set forth in complete detail in order that the rules

---

[19]Guide, pp. 238 to 240.

for each method may stand alone without need for footnotes or reference to other sections. If the methods are conscientiously implemented, good results should be routinely achieved.

Connections not Requiring Full Tensioning. In the Commentary, Section C6 of the previous edition of the Specification it was pointed out that "bearing" type connections need not be tested to assure that the specified pretension in the bolts had been provided, but specific provision permitting relaxation of the tensioning requirement was not contained in the body of the Specification. In this edition of the Specification, separate installation procedures are provided for bolts that are not within the slip-critical or direct tension category. The intent in making this change is to improve the quality of bolted steel construction and reduce the frequency of costly controversies by focusing attention, both during the installation and tensioning phase and during inspection, on the true slip-critical connections rather than diluting the effort through the requirement for costly tensioning and tension testing of the great many connections where such effort serves no useful purpose. The requirement for identification of connections on the drawings may be satisfied either by identifying the slip-critical and direct tension connections which must be fully tightened and inspected or by identifying the connections which need be tightened only to the snug tight condition.

In the Specification, snug tight is defined as the tightness that exists when all plies are in firm contact. This may usually be attained by a few impacts of an impact wrench or the full effort of a man using an ordinary spud wrench. In actuality, snug tight is a degree of tightness which will vary from joint to joint depending upon the thickness and degree of parallelism of the connected material. In most joints the plies will pull together; however, in some joints, it may not be possible at snug tight to have contact throughout the faying surface area.

Tension Calibrating Devices. At the present time, there is no device or economical means for determining the actual tension in a bolt that is installed in a connection. However, the actual tension in a bolt installed in a tension calibrator (hydraulic tension indicating device) is directly indicated by the dial of the device. Thus, such a device is an economical and valuable tool that should be readily available whenever high-strength bolts in slip critical joints or bolts subject to applied axial tension are to be tensioned to the pretension specified in Table 4. Although each element of a fastener assembly may conform to the minimum requirements of their separate ASTM specifications, their compatability in an assembly or the need for lubrication can only be assured by testing the assembly. Therefore, such devices are important for testing the complete fastener assembly as it will be used with any method of tightening to assure the suitability of bolts and nuts (probably produced by different manufacturers), as well as other elements, and the ability of the assembly to provide the specified tension using the selected method. Testing before starting to install fasteners in the work will also identify potential sources of problems, such as the need for lubrication, weakened nuts due to excessive over tapping in the case of galvanized nuts, or failure of bolts subject to combined torque and tension due to over strength load indicators, and to clarify, for the bolting crews and inspectors, the proper implementation of the selected installation method to be used. Such devices are essential to the specified procedure for the calibrated wrench method of installation and for specified procedure for arbitration inspection when such inspection is required.

Experience on many projects has shown that bolts and/or nuts not meeting the requirements of the applicable ASTM specification, but which were intended for installation by turn-of-nut method, would have been identified prior to installation; thus saving the controversy and great expense of replacing bolts installed in the structure when difficulties were discovered at a later date.

Hydraulic tension calibrating devices capable of indicating bolt tension undergo a slight deformation under load. Hence the nut rotation corresponding to a given tension reading may be somewhat larger than it would be if the same bolt were tightened against a solid steel abutment. Stated differently, the reading of the calibrating device tends to underestimate the tension which a given rotation of the turned element would induce in a bolt in an actual joint. This should be borne in mind when using such devices to establish a tension-rotation relationship.

Slip-critical Connections and Connections Subject to Direct Tension. Four methods for joint assembly and tightening are provided for slip-critical and direct tension connections. Regardless of the method used, it should be demonstrated prior to the commencement of work that the procedure to be used with the fasteners to be used and by the crews who will be doing the work that the specified pretension is achieved. For this reason, it is a requirement that a tension measuring device be provided at the job site.

With any of the four described tensioning methods, it is important to install bolts in all holes of the connection and bring them to an intermediate level of tension generally corresponding to snug tight in order to compact the joint. Even after being fully tightened, some thick parts with uneven surfaces may not be in contact over the entire faying surface. This is not detrimental to the performance of the joint. As long as the specified bolt tension is present in all bolts of the completed connection, the clamping force equal to the total of the tensions in all bolts will be transferred at the locations that are in contact and be fully effective in resisting slip through friction. If however, bolts are not installed in all holes and brought to an intermediate level of tension to compact the joint, bolts which are tightened first will be subsequently relaxed by the tightening of the adjacent bolts. Thus the total of the forces in all bolts will be reduced which will reduce the slip load whether there is uninterrupted contact between the surfaces or not.

With all methods, tightening should begin at the most rigidly fixed or stiffest point and progress toward the free edges, both in the initial snugging up and in the final tightening.

Turn-of-Nut-Tightening. When properly implemented, turn-of-nut method provides more uniform tension in the bolts than does torque controlled tensioning methods because it is primarily dependent upon bolt elongation slightly into the inelastic range.

Consistency and reliability is dependent upon assuring that the joint is well compacted and all bolts at a snug tight condition prior to application of the final required partial turn. Reliability is also dependent upon assuring that the turn that is applied is relative between the bolt and nut; thus the element not turned in tightening should be prevented from rotating while the required degree of turn is applied to the turned element. Reliability and inspectability of the method may be improved by having the outer face of the nut match-marked to the protruding end of the bolt after the joint has been snug tightened but prior to final tightening. Such marks may be applied by the wrench operator using a crayon or dab of paint. Such marks in their relatively displaced position after tightening will afford the inspector a means for noting the rotation that was applied.

Some problems with turn-of-nut tightening encountered with galvanized bolts have been attributed to especially effective lubricant applied by the manufacturer to meet ASTM Specification requirements. Job site tests in the tension indicating device demonstrated the lubricant reduced the coefficient of friction between the bolt and nut to the degree that "the full effort of a man using an ordinary spud wrench" to snug tighten the joint actually induced the full required tension. Because the nuts could be removed by an ordinary spud wrench they were erroneously judged improperly tightened by the inspector. Research[5] confirms that lubricated high-strength bolts may require only one-half as much torque to induce the specified tension. For such situations, use of a tension indicating device and the fasteners being installed may be helpful in establishing alternate criteria for snug tight at about one-half the tension required by Table 4.

Because reliability of the method is independent of the presence or absence of washers, washers are not required except for oversize and slotted holes in an outer ply. Thus, in the absence of washers, testing after the fact using a torque wrench method is highly unreliable. That is, the turn-of-nut method of installation, properly applied, is more reliable and consistent than the testing method. The best method for inspection of the method is for the inspector to observe any job site confirmation testing of the fasteners and the method to be used followed by monitoring of the work in progress to assure that the method is routinely properly applied.

Calibrated Wrench Method.   Research has demonstrated that scatter in induced tension is to be expected when torque is used as an indirect indicator of tension. Numerous variables, which are not related to tension, affect torque. For example, the finish and tolerance on bolt threads, the finish and tolerance on the nut threads, the fact that the bolt and nut may not be produced by the same manufacturer, the degree of lubrication, the job site conditions contributing to dust and dirt or corrosion on the threads, the friction that exists to varying degree between the turned element and the supporting surface, the variability of the air pressure on the torque wrenches due to length of air lines or number of wrenches operating from the same source, the condition and lubrication of the wrench which may change within a work shift and other factors all bear upon the relationship between torque and induced tension.

Recognition of the calibrated wrench method of tightening was removed from the Specification with the 1980 edition. This action was taken because it is the least reliable of all methods of installation and many costly controversies had occured. It is to be suspected that short cut procedures in the use of the calibrated wrench method of installation, not in accordance with the Specification provisions, were being used. Further, torque controlled inspection procedures based upon "standard" or calculated inspection torques rather than torques determined as required by the Specification were being routinely used. These incorrect procedures plus others had a compounding effect upon the uncertainty of the installed bolt tension, and were responsible for many of the controversies.

It is recognized, however, that if the calibrated wrench method is implemented without short cuts as intended by the Specification, that there will be a ninety percent assurance that the tensions specified in Table 4 will be equaled or exceeded. Because the Specification should not prohibit any method which will give acceptable results when used as specified, the calibrated wrench method of installation is reinstated in this edition of the Council Specification. However, to improve upon the previous situation, the 1985 version of the Specification has been modified to require better control. Wrenches must be calibrated daily. Hardened washers must be used. Fasteners must be protected from dirt and moisture at the job site. Additionally, to achieve reliable results attention should be given to the control, insofar as it is practical, of those controllable factors which contribute to variability. For example, bolts and nuts should be purchased from single sources, insofar as practical, to minimize the variability of the fit. Bolts and nuts should be adequately and uniformly lubricated. Water soluble lubricants should be avoided.

Installation of Alternate Design Fasteners. It is the policy of the Council to recognize only fasteners covered by ASTM Specifications, however, it cannot be denied that a general type of alternate design fastener, produced by several manufacturers, are used on a significant number of projects as permitted by Section 2(d). The bolts referenced involve a splined end extending beyond the threaded portion of the bolt which is gripped by a special design wrench chuck providing a means for turning the nut relative to the bolt until the splined end is sheared off. While such bolts are subject to many of the variables affecting torque mentioned in the preceding section, they are produced and shipped by the manufacturers as a nut-bolt assembly under good quality control which apparently minimizes some of the negative aspects of the torque controlled process.

While these alternate design fasteners have been demonstrated to consistently provide tension in the fastener meeting the requirements of Table 5 in controlled tests in tension indicating devices, it must be recognized that the fastener may be misused and provide results as unreliable as those with other methods. The requirements of this Specification and the installation requirements of the manufacturer's specification required by Section 2(d) must be adhered to.

As with other methods, a representative sample of the bolts to be used should be tested to assure that they do, in fact, when used in accordance with the manufacturer's instructions, provide tension as specified in Table 5. In actual joints, bolts must be installed in all holes of a connection and all fasteners tightened to an intermediate level of tension adequate to pull all material into contact. Only after this has been accomplished should the fasteners be fully tensioned in a systematic manner and the splined end sheared off. The sheared off splined end merely signifies that at some time the bolt has been subjected to a torque adequate to cause the shearing. If the fasteners are installed and tensioned in a single continuous operation, they will give a misleading indication to the inspector that the bolts are properly tightened. Therefore, the only way to inspect these fasteners with assurance is to observe the job site testing of the fasteners and installation procedure and then monitor the work while in progress to assure that the specified procedure is routinely followed.

Direct Tension Indicator Tightening. Proprietary load indicating devices, not yet covered by an ASTM Specification, but recognized under this Specification in Section 2(f) are being specified and used in a significant number of projects. The referenced device is a hardened washer incorporating several small formed arches which are designed to deform in a controlled manner when subjected to load. These load indicator washers are the single device known which is directly dependent upon the tension load in the bolt, rather than upon some indirect parameter, to indicate the tension in a bolt.

As with the alternate design load indicating bolts, load indicating washers are dependent upon the quality control of the producer and proper use in accordance with the manufacturer's installation procedures and these Specifications. Load indicator washers delivered for use in a specific application should be tested at the job site to demonstrate that they do, in fact, provide a proper indication of bolt tension, and that they are properly used by the bolting crews. Because the washers depend upon an irreversible mechanism (inelastic deformation of the formed arches) bolts together with the load indicator washer plus any other washers required by Specification should be installed in all holes of the connection and the bolts tightened to approximately one-half the specified tension. Only after this initial tightening operation should the bolts be fully tensioned in a systematic manner. If the bolts are installed and tensioned in a single continuous operation, the load indicator washers will give the inspector a misleading indication that bolts are properly tightened. Therefore, the only way to inspect fasteners with which load indicator washers are used with assurance is to observe the job site testing of the devices and installation procedure and then routinely monitor the work in progress to assure that the specified procedure is followed.

During installation care must be taken to assure that the indicator nubs are oriented to bear against the hardened bearing surface of the bolt head or against an extra hardened flat washer if used under the nut.

C9 Inspection

It is apparent from the commentary on installation procedures that the inspection procedures giving the best assurance that bolts are properly installed and tensioned is provided by inspector observation of the calibration testing of the fasteners using the selected installation procedure followed by monitoring of the work in progress to assure that the procedure which was demonstrated to provide the specified tension is routinely adhered to. When such a program is followed, no further evidence of proper bolt tension is required.

However, if testing for bolt tension using torque wrenches is conducted subsequent to the time the work of installation and tightening of bolts is performed, the test procedure is subject to all of the uncertainties of torque controlled calibrated wrench installation. Additionally, the absence of many of the controls necessary to minimize variablity of the torque to tension relationship, which are unnecessary for the other methods of bolt installation, such as, use of hardened washers, careful attention to lubrication and the uncertainty of the effect of passage of time and exposure in the installed condition all reduce the reliability of the arbitration inspection results. The fact that it may, of necessity, have to be based upon a job test torque determined by bolts only assumed to be representative of the bolts in the actual job or bolts removed from completed joints, in many cases, makes the test procedure less reliable than a properly implemented installation procedure it is used to verify. Other verification inspection procedures available at this time are more accurate but too costly and time consuming for all but the most critical structural applications. The arbitration inspection procedure contained in the Specification is provided, in spite of its limitations, as the most feasible available at this time.

SPECIFICATION FOR ALLOWABLE STRESS

DESIGN OF SINGLE-ANGLE MEMBERS

PREFACE

The intention of the AISC Specification is to cover the common everyday design criteria in routine design office usage. It is not feasible to also cover the many special and unique problems encountered within the full range of structural design practice. This separate Specification and Commentary addresses one such topic—single-angle members—to provide needed design guidance for this more complex structural shape under various load and support conditions.

The single-angle Allowable Stress Design criteria were developed through a consensus process by a balanced ad-hoc Committee on Single Angle Members:

Donald R. Sherman, Chairman
Hansraj G. Ashar
Wai-Fah Chen
Raymond D. Ciatto
Mohamed Elgaaly
Theodore V. Galambos
Nestor R. Iwankiw
Thomas G. Longlais
Leroy A. Lutz
William A. Milek
Raymond H. R. Tide

The assistance of the Structural Stability Research Council Task Group on Single Angles in the preparation and review of this document is acknowledged.

In addition, the full AISC Committee on Specifications has reviewed and endorsed this Specification.

The reader is cautioned that professional judgment must be exercised when data or recommendations in this Specification are applied. The publication of the material contained herein is not intended as a representation or warranty on the part of the American Institute of Steel Construction, Inc.—or any other person named herein—that this information is suitable for general or particular use, or freedom from infringement of any patent or patents. Anyone making use of this information assumes all liability arising from such use. The design of structures is within the scope of expertise of a competent licensed structural engineer, architect or other licensed professional for the application of principles to a particular structure.

1. SCOPE

This document contains allowable stress design criteria for hot-rolled, single-angle members with equal and unequal legs in tension, shear, compression, flexure and for combined stresses. It is intended to be compatible with, and a supplement to, the

1989 AISC Specification for Structural Steel Buildings—Allowable Stress Design (AISC ASD) and repeats some common criteria for ease of reference. For design purposes, several conservative simplifications and approximations were made in the Specification provisions for single-angles which can be refined through a more precise analysis.

The Specification for single-angle design supersedes any comparable but more general requirements of the AISC ASD. All other design, fabrication and erection provisions not directly covered by this document shall be in compliance with the AISC ASD. For design of slender, cold-formed steel angles, the current AISI *Specification for the Design of Cold-formed Steel Structural Members* is applicable.

2. TENSION

The allowable tension stress F_t shall not exceed $0.6F_y$ on the gross area A_g, nor $0.50F_u$ on the effective net area A_e.

   a. For members connected by bolting, the net area and effective net area shall be determined from AISC ASD Specification Sects. B1 to B3 inclusive.
   b. When the load is transmitted by longitudinal or a combination of longitudinal and transverse welds through just one leg of the angle, the effective net area shall be

$$A_e = 0.85A_g \qquad (2\text{-}1)$$

   c. When load is transmitted by transverse weld through just one leg of the angle, A_e is the area of the connected leg.

For members whose design is based on tensile force, the slenderness ratio L/r preferably should not exceed 300. Members which have been designed to perform as tension members in a structural system, but may experience some compression, need not satisfy the compression slenderness limits.

3. SHEAR

The allowable shear stress due to flexure and torsion shall be:

$$F_v = 0.4F_y \qquad (3\text{-}1)$$

4. COMPRESSION

The allowable compressive stress on the gross area of axially compressed members shall be:

when $Kl/r < C_c'$

$$F_a = \frac{Q\left[1 - \dfrac{(Kl/r)^2}{2C_c'^2}\right]F_y}{\dfrac{5}{3} + \dfrac{3}{8}\left(\dfrac{Kl/r}{C_c'}\right) - \dfrac{(Kl/r)^3}{8C_c'^3}} \qquad (4\text{-}1)$$

when $Kl/r > C_c'$

$$F_a = \frac{12\,\pi^2 E}{23\,(Kl/r)^2} \qquad (4\text{-}2)$$

where

Kl/r = largest effective slenderness ratio of any unbraced length as defined in AISC ASD Specification Sect. E1

$$C'_c = \sqrt{\frac{2\pi^2 E}{QF_y}}$$

The reduction factor Q shall be:

when $b/t \le 76/\sqrt{F_y}$

$$Q = 1 \qquad\qquad (4\text{-}3a)$$

when $76/\sqrt{F_y} < b/t < 155/\sqrt{F_y}$

$$Q = 1.340 - 0.00447 \, (b/t) \sqrt{F_y} \qquad\qquad (4\text{-}3b)$$

when $b/t \ge 155/\sqrt{F_y}$

$$Q = 15,500/[F_y \, (b/t)^2] \qquad\qquad (4\text{-}3c)$$

where

b = full width of the longest angle leg
t = thickness of angle

For short, thin or unequal leg angles, flexural-torsional buckling may produce a significant reduction in strength. In such cases, the allowable stress shall be determined by the previous equations substituting an equivalent slenderness ratio $(Kl/r)_{equiv}$ for Kl/r

$$(Kl/r)_{equiv} = \pi \sqrt{E/F_e} \qquad\qquad (4\text{-}4)$$

where F_e is the elastic buckling strength for the flexural-torsional mode.

For members whose design is based on compressive force, the largest effective slenderness ratio preferably should not exceed 200.

5. FLEXURE

The allowable bending stress limits of Sect. 5.1 shall be used as indicated in Sects. 5.2. and 5.3.

5.1. Allowable Bending Stress

The bending stress is limited to the minimum allowable value F_b determined from Sects. 5.1.1, 5.1.2 and 5.1.3, as applicable.

5.1.1. To prevent local buckling when the tip of an angle leg is in compression,

when $b/t \le 65/\sqrt{F_y}$:

$$F_b = 0.66F_y \qquad\qquad (5\text{-}1a)$$

when $65/\sqrt{F_y} < b/t \le 76/\sqrt{F_y}$:

$$F_b = 0.60F_y \qquad\qquad (5\text{-}1b)$$

when $b/t > 76/\sqrt{F_y}$:

$$F_b = 0.60Q \, F_y \qquad\qquad (5\text{-}1c)$$

where

b = full width of angle leg in compression

Q = stress reduction factor per Eq. (4-3a), (b) and (c)

An angle leg shall be considered to be in compression if the tip of the angle leg is in compression, in which case the calculated stress f_b at the tip of this leg is used.

5.1.2. For the tip of an angle leg in tension

$$F_b = 0.66 \, F_y \tag{5-2}$$

5.1.3. To prevent lateral-torsional buckling, the maximum compression stress shall not exceed:

when $F_{ob} \leq F_y$

$$F_b = [0.55 - 0.10 \, F_{ob}/F_y] \, F_{ob} \tag{5-3a}$$

when $F_{ob} > F_y$

$$F_b = [0.95 - 0.50 \, \sqrt{F_y/F_{ob}}] \, F_y \leq 0.66F_y \tag{5-3b}$$

where

F_b =allowable bending stress at leg tip, ksi

F_{ob} =elastic lateral-torsional buckling stress, from Sect. 5.2 or 5.3 as applicable, ksi

F_y =yield stress, ksi

5.2. Bending About Geometric Axes

5.2.1.

a. Angle bending members with lateral-torsional restraint of leg in compression along the length may be designed on the basis of geometric axis bending with stress limited by the provisions of Sects. 5.1.1 and 5.1.2.

b. For equal leg angles if the lateral-torsional restraint of leg in compression is only at the point of maximum moment, the stress, f_b, is calculated on the basis of geometric axis bending limited by F_b in Sect. 5.2.2b.

5.2.2. Equal leg angle members without lateral-torsional restraint subjected to flexure applied about one of the geometric axes may be designed considering only geometric axis bending provided:

a. The calculated compressive stress f_b, using the geometric axis section modulus, is increased by 25%.

b. For the angle leg tips in compression, the allowable bending stress F_b is determined according to Sect. 5.1.3, where

$$F_{ob} = \frac{85,900}{(\ell/b)^2} \, C_b \, [\sqrt{1 + 0.78 \, (\ell t/b^2)^2} - 1] \tag{5-4}$$

and by b/t provisions in Sect. 5.1.1. When the leg tips are in tension, F_b is determined only by Sect. 5.1.2.

ℓ =unbraced length, in.

C_b =1.75 + 1.05(M_1/M_2) + 0.3$(M_1/M_2)^2 \le$ 1.5 where M_1 is the smaller and M_2 the larger end moment in the unbraced segment of the beam; (M_1/M_2) is positive when the moments cause reverse curvature and negative when bent in single curvature. C_b shall be taken as unity when the bending moment at any point within an unbraced length is larger than at both ends of its length.

5.2.3. Unequal leg angle members without lateral-torsional restraint subjected to bending about one of the geometric axes shall be designed using 5.3.

5.3. Bending about Principal Axes

Angles without lateral-torsional restraint shall be designed considering principal-axis bending except for cases covered by Sect. 5.2.2. Bending about both of the principal axes shall be evaluated using the interaction equations in AISC ASD Specification Sect. H1.

5.3.1. Equal leg angles

a. Major axis bending

The principal bending compression stress f_{bw} shall be limited by F_b in Sect. 5.1.3, where

$$F_{ob} = C_b \frac{28,250}{(\ell/t)} \tag{5-5}$$

and by b/t provisions in Sect. 5.1.1.

b. Minor axis bending

The principal bending stress f_{bz} shall be limited by F_b in Sect. 5.1.1 when the leg tips are in compression, and by Sect. 5.1.2 when the leg tips are in tension.

5.3.2. Unequal leg angles

a. Major axis bending

The principal bending compression stress f_{bw} shall be limited by F_b in Sect. 5.1.3, where

$$F_{ob} = \frac{143,100 I_z}{S_w \ell^2} C_b [\sqrt{\beta_w^2 + 0.052 (\ell t/r_z)^2} + \beta_w] \tag{5-6}$$

and by b/t provisions in Sect. 5.1.1 for the compression leg.

S_w = section modulus to tip of leg in compression, in.[3]
I_z = minor principal axis moment of inertia, in.[4]
r_z = radius of gyration for minor principal axis, in.

$\beta_w = \left[\dfrac{1}{I_w} \int_A z(w^2 + z^2)dA\right] - 2z_o$, special section property for unequal leg angles, positive for short leg in compression and negative for long leg in compression, in. (see Commentary for values). If the long leg is in compression anywhere along the unbraced length of the member, use the negative values of β_w, in.

z_o = coordinate along z axis of the shear center with respect to centroid, in.

I_w = major principal axis moment of inertia, in.[4]

b. Minor axis bending

The principal bending stress f_{bz} shall be limited by F_b in Sect. 5.1.1 when leg tips are in compression and by Sect. 5.1.2 when the leg tips are in tension.

6. COMBINED STRESSES

6.1. Axial Compression and Flexure

Members subjected to both axial compression and bending shall satisfy the requirements of AISC ASD Specification Sect. H1, subject to the following conditions:

6.1.1. In evaluating AISC ASD Specification Eqs. (H1-1) or (H1-2), the maximum compression bending stresses due to each moment acting alone must be used even though they may occur at different cross sections of the member.

6.1.2. AISC ASD Specification Eq. (H1-2) is to be evaluated at the critical member support cross section and need not be based on the maximum moments along the member length.

6.1.3. For members constrained to bend about a geometric axis with compressive stress and allowable stress determined per Sect. 5.2.1a, the radius of gyration r_b for F'_e shall be taken as the geometric axis value.

6.1.4. For equal leg angles without lateral-torsional restraint along the length and with bending applied about one of the geometric axes, the provisions of Sect. 5.2.2 shall apply for the calculated and allowable bending stresses. If Sect. 5.2.1b or 5.2.2 is used for F_b, the radius of gyration about the axis of bending r_b for F'_e should be taken as the geometric axis value of r divided by 1.35 in the absence of a more detailed analysis.

6.1.5. For members that do not meet the conditions of Sect. 6.1.3 or 6.1.4, the evaluation shall be based on principal axis bending according to Sect. 5.3 and the subscripts x and y in AISC ASD Specification Sect. H1 shall be interpreted as the principal axes, w and z, in this Specification when evaluating the length without lateral-torsional restraint.

6.2. Axial Tension and Bending

Members subjected to both axial tension and bending stresses due to transverse loading shall satisfy the requirements of AISC ASD Specification Sect. H2. Bending stress evaluation shall be as directed by Sects. 6.1.3, 6.1.4 and 6.1.5 for compressive stresses.

COMMENTARY

C2. TENSION

The criteria for the design of tension members in AISC ASD Specification Sect. D1 have been adopted for angles with bolted connections. However, recognizing the effect of shear lag when the connection is welded, the criteria in Sect. B3 of the AISC Allowable Stress Design Specification have been applied.

The advisory upper slenderness limits are not due to strength considerations, but are based on professional judgement and practical considerations of economics, ease of handling and transportability. The radius of gyration about the z-axis will produce the maximum ℓ/r and, except for very unusual support conditions, the maximum $K\ell/r$. Since the advisory slenderness limit for compression members is less than for tension members, an accommodation has been made for members with $K\ell/r > 200$ that are always in tension, except for unusual load conditions which produce a small compression force.

C3. SHEAR

Shear stresses in a single angle member are the result of the gradient in the bending moment along the length (flexural shear) and the torsional moment.

The elastic stress due to flexural shear may be computed by

$$f_v = 1.5 V_b / bt \qquad \text{(C3-1)}$$

where

> V_b = component of the shear force parallel to the angle leg with length b, and thickness t, kips

The stress, which is constant through the thickness, should be determined for both legs to determine the maximum.

The 1.5 factor is the calculated elastic value for equal leg angles loaded along one of the principal axes. For equal leg angles loaded along one of the geometric axes (laterally braced or unbraced) the factor is 1.35. Constants between these limits may be calculated conservatively from $V_b Q/It$ to determine the maximum stress at the neutral axis.

Alternatively, a uniform flexural shear stress in the leg of V_b / bt may be used due to inelastic material behavior and stress redistribution.

If the angle is not laterally braced against twist a torsional moment is produced equal to the applied transverse load times the perpendicular distance e to the shear center, which is at the heel of the angle cross section. Torsional moments are resisted by two types of shear behavior: pure torsion (St. Venant) and warping torsion (AISC,

1983). If the boundary conditions are such that the cross section is free to warp, the applied torsional moment M_T is resisted by pure shear stresses as shown in Fig. C3.1a. Except near the ends of the legs, these stresses are constant along the length of the leg and the maximum value can be approximated by

$$f_v = M_T t/J = 3M_T/At \qquad\qquad \text{(C3-2)}$$

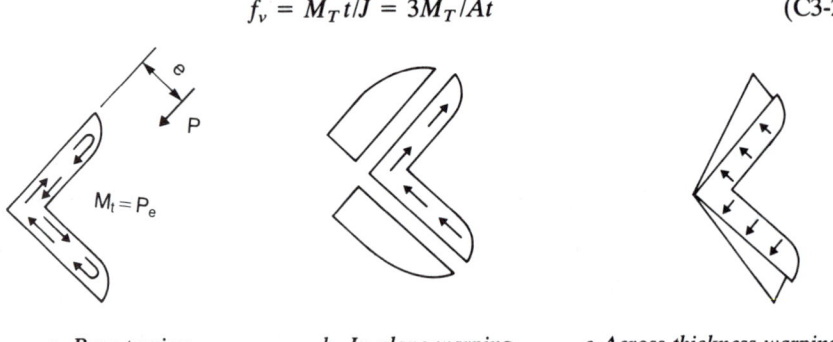

| a. Pure torsion | b. In-plane warping | c. Across-thickness warping |

Fig. C3.1.   Shear stresses due to torsion

where

 J = torsional constant, in.[4] (approximated by $\Sigma\, bt^3/3$ when precomputed value
   unavailable.)
 A = angle cross-sectional area, in.[2]

At section where warping is restrained, the torsional moment is resisted by warping shear stresses of two types (Gjelsvik, 1981). One type is in-plane (contour) as shown in Fig. C3.1b, which varies from zero at the toe to a maximum at the heel of the angle. The other type is across the thickness and is sometimes referred to as secondary warping shear. As indicated in Fig. C3.1c, it varies from zero at the heel to a maximum at the toe.

In an angle with typical boundary conditions and unrestrained load point, the torsional moment produces all three types of shear stresses (pure, in-plane warping and secondary warping) in varying proportions along its length. The total applied moment is resisted by a combination of three types of internal moments that differ in relative proportions according to the distance from the boundary condition. Using typical angle dimensions, it can be shown that the two warping shears are approximately the same order of magnitude and are less than 20% of the pure shear stress for the same torsional moment. Therefore, it is conservative to compute the torsional shear stress using the pure shear equation and total applied torsional moment M_T as if no warping restraint were present. This stress is added directly to the flexural shear stress to produce a maximum surface shear stress near the mid-length of a leg. Since this sum is a local maximum that does not extend through the thickness, applying the limit of $0.4F_y$ adds another degree of conservatism relative to the design of other structural shapes.

In general, torsional moments from laterally unrestrained transverse loads also produce warping normal stresses that are superimposed on bending stresses. However, since the warping strength for a single angle is relatively small, this additional bending effect is negligible and often ignored in design practice.

C4. COMPRESSION

The provisions for the allowable compression stress account for three possible failure modes that may occur in an angle depending on its proportions: general column flexural buckling, local buckling of thin legs and flexural-torsional buckling of the member. The Q factor in the equations for allowable stress accounts for the local buckling and the provisions are extracted from AISC ASD Specification Appendix B5. The F_e used for modification of the slenderness ratio for flexural-torsional buckling may be based on the provisions of Appendix E of the AISC LRFD Specification (AISC, 1986) while conservatively neglecting warping resistance, which is approximately less than 3% at unbraced lengths of 5 ft or more. The angle-end restraint conditions should be considered in determining the appropriate member effective length.

The equations for the elastic flexural-torsional buckling stress from AISC LRFD Appendix E with no warping resistance are:

For equal leg angles with w as the axis of symmetry:

$$F_e = \frac{F_{ew} + F_{ej}}{2H} \left(1 - \sqrt{1 - \frac{4 F_{ew} F_{ej} H}{(F_{ew} + F_{ej})^2}} \right) \tag{C4-1}$$

For unequal leg angles, F_e is the lowest root of the cubic equation:

$$(F_e - F_{ez})(F_e - F_{ew})(F_e - F_{ej}) - F_e^2(F_e - F_{ew})(z_o/\bar{r}_o)^2 - F_e^2(F_e - F_{ez})(w_o/\bar{r}_o)^2 = 0 \tag{C4-2}$$

where

E = modulus of elasticity, ksi

G = shear modulus, ksi

J = torsional constant = $\Sigma\, bt^3/3 = t^2 A/3$, in.^{4}

I_z, I_w = moment of inertia about *principal* axes, in.^{4}

z_o, w_o = coordinates of the shear center with respect to the centroid, in.

$\bar{r}_o^2$ = $z_o^2 + w_o^2 + (I_z + I_w)/A$, in.^{2}

H = $1 - (z_o^2 + w_o^2)/\bar{r}_o^2$

F_{ez} = $\dfrac{\pi^2 E}{(K_z \ell/r_z)^2}$, ksi

F_{ew} = $\dfrac{\pi^2 E}{(K_w \ell/r_w)^2}$, ksi

F_{ej} = $\dfrac{GJ}{A\bar{r}_o^2}$, ksi

A = cross-sectional area of member, in.^{2}

ℓ = unbraced length, in.

K_z, K_w = effective length factors in z and w directions

r_z, r_w = radii of gyration about the principal axes, in.

The coordinate axes are defined in Fig. C5.3. For equal leg angles, the flexural-torsional buckling stress will not control if

$$(K\ell/r)_{max} > 5.4(b/t)/Q \qquad \text{(C4-3)}$$

This limit can be derived for equal leg angles with $Q = 1$ by equating the equation for the elastic flexural-torsional stress to the Euler equation for column flexural buckling. For unequal leg angles, flexural-torsional buckling always controls though for higher slenderness ratios, it will be approximately equal to the minimum flexural buckling stress. Also, when $Q<1$, the limit cannot be derived because Q does not appear in the Euler equation. Numerical studies of the inelastic buckling strength of angles with a wide range of proportions indicate for members that exceed the $K\ell/r$ limit, the flexural-torsional buckling stress will be only a few percent less than the column buckling stress except when one leg is more than twice the length of the other. In the latter case, reductions as high as 10% may occur.

C5. FLEXURE

Flexural stress limits are established with consideration of local buckling and lateral-torsional buckling. In addition to addressing the general case of unequal leg single angles, the equal leg angle is treated as a special case. Furthermore, bending of equal leg angles about a geometric axis, an axis parallel to one of the legs, is addressed separately as it is a very common situation.

C5.1.1. These provisions follow typical AISC criteria for single angles under uniform compression. They are conservative when a leg is subjected to nonuniform compression stress if the maximum compression stress at the leg tip is used.

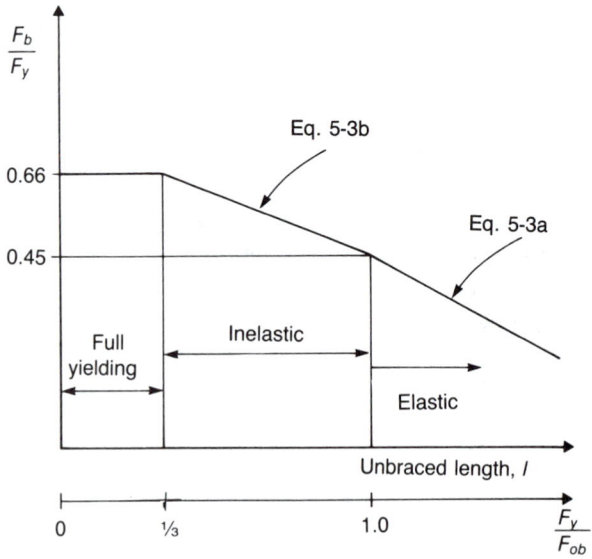

Fig. C5.1.   Lateral-torsional buckling of a single-angle beam

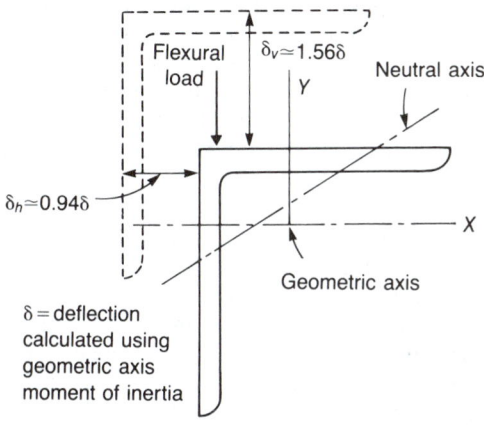

Fig. C5.2. Geometric axis bending of laterally unrestrained equal-leg angles

C5.1.2. Since the shape factor for angles is in excess of 1.5, the maximum allowable bending stress, $F_b = 0.66\ F_y$, for compact members is justified as long as instability does not control.

C5.1.3. Lateral-torsional instability may limit the allowable flexural stress of an unbraced single angle beam. As illustrated in Fig. C5.1, Eq. (5-3a) represents the elastic buckling portion with a variable factor of safety ranging from 2.22 to 1.82. Equation (5-3b) represents the inelastic buckling transition expression between $0.45F_y$ and $0.66F_y$. At F_{ob} greater than about $3F_y$, the unbraced length is adequate to develop the maximum beam flexural strength of $F_b = 0.66F_y$. These formulas were based on Australian research on single angles in flexure and on an analytical model consisting of two rectangular elements of length equal to actual angle leg width minus one-half the thickness (Leigh and Lay, 1984; Australian Institute of Steel Construction, 1975; Leigh and Lay, 1978; Madugula and Kennedy, 1985).

The familiar C_b moment gradient formula based on doubly symmetric wide flanges is used to correct lateral-torsional stability equations from the assumed most severe case of uniform moment throughout the unbraced length ($C_b = 1.0$). However, in lieu of a more detailed analysis, the reduced maximum limit of 1.5 is imposed for single angle beams to represent conservatively the lower envelope of this cross-section's non-uniform bending response.

C5.2.1. An angle beam loaded parallel to one leg will deflect and bend about that leg only if the angle is restrained laterally along the length. In this case simple bending occurs without any torsional rotation or lateral deflection and the geometric axis section properties should be used in the evaluation of the flexural stresses and deflection. If only the point of maximum moment is laterally braced, lateral-torsional buckling of the unbraced length under simple bending must also be checked.

C5.2.2. When bending is applied about one leg of a laterally unrestrained single angle, it will deflect laterally as well as in the bending direction and twist. Its behavior can be evaluated by resolving the load and/or moments into principal axis components and determining the sum of these principal axis flexural effects while neglecting the relatively minor torsional response. In order to simplify and expedite the design calculations for this common situation with equal leg angles, an alternate method may be used.

For such unrestrained bending of an equal leg angle, the resulting maximum normal stress at the angle tip (in the direction of bending) will be approximately 25% greater than calculated using the geometric axis section modulus. The deflection calculated using the geometric axis moment of inertia has to be increased 82% to approximate the total deflection. Deflection has two components, a vertical component (in the direction of applied load) 1.56 times the calculated value and a horizontal component of 0.94 of the calculated value. The resultant total deflection is in the general direction of the weak principal axis bending of the angle (see Fig. C5.2). These unrestrained bending deflections should be considered in evaluating serviceability.

The horizontal component of deflection being approximately 60% of the vertical deflection means that the lateral restraining force required to achieve purely vertical deflection (Sect. 5.2.1) must be 60% of the applied load value (or produce a moment 60% of the applied value) which is very significant.

The lateral-torsional buckling is limited by F_{ob} (Leigh and Lay, 1984 and 1978) in Eq. 5-4, which is based on

$$M_{cr} = \frac{2.33Eb^4t}{(1 + 3\ cos^2\Theta)\ (K\ell)^2} \times$$

$$\left[\sqrt{sin^2\Theta + \frac{0.162\ (1 + 3\ cos^2\Theta)\ (K\ell)^2t^2}{b^4}} + sin\ \Theta \right] \qquad \text{(C5-1)}$$

(the general expression for the critical moment of an equal leg angle) with $\Theta = -45°$ which is the most severe condition with the angle heel (shear center) in tension. Flexural loading which produces angle heel compression can be conservatively designed by Eq. (5-4) or more exactly by using the above general M_{cr} Equation with $\Theta = 45°$ (see Fig. C5.3).

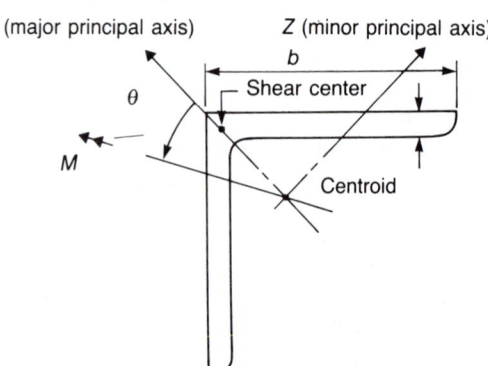

Fig. C5.3. Equal-leg angle with general moment loading

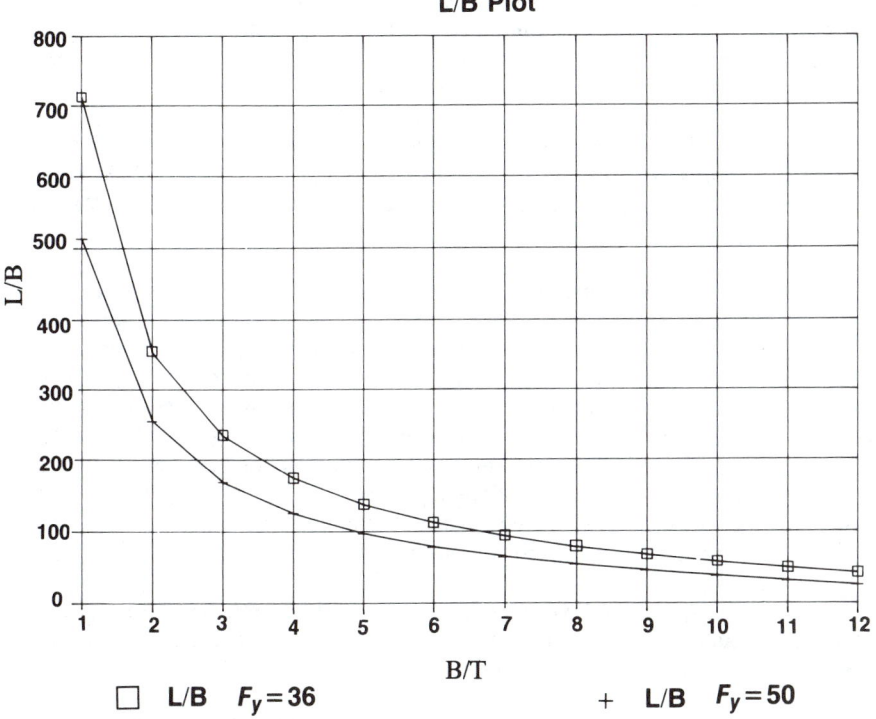

Fig. C5.4.   *Single-angle limits for $F_b = .66_y$*

Lateral-torsional buckling will reduce the stress limit only when ℓ/b is relatively large. If the $\ell t/b^2$ parameter (which is a ratio of ℓ/b over b/t) is less than 2.43 (with $C_b = 1$), there is no need to check lateral-torsional stability inasmuch as local buckling provisions of Sect. 5.1.1 will control the allowable flexural stress.

Lateral-torsional buckling will produce $F_b < 0.66F_y$ for equal leg angles only if F_{ob} by Eq. (5-4) is less than about $3F_y$, for $C_b = 1.0$. Limits for ℓ/b as a function of b/t are shown graphically in Fig. C5.4. Local buckling must be checked separately.

Stress at the tip of the angle leg parallel to the applied bending axis is of the same sign as the maximum stress at the tip of the other leg when the single angle is unrestrained. For an equal-leg angle this stress is about one third of the maximum stress. It is only necessary to check the stress condition at the tip of the angle leg with the maximum stress when evaluating such an angle. Since this maximum applied compressive stress per Sect. 5.2.1a represents combined principal axis stresses and Eq. (5-4) represents the design limit for this combined flexural stress, only a single flexural term needs to be considered when evaluating combined flexural and axial effects.

C5.2.3.   For unequal leg angles without lateral-torsional restraint the applied load or moment must be resolved into components along the two principal axis in all cases and designed for biaxial bending using the interaction equation.

C5.3.1. Under major axis bending of equal leg angles Eq. (5-5) in combination with (5-3a) or (5-3b) controls the flexural stress f_{bw} against overall lateral-torsional buckling of the angle. This is based on M_{cr}, given earlier with $\Theta = 0$.

Lateral-torsional buckling for this case will reduce the stress below 0.66 F_y only for $\ell/t \geq 9400/F_y$ ($F_{obw} = 3F_y$). If the $\ell t/b^2$ parameter is less than $1.42C_b$ for this case, local buckling will control the allowable flexural stress and F_b based on lateral-torsional buckling need not be evaluated. Local buckling must be checked using 5.1.1.

C5.3.2. Lateral-torsional buckling about the major principal W axis of an unequal leg angle is controlled by M_{ow} in Eq. (5-6). Section property β_w reflects the location of the shear center relative to the principal axis of the section and the bending direction under uniform bending. Positive β_w and maximum M_{ow} occurs when the shear center is in flexural compression while negative β_w and minimum M_{ow} occurs when the shear center is in flexural tension (see Fig. C5.5). This β_w effect is consistent with behavior of singly symmetric I- shaped beams which are more stable when the compression flange is larger than the tension flange. For principal W axis bending of equal leg angles, β_w is equal to zero due to symmetry and Eq. (5-6) reduces to Eq. (5-5) for this special case.

For reverse curvature bending, part of the unbraced length has positive β_w, while the remainder negative β_w, and conservatively, the negative value is assigned for that entire unbraced segment.

β_w is essentially independent of angle thickness (less than 1% variation from mean value) and is primarily a function of the leg widths. The average values shown in Table C5.1 may be used for design.

C6. COMBINED STRESSES

The stability and strength interaction equations of AISC ASD Specification Chap. H have been adopted with modifications to account for various conditions of bending

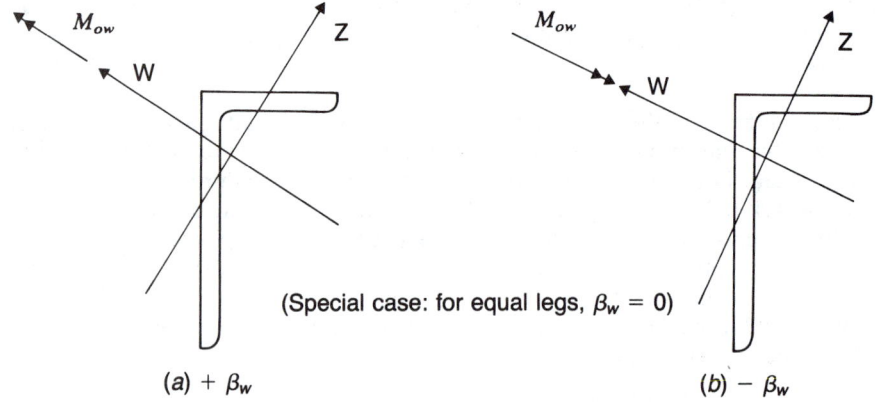

(Special case: for equal legs, $\beta_w = 0$)

(a) $+ \beta_w$ (b) $- \beta_w$

Fig. C5.5. Unequal-leg angle in bending

Table C5.1 β_w Values for Angles

Angle Size (in.)	β_w (in.)
9 × 4	6.54
8 × 6	3.31
8 × 4	5.48
7 × 4	4.37
6 × 4	3.14
6 × 3.5	3.69
5 × 3.5	2.40
5 × 3	2.99
4 × 3.5	0.87
4 × 3	1.65
3.5 × 3	0.87
3.5 × 2.5	1.62
3 × 2.5	0.86
3 × 2	1.56
2.5 × 2	0.85
Equal legs	0.00

that may be encountered. Bending will usually accompany axial loading in a single angle member since the axial load and connection along the legs are eccentric to the centroid of the cross section. Unless the situation conforms to Sect. 5.2.1 or 5.2.2 in that Sect. 6.1.3 or 6.1.4 may be used, the applied moment should be resolved about the principal axes for the interaction check.

C6.1.4. When the total maximum flexural stress is evaluated for a laterally unrestrained length of angle per Sect. 5.2, the bending axis is the inclined axis shown in Fig. C5.2. The radius of gyration modification for the moment amplification about this axis is equal to $\sqrt{1.82} = 1.35$ to account for the increased unrestrained bending deflection relative to that about the geometric axis for the laterally unrestrained length. The 1.35 factor is retained for angles braced only at the point of maximum moment to maintain a conservative calculation for this case. If the brace exhibits any flexibility permitting lateral movement of the angle, use of $r_b = r_x$ would not be conservative.

REFERENCES

American Institute of Steel Construction, Inc. (1983) Torsional Analysis of Steel Members *Chicago, IL.*

American Institute of Steel Construction, Inc. (1986) Load and Resistance Factor Design Specification for Structural Steel Buildings *Chicago, IL.*

Australian Institute of Steel Construction (1975) Australian Standard AS1250 *1975.*

Gjelsvik, A. (1981) The Theory of Thin-walled Bars *John Wiley and Sons, New York.*

Leigh, J.M. and M.G. Lay (1978) Laterally Unsupported Angles with Equal and Unequal Legs *Report MRL 22/2 July 1978, Melbourne Research Laboratories, Clayton.*

Leigh, J.M. and M.G. Lay (1984) "The Design of Laterally Unsupported Angles," in Steel Design Current Practice, Section 2, Bending Members *American Institute of Steel Construction, Inc., January, 1984.*

Madugula, M.K.S. and J.B. Kennedy (1985) Single and Compound Angle Members *Elsevier Applied Science New York.*

AISC
Quality Certification
Program

AMERICAN INSTITUTE OF STEEL CONSTRUCTION, INC.

Notes

AISC
Quality Certification
Program

In recent years, the quality of construction methods and materials has become the subject of increasing concern to building officials, highway officials, and designers. One result of this concern has been the enactment of ever more demanding inspection requirements intended to ensure product quality. In many cases, however, these more demanding inspection requirements have not been based upon demonstrated unsatisfactory performance of structures in service. Rather, they have been based upon the capacity of sophisticated test equipment, or upon standards developed for nuclear construction rather than conventional construction. Adding to the problem, arbitrary interpretation of specifications by inspectors has too often been made without rational consideration of the type of construction involved. The result has been spiraling increases in the costs of fabrication of structural steel and of inspection, which must be paid by owners without necessarily assuring that the product quality required has been improved.

Product inspection, although it has a valid place in the construction process, is not the most logical or practical way to assure that structural steelwork will conform to the requirements of contract documents and satisfy the intended use. A better solution can be found in the exercise of good quality control and quality assurance by the fabricator *throughout the entire production process.*

Recognizing this fact, and seeking some valid, objective method whereby a fabricator's capability for assuring a quality product could be evaluated, a number of code authorities have, in recent years, instituted steps to establish fabricator registration programs. However, these independent efforts resulted in extremely inconsistent criteria. They were developed primarily by inspectors or inspection agencies who were experienced in testing, but were not familiar with the complexities of the many steps, procedures, techniques, and controls required to assure quality throughout the fabricating process. Neither were these inspection agencies qualified to determine the various levels of quality required to assure satisfactory performance in meeting the service requirements of the many different types of steel structures.

Recognizing the need for a comprehensive national standard for fabricator certification, and concerned by the trend toward costly inspection requirements that could not be justified by rational quality standards, the American Institute of Steel Construction has developed and implemented a voluntary Quality Certification Program, whereby any structural steel fabricating plant—whether a member of AISC or not—can have its capability for assuring quality production evaluated on a fair and impartial basis.

THE AISC PROGRAM

The AISC Quality Certification Program does not involve inspection and/or judgment of product quality on individual projects. Neither does it guarantee the quality of specific fabricated steel products. Rather, the purpose of the AISC Quality Certification Program is to confirm to the construction industry that a Certified structural steel fabricating plant *has the personnel, organization, experience, procedures, knowledge, equipment, capability and commitment to produce fabricated steel of the required quality for a given category of structural steelwork.*

The AISC Quality Certification Program was developed by a group of highly qualified shop operation personnel from large, medium, and small structural steel fabricating plants throughout the United States. These individuals all had extensive experience and were fully aware of where and how problems can arise during the production process and of the steps and procedures that must be followed during fabrication to assure that the finished product meets the quality requirements of the contract.

The program was reviewed and strongly endorsed by an Independent Board of Review comprised of 17 prominent structural engineers from throughout the United States, who were not associated with the steel fabricating industry, but were well qualified in matters of quality requirements for reliable service of all types of steel structures.

CATEGORIES OF CERTIFICATION

A fabricator may apply for certification of a plant in one of the following categories of structural steelwork:

I: Conventional Steel Structures — Small Public Service and Institutional Buildings, (Schools, etc.), Shopping Centers, Light Manufacturing Plants, Miscellaneous and Ornamental Iron Work, Warehouses, Sign Structures, Low Rise, Truss Beam/Column Structures, Simple Rolled Beam Bridges.

II: Complex Steel Building Structures — Large Public Service and Institutional Buildings, Heavy Manufacturing Plants, Powerhouses (fossil, non-nuclear), Metal Producing/Rolling Facilities, Crane Bridge Girders, Bunkers and Bins, Stadia, Auditoriums, High Rise Buildings, Chemical Processing Plants, Petroleum Processing Plants.

III: Major Steel Bridges — All bridge structures other than simple rolled beam bridges.

MB: Metal Building Systems — Pre-engineered Metal Building Structures.

Supplement: Auxiliary and Support Structures for Nuclear Power Plants — This supplement, applicable to nuclear plant structures designed under the AISC Specification, but not to pressure-retaining structures, offers utility companies and designers of nuclear power plants a certification program that will eliminate the need for many of the more costly, conflicting programs now in use. A fabricator must hold certification in either Category I, II or III prior to application for certification in this category.

Certification in Category II automatically includes Category I. Certification in Category III automatically includes Categories I and II. Certification in Category MB is not transferable to any other Category.

INSPECTION-EVALUATION PROCEDURE

An outside, experienced, professional organization, ABS Worldwide Technical Services, Inc. (a subsidiary of American Bureau of Shipping) has been retained by AISC to perform the plant Inspection-Evaluation in accordance with a standard check list and rating procedure established by AISC for each certification category in the program. Upon completion of this Inspection-Evaluation, ABS Worldwide Technical Services (commonly known as ABSTECH) will recommend to AISC that a fabricator be approved or disapproved for certification. ABSTECH's Inspection-Evaluation is totally independent of the fabricator's and AISC's influence, and their evaluation is not subject to review by AISC.

At a time mutually agreed upon by the fabricator, AISC, and ABSTECH, the Inspection-Evaluation team visits the plant to investigate and rate the following basic plant functions directly and indirectly affecting quality assurance: General Management, Engineering and Drafting, Procurement, Shop Operations, and Quality Control. The Inspection-Evaluation team will perform the following:

1. Confirm data submitted with the Application for Certification.
2. Interview key supervisory personnel and subordinate employees.
3. Observe and rate the organization in operation, including procedures used in functions affecting quality assurance.
4. Inspect and rate equipment and facilities.
5. At an "exit interview," review with plant management the completed check list observations and evaluation scoring, including discussions of deficiencies and omissions, if any.

The number of days required for Inspection-Evaluation varies according to the size and complexity of the plant, but usually requires two to five days.

CERTIFICATION

Following recommendation for Certification by the Inspection-Evaluation team, AISC will issue a certificate identifying the fabricator, the plant, and the Category of Certification. The certificate is valid for a three year period, subject to annual review in the form of unannounced inspections early in the second and third year periods. The certificate is endorsed annually, provided there is successful completion of the unannounced second and third year inspection.

An annual self-audit, based on the standard check list, must be made by plant management during the 11th and 23rd months after initial Certification. This self-audit must be retained at the plant and made available to the Inspection-Evaluation team during the unannounced second and third year inspections.

At the end of the third year, the cycle begins again with a complete prescheduled Inspection-Evaluation and the issuance of a new certificate.

PRESENT STATUS

Two of the major Building Code bodies in the country have recognized that the AISC Quality Certification Program assures uniform minimum standards of quality in structural steel fabrication. AISC has been named a Quality Assurance Agency by Southern Building Code Congress International, Inc. in their report number Q.A. 7801-78 and by Building Officials and Code Administrators International, Inc. in their report RR 77-61.

For additional information on this program, write to: AISC Quality Certification Administrator, One East Wacker Drive, Suite 3100, Chicago, IL 60601-2001.

PART 6
Miscellaneous Data and Mathematical Tables

WIRE AND SHEET METAL GAGES
Equivalent thickness in decimals of an inch

Gage No.	U.S. Standard Gage for Uncoated Hot & Cold Rolled Sheets[b]	Galvanized Sheet Gage for Hot-Dipped Zinc Coated Sheets[b]	USA Steel Wire Gage	Gage No.	U.S. Standard Gage for Uncoated Hot & Cold Rolled Sheets[b]	Galvanized Sheet Gage for Hot-Dipped Zinc Coated Sheets[b]	USA Steel Wire Gage
7/0	—	—	.490	13	.0897	.0934	.092[a]
6/0	—	—	.462[a]	14	.0747	.0785	.080
5/0	—	—	.430[a]	15	.0673	.0710	.072
4/0	—	—	.394[a]	16	.0598	.0635	.062[a]
3/0	—	—	.362[a]	17	.0538	.0575	.054
2/0	—	—	.331	18	.0478	.0516	.048[a]
1/0	—	—	.306	19	.0418	.0456	.041
1	—	—	.283	20	.0359	.0396	.035[a]
2	—	—	.262[a]	21	.0329	.0366	—
3	.2391	—	.244[a]	22	.0299	.0336	—
4	.2242	—	.225[a]	23	.0269	.0306	—
5	.2092	—	.207	24	.0239	.0276	—
6	.1943	—	.192	25	.0209	.0247	—
7	.1793	—	.177	26	.0179	.0217	—
8	.1644	.1681	.162	27	.0164	.0202	—
9	.1495	.1532	.148[a]	28	.0149	.0187	—
10	.1345	.1382	.135	29	—	.0172	—
11	.1196	.1233	.120[a]	30	—	.0157	—
12	.1046	.1084	.106[a]				

[a] Rounded value. The steel wire gage has been taken from ASTM A510 "General Requirements for Wire Rods and Coarse Round Wire, Carbon Steel". Sizes originally quoted to 4 decimal equivalent places have been rounded to 3 decimal places in accordance with rounding procedures of ASTM "Recommended Practice" E29.

[b] The equivalent thicknesses are for information only. The product is commonly specified to decimal thickness, not to gage number.

AISI STANDARD NOMENCLATURE FOR
FLAT ROLLED CARBON STEEL

Thickness (In.)	Width (In.)					
	To 3½ incl.	Over 3½ To 6	Over 6 To 8	Over 8 To 12	Over 12 To 48	Over 48
0.2300 & thicker	Bar	Bar	Bar	Plate	Plate	Plate
0.2299 to 0.2031	Bar	Bar	Strip	Strip	Sheet	Plate
0.2030 to 0.1800	Strip	Strip	Strip	Strip	Sheet	Plate
0.1799 to 0.0449	Strip	Strip	Strip	Strip	Sheet	Sheet
0.0448 to 0.0344	Strip	Strip				
0.0343 to 0.0255	Strip	Hot rolled sheet and strip not generally produced in these widths and thicknesses				
0.0254 & thinner						

EFFECT OF HEAT ON STRUCTURAL STEEL

Short-time elevated-temperature tensile tests on the constructional steels permitted by the AISC Specification indicate that the ratios of the elevated-temperature yield and tensile strengths to their respective room-temperature strength values are reasonably similar at any particular temperature for the various steels in the 300 to 700° F. range, except for variations due to strain aging. (The tensile strength ratio may increase to a value greater than unity in the 300 to 700° F. range when strain aging occurs.) Above this range, the ratio of elevated-temperature to room-temperature strength decreases as the temperature increases.

The composition of the steels is usually such that the carbon steels exhibit strain aging with attendant reduced notch toughness. The high-strength low-alloy and heat-treated constructional alloy steels exhibit less-pronounced or little strain aging.

As examples of the decreased ratio levels obtained at elevated temperature, the yield strength ratios for carbon and high-strength low-alloy steels are approximately 0.77 at 800° F., 0.63 at 1000° F., and 0.37 at 1200° F.

FIRE-RESISTANT CONSTRUCTION

ASTM Specification E119, *Standard Methods of Fire Tests of Building Construction and Materials,* outlines the procedures of fire testing of structural elements located inside a building and exposed to fire within the compartment or room in which they are located. The temperature criterion used requires that the average of the temperature readings not exceed 1000° F. for columns and 1100° F. for beams. An individual temperature reading may not exceed 1100° F. for columns and 1200° F. for beams.

Steel buildings whose condition of exterior exposure and whose combustible contents under fire hazards will not produce a steel temperature greater than the foregoing criteria may therefore be considered fire-resistive without the provision of insulating protection for the steel.

A fire exposure of severity and duration sufficient to raise the temperature of the steel much above the fire test criteria temperature will seriously impair its ability to sustain loads at the unit stresses or plasticity load factors permitted by the AISC Specification. In such cases, the members upon which the stability of the structure depends should be insulated by fire-resistive materials or construction capable of holding the average temperature of the steel to not more than that specified for the fire test standard.

Under the E119 specification, each tested assembly is subjected to a standard fire of controlled extent and severity. The fire resistance rating is expressed as the time, in hours, that the assembly is able to withstand the fire exposure before the first critical point in its behavior is reached. These tests indicate the minimum period of time during which structural members, such as columns and beams, are capable of maintaining their strength and rigidity when subjected to the standard fire. They also establish the minimum period of time during which floors, roofs, walls or partitions will prevent fire spread by protecting against the passage of flame, hot gases and excessive heat.

Tables of fire resistance ratings for various insulating materials and constructions applied to structural elements are published in the AISI booklets *Fire Resistant Steel Frame Construction, Designing Fire Protection for Steel Columns* and *Designing Fire Protection for Steel Trusses.* Ratings may also be found in publications of the Underwriters' Laboratories, Inc.

A new rational fire-protection design procedure for exposed columns and beams at building exteriors has been developed by the American Iron and Steel Institute, and is described in AISI publication no. FS3, *Fire-safe Structural Steel — A Design Guide*. The Design Guide provides a step by step procedure which enables building designers to estimate the maximum steel temperature that would occur during a fire at any location on a structural member located outside a building. The design procedure is accepted by some building codes and is under study for adoption by others.

To judge the effect of a fire on structural steel, it is necessary to consider what happens in such an exposure. Peculiarities of this exposure are: (1) temperature attained by the steel can only be estimated, (2) time of exposure at any given temperature is unknown, (3) heating is uneven, (4) cooling rates vary and can only be estimated and (5) the steel is usually under load, and is sometimes restrained from normal expansion.

Carbon and high-strength low-alloy steels that show no evidence of gross damage from exposure to high temperatures, or from sudden cooling from high temperatures, can usually be straightened as necessary and be reused without reduction of working stress. Quenched and tempered alloy steels should not be heated to temperatures within 50° F. of the tempering temperature used in heat treatment. Thus, for the quenched and tempered constructional alloy steels approved by they AISC Specification, i.e., ASTM A514, for which the tempering temperature is 1150° F., the maximum steel temperature should be 1100° F.

Steel that has been exposed to very high temperatures can be identified by very heavy scale, pitting, and surface erosion. Such temperatures may not only cause a loss of cross section, but may also result in metallurgical changes. Normally these conditions will be accompanied by such severe deformation that the cost and difficulty of straightening such members, as compared to replacement, dictates that they be discarded.

Steel members that have suffered rapid cooling will usually be so severely distorted that straightening for reuse will seldom be considered practicable.

In some cases, there may be some deformation in members whose normal thermal expansion is inhibited or prevented by the nature of the construction. Such members may usually be straightened and reused.

Connections require special attention to make sure that the stresses induced by a fire, and by subsequent cooling after the fire, have not sheared or loosened bolts or rivets, or cracked welds.

COEFFICIENT OF EXPANSION

The average coefficient of expansion for structural steel between 70° F. and 100° F. is 0.0000065 for each degree. For temperatures of 100° F. to 1200° F. the coefficient is given by the approximate formula:

$$\epsilon = (6.1 + 0.0019t) \times 10^{-6}$$

in which ϵ is the coefficient of expansion for each degree Fahrenheit and t is the temperature in degrees Fahrenheit.

The modulus of elasticity of structural steel is approximately 29,000 ksi at 70° F. It decreases linearly to about 25,000 ksi at 900° F., and then begins to drop at an increasing rate at higher temperatures.

EFFECT OF HEAT DUE TO WELDING

Application of heat by welding produces residual stresses, which are generally accompanied by distortion of various amounts. Both the stresses and distortions are minimized by controlled welding procedures and fabrication methods. In normal structural practice, it has not been found necessary or desirable to use heat treatment (stress-relieving) as a means of reducing residual stresses. Procedures normally followed include: (1) proper positioning of the components of joints before welding, (2) selection of welding sequences determined by experience, (3) deposition of a minimum volume of weld metal with a minimum number of passes for the design condition and (4) preheating as determined by experience (usually above the specified minimums).

USE OF HEAT TO STRAIGHTEN, CAMBER, OR CURVE MEMBERS

With modern fabrication techniques, a controlled application of heat can be effectively used to either straighten or to intentionally curve structural members. By this process, the member is rapidly heated in selected areas; the heated areas tend to expand, but are restrained by adjacent cooler areas. This action causes a permanent plastic deformation or "upset" of the heated areas and, thus, a change of shape is developed in the cooled member.

"Heat straightening" is used in both normal shop fabrication operations and in the field to remove relatively severe accidental bends in members. Conversely, "heat cambering" and "heat curving" of either rolled beams or welded girders are examples of the use of heat to effect a desired curvature.

As with many other fabrication operations, the use of heat to straighten or curve will cause residual stresses in the member as a result of plastic deformations. These stresses are similar to those that develop in rolled structural shapes as they cool from the rolling temperature; in this case, the stresses arise because all parts of the shape do not cool at the same rate. In like manner, welded members develop residual stresses from the localized heat of welding.

In general, the residual stresses from heating operations do not affect the ultimate strength of structural members. Any reduction in column strength due to residual stresses is incorporated in the present design provisions.

The mechanical properties of steels are largely unaffected by heating operations, provided that the maximum temperature does not exceed 1100° F. for quenched and tempered alloy steels, and 1300° F. for other steels. The temperature should be carefully checked by temperature-indicating crayons or other suitable means during the heating process.

COEFFICIENTS OF EXPANSION

The coefficient of linear expansion (ϵ) is the change in length, per unit of length, for a change of one degree of temperature. The coefficient of surface expansion is approximately two times the linear coefficient, and the coefficient of volume expansion, for solids, is approximately three times the linear coefficient.

A bar, free to move, will increase in length with an increase in temperature and will decrease in length with a decrease in temperature. The change in length will be $\epsilon t l$, where ϵ is the coefficient of linear expansion, t the change in temperature, and l the length. If the ends of a bar are fixed, a change in temperature t will cause a change in the unit stress of $E\epsilon t$, and in the total stress of $AT\epsilon t$, where A is the cross sectional area of the bar and E the modulus of elasticity.

The following table gives the coefficient of linear expansion for 100°, or 100 times the value indicated above.

Example: A piece of medium steel is exactly 40 ft long at 60° F. Find the length at 90° F., assuming the ends free to move.

$$\text{Change of length} = \epsilon t l = \frac{.00065 \times 30 \times 40}{100} = .0078 \text{ ft}$$

The length at 90° F. is 40.0078 ft.

Example: A piece of medium steel is exactly 40 ft long and the ends are fixed. If the temperature increases 30° F., what is the resulting change in the unit stress?

$$\text{Change in unit stress} = E \epsilon t = \frac{29,000,000 \times .00065 \times 30}{100} = 5655 \text{ lbs. per sq. in.}$$

COEFFICIENTS OF EXPANSION FOR 100 DEGREES = 100ε

Materials	Linear Expansion		Materials	Linear Expansion	
	Centigrade	Fahrenheit		Centigrade	Fahrenheit
METALS AND ALLOYS			**STONE AND MASONRY**		
Aluminum, wrought	.00231	.00128	Ashlar masonry	.00063	.00035
Brass	.00188	.00104	Brick masonry	.00061	.00034
Bronze	.00181	.00101	Cement, portland	.00126	.00070
Copper	.00168	.00093	Concrete	.00099	.00055
Iron, cast, gray	.00106	.00059	Granite	.00080	.00044
Iron, wrought	.00120	.00067	Limestone	.00076	.00042
Iron, wire	.00124	.00069	Marble	.00081	.00045
Lead	.00286	.00159	Plaster	.00166	.00092
Magnesium, various alloys	.0029	.0016	Rubble masonry	.00063	.00035
Nickel	.00126	.00070	Sandstone	.00097	.00054
Steel, mild	.00117	.00065	Slate	.00080	.00044
Steel, stainless, 18-8	.00178	.00099			
Zinc, rolled	.00311	.00173			
TIMBER			**TIMBER**		
Fir ⎫	.00037	.00021	Fir ⎫	.0058	.0032
Maple ⎬ parallel to fiber	.00064	.00036	Maple ⎬ perpendicular to	.0048	.0027
Oak ⎪	.00049	.00027	Oak ⎪ fiber	.0054	.0030
Pine ⎭	.00054	.00030	Pine ⎭	.0034	.0019

EXPANSION OF WATER
Maximum Density = 1

C°	Volume	C°	Volume	C°	Volume	C°	Volume	C°	Volume	C°	Volume
0	1.000126	10	1.000257	30	1.004234	50	1.011877	70	1.022384	90	1.035829
4	1.000000	20	1.001732	40	1.007627	60	1.016954	80	1.029003	100	1.043116

WEIGHTS AND SPECIFIC GRAVITIES

Substance	Weight Lb. per Cu. Ft	Specific Gravity	Substance	Weight Lb. per Cu. Ft	Specific Gravity
ASHLAR MASONRY			**MINERALS**		
Granite, syenite, gneiss ...	165	2.3-3.0	Asbestos	153	2.1-2.8
Limestone, marble	160	2.3-2.8	Barytes	281	4.50
Sandstone, bluestone	140	2.1-2.4	Basalt	184	2.7-3.2
			Bauxite	159	2.55
MORTAR RUBBLE			Borax	109	1.7-1.8
MASONRY			Chalk	137	1.8-2.6
Granite, syenite, gneiss ...	155	2.2-2.8	Clay, marl	137	1.8-2.6
Limestone, marble	150	2.2-2.6	Dolomite	181	2.9
Sandstone, bluestone	130	2.0-2.2	Feldspar, orthoclase	159	2.5-2.6
			Gneiss, serpentine	159	2.4-2.7
DRY RUBBLE MASONRY			Granite, syenite	175	2.5-3.1
Granite, syenite, gneiss ...	130	1.9-2.3	Greenstone, trap	187	2.8-3.2
Limestone, marble	125	1.9-2.1	Gypsum, alabaster	159	2.3-2.8
Sandstone, bluestone	110	1.8-1.9	Hornblende	187	3.0
			Limestone, marble	165	2.5-2.8
BRICK MASONRY			Magnesite	187	3.0
Pressed brick	140	2.2-2.3	Phosphate rock, apatite ...	200	3.2
Common brick	120	1.8-2.0	Porphyry	172	2.6-2.9
Soft brick	100	1.5-1.7	Pumice, natural	40	0.37-0.90
			Quartz, flint	165	2.5-2.8
CONCRETE MASONRY			Sandstone, bluestone	147	2.2-2.5
Cement, stone, sand	144	2.2-2.4	Shale, slate	175	2.7-2.9
Cement, slag, etc.	130	1.9-2.3	Soapstone, talc	169	2.6-2.8
Cement, cinder, etc.	100	1.5-1.7			
			STONE, QUARRIED, PILED		
VARIOUS BUILDING			Basalt, granite, gneiss	96	—
MATERIALS			Limestone, marble, quartz .	95	—
Ashes, cinders	40-45	—	Sandstone	82	—
Cement, portland, loose ...	90	—	Shale	92	—
Cement, portland, set	183	2.7-3.2	Greenstone, hornblende ..	107	—
Lime, gypsum, loose	53-64	—			
Mortar, set	103	1.4-1.9	**BITUMINOUS SUBSTANCES**		
Slags, bank slag	67-72	—	Asphaltum	81	1.1-1.5
Slags, bank screenings ...	98-117	—	Coal, anthracite	97	1.4-1.7
Slags, machine slag	96	—	Coal, bituminous	84	1.2-1.5
Slags, slag sand	49-55	—	Coal, lignite	78	1.1-1.4
			Coal, peat, turf, dry	47	0.65-0.85
EARTH, ETC., EXCAVATED			Coal, charcoal, pine	23	0.28-0.44
Clay, dry	63	—	Coal, charcoal, oak	33	0.47-0.57
Clay, damp, plastic	110	—	Coal, coke	75	1.0-1.4
Clay and gravel, dry	100	—	Graphite	131	1.9-2.3
Earth, dry, loose	76	—	Paraffine	56	0.87-0.91
Earth, dry, packed	95	—	Petroleum	54	0.87
Earth, moist, loose	78	—	Petroleum, refined	50	0.79-0.82
Earth, moist, packed	96	—	Petroleum, benzine	46	0.73-0.75
Earth, mud, flowing	108	—	Petroleum, gasoline	42	0.66-0.69
Earth, mud, packed	115	—	Pitch	69	1.07-1.15
Riprap, limestone	80-85	—	Tar, bituminous	75	1.20
Riprap, sandstone	90	—			
Riprap, shale	105	—	**COAL AND COKE, PILED**		
Sand, gravel, dry, loose ...	90-105	—	Coal, anthracite	47-58	—
Sand, gravel, dry, packed .	100-120	—	Coal, bituminous, lignite ..	40-54	—
Sand, gravel, wet	118-120	—	Coal, peat, turf	20-26	—
			Coal, charcoal	10-14	—
EXCAVATIONS IN WATER			Coal, coke	23-32	—
Sand or gravel	60	—			
Sand or gravel and clay ...	65	—			
Clay	80	—			
River mud	90	—			
Soil	70	—			
Stone riprap	65	—			

The specific gravities of solids and liquids refer to water at 4°C., those of gases to air at 0°C. and 760 mm. pressure. The weights per cubic foot are derived from average specific gravities, except where stated that weights are for bulk, heaped or loose material, etc.

WEIGHTS AND SPECIFIC GRAVITIES

Substance	Weight Lb. per Cu. Ft	Specific Gravity	Substance	Weight Lb. per Cu. Ft	Specific Gravity
METALS, ALLOYS, ORES			**TIMBER, U.S. SEASONED**		
Aluminum, cast,			Moisture Content by		
hammered	165	2.55-2.75	Weight:		
Brass, cast, rolled	534	8.4-8.7	Seasoned timber 15 to 20%		
Bronze, 7.9 to 14% Sn	509	7.4-8.9	Green timber up to 50%		
Bronze, aluminum	481	7.7	Ash, white, red	40	0.62-0.65
Copper, cast, rolled	556	8.8-9.0	Cedar, white, red	22	0.32-0.38
Copper ore, pyrites	262	4.1-4.3	Chestnut	41	0.66
Gold, cast, hammered	1205	19.25-19.3	Cypress	30	0.48
Iron, cast, pig	450	7.2	Fir, Douglas spruce	32	0.51
Iron, wrought	485	7.6-7.9	Fir, eastern	25	0.40
Iron, spiegel-eisen	468	7.5	Elm, white	45	0.72
Iron, ferro-silicon	437	6.7-7.3	Hemlock	29	0.42-0.52
Iron ore, hematite	325	5.2	Hickory	49	0.74-0.84
Iron ore, hematite in bank	160-180	—	Locust	46	0.73
Iron ore, hematite loose	130-160	—	Maple, hard	43	0.68
Iron ore, limonite	237	3.6-4.0	Maple, white	33	0.53
Iron ore, magnetite	315	4.9-5.2	Oak, chestnut	54	0.86
Iron slag	172	2.5-3.0	Oak, live	59	0.95
Lead	710	11.37	Oak, red, black	41	0.65
Lead ore, galena	465	7.3-7.6	Oak, white	46	0.74
Magnesium, alloys	112	1.74-1.83	Pine, Oregon	32	0.51
Manganese	475	7.2-8.0	Pine, red	30	0.48
Manganese ore, pyrolusite	259	3.7-4.6	Pine, white	26	0.41
Mercury	849	13.6	Pine, yellow, long-leaf	44	0.70
Monel Metal	556	8.8-9.0	Pine, yellow, short-leaf	38	0.61
Nickel	565	8.9-9.2	Poplar	30	0.48
Platinum, cast, hammered	1330	21.1-21.5	Redwood, California	26	0.42
Silver, cast, hammered	656	10.4-10.6	Spruce, white, black	27	0.40-0.46
Steel, rolled	490	7.85	Walnut, black	38	0.61
Tin, cast, hammered	459	7.2-7.5	Walnut, white	26	0.41
Tin ore, cassiterite	418	6.4-7.0			
Zinc, cast, rolled	440	6.9-7.2	**VARIOUS LIQUIDS**		
Zinc ore, blende	253	3.9-4.2	Alcohol, 100%	49	0.79
			Acids, muriatic 40%	75	1.20
VARIOUS SOLIDS			Acids, nitric 91%	94	1.50
Cereals, oats bulk	32	—	Acids, sulphuric 87%	112	1.80
Cereals, barley bulk	39	—	Lye, soda 66%	106	1.70
Cereals, corn, rye bulk	48	—	Oils, vegetable	58	0.91-0.94
Cereals, wheat bulk	48	—	Oils, mineral, lubricants	57	0.90-0.93
Hay and Straw bales	20	—	Water, 4°C. max density	62.428	1.0
Cotton, Flax, Hemp	93	1.47-1.50	Water, 100°C.	59.830	0.9584
Fats	58	0.90-0.97	Water, ice	56	0.88-0.92
Flour, loose	28	0.40-0.50	Water, snow, fresh fallen	8	.125
Flour, pressed	47	0.70-0.80	Water, sea water	64	1.02-1.03
Glass, common	156	2.40-2.60			
Glass, plate or crown	161	2.45-2.72	**GASES**		
Glass, crystal	184	2.90-3.00	Air, 0°C. 760 mm.	.08071	1.0
Leather	59	0.86-1.02	Ammonia	.0478	0.5920
Paper	58	0.70-1.15	Carbon dioxide	.1234	1.5291
Potatoes, piled	42	—	Carbon monoxide	.07821	0.9673
Rubber, caoutchouc	59	0.92-0.96	Gas, illuminating	.028-.036	0.35-0.45
Rubber goods	94	1.0-2.0	Gas, natural	.038-.039	0.47-0.48
Salt, granulated, piled	84	—	Hydrogen	.00559	0.0693
Saltpeter	67	—	Nitrogen	.0784	0.9714
Starch	96	1.53	Oxygen	.0892	1.1056
Sulphur	125	1.93-2.07			
Wool	82	1.32			

The specific gravities of solids and liquids refer to water at 4°C., those of gases to air at 0°C. and 760 mm. pressure. The weights per cubic foot are derived from average specific gravities, except where stated that weights are for bulk, heaped or loose material, etc.

WEIGHTS OF BUILDING MATERIALS

Materials	Weight Lb. per Sq. Ft	Materials	Weight Lb. per Sq. Ft
CEILINGS		PARTITIONS	
Channel suspended		Clay Tile	
system	1	3 in.	17
Lathing and plastering	See Partitions	4 in.	18
Acoustical fiber tile	1	6 in.	28
		8 in.	34
FLOORS		10 in.	40
Steel Deck	See	Gypsum Block	
	Manufacturer	2 in.	9½
Concrete-Reinforced 1 in.		3 in.	10½
Stone	12½	4 in.	12½
Slag	11½	5 in.	14
Lightweight	6 to10	6 in.	18½
		Wood Studs 2 × 4	
Concrete-Plain 1 in.		12–16 in. o.c.	2
Stone	12	Steel partitions	4
Slag	11	Plaster 1 in.	
Lightweight	3 to 9	Cement	10
		Gypsum	5
Fills 1 in.		Lathing	
Gypsum	6	Metal	½
Sand	8	Gypsum Board ½ in.	2
Cinders	4		
		WALLS	
Finishes		Brick	
Terrazzo 1 in.	13	4 in.	40
Ceramic or Quarry Tile		8 in.	80
¾ in.	10	12 in.	120
Linoleum ¼ in.	1	Hollow Concrete Block	
Mastic ¾ in.	9	(Heavy Aggregate)	
Hardwood ⅞ in.	4	4 in.	30
Softwood ¾ in.	2½	6 in.	43
		8 in.	55
ROOFS		12½ in.	80
Copper or tin	1	Hollow Concrete Block	
Corrugated steel	See	(Light Aggregate)	
	Manufacturer	4 in.	21
3-ply ready roofing	1	6 in.	30
3-ply felt and gravel	5½	8 in.	38
5-ply felt and gravel	6	12 in.	55
		Clay tile	
Shingles		(Load Bearing)	
Wood	2	4 in.	25
Asphalt	3	6 in.	30
Clay tile	9 to 14	8 in.	33
Slate ¼	10	12 in.	45
		Stone 4 in.	55
Sheathing		Glass Block 4 in.	18
Wood ¾ in.	3	Windows, Glass, Frame	8
Gypsum 1 in.	4	& Sash	
		Curtain Walls	See
Insulation 1 in.			Manufacturer
Loose	½	Structural Glass 1 in.	15
Poured-in-place	2	Corrugated Cement	
Rigid	1½	Asbestos ¼ in.	3

For weights of other materials used in building construction, see pages 6- 7 and 6-8

WEIGHTS AND MEASURES
International System of Units (SI)[a]
(Metric practice)

BASE UNITS			SUPPLEMENTARY UNITS		
Quantity	Unit	Symbol	Quantity	Unit	Symbol
Length	Metre	m	Plane angle	Radian	rad
Mass	Kilogram	kg	Solid angle	Steradian	sr
Time	Second	s			
Electric current	Ampere	A			
Thermodynamic temperature	Kelvin	K			
Amount of substance	Mole	mol			
Luminous intensity	Candela	cd			

DERIVED UNITS (WITH SPECIAL NAMES)

Quantity	Unit	Symbol	Formula
Force	Newton	N	$kg\text{-}m/s^2$
Pressure, stress	Pascal	Pa	N/m^2
Energy, work, quantity of heat	Joule	J	$N\text{-}m$
Power	Watt	W	J/s

DERIVED UNITS (WITHOUT SPECIAL NAMES)

Quantity	Unit	Formula
Area	Square metre	m^2
Volume	Cubic metre	m^3
Velocity	Metre per second	m/s
Acceleration	Metre per second squared	m/s^2
Specific volume	Cubic metre per kilogram	m^3/kg
Density	Kilogram per cubic metre	kg/m^3

SI PREFIXES

Multiplication Factor		Prefix	Symbol
1 000 000 000 000 000 000	$=10^{18}$	exa	E
1 000 000 000 000 000	$=10^{15}$	peta	P
1 000 000 000 000	$=10^{12}$	tera	T
1 000 000 000	$=10^{9}$	giga	G
1 000 000	$=10^{6}$	mega	M
1 000	$=10^{3}$	kilo	k
100	$=10^{2}$	hecto[b]	h
10	$=10^{1}$	deka[b]	da
0.1	$=10^{-1}$	deci[b]	d
0.01	$=10^{-2}$	centi[b]	c
0.001	$=10^{-3}$	milli	m
0.000 001	$=10^{-6}$	micro	μ
0.000 000 001	$=10^{-9}$	nano	n
0.000 000 000 001	$=10^{-12}$	pico	p
0.000 000 000 000 001	$=10^{-15}$	femto	f
0.000 000 000 000 000 001	$=10^{-18}$	atto	a

[a]Refer to ASTM E380-79 for more complete information on SI.
[b]Use is not recommended.

WEIGHTS AND MEASURES
United States System

LINEAR MEASURE

Inches		Feet		Yards		Rods		Furlongs		Miles
1.0	=	.08333	=	.02778	=	.0050505	=	.00012626	=	.00001578
12.0	=	1.0	=	.33333	=	.0606061	=	.00151515	=	.00018939
36.0	=	3.0	=	1.0	=	.1818182	=	.00454545	=	.00056818
198.0	=	16.5	=	5.5	=	1.0	=	.025	=	.003125
7920.0	=	660.0	=	220.0	=	40.0	=	1.0	=	.125
63360.0	=	5280.0	=	1760.0	=	320.0	=	8.0	=	1.0

SQUARE AND LAND MEASURE

Sq. In.		Sq. Ft		Sq. Yds.		Sq. Rods		Acres		Sq. Miles
1.0	=	.006944	=	.000772						
144.0	=	1.0	=	.111111						
1296.0	=	9.0	=	1.0	=	.03306	=	.000207		
39204.0	=	272.25	=	30.25	=	1.0	=	.00625	=	.0000098
		43560.0	=	4840.0	=	160.0	=	1.0	=	.0015625
				3097600.0	=	102400.0	=	640.0	=	1.0

AVOIRDUPOIS WEIGHTS

Grains		Drams		Ounces		Pounds		Tons
1.0	=	.03657	=	.002286	=	.000143	=	.0000000714
27.34375	=	1.0	=	.0625	=	.003906	=	.00000195
437.5	=	16.0	=	1.0	=	.0625	=	.00003125
7000.0	=	256.0	=	16.0	=	1.0	=	.0005
14000000.0	=	512000.0	=	32000.0	=	2000.0	=	1.0

DRY MEASURE

Pints		Quarts		Pecks		Cubic Feet		Bushels
1.0	=	.5	=	.0625	=	.01945	=	.01563
2.0	=	1.0	=	.125	=	.03891	=	.03125
16.0	=	8.0	=	1.0	=	.31112	=	.25
51.42627	=	25.71314	=	3.21414	=	1.0	=	.80354
64.0	=	32.0	=	4.0	=	1.2445	=	1.0

LIQUID MEASURE

Gills		Pints		Quarts		U.S. Gallons		Cubic Feet
1.0	=	.25	=	.125	=	.03125	=	.00418
4.0	=	1.0	=	.5	=	.125	=	.01671
8.0	=	2.0	=	1.0	=	.250	=	.03342
32.0	=	8.0	=	4.0	=	1.0	=	.13378
						7.48052	=	1.0

SI CONVERSION FACTORS[a]

Quantity	Multiply	by	to obtain	
Length	Inch	[b]25.400	Millimetre	mm
	Foot	[b] 0.304 800	Metre	m
	Yard	[b] 0.914 400	Metre	m
	Mile (U.S. Statute)	1.609 347	Kilometre	km
	Millimetre	$39.370\ 079 \times 10^{-3}$	Inch	in
	Metre	3.280 840	Foot	ft
	Metre	1.093 613	Yard	yd
	Kilometre	0.621 370	Mile	mi
Area	Square inch	[b] $0.645\ 160 \times 10^{3}$	Square millimetre	mm²
	Square foot	[b] 0.092 903	Square metre	m²
	Square yard	0.836 127	Square metre	m²
	Square mile (U.S. Statute)	2.589 998	Square kilometre	km²
	Acre	$4.046\ 873 \times 10^{3}$	Square metre	m²
	Acre	0.404 687	Hectare	
	Square millimetre	$1.550\ 003 \times 10^{-3}$	Square inch	in²
	Square metre	10.763 910	Square foot	ft²
	Square metre	1.195 990	Square yard	yd²
	Square kilometre	0.386 101	Square mile	mi²
	Square metre	$0.247\ 104 \times 10^{-3}$	Acre	
	Hectare	2.471 044	Acre	
Volume	Cubic inch	[b]$16.387\ 06 \times 10^{3}$	Cubic millimetre	mm³
	Cubic foot	$28.316\ 85 \times 10^{-3}$	Cubic metre	m³
	Cubic yard	0.764 555	Cubic metre	m³
	Gallon (U.S. liquid)	3.785 412	Litre	l
	Quart (U.S. liquid)	0.946 353	Litre	l
	Cubic millimetre	$61.023\ 759 \times 10^{-6}$	Cubic inch	in³
	Cubic metre	35.314 662	Cubic foot	ft³
	Cubic metre	1.307 951	Cubic yard	yd³
	Litre	0.264 172	Gallon (U.S. liquid)	gal
	Litre	1.056 688	Quart (U.S. liquid)	qt
Mass	Ounce (avoirdupois)	28.349 52	Gram	g
	Pound (avoirdupois)	0.453 592	Kilogram	kg
	Short ton	$0.907\ 185 \times 10^{3}$	Kilogram	kg
	Gram	$35.273\ 966 \times 10^{-3}$	Ounce (avoirdupois)	oz av
	Kilogram	2.204 622	Pound (avoirdupois)	lb av
	Kilogram	$1.102\ 311 \times 10^{-3}$	Short ton	

[a]Refer to ASTM E380-79 for more complete information on SI.
[b]Indicates exact value.

SI CONVERSION FACTORS[a]

Quantity	Multiply	by	to obtain	
Force	Ounce-force	0.278 014	Newton	N
	Pound-force	4.448 222	Newton	N
	Newton	3.596 942	Ounce-force	
	Newton	0.224 809	Pound-force	lbf
Bending Moment	Pound-force-inch	0.112 985	Newton-Metre	N-m
	Pound-force-foot	1.355 818	Newton-metre	N-m
	Newton-metre	8.850 748	Pound-force-inch	lbf-in
	Newton-metre	0.737 562	Pound-force-foot	lbf-ft
Pressure, Stress	Pound-force per square inch	6.894 757	Kilopascal	kPa
	Foot of water (39.2 F)	2.988 98	Kilopascal	kPa
	Inch of Mercury (32 F)	3.386 38	Kilopascal	kPa
	Kilopascal	0.145 038	Pound-force per square inch	lbf/in^2
	Kilopascal	0.334 562	Foot of water (39.2 F)	
	Kilopascal	0.295 301	Inch of mercury (32 F)	
Energy, Work, Heat	Foot-pound-force	1.355 818	Joule	J
	[c]British thermal unit	1.055 056 × 10^3	Joule	J
	[c]Calorie	[b] 4.186 800	Joule	J
	Kilowatt hour	[b] 3.600 000 × 10^6	Joule	J
	Joule	0.737 562	Foot-pound-force	ft-lbf
	Joule	0.947 817 × 10^{-3}	[c]British thermal unit	Btu
	Joule	0.238 846	[c]Calorie	
	Joule	0.277 778 × 10^{-6}	Kilowatt hour	kW-h
Power	Foot-pound-force/second	1.355 818	Watt	W
	[c]British thermal unit per hour	0.293 071	Watt	W
	Horsepower (550 ft. lbf/s)	0.745 700	Kilowatt	kW
	Watt	0.737 562	Foot-pound-force/ second	ft-lbf/s
	Watt	3.412 141	[c]British thermal unit per hour	Btu/h
	Kilowatt	1.341 022	Horsepower (550 ft.-lbf/s)	hp
Angle	Degree	17.453 29 × 10^{-3}	Radian	rad
	Radian	57.295 788	Degree	
Temperature	Degree Fahrenheit	$t°C = (t°F - 32)/1.8$	Degree Celsius	
	Degree Celsius	$t°F = 1.8 × t°C + 32$	Degree Fahrenheit	

[a]Refer to ASTM E380-79 for more complete information on SI.
[b]Indicates exact value.
[c]International Table.

BRACING FORMULAS

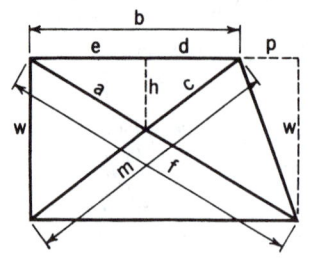
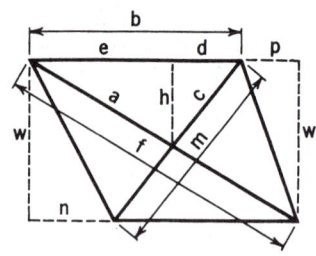

Given	To Find	Formula	Given	To Find	Formula
bpw	f	$\sqrt{(b + p)^2 + w^2}$	bpw	f	$\sqrt{(b + p)^2 + w^2}$
bw	m	$\sqrt{b^2 + w^2}$	bnw	m	$\sqrt{(b - n)^2 + w^2}$
bp	d	$b^2 \div (2b + p)$	bnp	d	$b(b - n) \div (2b + p - n)$
bp	e	$b(b + p) \div (2b + p)$	bnp	e	$b(b + p) \div (2b + p - n)$
bfp	a	$bf \div (2b + p)$	bfnp	a	$bf \div (2b + p - n)$
bmp	c	$bm \div (2b + p)$	bmnp	c	$bm \div (2b + p - n)$
bpw	h	$bw \div (2b + p)$	bnpw	h	$bw \div (2b + p - n)$
afw	h	$aw \div f$	afw	h	$aw \div f$
cmw	h	$cw \div m$	cmw	h	$cw \div m$

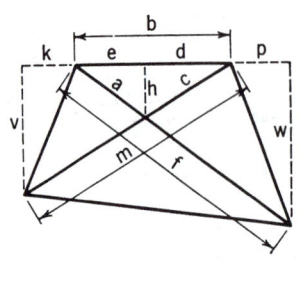

Given	To Find	Formula
bpw	f	$\sqrt{(b + p)^2 + w^2}$
bkv	m	$\sqrt{(b + k)^2 + v^2}$
bkpvw	d	$bw(b + k) \div [v(b + p) + w(b + k)]$
bkpvw	e	$bv(b + p) \div [v(b +p) + w(b + k)]$
bfkpvw	a	$fbv \div [v(b + p) + w(b + k)]$
bkmpvw	c	$bmw \div [v(b + p) + w(b + k)]$
bkpvw	h	$bvw \div [v(b + p) + w(b + k)]$
afw	h	$aw \div f$
cmw	h	$cw \div m$

PARALLEL BRACING

$k = (\log B - \log T) \div$ no. of panels. Constant k plus the logarithm of any line equals the log of the corresponding line in the next panel below.

$$a = TH \div (T + e + p)$$
$$b = Th \div (T + e + p)$$
$$c = \sqrt{(\tfrac{1}{2} T + \tfrac{1}{2} e)^2 + a^2}$$
$$d = ce \div (T + e)$$

$$\log e = k + \log T$$
$$\log f = k + \log a$$
$$\log g = k + \log b$$
$$\log m = k + \log c$$
$$\log n = k + \log d$$
$$\log p = k + \log e$$

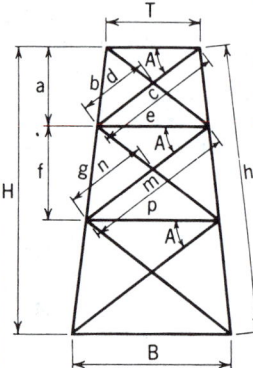

The above method can be used for any number of panels.
In the formulas for "a" and "b" the sum in parenthesis, which in the case shown in $(T + e + p)$, is always composed of all the horizontal distances

PROPERTIES OF PARABOLA AND ELLIPSE

PARABOLA	ELLIPSE

PARABOLA

Apex

Abscissa = x

0.6 H

Height = H

Ordinate = y

c. of g.

½ perimeter

.375 B

½ base = B

Parameter $P = B^2 \div H$ Area $= \frac{2}{3} HB$

$x = y^2 \div P$

$y = \sqrt{xP}$

Construction

a b c d e

H

B

1
2
3
4
5

ELLIPSE

$(x^2 \div H^2) + (y^2 \div B^2) = 1$

Major semi-axis = H

Abscissa = x

Ordinate = y

c. of g.

¼ Perimeter

.424 B

.424 H

Minor semi-axis = B

Area $= .7854\, Dd$

D

d

Construction

a

b

c

e

1
2
3
4

H

B

AREA BETWEEN PARABOLIC CURVE AND SECANT

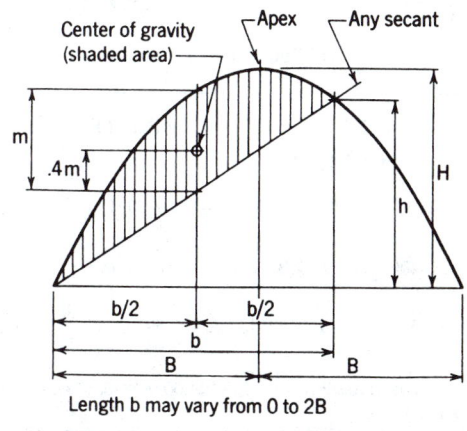

Center of gravity (shaded area)

Apex

Any secant

m

.4 m

H

h

b/2   b/2

b

B   B

Length b may vary from 0 to 2B

PROPERTIES OF THE CIRCLE

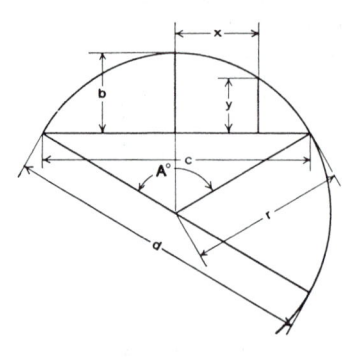

Circumference = 6.28318 r = 3.14159 d
Diameter = 0.31831 circumference
Area = 3.14159 r^2

Arc $a = \dfrac{\pi r A°}{180°} = 0.017453\ r\ A°$

Angle $A° = \dfrac{180°\ a}{\pi r} = 57.29578\ \dfrac{a}{r}$

Radius $r = \dfrac{4\ b^2 + c^2}{8\ b}$

Chord $c = 2\ \sqrt{2\ br - b^2} = 2\ r\ \sin\dfrac{A}{2}$

Rise $b = r - \tfrac{1}{2}\ \sqrt{4\ r^2 - c^2} = \dfrac{c}{2}\ \tan\dfrac{A}{4}$

$= 2\ r\ \sin^2\dfrac{A}{4} = r + y - \sqrt{r^2 - x^2}$

$y = b - r + \sqrt{r^2 - x^2}$
$x = \sqrt{r^2 - (r + y - b)^2}$

Diameter of circle of equal periphery as square = 1.27324 side of square
Side of square of equal periphery as circle = 0.78540 diameter of circle
Diameter of circle circumscribed about square = 1.41421 side of square
Side of square inscribed in circle = 0.70711 diameter of circle

CIRCULAR SECTOR

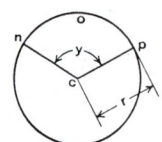

r = radius of circle y = angle ncp in degrees
Area of Sector ncpo = ½ (length of arc nop × r)

$= \text{area of circle} \times \dfrac{y}{360}$

$= 0.0087266 \times r^2 \times y$

CIRCULAR SEGMENT

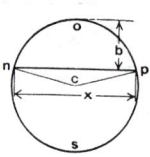

r = radius of circle x = chord b = rise
Area of Segment nop = area of Sector ncpo − area of triangle ncp

$= \dfrac{(\text{Length of arc nop} \times r) - x\ (r - b)}{2}$

Area of Segment nsp = area of circle − area of segment nop

VALUES FOR FUNCTIONS OF π

$\pi = 3.14159265359$, log = 0.4971499

$\pi^2 = 9.8696044$, log = 0.9942997 $\dfrac{1}{\pi} = 0.3183099$, log = $\overline{1}.5028501$ $\sqrt{\dfrac{1}{\pi}} = 0.5641896$, log = $\overline{1}.7514251$

$\pi^3 = 31.0062767$, log = 1.4914496 $\dfrac{1}{\pi^2} = 0.1013212$, log = $\overline{1}.0057003$ $\dfrac{\pi}{180} = 0.0174533$, log = $\overline{2}.2418774$

$\sqrt{\pi} = 1.7724539$, log = 0.2485749 $\dfrac{1}{\pi^3} = 0.0322515$, log = $\overline{2}.5085504$ $\dfrac{180}{\pi} = 57.2957795$, log = 1.7581226

Note: Logs of fractions such as $\overline{1}.5028501$ and $\overline{2}.5085500$ may also be written 9.5028501 − 10 and
8.5085500 − 10 respectively.

PROPERTIES OF GEOMETRIC SECTIONS

SQUARE
Axis of moments through center

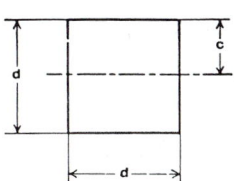

$A = d^2$

$c = \dfrac{d}{2}$

$I = \dfrac{d^4}{12}$

$S = \dfrac{d^3}{6}$

$r = \dfrac{d}{\sqrt{12}} = .288675\,d$

$Z = \dfrac{d^3}{4}$

SQUARE
Axis of moments on base

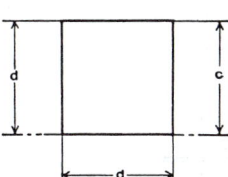

$A = d^2$

$c = d$

$I = \dfrac{d^4}{3}$

$S = \dfrac{d^3}{3}$

$r = \dfrac{d}{\sqrt{3}} = .577350\,d$

SQUARE
Axis of moments on diagonal

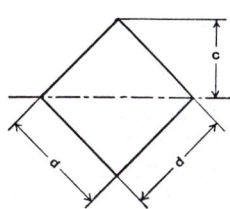

$A = d^2$

$c = \dfrac{d}{\sqrt{2}} = .707107\,d$

$I = \dfrac{d^4}{12}$

$S = \dfrac{d^3}{6\sqrt{2}} = .117851\,d^3$

$r = \dfrac{d}{\sqrt{12}} = .288675\,d$

$Z = \dfrac{2c^3}{3} = \dfrac{d^3}{3\sqrt{2}} = .235702\,d^3$

RECTANGLE
Axis of moments through center

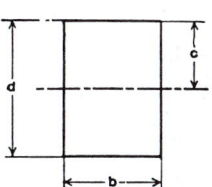

$A = bd$

$c = \dfrac{d}{2}$

$I = \dfrac{bd^3}{12}$

$S = \dfrac{bd^2}{6}$

$r = \dfrac{d}{\sqrt{12}} = .288675\,d$

$Z = \dfrac{bd^2}{4}$

PROPERTIES OF GEOMETRIC SECTIONS

RECTANGLE
Axis of moments on base

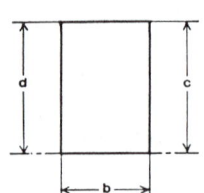

$$A = bd$$
$$c = d$$
$$I = \frac{bd^3}{3}$$
$$S = \frac{bd^2}{3}$$
$$r = \frac{d}{\sqrt{3}} = .577350\,d$$

RECTANGLE
Axis of moments on diagonal

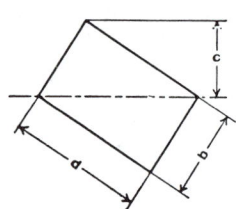

$$A = bd$$
$$c = \frac{bd}{\sqrt{b^2 + d^2}}$$
$$I = \frac{b^3d^3}{6\,(b^2 + d^2)}$$
$$S = \frac{b^2d^2}{6\,\sqrt{(b^2 + d^2}}$$
$$r = \frac{bd}{\sqrt{6\,(b^2 + d^2)}}$$

RECTANGLE
Axis of moments any line
through center of gravity

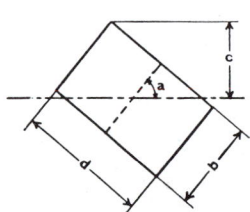

$$A = bd$$
$$c = \frac{b \sin a + d \cos a}{2}$$
$$I = \frac{bd\,(b^2 \sin^2 a + d^2 \cos^2 a)}{12}$$
$$S = \frac{bd\,(b^2 \sin^2 a + d^2 \cos^2 a)}{6\,(b \sin a + d \cos a)}$$
$$r = \sqrt{\frac{b^2 \sin^2 a + d^2 \cos^2 a}{12}}$$

HOLLOW RECTANGLE
Axis of moments through center

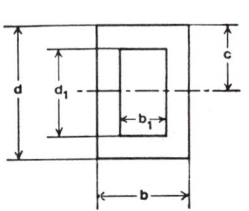

$$A = bd - b_1 d_1$$
$$c = \frac{d}{2}$$
$$I = \frac{bd^3 - b_1 d_1^3}{12}$$
$$S = \frac{bd^3 - b_1 d_1^3}{6d}$$
$$r = \sqrt{\frac{bd^3 - b_1 d_1^3}{12A}}$$
$$Z = \frac{bd^2}{4} - \frac{b_1 d_1^2}{4}$$

PROPERTIES OF GEOMETRIC SECTIONS

EQUAL RECTANGLES

Axis of moments through center of gravity

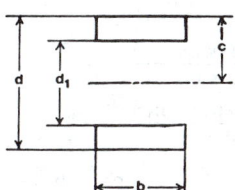

$$A = b(d - d_1)$$

$$c = \frac{d}{2}$$

$$I = \frac{b(d^3 - d_1^3)}{12}$$

$$S = \frac{b(d^3 - d_1^3)}{6d}$$

$$r = \sqrt{\frac{d^3 - d_1^3}{12(d - d_1)}}$$

$$Z = \frac{b}{4}(d^2 - d_1^2)$$

UNEQUAL RECTANGLES

Axis of moments through center of gravity

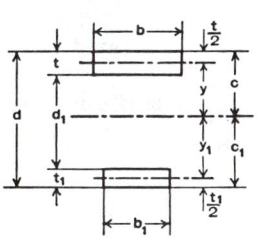

$$A = bt + b_1 t_1$$

$$c = \frac{\frac{1}{2}bt^2 + b_1 t_1 (d - \frac{1}{2}t_1)}{A}$$

$$I = \frac{bt^3}{12} + bty^2 + \frac{b_1 t_1^3}{12} + b_1 t_1 y_1^2$$

$$S = \frac{I}{c} \qquad S_1 = \frac{I}{c_1}$$

$$r = \sqrt{\frac{I}{A}}$$

$$Z = \frac{A}{2}\left[d - \left(\frac{t + t_1}{2}\right)\right]$$

TRIANGLE

Axis of moments through center of gravity

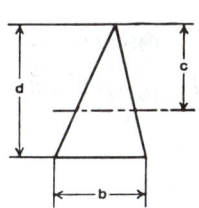

$$A = \frac{bd}{2}$$

$$c = \frac{2d}{3}$$

$$I = \frac{bd^3}{36}$$

$$S = \frac{bd^2}{24}$$

$$r = \frac{d}{\sqrt{18}} = .235702\, d$$

TRIANGLE

Axis of moments on base

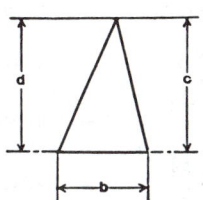

$$A = \frac{bd}{2}$$

$$c = d$$

$$I = \frac{bd^3}{12}$$

$$S = \frac{bd^2}{12}$$

$$r = \frac{d}{\sqrt{6}} = .408248\, d$$

PROPERTIES OF GEOMETRIC SECTIONS

TRAPEZOID
Axis of moments through
center of gravity

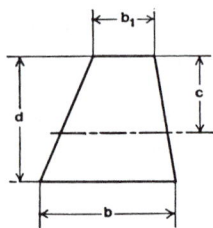

$$A = \frac{d(b + b_1)}{2}$$

$$c = \frac{d(2b + b_1)}{3(b + b_1)}$$

$$I = \frac{d^3(b^2 + 4bb_1 + b_1^2)}{36(b + b_1)}$$

$$S = \frac{d^2(b^2 + 4bb_1 + b_1^2)}{12(2b + b_1)}$$

$$r = \frac{d}{6(b + b_1)}\sqrt{2(b^2 + 4bb_1 + b_1^2)}$$

CIRCLE
Axis of moments
through center

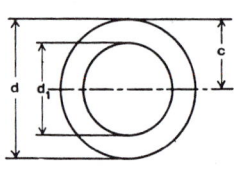

$$A = \frac{\pi d^2}{4} = \pi R^2 \quad .785398\, d^2 = 3.141593\, R^2$$

$$c = \frac{d}{2} = R$$

$$I = \frac{\pi d^4}{64} = \frac{\pi R^4}{4} = .049087\, d^4 = .785398\, R^4$$

$$S = \frac{\pi d^3}{32} = \frac{\pi R^3}{4} = .098175\, d^3 = .785398\, R^3$$

$$r = \frac{d}{4} = \frac{R}{2}$$

$$Z = \frac{d^3}{6}$$

HOLLOW CIRCLE
Axis of moments
through center

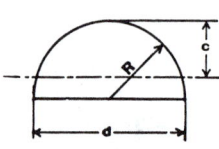

$$A = \frac{\pi(d^2 - d_1^2)}{4} = .785398\,(d^2 - d_1^2)$$

$$c = \frac{d}{2}$$

$$I = \frac{\pi(d^4 - d_1^4)}{64} = .049087\,(d^4 - d_1^4)$$

$$S = \frac{\pi(d^4 - d_1^4)}{32d} = .098175\,\frac{d^4 - d_1^4}{d}$$

$$r = \frac{\sqrt{d^2 + d_1^2}}{4}$$

$$Z = \frac{d^3}{6} - \frac{d_1^3}{6}$$

HALF CIRCLE
Axis of moments through
center of gravity

$$A = \frac{\pi R^2}{2} \qquad\qquad = 1.570796\, R^2$$

$$c = R\left(1 - \frac{4}{3\pi}\right) \qquad = .575587\, R$$

$$I = R^4\left(\frac{\pi}{8} - \frac{8}{9\pi}\right) \qquad = .109757\, R^4$$

$$S = \frac{R^3}{24}\frac{(9\pi^2 - 64)}{(3\pi - 4)} \qquad = .190687\, R^3$$

$$r = R\frac{\sqrt{9\pi^2 - 64}}{6\pi} \qquad = .264336\, R$$

PROPERTIES OF GEOMETRIC SECTIONS

PARABOLA

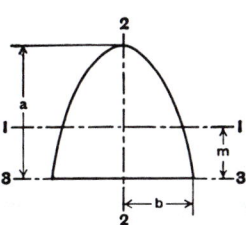

$$A = \frac{4}{3}ab$$

$$m = \frac{2}{5}a$$

$$I_1 = \frac{16}{175}a^3b$$

$$I_2 = \frac{4}{15}ab^3$$

$$I_3 = \frac{32}{105}a^3b$$

HALF PARABOLA

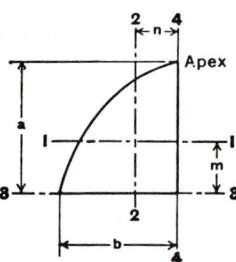

$$A = \frac{2}{3}ab$$

$$m = \frac{2}{5}a$$

$$n = \frac{3}{8}b$$

$$I_1 = \frac{8}{175}a^3b$$

$$I_2 = \frac{19}{480}ab^3$$

$$I_3 = \frac{16}{105}a^3b$$

$$I_4 = \frac{2}{15}ab^3$$

COMPLEMENT OF HALF PARABOLA

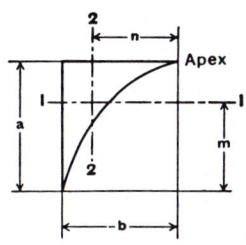

$$A = \frac{1}{3}ab$$

$$m = \frac{7}{10}a$$

$$n = \frac{3}{4}b$$

$$I_1 = \frac{37}{2100}a^3b$$

$$I_2 = \frac{1}{80}ab^3$$

PARABOLIC FILLET IN RIGHT ANGLE

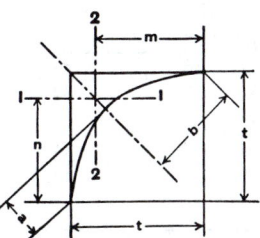

$$a = \frac{t}{2\sqrt{2}}$$

$$b = \frac{t}{\sqrt{2}}$$

$$A = \frac{1}{6}t^2$$

$$m = n = \frac{4}{5}t$$

$$I_1 = I_2 = \frac{11}{2100}t^4$$

PROPERTIES OF GEOMETRIC SECTIONS

*HALF ELLIPSE

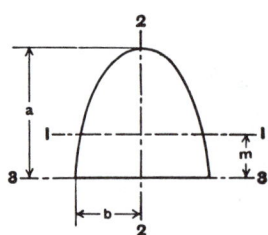

$$A = \frac{1}{2}\pi ab$$

$$m = \frac{4a}{3\pi}$$

$$I_1 = a^3b\left(\frac{\pi}{8} - \frac{8}{9\pi}\right)$$

$$I_2 = \frac{1}{8}\pi ab^3$$

$$I_3 = \frac{1}{8}\pi a^3b$$

*QUARTER ELLIPSE

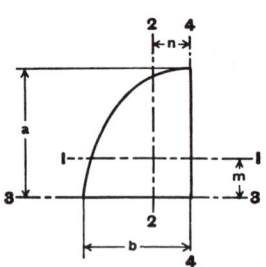

$$A = \frac{1}{4}\pi ab$$

$$m = \frac{4a}{3\pi}$$

$$n = \frac{4b}{3\pi}$$

$$I_1 = a^3b\left(\frac{\pi}{16} - \frac{4}{9\pi}\right)$$

$$I_2 = ab^3\left(\frac{\pi}{16} - \frac{4}{9\pi}\right)$$

$$I_3 = \frac{1}{16}\pi a^3b$$

$$I_4 = \frac{1}{16}\pi ab^3$$

*ELLIPTIC COMPLEMENT

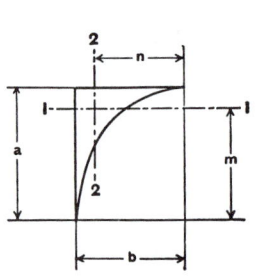

$$A = ab\left(1 - \frac{\pi}{4}\right)$$

$$m = \frac{a}{6\left(1 - \frac{\pi}{4}\right)}$$

$$n = \frac{b}{6\left(1 - \frac{\pi}{4}\right)}$$

$$I_1 = a^3b\left(\frac{1}{3} - \frac{\pi}{16} - \frac{1}{36\left(1 - \frac{\pi}{4}\right)}\right)$$

$$I_2 = ab^3\left(\frac{1}{3} - \frac{\pi}{16} - \frac{1}{36\left(1 - \frac{\pi}{4}\right)}\right)$$

*To obtain properties of half circle, quarter circle and circular complement substitute a = b = R.

PROPERTIES OF GEOMETRIC SECTIONS

REGULAR POLYGON
Axis of moments
through center

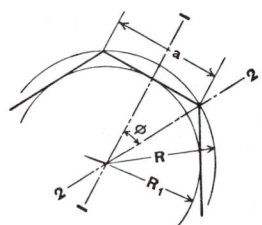

n = Number of sides

$\phi = \dfrac{180°}{n}$

$a = 2\sqrt{R^2 - R_1^2}$

$R = \dfrac{a}{2 \sin \phi}$

$R_1 = \dfrac{a}{2 \tan \phi}$

$A = \dfrac{1}{4} na^2 \cot \phi = \dfrac{1}{2} nR^2 \sin 2\phi = nR_1^2 \tan \phi$

$I_1 = I_2 = \dfrac{A(6R^2 - a^2)}{24} = \dfrac{A(12R_1^2 + a^2)}{48}$

$r_1 = r_2 = \sqrt{\dfrac{6R^2 - a^2}{24}} = \sqrt{\dfrac{12R_1^2 + a^2}{48}}$

ANGLE
Axis of moments through
center of gravity

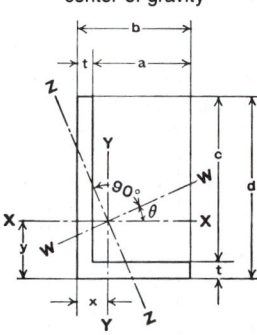

Z-Z is axis of minimum I

$\tan 2\theta = \dfrac{2 K}{I_Y - I_X}$

$A = t(b + c) \quad x = \dfrac{b^2 + ct}{2(b + c)} \quad y = \dfrac{d^2 + at}{2(b + c)}$

K = Product of Inertia about X-X & Y-Y

$= \mp \dfrac{abcdt}{4(b + c)}$

$I_X = \dfrac{1}{3} [t(d - y)^3 + by^3 - a(y - t)^3]$

$I_Y = \dfrac{1}{3} [t(b - x)^3 + dx^3 - c(x - t)^3]$

$I_Z = I_X \sin^2\theta + I_Y \cos^2\theta + K \sin2\theta$

$I_W = I_X \cos^2\theta + I_Y \sin^2\theta - K \sin2\theta$

K is negative when heel of angle, with respect to c. g., is in 1st or 3rd quadrant, positive when in 2nd or 4th quadrant.

BEAMS AND CHANNELS
Transverse force oblique
through center of gravity

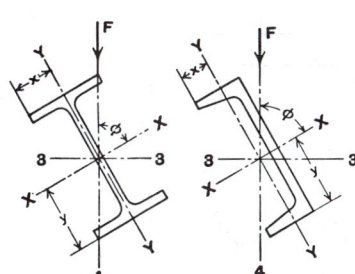

$I_3 = I_X \sin^2\phi + I_Y \cos^2\phi$

$I_4 = I_X \cos^2\phi + I_Y \sin^2\phi$

$f_b = M \left(\dfrac{y}{I_X} \sin\phi + \dfrac{x}{I_Y} \cos\phi \right)$

where Mj is bending moment due to force F.

TRIGONOMETRIC FORMULAS

TRIGONOMETRIC FUNCTIONS

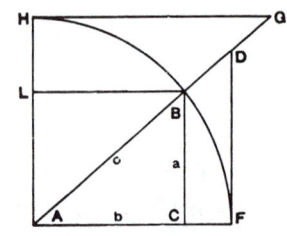

Radius AF $= 1$
$= \sin^2 A + \cos^2 A = \sin A \, \mathrm{cosec}\, A$
$= \cos A \sec A = \tan A \cot A$

Sine A $= \dfrac{\cos A}{\cot A} = \dfrac{1}{\mathrm{cosec}\, A} = \cos A \tan A = \sqrt{1-\cos^2 A} = BC$

Cosine A $= \dfrac{\sin A}{\tan A} = \dfrac{1}{\sec A} = \sin A \cot A = \sqrt{1-\sin^2 A} = AC$

Tangent A $= \dfrac{\sin A}{\cos A} = \dfrac{1}{\cot A} = \sin A \sec A \hspace{2cm} = FD$

Cotangent A $= \dfrac{\cos A}{\sin A} = \dfrac{1}{\tan A} = \cos A \, \mathrm{cosec}\, A \hspace{1cm} = HG$

Secant A $= \dfrac{\tan A}{\sin A} = \dfrac{1}{\cos A} \hspace{3cm} = AD$

Cosecant A $= \dfrac{\cot A}{\cos A} = \dfrac{1}{\sin A} \hspace{3cm} = AG$

RIGHT ANGLED TRIANGLES

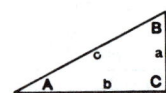

$$a^2 = c^2 - b^2$$
$$b^2 = c^2 - a^2$$
$$c^2 = a^2 + b^2$$

Known	Required					
	A	B	a	b	c	Area
a, b	$\tan A = \dfrac{a}{b}$	$\tan B = \dfrac{b}{a}$			$\sqrt{a^2 + b^2}$	$\dfrac{ab}{2}$
a, c	$\sin A = \dfrac{a}{c}$	$\cos B = \dfrac{a}{c}$		$\sqrt{c^2 - a^2}$		$\dfrac{a\sqrt{c^2 - a^2}}{2}$
A, a		$90° - A$		$a \cot A$	$\dfrac{a}{\sin A}$	$\dfrac{a^2 \cot A}{2}$
A, b		$90° - A$	$b \tan A$		$\dfrac{b}{\cos A}$	$\dfrac{b^2 \tan A}{2}$
A, c		$90° - A$	$c \sin A$	$c \cos A$		$\dfrac{c^2 \sin 2A}{4}$

OBLIQUE ANGLED TRIANGLES

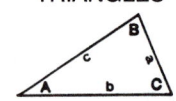

$$s = \dfrac{a + b + c}{2}$$

$$K = \sqrt{\dfrac{(s - a)(s - b)(s - c)}{s}}$$

$$a^2 = b^2 + c^2 - 2bc \cos A$$
$$b^2 = a^2 + c^2 - 2ac \cos B$$
$$c^2 = a^2 + b^2 - 2ab \cos C$$

Known	Required					
	A	B	a	b	c	Area
a, b, c	$\tan \dfrac{1}{2} A = \dfrac{K}{s-a}$	$\tan \dfrac{1}{2} B = \dfrac{K}{s-b}$	$\tan \dfrac{1}{2} C = \dfrac{K}{s-C}$			$\sqrt{s(s-a)(s-b)(s-c)}$
a, A, B			$180° - (A+B)$	$\dfrac{a \sin B}{\sin A}$	$\dfrac{a \sin C}{\sin A}$	
a, b, A		$\sin B = \dfrac{b \sin A}{a}$			$\dfrac{b \sin C}{\sin B}$	
a, b, C	$\tan A = \dfrac{a \sin C}{b - a \cos C}$				$\sqrt{a^2 + b^2 - 2ab \cos C}$	$\dfrac{ab \sin C}{2}$

DECIMALS OF AN INCH
For each 1/64 of an inch
With Millimeter Equivalents

Fraction	1/64	Decimal	Millimeters (Approx.)	Fraction	1/64	Decimal	Millimeters (Approx.)
—	1	.015625	0.397	—	33	.515625	13.097
1/32	2	.03125	0.794	17/32	34	.53125	13.494
—	3	.046875	1.191	—	35	.546875	13.891
1/16	4	.0625	1.588	9/16	36	.5625	14.288
—	5	.078125	1.984	—	37	.578125	14.684
3/32	6	.09375	2.381	19/32	38	.59375	15.081
—	7	.109375	2.778	—	39	.609375	15.478
1/8	8	.125	3.175	5/8	40	.625	15.875
—	9	.140625	3.572	—	41	.640625	16.272
5/32	10	.15625	3.969	21/32	42	.65625	16.669
—	11	.171875	4.366	—	43	.671875	17.066
3/16	12	.1875	4.763	11/16	44	.6875	17.463
—	13	.203125	5.159	—	45	.703125	17.859
7/32	14	.21875	5.556	23/32	46	.71875	18.256
—	15	.234375	5.953	—	47	.734375	18.653
1/4	16	.250	6.350	3/4	48	.750	19.050
—	17	.265625	6.747	—	49	.765625	19.447
9/32	18	.28125	7.144	25/32	50	.78125	19.844
—	19	.296875	7.541	—	51	.796875	20.241
5/16	20	.3125	7.938	13/16	52	.8125	20.638
—	21	.328125	8.334	—	53	.828125	21.034
11/32	22	.34375	8.731	27/32	54	.84375	21.431
—	23	.359375	9.128	—	55	.859375	21.828
3/8	24	.375	9.525	7/8	56	.875	22.225
—	25	.390625	9.922	—	57	.890625	22.622
13/32	26	.40625	10.319	29/32	58	.90625	23.019
—	27	.421875	10.716	—	59	.921875	23.416
7/16	28	.4375	11.113	15/16	60	.9375	23.813
—	29	.453125	11.509	—	61	.953125	24.209
15/32	30	.46875	11.906	31/32	62	.96875	24.606
—	31	.484375	12.303	—	63	.984375	25.003
1/2	32	.500	12.700	1	64	1.000	25.400

DECIMALS OF A FOOT
For each 1/32 of an inch

Inch	0	1	2	3	4	5
0	0	.0833	.1667	.2500	.3333	.4167
1/32	.0026	.0859	.1693	.2526	.3359	.4193
1/16	.0052	.0885	.1719	.2552	.3385	.4219
3/32	.0078	.0911	.1745	.2578	.3411	.4245
1/8	.0104	.0938	.1771	.2604	.3438	.4271
5/32	.0130	.0964	.1797	.2630	.3464	.4297
3/16	.0156	.0990	.1823	.2656	.3490	.4323
7/32	.0182	.1016	.1849	.2682	.3516	.4349
1/4	.0208	.1042	.1875	.2708	.3542	.4375
9/32	.0234	.1068	.1901	.2734	.3568	.4401
5/16	.0260	.1094	.1927	.2760	.3594	.4427
11/32	.0286	.1120	.1953	.2786	.3620	.4453
3/8	.0313	.1146	.1979	.2812	.3646	.4479
13/32	.0339	.1172	.2005	.2839	.3672	.4505
7/16	.0365	.1198	.2031	.2865	.3698	.4531
15/32	.0391	.1224	.2057	.2891	.3724	.4557
1/2	.0417	.1250	.2083	.2917	.3750	.4583
17/32	.0443	.1276	.2109	.2943	.3776	.4609
9/16	.0469	.1302	.2135	.2969	.3802	.4635
19/32	.0495	.1328	.2161	.2995	.3828	.4661
5/8	.0521	.1354	.2188	.3021	.3854	.4688
21/32	.0547	.1380	.2214	.3047	.3880	.4714
11/16	.0573	.1406	.2240	.3073	.3906	.4740
23/32	.0599	.1432	.2266	.3099	.3932	.4766
3/4	.0625	.1458	.2292	.3125	.3958	.4792
25/32	.0651	.1484	.2318	.3151	.3984	.4818
13/16	.0677	.1510	.2344	.3177	.4010	.4844
27/32	.0703	.1536	.2370	.3203	.4036	.4870
7/8	.0729	.1563	.2396	.3229	.4063	.4896
29/32	.0755	.1589	.2422	.3255	.4089	.4922
15/16	.0781	.1615	.2448	.3281	.4115	.4948
31/32	.0807	.1641	.2474	.3307	.4141	.4974

DECIMALS OF A FOOT
For each 1/32 of an inch

Inch	6	7	8	9	10	11
0	.5000	.5833	.6667	.7500	.8333	.9167
1/32	.5026	.5859	.6693	.7526	.8359	.9193
1/16	.5052	.5885	.6719	.7552	.8385	.9219
3/32	.5078	.5911	.6745	.7578	.8411	.9245
1/8	.5104	.5938	.6771	.7604	.8438	.9271
5/32	.5130	.5964	.6797	.7630	.8464	.9297
3/16	.5156	.5990	.6823	.7656	.8490	.9323
7/32	.5182	.6016	.6849	.7682	.8516	.9349
1/4	.5208	.6042	.6875	.7708	.8542	.9375
9/32	.5234	.6068	.6901	.7734	.8568	.9401
5/16	.5260	.6094	.6927	.7760	.8594	.9427
11/32	.5286	.6120	.6953	.7786	.8620	.9453
3/8	.5313	.6146	.6979	.7812	.8646	.9479
13/32	.5339	.6172	.7005	.7839	.8672	.9505
7/16	.5365	.6198	.7031	.7865	.8698	.9531
15/32	.5391	.6224	.7057	.7891	.8724	.9557
1/2	.5417	.6250	.7083	.7917	.8750	.9583
17/32	.5443	.6276	.7109	.7943	.8776	.9609
9/16	.5469	.6302	.7135	.7969	.8802	.9635
19/32	.5495	.6328	.7161	.7995	.8828	.9661
5/8	.5521	.6354	.7188	.8021	.8854	.9688
21/32	.5547	.6380	.7214	.8047	.8880	.9714
11/16	.5573	.6406	.7240	.8073	.8906	.9740
23/32	.5599	.6432	.7266	.8099	.8932	.9766
3/4	.5625	.6458	.7292	.8125	.8958	.9792
25/32	.5651	.6484	.7318	.8151	.8984	.9818
13/16	.5677	.6510	.7344	.8177	.9010	.9844
27/32	.5703	.6536	.7370	.8203	.9036	.9870
7/8	.5729	.6563	.7396	.8229	.9063	.9896
29/32	.5755	.6589	.7422	.8255	.9089	.9922
15/16	.5781	.6615	.7448	.8281	.9115	.9948
31/32	.5807	.6641	.7474	.8307	.9141	.9974

SYMBOLS

A	Cross-sectional area, in.^{2}
	Gross area of an axially loaded compression member, in.^{2}
A_b	Nominal body area of a fastener, in.^{2}
	Area of an upset rod based upon the major diameter of its threads, i.e., the diameter of a coaxial cylinder which would bound the crests of the upset threads, in.^{2}
A_{bc}	Planar area of web at beam-to-column connection, in.^{2}
A_c	Actual area of effective concrete flange in composite design, in.^{2}
A_{ctr}	Concrete transformed area in compression, in.^{2}
	$= \left(\dfrac{b}{n}\right) t$
A_e	Effective net area of an axially loaded tension member, in.^{2}
A_f	Area of compression flange, in.^{2}
A_{fe}	Effective tension flange area, in.^{2}
A_{fg}	Gross beam flange area, in.^{2}
A_{fn}	Net beam flange area, in.^{2}
A_g	Gross area of member, in.^{2}
A_n	Net area of an axially loaded tension member, in.^{2}
A_s	Area of steel beam in composite design, in.^{2}
A_s'	Area of compressive reinforcing steel, in.^{2}
A_{sr}	Area of reinforcing steel providing composite action at point of negative moment, in.^{2}
A_{st}	Cross-sectional area of a stiffener or pair of stiffeners, in.^{2}
A_t	Net tension area, in.^{2}
A_v	Net shear area, in.^{2}
A_w	Area of girder web, in.^{2}
A_1	Area of steel bearing concentrically on a concrete support, in.^{2}
A_2	Maximum area of the portion of the supporting surface that is geometrically similar to and concentric with the loaded area, in.^{2}
B	Bending coefficient dependent upon computed moment or stress at the ends of unbraced segments of a tapered member
	Allowable load per bolt, kips
B_c	Load per bolt, including prying action, kips
C	Coefficient for determining allowable loads in kips for eccentrically loaded connections
C_a	Coefficient used in Table 4 of Numerical Values
	Constant used in calculating moment for end-plate design: 1.13 for 36-ksi and 1.11 for 50-ksi steel
C_b	Bending coefficient dependent upon moment gradient
	$= 1.75 + 1.05 \left(\dfrac{M_1}{M_2}\right) + 0.3 \left(\dfrac{M_1}{M_2}\right)^2$
	Coefficient used in calculating moment for end-plate design
	$= \sqrt{b_f/b_p}$

C_c Column slenderness ratio separating elastic and inelastic buckling

C_c' Column slenderness ratio dividing elastic and inelastic buckling, modified to account for effective width of wide compression elements

$$= C_c \frac{1}{\sqrt{Q_a}} \text{ or } C_c \frac{1}{\sqrt{Q_s}} \text{ or } C_c \frac{1}{\sqrt{Q_a Q_s}}$$

C_h Coefficient used in Table 12 of Numerical Values

C_m Coefficient applied to bending term in interaction equation for prismatic members and dependent upon column curvature caused by applied moments

C_m' Coefficient applied to bending term in interaction equation for tapered members and dependent upon axial stress at the small end of the member

C_p Stiffness factor for primary member in a flat roof

C_s Stiffness factor for secondary member in a flat roof

C_v Ratio of "critical" web stress, according to the linear buckling theory, to the shear yield stress of web material

C_w Warping constant for a section, in.^{6}

C_1 Coefficient for web tear-out (block shear)

        Increment used in computing minimum spacing of oversized and slotted holes

C_2 Coefficient for web tear-out (block shear)

        Increment used in computing minimum edge distance for oversized and slotted holes

D Factor depending upon type of transverse stiffeners

        Outside diameter of tubular member, in.

        Number of $\frac{1}{16}$-inches in weld size

E Modulus of elasticity of steel (29,000 ksi)

E_c Modulus of elasticity of concrete, ksi

E_t Tangent modulus of elasticity, ksi

F_a Axial compressive stress permitted in a prismatic member in the absence of bending moment, ksi

$F_{a\gamma}$ Axial compressive stress permitted in a tapered member in the absence of bending moment, ksi

F_b Bending stress permitted in a prismatic member in the absence of axial force, ksi

F_b' Allowable bending stress in compression flange of plate girders as reduced for hybrid girders or because of large web depth-to-thickness ratio, ksi

$F_{b\gamma}$ Bending stress permitted in a tapered member in the absence of axial force, ksi

F_e' Euler stress for a prismatic member divided by factor of safety, ksi

$F_{e\gamma}'$ Euler stress for a tapered member divided by factor of safety, ksi

F_f Flange force due to moment in end-plate connections, kips

F_p Allowable bearing stress, ksi

$F_{s\gamma}$ St. Venant torsion resistance bending stress in a tapered member, ksi

F_t Allowable axial tensile stress, ksi

F_u Specified minimum tensile strength of the type of steel or fastener being used, ksi

F_v Allowable shear stress, ksi

$F_{w\gamma}$ Flange warping torsion resistance bending stress in a tapered member, ksi

F_y Specified minimum yield stress of the type of steel being used, ksi. As used in this Manual, "yield stress" denotes either the specified minimum yield point (for those steels that have a yield point) or specified minimum yield strength (for those steels that do not have a yield point)

F_y' The theoretical maximum yield stress (ksi) based on the width-thickness ratio of one-half the unstiffened compression flange, beyond which a particular shape is not "compact." See AISC Specification Sect. B5.1.

$$= \left[\frac{65}{b_f/2t_f} \right]^2$$

F_y''' The theoretical maximum yield stress (ksi) based on the depth-thickness ratio of the web below which a particular shape may be considered "compact" for any condition of combined bending and axial stresses. See AISC Specification Sect. B5.1.

$$= \left[\frac{257}{d/t_w} \right]^2$$

F_{yc} Specified minimum column yield stress, ksi

F_{yf} Specified minimum yield stress of flange, ksi

F_{yr} Specified minimum yield stress of the longitudinal reinforcing steel, ksi

F_{yst} Specified minimum stiffener yield stress, ksi

F_{yw} Specified minimum yield stress of beam web, ksi

G Shear modulus of elasticity of steel (11,200 ksi)

    Nomograph designation of end condition used in column design to determine the effective length

H_s Length of a stud shear connector after welding, in.

I Moment of inertia of a section, in.^{4}

I_d Moment of inertia of steel deck supported on secondary members, in.^{4}

I_{eff} Effective moment of inertia of composite sections for deflection computations, in.^{4}

I_p Moment of inertia of primary member in flat-roof framing, in.^{4}

    Polar moment of inertia, in.^{4}

I_s Moment of inertia of secondary member in flat-roof framing, in.^{4}

    Moment of inertia of steel beam in composite construction, in.^{4}

I_{tr} Moment of inertia of transformed composite section, in.^{4}

I_x Moment of inertia of a section about the X - X axis, in.^{4}

I_y Moment of inertia of a section about the Y - Y axis, in.^{4}

J Torsional constant of a cross-section, in.^{4}

K Effective length factor for a prismatic member

K_γ Effective length factor for a tapered member

L Span length, ft

    Length of connection angles, in.

    Unbraced length of tensile members, in.

    Unbraced length of member measured between centers of gravity of the bracing members, in.

    Plate length, in.

L_c Maximum unbraced length of the compression flange at which the allowable bending stress may be taken at $0.66F_y$ or as determined by AISC Specification Eq. (F1-3) or Eq. (F2-3), when applicable, ft

Unsupported length of a column section, ft

L_e Distance from free edge to center of the bolt, in.

L_g Unsupported length of a girder or other restraining member, ft

L_p Length of primary member in flat-roof framing, ft

L_s Length of secondary member in flat-roof framing, ft

L_u Maximum unbraced length of the compression flange at which the allowable bending stress may be taken at $0.6F_y$, ft

L_v Span for maximum allowable web shear of uniformly loaded beam, ft

M Moment, kip-ft

Maximum factored bending moment, kip-ft

M_1 Smaller moment at end of unbraced length of beam-column

Sum of moments due to lateral load and wind load on the leeward side of beam-to-column connections, kip-in.

Larger moment at one end of three-segment part of a tapered member

M_2 Larger moment at end of unbraced length of beam-column

Difference between the moments due to lateral load and gravity load on the windward side of beam-to-column connections, kip-in.

Maximum moment in three adjacent segments of a tapered member

M_D Moment produced by dead load

M_L Moment produced by live load

Moment produced by loads imposed after the concrete has achieved 75% of its required strength

M_e Extreme fiber bending moment in end-plate design, kip-in.

M_m Critical moment that can be resisted by a plastically designed member in the absence of axial load, kip-ft

M_p Plastic moment, kip-ft

N Length of base plate, in.

Length of bearing of applied load, in.

N_e Length at end bearing to develop maximum web shear, in.

N_r Number of stud shear connectors on a beam in one transverse rib of a metal deck, not to exceed 3 in calculations

N_1 Number of shear connectors required between point of maximum moment and point of zero moment

N_2 Number of shear connectors required between concentrated load and point of zero moment

P Applied load, kips

Force transmitted by a fastener, kips

Factored axial load, kips

Normal force, kips

P_R Beam reaction divided by the number of bolts in high-strength bolted connection, kips

P_b Plate bearing capacity in single-plate shear connections, kips

P_{bf}	Factored beam flange or connection plate force in a restrained connection, kips
P_{cr}	Maximum strength of an axially loaded compression member or beam, kips
P_e	Euler buckling load, kips
P_{ec}	Effective horizontal bolt distance used in end-plate connection design, in.
P_f	Distance between top or bottom of top flange to nearest bolt, in.
P_{fb}	Force, from a beam flange or moment connection plate, that a column will resist without stiffeners, as determined using Eq. (K1-1), kips
P_{wb}	Force, from a beam flange or moment connection plate, that a column will resist without stiffeners, as determined using Eq. (K1-8), kips
P_{wi}	Force, in addition to P_{wo}, that a column will resist without stiffeners, from a beam flange or moment connection plate of one inch thickness, as derived from Eq. (K1-9), kips
P_{wo}	Force, from a beam flange or moment connection plate of zero thickness, that a column will resist without stiffeners, as derived from Eq. (K1-9), kips
P_y	Plastic axial load, equal to profile area times specified minimum yield stress, kips
Q	Prying force per fastener, kips
	Full reduction factor for slender compression elements
Q_a	Ratio of effective profile area of an axially loaded member to its total profile area, Appendix B5.2
Q_f	Statical moment of flange, in.^{3}
Q_s	Axial stress reduction factor where width-thickness ratio of unstiffened elements exceeds noncompact section limits given in Specification Sect. B5.1
Q_w	Statical moment of cross section, in.^{3}
R	Maximum end reaction for 3½ in. of bearing, kips
	Reaction or concentrated load applied to beam or girder, kips
	Radius, in.
	Shear force in a single element at any given deformation, kips
R_1	A constant used in web yielding calculations, from Eq. (K1-3), kips $= 0.66\, F_y\, t_w\, (2.5k)$
R_2	A constant used in web yielding calculations, from Eq. (K1-3), kips/in. $= 0.66\, F_y\, t_w$
R_3	A constant used in web crippling calculations, from Eq. (K1-5), kips $= 34\, t_w^2\, \sqrt{F_{yw} t_f / t_w}$
R_4	A constant used in web crippling calculations, from Eq. (K1-5), kips/in. $= 34\, t_w^2 \left[3\left(\dfrac{1}{d}\right)\left(\dfrac{t_w}{t_f}\right)^{1.5} \right] \sqrt{F_{yw} t_f / t_w}$
R_{BS}	Resistance to web tear-out (block shear), kips
R_{PG}	Plate girder bending strength reduction factor
R_b	Bolt group capacity in single-plate shear connections, kips
R_e	Hybrid girder factor
R_i	Increase in reaction R in kips for each additional inch of bearing
R_o	Plate capacity in yielding in single-plate shear connections, kips
R_{ult}	Ultimate shear load of a single element

R_v Shear capacity of the net section of connection angles

S Elastic section modulus, in.^{3}

Spacing of secondary members in a flat roof, ft

Governing slenderness ratio of a tapered member

S' Additional section modulus corresponding to $\frac{1}{16}$-in. increase in web thickness for welded plate griders, in.^{3}

S_{eff} Effective section modulus corresponding to partial composite action, in.^{3}

S_s Section modulus of steel beam used in composite design, referred to the bottom flange, in.^{3}

S_t Section modulus of transformed composite cross-section, referred to the top of concrete, in.^{3}

$S_{t\text{-}eff}$ Section modulus relative to the top of the equivalent transformed steel section, in.^{3}

S_{tr} Section modulus of transformed composite cross section, referred to the bottom flange; based upon maximum permitted effective width of concrete flange, in.^{3}

S_w Warping statical moment at a point in the section, in.^{4}

S_x Elastic section modulus about the X - X axis, in.^{3}

T Horizontal force in flanges of a beam to form a couple equal to beam end moment, kips

Bolt force, kips

T_b Specified pretension of a high-strength bolt, kips

U Factor for converting bending moment with respect to Y - Y axis to an equivalent bending moment with respect to X - X axis

$$= \frac{F_{bx}S_x}{F_{by}S_y}$$

Reduction coefficient used in calculating effective net area

V Maximum web shear, kips

Statical shear on beam, kips

Shear produced by factored loading, kips

Friction force, kips

V_h Total horizontal shear to be resisted by connectors under full composite action, kips

V_h' Total horizontal shear provided by the connectors providing partial composite action, kips

V_s Story shear, kips

W Total uniform load, including weight of beam, kips

W_{no} Normalized warping function at a point at the flange edge, in.^{2}

Y Ratio of yield stress of web steel to yield stress of stiffener steel

Y_2 Distance from top of steel beam to centroid of concrete compressive area, in.

Z Plastic section modulus, in.^{3}

Z_x Plastic section modulus with respect to the major $(X\text{ -}X)$ axis, in.^{3}

Z_y Plastic section modulus with respect to the minor $(Y\text{ -}Y)$ axis, in.^{3}

a	Distance from bolt line to application of prying force Q, in.
	Clear distance between transverse stiffeners, in.
	Dimension parallel to the direction of stress, in.
a'	Distance beyond theoretical cut-off point required at ends of welded partial length cover plate to develop stress, in.
b	Actual width of stiffened and unstiffened compression elements, in.
	Dimension normal to the direction of stress, in.
	Fastener spacing vertically, in.
	Distance from the bolt centerline to the face of tee stem or angle leg in determining prying action, in.
	Effective concrete slab width based on AISC Specification Sect. I1, in.
b_e	Effective width of stiffened compression element, in.
b_f	Flange width of rolled beam or plate girder, in.
b_{fb}	Beam flange width in end-plate design, in.
b_p	End-plate width, in.
d	Depth of column, beam or girder, in.
	Diameter of a roller or rocker bearing, in.
	Nominal diameter of a fastener, in.
d_1	Depth of beam framing into a column on leeward side of connection, in.
d_2	Depth of beam framing into a column on windward side of connection, in.
d_L	Depth at the larger end of a tapered member, in.
d_b	Bolt diameter, in.
d_c	Web depth clear of fillets, in.
d_h	Diameter of hole, in.
d_l	Depth of the larger end of an unbraced segment of a tapered member, in.
d_o	Depth at the smaller end of a tapered member or unbraced segment thereof, in.
e	Base of natural logarithm ($\sim$2.718)
	Eccentricity or distance from point of load application to bolt line
e_o	Distance from outside face of web to the shear center of a channel section, in.
f	Axial compression stress on member based on effective area, ksi
f_a	Computed axial stress, ksi
f_{ao}	Computed axial stress at the smaller end of a tapered member or unbraced segment thereof, ksi
f_b	Computed bending stress, ksi
f_{b1}	Smallest computed bending stress at one end of a tapered segment, ksi
f_{b2}	Largest computed bending stress at one end of a tapered segment, ksi
f_{bl}	Computed bending stress at the larger end of a tapered member or unbraced segment thereof, ksi
f'_c	Specified compression strength of concrete, ksi
f_p	Actual bearing pressure on support, ksi
f_t	Computed tensile stress, ksi
f_v	Computed shear stress, ksi

f_{vs} — Shear between girder web and transverse stiffeners kips per linear inch of single stiffener or pair of stiffeners

g — Transverse spacing locating fastener gage lines, in.

h — Clear distance between flanges of a beam or girder at the section under investigation, in.

Total depth of composite beam, from bottom of steel beam to top of concrete, in.

h_r — Nominal rib height for steel deck, in.

h_s — Factor applied to the unbraced length of a tapered member

h_w — Factor applied to the unbraced length of a tapered member

k — Distance from outer face of flange to web toe of fillet of rolled shape or equivalent distance on welded section, in.

k_c — Compression element restraint coefficient

k_v — Shear buckling coefficient for girder webs

l — For beams, distance between cross sections braced against twist or lateral displacement of the compression flange, in.

For columns, actual unbraced length of member, in.

Unsupported length of a lacing bar, in.

Length of weld, in.

Largest laterally unbraced length along either flange at the point of load, in.

l_b — Actual unbraced length in plane of bending, in.

l_{cr} — Critical unbraced length adjacent to plastic hinge, in.

l_v — Distance from centerline of fastener hole to free edge of part in the direction of the force, in.

l_h — Distance from centerline of fastener hole to end of beam web, in.

m — Factor for converting bending to an approximate equivalent axial load in columns subjected to combined loading conditions

Cantilever dimensions of base plate, in.

n — Number of fasteners in one vertical row

Cantilever dimension of base plate, in.

Modular ratio (E/E_c)

q — Allowable horizontal shear to be resisted by a shear connector, kips

r — Governing radius of gyration, in.

r_T — Radius of gyration of a section comprising the compression flange plus ⅓ of the compression web area, taken about an axis in the plane of the web, in.

r_{To} — Radius of gyration at the smaller end of a tapered member or unbraced segment thereof, considering only the compression flange plus ⅓ of the compression web area, taken about an axis in the plane of the web, in.

r_b — Radius of gyration about axis of concurrent bending, in.

r_{bo} — Radius of gyration about axis of concurrent bending at the smaller end of a tapered member or unbraced segment thereof, in.

r_o — Radius of gyration at the smaller end of a tapered member, in.

r_v — Allowable shear or bearing value for one fastener, kips

r_x — Radius of gyration with respect to the X - X axis, in.

r_y — Radius of gyration with respect to the Y - Y axis, in.

r_y' — Radius of gyration with respect to Y - Y axis of double angle member, in.

s	Longitudinal center-to-center spacing (pitch) of any two consecutive holes, in.
t	Thickness of a connected part, in.
	Wall thickness of a tubular member, in.
	Angle thickness, in.
	Compression element thickness, in.
	Filler thickness, in.
	Thickness of concrete in compression, in.
t_b	Thickness of beam flange or moment connection plate at rigid beam-to-column connection, in.
t_f	Flange thickness, in.
t_{fb}	Thickness of beam flange in end-plate connection design, in.
t_o	Thickness of concrete slab above metal deck, in.
t_p	End-plate thickness, in.
t_s	Stiffener plate thickness, in.
t_w	Web thickness, in.
t_{wc}	Column web thickness, in.
w	Length of channel shear connectors, in.
	Plate width (distance between welds), in.
w_r	Average width of rib or haunch of concrete slab on formed steel deck, in.
x	Subscript relating symbol to strong axis bending
y	Subscript relating symbol to weak axis bending
$\bar{y}_b$	Distance from neutral axis of composite beam to bottom of steel beam, in.
$\bar{y}_{eff}$	Location of elastic neutral axis from bottom of steel beam, in.
z	Distance from the smaller end of a tapered member, in.
α	Ratio of sides of a flat plate with one edge fixed, one edge free, and the two short edges supported

$$= \frac{b_f - t_w}{2(d - 2t_f)}$$

Constant used in equation for hybrid girder factor R_e, Ch. G

$= 0.6\, F_{yw}/F_b \le 1.0$

Moment ratio used in prying action formula for end-plate design

β	Ratio S_{tr}/S_s or S_{eff}/S_s
Δ	Beam deflection, in.
	Displacement of the neutral axis of a loaded member from its position when the member is not loaded, in.
δ	Ratio of net area (at bolt line) to the gross area (at the face of the stem on angle leg)
γ	Tapering ratio of a tapered member or unbraced segment of a tapered member
	Subscript relating symbol to tapered members
μ	Coefficient of friction
ν	Poisson's ratio, may be taken as 0.3 for steel
kip	1,000 lbs.
ksi	Expression of stress in kips per sq. in.

INDEX